Handbook of
Neurochemistry

SECOND EDITION

Volume 3
METABOLISM IN
THE NERVOUS SYSTEM

Handbook of
Neurochemistry

SECOND EDITION

Edited by Abel Lajtha
Center for Neurochemistry, Wards Island, New York

Handbook of
Neurochemistry

SECOND EDITION

Volume 3
METABOLISM IN THE NERVOUS SYSTEM

Edited by
Abel Lajtha

Center for Neurochemistry
Wards Island, New York

SPRINGER SCIENCE+BUSINESS MEDIA, LLC

Library of Congress Cataloging in Publication Data

Main entry under title:

Handbook of neurochemistry.

Includes bibliographical references and index.
Contents: v. 1. Chemical and cellular architecture—v. 2. Experimental neuro-
chemistry—v. 3. Metabolism in the nervous system—v. 4. Enzymes in the nervous
system.
1. Neurochemistry—Handbooks, manuals, etc. 2. Neurochemistry. I. Lajtha,
Abel. [DNLM: WL 104 H235 1982]
QP356.3.H36 1982   612 .814   82-493
ISBN 978-1-4684-4369-1   ISBN 978-1-4684-4367-7 (eBook)
DOI 10.1007/978-1-4684-4367-7

© 1983 Springer Science+Business Media New York
Originally published by Plenum Press, New York 1983
Softcover reprint of the hardcover 1st edition 1983

A Division of Plenum Publishing Corporation
233 Spring Street, New York, N.Y. 10013

Contributors

Carl-David Agardh, Laboratory for Experimental Brain Research, E–Bocket, University Hospital of Lund, S-221 85 Lund, Sweden

M. H. Aprison, Institute of Psychiatric Research, and Departments of Psychiatry, Neurology and Biochemistry, Indiana University School of Medicine, Indianapolis, Indiana 46223

Giuseppe Arienti, Department of Biochemistry, The Medical School, University of Perugia, 06100 Perugia, Italy

Nicholas G. Bazan, LSU Eye Center, Louisiana State University Medical Center School of Medicine, New Orleans, Louisiana 70112

John P. Blass, Department of Neurology, Cornell University Medical College, Burke Rehabilitation Center, White Plains, New York 10605

Gerald P. Brierley, Department of Physiological Chemistry, College of Medicine, Ohio State University, Columbus, Ohio 43210

Graham LeM. Campbell, Department of Neurology, The Graduate Hospital, Philadelphia, Pennsylvania 19146; and Department of Physiology, Temple University School of Medicine, Philadelphia, Pennsylvania 19140

E. C. Daly, The Institute of Psychiatric Research and Departments of Psychiatry, Neurology and Biochemistry, Indiana University School of Medicine, Indianapolis, Indiana 46223

G. E. Gaull, Department of Human Development and Nutrition, New York State Institute for Basic Research in Developmental Disabilities, Staten Island, New York 10314; and Division of Human Genetics, Department of Pediatrics, Mount Sinai School of Medicine of the City University of New York, New York, New York 10029

Ezio Giacobini, Laboratory of Neuropsychopharmacology, Department of Biobehavioral Sciences, University of Connecticut, Storrs, Connecticut 06268.

Present address: Department of Pharmacology Southern Illinois University, School of Medicine, Springfield, Illinois 62708

Gary E. Gibson, Department of Neurology, Cornell University Medical College, Burke Rehabilitation Center, White Plains, New York 10605

Hubert W. Harder, Department of Physiological Chemistry, The Ohio State University, Columbus, Ohio 43210

Richard A. Hawkins, Departments of Anesthesia and Physiology, The Pennsylvania State University College of Medicine, Milton S. Hershey Medical Center, Hershey, Pennsylvania 17033

Lloyd A. Horrocks, Department of Physiological Chemistry, The Ohio State University, Columbus, Ohio 43210

Martin Ingvar, Laboratory for Experimental Brain Research, E-Bocket, University Hospital of Lund, S-221 85 Lund, Sweden

Dennis W. Jung, Department of Physiological Chemistry, College of Medicine, Ohio State University, Columbus, Ohio 43210

Yasuo Kishimoto, John F. Kennedy Institute and the Department of Neurology, Johns Hopkins University School of Medicine, Baltimore, Maryland 21205

Pirjo Kontro, Department of Biomedical Sciences, University of Tampere, Tampere, Finland

Elling Kvamme, Neurochemical Laboratory, Preclinical Medicine, Oslo University, Oslo, Norway

Ata A. Abdel-Latif, Department of Cell and Molecular Biology, Medical College of Georgia, Augusta, Georgia 30912

Robert W. Ledeen, The Saul A. Korey Department of Neurology and Department of Biochemistry, Albert Einstein College of Medicine, Bronx, New York 10461

Howard S. Maker, Department of Neurology, Mount Sinai School of Medicine of the City University of New York, New York, New York 10029; and The Bronx Veterans Administration Medical Center, Bronx, New York 10468

Anke M. Mans, Departments of Anesthesia and Physiology, The Pennsylvania State University, College of Medicine, The Milton S. Hershey Medical Center, Hershey, Pennsylvania 17033

S. S. Oja, Department of Biomedical Sciences, University of Tampere, Tampere, Finland

Giuseppe Porcellati, Department of Biochemistry, The Medical School, University of Perugia, 06100 Perugia, Italy

Norman S. Radin, Mental Health Research Institute, (Department of Psychiatry) and Department of Biological Chemistry, University of Michigan, Ann Arbor, Michigan 48109

D. K. Rassin, Division of Developmental Nutrition and Metabolism, Department of Pediatrics, University of Texas Medical Branch, Galveston, Texas 77550

William Sacks, New York State Rockland Research Institute, Orangeburg, New York 10962

Richard P. Shank, Department of Neurology, The Graduate Hospital, Philadelphia, Pennsylvania 19146; and Department of Physiology, Temple University School of Medicine, Philadelphia, Pennsylvania 19140

Bo K. Siesjö, Laboratory for Experimental Brain Research, E-Bocket, University Hospital of Lund, S-221 85 Lund, Sweden

Marion E. Smith, Department of Neurology, Veterans Administration Medical Center, Palo Alto, California 94304; and Stanford University School of Medicine, Stanford, California 94305

Louis Sokoloff, Laboratory of Cerebral Metabolism, National Institute of Mental Health, U.S. Public Health Service, Department of Health and Human Services, Bethesda, Maryland 20205

J. A. Sturman, Developmental Neurochemistry Laboratory, Department of Pathological Neurobiology, New York State Institute for Basic Research in Developmental Disabilities, Staten Island, New York 10314

H. H. Tallan, Department of Human Development and Nutrition, New York State Institute for Basic Research in Developmental Disabilities, Staten Island, New York 10314; and Division of Human Genetics, Department of Pediatrics, Mount Sinai School of Medicine of the City University of New York, New York, New York 10029

Ricardo Tapia, Department of Neurosciences, Center for Investigations in Cellular Physiology, Universidad National Autónoma de México, 04510 México, D.F., México

Simon N. Young, Department of Psychiatry, McGill University, Montreal, Quebec H3A 1A1, Canada

Preface

This volume is concerned with metabolic reactions occurring in the nervous system. Some time ago, it was thought that since most of the intermediary metabolism that can be observed in the brain is not specific to this organ, there is little justification in studying neural metabolism as such. Later it was realized that for an understanding of neural functions, the understanding of metabolism in the brain and its alterations is essential.

All aspects of the metabolism of a substrate in brain, or all metabolic reactions of the nervous system, could not be included in this volume; some will be dealt with in other volumes (such as the ones covering metabolic turn-over, alterations of metabolism, or pathology). Review of the aspects covered here clearly shows that the study of metabolic reactions in the nervous system is a very active field, producing important results. As in so many areas of research, as we learn more, new aspects become known, new questions emerge, and we see that in solving some problems we open areas with many additional problems to solve. But the accomplishments to date are impressive and indicate further important advances in the future. Brain metabolism is more active, more plastic, and more comprehensive than previously estimated. It is an essential part of brain function, and with its alteration, brain function will be altered. This shows the importance of more knowledge in this area. It is hoped that this volume will be of assistance in such further studies.

Abel Lajtha

Contents

Chapter 3

Ganglioslides

   R. W. Ledeen

Chapter 4

Metabolism of Phosphoinositides

    A. A. Abdel-Latif

Chapter 5

Metabolism of Phosphoglycerides

    G. Porcellati and G. Arienti

Chapter 6

Sulfatides

    N. S. Radin

Chapter 7

Ceramides and Cerebrosides

    Y. Kishimoto

Chapter 8

Peripheral Nervous System Myelin: Properties and Metabolism

    M. E. Smith

Chapter 9

Measurement of Local Glucose Utilization in the Central Nervous System and its Relationship to Local Functional Activity

    L. Sokoloff

Chapter 10

Intermediary Metabolism of Carbohydrates and Other Fuels

  R. A. Hawkins and A. M. Mans

Chapter 11

Oxidative Phosphorylation

D. W. Jung and G. P. Brierley

Chapter 12

Cerebral Metabolism in Vivo

W. Sacks

Chapter 15

Glutamine

E. Kvamme

Chapter 18

Taurine

    S. S. Oja and P. Kontro

Chapter 19

Methionine Metabolism in the Brain

H. H. Tallan, D. K. Rassin, J. A. Sturman and G. E. Gaull

Chapter 20

*The Significance of Tryptophan, Phenylalanine, Tyrosine, and Their
Metabolites in the Nervous System*

S. N. Young

Chapter 21

Imino Acids of the Brain

E. Giacobini

Chapter 22

Glutathione

   H. S. Maker

Chapter 23

Metabolism and Neurotransmission

   G. E. Gibson and J. P. Blass

Chapter 24

Blood Flow

   B. K. Siesjö and M. Ingvar

Fatty Acids and Cholesterol

Lloyd A. Horrocks and Hubert W. Harder

1. INTRODUCTION

Many new areas of research on fatty acids and less-polar lipids have developed since the first edition of this handbook. Among these are the studies on α oxidation of very-long-chain fatty acids by Kishimoto's group and on regulation of fatty acid and cholesterol synthesis by Volpe's group. It is now recognized that cholesterol in the brain is not inert but exchanges with plasma cholesterol at a slow rate. The exchange of free fatty acids between plasma and brain seems to be a rapid process. Ketone bodies are major sources of the carbon atoms for lipogenesis during development. The rapid metabolism and possible role of triacylglycerols in supplying acyl groups for glycerophospholipid synthesis have been recognized. Cultures of individual cell types, particularly oligodendroglia, are becoming valuable for studies of lipid metabolism. Normal brain contains very low levels of free fatty acids, but they are liberated rapidly by ischemia and other injuries.

The scope of this chapter includes the incorporation of fatty acids or acyl-CoA into triacylglycerols and cholesteryl esters but excludes incorporation into other lipids. The coverage is similar for hydrolases catalyzing the opposite reactions. Nearly all components of less-polar lipid fractions are included except tocopherol, quinones, prostaglandins, and lipoxygenase products. The metabolism of diacylglycerols is excluded except that related to triacylglycerols and monoacylglycerols.

2. OXIDATION

2.1. Ketone Bodies

The ketone bodies, acetoacetate and D-(−)-3-hydroxybutyrate, may be utilized for energy and for lipogenesis when available to the brain. They are pri-

---

Lloyd A. Horrocks and Hubert W. Harder • Department of Physiological Chemistry, The Ohio State University, Columbus, Ohio 43210.

mary sources of energy for the human brain during starvation[1] and development.[2] In the rat, the enzymes of ketone body utilization increase in activity during suckling, then decrease to adult values,[3,4] whereas the pyruvate dehydrogenase does not attain full activity until at least 30 days of age.[3] At 18 days of age, the rates of utilization are 148 nmol/min per ml for ketone bodies and 164 nmol/min per ml for glucose.[5]

The utilization of ketone bodies by brain and other organs was reviewed recently by Robinson and Williamson.[4] The entry of ketone bodies into brain is proportional to concentrations in the blood. Acetoacetate has a lower concentration than 3-hydroxybutyrate in blood, but the arteriovenous difference is greater for acetoacetate. A carrier membrane protein facilitates the transport of ketone bodies into brain cells. Permeability of the brain for ketone bodies is three to four times greater in the suckling than in the adult rat and human. For oxidation, the ketone bodies enter the mitochondria. The D-($-$)-3-hydroxybutyrate is oxidized to acetoacetate by a dehydrogenase (E.C. 1.1.1.30). Acetoacetate is converted to the CoA derivative by succinyl-CoA:3-oxoacid-CoA transferase (E.C. 2.8.3.5), then cleaved to two acetyl-CoA by the thiolase (E.C. 2.3.1.9). The acetyl-CoA may then be oxidized through the tricarboxylic acid cycle or be utilized for lipid synthesis (Section 3.2.1). The presence of ketone bodies decreases the rate of oxidation of glucose and pyruvate to carbon dioxide and increases the rate of formation of lactate from pyruvate. The role of ketone bodies as respiratory substrates and metabolic signals and the glucose–fatty acid–ketone body cycle were described by Stanley.[6]

2.2. β Oxidation

Only one-fourth or less of the radioactivity in labeled fatty acids can be recovered in the lipids after intracerebral injection. Typical recoveries are 14–23% for oleate and arachidonate at 1–40 min.[7,8] Much of this disappearance reflects transport into the circulation, but some may be the result of β oxidation of these fatty acids to acetyl-CoA and thence to carbon dioxide[9–12] in the mitochondria.[13] With rat brain homogenates, the activity of carnitine palmitoyltransferase (E.C. 2.3.1.2) and the capacity to oxidize fatty acids to CO_2 increase during development to a peak at 10–15 days of age.[14] These peak levels are five-fold and 2.5-fold the adult levels, respectively. The rate of fatty acid oxidation by newborn brain is approximately 20% of that observed in liver.[15] Although fatty acid oxidation can be detected, it is a minor source of energy for the adult brain. With a perfused cat brain preparation, the contribution of albumin-bound palmitate was 2.8% of the CO_2.[10] Slices of rabbit cerebral cortex have very little ability to oxidize palmitate to CO_2.[16] The yield of CO_2 (nmol/ h per g tissue) from fetal tissue was 0.64 for palmitate compared with 960 for glucose, with an increased disparity at later ages. Mitochondria from rabbit brain have a relatively low activity of palmitoyl-CoA synthetase.[17]

2.3. α Oxidation

Fatty acids with 2-hydroxy groups are abundant in cerebrosides and cerebroside sulfates.[18] These 2-hydroxy fatty acids include relatively large amounts of 22h:0, 23h:0, 24h:0, and 24h:1.[18] The 2-hydroxy fatty acids are derived directly from the corresponding nonhydroxy fatty acids according to *in vivo* results.[19–21] Further oxidation of the 2-hydroxy fatty acids gives a fatty acid with one less carbon.[19] The odd-chain fatty acids 23:0, 25:0, and 25:1 are made primarily by α oxidation. The hydroxylation reaction involves a direct replacement of a hydrogen atom by a hydroxyl group.[22] The activity with tetracosanoate is found in the heavy particulate fraction of rat brain and requires heat-stable and heat-labile cytoplasmic factors but does not require ATP or CoA.[23] All of the 2-hydroxy tetracosanoate produced *in vitro* was found in ceramide and cerebroside, indicating a close integration of α hydroxylation with their synthesis.[24] During these incubations, radioactivity from [1-^{14}C]tetracosanoate was found in glutamate, suggesting a unique type of fatty acid oxidation in brain.[25]

3. SOURCES OF FATTY ACIDS IN THE BRAIN

3.1. Transport

Polyunsaturated fatty acids of the (n-6) and (n-3) series are present in high concentrations in the glycerophospholipids of the nervous system. These fatty acids may be synthesized in the brain from the essential fatty acids linoleate and linolenate, may be synthesized in the liver and transported to the brain, or may be obtained from the diet and transported to the brain.[26] In any case, either the precursor essential fatty acids or the polyunsaturated fatty acids must be transported from the blood into the brain. Synthesis in the brain is relatively less important in the adult rat,[27] but even in the developing rat, the major sources of polyunsaturated fatty acids are synthesis in the liver and uptake from the diet.[26,28]

The composition of the polyunsaturated fatty acids of brain is more resistant to dietary manipulation than is the composition of other tissues.[26,29] In part, this may be because of the extensive reutilization of fatty acids by the brain.[30] With a sufficiently severe deficiency of essential fatty acids, a marked accumulation of 20:3 (n-9) can be found in brain[31] and peripheral nerve.[32] These changes in brain can be reversed within in a few weeks, even in myelin, indicating a substantial rate of turnover in adult animals.[33]

Direct uptake from the circulation and incorporation into complex lipids of rat brain have been shown for palmitic, oleic, linoleic, and linolenic acids.[34–40] Stearic acid administered by subcutaneous injection is incorporated into lipids of myelin.[41] This uptake process is active in developing and adult animals for both essential and nonessential fatty acids. Dhopeshwarkar and colleagues have shown that fatty acids are taken up intact as shown by their labeling patterns. In a thorough review, Dhopeshwarkar and Mead[42] state that the blood–brain barrier is not a barrier for lipids. The rapid labeling of brain lipids by circulating fatty acids shows transport and/or exchange. The uptake can be explained on the basis of metabolic requirements and turnover. Palmitic and oleic acids are taken up from the free fatty acid pool in plasma at a substantial rate that is probably balanced by a simultaneous release of free fatty acid.[43,44] A carrier mechanism exists for palmitic acid.[44] Uptake in the form of lysophosphatidylcholine[45] may account for a small part of the uptake. The alternative to uptake from the circulation is endogenous biosynthesis. Bourre *et al.*[46,47] believe that the latter is important but that the uptake from the circulation is particularly important during myelination.

3.2. Synthesis

De novo biosynthesis is primarily by a cytosolic multienzyme complex as in other mammalian tissues.[46,48,49] The small amounts of activity found in other subcellular fractions may reflect the cytosol in synaptosomes. Acetyl-CoA carboxylase (E.C. 6.4.1.2) catalyzes the reaction

$$\text{acetyl-CoA} + CO_2 + \text{ATP} \rightarrow \text{malonyl-CoA} + \text{ADP} + \text{Pi}$$

The malonyl-CoA is required for fatty acid synthetase

$$\text{acetyl-CoA} + 7 \text{ malonyl-CoA} + 14 \text{ NADPH} + 14 \text{ H}^+ \rightarrow$$
$$\text{palmitic acid} + 8 \text{ CoA} + 14 \text{ NADP}^+ + 7 CO_2 + 6 H_2O$$

Small amounts of other fatty acids may be produced.[50] Acetyl-CoA carboxylase is generally thought to be the rate-controlling enzyme.[51] The partially purified enzyme from C6 glial cells and neuroblastoma cells was markedly inhibited by 1 μM acyl-CoA, a concentration below the critical micelle concentration.[52] Thus, the uptake of exogenous fatty acids may regulate the short-term activity for endogenous fatty acid synthesis. Highest specific activities of both enzymes are found at relatively early stages of development. In rat brain, the specific activity of acetyl-CoA carboxylase was higher at 6 days than at 2 or 10 days.[51] The specific activity of fatty acid synthetase was higher in the 20 to 21-day embryo than at any age after birth.[49] However, if the activity was calculated for the total brain, it continued to rise during the first 3 weeks of life. The highest specific activity for endogenous fatty acid synthesis precedes myelination but coincides with cell multiplication.[17,46,53] However, fatty acid synthesis is 10 to 12-fold more active in C6 glial cells than in neuroblastoma cells.[52] Fatty acid synthesis has not been assayed in isolated nontumor cells. Fatty acid synthetase is also present in the peripheral nervous system.[54,55]

Fatty acid synthetase has a half-life of 1.9 days in brains of young rats and 6.4 days in adult rats.[56] The activity is not affected by fasting or short-term fat-free feeding.[56] A long-term fat-free diet increased the activity by 40%.[49] The specific activity was reduced by methimazole-induced hypothyroidism but was not affected by hydrocortisone, adrenalectomy, or triiodothyronine.[49]

The long-term regulation of endogenous fatty acid synthesis involves a close, coordinated relationship between the amounts synthesized of acetyl-CoA carboxylase and fatty acid synthetase.[57,58] Synthesis of these enzymes is inhibited by dibutyryl cyclic AMP, theophylline, hydrocortisone, and exogenous fatty acids.[57,58] Mechanisms of these changes in protein synthesis are not known. The rates of change in C6 glial cells indicate a half-life of less than 1 day for these enzymes.

3.2.1. Synthesis from Glucose and Acetate

Major fuels of the developing brain are glucose and ketone bodies. At physiological concentrations of these substances in the blood of suckling rats, ketone bodies are better precursors than glucose for the biosynthesis of cerebral lipid.[59] When both are present, glucose metabolism provides the reducing equivalents (NADPH) that are required for fatty acid synthesis. Using nearly pure gray matter from the first slice of rat cerebral cortex, Patel and Tonkonow[51] found that the oxidation of glucose was low with 2-day-old rats but was increased twofold during the second and third postnatal weeks. The maximum incorporation of glucose carbon into lipids was at day 6–7, the same time as the maximum specific activity of acetyl-CoA carboxylase. With slices of rabbit cerebral cortex, Carey[16] found the highest incorporation of glucose carbon into lipid to occur before birth.

Radioactivity from acetate is found in palmitate in complex lipids of the brain within a minute after injection into the carotid artery.[39] Dhopeshwarkar and Subramanian[60,61] compared the relative incorporation into lipids of various precursors administered by intracranial injection. The values were 2.75% for acetate, 0.97% for leucine, 0.87% for 3-hydroxybutyrate, 0.75% for glucose, 0.48% for octanoate, and 0.42% for isoleucine. The leucine and 3-hydroxybutyrate were both metabolized through acetoacetyl-CoA to give high proportions of sterols. The octanoate was metabolized via acetyl-CoA. The rats used in these experiments were 15–16 days old, an age at which incorporation from acetate is near the maximal rate[51] and at which the activity of acetyl-CoA synthetase is relatively high.[53]

Mechanisms of carbon transfer from mitochondrial pyruvate into acetyl-CoA in the cytosol include transport through citrate, N-acetylaspartate, and 2-oxoglutarate or glutamate.[62,63] The 2-oxoglutarate can be converted back to citrate in the cytosol. Citrate is cleaved by ATP-dependent citrate lyase to oxaloacetate and acetyl-CoA. This accounts for about 65% of the cytosolic acetyl-CoA from mitochondrial pyruvate in homogenates of 7-day-old rat brain. The remainder is apparently transported as N-acetylaspartate. Since cytosolic glutamate can be metabolized to citrate, and since glutamate is transported into neonatal rat brain, glutamate may be a source of acetyl-CoA for lipogenesis at very early ages.[63]

In brain, acetyl groups can be transported from the mitochondrion as N-acetylaspartate. These acetyl groups can then be used for fatty acid synthesis. D'Adamo et al.[64] found a 20% incorporation of radioactivity from N-[1-^{14}C]acetylaspartate into fatty acids after intracerebral injection into 8-day-old rats. Only 11% was incorporated at 16 days of age. However, the utilization at 16 days, 97 μg/g wet wt., was nearly twice that at 8 days. At 16 days, 35%

Conversions of polyunsaturated fatty acids in brain have been reviewed recently by Naughton[26] and Mead *et al.*[70] The central nervous system contains relatively large amounts of 20:4(n-6), 22:5(n-6), and 22:6(n-3) in the gray matter, very large amounts of 22:6(n-3) in the retina, and large amounts of 20:4(n-6) and 22:4(n-6) in the ethanolamine glycerophospholipids of white matter but very low amounts of 18:2(n-6) and 18:3(n-3).[82-84]

Brain seems to possess the same pathways as liver for the conversion of polyunsaturated acids. For the (n-6) family, 18:2 is Δ^6 desaturated to 18:3. After elongation to 20:3, it is Δ^5 desaturated to 20:4. Some is further elongated to 22:4 and Δ^4 desaturated to 22:5. For the (n-3) family, 18:3 is Δ^6 desaturated to 18:4. After elongation to 20:4, it is Δ^5 desaturated to 20:5, then elongated to 22:5, then Δ^4 desaturated to 22:6. During essential fatty deficiency, 18:1(n-9) can be Δ^6 desaturated to 18:2, elongated to 20:2, and Δ^5 desaturated to 20:3. Evidence for these pathways is from (1) analogy to liver, (2) the labeling after administration of labeled 18:2(n-6), 20:4(n-6), and 18:3(n-3), (3) *in vitro* assays of specific desaturases, and (4) evidence from the lack of a Δ^8 desaturase. Most steps have not been demonstrated directly in nervous tissue by injection of the appropriate labeled precursor. Recently, further evidence for the pathway of the (n-3) family and the presence of the appropriate Δ^6, Δ^5, and Δ^4 desaturases has been reported for retinoblastoma cells[84] and for rat brain.[85-87] Polyunsaturated fatty acids in brain may also be retroconverted to a fatty acid with 2 fewer carbons.[88,89]

The Δ^6 desaturase is the rate-controlling enzyme for conversion of 18:2(n-6) to 20:4(n-6) in the liver[90] and possibly also in the brain.[91] The very low amounts of 18:2(n-6) and 18:3(n-3) in brain suggest that transport might be rate limiting. Both Δ^6 and Δ^5 desaturase activities have been detected in perinatal rat brain homogenates.[27] The subcellular location of the conversion and retroconversion reactions in brain is not established. By analogy to liver, the conversion reactions should occur in endoplasmic reticulum, but brain microsomal fractions have less than half of the desaturase activity.[27] During development of rat brain, the activities of the Δ^6 and Δ^5 desaturases are maximal between 5 and 20 days of age.[27] At later ages, the liver becomes much more active in production of polyunsaturated fatty acids.

3.4. Fatty Alcohols and Alkanes

Long-chain alcohols comprise about 0.002% of the total lipids in bovine and porcine brain.[92] In porcine brain, the alcohols are primarily 16:0 (35%), 18:0 (30%), and 18:1 (27%). No free aldehydes were found.[93] After an intracerebral injection of labeled palmitate into young rats, free long-chain alcohols were isolated with a relatively very high specific radioactivity.[93] The alcohols are precursors of alkyl and alk-1-enyl groups in glycerophospholipids. Saturated and 18:1 alcohols are formed by microsomal fractions from the corresponding acyl-CoA by reduction with NADPH.[94,95] Long-chain aldehydes are intermediates in this reaction.[96]

Long-chain *n*-alkanes comprise 0.6 mg/g dry wt. of mouse brain.[97] They are most concentrated in myelin (7 mg/g dry wt.), with smaller amounts in

3.3. Elongation and Desaturation

Activation of fatty acids by formation of the CoA derivatives is required before further metabolism including elongation, desaturation, α hydroxylation, β oxidation, esterification, or amide formation. Long-chain fatty acid:CoA ligase (E.C. 6.2.1.3) has a relatively low activity in brain.[69] The activity is maximal at the peak of myelination[16,17] and is primarily located in the microsomal fraction.[17,69] Exogenous [1-^{14}C]palmitate is metabolized in rat brain to 18:0, 18:1, 20:4, 22:6, 24:0, and 24:1.[40] At 2 months after administration by intraperitoneal injection to 12- to 13-day-old rats, the specific radioactivities of these fatty acids were generally higher than that of the palmitate in phosphatidylcholine and cerebrosides.[40] Even at 1 h after an intracerebral injection of [^{14}C]glucose, label is found in 18:0 and 18:1 in addition to 16:0.[60] After intracerebral [^{14}C]acetate, more than 11% of the label was found in polyunsaturated fatty acids.[60]

Enzymes for elongation by two carbons are found in mitochondria and microsomes. The microsomes also contain desaturases acting at the 9, 6, 5, and 4 positions. The elongation of saturated and monounsaturated acyl-CoA in mitochondria requires acetyl-CoA, NADH, and NADPH.[46,70] Activities decrease with increasing chain length of the saturated acyl-CoA with little change in activity from 8 to 40 days of age.[71] The elongated products are not found in myelin but are utilized in the mitochondria.[46,71]

Microsomal elongation requires malonyl-CoA and NADPH.[46] A separate enzyme is present for elongation of 16-carbon acyl-CoA. The presence of two additional enzymes for 18-carbon and for longer chain lengths[72] has been supported by Murad and Kishimoto[71] but not by Bourre *et al.*[46] The elongation activity is fivefold greater for 16:0 than for 18:0.[73] The enzymes are somewhat more active with monounsaturated than with saturated acyl-CoA[46,71] and have peak activities during active myelination.[71,74] The very-long-chain products are utilized for synthesis of sphingolipids for myelin.[48,71] According to Bourre *et al.*,[75] at 18 days of age, the mouse brain incorporates 250 μg 24:0 into membranes, but the synthesis of only 160 μg can be demonstrated *in vitro*. Either the conditions of the *in vitro* assay are not optimal or exogenous 24:0 is required. The latter is unlikely. The microsomal fraction from mouse sciatic nerve can also elongate very-long-chain fatty acids.[55] The elongation of 16:1(n-7) to 18:1(n-7)[71] explains the appearance of the latter in acyl and alkenyl groups of ethanolamine glycerophospholipids.[76] In brain as in liver, the rate-limiting step for elongation of 16:0-CoA and 18:0-CoA by microsomes is the condensation reaction with malonyl-CoA to form the 3-ketoacyl-CoA.[73]

Desaturation of 18:0 to 18:1 was found at 1 min after intracerebral injection into the developing mouse brain.[77] Microsomal fractions desaturate palmitoyl-CoA and stearoyl-CoA at similar rates.[78] Desaturation is normally at the 9 position, but the microsomal 6-desaturase may also act on palmitoyl-CoA and stearoyl-CoA at an early age.[79] The specific activity of 9-desaturase decreases markedly during development, but the activity per brain is nearly constant.[77,78,80] Desaturation of lignoceroyl-CoA cannot be detected, but 24:1 can be formed by microsomal elongation of 22:1.[81]

microsomal, mitochondrial, and synaptosomal fractions. Even- and odd-number chain lengths are present in nearly equal amounts. Chain lengths range from 19 to 32 carbons. In whole brain, heptacosane and octacosane each account for 14% of the total *n*-alkanes. Sciatic nerve particulate fractions can synthesize alkanes from the corresponding fatty acid.[98]

4. FREE FATTY ACIDS

Free fatty acids are metabolic intermediates that are present at very low levels in normal brain.[99,100] Concentrations of free fatty acids increase rapidly within seconds after decapitation, ischemia, severe hypoxia, convulsions, or hypoglycemia.[99,101–103] Thus, a primary problem in the measurement of endogenous levels of free fatty acids in normal brain is to prevent the hydrolysis of esterified fatty acids. This has been approached by rapid dissection,[104] microwave irradiation,[103] and rapid freezing in liquid nitrogen.[102] All of these methods may involve up to 20 sec of ischemia for "normal" tissue, so reported "normal" values may be somewhat higher than true endogenous levels.

Free fatty acids from normal tissues are primarily 16:0, 18:0, 18:1, and 20:4. During ischemia or severe hypoxia, the largest relative increases in free fatty acids are in 18:0, 20:4, and 22:6.[102,103] Primary sources of these fatty acids are phosphatidylcholine, phosphatidylinositol, and ethanolamine plasmalogens[102,106] because of activation of lipases[99] and the reversal of glycerophospholipid synthetic reactions.[107] The released arachidonate may be converted to prostaglandins and related substances, causing tissue edema and cell damage.[100] The free fatty acids themselves inhibit oxidative phosphorylation in mitochondria,[108–110] thus decreasing ATP synthesis when it is most needed.

In peripheral nerve, the free fatty acids have a much different composition.[111] More than 75% are saturated, less than 5% are polyunsaturated, and 10–15% have chain lengths of 24–30 carbons. The metabolic significance is unknown.

5. DIACYLGLYCEROLS AND TRIACYLGLYCEROLS

Triacylglycerols account for about 0.1% of the total lipids of human, rat, and rabbit brain.[112–114] Smaller amounts of diacylglycerols and triacylglycerols are present in mouse brain.[115] These lipids are very active metabolically.[7,116] Diacylglycerols are key intermediates in lipid metabolism and are highly labeled within a few minutes after intracerebral injection. Triacylglycerols are labeled at a slower rate, reaching a peak between 10 and 20 min after injection. Presumably they are formed from diacylglycerols by an acyl-CoA transferase, but this has not been demonstrated. The metabolism of triacylglycerols is more rapid than that of glycerophospholipids.[7,116] This suggests that triacylglycerols may be localized to specialized regions such as choroid plexus[117] and/or that their function is to take up excesses of free fatty acids for later transfer to glycerophospholipids.[118] The latter seems to occur during *in vitro* incubations of primary cultures,[119] slices,[16] and isolated oligodendroglia.[120]

The acyl groups of mouse brain triacylglycerols are primarly 16:0, 18:1, 18:0, and 22:6.[115] The diacylglycerols of mouse, rat, and toad brain are enriched in 18:0 and 20:4, reflecting their metabolic relationship with inositol glycerophospholipid metabolism.[115,121,122] The diacylglycerol content of the retina is sevenfold greater than that of brain, with very high levels of 22:6.[121]

Lipases acting on triacylglycerols, diacylglycerols, and monoacylglycerols are present in brain tissue.[123–126] The lipases that are quuite active at pH 4.8 are presumably located in lysosomes.[123] Considerable lipase activity is also found at neutral pH. None of the lipases have been completely purified or characterized. Most of the fatty acids from triacylglycerols appear subsequently in glycerophospholipids.[119]

6. CHOLESTEROL

6.1. Content

The concentration of cholesterol in rat brain increases with age up to 63 days of age.[127] Even from 400 to 630 days, the cholesterol content increases from 37 to 42 mg per brain. Increases during early development of brain and spinal cord are most pronounced.[127,128] Small losses of cholesterol may take place very late (1400 to 2200 days) in the life-span of *Peromyscus leucopus*, a long-lived field mouse.[129] According to Rouser and Yamamoto,[114] the concentration of cholesterol in human brain is 66.2 mmol/kg wet wt. at 40 years of age. At 70 years of age, the concentration is 11% less. Concentrations in rat cerebrum and spinal cord at 127 days of age are 50 and 95 mmol/kg wet wt., respectively.[128] The differences in concentration result from different concentrations of myelin in which cholesterol is enriched. Within the myelin, cholesterol is symmetrically distributed between the two halves of the bilayer membrane.[130]

6.2. Synthesis

Labeled acetate readily labels brain sterols in weaning and adult rats.[131] Label is found in cholesterol from the brain less than 1 min after the injection of labeled acetate into the carotid artery of the adult rat.[37] The conventional pathway is followed to lanosterol. The alternative pathways from lanosterol to cholesterol are shown by Ramsey.[132]

Rates of cholesterol synthesis correlate well with the rates of cholesterol deposition. Very little or no cholesterol synthesis could be detected in slices of adult rat brain.[133] With acetate, synthesis per mouse brain is maximal between 7 and 20 days of age.[134] When measured with tritiated water, the labeling of sterols per rat brain was greater at 18 than at 12 or 24 days of age, but the labeling of sterols per gram of brain was greater at 12 than at 6 or 18 days of age.[5]

In slices of rat cerebral cortex, sterol synthesis from ketone bodies was greatest at 6 days of age.[59] During this period, concentrations of ketone bodies are relatively high, and that of glucose is relatively low. Also, the activities of enzymes involved in ketone body utilization are increased.[4] Acetoacetate and D-(−)-3-hydroxybutyrate are particularly good precursors of sterols because they can be converted directly to acetoacetyl-CoA in the cytosol,[5] whereas the acetyl-CoA from glucose must be transported from the mitochondria. The more direct utilization of acetoacetyl-CoA by 3-hydroxy-3-methylglutaryl-CoA synthase than by acetyl-CoA carboxylase also favors the utilization of ketone bodies for sterol synthesis rather than fatty acid synthesis.[5] This is particularly important in the brainstem during active myelination.[135] The relative rates of *in vivo* sterol synthesis in 18-day rat brain were 7 from glucose, 24 from acetoacetate, and 30 from D-(−)-3-hydroxybutyrate. The latter is not incorporated into sterols by the cytosolic pathway in isolated oligodendroglia.[66] These cells are more than 20-fold enriched in cytosolic acetoacetyl-CoA synthase activity relative to whole brain. Acetoacetate is a significant source of carbons for sterol synthesis by oligodendroglia in culture and presumably also for oligodendroglia that are producing myelin.[66] Leucine is also an efficient precursor for cholesterol synthesis.[136]

The activity of 3-hydroxy-3-methylglutaryl-CoA reductase, the rate-limiting enzyme for cholesterol biosynthesis, is elevated in developing brain.[137] In C6 glioma cells, the specific activities of HMG-CoA synthase and HMG-CoA reductase decreased markedly in conjunction with decreased sterol synthesis when serum was removed.[137] By disruption of microtubules, the activity of HMG-CoA reductase but not HMG-CoA synthase is sharply reduced because of a decreased content of the enzyme.[138] The half-life of HMG-CoA reductase in cultured oligodendroglia, as in other cultured cells, is less than 4 hr.[139] These oligodendroglia are sevenfold enriched in HMG-CoA reductase over whole white matter.[139]

Dolichol is an isoprenol involved in the synthesis of glycoproteins. The pathway for dolichol synthesis branches from that for cholesterol synthesis after HMG-CoA reductase. A sharp peak in the rate of dolichol synthesis in mouse brain was found at 3 to 7 days of age.[134] Since the rate of cholesterol synthesis does not correlate with the rate of dolichol synthesis, independent regulation was suggested, perhaps because of different cellular sites of synthesis.[134]

Desmosterol is normally present in developing brain but has virtually disappeared from rat brain by 50 days of age.[132] Several other precursor sterols are present in developing rabbit brain.[140] Various hypocholesterolemic drugs block sterol synthesis so that 7-dehydrodesmosterol, desmosterol, and/or 7-dehydrocholesterol accumulates.[132]

6.3. Metabolism

Cholesterol in the brain is derived from *de novo* synthesis within the brain and from transfer from plasma. Since the latter process in the rat occurs at a rate of 7.4 mg/month,[127] it cannot be detected by experiments utilizing intracarotid injection of labeled cholesterol incorporated into plasma lipoproteins.[43]

When cholesterol was adsorbed on albumin, 63% was taken up by brain during one pass. Exchange with plasma cholesterol is the only possibility for metabolism of cholesterol in the brain, because cholesterol degradation occurs only in the liver and some endocrine glands.

No pool of inert cholesterol can be found in brain.[127] Serougne *et al.*[127] have proposed a kinetic model for cholesterol turnover in the rat. The bulk of the cholesterol (94%) has a turnover flux of 4 mg/month and a turnover rate of 0.118/month (recalculated from values given in ref. [127]). This pool includes the cholesterol in myelin. Four other pools are included in the model. The model explains why the radioactivity in brain cholesterol decreases rapidly at first and then seems to reach a plateau value.

Cholesterol might be expected to turn over faster than any other component of membranes. Proteins, glycerophospholipids, and sphingolipids have relatively strong hydrogen bonding and often ionic bonding in addition to hydrophobic forces to hold them within a membrane, but cholesterol has only one hydroxyl group for a hydrogen bond besides the hydrophobic forces. However, when cholesterol does exchange with another membrane or with a carrier protein, very little can happen to it, since oxidative enzymes are missing. Thus, the rapid turnover of cholesterol leads only to isotopic equilibrium, and the only way for loss of labeled cholesterol from the brain is by exchange with unlabeled plasma cholesterol as described above.

Cholesterol can be transported in axons of optic nerve and peripheral nerves.[136,141] It is a component of the rapid phase of axonal transport.[141]

Cholesteryl esters are present in developing rat brain in quite small amounts of 20 μg/brain or less.[142] Major components of the acyl groups are 16:0, 18:1, 16:1, and 20:4.[142] Esterified cholesterol may be involved in the process of myelination.[142] This may reflect myelin disintegration as a part of normal postnatal maturation[143] and may relate to the somewhat higher content of cholesteryl esters in the brains of undernourished mice.[144] Quite high concentrations of esterified cholesterol are found in demyelinating conditions.[145,146]

A cholesterol-esterifying enzyme that incorporates free fatty acids into cholesteryl esters without participation of CoA is present in rat brain.[147] The activity of this enzyme is low at birth and reaches the adult level at 20 days of age.[148] It is present in human cerebrospinal fluid[149] and rat peripheral nerve.[150] An acyl-CoA-dependent enzyme is more active during development, then decreases markedly in activity.[151]

Three cholesteryl ester hydrolases (E.C. 3.1.1.13) are also found in brain.[148,152] Of these, one has an optimal pH of 4.2 and may be lysosomal. The microsomal hydrolase has optimal activity at pH 6.0, whereas the myelin hydrolase has a pH optimum of 7.2. Hydrolase activity is also found in human cerebrospinal fluid.[153]

ACKNOWLEDGMENTS. The writing of this review was supported in part by NIH Research Grants NS-08291 and NS-10165.

REFERENCES

1. Owen, O. E., Morgan, A. P., Kemp, H. G., Sullivan, J. M., Herrera, M. G., and Cahill, G. F., 1967, *J. Clin. Invest.* **46**:1589–1595.
2. Williamson, D. H., and Buckley, B. M., 1973, *Inborn Errors of Metabolism* (F. A. Hommes and C. J. van den Bergh, eds.), Academic Press, New York, pp. 81–96.
3. Land, J. M., Booth, R. F. G., Berger, R., and Clark, J. B., 1977, *Biochem. J.* **164**:339–348.
4. Robinson, A. M., and Williamson, D. H., 1980, *Physiol. Rev.* **60**:143–187.
5. Webber, R. J., and Edmond, J., 1979, *J. Biol. Chem.* **245**:3912–3920.
6. Stanley, J. C., 1981, *Br. J. Anaesth.* **53**:131–136.
7. Yau, T. M., and Sun, G. Y., 1974, *J. Neurochem.* **23**:99–104.
8. Sun, G. Y., and Yau, T.M., 1976, *J. Neurochem.* **26**:291–295.
9. Volk, M. E., Millington, R. H., and Weinhouse, S., 1952, *J. Biol. Chem.* **195**:493–501.
10. Allweis, C., Landau, T., Abeles, M., and Magnes, J., 1966, *J. Neurochem.* **13**:795–804.
11. Little, J. R., Hori, S., and Spitzer, J. J., 1969, *Am. J. Physiol.* **217**:919–922.
12. Spitzer, J. J., and Wolf, E. H., 1971, *Am. J. Physiol.* **221**:1426–1430.
13. Beattie, D. S., and Basford, R. E., 1966, *J. Biol. Chem.* **241**:1412–1418.
14. Warshaw, J. B., and Terry, M. L., 1976, *Dev. Biol.* **52**:161–166.
15. Warshaw, J. B., 1979, *Semin. Perinatol.* **3**:131–139.
16. Carey, E. M., 1975, *J. Neurochem.* **24**:237–244.
17. Cantrill, R. C., and Carey, E. M., 1975, *Biochim. Biophys. Acta* **380**:165–175.
18. Hoshi, M., and Kishimoto, Y., 1973, *J. Biol. Chem.* **248**:4123–4130.
19. Kishimoto, Y., and Radin, N. S., 1966, *Lipids* **1**:47–61.
20. Fulco, A. J., and Mead, J. F., 1961, *J. Biol. Chem.* **236**:2416–2420.
21. Mead, J. F., and Levis, G. M., 1963, *J. Biol. Chem.* **238**:1634–1636.
22. Tatsumi, K., Murad, S., and Kishimoto, Y., 1975, *Arch. Biochem. Biophys.* **171**:87–92.
23. Singh, I., and Kishimoto, Y., 1979, *J. Biol. Chem.* **254**:7698–7704.
24. Kishimoto, Y., Akanuma, H., and Singh, I., 1979, *Mol. Cell. Biochem.* **28**:93–105.
25. Uda, M., Singh, I., and Kishimoto, Y., 1981, *Biochemistry* **20**:1295–1300.
26. Naughton, J. M., 1981, *Int. J. Biochem.* **13**:21–32.
27. Cook, H. W., 1978, *J. Neurochem.* **30**:1327–1334.
28. Hassam, A. G., and Crawford, M. A., 1976, *J. Neurochem.* **27**:967–968.
29. Mohrhauer, H., and Holman, R. T., 1963, *J. Neurochem.* **10**:523–530.
30. Sun, G. Y., and Horrocks, L. A., 1973, *J. Lipid Res.* **14**:206–214.
31. McKenna, M. C., and Campagnoni, A. T., 1979, *J. Nutr.* **109**:1195–1204.
32. Evans, B. A., Yao, J. K., Holman, R. T., Brimijoin, W. S., Lambert, E. H., and Dyck, P. J., 1980, *J. Neuropathol. Exp. Neurol.* **39**:683–691.
33. Sun, G. Y., Winniczek, H., Go, J., and Sheng, S. L., 1975, *Lipids* **10**:365–373.
34. Dhopeshwarkar, G. A., and Mead, J. F., 1969, *Biochim. Biophys. Acta* **187**:461–467.
35. Dhopeshwarkar, G. A., and Mead, J. F., 1970, *Biochim. Biophys. Acta* **210**:250–256.
36. Dhopeshwarkar, G. A., Subramanian, C., and Mead, J. F., 1971, *Biochim. Biophys. Acta* **231**:8–14.
37. Dhopeshwarkar, G. A., Subramanian, C., and Mead, J. F., 1971, *Biochim. Biophys. Acta* **248**:41–47.
38. Dhopeshwarkar, G. A., Subramanian, C., and Mead, J. F., 1971, *Biochim. Biophys. Acta* **239**:162–167.
39. Dhopeshwarkar, G. A., Subramanian, C., McConnell, D. H., and Mead, J. F., 1972, *Biochim. Biophys. Acta* **255**:572–579.
40. Dhopeshwarkar, G. A., Subramanian, C., and Mead, J. F., 1973, *Biochim. Biophys. Acta* **296**:257–264.
41. Gozlan-Devilliere, N., Baumann, N., and Bourre, J. M., 1978, *Biochim. Biophys. Acta* **528**:490–496.
42. Dhopeshwarkar, G. A., and Mead, J. F., 1973, *Adv. Lipid Res.* **11**:109–142.
43. Pardridge, W. M., and Mietus, L. J., 1980, *J. Neurochem.* **34**:463–466.
44. Drewes, L. R., and Gilboe, D. D., 1977, *Fed. Proc.* **36**:166–177.

45. Illingworth, D. R., and Portman, O. W., 1972, *Biochem. J.* **130**:557–567.
46. Bourre, J. M., Pollet, S., Paturneau-Jouas, M., and Baumann, N., 1978, *Enzymes of Lipid Metabolism* (S. Gatt, L. Freysz, and P. Mandel, eds.), Plenum Press, New York, pp. 17–25.
47. Bourre, J. M., Gozlan-Devilliere, N., Morand, O., and Baumann, N., 1979, *Ann. Biol. Anim. Biochem. Biophys.* **19**(1B):173–180.
48. Bourre, J. M., Pollet, S., Paturneau-Jouas, M., and Baumann, N., 1977, *Function and Biosynthesis of Lipids* (N. G. Bazan, R. R. Brenner, and N. M. Giusto, eds.) Plenum Press, New York, pp. 103–109.
49. Volpe, J. J., and Kishimoto, Y., 1972, *J. Neurochem.* **19**:737–753.
50. Bourre, J. M., Daudu, O. L., and Baumann, N. A., 1975, *J. Neurochem.* **24**:1095–1097.
51. Patel, M. S., and Tonkonow, B. L., 1974, *J. Neurochem.* **23**:309–313.
52. Volpe, J. J., and Marasa, J. C., 1977, *Brain Res.* **129**:91–106.
53. Koeppen, A. H., Mitzen, E. J., and Papandrea, J. D., 1980, *J. Neurochem.* **34**:261–268.
54. Koeppen, A. H., Papandrea, J. D., and Mitzen, E. J., 1979, *Muscle Nerve* **2**:369–375.
55. Cassagne, C., Darriet, D., Boiron, F., Larrouquere-Regnier, S., and Bourre, J. M., 1980, *C. R. Soc. Biol.* (*Paris*) **174**:387–399.
56. Volpe, J. J., Lyles, T. O., Roncari, D. A. K., and Vagelos, P. R., 1973, *J. Biol. Chem.* **248**:2502–2513.
57. Volpe, J. J., and Marasa, J. C., 1976, *J. Neurochem.* **27**:841–845.
58. Volpe, J. J., and Marasa, J. C., 1976, *Biochim. Biophys. Acta* **431**:195–205.
59. Patel, M. S., and Owen, O. E., 1977, *J. Neurochem.* **28**:109–114.
60. Dhopeshwarkar, G. A., and Subramanian, C., 1977, *Lipids* **12**:762–764.
61. Dhopeshwarkar, G. A., and Subramanian, C., 1979, *Lipids* **14**:47–51.
62. Patel, T. B., and Clark, J. B., 1980, *Biochem. J.* **188**:163–168.
63. D'Adamo, A. F., Smith, J. C., and Frigyesi, G., 1975, *J. Neurochem.* **24**:597–600.
64. D'Adamo, A. F., Jr., Gidez, L. I., and Yatsu, F. M., 1968, *Exp. Brain Res.* **5**:267–273.
65. Patel, T. B., and Clark, J. B., 1979, *Biochem. J.* **184**:539–546.
66. Pleasure, D., Lichtman, C., Eastman, S., Lieb, M., Abramsky, O., and Silberberg, D., 1979, *J. Neurochem.* **32**:1447–1450.
67. Webber, R. J., and Edmond, J., 1977, *J. Biol. Chem.* **252**:5222–5226.
68. Yeh, Y. Y., Streuli, V. L., and Zee, P., 1977, *Lipids* **12**:957–964.
69. Murphy, M. G., and Spence, M. W., 1980, *J. Neurochem.* **34**:367–373.
70. Mead, J. F., Dhopeshwarkar, G. A., and Gan Elpano, M., 1977, *Adv. Exp. Med. Biol.* **83**:313–328.
71. Murad, S., and Kishimoto, Y., 1978, *Arch. Biochem. Biophys.* **185**:300–306.
72. Goldberg, I., Shechter, I., and Bloch, K., 1973, *Science* **182**:497–499.
73. Bernert, J. T., Jr., Bourre, J. M., Baumann, N. A., and Sprecher, H., 1979, *J. Neurochem.* **32**:85–90.
74. Brophy, P. J., and Vance, D. E., 1975, *Biochem. J.* **152**:495–501.
75. Bourre, J. M., Paturneau-Jouas, M. Y., Daudu, O. L., and Baumann, N. A., 1977, *Eur. J. Biochem.* **72**:41–47.
76. Pullarkat, R. K., and Reha, H., 1978, *J. Neurochem.* **31**:707–711.
77. Wise, R. W., MacQuarrie, R., and Sun, G. Y., 1979, *J. Neurochem.* **33**:351–354.
78. Carreau, J. P., Daudu, O., Mazliak, P., and Bourre, J. M., 1979, *J. Neurochem.* **32**:659–660.
79. Cook, H. W., 1979, *Lipids* **14**:763–767.
80. Pullarkat, R. K., and Reha, H., 1975, *J. Neurochem.* **25**:607–610.
81. Bourre, J. M., Daudu, O., and Baumann, N., 1976, *Biochim. Biophys. Acta* **424**:1–7.
82. Svennerholm, L., 1968, *J. Lipid Res.* **9**:570–579.
83. Sun, G. Y., and Horrocks, L. A., 1968, *Lipids* **3**:79–83.
84. Hyman, B. T., and Spector, A. A., 1981, *J. Neurochem.* **37**:60–69.
85. Dwyer, B. E., and Bernsohn, J., 1979, *Biochim. Biophys. Acta* **575**:309–317.
86. Dhopeshwarkar, G. A., and Subramanian, C., 1976, *Lipids* **11**:67–71.
87. Cohen, S. R., and Bernsohn, J., 1978, *J. Neurochem.* **30**:661–669.
88. Dhopeshwarkar, G. A., and Subramanian, C., 1976, *Lipids* **11**:689–692.
89. Dhopeshwarkar, G. A., and Subramanian, C., 1976, *J. Neurochem.* **26**:1175–1179.
90. Sprecher, H., 1977, *Adv. Exp. Biol. Med.* **83**:35–50.

91. Hassam, A. G., and Crawford, M. A., 1978, *Lipids* **13**:801–803.
92. Takahashi, T., and Schmid, H. H. O., 1970, *Chem. Phys. Lipids* **4**:243–246.
93. Schmid, H. H. O., and Takahashi, T., 1970, *J. Lipid Res.* **11**:412–419.
94. Natarajan, V., and Schmid, H. H. O., 1978, *Arch. Biochem. Biophys.* **187**:215–222.
95. Bourre, J. M., and Daudu, O., 1978, *Neurosci. Lett.* **7**:225–230.
96. Natarajan, V., and Sastry, P. S., 1976, *J. Neurochem.* **26**:107–113.
97. Darriet, D., Cassagne, C., and Bourre, J. M., 1978, *Neurosci. Lett.* **8**:77–81.
98. Cassagne, C., Darriet, D., and Bourre, J. M., 1977, *FEBS Lett.* **82**:51–54.
99. Bazan, N. G., Aveldano de Caldironi, M. I., Cascone de Suarez, G. D., and Rodriguez de Turco, E. B., 1980, *Prog. Clin. Biol. Res.* **39**:167–179.
100. Galli, C., Spagnuolo, C., Petroni, A., and Sautebin, L., 1980, *Ann. Nutr. Aliment.* **34**:415–422.
101. Agardh, C. D., Chapman, A. G., Nilsson, B., and Siesjö, B. K., 1981, *J. Neurochem.* **36**:490–500.
102. DeMedio, G. E., Goracci, G., Horrocks, L. A., Lazarewicz, J. W., Mazzari, S., Porcellati, G., Strosznajder, J., and Trovarelli, G., 1980, *Ital. J. Biochem.* **29**:412–432.
103. Gardiner, M., Nilsson, B., Rehncrona, S., and Siesjö, B. K., 1981, *J. Neurochem.* **36**:1500–1505.
104. Aveldano, M. I., and Bazan, N. G., 1975, *Brain Res.* **100**:99–110.
105. Cenedella, R. J., Galli C., and Paoletti, R., 1975, *Lipids* **10**:290–293.
106. Marion, J., and Wolfe, L. S., 1979, *Biochim. Biophys. Acta* **574**:25–32.
107. Dorman, R. V., Dabrowiecki, Z., DeMedio, G. E., Porcellati, G., and Horrocks, L. A., 1982, *Head Injury* (R. G. Grossman and P. L. Gildenberg, eds.), Raven Press, New York, pp. 93–101.
108. Ansevin, C. F., 1980, *Neurology (N.Y.)* **30**:160–166.
109. Lazarewicz, J. W., Strosznajder, J., and Gromek, A., 1972, *Bull. Acad. Pol. Sci.* **20**:599–606.
110. Ozawa, K., Seta, H., Araki, H., and Handa, H., 1967, *J. Biochem.* **62**:584–590.
111. Yao, J. K., Dyck, P. J., Van Loon, J. A., and Moyer, T. P., 1981, *J. Neurochem.* **36**:1211–1218.
112. Bazan, N. G., 1971, *J. Neurochem.* **18**:1379–1385.
113. Odutuga, A. A., Carey, E. M., and Prout, R. E. S., 1973, *Biochim. Biophys. Acta* **316**:115–123.
114. Rouser, G., and Yamamoto, A., 1969, *Handbook of Neurochemistry*, Volume 1 (A. Lajtha, ed.), Plenum Press, New York, pp. 121–169.
115. Sun, G. Y., 1970, *J. Neurochem.* **17**:445–446.
116. Sun, G. Y., and Horrocks, L. A., 1971, *J. Neurochem.* **18**:1963–1969.
117. Marinetti, G. V., Weindl, A., and Kelly, J., 1971, *J. Neurochem.* **18**:2003–2006.
118. Dwyer, B., and Bernsohn, J., 1979, *J. Neurochem.* **32**:833–838.
119. Yavin, E., and Menkes, J. H., 1973, *J. Neurochem.* **21**:901–912.
120. Carey, E. M., Stoll, U., and Carruthers, A., 1980, *Biochem. Soc. Trans.* **8**:368–369.
121. Aveldano, M. I., and Bazan, N. G., 1973, *Biochim. Biophys. Acta* **296**:1–9.
122. MacDonald, G., Baker, R. R., and Thompson, W., 1975, *J. Neurochem.* **24**:655–661.
123. Cabot, M. C., and Gatt, S., 1978, *Biochim. Biophys. Acta* **530**:508–512.
124. Mizobuchi, M., Shirai, K., Matsuoka, N., Saito, Y., and Kumagai, A., 1981, *J. Neurochem.* **36**:301–303.
125. Pendley, C., Singh, H., and Cox, J., 1981, *Fed. Proc.* **40**:1708.
126. Strosznajder, J., Singh, H., and Horrocks, L. A., 1982, *Metabolism of Phospholipids in the Nervous System*, Vol. I, Metabolism (L. A. Horrocks, G. B. Ansell, and G. Porcellati, eds.), Raven Press, New York, p. 343.
127. Serougne, G., Lefevre, C., and Chevallier, F., 1976, *Exp. Neurol.* **51**:229–240.
128. de Sousa, B. N., and Horrocks, L. A., 1979, *Dev. Neurosci.* **2**:122–128.
129. Torello, L. A., and Yates, A. J., cited by Horrocks, L. A., VanRollins, M., and Yates, A. J., 1981, *Molecular Basis of Neuropathology* (R. H. S. Thompson and A. N. Davison, eds.), Edward Arnold, London, pp. 601–630.
130. Scott, S. C., Bruckdorfer, K. R., and Worcester, D. L., 1980, *Biochem. Soc. Trans.* **8**:717.
131. Kabara, J. J., 1973, *Prog. Brain Res.* **40**:363–382.

132. Ramsey, R. B., 1977, *Lipids* **12:**841–846.
133. Andersen, J. M., and Dietschy, J. M., 1979, *J. Lipid Res.* **20:**740–752.
134. James, M. J., and Kandutsch, A. A., 1980, *Biochim. Biophys. Acta* **619:**432–435.
135. Yeh, Y. Y., 1980, *Lipids* **15:**904–907.
136. Wood, P. L., and Boegman, R. J., 1980, *FEBS Lett.* **115:**110–112.
137. Maltese, W. A., Reitz, B. A., and Volpe, J. J., 1980, *Biochem. J.* **192:**709–717.
138. Volpe, J. J., and Obert, K. A., 1981, *J. Biol. Chem.* **256:**2016–2021.
139. Pleasure, D., Abramsky, O., Silberberg, D., Quinn, B., Parris, J., and Saida, T., 1977, *Brain Res.* **134:**377–382.
140. Adamczewska-Goncerzewicz, Z., and Trzebny, W., 1981, *J. Neurochem.* **36:**1378–1382.
141. Blaker, W. D., Toews, A. D., and Morell, P., 1980, *J. Neurobiol.* **11:**243–250.
142. Eto, Y., and Suzuki, K., 1972, *J. Neurochem.* **19:**109–115.
143. Berthold, C. H., and Hildebrand, C., 1979, *J. Neurochem.* **32:**237–240.
144. Yusuf, H. K., and Mozaffar, Z., 1979, *J. Neurochem.* **32:**273–275.
145. Yao, J. K., Natarajan, V., and Dyck, P. J., 1980, *J. Neurochem.* **35:**933–940.
146. Norton, W. T., 1977, *Myelin* (P. Morell, ed.), Plenum Press, New York, pp. 383–413.
147. Choi, M. U., and Suzuki, K., 1978, *J. Neurochem.* **31:**879–885.
148. Eto, Y., and Suzuki, K., 1972, *J. Neurochem.* **19:**117–121.
149. Johnson, R. C., and Shah, S. N., 1979, *Brain Res.* **162:**353–357.
150. Yao, J. K., and Dyck, P. J., 1981, *J. Neurochem.* **37:**156–163.
151. Jagannatha, H. M., and Sastry, P. S., 1981, *J. Neurochem.* **36:**1352–1360.
152. Igarashi, M., and Suzuki, K., 1977, *J. Neurochem.* **28:**729–738.
153. Reiber, H., and Voss, W., 1980, *J. Neurochem.* **34:**1324–1326.

Metabolism of Phosphatidic Acid

Nicolas G. Bazan

1. INTRODUCTION

In the first edition of the *Handbook of Neurochemistry*, Rossiter and Strickland[1] reviewed the metabolism of phosphatidic acid in relation to the overall metabolism of phosphoglycerides, and Hawthorne and Kai[2] described the metabolic relationship between phosphatidic acid and phosphoinositides. The past 12 years have seen many advances in the area of phosphatidic acid biochemistry. A review of the existing information, therefore, is desirable. This chapter focuses primarily on work carried out in the nervous system, although background information obtained in nonneural tissues is also presented when relevant to the overall picture. Several aspects of phosphatidic acid composition and metabolism in the neural tissue remain unexplained because of the methodological difficulties of isolating this minor lipid from neural tissue.

During the second half of the 1970s, research on phospholipids of the nervous tissue was influenced greatly by an increased understanding of the metabolism of phosphatidylinositol[3-8] (A. A. Abdel-Latif, this volume). Hokin[9,10] found that during ^{32}P labeling of phosphatidylinositol in the pancreas, there was a concomitant stimulation of amylase secretion by acetylcholine. Phosphatidic acid was regarded as an intermediate in the cycle of phosphatidylinositol degradation and resynthesis.[4-8] Hypothetically, this cycle was thought to be involved in the control of Ca^{2+} influx across plasma cell membranes. However, phosphatidylinositol degradation was shown to follow calcium mobilization in several systems.[11-15]

New ideas and supporting data suggested that phosphatidic acid may be the Ca^{2+} ionophore in synaptic membranes linking depolarization with neurotransmitter release.[16-18] Moreover, the study of several physical chemical properties of phosphatidic acid in model membranes demonstrated that Ca^{2+} promotes lateral phase segregation and cluster formation in phosphatidic acid.[19]

Nicolas G. Bazan • LSU Eye Center, Louisiana State University Medical Center School of Medicine, New Orleans, Louisiana 70112.

In addition to high ^{32}P turnover, a high rate of *de novo* phosphatidic acid biosynthesis occurs *in vivo* in the nervous system. It is conceivable that phosphatidic acid has a functional role of its own in addition to serving as a key intermediate in glycerolipid synthesis. Recent reviews dealing with phospholipid metabolism have described several aspects of phosphatidic acid synthesis and degradation in nonneural tissues[3,5–8,20–25] and in the nervous system[26] (A. A. Abdel-Latif, this volume). A few overviews have been published on some aspects of phosphatidic acid metabolism in the nervous system[27–29] and in nonneural tissues.[30]

2. PROPERTIES, COMPOSITION, AND DISTRIBUTION OF PHOSPHATIDIC ACID IN ANIMAL TISSUES

Phosphatidic acid represents approximately 1.8% of the total lipid P in the rat brain.[27,31–33] At the subcellular level, a nonuniform distribution of this phospholipid has been observed in both brain[34,35] and retina.[36,37] Synaptosomal membranes contain 0.2 mol% of total lipids not including glycolipids.[38] Nonneural membranes, such as dog and sheep[39] erythrocytes, have been shown to display proportions of phosphatidic acid similar to those of synaptosomal membranes. However, lymphocytes[40] and pancreatic zymogen granules[41] contain about 2.0 and 3.8 mol%, respectively, total lipids.

2.1. Properties

The membrane organization, phase behavior, and other physical chemical properties of phosphatidic acid in the central nervous system are not known. This is also the case for other lipids. However, recently several studies have used model membranes to gather information on these properties of phosphatidic acid.[20] The growing interest in this subject has arisen mainly from investigations suggesting that phosphatidic acid may function as the calcium ionophore in the presynaptic membrane. This is discussed in Section 7 of this chapter.

At physiological pH, phosphatidic acid is anionic and carries a net negative charge in the polar moiety. Hence, this lipid is more properly "phosphatidate." The charged moiety allows phosphatidic acid to interact with other lipids and proteins and plays an important role in the conformational changes of artificial lipidic membranes. Phase separation leading to clustering and domain formation occurs in lipid membranes in addition to transition from a crystalline to a highly fluid liquid crystalline phase. Phosphatidic acid–phosphatidylcholine mixtures display phase separation when Ca^{2+} is added.[42] Phosphatidic acid conforms to a different structure after lateral lipid segregation.[43,44] The resulting structure may form electronegative domains or inverted micelles which, in turn, may be translocated to the other side of a bilayer.[42–45] Moreover, the affinity of phosphatidic acid for Ca^{2+} is much greater than for Na^+ or K^+ in an aqueous medium.[46] Basic proteins also promote the formation of inverted micelles of phosphatidic acid.[19,42]

When phospholipids were compared, phosphatidic acid and cardiolipin were found to be the most active in translocating Ca^{2+} across an organic phase in a Pressman cell. The activities were similar to those obtained with ionophore X537A.[47] Calcium salts of phosphatidic acid separate from monolayers, indicating the possibility of ionophoretic properties.[48]

Phosphatidic acid, as well as other acidic phospholipids, has high binding affinities with several substances of neurochemical interest. 5-Hydroxytryptamine in isobutanol displays a high affinity for acidic lipids.[49] Much lower values were found, however, when the same measurements were performed in an aqueous salt medium.[50]

Barbiturates and phenytoin enhance the binding of $^{45}Ca^{2+}$ to phosphatidic acid and other acidic lipids in organic solvent partition systems. Ethanol, however, does not modify $^{45}Ca^{2+}$ binding to these lipids. It appears that as a result of these physical chemical properties, phosphatidic acid may play a role not only as an ionophore but also as a regulator of divalent cation concentration in excitable membranes.[51]

2.2. Fatty Acid Composition

In whole rat brain, approximately 70% of the acyl chains of phosphatidic acid are composed of oleate, stearate, and palmitate.[31] Arachidonate, linolenate, and docosahexaenoate represent about 7.7%, 3.7%, and 1.2%, respectively.[31] During early postnatal development of the mouse brain, the level of arachidonate observed in phosphatidic acid was approximately 20% greater than later values.[27,34] Similar proportions of palmitate were observed in guinea pig cerebral hemispheres[32] and rat brain.[31] However, much higher stearate and lower oleate and arachidonate levels were seen in these animals. Docosahexaenoate concentration was 8.8 mol% in guinea pig cerebral hemispheres.[37] Much higher proportions of docosahexaenoate were observed in the phosphatidic acid from retinal microsomes[27,36,52] and rod outer segments.[27,29]

2.3. Distribution in Nervous Tissue

Information on the pool size, fatty acid composition, and metabolism of individual lipids is essential for an understanding of their physiological significance. This is true of all lipids whether one is interested in different cell types, subcellular fractions, neuroanatomic regions, or developmental profiles. Unfortunately, in the case of phosphatidic acid, scant information is available on these topics.

Brain tissue contains larger amounts of phosphatidic acid than retina,[28] and toad retina contains more phosphatidic acid than bovine retina.[27] The diglyceride pool is also larger in toad retina than in bovine retina.[53-55] The amount of phosphatidic acid increases during postnatal development.[27,34] In mouse brain, a severalfold enrichment in oleate, palmitate, and stearate, even higher than the rise in polyenoic acyl chains, takes place between birth and adulthood.[27,34]

Only trace amounts of eicosamonoate (20:1) are present in phosphatidic acid in newborn mammalian brain. This fatty acid appears in brain phosphatidic acid during myelination.[27,34] Retinal phosphatidic acid also contains trace amounts of eicosamonoate, which, however, is absent from liver phosphatidic acid.[27] Also of interest is the fact that oleic acid increases postnatally in brain phosphatidic acid.[27,34]

Large variations in energy metabolism, particularly in glucose uptake, have been observed in different brain regions.[56] Regional differences were also seen in the size of the phosphatidic acid pool in rat brain.[27] However, it is not known how this is related to the glucose carbon flow. The highest concentration of phosphatidic acid was seen in the brainstem, and the lowest in the hypothalamus.[27]

The largest amounts of brain phosphatidic acid have been recovered from microsomal preparations. However, other subcellular fractions contain varying amounts of this lipid.[34,35,57] Myelin subfractionation showed that the highest concentrations of phosphatidic acid, diphosphoinositol, and triphosphoinositol were present in the light myelin fraction.[35] Moreover, it was suggested that these lipids may be involved in the tightly packed central lamellae of the myelin sheath.[35]

A nonuniform distribution of phosphatidic acid was also observed in the retinal subcellular fraction.[36,37] A relatively larger amount of the lipid was found in rod outer segments, and a small pool was recovered from 133,000 × *g* postmicrosomal supernatant.[37]

Phosphatidic acid was distributed unevenly among neuronal perikaryonal subfractions.[57] The largest quantities were found in fractions containing large amounts of plasma membranes, lysosomes, and microsomes; the least amounts were found in fractions containing nuclei and mitochondria.[57]

3. PHOSPHATIDIC ACID BIOSYNTHESIS

Phosphatidic acid is the common precursor in the biosynthesis of phospholipids and neutral glycerides in the nervous tissue as well as in other tissues. During the biosynthesis of glycerolipids, phosphatidic acid is the first diacylated compound to evolve.

The glycerol backbone of phosphatidic acid in the neural tissue can be derived from (1) dihydroxyacetone phosphate, (2) glycerol, or (3) the hydrolysis of phospholipids synthesized in other tissues and transferred to the neural tissue.

3.1. De Novo Synthesis from Dihydroxyacetone Phosphate or Glycerol

Dihydroxyacetone phosphate, formed from glucose during glycolysis, yields glycerol-3-phosphate through a reaction catalyzed by glycerol-3-phosphate dehydrogenase (NAD$^+$) (E.C. 1.1.1.8):

Dihydroxyacetone phosphate + NADH → *sn*-glycerol-3-phosphate + NAD

[1]

In vivo,[31,58–62] *in vitro*,[63–68] and enzymatic[69,70] studies have indicated that neural tissue can utilize glycerol to produce glycerol-3-phosphate by means of glycerol kinase. It is not known, however, under what conditions the neural tissue uses glycerol for lipid synthesis or how much is used. Lipolysis, for instance, enhances the availability of free glycerol in the bloodstream,[71] and the mammalian brain can use glycerol as a metabolic substrate.[72]

Brain lipids can be readily formed from glycerol.[31,58,59,68,73] Some time after *in vivo* injection of [^{14}C]glycerol, increasing amounts of ^{14}C were found in the acyl moiety of brain lipids.[58,74] Eighteen hours after intracranial injection of [2-^{3}H]glycerol or [^{14}C]glycerol, similar labeling profiles were observed in brain lipids. However, [2-^{3}H]glycerol is a much more efficient and specific precursor of brain glycerolipids, since tritium at the 2 position is lost to form cellular ^{3}H$_2$O from any [2-^{3}H]glycerol channeled to dihydroxyacetone phosphate.[58]

The retina is a useful model of the central nervous system; when retinal tissue is incubated *in vitro* with radioactive glycerol, rapid uptake of the precursor and labeling of retinal lipids are seen.[68] Since labeled glycerol appears very rapidly in retinal phosphatidic acid, it has been suggested that the precursor is taken into the retina and readily phosphorylated and subjected to two acylation steps.[63,65,67,68] Alternate routes for glycerol utilization in nonneural tissues involve glyceraldehyde dehydrogenase and pyrophosphate:glycerol phosphotransferase, whose K_ms are 0.5 M and 3 M, respectively.[75,76] The K_m of glycerol kinase, on the other hand, varies between 0.035[77] and 0.003 mM.[78] Hence, at low glycerol concentrations, the preferred pathway may be through glycerol kinase. In retinal studies, 95 μM of labeled glycerol was used[68]; thus, essentially all the precursor present in phosphatidic acid may have been followed in reactions, at least during early incubation. In several nonneural tissues, glycerol is actively used for *de novo* synthesis of phosphatidic acid and other lipids.[79–84]

Glycerol kinase (ATP:glycerol-3-phosphotransferase, E.C. 2.7.1.30) catalyzes the phosphorylation of glycerol in rat brain[85]:

$$\text{Glycerol} + \text{ATP} \rightarrow \textit{sn}\text{-glycerol-3-phosphate} + \text{ADP} \qquad [2]$$

This enzyme seemed to be confined primarily to the mitochondrial fractions and displayed much less activity in the 100,000 × g supernatant. The K_m for glycerol was 70 μM at pH 7 for the mitochondrial enzyme. Enzyme activity was also found in brain cytosol, which phosphorylated dihydroxyacetone at much faster rates than glycerol.[85] The acylation of *sn*-glycerol-3-phosphate has been shown to result in phosphatidic acid formation in the brain[86–90]:

$$\text{Glycerol-3-phosphate} + \text{acyl-CoA} \rightarrow$$
$$\text{monoacylglycerol-3-phosphate} + \text{CoA} \quad [3]$$

$$\text{Monoacylglycerol-3-phosphate} + \text{acyl-CoA} \rightarrow$$
$$\text{diacylglycerol-3-phosphate (phosphatidic acid} + \text{CoA)} \quad [4]$$

Radioactive glucose has also been studied as a precursor of the glycerol backbone in brain lipids,[32,60,91,92] and differences in labeling were found in the phospholipids of different neuroanatomic regions.[92] Although glucose is a more physiological precursor than glycerol, recycling and labeling of moieties other than the glycerol backbone may occur.

The acylation of glycerol-3-phosphate (reactions 3 and 4) has been studied primarily in nonneural tissues.[93–97] Most of the fatty acids being introduced are long-chain saturated and lower unsaturated.[98] Polyunsaturated fatty acids are thought to be acylated by a retailoring mechanism after the phospholipids are formed (see Section 5.1). There are different acyltransferases located in mitochondria and endoplasmic reticulum.[99–122] These acyltransferases catalyze the introduction of fatty acids in positions 1 and 2. The order in which these enzymes acylate glycerol-3-phosphate and, under certain conditions, form 2-acyl-glycerol-3-phosphate and 1-acyl-glycerol-3-phosphate is not well understood. Kinetic studies designed to determine which molecular species of phospholipids arise from the acylation of glycerol-3-phosphate have proven useful.[109] Also, isoenzymes of glycerol phosphate acyltransferase have been found.[123]

In the nervous system, labeling of phosphatidic acid by means of radioactive fatty acids has shown that acylation is relatively active compared to other phospholipids.[124–126] The metabolism and composition of molecular species of brain phosphatidic acid and related metabolites have also been studied to shed some light on the dynamics of the acyl groups in this lipid.[32–36,127–152] In some *in vitro* studies, brain microsomes were used to follow the acylation of glycerol-3-phosphate.[131–186] Arachidonoyl- and docosahexanoyl-CoA acylate 1-alkyl-*sn*-glycero-3-phosphate much more efficiently than 1-acyl-*sn*-glycero-3-phosphate.[132]

3.2. Alternate Pathway via 1-Acyldihydroxyacetone-3-Phosphate

An alternate pathway for phosphatidic acid formation in liver and other tissues, including brain, involves the acylation of dihydroxyacetone-3-phosphate.[82,133] Acylation of dihydroxyacetone phosphate (DHAP) is catalyzed by an acyl-CoA:DHAP acyltransferase (E.C. 2.3.1.42)[133,134]:

Dihydroxyacetone phosphate + acyl-CoA →
$$\text{acyldihydroxyacetone phosphate} + \text{CoA} \quad [5]$$

Enzymatic reduction by NADPH through acyldihydroxyacetone phosphate:NADP oxidoreductase (E.C. 1.1.1.101)[135,136] results in formation of 1-acyl-*sn*-glycerol-3-phosphate (lysophosphatidic acid):

Acyldihydroxyacetone phosphate + $NADPH_2$ →
$$\text{glycerolphosphate} + \text{NADP} \quad [6]$$

Lysophosphatidic acid is subsequently acylated by acyl-CoA:1-acylglycerol-3-phosphate acyl transferase[132] as shown in reaction 4.

The enzymes of this alternate pathway for phosphatidic acid biosynthesis have been localized mainly in the microsomal fraction of brain tissue from 12- to 14-day-old rats.[69]

Acyl-CoA: 1-acyl-glycerol-3-phosphate acyl transferase is very specific for NADPH and displays an optimal pH between 5 and 9, with greatest activity at pH 5.4. The K_m for dihydroxyacetone phosphate is 100 μM.[69] It has been suggested that this particular enzyme differs from the one that catalyzes reaction 3 because acyl-CoA:*sn*-glycerol-3-phosphate acyl transferase shows an optimal pH of 7.6 and because sodium cholate differentially affects the enzymes and base in substrate competition studies. There are similarities between liver and brain acyl-CoA:DHAP acyl transferase[69,133,137]; however, the latter is mainly microsomal and the former largely mitochondrial in location. Since there are large amounts of the liver enzyme in the peroxisomal fraction,[138] the possibility has been raised that catalase-containing microperoxisomes may contain the brain enzyme. The acylation of dihydroxyacetone phosphate was preferentially active with palmitoyl CoA, although some activity toward oleoyl CoA and linoleoyl CoA was reported.[69] However, no acylation has been found to occur in the presence of either arachidonoyl-CoA, docosatetraenoyl-CoA, or docosahexaenoyl-CoA[132]:

Acyldihydroxyacetone phosphate + fatty aldehyde →
$$\text{alkyldihydroxyacetone phosphate + fatty acid} \quad [7]$$

Alkyldihydroxyacetone phosphate + $NADPH_2$ →
$$\text{alkylglycerol phosphate + NADP} \quad [8]$$

Alkylglycerol phosphate + acyl CoA →
$$\text{1-alkyl-2-acyl-glycerol phosphate + CoA} \quad [9]$$

Estimation of the rate of this pathway relative to that of glycerol-3-phosphate has been made in several tissues by means of D-[U-^{14}C,3-^{3}H]glucose. The ^{3}H:^{14}C ratio at position 2 of the glycerol backbone of lipids indicates the relative activity of the NADPH-specific acyldihydroxyacetone phosphate pathway. Significant lipid formation occurs by this route.[139–141] In microsomes from liver and other tissue, the acyltransferases that are active toward glycerol-3-phosphate and dihydroxyacetone phosphate have been shown to be dual catalytic functions of a single enzyme.[142,143]

3.3. Monoacylglycerol Pathway

After intracerebral injection of glycerol, the specific activity of monoacylglycerol is much higher than that of diglycerides and phosphatidic acid.[62] Thus, it is possible that the following reaction may engage monoacylglycerol as a precursor of monoacylglycerol phosphate and then, through reaction 4, lead to phosphatidic acid synthesis[144]:

Monoacylglycerol + ATP →
$$\text{monoacylglycerol phosphate + ADP} \qquad [10]$$

Active synthesis of phosphatidic acid through monoglycerides has been observed in nonneural tissues,[145] mainly in the intestine.[146,147] In the brain, monoacylglycerol kinase may give rise to the formation of lysophosphatidic acid,[148] after which phosphatidic acid may result from a subsequent acylation.

3.4. Diacylglycerol Kinase and Diacylglycerol as a Precursor for Phosphatidic Acid Synthesis

The phosphorylation of diacylglycerol in nervous tissue was first demonstrated by Hokin and Hokin[149] and Strickland[150] and explored further by others[151]:

$$\text{Diacylglycerol} + \text{ATP} \rightarrow \text{phosphatidic acid} + \text{ATP} \qquad [11]$$

Diacylglycerol, the precursor of phosphatidic acid in this reaction, is one of the most active centers of brain lipid metabolism.[64,73,152–157] Several molecular species may make up heterogenous pools in different neuronal membranes. Diacylglycerol metabolism is highly active in both brain[73,152,154,155] and retina.[29,61,64] Ischemia[154] and electroshock[155] trigger the production of brain diglycerides enriched in arachidonate and stearate.[155] In the toad retina, diacylglycerols were found to contain large quantities of docosahexaenoate.[53,54] The exact role of diacylglycerol kinase in the formation of phosphatidic acid in the nervous system and its quantitative importance as an alternate pathway have not been assessed.

It has been suggested that diacylglycerol kinase is an integral part of the cycle of the degradation and resynthesis of phosphatidylinositol (A. A. Abdel-Latif, this volume). Several drugs, hormones, neurohormones, and other stimuli enhance the metabolism of phosphatidylinositol in brain and other tissues. Diacylglycerol kinase, however, is not the site of cholinergic stimulation of the cycle.[153] A phosphatidic acid pool has been shown to give rise to CDP-DG:

Phosphatidic acid + cytidine triphosphate →
      CMP-phosphatidic acid (CDP-diglyceride) + pyrophosphate   [12]

Subsequently, phosphatidylinositol is formed (A. A. Abdel-Latif, this volume). The breakdown of this lipid may be the site of stimulation, resulting in diglyceride formation, which, in turn, completes the cycle with the resynthesis of phosphatidic acid through diglyceride kinase. Since brain phosphatidylinositol contains large amounts of the 1-stearoyl,2-arachidonoyl molecular species,[31] one would expect 1-stearoyl,2-arachidonoyl glycerol to be an efficient substrate for diglyceride kinase as part of the cycle. However, diglycerides with 18:1 and 18:3 were more effective substrates for diacylglycerol kinase than the tetraenoic molecular species.[148]

In brain, diacylglycerol can also be produced by the reversal of cytidine diphosphate choline and CDP-ethanolamine diacylglycerol phosphotransferases.[158]

3.5. Phospholipase D

Phosphatidic acid can be produced by enzymatic removal of choline or ethanolamine from phosphatidylcholine or phosphatidylethanolamine using phospholipase D:

$$\text{Phosphatidylcholine} \rightarrow \text{phosphatidic acid} + \text{choline} \qquad [13]$$

Only recently has this enzyme been described in mammalian tissues[159]; the greatest levels of activity were seen in brain and lung. Phospholipase D is membrane bound and exhibits an absolute requirement for detergent. Inhibition of the enzyme occurs above the optimal 6 mM concentration of taurodeoxycholate. The brain and lung enzyme seems to be microsomal and shows an optimum pH of 6.5.[149] The relative proportion of phosphatidic acid formed by phospholipase D in neural tissue is not known. A lysophospholipase D active on ether-linked lipids was also shown to be present in brain microsomes.[160] The physiological roles of these enzymes are not known.

4. PHOSPHATIDIC ACID: FURTHER METABOLIC CONVERSIONS

4.1. Phosphatidic Acid Phosphohydrolase

The primary step in the further metabolic conversion of phosphatidic acid involves phosphatidic acid phosphohydrolase:

$$\text{Phosphatidic acid} + H_2O \rightarrow \text{diacylglycerol} + \text{inorganic phosphate} \quad [14]$$

Most of our knowledge of this enzyme has been obtained in studies of liver tissue. In general, mammalian cells contain phosphatidate phosphohydrolase in both particulate and soluble fractions.[161–175] This enzymatic activity varies according to the form of substrate employed. When aqueous dispersions of phosphatidic acid are examined, the bulk of the activity is found in particulate fractions. However, when membrane-bound substrate is used in the assay, a large part of the activity is soluble.[163] The highest specific activity in the liver was found in the lysosomal fraction. Plasma membranes, microsomes, and mitochondria also displayed activity.[30] It is not possible to integrate the widely distributed enzyme with the metabolic pathways involving phosphatidic acid. The soluble enzyme, acting on the endoplasmic reticulum and the membrane-bound microsomal activity, may be part of the biosynthetic pathway from phosphatidic acid to diacylglycerol and subsequently to zwitterionic lipids. The enzyme present in plasma membrane may be concerned with the degradation of a phosphatidic acid pool involved in cycles other than *de novo* biosynthesis. Lysosomal phosphatidate phosphohydrolase from the liver may be engaged in the breakdown of phosphatidic acid as part of the phagolysosomal system.[30]

In the central nervous system, phosphatidate phosphohydrolase was concentrated in the plasma membranes; the samples were derived from synaptosomal fractions containing large amounts of synaptic thickenings.[176] In the same study, phosphatidic acid cytidyltransferase (E.C. 2.7.7.41) was found in synaptosomal mitochondria. These observations point to a heterogenous distribution of the enzymes that metabolize phosphatidic acid in subsynaptosomal structures.

In synaptosomes, acetylcholine-stimulated loss of ^{32}P from phosphatidate was explained by activation of phosphatidate phosphohydrolase.[177] At the same time, *de novo* synthesis was excluded, inasmuch as the labeling of phosphatidic acid with radioactive glucose or glycerol was not altered in the presence of acetylcholine.[177]

Some soluble factors, such as soluble proteins and unsaturated fatty acids, are required for phosphohydrolase activity. However, Ca^{2+} binds to the substrate and exerts an inhibitory effect on both soluble and membrane-bound enzymes.[161,172,178] Fluoride is also an inhibitor of phosphatidate phosphohydrolase.[161,169] Although there is no absolute requirement for Mg^{2+}, soluble phosphatidate phosphohydrolase is stimulated by this cation.[172]

It is not clear if phosphatidate phosphohydrolase exhibits selectivity on the acyl group composition of its substrate. *In vivo* experiments in liver argue against this possibility.[84]

Phosphatidate phosphohydrolase appears to act as a regulator in the synthesis of triglycerides in liver and adipose tissue.[30,120,162,165,166,171,172,179,180] The enzyme reaction rate is the lowest in the neobiosynthetic pathway of glycerolipids, and when triglyceride formation is enhanced by diets rich in fructose, there is also an increase in phosphatidate phosphohydrolase.[181] A similar increase in enzyme activity is seen when liver triglyceride increases, e.g., after partial hepatectomy or in the case of obese rats.[181]

The specific activity of phosphatidate phosphohydrolase in rat and rabbit lung[162] increases during perinatal life and after glucocorticoid treatment. Hence, it has been suggested that this hormonal effect is an underlying factor in enzyme induction and the further phosphatidylcholine synthesis of the surfactant membrane. The Mg^{2+}-dependent activity in adipocytes is inhibited by norepinephrine[166] and stimulated by insulin treatment.[165] It has been suggested that these modifications play an important role in the regulation of triglyceride synthesis.

In pineal gland,[182,183] retina,[27,36,37,64,65,68] brain,[184] lymphocytes,[185] iris muscle,[186,187] and liver,[188] amphiphilic cationic drugs tend to redirect the *de novo* biosynthetic pathway of glycerolipids toward phosphatidylinositol. This redirection is caused largely by the inhibitory effect of phosphatidate phosphohydrolase. This enzyme inhibits the flux of labeled precursors such as [^{14}C]- or [2-^{3}H]glycerol toward phosphatidylcholine, phosphatidylethanolamine, and triacylglycerol. Propranolol, phentolamine, chloropromazine, fenfluoramine and its derivatives, and other drugs having an ionizable amine group and hydrophobic moiety exert an inhibitory effect on phosphatidate phosphohydrolase.[189] It has been suggested that these drugs may remove the phosphatidate by means of their positively charged group, making the substrate unavailable for the enzyme.[188] Amphiphilic cations do interact with acidic lipids.[190] It has also been suggested that the amphiphilic cations are intercalated in the endoplasmic reticulum membrane and inhibit phosphatidate phosphohydrolase. However, these drugs seem to elicit other changes in lipid metabolism, not necessarily only a selective effect on phosphatidate phosphohydrolase.[64,65,191,192]

4.2. Conversion into CDP-Diacylglycerol

An important branch of the phospholipid pathway arises when phosphatidic acid is converted into CDP-diacylglycerol. The following acidic phospholipids are derived from this liponucleotide: phosphatidylinositol, diphosphatidylinositol, triphosphatidylinositol, phosphatidylglycerol, and diphosphatidylglycerol (cardiolipin). Other chapters in this volume are devoted to the description of the metabolism of these ,lipids. The enzyme CTP:phosphatidic acid cytidyltransferase catalyzes reaction 12. This enzyme was found mainly in brain microsomes, although mitochondrial fractions also showed some activity. CDP-diacylglycerol is the coenzyme that transfers the phosphatidic acid to *myo*-inositol to yield phosphatidylinositol.[127–129,193] Both phosphatidylinositol and CDP-diacylglycerol contain large quantities of 1-stearoyl-2-arachidonoyl-*sn*-glycerol backbone. However, the fatty acid composition of phosphatidic acid is unlike that of either CDP-diacylglycerol or phosphatidylinositol.[127] After the liponucleotide is formed, active acyl transfer reactions occur that are thought to be responsible for these differences.

Synthesis of CDP-diacylglycerol from phosphatidic acid and CMP has also been reported.[194] It has also been shown that 2'-deoxycytidine diphosphate diglyceride is synthesized in neuronal nuclei.[195–197]

4.3. Pyrophosphatidic Acid and the Metabolic Relationship of Phosphatidic Acid to Phosphatidylserine Biosynthesis

In neural and other tissues, phosphatidylserine biosynthesis takes place through a Ca^{2+}-dependent, energy-independent base-exchange reaction.[198] However, data gathered from labeling experiments in which the retina was used as a neural tissue model suggest that phosphatidic acid may be involved in an alternate route of phosphatidylserine synthesis.[64] Similar results were obtained in studies using other tissues,[199] although in rat heart histological preparations, the base-exchange reaction still appeared to be the pathway.[200] Further drug-induced modifications in the *de novo* biosynthesis of phosphatidylserine provided additional support for the possible involvement of phosphatidic acid as a precursor.[65,191,192]

Phosphatidic acid stimulates the incorporation of L-serine into phosphatidylserine through an energy-linked route in rat brain microsomes.[201] The following reactions describe a speculative pathway for the synthesis of phosphatidylserine from phosphatidic acid[201]:

$$2 \text{ Phosphatidic acid} \rightarrow \text{pyrophosphatidic acid} \qquad [15]$$

$$\text{Pyrophosphatidic acid} + \text{L-serine} \rightarrow \text{phosphatidylserine} \quad [16]$$

Reaction 16 requires ATP and Ni^{2+}.[201] It has been proposed that pyrophosphatidic acid [P,P'-bis(1,2-diacyl-*sn*-glycero-3-)pyrophosphate] is the immediate precursor of phosphatidylserine synthesis,[201] although it has not been detected in tissues, probably because of its high turnover rate.

4.4. Bis-Phosphatidic Acid and Related Compounds

Several tissues have been found to contain trace amounts of two-, three-, and four-fatty-acyl-chain phospholipids, frequently referred to, respectively, as lysobisphosphatidic acid, semilysobisphosphatidic acid, and bis-phosphatidic acid. These phospholipids are considered to be metabolically unrelated to phosphatidic acid, since their backbone is *sn*-1-glycerophospho-1'-*sn*-glycerol.[202] Lysobisphosphatidic acid and bis-monoacylglycerophosphate have been found in liver lysosomes and appear to be the products of phosphatidylglycerol and diphosphatidylglycerol metabolism.[203,204] Bis-phosphatidic acid plasmalogen has been found in the brain,[205] and in patients with storage diseases, lysobisphosphatidic acid has been shown to increase in the liver and brain.[206]

5. TURNOVER AND BREAKDOWN OF PHOSPHATIDIC ACID

5.1. Acyl Chains and Lysophosphatidic Acid

From the time of the classical bilayer model of cell membrane structure to the development of the fluid mosaic concept, the phospholipid components of cell membranes have been depicted as a single structure possessing a polar head and hydrophobic tails. Proper membrane fluidity is maintained by these lipids, and membrane proteins display their functions in a hydrophobic environment. Hence, no explanation or alternate role has been given to explain the wide variety of lipid classes and molecular species in such membranes. The heterogeneity of acyl chains with varying numbers of carbon atoms and double bonds has been studied in neural tissue membranes; however, the pathways by which these fatty acids are introduced and replaced are not well understood.

Neural tissues, particularly synaptic membranes[207-229] and the highly specialized photoreceptor membranes in the retina,[210-212] are endowed with relatively large quantities of a wide variety of phospholipids, all strikingly rich in polyenoic fatty acyl chains. Docosahexaenoic [22:6(n-3)] and arachidonic acids [20:4(n-6)] are fatty acids of this type that are derived from dietary linolenic and linoleic acids. It is thought that deacylation–reacylation reactions acting on phospholipids may alter the fatty acid composition by means of phospholipases A_1-A_2 and acyltransferases. In addition, the acylations that take place during the *de novo* biosynthesis of phosphatidic acid may play a role in modifying the fatty acid composition of membrane lipids.

The fatty acid composition of phospholipids in various membranes is very heterogeneous, and in neural tissue, within a given class (e.g., phosphatidylcholine), a wide variety of molecular species is found. In these lipids, most of the polyunsaturated acyl chains are esterified to the second carbon hydroxyl group, and the saturated chains to the first carbon glycerol. Recently it has been shown that small amounts of didocosahexaenoyl species are present in retina.[130,213] The pathways and control mechanisms of fatty acid acylation are not well understood. However, two major routes are thought to be responsible for acylation in neural and other tissue.

During the *de novo* synthesis of phosphatidic acid, two subsequent reactions take place (see reactions 3 and 4). The other pathway is known as the Lands cycle and is composed of deacylation and reacylation reactions in each phospholipid. The deacylation is catalyzed by either phospholipase A_1 or A_2. For instance, in the case of phosphatidylcholine, the reactions are:

Phosphatidylcholine + H_2O → free fatty acid
+ 1- or 2-monoacylglycerophosphatidylcholine (lysophospholipids)   [17]

A 1-monoacyl derivative is produced, in part, by a phospholipase A_2 enzyme, and a 2-monoacyl derivative is formed as a product of phospholipase A_1. Fatty acids are acylated by a two-step enzymatic system:

Free fatty acid + CoA → Acyl-CoA

Acyl-CoA + monoacylphosphatidylcholine →
diacylphosphatidylcholine + CoA   [18]

Acylation steps during phosphatidic acid synthesis produce saturated and unsaturated fatty acids.[99,121] Polyunsaturated fatty acids are thought to be acylated primarily by reaction 18.[3,24,214–216] Recently, it has been shown that the phosphatidic acid of retinal microsomes contains large quantities of docosahexaenoate[36,52] and that rapid labeling with docosapentaenoic acid and its metabolites takes place in retinal phosphatidic acid.[217] Taking into consideration the high rate of synthesis of this phospholipid, it has been suggested that a significant amount of docosahexaenoate may be acylated during the *de novo* biosynthesis of phosphatidic acid[36]:

Phosphatidic acid →
monoacylglycerol phosphate + free fatty acid   [19]

An active acylation–deacylation of arachidonate in phosphatidic acid has been seen in horse neutrophils and has been cited as a possible source of the arachidonic acid necessary for prostaglandin synthesis.[218]

Rapid phosphorylation of mono- and diglycerides was thought to be the underlying reason for the appearance of [32P]phosphatidic acid and [32P]lysophosphatidic acid in human platelets during aggregation induced by thrombin or phospholipase C treatment.[219]

5.2. Phosphate Moiety

An increase in ^{32}P labeling of phosphatidic acid and phosphatidylinositol occurs when muscarinic cholinergic and α-adrenergic receptors are activated. In addition to this lipid effect, activation of these cell surface receptors triggers an increase in cyclic GMP and assists Ca^{2+} in the mediation of physiological effects. Abdel-Latif reviews the effect of phosphoinositides in detail elsewhere (Chapter 4). Therefore, this discussion covers only those effects that relate directly to the turnover of phosphatidic acid.

Muscarinic cholinergic stimulation has been shown to increase ^{32}P labeling of phosphatidic acid in whole brain,[220] cerebral cortex slices,[221,222] and cerebral cortex synaptosomes.[223-225] Similar changes were also reported when adrenergic receptors were stimulated in whole brain[226] and brain slices.[91,227] Nonneural tissues—iris smooth muscle[228,229] adrenal medulla,[230-232] and avian salt gland[233]—have also shown increased ^{32}P turnover in phosphatidic acid in response to muscarinic cholinergic activation.

Receptor activation leads to an increased turnover of the polar moiety of several phospholipids, mainly phosphatidic acid and phosphatidylinositol; however, changes have also been reported in phosphatidylcholine and polyphosphoinositides (A. A. Abdel-Latif, this volume).

6. REGULATION OF PHOSPHATIDIC ACID METABOLISM

The ultimate goal of biochemists is to isolate and characterize the enzymes involved in a given pathway and to understand the regulatory mechanisms at the molecular level. The classic approaches to lipid metabolism enzymes have not been successful for several reasons, including the water insolubility of the substrate and the membrane-bound nature of many of the enzymes. Therefore, most of our knowledge of the regulation of phosphatidic acid metabolism comes from measuring enzyme activities in unpurified extracts, labeling with fatty acids and ^{32}P, and tracing the flux of radioactivity in the glycerol backbone. The retina is an integral part of the central nervous system and displays an exceptionally high level of de novo phosphatidic acid biosynthesis when radioactive glycerol is used as an in vivo precursor.[63,68]

Cytidine nucleotides alter the synthesis of brain phosphatidic acid[89,234]: CMP and CTP inhibit labeling with [^{14}C]glycerol-3-phosphate, whereas CDP-choline displays the opposite effect.[235] Hormones also influence the synthesis of phosphatidic acid. In sections of bovine thyroid, thyrotropin enhances the conversion of glycerophosphate to phosphatidic acid,[236] and ACTH, via cyclic AMP, stimulates the de novo biosynthesis of phosphatidic acid in adrenal sections and cells of the rat.[237] This activation is cycloheximide sensitive, Ca^{2+} dependent, and results in a rapid increase of adrenal phospholipids.[238]

Recently, other factors involved with the regulation of phosphatidic acid enzymes in several nonneural tissues have been reviewed.[30]

Phosphatidic acid is located at a branch of the glycerolipid biosynthetic pathway that has potential regulatory properties. The carbon flux can be directed by several conditions toward the acidic lipids or to the zwitterionic lipids. In fact, there is ample evidence indicating that the enzyme phosphatidate phosphohydrolase may act as a regulatory mechanism. Glucocorticoids stimulate this enzyme in the liver, increasing the formation of diglycerides and triglycerides.[180,238] This is particularly true when fatty acids are readily available.[239] Insulin and epinephrine are also involved in the synthesis of phosphatidic acid in fat cells.[240,241] Other factors that point to the regulatory role of this enzyme have been mentioned in Section 4.2.

7. INVOLVEMENT OF PHOSPHATIDIC ACID IN THE FUNCTIONING OF THE CENTRAL NERVOUS SYSTEM

Phosphatidic acid makes up a small fraction of the neural tissue phospholipids that display a high turnover rate in acyl groups and in the phosphate moiety. In addition, phosphatidic acid displays a high rate of *de novo* biosynthesis in neural tissue.[32,68] A net synthesis was observed in retinas incubated with cationic amphiphilic drugs for short periods.[192] The rate at which neural tissue synthesizes phosphatidic acid may be related to the needs of newly formed phospholipids of nerve cell membranes. Alternatively, an increase in the formation of phosphatidic acid may affect membrane-bound enzymes, other proteins, or ionophoretic properties activating or inhibiting certain physiological processes.

The functional significance of inducing a higher rate of phosphatidic acid and phosphatidylcholine ^{32}P labeling by muscarinic antagonists and α-adrenergic activation is not known. The functional implications of these metabolic modifications in terms of the "phosphoinositide effect" are discussed elsewhere (A. A. Abdel-Latif, this volume). Therefore, this section focuses on the functional significance of phosphatidic acid. However, in many instances, the proposed pathways involve both phosphatidic acid and phosphatidylinositol, and, therefore, considerable overlap may be seen.

Several roles have been suggested for phosphatidic acid in the transmembrane signaling of synaptic transmission. Some of these roles overlap, and many are still speculative.

Receptor activation of broken-cell preparations does not result in enhanced phospholipid turnover except in synaptosomes.[177] Hence, the exact subcellular site of this effect has been difficult to locate, and several hypotheses have placed it as either a presynaptic or a postsynaptic event.

Early proposals have linked phosphatidic acid and phosphatidylinositol changes with the deposition of proteins and catecholamines in (1) synaptic vesicles,[242] (2) a Na^+ pump of the salt gland,[8] and (3) the axonal membrane.[2]

In the cycle of phosphatidylinositol degradation and resynthesis, the activation of muscarinic cholinergic and α-adrenergic receptors involves phosphatidic acid as a metabolic intermediate. These phospholipid changes are envisaged as a reaction coupling the receptor activation to the calcium mobilization at the postsynapse.[5-7] When cytosolic calcium is released, physiological cellular responses are evoked.

Evidence that an increase in the turnover of phosphatidic acid and phosphatidylinositol occurs at a postsynaptic site was obtained earlier by means of sympathetic ganglia.[242,243] More recently, the α-receptor-mediated stimulation of phosphatidylinositol turnover in the pineal gland was also shown to be a postsynaptic event.[244] Moreover, a marked decrease in carbachol stimulation of phosphatidic acid and phosphatidylinositol labeling occurred in hippocampus nerve endings when ibotenic acid destroyed the nerve cells and muscarinic receptors.[245] These and other observations[225,246,247] suggest that the phospholipid effect is located in cholinoreceptive structures and not in cholinergic endings.

Phosphatidic acid has also been involved in presynaptic events. The action potential brings about membrane depolarization and, in turn, Ca^{2+} influx into the nerve ending. Consequently, neurotransmitter release takes place. Phosphatidic acid has been involved as a precursor in diglyceride formation. This neutral lipid may participate in membrane fusion of synaptic vesicles and presynaptic membranes during neurotransmitter exocytosis.[4,28,248]

Phosphatidic acid has been implicated as a Ca^{2+} ionophore, linking depolarization and neurotransmitter release. Through muscarinic receptors, acetylcholine may promote phosphatidic acid accumulation and, in turn, make the membrane permeable to Ca^{2+}. Exogenously added phosphatidic acid and ^{32}P labeling of phosphatidic acid correlate well with Ca^{2+} permeability and functioning in parotid cells,[17] smooth muscle,[18] and synaptosomes.[16] It has been suggested that phosphatidic acid changes regulate adenylate cyclase in fibroblasts, because phosphatidic acid inhibits prostaglandin PGE_1-stimulated cyclic AMP production.[249]

It has been suggested that deacylation of phosphatidic acid in platelets may result in production of free arachidonic acid and, later, formation of prostaglandins.[250,251] Thrombin, collagen, and divalent cation ionophores also promote increased ^{32}P turnover and changes in fatty acid composition in platelet phosphatidic acid and phosphatidylinositol; this precedes platelet aggregation and formation of eicosanoids via the cyclooxygenase and lipoxygenase pathways.[251–254]

Changes in membrane lipids have also been identified with specific functions such as the formation of diglycerides, the activation of protein kinase C,[255] and the methylation of phosphatidylethanolamine to phosphatidylcholine in membrane fluidity, β-adrenergic receptors, Ca^{2+} influx, and histamine release.[256] At the same time, this pathway has been linked to the deacylation of phospholipids through phospholipase A_2 and to arachidonic acid formation.[256] Brain free arachidonic acid and diglycerides containing large amounts of arachidonate and stearate are released at the onset of ischemia[257] and after electroconvulsive shock[257,258] or drug-induced convulsions.[156] These are the earliest changes known to occur in endogenous brain membrane lipids during seizures. It is likely that the diacylglycerols arise through phosphatidylinositol breakdown and not from phosphatidic acid, and exactly how phosphatidic acid is involved has not been clearly established. It is known that anoxia stimulates phosphatidic acid breakdown to diglycerides and a concurrent increase in free arachidonic acid in the newborn mouse brain.[27,259]

Lysophosphatidic acid also seems to exert potent effects on the arterial blood pressure of several animals species[260,261] and has also been involved as a Ca^{2+} ionophore.[263,264]

In addition to the enhanced ^{32}P turnover in phosphatidic acid discussed above, increases in stearate and arachidonate were found in mouse pancreas in response to acetylcholine stimulation.[265] Moreover, light[266] and divalent cations[267] enhance radioactive glycerol labeling in retinal phosphatidic acid, and drug-induced accumulation of phosphatidic acid has been observed in neural[192] and nonneural[268] tissues.

Phosphatidic acid synthesis can be detected within 2 s of platelet activation. This lipid modification is thought to result from phosphatidylinositol breakdown to diglycerides followed by the formation of phosphatidic acid by diglyceride kinase. Platelet aggregation may be triggered by increased intracellular Ca^{2+} brought about by phosphatidic acid acting on the flux of Ca^{2+} through the membrane. Hence, phosphatidic acid may act as the physiological mediator of ADP-induced Ca^{2+} movement.[260] Lysophosphatidic acid is also increased and displays aggregation and ultrastructural modifications similar to those seen when ADP is added and Ca^{2+} released.[262,263]

In summary, further elucidation of the relationships between receptor activation, drug action, and phosphatidic acid metabolism[269,270] should lead to the identification of specific pools of this lipid and their links to specific effects. To this end, enzyme studies and the use of antibodies may be instrumental in the documentation of the subcellular distribution of the different pools of phosphatidic acid and its molecular species in nervous tissue. Knowledge of the metabolism of molecular species of brain phosphatidic acid[271] has provided important information regarding metabolic interrelationships with other lipids.

In the 1980s, it is hoped that we will further our knowledge of the metabolism of phosphatidic acid in nervous tissue and will define more precisely how this acidic lipid is involved in the specific functions of excitable membranes and how its metabolism is coordinated among the phospholipids.

ACKNOWLEDGMENTS. This work was supported in part by a Research Manpower Award from Research to Prevent Blindness, Inc., New York City.

REFERENCES

1. Rossiter, R. J., and Strickland, K. P., 1970, *Handbook of Neurochemistry*, Volume 3 (A. Lajtha, ed.), Plenum Press, New York, pp. 467–489.
2. Hawthorne, J. N., and Kai, M., 1970, *Handbook of Neurochemistry*, Volume 3 (A. Lajtha, ed.), Plenum Press, New York, pp. 491–508.
3. Bell, R. M., and Coleman, R. A., 1980, *Annu. Rev. Biochem.* **49**:459–487.
4. Hawthorne, J. N., and Pickard, M. R., 1979, *J. Neurochem.* **32**:5–14.
5. Michell, R. H., 1975, *Biochim. Biophys. Acta* **415**:81.
6. Michell, R. H., 1979, *Companion to Biochemistry* (A. T. Bull, J. R. Lagnado, J. O. Thomas, and K. F. Tipton, eds.), Longmans, London, pp. 205–228.
7. Michell, R. H., 1980, *Cellular Receptors for Hormones and Neurotransmitters* (D. Schulster and A. Levitzki, eds.), Wiley, London, pp. 353–368.
8. Hokin-Neaverson, M., 1977, *Adv. Exp. Med. Biol.* **83**:429–446.

9. Hokin, M. R., and Hokin, L. E., 1953, *J. Biol. Chem.* **203:**967–977.
10. Hokin, L. E., and Hokin, M. R., 1958, *J. Biol. Chem.* **233:**818–821.
11. Broekman, M. J., Ward, J. W., and Marcus, A. J., 1980, *J. Clin. Invest.* **66:**275–283.
12. Cockcroft, S., and Gomperts, B. D., 1979, *Biochem. J.* **178:**681–687.
13. Cockcroft, S., Bennett, J. P., and Gomperts, B. D., 1980, *FEBS Lett.* **110:**115–118.
14. Cockcroft, S., and Gomperts, B. D., 1980, *Biochem. J.* **188:**789–798.
15. Egawa, K., Sacktor, B., and Takenawa, T., 1981, *Biochem. J.* **194:**129–136.
16. Harris, R. A., Schmidt, J., Hitzemann, B. A., and Hitzemann, R. J., 1981, *Science* **212:**1290–1291.
17. Putney, J. W., Weiss, S. J., Van De Walle. C. M., and Haddas, R. A., 1980, *Nature* **284:**345–347.
18. Salmon, D. M., and Honeyman, T. W., 1980, *Nature* **284:**344–345.
19. Boggs, J. M., 1980, *Can. J. Biochem.* **58:**755–769.
20. Bell, R. M., Ballas, L. M., and Coleman, R. A., 1981, *J. Lipid Res.* **22:**391–403.
21. McMurray, W. C., and Magee, W. L., 1972, *Annu. Rev. Biochem.* **41:**129–160.
22. Snyder, F., 1977, *Lipid Metabolism in Mammals,* Volumes 1, 2, Plenum Press, New York.
23. Thompson, G., 1973, *Form and Function of Phospholipids,* Elsevier, Amsterdam, p. 67.
24. van den Bosch, H., 1974, *Annu. Rev. Biochem.* **43:**243–277.
25. van Golde, L. M. G., and van den Bergh, S. G., 1977, *Lipid Metabolism in Mammals* (F. Snyder, ed.), Plenum Press, New York, pp. 1–33.
26. Wykle, R. L., and Brain, 1977, *Lipid Metabolism in Mammals* (F. Snyder, ed.), Plenum Press, New York, pp. 317–366.
27. Bazan, N. G., Aveldano de Caldironi, M. I., Giusto, N. M., and Rodriguez de Turco, E. B., 1982, *Prog. Lipid Res.* **20:**307–313.
28. Hawthorne, J. N., and Bleasdale, J. E., 1975, *Mol. Cell Biochem.* **8:**83–87.
29. Bazan, N. G., 1982, *Phospholipid Metabolism in the Nervous System* (L. A. Horrocks, G. B. Ansell and G. Porcellati, eds.), Raven Press, New York (in press).
30. Brindley, D. N., and Sturton, R. G., 1982, *Phospholipids* (J. N. Hawthorne and G. B. Ansell, eds.), Elsevier/North Holland Biomedical Press, Amsterdam (in press).
31. Baker, R. R., and Thompson, W., 1972, *Biochim. Biophys. Acta* **270:**489–503.
32. Luthra, M. G., and Sheltawy, A., 1976, *J. Neurochem.* **27:**1503–1511.
33. Friedel, R. O., and Schanberg, S. M., 1971, *J. Neurochem.* **18:**2191–2200.
34. Su, K. L., and Sun, G. Y., 1978, *J. Neurochem.* **31:**1043–1047.
35. Deshmukh, D. S., Kuizon, S., Bear, W. D., and Brockerhoff, H., 1980, *Lipids* **15:**14–18.
36. Bazan, N. G., and Giusto, N. W., 1980, *Control of Membrane Fluidity* (M. Kates and A. Kuksis, eds.), The Humana Press, Clifton, New Jersey, pp. 223–388.
37. Bazan, H. E. P., Careaga, M. M., and Bazan, N. G., 1981, *Biochim. Biophys. Acta* **666:**63–71.
38. Breckenridge, W. C., Gombos, G., and Morgan, I. G., 1972, *Biochim. Biophys. Acta* **266:**695–707.
39. Nelson, G. J., 1967, *Biochim. Biophys. Acta* **144:**221–232.
40. Gottfried, E. L., 1971, *J. Lipid Res.* **12:**531–537.
41. Meldolesi, J., Jamieson, J. D., and Palade, G. E., 1971, *J. Cell Biol.* **49:**130–149.
42. Galla, H. J., and Sackmann, E., 1975, *Biochim. Biophys. Acta* **401:**509–529.
43. Massari, S., and Pascolini, D., 1977, *Biochemistry* **16:**1189–1195.
44. Verkleij, A. J., Mombers, C., Leuniussen-Bijvelt, J., and Ververgaert, P. H. J. T., 1979, *Nature* **279:**162–163.
45. Krebecek, R., Gebhardt, C., Gruler, H., and Sackmann, E., 1979, *Biochim. Biophys. Acta* **554:**1–22.
46. Abramson, M. B., Katzman, R., Gregor, H., and Curci, R., 1966, *Biochemistry* **5:**2207–2213.
47. Tyson, C. A., Zande, H. V., and Green, D. E., 1976, *J. Biol. Chem.* **251:**1326–1332.
48. Jacobson, K., and Papahadjopoulos, D., 1975, *Biochemistry* **14:**152–161.
49. Johnson, D. A., Cho, T. M., and Loh, H. H., 1977, *J. Neurochem.* **29:**1101–1103.
50. Johnson, D. A., Merlone, S. C., Loh, H. H., and Ellman, G. L., 1978, *J. Neurochem.* **31:**713–717.
51. Chweh, A. Y., and Leslie, S. W., 1981, *J. Neurochem.* **36:**1865–1867.

52. Giusto, N. M., and Bazan, N. G. 1979, *Biochem. Biophys. Res. Commun.* **91:**791–794.
53. Aveldano, M. I., and Bazan, N. G., 1972, *Biochem. Biophys. Res. Commun.* **48:**689–694.
54. Aveldano, M. I., and Bazan, N. G., 1973, *Biochim. Biophys. Acta* **296:**1–9.
55. Wells, M. A., and Dittmer, J. C., 1967, *Biochemistry* **6:**3169–3175.
56. Sokoloff, L., Reivich, M., Kennedy, C., DeRosiers, M. H., Patlak, C. S., Pettigrew, K. D., Sakadura, O., and Shinohara, M., 1977, *J. Neurochem.* **28:**897–916.
57. Baker, R. R., 1979, *Brain Res.* **169:**65–81.
58. Benjamin, J. A., and McKhann, G. M., 1973, *J. Neurochem.* **20:**1111–1120.
59. O'Brien, J. F., and Geison, R. L., 1974, *J. Lipid Res.* **15:**44–49.
60. Brauning, C., and Gercken, G., 1976, *J. Neurochem.* **26:**1257–1261.
61. Careaga, M. M., and Bazan, H. E. P., 1981, *Neurochem. Res.* **6:**1169–1178.
62. Porcellati, G., and Binaglia, L., 1976, *Lipids* (R. Paoletti, G. Porcellati, and G. Jacini, eds.), Raven Press, New York, pp. 75–88.
63. Bazan, N. G., Aveldano, M. I., Pascual de Bazan, H. E., and Giusto, N. M., 1976, *Lipids* (R. Paoletti, G. Porcellati, and G. Jacini, eds.), Raven Press, New York, pp. 89–97.
64. Bazan, N. G., Ilincheta de Boschero, M. G., and Giusto, N. M., 1977, *Adv. Exp. Med. Biol.* **83:**377–388.
65. Bazan, N. G., Ilincheta de Boschero, M. G., Giusto, N. M., and Pascual de Bazan, H. E., 1976, *Adv. Exp. Med. Biol.* **72:**139–148.
66. Binaglia, L., Roberti, R., and Porcellati, G., 1977, *Biochem. Soc. Trans.* **5:**175–178.
67. Giusto, N. M., and Bazan, N. G., 1979, *Exp. Eye Res.* **29:**155–168.
68. Bazan, H. E. P., and Bazan, N. G., 1976, *J. Neurochem.* **27:**1051–1057.
69. Hajra, A. K., and Burke, C., 1978, *J. Neurochem.* **31:**125–134.
70. Tildon, J. T., Stevenson, J. H., and Ozand, P. T., 1976, *Biochem. J.* **157:**513–516.
71. Lin, E. C. C., 1977, *Annu. Rev. Biochem.* **46:**765–795.
72. Sloviter, H. A., Shimkin, P., and Suhara, K., 1966, *Nature* **210:**1334–1336.
73. Yavin, E., and Menkes, J. H., 1973, *J. Neurochem.* **21:**901–912.
74. Lapetina, E. G., Rodriguez de Lores Arnaiz, G., and DeRobertis, E., 1969, *Biochim. Biophys. Acta* **176:**643–646.
75. Stetten, M. R., and Rounbehler, D., 1968, *J. Biol. Chem.* **243:**1823–1829.
76. Himms-Hagen, J., 1968, *Can. J. Biochem.* **46:**1107–1112.
77. Grunnet, N., and Lundqvist, F., 1967, *Eur. J. Biochem.* **3:**78–82.
78. Robinson, J., and Newsholme, E. A., 1969, *Biochem. J.* **112:**455–502.
79. Sloviter, H. A., and Tanaka, S., 1967, *Biochim. Biophys. Acta* **137:**70–79.
80. Mims, L. C., and Kotas, R. V., 1973, *Biol. Neonate* **22:**436–443.
81. Hill, E. E., Husbands, D. R., and Lands, W. E. M., 1968, *J. Biol. Chem.* **243:**4440–4451.
82. Hajra, A. K., 1977, *Biochem. Soc. Trans.* **5:**34–36.
83. Clouet, E., Paris, R., and Clement, J., 1974, *Biochimie* **56:**145–152.
84. Akesson, B., Elovson, J., and Arvidson, G., 1970, *Biochim. Biophys. Acta* **210:**15–27.
85. Jenkins, B. T., and Hajra, A. K., 1976, *J. Neurochem.* **26:**377–385.
86. Kuwahara, S. S., 1972, *J. Neurochem.* **19:**641–651.
87. Martensson, E., and Kanfer, J., 1968, *J. Biol. Chem.* **243:**497–501.
88. McMurray, W. C., Strickland, K. P., Berry, J. F., and Rossiter, R. J., 1957, *Biochem. J.* **66:**634–644.
89. Possmayer, F., Meiners, B., and Mudd, J. B., 1973, *Biochem. J.* **132:**381–394.
90. Sanchez de Jimenez, E., and Cleland, W. W., 1969, *Biochim. Biophys. Acta* **176:**685–691.
91. Abdel-Latif, A. A., Yau, S. J., and Smith, J. P., 1974, *J. Neurochem.* **22:**383–393.
92. Barkai, A. I., 1981, *J. Neurosci. Res.* **6:**585–595.
93. Aas, M., and Daae, L. N. W., 1971, *Biochim. Biophys. Acta* **239:**208–216.
94. Abou-Issa, H. M., and Cleland, W. W., 1969, *Biochim. Biophys. Acta* **176:**692–698.
95. Bremer, J., Bjerve, K. S., Borrebaek, B., and Christiansen, R., 1976, *Mol. Cell. Biochem.* **12:**113–125.
96. Daae, L. N. W., 1973, *Biochim. Biophys. Acta* **306:**186–193.
97. Halder, D., Tso, W. W., and Pullman, K. M. E., 1979, *J. Biol. Chem.* **254:**4502–4509.
98. Hill, E. E., and Lands, W. E. M., 1968, *Biochim. Biophys. Acta* **152:**645–648.
99. Bates, E., and Saggerson, E. D., 1979, *Biochem. J.* **182:**751–762.

100. Bjerve, K. S., Daae, L. N., and Bremer, J., 1976, *Biochem. J.* **158**:249–254.
101. Daae, L. N. W., 1972, *FEBS Lett.* **27**:46–48.
102. Daae, L. N. W., 1972, *Biochim. Biophys. Acta* **270**:23–31.
103. Davidson, J. B., and Stanacev, N. Z., 1972, *Can. J. Biochem.* **50**:936–939.
104. Haldar, G., Carroll, M., Morris, P., Grosjean, C., and Anzalone, T., 1980, *Fed. Proc.* **39**:1992.
105. Kako, K. J., and Liu, M. S, 1974, *FEBS Lett.* **39**:243–246.
106. Kako, K. J., and Peckett, S. D., 1981, *Lipids* **16**:23–29.
107. Kelker, H. C., and Pullman, M. E., 1979, *J. Biol. Chem.* **254**:5364–5371.
108. Kinsella, J. E., 1976, *Lipids* **11**:680–684.
109. Kito, M., Ishinaga, M., Nishihara, M., Kanamoto, R., Yamamoto, K., and Hiromi, K., 1979, *J. Biochem. (Tokyo)* **85**:1527–1529.
110. Kornberg, A., and Pricer, W. E., 1953, *J. Biol. Chem.* **204**:345–358.
111. Liu, M. S., and Kako, K. J., 1974, *Biochem. J.* **138**:11–21.
112. Monroy, G., Kelker, H. C., and Pullman, M. E., 1973, *J. Biol. Chem.* **248**:2845–2852.
113. Monroy, G., Rola, F. J., and Pullman, M. E., 1972, *J. Biol. Chem.* **247**:6884–6894.
114. Nachbaur, J., Colbeau, A., and Vignais, P. M., 1971, *C. R. Acad. Sci. [D] (Paris)* **272**:1015–1018.
115. Yamashita, S., and Numa, S., 1972, *Eur. J. Biochem.* **31**:565–573.
116. Zaror-Behrens, G., and Kako, K. J., 1976, *Lipids* **11**:713–717.
117. Zaror-Behrens, G., and Kako, K. J., 1976, *Biochim. Biophys. Acta* **441**:1–13.
118. Ockner, R. K., Burnett, D. A., Lysenko, N., and Manning, J. A., 1979, *J. Clin. Invest.* **64**:172–181.
119. Okuyama, H., Yamada, K., and Ikezawa, H., 1975, *J. Biol. Chem.* **250**:1710–1713.
120. Lamb, R. G., Wyrick, S. D., and Piantadosi, C., 1977, *Atherosclerosis* **27**:147–154.
121. Tamai, Y., and Lands, W. E. M., 1974, *J. Biochem. (Tokyo)* **76**:847–860.
122. Stern, W., and Pullman, M. E., 1978, *J. Biol. Chem.* **253**:8047–8055.
123. Nimmo, H. G., 1979, *Biochem. J.* **177**:283–288.
124. Su, K. L., and Sun, G. Y., 1977, *J. Neurochem.* **29**:1059–1063.
125. Sun, G. Y., and Horrocks, L. A., 1973, *J. Lipid Res.* **14**:206–214.
126. Sun, G. Y., Su, K. L., Der, O. M., and Tang, W., 1979, *Lipids* **14**:229–235.
127. Thompson, W., 1977, *Adv. Exp. Med. Biol.* **83**:367–376.
128. Thompson, W., and Macdonald, G., 1975, *J. Biol. Chem.* **250**:6779–6785.
129. Thompson, W., and Macdonald, G., 1976, *Eur. J. Biochem.* **65**:107–111.
130. Aveldano de Caldironi, M. E., and Bazan, N. G., 1977, *Adv. Exp. Med. Biol.* **72**:387–404.
131. Kuwahara, S. S., 1972, *Physiol. Chem. Phys.* **4**:449–456.
132. Fleming, P. J., and Hajra, A. K., 1977, *J. Biol. Chem.* **252**:1663–1672.
133. Hajra, A. K., 1968, *Biochem. Biophys. Res. Commun.* **33**:929–935.
134. LaBelle, E. F., Jr., and Hajra, A. K., 1972, *J. Biol. Chem.* **247**:5835–5841.
135. Hajra, A. K., and Agranoff, B. W., 1968, *J. Biol. Chem.* **243**:3542–3543.
136. LaBelle, E. F., and Hajra, A. K., 1972, *J. Biol. Chem.* **247**:5825–5834.
137. LaBelle, E. F., and Hajra, A. K., 1972, *J. Biol. Chem.* **249**:6936–6944.
138. Jones, C. L., and Hajra, A. K., 1980, *J. Biol. Chem.* **255**:8289–8295.
139. Pollack, R. J., Hajra, A. K., and Agranoff, B. W., 1976, *J. Biol. Chem.* **251**:5149–5154.
140. Pollack, R. J., Hajra, A. K., and Agranoff, B. W., 1975, *Biochim. Biophys. Acta* **380**:421–436.
141. Pollack, R. J., Hajra, A. K., Folk, W. R., and Agranoff, B. W., 1975, *Biochem. Biophys. Res. Commun.* **65**:658–664.
142. Schlossman, D. M., and Bell, R. M., 1977, *Arch. Biochem. Biophys.* **182**:732–742.
143. Schlossman, D. M., and Bell, R. M., 1976, *J. Biol. Chem.* **251**:5738–5744.
144. Pieringer, R. A., and Hokin, L. E., 1962, *J. Biol. Chem.* **237**:653–658.
145. Dodds, P. F., Gurr, M. I., and Brindley, D. N., 1976, *Biochem. J.* **160**:693–700.
146. Paris, R., and Clement, G., 1969, *Proc. Soc. Exp. Biol. Med.* **131**:363–365.
147. Lamb, R. G., Gardner, T. G., and Fallon, H. J., 1980, *Biochim. Biophys. Acta* **619**:385–395.
148. Bishop, H. H., and Strickland, K. P., 1980, *Lipids* **15**:285–291.
149. Hokin, L. E., and Hokin, M. R., 1959, *J. Biol. Chem.* **234**:1381–1386.
150. Strickland, K. P., 1962, *Can. J. Biochem. Physiol.* **40**:247–259.

151. Lapetina, E. G., and Hawthorne, J. M., 1971, *Biochem. J.* **122**:171–179.
152. Sun, G. Y., and Horrocks, L. A., 1969, *J. Neurochem.* **16**:181–189.
153. Sun, G. Y., and Horrocks, L. A., 1971. *J. Neurochem.* **18**:1963–1969.
154. Aveldano, M. I., and Bazan, N. G., 1975, *J. Neurochem.* **25**:919–920.
155. Aveldano de Caldironi, M. I., and Bazan, N. G., 1979, *Neurochem. Res.* **4**:213–221.
156. Bazan, N. G., 1976, *Adv. Exp. Med. Biol.* **72**:317–335.
157. Geising, M., and Zilliken, F., 1980, *Neurochem. Res.* **5**:257–270.
158. Binaglia, L., Roberti, R., and Porcellati, G., 1978, *Adv. Exp. Med. Biol.* **101**:353–366.
159. Chalifour, R. J., and Kanfer, J. N., 1980, *Biochem. Biophys. Res. Commun.* **96**:742–747.
160. Wykle, R. L., and Schremmer, J. M., 1974, *J. Biol. Chem.* **249**:1742–1746.
161. Caras, I., and Shapiro, B., 1975, *Biochim. Biophys. Acta* **409**:201–211.
162. Casola, P. G., and Possmayer, R., 1981, *Biochim. Biophys. Acta* **665**:186–194.
163. Casola, P. G., and Possmayer, R., 1981, *Can. J. Biochem.* **59**:500–510.
164. Casola, P. G., and Possmayer, F., 1981, *Biochim. Biophys. Acta* **664**:298–315.
165. Cheng, C. H. K., and Saggerson, E. D., 1978, *FEBS Lett.* **93**:120–124.
166. Cheng, C. H. K., and Saggerson, E. D., 1978, *FEBS Lett.* **87**:65–68.
167. Jamdar, S. C., and Fallon, H. J., 1973, *J. Lipid Res.* **14**:517–524.
168. Sedgwick, B., and Hübscher, G., 1967, *Biochim. Biophys. Acta* **144**:397–408.
169. Smith, M. E., Sedgwick, B., Brindley, D. N., and Hübscher, G., 1967, *Eur. J. Biochem.* **3**:70–77.
170. Smith, S. W., Weiss, S. B., and Kennedy, E. P., 1957, *J. Biol. Chem.* **228**:915–922.
171. Sturton, R. G., and Brindley, D. N., 1977, *Biochem. J.* **162**:25–32.
172. Sturton, R. G., and Brindley, D. N., 1980, *Biochim. Biophys. Acta* **619**:494–505.
173. Sturton, R. G., and Brindley, D. N., 1978, *Biochem. J.* **171**:263–266.
174. Sturton, R. G., Pritchard, P. H., Han, L. Y., and Brindley, D. N., 1978, *Biochem. J.* **174**:667–670.
175. van Heusden, G. P., and van den Bosch, H., 1978, *Eur. J. Biochem.* **84**:405–412.
176. Cotman, C. W., McCaman, R. E., and Dewhurst, S. A., 1971, *Biochim. Biophys. Acta* **249**:395–405.
177. Schacht, J., Neale, E. A., and Agranoff, B. W., 1974, *J. Neurochem.* **23**:211–218.
178. Lamb, R. G., and Fallon, H. J., 1974, *Biochim. Biophys. Acta* **348**:166–178.
179. Bowley, M., Cooling, J., Burditt, S. L., and Brindley, D. N., 1977, *Biochem. J.* **165**:447–454.
180. Brindley, D. N., Cooling, J., Burditt, S. D., Pritchard, P. H., Pawson, S., and Sturton, R. G., 1979, *Biochem. J.* **180**:195–199.
181. Fallon, H. J., Lamb, R. G., and Jamdar, S. C., 1977, *Biochem. Soc. Trans.* **5**:37–43.
182. Eichberg, J., Gates, J., and Hauser, G., 1979, *Biochim. Biophys. Acta* **573**:90–106.
183. Hauser, G., and Eichberg, J., 1975, *J. Biol. Chem.* **250**:105–112.
184. Pappu, A. S., and Hauser, G., 1981. *J. Neurochem.* **37**:1006–1014.
185. Allan, D., and Michell, R. H., 1975. *Biochem. J.* **148**:471–478.
186. Abdel-Latif, A. A., 1976, *Function and Metabolism of Phospholipids in the Central and Peripheral Nervous System*, Plenum Press, New York, pp. 227–256.
187. Abdel-Latif, A. A., and Smith, J. P., 1976, *Biochem. Pharmacol.* **25**:1697–1704.
188. Brindley, D. N., and Bowley, M., 1975, *Biochem. J.* **148**:461–469.
189. Michell, R. H., Allan, D., Bowley, M., and Brindley, D. N., 1976, *J. Pharm. Pharmacol.* **28**:331–332.
190. Papahadjopoulos, D., Jacobson, K., Poste, G., and Shepherd, G., 1975, *Biochim. Biophys. Acta* **394**:504–519.
191. Ilincheta de Boschero, M. G., and Bazan, N. G., 1982, *Biochem. Pharmacol.* **31**:1049–1055.
192. Ilincheta de Boschero, M. G., Giusto, N. M., and Bazan, N. G., 1980, *Neurochemistry* **1**:17–28.
193. Bishop, H. H., and Strickland, K. P., 1976, *Can. J. Biochem.* **54**:249–260.
194. Hokin, M., Sadeghian, K., Harris, D. W., and Merrin, J. S., 1977, *Biochem. Biophys. Res. Commun.* **78**:364–371.
195. Thompson, R. J., 1977, *Biochem. Soc. Trans.* **5**:49–51.
196. Thompson, R. J., 1977, *J. Neurochem.* **29**:383–395.

197. Thompson, R. J., 1977, *J. Neurochem.* **29**:387–391.
198. Porcellati, G., Gaiti, A., Woelk, H., DeMedio, G. E., Brunetti, M., Francescangali, E., and Torovarelli, E., 1978, *Adv. Exp. Med. Biol.* **101**:301–318.
199. Itoh, T., and Kaneko, H., 1977, *Lipids* **12**:809–813.
200. Kiss, Z., 1976, *Eur. J. Biochem.* **67**:557–561.
201. Pullarkat, R. K., Sbaschnig-Alger, M., and Reha, H., 1981, *Biochim. Biophys. Acta* **663**:117–123.
202. Somerharju, P., Brotherus, J., Kahma, K., and Renkonen, O., 1977, *Biochim. Biophys. Acta* **487**:154–162.
203. Somerharju, P., and Renkonen, O., 1980, *Biochim. Biophys. Acta* **618**:407–419.
204. Bleistein, J., Heidrich, H. G., and Debuch, H., 1980, *Hoppe Seylers Z. Physiol. Chem.* **361**:595–597.
205. Hack, M. H., and Helmy, F. M., 1975, *Comp. Biochem. Physiol. [C]* **52**:139–145.
206. Kahma, I., Brotherus, J., Haltia, M., and Renkonen, O., 1976, *Lipids* **11**:539–544.
207. Cotman, C., Blank, M. L., Moehl, A., and Snyder, F., 1969, *Biochemistry* **8**:4606–4612.
208. Kishimoto, Y. L., Agranoff, B. W., Radin, N. S., and Burton, R. M., *J. Neurochem.* **16**:397–404.
209. Sun, G. Y., and Sun, A. Y., *Biochim. Biophys. Acta* **280**:306–318.
210. Anderson, R. E., and Maude, M. B., 1970, *Biochemistry* **9**:3624–3629.
211. Daemen, F. J. M., 1973, *Biochim. Biophys. Acta* **300**:255–288.
212. Aveldano, M. I., and Bazan, N. G., 1974, *J. Neurochem.* **23**:1127–1135.
213. Aveldano de Caldironi, M. I., and Bazan, N. G., 1980, *Neurochemistry* **1**:381–392.
214. Harmon, C. K., and Neiderhiser, D. H., 1978, *Biochim. Biophys. Acta* **530**:217.
215. Holub, B. J., MacNaughton, J. A., and Piekarski, J., 1979, *Biochim. Biophys. Acta* **572**:413.
216. Yamashita, S., Noriko, N., Miki, Y., and Numa, S., 1975, *Proc. Natl. Acad. Sci. U.S.A.* **72**:600.
217. Bazan, H. E., Careaga, M. M., Sprecher, H., and Bazan, N. G., 1982, *Biochim. Biophys. Acta* **712**:123–128.
218. Lapetina, E. G., Billah, M. M., and Cuatrecasas, P., 1980, *J. Biol. Chem.* **255**:10966–10970.
219. Mauco, G., Chap, H., Simon, M. F., and Douste-Blazy, L., 1978, *Biochimie* **60**:653–661.
220. Friedel, R. O., and Schanberg, S. M., 1972, *J. Pharmacol. Exp. Ther.* **183**:326–332.
221. Hokin, L. E., and Hokin, M. R., 1955, *Biochim. Biophys. Acta* **18**:102–10.
222. Hokin, L. E., and Hokin, M. R., 1955, *Biochim. Biophys. Acta* **16**:229–237.
223. Schacht, J., and Agranoff, B. W., 1972, *J. Biol. Chem.* **247**:771–777.
224. Yagihara, Y., and Hawthorne, J. N., 1972, *J. Neurochem.* **19**:355–367.
225. Fischer, S. K., and Agranoff, B. W., 1981, *J. Neurochem.* **37**:968–977.
226. Friedel, R. O., Johnson, J. R., and Schanberg, S. M., 1973, *J. Pharmacol. Exp. Ther.* **184**:583–589.
227. Hokin, M. R., 1969, *J. Neurochem.* **16**:127–134.
228. Abdel-Latif, A. A., Akhtar, R. A., and Hawthorne, J., 1977, *Biochem. J.* **162**:61–73.
229. Abdel-Latif, A. A., Green, K., Smith, J., McPherson, J. C., and Matheny, J. L., 1978, *J. Neurochem.* **30**:517–525.
230. Hokin, M. R., Benfy, B. G., and Hokin, L. E., 1958, *J. Biol. Chem.* **233**:814–817.
231. Trifaro, J. M., 1969, *Mol. Pharmacol.* **5**:382–393.
232. Adnan, N. A. M., and Hawthorne, J. N., 1981, *J. Neurochem.* **36**:1858–1860.
233. Hokin, M. R., and Hokin, L. E., 1967, *J. Gen. Physiol.* **50**:793–811.
234. Possmayer, F., 1974, *Biochem. Biophys. Res. Commun.* **61**:1415–1426.
235. Possmayer, F., and Mudd, J. B., 1971, *Biochim. Biophys. Acta* **239**:217–233.
236. Schneider, P. B., 1971, *J. Biol. Chem.* **247**:7910–7914.
237. Farese, R. V., Sabir, M. A., and Larson, R. E., 1981, *Endocrinology* **109**:1895–1901.
238. Glenny, H. P., and Brindley, D. N., 1978, *Biochem. J.* **176**:777–784.
239. Hülsmann, W. C., 1978, *Mol. Cell Endocrinol.* **12**:1–8.
240. Sooranna, S. R., and Saggerson, E. D., 1976, *FEBS Lett.* **64**:36–39.
241. Sooranna, S. R., and Saggerson, E. D., 1978, *FEBS Lett.* **90**:141–144.
242. Larrabee, M. G., 1968, *J. Neurochem.* **15**:803–808.
243. Larrabee, M. G., and Leicht, W. S., 1965, *J. Neurochem.* **12**:1–13.

244. Smith, T. L., Eichberg, J., and Hauser, G., 1979, *Life Sci.* **24**:2179–2184.
245. Fisher, S. K., Frey, K. A., and Agranoff, B. W., 1981, *J. Neurosci.* **1**:1407–1413.
246. Fisher, S. K., Boast, C. A., and Agranoff, B. W., 1980, *Brain Res.* **189**:284–288.
247. Fisher, S. K., and Agranoff, B. W., 1980, *J. Neurochem.* **34**:1231–1240.
248. Hawthorne, J. N., and Pickard, M. R., 1977, *Adv. Exp. Med. Biol.* **83**:419–427.
249. Clark, R. B., Salmon, D. M., and Honeyman, T. W., 1980, *J. Cyclic Nucleotide Res.* **6**:37–49.
250. Gerrard, J. M., Butler, A. M., Peterson, D. A., and White, J. G., 1978, *Prostagland. Med.* **1**:387–396.
251. Lapetina, E. G., Billah, M. M., and Cuatrecasas, P., 1981, *J. Biol. Chem.* **256**:5037–3040.
252. Broekman, M. F., Ward, J. W., and Marcus, A. J., 1981, *J. Biol. Chem.* **256**:8271–8274.
253. Rittenhouse-Simmons, S., 1981, *J. Biol. Chem.* **256**:287–298.
254. Bell, R. L., Kennerly, D. A., Stanford, N., and Majerus, P. W., 1979, *Proc. Natl. Acad. Sci. U.S.A.* **76**:3238–3241.
255. Kawahara, Y., Takai, Y., Minakuchi, R., Kimihiko, S., and Nishizuka, Y., 1980, *Biochem. Biophys. Res. Commun.* **97**:309–317.
256. Hirata, F., and Axelrod, J., 1980, *Science* **209**:1082–1090.
257. Bazan, N. G., 1970, *Biochim. Biophys. Acta* **218**:1–10.
258. Bazan, N. G., and Rakowski, H., 1970, *Life Sci.* **9**:501–507.
259. Rodriguez de Turco, E. G., Cascone, G. D., Pediconi, M. F., and Bazan, N. G., 1977, *Adv. Exp. Med. Biol.* **83**:389–396.
260. Tokumura, A., Akamatsy, Y., Tamada, S., and Tsukatani, H., 1978, *Agric. Biol. Chem.* **42**:515–521.
261. Tokimura, A., Fukuzawa, K., and Tsukatani, H., 1978, *Lipids* **13**:572–574.
262. Gerrard, J. M., Kindom, S. E., Peterson, D. A., Peller, J., Krantz, K. E., and White, J. G., 1979, *Am. J. Pathol.* **96**:423–436.
263. Gerrard, J. M., Kindon, S. E., Peterson, D. A., and White, J. G., 1979, *Am. J. Pathol.* **97**:531–543.
264. Lapetina, E. G., Motasim Billah, M., and Cuatrecasas, P., 1981, *J. Biol. Chem.* **256**:11984–11987.
265. Geison, R. L., Banschbach, M. W., Sadeghian, K., and Hokin-Neaverson, M., 1976, *Biochem. Biophys. Res. Commun.* **68**:343–349.
206. Bazan, H. E. P., and Bazan, N. G., 1977, *Adv. Exp. Med. Biol.* **83**:489–495.
267. Giusto, N. M., and Bazan, N. G., 1977, *Adv. Exp. Med. Biol.* **83**:481–488.
268. Allan, D., Watts, R., and Michell, H., 1976, *Biochem. J.* **156**:225.
269. Miller, J. C., 1977, *Biochem. J.* **168**:549–555.
270. Miller, J. C., and Leung, I., 1979, *Biochem. J.* **178**:9–13.
271. McDonald, G., Baker, R. R., and Thompson, W., 1975, *J. Neurochem.* **24**:655.

Gangliosides

Robert W. Ledeen

1. INTRODUCTION

Gangliosides occur primarily in the vertebrate phylum and have been detected in virtually all vertebrate tissues subjected to careful analysis. They constitute part of the glycoconjugate network extending from the membrane and contribute in a crucial manner to the surface properties of cells. In most instances, they comprise only a small portion of the lipid content of the plasma membrane. An important exception, however, is the CNS neuronal membrane in which gangliosides constitute approximately 10% of total lipid and two-thirds of glycoconjugate sialic acid. Organs such as intestine,[1] mammary gland,[1] and thyroid[2] have total sialic acid concentrations equal to or greater than that of brain, but in these and virtually every other extraneural tissue that has been studied, protein-bound sialic acid was found to predominate. Gangliosides of such tissues show characteristic patterns that differ from brain and usually from each other; additional variation is attributed to species and developmental state. Thyroid,[3] erythrocytes,[4] and intestinal mucosa[5] are examples of tissues that show marked ganglioside pattern differences among species. On the other hand, the intestinal muscular layer showed only small differences.[5] Brain displays significant variation when lower vertebrates are compared to higher,[6] but among mammals, the concentrations and patterns of the major brain gangliosides are relatively invariant.[3,7]

The history of ganglioside research may be considered to have passed through a number of identifiable (and overlapping) stages. The first, discovery, occurred in the 1930s and early 1940s in Klenk's laboratory during extensive investigation of normal and sphingolipidosis brains.[8-10] In retrospect, it is now clear that Landsteiner and Levene[11,12] had detected such substances even earlier in extracts from brain and kidney through reaction with p-dimethylaminobenzaldehyde and orcinol. Klenk carried out the first extensive purification and systematic compositional studies and eventually proposed the name ''gan-

Robert W. Ledeen • The Saul R. Korey Department of Neurology and Department of Biochemistry, Albert Einstein College of Medicine, Bronx, New York 10461.

glioside" in the belief they were concentrated in the "Ganglienzellen" (neurons) of gray matter.[10] Klenk also identified and named a new carbohydrate, neuraminic acid, which was responsible for their distinctive orcinol–hydrochloric acid reaction and which has become the identifying carbohydrate of this glycolipid class. Compositional analyses were followed by structural studies which continued over many years, leading by the early 1960s to full characterization of the four major brain gangliosides; this occurred primarily in the laboratories of Kuhn[13–15] and Klenk,[16–18] with many workers contributing subsequently. Some of the minor brain gangliosides were elucidated at that time, but several years passed before the great diversity of these structures was revealed. Chromatographic separation techniques and a now widely employed nomenclature system were described by Svennerholm.[19] Earlier in that period, Yamakawa's[20] discovery of G_{M3} in erythrocyte membranes demonstrated for the first time the existence of sialoglycolipids in extraneural tissues. The term he coined, "hematoside," has been used ever since to denote the subgroup of gangliosides lacking hexosamine.

The biochemical phase started in the 1960s when metabolic pathways began to be constructed. Roseman and co-workers[21,22] proposed a biosynthetic sequence that has proved generally valid in vertebrate systems. Several workers contributed to the elucidation of catabolic pathways during the same period, leading eventually to discovery of the primary metabolic defects in the gangliosidoses.[23–28] Cellular and subcellular distributional studies commenced during this biochemical period and continue to the present with certain key questions still unresolved (see below).

Current ganglioside research, although continuing to address structural and biochemical questions, has entered a new phase in which functional aspects are receiving greater emphasis. Experimentation in this area has pointed to a broad range of putative functions not restricted to the nervous system, but many of these are tentative and poorly defined at present. No unifying principles have yet emerged that would correlate the growing volume of data in these diverse areas, but important clues are beginning to appear. Progress in some of the current functional studies is surveyed in this chapter, along with new findings on metabolism, distribution, and structure. In general, an effort has been made to stress developments occurring since publication of the first edition of this series.[29] Methodologies for the isolation, separation, assay, and characterization of gangliosides have been described in several recent reviews[30–37] and are not considered here. Most of the above reviews also deal with other aspects of ganglioside biochemistry and function; to this list may be added a few comprehensive treatments[38–40] that appeared in the last few years.

2. STRUCTURE

Diversity of molecular structure is the hallmark of gangliosides, as of glycosphingolipids generally. Approximately 40 different oligosaccharide structures have been identified in gangliosides from vertebrate brain and extraneural tissues. The number increases to approximately 55 when variation in sialic acid type is taken into account. Such tabulations are necessarily provisional, since new structures continue to be discovered at a rapid rate. They range in complexity from two to ten (or more) sugar units. The carbohydrate common to all gangliosides—the one defining this family of glycolipids—is sialic acid, a generic term for the series of compounds derived from the parent neuraminic acid (5-amino-3,5-dideoxy-D-glycero-D-galacto-nonulosonic acid). Of the many different sialic acids known to exist, only a few have been found in gangliosides. N-Acetylneuraminic acid (NeuAc) is the major type in brain of most mammals and appears to be the exclusive type in human brain. N-Glycolylneuraminic acid (NeuGc) is a minor form in brain gangliosides of some species but is widely distributed, along with NeuAc, in extraneural gangliosides. Rabbit thymus was recently revealed[41] as a tissue with an unusually high proportion of NeuGc. Sialic acids with O-acyl groups have been found in hematosides of horse erythrocytes[42] and in brain gangliosides of lower vertebrates[43] and developing mice.[44,45]

Gangliosides occur primarily in two of the major glycosphingolipid families built up from lactosylceramide, the ganglio and neolacto series. A preliminary report[46] describes a ganglioside from chicken muscle belonging to the globo series—the first of its kind. Sialosylgalactosylceramide (G_{M4}) may be considered as belonging to the gala series and also to the hematosides, the subgroup lacking hexosamine. To date, gangliosides have not been found in the other series,[37] e.g., mucotetraose, isoglobotetraose, lactotetraose. Table I summarizes the glycolipid families in which gangliosides have been found to occur. In regard to nomenclature, it may be noted that the neolacto series lends itself to the derivative designations "neolactohexaose" and "neolactooctaose" because of the presence of repeating disaccharide units. Such homology has not yet been observed with the ganglio series.

2.1. The Ganglio Series

The large majority of brain gangliosides belong to this family, characterized by the presence of GalNAc in the form of $GgOse_3Cer$, $GgOse_4Cer$, or their

Table I
Glycosphingolipid Families that Contain Gangliosides (Vertebrate)

Family	Abbreviation	Structure
Galaose[a]	GaOseCer	Galβ1–1'Cer
Gangliotriaose	GgOse₃Cer	GalNAcβ1–4Galβ1–4Glcβ1–1'Cer
Gangliotetraose	GgOse₄Cer	Galβ1–3GalNAcβ1–4Galβ1–4Glcβ1–1'Cer
Globotetraose	GbOse₄Cer	GalNAcβ1–3Galα1–4Galβ1–4Glcβ1–1'Cer
Neolactotetraose	nLcOse₄Cer	Galβ1–4GlcNAcβ1–3Galβ1–4Glcβ1–1'Cer
Neolactohexaose[a]	nLcOse₆Cer	(Galβ1–4GlcNAcβ1–3)₂Galβ1–4Glcβ1–1'Cer
Neolactooctaose[a]	nLcOse₈Cer	(Galβ1–4GlcNAcβ1–3)₃Galβ1–4Glcβ1–1'Cer

[a] These terms, although not included in the recommendations of the IUPAC-IUB Commission on Biochemical Nomenclature[45a,b] are logical extensions of the systems adopted.

derivatives. The major gangliosides of mammalian brain are based on the GgOse$_4$Cer structure with one to three sialic acids (Fig. 1). These types also occur widely in extraneural tissues, though often as minor species in comparison to hematosides and the neolacto series. Over 30 different structures belonging to the ganglio series have been characterized to date. Table II represents an updating of recent compilations.[32–34,37,38,47–49]

A recently discovered ganglioside of some interest is G$_{M1b}$ (Table II, entry 7), isomeric to G$_{M1(a)}$ with sialic acid attached to terminal galactose. Previously it came to light[74] as a product of *in vitro* synthesis from GgOse$_4$Cer (G$_{A1}$) by rat brain homogenate and was thought to represent an artifact since the "natural" monosialoganglioside of the GgOse$_4$Cer type (G$_{M1}$) contained sialic acid on the inner galactose (Fig. 1). However, G$_{M1b}$ has now been isolated from natural sources such as erythrocytes,[54] rat ascites hepatoma,[55] and mouse myeloid leukemia M1 cells.[75] A derivative of this structure with an additional GalNAc (Table II, entry 9) was isolated from Tay–Sachs brain.[57] A preliminary report[64] describes yet another derivative of G$_{M1b}$ from frog brain, with the unusual feature of a NeuAc–GalNAc linkage (Table II, entry 21). Three related structures from the same source were tentatively identified as having additional NeuAc attached to one or both of the sialic acids shown.

Fucose-containing gangliosides belonging to the ganglio series were first discovered in extraneural tissues[51,76] and later in brain.[59,77] The three species that have been found in brain (Table II, entries 13,14,23) generally occur as minor components, although the mini-pig, type Gottingen, provides a notable exception with 17–32 mol%.[77]

A highly unusual derivative of G$_{M2}$ containing a sulfate group at the 8-position of NeuAc (Table II, entry 3) was recently isolated from bovine gastric mucosa.[52] The NeuGc analogue was also detected. It may be noted that pen-

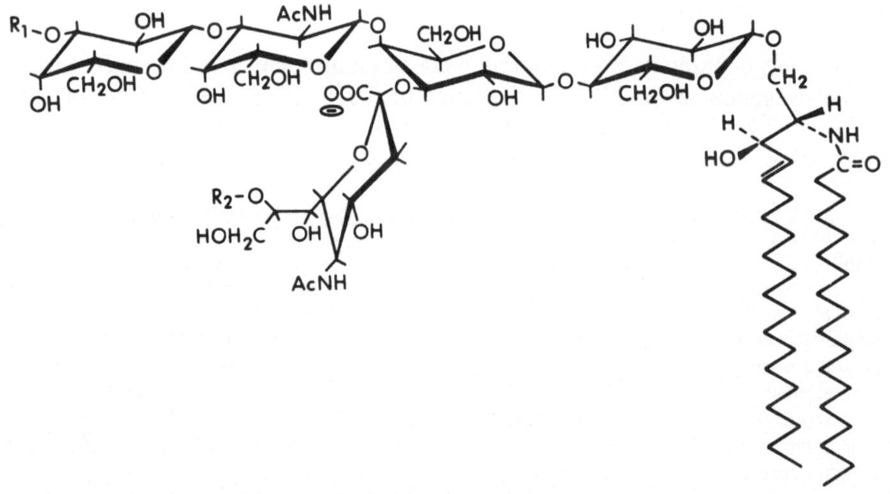

Fig. 1.   Structures of the four major gangliosides of mammalian brain. G$_{M1}$, R$_1$ = R$_2$ = H; G$_{D1a}$, R$_1$ = NeuAc, R$_2$ = H; G$_{D1b}$, R$_1$ = H, R$_2$ = NeuAc; G$_{T1b}$, R$_1$ = R$_2$ = NeuAc.

tasialoganglioside (Table II, entry 32) is the most polar to be characterized to date. However, a study of embryonic chick brain[78] has suggested the presence of even more polar types containing six or seven sialic acids. It thus appears likely that additional gangliosides will be found in this series.

2.2. Neolacto Series

Virtually all of the well-characterized glucosamine-containing gangliosides belong to the neolacto series (Table III). Some of these have the neolactote-traosyl (nLcOse$_4$Cer) structure, but a number have one or two additional Galβ1-4GlcNAcβ1-3- repeating disaccharide units giving rise to the nLcOse$_6$Cer and nLcOse$_8$Cer structures, respectively. One of the latter type, VIII^{3}NeuAc-nLcOse$_8$Cer (Table III, entry 14), has been isolated from human erythrocytes[89] and rabbit skeletal muscle.[19] Another erythrocyte ganglioside (Table III, entry 6), unusual in containing both GlcNAc and GalNAc, has the additional inter-esting feature of a NeuAcα2-3GalNAc linkage.[85] The structure IV^{3}NeuAc-nLcOse$_4$Cer (Table III, entry 1), bearing the common name "sialosylparag-loboside," was shown to be the major ganglioside in human peripheral nerve[81] and human erythrocytes.[79,80] It is a minor ganglioside of human brain and the only glucosamine-containing species found in that organ to date.[81] Gangliosides of the neolacto series are relatively more prominent than members of the gan-glio series in extraneural tissues. An extreme example is rabbit thymus[41] which was found to contain this family and hematosides but no members of the ganglio series.

2.3. Hematosides

This category of ganglioside is widespread and prominent in vertebrate tissues generally but occurs to only a minor extent in the nervous system. The majority of these gangliosides are based on lactosylceramide, G$_{M3}$ and G$_{D3}$ being the most prevalent (Table IV). They tend to become elevated in certain disease states characterized by gliosis. Ganglioside G$_{D3}$ occurs prominently in the retina of certain species.[91] A detailed study of G$_{M3}$ in several rabbit tissues revealed significant differences in concentration, sialic acid type, and lipophilic composition.[92] It demonstrated that although G$_{M3}$ is frequently the major gan-glioside in extraneural tissues, this is not invariably the case. Discovery of G$_{T3}$, first in fish brain[69,93,94] and later in pig kidney cortex,[95] marked the demon-stration of a trisialo form. Sialosylgalactosylceramide (G$_{M4}$) is a hematoside of the gala series that was shown to be the third most abundant ganglioside of human white matter.[96] It is primarily localized in CNS myelin (see below), although recent reports[97,98] indicate that it occurs to some extent in certain other tissues as well.

2.4. Lipophilic Components

Ceramide, the hydrophobic moiety that anchors the ganglioside in the membrane, consists of fatty acid joined in amide linkage to sphingosine or a

Table II

Carbohydrate Structures of Vertebrate Gangliosides: Ganglio Series

Structure[a]	Symbol[b]	Source
Monosialo		
1. GalNAcβ1-4Galβ1-4Glc- $\quad\quad\quad$3 $\quad\quad\quad\mid$ $\quad\quad\quad$NeuAcα2	II^{3}NeuAc-GgOse$_3$ G$_{M2}$	Human brain,[15] Tay–Sachs brain[50]
2. GalNAcβ1-4Galβ1-4Glc- $\quad\quad\quad$3 $\quad\quad\quad\mid$ $\quad\quad\quad$NeuGcα2	II^{3}NeuGc-GgOse$_3$ G$_{M2}$(NeuGc)	Bovine spleen[51]
3. $\quad\quad\quad$(8-0-SO$_4$)NeuAcα2 $\quad$GalNAcβ1-4Galβ1-4Glc- $\quad\quad\quad$3	II^3(8-0-SO$_4$)NeuAc-GgOse$_3$	Bovine gastric mucosa[52]
4. $\quad\quad\quad$(8-0-SO$_4$)NeuGcα2 $\quad$GalNAcβ1-4Galβ1-4Glc- $\quad\quad\quad$3	II^3(8-0-SO$_4$)NeuGc-GgOse$_3$	Bovine gastric mucosa[52]
5. Galβ1-3GalNAcβ1-4Galβ1-4Glc- $\quad\quad\quad\quad\quad$3 $\quad\quad\quad\quad\quad\mid$ $\quad\quad\quad\quad\quad$NeuAcα2	II^{3}NeuAc-GgOse$_4$ G$_{M1}$, G$_{M1a}$	Human brain,[13] G$_{M1}$-gangliosidosis,[53] bovine spleen[51]
6. Galβ1-3GalNAcβ1-4Galβ1-4Glc- $\quad\quad\quad\quad\quad$3 $\quad\quad\quad\quad\quad\mid$ $\quad\quad\quad\quad\quad$NeuGcα2	II^{3}NeuGc-GgOse$_4$ G$_{M1}$(NeuGc)	Bovine spleen, kidney, and liver[51]
7. NeuAcα2-3Galβ1-3GalNAcβ1-4Galβ1-4Glc-	IV^{3}NeuAc-GgOse$_4$ G$_{M1b}$	Human erythrocytes,[54] rat tumor[55]
8. GalNAcβ1-4Galβ1-3GalNAcβ1-4Galβ1-4Glc- $\quad\quad\quad\quad\quad\quad\quad\quad$3 $\quad\quad\quad\quad\quad\quad\quad\quad\mid$ $\quad\quad\quad\quad\quad\quad\quad\quad$NeuAcα2	IV^{4}βGalNAc, II^{3}NeuAc-GgOse$_4$ G$_{M1}$-GalNAc	Human brain[56]

9. GalNAcβ1-4Galβ1-3GalNAcβ1-4Galβ1-4Glc-
   $\quad\quad$ 3
   $\quad\quad$ |
   $\quad\quad$ NeuAcα2

   IV^3NeuAc, IV^4βGalNAc-GgOse$_4$
   G_{M1b}-GalNAc

   Tay–Sachs brain[57]

10. Galα1-3Galβ1-3GalNAcβ1-4Galβ1-4Glc-
    $\quad\quad$ 3
    $\quad\quad$ |
    $\quad\quad$ NeuAcα2

    IV^3αGal, II^3NeuAc-GgOse$_4$
    G_{M1}-Gal

    Frog fat body[58]

11. Galβ1-3Galα1-3Galβ1-3GalNAcβ1-4Galβ1-4Glc-
    $\quad\quad$ 3
    $\quad\quad$ |
    $\quad\quad$ NeuAcα2

    IV^3(Galβ1-3Gal-), II^3NeuAc-GgOse$_4$
    G_{M1}-Gal$_2$

    Frog fat body[58]

12. Galα1-3Galβ1-3Galα1-3Galβ1-3GalNAcβ1-4Galβ1-4Glc-
    $\quad\quad$ 3
    $\quad\quad$ |
    $\quad\quad$ NeuAcα2

    IV^3α(Galα1-3Gal-), II^3NeuAc-GgOse$_4$
    G_{M1}Gal$_3$

    Frog fat body[58]

13. Fucα1-2Galβ1-3GalNAcβ1-4Galβ1-4Glc-
    $\quad\quad$ 3
    $\quad\quad$ |
    $\quad\quad$ NeuAcα2

    IV^2αFuc, II^3NeuAc-GgOse$_4$
    G_{M1}-Fuc

    Bovine brain,[59] bovine thyroid,[60,61] pig adipose,[62] mini-pig brain[77]

14. Fucα1-2Galβ1-3GalNAcβ1-4Galβ1-4Glc-
    $\quad\quad$ 3
    $\quad\quad$ |
    $\quad\quad$ NeuGcα2

    IV^2αFuc, II^3NeuGc-GgOse$_4$
    G_{M1}-Fuc(NeuGc)

    Bovine liver,[51] pig adipose,[62] mini-pig brain[77]

15. Fucα1-3Galβ1-3GalNAcβ1-4Galβ1-4Glc-
    $\quad\quad$ 3
    $\quad\quad$ |
    $\quad\quad$ NeuAcα2

    IV^3αFuc, II^3NeuAc-GgOse$_4$

    Pig adipose[62]

Disialo

16. GalNAcβ1-4Galβ1-4Glc-
    $\quad\quad$ 3
    $\quad\quad$ |
    $\quad\quad$ NeuAcα2
    $\quad\quad$ 8
    $\quad\quad$ |
    $\quad\quad$ NeuAcα2

    II^3(NeuAc)$_2$-GgOse$_3$
    G_{D2}

    Human brain[15,18]

(*continued*)

Table II
(Continued)

Structure[a]	Symbol[b]	Source

17. NeuAcα2-3Galβ1-3GalNAcβ1-4Glc-
        |3
    NeuAcα2
	IV³NeuAc, II³NeuAc-GgOse₄
	G_{D1a}
		Human brain,[14,16] bovine adrenal medulla[63]

18. NeuAcα2-3Galβ1-3GalNAcβ1-4Glc-
        |3
    NeuGcα2
	IV³NeuAc, II³NeuGc-GgOse₄
	G_{D1a}(NeuAC/NeuGc)
		Bovine brain,[59] bovine adrenal medulla[63]

19. NeuGcα2-3Galβ1-3GalNAcβ1-4Glc-
        |3
    NeuAcα2
	IV³NeuGc, II³NeuAc-GgOse₄
	G_{D1a}(NeuGc/NeuAc)
		Bovine brain[59]

20. NeuGcα2-3Galβ1-3GalNAcβ1-4Glc-
        |3
    NeuGcα2
	IV³NeuGc, II³NeuGc-GgOse₄
	G_{D1a}(NeuGc)₂
		Bovine spleen, kidney[51]

21. NeuAcα2-3Galβ1-3GalNAcβ1-4Glc
        |6
    NeuGcα2
	IV³NeuAc, III⁶NeuAc-GgOse₄	Frog brain[64]

22. Galβ1-3GalNAcβ1-4Glc-
        |3
    NeuAcα2-8NeuAcα2
	II³(NeuAc)₂-GgOse₄
	G_{D1b}
		Human brain[14,17]

23. Fucα1-2Galβ1-3GalNAcβ1-4Glc-
        |3
    NeuAcα2-8NeuAcα2
	IV²αFuc, II³(NeuAc)₂-GgOse₄
	G_{D1b}-Fuc
		Human brain[65,66] pig cerebellum,[67] mini-pig brain[77]

24. GalNAcβ1-4Galβ1-3GalNAcβ1-4Galβ1-4Glc-
    |
    3
    |
    NeuAcα2

$IV^4\beta GalNAc, IV^3NeuAc, II^3NeuAc-GgOse_4$
G_{D1a}-GalNAc

Human brain[68]

Trisialo

25. NeuAcα2-8NeuAcα2-8NeuAcα2-3Galβ1-4Glc-
    |
    4
    |
    GalNAcβ1

$II^3(NeuAc)_3-GgOse_3$
G_{T2}

Fish brain[69]

26. NeuAcα2-3Galβ1-3GalNAcβ1-4Galβ1-4Glc-
    |
    3
    |
    NeuAcα2
    |
    8
    |
    NeuAcα2

$IV^3(NeuAc)_2, II^3NeuAc-GgOse_4$
G_{T1a}

Human brain[70]

27. NeuAcα2-3Galβ1-3GalNAcβ1-4Galβ1-4Glc-
    |
    3
    |
    NeuAcα2
    |
    8
    |
    NeuAcα2

$IV^3NeuAc, II^3(NeuAc)_2-GgOse_4$
G_{T1b}

Human brain[14,17]

28. NeuAcα2-3Galβ1-3GalNAcβ1-4Galβ1-4Glc-
    |
    3
    |
    (9-OAc)NeuAcα2-8NeuAcα2

$IV^3NeuAc, II^3(9-OAcNeuAcα2-8NeuAc-)GgOse_4$
G_{T1L}

Mouse brain[45]

29. Galβ1-3GalNAcβ1-4Galβ1-4Glc-
    |
    3
    |
    NeuAcα2
    |
    8
    |
    NeuAcα2
    |
    8
    |
    NeuAcα2

$II^3(NeuAc)_3-GgOse_4$
G_{T1c}

Fish brain[69]

Tetrasialo

30. NeuAcα2-3Galβ1-3GalNAcβ1-4Galβ1-4Glc-
    |
    3
    |
    NeuAcα2
    |
    8
    |
    NeuAcα2

$IV^3(NeuAc)_2, II^3(NeuAc)_2-GgOse_4$
G_{Q1b}

Human, bovine, and chicken brain[72,73]

(continued)

Table II
(Continued)

Structure[a]	Symbol[b]	Source
31. NeuAcα2-3Galβ1-3GalNAcβ1-4Galβ1-Glc- 　　　　　　　　　　　　　　　3 　　　　　　　　　　　　　　　| 　　NeuAcα2-8NeuAcα2-8NeuAcα2	IV^{3}NeuAc, II^3(NeuAc)$_3$-GgOse$_4$ G$_{Q1c}$	Fish brain[69,71]
Pentasialo 32.　NeuAcα2-3Galβ1-3GalNAcβ1-4Galβ1-4Glc- 　　　　8　　　　　　　　　　　　　3 　　　　|　　　　　　　　　　　　　| 　NeuAcα2　　NeuAcα2-8NeuAcα2 　　　　　　　　　8 　　　　　　　　　| 　　　　　　　NeuAcα2	IV^3(NeuAc)$_2$, II^3(NeuAc)$_3$-GgOse$_4$ G$_{P1c}$	Fish brain[71]

[a] Carbohydrate portions only are shown. The glucose moiety of each oligosaccharide chain is joined to the C-1 hydroxyl of ceramide with a β-glycosidic linkage.

[b] Symbols are in accordance with the recommendations of the IUPAC-IUB Commission on Biochemical Nomenclature.[45a,b] Symbols based on the Svennerholm system[19] are also given; additional symbols beyond those originally proposed have been added in a manner thought to be consistent with the system as a whole. Where more than one type of sialic acid is present in the same molecule, the first designated within parentheses is that most distal from ceramide.

Table III

Carbohydrate Structures of Vertebrate Gangliosides: Neolacto Series

	Structure[a]	Symbol[b]	Source
1.	NeuAcα2-3Galβ1-4GlcNAcβ1-3Galβ1-4Glc-	IV³NeuAc-nLcOse₄	Human erythrocytes,[79,80] peripheral nerve,[81] pig adipose[62]
2.	NeuGcα2-3Galβ1-4GlcNAcβ1-3Galβ1-4Glc-	IV³NeuGc-nLcOse₄	Bovine spleen and kidney[51]
3.	NeuAcα2-6Galβ1-4GlcNAcβ1-3Galβ1-4Glc-	IV⁶NeuAc-nLcOse₄	Bovine spleen and kidney,[51] human erythrocytes[54]

4.	Fucα1
	    \
	     3
	NeuAcα2-3Galβ1-4GlcNAcβ1-3Galβ1-4Glc-
		IV³NeuAc, III³αFuc-nLcOse₄	Human kidney[82]

5.	NeuAcα2-8NeuNAcα2-3Galβ1-4GlcNAcβ1-3Galβ1-4Glc-	IV³(NeuAc)₂-nLcOse₄	Human kidney[83]
6.	NeuAcα2-3GalNAcβ1-3Galβ1-4GlcNAcβ1-3Galβ1-4Glc-	IV³(NeuAcα2-3GalNAc-)LcOse₄	Human erythrocytes[84]
7.	NeuAcα2-3Galβ1-4GlcNAcβ1-3Galβ1-4GlcNAcβ1-3Galβ1-4Glc-	VI³NeuAc-nLcOse₆	Human spleen,[85] bovine erythrocytes[86]
8.	NeuGcα2-3Galβ1-4GlcNAcβ1-3Galβ1-4GlcNAcβ1-3Galβ1-4Glc-	VI³NeuGc-nLcOse₆	Bovine erythrocytes[86]
9.	NeuAcα2-6Galβ1-4GlcNAcβ1-3Galβ1-4GlcNAcβ1-3Galβ1-4Glc-	VI⁶NeuAc-nLcOse₆	Human erythrocytes[54]

10.	Galβ1-4GlcNAcβ1
	                        \
	                         6
	NeuAcα2-3Galβ1-4GlcNAcβ1-3Galβ1-4GlcNAcβ1-3Galβ1-4Glc-
		VI³NeuAc, IV⁶(Galβ1-4GlcNAcβ)-nLcOse₆	Human erythrocytes[54]

11.	Fucα1-2Galβ1-4GlcNAcβ1
	                                        \
	                                         6
	NeuAcα2-3Galβ1-4GlcNAcβ1-3Galβ1-4GlcNAcβ1-3Galβ1-4Glc-
		VI³NeuAc, IV⁶(Fucα1-2Galβ1-4GlcNAcβ1-)-nLcOse₆	Human erythrocytes[87]

12.	NeuAcα2-3Galβ1-4GlcNAcβ1
	                                          \
	                                           6
	NeuAcα2-3Galβ1-4GlcNAcβ1-3Galβ1-4GlcNAcβ1-3Galβ1-4Glc-
		VI³NeuAc, IV⁶(NeuAcα2-3Galβ1-4GlcNAcβ1-)-nLcOse₆	Human erythrocytes[88]

13.	Galα1-3Galβ1-4GlcNAcβ1
	                                        \
	                                         6
	Siaα2-3Galβ1-4GlcNAcβ1-3Galβ1-4GlcNAcβ1-3Galβ1-4Glc-
		VI³Sia, IV⁶(Galα1-3Galβ1-β1-)-nLcOse₆	Bovine erythrocytes[89]

14.	NeuAcα2-3Galβ1-4GlcNAcβ1-3Galβ1-4GlcNAcβ1-3Galβ1-4GlcNAcβ1-3Galβ1-4Glc-	VIII³NeuAc-nLcOse₈	Human erythrocytes[88,90]

[a] Carbohydrate portions only are shown. The glucose moiety of each oligosaccharide chain is joined to the C-1 hydroxl of ceramide with a β-glycosidic linkage.

[b] Recommendations of the IUPAC-IUB Commission on Biochemical Nomenclature.[45a,b]

Table IV

Carbohydrate Structures of Vertebrate Gangliosides: Hematoside Series

Structure[a]	Symbol[b]	Source
1. NeuAcα2-3Gal-	I³NeuAc-Gal $G_{M4}(G_7)$	Human brain,[15,96] avian myelin,[99] egg yolk[97]
2. NeuAcα2-3Galβ1-4Glc-	II³NeuAc-Lac G_{M3}	Human brain,[15] liver,[100] bovine liver, spleen, and kidney,[51] bovine adrenal medulla,[101] rabbit tissues[92]
3. NeuGcα2-3Galβ1-4Glc-	II³NeuGc-Lac G_{M3}(NeuGc)	Horse erythrocytes,[20,102] bovine adrenal medulla,[101] bovine liver, spleen, and kidney[51]
4. 4-O-AcNeuGcα2-3Galβ1-4Glc-	II³(4-O-AcNeuGc)-Lac	Horse erythrocytes[42]
5. NeuAcα2-8NeuAcα2-3Galβ1-4Glc-	II³(NeuAc)₂-Lac G_{D3}	Human brain,[15] bovine liver, spleen, and kidney,[51] bovine retina[103]
6. NeuAcα2-8NeuGcα2-3Galβ1-4Glc-	II³(NeuAcα2-8NeuGc)-Lac G_{D3}(NeuAc/NeuGc)	Bovine liver, spleen, and kidney[51]
7. NeuGcα2-8NeuAcα2-3Galβ1-4Glc-	II³(NeuGcα2-8NeuAc)-Lac G_{D3}(NeuGc/NeuAc)	Bovine liver, spleen, and kidney,[51] rabbit thymus[104]
8. NeuGcα2-8NeuGcα2-3Galβ1-4Glc-	II³(NeuGc)₂-Lac G_{D3}(NeuGc)₂	Bovine liver, spleen, and kidney[51] cat erythrocytes[105]
9. NeuAcα2-8NeuAcα2-8NeuAcα2-3Galβ1-4Glc-	II³(NeuAc)₃-Lac G_{T3}	Fish brain,[69,93,94] pig kidney[95]

[a] Carbohydrate portions only are shown. The glucose moiety of each oligosaccharide chain is joined to the C-1 hydroxyl of ceramide with a β-glycosidic linkage.

[b] Symbols are in accordance with the recommendations of the IUPAC-IUB Commission on Biochemical Nomenclature.[45a,b] Symbols based on the Svennerholm system[19] are also given; additional symbols beyond those originally proposed have been added in a manner thought to be consistent with the system as a whole. Where more than one type of sialic acid is present in the same molecule, the first designated within parentheses is that most distal from ceramide.

similar long-chain base. Stearate is the major fatty acid in the brain gangliosides, whereas a mixture ranging from C_{16} to C_{24} is usually found outside the CNS. Fatty acid and long-chain base patterns of peripheral nerve gangliosides proved to be a hybrid of CNS and extraneural properties.[106] The major individual gangliosides of brain had fatty acid patterns that gradually changed from a stearate composition of 92–94% in the newborn to 86–89% in old age.[107]

Brain gangliosides appear to be unique in containing C_{20} as well as C_{18} long-chain bases. Rat brain gangliosides contained only d18 sphingosine at birth but then showed rising levels of d20 sphingosine until the two were approximately equivalent at 6–8 weeks.[108] Human brain also showed increasing proportions of d20 sphingosine during the first 20 to 30 years of life, final values reaching 60–70% in individual gangliosides.[107,109] Ganglioside G_{M4} (G_7) from human[96] and avian[99] myelin was shown to have radically different lipophilic components compared to other brain gangliosides, the patterns for fatty acids and long-chain bases closely resembling those of myelin cerebrosides. Gangliosides containing 2-hydroxy fatty acids were also reported in human liver,[100] but such fatty acids are relatively rare in these sphingolipids.

2.5. Gangliosides of Invertebrates

Although sialoglycolipids are considered to be characteristic of vertebrates, a variety of such lipids have been found in one invertebrate phylum, Echinodermata. Their structures differ significantly from the vertebrate type, some containing glucose as the only carbohydrate besides sialic acid.[110,111] The simplest of these was NeuAc(2–6)Glc(1–1)Cer.[111] Disialogangliosides of this type containing one[110] and two[111,112] glucose units have been described. Structures with arabinose at the nonreducing end and sialic acid located internally in the oligosaccharide chain were isolated from *Asterina pectinifera*. One of these[113] had the structure:

$$\text{Ara}(1-6)\text{Gal}(\beta 1-4)\text{NeuGc}(\alpha 2-3)\text{Gal}(\beta 1-4)\text{Glc}(\beta 1-1)\text{Cer}$$

and two closely related forms[114] with methyl and β-galactosyl substituents on the 8-OH of sialic acid were also obtained. Additional structures with nonterminal sialic acid have been reported.[112,115]

3. DISTRIBUTION

3.1. Gross Compartmentalization

Early studies on the distribution of gangliosides[116,117] showed gray matter to have a substantially higher concentration than white matter, and from that arose the concept of neuronal specificity. That idea has been drastically revised in light of numerous studies demonstrating occurrence in all cell types and subcellular fractions of brain and extraneural tissues as well. However, it remains true that the bulk of brain ganglioside resides in neurons, although their mode of distribution within these complex cells is still controversial (see below).

Table V summarizes ganglioside concentrations reported for several human tissues, which are typical of mammalian tissues generally. Peripheral nerve contains a fraction of the level found in brain, and within the CNS, large differences are seen between brain and spinal cord. Variation of ganglioside concentration has been observed for different gray matter regions and also white matter tracts.[129–131] Extraneural tissues in general contain less than 10% of the gray matter level. Adrenal medulla, derived embryologically from the neural crest and considered a component of the autonomic nervous system, has somewhat more than the typical extraneural tissues.[101,125,126] Cerebrospinal fluid concentrations are extremely low[128] but nevertheless show some potential for disease diagnosis.[132,133]

Distributional patterns of gangliosides also vary in different regions of the nervous system.[130,131,134] Cerebellum, for example, differs from cerebral cortex in having higher G_{T1b} and reduced G_{D1a}.[130,134] White matter usually contains somewhat more G_{M1} than gray matter regions, reflecting the presence of myelin (see below). Spinal cord has elevated G_{M3} and G_{D3} and reduced G_{D1a} compared to cerebrum.[135] Peripheral nerve appears to have a pattern that is entirely different from brain, with a high proportion of glucosamine-containing gangliosides.[81,106] This tissue has not yet been thoroughly studied. Lumbar CSF had a pattern somewhat similar to brain but with relatively more tri- and tetrasialogangliosides; the pattern for ventricular CSF, however, was closer to that of plasma in the relative enrichment of G_{M3}.[128]

3.2. Brain Cells

Gangliosides of isolated neuronal and glial cells have been studied by several laboratories, and the results compiled in recent reviews.[7,32,38,47,48] Some of these findings are summarized in Table VI. The results vary considerably, possibly because of differences in cell isolation and assay procedures. The

Table V
Ganglioside Content of Human Tissues

Tissue	Concentration[a]	Reference
Brain gray matter	2850–3530	37,118–121
Brain white matter	900–1570	37,118–121
Spinal cord gray matter	751	37,121
Spinal cord white matter	430	37,121
Retina	366	122
Sciatic nerve	259	123
Adrenal medulla	407–757	125,126
Muscle	52	124
Liver	214	100
Plasma	11.3	127
CSF (lumbar)	0.841	128
CSF (ventricular)	0.259	128

[a] Tissue concentrations are expressed as nmoles of lipid-bound sialic acid per gram fresh weight; plasma and CSF are expressed as nmoles lipid-bound sialic acid per milliliter.

Table VI
Gangliosides of Isolated Brain Cells

Cell and source	Concentration[a]	Investigator	Ref.
Neurons			
Ox Deiter's nucleus	18.8[b]	Derry and Wolfe (1967)	136
Rat cortex	3.8	Norton and Poduslo (1971)	137
Rat brain	2.5	Skrivanek *et al.* (1978)	138
Rat brain	6.6	Maccioni *et al.* (1978)	139
Rabbit cortex	9.4	Hamburger and Svennerholm (1971)	140
Pig brainstem	6.5[b]	Tamai *et al.* (1971)	141
Human brain	4.2	Yu and Iqbal (1979)	142
Astroglia			
Rat cortex	10.3	Norton and Poduslo (1971)	137
Rat brain	6.8	Skrivanek *et al.* (1978)	138
Rabbit cortex	18.1	Hamburger and Svennerholm (1971)	140
Oligodendroglia			
Calf white matter	4.2	Poduslo and Norton (1972)	143
Human white matter	1.1	Yu and Iqbal (1979)	142

[a] Expressed as nmoles of lipid-bound sialic acid per milligram protein.
[b] Recalculated on the assumption of 50% protein in dry weight.

relatively high concentration in neurons from Deiter's vestibular nucleus may have been caused by ganglioside-rich processes contaminating the hand-dissected preparation. Astroglia, when isolated in bulk at the same time as neurons, have generally yielded twice the ganglioside concentration of the latter, although absolute values vary between laboratories. It has been pointed out[137] that astroglia have a much higher surface area to volume ratio, thus providing more plasma membrane per cell. Hence, ganglioside concentration may be considerably greater in the neuronal membrane. Although oligodendroglia were initially shown[143] to have a ganglioside concentration comparable to that of neurons, a recent study[142] indicated significantly lower concentration in the former.

Ganglioside distributional patterns of neurons and astroglia proved rather similar, again providing analyses were carried out in the same laboratory.[140,144,145] One study[141] found close correspondence between neurons and the gray matter pattern. Gangliosides labeled *in vivo* and subsequently extracted from isolated neurons and astroglia showed virtually identical patterns of major and minor species when subjected to two-dimensional TLC and autoradiography.[146] A recent study[142] of human oligodendroglia revealed a pattern distinctly different from neurons and astroglia, characterized by the presence of G_{M4} and relative enrichment of G_{M3}, G_{M2}, and G_{D3}.

3.3. Subcellular Distribution: Oligodendroglia

The CNS myelin membrane has a ganglioside concentration comparable to that of oligodendroglia on a dry weight basis and a pattern that is different from all other cellular and subcellular fractions. This includes a predominance

of G_{M1}[147] and the presence of sialosylgalactosylceramide (G_{M4}).[96] The latter
is relatively abundant in CNS myelin of primates and birds, less prominent
though detectable in CNS myelin of other mammals, and not detectable at all
in myelin of lower vertebrates.[99,148] Within the nervous system of mammals,
it appears to be a specific marker for oligodendroglia and CNS myelin. Peripheral nerve myelin was not found to contain G_{M4}.[106]

3.4. Subcellular Distribution: Neurons

Despite the high concentration in astroglia, neurons contain the bulk of
cortical ganglioside because of their greater number and surface area.[149–151]
Plasma membrane is believed to be their principal locus, although one study[152]
indicated that internal cell membranes and the nuclear membrane also bind
cholera toxin. The fact that gangliosides are synthesized by internal organelles
prior to entering the plasma membrane (see below) would suggest the likelihood
of some internal localization. Their mode of distribution within the neuronal
membrane is still a matter of controversy. The nerve ending has been considered the most likely zone of concentration, stemming from early reports of
unusually high ganglioside content in hand-dissected neuropil[136] and isolated
synaptic plasma membranes (SPM).[153,155] However, a survey of the literature
(Table VII) reveals that this has not been a universal finding; the majority of
studies report one-half to one-third as much ganglioside in SPM as those early
reports. These values fall in the approximate range of 40–70 nmol NeuAc/mg
protein. Because of the sixfold variation between the highest and lowest assays,
the true value remains uncertain.

Other neuronal subfractions have been widely studied in relation to ganglioside content. The variation in reported values for intact synaptosomes,
particularly in the rat, is much less than for SPM (Table VII), and this average
value, together with the reported estimate of the number of axon terminals per
unit of cortex, permitted calculation of the portion of total cortical ganglioside
present in nerve endings.[47] The fact that the large majority of synapses in the
cerebral cortex involve axon terminals rather than dendrodendritic
connections[175] lends credence to this type of calculation. The result suggested
that approximately 12% or less of total ganglioside in the rat cortex resides in
nerve-ending complexes.[47]

From this result and other considerations, it was postulated[47] that gangliosides are distributed over a major portion (perhaps all) of the neuronal
surface. Histochemical studies that bear on this question have been reviewed;[169] with few exceptions, these support the hypothesis. Biochemical
evidence favoring the hypothesis includes the finding[170] that axolemma-enriched fractions from brain have concentrations and patterns of ganglioside
roughly equivalent to that of SPM. There is also the observation that microsomes, which originate from a diversity of brain cell membranes and contain
only a small portion of synaptic origin, have a relatively high ganglioside content
(reviewed in ref. 169). Two recent studies[168,169] have reported brain microsomes and SPMs to have similar ganglioside concentrations and patterns, but
another[161] found a higher concentration in SPM. The latter study also calculated

Table VII
Gangliosides of Neuronal Subfractions

Fraction	Concentration[a]	Investigator	Ref.
Synaptosomes			
Ox	53	Wiegandt (1967)	154
Calf	26	Tettamanti *et al.* (1980)	161
Rabbit	45	Hamburger and Svennerholm (1971)	140
Rabbit	39	Tettamanti *et al.* (1973)	156
Rat	35	Yohe *et al.* (1980)	157
Rat	31	Dekirmenjian and Brunngraber (1969)	158
Rat	30	Avrova *et al.* (1973)	155
Rat	30	Seminario *et al.* (1964)	159
Rat	24	Maccioni *et al.* (1978)	139
Guinea pig	27	Eichberg *et al.* (1964)	160
Human	23	Kornguth *et al.* (1974)	162
Synaptic plasma membranes			
Rat	146, 60	Avrova *et al.* (1973)	155
Rat	144	Breckenridge *et al.* (1972)	163
Calf	90	Tettamanti *et al.* (1980)	161
Rat	68, 57	Hungund and Mahadik (1981)	164
Rat	62	Brunngraber *et al.* (1967)	165
Guinea pig	55	Whittaker (1969)	166
Rat	54, 24	Lapetina *et al.* (1968)	167
Rat	51	Cruz and Gurd (1981)	168
Rat	46, 38	Skrivanek *et al.* (1982)	169
Axolemma			
Rat	78	DeVries *et al.* (1981)	170
Bovine	49	DeVries *et al.* (1981)	170
Human	45	DeVries *et al.* (1981)	170
Synaptic vesicles			
Rat	16	Lapetina *et al.* (1968)	167
Rat	9.4	Breckenridge *et al.* (1973)	171
Soluble			
Rat neurons and glia	4.3	Sonnino *et al.* (1979)	172
Rat synaptosomes	4.8	Sonnino *et al.* (1979)	172
Rat synaptosomes	1.8	Lapetina *et al.* (1968)	167
Rat synaptosomes	1.1	Ledeen *et al.* (1976)	173
Axons			
Ox[b]	1.2	DeVries and Norton (1974)	174

[a] Expressed as nmoles of lipid-bound sialic acid per milligram protein. Some values have been recalculated.
[b] Primarily axoplasm with little or no axolemma.

a greater proportion of cortical ganglioside present in SPM. It was recently proposed,[176] on the basis of immuno-electron-microscopic observations, that gangliosides are evenly distributed over the neuronal surface except for a local increase in the synaptic junction. The latter aspect appears in conflict with other reports[173,177] claiming no enrichment of ganglioside in isolated junctional complexes, but since detergents were used in those isolation procedures, a definitive answer awaits further experimentation. For the present, it is clear that no concensus has yet been reached on the question of ganglioside localization in the neuron.

3.5. Soluble Gangliosides

Although the large bulk of brain gangliosides is undoubtedly membrane bound, reports from three laboratories have indicated low concentrations of soluble gangliosides in brain (Table VII). It may be noted that axoplasm (viz., axons isolated without their plasma membranes) also contains gangliosides at comparably low concentrations (Table VII). One study[172] has indicated that cytosolic gangliosides in calf brain occur complexed to protein. There is some indication that soluble gangliosides in nerve-ending cytoplasm may have different metabolic properties than membrane-bound gangliosides.[173]

4. METABOLISM

The main pathways for synthesis and degradation of gangliosides have been defined through a combination of *in vivo* turnover studies and *in vitro* enzymology. Ganglioside biosynthesis has been shown to proceed in stepwise fashion through transfer of individual sugars from their nucleotide conjugates to the growing oligosaccharide chain. The concept of "cooperative sequential specificity" was proposed[21] as a mechanism by which the glycolipid product of one glycosyltransferase serves as the substrate for the next enzyme in the sequence. There is now evidence to indicate that this coordinated multiglycosyltransferase is part of the Golgi apparatus (see below). According to one model,[178] each ganglioside is synthesized by a separate multienzyme complex. Enzymes involved in degradation of brain gangliosides may also function together in a multienzyme system analogous to that suggested for biosynthesis.[179] Several reviews on ganglioside metabolism have appeared in recent years.[35,36,38,180-185]

4.1. In Vivo Studies

The results of metabolic studies have proved quite variable, depending on the choice of precursor, age of the animals, and other factors. For example, rat brain studies in which glucosamine was injected gave a half-life of 24 days for the gangliosides,[186,187] whereas galactose[186] and glucose[187] gave values of 20 and 10 days, respectively. The shorter half-lives most likely reflected less reutilization of metabolized precursor and thus are considered closer to the true values. In another study,[188] sphingosine and stearate moieties, labeled by intracerebral injection of acetate, lost specific radioactivity simultaneously at the same rate in all four major gangliosides. However, the observation of half-lives of approximately 60 days indicated extensive reutilization. Younger animals generally showed faster turnover rates than older ones, although one of the earlier studies[187] reported slower turnover during the first 10 days postpartum than in the next 8-day period in the rat.

Evidence of possible metabolic interrelationships among gangliosides has been sought in several studies. The general finding has been that similar rates of labeling and apparent turnover prevail for individual gangliosides, thus indicating absence of precursor–product relationships. This would be consistent with the proposed model[178] of a separate multienzyme complex for each ganglioside. Although some of the studies have suggested that a small amount of G_{M1} may constitute a precursor pool leading to the oligosialogangliosides, it is clear that the whole pool of any one of the major gangliosides is not the precursor pool to any of the others. Catabolism also appears to involve maintenance of separate compartments. Thus, turnover studies in whole rat brain[187] showed that G_{M1} produced by degradation of multisialogangliosides does not mix with the whole pool of G_{M1} in brain.

Studies comparing turnover rates of sialidase-sensitive and sialidase-resistance NeuAc in brain gangliosides have given conflicting results. In one study,[189] double-labeling experiments with [^{3}H]glucosamine and [^{14}C]glucose injected into brains of 8-day-old rats produced the same ^{3}H/^{14}C ratio for the two forms of NeuAc in all of the gangliosides, indicating a single pool of CMP-NeuAc. Further evidence against selective turnover has been cited in reference to whole brain[188] and the specific pool of gangliosides reaching the nerve ending by axonal transport.[173] Earlier studies,[190,191] however, indicated differential labeling and turnover.

An important consideration in the study of ganglioside metabolism came to light in a recent finding[192] that free NeuAc, CMP-NeuAc, lipid-bound NeuAc, and protein-bound NeuAc all had similar specific radioactivities a few days after intraventricular injection of labeled ManNAc in the rat. That and the further observation that all four pools lost specific radioactivity at the same rate were interpreted as evidence for active recycling of sialic acid molecules or precursors. It was concluded that the calculated half-life of 3.5 weeks was not a true half-life of brain sialoglycoconjugates but merely the rate of leakage out of the brain of labeled NeuAc and/or precursors. Using the rate of incorporation of labeled ManNAc, the authors calculated approximate half-lives of 6–8 days and 2–3 days for the sialic acid residues of gangliosides and glycoproteins, respectively. Slower turnover of gangliosides compared to sialoglycoproteins has been observed elsewhere.[169]

4.2. Cell Cultures

Metabolic studies carried out *in vivo* are potentially complex because of the multiplicity of cell types and additional factors such as axonal transport. Use of whole brain can obscure possible differences between specific compartments, as revealed for example in a study[193] of myelin G_{M1}. The latter was found to turn over more slowly than G_{M1} of whole rat brain, approaching the rate of other myelin lipids. Such difficulties have caused investigators to turn to simpler systems such as cultured cells to gain further insight into ganglioside metabolism. Transformed and tumoral cells have provided one approach (ref. 194 for review). Studies with cultured glioblastoma and neuroblastoma cells revealed relatively simple ganglioside patterns in the former (mainly G_{M3} and G_{D3}) and somewhat more complex patterns for the latter.[195–198] These patterns were consistent with the presence of UDP-GalNAc:G_{M3} N-acetylgalactosaminyl transferase in neuronal but not glial clones. Several neuroblastoma clones were found to have a predominance of monosialogangliosides and also appreciable G_{D1a}.[195,196,199] Neuroblastoma cells grown *in vivo* differed from those grown in culture in having the capacity to produce G_{T1}.[200]

Morphological differentiation of neuroblastoma cells, induced by serum deprivation, dibutyryl cyclic AMP, or bromodeoxyuridine, caused variable changes but did not restore a true neuronal pattern.[199] The differentiated cells always exhibited increased levels of total ganglioside, larger increases corresponding to greater differentiation.[201] Changes in ganglioside pattern also accompanied such differentiation. Neurotumor cell lines are being employed to study the effect of opiates on ganglioside metabolism (Section 4.3.4).

Primary cultures of brain cells provide another potentially useful approach now that methods are available for culturing relatively pure neurons and glia.[202,203] Such cultures were shown to contain tri- and tetrasialogangliosides[203] and thus were closer to brain patterns then transformed cell lines. The fact that gangliosides of this type were detected in primary cultures of glial cells ruled out the possibility that their presence in bulk-isolated glia resulted from neuronal contamination.[203] It was of some interest that coculture of the two cell types produced an overall change in ganglioside pattern. The same study noted that even with transformed cells, coculture of glial and neuronal types caused appearance of low levels of G_{T1b} and G_{T1L}.[203]

4.3. Biosynthesis of Gangliosides

4.3.1. In Vitro Studies Establishing the Major Pathways

The major biosynthetic pathways leading to G_{D1a}, G_{T1b}, and other related species are now well defined (Fig. 2). Embryonic chick brain was the principal enzyme source employed in the early studies,[21,22] but many of the same reactions were later observed in other species. Detailed findings on the individual reactions were described in the first edition of this series and in several reviews appearing since (see above).

An important development has been recognition of the pivotal role of N-acetylgalactosaminyl transferase, responsible for conversion of G_{M3} to G_{M2}. It is seen as having a regulatory function, since only those tissues and cells containing this enzyme have appreciable levels of the more complex gangliosides.[183] Diminished activity is probably responsible for some of the altered ganglioside patterns in transformed cells. It has been suggested that activity of this enzyme may be controlled in part by the cellular content of cyclic AMP (section 4.3.4).

A rat brain preparation that converted G_{M3} to G_{M2} was also able to convert G_{D3} to G_{D2},[204] although it is not known whether a separate enzyme was responsible. The enzyme forming G_{M2} was reported[205] to differ from that which transfers GalNAc to lactosylceramide to form GgOse$_3$Cer (GA$_2$). The latter activity was not seen in the original chick embryo preparation[22] but appears to be part of the "aminoglycolipid pathway" observed in other systems. The early studies established that the major synthetic pathway involved addition of NeuAc prior to GalNAc, with the sequence Lac–Cer → G_{M3} → G_{M2} →

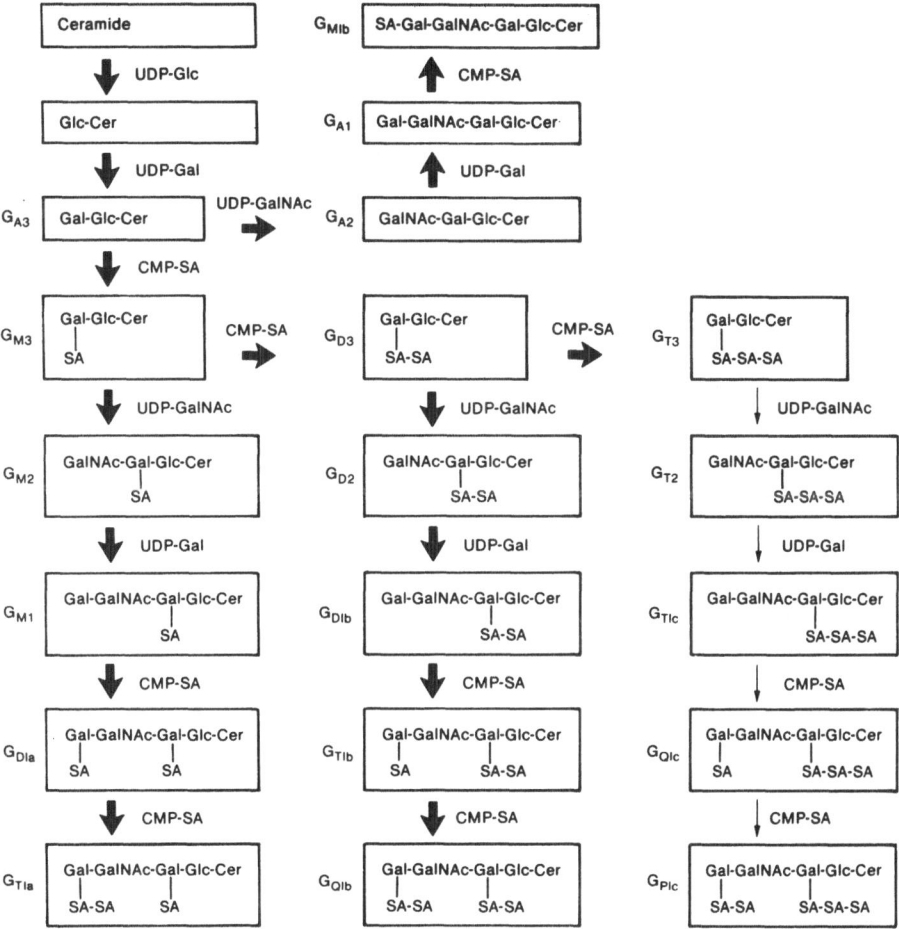

Fig. 2. Pathways for biosynthesis of gangliosides of the ganglio series. Heavy arrows indicate reactions demonstrated *in vitro*; thin arrows indicate proposed reactions not yet observed. Abbreviations are standard; SA, sialic acid.

G_{M1}. Confirmation of this was obtained in a study[206] utilizing double-label procedures and the endogenous glycolipid acceptors present in the enzyme preparation. Lactosylceramide is seen as one of the branch points in biosynthesis of gangliosides and blood group glycolipids (Figs. 2 and 3).

Further study of galactosyltransferase[207,208] revealed that the enzyme that synthesizes G_{M1} from G_{M2} also catalyzes the conversion $G_{D2} \rightarrow G_{D1b}$. This galactosyltransferase is different from the one that forms lactosylceramide, and both differ from that responsible for synthesis of galactosylceramide.[184] Another galactosyltransferase, which completes the synthesis of nLcOse$_4$Cer, has been characterized in embryonic chick brain[209] and rabbit bone marrow.[210] This enzyme has been separated from the $G_{M2} \rightarrow G_{M1}$ galactosyltransferase in embryonic chick brain.[211]

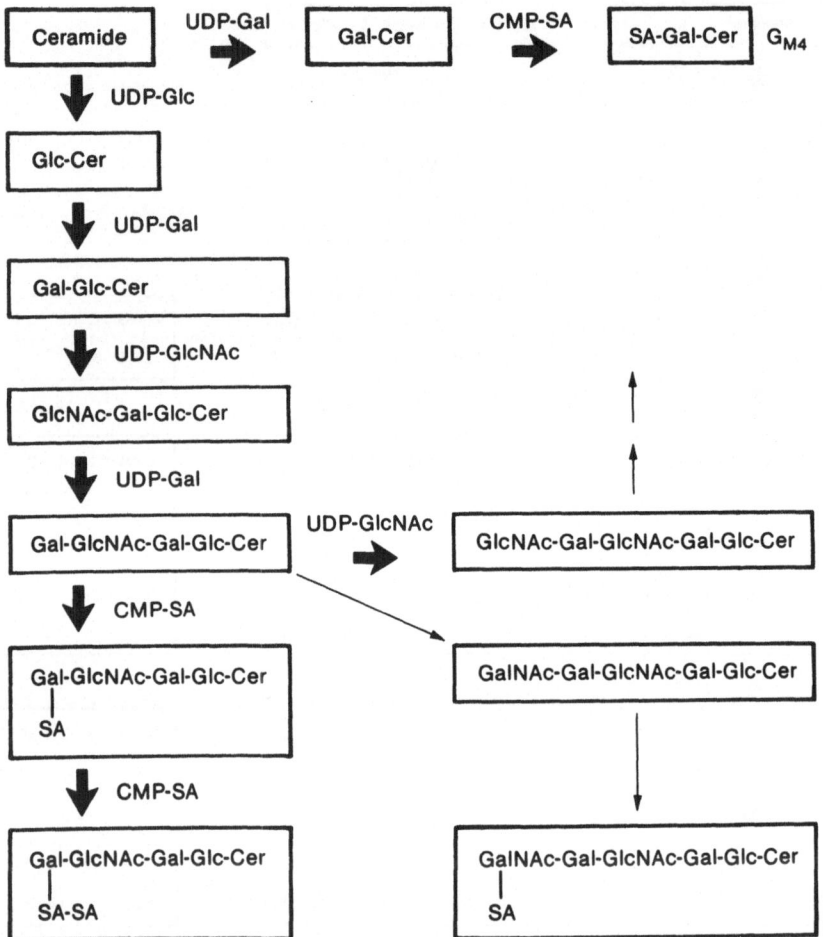

Fig. 3.   Pathways for biosynthesis of gangliosides of the neolacto series and G_{M4}. Heavy arrows indicate reactions demonstrated *in vitro*; thin arrows indicate proposed reactions not yet observed. Abbreviations are standard; SA, sialic acid.

New information has been acquired on the sialosyltransferases involved in ganglioside biosynthesis. The enzyme CMP-NeuAc : G_{M1} sialosyltransferase which converts G_{M1} to G_{D1a} was shown to differ from that forming G_{M3} in embryonic chick[212] and neonatal rat[213] brain. Biosynthesis of G_{D1b}, isomeric to G_{D1a}, is believed to proceed through G_{D3} and G_{D2} (Fig. 2). This pathway was suggested by discovery of a third sialosyltransferase in embryonic chick brain which converts G_{M3} to G_{D3}.[212] It was shown to be different, both in regard to enzymatic properties[212] and subcellular localization,[208] from the two sialosyltransferases that synthesize G_{M3} and G_{D1a}. Experiments with a rat brain preparation[204,207] demonstrated conversion of G_{D3} and G_{D2} to G_{D1b} when these were added as exogenous acceptors along with the appropriate sugar donors. The possibility of an alternate pathway involving G_{M1} as the immediate precursor to G_{D1b} was suggested by experiments utilizing endogenous glycolipid acceptors,[206] but the interpretation of those results has been questioned.[38]

Synthesis of trisialoganglioside G_{T1b} was accomplished by a sialosyltransferase in embryonic chick brain with CMP-NeuAc and G_{D1b} as exogenous substrates; G_{D1a} in the same system did not react.[214,215] In both embryonic chick brain[216] and bovine thyroid,[208] this sialosyltransferase was found to closely resemble the one that converts G_{M1} to G_{D1a}. The isomeric trisialoganglioside G_{T1a} was recently synthesized from G_{D1a} by a particulate fraction from embryonic chick brain,[217] and a similar preparation catalyzed the conversion of G_{T1b} to G_{Q1b}[218] (Fig. 2). Both reactions involve attachment of NeuAc via a $2{\rightarrow}8$ linkage to a terminal NeuAc$(2{\rightarrow}3)$Gal residue, but it is not known whether they are catalyzed by the same enzyme.

Details of the neolacto pathway (Fig. 3) are also being studied, and at this stage virtually all the reactions leading to $IV^3NeuAc\text{-}nLcOse_4Cer$ (sialosylparagloboside) have been carried out *in vitro*.[184] There is some indication[219] that the sialosyltransferase involved in synthesis of the latter substance may be the one forming G_{M1b} (see below). A disialosyl derivative of $nLcOse_4Cer$ has also been prepared *in vitro*, and competition studies with G_{M3} and NeuGc-$nLcOse_4Cer$ suggested that the same enzyme may catalyze formation of both that and G_{D3}.[219] These observations raise the intriguing possibility of metabolic interrelationships between gangliosides of the two glycolipid families. A pentaglycosyl homologue of $nLcOse_4Cer$, recently synthesized *in vitro* by an N-acetylglucosaminyltransferase from mouse T-lymphoma, appeared to have a terminal $\beta(1{-}6)$-linked GlcNAc residue.[220]

4.3.2. Additional Synthetic Pathways

In addition to the two established pathways in the ganglio series leading to G_{T1a} and G_{Q1b}, a third pathway has been proposed[69] leading to G_{Q1c} (Fig. 2). The latter ganglioside has been isolated from fish but not mammalian brain (see above), whereas its isomer, G_{Q1b}, shows the reverse distribution. It was suggested[69] that this may represent the major pathway for ganglioside biosynthesis in fish and thus constitute a phylogenetically older route. Essential links for this pathway were provided by isolation of three new trisialogangliosides[69] from fish brain: G_{T3}, G_{T2}, and G_{T1c} (Fig. 2, Tables II and IV). One enzyme in the pathway, a sialosyltransferase that converts G_{D3} to G_{T3}, has been detected in hog kidney,[95] and G_{T3} itself was isolated from the same source.

The above-mentioned aminoglycolipid pathway was originally postulated[221,222] as an alternative route to G_{M1} and higher gangliosides. This seemed to be supported by the observed conversion of lactosylceramide to $GgOse_3Cer$[222] by an enzyme that appeared different from that which converted G_{M3} to G_{M2},[205] and by the presence of UDP-Gal:$GgOse_3Cer$ galactosyltransferase in brain preparations of various species.[221] However, the sialosyltransferase in brain tissue and various cell lines that converted $GgOse_4Cer$ to a monosialoganglioside was shown to produce G_{M1b} (Fig. 2) rather than G_{M1}.[74,223] It is now clear that G_{M1} (G_{M1a}) is formed via G_{M3} and G_{M2} (Fig. 2) rather than $GgOse_4Cer$, but recent detection of several gangliosides containing the G_{M1b}

structure (see above) raises the possibility that the aminoglycolipid pathway (Fig. 2) may have functional significance in brain and elsewhere.

Conversion of galactosylceramide to G_{M4} appeared in one study[224] to be catalyzed by the same sialosyltransferase that synthesizes G_{M3}. However, G_{M4} synthesis was recently shown to occur in isolated oligodendroglia but not in astrocytes or neurons of rat brain.[225] All three cell types synthesized G_{M3} from lactosylceramide. This supplemented the list of cells in which activity for G_{M3} exists without a capability for G_{M4} synthesis, thus strengthening the case for two separate sialosyltransferases.

4.3.3. Site of Ganglioside Biosynthesis

Attempts have been made to identify the site of ganglioside synthesis at both the cellular and subcellular levels. Comparison of cell types revealed neurons to have much greater capability for biosynthesis of glucosylceramide, the first step in forming ganglioside oligosaccharide chains. On that basis it was suggested[226] that neurons are the primary sites of ganglioside synthesis in brain. That view received some support in the claim[225] that CMP-NeuAc:lactosylceramide sialosyltransferase activity is much higher in isolated neurons than in astrocytes or oligodendroglia. Another study[227] had shown that a neuron-enriched fraction incorporated serine and NeuAc into gangliosides much more actively than a glia-enriched fraction. The fact that glial cells, particularly astrocytes, contain a high level of gangliosides suggests either that the low levels of biosynthesis observed are sufficient to meet the needs of the cell or that gangliosides synthesized in the neuron are transferred to glia.[226]

Early subcellular localization studies suggested the nerve ending as the site of ganglioside biosynthetic activity (reviewed in ref. 38, p. 349), but subsequent work has tended to discount that theory. The accumulated evidence indicates that most of the glycosyltransferase activities detected in synaptosomal fractions and subfractions resulted from contamination with membranes of the Golgi apparatus and/or endoplasmic reticulum.[228–230] The Golgi apparatus has been shown to have a prominent role in ganglioside biosynthesis in several extraneural tissues, e.g., liver,[231,232] spleen,[233] kidney,[234] and mammary gland.[235] A recent study of thyroid[208] revealed most of the ganglioside biosynthetic activity to be located in the Golgi apparatus, although a different distribution was indicated for CMP-NeuAc:G_{M3} sialosyltransferase. A similar role for the Golgi apparatus in neurons and glia seems probable, but the evidence to data is only indirect.[236]

The idea of nerve-ending biosynthesis was recently revived to a limited extent in the claim[237] that two sialosyltransferase activities exist in synaptic plasma membranes (SPM), one acting on lactosylceramide and the other on glycoprotein acceptor. The former enzyme was found to have a specific activity in purified SPM over twice that of homogenate. A bimodal distribution of sialosyltransferase between SPM and intracellular structures was thus postulated. Other workers[139,236,238] did not find nerve-ending localization of this enzyme, although they did not look specifically at SPM. If this enzyme is indeed a component of SPM, its functional role will need to be clarified in view of the fact that none of the other ganglioside glycosyltransferase activities have been identified in that membrane. The possibility of a "sialylation–desialylation cycle" was suggested,[237] taking account of the demonstrated presence of sialidase in the same membrane (see below). It is perhaps significant that ganglioside sialosyltransferase has also been detected on the cell surface of neurons in culture.[91] However, the participation of these or other mammalian cell surface glycosyltransferases in membrane biogenesis *per se* has been questioned.[239]

4.3.4. Exogenous Factors That Affect Ganglioside Biosynthesis

Anesthesia, hypoxia, and various pharmacological agents such as chloropromazine and 2,4-dinitrophenol were reported to depress the rate of incorporation of labeled precursors into brain gangliosides (reviewed in ref. 38, p. 338). Puromycin injected into brains of neonatal rats caused a 70–80% decrease in incorporation of labeled sugars into gangliosides.[240,241] This drug reduced the activity of the enzyme UDP-Glc:Cer glucosyltransferase[242] which initiates ganglioside synthesis. However, a study with goldfish[243] found that incorporation of radiolabeled galactose into brain gangliosides was not affected by concentrations of puromycin known to block memory and protein synthesis. Hence, the putative linkage between ganglioside and protein synthesis may be dependent on species and on other factors not well understood.

Recent studies[244–246] revealed that exposure of opiate-receptor-positive mouse neuroblastoma cell lines to opiates, β-endorphin, or [D-Ala²,D-Leu⁵] enkephalin produced naloxone-reversible inhibition of ganglioside biosynthesis as measured by incorporation of radiolabeled precursors. Opiate-receptor-negative cells did not respond. Specific inhibition of the enzymes UDP-GalNAc:G_{M3} N-acetylgalactosaminyltransferase and UDP-Gal:G_{M2} galactosyltransferase was demonstrated, and evidence was obtained that the effect was mediated through reduction of cyclic AMP. A model was proposed[246] for activation of glycosyltransferases by various ligands (e.g., serotonin, cholera toxin) interacting with receptor linked to GTP regulatory protein which, in turn, is linked to adenylate cyclase. The resultant increase of intracellular cyclic AMP and activation of protein kinase were postulated to cause phosphorylation of glycosyltransferase. Opiates are believed to operate through the GTP regulatory protein in accordance with their ability to stimulate hydrolysis of GTP and thereby inhibit adenylate cyclase.[247] It was proposed[246] that inactivation of N-acetylgalactosaminyltransferase in virally transformed cells may be caused by the reduced levels of cyclic AMP usually seen in such cells.

4.4. Axonal Transport of Gangliosides

A study[248] employing the goldfish optic system indicated that gangliosides are synthesized in the perikarya of retinal ganglion cells and transported to the optic tectum in the wave of rapidly transported macromolecules. Subsequent work[249–251] gave results partially in agreement but not excluding local synthesis within the axon and nerve ending. The latter mechanism had been suggested from a study[252] with the rabbit optic system. Later work[230] with the chick optic system supported the concept of synthesis in the neuronal perikaryon followed by translocation to the nerve ending. It seems likely that the same mechanism applies to all vertebrate systems. Results of a recent study[146] indicated that both the axonal and synaptic membranes received gangliosides through the transport process, functioning as a unit in the uptake and turnover of these glycolipids. That study also demonstrated that virtually all gangliosides undergo transport simultaneously. Exposure of chicks to light increased the amount of radioactivity present in axonally transported gangliosides entering the optic tectum, although no difference was observed in the labeling of retinal gangliosides between light and dark.[253]

4.5. Catabolism of Gangliosides

4.5.1. Localization

Degradation of gangliosides proceeds in stepwise fashion with liberation of individual sugars in reverse order to biosynthesis. It has been postulated[179] that these reactions occur in a multienzyme complex analogous to that suggested for synthesis (see above). The majority of catabolic enzymes have features that identify them as lysosomal in origin, e.g., acidic pH optima, enhanced activity with detergents, and occurrence in particulate fractions sedimenting between 800 and 15,000 g. An important exception is sialidase which occurs in both lysosomal and nonlysosomal compartments. A study of the particle-bound sialidase from bovine brain provided evidence for the concentration of this enzyme in the synaptosomal plasma membrane.[254] Similar observations were made with SPM of rat[255] and calf[161] brain. Thus far, no other ganglioside hydrolase has been positively identified in this membrane.

Degradative enzymes appear to be present in all cell types. Two β-galactosidase activities along with other hydrolytic enzymes were detected in neurons, astrocytes, and oligodendroglia from bovine brain and in neurons and astrocytes of rat brain.[256] This confirmed an earlier report[257] on the presence of several lysosomal hydrolases in neuronal and astrocytic fractions of rat brain but contradicted an earlier claim[258] that β-galactosidase is a neuronal marker. Another study,[259] however, supported the latter claim in showing β-galactosidase activity to be considerably higher in neuronal cell bodies than in astroglial and oligodendroglial fractions of adult rabbit and bovine brain. The discrepancy was attributed to age differences[259] following an earlier demonstration[260] that several lysosomal hydrolases varied considerably in different cells of rat brain during development.

4.5.2. Sialidase

The first step in degradation of the oligosialogangliosides is cleavage of terminal sialic acid residues by sialidase. A particulate enzyme from human brain degraded G_{T1b} preferentially to G_{D1b} rather than G_{D1a}, indicating that the sialosylgalactosyl linkage is more reactive to sialidase than the sialosylsialosyl linkage.[261] The same phenomenon is observed with bacterial sialidase. However, it is not known whether the implied sequence $G_{T1b} \rightarrow G_{D1b} \rightarrow G_{M1}$ represents the actual *in vivo* pathway. In studying G_{T1a}, a ganglioside with a disialosyl grouping on terminal galactose, it was found that cleavage of the sialosylsialosyl bond proceeded more rapidly than for the same grouping in G_{D1b}, suggesting that the sluggishness of the latter may result from steric hindrance when it is situated on the internal galactose.[70]

Even more pronounced steric hindrance is responsible for the nearly complete resistance of G_{M1} and G_{M2} to bacterial sialidase in the absence of detergent. However, limited hydrolysis of G_{M1} can occur below the critical micelle concentration.[262] Steric hindrance has been depicted as resulting from an "oxygen cage" surrounding the NeuAc ketosidic linkage.[262,263] Sialidase from *Clostridium perfringens* was able to hydrolyze G_{M1} in the presence of cholate.[262,264] A lysosomal sialidase from brain was reported[265,266] to cleave NeuAc from G_{M2}, but the fact that detergent was employed leaves open the question of functional significance. The major, and possibly exclusive, catabolic pathway appears to be $G_{M1} \rightarrow G_{M2} \rightarrow G_{M3}$, the latter product being fully susceptible to sialidase.[267]

Sialidase in SPM has been shown to effect hydrolysis of endogenous gangliosides in the same membrane as well as exogenously added gangliosides.[156,254,255] It has been suggested[38] that this sialidase could initiate a sequence of events leading to return of the ganglioside to the cell body by retrograde axonal transport. It may be noted that the process is probably not restricted to the neuron, since fibroblasts were reported to contain an ectosialidase as well as a lysosomal sialidase,[268] and spleen cells appear to have sialidase activity in the plasma membrane/microsomal fraction.[269]

4.5.3. β-Galactosidase

Metabolic degradation of G_{M1} to G_{M2} involves action of a β-galactosidase. One such enzyme was first isolated from rat brain and found to be active toward G_{M1}, GgOse$_4$Cer, and lactosylceramide.[270,271] Subsequently β-galactosidase activity in liver was fractionated into three components by electrofocusing.[272,273] Present evidence indicates the presence of at least two genetically distinct acidic lysosomal β-galactosidases in mammalian brain.[274] One of these, G_{M1} β-galactosidase, hydrolyzes G_{M1} and GgOse$_4$Cer, whereas the other, galactosylceramide β-galactosidase, is primarily responsible for degradation of galactosylceramide. Both enzymes are capable of hydrolyzing lactosylceramide, and although the existence of a separate β-galactosidase for the latter substrate has been suggested,[275] that finding has not yet received independent confirmation. Metabolic degradation of lactosylceramide could occur through the two known β-galactosidases. Profound deficiency of G_{M1} β-galactosidase was shown[24] to be the genetic origin of "generalized gangliosidosis," later named G_{M1}-gangliosidosis. A protein activator of this enzyme was recently isolated from human liver.[276]

4.5.4. β-Hexosaminidase

Degradation of G_{M2} to G_{M3} is catalyzed by β-hexosaminidase A, the activity being greatly enhanced by a protein activator.[277] This activator was shown to be different from that activating β-galactosidase.[276] Absence of this enzyme is responsible for the classical form (variant B) of infantile G_{M2}-gangliosidosis (Tay–Sachs disease).[25,278] The B form of the enzyme exhibited little activity toward G_{M2}, even in the presence of protein activator. In the variant O of the disease, both A and B isozymes are deficient,[26,279] whereas the variant AB is characterized by normal or elevated levels of these two forms with deficiency of the activator protein.[280,281] A new variant of the type AB G_{M2}-gangliosidosis was recently reported[282] in which β-hexosaminidase A and B of brain showed normal activity toward synthetic substrates but could not hydrolyze G_{M2}. The G_{M2} activator in brain of that patient was elevated. The gangliosidoses comprise an increasingly complex group of biochemically and phenotypically variant diseases whose characteristics were recently summarized.[185,283,284]

4.5.5. β-Glucosidase

The final carbohydrate, Glc, is cleaved by glucosyl-ceramide β-glucosidase. The enzyme purified from rat and human spleen did not cleave galactosylceramide.[285] Ceramide is hydrolyzed to fatty acid and to sphingosine, which can undergo further degradation following phosphorylation of the primary hydroxyl.[286]

5. FUNCTIONS OF GANGLIOSIDES

Speculation concerning ganglioside function was initially based on the assumption of neuronal specificity. Current awareness of their widespread distribution has considerably broadened the range of potential functions now receiving consideration, although the neuron continues to command special interest because of the exceptionally high ganglioside concentration in its membrane. Aside from quantitative considerations, however, it is well to recall the virtual identity of ganglioside distributional patterns for neurons and astrocytes (see above) as a caution against overspeculation regarding neuron-specific functions of any particular ganglioside. The various theories to be outlined regarding ganglioside function are still in their formative stages and should be viewed only as very tentative hypotheses.

5.1. Ion Fluxes

The idea that gangliosides might influence ion permeability and flux in neural membranes was fostered by earlier reports[287–289] purporting to show that they maintain the excitability of nervous tissue *in vitro*. Brain slices responding to electrical pulses or potassium chloride with increased oxygen uptake lost this response in the cold or after addition of protamine or histones, whereas exogenous gangliosides in the medium restored the response. It was proposed that the native gangliosides offer acidic sites that function in active cation transport. However, reversal of the inhibitory effect of protamine and histones was also accomplished with sulfatides, phosphatidylserine, and other acidic substances.[290] Other findings also cast doubt on the physiological significance of the ganglioside effect.[291,292] The idea of a role in ion transport has been revived in recent years with evidence that gangliosides may influence surface potential and membrane enzymes that mediate sodium and potassium flux (see below).

5.2. Synaptic Transmission

A possible role in synaptic transmission was considered from the standpoint of calcium binding[293] following demonstration[294] of calcium–ganglioside complex formation. Gangliosides in this light were viewed as the extracellular calcium storage system of the synapse. Complex formation was then considered to cause the synaptic membrane to "close," whereas dissociation of calcium, purportedly initiated by K^+, would cause an "opening" of the presynaptic region.[295] The fact that such complexes are more stable with increasing sialic acid residues was used to explain an apparent compensatory increase of such gangliosides in brains of poikilotherm fish transferred to lower temperatures at which calcium–ganglioside complexes are thought to be destabilized.[293,296]

A similar view of gangliosides as carriers of calcium[176] pictures the latter as entering the nerve ending through specific calcium channels during neurotransmission and postulates replacement of such bound calcium by sodium with destabilization of the synaptic membrane in a manner that facilitates coalescence of the latter with the synaptic vesicle membrane. This model presumes (though does not necessarily require) enrichment of gangliosides at the synapse.

A somewhat different view of ganglioside function at the nerve ending was based on the observation that exogenous polysialogangliosides, but not monosialogangliosides, induced release of dopamine from synaptosomes in the presence of calcium.[297] Since membrane surface potential influences ionic permeabilities,[298,299] it was proposed that the polysialogangliosides function by causing a decrease in surface potential and modification of molecular packing. This property was shared with other fusogenic agents.

5.3. Receptors

One of the earliest suggested functions was that of serotonin receptor following the observation[300] that addition of gangliosides to sialidase-treated muscle restored the tissue's contractile response to serotonin. G_{D3} was the most active fraction. Although this idea was supported by some studies, other work[301] has cast doubt on that interpretation. Another possible interaction with serotonin was suggested by the report[302] that G_{D3} and other gangliosides greatly enhance the binding of this transmitter to a specific serotonin-binding protein which is localized intracellularly.

The idea that gangliosides may function as biological receptors has gained ground in recent years, in large part because of the successful demonstration of a receptor role toward cholera toxin and possibly other cell surface agents. The cholera toxin phenomenon has been well studied and is viewed as a possible model for high-affinity transport of polypeptide messages into the cell. The large variety of oligosaccharide structures occurring in gangliosides and other glycosphingolipids might thus provide the specificity required for recognition of a diverse group of biologically active peptides. At the present time, cholera toxin remains the only substance convincingly shown to function through mediation of a ganglioside receptor, although some evidence supports a similar function for tetanus toxin. This subject has been reviewed in recent years[181,183] and is only briefly surveyed here.

5.3.1. Cholera Toxin

It was first observed in 1971[303] that a crude ganglioside mixture could inactivate cholera toxin, and shortly thereafter three groups simultaneously reported G_{M1} to be the active species.[304–306] Subsequent studies in several laboratories provided strong proof that G_{M1} is the natural receptor.[183,307,308] This toxin is composed of five relatively small receptor-binding (B) subunits and a larger (A) subunit which penetrates the cell, giving rise to the A_1 effector fragment.[309] The latter activates adenylate cyclase by catalyzing an ADP–ribose transfer from NAD onto the GTP-binding regulatory component of adenylate cyclase. The G_{M1} receptor is believed to alter the conformation of the B subunit (during binding) in a way that permits entry of the A subunit into the cell. Various penetration models involving such conformational change have been proposed.[183,308] The high-affinity nature of the binding has made cholera toxin a useful tool for localization studies. However, caution is required in such work, since it has been claimed that several glycoproteins, e.g., in microvillus membranes,[310] lymphoid cells,[311] and Balb/c 3T3 cells,[311] also react with cholera toxin.

5.3.2. Tetanus Toxin

Interaction between tetanus toxin and gangliosides, first described over 20 years ago,[312,313] was observed with boundary electrophoresis and analytical ultracentrifugation. Later it was reported[314,315] that G_{D1b} and G_{T1b} were especially effective in both binding and neutralizing the toxin. This was confirmed in more recent studies[308] with a new assay in which plastic-attached gangliosides bind toxin which is then quantified by vapor condensation on surface or an immunoenzymic method. Strong reactions were obtained with G_{T1b}, G_{Q1b}, and G_{D1b}, closely followed by G_{T1a}; distinctly lower but still significant activity was seen with G_{M1}, G_{D1a}, and G_{D3}. Binding was thus less specific and also of lower affinity than with cholera toxin. The minimal structural requirement appeared to be one sialosyl residue linked to the inner galactose, whereas optimal activity required a disialosyl group at that position.

The evidence is not yet conclusive that any of these gangliosides are true functional receptors for tetanus toxin, responsible for its specific binding to presynaptic nerve endings and subsequent transfer to the neuron cell body via retrograde axonal transport.[316] It is similar to cholera toxin in having separate binding and effector regions but possesses one binding site per molecule. Tetanus toxin has become an accepted cytochemical agent for identification of neurons.[317] and, as such, binds to the entire cell rather than just to the nerve ending. An unresolved question is why this cytochemical specificity should be observed, considering that glial cells as well as neurons contain complex gangliosides of the G_{1b} type (see above). A related question is whether this phenomenon has any relation to the high-affinity binding that occurs at the nerve ending and leads to retrograde transport and disease induction. The possibility of a glycoprotein receptor for the latter has not been excluded.

5.3.3. Other Bacterial Toxins

Gangliosides were reported to inactivate botulinum toxin in proportion to the number of sialic acid residues.[318,319] However, a subsequent study[320] revealed that G_{T1b} was more effective than G_{T1a} or G_{Q1b} in both detoxification and inhibition of neurotoxin binding to synaptosomes, thus indicating the importance of location as well as number of sialic acids. It has been pointed out, in observing fixation and deactivation of various toxins, that a distinction should be made between specific and nonspecific phenomena.[321]

The *E. coli* enterotoxin, which produces diarrhea in a manner similar to cholera toxin, also interacts with G_{M1}.[322,323] Gangliosides were found to sensitize unresponsive fibroblasts to this toxin, as evidenced by increased cyclic AMP formation.[324] It has been suggested[325] that *Vibrio parahaemolyticus* haemolysin has a ganglioside receptor that has not been identified. The staphylococcal α toxin was claimed[326] to interact specifically with IV^3NeuAc-$nLcOse_4Cer$, showing coprecipitation on gel diffusion.

5.3.4. Glycoprotein Hormones

The receptors for thyrotropin (TSH) and related glycoprotein hormones are generally believed to be sialoglycoconjugates, although the precise nature of such conjugates has been a matter of controversy. Early evidence[327] pointed to a membrane glycoprotein as the active receptor for TSH, whereas the subsequent discovery[328] that gangliosides inhibit binding of the hormone to thyroid plasma membranes and themselves bind to TSH when incorporated into liposomes suggested a role for these substances. Further, those gangliosides that most effectively inhibited TSH binding (G_{D1b}, G_{T1b}) also produced the most marked conformational alteration of the hormone as measured by fluorescence changes. Thyrotropin and other glycoprotein hormones bind to specific receptors on the cell surface through the β subunit, releasing the α subunit intracellularly for activation of adenylate cyclase. It was claimed[329,330] that regions of amino acid sequence homology existed between the cholera toxin B fragment and the β chains of TSH, luteinizing hormone, human chorionic gonadotrophin, and follicle-stimulating hormone. Sequence homology was also noted between the A_1 subunit of cholera toxin and the α chain of these hormones[330] These and other findings pointed to a receptor role for gangliosides and to cholera toxin as a possible model for the mechanism of action of these several hormones.[331]

To take account of all of these results, a two-component model was proposed for the TSH receptor, according to which both ganglioside and glycoprotein act in concert.[331,332] Protein was believed to provide the high-affinity recognition site, and ganglioside the low-affinity component that induces the necessary conformation in the ligand for its entry into the lipid bilayer. However, not all the evidence has supported this model.[331] Thus, certain phospholipids mimicked gangliosides in preventing binding of TSH to thyroid plasma membranes, although these did not bind to TSH when incorporated into liposomes.[333,334] Various investigators have reported that under more physiological conditions gangliosides and cholera toxin did not block TSH binding or action.[335,336] A recent study[337] employing FTTL, a cloned line of normal rat thyroid cells with high-affinity functional receptors for TSH, found these cells to have none of the complex gangliosides postulated as essential components of the TSH receptor. From that and other evidence, it was concluded that gangliosides have no physiological role in the binding of TSH to thyroid cells. However, the sensitivity of ganglioside detection in that study was an unknown factor, and, hence, the question of whether gangliosides have a role in the receptor for TSH and other glycoprotein hormones is still in abeyance. It has been pointed out[49] that gangliosides in model systems could mimic glycoproteins with similar oligosaccharide chains, and, indeed, a group of such "ganglioproteins" has been isolated.[338]

5.3.5. Interferon

The evidence that interferon functions through attachment to a ganglioside receptor is largely circumstantial. Soluble interferon bound to sepharose–ganglioside beads and sepharose-bound interferon lost their antiviral activity after preincubation with gangliosides.[339] The most active species were G_{M2} and G_{T1}, gangliosides of very different structures. The inhibitory effect of gangliosides was shown to involve the antiviral and antigrowth activities of mouse fibroblast interferon but not mouse T-cell interferon.[340] Human fibroblast and leukocyte interferons were also neutralized by preincubation with gangliosides.[341] The reservations regarding glycoprotein hormones with respect to ganglioside receptors obviously apply to interferon as well.

5.3.6. Sendai Virus

The possibility of a receptor function for Sendai virus was first suggested by the observation[342] that ganglioside-containing liposomes inhibited the agglutination of erythrocytes by this virus. The idea was later extended by the demonstration[308,343] of virus binding to plastic-adsorbed gangliosides having the structure NeuAcα2→8NeuAcα2–3Galβ1→3GalNAc→ (e.g., G_{T1a},

G_{Q1b}, G_{P1c}). Lower-affinity binding was obtained with G_{D1a} and G_{T1b}. These gangliosides in the same relative potency were found to restore susceptibility of sialidase-treated Madin–Darby bovine kidney cells to Sendai virus infection.[344] Somewhat different results were obtained with HeLa cell gangliosides which, when incorporated into liposomes, did not bind the virus.[345] On the other hand, a glycoprotein fraction extracted from HeLa cells with chloroform–methanol showed avid binding of this virus. These apparently contradictory results were obtained with different assays applied to different cell types, and the cause of the discrepancy is therefore uncertain. It was suggested[342] that sialoglycoconjugates of both types may behave as receptors, but this possibility has not been rigorously tested.

5.3.7. Fibronectin

On the basis of their inhibition of fibronectin-mediated cell adhesion to collagen, gangliosides were proposed as possible receptors for fibronectin on the cell surface.[346] G_{T1} and G_{D1a} were the most effective inhibitors. However, cell attachment on a collagen or lectin layer was also inhibited, suggesting that the active gangliosides may have functioned in a nonspecific manner.

5.4. Protein Modification

In addition to their potential function as membrane receptors for exogenous proteins, gangliosides may also interact in various ways with protein components of the membrane itself. This was suggested by the ability of spin-labeled gangliosides, incorporated into both artificial and natural membranes, to cluster around and bind to glycoproteins in a reversible manner.[347] Such associations were postulated as particularly strong in the case of glycoproteins bearing carboxyl residues (e.g., on sialic acid) to which cross linking by divalent cations can occur. It was postulated[347] that intertwining oligosaccharide chains of this type, with extensive cross linking, could form stable yet potentially malleable units of the cell's glycocalyx. A similar idea was proposed in the dynamic annular model wherein gangliosides (and other glycolipids) would comprise part of the closely associated ring of annular lipids surrounding a functional membrane protein.[4,348] It is conceivable that certain proteins (e.g., enzymes) on the cell surface may require such interaction for optimal conformation and reactivity.

5.4.1. Enzyme Activation

Recent reports have suggested that gangliosides may alter the activity of membrane-bound enzymes. Thus, adenylate cyclase of rat cerebral membranes experienced 50–95% activation by exogenous gangliosides[349] Cyclic nucleotide phosphodiesterase, in the membrane-bound state, did not appear to be affected, but the soluble or solubilized forms were activated up to 130%.[350] A recent study of brain (Na^+,K^+)-ATPase showed enhancement of 26–43% by added gangliosides.[351] The four major gangliosides were equally effective, but other amphipathic lipids and free sialic acid were ineffective. The process was concentration dependent and required stable, irreversible insertion of ganglioside into the membrane bilayer. Previously observed[352,353] inhibition was attributed to the loose ganglioside interaction occurring at higher concentrations. The Mg^{2+}-ATPase was reported[352] to be activated by gangliosides, but the effect was smaller on Ca^{2+}-ATPase.

5.4.2. Myelin Proteins

Recent studies have suggested the possibility of *in situ* association of myelin gangliosides and the myelin basic protein (P_1) present in both the CNS and PNS. This protein released large quantities of entrapped glucose from liposomes containing G_{M4},[354] a ganglioside specific to CNS myelin (see above). The effect was relatively specific for G_{M4}. Another study[355] reported this ganglioside to be protected against sialidase when combined with the myelin basic protein, whereas G_{M3} under similar circumstances was hydrolyzed. Other studies[106,356] had pointed out that ganglioside sialic acid and myelin basic protein are approximately equimolar *in situ*, and evidence for association of G_{M1} with specific amino acids of the protein was obtained from NMR studies.[357] Extraction of myelin with organic solvents produced a protein–lipid complex containing basic protein, ganglioside, phosphatidyl serine, and sulfatide in molar ratio of $1:1:3:6$.[358] The ability of gangliosides to modulate the antigenicity of P_2 protein of the PNS in inducing EAN suggested a molecular association.[4]

5.5. Development and Differentiation

Ganglioside changes related to development have been studied in brain and retina of several species (for review, refs. 38,359). The pronounced changes observed in concentration and distributional pattern suggested developmentally related functions. Ganglioside pattern changes observed in the developing chick retina have been correlated with fluctuations in the activities of biosynthetic enzymes.[360] Developmentally related changes in the small pool of soluble gangliosides have also been described.[359] An important advance in this area has been the use of primary cultures of neuronal and glial cells from 8-day-old chick embryo hemispheres.[361] Neurons were found to have two periods of ganglioside accumulation: a first phase corresponding to cell division in which ganglioside content increased slightly, and a second period corresponding to cell maturation in which all of the major CNS gangliosides accumulated. The period of synapse formation was marked by decrease of G_{D3} and rapid increase of G_{D1a}. In a related approach, comparison of chick optic tectum and aggregating tectal cell cultures revealed similar changes in ganglioside patterns including the transient appearance of four new gangliosides early in development.[362]

The precise mechanism by which gangliosides might influence brain development is not known. Two basically different models have been considered for glycolipid-mediated cell–cell recognition: (1) the glycolipid arrangement on one cell surface is complementary to the same order and number of glycolipids on the counterpart cell surface (Steinberg's self–self interaction model[363]); (2) a "lock-and-key" interaction between a glycolipid on one cell and a protein on the other. The latter is thought to be a glycoprotein, either an enzyme (e.g., glycosyltransferase[21]) or a lectin.[364,365] A few attempts to apply these concepts to the nervous system are cited below.

5.5.1. Cell–Cell Interactions

In the nervous system, retinotectal specificity has been studied as a prototypic process of neuronal recognition during development. A ganglioside or related glycosphingolipid has been implicated in the preferential adhesion of chick retinal cells to the surfaces of intact optic tecta.[366] Adhesion of cells from the ventral retina to the dorsal tecta appeared to depend on recognition between a terminal β-GalNAc residue of a ganglioside in the dorsal tecta and a specific protein on the surface of cells from the ventral retina. A reverse recognition was observed between lipid-bound β-GalNAc on cells from the dorsal retina and a specific protein on the ventral tecta. Liposomes containing G_{M2} inhibited the adhesion, and this ganglioside was proposed as the recognition molecule. The adhesive protein was postulated to be the enzyme UDP-Gal:G_{M2} galactosyltransferase, and a double-gradient model involving these two complementary molecules was proposed to account for retinotectal specificity.[366,367]

Muscle cells (L6) at the aligning stage showed enhanced synthesis of a ganglioside[368] initially thought to be G_{D1a} but subsequently shown to be G_{D3}.[369] In mixed cultures of skeletal muscle and spinal cord, formation of neuromuscular junctions was specifically inhibited by high concentrations of globoside and G_{M1} but stimulated by lower concentrations of both glycolipids.[370] Differentiation of myeloid cells into macrophages was accompanied by marked increase of the unusual ganglioside G_{M1b}.[371] Interactions between neurons and glia are another aspect of development that may involve gangliosides.[203]

Numerous examples of glycolipid involvement in cellular interactions of nonneural systems could be cited. These include the areas of cell growth control and oncogenic transformation, processes characterized by major alterations in glycolipid pattern and content. Mediation of the immune response is another surface-related phenomenon in which gangliosides and other glycolipids have been strongly implicated. These aspects of cellular interaction, dealing primarily with nonneural systems, have been discussed at length in a recent review[49] and are not repeated here. However, an aspect of cell differentiation in the nervous system that may relate to membrane gangliosides is discussed below.

5.5.2. Neuritogenesis

A prospective role for gangliosides in neurite growth was suggested by the discovery of aberrant secondary neurites in pyramidal cells of brains afflicted with ganglioside storage disease.[372–374] These mature neurons, with well-developed axon and dendritic structures, produced new processes including occasional synapses from regions of the cell body (e.g., meganeurites) that were filled with stored ganglioside. Studies employing lectin labeling of sprouting neurons suggested that gangliosides may be the predominant glycoconjugate in the growing neurite.[375] Additional evidence has come from two general sources: (1) observation of ganglioside changes in cells reacting to differentiation inducers; (2) neurite growth as affected by exogenous gangliosides.

5.5.2a. Ganglioside Changes by Differentiation Inducers. Several lines of cultured cells show striking morphological changes, including extension of neuritelike processes, on exposure to millimolar concentrations of short-chain fatty acids, especially butyrate.[376] This was shown[377] to be accompanied by marked increase in the cellular concentration of G_{M3} and the sialosyltransferase that synthesizes it. G_{M1} also increased manyfold in HeLa cells, though still remaining at levels well below that of G_{M3}.[378] The AH7974 cells from rat ascites hepatoma showed a two- to threefold elevation of the sialosyltransferase and a reduction in the levels of four other glycolipid glycosyltransferases.[379] Transformed cells, in addition to morphological changes, showed reduced cell saturation density and restored contact inhibition.[380] Butyrate-induced neurite growth has been observed in a wide variety of vertebrate cells including neuroblastoma lines.[381]

Similar effects have been produced by the tumor promoter phorbol-12-myristate-13-acetate and other powerful differentiation inducers such as the retinoid compounds. Their mechanism of action as well as that of butyrate have yet to be revealed. The specific biochemical effect of butyrate, apparently manifested by all vertebrate cell types, is the hyperacetylation of nucleosomal histone proteins[382] believed to result from inhibition of the deacetylase activity[383] Butyrate is also known to inhibit DNA synthesis, induce new β-adrenergic receptors, and activate alkaline phosphatase, but the relationship of these events to stimulated G_{M3} biosynthesis and process formation is not known.

5.5.2b. Neuritogenesis through Exogeneous Gangliosides. In vitro studies have demonstrated a neuritogenic effect caused by exogenous gangliosides that are taken up from the medium by cells in culture. Use of Neuro-2a cells, an established neuroblastoma cell line, revealed that addition of a bovine brain ganglioside mixture to the culturing medium produced an increase in the length and number of cell processes[384] (Fig. 4). Media containing gangliosides from a feline G_{M1}-gangliosidosis brain showed the same enhancement of neurite growth.[385] Similar findings were obtained in another study[386] which reported sprouting of neurites in a dose-dependent manner and formation of a stable network on ganglioside addition. Biochemical alterations included an increase in cellular cyclic AMP content[386] and twofold elevation of ornithine decarboxylase activity.[384] The relatively rapid effect of the added gangliosides was said to resemble the sprouting that occurs *in vitro* and *in vivo* during regeneration.[386] A recent investigation of Neuro-2a cells revealed that in addition to increased cyclic AMP content, ganglioside–induced differentiation is accom-

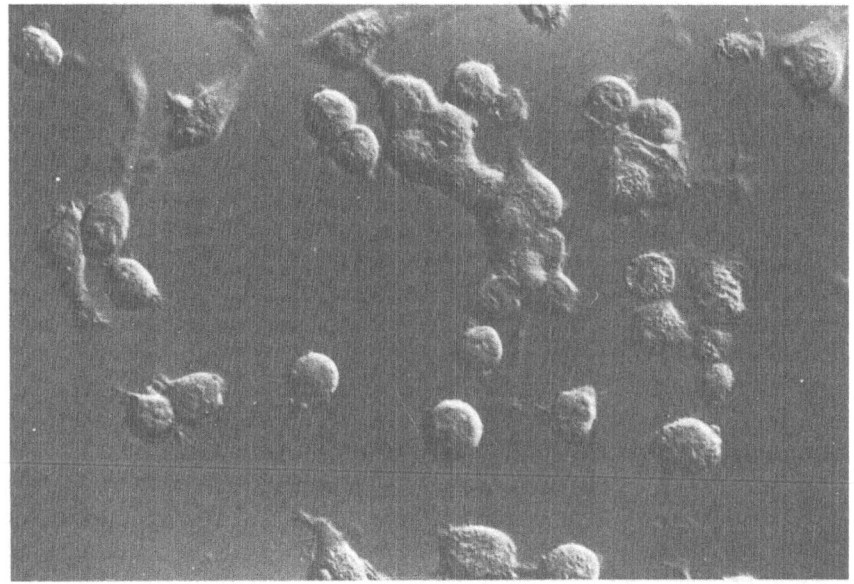

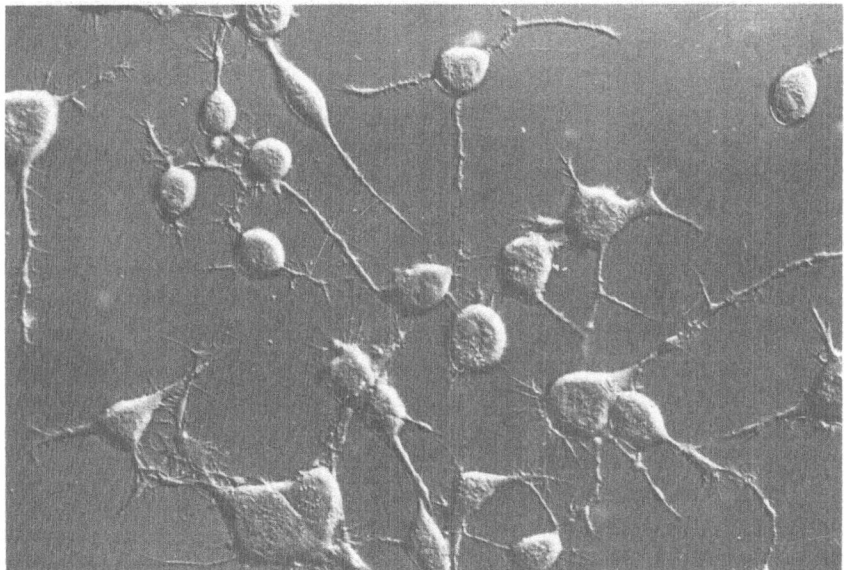

Fig. 4. Neurite growth induced by gangliosides. Upper: Photomicrograph of Neuro-2a cells grown for 24 h on a glass surface in standard medium: Minimum Essential Medium (Eagle) with Hanks' balanced salt solution (90%), heat-inactivated fetal calf serum (10%), nonessential amino acids, and gentamicin (10 mg%). Nomarski optics, 350 ×. Lower: Cells grown under the same conditions as above except that the standard medium was supplemented with 250 µg/ml bovine brain ganglioside mixture (27% GM_1, 40% G_{D1a}, 16% G_{D1b}, and 17% G_T). Nomarski optics, 350 ×. Kindly provided by Dr. Fred J. Roisen, Department of Anatomy, CMDNJ-Rutgers Medical School, Piscataway, NJ (cf. Ref. 384).

panied by a significant alongation of G_1 phase and a decrease in the rate of ^{3}H-thymidine incorporation.[386a]

It has been pointed out[384] that not all cell lines respond to exogenous gangliosides in this manner. Addition of several gangliosides to culture media reduced both the growth rate and saturation density of SV40-transformed and untransformed 3T3 cells.[387] In the latter study, no significant morphological alterations were observed, although trisialoganglioside caused cell damage and lysis in some experiments. The fate of added gangliosides in such systems has been of interest with regard to assessing their ability to assume the role of natural membrane constituents. Although some could be removed from the cell surface with trypsin,[387] another portion appeared to be oriented as true membrane lipid with ceramide embedded in the bilayer.[347,388,389] Incorporation into membranes in this manner was recently shown to be a concentration-dependent phenomenon.[351] The B104 neuronal cell line was used[390] to probe the mechanism by which gangliosides act as differentiation and survival factors. The evidence pointed to a revised model of cell development involving a differentiated (D) state in addition to the previously defined growing (P) and resting (R) states. This postulated D state, characterized by a long survival time, was believed to be controlled by gangliosides in contrast to the R state which may be regulated by cyclic nucleotides.

An alternative approach has been the use of primary cultures such as dorsal root ganglia from chick embryos[384] and thoracic spinal ganglia of guinea pig embryos.[391] The latter required much less ganglioside than the former for optimal increase in the number of neurites. Surprisingly, individual ganglioside fractions were not able to produce the same neuritogenic effect as the ganglioside mixture. The same study demonstrated that human fibroblast growth rates were also increased in the presence of ganglioside.

A number of *in vivo* models have also come into use. Nerve regeneration was studied in rats following denervation of the extensor digitorum longus muscle, and the enhanced recovery of animals injected with gangliosides was was explained by stimulation of axonal sprouting.[392] Exogenous gangliosides were also found to accelerate reinnervation of the cat nictitating membrane[393] and rat skeletal muscle.[394] The characteristic increase of the sensory threshold following peripheral denervation of the somatosensory pathway was significantly reduced in rats by ganglioside administration.[395] The axonal atrophy responsible for slowed conduction velocity in diabetic mutant mice was reportedly reversed by ganglioside therapy.[395a] Local application of ganglioside to the site of a crush in the rat sciatic nerve resulted in a 19 to 33% increase in the number of regenerating axons at 7 days after the lesion, although the maximum axonal outgrowth distance was not affected;[395b] these modest effects were contrasted to the enhanced regeneration achieved by exogenous gangliosides in goldfish optic axons.[395c,395d] Not all such studies have yielded positive results. Gangliosides administered by intraperitoneal injection were reported to have no effect on the rate of regeneration of sensory axons in rats after sciatic nerve crush[395e] and exogenous gangliosides applied to regenerating optic nerve of the carp appeared to have no beneficial effect.[396]

5.5.3. Treatment of Neurological Diseases

The beneficial effects of gangliosides observed in some nerve regeneration experiments have led to their use in treatment of certain human neurological conditions in some parts of the world. Patients with alcoholic neuropathy, treated by daily intramuscular injection of 20 mg of bovine brain ganglioside mixture over 4 weeks, showed improvement in the sensory but not the motor disorders.[397] Human diabetic peripheral neuropathy is also alleged to have responded favorably.[398] Insulin-dependent diabetics given the above dose for 40 consecutive days were reported to show significant improvement of the median nerve sensory action potential latency and the median nerve mixed conduction velocity. Similar improvements were found using animal models of diabetes;[399] a dose-dependent improvement in maximal nerve conduction velocity was found in both alloxan-induced diabetic mice and an inbred strain of diabetic mice. It is perhaps surprising that gangliosides were reported[395f] to exert a beneficial effect on patients with senile and presenile dementia when tested for immediate visual memory and with Raven's Coloured Progressive Matrices. It is obviously too early for an informed judgment on the real benefits and possible side effects of such therapy in man.

5.6. Antibodies to Gangliosides

Because of their location on the external portion of plasma membranes, gangliosides and other glocosphingolipids serve as a variety of specific antigens on the surface of cells.[400,401] Despite the many problems encountered in immunologic studies of gangliosides,[402] antibodies to these substances provide a potentially useful tool for the study of function. Application at the cellular level revealed that monovalent antibodies to G_{M3} inhibited growth of Balb/3T3 and NIL hamster fibroblasts but not that of their transformed counterparts.[403] However, other studies[203] employing cultures of a clone of spontaneously transformed hamster astrocytes (clone NN) and a clone from rat astrocytoma (C6) showed inhibition when treated with anti-G_{M1} or anti-G_{M3} antibodies during exponential growth. These antibodies were shown to produce an increase of adenylate cyclase and decrease of guanylate cyclase, changes that usually accompany a block of cell proliferation.[404] Although such studies probably reflected interaction of antibodies with cell surface antigens, antibodies have also been used to perturb biosynthesis in tissue homogenates of brain and retina.[91] Thus, anti-G_{M1} and anti-G_{M3} inhibited biosynthesis of G_{D1a} and G_{D3}, respectively, along with other gangliosides. The generality of the effect was interpreted as evidence for lack of antibody specificity.

Antiganglioside sera have recently been employed as interventive agents to study behavioral and pharmacological paradigms in rats. The initial experiments indicated recurrent epileptiform activity lasting several weeks following a single injection of antiserum to total brain gangliosides into the sensorimotor cortex.[405,406] Absorption of antibody with pure G_{M1} completely abolished the effect. Similar results were obtained with cholera toxin and choleragenoid,[407] suggesting the primary action to be the binding of ligand to G_{M1} ganglioside receptors. This was supported by the observation[407] that antibodies to G_{M1} purified by affinity chromatography also induced recurrent epileptiform discharges, although not as effectively as the native antiserum. It was pointed out[407] that since various convulsive agents are suggested to have a common mechanism involving the closing of chloride channels,[408] it is possible that G_{M1} receptor binding may affect such channels. Subsequent work indicated that the same antisera apparently caused inhibition of learning, in particular the consolidation and retrieval stages of passive avoidance.[409,410] Using as a control antiganglioside sera that had been absorbed with G_{M1} ganglioside, additional effects were observed,[411] including inhibition of morphine analgesia, blockage of resperine sedation, and blockage of cholinergic stimulation of drinking. Despite this astonishing variety of behavior changes, some selectivity was indicated in the absence of alteration of pattern discrimination, fixed-ratio conditioning, self-stimulation, or pain thresholds.[411] It was recognized[411] that the widespread distribution of G_{M1} in neuronal and glial membranes together with the high degree of accessibility of G_{M1} to the extracellular space in synaptic membranes poses special problems concerning possible mechanisms for these varied effects. Morphological alterations have also been reported with this interventive agent: dendritic spines were reduced in number and altered in appearance following injection of ganglioside antiserum in the cisterna magna of 5-day-old rats.[412]

ACKNOWLEDGMENTS. This work was supported by NIH grants NS-04834, NS-03356, and NS-16181, United States Public Health Service. For manuscript preparation the contributions of Ms. Renée Sasso and Ms. Marion Levine are gratefully acknowledged.

REFERENCES

1. Puro, K., Maury, P., and Huttunen, J. K., 1969, *Biochim. Biophys. Acta* **187**:230–235.
2. Warren, L., 1959, *J. Biol. Chem.* **234**:1971–1975.
3. Nagai, Y., and Iwamori, M., 1980. *Mol. Cell. Biochem.* **29**:81–90.
4. Yamakawa, T., and Nagai, Y., 1978, *Trends Biochem. Sci.* **3**:128–131.
5. Holmgren, J., Lönnroth, I., Månsson, J.-E., and Svennerholm, L., 1975, *Proc. Natl. Acad. Sci. U.S.A.* **72**:2520–2524.
6. Avrova, N. F., 1971, *J. Neurochem.* **18**:667–674.
7. Ledeen, R. W., and Yu, R. K., 1976, *Glycolipid Methodology* (L. A. Witting, ed.), American Oil Chemists Society, Champaign, Illinois, pp. 187–214.
8. Klenk, E., 1935, *Hoppe Seylers Z. Physiol. Chem.* **235**:24–36.
9. Klenk, E., 1939, *Hoppe Seylers Z. Physiol. Chem.* **262**:128–143.
10. Klenk, E., 1942, *Hoppe Seylers Z. Physiol. Chem.* **273**:76–86.
11. Landsteiner, K., and Levene, P. A., 1925, *J. Immunol.* **10**:731–733.
12. Levene, P. A., and Landsteiner, K., 1927, *J. Biol. Chem.* **75**:607–612.
13. Kuhn, R., and Wiegandt, H., 1963, *Chem. Ber.* **96**:866–880.
14. Kuhn, R., and Wiegandt, H., 1963, *Z. Naturforsch.* **18b**:541–543.
15. Kuhn, R., and Wiegandt, H., 1964, *Z. Naturforsch.* **19b**:256–257.
16. Klenk, E., and Gielen, W., 1963, *Hoppe Seylers Z. Physiol. Chem.* **330**:218–226.
17. Klenk, E., Hof, L., and Georgias, L., 1967, *Hoppe Seylers Z. Physiol. Chem.* **348**:149–166.

18. Klenk, E., and Naoi, M., 1968, *Hoppe Seylers Z. Physiol. Chem.* **349**:288–292.
19. Svennerholm, L., 1963, *J. Neurochem.* **10**:613–623.
20. Yamakawa, T., and Suzuki, S., 1951, *J. Biochem. (Tokyo)* **38**:199–212.
21. Roseman, S., 1970, *Chem. Phys. Lipids* **5**:270–297.
22. Kaufman, B., Basu, S., and Roseman, S., 1967, *Inborn Disorders of Sphingolipid Metabolism* (A. M. Aronson and B. W. Volk, eds.), Pergamon Press, Oxford, pp. 193–213.
23. Gatt, S., and Rapport, M. M., 1966, *Biochim. Biophys. Acta* **101**:680–686.
24. Okada, S., and O'Brien, J. S., 1968, *Science* **160**:1002–1004.
25. Okada, S., and O'Brien, J. S., 1969, *Science* **165**:698–700.
26. Sandhoff, K., Andreae, U., and Jatzkewitz, H., 1968, *Life Sci.* **7**:283–288.
27. Brady, R. O., Kanfer, J. N., Bradley, R. M., and Shapiro, D., 1966, *J. Clin. Invest.* **45**:1112–1115.
28. Suzuki, K., and Suzuki, Y., 1970, *Proc. Natl. Acad. Sci. U.S.A.* **66**:302–309.
29. Svennerholm, L., 1970, *Handbook of Neurochemistry*, Volume 3 (A. Lajtha, ed.), Plenum Press, New York, pp. 425–452.
30. Hakomori, S.-I., and Siddiqui, B., 1974, *Methods Enzymol.* **32**:345–367.
31. Witting, L. A. (ed.), 1976, *Glycolipid Methodology*, American Oil Chemists Society, Champaign, Illinois.
32. Sweeley, C. C., and Siddiqui, B., 1977, *The Glycoconjugates*, Volume I (M. I. Horowitz and W. Pigman, eds.), Academic Press, New York, pp. 459–540.
33. Macher, B. A., and Sweeley, C. C., 1978, *Methods Enzymol.* **50**:236–251.
34. Ledeen, R. W., and Yu, R. K., 1978, *Research Methods in Neurochemistry*, (N. Marks and R. Rodnight, eds.), Plenum Press, New York, pp. 371–410.
35. Sweeley, C. C. (ed.), 1980, *Cell Surface Glycolipids*, American Chemical Society, Washington.
36. Svennerholm, L., Mandel, P., Dreyfus, P., and Urban, P.-F., 1980, *Structure and Function of Gangliosides*, Plenum Press, New York.
37. Ledeen, R. W., and Yu, R. K., 1982, *Methods Enzymol.* **83**:139–191.
38. Brunngraber, E., 1979, *Neurochemistry of Aminosugars. Neurochemistry and Neuropathology of the Complex Carbohydrates*, Charles C. Thomas, Springfield, Illinois.
39. Margolis, R. U., and Margolis, R. K. (eds.), 1979, *Complex Carbohydrates of Nervous Tissue*, Plenum Press, New York.
40. Rapport, M. M., and Gorio, A. (eds.), 1981, *Gangliosides in Neurological and Neuromuscular Function, Development, and Repair*, Raven Press, New York.
41. Iwamori, M., and Nagai, Y., 1981, *Biochim. Biophys. Acta* **665**:205–213.
42. Hakomori, S., and Saito, T., 1969, *Biochemistry* **8**:5082–5088.
43. Ishizuka, I., Kloppenburg, M., and Wiegandt, H., 1970, *Biochim. Biophys. Acta* **210**:299–305.
44. Baumann, N., Pollet, S., and Harpin, M. L., 1976, *C. R. Acad. Sci. [D]* (Paris), **238**:1113.
45. Ghidoni, R., Sonnino, S., Tettamanti, G., Baumann, N. Reuter, G., and Schauer, R., 1980, *J. Biol. Chem.* **255**:6990–6995.
45a. IUPAC-IUB Commission on Biological Nomenclature, 1978, *J. Lipid Res.* **19**:114–129.
45b. IUPAC-IUB Commission on Biological Nomenclature, 1977, *Hoppe Seylers Z. Physiol. Chem.* **358**:617–631.
46. Chien, J.-L., and Hogan, E. L., 1980, *Fed. Proc.* **39**:2183.
47. Ledeen, R. W., 1978, *J. Supramol. Struct.* **8**:1–17.
48. Ledeen, R. W., 1979, *Complex Carbohydrates of Nervous Tissue* (R. U. Margolis and R. K. Margolis, eds.), Plenum Press, New York, pp. 1–23.
49. Hakomori, S.-I., 1981, *Annu. Rev. Biochem.* **50**:733–764.
50. Ledeen, R., and Salsman, K., 1965, *Biochemistry* **4**:2225–2233.
51. Wiegandt, H., 1973, *Hoppe Seylers Z. Physiol. Chem.* **354**:1049–1056.
52. Slomiany, B. L., Kojima, K., Banas-Gruszka, Z., Murty, V. L. N., Galicki, N. I., and Slomiany, A., 1981, *Eur. J. Biochem.* **119**:647–650.
53. Ledeen, R., Salsman, K., Gonatas, J., and Taghavy, A., 1965, *J. Neuropathol. Exp. Neurol.* **24**:341–351.
54. Watanabe, K., Powell, M., and Hakomori, S. I., 1979, *J. Biol. Chem.* **254**:8223–8229.

55. Hirabayashi, Y., Taki, T., and Matsumoto, M., 1979, *FEBS Lett.* **100**:253–257.
56. Iwamori, M., and Nagai, Y., 1978, *J. Biochem.* (*Tokyo*) **84**:1601–1608.
57. Itoh, T., Li, Y.-T., Li, S.-C., and Yu, R. K., 1981, *J. Biol. Chem.* **256**:165–169.
58. Ohashi, M., 1980, *J. Biochem.* (*Tokyo*) **88**:583–589.
59. Ghidoni, R., Sonnino, S., Tettamanti, G., Wiegandt, H., and Zambotti, V., 1976, *J. Neurochem.* **27**:511–515.
60. Macher, B. A., Pacuszka, T., Mullin, B. R., Sweeley, C. C. Brady, R. O., and Fishman, P. H., 1979, *Biochim. Biophys. Acta* **588**:35–43.
61. Van Dessel, G. A. F., Lagrou, A. R., Hilderson, H. J. J., Dierick, W. S. H., and Lauwers, W. F. J., 1979, *J. Biol. Chem.* **254**:9305–9310.
62. Ohashi, M., and Yamakawa, T., 1977, *J. Biochem.* (*Tokyo*) **81**:1675–1690.
63. Price, H., Kundu, S., and Ledeen, R., 1975, *Biochemistry* **14**:1512–1518.
64. Ohashi, M., 1981, *Proc. 6th Int. Symp. Glycoconjugates*, Japan Scientific Societies Press, Tokyo, p. 33–34.
65. Ando, S., and Yu, R. K., 1977, *Proc. 4th Int. Symp. Glycoconjugates*, p. 1.
66. Ando, S., and Yu, R. K., 1979, *Glycoconjugate Research*: Proceedings 4th International Symposium Glycoconjugates, (J. D. Gregory and R. W. Jeanloz, eds.) Academic Press, New York, pp. 79–82.
67. Sonnino, S., Ghidoni, R., Galli, G., and Tettamanti, G., 1978, *J. Neurochem.* **31**:947–956.
68. Svennerholm, L., Månsson, J.-E., and Li, Y.-T., 1973, *J. Biol. Chem.* **248**:740–742.
69. Yu, R. K., and Ando, S., 1980, *Adv. Exp. Med. Biol.* **125**:33–46.
70. Ando, A., and Yu, R. K., 1977, *J. Biol. Chem.* **252**:6247–6250.
71. Ishizuka, I., and Wiegandt, H., 1972, *Biochim. Biophys. Acta* **260**:279–289.
72. Ando, S., and Yu, R. K., 1979, *J. Biol. Chem.* **254**:12224–12229.
73. Fredman, P., Månsson, J.-E., Svennerholm, L., Karlsson, K.-A., Pascher, I., and Samuelsson, B. E., 1980, *FEBS Lett.* **110**:80–84.
74. Stoffyn, A., Stoffyn, P., and Yip, M. C. M., 1975, *Biochim. Biophys. Acta* **409**:97–103.
75. Saito, M., Nojiri, H., and Yamada, M., 1980, *Biochem. Biophys. Res. Commun.* **97**:452–462.
76. Suzuki, A., Ishizuka, I., and Yamakawa, T., 1975, *J. Biochem.* (*Tokyo*) **78**:947–954.
77. Fredman, P., Månsson, J.-E., Svennerholm, L., Samuelsson, B., Pascher, I., Pimlott, W., Karlsson, K.-A., and Klinghardt, G. W., 1981, *Eur. J. Biochem.* **116**:553–564.
78. Rösner, H., 1981, *J. Neurochem.* **37**:993–997.
79. Ando, S., Kon, K., Isobe, M., and Yamakawa, T., 1973, *J. Biochem.* (*Tokyo*) **73**:893–895.
80. Wherrett, J. R., 1973, *Biochim. Biophys. Acta* **326**:63–73.
81. Li, Y.-T., Månsson, J. E., Vanier, M.-T., and Svennerholm, L., 1973, *J. Biol. Chem.* **248**:2634–2636.
82. Rauvala, H., 1976, *FEBS Lett.* **62**:161–164.
83. Rauvala, H., Krusius, T., and Finne, J., 1978, *Biochim. Biophys. Acta* **531**:266–274.
84. Watanabe, K., and Hakomori, S.-I., 1979, *Biochemistry* **18**:5502–5504.
85. Wiegandt, H., 1974, *Eur. J. Biochem.* **45**:367–369.
86. Chien, J.-L., Li, S.-C., Laine, R. A., and Li, Y.-T., 1978, *J. Biol. Chem.* **253**:4031–4035.
87. Watanabe, K., Powell, M., and Hakomori, S.-I., 1978, *J. Biol. Chem.* **253**:8962–8967.
88. Kundu, S. K., Marcus, D. M., Pascher, I., and Samuelsson, B. E., 1981, *Fed. Proc.* **40**:1545.
89. Watanabe, K., Hakomori, S.-I., Childs, R. A., and Feizi, T., 1979, *J. Biol. Chem.* **254**:3221–3228.
90. Iwamori, M., and Nagai, Y., 1981, *J. Biochem.* (*Tokyo*) **89**:1253–1264.
91. Dreyfus, H., Urban, P. F., Harth, S., Preti, A., and Mandel, P., 1976, *Adv. Exp. Med. Biol.* **71**:163–188.
92. Iwamori, M., and Nagai, Y., 1978, *J. Biochem.* (*Tokyo*) **84**:1609–1615.
93. Yu, R. K., and Ando, S., 1978, *Trans. Am. Soc. Neurochem.* **9**:135.
94. Avrova, N. F., Li, Y.-T., and Obukhova, E. L., 1979, *J. Neurochem.* **32**:1807–1815.
95. Murakami-Murofushi, K., Tadano, K., Koyama, I., and Ishizuka, I., 1981, *J. Biochem.* (*Tokyo*) **90**:1817–1820.
96. Ledeen, R. W., Yu, R. K., and Eng, L. F., 1973, *J. Neurochem.* **21**:829–839.
97. Li, S.-C., Chien, J.-L., Wan, C. C., and Li, Y.-T., 1978, *Biochem. J.* **173**:697–699.
98. Hamanaka, S., Handa, S., and Yamakawa, T., 1979, *J. Biochem.* (*Tokyo*) **65**:1623–1626.

99. Cochran, F. B., Yu, R. K., Ando, S., and Ledeen, R. W., 1981, *J. Neurochem.* **36**:696–702.
100. Seyfried, T. N., Ando, S., and Yu, R. K., 1978, *J. Lipid Res.* **19**:538–543.
101. Ledeen, R., Salsman, K., and Cabrera, M., 1968, *Biochemistry* **7**:2287–2295.
102. Klenk, E., and Lauenstein, I., 1953, *Hoppe Seylers Z. Physiol. Chem.* **295**:164–173.
103. Handa, S., and Burton, R. M., 1969, *Lipids* **4**:205–208.
104. Iwamori, M., and Nagai, Y., 1978, *J. Biol. Chem.* **253**:8328–8331.
105. Handa, N., and Handa, S., 1965, *Jpn. J. Exp. Med.* **35**:331–341.
106. Fong, J. W., Ledeen, R. W., Kundu, S. K., and Brostoff, S. W., 1976, *J. Neurochem.* **26**:157–162.
107. Månsson, J.-E., Vanier, M.-T., and Svennerholm, L., 1973, *J. Neurochem.* **30**:273–275.
108. Rosenberg, A., and Stern, N., 1966, *J. Lipid Res.* **7**:122–131.
109. Kawamura, N., and Taketomi, T., 1975, *Jpn. J. Exp. Med.* **45**:489–500.
110. Kochetkov, N. K., Zhukova, I. G., Smirnova, G. P., and Glukhoded, I. S., 1973, *Biochim. Biophys. Acta* **326**:74–83.
111. Hoshi, M., and Nagai, Y., 1975, *Biochim. Biophys. Acta* **388**:152–162.
112. Prokazova, N. V., Mikhailov, A. T., Kocharov, S. L., Malchenko, L. A., Zvezdina, N. D., Buznikov, G., and Bergelson, L. D., 1981, *Eur. J. Biochem.* **115**:671–677.
113. Sugita, M., 1979, *J. Biochem.* (*Tokyo*) **86**:289–300.
114. Sugita, M., 1979, *J. Biochem.* (*Tokyo*) **86**:765–772.
115. Smirnova, G. P., and Kochetkov, N. K., 1980, *Biochim. Biophys. Acta* **618**:486–495.
116. Klenk, E., and Langerbeins, H., 1941, *Hoppe Seylers Z. Physiol. Chem.* **270**:185–193.
117. Svennerholm, L., 1957, *Biochim. Biophys. Acta* **24**:604–611.
118. Ledeen, R. W., and Yu, R. K., 1970, *J. Lipid Res.* **11**:506–516.
119. Tettamanti, G., Bonali, F., Marchesini, S., and Zambotti, V., 1973, *Biochim. Biophys. Acta* **296**:160–170.
120. Svennerholm, L., and Fredman, P., 1980, *Biochim. Biophys. Acta* **617**:97–109.
121. Yu, R. K., Ando, S., Takahashi, A., and Miyatake, T., 1980. *Proc. Jpn. Conf. Biochem. Lipids* **22**:332–334.
122. Holm, M., Månsson, J.-E., Vanier, M.-T., and Svennerholm, L., 1972, *Biochim. Biophys. Acta* **280**:356–364.
123. MacMillan, V. H., and Wherrett, J. R., 1969, *J. Neurochem.* **16**:1621–1624.
124. Svennerholm, L., Bruce, A., Månsson, J.-E., Rynmark, B.-M., and Vanier, M.-T., 1972, *Biochim. Biophys. Acta* **280**:626–636.
125. Price, H. C., and Yu, R. K., 1976, *Comp. Biochem. Physiol.* **54B**:451–454.
126. Ariga, T., Ando, S., Takahashi, A., and Miyatake, T., 1980, *Biochim. Biophys. Acta* **618**:480–485.
127. Yu, R. K., and Ledeen, R. W., 1972, *J. Lipid Res.* **13**:680–686.
128. Ledeen, R. W., and Yu, R. K., 1972, *Adv. Exp. Med. Biol.* **19**:77–93.
129. Lowden, J. A., and Wolfe, L. S., 1964, *Can. J. Biochem.* **42**:1587–1594.
130. Suzuki, K., 1967, *Inborn Disorders of Sphingolipid Metabolism* (A. M. Aronson and B. W. Volk, eds.), Pergamon Press, Oxford, pp. 215–230.
131. Dominick, V., and Gielen, W., 1968, *Hoppe Seylers Z. Physiol. Chem.* **349**:731–736.
132. Bernheimer, H., 1968, *Klin. Wochenschr.* **46**:258–261.
133. Ginns, E., French, J., Fleischman, A., and Cohen, S., 1980, *Pediatr. Res.* **14**:1276–1277.
134. Urban, P. F., Harth, S., Freysz, L., and Dreyfus, H., 1980, *Adv. Exp. Med. Biol.* **125**:149–157.
135. Ueno, K., Ando, S., and Yu, R. K., 1978, *J. Lipid Res.* **19**:863–871.
136. Derry, D. M., and Wolfe, L. S., 1967, *Science* **158**:1450–1452.
137. Norton, W. T., and Poduslo, S. E., 1971, *J. Lipid Res.* **12**:84–90.
138. Skrivanek, J., Ledeen, R., Norton, W., and Farooq, M., 1978, *Trans. Am. Soc. Neurochem.* **9**:133.
139. Maccioni, H. J. F., Defilpo, S. S., Landa, C. A., and Caputto, R., 1978, *Biochem. J.* **174**:673–680.
140. Hamburger, A., and Svennerholm, L., 1971, *J. Neurochem.* **18**:1821–1829.
141. Tamai, Y., Matsukawa, S., and Satake, M., 1971, *J. Biochem.* (*Tokyo*) **69**:235–238.
142. Yu, R. K., and Iqbal, K., 1979, *J. Neurochem.* **32**:293–300.

143. Poduslo, S. E., and Norton, W. T., 1972, *J. Neurochem.* **19:**727–736.
144. Abe, T., and Norton, W. T., 1974, *J. Neurochem.* **23:**1025–1036.
145. Robert, J., Freysz, L., Sensenbrenner, M., Mandel, P., and Rebel, G., 1975, *FEBS Lett.* **50:**144–146.
146. Ledeen, R. W., Skrivanek, J. A., Nuñez, J., Sclafani, J. R., Norton, W. T., and Farooq, M., 1981, *Gangliosides in Neurological and Neuromuscular Functions, Development, and Repair* (M. M. Rapport and A. Gorio, eds.), Raven Press, New York, pp. 211–224.
147. Suzuki, K., Poduslo, S. E., and Norton, W. T., 1967, *Biochim. Biophys. Acta* **144:**375–381.
148. Ledeen, R. W., Cochran, F. B., Yu, R. K., Samuels, F. G., and Haley, J. E., 1980, *Adv. Exp. Med. Biol.* **125:**167–176.
149. Gambetti, P., Autilio-Gambetti, L., Rizutto, N., Shafer, B., and Pfaff, L., 1974, *Exp. Neurol.* **43:**464–473.
150. Hess, H. H., Bass, N. H., Thalheimer, C., and Devarakonda, R., 1976, *J. Neurochem.* **26:**1115–1121.
151. Diamond, M. C., Johnson, R. E., and Gold, M. W., 1977, *Behav. Biol.* **20:**409–418.
152. Manuelidis, L., and Manuelidis, E. E., 1976, *J. Neurocytol.* **5:**575–589.
153. Morgan, I. G., Wolfe, L. S., Mandel, P., and Gombos, G., 1971, *Biochim. Biophys. Acta* **241:**737–751.
154. Wiegandt, H., 1967, *J. Neurochem.* **14:**671–674.
155. Avrova, N. F., Chenykaeva, E. Y., and Obukhova, E. L., 1973, *J. Neurochem.* **20:**997–1004.
156. Tettamanti, G., Preti, A., Lombardo, A., Bonali, F., and Zambotti, V., 1973, *Biochim. Biophys. Acta* **306:**466–477.
157. Yohe, H. C., Ueno, K., Chang, N.-C., Glaser, G. H., and Yu, R. K., 1980, *J. Neurochem.* **34:**560–568.
158. Dekirmenjian, H., and Brunngraber, E. G., 1969, *Biochim. Biophys. Acta* **177:**1–10.
159. Seminario, L. M., Hren, N., and Gomez, C. J., 1964, *J. Neurochem.* **11:**197–209.
160. Eichberg, J., Whittaker, V. P., and Dawson, R. M. C., 1964, *Biochem. J.* **92:**91–100.
161. Tettamanti, G., Preti, A., Cestaro, B., Venerando, B., Lombardo, A., Ghidoni, R., and Sonnino, S., 1980, *Adv. Exp. Med. Biol.* **125:**263–281.
162. Kornguth, S., Wannamaker, B., Kolodny, E., Geison, R., Scott, G., and O'Brien, J. F., 1974, *J. Neurol. Sci.* **22:**383–406.
163. Breckenridge, W. C., Gombos, G., and Morgan, I. G., 1972, *Biochim. Biophys. Acta* **266:**695–707.
164. Hungund, B. L., and Mahadik, S. P., 1981, *Neurochem. Res.* **6:**183–191.
165. Brunngraber, E. G., Dekirmenjian, H., and Brown, B. D., 1967, *Biochem. J.* **103:**73–78.
166. Whittaker, V. P., 1969, *Handbook of Neurochemistry*, Volume 2 (A. Lajtha, ed.), Plenum Press, New York, pp. 327–364.
167. Lapetina, E. G., Soto, E. F., and DeRobertis, E., 1968, *J. Neurochem.* **15:**437–445.
168. Cruz, T. F., and Gurd, J. W., 1981, *Biochim. Biophys. Acta* **675:**201–208.
169. Skrivanek, J. A., Ledeen, R. W., Margolis, R. U., and Margolis, R. K., 1982, *J. Neurobiol.* **13:**95–106.
170. DeVries, G. H., Payne, W., and Saul, R. G., 1981, *Neurochem. Res.* **6:**521–537.
171. Breckenridge, W. C., Morgan, I. G., Zanetta, J. P., and Vincendon, G., 1973, *Biochim. Biophys. Acta* **320:**681–686.
172. Sonnino, S., Ghidoni, R., Marchesini, S., and Tettamanti, G., 1979, *J. Neurochem.* **33:**117–121.
173. Ledeen, R. W., Skrivanek, J. A., Tirri, L. J., Margolis, R. K., and Margolis, R. U., 1976, *Adv. Exp. Med. Biol.* **71:**83–105.
174. DeVries, G. H., and Norton, W. T., 1974, *J. Neurochem.* **22:**259–264.
175. Shepherd, G. M., 1979, *The Synaptic Organization of the Brain*, 2nd ed., Oxford University Press, New York.
176. Svennerholm, L., 1980, *Adv. Exp. Med. Biol.* **125:**533–544.
177. Lapetina, E. G., and DeRobertis, E., 1968, *Life Sci.* **7:**203–208.
178. Caputto, R., Maccioni, H. J., Arce, A., and Cumar, F. A., 1976, *Adv. Exp. Med. Biol.* **71:**27–44.
179. Gatt, S., 1970, *Chem. Phys. Lipids* **5:**235–249.

180. Rosenberg, A., 1979, *Complex Carbohydrates of Nervous Tissue* (R. U. Margolis and R. K. Margolis, eds.), Plenum Press, New York, pp. 25–44.
181. Ledeen, R. W., and Mellanby, J., 1977, *Perspectives in Toxicology* (A. W. Bernheimer, ed.), John Wiley & Sons, New York, pp. 16–42.
182. Schachter, H., and Roseman, S., 1980, *The Biochemistry of Glycoproteins and Proteoglycans* (W. J. Lennerz, ed.), Plenum Press, New York, pp. 85–160.
183. Fishman, P. H., and Brady, R. O., 1976, *Science* 194:906–915.
184. Basu, S., and Basu, M., 1982. *The Glycoconjugates*, Volume 3 (M. Horowitz, ed.), Academic Press, New York, pp. 265–285.
185. Brady, R. O., 1978, *Annu. Rev. Biochem.* 47:687–714.
186. Burton, R. M., Balfour, Y. M., and Gibbons, J. M., 1964, *Fed. Proc.* 23:230.
187. Suzuki, K., 1967, *J. Neurochem.* 14:917–925.
188. Holm, M., and Svennerholm, L., 1972, *J. Neurochem.* 19:609–622.
189. Maccioni, H. J., Arce, A., and Caputto, R., 1971, *Biochem. J.* 125:1131–1137.
190. Suzuki, K., and Korey, S. R., 1963, *Biochim. Biophys. Acta* 78:388–389.
191. Suzuki, K., and Korey, S. R., 1964, *J. Neurochem.* 11:647–653.
192. Ferwerda, W., Blok, C. M., and Heijlman, J., 1981, *J. Neurochem.* 36:1492–1499.
193. Suzuki, K., 1970, *J. Neurochem.* 17:209–213.
194. Dawson, G., 1979, *Complex Carbohydrates of Nervous Tissue* (R. U. Margolis and R. K. Margolis, eds.), Plenum Press, New York, pp. 291–326.
195. Dawson, G., Kemp, S. F., Stoolmiller, A. C., and Dorfman, A., 1971, *Biochem. Biophys. Res. Commun.* 44:687–694.
196. Yogeeswaran, G., Murray, R. K., Pearson, M. L., Sanwal, B. D., McMorris, F. A., and Ruddle, F. A., 1973, *J. Biol. Chem.* 248:1231–1239.
197. Robert, J., Freysz, L., Sensenbrenner, M., Mandel, P., and Rebel, G., 1975, *FEBS Lett.* 50:144–146.
198. Duffard, R. O., Fishman, P. H., Bradley, R. M., Lauter, C. J., Brady, R. O., and Trams, E. G., 1977, *J. Neurochem.* 28:1161–1166.
199. Ciesielski-Treska, J., Robert, J., Rebel, G., and Mandel, P., 1977, *Differentiation* 8:31–37.
200. Dawson, G., and Stoolmiller, A. C., 1976, *J. Neurochem.* 26:225–226.
201. Rebel, G., Robert, J., and Mandel, P., 1980, *Adv. Exp. Med. Biol.* 125:159–166.
202. Yavin, E., and Yavin, Z., 1974, *J. Cell Biol.* 62:540–546.
203. Mandel, P., Dreyfus, H., Yusufi, A. N. K., Sarlieve, L., Robert, J., Neskovic, N., Harth, S., and Rebel, G., 1980, *Adv. Exp. Med. Biol.* 125:515–531.
204. Cumar, F. A., Fishman, P. H., and Brady, R. O., 1971, *J. Biol. Chem.* 246:5075–5084.
205. DiCesare, J. L., and Dain, J. A., 1972, *J. Neurochem.* 19:403–410.
206. Caputto, R., Maccioni, H. J., and Arce, A., 1974, *Mol. Cell. Biochem.* 4:97–106.
207. Cumar, F. A., Tallman, J. F., and Brady, R. O., 1972, *J. Biol. Chem.* 247:2322–2327.
208. Pacuszka, T., Duffard, R. O., Nishimura, R. N., Brady, R. O., and Fishman, P. H., 1978, *J. Biol. Chem.* 253:5839–5846.
209. Basu, S., Basu, M., Chien, J.-L., and Presper, K. A., 1980, *Adv. Exp. Med. Biol.* 125:213–226.
210. Basu, M., and Basu, S., 1972, *J. Biol. Chem.* 247:1489–1495.
211. Basu, S., Higashi, H., Basu, S., and Evans, C. H., 1980, *Fed. Proc.* 39:2184.
212. Kaufman, B., Basu, S., and Roseman, S., 1968, *J. Biol. Chem.* 243:5804–5807.
213. Ng, S.-S., and Dain, J. A., 1977, *J. Neurochem.* 29:1075–1083.
214. Arce, A., Maccioni, H. J., and Caputto, R., 1971, *Biochem. J.* 121:483–493.
215. Mestrallet, M. G., Cumar, F. A., and Caputto, R., 1974, *Biochem. Biophys. Res. Commun.* 59:1–7.
216. Mestrallet, M. G., Cumar, F. A., and Caputto, R., 1977, *Mol. Cell. Biochem.* 16:63–70.
217. Yohe, H. C., and Yu, R. K., 1980, *J. Biol. Chem.* 255:608–613.
218. Yohe, H. C., Macala, L. J., and Yu, R. K., 1982, *J. Biol. Chem.* 257:249–252.
219. Basu, S., Basu, M., and Higashi, H., 1981, *Proc. 6th International Symp. Glycoconj.*, Japan Scientific Societies Press, p. 41.
220. Basu, M., Kyle, J. W., and Basu, S., 1982, *Fed. Proc.* 41:3613.
221. Yip, M. C. M., and Dain, J. A., 1969, *Lipids* 4:270–277.

222. Handa, S., and Burton, R. M., 1969, *Lipids* **4**:589–598.
223. Stoffyn, P., and Stoffyn, A., 1980, *Carb. Res.* **78**:327–340.
224. Yu, R. K., and Lee, S. H., 1976, *J. Biol. Chem.* **251**:198–203.
225. Stoffyn, A., Stoffyn, P., Farooq, M., Snyder, D. S., and Norton, W. T., 1981, *Neurochem. Res.* **6**:1143–1151.
226. Radin, N. S., Brenkert, A., Arora, R., Sellinger, O. Z., and Flangas, A. I., 1972, *Brain Res.* **39**:163–169.
227. Jones, J. P., Ramsey, R. B., Aexel, R. T., and Nicholas, H. J., 1972, *Life Sci.* **11**:309–315.
228. Reith, M., Morgan, I. G., Gombos, G., Breckenridge, W. C., and Vincendon, G., 1972, *Neurobiology* **2**:169–175.
229. Raghupathy, E., Ko, G. K. W., and Peterson, N. A., 1972, *Biochim. Biophys. Acta* **286**:339–349.
230. Landa, C. A., Maccioni, H. J. F., and Caputto, R., 1979, *J. Neurochem.* **33**:825–838.
231. Keenan, T. W., Morré, D. J., and Basu, S., 1974, *J. Biol. Chem.* **249**:310–315.
232. Richardson, C. L., Keenan, T. W., and Morré, D. J., 1977, *Biochim. Biophys. Acta* **488**:88–96.
233. Basu, S., Basu, M., Moskal, J. R., and Chien, J.-L., 1976, *Glycolipid Methodology* (L. A. Witting, ed.), American Oil Chemists Society, Champaign, Illinois, pp. 123–139.
234. Fleischer, B., 1977, *J. Supramol. Struct.* **7**:79–89.
235. Keenan, T. W., 1974, *J. Dairy Sci.* **57**:187–192.
236. Ng, S.-S., and Dain, J. A., 1977, *J. Neurochem.* **29**:1085–1093.
237. Preti, A., Fiorilli, A., Lombardo, A., Caimi, L., and Tettamanti, G., 1980, *J. Neurochem.* **35**:281–296.
238. Van den Eijnden, D. H., and van Dijk, W., 1974, *Biochim. Biophys. Acta* **362**:136–149.
239. Keenan, T. W., and Morré, D. J., 1975, *FEBS Lett.* **55**:8–13.
240. Kanfer, J. N., and Richards, R. L., 1967, *J. Neurochem.* **14**:513–518.
241. Kanfer, J. N., Bradley, R. M., and Gal, A. E., 1967, *J. Neurochem.* **14**:1095–1098.
242. Shah, S. N., and Peterson, N. A., 1971, *Biochim. Biophys. Acta* **239**:126–131.
243. McCluer, R. H., and Agranoff, B. W., 1972, *J. Neurochem.* **19**:2307–2315.
244. Dawson, G., McLawhon, R., and Miller, R. J., 1980, *J. Biol. Chem.* **255**:129–137.
245. McLawhon, R. W., Schoon, G. S., and Dawson, G., 1981, *J. Neurochem.* **37**:132–139.
246. McLawhon, R. W., Schoon, G. S., and Dawson, G., 1981, *Eur. J. Cell Biol.* **25**:353–358.
247. Koski, G., and Klee, W. A., 1981, *Proc. Natl. Acad. Sci. U.S.A.* **78**:4185–4189.
248. Forman, D. S., and Ledeen, R. W., 1972, *Science* **177**:630–633.
249. Rösner, H., Wiegandt, H., and Rahmann, H., 1973, *J. Neurochem.* **21**:655–665.
250. Rahmann, H., and Breer, H., 1975, *Brain Res.* **85**:301–305.
251. Rösner, H., 1975, *Brain Res.* **97**:107–116.
252. Holm, M., 1972, *J. Neurochem.* **19**:623–629.
253. Caputto, B. L., Maccioni, A. H. R., Landa, C. A., and Caputto, R., 1979, *Biochem. Biophys. Res. Commun.* **86**:849–854.
254. Schengrund, C. L., and Rosenberg, A., 1970, *J. Biol. Chem.* **245**:6196–6200.
255. Tettamanti, G., Morgan, I. G., Gombos, G., Vincendon, G., and Mandel, P., 1972, *Brain Res.* **47**:515–518.
256. Abe, T., Miyatake, T., Norton, W. T., and Suzuki, K., 1979, *Brain Res.* **161**:179–182.
257. Ragahavan, S. S., Rhoads, D. B., and Kanfer, J. N., 1972, *Biochim. Biophys. Acta* **268**:755–760.
258. Sinha, A. K., and Rose, S. P. R., 1973, *J. Neurochem.* **20**:39–44.
259. Freysz, L., Farooqui, A. A., Adamczewska-Goncerzewicz, and Mandel, P., 1979, *J. Lipid Res.* **20**:503–508.
260. Arbogast, B. W., and Arsenis, C., 1974, *Neurobiology* **4**:21–37.
261. Ohman, R., Rosenberg, A., and Svennerholm, L., 1970, *Biochemistry* **9**:3774–3782.
262. Schauer, R., Veh, R. W., Sander, M., Corfield, A. P., and Wiegandt, H., 1980, *Adv. Exp. Med. Biol.* **125**:283–294.
263. Harris, P. L., and Thornton, E. R., 1978, *J. Am. Chem. Soc.* **100**:6738–6745.
264. Wenger, D. A., and Wardell, S., 1973, *J. Neurochem.* **20**:607–612.

265. Kolodny, E. H., Kanfer, J. N., Quirk, J. M., and Brady, R. O., 1971, *J. Biol. Chem.* **246:**1426–1431.
266. Tallman, J. F., and Brady, R. O., 1972, *J. Biol. Chem.* **247:**7570–7575.
267. Ledeen, R. W., 1970, *Chem. Phys. Lipids,* **5:**205–219.
268. Schengrund, C. L., Rosenberg, A., and Repman, M. A., 1976, *J. Cell Biol.* **70:**555–561.
269. Schengrund, C. L., Repman, M. A., and Nelson, J. T., 1979, *Biochim. Biophys. Acta* **568:**377–385.
270. Gatt, S., and Rapport, M. M., 1966, *Biochim. Biophys. Acta* **113:**567–576.
271. Miyatake, T., and Suzuki, K., 1974, *Biochim. Biophys. Acta* **337:**333–342.
272. Suzuki, Y., and Suzuki, K., 1974, *J. Biol. Chem.* **249:**2105–2108.
273. Suzuki, Y., and Suzuki, K., 1974, *J. Biol. Chem.* **249:**2113–2117.
274. Suzuki, K., Tanaka, H., Yamanaka, T., and Van Damme, O., 1980, *Adv. Exp. Med. Biol.* **125:**307–318.
275. Nishimura, K., and Amano, R., 1976, *J. Biochem. (Tokyo)* **80:**209–215.
276. Li, S. C., Nakamura, T., Ogamo, A., and Li, Y.-T., 1979, *J. Biol. Chem.* **254:**10592–10595.
277. Li, Y.-T., Mazzotta, M. Y., Wan, C. C., Orth, R., and Li, S. C., 1973, *J. Biol. Chem.* **248:**7512–7515.
278. Sandhoff, K., 1969, *FEBS Lett.* **4:**351–354.
279. Sandhoff, K., Harzer, K., Waessle, W., and Jatzkewitz, H., 1971, *J. Neurochem.* **18:**2469–2489.
280. Conzelmann, E., and Sandhoff, K., 1978, *Proc. Natl. Acad. Sci. U.S.A.* **75:**3979–3983.
281. Conzelmann, E., and Sandhoff, K., 1980, *Adv. Exp. Med. Biol.* **125:**295–306.
282. Li, S. C., Hirabayashi, Y., and Li, Y.-T., 1981, *Biochem. Biophys. Res. Commun.* **101:**479–485.
283. Sandhoff, K., and Christomanon, H., 1979, *Hum. Genet.* **50:**107–143.
284. Volk, B. W., and Schneck, L. (eds.), 1975, *The Gangliosidoses*, Plenum Press, New York.
285. Brady, R. O., Kanfer, J. N., and Shapiro, D., 1965, *J. Biol. Chem.* **240:**39–43.
286. Stoffel, W., and Sticht, G., 1967, *Hoppe Seylers Z. Physiol. Chem.* **348:**1345–1351.
287. Marks, N., and McIlwain, H., 1959, *Biochem. J.* **73:**401–410.
288. McIlwain, H., 1961, *Biochem. J.* **78:**24–32.
289. McIlwain, H., 1960, *Chemical Exploration of the Brain. A Study of Excitability and Ion Movement*, Elsevier, Amsterdam.
290. McIlwain, H., 1964, *Biochem. J.* **90:**442–448.
291. Evans, W. H., and McIlwain, H., 1967, *J. Neurochem.* **14:**35–44.
292. Yogeeswaran, G., Murray, R. K., Pearson, M. L., Sanwal, B. D., McMorris, F. A., and Ruddle, F. H., 1973, *J. Biol. Chem.* **248:**1231–1239.
293. Rahmann, H., 1976, *Adv. Exp. Med. Biol.* **71:**151–161.
294. Behr, J. P., and Lehn, J. M., 1974, *FEBS Lett.* **31:**297–300.
295. Rahmann, H., Rösner, H., and Breer, H., 1976, *J. Theor. Biol.* **57:**231–237.
296. Rahmann, H., 1980, *Adv. Exp. Med. Biol.* **125:**505–514.
297. Cumar, F. A., Maggio, B., and Caputto, R., 1978, *Biochem. Biophys. Res. Commun.* **84:**65–69.
298. Gingell, D., 1967, *J. Theor. Biol.* **17:**451–482.
299. Bangham, A. D., 1968, *Prog. Biophys. Mol. Biol.* **18:**29–95.
300. Wooley, D. W., and Gommi, B. W., 1963, *Proc. Natl. Acad. Sci. U.S.A.* **53:**959–963.
301. Carroll, P. M., and Sereda, D. D., 1968, *Nature* **217:**667–668.
302. Tamir, H., Brunner, W., Casper, D., and Rapport, M. M., 1980, *J. Neurochem.* **34:**1719–1724.
303. van Heyningen, W. E., Carpenter, C. C. J., Pierce, N. F., and Greenough, W. B., 1971, *J. Infect. Dis.* **124:**415–418.
304. King, C. A., and van Heyningen, W. E., 1973, *J. Infect. Dis.* **127:**639–647.
305. Holmgren, J., Lönnroth, I., and Svennerholm, L., 1973, *Infect. Immun.* **8:**208–214.
306. Cuatrecasas, P., 1973, *Biochemistry* **12:**3558–3566.
307. Fishman, P. H., and Henneberry, R. C., 1980, *Cell Surface Glycolipids* (C. C. Sweeley, ed.), American Chemical Society, Washington, pp. 223–240.

308. Holmgren, J., Elwing, H., Fredman, P., Strannegård, Ö., and Svennerholm, L., 1980, *Adv. Exp. Med. Biol.* **125:**453–470.
309. Lönnroth, I., and Holmgren, J., 1973, *J. Gen. Microbiol.* **76:**417–427.
310. Morita, A., Tsao, D., and Kim, Y. S., 1980, *J. Biol. Chem.* **255:**2549–2553.
311. Critchley, D. R., Ansell, S., Perkins, R., Dilks, S., and Ingram, J., 1979, *J. Supramol. Struct.* **12:**273–291.
312. van Heyningen, W. E., 1959, *J. Gen. Microbiol.* **20:**310–320.
313. van Heyningen, W. E., and Miller, P. A., 1961, *J. Gen. Microbiol.* **24:**107–119.
314. van Heyningen, W. E., and Mellanby, J., 1971, *Microbiological Toxins*, Volume 2A (S. Kadis, T. C. Montie, and S. J. Ajl, eds.), Academic Press, New York, pp. 69–108.
315. Ledley, F. D., Lee, G., Kohn, L. D., Habig, W. H., and Hardegree, M. C., 1977, *J. Biol. Chem.* **252:**4049–4055.
316. Price, D. L., Griffin, J. W., and Peck, K., 1977, *Brain Res.* **121:**379–384.
317. Mirsky, R., Wendon, L. M. B., Black, P., Stolkin, C., and Bray, D., 1978, *Brain Res.* **148:**251–259.
318. Simpson, L. L., and Rapport, M. M., 1971, *J. Neurochem.* **18:**1341–1343.
319. Simpson, L. L., and Rapport, M. M., 1971, *J. Neurochem.* **18:**1751–1759.
320. Kitamura, M., Iwamori, M., and Nagai, Y., 1980, *Biochim. Biophys. Acta* **628:**328–335.
321. van Heyningen, W. E., and Mellanby, J., 1973, *Naunyn Schmiedelbergs Arch. Pharmacol.* **276:**297–302.
322. Pierce, N. F., 1973, *J. Exp. Med.* **137:**1009–1023.
323. Holmgren, J., 1973, *Infect. Immun.* **8:**851–859.
324. Moss, J., Garrison, S., Fishman, P. H., and Richardson, S. H., 1979, *J. Clin. Invest.* **64:**381–384.
325. Takeda, Y., Takeda, T., Honda, T., Sakurai, J., Ohtomo, N., Miwatani, T., 1975, *Infect. Immun.* **12:**931–933.
326. Kato, I., and Naiki, M., 1976, *Infect. Immun.* **13:**289–291.
327. Winand, R. J., and Kohn, L. D., 1975, *J. Biol. Chem.* **250:**6534–6540.
328. Mullin, B. R., Aloj, S. M., Fishman, P. H., Lee, G., Kohn, L. D., Brady, R. O., 1976, *Proc. Natl. Acad. Sci. U.S.A.* **73:**1679–1683.
329. Mullin, B. R., Fishman, P. H., Lee, G., Aloj, S. M., Ledley, F. D., Winand, R. J., Kohn, L. D., and Brady, R. O., 1976, *Proc. Natl. Acad. Sci. U.S.A.* **73:**842–846.
330. Ledley, F. D., Mullin, B. R., Lee, G., Aloj, S. M., Fishman, P. H., Hunt, L. T., Dayhoff, M. O., and Kohn, L. D., 1976, *Biochem. Biophys. Res. Commun.* **69:**852–859.
331. Kohn, L. D., Consiglio, E., DeWolf, M. J. S., Grollman, E. F., Ledley, F. D., Lee, G., and Morris, N. P., 1980, *Adv. Exp. Med. Biol.* **125:**487–503.
332. Kohn, L. D., 1978, *Receptors and Recognition*, Series A, Volume 5 (P. Cuatrecasas and M. F. Greaves, eds.), Chapman Hall, London, pp. 134–212.
333. Aloj, S. M., Lee, G., Grollman, E. F., Beguinot, F., Consiglio, E., and Kohn, L. D., 1979, *J. Biol. Chem.* **254:**9040–9049.
334. Omodeo-Sale, F., Brady, R. O., and Fishman, P. H., 1978, *Proc. Natl. Acad. Sci. U.S.A.* **75:**5301–5305.
335. Pekonen, F., and Weintraub, B. D., 1979, *Endocrinology* **105:**352–359.
336. Holmes, S. D., Titus, G., Chou, M., and Field, J. B., 1980, *Endocrinology* **107:**2076–2081.
337. Beckner, S., Brady, R. O., and Fishman, P. H., 1981, *Proc. Natl. Acad. Sci. U.S.A.* **78:**4848–4852.
338. Tonegawa, Y., and Hakomori, S.-I., 1977, *Biochem. Biophys. Res. Commun.* **76:**9–17.
339. Besancon, F., and Ankel, H., 1974, *Nature* **252:**478–480.
340. Ankel, H., Besancon, F., and Krishnamurti, C., 1980, *Cell Surface Glycolipids* (C. C. Sweeley, ed.), American Chemical Society, Washington, pp. 391–405.
341. Vengris, V. E., Fernie, B. F., and Pitha, P. M., 1980, *Adv. Exp. Med. Biol.* **125:**479–486.
342. Haywood, A. M., 1974, *J. Mol. Biol.* **83:**427–436.
343. Holmgren, J., Svennerholm, L., Elwing, H., Fredman, P., and Strannegård, Ö., 1980, *Proc. Natl. Acad. Sci. U.S.A.* **77:**1947–1950.
344. Markwell, M. A. K., Svennerholm, L., and Paulson, J. C., 1981, *Proc. Natl. Acad. Sci. U.S.A.* **78:**5406–5410.

345. Wu, P.-S., Ledeen, R. W., Udem, S., and Isaacson, Y. A., 1980, *J. Virol.* **33**:304–310.
346. Kleinman, H. K., Martin, G. R., and Fishman, P. H., 1979, *Proc. Natl. Acad. Sci. U.S.A.* **76**:3367–3371.
347. Sharom, F. J., and Grant, C. W. M., 1978, *Biochim. Biophys. Acta* **507**:280–293.
348. Lee, A. G., 1977, *Trends Biochem. Sci.* **2**:231–233.
349. Partington, C. R., and Daly, J. W., 1979, *Mol. Pharmacol.* **15**:484–491.
350. Davis, C. W., and Daly, J. W., 1980, *Mol. Pharmacol.* **17**:206–211.
351. Leon, A., Facci, L., Toffano, G., Sonnino, S., and Tettamanti, G., 1981, *J. Neurochem.* **37**:350–357.
352. Caputto, R., Maccioni, A. H. R., and Caputto, B. L., 1977, *Biochem. Biophys. Res. Commun.* **74**:1046–1052.
353. Jeserich, G., Breer, H., and Düvel, M., 1981, *Neurochem. Res.* **6**:465–474.
354. Mullin, B. R., Decandis, F. X., Montanaro, A. J., and Reid, J. D., 1981, *Brain Res.* **222**:218–221.
355. Yohe, H. C., and Yu, R. K., 1982, *Trans. Am. Soc. Neurochem.* **13**:32.
356. Cochran, F. B., Ledeen, R. W., and Yu, R. K., 1983. *Developmental Brain Res.* **6**:27–32.
357. Littlemore, L. A. T., and Ledeen, R. W., 1979, *Aust. J. Chem.* **32**:2631–2636.
358. Koh, C.-S., Tsukada, N., Yanagisawa, N., Kunishita, T., Uemura, K., and Taketomi, T., 1981, *J. Neuroimmunol.* **1**:69–80.
359. Sonnino, S., Ghidoni, R., Masserini, M., Aporti, F., and Tettamanti, G., 1981, *J. Neurochem.* **36**:227–232.
360. Panzetta, P., Maccioni, H. J. F., and Caputto, R., 1980, *J. Neurochem.* **35**:100–108.
361. Dreyfus, H., Lonis, J. C., Harth, S., and Mandel, P., 1980, *Neuroscience* **5**:1647–1655.
362. Engel, E. L., Wood, J. G., and Byrd, F. I., 1979, *J. Neurobiol.* **10**:429–440.
363. Steinberg, M. S., 1963, *Science* **141**:401–408.
364. Barondes, S. H., and Rosen, S. D., 1976, *Neuronal Recognition* (S. H. Barondes, ed.), Plenum Press, New York, pp. 331–358.
365. Frazier, W., and Glaser, L., 1979, *Annu. Rev. Biochem.* **48**:491–523.
366. Marchase, R. B., 1977, *J. Cell Biol.* **75**:237–257.
367. Roth, S., and Marchase, R. B., 1976, *Neuronal Recognition* (S. H. Barondes, ed.), Plenum Press, New York, pp. 227–248.
368. Whattey, R., Ng, S. K.-C., Roger, J., McMurray, W. C., and Sanwal, B. D., 1976, *Biochem. Biophys. Res. Commun.* **70**:180–185.
369. McKay, J., 1980, *Glycolipids in Myogenesis*, Ph.D. Thesis, University of Washington, Seattle.
370. Obata, K., Oide, M., and Handa, S., 1977, *Nature* **266**:369–371.
371. Saito, M., Nojiri, H., and Yamada, M., 1980, *Biochem. Biophys. Res. Commun.* **97**:452–462.
372. Purpura, D. P., and Baker, H. J., 1978, *Brain Res.* **143**:13–26.
373. Purpura, D. P., Pappas, G. D., and Baker, H. J., 1978, *Brain Res.* **143**:1–12.
374. Walkley, S. U., Wurzelmann, S., and Purpura, D. P., 1981, *Brain Res.* **211**:393–398.
375. Pfenninger, K. H., and Maylie-Pfenninger, M.-F., 1978, *Neuronal Information Transfer* (A. Karlin, H. J. Vogen, and V. M. Tennyson, eds.), Academic Press, New York, pp. 373–386.
376. Fishman, P. H., and Henneberry, R. C., 1980, *Cell Surface Glycolipids* (C. C. Sweeley, ed.), American Chemical Society Washington, pp. 223–239.
377. Simmons, J. L., Fishman, P. H., Freese, E., and Brady, R. O., 1975, *J. Cell Biol.* **66**:414–424.
378. Fishman, P. H., and Atikkan, E. A., 1979, *J. Biol. Chem.* **254**:4342–4344.
379. Taki, T., Hirabayashi, Y., Kondo, R., Matsumoto, M., and Kojima, K., 1979, *J. Biochem. (Tokyo)* **86**:1395–1402.
380. Ginsburg, E., Salomon, D., Sreevalson, T., and Freese, E., 1973, *Proc. Natl. Acad. Sci. U.S.A.* **70**:2457–2461.
381. Sheu, C. W., Salomon, D., Simmons, J. L., Sreevalsan, T., and Freese, E., 1975, *Antimicrob. Agents Chemother.* **7**:349–363.
382. Riggs, M. G., Whittaker, R. G., Neumann, J. R., and Ingram, V. M., 1977, *Nature* **268**:462–464.
383. Sealy, L., and Chalkey, R., 1978, *Cell* **14**:115–121.
384. Roisen, F., Bartfeld, H., Nagele, R., and Yorke, G., 1981, *Science* **214**:577–578

385. Ledeen, R. W., Roisen, F. J., Byrne, M. C., Yorke, G., and Sclafani, J. R., 1982, *Trans. Am. Soc. Neurochem.* **13**:106.

386. Dimpfel, W., Möller, W., and Mengs, U., 1981, *Gangliosides in Neurological and Neuromuscular Function, Development, and Repair* (M. M. Rapport and A. Gorio, eds.), Raven Press, New York, pp. 119–134.

386a. Leon, A., Facci, L., Benvegnù, D., and Toffano, G., 1982, *Developmental Neurosci.* **5**:108–114.

387. Keenan, T. W., Schmid, E., Franke, W. W., and Wiegandt, H., 1975, *Exp. Cell Res.* **92**:259–270.

388. Callies, R., Schwarzmann, G., Radsak, K., Siegert, R., and Wiegandt, H., 1977, *Eur. J. Biochem.* **80**:425–432.

389. Radsak, K., Schwarzmann, G., and Wiegandt, H., 1982, *Hoppe–Seyler's Z. Physiol. Chem.* **363**:263–272.

390. Morgan, J. I., and Seifert, W., 1979, *J. Supramol. Struct.* **10**:111–124.

391. Hauw, J. J., Fenelon, S., Boutry, J.-M., Nagai, Y., and Escourolle, R., 1981, *Gangliosides in Neurological and Neuromuscular Function, Development, and Repair* (M. M. Rapport and A. Gorio, eds.), Raven Press, New York, pp. 171–175.

392. Gorio, A., Carmignoto, G., Facci, L., and Finesso, M., 1980, *Brain Res.* **197**:236–241.

393. Ceccarelli, B., Aporti, F., and Finesso, M., 1976, *Adv. Exp. Med. Biol.* **71**:275–293.

394. Ceccarelli, B., Aporti, F., and Finesso, M., 1977, *Adv. Exp. Med. Biol.* **83**:283–287.

395. Norido, F., Canella, R., and Aporti, F., 1981, *Experientia* **37**:301–302.

395a. Norido, F., Canella, R., and Gorio, A., 1982, *Muscle & Nerve* **5**:107–110.

395b. Sparrow, J. R., and Grafstein, B., 1982, *Experimental Neurol.* **77**:230–235.

395c. Grafstein, B., Yip, H. K., and Meiri, H., 1982, *Nervous System Regeneration* (A. M. Giuffrida-Stella, B. Haber, G. Hashim, J. R. Perez-Polo, eds.), Liss, New York, in press.

395d. Yip, H. K., and Grafstein, B., 1981, *Soc. Neurosci. Abstr.,* **7**:680.

395e. Verghese, J. P., Bradley, W. G., Mitsumoto, B. H., and Chad, D., 1982, *Exp. Neurol.* **77**:455–458.

395f. Miceli, G., Caltagirone, C., and Gainotti, G., 1977, *Acta Psychiat. Scand.* **55**:102–110.

396. Beuttler, H., Jeserich, G., and Rahmann, H., 1980, *Jpn. J. Exp. Med.* **50**:63–65.

397. Mamoli, B., Brunner, G., Mader, R., and Schanda, H., 1980, *Eur. Neurol.* **19**:320–326.

398. Pozza, G., Saibene, V., Comi, G., and Canal, N., 1981, *Gangliosides in Neurological and Neuromuscular Function, Development and Repair* (M. M. Rapport and A. Gorio, eds.), Raven Press, New York, pp. 253–257.

399. Gorio, A., Aporti, F., Norido, F., and Canella, R., 1981, *Gangliosides in Neurological and Neuromuscular Function, Development and Repair* (M. M. Rapport and A. Gorio, eds.), Raven Press, New York, pp. 259–266.

400. Marcus, D. M., and Schwarting, G. A., 1976, *Adv. Immunol.* **23**:203–240.

401. Hakomori, S., and Young, W. W., Jr., 1978, *Scand. J. Immunol.* **6**:97–117.

402. Marcus, D. M., and Kundu, S. K., 1980, *Adv. Exp. Med. Biol.* **125**:321–326.

403. Lingwood, C. A., and Hakomori, S., 1977, *Exp. Cell Res.* **108**:385–391.

404. Rasmussen, H., and Goodman, D. B. P., 1977, *Physiol. Rev.* **57**:421–509.

405. Karpiak, S. E., Graf, L., and Rapport, M. M., 1976, *Science* **194**:735–737.

406. Karpiak, S. E., Mahadik, S. P., Graf, L., and Rapport, M. M., 1981, *Epilepsia* **22**:189–196.

407. Karpiak, S. E., Mahadik, S. P., and Rapport, M. M., 1978, *Exp. Neurol.* **62**:256–259.

408. Pellmar, T. C., and Wilson, W. A., 1977, *Science* **197**:912–914.

409. Karpiak, S. E., Graf, L., and Rapport, M. M., 1978, *Brain Res.* **151**:637–640.

410. Karpiak, S. E., and Rapport, M. M., 1979, *Behav. Neural Biol.* **27**:146–156.

411. Rapport, M. M., 1981, *Gangliosides in Neurological and Neuromuscular Function, Development and Repair* (M. M. Rapport and A. Gorio, eds.), Raven Press, New York, pp. 91–97.

412. Kasarskis, E. J., Karpiak, S. E., Rapport, M. M., Yu, R. K., and Bass, N. H., 1981, *Dev. Brain Res.* **1**:25–35.

Metabolism of Phosphoinositides

Ata A. Abdel-Latif

1. INTRODUCTION

The literature on the metabolism of phosphoinositides, phosphatidylinositol (PI),* and the polyphosphoinositides, phosphatidylinositol-4-phosphate (PI-P; diphosphoinositide, DPI), and phosphatidylinositol-4,5-bisphosphate (PI-bis P; triphosphoinositide, TPI), prior to 1970 was reviewed by Hawthorne and Kai[1] in the earlier edition of the *Handbook of Neurochemistry*. Therefore, in this chapter, I limit my discussion to work that has been published since then, giving background summaries of earlier studies only when necessary to complete the overall picture.

In the last decade, there has been increased attention directed to the study of phosphinositides in several tissues by a number of investigators working in a variety of disciplines. This is evident from a recent conference on cyclitols and phosphoinositides[2] at which studies on the ubiquitous *myo*-inositol and its phospholipids in animals, plants, yeast, bacteria, and molds, and in brain, nerves, pineal, iris, retina, pancreas, testis, parotid, lymphocytes, synaptosomes, and lysosomes were discussed. Although phosphoinositides have been found to be minor constituents of all animal tissues examined, the present interest in their metabolism is derived from the fact that they may be involved in certain physiological processes. Thus, in addition to their contributing to membrane structure, suggestions have been made in the past that they may be involved in the following processes: (1) active transport of ions across cell membranes[3]; (2) permeability of the axonal membrane and conduction of impulses along nerve axons[1]; (3) ganglionic transmission[4]; (4) receptor function in a variety of tissues[2,5,6]; (5) activation of enzymes such as $(Na^+ - K^+)$-

---

* Abbreviations used: PI, phosphatidylinositol; PI-P, phosphatidylinositol-4-phosphate; PI-bis P, phosphatidylinositol-4,5-bisphosphate; PA, phosphatidic acid; PC, phosphatidylcholine; PS, phosphatidylserine. Nomenclature of phosphoinositides as recommended by IUPAC-IUB Commission on Biochemical Nomenclature is followed throughout this chapter (*Biochem. J.* **171**:1–19, 1978).

Ata A. Abdel-Latif • Department of Cell and Molecular Biology, Medical College of Georgia, Augusta, Georgia 30912.

ATPase[7] and tyrosine hydroxylase[8] and attachment of alkaline phosphatase to membranes[9]; (6) as a source of arachidonic acid for prostaglandin biosynthesis[10–12]; and (7) possibly certain neurological diseases.

In this chapter, first the structure and metabolism of phosphoinositides are discussed; then their possible role(s) in membrane physiological functions is reviewed. A few recent related reviews may be of interest to the reader. Hawthorne and his colleagues have discussed phospholipid metabolism in relation to transport across cell membranes,[13] metabolism and function of *myo*-inositol lipids,[14] and phospholipids in synaptic function.[6] Michell reviewed receptor activation, phosphoinositide metabolism[5], and inositol phospholipids in membrane function.[15] A recent symposium was devoted to cyclitols and phosphoinositides.[2]

2. CHEMISTRY AND DISTRIBUTION OF PHOSPHOINOSITIDES IN ANIMAL TISSUES

At one time *myo*-inositol was classified as an essential growth factor (in a vitamin B group) in animal tissues.[2] However, there is little evidence at this time to suggest that it acts as a coenzyme in enzymatic reactions, and, furthermore, it is well established that it can be generated by the irreversible isomerization of D-glucose-6-P to L-*myo*-inositol-1-P by L-*myo*-inositol-1-phosphate synthase.[16,17] *Myo*-inositol-1-P is dephosphorylated by *myo*-inositol-1-P phosphatase. Free *myo*-inositol is found in high concentrations in many tissues of the body, especially brain. *Myo*-inositol content of adult male rat brain was reported by Gaitonde and Griffiths[18] to be 5.3–5.7 μmol/g. Concentrations of *myo*-inositol (mmol/kg dry tissue) in molecular, granular, and medullary layers of the rat cerebellum were found to be 28.2, 14.1, and 17.6, respectively.[19] Its essential function seems to reside in its being a constituent of the phosphoinositides of all cell membranes. These phosphoinositides account for all the combined inositol in animal tissues.

In the early 1940s Folch and Woolley[20] reported that phosphoinositides in brain and spinal cord contained inositol. Later, Folch and his colleagues,[21–23], who coined the name phosphoinositides, isolated a diphosphoinositide fraction as Ca^{2+}–Mg^{2+} complex from brain. Since then, monophosphoinositide and the polyphosphoinositides have been isolated from several animal tissues, and their structures, cellular distributions, metabolism, and possible physiological functions have been investigated in detail.

2.1. Structures and Ionic Properties of Phosphoinositides

The structures and nomenclature of the phosphoinositides are given in Fig. 1. Although the polyphosphoinositides were first isolated from brain in 1949 by Folch,[22] their structures were established by Brockerhoff and Ballou[24] in 1961. There are nine possible isomers of hexahydroxycyclohexane, and the only inositol isomer found so far as a lipid constituent is *myo*-inositol.[25,26] In the preferred chair form of the ring (Fig. 2), the hydroxyl on carbon 2 has the

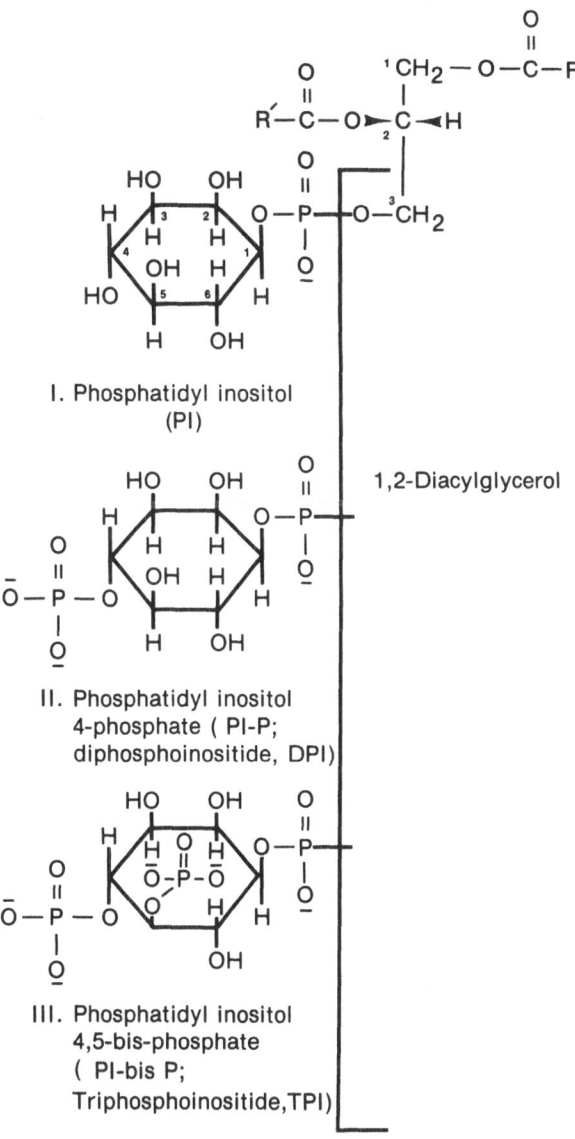

Fig. 1.   Structures and nomenclature of phosphoinositides.

axial configuration. The remaining five hydroxyls are equatorial. It is important to note that it is the 2-hydroxyl that takes part in cyclic ester formation when PI is hydrolyzed. In PI (Fig. 1, Structure I), 1,2-diacyl-*sn*-glycero-3-P (phosphatidic acid) is esterified to an equatorial hydroxyl of *myo*-inositol. In PI-P (Fig. 1, Structure II) and PI-bisP (Fig. 1, Structure III), one and two monoesterified phosphate groups, respectively, are attached to the appropriate hydroxyl groups of *myo*-inositol. Another structural feature that distinguishes the phosphoinositides from other phospholipids is that in some tissues, such as the

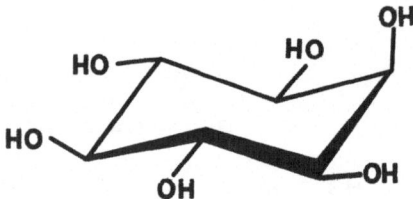

Fig. 2. *Myo*-inositol, chair form (reproduced with modification from ref. 27 with permission).

brain, their nonpolar side chains are comprised mainly of stearoylarachidonyl species,[28] the prevalent pattern being R = stearoyl and R' = arachidonyl (Fig. 1).

At physiological pH, the phosphoinositides are anionic, with PI, PI-P, and PI-bisP carrying one, three, and five net negative charges, respectively (Fig. 1). Phosphoinositides, especially PI-bisP, complex readily with divalent cations such as Ca^{2+} and Mg^{2+}.[29-33] In fact, Dawson[30] reported that PI-bisP has a greater affinity than EDTA for Ca^{2+}, resulting in a highly hydrophobic complex with Ca^{2+}. Hayashi *et al.*[31] reported on a Ca^{2+}-sensitive univalent cation channel formed by lyso-PI-bisP in bilayer lipid membranes. The lyso-PI-bisP could penetrate into the membrane and form a channel system that was selectively permeable to univalent cations. Moreover, the univalent cation conductance of the membrane was greatly influenced by the concentration of Ca^{2+} in the aqueous phase. The affinities of phosphoinositides for monovalent cations are much lower.

Phosphoinositides complex readily with a variety of proteins.[34] In this connection, Armitage *et al.*[35] reported that glycophorin preparations isolated from erythrocytes by a lithium diiodosalicylate extraction, phenol partition method contain appreciable quantities of polyphosphoinositides. This was confirmed by Buckley.[36] More recently, Shukla *et al.*[37] showed that glycophorin extracted by Triton X-100 showed no such enrichment. These workers concluded that the majority of the polyphosphoinositides are not associated with glycophorin. Wu *et al.*[38] showed that the proteolipid receptor fraction isolated from cerebral cortex contains PI-bisP. Later, this group suggested that this phosphoinositide may function as a binding component of the nicotinic cholinergic receptor.[39] These findings clearly suggest that there is a considerable affinity between polyphosphoinositides and tissue proteins.

When polyphosphoinositides are complexed with monovalent cations they are largely water soluble, and when they are complexed with divalent cations or proteins, they are water insoluble but are readily soluble in acidified chloroform–methanol.[30] These solubility properties constitute the basis for the common use of chloroform–methanol–HCl in the extraction of higher inositides from a variety of tissues.[14,40] In contrast to polyphosphoinositides, PI has a low affinity for Ca^{2+}, and it is readily soluble in neutral chloroform–methanol. Neomycin, an aminocyclitol antibiotic that has a high affinity for polyphosphoinositides and can competitively displace Ca^{2+} from membranes,[41] was recently employed in the isolation of polyphosphoinositides from brain.[42] In this method, neomycin is reductively coupled to reactive glass beads (Glyco-phase-CPG) and serves as the stationary phase in column chromatography. All lipids but the polyphosphoinositides are eluted from the column by 150 mM ammonium acetate in chloroform–methanol–water. The PI-P was eluted by increasing the salt concentration to 600 mM, and removal of PI-bisP was accomplished by addition of either ammonia or HCl to the eluent.

In view of these ionic and binding properties, several investigators have suggested that polyphosphoinositides could be involved in ion permeability in membranes.[1,33,43] They speculated that these phospholipids may have a role in the control of membrane permeability, possibly through binding and release of Ca^{2+}.

2.2. Distribution

Phosphoinositides have been demonstrated in all membranes of animal tissues that have been investigated. Phosphtidylinositol occurs widely in nature in amounts usually corresponding to 2–12% of the total phospholipids. It represents up to 20% of the phospholipids of certain yeasts (*Saccharomyces*) and some kinetoplastid protozoa.[44,45] The higher inosites, PI-P and PI-bisP, are minor lipid components of most animal tissues. They occur in greatest amounts in the nervous system and are especially enriched in the white matter or myelin sheath; however, significant quantities of these substances are also present in other subcellular fractions (Tables I and II). Thus, concentrations of the polyphosphoinositides increased considerably during myelination in both the central and peripheral nervous systems of the chicken (Table I). Eichberg and Hauser[48] have compared the subcellular distribution of PI-P and PI-bisP in brains of 7-day (unmyelinated) and 34-day (myelinated) rats. In the unmyelinated brain, distribution of the polyphosphoinositides resembled that of 5′-nucleotidase and acetylcholinesterase, indicating the presence of these phospholipids in neuronal and glial plasma membranes. In myelinated brains, over half of the PI-P and 75% of the PI-bisP were in myelin-rich fractions. The ratio PI-P/PI-bisP was 0.47 for the myelin fraction and 1.96 for synaptosomes. Because of the slow turnover of the lipids and proteins of the myelin sheath, it has been suggested that the highly metabolically active polyphosphoinositides

Table I
Levels of Phosphoinositides in Developing Chick Brain and Sciatic Nerve[a]

Tissue	Phosphoinositides (μmol/g wet wt.)								
	Embryo (21 days gestation)			8-Day old			Adult		
	PI	PI-P	PI-bisP	PI	PI-P	PI-bisP	PI	PI-P	PI-bisP
Brain	1800	197	235	2400	232	401	1800	202	707
Sciatic nerve	1700	31	293	2300	45	538	3600	67	875

[a] Data taken from reference 46.

Table II
Levels of Phosphoinositides in Rat Brain and Its Subcellular
Fractions

Tissue	Phosphoinositides		
	PI	PI-P	PI-bisP
Whole brain	2220[a]	250[a]	400[a]
Membranes			
Myelin	52[b]	35[c]	75[c]
Mitochondrial	17[b]	6[c]	4[c]
Synaptosomal	51[b]	45[c]	23[c]
Microsomal	18[b]	18[c]	11[c]

[a] Data from reference 40. Results are expressed as nmol/g wet wt. brain.
[b] Data from reference 47. Results are expressed as nmol/mg protein.
[c] Data from reference 48. Results are expressed as nmol/g wet brain.

and their enzymes might be located in myelin appurtenances such as the internal or external mesaxon or the membrane loops at the nodes of Ranvier.[49,50]

Recently Deshmukh et al.[50] separated rat brain myelin into three subfractions, heavy, medium, and light. The PI was evenly distributed among the fractions, and PA, PI-P, and PI-bisP occurred in highest concentrations in the light myelin. They concluded that these data could indicate that polyphosphoinositides may play an important role in the tightly packed central lamellae of the myelin sheath. Representative values for the concentrations of polyphosphoinositides in the nervous system are given in Tables I and II. Low polyphosphoinositide concentrations, coupled with their rapid breakdown and the difficulties encountered in their extractions, separation, and determination, have contributed to the difficulty of giving reliable values for their concentrations in certain organs. Values for phospholipid concentrations in a variety of tissues, including polyphosphoinositides, have been compiled by White.[51] There is little published information on the localization of phosphoinositides in biological membranes. In erythrocyte membranes, choline-containing phospholipids are located mainly in the outer layer, and amino phospholipids and PI are preferentially located in the inner leaflet.[52–55] There is evidence that this phospholipid asymmetry also occurs in other mammalian membranes.[52] From the localization of the enzymes involved in their metabolism, it appears that phosphoinositides are localized at the cytoplasmic side of the membranes. As for the polyphosphoinositides, Garrett and Redman[56] prepared inside-out and right-side-out vesicles from human erythrocytes that were impermeable to the γ-^{32}P-ATP substrate. Synthesis of polyphosphoinositides only occurs with the inside-out vesicles, suggesting that the phosphoinositide kinases are on the cytoplasmic side of the plasma membrane. Although the above studies suggest that PI and the polyphosphoinositides are located at the cytoplasmic side of cell membranes, it must be emphasized that the data on phospholipid asymmetry in biological membranes are still incomplete and, in many cases, controversial.

3. METABOLIC PATHWAYS AND PROPERTIES OF ENZYMES INVOLVED IN PHOSPHOINOSITIDE METABOLISM

3.1. Overview of Biosynthesis and Catabolism

A summary of the major pathways for phosphoinositide biosynthesis and catabolism is given in Fig. 3. CDP-Diacylglycerol, formed from PA and CTP in the presence of CTP-PA cytidyltransferace (Fig. 3, Step 1), is the key intermediate in phosphoinositide biosynthesis. In general, the biosynthesis of PI from CDP-diacylglycerol and *myo*-inositol (Fig. 3, Step 2) occurs at the endoplasmic reticulum, whereas its phosphorylation by ATP in the presence of specific kinase to PI-P (Fig. 3, Step 3) and PI-bisP (Fig. 3, Step 4) takes place at the plasma membrane and membranes of other cellular organelles. Phosphorylation of PI by ATP results in formation of the polyphosphoinositides in which the phosphodiester originates from PA and the monoester directly from ATP. Removal of the phosphomonoesters from the polyphosphoinositides is achieved via specific phosphomonoesterases (Fig. 3, Steps 5 and 6). The phosphomonoesterases are activated by Mg^{2+} and are found in the soluble and particulate fractions, the ratio of soluble to particulate being dependent on the tissue. The breakdown of phosphoinositides into 1,2-diacylglycerol and water-soluble inositol phosphates (Fig. 3, Steps 7, 8, and 9) is achieved by specific phosphodiesterases which are found in the soluble and particulate fractions

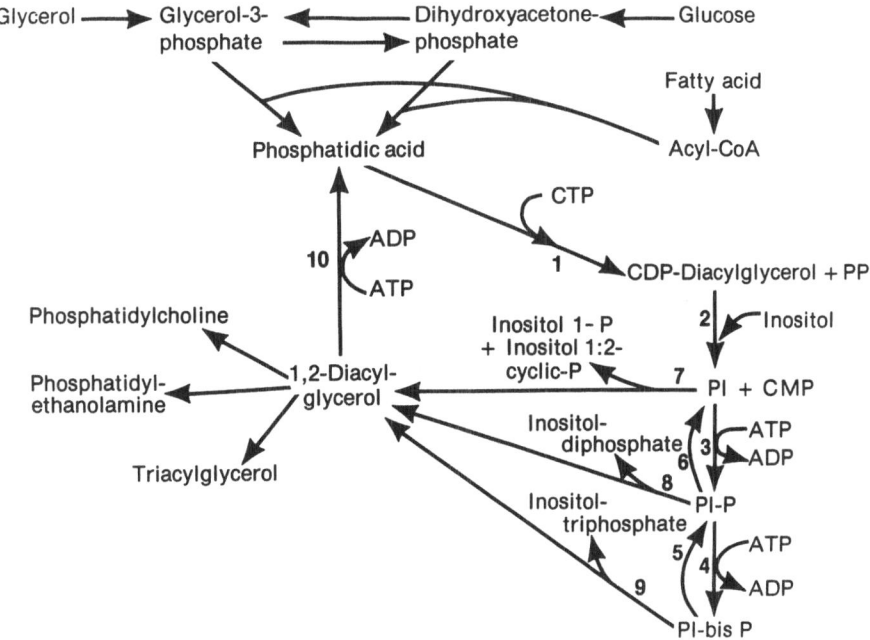

Fig. 3. Pathways of phosphoinositide metabolism.

and, in general, are activated by Ca^{2+}. Finally, diacylglycerol, the product of phosphoinositide breakdown, is phosphorylated by ATP (Fig. 3, Step 10) in the presence of the plasma membrane enzyme diacylglycerol kinase to regenerate PA, the precursor of CDP-diacylglycerol and subsequently the phosphoinositides. 1,2-Diacylglycerol also serves as a precursor for PC, phosphatidylethanolamine, and triacylglycerol (Fig. 3). The other products of phosphoinositide breakdown, namely, the water-soluble inositol phosphates, are degraded by inositolphosphatases into free *myo*-inositol and inorganic phosphate.

Although there has been a large amount of literature on the properties of the enzymes involved in the phosphoinositide metabolism, there is still a need to establish their exact localization in the cell and their ionic requirements. Much of the work on phosphoinositide turnover* in a variety of tissues is covered in Section 4, since it is related to changes in their metabolism in response to stimulation. The following is a concise summary of the more recent studies on the metabolism of CDP-diacylglycerol and the individual phosphoinositides.

3.1.1. CDP-Diacylgylcerol

3.1.1a Biosynthesis. One of the key features of PI biosynthesis is the prominent involvement of the liponucleotide CDP-diacylgylcerol which is formed from CTP and PA in the presence of the enzyme CTP–PA cytidyl transferase:

$$\text{Phosphatidic acid} + \text{CTP} \rightleftarrows \text{CDP-Diacylgylcerol} + \text{Pi} \qquad [1]$$

Net synthesis of CDP-diacylgylcerol is dependent on the addition of PA and CTP. This enzyme was studied in rat brain preparations[57] and other tissues.[58,59] In brain, the enzyme was mainly located in the microsomal fraction with small but significant activity in the mitochondrial fraction. Comparison of activities (nanomoles CTP incorporated per milligram protein per minute) among tissues showed the following order: brain, 1.87; liver, 1.32; lung, 1.19; small intestine, 1.0; kidney, 0.69; heart, 0.41; diaphragm, 0.07; and skeletal muscle, 0.02. The enzyme showed no selectivity with respect to the fatty acid distribution in phosphatidic acids.[57] A 550-fold purification of this enzyme was achieved from yeast mitochondria.[58] Phosphatidic acid accumulates in the membranes of *E. coli* mutants defective in CTP–PA cytidyl transferase.[60] In rat liver microsomes, the membrane-bound enzyme cannot utilize PA substrate present in heat-denatured membranes but is active on PA incorporated into membranes of phospholipid vesicles.[61]

In rat liver microsomes, GTP has been found to markedly enhance the formation of CDP-diacylglycerol. By stimulating the formation of the liponucleotide, it enhances the formation of PI from CTP, PA, and inositol. The concentrations of CDP-diacylgylcerol in microorganisms and mammalian tissues are extremely small, and thus, until a few years ago, very little work had been done on its characterization. Recently, Thompson and MacDonald[63,64]

---

* For more detail on the turnover of other phospholipids see A. Porcellati and G. Arienti (Chapter 5, this volume).

isolated the liponucleotide from bovine brain and liver in sufficient quantities to permit its complete chemical characterization. A procedure for the preparation of ^{32}P- or ^{3}H-labeled CDP-diacyglycerol from rat pineal glands has also been described.[65]

Yield of CDP-diacylglycerol from brain ranged from 9.2 to 15.5 μmol per kg of tissue, which corresponds to about 1% of the level of PA. Analysis of fatty acid composition showed that 1-steroyl,2-arachidonyl is the major species of brain CDP-diacylglycerol. This differs significantly from the composition of PA, whose predominant unsaturated fatty acid is oleate, but is very similar to that of PI. The presence in the tissues of CDP-diacylglycerol with high arachidonate content suggests that arachidonyl-rich PI can also be derived from the liponucleotide. Although the experimental evidence on how the high arachidonate content of CDP-diacylglycerol is achieved is inconclusive, the enrichment of arachidonate in PI could be achieved in large measure by a combination of deacylation and reacylation reactions.[66–70] A high selectivity for arachidonyl-CoA in the acylation of (1-acyl) lyso-PI has been demonstrated with rat brain and rat liver microsomes.[71] The isolation of a specific arachidonyl coenzyme A cytidine diphosphate monoacylglycerolacyltransferase from rat liver microsomes which shows strict specificity for liponucleotide as acyl group acceptor and arachidonyl-CoA as the sole acyl group donor has been reported[72,73]; however, this work has since been retracted (W. Thompson, personal communication, 1980).

The above studies could explain why brain phosphoinositides are enriched with arachidonate, the precursor of the prostaglandins.

3.1.1b. Catabolism. CDP-diacylglycerol is split quantitatively by a membrane-bound hydrolase from *E. coli* that is specific for the liponucleotide into PA and CMP.[74] A similar activity has been reported in a lysosomal fraction obtained from guinea pig cerebral cortex.[75]

3.1.2. Phosphatidylinositol

3.1.2a. Biosynthesis. The biosynthesis and catabolism of PI are now well established in brain and other tissues.[76–78] In general, biosynthesis of PI occurs at the endoplasmic reticulum, and its degradation occurs in both the cytoplasm and particulate fractions.[79,80] The newly synthesized PI is transported from the endoplasmic reticulum to other nonsynthetic membranes by carrier exchange proteins which are found in the cytoplasm of the cell.

Free *myo*-inositol is incorporated into PI of animal tissues by two distinct mechanisms. One is *de novo* synthesis of PI from CDP-diacylglycerol and free *myo*-inositol (equation 2); in the other, *myo*-inositol is incorporated into PI in the presence of Mn^{2+} as a result of an enzymatic exchange reaction with endogenous PI.[78] In the latter, *myo*-inositol incorporation is not mediated by CDP-diacylglycerol. In brain, reactions involved in the *de novo* biosynthesis of PI result in the formation of predominant 1-steroyl,2-arachidonyl molecular species.

$$\text{CDP-diacylglycerol} + \textit{myo}\text{-inositol} \rightleftarrows \text{phosphatidylinositol} + \text{CMP} \quad [2]$$

Holub[81] injected rats intraperitoneally with [³H]inositol and then determined the molecular species of PI produced in liver microsomes by the Mn^{2+}-stimulated entry of free inositol to assess the possible quantitative significance of this pathway. He concluded that a major proportion of the free inositol entering the phosphatidylinositols under physiological conditions does so during the *de novo* biosynthesis of these lipids. The PI–*myo*-inositol exchange enzyme from rat liver microsomes has recently been purified and characterized.[82]

The CDP-diacylglycerol:inositol transferase, which is responsible for the *de novo* biosynthesis of PI (equation 2), was investigated in brain[83,84] and other tissues.[85,86] In guinea pig brain, the enzyme showed a high selectivity for *myo*-inositol.[83] Comparison of activity (nanomoles of CMP released per milligram protein per hour) among tissues showed the following[83]: brain, 49; liver, 31; kidney, 31; heart, 76; lung, 76; and spleen, 64.

This enzyme was partially purified from rat brain microsomes[84] and rat liver microsomes.[87] In brain,[84] the enzyme loses activity when the lipids are removed, and addition of total microsomal lipids to lipid-depleted enzyme preparations restored 10–18% of the original enzyme activity. The reversal of CDP-diacylglycerol:inositol transferase activity has been reported in thyroid[88] and lungs,[85] and in rat brain synaptosomes, incorporation of [³H]inositol into PI appears to take place through this pathway.[89] Incorporation of [³H]inositol into PI of synaptosomes and purified mitochondria was about 16 and 11%, respectively, of that in microsomes.[89] The fact that of the nine possible isomers of hexahydroxycyclohexane only *myo*-inositol (Fig. 2) is incorporated rapidly into the inositol lipids probably results from the high degree of specificity the inositol transferase has for this isomer.[78,83]

3.1.2b. Catabolism. In animal tissues, PI is degraded by PI phosphodiesterase (PI-specific phospholipase C) into 1,2-diacylglycerol and a mixture of *myo*-inositol-1-P and *myo*-inositol 1:2-cyclic-P:

$$\text{phosphatidylinositol} + H_2O \rightarrow \text{1,2-diacylglycerol} +$$
$$\textit{myo}\text{-inositol 1:2-cyclic-P} + \textit{myo}\text{-inositol-1-P} \quad [3]$$

Phosphoinositides are the only phospholipids for which there is well-documented evidence of hydrolysis by a phospholipase C in mammalian tissues. Because of the possible role of this enzyme in the stimulated PI turnover in cell membranes, a number of investigators have investigated its properties in a variety of tissues, including brain,[90-97] liver,[98] intestinal mucosa,[99] kidney,[100] smooth muscle,[101,102] platelets,[103,104] lymphocytes,[105] and erythrocytes.[106] The properties of this enzyme have also been studied in microorganisms.[107-109]

In brain the enzyme has been reported to occur in both particulate[93] and soluble[85] forms; it is activated by Ca^{2+} and has a pH optimum of about 5.6. Irvine and Dawson,[95] working with rat brain, concluded that all of the properties of Ca^{2+}-dependent PI phosphodiesterase in this tissue can be explained by the existence of only the soluble cytoplasmic enzyme: no evidence confirming a distinct membrane-bound activity has been obtained. In smooth muscle, the enzyme does not appear to be exclusively cytosolic.[101,102] Thus, in the iris smooth muscle,[102] about 58% of the Ca^{2+}-sensitive enzyme activity was found in the soluble fraction, the remainder being particulate. Allan and Michell,[105] working with lymphocytes, reported that this enzyme has a pH optimum at 7.0 and is activated by 0.7 μM Ca^{2+}, a concentration similar to common intracellular values. More recently Hirasawa *et al.*[110] reported that the Ca^{2+}-dependent PI phosphodiesterose from the cytosolic supernatant of rat brain was active against exogenous [^{32}P] PI from pH 5.0 to pH 8.5. However, the activity in the range pH 7.0–8.5 could not be recovered after precipitation with $(NH_4)_2SO_4$; most of the enzyme activity was recovered in the 30–40% fraction and showed a single sharp pH optimum at 5.5. They concluded that the Ca^{2+}-dependent PI phosphodiesterose exhibits considerable heterogeneity, both with respect to pH optima of activity, and its isoelectric properties.

Another PI phosphodiesterase, which is distinct from the Ca^{2+}-sensitive enzyme, was reported in lysosomes from rat brain and liver.[111] The lysosomal enzyme has a pH optimum of 4.8; it is inhibited by Ca^{2+} but not by EDTA, and it appears to be exclusively localized in the lysosomal fraction. The PI phosphodiesterase is a cyclizing phosphotransferase[112] that catalyses the release from PI of *myo*-inositol-1:2-cyclic P (equation 3). The formation of the latter, which constitutes about 55% of the total water-soluble inositol phosphates, has now been confirmed in several tissues. A specific phosphodiesterase that converts the cyclic ester to *myo*-inositol-1-P has been reported in kidney, where it appears to be selectively localized in the brush border of the proximal tubules.[113] It is possible that the initial water-soluble product of PI hydrolysis is the cyclic ester which is then hydrolyzed by the phosphodiesterase into inositol-1-P. Michell and Lapetina[114] suggested that the cyclic ester might act as an intracellular "second messenger"; however, its physiological role remains enigmatic.[115]

Other pathways that have been reported recently on the catabolism of PI are deacylation at the 2-position to give lyso-PI and free fatty acids[116] and cleavage of glycerophosphoinositol by a phosphodiesterase into glycerophosphate and inositol.[117] The deacylase activity was demonstrated in the 12,000–106,000 *g* pellet and supernatant prepared from rat brain homogenate. The major activity had a pH optimum at pH 7.5 and was stimulated by Ca^{2+}. This study provides evidence for the deacylation component of a deacylation–reacylation cycle for the generation of specific molecular species of PI.

3.1.3. Polyphosphoinositides

3.1.3a. Biosynthesis. Biosynthesis of the polyphosphoinositides proceeds by a sequential kinase reactions from PI:

phosphatidyl inositol + ATP → phosphatidyl inositol 4-phosphate + ADP

[4]

phosphatidylinositol 4-phosphate + ATP →
                    phosphatidyl inositol 4,5-bisphosphate + ADP   [5]

These reactions were first demonstrated by incorporation of radioactive phospholipid precursors, e.g., ^{32}Pi, [^{14}C]glycerol, and [^{3}H]inositol, *in vivo* and *in vitro* in brain slices (for review see references 1 and 25). These studies have shown that whereas [^{14}C]glycerol and [^{3}H]inositol are incorporated in the order PI > PI-P > PI-bisP, ^{32}Pi is incorporated in the order PI-bisP > PI-P > PI. Phosphorylation of PI by ATP produces the polyphosphoinositides in which the phosphodiester originates from PA and the monoester directly from ATP. Phosphatidylinositol and PI-P kinases add phosphates to the 4 and 5 positions of inositol in PI; the structure of the latter remains intact during these inter-conversions. The properties of these enzymes have been studied in rat brain,[118-120] chick brain and sciatic nerve,[121] rabbit sciatic nerve myelin,[122] kidney cortex,[123] adrenal chromaffin vesicles,[124-127] parotids, [128] erythrocytes,[129] and yeast.[130] The PI-kinase (equation 4) has a subcellular distribution similar to that of 5'-nucleotide and (Na^{+}–K^{+})-ATPase in brain. It is present in brain microsomes,[120] kidney cortex microsomes,[123] plasma membranes of brain,[119,131] chromaffin granule membrane,[124-127] and all particulate subcellular fractions from rat parotid glands.[128] In brain, the PI-kinase requires Mg^{2+}, is inhibited by Ca^{2+}, and it is stimulated by certain detergents.[131] In erythrocytes, PI-kinase activity is increased more than tenfold by addition of mercaptoeth-anol and is inhibited by adenosine and ADP.[129]

The PI-P kinase (equation 5) has been shown to be cytoplasmic in rat brain[119] and parotids[128] and membraneous in rabbit sciatic nerve myelin[122] and rat kidney cortex.[123] The PI and PI-P kinases appear to be localized on the cytoplasmic surface of the erythrocyte membrane.[56] Both enzymes are acti-vated by Mg^{2+} and inhibited by Ca^{2+} when Mg^{2+} is present, and neither of the kinases is activated by Na^{+} or K^{+}. Developmental changes were the same for both kinase activities, being low in unmyelinated brain and sciatic nerve.[121]

3.1.3b Catabolism. The polyphosphoinositides are rapidly catabolized either to PI and Pi by monoesteratic cleavage of the phosphate groups by a polyphosphoinositide phosphomonoesterase (equations 6 and 7) or to diacyl-glycerol and the respective inositol phosphates by a polyphosphoinositide phos-phodiesterase (equations 8 and 9).

phosphatidylinositol 4,5-bisphosphate + H$_2$O →

$\qquad\qquad\qquad\qquad\qquad$ phosphatidylinositol 4-phosphate + Pi   [6]

phosphatidylinositol 4-phosphate + H$_2$O → phosphatidylinositol + Pi   [7]

phosphatidylinositol 4,5-bisphosphate + H$_2$O →

$\qquad\qquad\qquad\qquad\qquad$ 1,2-diacylglycerol + inositol triphosphate   [8]

phosphatidylinositol 4-phosphate + H$_2$O →

$\qquad\qquad\qquad\qquad\qquad$ 1,2-diacylglycerol + inositol diphosphate   [9]

The properties and subcellular distribution of these hydrolytic enzymes have been studied in brain,[132-135] kidney,[136,137] smooth muscle,[138] erythro-cytes,[139] and the protozoan *Crithidia fasciculata.*[140,141] Studies on their sub-cellular distribution revealed them to be localized in both the soluble and par-ticulate fractions. Polyphosphoinositide phosphomonoesterase was purified from rat brain cytosol.[134] The enzyme has a relatively low molecular weight and an isoelectric point of 6.8. It showed a high affinity for PI-bisP and a pH optimum between 7.4 and 7.6, and it was activated by Ca^{2+} and Mg^{2+}. In most tissues, the phosphomonoesterase and the phosphodiesterase are activated by Mg^{2+} and Ca^{2+}, respectively. Although the PI-bisP is a better substrate than PI-P for both enzymes, there is little evidence to suggest that these polyphosphoinositides are attacked by separate enzymes. The developmental pattern of both enzymes was investigated in the central and peripheral nervous systems of the chicken.[135] The PI-bisP phosphomonoesterase did not change with age, whereas the PI-bisP phosphodiesterase decreased steadily with no marked change during the period of most active myelination.

3.2. Phosphatidylinositol Exchange between Membranes

De novo biosynthesis of the major phospholipids, including PI, occurs at the endoplasmic reticulum, both rough and smooth. The newly synthesized phospholipids are then transported to other, nonsynthetic membrane sites by phospholipid exchange proteins which are found in the cytosol. These proteins have been investigated in a variety of animal tissues (for review see Wirtz[142]). They range from 12,500 to 25,000 daltons and catalyze the transfer of phospholipids from a donor such as endoplasmic reticulum or artificial membranes (liposomes) to an acceptor such as mitochondria, nerve terminals, or myelin. Thus, if two different membrane preparations, such as microsomes and mitochondria, are mixed and then incubated, one observes an exchange of phospholipid only if the soluble fraction from the tissue is added back to the incubation medium.

When unlabeled mitochondria or synaptosomes from guinea pig brain were incubated with [^{32}P]-labeled microsomal fraction in the presence of brain soluble fractions, there was a transfer of phospholipids, including PI, to the synaptosomes.[143] Soluble proteins have been shown to catalyze the transfer of phospholipids to the myelin membranes[144-146] and synaptosomes[144] *in vitro*. There is an increase in PI transfer activity in rat cerebral hemispheres during the stage of most active myelination.[147] Exchange of PI and PC between microsomal and myelin membranes has been demonstrated.[148] This exchange is reversible and is catalyzed by soluble proteins from the brain homogenate precipitated at pH 5.1. Specific proteins for the transfer of PI between membranes have been isolated from brain.[149,150] The molecular properties and phospholipid transfer specificity and activity of these proteins are distinct from those reported for phospholipid exchange proteins from bovine liver and heart.

Brammer and Sheltawy[151] purified true proteins from bovine brain, each of 22,000 daltons, that exchange PI and PC between mitochondria of bovine brain and PI/PC liposomes. A highly purified exchange protein from beef brain retained the ability to catalyze a net transfer of both PI and PC.[152] This protein showed a stronger affinity for PI, an eightfold higher transfer activity, than for PC. The distribution of PI and PC in sonicated vesicles has recently been determined by the use of an exchange protein from heart and PI phosphodiesterase from *Staphylococcus aureus*.[153] About 70% of PI and 65% of PC were accessible to the exchange protein, and 70–75% was accessible to the enzyme.

This finding could suggest that PI is not preferentially located in either surface of the phospholipid bilayer.

The above *in vitro* studies indicate that cells are capable of protein-mediated exchange of all of their phospholipids from membrane to membrane; however, the role of these specific exchange proteins *in vivo* is still unknown. It is also not known whether other mechanisms, which do not require the participation of these proteins, such as lateral diffusion, could be involved in phospholipid transport from the endoplasmic reticulum to other membranes.

In summary, PI is synthesized in the endoplasmic reticulum from which it is transported via specific proteins or other unknown mechanisms throughout the cell. Phosphorylation of PI to polyphosphoinositides occurs in other membranes of the cell. All of the enzymes involved in phosphoinositide catabolism are found in both the cytosol and other membranes of the cell. Although there has been a large amount of literature published on these enzymes in the last decade, there is still a need to establish their subcellular distributions, to purify them, and to investigate their properties. This is essential, since in order to understand the physiological significance of phosphoinositides in brain structure and function, it is necessary to know the characteristics of the enzymes involved in their synthesis and catabolism.

4. PHOSPHOINOSITIDE TURNOVER AND PLASMA MEMBRANE ACTIVATION AND PERMEATION

4.1. Introduction and Overview

Although phosphoinositides constitute a small percentage of membrane phospholipids (representing roughly 3–5% of the total lipid of whole brain), their metabolism and function have captured the interest and imagination of several investigators for the past 30 years. The continued interest in these phospholipids stems from the following findings on phosphoinositide metabolism in brain and other tissues. (1) Turnover (defined as simultaneous synthesis and degradation) of the polar head groups of phosphoinositides is much higher than that of any other membrane phospholipid. This can easily be seen from the rapidity with which ^{32}Pi is incorporated into the inositol-containing phospholipids, especially the polyphosphoinositides. This rapid labeling could be associated with plasma membrane function. (2) A variety of stimuli, including neurotransmitters, hormones, certain neuropharmacological agents, and electrical pulses, significantly increase the turnover of the polar head groups of phosphoinositides and their precursor, phosphatidic acid. Addition of ACh to brain slices incubated in an isosmotic medium containing ^{32}Pi alters considerably the ^{32}P labeling of phosphatidic acid and phosphoinositides. The ACh-stimulated ^{32}P labeling of these phospholipids is blocked by atropine, a muscarinic cholinergic blocking agent. These findings suggest a connection between the metabolism of phosphoinositides and receptor function. (3) Polyphosphoinositides have high affinity for binding Ca^{2+} and other cations (see Section 2.1). This property, coupled with the fact that they are rapidly dephosphorylated and rephosphorylated, suggest that they may be important in calcium binding and permeability changes related to axonal conduction. Such changes would affect Na^+ and K^+ permeability. More recent evidence suggests that they may also play an important role at the synapse.[155]

In the last decade, a large amount of literature has appeared on the role of phosphoinositides in receptor function; this includes a number of excellent reviews on the whole field.[2,5,6,14,154] Briefly, in 1953 Hokin and Hokin[156] described the first phospholipid effect (or phosphoinositide effect), which may be defined as a change in the rate of turnover of a phospholipid when the tissue in which it occurs is stimulated by various stimuli, when they discovered that ACh stimulated the release of digestive enzymes from pancreas and that, at the same time, there was an increased incorporation of ^{32}Pi into phospholipids. Later, these investigators described a similar phosphoinositide effect in brain slices[157-159] and furthermore showed that this phenomenon is largely attributable to an increase in the ^{32}P labeling of PI and its precursor, PA. Since then, the phosphoinositide effect has been demonstrated in several tissues (see below).

Studies on the phosphoinositide effect revealed the following common properties. (1) It is a property of whole cells and is lost when the cells are broken, e.g., by homogenization or freezing and thawing. With the exception of synaptosomes, which retain most of the properties of intact cells, attempts to demonstrate this effect in subcellular fractions have been unsuccessful so far. Synaptosomes that have been lysed osmotically, by freezing and thawing, or by detergents do not show a phosphoinositide effect. (2) In general, the effect is mostly confined to the acidic phospholipids, namely, PA and the phosphoinositides, although in some tissues alterations in the turnover of other phospholipids, e.g., PC and phosphatidyl glycerol, have been reported. (3) Only the turnover of the polar head groups of these phospholipids is affected, the diacylglycerol backbone being conserved. *De novo* synthesis, e.g., from [^{14}C]glycerol, does not appear to be involved. In fact, present evidence suggests an increase of phosphoinositide breakdown as the rate-limiting step that mediates this phenomenon. (4) In spite of the fact that various tissues respond to different stimuli, the effect is associated with membrane processes, mostly involving permeability changes across the plasma membrane. (5) The effect does not appear to be a result of increased labeling or transport of precursors. (6) A direct effect of neurotransmitters on any of the enzymes involved in phosphoinositide metabolism has not yet been unequivocally demonstrated. (7) In most tissues, this effect is mediated through muscarinic cholinergic and α-adrenergic receptors (see below). (8) Efforts to demonstrate an involvement of cyclic nucleotides as second messengers have been unsuccessful, although a report, which needs to be confirmed, has appeared recently showing effects of cyclic nucleotides and selected agents known to act via the cyclic nucleotide system on the ^{32}Pi labeling of polyphosphoinositides in rabbit renal cortex slices.[160] However, although this phenomenon appears to be Ca^{2+} independent in some tissues such as the parotids, insect salivary gland, and ileum muscle, it is Ca^{2+} dependent in others such as the iris smooth muscle, synaptosomes, chromaffin cell cultures, neutrophils, hepatocytes, and pancreas (see below).

4.2. Phosphoinositides and Receptor Function

4.2.1. Types of Receptors Involved

Activation of a variety of cell-surface receptors leads to an increase in ^{32}P labeling of PA and PI and a decrease in that of polyphosphoinositides. Both *in vivo* and *in vitro* studies have established that the phosphoinositide effect is associated with muscarinic cholinergic and α-adrenergic receptors (Table III). This conclusion is based on the finding that cholinergic and adrenergic stimulation of ^{32}P labeling of phosphoinositides are blocked, respectively, by the antimuscarinic antagonist atropine but not by the nicotinic antagonist *d*-tubocurarine, and by the α-adrenergic antagonist phentolamine but not the β-adrenergic antagonists propranolol and sotalol. There is agreement now, from work done in a variety of tissues, that the phosphoinositide effect is mediated through α1-adrenoreceptors. This conclusion is based on the finding that only prazosin, an α1-adrenergic antagonist, and not yohimbine, an α2-antagonist, inhibited the norepinephrine-stimulated ^{32}P labeling of PI. The possible role(s) of phosphoinositides in the functional activities of these receptors has been extensively investigated in several tissues (Table III). It is interesting to note that muscarinic cholinergic and α-adrenergic receptors have the following properties in common: (1) activation of both leads to an increase in phosphoinositide turnover; (2) activation of both results in an increase in cyclic GMP but not cyclic AMP synthesis; (3) calcium ion mediates the physiological functions of both types of receptors.

Other types of receptors that have been reported to be involved in the control of phosphoinositide metabolism in various tissues are: dopamine,[166,167] histamine,[204,205] serotonin,[199] pancreozymin,[206] substance P,[207] angiotensin,[208] vasopressin,[209–212] adrenocorticotropin (ACTH),[211,213,214] and GABA.[215] The association between these receptors and phosphoinositide metabolism has not yet been well documented in excitable tissues.

4.2.2. Phosphoinositide Effect in Excitable Tissues

Tissues in which the phosphoinositide effect* has been shown to be associated with activation of cholinergic and adrenergic receptors are listed in Table III. In excitable tissues, such as brain, nerve, and smooth muscle, *in vivo* or *in vitro* activation of muscarinic cholinergic and α-adrenergic receptors results in an increase in the turnover of the polar head groups of PA and PI of 70–300% of that of the control, and in that of PC by up to 30% of the control. Under the same experimental conditions, there are significant changes (up to 30% of that of the control) in the turnover of the polar head groups of the polyphosphoinositides. In addition, studies on the ^{32}P incorporation into phospholipids of rat pineal glands in organ culture revealed that both norepinephrine and α-adrenergic blocking agents markedly stimulated the labeling of PI and phosphatidylglycerol.[189]

---

* Includes both the phosphatidylinositol and polyphosphoinositide effects. The simplest method to assay for the phosphoinositide effect is to determine changes that occur in ^{32}P labeling of phosphatidic acid in response to addition of agonists and/or antagonists.

There is some confusion in the literature as to whether the observed effects on other phospholipids in some tissues (Table III) are in fact enhanced turnover or are increased net synthesis caused by the increased diacylglycerol levels that are formed as a result of the increase in phosphoinositide breakdown in response to receptor activation. The fact that in animal tissues only phosphoinositides are hydrolyzed by a specific phospholipase C-like enzyme favors the latter possibility. Both of these phosphatides have a common precursor, CDP-diacylglycerol. In early studies, most workers measured the phosphoinositide effect by incubating the tissue in an isosmotic medium containing ^{32}Pi or [^{3}H]inositol in the presence and absence of the neurotransmitter (usually about 10^{-4} M). The individual phosphatides were then separated, usually by TLC, and the radioactivity determined. These studies are of limited value in deciding which of the reactions involved in PI metabolism is under the control of the stimulus. Some investigators have chemically measured changes in PI levels in response to the neurotransmitter, and a few have done chase experiments to follow loss of radioactivity from prelabeled tissue.

Early in 1974, it became clear to a number of investigators that neurotransmitters act on membrane phospholipids by stimulating the breakdown, not synthesis, of PA or PI in the tissue stimulated. Thus, Schact and Agranoff[216] labeled synaptosomes with ^{32}Pi, then blocked further [γ-^{32}P]ATP formation by adding 2,4-dinitrophenol, and demonstrated a 60–70% increase in the hydrolysis of [^{32}P]PA on the addition of ACh. These authors suggested that PA hydrolysis is the initial muscarinic response in synaptosomes. In contrast, Durell and Garland,[217] on the basis of preliminary data, suggested that stimulation by ACh occurs by activation of PI breakdown. Acetylcholine does not affect partially purified diglyceride kinase, and there are conflicting reports about its action on PI phosphodiesterase.[92,176] The concept that PI breakdown could be the mechanism underlying the enhanced PI turnover that occurs in stimulated tissues gained support from the studies of Hokin-Neaverson and her collaborators[219,220] working with pancreas and Jones and Michell[193] working with rat parotid fragments who demonstrated a decrease in the level of PI in response to ACh. Furthermore, they demonstrated an increase in the concentrations of 1,2-diacylglycerol and PA that corresponded to the decrease in PI in response to ACh. Hokin-Neaverson[206] has reviewed the work from her laboratory on the PI response to ACh in mouse pancreas. A net loss of 40% of the tissue PI was measured chemically. There was no production of inositol phosphates but only of free inositol. In more recent studies on the effects of cholinergic stimulation on levels and fatty acid composition of diacylglycerol in mouse pancreas, these authors[222] concluded that in both the *in vivo* and *in vitro* experiments, the changes in levels of stearic and arachidonic acids in diacylglycerol indicated that only a small proportion of the total increase in diacylglycerol level could have been derived from the PI breakdown that occurs in response to cholinergic stimulation. The major effect of ACh in diacylglycerol metabolism in mouse pancreas is separate, therefore, from the effect on PI metabolism.

There is accumulating evidence from some but not all (e.g., the synaptosome) tissues that suggests that the reaction controlled by receptor activation is PI breakdown and that the diacylglycerol released is recycled back to PI

Table III

Tissues in Which the Phosphoinositide Effect Has Been Shown to Be Associated with Activation of Cholinergic and Adrenergic
Receptors

Tissue	Stimulus	Phospholipids affected	References
A. Excitable			
1. Nervous			
Whole brain[a]	Cholinergic	PI	161
Whole brain[a]	Muscarinic cholinergic	PI, PA, PC	162
Whole brain[a]	Cholinergic	Polyphosphoinositides	163
Whole brain[a]	α-Adrenergic	PI, PA	164
Cerebral cortex[a]	Cholinergic	PI	165
Cerebral cortex slices	Muscarinic cholinergic	PI, PA	157, 158
Cerebral cortex slices	Adrenergic	PI, PA, PC	166, 167
Rat superior cervical ganglion	Nicotinic cholinergic (electrical stimulation)	PI	168, 169
Rat superior cervical ganglion	Muscarinic cholinergic	PI	170, 171
Eel electroplax	Cholinergic	PI, PC	172
Electric organ or Torpedo[a]	Cholinergic (electrical stimulation)	PI, PA, PC	173
Crab nerve fibers	Cholinergic	Polyphosphoinositides	174
Cerebral cortex synaptosomes	Muscarinic cholinergic	PI, PA, polyphosphoinositides	175–179 (also see 2, 6)

2. Smooth Muscle			
Iris	Muscarinic cholinergic and α-adrenergic	PI, PA, polyphosphoinositides	155, 180 (see also 2)
Iris[a]	Sympathetic electrical stimulation	PI, PA, polyphosphoinositides	181
Vas deferens	α-Adrenergic and muscarinic cholinergic	PI	182, 183
Ileum	Muscarinic cholinergic	PI	184
B. Endocrine			
Adrenal medulla	Cholinergic	PI, PA, PC	185–187
Pineal	α-Adrenergic	PI and phosphatidylglycerol	188–191 (see also 2)
C. Exocrine			
Pigeon pancreas	Muscarinic cholinergic	PI, PA	159 (see also 2)
Parotids	Muscarinic cholinergic and α-adrenergic	PI	192–197
Avian salt gland	Muscarinic cholinergic	PI, PA	198
Blowfly salivary gland	Serotonin	PI and polyphosphoinositides	199
D. Others			
Liver	α-Adrenergic	PI	200–202
Adipose	α-Adrenergic	PI	203

[a] Carried out *in vivo*.

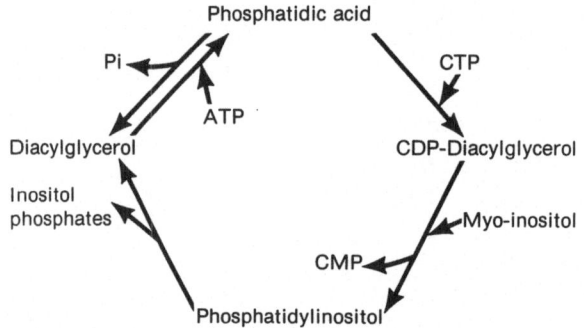

Fig. 4. The cycle of PI hydrolysis and resynthesis that appears to occur in response to receptor activation.

(Fig. 4). Although this concept has been accepted by many workers in the field, there is still a need to measure PI breakdown in response to receptor activation in other tissues in which the PI effect has been reported.

An effect of ACh on polyphosphoinositide metabolism has also been suggested as part of a model on the mechanism of action of ACh.[223] However, the experimental evidence reported by several investigators working on the effects of external stimuli on polyphosphoinositides in various tissues has been confusing and in many instances contradictory (for review see ref. 6). For instance, Yagihara and Hawthorne[176] found no effect of ACh on PI-bis P labeling in synaptosomes, and although Schacht and Agranoff[175] observed some reduction in labeling, it was not atropine sensitive. Strong evidence for a role of polyphosphoinositides in synaptic transmission came from work on the iris smooth muscle.[155] Thus, in this tissue, ACh and norepinephrine were found to stimulate the breakdown of PI-bis P, and these effects were blocked by atropine and phentolamine, respectively.[155,181] These observations suggest that the polyphosphoinositide effect is controlled through muscarinic-cholinergic and α-adrenergic receptors. More recently, Fisher and Agranoff[178] reported a polyphosphoinositide effect in synaptosomes that was also blocked by atropine. Gispen and his colleagues,[179] working with rat brain crude mitochondrial fractions, demonstrated a relationship between polyphosphoinositide metabolism and protein phosphorylation. Thus, the phosphorylation of proteins from [^{32}P]ATP was dose-dependently inhibited by corticotropin, whereas that of PI-bis P increased with the peptide concentration. Under the same experimental conditions, the ^{32}P labeling of PA was inhibited, and that of PI-P was unaffected by the peptide.

It can be concluded from the above that the initial response to receptor activation appears to be the breakdown of phosphoinositides.

4.2.3. Localization

Studies on the regional distribution of the phosphoinositide effect in brain revealed it to be associated with those regions that are enriched in muscarinic-cholinergic and adrenergic receptors. The muscarinic-cholinergic and α-adrenergic receptors responsible for the phosphoinositide effect are either presynaptic, as evidenced from the studies with synaptosomes, or postsynaptic, as has been concluded from denervation studies on sympathetic ganglia,[224] iris smooth muscle,[225] pineal gland,[226] and hippocampal synaptosomes.[227] Although activation of these receptors leads to phosphoinositide breakdown, there is no convincing evidence at this time to show that this phenomenon does indeed occur at the plasma membrane of the cell. Autoradiographic studies of pancreas,[228] sympathetic ganglia,[229] and subcellular fractions of cerebral cortex[167,230] and sympathetic ganglia[231] revealed that the stimulated labeling of phosphoinositides is distributed throughout the cell.

Using cerebral cortex slices, Abdel-Latif *et al.*[167] found increased ^{32}P labeling of PA, PI, and PC by ACh, norepinephrine, and other neurotransmitters to be widely distributed among subcellular fractions. Synaptosomes showed a marked effect, but stimulation was also seen in microsomal, mitochondrial, and nuclear fractions. Glial cells, as well as neuronal cells, exhibited increases. In synaptosomes, there is evidence that the ACh-stimulated PA labeling is localized in the synaptic vesicles[232,233] and synaptosomal plasma membranes.[234] Bleasdale *et al.*[173] showed that stimulation of the electric lobe of the brain of *Torpedo marmorata* increased PA labeling in the electric organ in fractions rich in synaptic vesicles. Stimulation *in vivo* of rat brain cortex with carbamylcholine and eserine resulted in a significant drop in the specific radioactivity of PI, and when the tissue was subjected to subcellular fractionation, the loss was chiefly from fractions rich in synaptic vesicles.[165] Preliminary studies on the distribution of the polyphosphoinositide effect in the iris muscle indicate that it is also distributed throughout the cell.[180]

It is clear from the above that the precise intracellular site of the phosphoinositide effect is far from clear. The fact that both hydrolysis and resynthesis of phosphoinositides proceed simultaneously in the stimulated and nonstimulated cell presents a technical difficulty. This is especially true since PI is synthesized at the endoplasmic reticulum, and its stimulated hydrolysis is presumably occurring at the plasma membrane. To understand the molecular mechanism(s) and physiological significance of this phenomenon, there is a need to localize it in the cell. More imaginative approaches armed with better methodology and more suitable tissues would be a step in the right direction.

4.2.4. Effects of Ca^{2+} and Other Cations and Possible Mechanisms and Physiological Significance

Until recently, most of the reports on the phosphoinositide effect have dealt with studies on its characteristics in a variety of tissues and some speculations on its molecular mechanism(s) and physiological significance. The latter studies were hampered by two main problems: (1) lack of experimental evidence showing the biochemical site of this phenomenon in the stimulated cell and (2) little information on how phosphoinositide turnover is actually coupled to receptor activation. The fact that this phenomenon is mediated through muscarinic-cholinergic and α-adrenergic receptors raises the possibility of an interrelationship between the enhanced phosphoinositide turnover and the ionic consequences of neurotransmitter–receptor interaction. Although activation of these receptors is known to lead to changes in the intracellular concentrations of a second messenger such as cyclic AMP or cyclic GMP or changes in relevant enzymes, only the possible involvement of cations has been investigated in detail. An effect brought about by activation of various receptors, including muscarinic-cholinergic and α-adrenergic receptors, is an increase in cell-surface Ca^{2+} permeability.[235–239] Calcium ion is known to play a key role as a regulatory ion controlling a wide variety of membrane and intracellular functions, and thus it has been of particular interest to many investigators to show if this cation has a role in this phenomenon. Thus, in the early 1970s, it was shown that in most systems that have been tested, enhanced PI turnover was found to be somewhat insensitive to omission of Ca^{2+} from the incubation medium (for review see ref. 5). Thus, it was reported that Ca^{2+} is not required for PI and/or PA turnover in response to various stimuli in adrenal medulla,[240] parotid,[241–243] guinea pig ileum,[244] and synaptosomes.[175,232] In contrast, Ca^{2+} was reported to be required for maximal PI and/or PA turnover in response to ACh in the pancreas,[245] thyroid-stimulating hormone in the thyroid,[246] ACh and norepinephrine in the iris muscle,[247] and electrical stimulation of the synaptosomal fraction.[248] There is also evidence that has been interpreted as indicating that in muscle, severe and prolonged depletion of Ca^{2+} can depress the "resting" rate of PI turnover.[249]

In 1978, Akhtar and Abdel-Latif[250] reported that in iris smooth muscle that was prelabeled with ^{32}Pi and had its Ca^{2+} content depleted with EGTA, the ACh-stimulated breakdown of polyphosphoinositides and ^{32}P labeling of PA and to a much lesser extent PI are dependent on the presence of extracellular Ca^{2+}. Later, it was found that in iris muscle that was prelabeled with [^{3}H]*myo*-inositol, ACh stimulated the release from tissue phosphoinositides of the water-soluble inositol monophosphate and inositol triphosphate.[251] In light of these findings, it was suggested that the interaction of the neurotranmitter with its receptor results in an activated complex that, in turn, could lead to the enhanced phosphoinositide breakdown via a Ca^{2+}-mediated step (Fig. 5) and consequently to muscle response.[250–252] In this tissue, ACh and norepinephrine act to increase the intracellular concentration of calcium ion,[253] and both PI[102] and polyphosphoinositide[138] phosphodiesterases are stimulated by this cation. Furthermore, kinetic studies on dose–phosphoinositide and dose–contraction responses revealed a close relationship between the biochemical and pharmacological responses.[252] The dependence of the phosphoinositide effect on the presence of extracellular Ca^{2+} has also been demonstrated in synaptosomes.[254–256] In both systems, the increase in labeling is blocked by either atropine or EGTA, and addition of the Ca^{2+} ionophone A23187 alone mimicked

Neurotransmitter + Receptor → Neurotransmitter-Receptor → Ca^{2+}_{int} ↑ →

Phosphoinositide breakdown → Diacylglycerol + Inositol phosphates → → Response

Fig. 5. A possible role for Ca^{2+} in the neurotransmitter-stimulated breakdown of phosphoinositides in the iris muscle (modified from ref. 251).

the action of the neurotransmitter; however, only the neurotransmitter effect is blocked by atropine.

In prelabeling experiments, addition of the Ca^{2+} ionophore to synaptosomes caused a marked loss of ^{32}P radioactivity from polyphosphoinositides and stimulated labeling of PA,[178,257] and the rate of loss was further augmented by addition of ACh.[178] Neither agent produced comparable effects on the breakdown of prelabeled PA and PI.

The effects of Ca^{2+} and ionophore A23187 on the release of water-soluble inositol phosphates from polyphosphoinositides have been reported in synaptosomes[257] and erythrocytes.[258] In erythrocytes polyphosphoinositides prelabeled with ^{32}Pi, decreases of 40% and 70% in the radioactivity associated with PI-bisP were observed in media containing 10 and 100 μM $CaCl_2$, respectively, in the presence of the ionophore, as compared to controls.[259] Ionophore A23187 increased ^{32}Pi incorporation into PA and phosphoinositides of polymorphonuclear leukocytes.[260]

The above studies suggest that stimulated polyphosphoinositide breakdown is dependent on the presence of extracellular Ca^{2+}. The increase in PA, and subsequently PI, labeling that results from receptor activation may result from an increased diacylglycerol availability as a result of breakdown of these phosphoinositides.[155,178,250]

Although there is agreement that Ca^{2+} may play an important role in the polyphosphoinositide effect, its role in the PI effect is unclear and even at times controversial. Thus, in the parotid,[241–243,261,262] the PI effect is neither dependent on influx of Ca^{2+} nor mimicked by the introduction of Ca^{2+} intracellularly with the cationophore A23187. In blowfly salivary glands, the breakdown of [^{32}P]-labeled PI was not increased by cyclic AMP or ionophore A23187, and the increase in response to 5-hydroxytryptamine was unaffected by the absence of Ca^{2+} from the bathing medium.[199]

α-Adrenergic activation of rat fat cells produces a marked increase in the incorporation of ^{32}Pi into PI which does not require the presence of extracellular calcium ions.[263] In contrast, situations in which the PI effect has been reported to be Ca^{2+} dependent are stimulation of rabbit neutrophils by the synthetic peptide f-Met-Leu-Phe,[264] stimulation of dispersed pancreatic fragments by ionophore A23187 and carbachol,[265] stimulation of rat pancreatic islets by glucose[266], and stimulation of rat hepatocytes by vasopressin and epinephrine.[212] In iris muscle labeled with [^{3}H]*myo*-inositol, ACh increased the production of inositol monophosphate and inositol triphosphate by 48% and 33%, respectively, and this was Ca^{2+} dependent. The marked increase observed in the production of inositol monophosphate could also result from Ca^{2+} activation of PI phosphodiesterase. Fisher *et al.*,[267] working with cultured bovine adrenal chromaffin cells, reported that although Ca^{2+} is essential for catecholamine secretion but not for stimulated phospholipid labeling, the latter is inhibited when endogenous Ca^{2+} is chelated with EGTA.

From the above studies, it can be concluded that although in some tissues the PI effect appears to be Ca^{2+} independent, there is accumulating evidence in more recent years to suggest that in certain tissues the cation could play a key role in this phenomenon.

Activation of muscarinic cholinergic and α-adrenergic receptors also leads to an increase in cell-surface permeability to Na$^+$ and K$^+$ in addition to Ca^{2+}. Brossard and Quastel,[268] working with rat brain slices, reported that ACh stimulation of ^{32}Pi incorporation into total phospholipids is dependent on the presence of Na$^+$. More recently, Keryer *et al.*[269] reported that in rat parotid glands cholinergic stimulation of [^{3}H]*myo*-inositol into PI is dependent on extracellular Na$^+$. In the iris muscle,[180,270,271] studies on the effects of monovalent cations on the phosphoinositide effect revealed the following. (1) The neurotransmitter-stimulated ^{32}P labeling of PA, PI, and PC is dependent on the presence of extracellular Na$^+$. (2) The monovalent cation requirement for Na$^+$ is specific. Of the monovalent cations Li$^+$, NH$_4^+$, K$^+$, choline$^+$, and Tris, only Li$^+$ partially substituted for Na$^+$. (3) A significant decrease in ^{32}P labeling of phospholipids in response to ACh was observed when Ca^{2+} and/or K$^+$ were added to an isomotic medium deficient in Na$^+$. (4) Ouabain, which blocks the Na$^+$ pump, inhibited the basal ^{32}Pi incorporation into PC and the ACh-stimulated ^{32}P labeling of PA, PI, and PC. It was suggested that phosphoinositide breakdown is associated with Ca^{2+} influx (Fig. 5) and that the enhanced ^{32}P-labeling of phosphoinositides could be associated with Na$^+$ efflux via the Na$^+$ pump mechanism (Fig. 6). Incubation of synaptosomes[248] or fragments of guinea pig ileum smooth muscle[272] in the presence of an elevated extracellular K$^+$ concentration, which causes an increase in cell-surface permeability, caused a marked increase in ^{32}Pi incorporation into Pi. The administration of LiCl (3.6 mg/kg per day) to adult male rats for 9 days resulted in an increase in the cerebral cortex level of *myo*-inositol-1-P to 4.43 ± 0.52 μmol/kg (dry wt.) compared to a control level of 0.24 ± 0.02 μmol/kg.[273] About 90% of the increase corresponds to the D-enantiomer, evidence that Li$^+$ largely produces this effect via phospholipase C-mediated phosphoinositide metabolism.

To summarize, on the basis of the requirement for Ca^{2+}, one can discern two categories of tissues, those that require Ca^{2+} for the phosphoinositide effect, such as the iris muscle,[250,251] synaptosomes,[178,254–257] bovine chromaffin cells,[267] neutrophils,[264] hepatocytes,[212] and pancreas,[265,266] and those in which these effects are Ca^{2+} independent, such as parotid gland,[241–243] ileum smooth muscle,[244] and insect salivary gland.[199] It remains to be established whether the molecular mechanisms underlying these effects in both types of tissues are the same or different. Conclusions in regard to the findings on a possible link between monovalent cations[268–271] and the enhanced phosphoinositide turnover should await further studies in other tissues.

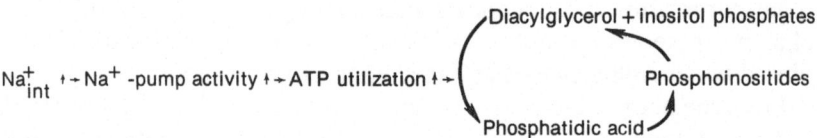

Fig. 6.   A possible role for Na$^+$ in the neurotransmitter-stimulated ^{32}P labeling of phosphoinositides in the iris muscle (taken from reference 270, with permission).

The proposed functions of the phosphoinositide effect are almost as numerous as the tissues in which this phenomenon has been investigated. Thus, because of the striking phospholipid effects in glands elicited by ACh and related secretagogues, Hokin[154] suggested that it may be involved in packaging of proteins or catecholamines into vesicles. Hokin-Neaverson[206] proposed that the phosphoinositide effect in the salt gland is associated with the Na^+ pump enzyme. For neural tissues, she speculates that phosphoinositides may control both protein synthesis in the perikaryon and plasma membrane transport processes. Hawthorne and Pickard[6] have suggested that the diacylglycerol produced from PI at the presynapse may have a role in exocytosis in nervous tissue. The rapid turnover of PI in frog rod outer segments, compared to other phospholipids, may be related to membrane fusion events associated with the assembly and/or turnover of outer segment membranes.[274]

Michell and his collaborators[5,15,272,275] proposed a hypothesis that envisages that at the postsynapse PI breakdown is a coupling reaction essential to the mechanism by which receptor activation leads to calcium mobilization. According to this hypothesis, the neurotransmitter-stimulated PI breakdown precedes Ca^{2+} entry (Fig. 7). More recently, Michell[277] has suggested that the polyphosphoinositides and not PI, as he had previously suggested, are involved in mobilization of Ca^{2+} in the cytosol. According to these authors, the disappearance of PI may be secondary to the breakdown of PI-bisP. Weiss and Putney[196] suggested that PI turnover may be a reaction more closely associated with Ca^{2+}-gating rather than a reaction directly involved in receptor activation or occupation.

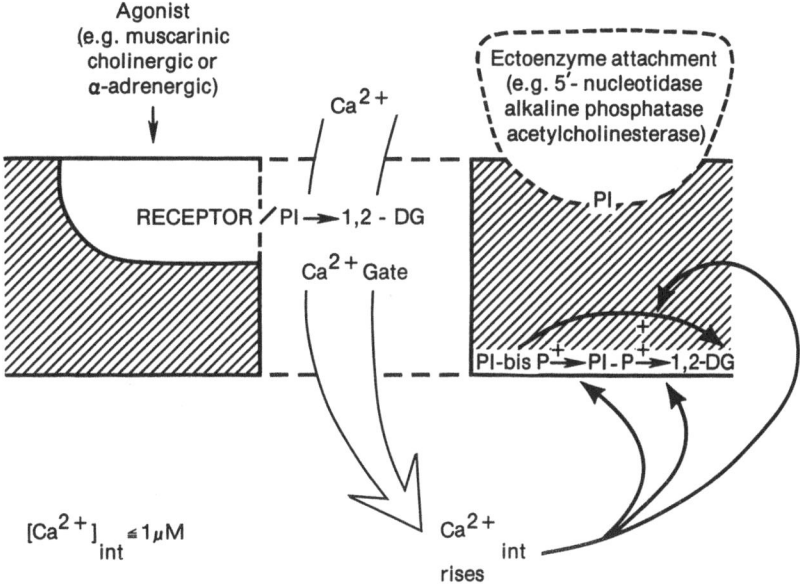

Fig. 7. A hypothetical plasma membrane at which PI breakdown is controlled by receptor activation: Ca^{2+} stimulates breakdown of PI-P and PI-bis P, and PI anchors ectoenzymes to the cell surface (reproduced from reference 15 with permission).

It has recently been suggested that it may be the net formation of PA that is most important in activating Ca^{2+} transport.[278,279] According to this concept, PA, which is synthesized from the diacylglycerol released during PI breakdown, might act as a natural Ca^{2+} ionophore in the stimulated tissue. Harris, et al.,[280] working with rat brain synaptosomes, showed that PA could function as a link between depolarization and neurotransmitter release. Thus, PA stimulated uptake of ^{45}Ca by synaptosomes and evoked the release of [^{3}H]dopamine from these particles.

The Ca^{2+}-gating hypothesis (Fig. 7) is based in part on the finding that in some tissues the phosphoinositide effect is Ca^{2+} independent. However, and in contrast, this phenomenon has been shown in a variety of tissues to be Ca^{2+} dependent (see above). It appears that in the latter tissues, phosphoinositide hydrolysis is a consequence and not a cause of Ca^{2+} entry. A plausible model of how Ca^{2+} and Na^+ could be involved in the phosphoinositide effect in the iris smooth muscle is given in Fig. 8. According to this hypothesis, the Ca^{2+} entering the cell in response to receptor activation could activate the enzymes, which are stimulated by Ca^{2+}, involved in phosphoinositide breakdown to diacylglycerol and inositol phosphates. Removal of the polar head groups from these phospholipids could facilitate the cationic fluxes through the Na^+-Ca^{2+} channels. The Na^+ entering the cell through passive flux and in response to ACh leads to an increase in intracellular Na^+ concentration which, in turn, stimulates Na^+ pump activity and ATP turnover. Restoration of the polar head groups to the diacylglycerol backbone at the Na^+-Ca^{2+} channels, which is reflected in the enhanced ^{32}P labeling of phosphoinositides from [^{32}P]ATP, could be associated with the extrusion of Na^+ and Ca^{2+}, presumably via the Na^+ and Ca^{2+} pumps.

It can be concluded from the above that in excitable tissues the turnover of the polar head groups of PA, PI, and the polyphosphoinositides is linked to

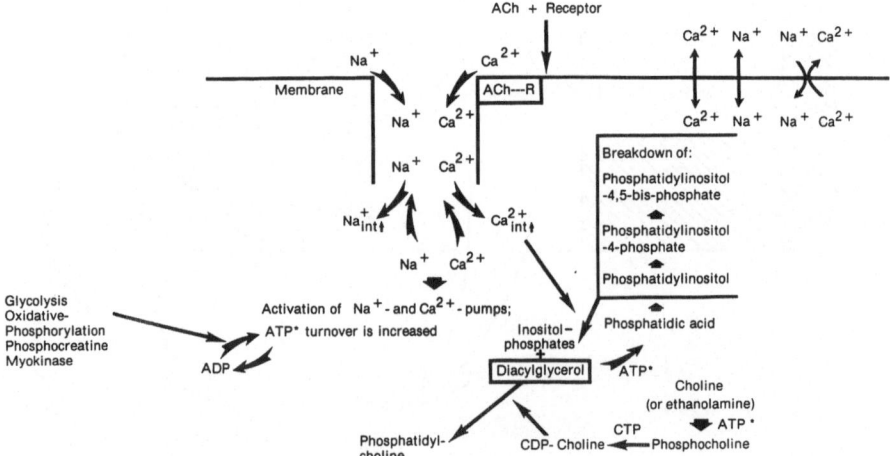

Fig. 8. Probable roles of Ca^{2+} and Na^+ in the neurotransmitter-stimulated phosphoinositide breakdown at muscarinic-cholinergic and α-adrenergic receptors of tissues in which a requirement for Ca^{2+} has been demonstrated.

the activation of muscarinic-cholinergic, α-adrenergic, and possibly other types of receptors and that the phosphoinositide effect takes place at presynaptic and postsynaptic sites. The initial reaction is probably a phospholipase C-catalyzed hydrolysis of minor pools of PI and polyphosphoinositides. The diacylglycerol formed is then utilized for resynthesis of phosphoinositides via PA and CDP-diacylglycerol. Although Ca^{2+} influx appears to precede and is required for the polyphosphoinositide effect, its requirement for the PI effect has been demonstrated in some but not all tissues. Although the restoration of the polar head group to diacylglycerol could be associated with Na^+ and Ca^{2+} effluxes via the pump mechanisms, the interrelationship between them remains to be established. The physiological significance of this phenomenon is still unclear. Among the possibilities mentioned are a mechanism for perturbation of cellular membranes and regulation of membrane-related metabolic and ionic changes, Ca^{2+}-gating, Ca^{2+} ionophore, exocytosis, neurotransmitter release, and production of arachidonic acid for prostaglandin synthesis. However, there is no conclusive experimental evidence at the present time to show that the phosphoinositide effect is an essential step in a physiological process.

4.3. Effects of Amphiphilic Drugs

A number of reports have recently appeared showing that a large variety of amphiphilic cationic drugs,[281–286] including chlorpromazine and other phenothiazine tranquilizers,[281] morphine,[287–289] nicotine,[290] fenfluramine and its derivatives,[282,285] and propranolol,[189,286,291–295] markedly alter phospholipid metabolism in a wide variety of tissues.* In general, these drugs interact with lipids, especially the acidic phospholipids. This association changes the properties of the membrane and displaces divalent cations such as Ca^{2+} and Mg^{2+}. This alters the physical properties of membrane lipids and their metabolic turnover.[296] The rate of conversion to diacylglycerol by PA phosphohydrolase is inhibited, whereas that to CDP-diacylglycerol by PA cytidyltransferase, is stimulated (Fig. 9). This redirects glycerolipid synthesis away from the production of PC, PE, and triacyglycerol and toward the production of phosphoinositides and cardiolipin. In addition, propranolol[293] and morphine[297] have been shown to increase the biosynthesis of phosphatidylserine by stimulating the Ca^{2+}-dependent base-exchange reaction. The effects that have been described are largely nonspecific effects of amphiphilic amines. Thus, they increase the incorporation of both ^{32}Pi and [^{14}C]glycerol into phosphoinositides. These effects can be distinguished from the receptor-mediated phosphoinositide effect described above in which there is an increased incorporation of ^{32}Pi and [^{3}H]inositol but not of [^{14}C]glycerol. More recently we observed that propranolol and mepacrine exert multiple effects on the enzymes involved in phospholipid synthesis in the iris muscle (A. A. Abdel-Latif, J. P. Smith and R. A. Akhtar, unpublished work). Thus in iris microsomes: (a) They stimulated the activities of PA phosphohydrolase, diacylgylcerolkinase, PI kinase and PI-P kinase; and (b) they inhibited the activity of diacylglycerocholine phospho-

* For more detail, see Chapter 2 by N. G. Bazan (this volume).

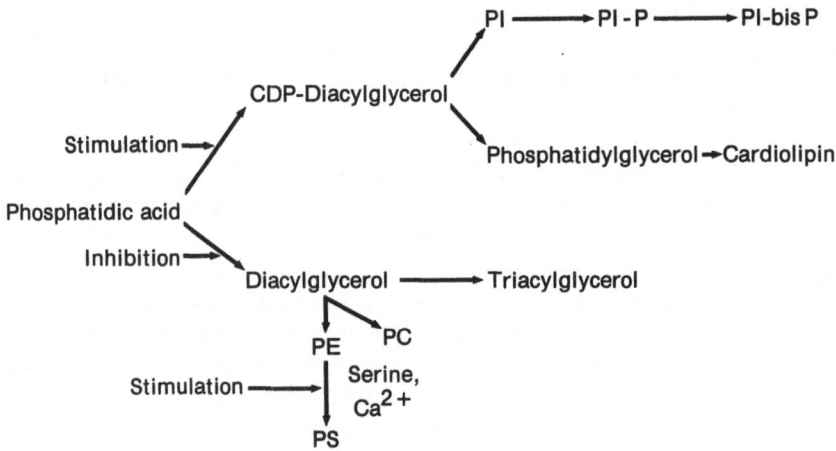

Fig. 9. Pathways of biosynthesis of phosphoinositides and other glycerolipids affected by propranolol and other amphiphilic drugs.

transferase. Their multiple effects on the enzymes of phospholipid metabolism could also underlie the drug-induced redirection of phospholipid synthesis.

5. INVOLVEMENT OF PHOSPHOINOSITIDES IN OTHER FUNCTIONAL ACTIVITIES

5.1. Precursor for Arachidonic Acid in Prostaglandin Biosynthesis

In animal tissues, there is no free arachidonic acid, and, furthermore, there are no free prostaglandins. Prostaglandins are synthesized as a result of membrane perturbations that cause the release of arachidonic acid from membrane phospholipids. This release can be brought about by a wide variety of stimuli, either directly or indirectly, which include neurotransmitters and hormones, inflammatory or immunologic stimuli, Ca^{2+} ionophores, UV, mechanical agitation, and electroconvulsive shock. A controlled release from phospholipids of the cell membranes by acylhydrolases to the prostaglandin synthetase is thought to be the rate-limiting step for prostaglandin biosynthesis. Stimulation of phospholipase A_2 within the cell membrane, e.g., by bradykinin, vaspressin, ACTH, Con A, chemotactic peptides, β-adrenergic agonists, and Ca^{2+} ionophore A23187, is the most likely reaction, but this has not yet been elucidated.

The release of arachidonic acid from phospholipids, followed by conversion to prostaglandins, is an important step in regulating a variety of physiological functions. Bazan,[298] working with rat brain, reported an increase in arachidonic acid release after electroconvulsive shock and during ischemia. In brain, PI is enriched with arachidonic acid (see Section 2.1). There is now accumulating evidence that in a variety of tissues, including platelets,[10,12] rabbit neutrophils,[299] cat adrenocortical cells,[300] mouse fibrosarcoma,[301,302] mouse pancreas,[303–305] platelets,[306,307] thyroid,[308] and smooth muscle,[309] the source of arachidonic acid for prostaglandin synthesis is PI.

Rubin and his colleagues,[300] working with cat adrenocortical cells, have demonstrated that ACTH induces a Ca^{2+}-dependent increase in phospholipase A_2 activity and prostaglandin synthesis in these cells. More recently, these authors[299] concluded that an early action of ACTH is a Ca^{2+}-dependent turnover of arachidonyl PI that is initiated by phospholipase A_2 and appears to be followed by a rapid, selective reacylation of lyso-PI. Hokin and his collaborators,[303] working with mouse pancreas, reported that the breakdown of PI provides arachidonic acid for prostaglandin synthesis and, furthermore, that prostaglandins are involved in stimulus–secretion coupling. More recently, they reported that the formation of prostaglandin E_2 due to PI breakdown is associated with the stimulation of enzyme secretion in this tissue.[305]

There are two possible pathways for arachidonate release from PI (Fig. 10). The first involves two enzymes: a PI phosphodiesterase (phospholipase C type) that is specific for PI (see Section 3.1.2b) and a diacylglycerol lipase that acts on the diglyceride liberated by the phospholipase. The other pathway is the deacylation at the 2-position, via phospholipase A_2, to give lyso-PI and free fatty acids[116] (see Section 3.1.2b). The activation of PI-hydrolyzing phospholipase A_2 activity by bradykinin, thrombin, and ionophore A23187 during prostaglandin biosynthesis was demonstrated in transformed mouse cells grown in tissue culture.[304] In thrombin-stimulated platelets, conversion of PI to PA precedes the production of arachidonic acid.[306,307] The PA produced is degraded by a specific phospholipase A_2, resulting in the production of arachidonic acid and lyso-PA.

There has been very little work done on the diacylglycerol lipase and phospholipase A_2, which is specific for PI, in brain and other tissues. The experimental evidence showing that PI is a source for arachidonic acid in prostaglandin biosynthesis is interesting and needs to be confirmed in other tissues.

5.2. Activation and Interaction with Enzymes

When phospholipids are removed from certain enzyme preparations, some of the enzymes are inactivated; however, under appropriate conditions, it is possible to reactivate these enzymes simply by mixing them with dispersions of specific phospholipids. Results of such studies have shown that PI is capable of significant reactivation. Thus, Mandersloot *et al.*[7] studied the phospholipid requirement of rabbit kidney microsomal Na^+–K^+ ATPase by selective hydrolysis of endogenous phospholipids with the aid of various phospholipases. They concluded that the enzyme has a specific requirement for PI as the essential endogenous phospholipid. Lloyd[310] investigated the effects of PI on tyrosine hydroxylase from rabbit adrenal glands and reported two potent effects on the enzyme *in vitro*: (1) a rapid reversible activation of the enzyme and (2) a slower, preincubation-dependent, irreversible inactivation of the enzyme. Heger and Peter[311] reported that PI is an essential constituent of the acetyl-CoA carboxylase from rat liver. The enzyme can be inactivated by removal of the phospholipids with phospholipases, and the lipid-depleted enzyme can be

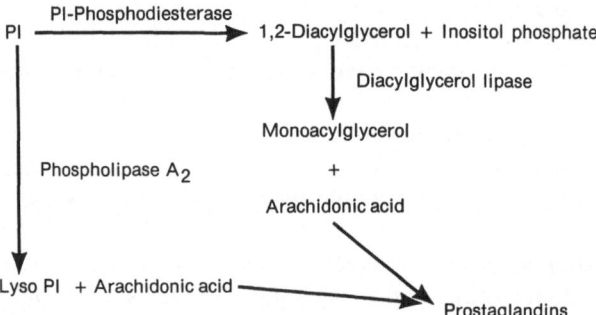

Fig. 10.   Proposed pathways for arachidonate release from PI in stimulated tissues.

reactivated to about 70% by incubation with PI. Levy[312] has shown that heart adenylate cyclase sensitivity to the β-adrenergic stimulus required the presence of PI.

Alkaline phosphatase, acetylcholinesterase, and 5'-nucleotidase are released from a wide variety of tissues by bacterial PI phosphodiesterase.[107,109,313–318] These observations suggest that PI may be responsible for the attachment of these enzymes to cellular membranes. Chymotrypsinogen A was almost quantitatively extracted from aqueous solution in the presence of inositol phosphatides at relatively low concentrations of both ligands, and this interaction was facilitated by Ca^{2+} (10^{-4}–10^{-5} M).[319]

The molecular mechanism(s) underlying the activation and interaction of PI with enzymes is unclear and at times is controversial. Thus, it has been reported that when phospholipids are totally depleted, reactivation of Na^+–K^+ ATPase is not possible.[320] DePont et al.,[321] working with highly purified Na^+–K^+ ATPase, concluded that there is no absolute requirement for either PI or phosphatidylserine. However, these authors also showed that the ATPase activity was markedly reduced when both PI and phosphatidylserine were removed, so the possibility remains that these negatively charged phospholipids are normally intimately associated with the ATPase in the cell membrane. The role of phospholipids in the functioning of the Na^+–K^+ ATPase has also been a subject of speculation. Among the suggestions as to the role of PI and other phospholipids is that they are involved in the interaction between the ATPase protein and the lipid bilayer in vivo.[322] They could create a negative charge near the active site, be components of the active site itself, provide ion-specific sites, or serve in some kind of recognition role to insure insertion of the protein both in the required part of the membrane and in the necessary alignment (for summary, see ref. 323).

In conclusion, phosphoinositides appear to play an important role in modulating enzyme activities in cellular membranes. Additional studies with purified enzymes and intact cell preparations, such as synaptosomes and cell culture, are required in order to evaluate their precise role and physiological significance.

6. ALTERATIONS IN PHOSPHOINOSITIDE METABOLISM AFTER DENERVATION AND IN CERTAIN TYPES OF NEUROLOGICAL DISORDERS

6.1. Effects of Denervation

In normal vertebrate skeletal muscle, application of ACh to any part of the muscle surface other than the synapse elicits no response. If the nerve to that muscle is cut, a phenomenon known as "denervation supersensitivity" may be observed. The denervated muscle now responds to the application of the neurotransmitter at any point of the plasma membrane surface. Several studies have revealed that a new population of extrajunctional receptors (10 to 40-fold increase in the total number of measurable cholinergic sites) is formed whose properties are in many cases identical with those found at the synapse. It is believed that a trophic substance that acts as a regulator of metabolic activity is continually being released from the nerve into the muscle cell.

The effect of denervation on PI turnover has been investigated in rat diaphragm,[324] rat skeletal muscle membranes,[325] and iris smooth muscle.[181,225,326] In the rat hemidiaphragm,[324] denervation depressed the incorporation of ^{32}Pi into PI. The effect of denervation on PI turnover was not entirely a direct consequence of the cessation of ACh release at the motor end plate, since the effect was seen in strips of muscle without motor end plates, but it was slightly greater in strips containing motor end plates. Appel *et al.*,[325] working on PI turnover in rat skeletal muscle membranes following denervation, reported that in all membrane subfractions the turnover of PI was enhanced in preparations derived from denervated muscles. Sympathetic denervation of the rabbit iris[326] caused small increases in the *in vitro* incorporation of ^{32}P, [^{14}C]glycerol, and [^{3}H]inositol into phosphoinositides. In these studies, it was found that the increase in PA and polyphosphoinositide turnover in response to norepinephrine is associated with denervation supersensitivity.

Studies on the stimulated turnover of PA and PI in normal and Duchenne-dystrophic human skin fibroblasts revealed that the pattern of ^{32}P incorporation into phospholipids is significantly altered in cells of the latter.[327] Concanavalin A, when added to ^{32}P-prelabeled fibroblasts, leads to increased ^{32}P incorporation into PA and a corresponding decrease into PI. The concanavalin A-induced increase in PA radioactivity and decrease in PI radioactivity was significantly greater in Duchenne-dystrophic cells than in normal ones. Fleming and his colleagues[328] investigated the relationship between the electrogenic Na^+-K^+ pump and postjunctional supersensitivity of the guinea pig vas deferens. They proposed a biochemical mechanism involving reduced Na^+-K^+ ATPase activity as the cause of the depolarization of the supersensitive vas deferens. In contrast to skeletal muscle, in smooth muscle there is no increase in extrajunctional receptors after denervation. It is interesting to note that ouabain, which blocks the Na^+ pump, also stimulates the ^{32}P labeling of PA and PI in smooth muscle.[270]

The mechanism(s) underlying the phenomenon of denervation supersensitivity in muscle and nerve is unclear at this time. This is also true of the alterations observed in phosphoinositide metabolism following denervation. Changes in availability of substrates or concentrations of the enzymes involved in phosphoinositide synthesis have been suggested.[325,327] Further studies are needed on the effects of denervation on the morphobiochemical properties, including phosphoinositide metabolism, of muscle and nerve. Such studies will undoubtedly throw more light on the role of phosphoinositides in membrane structure and receptor function.

6.2. Alterations in Hereditary Ataxia and in Animals with Streptozotocin-Induced Diabetes

Hereditary ataxia in rabbits progressively involves lateral brainstem and deep cerebellar nuclei. Glycogen deposits have been observed in the areas of the brain involved.[329] Eliasson et al.[330] found decreased levels of the phosphoinositides in the brainstem of ataxic rabbits; incorporation of inositol into these lipids was also decreased in brainstem slices in the affected animals in comparison with control slices from normal rabbits. Later, these authors[331] found that the decrease in phosphoinositide level in brainstem of ataxic rabbits results from increased breakdown of CDP-diacylglycerol, presumably by CDP-diglyceride hydrolase. The activity of the latter has recently been demonstrated in brain.[75]

Another neurological disorder in which alterations in phosphoinositide metabolism have been reported to occur is in diabetic neuropathy. Experimental diabetes is known to cause a decrease in inositol content of peripheral nerve and decreased motor nerve conduction velocity. Diabetes is associated with three types of nerve diseases[332]: (1) the autonomic neuropathy affecting peripheral cholinergic and noradrenergic nerves, (2) the peripheral or distal symmetric polyneuropathy affecting sensory and somatic nerves of the extremities, and (3) the mononeuropathy involving isolated peripheral or cranial nerves.

Changes in the metabolism of *myo*-inositol, phosphoinositides, and other phospholipids have been demonstrated in peripheral nerves in experimental diabetes.[333-339] In these studies, diabetes was induced in rats by intraperitoneal injection of streptozotocin. On a wet-weight basis, the nerves from the diabetic animals showed a 7% decrease in total phospholipid from those of controls and a relative decrease in PI.[338] Incubations of isolated sciatic nerves of diabetic rats in a medium containing ^{32}Pi gave decreased labeling of PI and PI-bis P and increased labeling of PI-P from that of controls. The alterations in ^{32}P labeling of phosphoniositides could be explained by changes in the enzymes involved in their metabolism. Thus, Whitting et al.[336] found a decrease in the specific activities of CDP-diacylglycerol–inositol phosphatidyl transferase and PI-P kinase of brain and sciatic nerve of diabetic rats. However, such a decrease cannot account for the ^{32}P-labeling pattern obtained from sciatic nerve incubated at longer time intervals.[338] Bell and Eichberg,[339] also working with ^{32}P incorporation into phospholipids of normal and diabetic sciatic nerves, reported a 31% increase and 27% decrease in the labeling of PI-bis P and PC, respectively.

The above studies indicate that changes in metabolism of phosphoinositides do occur in diabetic nerve. Such changes could contribute to human diabetic neuropathy.

7. CONCLUSIONS

That phosphoinositide metabolism plays an important role in the structure and physiological function of excitable membranes is now well established. How the metabolism of the polar head groups of these phospholipids is actually coupled to receptor activation in a variety of tissues continued to occupy the attention of many in the past decade, and undoubtedly, it will continue to do so in the future. Studies designed to determine the subcellular localization of the phosphoinositide effect, to characterize the enzymes involved in phosphoinositide metabolism, to determine the effects of the cationic environment on the metabolic reactions involved in their biosynthesis and degradation, and to demonstrate this effect in isolated membranes should contribute to our present understanding of this important phenomenon. The proposed functions of diacylglycerol, which results from the stimulated phosphatidylinositol breakdown, in endocytosis–exocytosis mechanisms and as a source for arachidonic acid in prostaglandin biosynthesis are feasible and ought to be pursued further. The findings that phosphatidylinositol is involved in the anchoring and activation of certain enzymes at the plasma membrane point to the importance of these phospholipids in membrane structure and function. The observations of a possible interrelationship between the phosphate turnover of polyphosphoinositides and phosphoproteins in neuronal membranes support early speculations that the metabolism of these phospholipids could be involved in the movement of cations across the plasma membrane.

As we learn more about these interesting phospholipids, we hope that a better understanding of their roles in nerve conduction, synaptic transmission, and, ultimately, in the molecular mechanisms underlying certain neurological disorders will be achieved.

ACKNOWLEDGMENTS. Work referred to in this chapter from the author's laboratory was supported by NIH grants EY-02918, EY-02181, and CEP-N01-NS-6-2340 from the United States Public Health Service. I am grateful to Dr. Rashid Akhtar and Mr. Jack P. Smith for suggestions and discussions and to my colleagues who kindly supplied manuscripts prior to publication. This paper is contribution No. 0671 from the Department of Cell and Molecular Biology, Medical College of Georgia.

REFERENCES

1. Hawthorne, J. N., and Kai, M., 1970, *Handbook of Neurochemistry* (A. Lajtha, ed.), Plenum Press, New York, pp. 491–508.
2. Wells, W. W., and Eisenberg, F., Jr., (eds.), 1978, *Cyclitols and Phosphoinositides*, Academic Press, New York.
3. Hokin, L. E., and Hokin, M. R., 1960, *Int. Rev. Neurobiol.* 2:99–136.
4. Larrabee, M. G., and Leicht, W. S., 1965, *J. Neurochem.* 12:1–13.

5. Michell, R. H., 1975, *Biochim. Biophys. Acta* **415**:81–147.
6. Hawthorne, J. N., and Pickard, M. R., 1979, *J. Neurochem.* **32**:5–14.
7. Mandersloot, J. G., Roelfsen, B., and DeGier, J., 1978, *Biochim. Biophys. Acta* **508**:478–85.
8. Lloyd, T., 1979, *J. Biol. Chem.* **254**:7247–7254.
9. Low, M. G., and Zilversmit, D. B., 1980, *Biochemistry* **19**:3913–3918.
10. Bell, R. L., Kennerly, D. A., Stanford, N., and Majerus, P. W., 1979, *Proc. Natl. Acad. Sci. U.S.A.* **76**:3237–3241.
11. Banschbach, M. W., and Hokin-Neaverson, M., 1980, *FEBS Lett.* **117**:131–133.
12. Billah, M. M., Lapetina, E. G., and Cuatrecasas, P., 1980, *J. Biol. Chem.* **255**:10227–10231.
13. Hawthorne, J. N., 1973, *Form and Function of Phospholipids* (G. B. Ansell, R. M. C. Dawson, and J. N. Hawthorne, eds.), Elsevier, Amsterdam, pp. 423–440.
14. Hawthorne, J. N., and White, D. A., 1975, *Vitam. Horm.* **33**:529–573.
15. Michell, R. H., 1979, *Trends Biochem. Sci.* **4**:128–131.
16. Eisenberg, F., Jr., 1967, *J. Biol. Chem.* **242**:1375–1382.
17. Maeda, T., and Eisenberg, F., Jr., 1980, *J. Biol. Chem.* **255**:8458–8464.
18. Gaitonde, M. K., and Griffiths, M., 1966, *Anal. Biochem.* **15**:532–535.
19. Allison, J. H., Boshans, R. L., Hallcher, L. M., Packman, P. M., and Sherman, W. R., 1980, *J. Neurochem.* **34**:456–458.
20. Folch, J., and Woolley, D. W., 1942, *J. Biol. Chem.* **142**:963–964.
21. Folch, J., and Sperry, W. M., 1948, *Annu. Rev. Biochem.* **17**:147–168.
22. Folch, J., 1949, *J. Biol. Chem.* **177**:497–504.
23. Folch, J., and LeBaron, 1956, *Can. J. Biochem. Physiol.* **34**:305–319.
24. Brockerhoff, H., and Ballou, C. E., 1961, *J. Biol. Chem.* **236**:1907–1911.
25. Hawthorne, J. N., and Kemp, P., 1964, *Adv. Lipid Res.* **2**:127–166.
26. Agranoff, B. W., 1978, *Trends Biochem. Sci.* **3**:N283–N285.
27. Hawthorne, J. N., 1960, *J. Lipid Res.* **1**:255–280.
28. Holub, B. J., Kuksis, A., and Thompson, W., 1970, *J. Lipid Res.* **11**:558–564.
29. Hendrickson, H. S., and Fullington, J. G., 1965, *Biochemistry* **4**:1599–1605.
30. Dawson, R. M. C., 1965, *Biochem. J.* **97**:134–138.
31. Hayashi, F., Sokabe, M., Takage, M., Hyashi, K., and Kishimoto, U., 1978, *Biochim. Biophys. Acta* **510**:305–315.
32. Hendrickson, H. S., and Reinertsen, J. L., 1969. *Biochemistry* **8**:4855–4858.
33. Hendrickson, H. S., and Reinertsen, J. L., 1971, *Biochem. Biophys. Res. Commun.* **44**:1258–1264.
34. Hendrickson, H. S., 1969, *Ann. N.Y. Acad. Sci.* **165**:668–676.
35. Armitage, I. M., Shapiro, D. L., Furthmayr, H., and Marchesi, V. R., 1977, *Biochemistry* **16**:1317–1320.
36. Buckley, J. T., 1978, *Can. J. Biochem.* **56**:349–351.
37. Shukla, S. D., Coleman, R., Finlan, J. B., and Michell, R. H., 1979, *Biochem. J.* **179**:441–444.
38. Wu, Y. C., Cho, T. M., and Loh, H. H., 1977, *J. Neurochem.* **29**:589–592.
39. Cho, T. M., Cho, J. S., and Loh, H. H., 1978, *Proc. Natl. Acad. Sci. U.S.A.* **75**:784–788.
40. Dittmer, J. C., and Douglas, M. G., 1969, *Ann. N.Y. Acad. Sci.* **165**:515–525.
41. Lodhi, S., Winer, N. D., and Schacht, J., 1979, *Biochim. Biophys. Acta* **557**:1–8.
42. Schacht, J., 1978, *J. Lipid Res.* **19**:1063–1067.
43. Kai, M., and Hawthorne, J. N., 1969, *Ann. N.Y. Acad. Sci.* **165**:761–773.
44. Lester, R. L., and Steiner, M. R., 1968, *J. Biol. Chem.* **243**:4889–4893.
45. Palmer, F. B. S. C., 1973, *Biochim. Biophys. Acta* **316**:296–304.
46. Shaikh, N. A., and Palmer, F. B. St. C., 1976, *J. Neurochem.* **26**:597–603.
47. Guarnieri, M., 1975, *Lipids* **10**:294–298.
48. Eichberg, J., and Hauser, G., 1973, *Biochim. Biophys. Acta* **326**:210–22.
49. Eichberg, J., and Dawson, R. M. C., 1965, *Biochem. J.* **96**:644–650.
50. Deshmukh, D. S., Bear, W. D., and Brockerhoff, H., 1978, *J. Neurochem.* **30**:1191–1193.
51. White, D. A., 1973, *Form and Function of Phospholipids* (G. B. Ansell, J. N. Hawthorne, and R. M. C. Dawson, eds.), Elsevier, Amsterdam, pp. 441–482.
52. OpdenKamp, J. A. F., 1979, *Annu. Rev. Biochem.* **48**:47–71.
53. Low, M. G., and Finean, J. B., 1977, *Biochem. J.* **162**:235–240.

54. Billington, D., Coleman, R., and Lusak, Y. A., 1977, *Biochim. Biophys. Acta* **466:**526–530.
55. Chap, H. J., Zwall, R. F., and Van Deenen, L. L. M., 1977, *Biochim. Biophys. Acta* **467:**146–164.
56. Garrett, R. J. B., and Redman, C. M., 1975, *Biochim. Biophys. Acta* **382:**58–64.
57. Bishop, H. H., and Strickland, K. P., 1976, *Can. J. Biochem.* **54:**249–260.
58. Belendiuk, G., Mangnall, D., Tung, B., Westley, J., and Getz, G. S., 1978, *J. Biol. Chem.* **253:**4555–4565.
59. Longmuir, K. J., and Johnson, J. M., 1980, *Biochim. Biophys. Acta* **620:**500–508.
60. Ganoug, B. R., Leonard, J. M., and Raetz, C. R. H., 1980, *J. Biol. Chem.* **255:**1623–1629.
61. Van Huesden, G. P. H., and van den Bosch, H., 1978, *Eur. J. Biochem.* **84:**405–412.
62. Liteplo, R. A., and Sribney, M., 1980, *Can. J. Biochem.* **58:**871–877.
63. Thompson, W., and MacDonald, G., 1975, *J. Biol. Chem.* **25:**6779–6785.
64. Thompson, W., and MacDonald, G., 1976, *Eur. J. Biochem.* **65:**107–111.
65. Eichberg, J., and Hauser, G., 1978, *J. Lipid Res.* **19:**778–783.
66. MacDonald, G., Baker, R. R., and Thompson, W., 1975, *J. Neurochem.* **24:**655–661.
67. Bishop, H. H., and Strickland, K. P., 1976, *Can. J. Biochem.* **54:**249–260.
68. Holub, B. J., and Piekarski, J., 1976, *Lipids* **11:**251–257.
69. Sue, K. L., and Sun, G. H., 1978, *J. Neurochem.* **31:**1043–1047.
70. Baker, R. R., and Thompson, W., 1973, *J. Biol. Chem.* **248:**7060–7065.
71. Holub, B. J., 1976, *Lipids* **11:**1–5.
72. Thompson, W., and MacDonald, G., 1978, *J. Biol. Chem.* **253:**2712–2715.
73. Thompson, W., and MacDonald, G., 1979, *J. Biol. Chem.* **254:**3311–3314.
74. Raetz, C. R. H., Hirschberg, C. B., Dohan, W., Wickner, W. T., and Kennedy, E. P., 1972, *J. Biol. Chem.* **247:**2245–2247.
75. Rittenhouse, H. G., Seguin, E. B., Fisher, S. K., and Agranoff, B. W., 1981, *J. Neurochem.* **36:**991–999.
76. Agranoff, B. W., Bradley, R. M., and Brady, R. O., 1958, *J. Biol. Chem.* **233:**1077–1083.
77. Paulus, H., and Kennedy, E. P., 1958, *J. Am. Chem. Soc.* **80:**6689–6691.
78. Paulus, H., and Kennedy, E. P., 1960, *J. Biol. Chem.* **235:**1303–1311.
79. Thompson, W., Strickland, K. P., and Rossiter, R. J., 1963, *Biochem. J.* **87:**136–142.
80. Prottey, C., and Hawthorne, J. N., 1967, *Biochem. J.* **105:**379–392.
81. Holub, B. J., 1974, *Biochim. Biophys. Acta* **369:**111–122.
82. Takenawa, T., and Egawa, K., 1980, *Arch. Biochem. Biophys.* **202:**601–607.
83. Benjamins, J. A., and Agranoff, B. W., 1969, *J. Neurochem.* **16:**513–527.
84. Rao, R. H., and Strickland, K. P., 1974, *Biochim. Biophys. Acta* **348:**306–314.
85. Bleasdale, J. E., Wallis, P., MacDonald, P. C., and Johnson, J. M., 1979, *Biochim. Biophys. Acta* **575:**135–147.
86. Daniels, C. J., and Palmer, F. B. St. C., 1980, *Biochim. Biophys. Acta* **618:**263–272.
87. Takenawa, T., and Egawa, K., 1977, *J. Biol. Chem.* **252:**5419–5423.
88. Jungalwala, F. B., Frienkel, N., and Dawson, R. M. C., 1971, *Biochem. J.* **123:**19–33.
89. Abdel-Latif, A. A., Roberts, M. B., Karp, W. B., and Smith, J. P., 1973, *J. Neurochem.* **20:**189–202.
90. Thompson, W., 1967, *Can. J. Biochem.* **45:**853–861.
91. Friedel, R. O., Brown, J. D., and Durell, J., 1967, *Biochim. Biophys. Acta* **144:**686–689.
92. Canessa de Scarnati, O., and Rodriguez de Lores Arnaiz, G., 1972, *Biochim. Biophys. Acta* **270:**218–225.
93. Lapetina, E. G., and Michell, R. H., 1973, *Biochem. J.* **131:**433–442.
94. Irvine, R. F., Hemington, N., and Dawson, R. M. C., 1979, *Eur. J. Biochem.* **99:**525–530.
95. Irvine, R. F., and Dawson, R. M. C., 1978, *J. Neurochem.* **31:**1427–1434.
96. Hirasawa, K., Irvine, R. F., and Dawson, R. M. C., 1981, *Biochem. J.* **193:**607–614.
97. Der, O. M., and Sun, G. Y., 1981, *J. Neurochem.* **36:**355–362.
98. Kemp, P., Hubscher, G., and Hawthorne, J. N., 1961, *Biochem. J.* **79:**193–200.
99. Atherton, R. S., and Hawthorne, J. N., 1978, *Eur. J. Biochem.* **4:**68–75.
100. Lapetina, E. G., Seguin, E. B., and Agranoff, B. W., 1976, *Biochim. Biophys. Acta* **398:**118–124.
101. Lapetina, E. G., Growman, M., and Canessa De Scarnati, O., 1976, *Int. J. Biochem.* **7:**507–513.

102. Abdel-Latif, A. A., Luke, B., and Smith, J. P., 1980, *Biochim. Biophys. Acta* **614:**424–434.
103. Mauco, G., Chap, H., and Douste-Blazy, L., 1979, *FEBS Lett.* **100:**367–370.
104. Billah, M. M., Lapetina, E. G., and Cuatrecasas, P., 1979, *Biochem. Biophys. Res. Commun.* **90:**92–98.
105. Allan, D., and Michell, R. H., 1974, *Biochem. J.* **142:**599–604.
106. Allan, D., and Michell, R. H., 1976, *Biochim. Biophys. Acta* **455:**824–830.
107. Ikezawa, H., Yamanegi, M., Taguchi, R., Miyashita, T., and Ohyabu, T., 1976, *Biochim. Biophys. Acta* **450:**154–164.
108. Sundler, R., Alberts, A. W., and Vagelos, P. R., 1978, *J. Biol. Chem.* **253:**4175–4179.
109. Ohyabu, T., Taguchi, R., and Ikezawa, H., 1978, *Arch. Biochem. Biophys.* **190:**1–7.
110. Hirasawa, Irvine, R. F., and Dawson, R. M. C., 1982, *Biochem. J.* **205:**437–442.
111. Irvine, R. F., Hemington, N., and Dawson, R. M. C., 1978, *Biochem. J.* **176:**475–484.
112. Dawson, R. M. C., Freinkel, N., Jugalawala, F. B., and Clarke, N., 1971, *Biochem. J.* **122:**605–607.
113. Dawson, R. M. C., and Clarke, N. G., 1972, *Biochem. J.* **127:**113–118.
114. Lapetina, E. G., and Michell, R. H., 1973, *FEBS Lett.* **31:**1–10.
115. Freinkel, N., and Dawson, R. M. C., 1973, *Nature* **243:**535–537.
116. Shum, T. Y. P., Gray, N. C. C., and Strickland, K. P., 1979, *Can. J. Biochem.* **57:**1359–1367.
117. Dawson, R. M. C., Hemington, N., Richards, D. E., and Irvine, R. F., 1979, *Biochem. J.* **182:**39–45.
118. Kai, M., White, G. L., and Hawthorne, J. N., 1966, *Biochem. J.* **101:**328–337.
119. Kai, M., Salway, J. G., and Hawthorne, J. N., 1968, *Biochem. J.* **106:**791–801.
120. Colodzin, M., and Kennedy, E. P., 1965, *J. Biol. Chem.* **240:**3771–3780.
121. Shaikh, N. A., and Palmer, F. B. St. C., 1977, *J. Neurochem.* **28:**395–402.
122. Iacobelli, S., 1969, *J. Neurochem.* **16:**909–911.
123. Tou, J.-S., Hurst, M. W., and Huggins, C. G., 1968, *Arch. Biochem. Biophys.* **131:**596–602.
124. Phillips, J. H., 1973, *Biochem. J.* **136:**579–587.
125. Muller, T. W., and Kirschner, N., 1975, *J. Neurochem.* **24:**1155–1161.
126. Trifaro, J. M., and Dworkind, J., 1975, *Can. J. Physiol. Pharmacol.* **53:**479–492.
127. Lefebvre, Y. A., White, D. A., and Hawthorne, J. N., 1976, *Can. J. Biochem.* **54:**746–753.
128. Oron, Y., Sharoni, Y., Lefkovitz, H., and Selinger, Z., 1978, *Cyclitols and Phosphoinositides* (W. W. Wells and F. Eisenberg, Jr., eds.), Academic Press, New York, pp. 383–397.
129. Buckley, J. T., 1977, *Biochim. Biophys. Acta* **498:**1–9.
130. Talwalker, R. T., and Lester, R. L., 1974, *Biochim. Biophys. Acta* **360:**306–311.
131. Harwood, J. L., and Hawthorne, J. N., 1969, *J. Neurochem.* **16:**1377–1387.
132. Thompson, W., and Dawson, R. M. C., 1964, *Biochem. J.* **91:**233–236, 237–243, 244–250.
133. Sheltawy, A., Brammer, M., and Borrill, D., 1972, *Biochem. J.* **128:**579–586.
134. Nijjar, M. S., and Hawthorne, J. N., 1977, *Biochim. Biophys. Acta* **480:**390–402.
135. Shaikh, N. A., and Palmer, F. B. St. C., 1977, *Brain Res.* **137:**333–342.
136. Lee, T.-C., and Huggins, C. G., 1968, *Arch. Biochem. Biophys.* **126:**214–220.
137. Cooper, P. H., and Hawthorne, J. N., 1975, *Biochem. J.* **150:**537–551.
138. Akhtar, R. A., and Abdel-Latif, A. A., 1978, *Biochim. Biophys. Acta* **527:**159–170.
139. Garrett, N. E., Burriss Garrett, R. J., Talwalker, R. T., and Lester, R. L., 1976, *J. Cell. Physiol.* **87:**63–69.
140. Palmer, F. B. St. C., 1973, *Biochim. Biophys. Acta* **326:**1942–200.
141. Palmer, F. B. St. C., 1976, *Biochim. Biophys. Acta* **441:**477–487.
142. Wirtz, K. W. A., 1974, *Biochim. Biophys. Acta* **344:**95–117.
143. Miller, E. K., and Dawson, R. M. C., 1972, *Biochem. J.* **126:**823–835.
144. Wirtz, K. W. A., Jolles, J., Westerman, J., and Neys, F., 1976, *Nature* **260:**354–355.
145. Carey, E. M., and Foster, P. C., 1977, *Biochem. Soc. Trans.* **5:**1412–1414.
146. Bramer, M. J., 1978, *J. Neurochem.* **31:**1435–1440.
147. Brophys, P. J., and Aitken, J. W., 1979, *J. Neurochem.* **33:**355–356.
148. Ruenwongsa, P., Singh, H., and Jungalwala, F. B., 1979, *J. Biol. Chem.* **254:**9385–9393.
149. Helmkamp, G. M., Harvey, M. S., Wirtz, K. W. A., and van Deenen, L. L. M., 1974, *J. Biol. Chem.* **249:**6382–6389.
150. Possmayer, F., 1974, *Brain Res.* **74:**167–174.

151. Brammer, M. J., and Sheltawy, A., 1975, *J. Neurochem.* **25**:699–705.
152. Demel, R. A., Kalsbeek, R., Wirtz, K. W. A., and van Deenen, L. L. M., 1977, *Biochim. Biophys. Acta* **466**:10–22.
153. Low, M. G., and Zilversmit, D. B., 1980, *Biochim. Biophys. Acta* **596**:223–234.
154. Hokin, L. E., 1969, *Structure and Function of Nervous Tissue* Volume 3 (G. B. Bourne, ed.), Academic Press, New York, pp. 161–184.
155. Abdel-Latif, A. A., Akhtar, R. A., and Hawthorne, J., 1977, *Biochem. J.* **162**:61–73.
156. Hokin, M. R., and Hokin, L. E., 1953, *J. Biol. Chem.* **203**:967–977.
157. Hokin, L. E., and Hokin, M. R., 1955, *Biochim. Biophys. Acta* **16**:229–237.
158. Hokin, L. E., and Hokin, M. R., 1955, *Biochim. Biophys. Acta* **18**:102–110.
159. Hokin, L. E., and Hokin, M. R., 1958, *J. Biol. Chem.* **233**:805–826.
160. Baricos, W. H., Hurst, M. W., and Huggins, C. G., 1979, *Arch. Biochem. Biophys.* **196**:227–232.
161. Margolis, R. U., and Heller, A., 1966, *J. Pharmacol. Exp. Ther.* **151**:307–312.
162. Friedel, R. O., and Schanberg, S. M., 1972, *J. Pharmacol. Exp. Ther.* **183**:326–332.
163. Soukup, J. F., Friedel, R. O., and Schanberg, S. M., 1978, *Biochem. Pharmacol.* **27**:1239–1243.
164. Friedel, R. O., Johnson, J. R., and Schanberg, S. M., 1973, *J. Pharmacol. Exp. Ther.* **184**:583–589.
165. Lunt, G. G., and Pickard, M. R., 1975, *J. Neurochem.* **24**:1203–1208.
166. Hokin, M. R., 1969, *J. Neurochem.* **16**:127–134.
167. Abdel-Latif, A. A., Yau, S.-J., and Smith, J. P., 1974, *J. Neurochem.* **22**:383–393.
168. Larrabee, M. G., and Leicht, W. S., 1965, *J. Neurochem.* **12**:1–13.
169. Larrabee, M. G., Klingman, J. D., and Leicht, W. S., 1963, *J. Neurochem.* **10**:549–560.
170. Lapetina, E. G., Brown, W. E., and Michell, R. H., 1976, *J. Neurochem.* **26**:649–651.
171. Pickard, M. R., Hawthorne, J. N., Hayashi, E., and Yamada, S., 1977, *Biochem. Pharmacol.* **26**:448–450.
172. Rosenberg, P., 1973, *J. Pharm. Sci.* **62**:1552–1554.
173. Bleasdale, J. E., Hawthorne, J. N., Widlund, L., and Heilbronn, E., 1976, *Biochem. J.* **158**:557–565.
174. Tret'jak, A. G., Limarenko, I. M., Kossora, G. V., Gulak, P. V., and Kozlov, Y. P., 1977, *J. Neurochem.* **28**:199–205.
175. Schacht, J., and Agranoff, B. W., 1972, *J. Biol. Chem.* **247**:771–777.
176. Yagihara, Y., and Hawthorne, J. N., 1972, *J. Neurochem.* **19**:355–367.
177. Aly, M. I., and Abdel, Latif, A. A., 1982, *Neurochem. Res.* **7**:159–169.
178. Fisher, S. K., and Agranoff, B. W., 1981, *J. Neurochem.* **37**:968–977.
179. Jolles, J., Zwiers, H., Dekker, A., Wirtz, J. W. A., and Gispen, W. H., 1981, *Biochem. J.* **194**:283–291.
180. Abdel-Latif, A. A., and Akhtar, R. A., 1982, *Phospholipids in the Nervous System*, (Horrocks, L. A., Ansell, G. B., and Porcellati, G., eds.) Raven Press, New York, pp. 251–264.
181. Abdel-Latif, A. A., Green, K., Smith, J., McPherson, J. C., and Matheny, J. L., 1978, *J. Neurochem.* **30**:517–525.
182. Canessa, De Scarnati, O., and Lapetina, E. G., 1974, *Biochim. Biophys. Acta* **360**:298–305.
183. Egawa, K., Sacktor, B., and Takenawa, T., 1981, *Biochem. J.* **194**:129–136.
184. Jafferji, S. S., and Michell, R. H., 1976, *Biochem. J.* **154**:643–657.
185. Hokin, M. R., Benfy, B. G., and Hokin, L. E., 1958, *J. Biol. Chem.* **233**:814–817.
186. Trifaro, J. M., 1969, *Mol. Pharmacol.* **5**:382–393.
187. Adnan, N. A. M., and Hawthorne, J. N., 1981, *J. Neurochem.* **36**:1858–1860.
188. Berg, G. R., and Klein, D. C., 1972, *J. Neurochem.* **19**:2519–2532.
189. Eichberg, J., Shein, G., Schwartz, M., and Hauser, G., 1973, *J. Biol. Chem.* **248**:3615–3622.
190. Smith, T. L., Eichberg, J., and Hauser, G., 1979, *Life Sci.* **24**:2719–2184.
191. Basinka, J., Sastry, P. S., and Stancer, H. C., 1973, *Endocrinology* **92**:1588–1595.
192. Michell, R. H., and Jones, L. M., 1974, *Biochem. J.* **138**:47–52.
193. Jones, L. M., and Michell, R. H., 1974, *Biochem. J.* **142**:583–590.
194. Oron, Y., Lowe, M., and Selinger, Z., 1974, *FEBS Lett.* **34**:198–200.
195. Oron, Y., Lowe, M., and Selinger, Z., 1975, *Mol. Pharmacol.* **11**:79–86.
196. Weiss, S. G., and Putney, J. W., 1981, *Biochem. J.* **194**:463–468.

197. Eggman, L. D., and Hokin, L. E., 1960, *J. Biol. Chem.* **235**:2569–2571.
198. Hokin, M. R., and Hokin, L. E., 1967, *J. Gen. Physiol.* **50**:793–811.
199. Fain, J. N., and Berridge, M. J., 1979, *Biochem. J.* **178**:45–58.
200. De Torrentegui, G., and Berthet, J., 1966, *Biochim. Biophys. Acta* **116**:467–476.
201. Billah, M. M., and Michell, R. H., 1979, *Biochem. J.* **182**:661–668.
202. Tolbert, M. E. M., White, A. C., Aspry, K., Cutts, J., and Fain, J. N., 1980, *J. Biol. Chem.* **255**:1938–1944.
203. Garcia Sainz, J. A., Hasfer, A. K., and Fain, J. N., 1980, *Biochem. Pharmacol.* **29**:3330–3333.
204. Jones, L. M., Cockcroft, S., and Michell, R. H., 1979, *Biochem. J.* **182**:669–676.
205. Subramanian, N., Whitmore, W. L., Seidler, F. J., and Slotkin, T. A., 1981, *J. Neurochem.* **36**:1137–1141.
206. Hokin-Neaverson, M. R., 1977, *Function and Biosynthesis of Lipids* (N. A. Bazan, R. R. Brenuer, and N. M. Aiusto, eds.), Plenum Press, New York, pp. 429–446.
207. Hanley, M. R., Lee, C. M., Jones, L. M., and Michell, R. H., 1980, *Mol. Pharmacol.* **18**:78–83.
208. Billah, M. M., and Michell, R. H., 1979, *Biochem. J.* **182**:661–668.
209. Takhar, A. P. S., and Kirk, C. J., 1981, *Biochem. J.* **194**:167–172.
210. Tolbert, M. E. M., White, A. C., Aspry, K., Cutts, J., and Fain, J. M., *J. Biol. Chem.* **255**:1938–1944.
211. Jolles, J., Wirtz, W. A., Schotman, P., and Gispen, W. H., 1979, *FEBS Lett.* **105**:110–114.
212. Prpic, V., Blackmore, P. F., and Exton, J. H., 1982, *J. Biol. Chem.,* **257**:11323–11331.
213. Farese, R. V., Sabir, A. M., and Vandor, S. L., 1979, *J. Biol. Chem.* **254**:6842–6844.
214. Farese, R. V., Sabir, A. M., Vandor, S. L., and Larson, R. E., 1980, *J. Biol. Chem.* **255**:5728–5734.
215. Friedel, R. O., Slotnick, R. N., and Bombardt, P. A., 1977, *Life Sci.* **20**:235–242.
216. Schacht, J., and Agranoff, B. W., 1974, *J. Biol. Chem.* **249**:1551–1557.
217. Durell, J., and Garland, J. T., 1969, *Ann. N.Y. Acad. Sci.* **165**:743–754.
218. Lapetina, E. A., and Hawthorne, J. N., 1971, *Biochem. J.* **122**:171–179.
219. Hokin-Neaverson, M., 1974, *Biochem. Biophys. Res. Commun.* **58**:763–768.
220. Bansbach, M. W., Aeison, R. L., and Hokin-Neaverson, M., 1974, *Biochem. Biophys. Res. Commun.* **58**:714–718.
221. Hokin-Neaverson, M., 1977, *Adv. Exp. Med. Biol.* **83**:429–446.
222. Bansbach, M. W., Aeison, R. L., and Hokin-Neaverson, M., 1981, *Biochim. Biophys. Acta* **663**:34–45.
223. Durell, J., Garland, J. T., and Friedel, R. O., 1969, *Science* **165**:862–866.
224. Hokin, L. E., 1966, *J. Neurochem.* **13**:179–184.
225. Abdel-Latif, A. A., Green, K., Matheny, J. L., McPherson, J. C., and Smith, J. P., 1975, *Life Sci.* **17**:1821–1828.
226. Smith, T. L., Eichberg, J., and Hauser, G., 1979, *Life Sci.* **24**:2179–2184.
227. Fisher, S. K., Boast, C. A., and Agranoff, B. W., 1980, *Brain Res.* **189**:284–288.
228. Hokin, L. E., and Huebuer, D., 1969, *J. Cell. Biol.* **33**:521–530.
229. Hokin, L. E., 1965, *Proc. Natl. Acad. Sci. U.S.A.* **53**:1369–1376.
230. Lapetina, E. A., and Michell, R. H., 1972, *Biochem. J.* **126**:1141–1146.
231. Burt, D. R., and Larrabee, M. A., 1973, *J. Neurochem.* **21**:255–272.
232. Yagihara, Y., Bleasdale, J. E., and Hawthorne, J. N., 1973, *J. Neurochem.* **21**:173–190.
233. Pickard, M. R., and Hawthorne, J. N., 1978, *J. Neurochem.* **30**:145–155.
234. Schacht, J., Neale, E. A., and Agranoff, B. W., 1974, *J. Neurochem.* **23**:211–218.
235. Berridge, M. J., 1975, *Adv. Cyclic Nucleotide Res.* **6**:1–98.
236. Hurwitz, L., and Suria, A., 1971, *Annu. Rev. Pharmacol.* **11**:303–326.
237. Triggle, D. J., 1972, *Prog. Surf. Memb. Sci.* **5**:267–331.
238. Triggle, D. J., and Triggle, C. R., 1976, *Chemical Pharmacology of the Synapse*, Academic Press, New York.
239. Weiss, G. B., (ed.), 1978, *Calcium in Drug Action*, Plenum Press, New York.
240. Trifaro, J. M., 1969, *Mol. Pharmacol.* **5**:424–427.
241. Oron, Y., Lowe, M., and Selinger, Z., 1975, *Mol. Pharmacol.* **11**:79–86.

242. Jones, L. M., and Michell, R. H., 1975, *Biochem. J.* **148**:479–485.
243. Jones, L. M., and Michell, R. H., 1976, *Biochem. J.* **158**:505–507.
244. Jafferji, S. S., and Michell, R. H., 1976, *Biochem. J.* **160**:163–169.
245. Hokin, L. E., 1966, *Biochim. Biophys. Acta* **115**:219–221.
246. Zor, U., Lowe, I. P., Bloom, G., and Field, J. B., 1968, *Biochem. Biophys. Res. Commun.* **33**:649–658.
247. Abdel-Latif, A. A., 1976, *Function and Metabolism of Phospholipids in the Central and Peripheral Nervous Systems* (G. Porcellati, L. Amaducci, and C. Galli, eds.), Plenum Press, New York, pp. 227–256.
248. Hawthorne, J. N., and Bleasdale, J. E., 1975, *Mol. Cell. Biochem.* **8**:83–87.
249. Lennon, A. M., and Steinberg, H. R., 1973, *J. Neurochem.* **20**:337–345.
250. Akhtar, R. A., and Abel-Latif, A. A., 1978, *J. Pharmacol. Exp. Ther.* **204**:655–668.
251. Akhtar, R. A., and Abdel-Latif, A. A., 1980, *Biochem. J.* **192**:785–791.
252. Grimes, M. J., Abdel-Latif, A. A., and Carrier, G. O., 1979, *Biochem. Pharmacol.* **28**:3213–3219.
253. Akhtar, R. A., and Abdel-Latif, A. A., 1979, *Gen. Pharmacol.* **10**:445–450.
254. Griffin, H. D., Hawthorne, J. N., and Sykes, M., 1979, *Biochem. Pharmacol.* **28**:1143–1147.
255. Miller, J. C., and Leung, I., 1979, *Biochem. J.* **178**:9–13.
256. Fisher, S. K., and Agranoff, B. W., 1980, *J. Neurochem.* **34**:1231–1240.
257. Griffin, H. D., and Hawthorne, J. N., 1978, *Biochem. J.* **176**:541–552.
258. Allan, D., and Michell, R. H., 1978, *Biochim. Biophys. Acta* **508**:277–286.
259. Ponnappa, B. C., Greenquist, A. C., and Shohet, S. B., 1980, *Biochim. Biophys. Acta* **598**:494–501.
260. Tou, J.-S., 1978, *Biochim. Biophys. Acta* **531**:167–178.
261. Rossignol, B., Herman, G., Chambaut, A. M., and Keryer, G., 1974, *FEBS Lett.* **43**:241–246.
262. Jone, L. M., and Michell, R. H., 1978, *Biochim. Soc. Trans.* **6**:1035–1037.
263. Garcia-Sainz, J. A., and Fain, J. N., 1980, *Biochem. J.* **186**:781–789.
264. Cockroft, S., Bennett, J. P., and Gomperts, B. D., 1980, *FEBS Lett.* **110**:115–118.
265. Farese, R. V., Larson, R. E., and Sabir, M. A., 1980, *Biochim. Biophys. Acta* **633**:479–484.
266. Clements, R. S., Evans, M. H., and Pace, C. S., 1981, *Biochim. Biophys. Acta* **674**:1–9.
267. Fisher, S. K., Holz, R. W., and Agranoff, B. W., 1981, *J. Neurochem.* **37**:491–497.
268. Brossard, M., and Quastel, J. H., 1963, *Can. J. Biochem. Physiol.* **41**:1243–1256.
269. Keryer, G., Herman, G., and Rossignol, G., 1979, *FEBS Lett.* **102**:4–8.
270. Abdel-Latif, A. A., and Luke, B., 1981, *Biochim. Biophys. Acta* **673**:64–74.
271. Akhtar, R. A., and Abdel-Latif, A. A., 1982, *J. Neurochem.* **39**:1374–1380.
272. Jafferji, S. S., and Michell, R. H., 1976, *Biochem. J.* **160**:397–399.
273. Sherman, W. R., Leavitt, A. L., Honchar, M. P., Hallcher, L. M., and Phillips, B. E., 1981, *J. Neurochem.* **36**:1947–1951.
274. Anderson, R. E., Maude, M. B., and Kelleher, P. A., 1980, *Biochim. Biophys. Acta* **620**:236–246.
275. Michell, R. H., Jafferji, S. S., and Jones, L. M., 1977, *Adv. Exp. Med. Biol.* **83**:447–464.
276. Michell, R. H., and Kirk, C. J., 1981, *Trends Pharmacol. Sci.* **2**:86–89.
277. Michell, R. H., Kirk, C. J., Jones, L. M., Downes, C. P., and Cerba, J. A., 1981, *Phil. Trans R. Soc. Lond. B.* **296**:123–137.
278. Salmon, D. M., and Honeyman, T. W., 1980, *Nature* **284**:344–345.
279. Putney, J. W., Weiss, S. J., Van De Walle, C. M., and Haddas, R. A., 1980, *Nature* **284**:345–347.
280. Harris, R. A., Schmidt, J., Hitzemann, B. A., and Hitzemann, R. J., 1981, *Science* **212**:1290–1291.
281. Magee, W. L., Berry, J. G., Strickland, K. P., and Rossiter, R. J., 1963, *Biochem. J.* **88**:45–52.
282. Brindley, D. N., and Bowley, M., 1975, *Biochem. J.* **148**:461–469.
283. Allan, D., and Michell, R. H., 1975, *Biochem. J.* **148**:471–478.
284. Salway, J. G., and Hughes, I. E., 1972, *J. Neurochem.* **19**:1233–1240.
285. Brindley, D., and Bowley, M., 1975, *Postgrad. Med. J.* **51**(Suppl):91–95.

286. Eichberg, J., and Hauser, G., 1974, *Biochem. Biophys. Res. Commun.* **60:**1460–1467.
287. Mule, S., 1967, *J. Pharmacol. Exp. Ther.* **156:**92–100.
288. Mule, S. J., 1970, *Biochem. Pharmacol.* **19:**581–593.
289. Natsuki, R., Hitzemann, R. J., and Loh, H. H., 1979, *Res. Commun. Chem. Pathol. Pharmacol.* **24:**233–250.
290. Hitzemann, R. J., Natsuki, R., and Loh, H. H., 1978, *Biochem. Pharmacol.* **27:**2519–2523.
291. Stein, J. M., and Hales, C. N., 1972, *Biochem. J.* **128:**531–541.
292. Hauser, G., and Eichberg, J., 1975, *J. Biol. Chem.* **250:**104–112.
293. Abdel-Latif, A. A., and Smith, J. P., 1976, *Biochem. Pharmacol.* **25:**1697–1704.
294. Bazan, N. G., Ilincheta, de Boschero, M. G., and Guisto, N. M., 1977, *Adv. Exp. Med. Biol.* **83:**429–446.
295. Eichberg, J., Gates, J., and Hauser, G., 1979, *Biochim. Biophys. Acta* **573:**90–106.
296. Brindley, D. N., Bowley, M., Sturton, T. G., Pritchard, P. H., Burditt, S. L., and Cooling, J., 1978, *Central Mechanisms of Anorectic Drugs* (S. Garattini and R. Samanin, eds.), Raven Press, New York, pp. 301–317.
297. Natsuki, R., Hitzemann, R., and Loh, H., 1978, *Mol. Pharmacol.* **14:**448–453.
298. Bazan, N. G., 1970, *Biochim. Biophys. Acta* **218:**1–10.
299. Rubin, R., Sink, L. E., and Freer, R. J., 1981, *Biochem. J.* **194:**479–505.
300. Schrey, M. P., and Rubin, R. P., 1979, *J. Biol. Chem.* **254:**11234–11241.
301. Bell, R. L., Baenziger, N. L., and Majerus, P. W., 1980, *Prostaglandins* **20:**269–274.
302. Hong, S. L., and Deykin, D., 1981, *J. Biol. Chem.* **256:**5215–5219.
303. Bansbach, M. W., and Hokin-Neaverson, M., 1980, *FEBS Lett.* **117:**131–133.
304. Marshall, P. J., Dixon, J. F., and Hokin, L. E., 1980, *Proc. Natl. Acad. Sci. U.S.A.* **77:**3292–3296.
305. Marshall, P. J., Boatman, D. E., and Hokin, L. E., 1981, *J. Biol. Chem.* **256:**844–847.
306. Lapetina, E. G., Billah, M. M., and Cuatrecasas, P., 1981, *J. Biol. Chem.* **256:**5037–5040.
307. Billah, M. M., Lapetina, E. G., and Cuatrecasas, P., 1981, *J. Biol. Chem.* **256:**5399–5403.
308. Haye, B., Champion, S., and Jacquemin, C., 1974, *FEBS Lett.* **41:**89–93.
309. Coburn, R. F., Cunningham, M., and Strauss, J. F., 1981, *Biochim. Biophys. Acta* **664:**188–199.
310. Lloyd, T., 1979, *J. Biol. Chem.* **254:**7247–7254.
311. Heger, H. W., and Peter, H. W., 1977, *Int. J. Biochem.* **8:**841–846.
312. Levy, G. S., 1971, *J. Biol. Chem.* **246:**7405–7407.
313. Slein, M. W., and Logan, G. F., 1965, *J. Bacteriol.* **90:**69–81.
314. Low, M. G., and Finean, J. B., 1976, *Biochem. J.* **154:**203–208.
315. Low, M. G., and Finean, J. B., 1977, *Biochem. J.* **167:**281–284.
316. Low, M. G., and Finean, J. B., 1978, *Biochim. Biophys. Acta* **508:**565–570.
317. Low, M. G., and Zilversmit, D. B., 1980, *Biochemistry* **19:**3913–3918.
318. Panagia, V., Heyliger, C. E., Choy, P. C., and Dhalla, N. S., 1981, *Biochim. Biophys. Acta* **640:**802–806.
319. Rothman, S. S., 1978, *Biochim. Biophys. Acta* **509:**374–383.
320. Goldman, S. S., and Albers, R. W., 1973, *J. Biol. Chem.* **248:**867–874.
321. DePont, J. J. H. H. M., Van Prooijen-Van Eeden, A., and Bonting, S. L., 1978, *Biochim. Biophys. Acta* **508:**464–477.
322. Isern De Caldentey, M., and Wheeler, K. P., 1979, *Biochem. J.* **177:**265–273.
323. Dahl, J. L., and Hokin, L. E., 1974, *Annu. Rev. Biochem.* **43:**327–356.
324. Steinberg, H. R., and Durell, J., 1971, *J. Neurochem.* **18:**277–286.
325. Appel, S. H., Andrews, C. A., and Almon, R. R., 1974, *J. Neurochem.* **23:**1077–1080.
326. Abdel-Latif, A. A., Green, J., and Smith, J. P., 1979, *J. Neurochem.* **32:**225–228.
327. Rounds, P. S., Jepson, A. B., McAllister, D. J., and Howland, J. L., 1980, *Biochem. Biophys. Res. Commun.* **97:**1384–1390.
328. Gerthoffer, W. T., Fedan, J. S., Westfall, D. P., Goto, K., and Fleming, W. W., 1979, *J. Pharmacol. Exp. Ther.* **210:**27–36.
329. O'Leary, J. L., Harris, A. B., and Fox, R. R., 1965, *Arch. Neurol.* **13:**238–262.
330. Eliasson, S. G., Scarpellini, J. D., and Fox, R. R., 1967, *Arch. Neurol.* **17:**661–665.
331. Eliasson, S. G., Scarpellini, J. D., and Fox, R. R., 1972, *Arch. Neurol.* **27:**535–539.

332. Giachetti, A., 1981, *Pharmacol. Res. Commun.* **13**:101–119.
333. Eliasson, S. G., 1965, *Lipids* **1**:237–240.
334. Greene, D. A., De Jesus, P. V., and Winegrad, A. I., 1975, *J. Clin. Invest.* **55**:1326–1336.
335. Palmato, K. P., Whitting, P. H., and Hawthorne, J. N., 1977, *Biochem. J.* **167**:229–235.
336. Whitting, P. H., Palmato, K. P., and Hawthorne, J. N., 1979, *Biochem. J.* **179**:549–553.
337. Hothersall, J. S., and McLean, P., 1979, *Biochem. Biophys. Res. Commun.* **88**:477–484.
338. Natarajan, V., Dyck, P. J., and Schmid, H. P., 1981, *J. Neurochem.* **36**:413–419.
339. Bell, M. E., and Eichberg, J., 1981, *Trans. Am. Soc. Neurochem.* **12**:240.

Metabolism of Phosphoglycerides

Giuseppe Porcellati and Giuseppe Arienti

1. INTRODUCTION

Glycerophospholipids are present in all living organisms and are necessary for membrane assembly and function. Their presence in nervous tissue has been recognized for many years,[1] and, with the development of suitable analytical techniques, their metabolism has been amply studied and is being investigated in different aspects in several laboratories.

The study of phospholipid metabolism aims to clarify biosynthetic and catabolic pathways and the relative importance and significance of the various metabolic processes. The mechanism by which lipid metabolism is influenced by physiological and pathological situations is also a very important topic, as is the action of lipid metabolism on the properties of nervous tissue.

This chapter deals with the anabolic and catabolic pathways of lipid metabolism in nervous tissue. Not all lipids and not every aspect of lipid metabolism are treated here, because other chapters of this *Handbook* cover particular aspects of phospholipid metabolism in nervous tissue.[2,3] (L. A. Horrocks and H. W. Harder, this volume; N. G. Bazan, this volume; A. A. Abdel-Latif, this volume).

2. CHEMISTRY AND DISTRIBUTION IN NERVOUS TISSUE

2.1. Structure and Properties

Phosphoglycerides are derivatives of *sn*-glycero-3-phosphoric acid containing at least one O-acyl or O-alkyl or 1-alk-1'-enyl residue bound to the glycerol moiety.[4–6] Lipids containing phosphorus in the phosphonate form have also been described[7] and represent the phosphonolipids.

A problem of designation of glycerol derivatives is their asymmetry. The most satisfactory way to indicate the stereochemical characteristics of glycerol

Giuseppe Porcellati and Giuseppe Arienti • Department of Biochemistry, The Medical School, University of Perugia, 06100 Perugia, Italy.

lipids is the "stereospecific numbering" (*sn*) recommended by the IUPAC–IUB Commission.[4–6]

Phosphoglycerides are usually divided into classes according to the substituent bound to the phosphoric acid moiety. There are several classes of glycerophospholipids, which can be further divided into subclasses according to the number and the type of chemical bond of the substituent(s) attached to position 1 and/or 2 of the *sn*-glycero-3-phosphate derivatives (Fig. 1). Each subclass can be divided into molecular species according to the nature of the fatty acid(s) (fatty aldehyde or fatty alcohol) present in the molecule. The interested reader may find more information on this subject in Ansell *et al.*,[8] whose book will soon appear in a new edition, and in Gaiti *et al.*[9] It is clear that many phosphoglycerides exist in nervous tissue.

Phospholipids are one of the major constituents of biological membranes; to a great extent they determine the properties of the membrane itself. The presence of one class or subclass or even a molecular species instead of another could be of great relevance to membrane function. The means by which the composition of membrane lipids is created and maintained is one of the most interesting topics in phospholipid metabolism.

Glycerophospholipids possess a hydrophilic head and a hydrophobic tail, permitting an amazing variety of interactions with both the nonpolar milieu of the membrane and the surrounding aqueous medium containing proteins and ions. Therefore, lipid composition can determine most of the chemicophysical properties of the membrane, such as fluidity and permeability to ions. Moreover, several membrane-bound enzymes require phospholipid to be active. The roles of phosphoglycerides are important in every cell, particularly in the nervous system, whose ability to transfer information by electrical waves depends essentially on membrane properties and integrity.

2.2. Distribution in Nervous Tissue

A study of phosphoglyceride distribution is obviously a difficult task, requiring the determination of several hundred molecules. An additional difficulty arises when nervous tissue is the object of such an investigation because different cell types and subcellular structures must be taken into account. Moreover, lipid composition is influenced by the animal species, by the age of the animal, and by physiological and pathological events.

Many studies have been performed on the lipid composition of central and peripheral nervous systems, and both vertebrate and invertebrate animals have

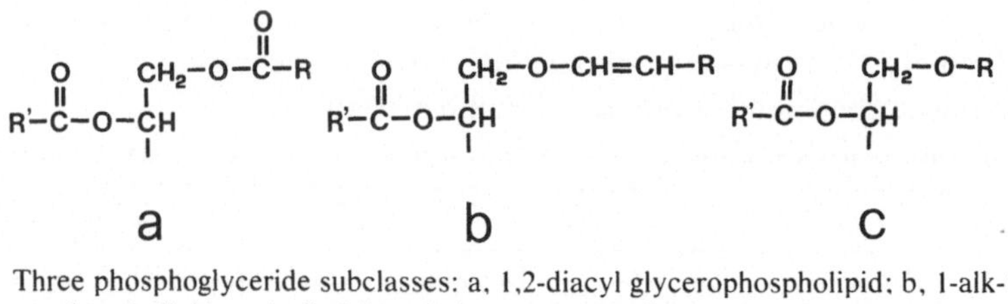

Fig. 1. Three phosphoglyceride subclasses: a, 1,2-diacyl glycerophospholipid; b, 1-alk-1′-enyl-2-acyl glycerophospholipid; c, 1-alkyl-2-acyl glycerophospholipid.

Table I
The Total Phospholipid Phosphorus Content of Brain and Peripheral Nerve

Tissue	Animal species	Lipid phosphorus content[a]	Reference[b]
Brain	Sheep	2.14	10
	Lamb	1.07	10
	Rat	1.85	11
	Rabbit	2.35	12
	Guinea pig	1.96	13
Peripheral nerve	Sheep (sciatic)	1.67	14
	Bovine (splenic)	0.42	14
	Rabbit (sciatic)	1.95	14
	Monkey (sciatic)	1.67	14
	Hen (sciatic)	1.66	14
	Crab, claw	0.38	14
	Crab, leg	0.35	14
	Lobster, claw	0.41	14
	Lobster, leg	0.52	14

[a] Results are expressed as mg lipid P/g wet wt. and represent mean values from various estimations.
[b] For more data see also reference 15.

been considered. Glycerophospholipids are abundant both in brain and peripheral nerve (Table I). In the latter case, the amount of myelin is important in this connection; indeed, the lipid phosphorus content is not less in well-myelinated peripheral nerves than in the brain.

The distribution of lipid phosphorus among lipid classes in brain and nerve is indicated in Table II, which presents only a few data from those available in the literature. The most abundant glycerophospholipid classes are ethanolamine and choline phosphoglycerides, followed by the serine lipids. Diacyl subclasses are the most represented in nervous tissue. An exception is ethanolamine plasmalogen which accounts for more than 50% of the total ethanolamine glycerophospholipids.

The lipid composition of different brain areas of the rabbit is reported in Table III. Peripheral nerve appears to contain more sphingomyelin than does brain white matter, whereas the phospholipid composition does not vary very much among cerebrum, cerebellum, and lower brainstem.

Brain tissue can be homogenized, and subcellular fractions can be obtained by centrifuging the homogenate. The purity of cellular subfractions is usually tested by microscopic and biochemical examinations. These preparations have been employed widely in neurochemistry, and their phospholipid composition is shown in Table IV. Choline glycerophospholipid is the most abundant phosphoglyceride in almost all subfractions except myelin, in which ethanolamine derivatives predominate. Serine glycerophospholipid and sphingomyelin are also well represented in most subfractions, whereas the content of other phospholipid classes is rather low.

Table II
Phospholipid Composition of Brain and Nerve[a]

	Human brain (55-year-old male)		Sheep brain (10)	Rat brain (17)	Bovine brain White, frontal		Optic nerve (19)	Rabbit Brain (12)	Sciatic nerve (14)
	Gray (16)	White (16)			Cortex (18)	Medulla (18)			
Phosphatidylcholine	39.1	30.7	37.3	36.8	24.1[b]	24.1[b]	17.5	31.2	13.3
Choline plasmalogen			0.9					1.0	
Lysophosphatidylcholine								0.4	
Phosphatidylethanolamine	40.0[b]	34.1[b]	7.7	36.4[b]	42.7[b]	35.4[b]	8.4	12.7	5.4
Ethanolamine plasmalogen			16.5					22.6	24.8
Phosphatidylinositol			2.1	3.1	12.5[c]	3.0		3.0	3.7[d]
Phosphatidylserine	12.6	15.7	9.2	11.8		16.6	9.5	15.8[e]	11.8
Diphosphatidylglycerol			2.0	2.2		2.1		2.0	
Phosphatidic acid			2.6	1.2					1.3
Sphingomyelin	8.3	19.5	12.8	5.7	16.9	16.6		12.4	19.7

[a] Results are expressed as percent of lipid phosporus content. Numbers in parentheses indicate references.
[b] Includes plasmalogen.
[c] Includes serine glycerophospholipids.
[d] Includes triphosphoinositides.
[e] Includes 1% serine plasmalogen.

Table III
Regional Differences of Phospholipid Composition in the White
Matter of CNS and in Peripheral and Cranial Nerves of the
Rabbit[a]

Tissue	Total phospholipid	CPG[b]	EPG[c]	SM[d]	SPG[e]
White matter of CNS					
Cerebrum	49.1	11.75	20.08	5.96	7.37
Pons + medulla	47.7	10.30	18.32	6.30	6.25
Cerebellum	51.9	11.11	20.76	6.80	7.89
Spinal cord					
Motor	50.3	8.95	20.32	9.66	7.85
Sensory	50.2	9.69	20.18	9.46	7.58
Spinal nerve					
Root nerve					
Anterior	60.6	8.48	20.18	17.82	9.51
Posterior	60.4	8.34	20.17	17.46	9.85
Sciatic nerve					
Proximal	56.3	8.28	19.48	16.05	8.84
Distal	61.0	9.85	20.86	16.23	9.64
Cranial nerves					
Optic	48.6	11.86	18.37	6.61	7.48
Oculomotor	50.5	7.42	17.52	13.43	8.13
Trigeminal	50.8	8.48	19.51	11.07	8.28
Vagal	56.0	10.02	19.04	13.66	9.02

[a] Results are expressed as μmol/100 mg of total lipids and represent mean values. Data are from reference 20.
[b] CPG, choline phosphoglycerides.
[c] EPG, ethanolamine phosphoglycerides.
[d] SM, sphingomyelin.
[e] SPG, serine phosphoglycerides.

The phospholipid composition of neuron- and glia-enriched fractions has also been studied and recently reviewed.[25] The data reported in Table V refer only to cell bodies, because neurites and dendrites are almost all lost during cell separation.

The fatty acid composition of cerebral lipids has been studied under physiological and pathological conditions. Table VI shows the fatty acid composition of choline and ethanolamine glycerophospholipids of 6-day-old rat brain. From this table, it can be assumed that choline diacyl lipids are rich in 16:0 and 18:1, whereas ethanolamine diacyl lipids are rich in 18:0, 20:4, and 22:6.

Much work has been performed on phospholipid composition during development.[30-32] In this case, the differences result mainly from cell proliferation and myelin deposition. The latter process implies an increase of ethanolamine phosphoglyceride and sphingomyelin, whereas choline phosphoglyceride does not increase during myelin formation.[31]

Another interesting aspect of phospholipid distribution in nervous tissue is the asymmetry of phospholipid across the membrane. Not much work has yet been done on nervous membranes; however, about 10–15% of ethanolamine

Table IV
Phospholipid Composition of Brain Subcellular Fractions[a]

	Nuclei		Mitochondria		Microsomes		Supernatant, rat (23)	White matter myelin, rat (18)
	Bovine (21)	Human (21)	Bovine (22)	Rat (17)	Rat (18)	Human (18)		
Phosphatidylcholine	53.1	47.6	28.5	50.0	34.7	43.3	53.6	26.2
Lysophosphatidylcholine		2.5	0.2					
Phosphatidylethanolamine	20.8	17.8	18.5	32.1	33.6	20.2	17.8	44.9
Ethanolamine plasmalogen			8.7					
Phosphatidylinositol	7.9	5.0	2.9	13.4	3.3	7.3	14.3	5.1
Phosphatidylserine	5.0	9.3	11.3		10.7	12.8	10.7	15.6
Sphingomyelin	3.0	10.7	17.3		7.3	14.1	3.6	8.6
Phosphatidylglycerol		0.6	0.5					
Diphosphatidylglycerol		0.3	4.6	1.6	3.3	2.3		2.7
Phosphatidic acid	0.6	1.2	0.7					

[a] Results are expressed as percent of total lipid phosphorus content. Numbers in parentheses indicate references.

Table V
Phospholipid Composition of Neuronal and Glial Cell-Enriched Fractions[a]

	Rabbit[b]		Rat[c]		Chicken[d]		Calf oligodendroglia[e]
	Neurons	Glia	Neurons	Glia	Neurons	Glia	
Diacyl-GPC[f]	49.5	37.8	37.0	37.0	41.6[g]	40.0[g]	47.6[g]
Alkenylacyl-GPC[f]		1.5	1.6	1.6			
Diacyl-GPE[h]	16.8	18.7	17.1	19.0	31.0[g]	32.0[g]	24.3[g]
Alkenylacyl-GPE[h]	14.9	19.2	13.9	14.2			
Diacyl-GPS[i]	7.2	11.1	9.8	9.9	8.1	8.8	6.8[g]
Diacyl-GPI[j]	3.9	4.8	3.9	3.0	4.9	4.8	7.7
Sphingomyelin	4.9	6.7	5.6	5.2	8.1	9.3	5.4

[a] Data are expressed as % of total lipid phosphorus content.
[b] From reference 26.
[c] From reference 27.
[d] From reference 28.
[e] From reference 29.
[f] GPC, *sn*-glycero-3-phosphorylcholine.
[g] Including ether lipids.
[h] GPE, *sn*-glycero-3-phosphorylethanolamine.
[i] GPS, *sn*-glycero-3-phosphorylserine.
[j] GPI, *sn*-glycero-3-phosphorylinositol.

glycerophospholipid and 20% of serine glycerophospholipid appear to be in the outer monolayer of synaptosomal membrane.[33] Somewhat different results have also been reported in the literature,[34] probably because of the different techniques that have been employed for this purpose.

3. GLYCEROPHOSPHOLIPID SYNTHESIS

3.1. In Vivo Studies: Assembly of Glycerophospholipid Molecules

Nervous tissue is able to synthesize glycerophospholipids from simpler precursors such as glycerol, fatty acids, nitrogenous bases, and so on.

Using ^{32}P-labeled phosphate, it has been possible to show that phosphatidylinositol and choline plasmalogen have the fastest turnover in the brain, whereas diphosphatidylglycerol has the lowest.[35] Moreover, as a general rule, lipids turn over faster in neurons than in astroglia.[3,25] On the other hand, the turnover rate of practically all phospholipids is slower in whole brain than in astroglial or neuronal cells alone, which means that lipids far from cell bodies turn over very slowly. This difference is particularly evident for plasmalogens which have been shown to turn over much faster in microsomes and mitochondria than in myelin.[36] On the other hand, each phospholipid tends to reach the same specific radioactivity in all subfractions provided enough time is allowed.[37] This would indicate that these molecules are formed in one place and then carried to other places. The time elapsed between the injection of the labeled precursor and the sacrifice of the animal is extremely important and may help to explain discrepancies reported in the literature.

Table VI
Fatty Acid Composition of 1,2-Diacyl-sn-glycero-
3-phosphorylcholine and of 1,2-Diacyl-sn-glycero-
3-phosphorylethanolamine of 6-Day-Old Rat
Brain[a]

Fatty acid[b]	Choline lipids	Ethanolamine lipids
14:0	4.2	1.4
16:0	51.1	14.5
16:1	7.1	1.8
18:0	5.2	24.7
18:1	18.1	9.5
18:2	1.9	1.1
20:4	6.3	19.0
22:4[c]	0.9	3.0
22:5	—	3.0
22:6	2.8	18.1

[a] Results are expressed as mol% (from reference 30).
[b] Fatty acids comprising less than 0.5% are omitted.
[c] Tentative identification.

Miller and Morell,[38] using different labeled precursors, found that the choline and acyl moieties are reutilized better than glycerol (whose reutilization should be minimal). The reutilization of a lipid precursor is a fact that further complicates the calculation of lipid turnover. The age of the animals is also relevant when turnover rates are to be estimated,[38,39] because cell proliferation and myelin deposition take place in young animals. For all these reasons, great discrepancies on this subject have been reported in the literature; for instance, the half-life of myelin phosphatidylcholine has been found to range between 10 days[40] and 167 days.[41]

The studies on phospholipid turnover can be further complicated by the existence of separate metabolic pools of lipids and their precursors.[42] The localization of the pools is sometimes very difficult. They could indeed reflect different metabolic rates of cells or organelles or even metabolic compartmentation in the same membrane. Phospholipids are synthesized mainly at the microsomal level, and it is possible that the rapid-turnover pool of phosphatidylcholine represents lipid molecules exchanging between the cytosol and the cytoplasmic side of the membrane, whereas the slow pool would be the true expression of the renewal of membrane segments.

Phospholipid turnover has also been studied in separated neuronal and astroglial cells and has been found to be higher in neurons.[25,43] On the other hand, it is possible that oligodendroglia differ very much from astroglia in this respect.[25,44] As a general rule, the turnover of glycerophospholipids decreases from neurons to glia to myelin.

The sciatic nerve of the hen is able to incorporate labeled phosphate into lipids *in vitro*.[45] The specific radioactivity of ^{32}P in the synthesized lipids decreases in the following order: phosphatidic acid, phosphatidylinositol, choline glycerophospholipid, cardiolipin, serine glycerophospholipid, and ethanolamine glycerophospholipid. A major problem of lipid synthesis in peripheral nerve is the localization of the biosynthetic reactions that could be present in the axon and/or in the Schwann cells. Choline, ethanolamine, and glycerol are not incorporated into lipids within the axons,[46,47] and choline phosphotransferase has been found mostly in Schwann cells.[48] Similarly, phospholipid synthesis in invertebrate axons appears to be limited in extent.[49]

3.2. Mechanisms of Synthesis

3.2.1. Choline Glycerophospholipids

3.2.1a. Net Synthesis. Choline glycerophospholipids can be synthesized in the brain by several pathways which probably have different metabolic and physiological meanings. However, only net synthesis can account for the assembly of choline glycerophospholipids from simpler precursors, because other pathways require a preformed lipid to be transformed into choline phosphoglyceride.

The net synthesis involves the transfer of choline phosphate (or phosphorylcholine) from cytidine-5'-diphosphate choline (CDP-choline) to 1,2-diacyl-*sn*-glycerol (diglyceride).[50] The syntheses of phosphatidic acid and of diglyceride, which are clearly a part of this pathway, are not treated here, since they are considered in other chapters of this *Handbook* (N. G. Bazan, this volume; A. A. Abdel-Latif, this volume). This chapter is primarily concerned with the formation of CDP-choline and the transfer of its choline phosphate moiety to diacylglycerol. This part of the pathway was first described in brain by McMurray *et al.*[50] and requires the sequential action of three enzymes: (1) choline kinase (ATP:choline phosphotransferase; E.C. 2.7.1.32) forming choline phosphate from free choline and ATP; (2) choline phosphate cytidylyltransferase (CTP:choline phosphate cytidylyltransferase; E.C. 2.7.7.15) forming CDP-choline from choline phosphate and CTP; and (3) choline phosphotransferase (CDP-choline:1,2-diacyl glycerol choline phosphotransferase; E.C. 2.7.8.2)* transferring the choline phosphate moiety from CDP-choline to diglyceride (Fig. 2).

Choline kinase is present in the cell supernatant, requires the presence of Mg^{2+}, and, at its pH optimum, is 20 times as active as the cytidylyltransferase reaction.[51] This and other findings led Porcellati and Arienti[52] to postulate for the first time that the cytidylyltransferase reaction constitutes a rate-limiting step in phosphatidylcholine biosynthesis. This kinase appears to be different from the one catalyzing the phosphorylation of ethanolamine and is present in the cellular supernatant.[53] The measurement of this enzymic activity can give lower values in the presence of microsomal membranes because of the action

---

* The classical name for choline phosphotransferase has been reported here. More correctly, the enzyme should be called CDP-choline:1,2-diradyl glycerol choline phosphotransferase. The same observation should also be valid for ethanolamine phosphotransferase.

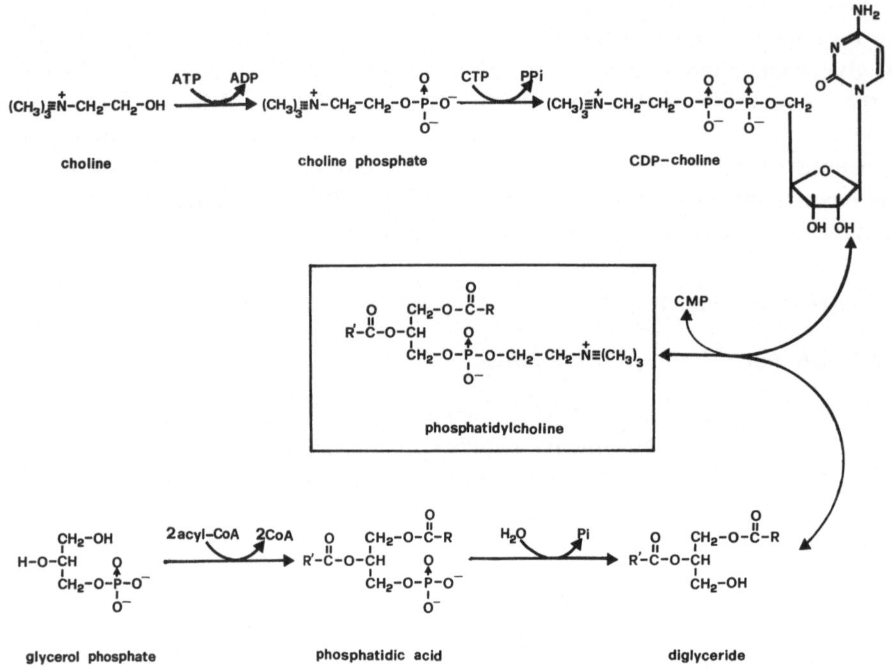

Fig. 2.   Net synthesis of phosphatidylcholine.

of microsomal alkaline phosphatase. The cytosolic V_{max} of the enzyme is 1.18 ± 0.11 μmol/g fresh tissue per h [54] with a K_M for choline of 2.2 mM, whereas the V_{max} in intact synaptosomes is 0.14–0.23 μmol/g fresh tissue per h. However, if cysteine (which inhibits alkaline phosphatase) is added, the activity in intact synaptosomes becomes similar to that of the cellular supernatant.[54]

The cytidylyltransferase is loosely bound to microsomal membranes and is easily released in the supernatant during the homogenization of the tissue and the preparation of subcellular fractions. This enzyme can be separated from the analogous one that forms CDP-ethanolamine.[55] In vitro, the formation of CDP-choline is the slowest step in the entire pathway,[52] and it is also possible that in vivo the rate of CDP-choline formation controls the velocity of the net synthesis of choline glycerophospholipid. The reaction can be stimulated in peripheral nerve and brain by the addition of some phospholipids (especially lysophospholipids) to the incubation mixture.[52,56] An example of this stimulation in reported in Table VII.

The final step in the biosynthesis of choline glycerophospholipid is the Mg^{2+}- or Mn^{2+}-dependent transfer of choline phosphate to the lipid acceptor.[57] The reaction is membrane bound, and the addition of diglyceride stimulates it[56]; moreover, its time dependence is biphasic if no diglyceride is added to the incubation mixture.[58] This finding means that there is a pool of readily disposable diglyceride at membrane level which is consumed in the first minutes of the reaction. The availability and/or mobility of diglyceride within membranes can therefore be another regulatory point of this pathway.

The preference of choline phosphotransferase for some diglyceride molecular species is debated. Technical difficulties add to the problem because when diglyceride is added to the incubation mixtures, the physical properties of the suspension (which depend also on diglyceride molecular species) could critically influence the availability of the lipid substrate to the enzyme. Therefore, a misleading assumption on enzyme specificity can be obtained from adding diglyceride to the incubation mixture. After the preincubation of the membranes with labeled glycerol-3-phosphate, no specificity towards diglyceride molecular species could be observed[59] on the basis of the specific radioactivity of diglyceride and choline glycerophospholipid. On the other hand, tetraenoic molecular species are the most abundant ones in diglyceride but not in phosphatidylcholine. It is therefore probable that the fatty acid composition of phosphatidylcholine is adjusted by means of interconversion reactions.[60] This point has not been completely cleared up.

Synaptic vesicles and synaptic microsomes possess a choline phosphotransferase activity that is influenced by a variety of neurotransmitters and by cyclic AMP.[61]

The net synthesis of choline glycerophospholipids has been studied in separated neuronal and astroglial cells and is present in both cell populations,[62]

Table VII
The Stimulation of CDP-Choline Synthesis in
Chicken Nerve Supernatant Fraction by Lipid
Material[a]

Lipid added	Activity[b]
None	15.4
Peripheral nerve lipids	
1,2-Diacyl-GPE[c]	14.6
1,2-Diacyl-GPC[d]	22.6
1-Acyl-GPC[d]	82.1
1,2-Diacyl-GPS[e]	16.4
Natural phospholipid preparations	
1,2-Diacyl-GPE[c] (egg lipids)	15.1
1,2-Diacyl-GPC[d] (egg lipids)	25.9
1,2-Diacyl-GPC[d] (soybeans lipids)	28.6
1-Acyl-GPC[d] (egg lipids)	59.1
Synthetic phospholipids	
1,2-Dipalmitoyl-GPE[c]	14.7
1,2-Distearoyl-GPE[c]	15.7
1,2-Dioleoyl-GPC[d]	28.5
1-Stearoyl-GPE[c]	60.5
1-Myristoyl-GPC[d]	59.4

[a] From reference 52.
[b] Activity is expressed as nmol CDP-choline/mg supernatant protein per hr.
[c] GPE, *sn*-glycero-3-phosphorylethanolamine.
[d] GPC, *sn*-glycero-3-phosphorylcholine.
[e] GPS, *sn*-glycero-3-phosphorylserine.

although neurons are more active than glia, and the kinetic properties of the enzyme appear to be different in the two cases.[63]

A decrease of choline phosphotransferase activity has been noticed in the neurons of aged rats.[64] The activity of choline phosphotransferase is lower in aged rats if no diglyceride is added to the incubation mixture; however, only the molecular species of diglyceride vary with aging, the total amount of these lipids being unaffected. It is therefore possible that the fatty acid pattern of the lipid acceptor influences the activity.[65]

An interesting aspect of choline phosphotransferase reaction is its reversibility,[66] a property shared also by ethanolamine phosphotransferase.[67,68] The addition of CMP to incubation mixtures produces CDP-choline and 1,2-diacyl glycerol which can be hydrolyzed by a lipase.[69] This mechanism could be implicated in the release of free fatty acids; moreover, the diglyceride formed by the reverse action of choline phosphotransferase could also be utilized for the formation of new lipid molecules, thus representing a system to obtain glycerophospholipid interconversions at membrane level (see also Section 5).

The regulation of choline phosphotransferase activity and its specificity towards the lipid acceptor are still little understood, and the presence of two isoenzymes differently affected by modulators[61] may further complicate this problem.

Choline phosphotransferase has been reported to be stimulated by oleic acid.[70] The addition of diacyl glycerols or alkylacyl glycerols to the incubation mixture and the use of different fatty acids can, however, inhibit the reaction[71]; moreover, ethanolamine phosphotransferase is activated by stearic, oleic, and linolenic acids when no lipid acceptors are added to the incubation mixture and is inhibited in the opposite case.[71]

Net synthesis of choline glycerophospholipid also occurs in peripheral nerve[72]; however, this process is limited to the myelin sheath surrounding the axon.[46–48] Choline phosphotransferase is present in Schwann cells and is not transported by axonal flow.[48] On the other hand, choline glycerophospholipids are transported by axonal flow.[73]

3.2.1b. Base Exchange. Base-exchange reactions form choline glycerophospholipid through a substitution of a lipid-bound base with the free choline present in the incubation medium (Fig. 3).[74]

The reaction is Ca^{2+} dependent and does not require energy; it is therefore noticeably different from net synthesis which is Mg^{2+} dependent and requires

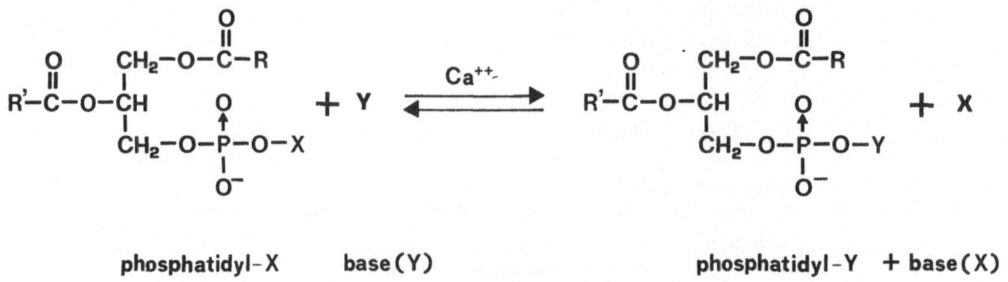

phosphatidyl-X   base(Y)   phosphatidyl-Y + base(X)

Fig. 3.   Base-exchange reaction.

ATP and CTP. The rate of the base-exchange reaction is estimated by measuring the labeling of membrane phospholipids after incubation with a free base. Base exchange has been thought to represent a reversal of phospholipase D action; on the other hand, base exchange and phospholipase D have different properties,[75,76] and the addition of cabbage phospholipase D, phosphatidic acid, and diglyceride to the incubation mixture[77] does not affect its rate, at least for ethanolamine and serine.

Ca^{2+} is an essential factor for the reaction,[78] but its optimal concentration depends on the pH; conversely, the optimal pH depends on the concentration of Ca^{2+}. Base exchange is a membrane-bound activity.[77,79] Microsomes are the most active cellular subfraction, but synaptosomes are also able to base exchange.[80] This system is probably present only in neurons,[80] at least for serine and ethanolamine, although an oligodendroglial activity has also been reported.[81]

De Medio *et al.*[82] have studied the fatty acid composition of the 1,2-diacyl-*sn*-glycero-3-phosphorylcholines that become labeled after the incubation of microsomes with labeled free choline. The incorporation of the base into tetraenoic species is six times as fast as into saturated species. Intermediate values were obtained for other molecular species, but these values depend on both pH and Ca^{2+} concentration.

Two types of base exchange are possible: homologous and nonhomologous. In the first case, free labeled choline would replace a nonlabeled lipid-bound choline; in the second case, free labeled choline would replace another lipid-bound base such as ethanolamine or serine. If only homologous base exchange were present, membrane lipids would not be modified by the reaction. Gaiti *et al.*[83] labeled microsomal membranes by preincubating them with labeled ethanolamine and serine in base-exchange conditions. These authors could successively displace the label with cold choline, demonstrating that nonhomologous base exchange can occur. The same authors[83] could also observe that not all membrane phosphoglycerides are substrates for base-exchange reactions, but only a small pool of them (about 5% of total glycerophospholipids).

The study of base-exchange reaction *in vivo* is particularly difficult because of the presence of other biosynthetic pathways. This problem has been approached with different techniques.[84,85] Sacrificing rats a few seconds after the intracerebral injection of labeled choline, Arienti *et al.*[84] observed that the specific radioactivity of CDP-choline increased steadily with time, whereas choline phosphoglyceride labeling was biphasic, showing a very fast incorporation of label in the first seconds. These authors[84] concluded that some choline glycerophospholipid became labeled without the intervention of CDP-choline. These findings have later been confirmed by Orlando *et al.*[85] who calculated the velocity constants for the incorporation of choline phosphate into choline glycerophospholipids after the intrathecal injection of labeled choline and labeled phosphate. The constant calculated from ^3H data exceeded, at least in the first minutes after the injection, the constant calculated from ^{32}P data. This fact indicates that some choline glycerophospholipid does not require the formation of CDP-choline for its synthesis. The same authors[85] calculated that base exchange *in vivo* is some five times as fast as net synthesis for the incorporation of choline into choline glycerophospholipids. However, it should be noticed that base exchange is probably underestimated because of the contemporary presence of net synthesis.

Base exchange is regulated by the pH and the concentrations of Ca^{2+} and free bases; however, it is possible that other factors influence the reaction. Calmodulin, for instance, has a stimulatory effect on the exchange of choline.[86]

Glycerophospholipids are asymmetrically distributed across the membranes,[33] and base exchange might be operative on one side of the membrane or on both. Buchanan and Kanfer[79] have found that choline exchange takes place predominantly on the cytoplasmic side of microsomal vesicles.

3.2.1c. N-Methylation. Choline glycerophospholipids can be formed by the N-methylation of ethanolamine lipids in nervous tissues.[87,88] This pathway involves the sequential transfer of three methyl groups from S-adenosyl-L-methionine to ethanolamine phosphoglyceride with the intermediate formation of monomethyl- and dimethylethanolamine glycerophospholipids.

The methyltransferase activity is localized in microsomes and synaptosomes.[89] The reaction is catalyzed by more than one enzyme, at least in synaptosomes.[89] The first enzyme forms N-monomethylethanolamine lipids, has a pH optimum of 7.5, a low K_M for S-adenosylmethionine, a partial requirement for Mg^{2+}, and is tightly bound to membranes. The second methyltransferase catalyzes the methylation of monomethyl- and dimethylethanolamine glycerophospholipids to the corresponding choline phosphoglycerides. It has a pH optimum of 10.5, a high K_M for S-adenosylmethionine, does not require Mg^{2+}, and is solubilized by sonication.[89]

The N-methylation is a rather slow pathway in the brain, and its detection has been a recent accomplishment.[87-89] Although this pathway does not constitute a major system for choline glycerophospholipid biosynthesis in nervous tissues, it may be important for the modification of restricted pools of membrane phospholipids. Choline and ethanolamine glycerophospholipids are differently localized across plasma membranes, ethanolamine phosphoglycerides (together with serine phosphoglycerides) largely facing the cytoplasmic side and choline phosphatides being mainly oriented towards the outside of the membrane surface.[90] The two methyltransferases are differently localized across synaptosomal plasma membranes,[91] and the methylation of ethanolamine lipids begins on the cytoplasmic side. The conversion of ethanolamine lipids to choline phosphoglycerides can alter some properties of the membrane that may be important for the functional pleiomorphism associated with nerve cell plasma membrane. It is possible that this reaction is implicated in the transmission of signals at synaptic level.

3.2.2. Ethanolamine Glycerophospholipids

3.2.2a. Net Synthesis. Ethanolamine glycerophospholipids are one of the major phospholipids in brain. In contrast to choline glycerophospholipids, whose plasmalogens are scarcely represented in nervous tissues, ethanolamine plasmalogens are abundant, especially in the myelin sheath. This section is mainly concerned with the biosynthesis of diacyl species, since plasmalogens and alkylacyl ethanolamine glycerophospholipids are treated separately in another section (Section 3.2.4).

The net synthesis of ethanolamine phosphoglyceride closely resembles that already described for the analogous choline-containing compounds (Section 3.2.1a and Fig. 2). In this case as well, the pathway consists of three reactions: (1) the formation of ethanolamine phosphate (or phosphorylethanolamine) from ethanolamine and ATP (ethanolamine:ATP phosphotransferase; E.C. 2.7.1.82); (2) the synthesis of cytidine-5'-diphosphate ethanolamine (CDP-ethanolamine) from ethanolamine phosphate and CTP (CTP:ethanolamine phosphate cytidylyltransferase; E.C. 2.7.7.14), and (3) the final transfer of ethanolamine phosphate from CDP-ethanolamine to the 1,2-diacyl glycerol (CDP-ethanolamine:1,2-diacylglycerol ethanolamine phosphotransferase; E.C. 2.7.8.1). Although this pathway is similar to that forming choline glycerophospholipids, it should be noticed that all the enzymes forming ethanolamine phosphoglycerides are different from those synthesizing choline lipids.

The pathway was first detected in brain *in vitro* by McMurray *et al.*[50] and *in vivo* by Ansell and Spanner.[92] A detailed study of its subcellular localization was first carried out by Porcellati *et al.*[93,94] The overall transfer of ethanolamine to ethanolamine glycerophospholipids is rather low, and the limiting step appears to be the formation of CDP-ethanolamine through the cytidylyltransferase reaction[93,94] as in the case of choline phospholipid synthesis.

The phosphorylation of ethanolamine has been particularly well studied in the synaptosomal fraction.[95] For optimal activity, the enzyme requires 30 mM Mg^{2+}-ATP and is not inhibited by hemicholinium-3 which inhibits the choline kinase. The enzyme is localized in the soluble portion of the cell. Interestingly, the cytidylyltransferase catalyzing the formation of CDP-ethanolamine appears to be loosely bound to microsomal membranes,[93] and its activity is low compared to that of other enzymes of this pathway.[96]

The transfer of ethanolamine phosphate to the lipid acceptor (ethanolamine phosphotransferase reaction) is the last step of ethanolamine glycerophospholipid net synthesis. The enzyme is microsomal and Mg^{2+} dependent[94]; however, some activity is also present in myelin[97] and disrupted synaptosomes.[61] The presence of ATP, cyclic AMP, and some neurotransmitters can differently influence the formation of ethanolamine phosphoglyceride subclasses in synaptosomal membranes.[61] The K_M for CDP-ethanolamine is 2.5×10^{-4} M and for the lipid acceptor approximately 1.6×10^{-3} M.[98] However, the determination of the K_M for diradyl glycerols is far from simple: the K_Ms for the lipid acceptors are similar with the diacyl, alkylacyl, and 1-alk-1'-enyl-2-acyl subclasses, indicating a lack of enzymic specificity.[99] The specificity of ethanolamine phosphotransferase towards diglyceride molecular species has been studied by Roberti *et al.*[59] who found that hexaenoic species are converted into ethanolamine phosphoglycerides at the highest rate[59]; moreover, diacyl glycerols inhibit the formation of ethanolamine plasmalogen.[98]

The net synthesis of ethanolamine phosphoglyceride has also been studied using preparations enriched in astroglial and neuronal cells.[63,100] Neurons are more active than astroglia.

The ethanolamine and choline phosphotransferases of chicken brain microsomes are inhibited by pretreatment of the membranes with phospholipase A_2,[101] suggesting a lipid requirement for their activity.

3.2.2b. Base Exchange. Ethanolamine is incorporated into ethanolamine glycerophospholipids by a base-exchange reaction.[77] Ethanolamine exchange possesses a number of properties common to choline exchange (Section 1.2.1b). However, the pH optimum is lower for ethanolamine incorporation into lipids.[78] The K_M and the V_{max} of the reaction appear to be influenced by the concentrations of Ca^{2+} [78]; moreover, the enzyme is saturated by two different concentrations of ethanolamine (0.2 and 2 mM).

Ethanolamine is incorporated into 1,2-diacyl-, 1-alkyl-2-acyl-, and 1-alk-1'-enyl-*sn*-glycero-3-phosphorylethanolamine. Diacyl species incorporate more radioactivity than other species after the incubation with labeled ethanolamine.[78] This difference is more evident at low ethanolamine concentrations.[78]

Ethanolamine exchange involves only a small pool of membrane glycerophospholipids (5–10%)[102] which may open interesting questions about the physiological significance of the reaction. Moreover, base exchange could be present on only one side of microsomal membranes.[79] Buchanan and Kanfer[79] proposed that this activity is localized on the luminal side of membranes because a preincubation with trypsin did not affect ethanolamine exchange, whereas choline exchange was inhibited by the same treatment. More complete data are, however, necessary to ascertain whether or not the activity is localized on one side of microsomal vesicles because the ethanolamine base-exchange enzyme could be a poor substrate for trypsin.

Ethanolamine exchange, similarly to serine exchange, is more active in neurons than in astroglia, and it has been proposed that the glial activity reflects neuronal contamination.[80]

The presence of base exchange *in vivo* has been studied using different approaches. After the intracerebral injection of free, labeled ethanolamine, the radioactivity of ethanolamine glycerophospholipids increased with a steeper slope in the first seconds than afterwards, whereas the labeling of CDP-ethanolamine increased almost linearly with time.[103] This finding would indicate that ethanolamine is incorporated into lipids through a pathway that does not imply the intermediate formation of CDP-ethanolamine. Moreover, the specific radioactivity of the molecular species that become labeled varied with the time elapsed from the injection of the base.[103] Orlando *et al.*[104] calculated the kinetic constants for the incorporation of ethanolamine phosphate into ethanolamine phosphoglyceride after the intracisternal injection of ^{32}P-labeled phosphate and [^{3}H]ethanolamine. In the first minutes after the administration of the precursors, the constants calculated from ^{3}H data exceeded the ones calculated from ^{32}P data, indicating that ethanolamine can be incorporated into rat brain lipids *in vivo* through a pathway that is not net synthesis. For these reasons, ethanolamine base exchange seems to occur *in vivo*.

3.2.2c. Other Pathways. In some tissues, ethanolamine phosphoglyceride can be obtained from the decarboxylation of serine phosphoglyceride. Frog retina seems able to produce ethanolamine glycerophospholipids by this pathway,[105] which appears to be very active in this tissue. A decarboxylation of serine phosphoglyceride has also been reported to occur in rat brain cells in culture.[106]

3.2.3. Serine Glycerophospholipids

A net synthesis of serine glycerophospholipid has never been reported for animal tissues. Both CDP-serine and CDP-diglyceride are not intermediates of serine glycerophospholipid biosynthesis. However, the presence of an ATP-stimulated incorporation of L-serine into serine glycerophospholipid has recently been claimed to occur in rat brain microsomes.[107] This system is stimulated by the addition of Ni^{2+} and is inhibited by sulfhydryl reagents that do not affect base-exchange reactions. Pyrophosphatidic acid [p-p'-bis'4-(1,2-diacyl-*sn*-glycero-3-)pyrophosphate] has been proposed as an intermediate of the reaction.

A pathway by which serine phosphoglyceride can be synthesized in nervous tissue is base exchange.[77] Serine exchange shares many properties with choline and ethanolamine exchanges. The solubilized enzyme has been studied to some extent by Taki and Kanfer[108] who found that ethanolamine lipids (especially plasmalogen) are the best lipid acceptors for serine, thus confirming the biosynthetic role of serine exchange. The K_M for ethanolamine glycerophospholipid ranged from 0.25 to 0.66 mM, depending on the source and the type of ethanolamine lipid employed.[108]

Serine exchange is more abundant in neurons than in astroglia[80] and should be present on the luminal side of membranes.[79]

3.2.4. Ether Lipids

3.2.4a. Plasmalogens. The most concentrated source of plasmalogen in CNS is the myelin sheath. The value of the mole ratio of alk-1'-enyl groups to lipid phosphorus is in the range of 0.31–0.37, whereas in the gray matter, this ratio is about 0.17.[109] Almost all of alk-1'-enyl groups belong to ethanolamine plasmalogen; serine and choline plasmalogens are indeed rather scarce in nervous tissues. 1-Alkyl-2-acyl-*sn*-glycero-3-phosphorylethanolamine represents only 2–3% of total phospholipids in mammalian brain.

Both base exchange and net synthesis are able to form ether lipids in brain, provided that the ether linkage is contained into the lipid acceptor.[110] The synthesis of both plasmalogens and alkylacyl glycerophospholipids by the phosphotransferase reaction is stimulated by the addition of 1-alkyl-2-acyl-*sn*-glycerol to the incubation mixtures.[111] Choline phosphotransferase is very active towards plasmalogenic diglyceride, which could explain the high turnover rate of choline plasmalogen.[62,111] Neuronal perikarya are more active than synaptosomes and astroglia in the synthesis of ether lipids through phosphotransferase reactions. Choline plasmalogen can be synthesized by the sequential methylation of the homologous ethanolamine compounds.[112]

One of the major problems encountered in the explanation of ether lipid biosynthesis is the formation of the ether bond. Radominska-Pyrék and Horrocks[99] suggested that plasmalogen could be formed through the oxidation of alkylacyl lipids. The biosynthesis of ethanolamine plasmalogen requires the presence of 1-alkyl-2-acyl-*sn*-glycero-3-phosphorylethanolamine, molecular oxygen, and NADH and is inhibited by KCN.[113] This system is a mixed-function oxidase, and cytochrome b_5 participates in the reaction[114] which is specific for the *sn*-1 position of glycerol.[115] It is apparently active only on ethanolamine lipids; choline compounds are not substrates for the reaction.[116] Therefore, choline plasmalogen should be formed either from the plasmalogenic diglyceride released from ethanolamine plasmalogen or through other mechanisms (base exchange or N-methylation).

The biosynthesis of choline plasmalogen through the phosphotransferase reaction remains, in addition, somewhat obscure because the presence of 1-alk-1'-enyl-2-acyl-*sn*-glycerol in biological membranes is open to question.

It is also possible that the lipid on which the vinyl–ether linkage is formed is not the 1-alkyl-2-acyl-*sn*-glycero-3-phosphorylethanolamine but the 1-alkyl-*sn*-glycero-3-phosphorylethanolamine.[117]

Although the desaturation of ethanolamine alkylacyl lipids is a microsomal reaction, a factor able to stimulate the synthesis of this lipid has been found in the cell supernatant of various organs including brain.[118]

3.2.4b. Alkylacyl Lipids. It is now well established that fatty alcohol is the precursor of the alkyl and alk-1'-enyl groups of mammalial phosphoglyceride.[119] Fatty alcohols can be obtained from the reduction of the corresponding fatty acids and are incorporated into alkyl dihydroxyacetone phosphate[120] through a Mg^{2+}-dependent substitution of the fatty acid of acyl dihydroxyacetone phosphate with a fatty alcohol.[121] The pathway continues with the reduction of the carbonyl group, yielding 1-alkyl-glycerol-3-phosphate which is subsequently acylated to 1-alkyl-2-acyl-glycerol-3-phosphate. The dephosphorylation of this compound produces 1-alkyl-2-acyl-glycerol[122] which can function as lipid acceptor for the ethanolamine phosphotransferase reaction (Fig. 4).

Although many different fatty alcohols can be incorporated into alkylacyl glycerophospholipids when they are administered to the animal, only saturated or monoenoic alcohols are present in natural lipids. This fact seems to depend on the specificity of the system that desaturates fatty acids and not on the subsequent reactions introducing fatty alcohols into alkyl dihydroxyacetone phosphate.[123] Indeed, in the brain, only palmitic, stearic, and oleic acids are reduced to the corresponding alcohols, whereas the polyunsaturated fatty acids (linoleic, linolenic, and arachidonic) and the short-chain fatty acids (myristic and lauric) are not.[124]

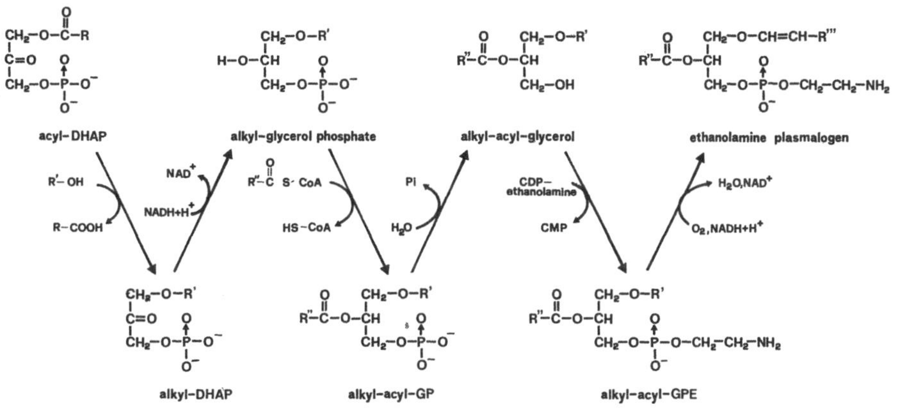

Fig. 4. Biosynthesis of ethanolamine plasmalogen. DHAP, dihydroxyacetone phosphate; GPE, glycero-3-phosphorylethanolamine; GP, glycerol phosphate.

From studies employing ^{3}H-labeled glucose in various positions, it has been concluded that ether lipids are synthesized only by the acyl dihydroxyacetone phosphate pathway and not through the glycerol-3-phosphate one.[124] The alkyl dihydroxyacetone phosphate synthase has been found associated with microsomal or mitochondrial membranes depending on the organ examined. Hajra *et al.*[124] found that this activity in the liver is associated with microbodies and postulated that the enzyme is microbody bound in every organ, and its appearance in microsomal or mitochondrial fractions depends on the different sedimentation characteristics of microbodies from various organs.

4. GLYCEROPHOSPHOLIPID BREAKDOWN

4.1. Phospholipase A_1

Glycerophospholipids can be hydrolyzed in nervous tissues where several phospholipases are present. Their action covers at least two aspects of glycerophospholipid metabolism: (1) the interconversion of phosphoglyceride molecules (in this case, the hydrolytic removal of a part of the lipid molecule is followed by resynthesis) and (2) a true catabolic aspect (in this case, the action of a phospholipase is followed by a further hydrolysis of the phospholipid). A review on intracellular phospholipases A (E.C. 3.1.1.4) has recently been published.[125] In this section, only those aspects peculiar to nervous system are considered.

The presence of phospholipase A activity in brain was recognized several years ago,[126] and both the A_1 and A_2 activities have been adequately demonstrated.[127] Brain appears to contain several types of phospholipase A_1. Cooper and Webster,[128] studying brain extracts, described a phospholipase A_1 that differs from the A_2 with respect to the fatty acid released (the A_1 cleaves the fatty acid from the position 1-*sn*, the A_2 from the position 2*sn*), pH optimum, stability to acidic conditions, and thermolability. The A_1 enzyme is stimulated by detergents and has an optimum in the range of pH 4–5. It does not require Ca^{2+},[129] and its activity is higher in human cerebral cortex and lowest in sciatic nerve.[128] Cerebral gray matter is slightly more active than white matter.[128]

A phospholipase A_1 that has a pH optimum of 6.4 and is bound to microsomal membranes has also been described.[130,131] This enzyme does not require Ca^{2+} and hydrolyzes choline better than ethanolamine or serine glycerophospholipids.[130]

A third phospholipase A_1 has a pH optimum of 9.3, is inhibited by detergents, and is activated by Ca^{2+}.[132] This enzyme is present in the soluble portion of the cell or is loosely bound to membranes from which it can be detached on homogenization of the tissue.[133]

Phospholipase A_1 has also been studied in separated neuronal and astroglial cells.[134,135] Neuronal phospholipase A_1 is more active than the corresponding glial enzyme; moreover, optimal activities were found at pH 7.2 in neurons and pH 5.4 in the glia, which probably indicates the presence of different phospholipases A_1 in the two cellular types. The nature of the fatty acid bound to the *sn*-1 position influences the rate of hydrolysis by neuronal phospholipase A_1, with saturated fatty acids being better released.[134] However, the fatty acid bound to the position *sn*-2 does not affect the rate of hydrolysis, whereas plasmalogens and alkylacyl lipids competitively inhibit the enzymic activity.[135]

Neuronal and glial plasma membranes contain a phospholipase A_1 activity.[136] In this case too, neurons are more active than astroglia, although the pH optima are similar (8.0–9.0); however, the enzymes from glia and neurons have different specificities towards the substrate.

4.2. Phospholipase A_2

Phospholipase A_2 catalyzes the removal of the fatty acid from the position *sn*-2 of glycerophospholipids. The enzyme is present in brain mitochondria,[136,137] and its pH optimum is 8.4. Therefore, the phospholipases A_1 and A_2 differ in their subcellular distribution, because the A_1 activity has been found in microsomes and cytosol.[130,133] Moreover, synaptosomes also contain an A_2 activity localized in synaptic vesicles and intraterminal mitochondria,[130] whereas the A_1 is localized in synaptic microsomes and ghosts.[131]

The activity of phospholipase A_2 is two- to threefold higher in neurons than in astroglia, and in both cases the enzyme(s) more rapidly hydrolyzes diacyl than ether glycerophospholipids.[137] Glial and neuronal activities differ in their pH optima which are 5.4 and 8.0, respectively.[134] This finding may suggest the presence of more than one phospholipase A_2 in the brain.

The activity of phospholipase A_2 can be connected with the release of polyunsaturated fatty acids necessary for the synthesis of prostaglandins. Neuronal phospholipase A_2 seems to release arachidonic acid faster than linoleic acid from both choline and ethanolamine glycerophospholipids, whereas arachidonic acid is removed less actively from ethanolamine glycerophospholipids by the glial enzyme.[137]

The action of norepinephrine on cerebral phospholipase A_2 is not completely clear, because the neurotransmitter injected intraventricularly into rabbit brain stimulates the hydrolysis of all glycerophospholipid classes in both glia and neurons.[137]

4.3. Other Phospholipases

Nervous tissues are claimed to contain phospholipases acting on phosphodiester bonds. Phospholipase C releases the phosphoryl base and the diglyceride, whereas phospholipase D liberates the free base.

Brain phospholipase(s) C (E.C. 3.1.4.3) is active towards several lipids such as phosphatidylinositol,[138] sphingomyelin,[139] phosphatidylethanolamine,[140] and phosphatidylcholine.[141]

The release of ethanolamine phosphate from ethanolamine glycerophospholipids has an optimum at pH 6.5 and does not require divalent cations.[140] The properties of this enzyme distinguish its action from a release of diglyceride via reversal of the phosphotransferase reaction.[67,68] The enzyme is present in both microsomes and mitochondria.[140] A phospholipase C active towards choline glycerophospholipids has been claimed to be active in several rat tissues including brain.[141]

Phospholipase D catalyzes the removal of the free base from a glycerophospholipid-forming phosphatidic acid. Its detection in nervous tissues is a rather recent accomplishment.[142] The enzyme is membrane bound and has been solubilized from rat brain.[142,143] The solubilized activity is stimulated by both Ca^{2+} and Mg^{2+} and has different pH optima for the hydrolytic and the transphosphatidylation reactions.[75] The enzyme is active on choline and ethanolamine glycerophospholipids with K_Ms of 0.75 and 0.91 mM, respectively.

Similarities exist between phospholipase D and base-exchange reactions. Both systems are stimulated by Ca^{2+}, are membrane bound, and act on the diester linkage between the base and the phosphoric acid moiety of glycerophospholipids. Moreover, phospholipase D can catalyze transphosphatidylation reactions.[144] However, several differences exist between base exchange and phospholipase D. The phospholipase D-catalyzed transphosphatidylations have acidic pH optima and require higher Ca^{2+} and base concentrations than base exchange; moreover, a solubilized preparation of phospholipase D has been reported to be devoid of base-exchange activity.[145] Brain phospholipase D has been partially purified and separated from base exchange.[145] Its molecular weight is about 200,000, and the pH optimum 6. Both Ca^{2+} and Fe^{2+} have a stimulatory action on the reaction, but the activity does not depend on the presence of divalent cations.

4.4. Lysophospholipase

Lysophospholipases hydrolyze lysophospholipids (1-, or 2-acyl glycerophospholipid) to fatty acid and glycerophosphoryl base. Their action usually follows the action of a phospholipase A and allows the further breakdown of phospholipids.

Lysophospholipase A (E.C. 3.1.1.5) is present in brain[146] and is distributed among several subcellular fractions. The enzyme(s) of different subfractions can be distinguished by different kinetic parameters and sensitivities to the activating action of albumin. The mitochondrial and microsomal enzyme is inhibited by an excess of substrate.[147] Probably, this is because of absorption of lysophosphatidylcholine molecules or micelles onto the particles or microsomes.[147] The enzyme is not active towards lysoplasmalogens because it cannot hydrolyze the vinyl–ether linkage of these lipids. However, these compounds can be hydrolyzed by a lysophospholipase D which releases the free base.[148,149] This reaction is inhibited by Ca^{2+} and activated by Mg^{2+}, which distinguishes it from base exchange. The enzymic activity is minimal when the sn-2 position is acylated, and it has a pH optimum of 7.2.

4.5. Plasmalogenase

Phospholipase A_1 and lysophospholipase A cannot hydrolyze the vinyl–ether bond of plasmalogens. On the other hand, ethanolamine plasmalogens have a substantial rate of turnover in the brain.[111] The first description of a specific enzyme (E.C. 3.3.2.2) capable of hydrolyzing the vinyl–ether linkage was reported by Warner and Lands.[150] In the brain, the enzyme is particularly active in oligodendroglia,[151] choline and ethanolamine plasmalogens are both substrates of the reaction,[152] although ethanolamine lipids are better hydrolyzed.

Diacyl glycerophospholipids are competitive inhibitors of plasmalogenase which has a pH optimum of 7.4[153] and is preferentially localized in mitochondria and microsomes.[154] In contrast to other enzymes of lipid metabolism, plasmalogenase is much more abundant in glia than in neurons. Astroglia are almost two times as active as neurons, and oligodendroglia are about seven times as active as astroglia.[151] On the other hand, myelin has no plasmalogenase activity.

5. GLYCEROPHOSPHOLIPID INTERCONVERSION

Although glycerophospholipids can be synthesized *de novo* from simpler precursors in nervous tissues, an already formed phospholipid can be modified by the so-called interconversion reactions. These reactions have already been partly described in this chapter in Sections 3 and 4. Interconversion reactions are responsible for the different turnover rates of the various portions of glycerophospholipids.[3]

Lysophosphatidylcholine generated by phospholipase A action can be reacylated in brain.[155] The reaction requires Mg^{2+}, CoA, and ATP and is present in the microsomal fraction.[156] The pathway consists of two steps: (1) the fatty acid activation to form an acyl-CoA (acyl-CoA synthetase; E.C. 6.2.1.3) and (2) the acyltransferase reaction (acyl-CoA:1-acylglycero-3-phosphorylcholine O-acyl transferase; E.C. 2.3.1.23). The acyl-CoA synthetase appears to be the limiting step of the pathway *in vitro*.[157] The acylation of lysophosphatidylcholine may occur at either the 1 or 2 position, depending on the lysophospholipid present in the incubation mixtures. However, the acylations in positions 1 and 2 differ somewhat in specificity. The system acylating position 1 does not distinguish between oleoyl-CoA and *cis*-vaccenoyl-CoA as acyl group donors, whereas the acylation in position 2 prefers oleoyl-CoA.[157] It is impossible to say if this fact implies the presence of more than one acylating enzyme.

The acylation rates of 1-acyl-*sn*-glycero-3-phosphorylcholine are quite comparable with acyl-CoA possessing two to four double bonds.[158] No special selectivity towards arachidonoyl-CoA is present *in vitro*, and this may be consistent with the low levels of arachidonate in brain choline phosphoglycerides.[60] However, although oleic acid is one of the major fatty acids of choline glycerophospholipids, oleoyl-CoA is a poor acyl group donor.

The acylation of lysophosphatidylserine is known to occur in brain[159] and to have an optimum of activity at pH 9. This system seems to possess some specificity towards long-chain unsaturated acyl-CoA and may be active in the regulation of the fatty acid profile of serine phosphoglyceride. However, although docosahexaenoic acid is one of the major unsaturated fatty acids of brain phosphatidylserine, its thioester is a poor acyl donor.[159]

Deacylation–reacylation reactions are generally believed to be a system for the regulation of the fatty acid composition of brain glycerophospholipid, especially in the case of choline phosphoglyceride.[60] However, it is difficult to establish how the fatty acid profile of brain phospholipid is regulated, because many reactions can affect it. Base exchange reactions, N-methylation of ethanolamine glycerophospholipids, the reverse of phosphotransferase reactions, and so on are all systems capable of influencing the fatty acid composition of brain glycerophospholipid classes. For this reason, this subject still awaits precise answers.

6. GLYCEROPHOSPHOLIPID EXCHANGE

Subcellular fractions often used to investigate glycerophospholipid metabolism in the brain are never absolutely pure; indeed, every type of separated membrane always contains some contaminants from other membranes. Moreover, brain is a complex structure formed by several cell types, and therefore, terms such as "microsomal membranes" indicate a fraction enriched in membranes of the endoplasmic reticulum of various cell types containing, as impurities, proportions of other membranes such as myelin or plasma membranes.

The specific activities of several enzymes of lipid metabolism increase in microsomal membranes. The question arises whether only microsomes can form glycerophospholipids or whether this property is shared by other membranes. Phosphatidic acid can be formed in either mitochondria or microsomes,[160] whereas the data reported in the literature disagree about the site of ethanolamine and choline glycerophospholipid biosynthesis.

Miller and Dawson,[161] mixing mitochondria with microsomes and extrapolating at zero microsomal concentration, concluded that only microsomes are able to form *de novo* choline and ethanolamine glycerophospholipids; on the other hand, Pasquini *et al.*[162] reported that sulfatides are transported intracellularly among organelles, whereas ethanolamine and choline glycerophospholipids are not. Clearly, this is a controversial field.

A cytoplasmic factor able to carry phospholipids and to redistribute them among subcellular organelles has been described by Wirtz and Zilversmit,[163] and this protein has been detected also in brain.[164]

An exchange of phosphatidylcholine between microsomes and myelin has been reported using the so-called 5.1 supernatant protein.[165] The exchange protein is probably the same for choline and inositol lipids; however, in brain there is a preference for phosphatidylinositol exchange, whereas in liver, phosphatidylcholine is preferentially exchanged.

7. REGULATION OF GLYCEROPHOSPHOLIPID METABOLISM

One of the major problems of phospholipid metabolism in nervous tissues is its regulation; unfortunately, little is known on this subject. The pathways leading to glycerophospholipid formation and breakdown are several, and the contribution of any single metabolic reaction is sometimes little understood. Another aspect that hinders a clear comprehension of the regulation of lipid metabolism is that glycerophospholipids might participate, through metabolic modifications, in functional events in nervous system. For this reason, metabolic pathways may have completely different meanings, and thus a pathway that is very important for some nervous function is of scarce relevance in the formation or hydrolysis of lipid molecules.

The net synthesis of ethanolamine and choline glycerophospholipids could be regulated at several levels. The formation of the CDP derivatives is the limiting step of the pathway *in vitro*[52] and is stimulated by lysophospholipids. Choline and ethanolamine phosphotransferases are adversely affected by CMP[67,68]; moreover, CDP-choline inhibits the formation of ethanolamine lipids, and CDP-ethanolamine the synthesis of choline glycerophospholipids. Various fatty acids may also influence this reaction,[71] and in synaptosomes, neurotransmitters and cyclic AMP can modify the rates of the phosphotransferases.[61] It is possible that either the lipid composition of the membrane or stimulation of the cell influences the rate of choline and ethanolamine glycerophospholipid biosynthesis. The breakdown of glycerophospholipids can also be influenced by neurotransmitters, with phospholipase A_2 being affected by norepinephrine.[136]

The availability of diglyceride can also be an important point in the regulation of phosphoglyceride metabolism. 1,2-Diacyl glycerols can be formed by several reactions, such as phospholipase C,[141] the reverse action of ethanolamine and choline phosphotransferases,[67,68] and the hydrolysis of phosphatidic acid.[166] Diglyceride can also be removed to form phospholipids, triglycerides, and phosphatidic acid[167] or hydrolyzed by diglyceride lipases.[168] Of course, the levels of membrane diglyceride could be determined by factors influencing any reactions of its metabolism.

Base-exchange reactions are apparently regulated by the concentration of the free bases; on the other hand, the lipid composition of membranes could also be important.[169] The availability of Ca^{2+} should also be relevant in this connection.

8. CONCLUSION

Glycerophospholipid is present in all living cells and is particularly abundant in nervous tissue (especially ethanolamine and choline phosphoglycerides). Its presence is necessary for the formation and functioning of biological membranes.

Mammalian nervous system can form glycerophospholipid from simpler molecules and hydrolyze it completely. Moreover, already formed lipids can be modified by interconversion reactions which make possible the transformation of a phospholipid molecule into another, changing or modifying its nitrogenous base or fatty acids.

Glycerophospholipid metabolism is probably very carefully regulated in nervous tissues; however, regulation mechanisms are not well known. It is possible that reactions of phospholipid metabolism play a part in physiological and pathological events occurring in the nervous system, and it is probable that this point will be the object of future research.

REFERENCES

1. Vaquelin, L. M., 1812, *Ann. Chim. (Paris)* **81**:37–45.
2. Brammer, M. J., 1982, *Handbook of Neurochemistry*, 2nd ed., Volume 5 (A. Lajtha, ed.), Plenum Press, New York (in press).
3. Porcellati, G., 1982, *Handbook of Neurochemistry*, 2nd ed., Volume 5 (A. Lajtha, ed.), Plenum Press, New York (in press).
4. IUPAC–IUB, 1968, *Biochim. Biophys. Acta* **152**:1–9.
5. IUPAC–IUB, 1970, *J. Lipid Res.* **11**:173.
6. IUPAC–IUB, 1967, *Biochem. J.* **105**:897–902.
7. Baer, E., and Stanacev, N. Z., 1964, *J. Biol. Chem.* **239**:3209–3214.
8. Ansell, G. B., Dawson, R. M. C., and Hawthorne, J. N. (eds.), 1973, *Form and Function of Phospholipids*, Elsevier, London.
9. Gaiti, A., Arienti, G., and Porcellati, G., 1977, *I Fosfolipidi*, Liviana, Padova.
10. Scott, T. W., Setchell, B. P., and Basset, J. N., 1967, *Biochem. J.* **104**:1040–1047.
11. Cuzner, M. L., and Davison, A. N., 1968, *Biochem. J.* **106**:29–34.
12. Owens, J., 1966, *Biochem. J.* **100**:354–361.
13. Eichberg, J., Whittaker, V. P., and Dawson, R. M. C., 1964, *Biochem. J.* **92**:91–100.
14. Sheltawy, A., and Dawson, R. M. C., 1966, *Biochem. J.* **100**:12–18.
15. Porcellati, G., 1969, *Handbook of Neurochemistry*, Volume 2 (A. Lajtha, ed.), Plenum Press, New York, pp. 393–342.
16. O'Brien, J. S., and Sampson, E. L, 1965, *J. Lipid Res.* **6**:537–544.
17. Wuthier, R. E., 1966, *J. Lipid Res.* **7**:544–550.
18. Cuzner, M. L., Davison, A. M., and Gregson, M. A., 1965, *J. Neurochem.* **12**:469–481.
19. Gurr, M. I., Prottey, C., and Hawthorne, J. N., 1965, *Biochim. Biophys. Acta* **106**:357–370.
20. Ishibe, T., and Yamamoto, A., 1979, *J. Neurochem.* **32**:1665–1670.

21. Siakotos, A. N., Rouser, G., and Fleischer, S., 1969, *Lipids* **4**:234–242.
22. Parsons, P., and Basford, R. E., 1967, *J. Neurochem.* **14**:823–840.
23. Biran, L. A., and Bartley, W., 1961, *Biochem. J.* **79**:159–176.
24. Herschowitz, M., McKhan, G. M., and Shooter, E. M., 1968, *J. Neurochem.* **15**:161–168.
25. Arienti, G., Goracci, G., and Porcellati, G., 1981, *Neurochem. Res.* **6**:729–742.
26. Porcellati, G., and Goracci, G., 1976, *Lipids* (R. Paoletti, G. Porcellati, and G. Iacini, eds.), Raven Press, New York, pp. 203–214.
27. Freysz, L., Bieth, R., Judes, G., Sensenbrenner, M., Jacob, M., and Mandel, P., 1968, *J. Neurochem.* **15**:307–313.
28. Freysz, L., and Mandel, P., 1970, *FEBS Lett.* **40**:110–113.
29. Farooq, M., Cammer, W., Synder, D. S., Raine, C. S., and Norton, W. T., 1981, *J. Neurochem.* **36**:431–440.
30. Crawford, C. G., and Wells, M. A., 1979, *Lipids* **14**:757–762.
31. Odutuga, A. A., Carey, E. M., and Prout, R. E. S., 1973, *Biochim. Biophys. Acta* **316**:115–123.
32. Gonzales-Ros, J. M., and Ribera, A., 1980, *Lipids* **15**::279–284.
33. Fontaine, R. N., Harris, R. A., and Schroeder, F., 1979, *Life Sci.* **24**:395–400.
34. Smith, A. P., and Loh, H. H., 1976, *Proc. West. Pharmacol. Soc.* **19**:147–151.
35. Freysz, L., Bieth, R., and Mandel, P., 1969, *J. Neurochem.* **16**:1417–1424.
36. Smith, M. E., and Eng. L. E., 1965, *J. Am. Oil Chem. Soc.* **42**:1013–1020.
37. Abdel-Latif, A., and Abood, L. G., 1965, *J. Neurochem.* **12**:157–166.
38. Miller, S., and Morell, P., 1978, *J. Neurochem.* **31**:771–777.
39. Horrocks, L. A., 1973, *Prog. Brain Res.* **40**:385–395.
40. Sun, G. Y., 1973, *J. Neurochem.* **21**:1083–1092.
41. Lapetina, E. G., Lunt, C. G., and De Robertis, E., 1970, *J. Neurobiol.* **1**:295–302.
42. Pasquini, J. M., Krawiec, L., and Soto, E. F., 1973, *J. Neurochem.* **21**:647–653.
43. Freysz, L., Farooqui, A. A., Adamkzewska-Goucerzewicz, Z., and Mandel, P., 1979, *J. Lipid Res.* **20**:503–508.
44. Chao, S. W., and Rumsby, M. G., 1978, *Proceedings of the European Society for Neurochemistry*, Volume 1 (V. Neuhoff, ed.), Verlag Chemie, Weinheim, p. 126.
45. Porcellati, G., and Mastrantonio, M. A., 1964, *Ital. J. Biochem.* **13**:332–352.
46. Droz, B., and Boyenval, J., 1975, *J. Microsc. Biol. Cell.* **23**:45–46a.
47. Gould, R. M., 1976, *Brain Res.* **117**:169–174.
48. Kumara-Siri, M. H., and Gould, R. M., 1980, *Brain Res.* **186**:315–330.
49. Larrabee, M. G., and Brinley, F. J., Jr., 1968, *J. Neurochem.* **15**:533–545.
50. McMurray, W. C., Strickland, K. P., Berry, J. F., and Rossiter, R. J., 1957, *Biochem. J.* **66**:634–644.
51. Ansell, G. B., and Spanner, S., 1972, *Current Trends in the Biochemistry of Lipids* (J. Ganguly and R. M. S. Smellie, eds.), Academic Press, London, pp. 151–159.
52. Porcellati, G., and Arienti, G., 1970, *Brain Res.* **19**:451–464.
53. Spanner, S., and Ansell, G. B., 1977, *Biochem. Soc. Trans.* **5**:164–165.
54. Miller, S. L., Benjamins, J. A., and Morell, P., 1977, *J. Biol. Chem.* **252**:4025–4037.
55. Chojnacki, T., Radominska-Pyrék, A., and Korzybski, T., 1967, *Acta Biochim. Pol.* **14**:383–388.
56. Porcellati, G., 1972, *Adv. Enzyme Regul.* **10**:83–100.
57. McCaman, R. E., and Cook, K., 1966, *J. Biol. Chem.* **241**:3390–3394.
58. Binaglia, L., Roberti, R., Vecchini, A., and Porcellati, G., 1980, *Ital. J. Biochem.* **29**:360–362.
59. Roberti, R., Binaglia, L., and Porcellati, G., 1980, *J. Lipid Res.* **21**:449–454.
60. Baker, R. R., and Thompson, W., 1972, *Biochim. Biophys. Acta* **270**:489–503.
61. Strosznajder, J., Radominska-Pyrék, A., and Horrocks, L. A., 1979, *Biochim. Biophys. Acta* **574**:48–56.
62. Francescangeli, E., Goracci, G., Piccinin, G. L., Mozzi, R., Woelk, H., and Porcellati, G., 1977, *J. Neurochem.* **28**:171–176.
63. Binaglia, L., Goracci, G., Porcellati, G., Roberti, R., and Woelk, H., 1973, *J. Neurochem.* **21**:1067–1082.

64. Gaiti, A., Sitkiewicz, D., Brunetti, M., and Porcellati, G., 1981, *Neurochem. Res.* **6**:11–20.
65. Brunetti, M., Gaiti, A., and Porcellati, G., 1979, *Lipids* **14**:925–931.
66. Strosznajder, J., Radominska-Pyrék, A., Lazarewicz, J., and Horrocks, L. A., 1977, *Bull. Acad. Pol. Sci. [Biol.]* **25**:363–370.
67. Goracci, G., Horrocks, L. A., and Porcellati, G., 1977, *FEBS Lett.* **80**:41–44.
68. Goracci, G., Francescangeli, E., Horrocks, L. A., and Porcellati, G., 1981, *Biochim. Biophys. Acta* **664**:373–379.
69. Cabot, M. C., and Gatt, S., 1977, *Biochemistry* **16**:2330–2334.
70. Sbriney, M., and Layman, E. M., 1973, *Can. J. Biochem.* **51**:1479–1486.
71. Radominska-Pyrék, A., Strosznajder, J., Dabrowiecki, Z., Chojnacki, T., and Horrocks, L. A., *J. Lipid Res.* **17**:657–662.
72. Porcellati, G., 1967, *Prog. Biochem. Pharmacol.* **3**:49–58.
73. Abe, T., Tatsuya, H., and Kurokawa, M., 1973, *Biochem. J.* **136**:731–740.
74. Kanfer, J. N., 1972, *J. Lipid Res.* **13**:468–474.
75. Saito, M., Bourque, E., and Kanfer, J. N., 1975, *Arch. Biochem. Biophys.* **169**:318–323.
76. Saito, M., and Kanfer, J. N., 1975, *Arch. Biochem. Biophys.* **169**:304–317.
77. Porcellati, G., Arienti, G., Pirotta, M. G., and Giorgini, D., 1971, *J. Neurochem.* **18**:1395–1417.
78. Gaiti, A., De Medio, G. E., Brunetti, M., Amaducci, L., and Porcellati, G., 1974, *J. Neurochem.* **23**:1153–1159.
79. Buchanan, A. G., and Kanfer, J. N., 1980, *J. Neurochem.* **34**:720–725.
80. Goracci, G., Blomstrand, C., Arienti, G., Hamberger, A., and Porcellati, G., 1973, *J. Neurochem.* **20**:1445–1460.
81. Brammer, M. J., and Carey, S. G., 1980, *J. Neurochem.* **35**:873–879.
82. De Medio, G. E., Woelk, H., Gaiti, A., Porcellati, G., and Fratini, F., 1975, *Ital. J. Biochem.* **24**:335–350.
83. Gaiti, A., Brunetti, M., and Porcellati, G., 1975, *FEBS Lett.* **49**:361–364.
84. Arienti, G., Corazzi, L., Woelk, H., and Porcellati, G., 1976, *J. Neurochem.* **27**:203–210.
85. Orlando, P., Arienti, G., Cerrito, F., Massari, P., and Porcellati, G., 1977, *Neurochem. Res.* **2**:191–201.
86. Buchanan, A. G., and Kanfer, J. N., 1980, *J. Neurochem.* **35**:814–822.
87. Mozzi, R., and Porcellati, G., 1979, *FEBS Lett.* **100**:363–366.
88. Blusztajn, J. K., Zeiseland, S. H., and Wurtman, R. J., 1979, *Brain Res.* **179**:315–327.
89. Crews, F. T., Hirata, F., and Axelrod, J., 1980, *J. Neurochem.* **34**:1491–1498.
90. Bergelson, L. D., and Barnskov, L. I., 1977, *Science* **197**:224–230.
91. Crews, F. T., Hirata, F., and Axelrod, J., 1980, *Neurochem. Res.* **5**:983–991.
92. Ansell, G. B., and Spanner, S., 1967, *J. Neurochem.* **14**:873–885.
93. Porcellati, G., Biasion, M. G., and Arienti, G., 1970, *Lipids* **5**:725–733.
94. Porcellati, G., Biasion, M. G., and Pirotta, M. G., 1970, *Lipids* **5**:734–742.
95. Spanner, S., and Ansell, G. B., 1978, *Enzymes of Lipid Metabolism* (S. Gatt, L. Freysz, and P. Mandel, eds.), Plenum Press, New York, pp. 237–245.
96. Porcellati, G., and Pirotta, M. G., 1970, *Enzymologia* **38**:351–369.
97. Wu, P.-S., and Ledeen, R. W., 1980, *J. Neurochem.* **35**:659–666.
98. Ansell, G. B., and Metcalfe, R. F., 1971, *J. Neurochem.* **18**:647–665.
99. Radominska-Pyrék, A., and Horrocks, L. A., 1972, *J. Lipid Res.* **13**:580–587.
100. Roberti, R., Binaglia, L., Francescangeli, E., Goracci, G., and Porcellati, G., 1975, *Lipids* **10**:121–127.
101. Freysz, L., Horrocks, L. A., and Mandel, P., 1978, *Enzymes of Lipid Metabolism* (S. Gatt, L. Freysz, and P. Mandel, eds.), Plenum Press, New York, pp. 253–268.
102. Gaiti, A., Brunetti, M., and Porcellati, G., 1975, *FEBS Lett.* **49**:361–364.
103. Arienti, G., Corazzi, L., Woelk, H., and Porcellati, G., 1977, *Brain Res.* **124**:317–329.
104. Orlando, P., Arienti, G., Massari, P., Porcellati, P., and Roberti, S., 1979, *Neurochem. Res.* **4**:595–603.
105. Anderson, R. E., Kelleher, P., and Maude, M. B., 1980, *Biochim. Biophys. Acta* **620**:227–235.
106. Yavin, E., and Leigler, B. P., 1977, *J. Biol. Chem.* **252**:260–267.

107. Pullakart, R. F., Sbasching-Aglez, M., and Reha, H., 1981, *Biochim. Biophys. Acta* **664**:117–123.
108. Taki, T., and Kanfer, J. N., 1978, *Biochim. Biophys. Acta* **528**:309–317.
109. Horrocks, L. A., 1972, *Ether Lipids* (F. Snyder, ed.), Academic Press, New York, pp. 177–272.
110. Gaiti, A., Goracci, G., De Medio, G. E., and Porcellati, G., 1972, *FEBS Lett.* **27**:116–120.
111. Horrocks, L. A.; and Chun Fu, S., 1978, *Enzymes of Lipid Metabolism* (S. Gatt, L. Freysz, and P. Mandel, eds.), Plenum Press, New York, pp. 397–406.
112. Mozzi, R., Siepi, D., Andreoli, V., and Porcellati, G., 1981, *FEBS Lett.* **131**:115–118.
113. Wykle, R. L., and Schremmer-Lockmiller, J. M., 1975, *Biochim. Biophys. Acta* **380**:291–298.
114. Paltauf, F., Prough, R. A., Masters, B. B. S., and Johnston, J. M., 1974, *J. Biol. Chem.* **249**:2661–2662.
115. Paltauf, F., and Holasek, A., 1973, *J. Biol. Chem.* **248**:1609–1615.
116. Schmidt, H. H. O., Bandi, P. C., Madson, T. H., and Baumann, W. J., 1977, *Biochim. Biophys. Acta* **488**:172–178.
117. Paltauf, F., 1978, *Enzymes of Lipid Metabolism* (S. Gatt, L. Freysz, and P. Mandel, eds.), Plenum Press, New York, pp. 387–396.
118. Gunawan, J., and Debuch, H., 1977, *Hoppe Seylers Z. Physiol. Chem.* **358**:537–542.
119. Wykle, R. L., and Snyder, F., 1976, *The Enzymes of Biological Membranes*, Volume 2 (A. Martonosi, ed.), Plenum Press, New York, pp. 87–117.
120. Snyder, F., 1972, *Ether Lipids* (F. Snyder, ed.), Academic Press, New York, pp. 121–156.
121. Wykle, R. L., Piantadosi, C., and Snyder, F., 1972, *J. Biol. Chem.* **247**:2944–2953.
122. Wykle, R. L., and Snyder, F., 1970, *J. Biol. Chem.* **245**:3047–3055.
123. Natarajan, V., and Schmidt, H. H. O., 1977, *Biochem. Biophys. Res. Commun.* **79**:411–416.
124. Hajra, A. K., Burke, C. L., and Jones, C. L., 1979, *J. Biol. Chem.* **254**:10896–10900.
125. van den Bosch, H., 1980, *Biochim. Biophys. Acta* **604**:191–246.
126. Gallai-Hatchard, J., Magee, W., Thompson, R. H. S., and Webster, G. R., 1962, *J. Neurochem.* **9**:545–554.
127. Goracci, G., Porcellati, G., and Woelk, H., 1978, *Advances in Prostaglandin and Thromboxane Research*, Volume 3 (C. Galli, G. Galli, and G. Porcellati, eds.), Raven Press, New York, pp. 55–68.
128. Cooper, M. F., and Webster, G. R., 1970, *J. Neurochem.* **17**:1543–1554.
129. Gatt, S., 1968, *Biochim. Biophys. Acta* **159**:304–316.
130. Woelk, H., and Porcellati, G., 1973, *Hoppe Seylers Z. Physiol. Chem.* **354**:90–100.
131. Woelk, H., Peiler-Ichikawa, K., Binaglia, L., Goracci, G., and Porcellati, G., 1974, *Hoppe Seylers Z. Physiol. Chem.* **355**:1535–1542.
132. Rooke, J. A., and Webster, G. R., 1976, *J. Neurochem.* **27**:613–620.
133. Doherty, F. J., and Rowe, C. E., 1980, *Brain Res.* **197**:113–122.
134. Woelk, H., Goracci, G., Gaiti, A., and Porcellati, G., 1973, *Hoppe Seylers Z. Physiol. Chem.* **354**:729–736.
135. Woelk, H., Porcellati, G., and Gaiti, A., 1979, *Neurochem. Res.* **4**:535–543.
136. Woelk, H., Rubly, N., Arienti, G., Gaiti, A., and Porcellati, G., 1981, *J. Neurochem.* **36**:875–880.
137. Woelk, H., Arienti, G., Gaiti, A., Kanig, K., and Porcellati, G., 1981, *Neurochem. Res.* **6**:23–32.
138. Friedel, R. O., Brown, J. D., and Durell, J., 1969, *J. Neurochem.* **16**:371–378.
139. Parenholtz, Y., Roitman, A., and Gatt, S., 1966, *J. Biol. Chem.* **241**:3731–3737.
140. Williams, J. D., Spanner, S., and Ansell, G. B., 1973, *Biochem. Soc. Trans.* **1**:466–467.
141. Hostetler, K. Y., and Hall, L. B., 1980, *Biochem. Biophys. Res. Commun.* **96**:388–393.
142. Saito, M., and Kanfer, J. N., 1973, *Biochem. Biophys. Res. Commun.* **53**:391–398.
143. Chalifour, R. J., and Kanfer, J. N., 1980, *Biochem. Biophys. Res. Commun.* **96**:742–747.
144. Kanfer, J. N., 1980, *Can. J. Biochem.* **58**:1370–1380.
145. Taki, T., and Kanfer, J. N., 1979, *J. Biol. Chem.* **254**:9761–9765.
146. Leibovitz Ben Gershon, Z., Kobiler, I., and Gatt, S., 1972, *J. Biol. Chem.* **247**:6840–6847.
147. Leibovitz Ben Gershon, Z., and Gatt, S., 1974, *J. Biol. Chem.* **249**:1525–1529.
148. Wykle, R. L., and Schremmer, J. M., 1974, *J. Biol. Chem.* **249**:1742–1746.

149. Wykle, R. L., Kraemer, W. F., and Schremmer, J. M., 1977, *Arch. Biochem. Biophys.* **184**:149–155.
150. Warner, H. R., and Lands, W. E. M., 1961, *J. Biol. Chem.* **236**:2404–2409.
151. Dorman, R. V., Toews, A. D., and Horrocks, L. A., 1977, *J. Lipid Res.* **18**:115–117.
152. D'Amato, R. A., Horrocks, L. A., and Richardson, K. E., 1975, *J. Neurochem.* **24**:1251–1255.
153. Ansell, G. B., and Spanner, S., 1965, *Biochem. J.* **94**:252–258.
154. Ansell, G. B., and Spanner, S., 1968, *Biochem. J.* **108**:207–209.
155. Webster, G. R., and Alpern, R. J., 1964, *Biochem. J.* **90**:35–42.
156. Fisher, S. K., and Rowe, C. E., 1980, *Biochim. Biophys. Acta* **618**:231–241.
157. Murphy, M. G., and Spencer, M. W., 1977, *J. Neurochem.* **29**:251–259.
158. Baker, R. R., and Thompson, W., 1973, *J. Biol. Chem.* **248**:7060–7065.
159. James, O. A., McDonald, G., and Thompson, W., 1979, *J. Neurochem.* **33**:1061–1066.
160. Mc Murray, W., and Dawson, R. M. C., 1969, *Biochem. J.* **112**:91–108.
161. Miller, E. K., and Dawson, R. M. C., 1972, *Biochem. J.* **126**:805–821.
162. Pasquini, J. N., Gomez, C. J., Najle, R., and Soto, E. F., 1975, *J. Neurochem.* **24**:439–443.
163. Wirtz, K. W. A., and Zilversmit, D. B., 1968, *J. Biol. Chem.* **243**:3569–3602.
164. Harvey, M. S., Wirtz, K. W. A., Kamp, H. H., Zegers, B. J. M., and van Deenen, L. L. M., 1973, *Biochim. Biophys. Acta* **323**:234–239.
165. Helmkamp, G. M., Nelemans, S. A., and Wirtz, K. W. A., 1976, *Biochim. Biophys. Acta* **424**:168–182.
166. Agranoff, B. W., 1962, *J. Lipid Res.* **3**:190–196.
167. Holub, B. J., and Pierkarski, J., 1978, *J. Neurochem.* **31**:903–908.
168. Rousseau, A., and Gatt, S., 1979, *J. Biol. Chem.* **254**:7741–7745.
169. Binaglia, L., Roberti, R., and Porcellati, G., 1978, *Bull. Mol. Biol. Med.* **3**:136s–149s.

Sulfatides

Norman S. Radin

1. PHYSICAL AND CHEMICAL MATTERS

1.1. Types of Sulfatides

The name sulfatide ordinarily refers to galactocerebroside-3-sulfate (also called 3-sulfogalactosyl ceramide or cerebroside sulfate and abbreviated as CS here). It is the half ester of sulfuric acid formed from galactosyl ceramide, and one ought to specify the cationic group that occupies the anionic site of the sulfate. Few researchers and suppliers, however, even raise the subject.

$$C_{13}H_{27}-CH{=}CH-\underset{\underset{\displaystyle OH}{|}}{CH}-\underset{\underset{\displaystyle NH}{|}}{CH}-CH_2-O-CH \cdots CH-CH_2OH$$

Like galactosyl ceramide, its metabolic precursor, sulfatide contains assorted long-chain bases and fatty acids, including the very long acids and acids bearing a 2-D-hydroxy group. In recent years, additional "sulfolipids," containing additional sugars, have been isolated, and it is not clear whether the name "sulfatide" should be extended to include them. Probably "sulfolipid" should be used as the most general term, to include steroidal and other lipoidal sulfate esters as well as the sulfonic acids. "Sulfatide" should be restricted to the sulfolipids derived from sphingosine and related long-chain bases.

Sulfatide occurs in nerve tissue in relatively high concentrations, primarily because of its localization in myelin. However it has also been found in preparations of neurons, axons, and axolemmas in unexpectedly high concentra-

Norman S. Radin • Mental Health Research Institute (Department of Psychiatry) and Department of Biological Chemistry, University of Michigan, Ann Arbor, Michigan 48109.

tions. It has been estimated that myelin contains only about 45% of the CS in brain, but another study found 95% in the myelin of sheep brain.

Glucocerebroside sulfate has been reported to be present in the brain of an infant with Gaucher's disease. Since glucosyl ceramide accumulates in this disease in the peripheral organs but not in brain, sulfation of the rare glycolipid seems unlikely, and the report needs confirmation.

The glyceryl analogue of CS has been isolated from testes and is called seminolipid. It is 1-O-alkyl-2-O-acyl-3-O-β-**D**-(3'-sulfo)galactopyranosyl glycerol. The hydrolase that releases the sulfate group seems to be the same one that acts on CS. Seminolipid and its diacyl analogue have been found in brain and retina, and both accumulate with age, as does CS.[1] In adult chick retina, the molar ratio of CS to sulfoglyceride is 14:1.

The sulfated glyceroglycolipids occur in rat and fish brain but could not be found in human brain, even in a person with metachromatic leukodystrophy. Perhaps they hydrolyze during the interval before the brain can be removed. The amount in rat brain was found to be 7.5 nmol at 14 days of age and 248 nmol at 48 days, at which point it leveled off, although CS continued to accumulate. The diacyl form predominated at first; then the ether form predominated.

Lactoside sulfate (3-sulfogalactosylglucosyl ceramide or ceramide dihexoside sulfate) was isolated from liver of humans with metachromatic leukodystrophy but could not be found in normal liver. Of interest is the occurrence of phytosphingosine as one of the sphingoid bases, presumably derived from the diet. This sulfatide has also been isolated from human kidney and erythrocytes and from hog gastric mucosa. The sulfate esters of ceramide tri- and tetrasaccharide have been isolated from hog gastric mucosa.

1.2. Physical Properties

Most studies of the physical properties have been made by partitioning CS between two solvents. Breyer[2] showed with a Folch/Lees/Sloane-Stanley mixture (C–M–W) that Ca^{2+} and Mg^{2+} compete effectively against Na^+ and K^+ at low ionic concentrations, apparently because CS exists in the chloroform-rich layer as the M^{2+} $(CS^-)_2$ salt. At high ionic concentrations, more of the monovalent salt forms, and some CS enters the upper layer. There is nothing unexpected in these results, which undoubtedly apply to other monoanionic lipids, but they could indicate that CS exists in the lipoidal regions of membranes as the Ca^{2+} and Mg^{2+} salts rather than as salts of hydrophilic cations.

A different kind of study, based on dispersions of the CS Na^+ salt in water with measurements of turbidity and coagulation, showed the relative affinities of CS for cations to be in the series $Ca^{2+} > Mg^{2+} > K^+ > Na^+ > Li^+$. However, salting out and solvation effects may have exerted a confusing influence on these measurements, and some of the data indicated that Li^+ binds more firmly than Na^+ or K^+. The inclusion of lecithin in the CS dispersions reduced the turbidity induced by monovalent cations but not by Ca^{2+} and Mg^{2+}. A finding of interest of preparative value is that the Na^+ salt could be converted to the K$^+$ salt simply by adding 0.15 M KCl to the dispersed lipid, which then precipitated the K$^+$ salt.

Partition experiments have also been done with organic cations of neurochemical interest, since CS could occur in nerve tissue (in part, at least) as a salt with neuroactive amines. Sulfatide was able to pull norepinephrine, histamine, acetylcholine, and serotonin into the chloroform layer of a C–M–W system. In a similar system containing salts, the partitioning of Ca^{2+} into the chloroform layer was enhanced by barbiturates but not by phenytoin.

Dialysis of CS against water with the lipid dissolved in C–M allows one to remove small molecules without loss of the lipid. This property could be used to exchange the cations.

Because of its amphipathic structure, CS ought to possess some detergent properties, but the lack of a good preparative method (until recently) has apparently discouraged studies of this sort. We have used it as part of a liposome preparation (with ceramide and lecithin) to stabilize the suspension.

The infrared absorption peak at 8.02 μm has been interpreted to mean that the sulfate ester is attached at an equatorial position of galactose. Another report assigned sulfate absorption peaks to 820 and 1240 cm^{-1} (12.20 and 8.06 μm).

1.3. Chemical Properties; Synthesis of Labeled Sulfatide

Sulfatide can be permethylated without loss of the sulfate group by the Hakomori procedure. Cleavage of the product with methanolic HCl yields the 2,4,6-trimethyl galactoside which can be hydrolyzed to the free hexose or converted to the anilide for further identification.

LysoCS can be prepared by alkaline hydrolysis of the amide group in CS, but several side reactions occur.[3] Refluxing for 1 h in *n*-butanol:70% KOH (9:1) yields lysoCS, ceramide, sphingosine, and 3,6-anhydrogalactosyl sphingosine. Psychosine sulfate is another name for lysoCS, but it is also the name that has been used for the salt formed from galactosyl sphingosine and sulfuric acid. Unfortunately, some researchers and one company that offered it for sale confused this compound with the sulfate ester of psychosine. It would be better to call the salt psychosinium sulfate in analogy with ammonium sulfate.

LysoCS, like many other amines, is readily acylated with fatty acid chlorides in aqueous sodium acetate and tetrahydrofuran. The reaction has been used to prepare CS with labeled lignoceric acid.

The amide group was apparently cleaved by heating CS in 6 M HCl at 100°C for 30 min, but the products were not identified.

The sulfate ester link can be cleaved under very mild conditions: 4 h at room temperature in 50 mM methanolic HCl. The reaction produces cerebroside and, perhaps, monomethyl sulfuric acid. It has also been cleaved faster with the same reagent by means of a higher temperature: 80 min at 40°C. Karlsson, Samuelsson, and Steen used 0.1 M HCl at room temperature for 4 h. Hara and Radin reported that a higher acid concentration (1 M HCl in methanol) is necessary for complete cleavage of the ester linkage, but then some degradation of the cerebroside occurs too. The reaction is interesting when followed by TLC, as several unknown intermediates form during the cleavage.[4]

Solvolysis of the sulfate group in sulfolipids can be accomplished by heating the lipids in dry dioxan for 20 min at 100°.[1] Ether has been used as a solvolytic reagent for other sulfate esters.

Sulfatide can be benzoylated with warm pyridine/benzoyl chloride, and the sulfate group can then be removed by acid-catalyzed solvolysis. A small loss of the sulfate group occurs during the acylation, and additional losses occur during washing to removing the reagents, so the yield of the benzoylated sulfate is about 85%.[5]

This sequence of reactions has been used to make [^{35}S]CS by sulfation of the benzoate with [^{35}S]sulfuric acid followed by removal of the benzoate groups by alkali catalysis.[6]

Labeled CS has also been prepared by treating CS in tetrahydrofuran with aqueous $PdCl_2$ and $NaBT_4$ in NaOH. This product, which probably contains the tritium where the double bonds had been, is useful in the assay for sulfatidase.[7] Catalytic (Pt) hydrogenation with tritium gas has also been used.

Sulfatide is not attacked by galactose oxidase or periodate. The periodate reaction is a simple way of ruling out the presence of sulfate on a position other than the C-3 of galactose.

1.4. Analytical Methods

Sulfatide has been detected qualitatively in urine sediments of children with metachromatic leukodystrophy by formalin fixation and staining with cresyl fast violet. The epithelial cells from the kidney, which form the sediment, yield a brown ("metachromatic") stain in the cytoplasm. Earlier studies by James Austin had shown similar metachromatic staining with toluidine blue in brain and kidney, the latter reacting more strongly in metachromatic leukodystrophy victims. Austin also used infrared spectroscopy at 8020 nm (a sulfate ester absorption peak) to make quantitative measurements. Phospholipids interfere at this wavelength but can be removed very readily by alkali-catalyzed methanolysis, which does not affect CS.

A very sensitive assay for CS involves high-performance liquid chromatography of the benzoylated, desulfated derivative described above.[5] The ester is monitored and quantitated by its absorbance at 230 nm. By use of a reverse-phase column, one can separate the different homologues somewhat.

Sulfoglycerides have been determined by removing the sulfate group (acidic solvolysis) and fatty acids; the residual structures were then converted to the trimethylsilyl ethers and analyzed quantitatively by gas chromatography.[1] If applied to CS, this method would yield cerebrosides the silyl ethers of which have been quantitated by gas chromatography. A short column is necessary because of the high molecular weight of the lipid.

The sulfoglycerides were determined in testes by benzoylation of total lipids, purification by Florisil® chromatography, and liquid chromatography with an octadecyl silica high-performance column.[8] An interesting feature of the reverse-phase separation was the use of a paired-ion solvent containing dilute tetrabutylammonium hydroxide in M–W.

The technique of pairing with a cationic dye has been applied to the problem of determining CS spectrophotometrically.[9] The dye used was azure blue A, which is partitioned by CS into the chloroform-rich layer of a biphasic solvent system made from C–M–W. The lower limit of determination was about 2 nmol. Other acidic lipids act similarly, so they must be removed first (most readily by treatment with alkali). In some laboratories, the solvent mixture was found to be sensitive to temperature (water can precipitate out of the lower layer after centrifugation), and recentrifuging the separated lower layer solved the problem. Perhaps drying the chloroform layer briefly with Na_2SO_4 after the first centrifugation would eliminate the clouding problem. This partitioning approach could probably be extended in sensitivity by using a fluorescent cationic material.

Mass spectrometry of CS can be done by first acetylating the hydroxyl groups with pyridine/acetic anhydride, then heating with trimethylchlorosilane and hexamethyldisilazane in pyridine. The silylation step replaces the sulfate group with a trimethylsilyl group. The galactose-derived fragments in the mass spectrum showed that the sulfate group was originally on the C-3 hydroxyl.

Tissues can be stained semispecifically for CS by cresyl violet, acriflavine, pseudoisocyanin, or Alcian blue/$MgCl_2$.

Thin-layer chromatography with CS has been done with silica gel and a great variety of solvent mixtures. Ordinarily, it migrates below cerebroside, but the inclusion of ammonia moves it to a position above cerebroside. Phosphatidyl glycerol, a trace lipid, tends to overlap CS; they can be separated with C–M–W–HOAc 50:15:1:15. An improved separation from other sulfolipids can be obtained with C–acetone–M–HOAc–W 50:20:10:5. In my experience with several solvents, the resolution of hydroxy and nonhydroxy fatty acid CS is improved by reducing the sample weight; the R_f values depend to some extent on the sample weight.

1.5. Isolation Methods

The methods thus far described have involved ion exchange (usually DEAE-cellulose), adsorption chromatography (silica gel, Florisil®, charcoal), or precipitation. The last of these methods has been described for use with brain extracts.[4] It is readily run at widely differing operating scales. The procedure involves extraction of the lipids with hexane–isopropanol and treatment of the extract with methanolic 5 M NaOH. The alkali cleaves the ester lipids within 15 min and also precipitates virtually all the CS while forming a small amount of a lower solvent layer. The lower layer, plus its interface, is neutralized with HCl and partitioned between C–M–W. The new lower layer contains the CS together with part of the cerebrosides and small amounts of other lipids. Although chromatographic isolation of CS from this mixture is feasible, the column separation is made much more effective by treating the mixture with periodic acid to reduce the polarity of the cerebroside fraction. This procedure yields the Na^+ salt. The product seems to be a hydrate or solvate, since it loses about 2.4% of its weight on storage at 100°C *in vacuo* over P_2O_5.

It should be noted that commercial analytical laboratories do not always produce accurate elemental values for sulfur and nitrogen. The published elemental analyses for some reported CS preparations have not been very satisfactory, but few workers have done enough evaluations to clarify the discrepancies. Tests of purity are TLC examination after solvolytic removal of the sulfate group and treatment with periodate, the amount of ash after dry combustion, the absence of nonacidic lipids in the effluent of an anion-exchange column, and the presence of the correct amount of a cation, determined chemically. Heavy samples (0.1–0.2 mg) should be examined by TLC. The two infrared peaks for sulfate should be easily seen, as well as the bands characteristic of cerebroside.

A very sensitive test of purity (for any compound) is examination by the above methods of the different fractions obtained from a chromatographic column. Chromatographic columns tend to produce CS containing related sulfolipids such as cholesterol sulfate, seminolipid, and sulfolactosyl ceramide. If one collects small fractions and examines each at a high sample level by TLC, one can see extra spots in the early and late fractions; this is especially noticeable if a brain prelabeled by injection with labeled sulfate is used as the source. Seminolipid elutes just ahead of CS with silica gel columns, but it would be removed by an alkaline pretreatment.

The sulfoglyceroglycolipids can be isolated by silica gel chromatography with bead-shaped silica gel (high-resolution silica) after initial purification by Florisil® and DEAE-Sephadex A-25 columns.

Acid-washed charcoal adsorbs CS with some specificity. The lipid can be eluted with C–M–ammonia. However, additional purification is necessary.

The isolation of labeled CS from brain after intracranial injection of labeled sulfate has been described for use in sulfatase assays.[10]

According to Pritchard, the extraction of brain with C–M does not solubilize all of the CS. An additional 5–10% radioactive CS was obtained by a subsequent extraction with acidified C–M. This observation does not seem to have received adequate attention.

1.6. Concentration in Tissues

Rat brain has been reported to contain 0.17 μmol at 10 days of age, 0.45 at 17 days, and 0.60 at 21 days. The concentrations at the same ages were 0.37, 0.57, and 0.99 μmol/g. A concentration of 2.2 mg/g was reported for "adult" rat brain. A more detailed analysis of young rat brain[11] showed the following concentrations (in nmol/g) for 5, 10, and 14 days of age: in the brainstem, the hydroxy acid CS content was 16, 81, and 285; the nonhydroxy CS content was 18, 202, and 724. In the cerebellum, the corresponding values were 12, 28, and 112 for hydroxy and 6, 50, and 244 for nonhydroxy. In the cerebral hemispheres, they were 6, 20, and 63; 2, 30, and 120. In the diencephalon, they were 7, 18, and 151; 4, 26, and 272. In all regions, the nonhydroxy acid type of CS dominated the hydroxy acid type except for two regions in the 5-day age rats.

The same workers reported on subcellular CS distribution in sheep brain, finding differences between gray and white matter.

Rat brain (140–400 days old) contained 2.8 mg/g (wet weight) in the forebrain, 0.83 mg/g in gray matter, none in astrocytoma, 4.8 mg/g in white matter, 3.6 in spinal cord, 2.3 in trigeminal nerve, 3.1 in sciatic nerve, and varying amounts in tumors of the nervous system.[12]

As the result of a very detailed study of human brains, Rouser and Yamamoto[13] reported the CS concentration in whole brain in the form of coefficients for the equation, $[CS] = k_1 + k_2 \log A$, where A is the age of the individual (in years), and the concentration of CS is in nmol/g wet tissue. From their data, we see that the concentration of CS increases throughout the first 40 years of life (about 50-fold compared with the 3-week level) and then starts to undergo a decrease.

In myelin of rat brain isolated at different ages, CS constituted 4.2 to 7.9% of the total myelin lipids, with the values rising at every age. It has been found at a very low concentration bound to cytosolic protein.

The concentrations in human axolemma-enriched fractions and in myelin were found to be about 5.2 and 8.0% of the total lipids or 2.7 and 5.5% of the dry weight, respectively. Axons prepared from bovine CNS were found to contain (and to be able to synthesize) CS. Thus, there is no need to postulate transfer of CS from oligodendroglial myelin to axons.

Sulfatide in bovine oligodendroglia preparations was found to be about 3.3% of the total lipids.

A qualitative study of a 30-day-old mouse with CS antibodies and an immunofluorescence binding technique[14] showed the lipid to be present in myelin, myelinating oligodendrocytes, ciliated ependymal cells, and subpial astrocytic processes. The lipid could not be seen in cortical oligodendrocytes, some of the subependymal cells, or astrocyte cell bodies. It was detected in periaqueductal gray matter and the nucleus interpeduncularis.

2. BIOSYNTHESIS

Firm proof for the utilization of galactosyl ceramide came from the use of hydroxy [^{14}C]cerebroside labeled in the galactose moiety. Sulfatide was isolated from the incubation mixture and shown, by methylation and degradation, to contain all the sulfate on the C-3 of the sugar. Thus, it was not formed by a nonspecific (or "other") sulfotransferase.

Cerebroside sulfotransferase was originally found in rat brain microsomes. It could be solubilized by aqueous Triton X-100, and it acted on galactocerebrosides as well as on lactosyl ceramide and galactosyl sphingosine but not on glucosyl ceramide. More recently, it was found that the enzyme is actually in the Golgi apparatus.[15] There is no enzymatic activity in myelin. The molecular weight of the enzyme is estimated to be 43,500.

The apparent K_m for the galactolipids as sulfate acceptors was found to be about 50 μM in rat brain at pH 6.8–7.0. For this assay, the galactosyl ceramide was dispersed with the detergents Brij® 96 or Triton X-100®. Some authors have found Mg^{2+} to be stimulatory; others have added ATP, KCl, and K_2SO_4.

The specific activity of the sulfotransferase was found to be higher in white matter than in gray (as might be expected from the location of CS), and it reached a peak value at 5 years in human brain.

Cultured Schwann cells from the rat could synthesize CS, even in the absence of neuronal cells, but fibroblasts could not. A rat cell line (RN2, derived from neurons via carcinogenic transformation) synthesized a different, unknown sulfolipid which migrated just ahead of CS on TLC plates.

Isolated oligodendroglial cells from calf brain were eight times more active in CS synthesis than neuronal cells. The ability of neuronal cells from rat brain to synthesize CS increased with age during the period of active myelination.

A solubilized preparation from boar testis was able to sulfate galactocerebroside, lactosyl ceramide, and desulfoseminolipid (1-alkyl-2-acyl-3-galactosyl glycerol). Apparently the same enzyme was involved in all three reactions. Digalactosyl ceramide, trihexosyl ceramide, glucosyl ceramide, and cholesterol were not sulfated. Detergents and EDTA were stimulatory, whereas Mg^{2+} and ATP were inhibitory.

Cerebroside sulfotransferase has been partially purified from rat brain.[16] The detergent Cemulsol NPT 12® was used to solubilize the enzyme.

A recently described assay for sulfotransferases acting on lipids utilized TLC with a solvent that brings the sulfolipid close to the solvent front while leaving residual radioactive phosphoadenosyl phosphosulfate and sulfate behind. Galactosyl sphingosine (psychosine) can also act as a sulfate acceptor, and evidence has been offered indicating that a different enzyme is involved. This raises the possibility that an acyltransferase could act on psychosine sulfate, forming CS by a second route. Indeed, if such a transferase existed, it could explain the claim in the literature that psychosine can be acylated to form cerebroside. Perhaps the sequence involves sulfation, acylation, and desulfation. However, endogenous levels of psychosine are extremely low.

The specific activity of cerebroside sulfotransferase is quite low in the brains from the mutants jimpy and quaking mice. This finding is typical of other major myelin-synthesizing enzymes. The enzyme does not seem to be abnormal, judging by its K_m and pH optimum. However, data from Herschkowitz's laboratory indicate that the abnormality is, to a considerable extent, due to the low lipid content of jimpy brain.[16a] Acetone extraction of the microsomes lowered the activity more in normal than in jimpy membranes, and adding back lipids brought the jimpy enzyme to a level that was 80% of normal. Cholesterol and lecithin were stimulatory. The need for lipid by sphingolipid synthetases is probably a general phenomenon for membrane enzymes, and the other enzyme deficiencies noticed in the jimpy mutant ought to be studied in a similar fashion.

Curiously, the same researchers have found an apparently opposite correlation in cultured glioblastoma cells. Here, estradiol doubled the sulfotransferase specific activity while lowering the cholesterol concentration. They concluded that the enzyme's activity is stimulated by lowering the cholesterol content.

The synthesis of CS by cultured oligodendroglioma cells (G-26) from mouse brain was shown to be enhanced considerably by addition of cortisol to the medium.[17] Dexamethasone was also effective, cortisone less so, and several other steroid hormones—including estradiol—were inactive. Other cell lines did not respond to the cortisol.

When Wallerian degeneration was induced in rabbit optic nerves by removal of the retina, cerebroside sulfotransferase disappeared from the nerve faster than the other components studied. Its activity was reduced by 40% within 24 h, from which it might be concluded that this enzyme normally has a high turnover rate.

3. ENZYMATIC HYDROLYSIS

The enzyme that removes the sulfate group is called aryl sulfatase A or sulfatidase. The former name is used when the assay substrate is a phenolic sulfate such as p-nitrocatechol sulfate (2-hydroxy-5-nitrophenyl sulfate). Aryl sulfatase B (N-acetylgalactosamine-4-sulfate sulfatase), which is often present together with A, acts on the same substrate and has to be blocked by adding 0.75 M NaCl. The use of methylumbelliferone sulfate as substrate makes it possible to use less enzyme because it forms a strongly fluorescent product. Unfortunately, the use of unnatural substrates instead of CS has yielded some published data that are difficult to interpret and are probably misleading. (This topic is too long to expound on here.)

The enzyme, studied with one of the unnatural substrates, exhibits odd characteristics. The sulfatase from rabbit liver has two pH optima (5.1 and 5.6), but the proportion observed at each optimum changes with enzyme concentration.[18] Apparently this is because of the presence of two active forms with two different optima; the polymerized form dissociates into the monomeric form on dilution. At pH 5.5, the enzyme quickly changes to a less active form, and the addition of dilute inorganic sulfate stimulates the enzyme. The monomeric enzyme can be bound to Sepharose, still retaining its activity, and it can then act as an affinity column, binding other monomeric sulfatase molecules and thereby effecting a considerable purification ("subunit affinity chromatography").

Use of labeled CS as substrate with human fibroblasts showed that the pH optimum was 4.5 in acetate buffer (very sharp), and the enzyme was stimulated by Mn^{2+} and Cl^-. Bile salt, such as taurodeoxycholate, was essential (also a sharp optimal concentration), and sulfate ions were inhibitory. Human urine yielded two forms of arylsulfatase A on fractionation by reverse-gradient solubilization chromatography, and the ratio of the two forms depended on the individual.

A similar study with [^{3}H]CS and human leukocytes called for pH 5 acetate buffer, taurodeoxycholate, and Mn^{2+}.[7] Removal of interfering materials by dialysis greatly improved the activity of the enzyme and the relationship between enzyme concentration and observed activity.

Various studies have reported inhibition by sulfate, sulfite, phosphate, borate, Cu^{2+}, Fe^{3+}, and Ag^+. Presumably it contains a vital thiol group, a probable characteristic of all lysosomal enzymes.

Sulfatase A has been isolated from a variety of sources. An interesting procedure starting with human liver involved some conventional steps (as well as the reverse ammonium sulfate solubilization column mentioned above) but also sequential gel permeation chromatography at two different pHs.[19] At pH 8.1, the enzyme eluted in a low-molecular-weight form, but at pH 5.0, it eluted as an aggregated form (probably a tetramer). The aggregation was reversible.

Another interesting aspect of aryl sulfatase is that its activity was reported to be enhanced when it was mixed with antibodies raised against the enzyme. The antibodies may have acted by counteracting the inhibitory action of the substrate, which is distinctly inhibitory at higher concentrations.

Some clarification of the many puzzling features of sulfatase activity comes from the observation that the enzyme binds sulfate covalently as the result of action on substrate and consequently undergoes a deleterious conformational change. Some protection against the change is produced if sulfate, phosphate, or pyrophosphate ions are present.

Probably many of the curious properties of the enzyme would disappear if the natural substrate and natural stimulating protein were to be used. Activity against CS, if a bile salt is omitted from the assay medium, requires a second protein.[20] This protein ought to be called cohydrolase-S in analogy with the name "colipase" assigned to the protein that helps pancreatic lipase act against triglycerides. A number of other hydrolases show a similar requirement for "helper" proteins. Cohydrolase-S is a heat-stable glycoprotein, a contaminant of sulfatase in some isolation procedures, separable from the enzyme by size exclusion chromatography, has an M_r of about 21,500 and an isoelectric point of 4.3 (human liver), and it binds to CS in a $1:1$ molar ratio. Although the enzyme can be stimulated more highly by taurodeoxycholate, the activator is more effective on a molar basis. Apparently, it acts by solubilizing CS (i.e., rendering it more hydrophilic), and it thus does not stimulate attack on the more soluble nitrocatechol sulfate. Cohydrolase-S also binds to cerebroside, but not as strongly as to CS. (This competition for the helper protein may explain why galactosyl ceramide inhibits CS hydrolysis.) Both the sulfatase and cohydrolase have been found to be concentrated in the lysosomal fraction.

The importance of the activating protein is indicated by the finding of a patient with metachromatic leukodystrophy (see Section 6.1) whose genetic defect is not in the enzyme but in cohydrolase-S.

Sulfogalactosyl sphingosine (lysoCS) can be hydrolyzed by the same enzyme that acts on CS. However, there is no evidence for the existence in nature of this amino acid or for its formation from CS by an amidase.

A recently described method for sulfatidase assay involves incubation of $[^3H]CS$ with Mn^{2+} and Na taurodeoxycholate at pH 5, then removal of unreacted substrate with a small column packed with DEAE-cellulose.[7] The enzymatic product, labeled cerebroside, is measured in the column effluent, and the bound CS is salvaged for reuse later.

The concentration of sulfatidase, like that of cerebroside sulfotransferase, was found to peak at 5 years of age in human brain and to be higher in white matter than in gray. In mouse brain, sulfatase specific activity peaked at 18 days (as for myelin-forming enzymes). In similar mice bearing the jimpy mutation, the specific activity curves followed the same pattern up to 15 days of age but then started to decline prematurely. The difference seems decidedly secondary to the other brain changes.

4. METABOLISM

Injection into rat brain of labeled sulfate gave rise to labeled CS, the maximal activity being reached in about 3 days. This activity remained constant for about 13 days and then slowly declined. The slow-rise period is probably caused by relatively rapid incorporation into oligosaccharides followed by relatively rapid release and competitive uptake by cerebroside ("recycling"). Inorganic sulfate and phosphoadenosine phosphosulfate exist at very low levels and could not act as a radioactive reservoir for 3 days. The long plateau could be caused by incorporation of the labeled CS into a membrane that has a specific life-span (such as red blood cells). However, interpreting turnover data from whole organs has distinct limits.

The turnover of CS in rat brain has also been followed with [^{14}C]glucose, from which a half-life of 315 days was adduced for the total carbon atoms and 46 days for the galactose moiety. These were the longest half-lives of the five sphingolipids studied.

Intracerebral injection of [^{3}H,^{35}S]CS into rats gave rise to labeled cerebrosides and other lipids, presumably derived by hydrolysis of the cerebrosides.[6] The sulfate group was lost preferentially but was recycled for the synthesis of CS, seminolipid, cholesterol sulfate, and water-soluble materials. It is an interesting question as to whether the reuse of sulfate occurs via transulfation (catalyzed by sulfatase) or via sulfation with phosphoadenosine phosphosulfate.

Rat brain, after injection of rats with [^{35}S]sulfate, was found to contain CS in the cytosol bound to proteins.[21] Judging by the time studies, newly synthesized CS entered this pool first, then moved to myelin. The cytosolic protein complex may indicate the existence of a carrier protein.

The injection of corticosterone into newborn rats resulted in a decrease in the concentration of CS in the brain, possibly because of suppression of the formation of glial cells (preferentially).

The incorporation of [^{35}S]sulfate into rat brain CS was found to be reduced by injection of chlorpromazine, trifluoperazine, and promethazine.[22] The sulfate was injected 24 h after the last tranquilizer injection. Chlorpromazine has been found in other studies to inhibit the synthesis of various kinds of lipids. Puromycin, usually thought of as an inhibitor of protein synthesis, was also found to inhibit CS formation from [^{35}S]sulfate by spinal cord slices.[23]

5. PHYSIOLOGICAL FUNCTIONS

5.1. Salt Transport

The outer part of the kidney medulla contains a high concentration of CS. Since this region is very active in Na$^+$ transport, the suggestion was made that the two phenomena are related. Further evidence came from the finding that the salt gland of the herring gull had a much higher concentration of CS than of cerebroside.[24] This is the gland that helps the bird eliminate excess salt, a need that arises when the birds scoop fish (and sea water) out of the sea. The gland contained 6 mg/g dry weight, which can be compared with the sphingomyelin content (14 mg/g) and cerebroside content (0.8 mg/g).

Evidence against involvement in Na^+ pumping comes from an immunostaining localization study in kidney with antibodies against CS.[25] This showed that CS is sharply localized in the thick ascending limb of the loop of Henle and in the initial portion of the distal convoluted tubule. In the latter region, the lipid is confined to the luminal membrane, whereas Na^+–K^+ ATPase is located at other sites. It was also noted[26] that there is a striking similarity in the distributions of CS and the Tamm–Horsfall protein, from which it was suggested that the lipid is involved in Cl^- transport.

An important point eliminating a simple relationship between CS and ATPase is that the antibodies to CS did not inhibit the kidney ATPase.

5.2. Opiate Receptor

Various types of evidence for a specific role of CS in the morphine receptor site have been offered by Horace Loh and colleagues.[26a] A partially purified opiate receptor preparation was found to have a high CS content. A variety of liquid/liquid partitioning experiments seemed to demonstrate a special affinity between opiates (and related compounds) and CS. Opiates and opiate agonists are basic lipids, and it is not surprising that they should show competitive affinity for an acidic lipid. Of the different acidic lipids tested, CS showed the highest affinity in various physical measurements.

The strongest evidence has been the demonstration that antibodies to CS injected into the brains of intact rats counteracted the narcotic action of morphine.[27] The antibodies were prepared by repeated injection of rabbits with CS mixed with lecithin, cholesterol, and methylated bovine serum albumin and then purified by immunoabsorption. Additional evidence came from the observation that fluorescein-labeled antibodies to CS could bind to brain regions known to be high in opiate receptors. As expected, the binding in these regions was blocked by opiate agonists, which did not block binding to myelinated regions.

5.3. Other Possibilities

A long time ago, this author suggested that CS may function as a memory deposit, much like the ink in a book. This idea was based on the high metabolic inertness of the lipid (shown with ^{35}S), the fact that it accumulates with age, and the possibility that memory is deposited in specific locations in a coded form, to be read out the way we read printed information. Perhaps it is a "memocule."

Perhaps the basic protein of myelin is bound to CS in salt linkage. Some support for this idea comes from the observation that one can extract a complex consisting of basic protein, CS, gangliosides, and phosphatidylserine[28] (see also Section 6.2).

One of the steps in clotting, the activation of factor XII by plasma kallikrein, is accelerated by CS. However, CS does not seem to have been found in plasma.

6. PATHOLOGY

6.1. Metachromatic Leukodystrophy or Sulfatidosis

Metachromatic leukodystrophy is a human genetic disorder in which sulfatase A is defective, CS accumulates in tissues, and various nervous system functions fail. The CS occurs in metachromatic deposits in tissues and urine. The defective hydrolase cross reacts with antibodies to the normal enzyme, as might be expected for a protein bearing a mutated locus. However, there is evidence for different kinds of mutations in different patients.

6.2. Experimental Allergic Encephalomyelitis

Sulfatide metabolism has been implicated in several ways in the development of EAE as a result of inoculation with white matter in Freund's adjuvant. The concentration of galactocerebroside and CS in rat brain decreased somewhat during development of the disorder and returned to normal during the recovery stage. The ability to form radioactive phosphoadenosine phosphosulfate and CS *in vitro* increased with severity of the symptoms. However the CS synthesis assays were run using endogenous cerebroside, which may have given misleading results.

Large decreases in CS content of guinea pig brain have been produced by inoculating the animals with Freund's adjuvant containing central nervous tissue, basic protein, basic protein mixed with spinal cord lipids, bovine serum albumin mixed with spinal cord lipids, polylysine, or polylysine mixed with spinal cord lipids.[29] The latter three treatments did not produce characteristic clinical symptoms of EAE, yet caused similar decreases in CS levels. Only the inocula containing lipids produced a marked decrease in levels of cerebroside.

Cultures of cerebellum from newborn mice and spinal cord from embryonic mice develop somewhat normally, with eventual formation of myelin. The rate of CS synthesis was found to increase with age, especially sharply just before myelin became visible. Both myelin formation and sulfotransferase increase could be inhibited by adding the serum of rabbits exhibiting EAE.[30] The inhibitory action of the serum on the enzyme developed rapidly, and it might therefore be the primary event in the development of the encephalomyelitis in intact animals. This intriguing observation ought to be checked with purified antibodies to see if the active component is antibasic protein or possibly antisulfotransferase. A similar interference in CS biosynthesis by antibodies against myelin basic protein has been demonstrated.[23] Here too there is evidence of a linkage between the basic protein and CS (see Section 5.3, paragraph 2). It would be of interest to check the inhibitory effects with subcellular preparations or purified sulfotransferase.

6.3. Miscellaneous Aspects of Pathology

A suggestion by Wago[31] that CS could be useful in the therapy of atherosclerosis deserves thought. He showed that the addition of CS slowed the clotting of plasma (initially protected against clotting with oxalate, then treated with Ca^{2+}). Thromboplastin formation was slowed at concentrations of 40 $\mu g/$ml. It was found possible to inject rabbits intravenously every other day for long periods of time with 10 mg of CS (not quite pure) dispersed in saline. The rabbits were maintained on a cholesterol-supplemented diet and developed atherosclerotic lesions. The CS treatment significantly reduced serum lipid and cholesterol concentrations and the α/β lipoprotein index. This ought to be repeated with purer CS.

Vitamin A may be involved in the synthesis of sulfate esters. An initial report indicating that its lack blocked CS synthesis in rat brain was contradicted by a study in which the control animals were pair fed. It would seem from this that mere depletion of food intake could produce a slowing of CS synthesis, even in brain. A more recent study of seminolipid in rat testis seemed to show a specific need for vitamin A. However, here too the control rats were not pair fed, and there was a marked difference in organ weights. Other studies with various kinds of depletion diets have shown that young animals are quite sensitive with respect to CS deposition. It is not clear from the depletion experiments whether the sulfation step or galactosylation step or an earlier step is the critical one. It is possible that the level of phosphoadenosine phosphosulfate is readily lowered and that other sulfation reactions suffer too.

An animal model of phenylketonuria was produced in rats by administration of phenylalanine and an inhibitor of phenylalanine metabolism, 2-methylphenylalanine.[32] This produced high blood levels of the amino acid as well as impairment of body and brain growth and myelin deposition. The CS content of the myelin was preferentially reduced.

REFERENCES

1. Ishizuka, I., Inomata, M., Ueno, K., and Yamakawa, T., 1978. *J. Biol. Chem.* **253**:898–907.
2. Breyer, U., 1965, *J. Neurochem.* **12**:131–133.
3. Nonaka, G., Kishimoto, Y., Seyama, Y., and Yamakawa, T., 1979, *J. Biochem.* **85**:511–518.
4. Hara, A., and Radin, N. S., 1979, *Anal. Biochem.* **100**:364–370.
5. Nonaka, G., and Kishimoto, Y., 1979, *Biochim. Biophys. Acta* **572**:423–431.
6. Nonaka, G., and Kishimoto, Y., 1979, *J. Neurochem.* **33**:23–27.
7. Raghavan, S. S., Gajewski, A., and Kolodny, E. H., 1981, *J. Neurochem.* **36**:724–731.
8. Suzuki, A., Sato, M., Handa, S., Muto, Y., and Yamakawa, T., 1977, *J. Biochem.* (*Tokyo*) **82**:461–467.
9. Kean, E. L., 1968, *J. Lipid Res.* **9**:319–327.
10. Fluharty, A. L., Davis, M. L., Kihara, H., and Kritchevsky, G., 1974, *Lipids* **9**:865–869.
11. Nonaka, G., and Kishimoto, Y., 1979, *Biochim. Biophys. Acta* **572**::432–441.
12. Chou, K.-H., and Jungalwala, F. B., 1981, *J. Neurochem.* **36**:394–401.
13. Rouser, G., and Yamamoto, A., 1968, *Lipids* **3**:284–287.
14. Zalc, B., Monge, M., Dupouey, P., Hauw, J. J., and Baumann, N., 1981, *Brain Res.* **211**:341–354.
15. Siegrist, H. P., Burkart, T., Wiesmann, U. N., Herschkowitz, N. N., and Spycher, M. A., 1979, *J. Neurochem.* **33**:497–504.

16. Sarlieve, L. L., Neskovic, N. M., Rebel, G., and Mandel, P., 1976, *J. Neurochem.* **26**:211–215.

16a. Siegist, H. P., Burkart, T., Wiesmann, U., and Herschkowitz, N. N., 1976, *J. Neurochem.* **27**::599–604.

17. Dawson, G., and Kernes, S. M., 1979, *J. Biol. Chem.* **254**:163–167.

18. Van Etten, R. L., and Waheed, A., 1980, *Arch. Biochem. Biophys.* **202**:366–373.

19. Draper, R. K., Fiskum, G. M., and Edmond, J., 1976, *Arch. Biochem. Biophys.* **177**::525–538.

20. Fischer, G., and Jatzkewitz, H., 1977, *Biochim. Biophys. Acta* **481**:561–572.

21. Herschkowitz, N., McKhann, G. M., Saxena, S., and Shooter, E. M., 1968, *J. Neurochem.* **15**:1181–1188.

22. Glende, E. A., and Cornatzer, W. E., 1963, *J. Pharmacol. Exp. Ther.* **139**::377–382.

23. Pellkofer, R., and Jatzkewitz, H., 1976, *J. Neurochem.* **27**::351–354.

24. Karlsson, K.-A., Samuelsson, B. E., and Steen, G. O., 1974. *Eur. J. Biochem.* **46**:243–258.

25. Zalc, B., Helwig, J. J., Ghandour, M. S., and Sarlieve, L., 1978, *FEBS Lett.* **92**:92–96.

26. Zalc, B., Ollier-Hartmann, M. P., and Baumann, N., 1981, Glycoconjugates, in: *Proceedings of the Sixth International Symposium on Glycoconjugates (Tokyo)* (T. Yamakawa, T. Olsawa, and S. Handa, eds.), Japan Scientific Societies Press, Tokyo, pp. 514–515.

26a. Loh, H. H., Law, P. Y., Ostwald, T., Cho, T. M., and Way, E. L. 1978, *Federation Proc.* **37**:147–152.

27. Craves, F. B., Zalc, B., Leybin, L., Baumann, N., and Loh, H. H., 1980, *Science* **207**:75–76.

28. Kunishita, T. K., Uemura, A., Okano, A., and Taketomi, T., 1979, *Jpn. J. Exp. Med.* **49**:391–396.

29. Maggio, B., and Cumar, F. A., 1975, *Nature* **243**::364–365.

30. Fry, J. M., Lehrer, G. M., and Bornstein, M. B., 1972, *Science* **175**:192–194.

31. Wago, K., 1961, *Jpn. Heart J.* **2**:354–367.

32. Johnson, R. C., and Shah, S. N., 1980, *Neurochem. Res.* **5**::709–718.

GENERAL REVIEWS

1. Farooqui, A. A., Rebel, G., and Mandel, P., 1977, *Life Sci.* **20**:569–584.

2. Balasubramanian, A. S., and Bachhawat, B. K., 1970, *Brain Res.* **20**:341–360.

3. Dulaney, J. T., and Moser H. W., 1978, *The Metabolic Basis of Inherited Disease* (J. Stanbury, J. B. Wyngaarden, and D. S. Fredrickson, eds.), McGraw-Hill, New York, pp. 770–809.

4. Murray, R. K., Levine, M., and Kornblatt, M. J., 1976, *Glycolipid Methodology* (L. A. Witting, ed.), American Oil Chemists' Society, Champaign, Illinois, pp. 305–327.

5. Farooqui, A. A., 1980, *Clin. Chim. Acta* **100**:285–299.

Ceramides and Cerebrosides

Yasuo Kishimoto

1. CERAMIDES

1.1. Structure and Distribution

1.1.1. Structure and Types of Ceramides

Ceramide* is the basic component of sphingolipids and a key intermediate of both synthesis and degradation of all sphingolipids.[1] Major brain sphingolipids are cerebrosides, sulfatides, sphingomyelins, and gangliosides. These are unique to and abundant in the nervous system. In addition, there is a small quantity of free ceramide in brain. As shown in Fig. 1, ceramide is a fatty acid amide of a sphingoid base. The major brain sphingoid base (over 80% of total) is sphingosine [(2S,3R,4E)-2-amino-4-octadecene-1,3-diol, with a D-*erythro* configuration].[2] Up to 15% of all sphingoid base in brain sphingolipids is sphinganine (dihydrosphingosine). Other minor sphingoid bases include icosasphingosine and hexadecasphingosine. 4-D-Hydroxysphinganine (phytosphingosine; the extra hydroxyl group has an R configuration), which is abundant in plant and yeast, is not found in the nervous system.

The fatty acids in brain ceramide are more complex. They can be divided into four groups: unsubstituted C_{14}–C_{20} carbon chain (nonhydroxy long chain), unsubstituted C_{22}–C_{27} (nonhydroxy very long chain), α-hydroxylated C_{14}–C_{20} (hydroxy long chain), and α-hydroxylated C_{22}–C_{27} (hydroxy very long chain). All of these fatty acids have straight chains, generally either saturated or monounsaturated, although there is a very small proportion of diunsaturated

---

* We followed the IUPAC-IUB nomenclature of lipids found in *Mol. Cell. Biochem.* **17**:158–171. Cerebrosides containing galactose and glucose were specifically called galactocerebroside and glucocerebroside, respectively. Fatty acids were abbreviated according to their carbon number followed by the number of their double bond; e.g., 18:0 and 24:1 are stearic acid and nervonic acid, respectively. Fatty acids containing an α-hydroxyl group are designated by h followed by the carbon number; thus, 24h:0 is α-hydroxylignoceric acid (cerebronic acid).

Yasuo Kishimoto • John F. Kennedy Institute and the Department of Neurology, Johns Hopkins University School of Medicine, Baltimore, Maryland 21205

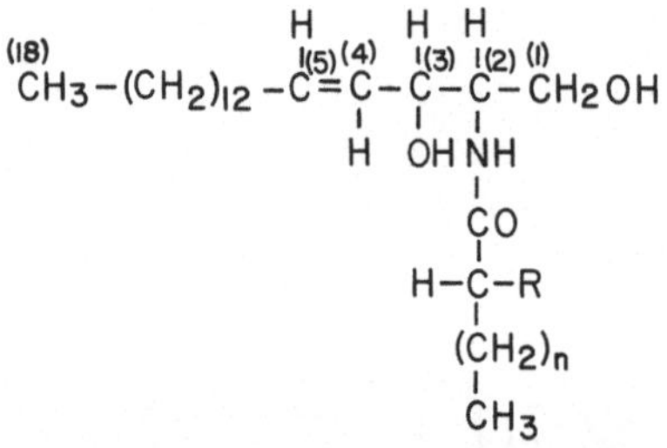

$$CH_3-(CH_2)_{12}-\overset{(18)}{\underset{H}{\overset{H}{\underset{|}{C}}}}=\overset{(5)}{\underset{|}{\overset{(4)}{C}}}-\overset{H}{\underset{OH}{\overset{(3)}{\underset{|}{C}}}}-\overset{H}{\underset{NH}{\overset{(2)}{\underset{|}{C}}}}-\overset{(1)}{CH_2}OH$$

$$\underset{\underset{\underset{\underset{CH_3}{|}}{(CH_2)_n}}{\underset{H-C-R}{|}}}{CO}$$

Fig. 1. Basic structure of ceramide. Structure of N-acyl sphingosine is shown. R is either –H (nonhydroxy) or –OH (hydroxy); n varies from 11 to 24. In ceramide containing sphinganine, the double bond in C_4–C_5 is saturated.

carbon chains.[3] All of the unsaturated bonds have a *cis* configuration and most commonly are situated either on the seventh or ninth carbon from the methyl end. In addition, there are a significant number of double bonds situated in the other positions. With these variations, it is assumed that brain sphingolipid fatty acids include more than 100 homologues and isomers, and when different sphingoid bases are taken into account, the variety of ceramides numbers several hundred.

Each brain sphingolipid contains a somewhat different ceramide species. The ceramide moieties of galactocerebrosides and sulfatides are similar to each other, reflecting their close metabolic relationship. Although both hydroxy and nonhydroxy very-long-chain fatty acids are the dominant fatty acids in cerebrosides (and less so in sulfatides), these sphingolipids also contain shorter-chain homologues. Ganglioside fatty acid has a simple composition, about 90% stearic acid (18:0) with varying amounts of palmitic (16:0) and eicosanoic (20:0) acids.[4] There is no α-hydroxy fatty acid. The sphingoid base, on the other hand, is more complex than in cerebrosides. There is a significant proportion of icosasphingosine.[5] Most cerebroside and sulfatide sphingoid bases are sphingosine (major) and sphinganine (minor). The variation in the ceramide moiety of sphingomyelin falls somewhere between cerebrosides and gangliosides; however, there are no hydroxy fatty acids.[6] In short, the ceramide composition in these sphingolipids probably reflects the site of synthesis, either neuron (shorter-chain, nonhydroxy fatty acids with more variation in sphingoid bases) or glia (longer chain with an abundance of hydroxy acids and a simple sphingoid base composition[7]). Molecular species of free ceramides in brain are discussed in Sections 1.1.5 and 1.1.6.

1.1.2. Chemistry

1.1.2a. Chemical and Physical Characteristics. Because it contains two long hydrocarbon chains, ceramide is extremely nonpolar in nature. Barely soluble in water, it is more soluble in organic solvents. However, ceramide also contains a small but quite active polar center in carbons 1 through 5 (Fig. 1). The C_1 hydroxyl group appears to be quite accessible to various transferase groups. Thus, galactose, glucose (Section 2.2.1a), and phosphorylcholine transferases[8] attach their moieties to the hydroxyl group. The C_3 hydroxyl group in sphingosine is allylic, but the one in sphinganine is not. The allylic

C_3 hydroxyl group can be readily oxidized to a ketone by a mild dehydrogenating reagent such as dichlorodicyanobenzoquinone or dichromate, which does not affect other hydroxyl groups.[9,10] In addition, the allylic hydroxyl group is involved in side reactions during hydrolysis of ceramide (Section 1.1.2d).

1.1.2b. Chemical Synthesis. Ceramides are synthesized by the condensation of sphingosine and fatty acid. Sphingosine can be prepared by hydrolysis of sphingolipids (see Section 1.1.2d) or by total chemical synthesis.[11] Although the chemically synthesized sphingosine is racemic, D- and L-sphingosine can be separated as the 0-acetyl-L-mandelate.[12] α-Hydroxy fatty acids prepared by chemical synthesis are also racemic, but these too can be separated as the acetyl-mandelate esters.[13]

Fatty acids are usually activated to acyl chloride[14] or N-hydroxysuccinimide ester[15] or by carbodiimide[16] for the condensation. Recently, a new synthesis directly condensing sphingosine and free fatty acid by an oxidation–reduction reagent has been developed.[17] For the synthesis of hydroxyceramide, the hydroxyl group in the fatty acid is covered by an acetyl group in all synthesis methods. The acetyl group is removed by mild alkaline methanolysis. Either sphingosine or the α-hydroxy fatty acid for hydroxyceramide need not be stereochemically pure, because the diastereomers that are formed can be separated by TLC.[17,18]

1.1.2c. Synthesis of Radioactively Labeled Ceramide. There are at least four different ways of labeling ceramide:

1. Catalytic reduction of the (4*E*)-double bond of the sphingosine moiety by tritium.[19,20] A labeled ceramide containing only sphinganine can be prepared by this method. If natural ceramide (either from free ceramide or from sphingomyelin or cerebrosides) is used as the starting material, part of the tritium is incorporated into the fatty acid in addition to the sphingosine moiety.
2. Acylation of sphingosine or sphinganine with a labeled fatty acid (see Section 1.1.2b for the synthesis). The preparation can be done with a labeled sphingoid base and a nonradioactive fatty acid or vice versa. The preparation of specifically labeled sphingoid bases[12,21,22] and fatty acids[23–25] has been reported.
3. Oxidation of the 3-hydroxyl group of sphingosine by dichlorodicyanobenzoquinone and the reduction of the 3-ketoceramide with Na-B^3H_4.[26] In addition to the labeled ceramide containing sphingosine, which is the major product, small portions of the D-*threo* isomer and sphinganine-containing labeled ceramide will be obtained.
4. The reduction of D-*erythro*-2-fatty acylamino-3-hydroxy-4*E*-octadecenoate by NaB^3H_4.[12] The starting material can be totally synthesized from myristoyl aldehyde.

With these methods, it is possible to prepare ceramide labeled with tritium either at the 1 or 3 position of sphingosine, 4 or 5 position of sphinganine, or the fatty acid. In addition, ceramide can also be labeled with ^3H or ^{14}C in the fatty acid. It is also possible that the C_3 carbon of sphingosine can be labeled by utilizing the total synthesis.[11] By a combination of these, a double-labeled ceramide can readily be prepared for metabolic studies.

1.1.2d. Hydrolysis. Hydrolysis involves the cleavage of the amide bond, which is somewhat difficult to achieve. This cleavage can be accomplished with a strong acid such as by boiling with methanolic HCl[27] or methanolic BF_3[28] for a prolonged time. One severe problem with acid hydrolysis is the instability of the allylic hydroxyl group at C_3 of sphingosine. Because of the ready mobility of this group under these conditions, isomerization at the C_3 chiral center to form D-*threo*-sphingosine, replacement of the hydroxyl group by a methoxyl group to form 3-O-methyl (both D and L) sphingosine, and migration of the double bond to the $C_{3,4}$ position accompanied by the migration of the hydroxyl group to the 5 position can occur.[29,30] These isomerization reactions can be minimized by the addition of water to the methanolic solution.

The amide cleavage can be achieved in a much shorter time when a strong alkali is used at higher temperatures, such as boiling with KOH in 1,2-propanediol.[31] The quantitative recovery of the sphingosine is, however, not known.

1.1.3. Analytical Methods

1.1.3a. Qualitative Analysis. Qualitative analysis of brain ceramide has been most conveniently done by TLC. Many commonly used TLC solvents separate ceramides into two bands according to chain length provided one applies the sample as a narrow band at the origin. The ceramides containing saturated acids and those containing unsaturated fatty acids separate well on a silica gel G plate by using the chloroform–methanol–acetic acid system.[32] By using borate-, arsenite-, and silver ion-impregnated as well as regular silica gel plates, the separation of ceramides according to the number of hydroxyl groups and unsaturation has been reported.[33,34] There are, however, no reagents that give a specific color to the spot of ceramide.

Fatty acid composition is most conveniently determined by GLC, as described in Section 1.1.3b. Sphingoid composition is determined by GLC of the trimethylsilyl ether[35,36] or acetate-trimethylsilyl ether.[37] In addition, reverse-phase HPLC of perbenzoylated ceramide[38] and mass spectrometric analysis of the trimethylsyl derivative[36,39–42] provide information on the structure; PMR spectra of ceramide and sphingoid base have been published.[43]

1.1.3b. Quantitative Analysis. Quantitative analysis can be done in three ways, by determining (1) the sphingoid moiety, (2) the fatty acid moiety, or (3) the whole molecule. The first two methods require prior hydrolysis of the amide bond as discussed in Section 1.1.2d. The amount of sphingoid base can be determined by the colorimetric[44,45] or fluorimetric method.[46–49] The amount of fatty acid can also be determined by colorimetry[50] or GLC as methyl esters.[4,51] Usually an internal standard such as 19:0 or 21:0 fatty acid ester is added for quantitation. The hydroxyl group should be converted to methyl ether[52] or trimethyl silyl ether.[53]

Quantitation of the whole molecule can be achieved by HPLC, GLC, or TLC. For HPLC analysis, the ceramide is converted to a perbenzoyl derivative.[54–56] Depending on the benzoylation method used, the nitrogen of the amide group of nonhydroxy ceramide can be benzoylated. The benzoylated nonhydroxyceramide and hydroxyceramide can be separately quantitated in picomole amounts on a silica column by monitoring the effluent by measuring its absorption at 230 nm. The benzoylated ceramides can also be analyzed by reverse-phase HPLC for the determination of individual homologues.[38] Analysis of ceramides by HPLC can also be performed on the 3-keto derivative with nearly equal sensitivity.[57] For GLC or GLC–MS analysis, ceramide is converted to a trimethylsilyl ether[39–41] or to a boronate derivative.[58] A 2- to 20-μg quantity of ceramide can also be determined by TLC. The plate is charred, and the quantity is determined by measuring scintillation quenching.[59]

1.1.4. Isolation Methods

Thin-layer chromatographic preparation is the most convenient and reliable method. Since ceramide is usually a very minor component in brain, prefractionation is necessary. The most effective method is mild alkaline methanolysis followed by silica gel column chromatography.[60] The column is first rinsed with chloroform, and then the ceramide is eluted with chloroform–methanol 98:2. Alternatively, the total lipids can be extracted with a hexane–isopropanol mixture, the extract washed with sodium borate solution, and the lipid recovered in upper layer applied directly on TLC.[59] Ceramide homologues separate slightly on preparative TLC; thus, it is advisable to scrape an area extending outside the visible band.

Ceramides can also be prepared as conveniently from more complex sphingolipids. Cerebrosides are first oxidized by periodate followed by $NaBH_4$ reduction.[61] The product is then hydrolyzed by mild acid hydrolysis. Direct acid hydrolysis causes isomerization of the ceramide moiety as discussed in Section 1.1.2d. Nonhydroxyceramides can also be prepared from sphingomyelin with phospholipase C.[33]

1.1.5. Concentration in Brain

The nonhydroxyceramide concentration in human fetal brain is about 1% of the total lipids and 0.25% of the dry weight.[62] The concentration slightly increases to 1.3% of the total lipids in a 6-month-old and then decreases to 0.3% in a 55-year-old human brain. This change during aging was later confirmed by HPLC analysis of the ceramide level during rat brain development.[55] The levels of nonhydroxyceramide remained stable at 0.08–0.09% of the dry weight. Difference in concentration in human autopsied brain and rat brain may be caused by the autolysis of complex sphingolipids.[59]

The amount of hydroxyceramide in brain is very small. Although its presence was reported in human brain,[63] many investigators failed to find hydroxyceramide in brains of experimental animals.[62,64] Therefore, it has been assumed that the hydroxyceramide detected in human brain was a product of autolysis. However, the presence of hydroxyceramide in normal brain was confirmed recently by HPLC,[55] and its concentration was reported to be about 1% of the nonhydroxyceramide and 0.001% of the brain dry weight. The presence of a large amount of hydroxyceramide as well as an increased amount of nonhydroxyceramide in the brain of a patient with Farber's disease which is characterized by the lack of acid ceramidase was reported.[64]

1.1.6. Composition of Brain Ceramides

As discussed above, nearly all ceramides in brain contain nonhydroxy fatty acids. The fatty acid composition appears to change during development, as a reflection of the active cerebroside and sphingomyelin synthesis associated with increased myelination. For example, in a 14-day-old mouse brain, nearly all the fatty acid in ceramide is C_{14}–C_{20} (I. Singh and Y. Kishimoto, unpublished result). On the other hand, a significant proportion of very-long-chain fatty acids are reported in more mature brain.[55,65] Probably, the ceramide synthesized in neuronal cells contains mostly long-chain fatty acids and that made in oligodendroglial cells contains mostly very-long-chain fatty acids.

As described in Section 1.1.1, brain gangliosides contain significant proportions of icosasphingosine as the sphingoid base component. However, free ceramide isolated from rat brain at the age when active ganglioside synthesis is taking place does not contain detectable icosasphingosine.[9,64] This observation indicates that there is more than one pool of nonhydroxyceramide in brain.

1.2. Metabolism

1.2.1. Biosynthesis of Ceramide Precursors

1.2.1a. Fatty Acids. Although some fatty acids are taken up by the brain from blood and used for sphingolipid synthesis,[66] most fatty acids appear to be made *in situ*.[3] Most C_{18} or longer fatty acids of the sphingolipids are made from palmitic acid (16:0) by the chain-elongation system present in microsomes[67,68]; this acid is made by cytosolic *de novo* fatty acid synthetase.[69] α-Hydroxy fatty acids are made from the nonhydroxy fatty acid of corresponding chain length by α-hydroxylase, also present in brain microsomes.[70] Most of the unsaturation appears to occur on palmitic and stearic acids whose products are then elongated.[3]

1.2.1b. Sphingoid Bases. The first step is the condensation of palmitoyl CoA and serine with decarboxylation to produce 3-ketosphinganine.[71,72] This key intermediate can be reduced to sphinganine and then dehydrogenated to yield sphingosine.[73–75] The sequence of the last two steps is still ambiguous and needs clarification. The simultaneous occurrence of the reduction of C_3-ketone and dehydrogenation at C_4 and C_5 by simple migration of hydrogens was proposed.[76] At least a part of the last step of dehydrogenation was shown to occur after the formation of ceramide containing sphinganine.[22,77] The possibility of the first condensation of 3-ketosphinganine with fatty acyl CoA followed by dehydrogenation and then by the reduction of ketone has also been suggested.[78,79]

1.2.2. Biosynthesis of Ceramide

There are at least three different pathways of ceramide synthesis in brain.[1] First, ceramide is synthesized by the condensation of sphingosine and fatty acyl CoA.[32,72,80,81] Second, the synthesis is catalyzed by a reverse reaction of both acid and alkaline ceramidase and involves condensation of free fatty acid and sphingoid base.[82–84] Physiological significance *in vivo* of this reaction is not apparent. The third pathway is now under study in our laboratory.[87,88] This enzymatic reaction also involves free fatty acid as a substrate and, unlike the reverse ceramidase reaction, requires a pyridine nucleotide and two cytosolic factors, one heat stable and the other heat labile. It appeared to be present only in brain, and a very-long-chain fatty acid is the preferred substrate. Kinetic studies indicated that the free acid may be first converted to an activated form before being transferred to sphingosine. Coenzyme A, however, did not appear to be required. Later studies indicated that the addition of an excess of ATP can eliminate the effect of pyridine nucleotide and cytosolic factors.[89]

In the last system, part of the lignoceric acid is also converted to hydroxyceramide when NADPH and cytosolic factors are added.[90] Apparently, α-hydroxylation and ceramide synthesis are occurring simultaneously. However, free cerebronic acid is not converted to hydroxyceramide in this system.[25] The CoA esters of α-hydroxyl fatty acids were, however, converted to hydroxyceramide by mouse brain microsomes.[81] It has been proposed that there are four different ceramide-synthesizeing enzymes, one each for ceramide containing regular-chain fatty acids (C_{16}–C_{20}), very-long-chain fatty acids (C_{22}–C_{26}), α-hydroxy regular-chain fatty acids, and α-hydroxy very-long-chain fatty acids.[81]

1.2.3. Hydrolysis

Two different hydrolytic enzymes are reported. One is an acidic lysosomal enzyme, which has an optimal pH of 4.5.[82,83] Another has an alkaline or neutral pH optimum.[84] The former enzyme has been partially purified and characterized. Its apparent physiological significance was demonstrated when it was discovered that Farber's disease is caused by its deficiency (see Section 1.2.6.). The significance of the alkaline ceramidase is not apparent. As mentioned above (Section 1.2.2.), both enzymes also catalyze the reverse reaction.

1.2.4. In Vivo Metabolism

The *in vivo* conversion of the precursors of ceramide synthesis, such as serine[91] or sphingosine,[75,92] into nonhydroxyceramide was reported. A rapid exchange of the fatty acid moiety of ceramide was demonstrated in rat brain.[93] A further conversion of ceramide into sphingomyelins and cerebrosides was also demonstrated.[14,21,93]

1.2.5. Physiological Significance

The importance of ceramide as the backbone structure of all sphingolipids is apparent. It is generally accepted that the ceramide portion of the complex sphingolipid serves as an anchor for a more polar region, such as carbohydrates, and, by burying itself in a membrane, provides a polar environment on the cell surface. The structure of ceramide resembles that of diglyceride which is the anchor backbone for phospholipids. Both structures have two long aliphatic chains to form a basic membrane structure. The significant difference between ceramide and diglyceride is that ceramide contains an extra hydroxyl group which is allylic in nature. This hydroxyl group may provide extra stability for the membrane by providing a hydrogen bond.[94,95] The α-hydroxyl group of the fatty acid in hydroxyceramide and the extra proton in the amide also provide a hydrogen bond. In addition, the extra-long and more saturated fatty acid chain in ceramide should provide extra rigidity. These features must contribute structural stability to myelin, in which sphingolipids are abundant.

1.2.6. Ceramides in Pathological States

The deficiency of acid ceramidase was identified as the cause of Farber's disease.[96] Farber's disease is an autosomal recessive inherited disorder and is characterized by the accumulation of ceramide-laden macrophases in all of the patient's tissue, including the cerebellum.[97]

A decrease in the synthesis of hydroxyceramide[98] and ceramide containing very-long-chain nonhydroxy fatty acids[99] was reported in the brain of quaking mice, which have a moderate deficiency in myelination. Although the synthesis of hydroxyceramide is severely affected in jimpy mouse brain,[98] another myelin-deficient mutant, the synthesis of nonhydroxyceramide appears to be normal (I. Singh and Y. Kishimoto, unpublished observation). The observed deficiency in hydroxyceramide synthesis may result from diminished α-hydroxylation activity in this mutant. Decreased levels of ceramides containing nonhydroxy very-long-chain fatty acids were reported.[100] The levels of regular-chain fatty acid ceramides remain normal in this mutant. A higher than normal level of free ceramide in a brain tumor caused by treating rats with N-methylnitrosourea was reported.[56] The cause of the high ceramide level is not known but is probably a perturbation of sphingolipid metabolism in the tumor cell.

2. CEREBROSIDES

2.1. Structure and Distribution

2.1.1. Types of Cerebrosides

Cerebroside is a derivative of ceramide to which a single carbohydrate is attached to the primary alcohol of the sphingoid base (Fig. 2). In brain, the carbohydrate is either β-D-galactose (galactosyl ceramide or galactocerebroside) or α-D-glucose (glucosyl ceramide or glucocerebroside). Galactocerebroside is a major component of the myelin sheath[101] and contains sphingosine as

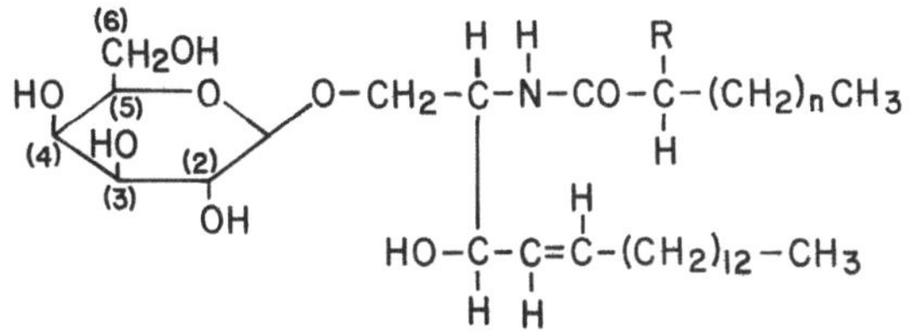

Fig. 2. Basic structure of cerebroside. The structure of galactosyl ceramide is shown. Glucosyl ceramide is a diastereomer of galactosyl ceramide in which the configuration of the 3-hydroxyl group is reversed. Sulfatide is the 3-O-sulfate ester of galactosyl ceramide. R is either –H (nonhydroxy) or –OH (hydroxy), and *n* varies from 11 to 24 or higher.

the major base component and sphinganine as the minor base component.[102] On the other hand, glucocerebroside is rich in neuronal cells and contains small amounts of icosasphingosine in addition to sphingosine and sphinganine.[103,104]

The fatty acids in these two cerebrosides are very different. Galactocerebrosides contain very-long-chain and regular long-chain fatty acids, both nonhydroxy and α-hydroxy.[3] Glucocerebrosides appear to contain a simple fatty acid composition. Although different fatty acid compositions of brain glucocerebroside have been reported from time to time, apparently the purest preparation contains stearic acid as the major component and palmitic and icosanoic acids as minor components.[104] This probably reflects its role as a precursor of ganglioside. Various techniques involved in cerebroside investigations have recently been compiled.[105]

Brain also contains a small amount of esterified cerebroside,[106] mostly in white matter. Both nonhydroxy and hydroxy cerebrosides are found in the ester form.[107] The ester-linked fatty acids are regular long chain and are attached to cerebrosides at galactose-6, galactose-3, or sphingosine-3.[107–109]

2.1.2. Physical and Chemical Properties

2.1.2a. Physical Properties. Although cerebrosides contain a carbohydrate, the nature of this lipid is determined by the two long, highly saturated alkyl chains of the ceramide moiety. Consequently, cerebrosides are much more soluble in organic solvents than in aqueous solvents and have a relatively low melting point. However, the hexose moiety gives the molecule its polar nature. Having the lipophilic ceramide moiety and the hydrophilic hexose, cerebroside is an ideally stable membrane component. Mass spectra of the trimethylsialyl and permethylated derivatives have been used for the elucidation of the fatty acid and sphingoid compositions.[110–114] The NMR spectra of cerebrosides were studied in connection with localizing it on the cell surface.[115–117]

2.1.2b. Chemical Synthesis. The only practical total synthesis of cerebroside consists of the condensation of 3-O-benzoylated ceramide and acetobromohexose[118] followed by mild alkaline saponification. The condensation was originally catalyzed by mercuric cyanide, but the use of stannous chloride[119] or ferric chloride[120] also appears to be satisfactory.

Various partial syntheses of radioisotopically labeled cerebrosides have been reported. Tritium can be introduced at the 6-carbon of the galactose moiety of galactocerebroside by first oxidizing galactocerebroside with microbial galactose oxidase.[121] 6-Dehydrocerebroside is then reduced with NaB^3H_4. Glucocerebroside is not this easy to label because the corresponding enzyme is not known. To label glucose with tritium, the primary alcohol of glucocerebroside is first converted to the trityl derivative, and then other hydroxyl groups are acetylated. The trityl group is removed, oxidized with chromic acid, and then reduced with NaB^3H_4. This is followed by alkaline methanolysis which yielded glucocerebroside labeled with ^3H at C_6.[122]

Tritium can also be placed on the C_3 of sphingosine. Cerebroside is first oxidized to 3-ketocerebroside by reacting it with dichlorodicyanobenzoquinone and then reducing it with NaB^3H_4. Unfortunately, the reduction yields a small portion of the *threo* isomer. The labeled cerebroside is saponified, and the *erythro*-psychosine separated from the *threo* isomer by column chromatography and reacylated.[26]

The third synthesis method is the acylation of psychosine with radioactive fatty acyl chloride.[14] Cerebroside is saponified with aqueous butanolic KOH at the boiling point to obtain psychosine.[123,124] The final and probably easiest method to prepare radioactive cerebroside is by catalytic reduction with tritium gas.[19,125] However, this method has the disadvantage that the product contains sphinganine, which is a minor component of brain cerebroside, and the labeling occurs not only in the sphingoid moiety but also in the fatty acid moiety.

2.1.2c. Chemical Cleavage. Acidic methanolysis has been most commonly used to study the fatty acid, hexose, and base compositions. Hydrochloric acid, sulfuric acid, and boron trifluoride have been used as the acidic component; however, the first acid is most frequently used.[27] Unfortunately, the cleavage has never been studied stoichiometrically. The recovery of the fatty acid methyl esters is usually lower than the calculated value. It is suspected that the hexose, sphingoid base, and fatty acid released by the methanolysis interact at a high temperature and thus yield uncharacterized side products. As described in Section 1.1.2d, the methanolysis causes isomerization and methylation of the sphingoid moiety. Therefore, it is recommended that aqueous methanolic HCl be used for the study of the sphingoid moiety.[35]

The fatty acid moiety can be cleaved without hydrolyzing the glycosidic linkage by alkaline saponification as described in Section 1.1.2b. Refluxing the KOH solution in 1,2-propanediol was used to quantitatively release the fatty acid moiety in a short time.[31]

The glycosidic linkage of the cerebroside can be released without causing sphingoid isomerization by removing the hexose moiety step by step,[61] first with periodate oxidation, then reduction and mild acid hydrolysis.

2.1.3. Analysis of Cerebrosides

2.1.3a. Qualitative Analysis. Thin-layer chromatography is still the most convenient method for the qualitative detection of cerebrosides. The total lipid extract or cerebroside-rich fraction can be applied to a TLC plate of silica gel.

The plate is developed by a neutral solvent such as chloroform–methanol–water (24:7:1), a basic solvent such as chloroform–methanol–2N NH$_4$OH (40:10:1), or an acidic solvent such as chloroform–methanol–acetic acid–water (80:13:8:1). Cerebroside spots can be detected either by a universal spray reagent such as bromothymol blue, sulfuric acid charring, or a specific coloring reagent such as α-naphthol–sulfuric acid.[126] Generally, cerebrosides give two or three spots. The top spot is nonhydroxycerebrosides, and the lower is hydroxycerebrosides. When the cerebrosides contain a considerable amount of regular long-chain α-hydroxy acids (such as 18h:0), these homologues give a third spot which is located below the hydroxycerebroside spot. The separation also appears to be influenced by number of double bonds and base composition.[127] Glucocerebrosides can be separated from galactocerebrosides on a silica gel G plate coated with sodium borate.[128]

For the analysis of the fatty acid and sphingoid base composition, see Section 1.1.3. Hexose can be identified by GLC as trimethylsilyl derivative.[129]

2.1.3b. Quantitative Determination. Cerebrosides can be quantitated in several ways. The quantitation of fatty acids by GLC and sphingoids by GLC or fluorometry, as discussed in Section 1.1.3b, can be used for cerebroside determination. These methods require the prior hydrolysis of the cerebrosides. In addition, cerebrosides can be quantitated from their sugar content. Various colorimetric methods such as those using anthrone-sulfuric acid[130] and GLC[129] have been developed. These procedures cannot be applied unless the cerebrosides are pure. Silica gel and Florisil® column chromatography, as discussed below, and TLC fractionation, as described above (Section 2.1.3a), can be used to purify the cerebrosides. With TLC fractionation, cerebrosides can be eluted from the powder or determined without elution. Some TLC powders may cause false peaks or color, and care should be taken to select the appropriate powder or to prewash the plate.

The recently developed HPLC method is more sensitive, simpler, and more specific.[131–133] The total lipid extract can be directly benzoylated and analyzed by HPLC. A column packed with microparticles of silica of 5-μm diameter is most suitable at the present time. A mixture of hexane–isopropanol is the best eluting solvent. The peak in the effluent can be determined by measuring its absorption at 230 nm. A gradient elution, which increases the solvent polarity, will shorten the run time and improve the shapes of the peaks. The benzoylated nonhydroxy- and hydroxycerebrosides give separate peaks. When the simultaneous determination of sulfatides is desirable, the benzoylated total lipids can be subjected to solvolysis. The solvolyzed products of perbenzoylated sulfatides yield peaks that are separated from perbenzoylated cerebrosides. Other minor sphingolipids such as glucocerebrosides and monogalactosyl diglyceride can also be separately determined under these conditions.[133] The effluents from cerebroside peaks can be collected and then studied further. Alternatively, these benzoylated cerebrosides can be separated by TLC.[134] Because of strong UV absorption, the spots can be easily detected by UV light on silica gel GF plates. These materials can be analyzed by reverse-phase HPLC to determine their homologue compositions.[38,135]

2.1.4. Isolation Methods

The most frequently used method for extracting cerebrosides from brain tissue is with chloroform–methanol, 2:1, followed by washing with a diluted salt solution.[136] Since chloroform has been identified as a potential carcinogen and is expensive, a new procedure which uses a hexane–isopropanol mixture has been developed.[137] The most effective method to purify cerebrosides uses mild alkaline methanolysis. The methanolyzed product can be further fractionated by either silica gel or Florisil® column chromatography.[3] Generally, the column is first rinsed with a nonpolar solvent such as chloroform to remove cholesterol and the fatty acid methyl esters formed by the mild alkaline methanolysis. A cerebroside-rich fraction can be eluted with either the chloroform–methanol mixture or acetone. A step-by-step method for eluting ceramide, cerebrosides, and sulfatides from the silica gel column has been developed.[105] The final purification of cerebrosides, however, usually requires preparative TLC. For a small-scale separation, the methanolyzed product can be placed on TLC without prior column fractionation.

The above method would be impractical if a large quantity of cerebroside is required because of the volume of solvent necessary. A new procedure was developed to minimize the amount of solvent and number of steps required.[138]

As described above (Section 2.1.3a), glucocerebrosides can be separated from galactocerebrosides by TLC on a borate-impregnated plate. A preparative method using column chromatography with a borate-impregnated silicic acid has recently been developed.[104]

2.1.5. Distribution and Concentration of Cerebrosides

2.1.5a. Distribution. Approximately 95% of the galactocerebrosides in rat brain appears to be associated with the myelin sheath.[139] The myelin fraction contains 90 nmol nonhydroxy- and 125 nmol hydroxycerebroside/mg total lipids. Accordingly, most cerebrosides are localized in the white matter, and the regional distribution also reflects the amount of myelin (or white matter) present. In young rat brain, the brainstem contains the highest amount of cerebrosides (25 μmol/g tissue), followed by cerebellum (14 μmol/g), diencephalon (10 μmol/g), and cerebral hemisphere (7 μmol/g). The ratio of nonhydroxy- to hydroxycerebrosides is almost constant in all regional areas and subcellular fractions. In contrast, glucocerebrosides appear to be predominant in gray matter and neurons. The distribution of this lipid has not been well studied because of the lack of adequate analysis methods.

2.1.5b. Homologue and Isomer Composition. As described briefly above, cerebrosides can be classified according to their hexose and fatty acid compositions. Galactocerebrosides generally contain more hydroxy fatty acids than nonhydroxy, whereas glucocerebrosides contain only nonhydroxy. The chain lengths of the nonhydroxy fatty acids of galactocerebrosides range from C_{14} to C_{27} or more. Lignoceric (24:0), nervonic (24:1), docosanoic (22:0), and stearic (18:0) are the major fatty acids, and 24:1 is generally the most predominant.[3] The chain lengths of the α-hydroxy fatty acids generally resemble those of the nonhydroxy fatty acids; however, α-hydroxystearic acid (18h:0) is only a minor component, whereas cerebronic acid (24h:0) is predominant. Beef hydroxycerebrosides are the exception and contain a significant amount of 18h:0. There are conflicting reports on the fatty acid composition of glucocerebrosides. This may be because glucocerebrosides are only a minor component and difficult to purify. The most recent report on the lipid prepared by borate silicic acid column chromatography indicated a simple fatty acid composition, predominantly 18:0 with much smaller portions of 16:0 and 20:0.[104] This fatty acid composition is similar to gangliosides which are the major lipids derived from glucocerebroside in brain.

2.1.5c. Change with Age. Since galactocerebrosides reside primarily in myelin, the concentration of this lipid in brain reflects the state of myelination. Thus, cerebrosides are barely detectable in rat brain before myelination occurs (20–50 nmol/g tissue).[139] The small amount of cerebroside found before myelination probably originates from oligodendroglia cell membranes.[7] The concentration starts increasing when the animal becomes 9–10 days old, and the increase is rapid during myelination (14–30 days) reaching 7–25 μmol/g tissue depending on the area of the brain (Section 2.1.5a). After myelination is complete, the concentration increases at a much slower rate as the animal matures.

The nonhydroxycerebroside concentration is usually higher than that of the hydroxy isomers in immature brain, but the ratio reverses as soon as myelination begins. The ratio remains almost constant thereafter, throughout the life of the animal. The fatty acid composition changes with the age of the animal.[140] For nonhydroxy fatty acids, regular long chain (mainly 18:0) predominate before 6 days of age. Very-long-chain fatty acids start appearing at the onset of myelination and predominate in mature brain. Thus, the proportion of very-long-chain fatty acids changes from zero at the third day after birth to 85% in adult rat brain. The changes in hydroxy fatty acids are less drastic. The proportion of very-long-chain fatty acids in 3-day-old rat brain is 55% and increases to 90% as soon as myelination begins.

Another characteristic change in the fatty acid pattern is the increase in odd-numbered very-long-chain fatty acids, both nonhydroxy and hydroxy.[3] There are no significant changes in the sphingoid base composition.

The concentration of glucocerebrosides is always low throughout brain development.[104,139] There are small increases in 5- to 10-postnatal-day-old rat brain, but the concentration appears to fall as the brain develops. There has not been a systematic study of the developmental changes of the fatty acid or sphingoid compositions of this lipid.

2.2. Metabolism of Cerebrosides

2.2.1. Biosynthesis

2.2.1a. Enzyme Characterization. Two pathways were proposed for the synthesis of galactocerebroside. One is the condensation of fatty acyl CoA and psychosine, and the other is the transferral of galactose from UDP-galactose to ceramide. Two studies support the first pathway: the synthesis of psychosine from sphingosine and UDP-galactose[141] and the condensation of psychosine and fatty acyl CoA by brain microsomes.[142] Although this pathway has been supported by two other investigators,[143,144] the condensation step could occur by a nonenzymatic reaction.[23,145] Further investigation is necessary to determine whether this pathway is physiologically significant.

The second pathway, the synthesis from ceramide by brain microsomes,[127,146–148] has been widely accepted. The enzyme activity is readily demonstrated with hydroxyceramide as a substrate but not with nonhydroxyceramide. For the latter substrate, the enzyme reaction has to be forced by a high concentration of phospholipids. It was proposed that this step is the rate-determining step of nonhydroxycerebroside synthesis.[23]

The rate-determining step of hydroxycerebroside synthesis appears to be α-hydroxylation of the fatty acid.[70] This conclusion was reached after the substrate specificity and deuterium isotope effect for α-hydroxylation were studies.[149] The V_{max} for each fatty acid was closely related to the abundance of the α-hydroxy fatty acid in brain cerebrosides. In addition to these two pathways, rat brain β-galactosidase was implicated in the transgalactosylation between ceramide and galactose.[150]

Glucocerebroside is also synthesized in brain microsomes from ceramide and UDP-glucose.[151,152] Oddly, although brain glucocerebroside does not contain any α-hydroxy fatty acids, the enzyme activity is more easily shown with hydroxyceramide.[152,153] This observation indicates that there is a lack of specificity for this enzyme.

The UDP-galactose:ceramide galactosyltransferase has been purified more than 100-fold from rat brain microsomes.[154] The enzyme was extracted with a detergent, cemolsol NPT-12, and the extract was purified by DEAE-cellulose chromatography, solvent extraction, pronase treatment, ammonium sulfate precipitation, and DEAE-Sephadex and CM-cellulose chromatography. The yield was 12.6%.

2.2.1b. Enzyme Localization. Since most brain galactocerebrosides are present in myelin, it is natural that most of the synthesis activity is located in oligodendroglia.[153,155] However, the enzyme is apparently present in neurons also.[155] The enzyme activity was originally found and characterized in the microsomal fraction,[127] especially from rough endoplasmic reticulum (ER).[156] However, the enzyme is apparently more widely distributed. A large amount of the ceramide:UDP-galactose galactosyltransferase was found in myelin.[156,157] Although the developmental changes of the myelin transferase are somewhat different, the nature of this enzyme is identical to the microsomal enzyme.[158] There is also an indication that axolemma contains the same enzyme[159]; a significant portion of the enzyme is also located in the Golgi apparatus.[160]

The function of this enzyme in the different locations has not been defined. It has been suggested that the myelin enzyme is used for the maintenance and normal turnover of cerebrosides in myelin. The mechanism that is involved in transferring the cerebrosides newly synthesized in microsomes or Golgi apparatus to the myelin assembly site is not known. The small amount of cerebrosides found in brain cytosol has been indicated as a possible intermediate in the transfer.[161]

Unlike galactocerebroside, glucocerebroside is a neuronal lipid, and the enzyme for its synthesis appears to be located mostly in the neurons.[155] This enzyme activity appears to be localized in microsomes, especially from smooth ER.[156,162]

2.2.1c. Control of Synthesis. Because galactocerebroside is a myelin lipid, the synthesis activity increases and decreases according to myelination.[152,163] Glucocerebroside is mainly a neuronal lipid and serves as a precursor for ganglioside; thus, the peak of synthesis appears earlier.[152,156,163]

The synthesis of galacto- as well as glucocerebroside is inhibited *in vitro* by an endogenous material that is extracted from rat brain microsomes.[164] This inhibitor is a nondialyzable heat-stable protein mixture. The synthesis is also inhibited by brain cytosol.[165] This inhibition is apparently caused by nucleotide pyrophosphatase which competes for UDP-galactose. The synthesis is also affected by a number of synthetic compounds whose structures are similar to ceramide.[166] The inhibition appears to be noncompetitive. Injection of cycloheximide or puromycin into rat brain reduced UDP-galactose:ceramide glucosyltransferase but not UDP-galactose:ceramide galactosyltransferase.[167] This finding implies that the glucosyltransferase has a faster turnover rate than galactosyltransferase.

The transferases for the synthesis of gluco- and galactocerebrosides both appear to have rather low substrate specificities. Although brain glucocerebrosides do not contain α-hydroxy fatty acids, the glucosyltransferase has nearly as high activity with hydroxyceramide as with nonhydroxyceramide.[168] The galactosyltransferase activity is also much higher with hydroxyceramide and shows highest activity with short-chain hydroxyceramides, which have not been detected in brain.[168]

2.2.2. Enzymatic Hydrolysis

2.2.2a. Enzyme Characterization. Both gluco- and galactocerebrosides are hydrolyzed to ceramides and hexose by brain enzymes. Both the β-glucosidase and β-galactosidase that have been characterized thus far are of lysosomal origin and have an acidic optimal pH.[169] These lysosomal enzymes are generally highly stimulated by detergents, which appear not only to help solubilize substrates not soluble in water but also to provide a suitable environment for the enzyme.

The study of β-galactosidase which hydrolyzes galactocerebroside in brain[170] was highly accelerated by the discovery of the enzyme defect in Krabbe's disease.[171] After a period of confusion, it is now accepted that the enzyme responsible for the hydrolysis of galactocerebroside is different from other β-galactosidases that also occur in brain and other tissue and hydrolyze galactose from lactosylceramide and G_{M1b}-gangliosides.[172,173] A 50-fold purification of rat brain galactocerebrosidase has been reported.[173] β-Glucosidase which hydrolyzes glucocerebroside has also been extensively studied, especially in connection with Gaucher's disease.[174] This enzyme has been extensively purified from rat intestine and human placenta.

Cerebrosides can be hydrolyzed by cleaving the amide linkage, thus producing psychosine and fatty acids. However, this pathway is not significant in normal tissue.[175] In tissue from patients with Krabbe's and Gaucher's diseases, a portion of the excess cerebrosides could be disposed of by the amide cleavage. In fact, a small amount of glucosyl sphingosine and galactosyl sphingosine has been reported in the tissue from patients with these diseases and has been suggested as the cause of pathogenesis.[176,177]

2.2.2b. Control Mechanism. As described above, the hydrolytic enzymes for cerebrosides are of lysosomal origin and do not seem to be affected by the age of the animal or brain development including myelination.[152] Many compounds resembling ceramides and cerebrosides were chemically synthesized and found to inhibit the cerebrosidase activity.[18,166] These inhibitors and stimulators should be useful in making an animal model of Gaucher's disease. Conduritol B epoxide, an analogue of gluconolactone, is effective in reducing glucocerebrosidase activity in rat *in vivo*.[178] Phenylhydrazine was shown to increase the specific activity of glucocerebrosidase in liver and spleen of mouse and to accelerate the reappearance of glucocerebrosidase activity after conduritol B epoxide treatment.[179] Glucocerebrosidase is also inhibited by glycoamides.[180]

Several natural compounds stimulate cerebrosidases. A thermostable glycoprotein, "P factor," and an unknown enzyme, "C factor," were found to be necessary to reconstitute glucocerebrosidase activity.[181] C factor is largely absent in Gaucher's tissues. The physiological significance of the activating factors has been questioned, however.[182,183] A relatively small heat-stable protein that stimulates galactocerebrosidase as well as glucocerebrosidase activity was isolated from the spleen of a patient with Gaucher's disease[184] and from bovine spleen.[185] The protein was shown to form a complex with glucocerebrosidase.[186] The stimulation occurs in the absence of the detergent that helps to solubilize substrates but is optimal when phosphatidyl serine is added. The stimulation of galactocerebrosidase activity by phosphatidyl serine was previously reported.[187]

2.3. In Vivo Metabolism

Numerous reports have been made of the incorporation and disappearance of radioactivity in brain cerebrosides of experimental animals after radioactive precursors were injected.[91,188] The precursors included glucose, galactose, fatty acids, acetate, sphingosine, L-serine, and ceramide. Most of these studies focused on galactocerebrosides. All of these experiments indicated a relatively quick synthesis of cerebroside from the precursors and slow turnover. Studies of turnover in individual subcellular organelles indicated that most cerebrosides are synthesized in microsomal or Golgi apparatus membranes and then transferred to myelin, possibly with the involvement of a cytosol component.[161] The turnover rate is generally high when the animal is young and in the stage of active myelination. The rate becomes much slower when the animal gets older. The turnover of glucocerebroside in brain is less well understood because the quantity of this lipid is limited.

2.4. Physiological Significance

The most apparent physiological role of galactocerebroside is as the major component of myelin.[101] As discussed in Section 1.2.5, the ceramide portion of cerebrosides provides the stability and inertness of the myelin sheath. The galactose moiety is supposed to be on the surface of the individual layer, whether it is facing the cytoplasmic or external side, and serves as an adhesive between the two membranes. Recently, over 50% of the myelin cerebroside was reported to be on the external surface.[189] This conclusion was made by treating myelin with galactose oxidase and measuring the unaltered remaining cerebrosides. However, our recent study indicated that galactose oxidase cannot react with cerebroside in myelin, probably because of steric inhibition.[134] Besides myelin, axons, axolemma, and oligodendroglia contain relatively high concentrations of galactocerebrosides.[7] The physiological role of this lipid in these locations is not understood but is assumed to be for membrane recognition and adhesiveness.

2.5. Pathological Aspects

As discussed above (Section 2.2.2a), the deficiency of galactocerebrosidase and glucocerebrosidase is the primary cause of Krabbe's and Gaucher's diseases, respectively. Although the enzyme defects are apparent in these diseases, their pathogenesis is not yet understood. Many acquired demyelinating diseases may be caused by autoimmune reactions, and cerebrosides are believed to play a critical role acting as the surface antigen, either on oligodendroglia[190] or myelin.[191] This aspect was recently demonstrated by injecting anticerebrosides into rat sciatic nerve which caused demyelination.[192] Similar antibodies also cause demyelination in myelinating organotypic culture of mouse cerebellum and spinal cord.[193,194] Since cerebroside is the major characteristic lipid of myelin sheath, it is natural to observe the significant reduction of cerebrosides in brains of patients with various demyelinating diseases.[195] In brains of myelin-deficient mouse mutants, such as jimpy and quaking, a significant but not specific decrease of cerebroside synthetic activity was reported.[100]

ACKNOWLEDGMENTS. The author is grateful to Dr. Norman S. Radin, University of Michigan, for his critical reading and to Ms. Geneva R. Davis and Ms. Janice White for skillful editorial and typing assistance.

REFERENCES

1. Kishimoto, Y., and Kawamura, N., 1979, *Mol. Cell. Biochem.* **23**:17–25.
2. Karlsson, K.-A., 1970, *Chem. Phys. Lipids* **5**:6–43.

3. Kishimoto, Y., and Radin, N. S., 1966, *Lipids* **1**:47–61.

4. Kishimoto, Y., and Radin, N. S., 1965, *J. Lipid Res.* **7**:141–145.

5. Klenk, E., and Huang, R. T. C., 1970, *Z. Physiol. Chem.* **351**:839–842.

6. O'Brien, J. S., and Rouser, G., 1964, *J. Lipid Res.* **5**:339–342.

7. Norton, W. T., Abe, T., Poduslo, S. E., and DeVries, G. H., 1975, *J. Neurosci. Res.* **1**:57–75.

8. Ullman, M. D., and Radin, N. S., 1974, *J. Biol. Chem.* **249**:1506–1512.

9. Kishimoto, Y., and Mitry, M. T., 1974, *Arch. Biochem. Biophys.* **161**:426–434.

10. Mendershausen, P. B., and Sweeley, C. C., 1969, *Biochemistry* **8**:2633–2635.

11. Shapiro, D., Segal, H., and Flowers, H. W., 1957, *J. Am. Chem. Soc.* **80**:1194–1197.

12. Shoyama, Y., Okabe, H., Kishimoto, Y., and Costello, C., 1978, *J. Lipid Res.* **19**:250–259.

13. Tatsumi, K., Kishimoto, Y., and Hignite, C., 1974, *Arch. Biochem. Biophys.* **165**:656–664.

14. Kopaczyk, K. C., and Radin, N. S., 1965, *J. Lipid Res.* **6**:140–145.

15. Ong, D. E., and Brady, R. N., 1972, *J. Lipid Res.* **13**:819–822.

16. Hammarström, S., 1971, *J. Lipid Res.* **12**:760–765.

17. Kishimoto, Y., 1975, *Chem. Phys. Lipids* **15**:33–36.

18. Arora, R. C., Lin, Y.-N., and Radin, N. S., 1973, *Arch. Biochem. Biophys.* **156**:77–83.

19. DiCesare, J. L., and Rapport, M. M., 1974, *Chem. Phys. Lipids* **13**:447–452.

20. Schwarzmann, G., 1978, *Biochim. Biophys. Acta* **529**:106–114.

21. Stoffel, W., and Sticht, G., 1967, *Z. Physiol. Chem.* **348**:1561–1569.

22. Ong, D. E., and Brady, R. N., 1973, *J. Biol. Chem.* **248**:3884–3888.

23. Morell, P., Costantino-Ceccarini, E., and Radin, N. S., 1970, *Arch. Biochem. Biophys.* **141**:738–748.

24. Hoshi, M., and Kishimoto, Y., 1973, *J. Biol. Chem.* **248**:4123–4130.

25. Murad, S., and Kishimoto, Y., 1977, *Biochem. Biophys. Acta* **488**:102–111.

26. Iwamori, M., Moser, H. W., and Kishimoto, Y., 1975, *J. Lipid Res.* **16**:332–336.

27. Kishimoto, Y., and Radin, N. S., 1965, *J. Lipid Res.* **6**:435–436.

28. Morrison, W. R., and Smith, L. M., 1964, *J. Lipid Res.* **5**:600–608.

29. Weiss, B., 1964, *Biochemistry* **3**:1288–1293.

30. Taketomi, T., and Kawamura, N., 1972, *J. Biochem.* **72**:189–193.

31. Kishimoto, Y., and Radin, N. S., 1963, *J. Lipid Res.* **4**:130–138.

32. Morell, P., and Radin, N. S., 1970, *J. Biol. Chem.* **245**:342–350.

33. Samuelsson, B., and Samuelsson, K., 1969, *J. Lipid Res.* **10**:47–55.

34. Karlsson, K. A., and Pascher, I., 1971, *J. Lipid Res.* **12**:466–472.

35. Gaver, R. C., and Sweeley, C. C., 1965, *J. Am. Oil Chem. Soc.* **42**:294–298.

36. Hammarström, S., 1970, *Eur. J. Biochem.* **15**:581–591.

37. Carter, H. E., and Gaver, R. C., 1967, *J. Lipid Res.* **8**:391–395.

38. Yahara, S., Moser, H. W., Kolodny, E. H., and Kishimoto, Y., 1980, *J. Neurochem.* **34**:694–699.

39. Samuelsson, K., and Samuelsson, B., 1970, *Chem. Phys. Lipids* **5**:44–79.

40. Hammarström, S., Samuelsson, B., and Samuelsson, K., 1970, *J. Lipid Res.* **11**:150–157.

41. Horning, M. G., Murakami, S., and Horning, E. C., 1971, *Am. J. Clin. Nutr.* **24**:1086–1096.

42. Oshima, M., Ariga, T., and Murata, T., 1977, *Chem. Phys. Lipids* **19**:289–299.

43. Ando, S., 1973, *Yukagaku* **21**:243–256.

44. Lauter, C. J., and Trams, E. G., 1961, *J. Lipid Res.* **3**:136–138.

45. Siakotos, A. N., Kulkarni, S., and Passo, S., 1971, *Lipids* **6**:254–259.

46. Coles, L., and Gray, G. M., 1970, *J. Lipid Res.* **11**:164–166.

47. Choi, Y. S., and Egawa, K., 1975, *Jpn. J. Exp. Med.* **45**:113–116.

48. Naoi, M., Lee, Y. C., and Roseman, S., 1974, *Anal. Biochem.* **58**:571–577.

49. Kisic, A., and Rapport, M. M., 1974, *J. Lipid Res.* **15**:179–180.

50. Rapport, M. M., and Alonzo, N., 1955, *J. Biol. Chem.* **217**:193–198.

51. Kishimoto, Y., Davies, W. E., and Radin, N. S., 1965, *J. Lipid Res.* **6**:525–531.

52. Kishimoto, Y., and Radin, N. S., 1959, *J. Lipid Res.* **1**:72–78.

53. Wood, R. D., Raju, P. K., and Reiser, R., 1965, *J. Am. Oil Chem. Soc.* **42**:81–85.

54. Sugita, M., Iwamori, M., Evans, J., McCluer, R. H., Dulaney, J. T., and Moser, H., 1974, *J. Lipid Res.* **15**:223–226.

55. Iwamori, M., Costello, C., and Moser, H. W., 1979, *J. Lipid Res.* **20**:86–96.

56. Chou, K. H., and Jungalwala, F. B., 1981, *J. Neurochem.* **36**:394–401.
57. Iwamori, M., Moser, H. W., McCluer, R. H., and Kishimoto, Y., 1975, *Biochim. Biophys. Acta* **380**:308–319.
58. Gaskell, S. J., Edmonds, C. G., and Brooks, C. J., 1976, *J. Chromatogr.* **126**:591–599.
59. Selvam, R., and Radin, N. S., 1981, *Anal. Biochem.* **112**:338–345.
60. Kishimoto, Y., 1971, *J. Neurochem.* **18**:1365–1368.
61. Carter, H. E., Rothfus, J. A., and Gigg, R., 1961, *J. Lipid Res.* **2**:228–234.
62. Rouser, G., and Yamamoto, A., 1972, *Lipids* **7**:561–563.
63. Klenk, E., and Huang, R. T. C., 1968, *Z. Physiol. Chem.* **349**:451–454.
64. Sugita, M., Connolly, P., Dulaney, J. T., and Moser, H. W., 1973, *Lipids* **8**:401–406.
65. Gerstl, B., Eng, L. F., Hayman, R. B., Tavaststjerna, M. G., and Bond, P. R., 1967, *J. Neurochem.* **14**:661–670.
66. Dhopeskwarkar, G. A., and Mead, J. F., 1973, *Adv. Lipid Res.* **11**:109–142.
67. Bourre, J.-M., Paturneau-Jouas, M. Y., Daudu, O. L., and Baumann, N. A., 1977, *Eur. J. Biochem.* **72**:41–47.
68. Murad, S., and Kishimoto, Y., 1978, *Arch. Biochem. Biophys.* **185**:300–306.
69. Volpe, J. J., and Kishimoto, Y., 1972, *J. Neurochem.* **19**:737–753.
70. Kishimoto, Y., Akanuma, H., and Singh, I., 1979, *Mol. Cell. Biochem.* **28**:93–105.
71. Braun, P. E., and Snell, E. E., 1967, *Proc. Natl. Acad. Sci. U.S.A.* **58**:298–303.
72. Braun, P. E., Morell, P., and Radin, N. S., 1970, *J. Biol. Chem.* **245**:335–341.
73. Brady, R. O., and Koval, G. J., 1958, *J. Biol. Chem.* **233**:26–31.
74. Karlsson, K.-A., 1970, *Lipids* **5**:878–891.
75. Stoffel, W., 1973, *Chem. Phys. Lipids* **11**:318–334.
76. Nakano, M., and Fujino, Y., 1973, *Biochim. Biophys. Acta* **296**:457–460.
77. Stoffel, W., and Bister, K., 1974, *Z. Physiol. Chem.* **355**:911–923.
78. Shoyama, Y., and Kishimoto, Y., 1976, *Biochem. Biophys. Res. Commun.* **70**:1035–1041.
79. Shoyama, Y., and Kishimoto, Y., 1978, *J. Neurochem.* **30**:377–382.
80. Sribney, M., 1966, *Biochim. Biophys. Acta* **125**:342–350.
81. Ullman, M. D., and Radin, N. S., 1972, *Arch. Biochem. Biophys.* **152**:767–777.
82. Gatt, S., 1966, *J. Biol. Chem.* **241**:3724–3730.
83. Yavin, E., and Gatt, S., 1969, *Biochemistry* **8**:1692–1697.
84. Sugita, M., Williams, M., Dulaney, J. T., and Moser, H. W., 1975, *Biochim. Biophys. Acta* **398**:125–131.
85. Stoffel, W., and Melzner, I., 1980, *Z. Physiol. Chem.* **361**:755–771.
86. Stoffel, W., Kruger, E., and Melzner, I., 1980, *Z. Physiol. Chem.* **361**:773–779.
87. Singh, I., and Kishimoto, Y., 1978, *Biochem. Biophys. Res. Commun.* **82**:1287–1293.
88. Singh, I., and Kishimoto, Y., 1980, *Arch. Biochem. Biophys.* **202**:93–100.
89. Singh, I., 1981, *Trans. Am. Soc. Neurochem.* **12**:120.
90. Singh, I., and Kishimoto, Y., 1979, *J. Biol. Chem.* **254**:7698–7704.
91. Carter, T. P., and Kanfer, J., 1974, *J. Neurochem.* **23**:589–594.
92. Kanfer, J. N., and Gal, A. E., 1966, *Biochem. Biophys. Res. Commun.* **22**:442–446.
93. Okabe, H., and Kishimoto, Y., 1977, *J. Biol. Chem.* **252**:7068–7073.
94. Pascher, I., 1976, *Biochim. Biophys. Acta* **455**:431–451.
95. Löfgren, H., and Pascher, I., 1977, *Chem. Phys. Lipids* **20**:273–284.
96. Sugita, M., Dulaney, J. T., and Moser, H. W., 1972, *Science* **178**:1100–1102.
97. Dulaney, J. T., and Moser, H. W., 1978, *The Metabolic Basis of Inherited Disease*, 3rd ed. (J. B. Stanbury, J. B. Wyngaarden, and D. S. Fredrickson, eds.), McGraw-Hill, New York, pp. 770–809.
98. Murad, S., and Kishimoto, Y., 1975, *J. Biol. Chem.* **250**:5841–5846.
99. Zalc, B., Pollet, S. A., Harpin, M.-L., and Baumann, N. A., 1974, *Brain Res.* **81**:511–518.
100. Hogan, E. L., 1977, *Myelin* (P. Morell, ed.), Plenum Press, New York, pp. 489–520.
101. Norton, W. T., 1977, *Myelin* (P. Morell, ed.), Plenum Press, New York, pp. 161–199.
102. Sweeley, C. C., and Moscatelli, E. A., 1959, *J. Lipid Res.* **1**:40–47.
103. Abe, T., and Norton, W. T., 1974, *J. Neurochem.* **23**:1025–1036.
104. Korniat, E. K., and Hof, L., 1978, *J. Neurochem.* **30**:557–562.
105. Kishimoto, Y., 1978, *Research Methods in Neurochemistry*, Volume 4 (N. Marks and R. Rodnight, eds.), Plenum Press, New York, pp. 411–436.

106. Norton, W. T., and Brotz, M., 1963, *Biochem. Biophys. Res. Commun.* **12**:198–203.

107. Kishimoto, Y., Wajda, M., and Radin, N. S., 1968, *J. Lipid Res.* **9**:27–33.

108. Klenk, E., and Löhr, J. P., 1967, *Z. Physiol. Chem.* **348**:1712–1714.

109. Tamai, Y., 1968, *Jpn. J. Exp. Med.* **38**:65–73.

110. Sweeley, C. C., and Dawson, G., 1969, *Biochem. Biophys. Res. Commun.* **37**:6–37.

111. Karlsson, K. A., Pascher, I., Samuelsson, B. E., and Steen, G. O., 1972, *Chem. Phys. Lipids* **9**:230–246.

112. Ledeen, R. W., Kundu, S. K., Price, H. C., and Fong, J. W., 1974, *Chem. Phys. Lipids* **13**:429–446.

113. Egge, H., 1978, *Chem. Phys. Lipids* **21**:349–360.

114. Murata, T., Ariga, T., Oshima, M., and Miyatake, T., 1978, *J. Lipid Res.* **19**:370–374.

115. Koerner, T. A. W., Jr., Cary, L. W., Li, S.-C., and Li, Y.-T., 1979, *J. Biol. Chem.* **254**:2326–2328.

116. Tkaczuk, P., and Thornton, E. R., 1979, *Biochem. Biophys. Res. Commun.* **91**:1415–1422.

117. Dabrowski, J., Egge, H., and Hanfland, P., 1980, *Chem. Phys. Lipids* **26**:187–196.

118. Shapiro, D., and Flowers, H. M., 1961, *J. Am. Chem. Soc.* **83**:3327–3332.

119. Auge, C., and Veyrières, A., 1979, *J. Chem. Soc.* **1979**:1825–1832.

120. Kiso, M., Nishiguchi, H., and Hasegawa, A., 1980, *Carbohydr. Res.* **81**:C13–C15.

121. Radin, N. S., 1972, *Methods Enzymol.* **28**:300–304.

122. McMaster, M. C., and Radin, N. C., 1976, *J. Labelled Comp. Radiopharm.* **13**:353–357.

123. Taketomi, T., and Yamakawa, T., 1963, *J. Biochem.* **54**:444–451.

124. Radin, N. S., 1974, *Lipids* **9**:358–360.

125. Seyama, Y., Yamakawa, T., and Komai, T., 1968, *J. Biochem.* **64**:487–493.

126. Skipski, V. P., and Barclay, M., 1969, *Methods Enzymol.* **14**:530–598.

127. Morell, P., and Radin, N. S., 1969, *Biochemistry* **8**:506–512.

128. Young, O. M., and Kanfer, J. N., 1965, *J. Chromatogr.* **19**:611–613.

129. Sweeley, C. C., and Walker, B., 1964, *Anal. Chem.* **36**:1461–1466.

130. Radin, N. S., 1958, *Methods Biochem. Anal.* **6**:163–189.

131. Ullman, M. D., and McCluer, R. H., 1977, *J. Lipid Res.* **19**:910–913.

132. Jungalwala, F. B., Hayes, L., and McCluer, R. H., 1977, *J. Lipid Res.* **18**:285–292.

133. Nonaka, G., and Kishimoto, Y., 1979, *Biochim. Biophys. Acta* **572**:423–431.

134. Yahara, S., Kishimoto, Y., and Poduslo, J., 1980, *Cell Surface Glycolipids* (C. C. Sweeley, ed.), American Chemical Society, Washington, pp. 15–33.

135. Suzuki, A., Handa, S., Ishizuka, I., and Yamakawa, T., 1977, *J. Biochem.* **81**:127–134.

136. Folch-Pi, J., Lees, M., and Sloane-Stanley, G. H., 1957, *J. Biol. Chem.* **226**:497–509.

137. Hara, A., and Radin, N. S., 1978, *Anal. Biochem.* **90**:420–426.

138. Radin, N. S., 1976, *J. Lipid Res.* **17**:290–293.

139. Nonaka, G., and Kishimoto, Y., 1979, *Biochim. Biophys. Acta* **572**:432–441.

140. Hoshi, M., Williams, M., and Kishimoto, Y., 1973, *J. Neurochem.* **21**:709–712.

141. Cleland, W. W., and Kennedy, E. P., 1960, *J. Biol. Chem.* **235**:45–51.

142. Brady, R. O., 1962, *J. Biol. Chem.* **237**:PC2416–2417.

143. Curtino, J. A., and Caputto, R., 1974, *Biochem. Biophys. Res. Commun.* **56**:142–147.

144. Hammarström, S., 1971, *Biochem. Biophys. Res. Commun.* **45**:459–467.

145. Hammarström, S., 1972, *FEBS Lett.* **21**:259–263.

146. Basu, S., Schultz, A. M., Basu, M., and Roseman, S., 1971, *J. Biol. Chem.* **246**:4272–4279.

147. Hammarström, S., 1971, *Biochem. Biophys. Res. Commun.* **45**:468–475.

148. Hammarström, S., and Samuelsson, B., 1972, *J. Biol. Chem.* **247**:1001–1011.

149. Murad, S., Chen, R. H. K., and Kishimoto, Y., 1977, *J. Biol. Chem.* **252**:5206–5210.

150. Carter, T. P., and Kanfer, J. N., 1976, *J. Neurochem.* **27**:53–62.

151. Basu, S., Kaufman, B., and Roseman, S., 1968, *J. Biol. Chem.* **243**:5802–5804.

152. Brenkert, A., and Radin, N. S., 1972, *Brain Res.* **36**:183–193.

153. Shah, S. N., 1971, *J. Neurochem.* **18**:395–402.

154. Neskovic, N. M., Sarlieve, L. L., and Mandel, P., 1976, *Biochem. Biophys. Acta* **429**:342–351.

155. Radin, N. S., Brenkert, A., Arora, R. C., Sellinger, O. Z., and Flangas, A. L., 1972, *Brain Res.* **39**:163–169.

156. Neskovic, N. M., Sarlieve, L. L., and Mandel, P., 1973, *J. Neurochem.* **20**:1419–1430.

157. Costantino-Ceccarini, E., and Suzuki, K., 1975, *Brain Res.* **93**:358–362.
158. Koul, O., Chou, K. H., and Jungalwala, F. B., 1980, *Biochem. J.* **186**:959–969.
159. Costantino-Ceccarini, E., Cesteli, A., and DeVries, G. H., 1979, *J. Neurochem.* **32**:1175–1182.
160. Siegrist, H. P., Burkart, T., Wiesmann, U. N., Herschkowitz, N. N., and Spycher, M. A., 1979, *J. Neurochem.* **33**:497–504.
161. Yahara, S., Singh, I., and Kishimoto, Y., 1980, *Biochim. Biophys. Acta* **619**:177–185.
162. Shah, S. N., 1973, *Arch. Biochem. Biophys.* **159**:143–150.
163. Costantino-Ceccarini, E., and Morell, P., 1972, *Lipids* **7**:656–659.
164. Costantino-Ceccarini, E., and Suzuki, K., 1978, *J. Biol. Chem.* **253**:340–342.
165. Satomi, D., and Kishimoto, Y., 1981, *J. Neurochem.* **36**:476–482.
166. Radin, N. S., Warren, K. R., Arora, R. C., Hyun, J. C., and Misra, R. S., 1975, *Modification of Lipid Metabolism* (E. G. Perkins and L. A. Whitting, eds.), Academic Press, New York, pp. 87–104.
167. Shah, S. N., and Peterson, N. A., 1971, *Biochim. Biophys. Acta* **239**:126–131.
168. Warren, K. R., Misra, R. S., Arora, R. C., and Radin, N. S., 1976, *J. Neurochem.* **26**:1063–1072.
169. Ho, M. W., Norden, A. G. W., Alhadeff, J. A., and O'Brien, J. S., 1977, *Mol. Cell. Biochem.* **17**:125–140.
170. Bowen, D. M., and Radin, N. S., 1968, *Biochim. Biophys. Acta* **152**:587–598.
171. Suzuki, K., and Suzuki, Y., 1978, *The Metabolic Basis of Inherited Disease*, 3rd ed. (J. B. Stambury, J. B. Wyngaaden, and D. S. Frederickson, eds.), McGraw-Hill, New York, pp. 747–769.
172. Wenger, D. A., Sattler, M., and Hiatt, W., 1974, *Proc. Natl. Acad. Sci. U.S.A.* **71**:854–857.
173. Miyatake, T., and Suzuki, K., 1975, *J. Biol. Chem.* **250**:585–592.
174. Brady, R. O., 1978, *The Metabolic Basis of Inherited Disease*, 4th ed. (J. B. Stanbury, J. B. Wyngaarden, and D. S. Fredrickson, eds.), McGraw-Hill, New York, pp. 731–746.
175. Lin, Y. N., and Radin, N. S., 1973, *Lipids* **8**:732–736.
176. Raghavan, S. S., Mumford, R. A., and Kanfer, J. N., 1974, *J. Lipid Res.* **15**:484–490.
177. Vanier, M. T., and Svennerholm, L., 1975, *Acta Paediatr. Scand.* **64**:641–648.
178. Kanfer, J. N., Legler, G., Sullivan, J., Raghavan, S. S., and Mumford, R. A., 1975, *Biochem. Biophys. Res. Commun.* **67**:85–90.
179. Hara, A., and Radin, N. S., 1979, *Biochim. Biophys. Acta* **528**:412–422.
180. Chiao, Y. B., Moffitt, K., Smallwood, Y., Glew, R. H., Naoi, M., and Lee, Y. C., 1979, *Arch. Biochem. Biophys.* **192**:1–9.
181. Ho, H. W., O'Brien, J. S., Radin, N. S., and Erickson, J. S., 1973, *Biochem. J.* **131**:173–176.
182. Pentchev, P. G., and Brady, R. O., 1973, *Biochim. Biophys. Acta* **297**:491–496.
183. Peters, S. P., Coyle, P., Coffee, C. J., Glew, R. H., Kuhlenschmidt, M. S., Rosenfeld, L., and Lee, Y. C., 1977, *J. Biol. Chem.* **252**:563–573.
184. Wenger, D. A., Sattler, M., and Roth, S., 1981, *Trans. Am. Soc. Neurchem.* **12**:210.
185. Berent, S. L., and Radin, N. S., 1981, *Arch. Biochem. Biophys.* **208**:248–260.
186. Berent, S. L., and Radin, N. S., 1981. *Biochim. Biophys. Acta* **664**:572–582.
187. Hanada, E., and Suzuki, K., 1979, *Biochim. Biophys. Acta* **575**:410–420.
188. Stoffel, W., 1971, *Annu. Rev. Biochem.* **40**:57–82.
189. Linington, C., and Rumsby, M. G., 1980, *J. Neurochem.* **35**:983–992.
190. Lisak, R. P., Abramsky, O., Dorfman, S. H., George, J., Manning, M. C., Pleasure, D. E., Saida, T., and Silberberg, D. H., 1979, *J. Neurol. Sci.* **40**:65–73.
191. Dupouey, P., Zalc, B., Lefroit-Joly, M., and Gomes, D., 1979, *Cell. Mol. Biol.* **25**:269–272.
192. Saida, T., Saida, K., Silberberg, D. H., Sumner, A. J., Manning, M. C., Lisak, R. P., and Brown, M. J., 1979, *Science* **204**:1103–1106.
193. Fry, J. M., Weissbarth, S., Lehrer, G. M., and Bornstein, M. B., 1974, *Science* **183**:540–542.
194. Dorfman, S. H., Fry, J. M., Silberberg, D. H., Grose, C., Manning, M. C., 1978, *Brain Res.* **147**:410–415.
195. Norton, W. T., 1977, *Myelin* (P. Morell, ed.), Plenum Press, New York, pp. 383–413.

8

Peripheral Nervous System Myelin Properties and Metabolism

Marion Edmonds Smith

1. INTRODUCTION

Available knowledge about peripheral nervous system (PNS) myelin has increased considerably since the first edition of the *Handbook of Neurochemistry*. At that time, CNS myelin was a subject of intense investigation in regard to both composition and metabolism, but very few laboratories had isolated peripheral nervous system myelin, and only the lipid composition had been measured. In this edition of the *Handbook*, aspects of PNS myelin and its pathology are a major subject of a number of chapters including those on the Schwann cell, peripheral nerve, nerve regeneration, Wallerian degeneration, and allergic neuritis. The interested reader may also refer to a chapter on CNS myelin for comparison of the characteristics and metabolism of CNS and PNS myelin.

Peripheral nervous system myelin has received less attention than its CNS counterpart for several reasons. Early work on the lipid composition revealed differences only in proportions, but not in species, of the various lipids in the two kinds of myelin, and it seemed likely that this would also be true of the protein constituents. It was not until the use of sodium dodecyl sulfate polyacrylamide gels was developed that it was fully appreciated that the proteins were quite different. Workers in PNS myelin research have encountered the difficulty of separating this membrane from the tough fibrous connective tissue of the sciatic nerve, the most abundant source material. The problem has been alleviated, however, by the development of high-speed homogenizers which can disintegrate the collagenous tissue; and, indeed, once the nerves are adequately disrupted, purification of PNS myelin is more easily accomplished than that of the CNS. Research on CNS myelin has been well supported for

Marion Edmonds Smith • Department of Neurology, Veterans Administration Medical Center, Palo Alto, California 94304; and Stanford University School of Medicine, Stanford, California 94305.

some years because of the need for investigations of demyelination related to multiple sclerosis. A recent event, that of the swine flu vaccine immunizations and the associated outbreak of Guillain–Barré syndrome cases, has stimulated interest in PNS myelin, and a number of investigators have turned to this area, resulting in a rapid increase in numbers of reports on this subject. In this chapter, the early work is reviewed, and the new findings on composition, metabolism, and pathology are discussed.

2. CHARACTERISTICS OF PNS MYELIN

2.1. Axon–Glial Interactions

The relationships between PNS myelin and the Schwann cell are discussed in detail in the chapter on the Schwann cell (Volume 1, this *Handbook*). The events leading to the ensheathment of the axon by myelin are mentioned only briefly here because of full treatment in the earlier volume and in reviews elsewhere.[1–3] Generally, myelination of peripheral nerves begins relatively early in the development of the animal and is well under way before any myelin is seen in the CNS. Both axons and Schwann cells are required for myelin formation, but the nature of the message from the axon to the Schwann cell to commence the myelination process is unknown. Several theories involving chemical communication have been proposed that implicate glycoprotein interactions, possibly mediated by cyclic nucleotide.[4]

2.1.1. Influence of the Axon

Although the axon contributes little or none of the actual metabolic activity to myelin formation, its role in the process is far from passive. The axon acts as a regulator in determining, by its size, the thickness of the myelin sheath and maintains an approximately linear relationship between axon circumference and number of lamellae.[5] Furthermore, there is evidence that the axon also specifies where myelin will be deposited, most likely by early differentiation of the sites of the nodes.[6] That a signal between the axon and Schwann cell is maintained in the adult animal is evident from the prompt dissolution of myelin and the Schwann cell reaction when the axon is severed from the neuronal cell body.

2.1.2. Role of the Schwann Cell

Small amounts of lipid and protein precursors appear to be derived from axonally transported material,[7] but the bulk of the myelin sheath components are synthesized by the Schwann cell. In a study of myelin formation in the sciatic nerve of the rat, Friede and Samorajski[5] noted that the mitochondrial density of axons being myelinated was lower than that of unmyelinated axons, whereas Schwann cells forming myelin showed a higher mitochondrial density than those around nonmyelinated fibers. They concluded that the large concentration of ribosomes and rough endoplasmic reticulum in Schwann cells was an indicator that these cells supplied the metabolic requirement for myelin formation. Little is known as yet about the metabolic properties of Schwann cells because of the difficulty of their isolation. Several new methods have been described recently for purification of these cells in viable functional form,[8-10] and studies of their characteristics *in vitro* will undoubtedly be forthcoming. Only those Schwann cells that have already been induced to myelinate, i.e., from 5 to 6-day-old rats, contain detectable amounts of the myelin components galactocerebroside sulfatide, myelin basic protein, and P_0 protein in culture, whereas cells from 1-day-old rats do not. Mirsky *et al.*[11] have interpreted these findings as indicating that synthesis of myelin lipids and proteins by Schwann cells need a continuous signal from the axon, in contrast to the oligodendroglia which are more autonomous.

The actual process of myelination by Schwann cells is well documented, and the sequence of events in which the Schwann cells, after migration and alignment along the axon, sort out the axons and ensheath them in multiple layers of myelin is described elsewhere[1] (see Chapter 15, Volume 1, this *Handbook*).

2.2. Composition of PNS Myelin

2.2.1. Isolation

Methods devised for the isolation of CNS myelin have generally been applicable to PNS myelin once the problem of homogenization of the nerve tissue is solved. Peripheral nervous system myelin is purified by most investigators according to the standard method of Norton and Poduslo[12] except that a single density gradient followed by several washes has been found sufficient to yield myelin with relatively little contamination when examined by ultrastructural microscopy.[13,14] In view of the lower density of PNS myelin compared to CNS, Wiggins *et al.*[14] have modified the procedure by using 0.29 M sucrose as a homogenizing medium rather than the 0.32 M sucrose used in the Norton–Poduslo method.

2.2.2. Lipids

The lipids contained in PNS myelin are qualitatively similar to those of CNS but are present in slightly different proportions. Although a vast amount of data on the lipid composition of CNS myelin is available, relatively fewer measurements of PNS myelin lipid have been published. Lipid compositions have been determined for myelin from monkey brachial plexus,[15] human femoral nerve,[16] rabbit sciatic nerve,[16] ox intradural roots,[17] and sciatic nerves from beef,[18] guinea pig,[19] rat,[19,20] frog (*Xenopus* and *Rana*),[20] and chick.[21] Although most lipids are found in comparable amounts in the two kinds of myelin, most data indicate that that of PNS contains more phosphatidyl serine and sphingomyelin and less cerebroside than CNS myelin [Table I for rat, beef, and frog (*Rana*)]. A comparison of the lipid compositions for different species

Table I
Comparison of Composition of CNS and PNS Myelin

	PNS Myelin			CNS Myelin		
	Rat (20)	Beef[b] (18)	Frog (Rana) (20)	Rat (20)	Beef[b] (18)	Frog (Rana) (20)
Total Phospholipid	50.6[a]	56.2	49.5	38.3	46.4	49.6
SGP	11.4	11.0	9.6	6.5	11.4	8.2
IGP	2.7 ⎤		2.3	1.3 ⎤		2.0
Sphingomyelin	8.8 ⎦ 17.1		5.8	3.0 ⎦ 7.4		2.9
CGP	10.0	11.0	14.6	10.2	11.0	15.4
EGP	17.7	17.1	17.2	17.2	16.6	21.1
Cerebroside	15.8 ⎤		16.7	28.9 ⎤		18.0
Cerebroside Sulfate	5.7 ⎦ 20.7		4.2	8.2 ⎦ 28.5		7.0
Cholesterol	27.2	23.2	24.8	25.6	25.1	25.3

[a] Numbers are percent dry weight of total lipid.
[b] Recalculated to percent dry weight using molecular weights of O'Brien and Sampson[22]. SGP = serine glycerophosphatide; IGP = inositol glycerophosphatide; CGP = choline glycerophosphatide; EGP = ethanolamine glycerophosphatide.

(Table II) shows some differences in galactolipid (lower in bovine roots and rabbit sciatic nerve), in sphingomyelin (lower in frog and rat), and in choline glycerophosphatides (lower in monkey, human, and chick). Whether myelin purified from different regions of the PNS (i.e., roots vs. nerves) might be of a different composition is unknown. Other lipids reported to be present include gangliosides, which occur in PNS myelin in less than half the amount found in CNS myelin, about equivalent to the amount of P_1,[23] and alkanes which have been detected in mouse sciatic nerves.[24]

Only a few determinations of the fatty acids in PNS myelin are to be found in the literature. The content of fatty aldehyde and the unsubstituted and hydroxy fatty acids in the major myelin lipids was measured in ox spinal root myelin[17] and in cerebrosides from chick myelin at 1 and 7 days of age.[21] O'Brien et al.[17] noted that linoleate comprised 5.6% of the fatty acids in choline glycerophosphatide of PNS myelin, about ten times that of CNS. A lower proportion of 25:0 and 26:0 in spinal root sphingolipids was balanced by an increased content of 22:0. The predominant fatty acid in cerebroside of chick sciatic nerve myelin (7 days) was 22:0, whereas that of beef was 24:0.[21]

Evidence for changes in lipid composition of PNS myelin during development are discussed in Section 3.3.1.

2.2.3. Proteins

Protein in PNS myelin is present in comparable amounts to that of CNS. Values for percent dry weight reported are 24.2% (ox[17]), 30.5% (squirrel monkey[15]), 22.3% (rabbit[16]), and 28.7% (human[16]). The proteins of PNS myelin are distinctive in that glycoproteins comprise at least 60% of the total protein, whereas CNS myelin contains much smaller amounts, with none detectable in

Table II
Lipid Composition of PNS Myelin from Different Species

	Ox (Roots, 17)	Beef (18)	Squirrel Monkey (Brachial Plexus) (15)	Human (Femoral, 16)	Rabbit (16)	Frog (Xenopus) (20)	Frog (Rana) (20)	Rat (20)	Rat (19)	Chick (Adult, 21)
Total Phospholipid	52.8[a]	56.2	48.9	54.9	57.6	55.4	49.5	50.6	56.2	48.2
SGP	8.8	11.0	9.8	9.5	7.4	11.6	9.6	11.4	11.1	9.2
IGP	—	17.1				2.2	2.3	2.7		—
Sphingomyelin	17.7	11.0	14.0	18.5	15.8	6.2	5.8	8.8	11.3	14.2
CGP	13.8	17.1	6.5	8.1	14.9	15.6	14.6	10.0	13.6	5.7
EGP	17.7		18.4	18.5	19.3	19.8	17.2	17.7	20.4	17.3
Total Galactolipid	16.6	20.7	24.5	22.1	15.1	21.1	20.9	21.5	20.0	32.8
Cholesterol	24.7	23.2	26.6	23.0	27.2	24.5	24.8	27.2	23.7	20.7

[a] Numbers represent percent dry weight of total lipid. Abbreviations are as on Table I. In some cases the original data are recalculated to common units using molecular weights of O'Brien and Sampson[22]. Unless indicated otherwise, myelin is prepared from sciatic nerve.

Table III
Distribution of Proteins in PNS Myelin of Different Species Estimated by Dye
Binding

	$P_0 + Y$	X	P_1	P_2
Human (sciatic)[b]	63.1	7.0	6.1	5.1
Rabbit (sciatic)[b]	57.1	9.3	11.6	5.5
Beef (roots)[b]	54	11	7	15
Guinea pig (sciatic)[b]	47.6	8.5	16.3	<2
Rat (sciatic)[b]	57.0	5.3	6.1	11.6[a]

	P_0	23K	19K (Contains P_1)	P_2
Human (sciatic)[c]	45.6	13.0	24.7	14.3
Frog (sciatic)[c]	73.1	3.7	16.6	6.2
Guinea pig (sciatic)[c]	69.6	3.8	22.0	3.8
Beef (roots)[c]	49.8	11.3	20.8	18.1
Rat (sciatic)[c]	67.8	4.5	20.2	8.1[a]
Rabbit (sciatic)[c]	56.9	10.3	22.7	9.0

[a] Contains mostly the small basic protein of CNS[31].
[b] Data of Greenfield et al.[28]
[c] Data of Smith et al.[30]

compact myelin.[25] The Schwann cell would appear, therefore, to invest a large proportion of its protein-synthesizing capacity in glycoprotein synthesis.

Earlier work on PNS myelin proteins has been reviewed by Braun and Brostoff[26] and by Benjamins and Morell,[27] but progress has been rapid in the past several years. The nomenclature of these proteins is confused, partly because the different proteins migrate somewhat differently on various polyacrylamide gel systems. Greenfield et al., who first described the complete protein composition,[28] named the three glycoproteins P_0, Y, and X and the two basic proteins P_1 and P_2. More recently, Eylar et al.[29] have proposed that the molecular weights 23K and 19K be used to identify the two glycoproteins migrating ahead of P_0 on polyacrylamide gels. The proportions of different PNS myelin proteins have been estimated, mostly by dye binding on polyacrylamide gels, and data from two laboratories for several species are shown in Table III. Comparisons between different analyses are difficult because of the varying methods of preparation of myelin, polyacrylamide gel separation methods, and nomenclature.

2.2.3a. P_0 Protein. The molecular weight of the main structural protein, P_0, has been estimated to be about 28,000. It comprises over 50% of the total protein and was early recognized as a glycoprotein[32,33] unique to the peripheral nervous system. Because of its insolubility, it has been difficult to purify, but this was eventually accomplished by gel filtration in sodium dodecyl sulfate by Kitamura et al.[34] and Roomi et al.[35] The carbohydrate that comprises about 6.3% of the total weight exists as a single nine-sugar chain containing three mannose, three N-acetylglucosamine, one sialic acid, one galactose, and one fucose moieties. It lacks N-acetylgalactosamine,[34,35] a characteristic of N-asparaginyl-linked glycoproteins, and determination of the sequence of the five-amino-acid peptide containing the carbohydrate chain has verified the asparaginyl linkage.[36] A high content of nonpolar amino acids, 32 mol%, partially accounts for the insolubility in aqueous solvents.[35]

2.2.3b. Other Glycoproteins. The protein designated as Y by Greenfield *et al.*[28] often appears as a shoulder to the P_0 protein on the SDS polyacrylamide gel. Appearance of the Y protein has been attributed to oxidation of P_0, and under adequate reducing conditions, the Y moves into the P_0 band.[37] The correspondence of the Y and X proteins of Greenfield *et al.* with the 23K and 19K proteins of Eylar *et al.* is difficult to ascertain because of the different gel systems by which they are defined. Both the 23K and 19K are glycoproteins, as are Y and X, but the 23K protein, as visualized on nongradient SDS polyacrylamide gels, persists in spite of adequate reducing conditions (unpublished observations). The 19K protein appears to be related to P_0 since trypsin treatment of P_0 yields a 19K polypeptide (TP_0) with similar properties to the naturally occurring 19K protein.[38] The products resulting from trypsin treatment and the 19K glycoprotein have similar carbohydrate chains, amino acid compositions, and partial amino acid sequences.[39]

Other glycoproteins described in the literature include a 16K protein[40] and a 13K protein described by Kitamura *et al.*[34] The latter two proteins have not been mentioned by most other investigators, but six glycoproteins ranging from 35,500 to 16,000 daltons have been detected in PNS myelin with the use of lectins,[41] and other glycoproteins are present in the high-molecular-weight fraction. Clearly, further study is needed to better define the interrelationships of the different glycoproteins.

2.2.3c. Basic Proteins. Two basic proteins (three in rodents) are present in PNS myelin and have been termed as P_1 and P_2. The P_1 protein (mol. wt. 18,000), which in some polyacrylamide gel systems migrates with 19K glycoprotein, is similar, if not identical, to the CNS myelin basic protein.[42] The P_2 protein is unique to PNS myelin,[43] has a molecular weight of about 15,000,[44] and is the agent that induces experimental allergic neuritis when whole nerve or PNS myelin is injected into animals under appropriate conditions.[45] The complete amino acid sequence of the P_2 protein has been published recently.[44] The fast-moving P_2 protein band has been found to consist of two components in rodent PNS myelin. A small amount of P_2 protein is present, but most of the material consists of the small myelin basic protein found in rodent CNS myelin.[31] This finding explains the observation by Smith *et al.*[30] that rat myelin induces only very mild experimental allergic neuritis in Lewis rats.

2.2.3d. High-Molecular-Weight Components. As indicated by labeling experiments, a number of glycoproteins of high molecular weight (greater than P_0) appear to be present in isolated PNS myelin.[46,47] Only one of these has been identified, and it appears to be similar to the myelin-associated glycoprotein (MAG) of CNS.[46] Antibody to CNS MAG reacts immunocytochemically with PNS myelin at the rim of the axon.[48] Other high-molecular-weight components may be related to axonal or Schwann cell membranes.[47]

2.2.4. Enzymes in Myelin

Earlier, myelin was considered to be a metabolically inert substance, and the absence of enzymes was sometimes used as a criterion of purity. More recently, a number of enzymes have been found to be associated with CNS myelin, and these include 2',3'-cyclic nucleotide 3'-phosphohydrolase (CNP) and pH 7.2 cholesterol hydrolase, both of which are believed to be myelin specific, as well as Na^+-K^+ ATPase, 5'-nucleotidase, carbonic anhydrase, and enzymes involved in lipid synthesis, among others.[49]

The PNS myelin has not been explored for enzyme activity as extensively, but those enzymes found to be present appear to be less active than in the CNS. CNP, frequently used as a CNS myelin marker, is much less active in sciatic nerve than white matter by a factor of 30,[50] and rigorous investigation of its subcellular localization indicates little enrichment of the myelin enzyme over the homogenate; it may be localized in Schwann cell plasma membranes.[51] This appears to be true also of 5'-nucleotidase.[52] Carbonic anhydrase is present in PNS myelin at a specific activity about 17% that of CNS myelin, but about 25% of the total is recovered in purified CNS myelin; therefore, it is probably an actual myelin component.[53] Other enzymes reported to be present in PNS myelin include protein kinases, both cyclic AMP-dependent[54] and -independent,[55] which phosphorylate P_0, Y, X, and, in the case of the cyclic AMP-dependent enzyme, also P_1. In view of the protein differences between CNS and PNS myelin, further investigations of enzymes in Schwann cells and PNS myelin may reveal specific marker enzymes that would be useful for studies of peripheral neuropathies involving demyelination.

2.3. Structural Relationships

For a more complete treatment of the molecular architecture of myelin, the reader is referred to a recent review.[56] Both PNS myelin and CNS myelin consist of a series of lipid bilayers within which the myelin proteins are embedded with various degrees of interaction with the lipids. The site of P_0 localization has been most investigated because it comprises the largest proportion of the total protein. Evidence obtained by various techniques including concanavalin A binding studies,[57] electron microscopy,[58] lactoperoxidase iodination,[59] and X-ray diffraction[60] indicated that this protein is localized in the intraperiod line or on the extracellular surface of the Schwann cell. The X protein also appears in the intraperiod line, whereas the basic proteins appear to be localized in the main period band, as determined by electron microscopic examination of ammonium acetate–Triton-X-extracted myelin,[59] and therefore are on the cytoplasmic surface. Ishaque *et al.*[39] have proposed that the 19K glycoprotein as prepared by trypsin treatment of isolated myelin is the hydrophobic core of P_0 which penetrates through the lipid bilayer, whereas the 30% of the molecule that is removed by trypsin is hydrophilic and therefore extends into the polar phase.

The location of the glycoprotein at the apposition of the external surfaces may be related to incomplete fusion of the two layers, since a 2-nm gap intervenes between the two surfaces, whereas no such gap appears in CNS myelin.[57] Unfortunately, studies of the localization of the various proteins have not been done on the Schwann cell membrane. Basic protein (P_1) is absent in PNS myelin of the mutant shiverer mouse, although the gross morphology appears to be normal. This protein, therefore, must have little importance in maintaining normal myelin structure in peripheral nerves. Quarles has devised a working model of the molecular architecture of PNS myelin that takes into account the accumulated findings.[25]

3. METABOLISM

3.1. Myelination: Dynamic Changes in Composition

In view of the extensive investigations on synthesis and turnover of CNS myelin, surprisingly little information is available about the metabolism of PNS myelin. The scarcity of data on the lipids is especially surprising since the methodology has been available for a number of years. Only one study has been made on changes of myelin lipid composition during development. Oulton and Mezei[21] prepared myelin from chick sciatic nerve at the 18-day embryo stage, on days 1, 4, and 7 after hatching, and from adults and were able to show an increase in cerebroside from 9.1 to 21.0, a decrease in phosphatidyl choline from 12.0 to 4.8, and an increase in phosphatidal ethanolamine from 5.1 to 10.4 mol % with increasing age. Longer-chain fatty acids in cerebroside also increased. These changes are similar to those found in CNS myelin (reviewed by Norton[62]).

Several studies are available on the appearance of PNS myelin protein in the developing animal. In a study in which myelin was purified from sciatic nerves of rats between 5 and 45 days of age, only minor changes in protein composition were seen, in contrast to marked changes in whole sciatic nerve in the same time interval when P_0 increased from 3 to 15% of the total nerve myelin.[63] This period of greatest accumulation of myelin protein also corresponded to the period of maximum incorporation of [^{35}S]sulfate into sulfatide. Very little myelin has been found in sciatic nerves of rats younger than 5 days of age, although it is detectable by electron microscopy as early as 2 days after birth.[5] Uyemura *et al.* examined the appearance of myelin proteins in the chick sciatic nerve.[64] Two glycoproteins, P_0 and PAS II, as well as the basic protein (BP) increased rapidly from 14 to 18 embryonic days. High-molecular-weight proteins were predominant at 14 embryonic days, and these decreased with development as P_0, PAS II, and BP increased. P_2 protein was not detectable in chick myelin.

3.2. Myelination: In Vivo Studies

The route of entrance of PNS myelin components was first demonstrated by Rawlins who injected labeled cholesterol into 10-day mice and followed the localization of label by electron microscopic autoradiography.[65] Radioactive silver grains appeared at the outer and adaxonal edges of the myelin sheath within 20 min after injection, and by 3 hr after injection, the grains were homogeneously distributed. An elegant series of experiments by Gould followed, using radioactive lipid precursors including choline[66] and inositol.[67] He demonstrated that phospholipids migrate from Schwann cell cytoplasm into myelin and gradually diffuse into the interior layers to become evenly distributed over a period of several days. In other autoradiographic studies, radioactive fucose was localized first over juxtanuclear Schwann cell cytoplasm in regions rich in Golgi apparatus; then the label gradually moved into the myelin sheath.[68] In developing animals, the label gradually became enveloped and concentrated over inner myelin layers, whereas in adult animals, the radioactivity remained confined to outer layers and was shown on gels to be associated exclusively with P_0 protein. These studies, which are discussed more authoritatively in Volume 1 of this *Handbook* (Chapter 15), indicate that PNS myelin lipids are mobile and diffuse throughout the membranous layers, whereas the glycoprotein tends to remain concentrated at the site of deposition.

3.3. Metabolic Studies

The synthesis and degradation of a complex membrane such as myelin will necessarily involve not only the formation and breakdown of the lipid and protein components but also the rate and mechanism of transfer of components into the myelin and the sequence of assembly and disassembly. Although, as mentioned earlier, the interactions between the components are largely unknown, some investigations described below are beginning to yield some insight into the assembly sequence.

3.3.1. Metabolic Changes with Age

Peripheral nerves dissected from the animal and incubated in a suitable medium with radioactive precursors are capable of synthesizing myelin components in a pattern resembling that occurring *in vivo*. In a study of myelin synthesis *in vitro*, Rawlins and Smith compared the rates of incorporation of [^{14}C]acetate into lipids and [^{14}C]leucine into total protein of CNS and PNS myelin from 25 days to 18 months of age.[13] Slices of rat brain, spinal cord, and segments of sciatic nerve were incubated with the precursor; then the myelin was isolated. The uptake of radioactive precursors of all three tissues into myelin decreased with increasing age, most markedly between 30 and 60 days of age, reaching a plateau at about 6 months of age. Lecithin showed the highest rate of uptake of [^{14}C]acetate at all ages, whereas cholesterol incorporated acetate very actively at 25 days but contained very little radioactivity in myelin from 18-month nerves. At all ages, however, incorporation of both lipid and protein precursors into PNS myelin components was several times higher than into those of CNS, possibly indicating a higher rate of PNS myelin metabolism.

3.3.2. Myelin Protein: Synthesis in Vitro

More recently, a series of investigations of synthesis of the individual PNS myelin proteins was undertaken using chopped spinal roots and sciatic nerves and measuring *in vitro* uptake of a [^{3}H]amino acid mixture into myelin in the 24-day rat.[47] Labeled amino acids were well incorporated into P_0, 23K, 19K + P_1, and small BP + P_2, as shown by separation of the myelin proteins on polyacrylamide gel electrophoresis, slicing of the gels, and counting of the slices (Fig. 1). The proteins of molecular weight higher than P_0 incorporated much of the radioactivity, over 40% of the total, although very little stained material was observed in this region. The major myelin proteins (P_0, 23K, P_1 + 19K, and small BP + P_2), containing about 87% of the fast-green-stained material, represented only about 60% of the incorporated radioactivity. In old rats (20–

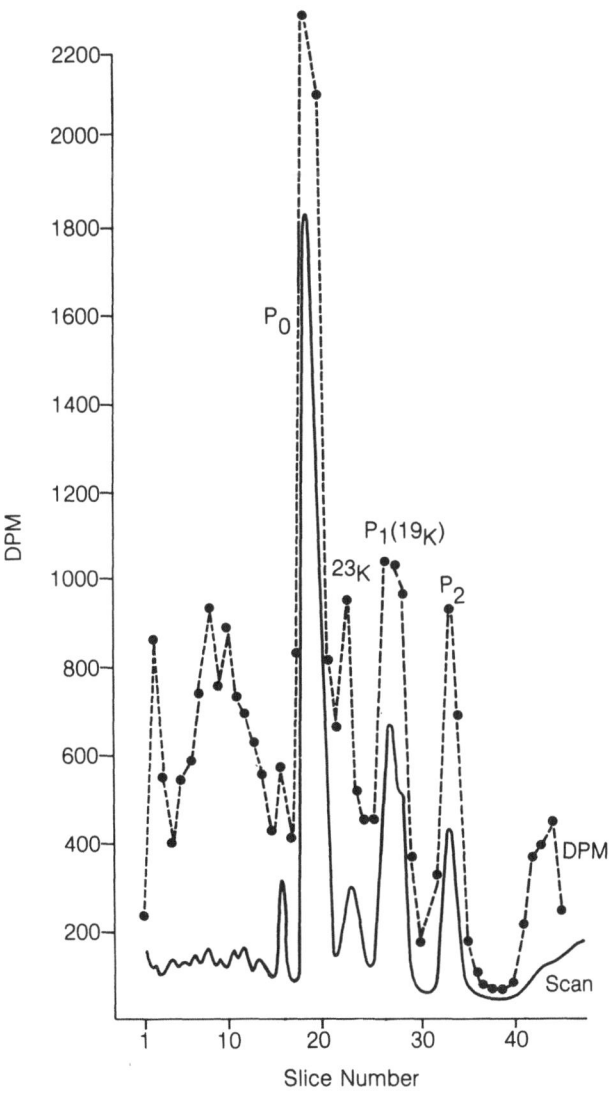

Fig. 1. Uptake of [^{3}H]amino acids into PNS myelin proteins. Solid lines indicate the spectrophotometric scan of the stained gell, dotted lines the radioactivity of each 2-mm slice. Gels were loaded with 70 g total myelin protein prepared from spinal roots of 24-day rats incubated with [^{3}H]amino acid mixture. From Smith,[47] reprinted with permission.

24 months), this discrepancy was even greater; the major myelin proteins represented 90% of the stained material but only 35–40% of the incorporated radioactivity. These data were interpreted as indicating that the high-molecular-weight proteins are more metabolically active than the prominent myelin proteins and may represent Schwann cell membranes or axonal material.

3.3.3. Specific Protein Markers

Radioactive labeling has been a useful technique for identification of myelin proteins, some of which are glycosylated, phosphorylated, or sulfated. Everly et al.[69] early identified P_0 as a glycoprotein when they injected labeled fucose directly into the nerve, isolated myelin 16 hr later, and found that the major myelin protein contained most of the label. By a similar technique, Matthieu et al. demonstrated that P_0 contains most of the injected [^{35}S]sulfate, whereas labeled fucose was present both on the major myelin glycoprotein and in a second smaller protein, probably 19K.[70]

Wiggins and Morell[71] incubated sciatic nerves in vitro and found a high amount of [^{3}H]fucose labeling in the 5-day rat, but this incorporation declined over 80% in the next 5 days as it was diluted by accumulating myelin. The glycoproteins P_0, Y, and X proteins were fucosylated, whereas incubation with ^{32}P labeled all myelin proteins including P_0, Y, X, P_1, and P_2. Again, labeling was highest at 5 days of age, followed by a sharp decline. P_0, the major myelin glycoprotein, appears to be phosphorylated at the serine residues as shown by Singh and Spritz[55] as well as sulfated, most likely on the carbohydrate chain.[70] Figlewicz et al. have shown by [^{3}H]fucose injection into sciatic nerve that a number of high-molecular-weight protein components are labeled, one of which has been identified as the myelin-associated glycoprotein similar to that in the CNS.[46]

3.3.4. Glycoprotein Synthesis in Vitro

In agreement with other investigators, we have found [^{3}H]fucose to be well incorporated in vitro into sciatic nerve and spinal root myelin proteins including P_0, 23K, and 19K in the 24-day rat.[72,73] Almost no fucose was incorporated into myelin proteins of old rats (20–24 months), which may be a result of protein stability in PNS myelin.[68] [^{3}H]Mannose and [^{3}H]glucosamine also served as effective precursors of myelin glycoproteins. Labeling of P_0 and 19K proteins was definite and predictable with all three labeled sugars, whereas labeling of 23K was uncertain and sometimes absent, perhaps as a result of its reported derivation from P_0. Incorporation of the three sugars in vitro was inhibited by the presence of tunicamycin,[73] an inhibitor of N-glycosylation by which the carbohydrate chain is attached to asparagine in the polypeptide chain (reviewed by Waechter and Scher[74]).

All three radioactive sugars also heavily labeled several components of high molecular weight, and it is evident from these studies and others[46] that a number of these glycoproteins in addition to the myelin-associated glycoprotein are metabolically active. Tunicamycin also inhibited substantially the labeling of these proteins, indicating that some of the high-molecular-weight proteins, including the MAG, contain carbohydrate chains attached by N-asparaginyl linkage. Inhibition of incorporation of carbohydrate chain precursor was not through inhibition of polypeptide synthesis.

3.4. Assembly and Relationship to Structure

Systems utilizing *in vitro* incubation of peripheral nerve slices or minces with radioactive precursors and subsequent purification and analysis of myelin have been useful in investigating mechanisms of myelin assembly. In a recent study with sciatic nerve slices from 9-day rats, Rapaport and Benjamins[75] have shown that [^{3}H]fucose incorporation into nerve homogenate is linear after 30 min but, in the isolated myelin, is linear only after an initial lag of 60 min. Using cycloheximide and cold-chase techniques, they determined that about 33 min elapsed between protein synthesis and appearance of P_0 in myelin and that 21 min elapsed between fucosylation and appearance of myelin. This time interval presumably corresponds to the length of time P_0 requires to move from the ribosomal site of translation to the Golgi (12 min) and then to the myelin sheath (21 min), although subcellular sites of translation and posttranslational modifications have not yet been isolated in this tissue.

In another recent study, the effect of tunicamycin on uptake of [^{3}H]amino acids into the individual myelin proteins was examined. Although there was little inhibition of short-term uptake (3 h), a new radioactive peak between positions occupied by P_0 and 23K appeared on the polyacrylamide gels, with a concomitant decrease in P_0 radioactivity.[72,73] The post-P_0 peak was not labeled with fucose and appeared in increasing proportions with time of incubation of nerves in the presence of tunicamycin. The new protein cross reacted with P_0 antiserum and was tentatively identified as P_0 without the carbohydrate chain, which appeared to be assembled into the myelin sheath almost as well as the normal P_0. Although it is possible that the post-P_0 protein did not actually become assembled into myelin but may have become associated with the myelin fraction after homogenization, the absence of the post-P_0 in the nonmyelin fraction may indicate that the carbohydrate chain on P_0 is not required for myelin component recognition and assembly, at least on a short-term basis. Other glycoproteins, especially those of high molecular weight such as the MAG protein localized in a periaxonal position,[48] may be involved in recognition between the axon and the myelin sheath. An approximate doubling of the amino acid uptake in lower-molecular-weight proteins in the presence of tunicamycin indicates that the unglycosylated P_0 may be more susceptible to degradation; thus, the carbohydrate chain may have a stabilizing effect.[73]

3.5. Maintenance and Turnover

Although uptake of lipid and protein precursors into PNS myelin *in vitro* has been found to be more rapid in the PNS than in the CNS,[13] no published studies have appeared on the actual turnover rate of PNS myelin components. From uptake data and unpublished experiments from the author's laboratory, it appears likely that, as in the CNS, certain myelin lipids and proteins and their immediate breakdown products are recycled during breakdown and resynthesis. The geometry and physical relationship of PNS myelin to the Schwann cell in a 1:1 relationship would make an efficient recycling mechanism, with the Schwann cell acting as a reservoir for metabolized or precursor materials. Gould, in his autoradiographic studies, demonstrated stability of fucose-labeled glycoproteins, and we have been unable to measure any uptake of [^{3}H]fucose into myelin of old rat sciatic nerves *in vitro*, although some amino acid uptake was measureable in the same animals.[47] In demyelination–remyelination studies (discussed in Section 4.4), the evidence for recycling of fucose and cholesterol within the Schwann cell–myelin complex is discussed.

4. DEMYELINATION

Metabolic studies of experimental demyelination may provide some clues regarding the mechanism of myelin destruction and the nature of myelin–axonal and Schwann cell–axonal interrelationships during breakdown and resynthesis. Such investigations are especially valuable in peripheral nerves because removal of the injurious agent or regeneration of the axon often results in rapid remyelination. Thus, within a comparatively short time, one can follow the metabolic correlation of myelin destruction, removal, and resynthesis compared with the much slower course of events in the CNS.

A number of agents are available for study of different kinds of peripheral neuropathies. Several industrial chemicals and environmental toxins have been found to result in peripheral neuropathies of the "dying back" type or in neuronopathies in which, if demyelination occurs, it is secondary to axonal or neuronal damage.[76] In a few conditions, the toxic or injurious agent is directed primarily to the myelin sheath. The Schwann cell may react to nerve damage in several ways: it often becomes reactive, ingests its own damaged myelin, proliferates, lines up along the denuded or regenerated axon, and begins the process of remyelination; or it may be uninvolved in the demyelinative process as in experimental allergic neuritis, in which the mononuclear hematogenous cells push the Schwann cell aside and attack the myelin sheath.[77] In at least one condition, that of leprosy, the Schwann cell may be the primary site of infection and injury.[78]

4.1. Vulnerability of PNS Myelin

The main structural protein of PNS myelin, P_0, unlike that of CNS myelin, is easily hydrolyzed by proteolytic enzymes, and the vulnerability of this protein may account partially for the rapid disappearance of myelin in Wallerian degeneration and other kinds of PNS demyelination. In a study of Wallerian degeneration in the rat, Wood and Dawson showed that the main glycoprotein disappeared at a steady rate with little remaining after 8 days, although other myelin proteins were undiminished.[79] Cholesterol ester also appeared within

3 days with cholesterol reduced, whereas dissolution of phospholipids was a later event.

We have studied the effects of several proteolytic enzymes on PNS myelin of rat[80] and beef.[81] Purified myelin incubated with acid proteinase or elastase or collagenase lost a large proportion of the P_0 within an hour, with very little effect on P_2. Acid proteinase hydrolyzed P_0 into peptides of molecular weight 23K, 19K, and several smaller peptides. Both elastase and collagenase degraded P_0 to peptides of molecular weight 23K and 19K, and the breakdown pattern of P_0 by lymphocytes and peritoneal exudate cells was similar. It was interesting that these enzymes incubated with CNS myelin hydrolyzed the basic protein with very little effect on the proteolipid protein, whereas in PNS myelin, the basic proteins, especially P_2, appear to be more protected, and P_0 is most vulnerable.

Cammer *et al.*[82] have shown that plasmin and conditioned medium from macrophage cultures plus plasminogen will hydrolyze P_0 to a series of smaller peptides, whereas P_2 is resistant to breakdown. The destruction of P_0 was inhibited by *p*-nitrophenylguanidinobenzoate, an inhibitor of plasminogen activator, and these investigators have proposed that the macrophage secretion products that contain plasminogen activator may participate in the inflammatory demyelination of experimental allergic neuritis.

As mentioned earlier, Roomi and Eylar[38] found that trypsin degrades P_0 to a 19K peptide (TP$_0$) which appears to be identical to the 19K protein found in PNS myelin. They have suggested that an endogenous neutral protease may cleave P_0 to 19K and that P_0 extends through the membrane and into the polar outer membrane surface where it is exposed to proteolytic enzymes. This theory, however, does not explain the finding of a 23K protein as a product of proteolytic enzyme hydrolysis of P_0.

4.2. Experimental Demyelination

Wallerian degeneration and experimental allergic neuritis are topics of separate chapters in this *Handbook* and are not further discussed here. A number of agents, for example, isonicotinic acid hydrazide, act primarily on the axon, causing degeneration of the Wallerian type. Some examples of experimental conditions in which myelin is primarily involved may have some bearing on certain kinds of human disease, and these are commented on below.

4.2.1. Dissolution by Lysolecithin

The possibility that phospholipases and lysolecithin (LPC) may be involved in certain kinds of demyelinative processes has been considered from time to time, especially in view of the properties of lysolecithin to "clear" brain homogenates[83] and myelin.[84] Banik *et al.* found that myelin incubated with snake venom or phospholipase A would convert phosphatidylcholine, phosphatidylethanolamine, and phosphatidylserine into the corresponding lyso compounds.[85] Furthermore, the presence of phospholipase increased the effects of trypsin on basic protein and proteolipid protein, and the authors proposed that cooperative effects of phospholipases and proteinases might be operative in myelin breakdown in degenerative disease. These experiments have not been repeated with PNS myelin, but a series of elegant experiments by Hall and Gregson[86-88] have shown that phospholipase A and LPC injected into the mouse sciatic nerve will induce demyelination within 30 min, with disappearance of the myelin sheath by 96 h. Remyelination was preceded by Schwann cell proliferation and contact with the axons, with remyelination beginning by 14 days after LPC injection.

We have carried out a metabolic study of the demyelination and remyelination process *in vitro* by dissecting the portion of rat sciatic nerve containing the site of LPC injection at different time intervals, incubating the chopped nerves with [^{3}H]amino acid, and measuring uptake into myelin proteins.[89] As early as 1 day after injection, the amino acid uptake into all myelin proteins was increased compared to that of the control, as would be expected by the greater permeability resulting from myelin disruption, but the increase in radioactivity of the high-molecular-weight proteins was much greater than in myelin proteins, defined as P_0, 19K, P_1, and P_2. This relationship was maintained up to 8 days after LPC injection; thereafter the myelin proteins became more active while the activity in the high-molecular-weight proteins subsided. By 30 days, the differential between high-molecular-weight and myelin proteins had disappeared.

We have shown in earlier studies that the high-molecular-weight proteins include the myelin-associated glycoprotein as well as other glycoproteins and are more active metabolically than those associated with the compact myelin sheath. This mixture of larger proteins is very likely associated with the axon and Schwann cell, possibly involved in myelin–axon recognition. The increased metabolism in the myelin disruption stage may represent activation of recognition signals and Schwann cells; then, at about 8–10 days after the insult, myelin protein synthesis begins in the remyelination process. The morphological observations by electron microscopy have correlated with these interpretations of metabolic changes.

4.2.2. Diphtheria Toxin

Peripheral neuropathy is a major feature of the clinical symptoms of diphtheria infection and is a result of segmental demyelination, especially in the ganglia of the peripheral nervous system.[90] Diphtheria toxin administered to experimental animals resulted in similar neuropathies, and direct injection of toxin into sciatic nerves of young rats caused separation of the outer myelin lamellae at the nodes of Ranvier.[91] When guinea pigs were injected with incompletely neutralized toxin–antitoxin mixtures, no damage was seen in peripheral nerves until 6–7 days, at which time some fibers showed segmental demyelination. When the nerves were incubated with radioactive acetate, however, the rate of incorporation into total lipids was profoundly depressed as early as 24 h after toxin injection. No such effect was found in liver slices of the same animals.[92]

Two metabolic studies have indicated an effect of diphtheria toxin on protein synthesis. Matheson found a concentration-dependent inhibition of uptake of [^{14}C]glycine into protein of rat peripheral nerve incubated with diphtheria toxin *in vitro*,[93] and Pleasure *et al.* observed a 79–88% inhibition of myelin proteins when chick sciatic nerves were incubated with diphtheria toxin.[94] Inhibition of [^{35}S]sulfate incorporation into sulfatide was delayed; therefore, this may have been a secondary effect of the altered protein synthesis. In this study, the differences between PNS and CNS proteins were not yet evident, and the greatest inhibition of incorporation was in proteins labeled "proteolipid protein" and "basic protein." From examination of the gels, however, it is evident that diphtheria toxin exerted a profound inhibition of synthesis of P_0 as well as of smaller myelin proteins. Diphtheria toxin inhibits protein synthesis by inactivating the translocating enzyme, elongation factor-2,[95] but it is not clear why sciatic nerve should be most affected. The experiments of Pleasure *et al.*[94] show that diphtheria toxin does not have a direct effect on myelin but probably inhibits the synthetic process, which leads to separation of lamellae and degradation.

4.2.3. Toxins Causing Myelin Edema

At least three toxic compounds, triethyl tin, hexachlorophene, and acetyl ethyl tetramethyl tetralin (AETT), cause severe edema and vacuolation of the CNS myelin sheath, and although the effects are less acute, PNS myelin is also affected, with the appearance of intramyelinic vacuolation and splitting of the intraperiod line. Triethyl tin was originally thought to affect only CNS, but further investigation with chronic administration in the drinking water revealed some degree of vacuolation in PNS[96] as well as reduction in motor nerve conduction velocity which paralleled an increase in number of axonal neurofilaments and neurotubules.[97] Hexachlorophene feeding to rats produced similar neuropathological findings and motor deficits.[98] Administration of this substance to developing rats caused severe vacuolation in roots, but the involvement was not as great as in optic nerve.[99] More recently, a neurotoxic fragrance was also found to split the CNS and PNS myelin sheath at the intraperiod line, resulting in bubbling.[100]

The metabolic effects of these substances have not been investigated in peripheral nerves, but in the CNS, triethyl tin inhibits a number of enzymes and uncouples oxidative phosphorylation, probably by acting as a carrier for chloride ion. Hexachlorophene and AETT also uncouple oxidative phosphorylation, and hexachlorophene inhibits carbonic anhydrase and a number of other enzymes.[101] These compounds appear to interfere with energy metabolism, perhaps directly within the myelin sheath, but as yet there is no evidence that myelin contains innate metabolic pumps apart from those of its supporting cell.

The effects of these inhibitors indicate that maintenance of the hydrophobic interior of myelin depends directly on intact metabolic processes which, when disrupted, allow fluid accumulation within the lamellae.

4.2.4. Diabetic Neuropathy

It is not certain whether the segmental demyelination in diabetic neuropathy is a primary or secondary effect of axonal damage. Slowing of motor and sensory nerve conduction has long been recognized in diabetic patients as well as in experimental animals made diabetic with alloxan, and this deficit has often been attributed to segmental demyelination and remyelination.[102] More recently, axonal changes have been observed in diabetic patients, and the axonal dysfunction appears to parallel the severity of the demyelinative lesion.[103] Tested fibers from biopsied sural nerves of diabetic patients showed evidence of segmental remyelination with shortening of internodal distances, indicating that demyelination and axonal loss are independent processes.[104]

Although most biochemical studies have been done on whole nerves, several reports indicate abnormal metabolism of the myelin. Although myelin isolated from sciatic nerves of rabbits made diabetic with alloxan was normal in composition, a significant decrease in total myelin was noted after diabetes of 3 to 11 months' duration.[105] Eliasson has shown decreased incorporation of [^{14}C]acetate into whole-nerve lipid, with intake into cerebroside, a characteristic myelin lipid, most depressed.[106] Spritz *et al.* reported a depressed *in vitro* incorporation of [^{14}C]leucine into sciatic nerve myelin proteins of rats made diabetic with streptozotocin.[107] Incorporation of [^{14}C]acetate and [^{3}H]water into myelin lipids was also decreased in diabetic rats and rabbits. Insulin stimulated the *in vitro* uptake of [^{14}C]leucine into myelin proteins of control sciatic nerves but had no effect on diabetic nerve.

The authors concluded that the Schwann cell may be affected in diabetes with a resultant change in myelin metabolism. In other studies, comparisons of precursor uptake *in vitro* into sciatic nerve myelin of streptozotocin and alloxan diabetic rats yielded significant increases in the ratio [^{3}H]fucose/[^{14}C]leucine, with the ratio change mostly representing a relative decrease in [^{14}C]leucine uptake.[108] Thus, definite effects on the Schwann cell/myelin unit which are not apparent in the CNS are implicated in diabetes.

Recent investigations have emphasized the role of *myo*-inositol in maintaining the ion flux in the nerve impulse and have suggested that a decrease in peripheral nerves of diabetic subjects may be a causative factor in the conduction deficit. Whereas normal rat sciatic nerves maintain a concentration of free *myo*-inositol 60–90 times that of plasma, the levels in diabetic sciatic nerves are considerably lower. Both insulin and a dietary supplement of 1% *myo*-inositol abolished the difference in *myo*-inositol concentrations between control and diabetic animals and prevented the impaired nerve conduction velocity in diabetics.[109] After streptozotocin treatment, a progressive decrease in both sciatic nerve inositol and lipid inositol has been measured.[110] These findings suggest that *myo*-inositol may be a likely therapeutic agent for treatment of diabetic neuropathies, although the role of this substance is obscure. *Myo*-inositol is mainly a precursor for lipid synthesis and is rapidly incorporated into axonal lipids.[67]

Three hypotheses as to the cause of diabetic neuropathy are currently emphasized. These include a vascular insufficiency, a direct effect on the

Schwann cell, thus affecting the myelin, and primary damage to the axon as a cause of the nerve conduction deficit. These theories are discussed in detail in a recent review.[111] Diabetic neuropathy, however, may be regarded as a class of disorders resulting from a number of insufficiencies.

4.3. Remyelination

Some evidence is available indicating that Schwann cells may derive certain components used in remyelination from previously phagocytosed myelin. Rawlins *et al.* injected [³H]cholesterol into suckling mice to label the sciatic nerve myelin and then, 5 weeks later, cut the nerve to induce Wallerian degeneration.[112] During degeneration, radioactivity was localized by autoradiography over degrading myelin; it was later found predominantly over regenerated myelin sheaths. The authors concluded that some of the cholesterol from the degenerated myelin was retained within the nerves in an exchangeable pool, possibly as cholesterol ester, and then reutilized for myelin regeneration.

In our studies of myelin degeneration from lysolecithin injection, uptake of labeled amino acids into myelin proteins, especially P_0, doubled 8 days after treatment, but incorporation of labeled fucose was augmented only slightly. Such a differential in uptake into polypeptide versus into the carbohydrate chain of the same protein can best be explained by reutilization of fucose from myelin previously disrupted by lysolecithin and retained within the nerve, possibly within the Schwann cell. This cell plays an instrumental role in myelin destruction in certain kinds of demyelination, including Wallerian degeneration, by phagocytosing the disrupted myelin. Certain components of the degraded myelin may then be recycled in regeneration of new myelin lamellae.

5. SUMMARY AND CONCLUSIONS

Peripheral nervous system myelin and its metabolism have recently become active areas of investigation, and we can now make some generalizations as to its nature. It is different from that of the CNS in regard to protein content and composition, although certain protein constituents are found in both. The PNS myelin protein is notable for its high content of glycoproteins. The lipid compositions of CNS and PNS myelin are qualitatively similar, but the proportions of certain lipids, especially cerebroside and sphingomyelin, are somewhat different. These differences are a result of the properties of the myelin-forming cells, the Schwann cells in the PNS and the oligodendroglia in the CNS, and these characteristics probably evolved from the more primitive animals in which the nervous system is predominantly peripheral to those with a more highly developed CNS.

Little is presently known about PNS myelin metabolism and turnover, but current information indicates that PNS myelin may be more active metabolically than that of CNS. This might be expected in view of the Schwann cell–axonal relationship of one cell/unit axon. Schwann cells begin to form myelin earlier, and throughout life, the uptake rate of radioactive myelin precursors is several times higher than in the CNS where one cell must maintain a large number of myelinated axons. It is interesting, however, that the myelin enzymes that have been found to be constituents of CNS myelin are generally less active in the PNS. In view of the closer relationships of the Schwann cell to PNS myelin, perhaps these enzymes are not needed in the myelin substance *per se.*

In the event of injury, the Schwann cell responds rapidly, ingests the damaged myelin, reorganizes, and remyelinates the demyelinated or regenerated nerve, whereas this course of events occurs much more slowly, if at all, in the CNS. The rapid reaction may be facilitated by the high glycoprotein content in PNS myelin which may lead to early recognition of disrupted contacts. After ingesting the myelin debris, the Schwann cell may reutilize the materials for remyelination. In view of the interdependencies in the morphology and metabolism of axons, Schwann cells, and myelin, injury to one of these elements may lead to rapid reactions or changes in the others. As a result of the interrelated reactions, it has been difficult to determine whether disease and injurious agents may act on a primary target.

The different characteristics of Schwann cells and oligodendroglia are being intensively investigated, and a comparison of their properties may help to explain the differences between CNS and PNS myelin in development and in diseased states.

ACKNOWLEDGMENT. I am grateful to Dr. Stephen G. Waxman for helpful discussions and the use of his library and to Drs. Wendy Cammer and Robert Gould for a critical reading of the manuscript. This work is supported by the Medical Research Service of the Veterans Administration and by Grant NS-02785 from the NIH.

REFERENCES

1. Webster, H. deF, 1975, *Peripheral Neuropathy*, Volume 1 (P. J. Dyck, P. K. Thomas, and E. H. Lambert, eds.), W. B. Saunders, Philadelphia, pp. 37–61.
2. Landon, D. N., and Hall, S., 1976, *Peripheral Nerve* (D. M. London, ed.), John Wiley & Sons, New York, pp. 1–105.
3. Raine, C. S., 1977, *Myelin* (P. Morell, ed.), Plenum Press, New York, pp. 1–49.
4. Spencer, P., and Weinberg, H. J., 1978, *Physiology and Pathobiology of Axons* (S. G. Waxman, ed.), Raven Press, New York, pp. 389–405.
5. Freide, R. L., and Samorajski, T., 1968, *J. Neuropathol. Exp. Neurol.* 27:546–570.
6. Waxman, S. G., and Foster, R. E., 1980, *Brain Res. Rev.* 2:205–234.
7. Brunetti, M., Di Giamberardino, L., Porcellati, G., and Droz, B., 1981, *Brain Res.* 219:73–84.
8. Spencer, P. S., Weinberg, H. J., Krygier-Brevart, V., and Zabrenetzky, V., 1979, *Brain Res.* 165:119–126.
9. Manthorpe, M., Skaper, S., and Varon, S., 1980, *Brain Res.* 196:467–482.
10. Kreider, B. Q., Messing, A., Doan, H., Kim, S. U., Lisak, R. P., and Pleasure, D. E., 1981, *Brain Res.* 207:433–444.
11. Mirsky, R., Winter, J., Abney, E. R., Pruss, R. M., Gavrilovic, J., and Raff, M. C., 1980, *J. Cell Biol.* 84:483–494.

12. Norton, W. T., and Poduslo, S. E., 1973, *J. Neurochem.* **21**:749–757.
13. Rawlins, F. A., and Smith, M. E., 1971, *J. Neurochem.* **18**:1861–1870.
14. Wiggins, R. C., Benjamins, J. A., and Morell, P., 1975, *Brain Res.* **89**:99–106.
15. Horrocks, L. A., 1967, *J. Lipid Res.* **8**:569–576.
16. Spritz, N., Singh, H., and Geyer, B., 1973, *J. Clin. Invest.* **52**:520–523.
17. O'Brien, J. S., Sampson, E. L., and Stern, M. B., 1967, *J. Neurochem.* **14**:357–365.
18. Eng, L. F., Chao, F.-C., Gerstl, B., Pratt, D., and Tavaststjerna, M. G., 1968, *Biochemistry* **7**:4455–4465.
19. Evans, M. J., and Finean, J. B., 1965, *J. Neurochem.* **12**:729–734.
20. Smith, M. E., and Curtis, B. M., 1979, *J. Neurochem.* **33**:447–452.
21. Oulton, M. R., and Mezei, C., 1976, *J. Lipid Res.* **17**:167–175.
22. O'Brien, J. S., and Sampson, E. L., 1965, *J. Lipid Res.* **6**:537–544.
23. Fong, J. W., Ledeen, R. W., Kundu, S. K., and Brostoff, S. W., 1976, *J. Neurochem.* **26**:157–162.
24. Darriet, D., Cassagne, C., and Bourre, J. M., 1978, *J. Neurochem.* **31**:1541–1543.
25. Quarles, R. H., 1979, *Complex Carbohydrates of Nervous Tissue* (R. U. Margolis and R. K. Margolis, eds.), Plenum Press, New York, pp. 209–233.
26. Braun, P. E., and Brostoff, S. W., 1977, *Myelin* (P. Morell, ed.), Plenum Press, New York, pp. 201–231.
27. Benjamins, J. A., and Morell, P., 1978, *Neurochem. Res.* **3**:137–174.
28. Greenfield, S., Brostoff, S., Eylar, E. H., and Morell, P., 1973, *J. Neurochem.* **20**:1207–1216.
29. Eylar, E. H., Uyemura, K., Brostoff, S. W., Kitamura, K., Ishaque, A., and Greenfield, S., 1979, *Neurochem. Res.* **4**:289–293.
30. Smith, M. E., Forno, L. S., and Hofmann, W. W., 1979, *J. Neuropathol. Exp. Neurol.* **38**:377–391.
31. Greenfield, S., Brostoff, S. W., and Hogan, E. L., 1980, *J. Neurochem.* **34**:453–455.
32. Everly, J. L., Brady, R. O., and Quarles, R. H., 1973, *J. Neurochem.* **21**:329–334.
33. Wood, J. G., and Dawson, R. M. C., 1973, *J. Neurochem.* **21**:717–719.
34. Kitamura, K., Suzuki, M., and Uyemura, K., 1976, *Biochim. Biophys. Acta* **455**:806–816.
35. Roomi, M. W., Ishaque, A., Khan, N. R., and Eylar, E. H., 1978, *Biochim. Biophys. Acta* **536**:112–121.
36. Kitamura, K., Suzuki, A., Suzuki, M., and Uyemura, K., 1979, *FEBS Lett.* **100**:67–70.
37. Cammer, W., Sirota, S. R., and Norton, W. T., 1980, *J. Neurochem.* **34**:404–409.
38. Roomi, M. W., and Eylar, E. H., 1978, *Biochim. Biophys. Acta* **536**:122–133.
39. Ishaque, A., Roomi, M. W., Szymanska, I., Kowalski, S., and Eylar, E. H., 1980, *Can. J. Biochem.* **58**:913–921.
40. Roomi, M. W., Ishaque, A., Khan, N. R., and Eylar, E. H., 1978, *J. Neurochem.* **31**:375–379.
41. Linington, C., and Waehneldt, T. V., 1981, *J. Neurochem.* **36**:1528–1535.
42. Brostoff, S. W., and Eylar, E. H., 1972, *Arch. Biochem. Biophys.* **153**:590–598.
43. Brostoff, S. W., Burnett, P., Lampert, P., and Eylar, E. H., 1972, *Nature [New Biol.]* **235**:210–212.
44. Kitamura, K., Suzuki, M., Suzuki, A., and Uyemura, K., 1980, *FEBS Lett.* **115**:27–30.
45. Kadlubowski, M., and Hughes, R. A. C., 1979, *Nature* **277**:140–141.
46. Figlewicz, D. A., Quarles, R. H., Johnson, D., Barbarash, G. R., and Sternberger, N. H., 1981, *J. Neurochem.* **37**:749–758.
47. Smith, M. E., 1980, *J. Neurochem.* **35**:1183–1189.
48. Sternberger, N. H., Quarles, R. H., Itoyama, Y., and Webster, H. DeF., 1979, *Proc. Natl. Acad. Sci. U.S.A.* **76**:1510–1514.
49. Norton, W. T., 1980, *The Search for the Cause of Multiple Sclerosis and Other Chronic Diseases and the Central Nervous System* (A. Boese, ed.), Verlag Chemie, Weinheim, pp. 64–75.
50. Kurihara, T., and Tsukada, Y., 1967, *J. Neurochem.* **14**:1167–1174.
51. Matthieu, J.-M., Costantino-Ceccarini, E., Beny, M., and Reigner, J., 1980, *J. Neurochem.* **35**:1345–1350.

52. Cammer, W., and Zimmerman, T. R., 1981, *Trans. Am. Soc. Neurochem.* **12:**81.
53. Cammer, W., 1979, *J. Neurochem.* **32:**651–654.
54. Krygier-Brevart, V., Zabrenetsky, V. S., and Spencer, P. S., 1977, *Trans. Am. Soc. Neurochem.* **8:**262.
55. Singh, H., and Spritz, N., 1976, *Biochem. Biophys. Acta* **448:**325–337.
56. Braun, P. E., 1977, *Myelin* (P. Morell, ed.), Plenum Press, New York, pp. 91–115.
57. Wood, J. G., and McLaughlin, B. J., 1975, *J. Neurochem.* **24:**233–235.
58. Sea, C. P., and Peterson, R. G., 1975, *Exp. Neurol.* **48:**252–260.
59. Peterson, R. G., and Gruener, R. W., 1978, *Brain Res.* **152:**17–29.
60. Blaurock, A. E., and Nelander, J. C., 1979, *J. Neurochem.* **32:**1753–1760.
61. Kirschner, D. A., and Ganser, A. L., 1980, *Nature* **283:**207–210.
62. Norton, W. T., 1977, *Myelin* (P. Morell, ed.), Plenum Press, New York, pp. 161–199.
63. Wiggins, R. C., Benjamins, J. A., and Morell, P., 1975, *Brain Res.* **89:**99–106.
64. Uyemura, K., Horie, K., Kitamura, K., Suzuki, M., and Uehara, S., 1979, *J. Neurochem.* **32:**779–788.
65. Rawlins, F. A., 1973, *J. Cell Biol.* **58:**42–53.
66. Gould, R. M., and Dawson, R. M. C., 1976, *J. Cell Biol.* **68:**480–496.
67. Gould, R. M., 1976, *Brain Res.* **117:**169–174.
68. Gould, R. M., 1977, *J. Cell Biol.* **75:**326–338.
69. Everly, J. L., Brady, R. O., and Quarles, R. H., 1973, *J. Neurochem.* **21:**329–334.
70. Matthieu, J.-M., Everly, J. L., Brady, R. O., and Quarles, R. H., 1975, *Biochim. Biophys. Acta* **392:**167–174.
71. Wiggins, R. C., and Morell, P., 1980, *J. Neurochem.* **34:**627–634.
72. Smith, M. E., 1980, *Abstr. Soc. Neurosci.* **10:**20.
73. Smith, M. E., and Sternberger, N. H., 1982, *J. Neurochem.* **38:**1044–1049.
74. Waechter, C. J., and Scher, M. G., 1979, *Complex Carbohydrates of Nervous Tissue* (R. U. Margolis and R. K. Margolis, eds.), Plenum Press, New York, pp. 75–102.
75. Rapaport, R. N., and Benjamins, J. A., 1981, *J. Neurochem.* **37:**164–171.
76. Spencer, P. S., and Schaumburg, H. F., 1978, *Physiology and Pathobiology of Axons* (S. G. Waxman, ed.), Raven Press, New York, pp. 265–282.
77. Lampert, P. W., 1969, *Lab. Invest.* **20:**127–138.
78. Asbury, A. K., and Johnson, P. C., 1978, *Pathology of Peripheral Nerve,* W. B. Saunders, Philadelphia, pp. 184–189.
79. Wood, J. G., and Dawson, R. M. C., 1974, *J. Neurochem.* **22:**631–635.
80. Smith, M. E., 1980, *Neurochemistry and Clinical Neurology* (L. Battistin, G. Hashim, and A. Lajtha, eds.), Alan R. Liss, New York, pp. 1–10.
81. Smith, M. E., 1980, *Trans. Am. Soc. Neurochem.* **11:**213.
82. Cammer, W., Brosnan, C. F., Bloom, B. R., and Norton, W. T., 1981, *J. Neurochem.* **36:**1506–1514.
83. Thompson, R. H. S., 1964, *Metabolism and Physiological Significance of Lipids* (R. M. C. Dawson and D. Rhodes, eds.), John Wiley & Sons, London, pp. 541–551.
84. Gent, W. L. G., Gregson, N. A., Gammack, D. B., and Raper, J. H., 1964, *Nature* **204:**553–555.
85. Banik, N. L., Gohil, K., and Davison, A. N., 1976, *Biochem. J.* **159:**273–277.
86. Hall, S. M., and Gregson, N. A., 1971, *J. Cell Sci.* **9:**769–789.
87. Gregson, N. A., and Hall, S. M., 1973, *J. Cell Sci.* **13:**257–277.
88. Hall, S. M., 1973, *J. Cell Sci.* **13:**461–477.
89. Smith, M. E., Kocsis, J. D., and Waxman, S. G., 1981, *Abstr. Int. Soc. Neurochem.* **8:**153.
90. McDonald, W. I., and Kocen, R. S., 1975, *Peripheral Neuropathy* (P. J. Dyck, P. K. Thomas, and E. H. Lambert, eds.), W. B. Saunders, Philadelphia, pp. 1281–1300.
91. Allt, G., and Cavanagh, J. B., 1969, *Brain* **92:**459–468.
92. Majno, G., Waksman, B. H., and Karnovsky, M. L., 1960, *J. Neuropathol. Exp. Neurol.* **19:**7–24.
93. Matheson, D. F., 1968, *J. Neurochem.* **15:**179–185.
94. Pleasure, D. E., Feldmann, B., and Prockop, D. J., 1973, *J. Neurochem.* **20:**81–90.
95. Pappenheimer, A. M., Jr., 1977, *Annu. Rev. Biochem.* **46:**69–94.

96. Graham, D. I., and Gonatas, N. K., 1973, *Lab. Invest.* **29:**628–632.
97. Graham, D. I., deJesus, P. V., Pleasure, D. E., and Gonatas, N. K., 1976, *Arch. Neurol.* **33:**40–48.
98. deJesus, P. V., and Pleasure, D. E., 1973, *Arch. Neurol.* **29:**180–182.
99. Towfighi, J., Gonatas, N. K., and McCree, L., 1975, *Lab. Invest.* **31:**712–721.
100. Spencer, P. S., Sterman, A. B., Haroupian, D. S., and Foulds, M. M., 1979, *Science* **204:**633–635.
101. Cammer, W., 1980, *Experimental and Clinical Neurotoxicology,* (P. S. Spencer and H. H. Schaumburg, eds.), Williams & Wilkins, Baltimore, pp. 239–256.
102. Thomas, P. K., and Eliasson, S. G., 1975, *Peripheral Neuropathy,* Volume II (P. J. Dyck, P. K. Thomas, and E. H. Lambert, eds.), W. B. Saunders, Philadelphia, pp. 956–981.
103. Hansen, S., and Ballantyne, J. P., 1977, *J. Neurol. Neurosurg. Psychiatry* **40:**555–564.
104. Behse, F., Buchthal, F., and Carlsen, F., 1977, *J. Neurol. Neurosurg. Psychiatry* **40:**1072–1082.
105. Spritz, N., Singh, H., and Marinan, B., 1975, *Diabetes* **24:**680–683.
106. Eliasson, S., 1966, *Lipids* **1:**237–240.
107. Spritz, N., Singh, H., and Marinan, B., 1975, *J. Clin. Invest.* **55:**1049–1056.
108. Baughman, S., Felten, S. Y., Lee, W., Moore, S. A., O'Conner, B. L., and Peterson, R. G., 1981, *Horm. Metab. Res.* **13:**331–335.
109. Greene, D. A., deJesus, P. V., Jr., and Winegrad, A. I., 1975, *J. Clin. Invest.* **55:**1326–1336.
110. Palmano, K. P., Whiting, P. H., and Hawthorne, J. N., 1977, *Biochem. J.* **167:**229–235.
111. Clements, R. S., Jr., 1979, *Diabetes* **28:**604–611.
112. Rawlins, F. A., Villegas, G. M., Hedley-Whyte, E. T., and Uzman, B. G., 1972, *J. Cell Biol.* **52:**615–625.

Measurement of Local Glucose Utilization in the Central Nervous System and Its Relationship to Local Functional Activity

Louis Sokoloff

1. INTRODUCTION

Tissues that do physical and/or chemical work, such as heart, kidney, and skeletal muscle, exhibit a close relationship between energy metabolism and the level of functional activity. The existence of a similar relationship in the tissues of the central nervous system has been more difficult to prove, partly because of uncertainty about the nature of the work associated with nervous functional activity but mainly because of the difficulty in assessing the levels of functional and metabolic activities in the same functional component of the brain at the same time.

Much of our present knowledge of cerebral energy metabolism *in vivo* has been obtained by means of the nitrous oxide technique of Kety and Schmidt[1] and its modifications,[2–5] which measure the average rates of energy metabolism in the brain as a whole. These methods have demonstrated changes in cerebral metabolic rate in association with gross or diffuse alterations of cerebral function and/or structure, as, for example, those that occur during postnatal development, aging, senility, anesthesia, disorders of consciousness, and convulsive states.[6–10] They have not detected changes in cerebral metabolic rate in a number of conditions with, perhaps, more subtle alterations in cerebral functional activity, for example, deep slow-wave sleep, performance of mental arithmetic, sedation and tranquilization, schizophrenia, and LSD-induced psychosis.[6,8,11] It is possible that there are no changes in cerebral energy metabolism in these conditions. The apparent lack of change could also be explained by either a redistribution of local levels of functional and metabolic activity

Louis Sokoloff • Laboratory of Cerebral Metabolism, National Institute of Mental Health, U.S. Public Health Service, Department of Health and Human Services, Bethesda, Maryland 20205.

without significant change in the average of the brain as a whole or the restriction of altered metabolic activity to regions too small to be detected in measurements of the brain as a whole.

Local cerebral blood flow is generally believed to be adjusted to local energy metabolism, and the blood flow might then be considered to reflect the local metabolic rate. Kety and his associates,[12-15] therefore, developed a method to measure local cerebral blood flow that was based on a quantitative autoradiographic technique to measure the local tissue concentrations of chemically inert, diffusible, radioactive tracers. Blood flow could be determined from the history of the arterial concentration of the tracer and the final tissue concentration in all the structural components visible and identifiable in the autoradiographs of serial sections of the brain. The application of this quantitative autoradiographic technique to the determination of local cerebral metabolic rate has proved to be more difficult because of the inherently greater complexity of the problem and the unsuitability of the labeled species of the normal substrates of cerebral energy metabolism, oxygen and glucose. The radioisotopes of oxygen have too short a physical half-life. Both oxygen and glucose are too rapidly converted to carbon dioxide, and CO_2 is too rapidly cleared from the cerebral tissues. Sacks,[16] for example, has found in man significant losses of $^{14}CO_2$ from the brain within 2 min after the onset of an intravenous infusion of [^{14}C]glucose, labeled either uniformly, or in the C-1, C-2, or C-6 positions.

These limitations of [^{14}C]glucose have been avoided by the use of 2-deoxy-D-[^{14}C]glucose, a labeled analogue of glucose with special properties that make it particularly appropriate for this application.[17] It is metabolized through part of the pathway of glucose metabolism at a definable rate relative to that of glucose. Unlike glucose, however, its product, [^{14}C]deoxyglucose-6-phosphate, is essentially trapped in the tissues, allowing the application of the quantitative autoradiographic technique. The use of radioactive 2-deoxyglucose to trace glucose utilization and the autoradiographic technique to achieve regional localization has recently led to the development of a method that measures the rates of glucose utilization simultaneously in all components of the central nervous system in the normal conscious state and during experimental physiological, pharmacological, and pathological conditions.[17] Because the procedure is so designed that the concentrations of radioactivity in the tissues during autoradiography are more or less proportional to the rates of glucose utilization, the autoradiographs provide pictorial representations of the relative rates of glucose utilization in all the cerebral structures visualized. Numerous studies with this method have established that there is a close relationship between functional activity and energy metabolism in the central nervous system,[18,19] and the method has become a potent new tool for mapping functional neural pathways on the basis of evoked metabolic responses.

2. THEORY

The method is based on a model of the biochemical properties of glucose and 2-deoxyglucose in brain (Fig. 1A).[17] 2-Deoxyglucose is transported bidi-

Table I

Values of Rate Constants in the Normal Conscious Albino Rat[a]

Structure	Rate constants (min^{-1})			Distribution volume (ml/g) $[k_1^*/(k_2^* + k_3^*)]$	Half-life of precursor pool (min) $[\log_e 2/(k_2^* + k_3^*)]$
	k_1^*	k_2^*	k_3^*		
Gray matter					
Visual cortex	0.189 ± 0.048	0.279 ± 0.176	0.063 ± 0.040	0.553	2.03
Auditory cortex	0.226 ± 0.068	0.241 ± 0.198	0.067 ± 0.057	0.734	2.25
Parietal cortex	0.194 ± 0.051	0.257 ± 0.175	0.062 ± 0.045	0.608	2.17
Sensory–motor cortex	0.193 ± 0.037	0.208 ± 0.112	0.049 ± 0.035	0.751	2.70
Thalamus	0.188 ± 0.045	0.218 ± 0.144	0.053 ± 0.043	0.694	2.56
Medial geniculate body	0.219 ± 0.055	0.259 ± 0.164	0.055 ± 0.040	0.697	2.21
Lateral geniculate body	0.172 ± 0.038	0.220 ± 0.134	0.055 ± 0.040	0.625	2.52
Hypothalamus	0.158 ± 0.032	0.226 ± 0.119	0.043 ± 0.032	0.587	2.58
Hippocampus	0.169 ± 0.043	0.260 ± 0.166	0.056 ± 0.040	0.535	2.19
Amygdala	0.149 ± 0.028	0.235 ± 0.109	0.032 ± 0.026	0.558	2.60
Caudate–putamen	0.176 ± 0.041	0.200 ± 0.140	0.061 ± 0.050	0.674	2.66
Superior colliculus	0.198 ± 0.054	0.240 ± 0.166	0.046 ± 0.042	0.692	2.42
Pontine gray matter	0.170 ± 0.040	0.246 ± 0.142	0.037 ± 0.033	0.601	2.45
Cerebellar cortex	0.225 ± 0.066	0.392 ± 0.229	0.059 ± 0.031	0.499	1.54
Cerebellar nucleus	0.207 ± 0.042	0.194 ± 0.111	0.038 ± 0.035	0.892	2.99
Mean ± S.E.M.	0.189 ± 0.012	0.245 ± 0.040	0.052 ± 0.010	0.647 ± 0.073	2.39 ± 0.40
White matter					
Corpus callosum	0.085 ± 0.015	0.135 ± 0.075	0.019 ± 0.033	0.552	4.50
Genu of corpus callosum	0.076 ± 0.013	0.131 ± 0.075	0.019 ± 0.034	0.507	4.62
Internal capsule	0.077 ± 0.015	0.134 ± 0.085	0.023 ± 0.039	0.490	4.41
Mean ± S.E.M.	0.079 ± 0.008	0.133 ± 0.046	0.020 ± 0.020	0.516 ± 0.171	4.51 ± 0.90

[a] From Sokoloff et al.[17]

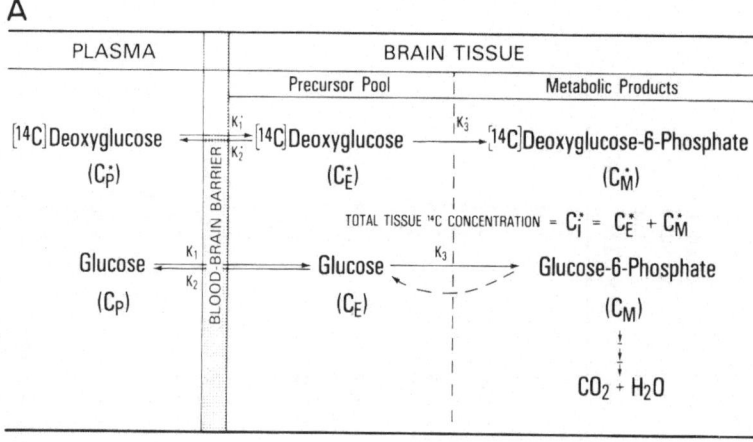

B

General Equation for Measurement of Reaction Rates with Tracers:

$$\text{Rate of Reaction} = \frac{\text{Labeled Product Formed in Interval of Time, 0 to T}}{\begin{bmatrix}\text{Isotope Effect} \\ \text{Correction Factor}\end{bmatrix}\begin{bmatrix}\text{Integrated Specific Activity} \\ \text{of Precursor}\end{bmatrix}}$$

Operational Equation of $[^{14}C]$ Deoxyglucose Method:

$$R_i = \frac{\overbrace{C_i^*(T) \quad - \quad k_1^* e^{-(k_2^*+k_3^*)T}\int_0^T C_p^* e^{(k_2^*+k_3^*)t}\,dt}^{\substack{\text{Total }^{14}\text{C in Tissue}\qquad ^{14}\text{C in Precursor Remaining in Tissue at Time, T}\\ \text{at Time, T}}}}{\underbrace{\left[\frac{\lambda\cdot V_m^*\cdot K_m}{\Phi\cdot V_m\cdot K_m^*}\right]}_{\substack{\text{"Isotope Effect"}\\\text{Correction}\\\text{Factor}}}\underbrace{\left[\underbrace{\int_0^T\left(\frac{C_p^*}{C_p}\right)dt}_{\substack{\text{Integrated Plasma}\\\text{Specific Activity}}} - \underbrace{e^{-(k_2^*+k_3^*)T}\int_0^T\left(\frac{C_p^*}{C_p}\right)e^{(k_2^*+k_3^*)t}\,dt}_{\substack{\text{Correction for Lag in Tissue}\\\text{Equilibration with Plasma}}}\right]}}$$

Labeled Product Formed in Interval of Time, 0 to T

Integrated Precursor Specific Activity in Tissue

Fig. 1. (A) Diagrammatic representation of the theoretical model. C_i^* represents the total ^{14}C concentration in a single homogeneous tissue of the brain. C_P^* and C_P represent the concentrations of $[^{14}C]$deoxyglucose and glucose in the arterial plasma, respectively; C_E^* and C_E represent their respective concentrations in the tissue pools that serve as substrates for hexokinase. C_M^* represents the concentration of $[^{14}C]$deoxyglucose-6-phosphate in the tissue. The constants k_1^*, k_2^*, and k_3^*, represent the rate constants for carrier-mediated transport of $[^{14}C]$deoxyglucose from plasma to tissue, for carrier-mediated transport back from tissue to plasma, and for phosphorylation by hexokinase, respectively. The constants k_1, k_2, and k_3 are the equivalent rate constants for glucose. $[^{14}C]$Deoxyglucose and glucose share and compete for the carrier that transports both between rectionally between blood and brain by the same carrier that transports glucose across the blood–brain barrier.[20–22] In the cerebral tissues, it is phosphorylated by hexokinase to 2-deoxyglucose-6-phosphate.[23] Deoxyglucose (DG) and glucose are, therefore, competitive substrates for both blood–brain transport and hexokinase-catalyzed phosphorylation. Unlike glucose-6-phosphate (G-6-P), however, which is metabolized further, eventually to CO_2 and water and to a lesser degree via the hexosemonophosphate shunt, deoxyglucose-6-phosphate cannot be converted to fructose-6-phosphate and is not a substrate for glucose-6-phosphate dehydrogenase.[23] There is very little glucose-6-phosphatase activity in brain[24] and even less deoxyglucose-6-phosphatase activity.[17] Deoxyglucose-6-phosphate (DG-6-P), once formed, is, therefore, essentially trapped in the cerebral tissues, at least long enough for the duration of the measurement.

If the interval of time is kept short enough, for example, less than 1 h, to allow the assumption of negligible loss of [14C]DG-6-P from the tissues, then the quantity of [14C]DG-6-P accumulated in any cerebral tissue at any given time following the introduction of [14C]DG into the circulation is equal to the integral of the rate of [14C]DG phosphorylation by hexokinase in that tissue during that interval of time. This integral is in turn related to the amount of glucose that has been phosphorylated over the same interval, depending on the time courses of the relative concentrations of [14C]DG and glucose in the precursor pools and the Michaelis–Menten kinetic constants for hexokinase with respect to both [14C]DG and glucose. With cerebral glucose consumption in a steady state, the amount of glucose phosphorylated during the interval of time equals the steady-state flux of glucose through the hexokinase-catalyzed step times the duration of the interval, and the net rate of flux of glucose through this step equals the rate of glucose utilization.

These relationships can be mathematically defined and an operational equation derived if the following assumptions are made: (1) a steady state for glucose (i.e., constant plasma glucose concentration and constant rate of glucose consumption) throughout the period of the procedure; (2) homogeneous tissue compartment within which the concentrations of [14C]DG and glucose are uniform and exchange directly with the plasma; and (3) tracer concentrations of [14C]DG (i.e., molecular concentrations of free [14C]DG essentially equal to zero). The operational equation that defines R_i, the rate of glucose consumption per unit mass of tissue, i, in terms of measurable variables is presented in Fig. 1B.

The rate constants are determined in a separate group of animals by a nonlinear, iterative process that provides the least-squares best fit of an equa-

---

plasma and tissue and for hexokinase which phosphorylates them to their respective hexose-6-phosphates. The dashed arrow represents the possibility of glucose-6-phosphate hydrolysis by glucose-6-phosphatase activity, if any.[17,95] (B) Operational equation of radioactive deoxyglucose method and its functional anatomy. T represents the time at the termination of the experimental period; λ equals the ratio of the distribution space of deoxyglucose in the tissue to that of glucose; Φ equals the fraction of glucose that, once phosphorylated, continues down the glycolytic pathway; and K_m^* and V_m^* and K_m and V_m represent the familiar Michaelis–Menten kinetic constants of hexokinase for deoxyglucose and glucose, respectively. The other symbols are the same as those defined in A.[95]

Table II
Values of the Lumped Constant in the Albino Rat, Rhesus Monkey, Cat, and Dog[a]

Animal	No. of animals	Mean ± S.D.	S.E.M.
Albino rat			
Conscious	15	0.464 ± 0.099^b	±0.026
Anesthetized	9	0.512 ± 0.118^b	±0.039
Conscious (5% CO_2)	2	0.463 ± 0.122^b	±0.086
Combined	26	0.481 ± 0.119	±0.023
Rhesus monkey			
Conscious	7	0.344 ± 0.095	±0.036
Cat			
Anesthetized	6	0.411 ± 0.013	±0.005
Dog (beagle puppy)			
Conscious	7	0.558 ± 0.082	±0.031

[a] The values were obtained as follows: rat, Sokoloff *et al.*[17]; monkey, Kennedy *et al.*[25]; cat, M. Miyaoka, J. Magnes, C. Kennedy, M. Shinohara, and L. Sokoloff (unpublished data); dog, Duffy *et al.*[93] From Sokoloff.[94]
[b] No statistically significant difference between normal conscious and anesthetized rats ($0.3 < P < 0.4$) and conscious rats breathing 5% CO_2 ($P > 0.9$).

tion that defines the time course of total tissue ^{14}C concentration in terms of the time, the history of the plasma concentration, and the rate constants to the experimentally determined time courses of tissue and plasma concentrations of ^{14}C.[17] The rate constants have thus far been completely determined only in normal conscious albino rats (Table I). Partial analyses indicate that the values are quite similar in the conscious monkey.[25]

The λ, Φ, and the enzyme kinetic constants are grouped together to constitute a single, lumped constant (Fig. 1B). It can be shown mathematically that this lumped constant is equal to the asymptotic value of the product of the ratio of the cerebral extraction ratios of $[^{14}C]DG$ and glucose and the ratio of the arterial blood to plasma specific activities when the arterial plasma $[^{14}C]DG$ concentration is maintained constant.[17] The lumped constant is also determined in a separate group of animals from arterial and cerebral venous blood samples drawn during a programmed intravenous infusion which produces and maintains a constant arterial plasma $[^{14}C]DG$ concentration.[17] Thus far, the lumped constant has been determined only in the albino rat, monkey, cat, and dog (Table II). The lumped constant appears to be characteristic of the species and does not appear to change significantly in a wide range of physiological conditions (Table II).[17]

Despite its complex appearance, the operational equation is really nothing more than a general statement of the standard relationship by which rates of enzyme-catalyzed reactions are determined from measurements made with radioactive tracers (Fig. 1B). The numerator of the equation represents the amount of radioactive product formed in a given interval of time; it is equal to C_i^*, the combined concentrations of $[^{14}C]DG$ and $[^{14}C]DG$-6-P in the tissue at time T, measured by the quantitative autoradiographic technique, less a term that represents the free unmetabolized $[^{14}C]DG$ still remaining in the tissue. The denominator represents the integrated specific activity of the precursor pool times a factor, the lumped constant, which is equivalent to a correction factor for an isotope effect. The term with the exponential factor in the denominator takes into account the lag in the equilibration of the tissue precursor pool with the plasma.

3. EXPERIMENTAL PROCEDURE FOR MEASUREMENT OF LOCAL CEREBRAL GLUCOSE UTILIZATION

3.1. Theoretical Considerations in the Design of the Procedure

The operational equation of the method specifies the variables to be measured in order to determine R_i, the local rate of glucose consumption in the brain. The following variables are measured in each experiment: (1) the entire history of the arterial plasma [^{14}C]deoxyglucose concentration, C_P^*, from zero time to the time of killing, T; (2) the steady-state arterial plasma glucose level, C_P, over the same interval; and (3) the local concentration of ^{14}C in the tissue at the time of killing, $C_i^*(T)$. The rate constants, k_1^*, k_2^*, and k_3^*, and the lumped constant, $\lambda V_m^- K_m / \Phi V_m K_m^-$, are not measured in each experiment; the values for these constants that are used are those determined separately in other groups of animals as described above and presented in Tables I and II.

The operational equation is generally applicable with all types of arterial plasma [^{14}C]DG concentration curves. Its configuration, however, suggests that a declining curve approaching zero by the time of killing is the choice to minimize certain potential errors. The quantitative autoradiographic technique measures only total ^{14}C concentration in the tissue and does not distinguish between [^{14}C]DG-6-P and [^{14}C]DG. It is, however, [^{14}C]DG-6-P concentration that must be known to determine glucose consumption. The [^{14}C]DG-6-P concentration is calculated in the numerator of the operational equation, which equals the total tissue ^{14}C content, $C_i^*(T)$, minus the [^{14}C]DG concentration present in the tissue, estimated by the term containing the exponential factor and rate constants. In the denominator of the operational equation, there is also a term containing an exponential factor and rate constants. Both of these terms have the useful property of approaching zero with increasing time if C_P^* is also allowed to approach zero. The rate constants, k_1^*, k_2^*, and k_3^*, are not measured in the same animals in which local glucose consumption is being measured. It is conceivable that the rate constants in Table I are not equally applicable in all physiological, pharmacological, and pathological states. One possible solution is to determine the rate constants for each condition to be studied. An alternative solution, and the one chosen, is to administer the [^{14}C]DG as a single intravenous pulse at zero time and to allow sufficient time for the clearance of [^{14}C]DG from the plasma and the terms containing the rate constants to fall to levels too low to influence the final result. To wait until these terms reach zero is impractical because of the long time required and the risk of effects of the small but finite rate of loss of [^{14}C]DG-6-P from the tissues. A reasonable time interval is 45 min; by this time, the plasma level has fallen to very low levels, and, on the basis of the values of $(k_2^* + k_3^*)$ in Table I, the exponential factors have declined through at least ten half-lives.

3.2. Experimental Protocol

The animals are prepared for the experiment by the insertion of polyeth-
ylene catheters in an artery and vein. Any convenient artery or vein can be
used. In the rat, the femoral or the tail arteries and veins have been found
satisfactory. In the monkey and cat, the femoral vessels are probably most
convenient. The catheters are inserted under anesthesia, and anesthetic agents
without long-lasting aftereffects should be used. Light halothane anesthesia
with or without supplementation with nitrous oxide has been found to be quite
satisfactory. At least 2 h is allowed for recovery from the surgery and anesthesia
before initiation of the experiment.

The design of the experimental procedure for the measurement of local
cerebral glucose utilization is based on the theoretical considerations discussed
above. At zero time, a pulse of 125 μCi (no more than 2.5 μmol) of
[^{14}C]deoxyglucose per kg of body weight is administered to the animal via the
venous catheter. Arterial sampling is initiated with the onset of the pulse, and
timed 50- to 100-μl samples of arterial blood are collected consecutively as
rapidly as possible during the early period so as not to miss the peak of the
arterial curve. Arterial sampling is continued at less frequent intervals later in
the experimental period but at sufficient frequency to define fully the arterial
curve. The arterial blood samples are immediately centrifuged to separate the
plasma, which is stored on ice until assayed for [^{14}C]DG by liquid scintillation
counting and glucose concentrations by standard enzymatic methods. At approximately 45 min, the animal is decapitated, and the brain is removed and
frozen in Freon XII® or isopentane maintained between $-50°$ and $-75°C$ with
liquid nitrogen. When fully frozen, the brain is stored at $-70°C$ until sectioned
and autoradiographed. The experimental period may be limited to 30 min. This
is theoretically permissible and may sometimes be necessary for reasons of
experimental expediency, but greater errors because of possible inaccuracies
in the rate constants may result.

3.3. Autoradiographic Measurement of Tissue ^{14}C Concentration

The ^{14}C concentrations in localized regions of the brain are measured by
a modification of the quantitative autoradiographic technique previously described.[15] The frozen brain is coated with chilled embedding medium (Lipshaw
Manufacturing Co., Detroit, MI) and fixed to object holders appropriate to the
microtome to be used.

Brain sections precisely 20 μm in thickness are prepared in a cryostat
maintained at $-21°C$ to $-22°C$. The brain sections are picked up on glass cover
slips, dried on a hot plate at 60°C for at least 5 min, and placed sequentially
in an X-ray cassette. A set of [^{14}C]methyl methacrylate standards (Amersham
Corp., Arlington Heights, IL), which include a blank and a series of progressively increasing ^{14}C concentrations, are also placed in the cassette. These
standards must previously have been calibrated for their autoradiographic
equivalence to the ^{14}C concentrations in brain sections 20 μm in thickness
prepared as described above. The method of calibration has been previously
described.[15]

Autoradiographs are prepared from these sections directly in the X-ray cassette with Kodak single-coated, blue-sensitive medical X-ray film, type SB-5 (Eastman Kodak Co., Rochester, NY). The exposure time is generally 5–6 days with the doses used as described above, and the exposed films are developed according to the instructions supplied with the film. The SB-5 X-ray film is rapid but coarse grained. For finer-grained autoradiographs and, therefore, better defined images with higher resolution, it is possible to use mammographic films such as DuPont LoDose® or Kodak MR-1 films or fine-grain panchromatic film such as Kodak Plus-X®, but the exposure times are 2–3 times longer. The autoradiographs provide a pictorial representation of the relative ^{14}C concentrations in the various cerebral structures and the plastic standards. A calibration curve of the relationship between optical density and tissue ^{14}C concentration for each film is obtained by densitometric measurements of the portions of the film representing the various standards. The local tissue concentrations are then determined from the calibration curve and the optical densities of the film in the regions representing the cerebral structures of interest. Local cerebral glucose utilization is calculated from the local tissue concentrations of ^{14}C and the plasma [^{14}C]DG and glucose concentrations according to the operational equation.

3.4. Computerized Color-Coded Image Processing

The autoradiographs provide pictorial representations of only the relative concentrations of the isotope in the various tissues. Because of the use of a pulse followed by a long period before killing, the isotope is contained mainly in deoxyglucose-6-phosphate, which reflects the rate of glucose metabolism. The autoradiographs are, therefore, pictorial representations also of the relative but not the actual rates of glucose utilization in all the structures of the nervous system. Furthermore, the resolution of differences in relative rates is limited by the ability of the human eye to recognize differences in shades of gray.

Manual densitometric analysis permits the computation of actual rates of glucose utilization with a fair degree of resolution, but it generates enormous tables of data which fail to convey the tremendous heterogeneity of metabolic rates, even within anatomic structures, or the full information contained within the autoradiographs. Goochee *et al.*[26] have developed a computerized image-processing system to analyze and transform the autoradiographs into color-coded maps of the distribution of the actual rates of glucose utilization exactly where they are located throughout the central nervous system. The autoradiographs are scanned automatically by a computer-controlled scanning microdensitometer. The optical density of each spot in the autoradiograph, from 25 to 100 μm as selected, is stored in a computer, converted to ^{14}C concentration on the basis of the optical densities of the calibrated ^{14}C plastic standards, and then converted to local rates of glucose utilization by solution of the operational equation of the method. Colors are assigned to narrow ranges of the rates of glucose utilization, and the autoradiographs are then displayed in a color TV monitor in color along with a calibrated color scale for identifying the rate of glucose utilization in each spot of the autoradiograph from its color. These color maps add a third dimension, the rate of glucose utilization on a color scale, to the spatial dimensions already present on the autoradiographs.

4. RATES OF LOCAL CEREBRAL GLUCOSE UTILIZATION IN THE NORMAL CONSCIOUS STATE

Thus far, quantitative measurements of local cerebral glucose utilization have been reported only for the albino rat[17] and monkey.[25] These values are presented in Table III. The rates of local cerebral glucose utilization in the normal conscious rat vary widely throughout the brain. The values in white structures tend to group together and are always considerably below those of gray structures. The average value in gray matter is approximately three times

Table III
Representative Values for Local Cerebral Glucose Utilization in the Normal Conscious Albino Rat and Monkey ($\mu mol/100$ g per min)[a]

Structure	Albino rat[b] (10)	Monkey[c] (7)
Gray matter		
Visual cortex	107 ± 6	59 ± 2
Auditory cortex	162 ± 5	79 ± 4
Parietal cortex	112 ± 5	47 ± 4
Sensory–motor cortex	120 ± 5	44 ± 3
Thalamus: Lateral nucleus	116 ± 5	54 ± 2
Thalamus: Ventral nucleus	109 ± 5	43 ± 2
Medial geniculate body	131 ± 5	65 ± 3
Lateral geniculate body	96 ± 5	39 ± 1
Hypothalamus	54 ± 2	25 ± 1
Mammillary body	121 ± 5	57 ± 3
Hippocampus	79 ± 3	39 ± 2
Amygdala	52 ± 2	25 ± 2
Caudate–putamen	110 ± 4	52 ± 3
Nucleus accumbens	82 ± 3	36 ± 2
Globus pallidus	58 ± 2	26 ± 2
Substantia nigra	58 ± 3	29 ± 2
Vestibular nucleus	128 ± 5	66 ± 3
Cochlear nucleus	113 ± 7	51 ± 3
Superior olivary nucleus	133 ± 7	63 ± 4
Inferior colliculus	197 ± 10	103 ± 6
Superior colliculus	95 ± 5	55 ± 4
Pontine gray matter	62 ± 3	28 ± 1
Cerebellar cortex	57 ± 2	31 ± 2
Cerebellar nuclei	100 ± 4	45 ± 2
White matter		
Corpus callosum	40 ± 2	11 ± 1
Internal capsule	33 ± 2	13 ± 1
Cerebellar white matter	37 ± 2	12 ± 1

[a] The values are the means ± standard errors from measurements made in the number of animals indicated in parentheses.
[b] From Sokoloff et al.[17]
[c] From Kennedy et al.[25]

that of white matter, but the individual values vary from approximately 50 to 200 μmol of glucose/100 g per min. The highest values are in the structures involved in auditory functions with the inferior colliculus clearly the most metabolically active structure in the brain.

The rates of local cerebral glucose utilization in the conscious monkey exhibit similar heterogeneity, but they are generally one-third to one-half the values in corresponding structures of the rat brain (Table III). The differences in rates in the rat and monkey brain are consistent with the different cellular packing densities in the brains of these two species.

5. EFFECTS OF BARBITURATE ANESTHESIA

General anesthesia produced by thiopental reduces the rates of glucose utilization in all structures of the rat brain (Table IV).[17] The effects are not

Table IV
Effects of Thiopental Anesthesia on Local Cerebral Glucose Utilization in the Rat[a]

Structure	Local cerebral glucose utilization (μmol/100 g per min)		
	Control[b] (6)	Anesthetized[b] (8)	Effect (%)
Gray matter			
Visual cortex	111 ± 5	64 ± 3	−42
Auditory cortex	157 ± 5	81 ± 3	−48
Parietal cortex	107 ± 3	65 ± 2	−39
Sensory–motor cortex	118 ± 3	67 ± 2	−43
Lateral geniculate body	92 ± 2	53 ± 3	−42
Medial geniculate body	126 ± 6	63 ± 3	−50
Thalamus: Lateral nucleus	108 ± 3	58 ± 2	−46
Thalamus: Ventral nucleus	98 ± 3	55 ± 1	−44
Hypothalamus	63 ± 3	43 ± 2	−32
Caudate–putamen	111 ± 4	72 ± 3	−35
Hippocampus: Ammon's horn	79 ± 1	56 ± 1	−29
Amygdala	56 ± 4	41 ± 2	−27
Cochlear nucleus	124 ± 7	79 ± 5	−36
Lateral lemniscus	114 ± 7	75 ± 4	−34
Inferior colliculus	198 ± 7	131 ± 8	−34
Superior olivary nucleus	141 ± 5	104 ± 7	−26
Superior colliculus	99 ± 3	59 ± 3	−40
Vestibular nucleus	133 ± 4	81 ± 4	−39
Pontine gray matter	69 ± 3	46 ± 3	−33
Cerebellar cortex	66 ± 2	44 ± 2	−33
Cerebellar nucleus	106 ± 4	75 ± 4	−29
White matter			
Corpus callosum	42 ± 2	30 ± 2	−29
Genu of corpus callosum	35 ± 5	30 ± 2	−14
Internal capsule	35 ± 2	29 ± 2	−17
Cerebellar white matter	38 ± 2	29 ± 2	−24

[a] Determined at 30 min following pulse of [14C]deoxyglucose. From Sokoloff *et al.*[17]

[b] The values are the means ± standard errors obtained in the number of animals indicated in parentheses. All differences are statistically significant at the $P < 0.05$ level.

uniform, however. The greatest reductions occur in the gray structures, particularly those of the primary sensory pathways. The effects in white matter, though definitely present, are relatively small compared to those of gray matter. These results are in agreement with those of previous studies in which anesthesia has been found to decrease the cerebral metabolic rate of the brain as a whole.[6,8,10]

6. RELATIONSHIP BETWEEN LOCAL FUNCTIONAL ACTIVITY AND ENERGY METABOLISM

The results of a variety of applications of the method demonstrate a clear relationship between local cerebral functional activity and glucose consumption. The most striking demonstrations of the close coupling between function and energy metabolism are seen with experimentally induced local alterations in functional activity that are restricted to a few specific areas in the brain. The effects on local glucose consumption are then so pronounced that they are not only observed in the quantitative results but can be visualized directly on the autoradiographs which are really pictorial representations of the relative rates of glucose utilization in the various structural components of the brain.

6.1. Effects of Increased Functional Activity

6.1.1. Effects of Sciatic Nerve Stimulation

Electrical stimulation of one sciatic nerve in the rat under barbiturate anesthesia causes pronounced increases in glucose consumption (i.e., increased optical density in the autoradiographs) in the ipsilateral dorsal horn of the lumbar spinal cord.[27]

6.1.2. Effects of Experimental Focal Seizures

The local injection of penicillin into the hand–face area of the motor cortex of the rhesus monkey has been shown to induce electrical discharges in the adjacent cortex and to result in recurrent focal seizures involving the face, arm, and hand on the contralateral side.[28] Such seizure activity causes selective increases in glucose consumption in areas of motor cortex adjacent to the penicillin locus and in small discrete regions of the putamen, globus pallidus, caudate nucleus, thalamus, and substantia nigra of the same side (Fig. 2).[27] Similar studies in the rat have led to comparable results and provided evidence on the basis of an evoked metabolic response of a "mirror" focus in the motor cortex contralateral to the penicillin-induced epileptogenic focus.[29]

6.2. Effects of Decreased Functional Activity

Decrements in functional activity result in reduced rates of glucose utilization. These effects are particularly striking in the auditory and visual systems of the rat and the visual system of the monkey.

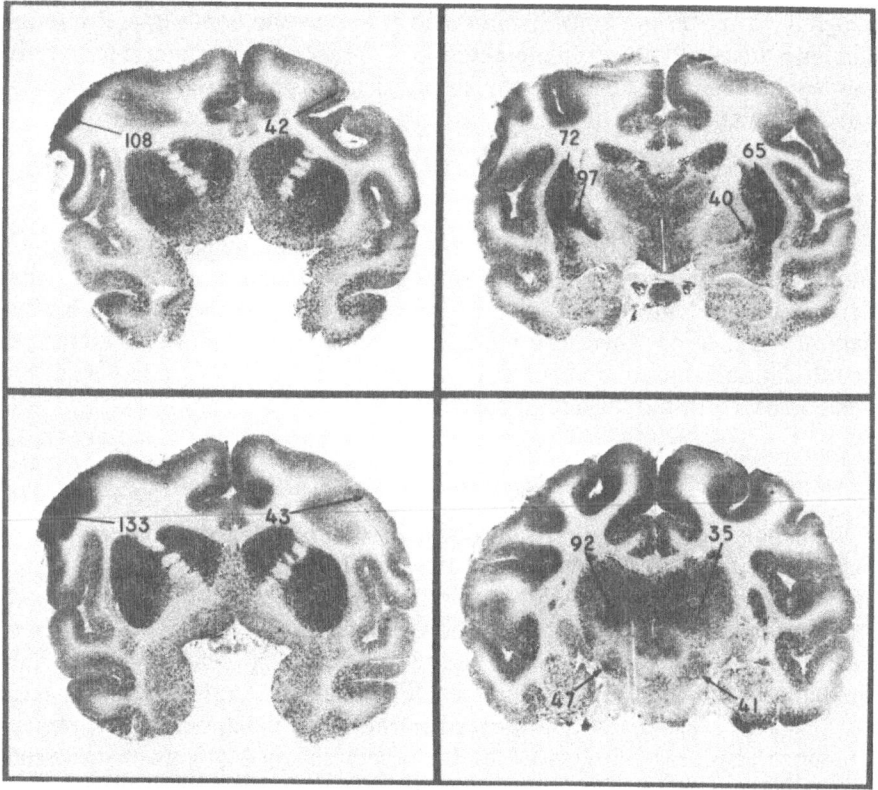

Fig. 2. Effects of focal seizures produced by local application of penicillin to motor cortex on local cerebral glucose utilization in the rhesus monkey. The penicillin was applied to the hand and face area of the left motor cortex. The left side of the brain is on the left in each of the autoradiographs in the figure. The numbers are the rates of local cerebral glucose utilization in μmol/100 g tissue per min. Note the following: upper left, motor cortex in region of penicillin application and corresponding region of contralateral motor cortex; lower left, ipsilateral and contralateral motor cortical regions remote from area of penicillin applications; upper right, ipsilateral and contralateral putamen and globus pallidus; lower right, ipsilateral and contralateral thalamic nuclei and substantia nigra.[18]

6.2.1. Effects of Auditory Deprivation

In the albino rat some of the highest rates of local cerebral glucose utilization are found in components of the auditory system, i.e., auditory cortex, medial geniculate ganglion, inferior colliculus, lateral lemniscus, superior olive, and cochlear nucleus (Table III). Bilateral auditory deprivation by occlusion of both external auditory canals with wax markedly depresses the metabolic activity in all of these areas.[18] The reductions are symmetrical bilaterally and range from 35 to 60%. Unilateral auditory deprivation also depresses the glucose consumption of these structures but to a lesser degree, and some of the structures are asymmetrically affected. For example, the metabolic activity of the ipsilateral cochlear nucleus equals 75% of the activity of the contralateral nucleus. The lateral lemniscus, superior olive, and medial geniculate ganglion are slightly lower on the contralateral side, whereas the contralateral inferior colliculus is markedly lower in metabolic activity than the ipsilateral structure. These results demonstrate that there is some degree of lateralization and crossing of auditory pathways in the rat.

6.2.2. Visual Deprivation in the Rat

In the rat, the visual system is 80 to 85% crossed at the optic chiasma,[30,31] and unilateral enucleation removes most of the visual input to the central visual structures of the contralateral side. In the conscious rat studied 2–24 h after unilateral enucleation, there are marked decrements in glucose utilization in the contralateral superior colliculus, lateral geniculate ganglion, and visual cortex as compared to the ipsilateral side.[27]

6.2.3. Visual Deprivation in the Monkey

In animals with binocular visual systems, such as the rhesus monkey, there is only approximately 50% crossing of the visual pathways, and the structures of the visual system on each side of the brain receive equal inputs from both retinas. Although each retina projects more or less equally to both hemispheres, their projections remain segregated and terminate in six well-defined laminae in the lateral geniculate ganglia, three each for the ipsilateral and contralateral eyes.[32–35] This segregation is preserved in the optic radiations which project the monocular representations of the two eyes for any segment of the visual field to adjacent regions of layer IV of the striate cortex.[32,33] The cells responding to the input of each monocular terminal zone are distributed transversely through the thickness of the striate cortex, resulting in a mosaic of columns, 0.3–0.5 mm in width, alternately representing the monocular inputs of the two eyes. The nature and distribution of these ocular dominance columns have previously been characterized by electrophysiological techniques,[32] Nauta degeneration methods,[33] and by autoradiographic visualization of axonal and transneuronal transport of [3H]proline- and [3H]fucose-labeled protein and/or glycoprotein.[34,35] Bilateral or unilateral visual deprivation, either by enucleation or by the insertion of opaque plastic disks, produces consistent changes in the pattern of distribution of the rates of glucose consumption, all clearly visible in the autoradiographs, which coincide closely with the changes in functional activity expected from known physiological and anatomic properties of the binocular visual system.[36]

In animals with intact binocular vision, no bilateral asymmetry is seen in the autoradiographs of the structures of the visual system (Figs. 3A, 4A). The lateral geniculate ganglia and oculomotor nuclei appear to be of fairly uniform density and essentially the same on both sides (Fig. 3A). The visual cortex is also the same on both sides (Fig. 4A), but throughout all of area 17, there is heterogeneous density distributed in a characteristic laminar pattern. These observations indicate that in animals with binocular visual input, the rates of glucose consumption in the visual pathways are essentially equal on both sides of the brain and relatively uniform in the oculomotor nuclei and lateral geniculate ganglia but markedly different in the various layers of the striate cortex.

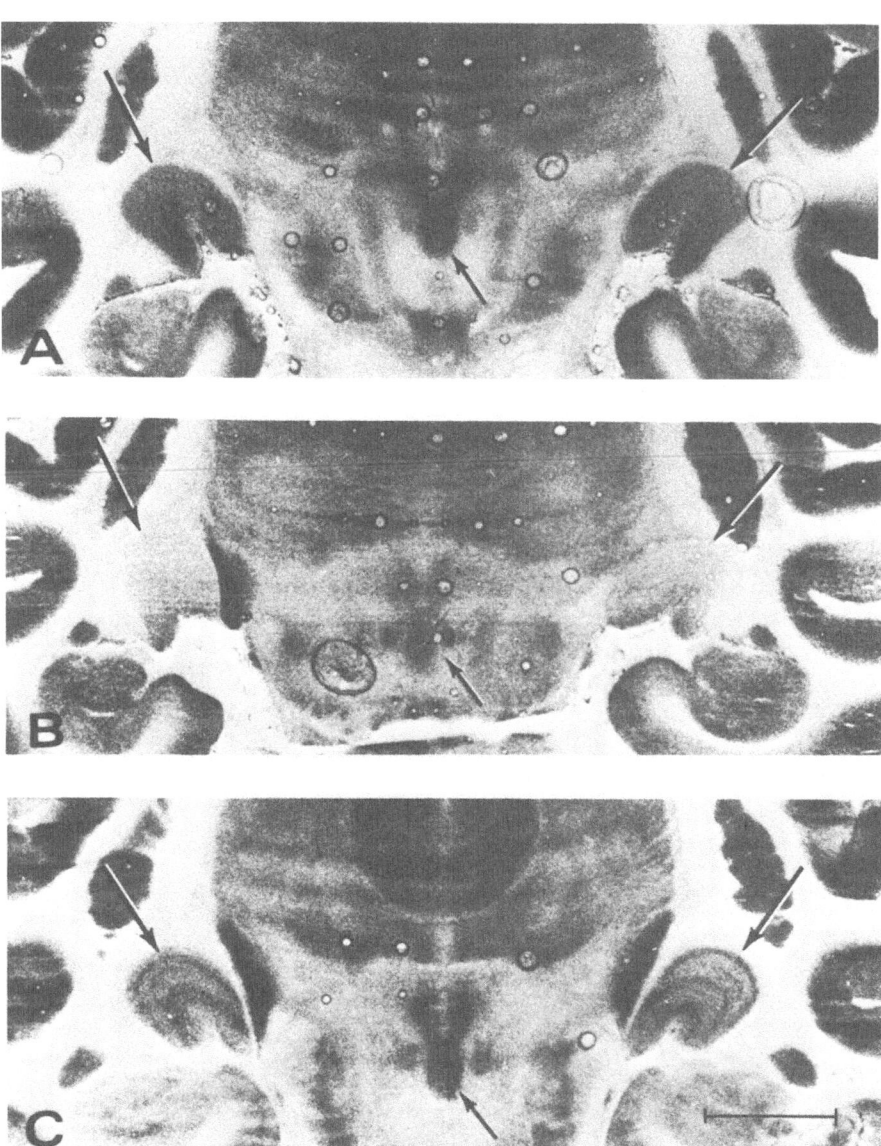

Fig. 3.   Autoradiography of coronal brain sections of monkey at the level of the lateral geniculate bodies. Large arrows point to the lateral geniculate bodies; small arrows point to oculomotor nuclear complex. (A) Animal with intact binocular vision. Note the bilateral symmetry and relative homogeneity of the lateral geniculate bodies and oculomotor nuclei. (B) Animal with bilateral visual occlusion. Note the reduced relative densities, the relative homogeneity, and the bilateral symmetry of the lateral geniculate bodies and oculomotor nuclei. (C) Animal with right eye occluded. The left side of the brain is on the left side of the photograph. Note the laminae and the inverse order of the dark and light bands in the two lateral geniculate bodies. Note also the lesser density of the oculomotor nuclear complex on the side contralateral to the occluded eye.[36] Scale bar = 5.0 mm.

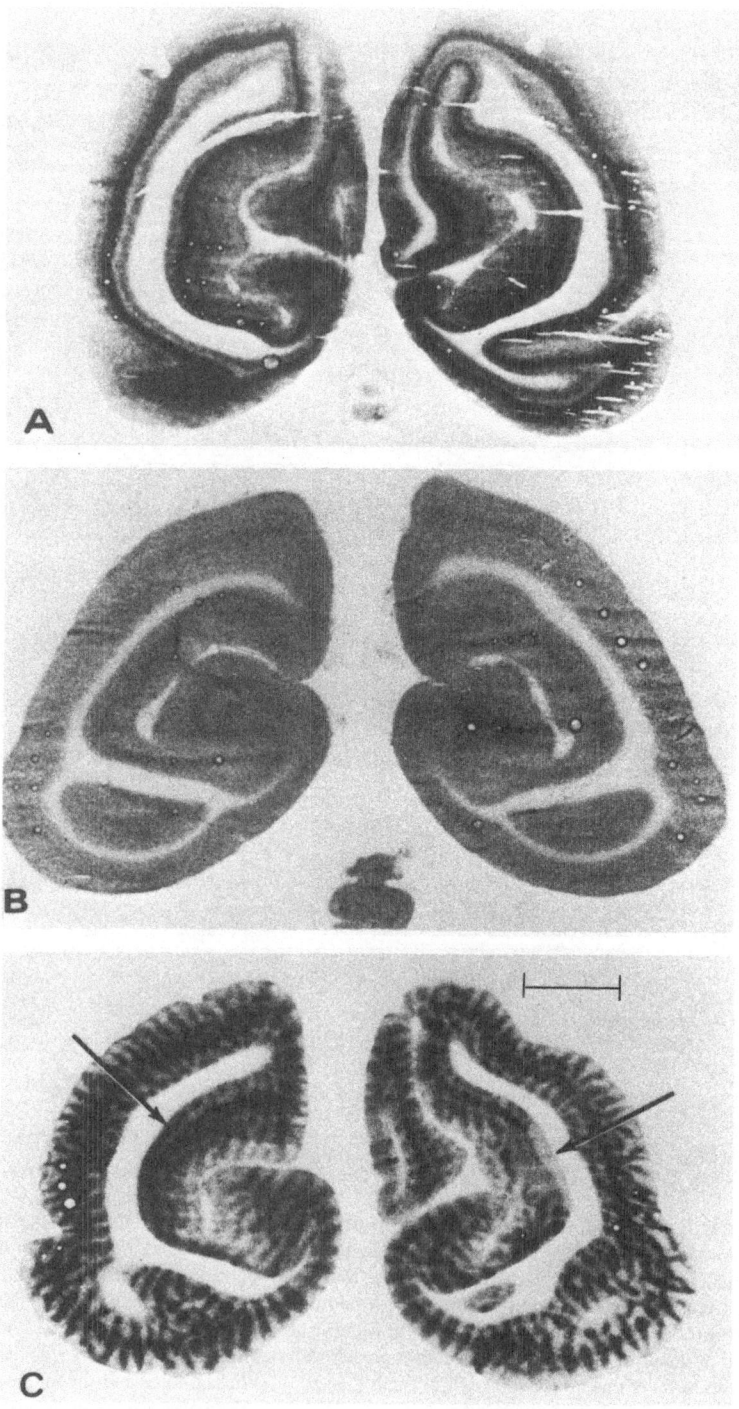

Autoradiographs from animals with both eyes occluded exhibit generally decreased labeling of all components of the visual system, but the bilateral symmetry is fully retained (Figs. 3B, 4B), and the density within each lateral geniculate body is for the most part fairly uniform (Fig. 3B). In the striate cortex, however, the marked differences in the densities of the various layers seen in the animals with intact bilateral vision (Fig. 4A) are virtually absent so that, except for a faint delineation of a band within layer IV, the concentration of the label is essentially homogeneous throughout the striate cortex (Fig. 4B).

Autoradiographs from monkeys with only monocular input because of unilateral visual occlusion exhibit markedly different patterns from those described above. Both lateral geniculate bodies exhibit exactly inverse patterns of alternating dark and light bands corresponding to the known laminae representing the regions receiving the different inputs from the retinas of the intact and occluded eyes (Fig. 3C). Bilateral asymmetry is also seen in the oculomotor nuclear complex; a lower density is apparent in the nuclear complex contralateral to the occluded eye (Fig. 3C). In the striate cortex, the pattern of distribution of the [^{14}C]DG-6-P appears to be a composite of the patterns seen in the animals with intact and bilaterally occluded visual input. The pattern found in the former regularly alternates with that of the latter in columns oriented perpendicularly to the cortical surface (Fig. 4C).

The dimensions, arrangement, and distribution of these columns are identical to those of the ocular dominance columns described by Hubel and Wiesel.[32-34] These columns reflect the interdigitation of the representations of the two retinas in the visual cortex. Each element in the visual fields is represented by a pair of contiguous bands in the visual cortex, one for each of the two retinas or their portions that correspond to the given point in the visual fields. With symmetrical visual input bilaterally, the columns representing the two eyes are equally active and, therefore, not visualized in the autoradiographs (Fig. 4A). When one eye is blocked, however, only those columns representing the blocked eye become metabolically less active, and the autoradiographs then display the alternate bands of normal and depressed activities corresponding to the regions of visual cortical representation of the two eyes (Fig. 4C).

There can be seen in the autoradiographs from the animals with unilateral visual deprivation a pair of regions in the folded calcarine cortex that exhibit

Fig. 4. Autoradiographs of coronal brain sections from rhesus monkeys at the level of the striate cortex. (A) Animal with normal binocular vision. Note the laminar distribution of the density; the dark band corresponds to layer IV. (B) Animal with bilateral visual deprivation. Note the almost uniform and reduced relative density, especially the virtual disappearance of the dark band corresponding to layer IV. (C) Animal with right eye occluded. The half-brain on the left side of the photograph represents the left hemisphere contralateral to the occluded eye. Note the alternate dark and light striations, each approximately 0.3–0.4 mm in width, that represent the ocular dominance columns. These columns are most apparent in the dark band corresponding to layer IV but extend through the entire thickness of the cortex. The arrows point to regions of bilateral asymmetry in which the ocular dominance columns are absent. These are presumably areas that normally receive only monocular input. The one on the left, contralateral to occluded eye, has a continuous dark lamina corresponding to layer IV which is completely absent on the side ipsilateral to the occluded eye. These regions are believed to be the loci of the cortical representations of the blind spots.[36]

bilateral asymmetry (Fig. 4C). The ocular dominance columns are absent on both sides, but on the side contralateral to the occluded eye, this region has the appearance of visual cortex from an animal with normal bilateral vision, and on the ipsilateral side, this region looks like cortex from an animal with both eyes occluded (Fig. 4). These regions are the loci of the cortical representation of the blind spots of the visual fields and normally have only monocular input.[27,36] The area of the optic disk in the nasal half of each retina cannot transmit to this region of the contralateral striate cortex which, therefore, receives its sole input from an area in the temporal half of the ipsilateral retina. Occlusion of one eye deprives this region of the ipsilateral striate cortex of all input while the corresponding region of the contralateral striate cortex retains uninterrupted input from the intact eye. The metabolic reflection of this ipsilateral monocular input is seen in the autoradiograph in Fig. 4C.

The results of these studies with the [^{14}C]deoxyglucose method in the binocular visual system of the monkey represent the most dramatic demonstration of the close relationship between physiological changes in functional activity and the rate of energy metabolism in specific components of the central nervous system.

7. APPLICATIONS OF THE DEOXYGLUCOSE METHOD

The results of studies like those described above on the effects of experimentally induced focal alterations of functional activity on local glucose utilization have demonstrated a close coupling between local functional activity and energy metabolism in the central nervous system. The effects are often so pronounced that they can be visualized directly on the autoradiographs, which provide pictorial representations of the relative rates of glucose utilization throughout the brain. This technique of autoradiographic visualization of evoked metabolic responses offers a powerful tool to map functional neural pathways simultaneously in all anatomic components of the central nervous system, and extensive use has been made of it for this purpose.[19] The results have clearly demonstrated the effectiveness of metabolic responses, either positive or negative, in identifying regions of the central nervous system involved in specific functions.

The method has been used most extensively in qualitative studies in which regions of altered functional activity are identified by the change in their visual appearance relative to other regions in the autoradiographs. Such qualitative studies are effective only when the effects are lateralized to one side or when only a few discrete regions are affected; other regions serve as the controls. Quantitative comparisons cannot, however, be made for equivalent regions between two or more animals. To make quantitative comparisons between animals, the fully quantitative method must be used, which takes into account the various factors, particularly the plasma glucose level, that influence the magnitude of labeling of the tissues. The method must be used quantitatively when the experimental procedure produces systemic effects and alters metabolism in many regions of the brain.

A comprehensive review of the many qualitative and quantitative applications of the method is beyond the scope of this chapter. Only some of the many neurophysiological, neuroanatomic, pharmacological, and pathophysiological applications of the method are briefly noted merely to illustrate the broad extent of its potential usefulness.

7.1. Neurophysiological and Neuroanatomic Applications

Many of the physiological applications of the [^{14}C]deoxyglucose method were in studies designed to test the method and to examine the relationship between local cerebral functional and metabolic activities. These applications have been described above. The most dramatic results have been obtained in the visual systems of the monkey and the rat. The method has, for example, been used to define the nature, conformation, and distribution of the ocular dominance columns in the striate cortex of the monkey (Fig. 4C).[36] It has been used by Hubel *et al.*[37] to do the same for the orientation columns in the striate cortex of the monkey. A byproduct of the studies of the ocular dominance columns was the identification of the loci of the visual cortical representation of the blind spots of the visual fields (Fig. 4C).[36]

Studies are currently in progress to map the pathways of higher visual functions beyond the striate cortex; the results thus far demonstrate extensive areas of involvement of the inferior temporal cortex in visual processing.[38] Des Rosiers *et al.*[39] have used the method to demonstrate functional plasticity in the striate cortex of the infant monkey. The ocular dominance columns are already present on the first day of life, but if one eye is kept patched for 3 months, the columns representing the open eye broaden and completely take over the adjacent regions of cortex containing the columns for the eye that had been patched. Inasmuch as there is no longer any cortical representation for the patched eye, the animal becomes functionally blind in one eye. This phenomenon is almost certainly the basis for the cortical blindness or amblyopia that often occurs in children with uncorrected strabismus.

There have also been extensive studies of the visual system of the rat. This species has little if any binocular vision and, therefore, lacks the ocular dominance columns. Batipps *et al.*[40] have compared the rates of local cerebral glucose utilization in albino and Norway brown rats during exposure to ambient light. The rates in the two strains were essentially the same throughout the brain except in the components of the primary visual system. The metabolic rates in the superior colliculus, lateral geniculate, and visual cortex of the albino rat were significantly lower than those in the pigmented rat.

Miyaoka *et al.*[41] have studied the influence of the intensity of retinal stimulation with randomly spaced light flashes on the metabolic rates in the visual systems of the two strains. In dark-adapted animals, there is relatively little difference between the two strains. With increasing intensity of light, the rates of glucose utilization first increase in the primary projection areas of the retina, e.g., superficial layer of the superior colliculus and lateral geniculate body, and the slopes of the increase are steeper in the albino rat (Fig. 5). At 7 lux, however, the metabolic rates peak in the albino rat and then decrease with increasing

SUPERIOR COLLICULUS

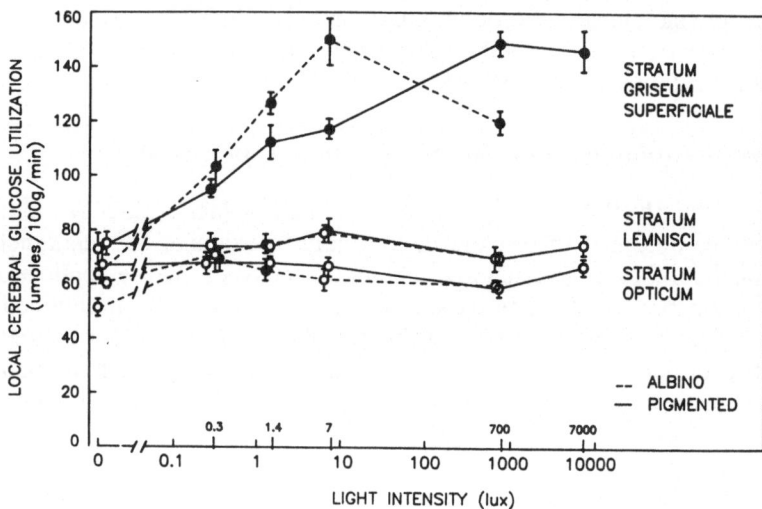

POSTEROLATERAL NUCLEUS OF THE THALAMUS

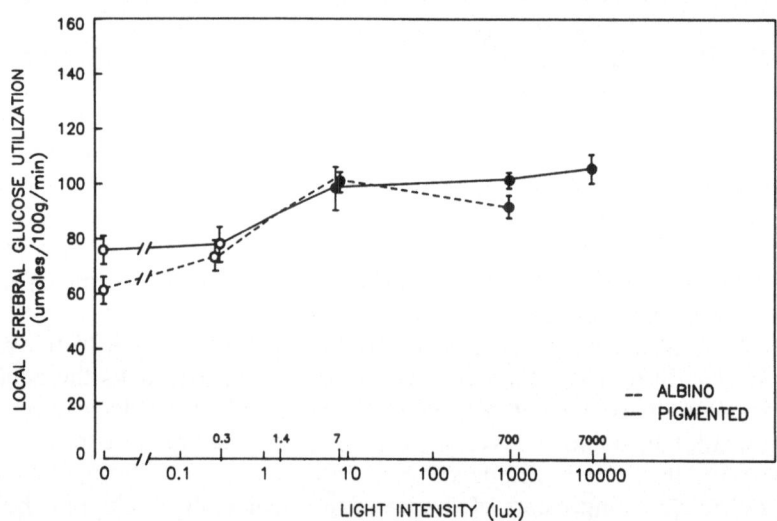

Fig. 5. Effects of intensity of retinal illumination with randomly spaced light flashes on local cerebral glucose utilization in components of the visual system of the albino and Norway brown rat. Note that the local glucose utilization is proportional to the logarithm of the intensity of illumination, at least at lower levels of intensity, in the primary projection areas of the retina.[41]

light intensity. In contrast, the metabolic rates in the pigmented rat rise until they reach a plateau at about 700 lux, approximately the ambient light intensity in the laboratory. At this level, the metabolic rates in the visual structures of the albino rat are considerably below those of the pigmented rat. These results are consistent with the greater intensity of light reaching the visual cells of the

LATERAL GENICULATE NUCLEUS

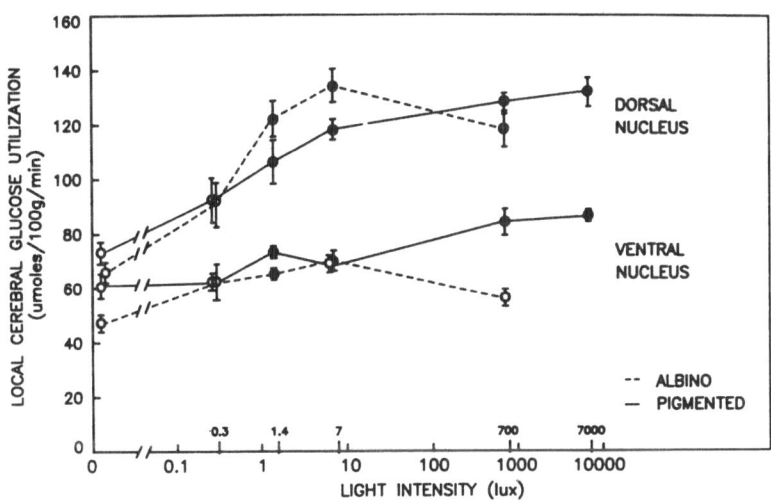

VISUAL CORTEX

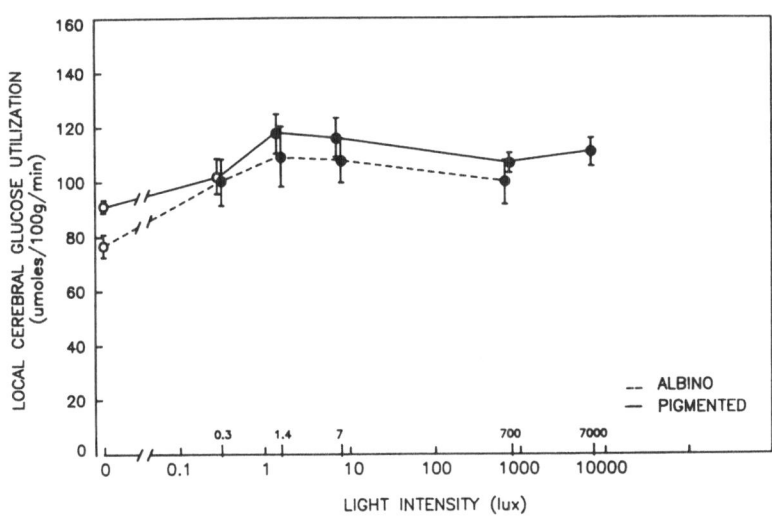

Fig. 5. (*continued*)

retina in the albino rats because of lack of pigment and the subsequent damage to the rods at higher light intensities. It is of considerable interest that the rates of glucose utilization in these visual structures obey the Weber–Fechner Law; i.e., the metabolic rate is directly proportional to the logarithm of the intensity of stimulation.[41] Inasmuch as this law was first developed from behavioral manifestations, these results imply that there is a quantitative relationship between behavioral and metabolic responses.

Although less extensive, there have also been applications of the method to other sensory systems. In studies of the olfactory system, Sharp *et al.*[42]

have found that olfactory stimulation with specific odors activates glucose utilization in localized regions of the olfactory bulb. In addition to the experiments in the auditory system described above, there have been studies of tonotopic representation in the auditory system. Webster *et al.*[43] have obtained clear evidence of selective regions of metabolic activation in the cochlear nucleus, superior olivary complex, nuclei of the lateral lemnisci, and the inferior colliculus in cats in response to different frequencies of auditory stimulation. Similar results have been obtained by Silverman *et al.*[44] in the rat and guinea pig. Studies of the sensory cortex have demonstrated metabolic activation of the "whisker barrels" by stimulation of the vibrissae in the rat.[45,46] Each vibrissa is represented in a discrete region of the sensory cortex; their precise location and extent have been elegantly mapped by Hand *et al.*[46] and Hand[47] by means of the [^{14}C]deoxyglucose method.

Thus far, there has been relatively little application of the method to the physiology of motor functions. Kennedy *et al.*[48] have studied monkeys that were conditioned to perform a task with one hand in response to visual cues; in the monkeys that were performing, they observed metabolic activation throughout the appropriate areas of the motor as well as sensory systems from the cortex to the spinal cord.

An interesting physiological application of the [^{14}C]deoxyglucose method has been to the study of circadian rhythms in the central nervous system. Schwartz and his co-workers[49,50] found that the suprachiasmatic nucleus in the rat exhibits circadian rhythmicity in metabolic activity that is high during the day and low during the night. None of the other structures in the brain that they examined showed rhythmic activity. The normally low activity present in the nucleus in the dark could be markedly increased by light, but darkness did not reduce the glucose utilization during the day. The rhythm is entrained to light; reversal of the light–dark cycle leads not only to reversal of the rhythm in running activity but also in the cycle of metabolic activity in the suprachiasmatic nucleus. These studies lend support to a role of the suprachiasmatic nucleus in the organization of circadian rhythms in the central nervous system.

Much of our knowledge of neurophysiology has been derived from studies of the electrical activity of the nervous system. Indeed, from the heavy emphasis that has been placed on electrophysiology, one might gather that the brain is really an electric organ rather than a chemical one that functions mainly by the release of chemical transmitters at synapses. Nevertheless, electrical activity is unquestionably fundamental to the process of conduction, and it is appropriate to inquire how the local metabolic activities revealed by the [^{14}C]deoxyglucose method are related to the electrical activity of the nervous system.

This question has been examined by Yarowsky and his co-workers[51] in the superior cervical ganglion of the rat. The advantage of this structure is that its preganglionic input and postganglionic output can be isolated and electrically stimulated and/or monitored *in vivo*. The results thus far indicate a clear relationship between electrical input to the ganglion and its metabolic activity. In normal conscious rats, its rate of glucose utilization equals approximately 35 μmol/100 g per min. This rate is markedly depressed by anesthesia or denervation and enhanced by electrical stimulation of the afferent nerves. The metabolic activation is frequency dependent in the range of 5 to 15 Hz, increasing linearly in magnitude with increasing frequency of stimulation.

Similar effects of electrical stimulation on the oxygen and glucose consumption of the excised ganglion studied *in vitro* have been observed.[52–54] Recent studies have also shown that antidromic stimulation of the postganglionic efferent pathways from the ganglion has similar effects; stimulation of the external carotid nerve antidromically activates glucose utilization in the region of distribution of the cell bodies of this efferent pathway, indicating that not only the preganglionic axonal terminals are metabolically activated, but the postganglionic cell bodies as well.[55] As already demonstrated in the neurohypophysis,[56] the effects of electrical stimulation on energy metabolism in the superior cervical ganglion are also probably caused by the ionic currents associated with the spike activity and the consequent activation of the Na^+–K^+ ATPase activity to restore the ionic gradients. Electrical stimulation of the afferents to sympathetic ganglia have been shown to increase extracellular K^+ concentration.[54,57] Each spike is normally associated with a sharp transient rise in extracellular K^+ concentration which then rapidly falls and transiently undershoots before returning to the normal level[57]; ouabain slows the decline in K^+ concentration after the spike and eliminates the undershoot. Continuous stimulation at a frequency of 6 Hz produces a sustained increase in extracellular K^+ concentration.[57] It is likely that the increased extracellular K^+ concentration and, almost certainly, increased intracellular Na^+ concentration activate the Na^+–K^+ ATPase, which in turn leads to the increased glucose utilization.

7.2. Pharmacological Applications

The ability of the deoxyglucose method to map the entire brain for localized regions of altered functional activity on the basis of changes in energy metabolism offers a potent tool to identify the neural sites of action of agents with neuropharmacological and psychopharmacological actions. It does not, however, discriminate between the direct and indirect effects of the drug. An entire pathway may be activated even though the direct action of the drug may be exerted only at the origin of the pathway. This is of advantage for relating behavioral effects to central actions, but it is a disadvantage if the goal is to identify the primary site of action of the drug. To discriminate between direct and indirect actions of a drug, the [^{14}C]deoxyglucose method must be combined with selectively placed lesions in the CNS that interrupt afferent pathways to the structure in question. If the metabolic effect of the drug then remains, it indicates direct action; if it is lost, the effect is likely to be indirect and mediated via the interrupted pathway. Nevertheless, the method has proved to be useful in a number of pharmacological studies.

7.2.1. Effects of γ-Butyrolactone

γ-Hydroxybutyrate and γ-butyrolactone, which is hydrolyzed to γ-hydroxybutyrate in plasma, produce trancelike behavioral states associated with marked suppression of electroencephalographic activity.[58] These effects are reversible, and these drugs have been used clinically as anesthetic adjuvants. There is evidence that these agents lower neuronal activity in the nigrostriatal pathway and may act by inhibition of dopaminergic synapses.[59] Studies in rats with the [^{14}C]deoxyglucose technique have demonstrated that γ-butyrolactone produces profound dose-dependent reductions of glucose utilization throughout the brain.[60] At the highest doses studied, 600 mg/kg of body weight, glucose utilization was reduced by approximately 75% in gray matter and 33% in white matter, but there was no obvious further specificity with respect to the local cerebral structures affected. The reversibility of the effects and the magnitude and diffuseness of the depression of cerebral metabolic rate suggest that this drug might be considered as a chemical substitute for hypothermia in conditions in which profound reversible reduction of cerebral metabolism is desired.

7.2.2. Effects of D-Lysergic Acid Diethylamide

The effects of the potent psychotomimetic agent D-lysergic acid diethylamide have been examined in the rat.[61] In doses of 12.5 to 125 μg/kg, it caused dose-dependent reductions in glucose utilization in a number of cerebral structures. With increasing dosage, more structures were affected and to a greater degree. There was no pattern in the distribution of the effects, at least none discernible at the present level of resolution, that might contribute to the understanding of the drug's psychotomimetic actions.

7.2.3. Effects of Morphine Addiction and Withdrawal

Acute morphine administration depresses glucose utilization in many areas of the brain, but the specific effects of morphine could not be distinguished from those of the hypercapnia produced by the associated respiratory depression.[62] In contrast, morphine addiction, produced within 24 hr by a single subcutaneous injection of 150 mg/kg of morphine base in an oil emulsion, reduces glucose utilization in a large number of gray structures in the absence of changes in arterial P_{CO_2}. White matter appears to be unaffected. Naloxone (1 mg/kg subcutaneously) reduces glucose utilization in a number of structures when administered to normal rats but, when given to the morphine-addicted animals, produces an acute withdrawal syndrome and reverses the reductions of glucose utilization in several structures, most strikingly in the habenula.[62]

7.2.4. Pharmacological Studies of Dopaminergic Systems

The most extensive applications of the deoxyglucose method to pharmacology have been in studies of dopaminergic systems. Ascending dopaminergic pathways appear to have a potent influence on glucose utilization in the forebrain of rats. Electrolytic lesions placed unilaterally in the lateral hypothalamus or pars compacta of the substantia nigra caused marked ipsilateral reductions of glucose metabolism in numerous forebrain structures rostral to the lesion, particularly the frontal cerebral cortex, caudate–putamen, and parts of the thalamus.[63,64] Similar lesions in the locus coeruleus had no such effects.

Table V

Effects of d-Amphetamine and l-Amphetamine on Local Cerebral Glucose Utilization in the Conscious Rat[a]

Structure	Control	d-Amphetamine	l-Amphetamine
Gray matter			
Visual cortex	102 ± 8	135 ± 11[b]	105 ± 8
Auditory cortex	160 ± 11	162 ± 6	141 ± 6
Parietal cortex	109 ± 9	125 ± 10	116 ± 4
Sensory–motor cortex	118 ± 8	139 ± 9	111 ± 4
Olfactory cortex	100 ± 6	93 ± 5	94 ± 3
Frontal cortex	109 ± 10	130 ± 8	105 ± 4
Prefrontal cortex	146 ± 10	166 ± 7	154 ± 4
Thalamus			
Lateral nucleus	97 ± 5	114 ± 8	117 ± 6
Ventral nucleus	85 ± 7	108 ± 6[b]	96 ± 4
Habenula	118 ± 10	71 ± 5[c]	82 ± 2[c]
Dorsomedial nucleus	92 ± 6	111 ± 8	106 ± 6
Medial geniculate	116 ± 5	119 ± 4	116 ± 4
Lateral geniculate	79 ± 5	88 ± 5	84 ± 4
Hypothalamus	54 ± 5	56 ± 3	52 ± 3
Suprachiasmatic nucleus	94 ± 4	75 ± 4[c]	67 ± 1[c]
Mamillary body	117 ± 8	134 ± 5	142 ± 5[b]
Lateral Olfactory Nucleus[d]	92 ± 6	95 ± 5	99 ± 6
A_{13}	71 ± 4	91 ± 4[c]	81 ± 4
Hippocampus			
Ammon's horn	79 ± 5	73 ± 2	81 ± 6
Dentate gyrus	60 ± 4	55 ± 3	67 ± 7
Amygdala	46 ± 3	46 ± 3	44 ± 2
Septal nucleus	56 ± 3	55 ± 2	54 ± 3
Caudate nucleus	109 ± 5	132 ± 8[b]	127 ± 3[b]
Nucleus accumbens	76 ± 5	80 ± 3	78 ± 3
Globus pallidus	53 ± 3	64 ± 2[b]	65 ± 3[b]
Subthalamic nucleus	89 ± 6	149 ± 10[c]	107 ± 2
Substantia nigra			
Zona reticulata	58 ± 2	105 ± 4[c]	72 ± 4
Zona compacta	65 ± 4	88 ± 6[c]	72 ± 3
Red nucleus	76 ± 5	94 ± 5[b]	86 ± 2
Vestibular nucleus	121 ± 11	137 ± 5	130 ± 4
Cochlear nucleus	139 ± 6	126 ± 1	141 ± 5
Superior olivary nucleus	144 ± 4	143 ± 4	147 ± 6
Lateral lemniscus	107 ± 3	96 ± 5	98 ± 3
Inferior colliculus	193 ± 10	169 ± 5	150 ± 8[c]
Dorsal tegmental nucleus	109 ± 5	112 ± 7	122 ± 6
Superior colliculus	80 ± 5	89 ± 3	91 ± 3
Pontine gray	58 ± 4	65 ± 3	60 ± 1
Cerebellar flocculus	124 ± 10	146 ± 15	153 ± 10
Cerebellar hemispheres	55 ± 3	68 ± 6	64 ± 2
Cerebellar nuclei	102 ± 4	105 ± 8	110 ± 3
White matter			
Corpus callosum	23 ± 3	24 ± 2	23 ± 1
Genu of corpus callosum	29 ± 2	30 ± 2	26 ± 2
Internal capsule	21 ± 1	24 ± 2	19 ± 2
Cerebellar white	28 ± 1	31 ± 2	31 ± 2

[a] All values are the means ± standard error of the mean for five animals. From Wechsler *et al.*[66]

[b] Significant difference from the control at the $P < 0.05$ level.

[c] Significant difference from the control at the $P < 0.01$ level.

[d] It was not possible to correlate precisely this area on autoradiographs with a specific structure in the rat brain. It is, however, most likely the lateral olfactory nucleus.

Enhancement of dopaminergic synaptic activity by administration of the agonist of dopamine, apomorphine,[65] or of amphetamine,[66] which stimulates release of dopamine at the synapse, produces marked increases in glucose consumption in some of the components of the extrapyramidal system known or suspected to contain dopamine-receptive cells. With both drugs, the greatest increases noted were in the zona reticulata of the substantia nigra and the subthalamic nucleus. Surprisingly, none of the components of the dopaminergic mesolimbic system appeared to be affected.

The studies with amphetamine[66] were carried out with the fully quantitative [^{14}C]deoxyglucose method. The results in Table V illustrate the comprehensiveness with which this method surveys the entire brain for sites of altered activity caused by actions of the drug. It also allows for quantitative comparison of the relative potencies of related drugs. For example, in Table V, the comparative effects of d-amphetamine and the less potent dopaminergic agent l-amphetamine are compared; the quantitative results clearly reveal that the effects of l-amphetamine on local cerebral glucose utilization are more limited in distribution and of lesser magnitude than those of d-amphetamine.

Indeed, in similar quantitative studies with apomorphine, McCulloch *et al.*[67,68] have been able to generate complete dose–response curves for the effects of the drug on the rates of glucose utilization in various components of dopaminergic systems. They have also demonstrated metabolically the development of supersensitivity to apomorphine in rats maintained chronically on the dopamine antagonist, haloperidol (J. McCulloch, H. E. Savaki, A. Pert, W. Bunney, and L. Sokoloff, unpublished observations). In the course of these studies with apomorphine, McCulloch *et al.*[69] obtained evidence of a retinal dopaminergic system that projects specifically to the superficial layer of the superior colliculus in the rat. Apomorphine administration activated metabolism in the superficial layer of the superior colliculus as well as in other structures, but the effect in the superficial layer was prevented by prior enucleation. M. Miyaoka (unpublished observations) subsequently observed that intraocular administration of minute amounts of apomorphine caused increased glucose utilization only in the superficial layer of the superior colliculus of the contralateral side.

7.2.5. Effects of α- and β-Adrenergic Blocking Agents

Savaki *et al.*[70] have studied the effects of the α-adrenergic blocking agent phentolamine and the β-adrenergic blocking agent propranolol. Both drugs produced widespread dose-dependent depressions of glucose utilization throughout the brain but exhibit particularly striking and opposite effects in the complete auditory pathway from the cochlear nucleus to the auditory cortex. Propranolol markedly depressed and phentolamine markedly enhanced glucose utilization in this pathway. The functional significance of these effects is unknown, but they seem to correlate with corresponding effects on the electrophysiological responsiveness of this sensory system. Propranolol depresses and phentolamine enhances the amplitude of all components of evoked auditory responses (T. Furlow and J. Hallenbeck, personal communication).

7.3. Pathophysiological Applications

The application of the deoxyglucose method to the study of pathological states has been limited because of uncertainties about the values for the lumped and rate constants to be used. There are, however, pathophysiological states in which there is no structural damage to the tissue, and the standard values of the constants can be used. Several of these conditions have been and are continuing to be studied by the [14C]deoxyglucose technique, both qualitatively and quantitatively.

7.3.1. Convulsive States

The local injection of penicillin into the motor cortex produces focal seizures manifested in specific regions of the body contralaterally. The [14C]deoxyglucose method has been used to map the spread of seizure activity within the brain and to identify the structures with altered functional activity during the seizure. The partial results of one such experiment in the monkey are illustrated in Fig. 2. Discrete regions of markedly increased glucose utilization, sometimes as much as 200%, are observed ipsilaterally in the motor cortex, basal ganglia, particularly the globus pallidus, thalamic nuclei, and contralaterally in the cerebellar cortex.[27]

Kato et al.,[71] Caveness et al.,[72] Hosokawa et al.,[73] and Caveness[74] have carried out the most extensive studies of the propagation of the seizure activity in newborn and pubescent monkeys. The results indicate that the brain of the newborn monkey exhibits similar increases of glucose utilization in specific structures, but the pattern of distribution of the effects is less well defined than in the pubescent monkeys. Collins et al.[29] have carried out similar studies in the rat with similar results but also obtained evidence on the basis of a local stimulation of glucose utilization of a "mirror focus" in the motor cortex contralateral to the side with the penicillin-induced epileptogenic focus.

Engel et al.[75] have used the [14C]deoxyglucose method to study seizures kindled in rats by daily electroconvulsive shocks. After a period of such treatment, the animals exhibit spontaneous seizures. Their results show marked increases in the limbic system, particularly the amygdala. The daily administration of the local anesthetic lidocaine kindles similar seizures in rats; Post et al.[76] have obtained similar results in such seizures with particularly pronounced increases in glucose utilization in the amygdala, hippocampus, and the enterorhinal cortex.

7.3.2. Spreading Cortical Depression

Shinohara et al.[77] studied the effects of local applications of KCl on the dura overlying the parietal cortex of conscious rats or directly on the pial surface of the parietal cortex of anesthetized rats in order to determine if K^+ stimulates cerebral energy metabolism *in vivo* as it is well known to do *in vitro*. The results demonstrate a marked increase in cerebral cortical glucose utilization in response to the application of KCl; NaCl has no such effect. Such application of KCl, however, also produces the phenomenon of spreading cortical depression. This condition is characterized by a spread of transient intense neuronal activity followed by membrane depolarization, electrical depression, and a negative shift in the cortical DC potential in all directions from the site of initiation at a rate of 2–5 mm/min. The depressed cortex also exhibits a number of chemical changes including an increase in extracellular K^+, presumably lost from the cells. At the same time, when the cortical glucose utilization is increased, most subcortical structures that are functionally connected to the depressed cortex exhibit decreased rates of glucose utilization. During recovery from the spreading cortical depression, the glucose utilization in the cortex is still increased, but it is distributed in columns oriented perpendicularly through the cortex. This columnar arrangement may reflect the columnar functional and morphological arrangement of the cerebral cortex. It is likely that the increased glucose utilization in the cortex during spreading cortical depression is the consequence of the increased extracellular K^+ and activation of the Na^+–K^+ ATPase.

7.3.3. Opening of Blood–Brain Barrier

Unilateral opening of the blood–brain barrier in rats by unilateral carotid injection with a hyperosmotic mannitol solution leads to widely distributed discrete regions of intensely increased glucose utilization in the ipsilateral hemisphere.[78] These focal regions of hypermetabolism may reflect local regions of seizure activity. The prior administration of diazepam in most cases prevents the appearance of these areas of increased metabolism,[78] and electroencephalographic recordings under similar experimental conditions reveal evidence of seizure activity (C. Fieschi, personal communication).

7.3.4. Hypoxemia

Pulsinelli and Duffy[79] have studied the effects of controlled hypoxemia on local cerebral glucose utilization by means of the qualitative [^{14}C]deoxyglucose method. Hypoxemia was achieved by artificial ventilation of the animals with a mixture of N_2, N_2O, and O_2 adjusted to maintain the arterial P_{O_2} between 28 and 32 mm Hg. All the animals had had one common carotid artery ligated to limit the increase in cerebral blood flow and the amount of O_2 delivered to the brain. Their autoradiographs provide striking evidence of marked and disparate changes in glucose utilization in the various structural components of the brain. The hemisphere ipsilateral to the carotid ligation was not unexpectedly more severely affected. The most striking effects were markedly higher increases in glucose utilization in white matter than in gray matter, presumably because of the Pasteur effect, and the appearance of transverse cortical columns of high activity alternating with columns of low activity. By studies with black plastic microspheres, they were able to show that the cortical columns were anatomically related to penetrating cortical arteries, with the columns of high metabolic activity lying between the arteries.

Miyaoka et al.[80] have also studied the effects of moderate hypoxemia in normal, spontaneously breathing conscious rats without carotid ligation. The hypoxemia was produced by lowering the O_2 in the inspired air to approximately 7%. Although this procedure reduced arterial P_{O_2} to approximately 30 mm Hg, the cerebral hypoxia was probably less than in the studies of Pulsinelli and Duffy[79] because of the intact cerebral circulation. The animals remained fully conscious under these experimental conditions, although they appeared subdued and less active. The quantitative [^{14}C]deoxyglucose method was employed, and rates of glucose utilization were determined.

The results revealed many similarities to those of Pulsinelli and Duffy.[79] There was a complete redistribution of the local rates of glucose utilization from the normal pattern. Metabolism in white matter was markedly increased. Many areas showed decreased rates of metabolism. Columns were seen in the cerebral cortex, and the caudate nucleus exhibited a strange lacelike heterogeneity quite distinct from its normal homogeneity. Despite the widespread changes, however, overall average glucose utilization remained unchanged. These results are of relevance to the studies by Kety and Schmidt,[81] who found in man that the breathing of 10% O_2 produced a wide variety of mental symptoms without altering the average O_2 consumption of the brain as a whole. The mental symptoms were probably the result of metabolic and functional changes in specific regions of the brain detectable only by methods such as the deoxyglucose method that measure metabolic rate in the structural components of the brain.

7.3.5. Normal Aging

Although, strictly speaking, aging is not a pathophysiological condition, many of its behavioral consequences are directly attributable to decrements in functions of the central nervous system.[82] Normal human aging has been found to be associated with a decrease in average glucose utilization of the brain as a whole.[83] Smith *et al.*[84] have employed the quantitative [^{14}C]deoxyglucose method to study normal aging in Sprague–Dawley rats between 5–6 and 36 months of age. Their results show widespread but not homogeneous reductions of local cerebral glucose utilization with age. The sensory systems, particularly auditory and visual, are particularly severely affected. The caudate nucleus is metabolically depressed, and preliminary experiments indicate that it loses responsivity to dopamine agonists such as apomorphine with age (C. Smith and J. McCulloch, unpublished observations). A striking effect was the loss of metabolically active neuropil in the cerebral cortex; layer IV is markedly decreased in metabolis activity and extent. Some of these changes may be related to specific functional disabilities that develop in old age.

8. MICROSCOPIC RESOLUTION

The resolution of the present [^{14}C]deoxyglucose method is at best approximately 100 μm. The use of [^{3}H]deoxyglucose does not greatly improve the resolution when the standard autoradiographic procedure is used. The limiting factor is the diffusion and migration of the water-soluble labeled compound in the tissue during the freezing of the brain and the cutting of the brain sections. Des Rosiers and Descarries[85] have been working to extend the resolution of the method to the light and electron microscopic levels. They use [^{3}H]deoxyglucose and dipping emulsion techniques. They have reported that fixation, postfixation, dehydration, and embedding of the brain by perfusion *in situ* results in negligible loss or migration of the label in the tissue. They can localize grain counts over individual cells or portions of them. Although the method is at present only qualitative, it is likely that it can eventually be adopted for quantitative use. An alternative promising approach to microscopic resolution is the use of freeze-substitution techniques.[86,87]

9. THE [^{18}F]FLUORODEOXYGLUCOSE TECHNIQUE

Because the deoxyglucose method requires the measurement of local concentrations of radioactivity in the individual components of the brain, it cannot be applied as originally designed to man. Recent developments in computerized emission tomography, however, have made it possible to measure local concentrations of labeled compounds *in vivo* in man. Emission tomography requires the use of γ-radiation, preferably annihilation γ-rays derived from positron emission. A positron-emitting derivative of deoxyglucose, 2-[^{18}F]fluoro-2-deoxy-D-glucose, has been synthesized and found to retain the necessary biochemical properties of 2-deoxyglucose.[88] The method has, therefore, been adapted for use in man with [^{18}F]fluorodeoxyglucose and positron-emission tomography.[88,89] The resolution of the method is still relatively limited, approximately 1 cm, but it is already proving to be useful in studies of the human visual system[90] and of clinical conditions such as focal epilepsy.[91,92] This technique is of immense potential usefulness for studies of human local cerebral energy metabolism in normal states and in neurological and psychiatric disorders.

10. SUMMARY

The deoxyglucose method provides the means to determine quantitatively the rates of glucose utilization simultaneously in all structural and functional components of the central nervous system and to display them pictorially superimposed on the anatomic structures in which they occur. Because of the close relationship between local functional activity and energy metabolism, the method makes it possible to identify all structures with increased or decreased functional activity in various physiological, pharmacological, and pathophysiological states.

The images provided by the method do resemble histological sections of nervous tissue, and the method is, therefore, sometimes misconstrued to be a neuroanatomic method and contrasted with physiological methods, such as electrophysiological recording. This classification obscures the most significant and unique feature of the method. The images are not of structure but of a dynamic biochemical process, glucose utilization, which is as physiological as electrical activity. In most situations, changes in functional activity result in changes in energy metabolism, and the images can be used to visualize and identify the sites of altered activity. The images are, therefore, analogous to infrared maps; they record quantitatively the rates of a kinetic process and display them pictorially exactly where they exist. The fact that they depict the anatomic structures is fortuitous; it indicates that the rates of glucose utilization are distributed according to structure and that specific functions in the nervous system are associated with specific anatomic structures.

The deoxyglucose method represents, therefore, in a real sense, a new type of encephalography, metabolic encephalography. At the very least, it should serve as a valuable supplement to more conventional types, such as electroencephalography. Because, however, it provides a new means to examine another aspect of function simultaneously in all parts of the brain, it is hoped that it and its derivative, the [18F]fluorodeoxyglucose technique, will open new roads to the understanding of how the brain works in health and disease.

ACKNOWLEDGMENT. The author wishes to express his appreciation to Mrs. Ruth Bower for her excellent editorial and bibliographic assistance.

REFERENCES

1. Kety, S. S., and Schmidt, C. F., 1948, *J. Clin. Invest.* **27:**476–483.
2. Scheinberg, P., and Stead, E. A., Jr., 1949, *J. Clin. Invest.* **28:**1163–1171.
3. Lassen, N. A., and Munck, O., 1955, *Acta Physiol. Scand.* **33:**30–49.
4. Eklöf, B., Lassen, N. A., Nilsson, L., Norberg, K., and Siesjö, B. K., 1973, *Acta Physiol. Scand.* **88:**587–589.
5. Gjedde, A., Caronna, J. J., Hindfelt, B., and Plum, F., 1975, *Am. J. Physiol.* **229:**113–118.
6. Kety, S. S., 1950, *Am. J. Med.* **8:**205–217.
7. Kety, S. S., 1957, *Metabolism of the Nervous System* (D. Richter, ed.), Pergamon Press, London, pp. 221–237.
8. Lassen, N. A., 1959, *Physiol. Rev.* **39:**183–238.
9. Sokoloff, L., 1960, *Handbook of Physiology–Neurophysiology*, Volume 3 (J. Field, H. W. Magoun, and V. E. Hall, eds.), American Physiological Society, Washington, pp. 1843–1864.
10. Sokoloff, L., 1976, *Basic Neurochemistry*, 2nd ed., (G. J. Siegel, R. W. Albers, R. Katzman, and B. W. Agranoff, eds.), Little, Brown, Boston, pp. 388–413.
11. Sokoloff, L., 1969, *Psychochemical Research in Man* (A. J. Mandell and M. P. Mandell, eds.), Academic Press, New York, pp. 237–252.
12. Landau, W. M., Freygang, W. H., Jr., Rowland, L. P., Sokoloff, L., and Kety, S. S., 1955, *Trans. Am. Neurol. Assoc.* **80:**125–129.
13. Freygang, W. H., Jr., and Sokoloff, L., 1958, *Adv. Biol. Med. Phys.* **6:**263–279.
14. Kety, S. S., 1960, *Methods Med. Res.* **8:**228–236.
15. Reivich, M., Jehle, J., Sokoloff, L., and Kety, S. S., 1969, *J. Appl. Physiol.* **27:**296–300.
16. Sacks, W., 1957, *J. Appl. Physiol.* **10:**37–44.
17. Sokoloff, L., Reivich, M., Kennedy, C., Des Rosiers, M. H., Patlak, C. S., Pettigrew, K. D., Sakurada, O., and Shinohara, M., 1977, *J. Neurochem.* **28:**897–916.
18. Sokoloff, L., 1977, *J. Neurochem.* **29:**13–26.
19. Plum, F., Gjedde, A., and Samson, F. E., 1976, *Neurosci. Res. Prog. Bull.* **14:**457–518.
20. Bidder, T. G., 1968, *J. Neurochem.* **15:**867–874.
21. Bachelard, H. S., 1971, *J. Neurochem.* **18:**213–222.

22. Oldendorf, W. H., 1971, *Am. J. Physiol.* **221**:1629–1638.
23. Sols, A., and Crane, R. K., 1954, *J. Biol. Chem.* **210**:581–595.
24. Hers, H. G., 1957, *Le Métabolisme du Fructose,* Editions Arscia, Brussels, p. 102.
25. Kennedy, C., Sakurada, O., Shinohara, M., Jehle, J., and Sokoloff, L., 1978, *Ann. Neurol.* **4**:293–301.
26. Goochee, C., Rasband, W., and Sokoloff, L., 1980, *Ann. Neurol.* **7**:359–370.
27. Kennedy, C., Des Rosiers, M., Jehle, J. W., Reivich, M., Sharp, F., and Sokoloff, L., 1975, *Science* **187**:850–853.
28. Caveness, W. F., 1969, *Basic Mechanisms of the Epilepsies* (H. H. Jasper, A. A. Ward, and A. Pope, eds.), Little, Brown, Boston, pp. 517–534.
29. Collins, R. C., Kennedy, C., Sokoloff, L., and Plum, F., 1976, *Arch. Neurol.* **33**:536–542.
30. Lashley, K. S., 1934, *J. Comp. Neurol.* **59**:341–373.
31. Montero, V. M., and Guillery, R. W., 1968, *J. Comp. Neurol.* **134**:211–242.
32. Hubel, D. H., and Wiesel, T. N., 1968, *J. Physiol. (Lond.)* **195**:215–243.
33. Hubel, D. H., and Wiesel, T. N., 1972, *J. Comp. Neurol.* **146**:421–450.
34. Wiesel, T. N., Hubel, D. H., and Lam, D. M. K., 1974, *Brain Res.* **79**:273–279.
35. Rakic, P., 1976, *Nature* **261**:467–471.
36. Kennedy, C., Des Rosiers, M. H., Sakurada, O., Shinohara, M., Reivich, M., Jehle, J. W., and Sokoloff, L., 1976, *Proc. Natl. Acad. Sci. U.S.A.* **73**:4230–4234.
37. Hubel, D. H., Wiesel, T. N., and Stryker, M. P., 1978, *J. Comp. Neurol.* **177**:361–380.
38. Macko, K. A., Jarvis, C. D., Kennedy, C., Miyaoko, M., Shinohara, M., Sokoloff, L., and Mishkin, M., 1982, *Science* **218**:394–397.
39. Des Rosiers, M. H., Sakurada, O., Jehle, J., Shinohara, M., Kennedy, C., and Sokoloff, L., 1978, *Science* **200**:447–449.
40. Batipps, M., Miyaoka, M., Shinohara, M., Sokoloff, L., and Kennedy, C., 1981, *Neurology (N.Y.)* **31**:58–62.
41. Miyaoka, M., Shinohara, M., Batipps, M., Pettigrew, K. D., Kennedy, C., and Sokoloff, L., 1979, *Acta Neurol. Scand. [Suppl.]* **72**:16–17.
42. Sharp, F. R., Kauer, J. S., and Shepherd, G. M., 1975, *Brain Res.* **98**:596–600.
43. Webster, W. R., Serviere, J., Batini, C., and LaPlante, S., 1978, *Neurosci. Lett.* **10**:43–48.
44. Silverman, M. S., Hendrickson, A. E., and Clopton, B. M., 1977, *Neurosci. Abstr.* **3**:11.
45. Durham, D., and Woolsey, T. A., 1977, *Brain Res.* **137**:169–174.
46. Hand, P. J., Greenberg, J. H., Miselis, R. R., Weller, W. L., and Reivich, M., 1978, *Neurosci. Abstr.* **4**:553.
47. Hand, P. J., 1981, *Neuroanatomical Tract-Tracing Methods* (L. Heimer and M. J. Robards, eds.), Plenum Press, New York pp. 511–538.
48. Kennedy, C., Miyaoka, M., Suda, S., Macko, K., Jarvis, C., Mishkin, M., and Sokoloff, L., 1980, *Trans. Am. Neurol. Assoc.* **105**:13–17.
49. Schwartz, W. J., and Gainer, H., 1977, *Science* **197**:1089–1091.
50. Schwartz, W. J., Davidsen, L. C., and Smith, C. B., 1980, *J. Comp. Neurol.* **189**:157–167.
51. Yarowsky, P. J., Jehle, J., Ingvar, D. H., and Sokoloff, L., 1979, *Neurosci. Abstr.* **5**:421.
52. Larrabee, M. G., 1958, *J. Neurochem.* **2**:81–101.
53. Horowicz, P., and Larrabee, M. G., 1958, *J. Neurochem.* **2**:102–118.
54. Friedli, C., 1978, *Adv. Exp. Med. Biol.* **94**:747–754.
55. Yarowsky, P., Crane, A. M., and Sokoloff, L., 1980, *Neurosci. Abstr.* **6**:340.
56. Mata, M., Fink, D. J., Gainer, H., Smith, C. B., Davidsen, L., Savaki, H., Schwartz, W. J., and Sokoloff, L., 1980, *J. Neurochem.* **34**:213–215.
57. Galvan, M., Ten Bruggencate, G., and Senekowitsch, R., 1979, *Brain Res.* **160**:544–548.
58. Roth, R. H., and Giarman, N. J., 1966, *Biochem. Pharmacol.* **15**:1333–1348.
59. Roth, R. H., 1976, *Pharmacol. Ther.* **2**:7188.
60. Wolfson, L. I., Sakurada, O., and Sokoloff, L., 1977, *J. Neurochem.* **29**:777–783.
61. Shinohara, M., Sakurada, O., Jehle, J., and Sokoloff, L., 1976, *Neurosci. Abstr.* **2**:615.
62. Sakurada, O., Shinohara, M., Klee, W. A., Kennedy, C., and Sokoloff, L., 1976, *Neurosci. Abstr.* **2**:613.
63. Schwartz, W. J., Sharp, F. R., Gunn, R. H., and Evarts, E. V., 1976, *Nature* **261**:155–157.
64. Schwartz, W. J., 1978, *Brain Res.* **158**:129–147.

65. Brown, L., and Wolfson, L., 1978, *Brain Res.* **148**:188–193.
66. Wechsler, L. R., Savaki, H. E., and Sokoloff, L., 1979, *J. Neurochem.* **32**:15–22.
67. McCulloch, J., Savaki, H. E., McCulloch, M. C., and Sokoloff, L., 1979, *Nature* **282**:303–305.
68. McCulloch, J., Savaki, H. E., and Sokoloff, L., 1980, *Brain Res.* **194**:117–124.
69. McCulloch, J., Savaki, H. E., McCulloch, M. C., and Sokoloff, L., 1980, *Science* **207**:313–315.
70. Savaki, H. E., Kadekaro, M., Jehle, J., and Sokoloff, L., 1978, *Nature* **276**:521–523.
71. Kato, M., Malamut, B. L., Caveness, W. F., Hosokawa, S., Wakisaka, S., and O'Neill, R. R., 1980, *Ann. Neurol.* **7**:204–212.
72. Caveness, W. F., Kato, M., Malamut, B. L., Hosokawa, S., Wakisaka, S., and O'Neill, R. R., 1980, *Ann. Neurol.* **7**:213–221.
73. Hosokawa, S., Iguchi, T., Caveness, W. F., Kato, M., O'Neill, R. R., Wakisaka, S., and Malamut, B. L., 1980, *Ann. Neurol.* **7**:222–229.
74. Caveness, W. F., 1980, *Ann. Neurol.* **7**:230–237.
75. Engel, J., Jr., Wolfson, L., and Brown, L., 1978, *Ann. Neurol.* **3**:538–544.
76. Post, R. M., Kennedy, C., Shinohara, M., Squillace, K., Miyaoka, M., Suda, S., Ingvar, D. H., and Sokoloff, L., 1979, *Neurosci. Abstr.* **5**:196.
77. Shinohara, M., Dollinger, B., Brown, G., Rapoport, S., and Sokoloff, L., 1979, *Science* **203**:188–190.
78. Pappius, H. M., Savaki, H. E., Fieschi, C., Rapoport, S. I., and Sokoloff, L., 1979, *Ann. Neurol.* **5**:211–219.
79. Pulsinelli, W. A., and Duffy, T. E., 1979, *Science* **204**:626–629.
80. Miyaoka, M., Shinohara, M., Kennedy, C., and Sokoloff, L., 1979, *Trans. Am. Neurol. Assoc.* **104**:151–154.
81. Kety, S. S., and Schmidt, C. F., 1948, *J. Clin. Invest.* **27**:484–492.
82. Birren, J. E., Butler, R. N., Greenhouse, S. W., Sokoloff, L., and Yarrow, M. R. (eds.), 1963, Public Health Service Publication No. 986, United States Government Printing Office, Washington.
83. Sokoloff, L., 1966, *Res. Publ. Assoc. Nerv. Ment. Dis.* **41**:237–254.
84. Smith, C. B., Goochee, C., Rapoport, S. I., and Sokoloff, L., 1980, *Brain* **103**:351–365.
85. Des Rosiers, M. H., and Descarries, L., 1978, *C. R. Acad. Sci. Paris [D]* **287**:153–156.
86. Ornberg, R. L., Neale, E. A., Smith, C. B., Yarowsky, P., and Bowers, L. M., 1979, *J. Cell Biol. [Abstr.]* **83**:CN142A.
87. Sejnowski, T. J., Reingold, S. C., Kelley, D. B., and Gelperin, A., 1980, *Nature* **287**:449–451.
88. Reivich, M., Kuhl, D., Wolf, A., Greenberg, J., Phelps, M., Ido, T., Cassella, V., Fowler, J., Hoffman, E., Alavi, A., Som, P., and Sokoloff, L., 1979, *Circ. Res.* **44**:127–137.
89. Phelps, M. E., Huang, S. C., Hoffman, E. J., Selin, C., Sokoloff, L., and Kuhl, D. E., 1979, *Ann. Neurol.* **6**:371–388.
90. Phelps, M. E., Mazziotta, J. C., and Kuhl, D. E., 1981, *Science* **211**:1445–1448.
91. Kuhl, D., Engel, J., Phelps, M., and Selin, C., 1979, *Acta Neurol. Scand. [Suppl.]* **72**:538–539.
92. Kuhl, D. E., Engel, J., Jr., Phelps, M. E., and Selin, C., 1980, *Ann. Neurol.* **8**:348–360.
93. Duffy, T. E., Cavazzuti, M., Cruz, M. S., and Sokoloff, L., 1982, *Ann. Neurol.* **11**:233–246.
94. Sokoloff, L., 1979, *Acta Neurol. Scand. [Suppl.]* **70**:640–649.
95. Sokoloff, L., 1978, *Trends Neurosci.* **1**(3):75–79.

Intermediary Metabolism of Carbohydrates and Other Fuels

Richard A. Hawkins and Anke M. Mans

1. INTRODUCTION

The central nervous system (CNS) is composed of many cell types, each with different functions and metabolic requirements. These are combined into a variety of structures which may be just as different metabolically as they are anatomically and functionally. Although considerable progress has been made in the separation and study of these different cell types (see Volume 1 of this *Handbook*), results from such studies may not necessarily reflect cerebral metabolism *in vivo*. It is, after all, the interaction of the various cell types that determines the unique character of cerebral function and metabolism. When the cells are separated, this aspect is lost to a greater degree in brain than in other organs which do not rely as heavily on intercellular communication. In view of these considerations, the emphasis of this chapter is on studies conducted *in vivo*.

Although the CNS is highly specialized, it shares many metabolic pathways with other organs, including the ability to synthesize and degrade glycogen, the glycolytic pathway, the Krebs cycle, the pentose phosphate pathway, gluconeogenic reactions, CO_2 fixation (anaplerotic reactions), and the ability to synthesize and oxidize lipids. In addition, the brain has the γ-aminobutyrate pathway which is not found in most other organs. Because of the heavy reliance of the brain on glucose for energy metabolism, close coordination of the glycolytic pathway and the Krebs cycle takes on particular importance.

To obtain a better perspective of intermediary metabolism in the brain, it is useful to consider several functional and anatomic features that define the routes by which the brain meets its metabolic requirements.

Richard A. Hawkins and Anke M. Mans • Departments of Anesthesia and Physiology, The Pennsylvania State University College of Medicine, The Milton S. Hershey Medical Center, Hershey, Pennsylvania 17033.

2. ENERGY REQUIREMENTS

Although nervous tissue does not participate in processes that require large amounts of energy, such as mechanical work, osmotic work, or extensive biosynthesis, it has almost as high a rate of oxidative metabolism as some tissues that do. For example, the rate of O_2 consumption of rat brain (about 4.4 $\mu mol \cdot min^{-1} \cdot g^{-1}$)[1] is comparable to that of the heart (8.7 $\mu mol \cdot min^{-1} \cdot g^{-1}$ working at 100 Torr aortic pressure),[2] kidney (7–13 $\mu mol \cdot min^{-1} \cdot g^{-1}$),[3] or liver (5.5 $\mu mol \cdot min^{-1} \cdot g^{-1}$).[4] In newborn animals, the rate of O_2 consumption is less than half the adult rate.[5,6] In general, there is an almost constant relationship between CNS size and whole-animal caloric consumption. Most vertebrates use 2–8% of their total resting metabolism for the CNS, except higher apes and man who use considerably more.[7] In humans, the rate is maximal at 5 years of age and declines throughout life.[8] Because of the large brain-to-body weight ratio, the human brain requires a much larger proportion of the whole-body energy requirement than does the brain of any other species.

The energy-utilizing reactions of brain can be divided into two major categories: biosynthesis and transport. Biosynthesis appears to occur mostly in cell bodies (proteins, polypeptides, lipids, etc.) or in nerve terminals (neurotransmitters). Transport processes are especially active in axons and dendrites and include the acquisition of essential nutrients, reuptake of neurotransmitters, and the transport of Na^+ and K^+ accomplished by means of Na^+–K^+ ATPases which are responsible for the maintenance of transmembrane electrical potentials. Some evidence suggests that the major proportion of energy is used for ion transport.[9-11] In barbiturate anesthesia deep enough to cause near isoelectric encephalographic activity, metabolism may be depressed by about 56%.[9] If it is assumed that biosynthetic and other energy-consuming processes were not affected by the anesthetic, this suggests that at least 56% of the normal cerebral activity was associated with electrical events, i.e., the maintenance of Na^+ and K^+ gradients. This must be regarded as an approximation because neither the residual ion transport activity nor the effect on biosynthetic processes was taken into account. However, in view of the continuous electrical activity in the normal CNS, it appears likely that a large proportion of the energy requirement is devoted to ion transport.

3. THE BLOOD–BRAIN BARRIER

The blood–brain barrier has a strong influence on the availability of substrates to brain. It may in fact be the limiting factor for the metabolism of ketone bodies and some free fatty acids,[12] for glucose during hypoglycemia,[13,14] and possibly for amino acids.[15] (See chapters by H. S. Bachelard and by W. H. Oldendorf, Volume 10, and refs. 16, 17 for comprehensive discussions of the blood–brain barrier and its transport systems.) The endothelial cells of most capillaries throughout the body have narrow gaps between them, allowing small molecules to pass relatively rapidly into the extracellular fluid while retaining most of the larger circulating molecules such as proteins. Nerve cells, however, require a strictly controlled environment for proper functioning. The endothelial cells of cerebral capillaries are joined to contiguous cells by tight junctions, which are in effect a fusion of the membranes.[18,19] Thus, molecules cannot pass between the cells but must penetrate the luminal and antiluminal membranes. This barrier constitutes the initial point of resistance to the entry of substrates into brain. Although lipid-soluble molecules, such as the gases O_2 and CO_2, may pass readily through the membranes, most nutrient molecules are hydrophilic, and their passage must be mediated by specific transport systems. Such systems have been described for glucose, amino acids, monocarboxylic acids, and other essential nutrients. After crossing the endothelial membranes, substrates must pass through a basement membrane consisting of glycoproteins, whereupon they encounter an almost complete layer formed by astrocytic processes. In order to reach neurons, substrates must either pass through the two membranes of the astrocytes or take a relatively circuitous route around them in the interstitual fluid. Therefore, before a substrate is in a position to be metabolized by neuronal cells, it must cross at least three and possibly five membranes in addition to the basement membrane layer.*

4. ANATOMIC AND METABOLIC COMPARTMENTS

It has been established that metabolic compartmentation exists in brain and may have, at least in part, an anatomic basis. Three major interlocking cellular compartments comprise the CNS: the vascular, the neuronal, and the glial compartments.[20] The capillaries are almost completely ensheathed by glial cells which also line the external surfaces of the brain adjacent to the pia-arachnoid. The neurons are therefore isolated by being suspended, as it were, in a framework of glia. The capillary endothelial cells, which comprise only a few percent of the total cellular volume of brain, contain five to seven times as many mitochondria as endothelial cells of systemic capillaries[21] and appear to be more active metabolically. The neuroglia occupy as much as 50% of the vertebrate CNS.[22] Of these, the astrocytes occupy 22–27% and are characterized by the presence of relatively few mitochondria and many glycogen granules,[23] although studies of astrocytes in culture suggest a more active rate of oxidative metabolism.[24] The oligodendrocytes, on the other hand, contain many more mitochondria and relatively little glycogen and appear to be much more active cells. Studies of the relative rates of oxidative metabolism of CNS cells are confusing. In general, as judged by morphological, neuropathological, and quantitative measurements made *in vitro,* the rates decrease in the order: neurons > oligodendrocytes > astrocytes (= capillary endothelial cells?) > microglia. After a careful review, Siesjö[25] concluded that neurons were probably responsible for about 75% of the O_2 consumed by the CNS. It would seem

---

* There are several areas in the brain, the so-called circumventricular organs, where there is no blood–brain barrier. However, in these areas, there appears to be a permeability barrier between the cerebrospinal fluid and the interstitial fluid which prevents entry of material into the cerebrospinal fluid cavities, thus isolating the area of leakage from the rest of the brain.[269]

that these cell groups with their different metabolic properties are at least partially responsible for the phenomenon of metabolic compartmentation within the CNS.

Metabolic compartmentation was defined[26] as the presence of two or more distinct pools of a given metabolite which do not rapidly exchange this metabolite but maintain their own integrity, turnover rates, and flux rates. Such pools could be located in different cell populations or be maintained within a single cell group providing that adequate separation exists. Several biochemical compartments have been recognized in brain on the basis of radioisotope precursor–product relationships of the amino acids glutamine, glutamate, and aspartate.

To be incorporated into these amino acids from an oxidizable substrate, label must first be converted into acetyl-CoA (or succinyl-CoA in the case of propionate) and then be further oxidized in the Krebs cycle to α-oxoglutarate before exchanging with glutamate. Normally the specific radioactivity of glutamate is expected to exceed that of glutamine. This is indeed the case with many substrates, including glucose, ketone bodies, lactate, and glycerol.[27] However, several other oxidizable substrates including the amino acids leucine, phenylalanine, proline, glutamate, aspartate, and γ-aminobutyrate, as well as the short- to medium-chain-length fatty acids, acetate, propionate, butyrate, and octanoate, label glutamine to a greater specific activity than glutamate.[27–31,35]

These results can be explained by postulating at least two separate pools of Krebs cycle intermediates, one containing a large pool of glutamate with relatively little conversion to glutamine, and the other containing a small glutamate pool which is rapidly converted to glutamine.[32] On the basis of considerable evidence, these major pools can be assigned to the neurons and glial cells, respectively.[30,33,34] (Compartmentation is almost absent at birth and develops in parallel with glial cells[35].) The nonneuronal compartment is probably responsible for most of the oxidation of amino acids and fatty acids that occurs in the CNS. In fact, it has been suggested that this compartment uses relatively little glucose.[28,30] It would follow that the neurons are responsible for the large glucose requirement of brain. The neuronal pool can be further subdivided into at least two pools, one comprised of nerve terminals and the other of neuronal perikarya and dendrites.[30] Evidence based on differential labeling patterns of leucine and fatty acids indicates that the glial compartment must likewise consist of at least two pools.[29] Undoubtedly, other metabolic pools associated with the Krebs cycle will be discovered in the future.

5. SUBSTRATES

5.1. Glucose

Glucose is the major but not the only source of cerebral energy (for review see ref. 8). Earlier beliefs were based on the observation that measured arteriovenous differences were sufficient to supply the energy requirement of brain.

The respiratory quotient has been measured to be close to 1 (0.92–0.99),[36] which is compatible with oxidation of carbohydrate. In addition, when circulating fuels (including glucose, free fatty acids, ketone bodies, and triglycerides) are decreased by insulin administration, the coma that results can be completely reversed only by glucose and mannose. Mannose is presumably phosphorylated by hexokinase and then reacts with phosphomannoisomerase to form fructose 6-phosphate and enters the glycolytic pathway.[37] However, mannose does not normally occur in the circulation in significant concentrations.

Although it is clear that normal cerebral function requires glucose, the widely held concept that it is the only fuel source requires closer examination. Respiratory quotients are notoriously difficult to measure and are relatively insensitive indicators of the substrates being oxidized. For instance, if brain were to oxidize a mixture of 25% free fatty acids and 75% glucose, the respiratory quotient would be 0.925. A respiratory quotient of 0.92 has been observed during diabetic ketoacidosis (when plasma free fatty acids and ketone bodies are expected to be elevated), but this was taken to be within normal limits of carbohydrate oxidation.[36] The measurement of arteriovenous differences is likewise technically difficult. Arteriovenous differences are usually very small (the net uptake of glucose is only about 10%), so that highly sensitive techniques must be used. No arteriovenous differences could be found for the ketone bodies during diabetic ketoacidosis in early studies,[36] but it is now known that they can in fact supply more than half of the metabolic requirement in this condition.[38] Thus, data based only on respiratory quotients or arteriovenous differences using older techniques must be interpreted with caution. There is considerable evidence that the oxidation of substrates besides glucose may be important in the CNS, especially for nonneuronal cells. In fact, some evidence suggests that certain cells in the CNS do not use substantial amounts of glucose.[28,30] Certainly, it has been well established that ^{14}C from a variety of noncarbohydrate compounds can rapidly appear in CNS amino acids, and this can only occur following metabolism by the Krebs cycle.

5.2. Ketone Bodies

Nervous tissue has the capability of oxidizing the ketone bodies 3-hydroxybutyrate and acetoacetate. Although these substrates are normally present in low concentrations in the circulation, there are several physiological and pathological conditions in which their concentrations rise and they become an important respiratory fuel. In addition to elevation during fasting and diabetes, increased concentrations occur during pregnancy, during prolonged exercise, in persons eating high-fat diets, in uremia, during the neonatal period, and during infancy (for review see ref. 39). Consumption of ketone bodies diminishes the demand for glucose, reducing the need for gluconeogenesis and concomitant degradation of protein.

The ability to use ketone bodies varies in different species. Rats can derive 5–20% of the total brain energy requirement from ketone bodies.[40–42] Adult dogs do not develop significant ketosis. Interestingly, neither do bears in winter sleep. Bears can synthesize enough glucose from the glycerol that is continuously released from triglycerides to supply the entire cerebral requirement.[43] Humans, on the other hand, may depend on ketone bodies for about 60% of the energy requirement after prolonged fasting.[38] Perhaps ketone bodies are of the greatest importance for children, in whom brain oxygen consumption may comprise 50% or more of the basal metabolic rate. Young children may develop hypoglycemia and ketoacidosis within several hours, and the ability to use fat-derived substrates may become necessary for survival. Although there is circumstantial evidence that ketone bodies can be almost the sole substrate for cerebral metabolism in humans,[43] controlled experiments in animals indicate that some glucose is also necessary.[44–47] The utilization of ketone bodies may decrease cerebral glycolysis (perhaps by indirectly inhibiting phosphofructokinase), increase brain glycogen content, and inhibit the oxidation of pyruvate at the pyruvate dehydrogenase step.[48,49]

Flux through the pathway of ketone body metabolism leading to the Krebs cycle appears to be determined by substrate availability. The metabolism of ketone bodies by human and rat brain is linearly related to arterial concentrations, and the uptake of acetoacetate generally exceeds that of 3-hydroxybutyrate at comparable arterial concentrations.[40,50] Thus, transport into brain is the rate-limiting step in the utilization of ketone bodies; i.e., they are metabolized as rapidly as they enter.[42,48,51] Ketone bodies enter brain by a transport system that also carries other monocarboxylic acids including short-chain fatty acids, lactate, and pyruvate.[52–54] Transport activity into brain varies from region to region, and glucose and ketones do not nourish the same areas proportionately.[12] Those areas of brain that have no blood–brain barrier consume many more ketone bodies than those areas with an intact blood–brain barrier. Ketone body transport activity in rats rises throughout the suckling period, reaching a peak before weaning and then declining to adult levels.[42,53] In contrast to the activity of the enzymes of the metabolic pathway in the brain, activity of the transport system is increased by starvation, diabetes, and a high-fat diet.[41,42,47,55,56]

5.3. Lactate

Because of the relative impermeability of the blood–brain barrier to circulating lactate, it cannot be considered a significant fuel for brain in adult mammals. However, in young mammals, the blood–brain barrier is much more permeable to monocarboxylic acids; the V_{max} of lactate entry into brain of suckling rats is many times greater than that of adults.[54] When the arterial concentration of lactate exceeds 2 mM in newborn dogs, net uptake occurs into the brain, whereas below 2 mM, the brain produces lactate.[57,58] At 10 mM lactate, a concentration that may result from intense exercise, a short period of hypoxia, or excessive insulin administration, it was found that lactate could supply more than 50% of the total metabolic fuel required.[58] Therefore, under abnormal circumstances, lactate can be oxidatively metabolized in lieu of glucose by brains of young mammals when the circulating concentration rises. In adults, it is unlikely that lactate ever becomes important as a fuel for brain regardless of the circulating concentrations.

5.4. Glycerol

Glycerol produced from fats can be converted by mammalian tissues to glucose. There is also some evidence that glycerol can be used directly by brain.[59] In the rabbit, the abnormal EEG produced by hypoglycemia could be altered toward normal by internal carotid infusion of glycerol, although ten times as much glycerol as glucose was required to produce a noticeable effect.[59,60]

5.5. Amino Acids

The combined unidirectional influx of all the amino acids normally present in the circulation is about 40–45 nmol $\cdot$ min^{-1} $\cdot$ g^{-1} in adult rats.[15,61] Since the normal rate of glucose utilization is between 700 and 800 nmol $\cdot$ min^{-1} $\cdot$ g^{-1}, it is obvious that even the complete oxidation of all entering amino acids would not contribute more than a few percent to the total energy requirement. Several studies in experimental animals could not demonstrate significant arteriovenous differences of amino acids.[15] This may have been because of the difficulty of measuring amino acids in small samples. Arteriovenous determinations in humans showed a consistent net uptake of amino acids after an overnight fast[62] and during prolonged starvation.[38] Interestingly, the arteriovenous differences (i.e., net influx) in humans are very similar to the rates of unidirectional influx determined in experimental animals, suggesting that net oxidative metabolism occurs. Certainly, it has been shown that ^{14}C or ^{3}H from leucine, proline, and phenylalanine can enter aspartate, glutamate, and glutamine.[28,29,63–65] Thus, label from these amino acids was converted to acetyl CoA and further oxidized in the Krebs cycle. The major site of oxidation of the amino acids is believed to be the "small" glutamate compartment, because administration of label results in a higher specific activity of glutamine than glutamate. The findings are compatible with the suggestion that the oxidative metabolism of these amino acids occurs in glial cells, whereas incorporation into protein (the major fate of amino acids entering brain) occurs in neuronal perikarya.[30]

5.6. Free Fatty Acids

Since fatty acids are the preferred fuel of oxidative metabolism in many tissues, it is surprising that the CNS is not more dependent on them. Brain mitochondrial preparations oxidize long-, medium-, and short-chain fatty acids at rates comparable to mitochondria from heart and liver.[66,67] The belief that fatty acids are not used by the CNS comes from the observation that the respiratory quotient is close to 1[36] under a variety of circumstances, and attempts to measure arteriovenous differences of free fatty acids, which are difficult to measure, have been unsuccessful.[8,38] On the other hand, there are several reports that indicate that contributions by free fatty acids to cerebral energy metabolism may be significant.

Long-chain fatty acids are acylated to fatty acyl-CoA in the cytoplasm, converted to acyl-carnitine by carnitine palmityltransferase, and may then enter the mitochondria where oxidation occurs. The carnitine palmityltransferase activity of brain is comparable to that of other tissues,[68] and the brain content of carnitine is high enough for activity (M. Babcock and R. A. Hawkins, unpublished observations). Several observations suggest that the blood–brain barrier may limit the oxidation of long-chain fatty acids. When a fatty acid–albumin complex was introduced into dog brain by ventriculocisternal perfusion, it was readily oxidized.[69] Also, it has been observed that the incorporation of ^{14}C from palmitate was relatively rapid in areas of the brain that lack a blood–brain barrier, such as the pineal and pituitary glands.[51] Direct measurement of albumin-bound palmitate uptake showed that about 5% was extracted on a single pass through the brain in adult rats.[70] This amount of palmitate could account for about 10–15% of the normal fuel requirement at a blood flow of 1 $ml \cdot min^{-1} \cdot g^{-1}$ and circulating free fatty acid concentrations of 0.5 mEq/liter. This agrees well with studies of the isolated perfused cat brain which was shown to derive about 10% of its energy requirements from albumin-bound palmitate.[71]* Fasting puppies, which do not readily become ketotic, can derive about 24% of their brain oxidative fuel from circulating fatty acids,[72] an amount that is certainly significant.

Short- and medium-chain fatty acid concentrations in the plasma are normally so low (with the possible exception of acetate[73]) that they cannot be considered significant fuels except perhaps in ruminants. The short-chain fatty acids, i.e, acetate and butyrate, enter the brain by the monocarboxylic acid transport system, which is relatively inactive.[73,74] Medium-chain fatty acids such as octanoate are, however, much more lipid soluble, and their entry into brain is relatively unrestricted.[74] In contrast to long-chain fatty acids, short- and medium-chain fatty acids are activated to acyl-CoA esters directly within the mitochondria where they are further metabolized. In view of the fact that the monocarboxylic acid transport system in young mammals has a capacity several times that of adults[53,54] and the fact that very-short-chain fatty acids (i.e., acetate) may be present in millimolar concentrations,[73] it seems possible that short-chain fatty acid oxidation by brain may be of greater significance in young mammals.

Oxidation of long-, medium-, and short-chain fatty acids results in a higher specific activity of label in glutamine than in glutamate. Therefore, their oxidation, like that of the amino acids, most likely occurs in nonneuronal cells. This is in contrast to the other substrates (glucose, ketone bodies, lactate, and glycerol) which may be considered to be more general fuels.

In summary, although the major fuel of respiration in the CNS is glucose, the role of other fuels cannot be ignored. In some circumstances, their combined role may be greater than that of glucose. The pineal and pituitary glands, which lack a blood–brain barrier, may rely more heavily on noncarbohydrate fuels of respiration.

---

* Cerebral capillary endothelial cells are metabolically active and associated with ion transport.[21,266] They are dependent on a supply of long-chain free fatty acids for optimal functioning and derive about 25% of their energy requirement from this source.[267]

6. *ENDOGENOUS SUBSTRATES*

The glycogen content of brain is relatively small (see Fig. 1) compared to other tissues. It has been suggested[43] that, since 3–4 g of water is required for each gram of glycogen, larger quantities of glycogen would result in volume fluctuations unacceptable in the brain because of the rigidity of the cranium. Glycogen granules are much more obvious in astrocytes than in other CNS cells, and hence glycogen may not be readily available as a reserve in other cells. In any event, the amount present (2.8 μmol/g in rats) is only enough to sustain brain glucose consumption (0.8 μmol $\cdot$ min^{-1} $\cdot$ g^{-1}) for 3.5 min when the supply of O_2 is sufficient.

There is considerable evidence that the CNS can use other, largely unidentified fuels for a large portion of its oxidative needs when the supply of circulating substrates is insufficient. It was found[75] that during severe insulin-induced coma, when the circulating concentrations of all known substrates are reduced, the rate of cerebral glucose utilization was almost negligible, but O_2 consumption was only halved. The isolated cat brain perfused with a glucose-free medium continued to use O_2 at near normal rates of 2.2–3.1 μmol $\cdot$ min^{-1} $\cdot$ g^{-1} for 60–90 min, while the respiratory quotient decreased from 1 to about 0.5.[76] Furthermore, it was found that O_2 consumption could be increased two- to threefold by pentylenetetrazole-induced seizures. Barbiturate-anesthetized sheep and rabbits used O_2 at normal rates for more than 2 h during severe hypoglycemia.[77] Under these circumstances, the molar ratio of glucose used to O_2 consumed decreased from 0.93 to less than 0.4. It was suggested that the O_2 was used for metabolism of lipids stored in the brain; less than 0.1 g of lipid/100 g brain would suffice to account for 3 h of nonglucose oxidative metabolism at the observed rate. Attempts to identify the endogenous substrate(s) have not produced convincing evidence.[78] A significant breakdown of brain proteins and nucleic acids in insulin-induced hypoglycemic animals could not be found.[79] Some evidence was produced that breakdown of phospholipid could provide substrate for fuel,[79,80] but others found no change in phospholipid content.[81] The amino acids glutamate, glutamine, and aspartate have also been suggested as endogenous fuels,[82] but they are sufficient to maintain normal oxidative metabolism for only about 10 min.

7. *CONTROL OF GLYCOLYSIS AND THE KREBS CYCLE*

The control of glycolysis and the Krebs cycle is necessary to insure that glucose is metabolized at a sufficient rate to provide for the needs of high-energy phosphate synthesis, production of reducing equivalents, and replenishment of Krebs cycle intermediates used for the synthesis of neurotransmitter amino acids. This requires precise synchrony within and between the various pathways. Although it is clear that there is a close correspondence between energy requirements and flux through the glycolytic pathway and Krebs cycle, the question may be asked, "Precisely what is it that cells regulate?" The major factors are probably the cytoplasmic phosphorylation state, the pyridine nu-

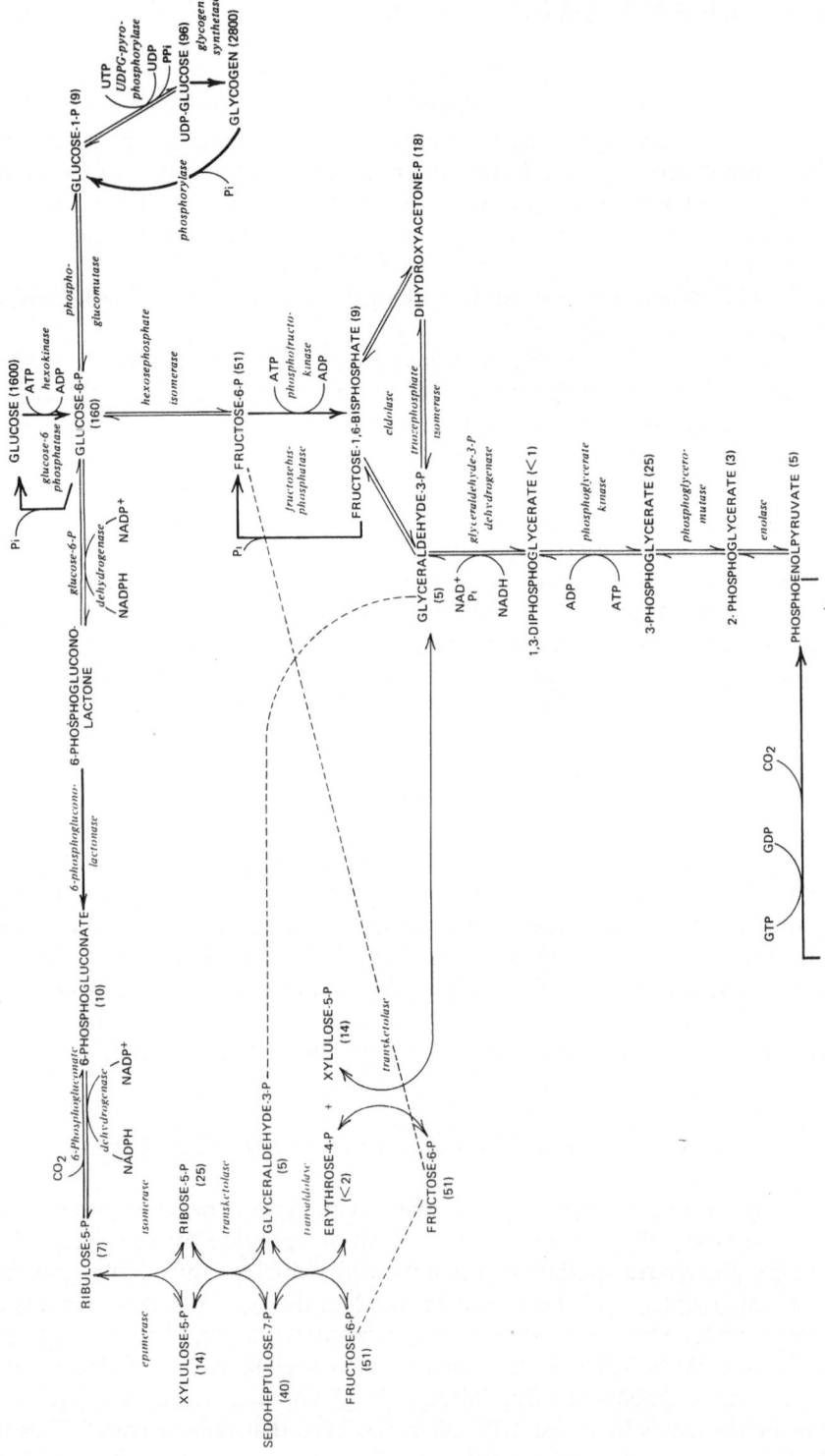

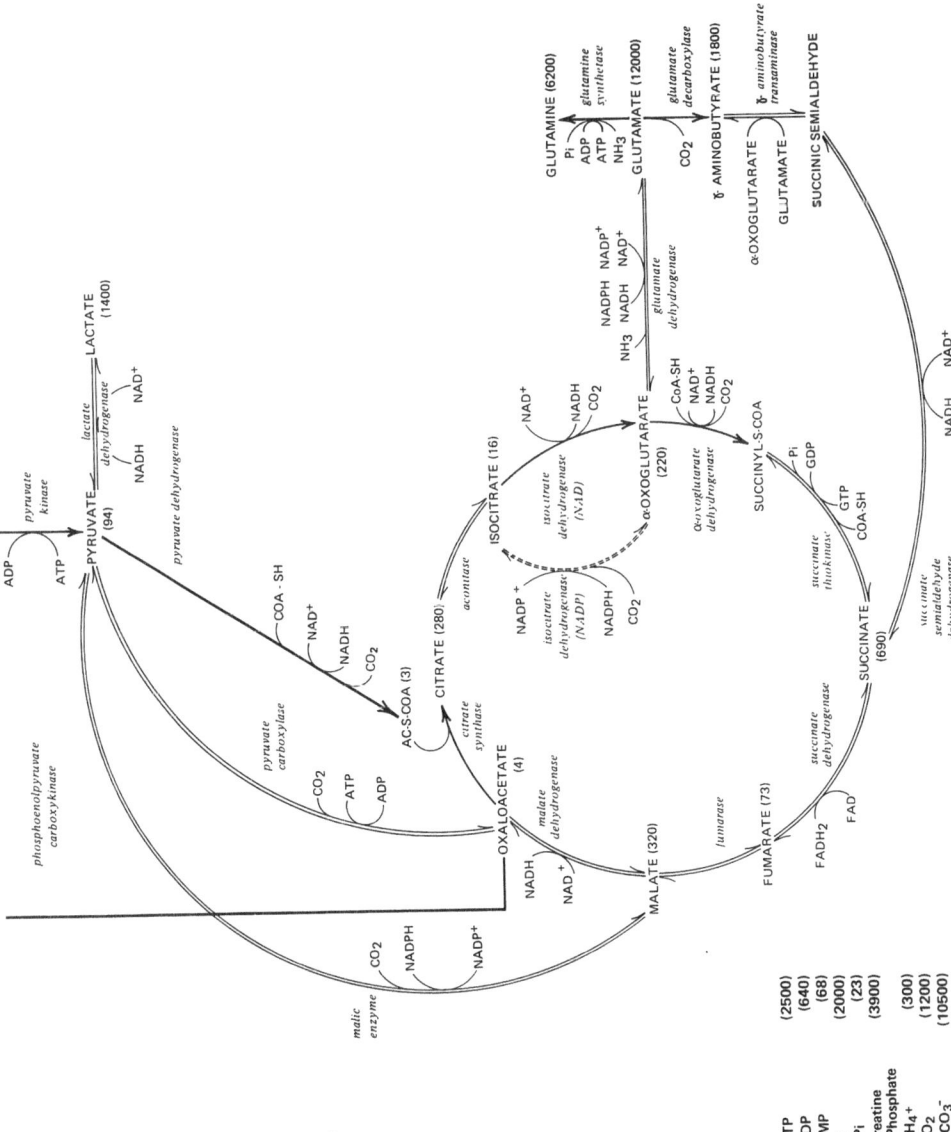

Fig. 1.   Metabolic pathways. Numbers in parentheses indicate the whole-brain content of substrate in nmol/g fresh weight.
From refs. 48, 149, 184, 196, 210, 282–290.

cleotide redox state, and the level of key intermediary metabolites. In addition, the creatine phosphate phosphorylation state may also be important. Thus, mechanisms sensitive to these parameters must be present.

The cytoplasmic phosphorylation state or the ratio (ATP)/[(ADP) · (Pi)] determines the amount of energy available for all cytoplasmic reactions that utilize the terminal phosphate of ATP. The most important synthetic and ion transport reactions occur in the cytoplasmic compartment, and the energy available to them will be determined by the relationship $G = G' + RT \ln \{(\text{ATP})/[(\text{ADP}) \cdot (\text{Pi})]\}$. There is considerable evidence that the cytoplasmic phosphorylation state is much higher than that of the mitochondria or that obtained by direct measurement in whole tissue.[83-85] It has been calculated that the free cytoplasmic phosphorylation state is about 30,000 M^{-1} compared to the measured cell content of 1,320 M^{-1}.[83]

Interestingly, the cytoplasmic phosphorylation potential of brain is similar to that calculated for muscle.* Both muscle and brain are subject to marked, transient changes in the rate of energy utilization, and both contain creatine phosphate, which may play a special role in transfering energy from the mitochondria directly to specialized sites of action. In muscle, the creatine phosphokinase system transports energy from mitochondria to myofibrils. There are two separate isoenzymes of creatine phosphokinase in heart, brain, and skeletal muscle.[86] One is located in the intermembrane space of the mitochondria. In muscle, the other isoenzyme is located on the meromyosin head, primarily in the M band. Thus, creatine phosphate is formed in the intermembrane space and transports energy exclusively to the myofibrils where the terminal phosphate is transferred to ADP. Because creatine phosphate is not a substrate for other enzymes, its energy is available for the exclusive use of myofibrils.[87-89] It is tempting to conceive of a similar situation in brain, with the difference that the energy supply is reserved for Na^+-K^+ ATPase, whose continued activity is essential for normal nerve cell function. There are two isoenzymes of creatine phosphokinase in brain,[86] but the location of the extramitochondrial enzyme has not been definitively established.[90,91] Whether the creatine phosphokinase system in the brain serves to direct energy from the mitochondria to specific sites of utilization as has been suggested[89] or simply serves as a high-energy phosphate buffer remains to be clarified.

It is now apparent that normal physiological circumstances demand rapid and marked alterations in activity and metabolism in specific CNS structures. For instance, there are focal increases in cortical blood flow of 40–50% in response to such normal activities as reading, finger movements, speaking, listening to music, etc.[92,93] The rate of accumulation of deoxyglucose (a glucose analogue that accumulates at similar rates to glucose) was nearly doubled in normal humans who were viewing an ordinary scene.[94] These were sharply localized alterations in specific cerebral areas, and it is quite possible that, if

---

* The free energy of hydrolysis of glucose in CNS is about 2847 kJ/mol, whereas the free energy of cytoplasmic ATP hydrolysis is about 58.85 kJ/mol.[83] Assuming that 38 ATP are generated per mole of glucose, this represents a recovery of 2236 kJ/mol or 79% of the available energy. The remaining energy, presumably dissipated as heat by the nonequilibrium reactions, probably represents the net cost of metabolic control.

the techniques had greater resolution, even larger changes in the rate of metabolism would be detected. To deal with these changes in energy metabolism, the enzyme activities of the glycolytic pathway and Krebs cycle must be closely coordinated. The principles involved in the control of glucose metabolism have been rigorously developed by Newsholme whose writings should be consulted for details.[95,96] Although many studies leading to discoveries about regulation were done on muscle, there is little doubt that the principles also apply to other tissues, including brain. Newsholme points out the importance of understanding the roles of flux-generating and other nonequilibrium reactions on the one hand and equilibrium reactions on the other hand. It is through the nonequilibrium reactions that the rate and direction of substrate utilization are controlled.

7.1. Near-Equilibrium Reactions

The forward and reverse reaction rates of a near-equilibrium reaction are similar, and both rates are much greater than flux through the reaction. Thus, near-equilibrium reactions respond rapidly and efficiently to changes in flux through substrate and product concentrations, and it is most unlikely that control is exerted through these reactions. Most reactions are of the near-equilibrium type (See Table I and Fig. 1).

7.2. Flux-Generating Reactions

A flux-generating reaction is situated at the beginning of a pathway. The K_m of the enzyme is considerably below the substrate concentration *in vivo*; i.e., the enzyme is saturated, and changes in substrate concentration alone have no effect on the rate. The reaction is, therefore, nonequilibrium, and all other reactions in the pathway must be regulated to respond to the flux generated by this reaction.

7.3. Other Nonequilibrium Reactions

Other nonequilibrium reactions are catalyzed by regulatory enzymes situated at strategic branch points in a pathway in order to direct flux in the proper direction. The enzymes involved are not saturated with substrate and thus respond to some degree to changes in substrate concentration. More importantly, they are controlled by allosteric factors.

7.4. Regulatory Enzymes

Identification of regulatory enzymes is based on the principle that their activities are affected by factors other than substrate availability. If the concentrations of substrates and products are expressed as the mass action ratio (products:substrates), the resulting number will be much less than the equilibrium constant of the reaction. Another approach, first suggested by Krebs,[97] relies on the fact that when a controlled reaction is stimulated, the substrates will decrease in concentration while the concentrations of the products stay

Table I
Metabolic Pathways in Brain[a]

Enzyme	E.C.#	Reaction	Activity[b] (μmol·min^{-1}·g^{-1})
Glycolysis			
Hexokinase	2.7.1.1	Glucose + ATP = Glucose 6-phosphate + ADP	28.[270]
Phosphoglucoisomerase	5.3.1.9	Glucose 6-phosphate = Fructose 6-phosphate	230.[270]
Phosphofructokinase	2.7.1.11	Fructose 6-phosphate + ATP = Fructose bisphosphate + ADP	36.[270]
Aldolase	4.1.2.13	Fructose bisphosphate = Glyceraldehyde 3-phosphate + Dihydroxyacetone phosphate	12.[270]
Triosephosphate isomerase	5.3.1.1	Dihydroxyacetone phosphate = Glyceraldehyde 3-phosphate	990.[271]
Glyceraldehyde phosphate dehydrogenase	1.2.1.12	Glyceraldehyde 3-phosphate + NAD$^+$ + Pi = 1,3-Diphosphoglycerate + NADH + H$^+$	130.[270]
Phosphoglycerate kinase	2.7.2.3	1,3-Diphosphoglycerate + ADP = 3-Phosphoglycerate + ATP	1100.[270]
Phosphoglyceromutase	2.7.5.3	3-Phosphoglycerate = 2-Phosphoglycerate	220.[270]
Enolase	4.2.1.11	2-Phosphoglycerate = Phosphoenolpyruvate + H$_2$O	62.[270]
Pyruvate kinase	2.7.1.40	Phosphoenolpyruvate + ADP = Pyruvate + ATP	170.[270]
Lactate dehydrogenase	1.1.1.27	Pyruvate + NADH + H$^+$ = Lactate + NAD$^+$	140.[270]
Krebs cycle			
Pyruvate dehydrogenase	1.2.4.1[c]	Pyruvate + CoA + NAD$^+$ = Acetyl-CoA + CO$_2$ + NADH + H$^+$	1.1–3.1[159–161]
Citrate synthase	4.1.3.7	Oxaloacetate + Acetyl-CoA + H$_2$O = Citrate + CoA	41.–58.[173,276]
Aconitase	4.2.1.3	Citrate = cis-Aconitate + H$_2$O = Isocitrate	2.0[213]
Isocitrate dehydrogenase (NAD)	1.1.1.41	Isocitrate + NAD$^+$ = α-Oxoglutarate + CO$_2$ + NADH + H$^+$	8.0[213]
α-Oxoglutarate dehydrogenase	1.2.4.2[c]	α-Oxoglutarate + NAD$^+$ + CoA = Succinyl-CoA + CO$_2$ + NADH + H$^+$	3.4[277]
Succinate thiokinase	6.2.1.4	Succinyl-CoA + GDP + Pi = Succinate + CoA + GTP	—
Succinate dehydrogenase	1.3.99.1	Succinate + FAD = Fumarate + FADH$_2$	19.[262]
Fumarase	4.2.1.2	Fumarate + H$_2$O = Malate	38.[278,d]
Malate dehydrogenase	1.1.1.37	Malate + NAD$^+$ = Oxaloacetate + NADH + H$^+$	47.–350.[173,215]
Pentose phosphate pathway			
Glucose 6-phosphate dehydrogenase	1.1.1.49	Glucose 6-phosphate + NADP$^+$ = 6-Phosphogluconolactone + NADPH + H$^+$	0.6–0.8[272,273]
6-Phosphogluconate dehydrogenase	1.1.1.43	6-Phosphogluconate + NADP$^+$ = Ribulose 5-phosphate + CO$_2$ + NADPH + H$^+$	0.4–1.3[272,273]
Transketolase	2.2.1.1	Ribulose 5-phosphate + Xylulose 5-phosphate = Sedoheptulose	0.5–0.7[272,273]

Enzyme	EC number	Reaction	Activity[b]
Transaldolase	2.2.1.2	Sedoheptulose 7-phosphate + Glyceraldehyde 3-phosphate = Fructose 6-phosphate + Erythrose 4-phosphate	0.6[272]
Glycogen metabolism			
Glycogen synthetase	2.4.1.11	$(Glucosyl)_n$ + UDPG = $(Glucosyl)_{n+1}$ + UDP	0.2[274]
Glycogen phosphorylase	2.4.1.1	$(Glucosyl)_n$ + Pi = $(Glucosyl)_{n-1}$ + Glucose 1-phosphate	2.0[274]
Phosphoglucomutase	2.7.5.1	Glucose 1-phosphate = Glucose 6-phosphate	11.[275]
Gluconeogenesis + CO_2 fixation			
Glucose 6-phosphatase	3.1.3.9	Glucose 6-phosphate + H_2O = Glucose + Pi	0.3–1.6[223,226]
Fructose bisphosphatase	3.1.3.11	Fructose bisphosphate + H_2O = Fructose 6-phosphate + Pi	0.04[236]
Malic enzyme	1.1.1.40	Pyruvate + CO_2 + NADPH + H^+ = Malate + $NADP^+$	0.9–1.7[212,213]
Phosphoenolpyruvate carboxykinase	4.1.1.32	Phosphoenolpyruvate + GDP + CO_2 = Oxaloacetate + GTP	0.06–0.3[213,214]
Pyruvate carboxylase	6.4.1.1	Pyruvate + CO_2 + ATP + H_2O = Oxaloacetate + ADP + Pi	0.4–1.4[212,213,215]
Isocitrate dehydrogenase (NADP)	1.1.1.42	α-Oxoglutarate + CO_2 + NADPH + H^+ = Isocitrate + $NADP^+$	1.2–3.5[212,213,215]
Aminotransferases			
Aspartate aminotransferase	2.6.1.1	Glutamate + Oxaloacetate = α-Oxoglutarate + Aspartate	33.–190.[215,279]
Alanine aminotransferase	2.6.1.2.	Glutamate – Pyruvate = α-Oxoglutarate + Alanine	5.[279]
Ammonium metabolism			
Glutamate dehydrogenase (NAD)	1.4.1.2	Glutamate + H_2O + NAD^+ = α-Oxoglutarate + NH_3 + NADH + H^+	16.–23.[173,280]
Glutamate dehydrogenase (NADP)	1.4.1.4	Glutamate + H_2O + $NADP^+$ = α-Oxoglutarate + NH_3 + NADPH – H^+	20.[173]
Glutamine synthetase	6.3.1.2	Glutamate + ATP + NH_3 = Glutamine + ADP + Pi	0.7[292]
γ-Aminobutyrate pathway			
Glutamate decarboxylase	4.1.1.15	Glutamate = γ-Aminobutyrate + CO_2	0.7–0.8[257,280]
γ-Aminobutyrate transaminase	2.6.1.19	γ-Aminobutyrate. + α-Oxoglutarate = Succinate semialdehyde + Glutamate	0.3[281]
Succinate semialdehyde dehydrogenase	1.2.1.16	Succinate semialdehyde + $NAD(P)^+$ + H_2O = Succinate + $NAD(P)H$ + H^+	0.2[281]

[a] Adapted from ref. 291.

[b] Activities are given per gram fresh weight and were corrected for 38°; Q_{10} = 1.7. It was assumed that 1 g of brain contains 100 mg total protein or 20 mg mitochondrial protein.[213] In some cases, a range of activities from different authors is given. Except where indicated, the data are for rat or mouse.

[c] Covers only part of reaction.

[d] Rabbit.

the same or increase (this is essentially the same as the "crossover" theorem often applied in mitochondrial studies). The application of these principles has helped to identify most of the important controlled reactions in the cytoplasm. It is not as useful for reactions occurring in the mitochondria because there is greater uncertainty in the determination of intramitochondrially located substrates. In subsequent sections, current knowledge regarding controlling enzymes is summarized. Near-equilibrium reactions or reactions controlled solely by substrate availability are not discussed (except glucose transport) because they are not considered to be of major importance in the control of energy metabolism.

8. GLYCOLYSIS AND GLUCOSE TRANSPORT

8.1. Glucose Transport

To reach the nerve cells from the blood, glucose must pass through several cell membranes (see Section 3). The primary resistance to movement probably resides at the capillary endothelial cell layer. Glucose transport has been the subject of many comprehensive articles, to which the reader is referred.[98–102] Transport of glucose across the endothelial cell membranes is believed to occur by an energy-independent, stereospecific, saturable, and symmetrical transport system which seems to be equilibrative, not concentrative. Most investigators have found that the process could be described by Michaelis–Menten kinetics with a K_m of 5–9 mM (although values ranging from 2 to 28 have been reported[98] and a V_{max} of 1.6–2.8 μmol $\cdot$ min^{-1} $\cdot$ g^{-1} in experimental animals. These constants should be considered as a purely functional description of the transport kinetics, because at least two cell membranes as well as an intervening layer of cytoplasm are involved.

Studies of glucose transport across glial and nerve cell membranes are more difficult to carry out, but results obtained so far indicate that the systems involved are likewise equilibrative, energy-independent, stereospecific mechanisms.[98] Their activity in relation to cellular requirements for glucose is relatively high, thereby insuring near equilibrium between intra- and extracellular glucose concentrations.[98] (For a different viewpoint, see ref. 101.)

Under normal circumstances, when plasma glucose concentration is 6–8 mM, glucose influx is more than sufficient to meet brain demands. However, when the plasma concentration drops below 3 mM, the rates of influx and utilization are approximately equal.[13,14] Below this concentration, the rate of utilization is limited by the rate of influx, brain content of glucose becomes nil, and the brain must consume alternative endogenous substrates (see above). At this point, the flux-limiting step of glycolysis ceases to be the hexokinase reaction, and the site of control moves to the liver which maintains circulating glucose concentrations.

Regional differences in the rate of glucose transport have not been described, but there is good reason for believing that they exist. The glucose content of various brain structures is about the same,[103,104] yet it is well established that glucose metabolism is substantially different in different regions (see L. Sokoloff, this volume). This apparent discrepancy could be explained

decreases the level of an effector in the liver, i.e., that phosphorylation is not involved.[133,140] This effector, which activates phosphofructokinase, has been identified as fructose 2,6-bisphosphate.[141–143] So far, these studies have been carried out only in the liver; whether similar mechanisms operate in brain remains to be established.

8.4. Pyruvate Kinase

Pyruvate kinase is known to be a regulatory enzyme in brain.[46] The flux through pyruvate kinase, expressed on a molar equivalent basis, can exceed that through phosphofructokinase, depending on the activity of the pentose phosphate pathway which produces glyceraldehyde phosphate without mediation by phosphofructokinase. Similarly, when glycogen is degraded to produce glucose, flux through pyruvate kinase can be greater than that through hexokinase. A controlled step at this point in the glycolytic pathway is necessary when gluconeogenic activity occurs with net production of PEP from pyruvate. Thus, pyruvate kinase must be adjusted appropriately to prevent reconversion of PEP to pyruvate and the occurrance of a "futile" cycle (Section 14).

There are several isozymes of pyruvate kinase,[131,144] one of which seems to be unique to brain.[145,146] ATP inhibits brain pyruvate kinase competitively with ADP but not PEP, whereas with the muscle enzyme, the inhibition is competitive with both substrates.[145] Brain pyruvate kinase is also inhibited by phenylalanine, a process that may be reversed by alanine, serine, and cystine.[147,148] Such inhibition was suggested to be involved in the brain damage observed in phenylketonuria,[149,150] but this has been disputed by others who showed no inhibition of brain pyruvate kinase *in vivo*.[146,151] Creatine phosphate was believed to be inhibitory to muscle pyruvate kinase and to act synergistically with ATP[152]; however, this effect is now thought to be caused by a contaminant.[153,154] Fructose 1,6-bisphosphate, which activates liver pyruvate kinase,[144] appears to have little effect on brain pyruvate kinase.[145,146] As with phosphofructokinase, pyruvate kinase may also be regulated by phosphorylation. In liver, glucagon decreases pyruvate kinase activity by stimulating enzyme phosphorylation.[134,155–158]

9. THE KREBS CYCLE

The conversion of glucose or glycogen to lactate, whether by glycolysis or the pentose phosphate pathway, occurs entirely within the cytoplasm. Only a fraction (about 5%) of the available energy is extracted, in the form of two molecules of ATP per molecule of glucose or three per molecule of glycogen. The complete oxidation of pyruvate occurs in the Krebs cycle within the mitochondria, resulting in the production of four molecules of NADH, one molecule of $FADH_2$, and one molecule of substrate-linked ATP. It is the oxidation of the reduced pyridine and flavin nucleotides by the respiratory chain that yields the major portion of energy. The primary function of the Krebs cycle is therefore the orderly oxidation of pyruvate.

A secondary function is the production of suitable carbon skeletons for the synthesis of certain amino acids, notably glutamate, glutamine, γ-aminobutyric acid, and aspartate. This requires that the Krebs cycle be connected with the glycolytic pathway not only by oxidative enzymes such as pyruvate dehydrogenase but also by anaplerotic reactions such as that mediated by pyruvate carboxylase (see below).

9.1. Pyruvate Dehydrogenase

Pyruvate dehydrogenase catalyzes the flux-generating step of the Krebs cycle. The total activity of pyruvate dehydrogenase is considerably less than that of any of the glycolytic enzymes[159,160] (see Table I), and this is true at all ages.[161] There are pathological circumstances (e.g., seizures) in which the energy requirement is so great that the rate of glycolysis may exceed the capacity of pyruvate dehydrogenase, and any additional energy must then be obtained through aerobic glycolysis or the oxidation of other substrates.

Pyruvate dehydrogenase is located within the inner mitochondrial membrane[162] and therefore responds to changes in concentrations of substrates and effectors in the mitochondrial matrix rather than cytoplasmic concentrations. This multienzyme complex requires thiamine pyrophosphate, lipoic acid, coenzyme A, FAD, and NAD for activity. The first step of the reaction, which involves decarboxylation, has the greatest free energy change and is probably rate limiting. The overall reaction is irreversible.

The regulation of pyruvate dehydrogenase has been extensively investigated in mammalian tissues other than brain (for reviews see refs. 162,163). Regulation occurs by interconversion between the inactive phosphorylated and an active nonphosphorylated form.[164-167] Phosphorylation is inhibited by several substances acting on the kinase that catalyzes the reaction. These include pyruvate and other carboxylic acids, PEP, thiamine pyrophosphate, and pyrophosphate, all of which inhibit the kinase, thereby stimulating pyruvate dehydrogenase activity.[163] ADP is a relatively weak kinase inhibitor by competition with ATP.[163] Dephosphorylation (stimulation) by the phosphatase is increased by Ca^{2+} and Mg^{2+} which also inhibit the kinase.[159,168-170] Pyruvate dehydrogenase complex in brain has been studied in purified form, and its properties appear to be similar to those of other mammalian tissues, although minor differences were found.[159-161,168,171-173] Activity of the brain enzyme, unlike that in other tissues, does not decrease during starvation,[168,172] probably reflecting the dependence of the CNS on glucose utilization.

9.2. Citrate Synthase

There are relatively few studies of CNS citrate synthase other than determination of its activity in homogenates[174] and mitochondria.[173] The reaction is readily reversible, and whether it is truly regulatory remains open. Citrate synthase has been extensively studied in other tissues where it has been shown that its activity is affected by its substrates oxaloacetate and acetyl CoA[175-179] (for review see ref. 180). Physiological concentrations of NAD, NADPH, and succinyl CoA are inhibitory.[176-178,181] Although the various factors listed above may alter the activity of citrate synthase, it seems most probable that it responds primarily to variations in substrate concentration.[4]

9.3. Isocitrate Dehydrogenase

Mammalian tissues contain two isocitrate dehydrogenases, an NAD-dependent enzyme located in the mitochondria and an NADP-dependent enzyme located in the mitochondria and cytoplasm. In the rat, the NAD isocitrate dehydrogenase predominates in brain, whereas the NADP form predominates in other tissues.[182] The two enzymes have different catalytic, physical, and regulatory properties.[183] Evidence has been provided that the reaction catalyzed by these enzymes in the CNS is far from equilibrium,[174] and in a study of the effect of altering mouse brain metabolic activity by various means, the results were consistent with possible control at this reaction.[184] The kinetic properties of NAD isocitrate dehydrogenase suggest that modulation of its activity may influence the *in vivo* activity of the Krebs cycle.[179,182,185–189] The properties of rat brain NAD isocitrate dehydrogenase are similar to those of the enzyme from other tissues.[190] Therefore, results obtained using other tissues are probably applicable to the brain. The activity of NAD-dependent isocitrate dehydrogenase is allosterically affected by several metabolites including ADP, ATP, NADH, and NADPH.[182,183,185,190,191] ADP activates the enzyme, whereas NADH, NADPH, and ATP are inhibitory.

9.4. 3-Oxoglutarate Dehydrogenase

3-Oxoglutarate dehydrogenase does not seem to have been examined in brain tissue. The enzyme complex appears to have similar properties to pyruvate dehydrogenase, and the reaction it catalyzes is therefore expected to be irreversible. Its position enables control over the direction of 3-oxoglutarate metabolism, either towards the synthesis of γ-aminobutyrate or glutamine or further oxidation in the Krebs cycle. There are three components, a dehydrogenase, lipoate succinyl transferase, and lipoamide dehydrogenase, arranged in a high-molecular-weight complex.[192,193] As with pyruvate dehydrogenase, thiamine pyrophosphate is needed for activity. Although α-oxoglutarate dehydrogenase and pyruvate dehydrogenase are similar, their regulation may occur through different mechanisms. As yet, there is no evidence that the 3-oxoglutarate dehydrogenase complex is controlled by a phosphorylation–dephosphorylation mechanism.[167] The activity of 3-oxoglutarate dehydrogenase is inhibited by NADH, succinyl CoA, and ATP and stimulated by ADP and Ca^{2+} (in contrast to the situation with pyruvate dehydrogenase, Ca^{2+} does not appear to act through a kinase; rather, it decreases the apparent K_m for 3-oxoglutarate).[194] It has been suggested that the primary regulatory factors are the NADH/NAD$^+$ and succinyl CoA/CoA-SH ratios rather than the ATP/ADP ratio.[181,195]

10. FLUX THROUGH THE GLYCOLYTIC PATHWAY AND KREBS CYCLE IN VIVO

Normally there is a near stoichiometric relationship between the glycolytic pathway and the Krebs cycle. The rate of glycolysis slightly exceeds that of the Krebs cycle, resulting in a small net production of lactate which leaves the brain across the blood–brain barrier. In a study of arteriovenous differences of glucose and lactate in overnight-fasted healthy young men, about 17% of the total glucose was converted to lactate. The discrepancy between the rates of flux through the two pathways appears to be a function of the overall rate of oxidative metabolism. For example, in barbiturate-anesthetized rats, arteriovenous differences of glucose and lactate demonstrated that only 2–4% of the glucose taken up appeared as lactate in circulation.[40,48] Under these circumstances, when glucose utilization is reduced by 50–60%,[103] the brain content of lactate decreases markedly.[196] Conversely, when oxidative metabolism is stimulated as during experimentally induced seizures, the glycolytic rate exceeds the capacity of pyruvate dehydrogenase and the Krebs cycle, and brain lactate accumulates rapidly.[197,198] This occurs even when oxygen supply is maintained and tissue oxygenation is not believed to be limiting.

Accumulation of lactate in the brain under these circumstances is primarily a consequence of the low capacity of the blood–brain barrier for the removal of lactate. The transport mechanism responsible has a V_{max} of 0.19 $\mu mol \cdot min^{-1} \cdot g^{-1}$.[54,199] Assuming a rate of glucose utilization of 0.7 $\mu mol \cdot min^{-1} \cdot g^{-1}$ and production of pyruvate of 1.4 $\mu mol \cdot min^{-1} \cdot g^{-1}$, it can be seen that when diversion of glycolytic flux into lactate exceeds 14% of normal flux, lactate will begin to accumulate. During bicuculline-induced seizures, when glucose utilization reaches 2.1 $\mu mol \cdot min^{-1} \cdot g^{-1}$,[197,200] brain lactate content may rise to about 10 $\mu mol/g$. A serious consequence of lactate accumulation is the resulting reduction of the cytoplasmic redox potential and lowering of the intracellular pH. It is perhaps to prevent dangerous increases in lactate content that the enzymes necessary for gluconeogenesis are present in brain (see below).

When there is an excessive demand for energy or when oxidative metabolism is restricted because of an insufficient supply of oxygen, glycolysis can provide supplementary energy. This is possible because of the very high potential for flux through the pathway, amounting to several times the rate of normal flux and greatly exceeding the capacity of the Krebs cycle (see Table I). Anaerobic glycolysis can supply about 40–70% of the normal rate of ATP production under conditions of near-maximal stimulation.*

---

* The proportion of the total energy requirement that can be supplied by glycolysis can be estimated from the "Meyerhoff quotient," which resulted from the observation by Meyerhoff and Warburg that consumption of one molecule of oxygen prevented the formation of two molecules of lactate in most tissues: Meyerhoff quotient = (anaerobic glycolysis − aerobic glycolysis)/oxygen consumption. Because one molecule of oxygen yields up to six molecules of ATP compared to one from lactate, a Meyerhoff quotient of 6 indicates that glycolysis alone can provide as much energy as the complete oxidative pathway. A Meyerhoff quotient of 2 means that total energy production is decreased by two-thirds. The ability of glycolysis to meet the cerebral energy requirement in extreme conditions can be calculated from the data of Lowry et al.,[196] who measured the initial rate of lactate production immediately after decapitation. They found that the rate of glycolysis rose from an estimated resting rate of 0.76 $\mu mol \cdot min^{-1} \cdot g^{-1}$ to a value between 5.6 and 8.7 $\mu mol \cdot min^{-1} \cdot g^{-1}$, corresponding to an increase of 7.4 to 11.4 times the normal rate. Taking oxygen consumption to be 3.89 $\mu mol \cdot min^{-1} \cdot g^{-1}$ (assuming that 85% of the normal rate of glucose utilization supplies oxidative metabolism), the Meyerhoff quotient is between 2.5 and 4.1.

The ability of anaerobic glycolysis to substitute for oxidative metabolism has, of course, limitations. For instance, it was shown[201] that when arterial oxygen tension was decreased to 50 torr, cerebral lactate content rose, indicating that glycolysis was stimulated above the rate required for oxidative metabolism. At 40 torr, brain creatine phosphate decreased, perhaps reflecting a lower intracellular pH or decreased oxidative phosphorylation. There was no change in adenine nucleotide content until oxygen tension dropped to 20 torr and brain lactate content had risen to 20 μmol/g or more.[201] These experiments demonstrate the important role that glycolysis can assume in the maintenance of a normal energy supply, as well as its limitations. Because of the relative impermeability of the blood–brain barrier to lactate, lactate will eventually accumulate in brain and lead to H^+ concentrations incompatible with normal function. Thus, anaerobic glycolysis must be viewed as a supplement to oxidative metabolism useful only for limited periods of time.

11. THE CRABTREE EFFECT

The inhibition of oxidative respiration by high concentrations of glucose, referred to as the Crabtree effect, is found in cells with a high glycolytic capacity such as embryonic tissue, tumor cells, and leukocytes. The effect has been likened by Krebs[202] to a competition between glycolysis and the respiratory chain for ADP and phosphate. Evidence for the Crabtree effect in nervous tissue is scanty. It has been observed with isolated retinal tissue[203] where it decreases with increasing age. In humans, intravenous glucose infusion increased CMRglucose slightly but without affecting $CMRO_2$.[204] No evidence for the effect could be found in barbiturate-anesthetized sheep and rabbits.[77] In this study, cerebral oxygen and glucose consumption were found to be independent of the glucose concentration over a range of 1–30 mM. However, in another study of rats, the rate of glucose consumption rose, and $CMRO_2$ fell when plasma glucose was increased from 14 to 44 mM.[205] This implies that inhibition of the hexokinase and phosphofructokinase reactions was partially overcome by increasing the substrate concentration. Even if it is present, the currently available evidence suggests that the Crabtree effect is not of great importance in the CNS.

12. CO₂ FIXATION

Anaplerotic reactions are needed in the brain to replenish Krebs cycle intermediates lost by synthesis of amino acids (aspartate, glutamate, glutamine, γ-aminobutyric acid) as well as the spontaneous decarboxylation of oxaloacetate to pyruvate and CO_2 which occurs continuously at a finite rate. Such anaplerotic reactions are believed to occur by CO_2 fixation which has been demonstrated in brain.[206–209] The ability of brain to replenish Krebs cycle intermediates appears to be considerable. For example, after an acute rise in circulating ammonia, net brain glutamine synthesis was measured at 0.33

$\mu mol \cdot min^{-1} \cdot g^{-1}$ for 5 min without detectably changing the cerebral content of Krebs cycle intermediates.[210] Without anaplerotic reactions, Krebs cycle intermediates would have been depleted within 2 or 3 min under such circumstances.

Theoretically, three reversible reactions could significantly contribute to CO_2 fixation:

$$\text{pyruvate} + HCO^{3-} + ATP \underset{\text{pyruvate carboxylase}}{\rightleftharpoons} \text{Oxaloacetate} + ADP + Pi \quad [1]$$

$$PEP + CO_2 + GDP \underset{\text{phosphoenolpyruvate carboxykinase}}{\rightleftharpoons} \text{Oxaloacetate} + GTP \quad [2]$$

$$\text{pyruvate} + CO_2 + NADPH \underset{\text{malic enzyme}}{\rightleftharpoons} \text{Malate} + NADP^+ \quad [3]$$

The finding that oxaloacetate was the primary reaction product of CO_2 fixation in brain suggested that phosphoenolpyruvate carboxykinase or pyruvate carboxylase was involved.[211] All three enzymes have been found in brain tissue.[212–214] However, the evidence suggests that pyruvate carboxylase is the major anaplerotic enzyme in the brain. Fixation of CO_2 by rat brain mitochondria by either phosphoenolpyruvate carboxykinase or malic enzyme was found to be negligible,[213,215] and the kinetics for carboxylation by these two enzymes appears to be very unfavorable compared to decarboxylation.[209,214,216]

It has also been suggested that cytosolic NADP-dependent isocitrate dehydrogenase may play a role in the formation of citrate from α-oxoglutarate and bicarbonate in nervous tissue.[208,217] However, carboxylation activity by this enzyme in brain tissue was found to be very low compared to that of pyruvate carboxylase.[213,215]

12.1. Pyruvate Carboxylase

Pyruvate carboxylase from rat brain has been purified and some of its properties investigated.[215,218] It is exclusively mitochondrial[212] and appears to be similar to that found in the liver. Acetyl CoA, ATP, Mg^{2+}, and biotin are needed for activity, whereas ADP and PEP are inhibitory. The K_m for pyruvate is about 0.5 mM; thus, it has been suggested that pyruvate concentrations may control the rate of carboxylation by brain.[215] The ATP/ADP ratio may also have a regulatory function.

12.2. Malic Enzyme

Two distinct forms of malic enzyme have been separated from rat and bovine brain,[212,213,216] one cytosolic and one mitochondrial. The two forms have different kinetic as well as electrophoretic and chromatographic properties.[216] With both forms, the rate of decarboxylation was about ten times the rate of carboxylation in preparations from rat brain.[213] Properties of malic enzyme from other tissues have been extensively investigated (for review see ref. 219).

12.3. Phosphoenolpyruvate Carboxykinase

Phosphoenolpyruvate carboxykinase in brain is found only in the mitochondria, unlike the liver where it is present in the cytosol.[213,214] Distribution in other tissues also appears to differ among different species.[220] The brain enzyme has been obtained in a partially purified form, and some of its properties described.[214] In general, its properties were similar to those of the liver enzyme. The large Michaelis constant for PEP in the direction of CO_2 fixation suggested that it would be more important for the decarboxylation of oxaloacetate and formation of PEP, i.e., gluconeogenesis, than for the reverse reaction.

13. GLUCONEOGENESIS

All of the enzymes necessary for gluconeogenesis are present in the CNS in significant amounts. However, the brain cannot be considered seriously as a gluconeogenic organ since the activities of most of the enzymes involved are relatively low compared to, for example, those in the liver of fasted mammals. The reason for the presence of gluconeogenic enzymes is not entirely clear. One function may be to remove the lactate that accumulates during periods of high metabolic activity. In extreme circumstances, lactate concentrations can reach 10 μmol/g or more in the CNS,[197,198] presumably because transport of lactate by the monocarboxylic acid transport system of the blood–brain barrier is relatively slow. During recovery, the synthesis of glucose would help to remove lactate and restore the intracellular pH. Conversion of lactate to glycogen may not be desirable because glycogen must be stored in the hydrated form which occupies considerable volume.

The complete gluconeogenic pathway uses pyruvate carboxylase, phosphoenolpyruvate carboxykinase, fructose 1,6-bisphosphatase, and glucose-6-phosphatase. The characteristics of the first two enzymes are discussed in the previous section on CO_2 fixation.

13.1. Glucose-6-Phosphatase

Glucose-6-phosphatase is a multifunctional enzyme capable of many other reactions in addition to hydrolyzing glucose-6-phosphate.[221] Its primary function in brain seems to be as a phosphohydrolase.[222] Glucose-6-phosphatase was long believed to be primarily an enzyme of liver and kidney, but it is now known to be widely distributed.[223,224] Its presence in brain has been demonstrated.[223,225–229] Among the enzymes involved in carbohydrate metabolism, glucose-6-phosphatase is unique in that it appears to be associated exclusively with intracellular membranes.[225,229] Glucose-6-phosphatase has been found in all parts of brain and all cell types; however, activity was most prominent in neurons[222] and in large cells such as parenchymal and pyramidal cells of the cerebral cortex, possibly because of their high content of smooth and rough endoplasmic reticulum.[227,229] Although the histochemical properties of the CNS enzyme are similar to those of liver and kidney,[227,230] the activity in rat cerebral cortical slices is about one-fifth that of the liver.[226]

The liver and kidney enzymes are affected by hormonal and dietary factors, but no specific mechanism for intracellular regulatory control has been identified. Suggestions have included regulation by membrane phospholipid environment[231] and possibly a phosphorylation–dephosphorylation mechanism.[232]

13.2. Fructose 1,6-Bisphosphatase

It was originally thought that fructose bisphosphatase was not present in the CNS,[233] but its presence in brain has been clearly demonstrated, and it has been obtained in purified form.[234-236] The activity of the enzyme is much lower than that of liver, but conversion of [^{14}C]fructose-6-phosphate has been demonstrated to occur in brain.[236] Unlike fructose bisphosphatase from liver, kidney, and muscle, brain fructose bisphosphatase was not inhibited by 5'-AMP (in this respect, the brain enzyme was similar to that from the bumblebee flight muscle, which participates in substrate cycling), and it does not need a divalent cation for activity.[236] Fructose bisphosphatase reacts to many effectors that regulate phosphofructokinase but in an opposing fashion. Since phosphofructokinase and fructose bisphosphatase catalyze opposing nonequilibrium reactions, the existence of substrate cycling with concomitant hydrolysis of ATP is a possibility. Liver fructose bisphosphatase is completely inhibited by fructose 2,6-bisphosphate which also affects phosphofructokinase.[237] Liver and kidney fructose bisphosphatase have been shown to be phosphorylated by a 3',5'-cyclic AMP-dependent protein kinase.[238-240] Although it has been suggested that there is a direct protein–protein interaction between rabbit liver phosphofructokinase and fructose bisphosphatase causing inhibition of phosphofructokinase,[129,241] this interaction may occur only *in vitro* by alteration of fructose bisphosphate concentrations.[242]

14. SUBSTRATE (FUTILE) CYCLES

There are at least three nonequilibrium steps in the metabolism of glucose at which so-called futile cycling may occur.[243,244] These consist of two opposing pathways in which the forward and reverse directions are catalyzed by separate enzymes. The reaction in one direction requires ATP (or GTP), whereas the other does not. There is no change in substrate or product concentration; thus, the only net change is the hydrolysis of ATP. Although this process appears to be without obvious benefit to the cell (hence the term "futile" cycling), it has been suggested that such cycles may be useful in the production of heat (not necessarily in the brain) and in the control of metabolic regulation.[96,245] Newsholme has therefore suggested that "substrate" cycling may be a better term to describe the process.[96] The three cycles that may exist in brain are: the glucose cycle, glucose → glucose-6-phosphate → glucose; the fructose 6-phosphate cycle, fructose 6-phosphate → fructose 1,6-bisphosphate → fructose 6-phosphate; and the phosphoenolpyruvate cycle, phosphoenolpyruvate → pyruvate → oxaloacetate → phosphoenolpyruvate.

There is good evidence that at least the glucose cycle operates in brain. Glucose-6-phosphatase is present in the CNS with sufficient activity to hydrolyze about 6% of the glucose-6-phosphate pool per minute.[246] In awake, undisturbed rats and humans, the rate of glucose 6-phosphate hydrolysis seems to be between 0.5 and 1% per minute.[246,247] This may be substantially increased during sleep and perhaps after surgical stress.[248] The effects of sleep were studied by injecting [^{14}C]2-deoxyglucose. Cerebral deoxyglucose phosphate was allowed to reach a virtual plateau at 45 min after infusion, whereupon the rats were allowed to sleep for 15 min. A 30% reduction in brain deoxyglucose phosphate content resulted, compared to the awake controls, indicating a stimulation of glucose-6-phosphatase activity during sleep.[248]*

An interesting hypothesis is that substrate cycling may be an effective method of amplifying the metabolic response of a controlling reaction. This may occur as follows. If the forward and reverse reactions occur simultaneously, the rate of flux through the reaction and its direction will be determined by the difference in the two rates. Therefore, a signal that stimulates one reaction and inhibits the other will have a greater effect than one that worked only on one reaction. Naturally, the amplification possible by such a mechanism is a function of the rate of cycling in relation to the rate of flux. The greater this ratio, the greater the potential amplification. The disadvantage of such a mechanism is of course, the expense incurred in unrecoverable energy (see refs. 96,245,249 for details). It does not appear likely that such a cycling process would be important in the regulation of the hexokinase and phosphofructokinase reactions because of the relatively low activity of the enzymes controlling the reverse reactions.

15. GLUTAMATE METABOLISM

15.1. Ammonium Removal

Glutamate may be metabolized to glutamine, thereby effectively removing ammonium[33] which may have toxic properties in the CNS. Glutamine synthetase is located exclusively within astrocytes.[34,214] The synthesis of glutamine removes intermediates from the Krebs cycle which must be replenished by anaplerotic reactions. Under extreme circumstances, such as acute ammonium toxicity, flux through glutamine synthesis and concomitant anaplerotic reactions may amount to as much as 30% of the flux through the glycolytic pathway.[210]

15.2. γ-Aminobutyrate Synthesis (GABA Pathway)

The metabolism of GABA in the CNS has been extensively studied.[35,251–255] The GABA pathway could be considered as an alternative route for the

---

* The hydrolysis of deoxyglucose phosphate by glucose 6-phosphatase, although slow, must be taken into account when one uses deoxyglucose for the quantitative determination of glucose utilization.[246,247]

oxidation of 3-oxoglutarate, which otherwise occurs in the Krebs cycle. This route is energetically less favorable, since the substrate-linked phosphorylation step (succinyl-CoA thiokinase) is bypassed. However, there is considerable evidence that GABA is synthesized and catabolized in structurally distinct regions,[253] indicating that its synthesis is necessary for specific functions. GABA is synthesized from glutamate by glutamate decarboxylase, found only in the CNS and retina, in an irreversible reaction. Both GABA and glutamate decarboxylase are found in higher concentrations in inhibitory GABAergic neurons; estimates of GABA content range from 12 to 150 μmol/g in whole brain.[256] Synthesis of GABA occurs from a "small glutamate compartment" at the expense of Krebs cycle intermediates. The CO_2 fixation reactions that must occur as a result likewise show evidence of similar compartmentation.[30,35,206] On release from nerve terminals, GABA is removed from extracellular fluid by high-affinity transport systems located in glial cells and possibly also by presynaptic and postsynaptic membranes (see L. Herz, Volume 2). Further metabolism of GABA occurs by transamination with α-oxoglutarate to regenerate glutamate and yield succinic semialdehyde, which is then oxidized to succinate, thereby reentering the Krebs cycle. These steps occur largely within glial cells and in neuronal perikarya and dendrites.[30,35] Whether some cells or cell compartments have the complete GABA pathway remains to be clarified. Interestingly, structures such as the pineal and posterior pituitary glands lack GABA inhibitory neurons but contain measurable amounts of endogenous glutamate decarboxylase and GABA transaminase, perhaps indicating another role for GABA besides neurotransmission.[256]

The activity of glutamate decarboxylase in the CNS at birth is about 10% of that in the adult.[257] Labeling studies have demonstrated that about 10% of the Krebs cycle flux proceeds through GABA.[35] However, it is conceivable that in some cerebral compartments a very large portion of the Krebs cycle intermediates are converted to GABA, which is then released. As a result, flux through the Krebs cycle between α-oxoglutarate and oxaloacetate may be greatly diminished in these cells.

16. PENTOSE PHOSPHATE PATHWAY

Metabolism of glucose can occur through the pentose phosphate pathway which operates in brain with a relatively limited capacity compared to glycolysis (see Table I). The pentose phosphate pathway cannot be considered an alternative to glycolysis since their functions are entirely different. The glycolytic pathway, in conjunction with the Krebs cycle, supplies energy and provides carbon skeletons for cellular constituents such as amino acids. The function of the pentose phosphate pathway is to maintain an effective NADPH redox state needed for many biosynthetic reactions and to maintain the glutathione redox state and the supply of ribose phosphate for nucleic acid synthesis. In adult animals, flux through the pentose phosphate pathway is only a few percent of the total glycolytic flux.[258-260] The relative activity of the pathway is greater in developing brain.[261] Perhaps as much as 50% of the glucose carbon may be metabolized via the pentose phosphate pathway during early development, when cell division and myelination are rapid.[262]

The flux-generating step of the pentose phosphate pathway is the reaction catalyzed by glucose 6-phosphate dehydrogenase. The product of this reaction, 6-phosphogluconolactone, is rapidly and irreversibly hydrolyzed to 6-phosphogluconate by gluconolactonase. All reactions past this point are freely reversible. Glucose 6-phosphate dehydrogenase is strongly inhibited by NADPH; physiological concentrations of NADPH are sufficient to almost completely inhibit enzyme action.[263] Inhibition of glucose 6-phosphate dehydrogenase can be overcome by oxidized glutathione.[263] Reduction of glutathione is catalyzed by glutathione reductase which is specific for NADPH. This therefore completes the feedback control loop. Glutathione is probably important in maintaining cellular sulfhydryl groups in a reduced state in addition to possibly acting as a free radical scavenger.[264] Thus, when spontaneous oxidation of cell constituents or production of free radicals threatens orderly cell function, reducing power generated by the pentose phosphate pathway may restore normal conditions.[265]

REFERENCES

1. Norberg, K., and Siesjö, B. K., 1974, *Acta Physiol. Scand.* **91**:154–164.
2. Neely, J. R., Bowman, R. H., and Morgan, H. E., 1969, *Am. J. Physiol.* **216**:804–811.
3. Ross, B. D., 1978, *Clin. Sci. Mol. Med.* **55**:513–521.
4. Krebs, H. A., 1970, *Advances in Enzyme Regulation,* Volume 8 (G. Weber, ed.), Pergamon Press, Oxford, pp. 335–353.
5. Cremer, J. E., and Heath, D. F., 1974, *Biochem. J.* **142**:527–544.
6. Hernandez, M. J., Brennan, R. W., Vannucci, R. C., and Bowman, G. S., 1978, *Am. J. Physiol.* **234**:R209–R215.
7. Mink, J. W., Blumenschine, R. J., and Adams, D. B., 1981, *Am. J. Physiol.* **241**:R203–R212.
8. Sokoloff, L., 1977, *Nutrition and the Brain,* Volume 1 (R. J. Wurtman and J. J. Wurtman, eds.), Raven Press, New York, pp. 87–139.
9. Crane, P. D., Braun, L. D., Cornford, E. M., Cremer, J. E., Glass, J. M., and Oldendorf, W. H., 1978, *Stroke* **9**:12–18.
10. Mata, M., Fink, D. J., Gainer, H., Smith, C. B., Davidsen, L., Savaki, H., Schwartz, W. J., and Sokoloff, L., 1980, *J. Neurochem.* **34**:213–215.
11. Whittam, R., and Blond, D. M., 1964, *Biochem. J.* **92**:147–158.
12. Hawkins, R. A., and Biebuyck, J. F., 1979, *Science* **205**:325–327.
13. Lewis, L. D., Ljunggren, B., Ratcheson, R. A., and Siesjö, B. K., 1974, *J. Neurochem.* **23**:673–679.
14. Lewis, L. D., Ljunggren, B., Norberg, K., and Siesjö, B. K., 1974, *J. Neurochem.* **23**:659–671.
15. Pardridge, W. M., 1977, *Nutrition and the Brain,* Volume 1, (R. J. Wurtman and J. J. Wurtman, eds.), Raven Press, New York, pp. 141–204.
16. Rapoport, S. I., 1976, *Blood–Brain Barrier in Physiology and Medicine*, Raven Press, New York.
17. Bradbury, M., 1979, *The Concept of a Blood–Brain Barrier*, John Wiley & Sons, New York.
18. Reese, T. S., and Karnovsky, M. J., 1967, *J. Cell. Biol.* **34**:207–217.
19. Brightman, M. W., and Reese, T. S., 1969, *J. Cell. Biol.* **40**:648–677.
20. Glees, P., 1972, *Metabolic Compartmentation in the Brain* (R. Balázs and J. E. Cremer, eds.), John Wiley & Sons, New York, pp. 209–234.
21. Oldendorf, W. H., Cornford, M. E., and Brown, W. J., 1977, *Ann. Neurol.* **1**:409–417.

22. Friede, R. L., 1970, *Triangle* **9**:165–173.
23. Wolff, J. R., 1970, *Triangle* **9**:153–164.
24. Hertz, L., 1981, *Glial and Neuronal Cell Biology* (E. A. Vidrio and S. Federoff, eds.), Alan R. Liss, New York, pp. 45–58.
25. Siesjö, B. K., 1978, *Brain Energy Metabolism* (B. K. Siesjö, ed.), John Wiley & Sons, New York, pp. 131–150.
26. Berl, S., and Clarke, D. D., 1975, *Metabolic Compartmentation and Neurotransmission. Relation to Brain Structure and Function* (S. Berl, D. D. Clarke, and D. Schneider, eds.), Plenum Press, New York, pp. xiii–xvii.
27. O'Neal, R. M., and Koeppe, R. E., 1966, *J. Neurochem.* **13**:835–847.
28. Cremer, J. E., Heath, D. F., Patel, A. J., Balazs, R., and Cavanagh, J. B., 1975, *Metabolic Compartmentation and Neurotransmission. Relation to Brain Structure and Function* (S. Berl, D. D. Clarke, and D. Schneider, eds.), Plenum Press, New York, pp. 461–478.
29. Cremer, J. E., Teal, H. M., Heath, D. F., and Cavanagh, J. B., 1977, *J. Neurochem.* **28**:215–222.
30. Balázs, R., Patel, A. J., and Richter, D., 1972, *Metabolic Compartmentation in the Brain* (R. Balázs and J. E. Cremer, eds.), John Wiley & Sons, New York, pp. 167–186.
31. Van den Berg, C. J., Matheson, D. F., Ronda, G., Reijnierse, G. L. A., Blockhuis, G. G. D., Kroon, M. C., Clarke, D. D., and Garfinkel, D., 1975, *Metabolic Compartmentation and Neurotransmission. Relation to Brain Structure and Function* (S. Berl, D. D. Clarke, and D. Schneider, eds.), Plenum Press, New York, pp. 515–543.
32. Berl, S., 1972, *Metabolic Compartmentation in the Brain* (R. Balázs and J. E. Cremer, eds.), John Wiley & Sons, New York, pp. 3–17.
33. Cooper, A. J. L., McDonald, J. M., Gelbard, A. S., Gledhill, R. F., and Duffy, T. E., 1979, *J. Biol. Chem.* **254**:4982–4992.
34. Martinez-Hernandez, A., Bell, K. P., and Norenberg, M. D., 1977, *Science* **195**:1356–1358.
35. Balázs, R., Machiyama, Y., and Patel, A. J., 1972, *Metabolic Compartmentation in the Brain* (R. Balázs and J. E. Cremer, eds.), John Wiley & Sons, New York, pp. 57–70.
36. Sokoloff, L., 1960, *Handbook of Physiology–Neurophysiology,* Volume 3, (J. Field, H. W. Magoun, and V. E. Hall, eds.), American Physiological Society, Washington, pp. 1843–1864.
37. Sloviter, H. A., and Kamimoto, T., 1970, *J. Neurochem.* **17**:1109–1111.
38. Owen, O. E., Morgan, A. P., Kemp, H. G., Sullivan, J. M., Herrera, M. G., and Cahill, G. F., Jr., 1967, *J. Clin. Invest.* **46**:1589–1595.
39. Robinson, A. M., and Williamson, D. M., 1980, *Physiol. Rev.* **60**:143–187.
40. Hawkins, R. A., Williamson, D. H., and Krebs, H. A., 1971, *Biochem. J.* **122**:13–18.
41. Mans, A. M., Biebuyck, J. F., and Hawkins, R. A., 1981, *J. Cereb. Blood Flow Metab.* **1**(Suppl.):90–91.
42. Moore, T. J., Lione, A. P., Sugden, M. C., and Regen, D. M., 1976, *Am. J. Physiol.* **230**:619–630.
43. Cahill, G. F., and Aoki, T. T., 1980, *Cerebral Metabolism and Neural Function* (J. V. Passonneau, R. A. Hawkins, W. D. Lust, and F. A. Welsh, eds.), Williams & Wilkins, Baltimore, pp. 234–242.
44. Ide, T., Steinke, J., and Cahill, G. F., Jr., 1969, *Am. J. Physiol.* **217**:784–792.
45. Patel, M. S., and Owen, O. E., 1977, *J. Neurochem.* **28**:109–114.
46. Rolleston, F. S., and Newsholme, E. A., 1967, *Biochem. J.* **104**:524–533.
47. Ziven, J. A., and Snarr, J. F., 1972, *J. Appl. Physiol.* **32**:664–667.
48. Ruderman, N. B., Ross, P. S., Berger, M., and Goodman, M. N., 1974, *Biochem. J.* **135**:1–10.
49. DeVivo, D. C., Leckie, M. P., Ferrendelli, J. S., and McDougal, D. B., Jr., 1978, *Ann. Neurol.* **3**:331–337.
50. Gottstein, U., Müller, W., Berghoff, W., Gärtner, H., and Held, K., 1971, *Klin. Wochenschr.* **49**:406–411.
51. Hawkins, R. A., and Biebuyck, J. F., 1980, *Cerebral Metabolism and Neural Function* (J. V. Passonneau, R. A. Hawkins, W. D. Lust, and F. A. Welsch, eds.), Williams & Wilkins, Baltimore, pp. 255–263.
52. Oldendorf, W. H., 1973, *Am. J. Physiol.* **224**:1450–1453.

53. Cremer, J. E., Braun, L. D., and Oldendorf, W. H., 1976, *Biochim. Biophys. Acta* **448**:633–637.
54. Cremer, J. E., Cunningham, V. J., Pardridge, W. M., Braun, L. D., and Oldendorf, W. H., 1979, *J. Neurochem.* **33**:439–445.
55. Pollay, M., and Stevens, F. A., 1980, *J. Neurosci. Res.* **5**:163–172.
56. Gjedde, A., and Crone, C., 1975, *Am. J. Physiol.* **229**:1165–1169.
57. Vannucci, R. C., Hellmann, J., Hernandez, M. J., and Vannucci, S. J., 1980, *Cerebral Metabolism and Neural Function* (J. V. Passonneau, R. A. Hawkins, W. D. Lust, and F. A. Welch, eds.), Williams & Wilkins, Baltimore, pp. 264–270.
58. Hernandez, M. J., Brennan, R. W., and Hawkins, R. A., 1980, *Cerebral Metabolism and Neural Function* (J. V. Passoneau, R. A. Hawkins, W. D. Lust, and F. A. Welsh, eds.), Williams & Wilkins, Baltimore, pp. 196–201.
59. Sloviter, H. A., Shimkin, P., and Suhara, K., 1966, *Nature* **210**:1334–1336.
60. Sloviter, H. A., and Suhara, K., 1967, *J. Appl. Physiol.* **23**:792–797.
61. Baños, G., Daniel, P. M., Moorhouse, S. R., and Pratt, O. E., 1973, *Proc. R. Soc. Lond.* [*Biol*] **183**:59–70.
62. Felig, P., Wahren, J., and Ahlborg, G., 1973, *Proc. Soc. Exp. Biol. Med.* **142**:230–231.
63. Van den Berg, C. J., and Van den Velden, J., 1970, *J. Neurochem.* **17**:985–991.
64. Berl, S., and Frigyesi, T. L., 1968, *J. Neurochem.* **15**:965–970.
65. Roberts, S., and Morelos, B. S., 1965, *J. Neurochem.* **12**:373–387.
66. Vignais, P. M., Gallagher, C. H., and Zabin, I., 1958, *J. Neurochem.* **2**:283–287.
67. Beattie, D. S., and Basford, R. E., 1966, *J. Biol. Chem.* **241**:1412–1418.
68. Bremer, J., 1963, *J. Biol. Chem.* **238**:2774–2779.
69. Wiener, R., Hirsch, H. J., and Spitzer, J. J., 1971, *Am. J. Physiol.* **220**:1542–1546.
70. Pardridge, W. M., and Mietus, L. J., 1980, *J. Neurochem.* **34**:463–466.
71. Allweis, C., Landau, T., Abeles, M., and Magnes, J., 1966, *J. Neurochem.* **13**:795–804.
72. Spitzer, J. J., 1973, *Physiologist* **16**:55–68.
73. Sarna, G. S., Bradbury, M. W. B., Cremer, J. E., Lai, J. C. K., and Teal, H. M., 1979, *Brain Res.* **160**:69–83.
74. Oldendorf, W. H., 1972, *Eur. Neurol.* **6**:49–55.
75. Kety, S. S., Woodford, R. B., Harmel, M. H., Freyhan, F. A., Appel, K. E., and Schmidt, C. F., 1948, *Am. J. Psychiatry.* **104**:765–770.
76. Geiger, A., Magnes, J., and Geiger, R. S., 1952, *Nature* **170**:754–755.
77. Pappenheimer, J. R., and Setchell, B. P., 1973, *J. Physiol. (Lond.)* **233**:529–551.
78. Agardh, C. D., Chapman, A. G., Nilsson, B., and Siesjö, B. K., 1981, *J. Neurochem.* **36**:490–500.
79. Knauff, H. G., Mark, D., and Mayer, G., 1961, *Hoppe Seylers Z. Physiol. Chem.* **326**:227–234.
80. Hinzen, D. H., Becker, P., and Müller, U., 1970, *Pfluegers Arch.* **321**:1–14.
81. Petersen, V. P., and Schou, M., 1955, *Acta Physiol. Scand.* **33**:309–315.
82. Miller, A. L., Hawkins, R. A., and Veech, R. L., 1975, *J. Neurochem.* **25**:553–558.
83. Veech, R. L., Lawson, J. W. R., Cornell, N. W., and Krebs, H. A., 1979, *J. Biol. Chem.* **254**:6538–6547.
84. Davis, E. J., and Lumeng, L., 1975, *J. Biol. Chem.* **250**:2275–2282.
85. Heldt, H. W., Klingenberg, M., and Milovancev, M., 1972, *Eur. J. Biochem.* **30**:434–440.
86. Jacobs, H., Heldt, H. W., and Klingenberg, M., 1964, *Biochem. Biophys. Res. Commun.* **16**:516–521.
87. Jacobus, W. E., and Lehninger, A. L., 1973, *J. Biol. Chem.* **248**:4803–4810.
88. Saks, V. A., Rosenshtraukh, L. V., Smirnov, V. N., and Chazov, E. I., 1978, *Can. J. Physiol. Pharmacol.* **56**:691–706.
89. Bessman, S. P., and Geiger, P. J., 1981, *Science* **211**:448–452.
90. Wood, T., and Swanson, P. D., 1964, *J. Neurochem.* **11**:301–307.
91. Swanson, P. D., 1967, *J. Neurochem.* **14**:343–356.
92. Larsen, B., Skinhoj, E., Soh, K., Endo, H., and Lassen, N. A., 1977, *Acta Neurol. Scand.* **56**(Suppl. 64):268–269, 280–281.
93. Lassen, N. A., Roland, P. E., Larsen, B., and Melamed, S., Soh, K., 1977, *Acta Neurol. Scand.* **56**(Suppl. 64):262–263, 274–275.

94. Phelps, M. E., Kuhl, D. E., and Mazziotta, J. C., 1981, *Science* **211**:1445–1448.
95. Newsholme, E. A., 1980, *FEBS Lett.* **117**:K121–K134.
96. Newsholme, E. A., and Start, C., 1973, *Regulation in Metabolism,* John Wiley & Sons, New York.
97. Krebs, H. A., 1957, *Endeavour* **16**:125–132.
98. Lund-Andersen, H., 1979, *Physiol. Rev.* **59**:305–352.
99. Lund-Andersen, H., 1980, *Cerebral Metabolism and Neural Function* (J. V. Passonneau, R. A. Hawkins, W. D. Lust, and F. A. Welsh, eds.), Williams & Wilkins, Baltimore, pp. 120–126.
100. Bachelard, H. S., 1975, *Brain Work* (D. H. Ingvar and N. A. Lassen, eds.), Munskgaard, Copenhagen, pp. 126–141.
101. Bachelard, H. S., 1980, *Cerebral Metabolism and Neural Function* (J. V. Passonneau, R. A. Hawkins, W. D. Lust, and F. A. Welsh, eds.), Williams & Wilkins, Baltimore, pp. 106–119.
102. Crone, C., 1980, *Microvasc. Res.* **20**:133–149.
103. Hawkins, R. A., Hass, W. K., and Ransohoff, J., 1979, *Stroke* **10**:690–703.
104. Shimada, M., Kihara, T., Watanabe, M., Kurimoto, K., 1977, *Neurochem. Res.* **2**:595–603.
105. Craigie, E. H., 1920, *J. Comp. Neurol.* **31**:429–464.
106. Sacks, W., 1969, *Handbook of Neurochemistry,* Volume 1 (A. J. Lajtha, ed.), Plenum Press, New York, pp. 301–321.
107. Grossbard, L., and Schimke, R. T., 1966, *J. Biol. Chem.* **241**:3546–3560.
108. Crane, R. K., and Sols, A., 1954, *J. Biol. Chem.* **210**:597–606.
109. Weil-Malherbe, H., and Bone, A. D., 1951, *Biochem. J.* **49**:339–347.
110. Ning, J., Purich, D. L., and Fromm, H. J., 1969, *J. Biol. Chem.* **244**:3840–3846.
111. Casazza, J. P., and Fromm, H. J., 1976, *Arch. Biochem. Biophys.* **177**:480–487.
112. Kosow, D. P., and Rose, I. A., 1970, *J. Biol. Chem.* **245**:198–204.
113. Uyeda, K., and Racker, E., 1965, *J. Biol. Chem.* **240**:4682–4688.
114. Rose, I. A., and Warms, J. V. B., 1967, *J. Biol. Chem.* **242**:1635–1645.
115. Ellison, W. R., Lueck, J. D., and Fromm, H. J., 1975, *J. Biol. Chem.* **250**:1864–1871.
116. Purich, D. L., and Fromm, H. J., 1971, *J. Biol. Chem.* **246**:3456–3463.
117. Wilson, J. E., 1980, *Curr. Top. Cell Regul.* **16**:1–44.
118. Knull, H. R., Taylor, W. F., and Wells, W. W., 1974, *J. Biol. Chem.* **249**:6930–6935.
119. Thompson, M. F., and Bachelard, H. S., 1970, *Biochem. J.* **118**:25–34.
120. Hernandez, A., and Crane, R. K., 1966, *Arch. Biochem. Biophys.* **113**:223–229.
121. Wilson, J. E., 1978, *Arch. Biochim. Biophys.* **185**:88–99.
122. Wilson, J. E., 1968, *J. Biol. Chem.* **243**:3640–3647.
123. Gots, R. E., Gorin, F. A., and Bessman, S. P., 1972, *Biochem. Biophys. Res. Commun.* **49**:1249–1255.
124. Felgner, P. L., 1973, *Fed. Proc.* **32**:488.
125. Kropp, E. S., and Wilson, J. E., 1970, *Biochem. Biophys. Res. Commun.* **38**:74–79.
126. Bielicki, L., Krieglstein, J., and Wever, K., 1980, *Arzneim. Forsch.* **30**:594–597.
127. Lusk, J. A., Manthorpe, C. M., Kao-Jen, J., and Wilson, J. E., 1980, *J. Neurochem.* **34**:1412–1420.
128. Uyeda, K., 1979, *Adv. Enzymol.* **48**:193–244.
129. Uyeda, K., and Luby, L. J., 1974, *J. Biol. Chem.* **249**:4562–4570.
130. Goldhammer, A. R., and Paradies, H. H., 1979, *Curr. Top. Cell. Regul.* **15**:109–141.
131. Tornheim, K., 1980, *J. Theor. Biol.* **85**:199–222.
132. Kagimoto, T., and Uyeda, K., 1980, *Arch. Biochem. Biophys.* **203**:792–799.
133. Claus, T. H., Schlumpf, J. R., El-Maghrabi, M. R., Pilkis, J., and Pilkis, S. J., 1980, *Proc. Natl. Acad. Sci. U.S.A.* **77**:6501–6505.
134. Nieto, A., and Castano, J. G., 1980, *Biochem. J.* **186**:953–957.
135. Furuya, E., and Uyeda, K., 1980, *J. Biol. Chem.* **255**:11656–11659.
136. Furuya, E., and Uyeda, K., 1980, *Proc. Natl. Acad. Sci. U.S.A.* **77**:5861–5864.
137. Kagimoto, T., and Uyeda, K., 1979, *J. Biol. Chem.* **254**:5584–5587.
138. Castano, J. G., Nieto, A., and Felia, J. E., 1979, *J. Biol. Chem.* **254**:5576–5579.
139. Pilkis, S. J., Schlumpf, J., Pilkis, J., and Claus, T. H., 1979, *Biochem. Biophys. Res. Commun.* **88**:960–967.

140. Van Schaftingen, E., Hue, L., and Hers, H.-G., 1980, *Biochem. J.* **192**:887–895.
141. Van Schaftingen, E., Hue, L., and Hers, H.-G., 1980, *Biochem. J.* **192**:897–901.
142. Van Schaftingen, E., and Hers, H.-G., 1980, *Biochem. Biophys. Res. Commun.* **96**:1524–1531.
143. Pilkis, S. J., El-Maghrabi, M. R., Pilkis, J., Claus, T. H., and Cumming, D. A., 1981, *J. Biol. Chem.* **256**:3171–3174.
144. Kayne, F. J., 1973, *The Enzymes,* Volume VIII (P. D. Boyer, ed.), Academic Press, New York, pp. 353–382.
145. Nicholas, P. C., and Bachelard, H. S., 1974, *Biochem. J.* **141**:165–171.
146. Tsao, M. U., 1979, *Moll. Cell. Biochem.* **24**:75–81.
147. Weber, G., 1969, *Proc. Natl. Acad. Sci. U.S.A.* **63**:1365–1369.
148. Schwark, W. S., Singhal, R. L., and Ling, G. M., 1971, *J. Neurochem.* **18**:123–134.
149. Miller, A. L., Hawkins, R. A., and Veech, R. L., 1973, *Science* **173**:904–906.
150. Wapnir, R. A., Moak, S. A., and Lifshitz, F., 1977, *Nature* **265**:647–648.
151. Akeson, A. L., Berry, H. K., Brunner, R. L., and Vorhees, C. V., 1979, *J. Neurochem.* **32**:233–235.
152. Kemp, R. G., 1973, *J. Biol. Chem.* **248**:3963–3967.
153. Tornheim, K., and Lowenstein, J. M., 1979, *J. Biol. Chem.* **254**:10586–10587.
154. Fitch, C. D., Chevli, R., and Jellinek, M., 1979, *J. Biol. Chem.* **254**:11357–11359.
155. Ishibashi, H., and Cottam, G. L., 1978, *J. Biol. Chem.* **253**:8767–8771.
156. Ljungström, O., Hjelmquist, G., and Engström, L., 1974, *Biochim. Biophys. Acta* **358**:289–298.
157. Ekman, P., Dahlqvist, U., Humble, E., and Engström, L., 1976, *Biochim. Biophys. Acta* **429**:374–382.
158. Engström, L., 1978, *Curr. Top. Cell Regul.* **13**:29–51.
159. Jope, R., and Blass, J. P., 1975, *Biochem. J.* **150**:397–403.
160. Booth, R. F. G., and Clark, J. B., 1978, *J. Neurochem.* **30**:1003–1008.
161. Cremer, J. E., and Teal, H. M., 1974, *FEBS Lett.* **39**:17–20.
162. Denton, R. M., Randle, P. J., Bridges, B. J., Cooper, R. H., Kerbey, A. L., Pask, H. T., Severson, D. L., Stansbie, D., and Whitehouse, S., 1975, *Mol. Cell. Biochem.* **9**:27–53.
163. Hucho, F., 1975, *Angew. Chem. [Engl.]* **14**:591–601.
164. Linn, T. C., Pettit, F. H., Hucho, F., and Reed, L. J., 1969, *Proc. Natl. Acad. Sci. U.S.A.* **62**:234–241.
165. Linn, T. C., Pelley, J. W., Pettit, F. H., Hucho, F., Randall, D. D., and Reed, L. J., 1972, *Arch. Biochem. Biophys.* **148**:327–342.
166. Pratt, M. L., Roche, T. E., Dyer, D. W., and Cate, R. L., 1979, *Biochem. Biophys. Res. Commun.* **91**:289–296.
167. Reed, L. J., Linn, T. C., Pettit, F. H., Oliver, R. M., Hucho, F., Pelley, J. W., Randall, D. D., and Roche, T. E., 1972, *Energy Metabolism and the Regulation of Metabolic Processes in Mitochondria* (M. A. Mehlman and R. W. Hanson, eds.), Academic Press, New York, pp. 253–270.
168. Siess, E., Wittmann, J., and Wieland, O., 1971, *Hoppe Seylers Z. Physiol. Chem.* **352**:447–452.
169. Randle, P. J., Denton, R. M., Pask, H. T., and Severson, D. L., 1974, *Biochem. Soc. Symp.* **39**:75–88.
170. Severson, D. L., Denton, R. M., Pask, H. T., and Randle, P. J., 1974, *Biochem. J.* **140**:225–237.
171. Ngo, T. T., and Barbeau, A., 1978, *J. Neurochem.* **31**:69–75.
172. Jope, R., and Blass, J. P., 1976, *J. Neurochem.* **26**:709–714.
173. Land, J. M., Booth, R. F. G., Berger, R., and Clark, J. B., 1977, *Biochem. J.* **164**:339–348.
174. Sugden, P. H., and Newsholme, E. A., 1975, *Biochem. J.* **150**:105–111.
175. LaNoue, K., Nicklas, W. J., and Williamson, J. R., 1970, *J. Biol. Chem.* **245**:102–111.
176. LaNoue, K. F., Walajtys, E. I., and Williamson, J. R., 1973, *J. Biol. Chem.* **248**:7171–7183.
177. Srere, P. A., Matsuoka, Y., and Mukherjee, A., 1973, *J. Biol. Chem.* **248**:8031–8035.
178. Smith, C. M., and Williamson, J. R., 1971, *FEBS Lett.* **18**:35–38.
179. Williamson, J. R., and Cooper, R. H., 1980, *FEBS Lett.* **117**:K73–K85.
180. Weitzman, P. D. J., and Danson, M. J., 1976, *Curr. Top. Cell. Regul.* **10**:161–204.

181. Williamson, J. R., Smith, C. M., LaNoue, K. F., and Bryla, J., 1972, *Energy Metabolism and the Regulation of Metabolic Processes in Mitochondria* (M. A. Mehlman and R. W. Hanson, eds.), Academic Press, New York, pp. 185–210.

182. Stein, A. M., Stein, J. H., and Kirkman, S. K., 1967, *Biochemistry* **6:**1370–1379.

183. Plaut, G. W. E., 1970, *Curr. Top. Cell. Regul.* **2:**1–27.

184. Goldberg, N. D., Passonneau, J. V., and Lowry, O. H., 1966, *J. Biol. Chem.* **241:**3997–4003.

185. Chen, R. F., and Plaut, G. W. E., 1963, *Biochemistry* **2:**1023–1032.

186. Sanwal, B. D., and Stachow, C. S., 1965, *Biochim. Biophys. Acta* **96:**28–44.

187. Shen, W.-C., Mauck, L., and Colman, R. F., 1974, *J. Biol. Chem.* **249:**7942–7949.

188. Colman, R. F., 1975, *Adv. Enzyme Regul.* **13:**413–433.

189. Dalziel, K., 1980, *FEBS Lett.* **117:**K45–55.

190. Willson, V. J. C., and Tipton, K. F., 1980, *J. Neurochem.* **34:**793–799.

191. Nicholls, D. G., and Garland, P. B., 1969, *Biochem. J.* **114:**215–225.

192. Tanaka, N., Koike, K., Hamada, M., Otsuka, K.-I., Suematsu, T., and Koike, M., 1972, *J. Biol. Chem.* **247:**4043–4049.

193. Koike, K., Hamada, M., Tanaka, N., Otsuka, K.-I., Ogasahara, K., and Koike, M., 1974, *J. Biol. Chem.* **249:**3836–3842.

194. McCormack, J. G., and Denton, R. M., 1981, *Biochem. J.* **196:**619–624.

195. Smith, C. M., Bryla, J., and Williamson, J. R., 1974, *J. Biol. Chem.* **249:**1497–1505.

196. Lowry, O. H., Passonneau, J. V., Hasselberger, F. X., and Schulz, D. W., 1964, *J. Biol. Chem.* **239:**18–30.

197. Chapman, A. G., Meldrum, B. S., and Siesjö, B. K., 1977, *J. Neurochem.* **28:**1025–1035.

198. Duffy, T. E., Howse, D. C., and Plum, F., 1975, *J. Neurochem.* **24:**925–934.

199. Pardridge, W. M., and Oldendorf, W. H., 1977, *J. Neurochem.* **28:**5–12.

200. Borgström, L., Chapman, A. G., and Siesjö, B. K., 1976, *J. Neurochem.* **27:**971–973.

201. Siesjö, B. K., and Nilsson, L., 1971, *Scan. J. Clin. Lab. Invest.* **27:**83–96.

202. Krebs, H. A., 1972, *Essays Biochem.* **8:**1–34.

203. Cohen, L. H., and Noell, W. K., 1965, *Biochemistry of the Retina* (C. N. Graymore, ed.), Academic Press, New York, pp. 36–50.

204. Gottstein, U., 1975, *Brain Work. The Coupling of Function, Metabolism and Blood Flow* (D. H. Ingvar and N. A. Lassen, eds.), Munksgaard, Copenhagen, pp. 144–148.

205. Gjedde, A., Hansen, A. J., and Siemkowicz, E., 1981, *J. Cereb. Blood Flow Metab.* **1:**172–173.

206. Berl, S., Takagaki, G., Clark, D. D., and Waelsch, H., 1962, *J. Biol. Chem.* **237:**2570–2573.

207. Felicioli, R. A., Gabrielli, F., and Rossi, C. A., 1967, *Life Sci.* **6:**133–143.

208. Waelsch, H., Cheng, S.-C., Cote, L. J., and Naruse, H., 1965, *Proc. Natl. Acad. Sci. U.S.A.* **54:**1249–1253.

209. Cheng, S.-C., 1971, *Int. Rev. Neurobiol.* **14:**125–157.

210. Hawkins, R. A., Miller, A. L., Nielsen, R. C., and Veech, R. L., 1973, *Biochem. J.* **134:**1001–1008.

211. Naruse, H., Cheng, S.-C., and Waelsch, H., 1966, *Exp. Brain Res.* **1:**291–298.

212. Salganicoff, L., and Koeppe, R. E., 1968, *J. Biol. Chem.* **243:**3416–3420.

213. Patel, M. S., 1974, *J. Neurochem.* **22:**717–724.

214. Cheng, S.-C., and Cheng, R. H. C., 1972, *Arch. Biochem. Biophys.* **151:**501–511.

215. Patel, M. S., and Tilghman, S. M., 1973, *Biochem. J.* **132:**185–192.

216. Frenkel, R., 1972, *Arch. Biochem. Biophys.* **152:**136–143.

217. D'Adamo, A. F., Jr., and D'Adamo, A. P., 1968, *J. Neurochem.* **15:**315–323.

218. Mahan, D. E., Mushahwar, I. K., and Koeppe, R. E., 1975, *Biochem. J.* **145:**25–35.

219. Frenkel, R., 1975, *Curr. Top. Cell Regul.* **9:**157–181.

220. Denton, R. M., and Halestrap, A. P., 1979, *Essays Biochem.* **15:**37–77.

221. Nordlie, R. C., 1974, *Curr. Top. Cell Regul.* **8:**33–116.

222. Anchors, J. M., Haggerty, D. F., and Karnovsky, M. L., 1977, *J. Biol. Chem.* **252:**7035–7041.

223. Colilla, W., Jorgenson, R. A., and Nordlie, R. C., 1975, *Biochim. Biophys. Acta* **377:**117–225.

224. Nordlie, R. C., 1971, *Enzymes* **4:**534–610.

225. Tewari, H. B., and Bourne, G. H., 1963, *J. Histochem. Cytochem.* **11:**121–122.
226. Prasannan, K. G., and Subrahmanyam, K., 1968, *Endocrinology* **82:**1–6.
227. Rosen, S. I., 1970, *Acta Histochem. (Jena)* **36:**44–53.
228. Anchors, J. M., and Karnovsky, M. L., 1975, *J. Biol. Chem.* **250:**6408–6416.
229. Stephens, H. R., and Sandborn, E. B., 1976, *Brain Res.* **113:**127–146.
230. Rosen, S. I., 1974, *Acta Histochem. (Jena)* **50:**1–18.
231. Zakim, D., 1970, *J. Biol. Chem.* **245:**4953–4961.
232. Burchell, A., and Burchell, B., 1980, *FEBS Lett.* **118:**180–184.
233. Scrutton, M. C., and Utter, M. F., 1968, *Annu. Rev. Biochem.* **37:**249–302.
234. Phillips, M. E., and Coxon, R. V., 1975, *Biochem. J.* **146:**185–189.
235. Hevor, T., and Gayet, J., 1981, *J. Neurochem.* **36:**949–958.
236. Majumder, A. L., and Eisenberg, F., Jr., 1977, *Proc. Natl. Acad. Sci. U.S.A.* **74:**3222–3225.
237. Pilkis, S. J., El-Maghrabi, M. R., Pilkis, J., and Claus, T., 1981, *J. Biol. Chem.* **256:**3619–3622.
238. Rious, J.-P., Claus, T. H., Flockhart, D. A., Corbin, J. D., and Pilkis, S. J., 1977, *Proc. Natl. Acad. Sci. U.S.A.,* **74:**4615–4619.
239. Pilkis, S. J., El-Maghrabi, M. R., Coven, B., Claus, T. H., Tager, H. S., Steiner, D. F., Keim, P. S., and Heinrikson, R. L., 1980, *J. Biol. Chem.* **255:**2770–2775.
240. Mendicino, J., Leibach, F., and Reddy, S., 1978, *Biochemistry* **17:**4662–4669.
241. Proffitt, R. T., Sankaran, L., and Pogell, B. M., 1976, *Biochemistry* **15:**2918–2925.
242. Soling, H. D., Bernhard, G., Kuhn, A., and Lück, H.-J., 1977, *Arch. Biochem. Biophys.* **182:**563–572.
243. Katz, J., and Rognstad, R., 1976, *Curr. Top. Cell Regul.* **10:**237–289.
244. Katz, J., and Rognstad, R., 1978, *Trends Biochem. Sci.* **3:**171–174.
245. Newsholme, E. A., and Crabtree, B., 1976, *Biochem. Soc. Symp.* **41:**61–109.
246. Hawkins, R. A., and Miller, A. L., 1978, *Neuroscience* **3:**251–258.
247. Huang, S. C., Phelps, M. E., Hoffman, E. J., Sideris, K., Selin, C. J., and Kuhl, D. E., 1980, *Am. J. Physiol.* **238:**E69–E82.
248. Karnovsky, M. L., Burrows, B. L., and Zoccoli, M. A., 1980, *Cerebral Metabolism and Neural Function* (J. V. Passonneau, R. A. Hawkins, W. D. Lust, and F. A. Welsh, eds.), Williams & Wilkins, Baltimore, pp. 359–366.
249. Newsholme, E. A., and Gevers, W., 1967, *Vitam. Horm.* **25:**1–87.
250. Norenberg, M. D., and Martinez-Hernandez, A., 1979, *Brain Res.* **161:**303–310.
251. Davidson, N., 1976, *Neurotransmitter Amino Acids,* Academic Press, New York, pp. 57–111.
252. Berl, S., Clark, D. D., and Schneider, D. (eds.), 1975, *Metabolic Compartmentation and Neurotransmission,* Plenum Press, New York.
253. Roberts, E., 1974, *Biochem. Pharmacol.* **23:**2637–2649.
254. Roberts, E., Chase, T. N., and Tower, D. B. (eds.), 1976, *GABA in Nervous System Function,* Raven Press, New York.
255. Baxter, C. F., 1970, *Handbook of Neurochemistry,* Volume 3 (A. Lajtha, ed.), Plenum Press, New York, pp. 289–356.
256. Iversen, L. L., Dick, F., Kelly, J. S., and Schon, F., 1975, *Metabolic Compartmentation and Neurotransmission* (S. Berl, D. D. Clarke, and D. Schneider, eds.), Plenum Press, New York, pp. 65–90.
257. Sims, K. L., and Pitts, F. N., 1970, *J. Neurochem.* **17:**1607–1612.
258. Hostetler, K. Y., Landau, B. R., White, R. J., Albin, M. S., and Yashon, D., 1970, *J. Neurochem.* **17:**33–39.
259. Gaitonde, M. K., and Arnfred, T., 1971, *J. Neurochem.* **18:**1971–1987.
260. Thoesen-Coleman, M. T., and Allen, N., 1978, *J. Neurochem.* **30:**83–90.
261. Maker, H. S., Clarke, D. D., and Lajtha, A. L., 1976, *Basic Neurochemistry,* 2nd ed. (G. J. Siegel, R. W. Albers, R. Katzman, B. W. Agranoff, eds.), Little, Brown, Boston, pp. 279–307.
262. McIlwain, H., and Bachelard, H. S., 1971, *Biochemistry and the Central Nervous System,* 4th ed., Churchill, London.
263. Krebs, H. A., and Eggleston, L. V., 1974, *Adv. Enzyme Regul.* **12:**421–434.

264. Siesjö, B. K., 1981, *J. Cereb. Blood Flow Metab.* **1**:155–186.
265. Hotta, S. S., 1962, *J. Neurochem.* **9**:43–51.
266. Eisenberg, H. M., Suddith, R. L., and Crawford, J. S., 1980, *Adv. Exp. Med. Biol.* **131**:57–68.
267. Goldstein, G. W., 1979, *J. Physiol. (Lond.)* **286**:185–195.
268. Craven, P. A., and Basford, R. E., 1974, *Biochim. Biophys. Acta* **354**:49–56.
269. Reese, T. S., and Brightman, M. W., 1968, *Anat. Rec.* **160**:414.
270. Lowry, O. H., and Passonneau, J. V., 1964, *J. Biol. Chem.* **239**:31–42.
271. Knull, H. R., 1978, *Biochim. Biophys. Acta* **522**:1–9.
272. Novello, F., and McLean, P., 1968, *Biochem. J.* **107**:775–791.
273. Soothill, P. W., Kouseibati, F., Watts, R. L., and Watts, D. C., 1981, *J. Neurochem.* **37**:506–510.
274. Watanabe, H., and Passonneau, J. V., 1973, *J. Neurochem.* **20**:1543–1554.
275. Wilson, J. E., and Felgner, P. L., 1977, *Mol. Cell. Biochem.* **18**:39–47.
276. Booth, R. F. G., Patel, T. B., and Clark, J. B., 1980, *J. Neurochem.* **34**:17–25.
277. Holowach, J., Kauffman, F., Ikossi, M. G., Thomas, C., and McDougal, D. B., Jr., 1968, *J. Neurochem.* **15**:621–631.
278. Lowry, O. H., Roberts, N. R., Wu, M. L., Hixon, W. S., and Crawford, E. J., 1954, *J. Biol. Chem.* **207**:19–37.
279. Balázs, R., 1965, *J. Neurochem.* **12**:63–76.
280. MacDonnell, P. C., and Greengard, O., 1974, *Arch. Biochem. Biophys.* **163**:644–655.
281. Passonneau, J. V., Lust, W. D., and Crites, S. K., 1977, *Neurochem. Res.* **2**:605–617.
282. Passonneau, J. V., Lowry, O. H., Schulz, D. W., and Brown, J. G., 1969, *J. Biol. Chem.* **244**:902–909.
283. Kauffman, F. C., Brown, J. G., Passonneau, J. V., and Lowry, O. H., 1969, *J. Biol. Chem.* **244**:3647–3653.
284. Veech, R. L., 1980, *Cerebral Metabolism and Neural Function* (J. V. Passonneau, R. A. Hawkins, W. D. Lust, and F. A. Welsh, eds.), Williams & Wilkins, Baltimore, pp. 34–41.
285. Bachelard, H. S., 1976, *Handbook of Clinical Neurology. Metabolic and Deficiency Diseases of the Nervous System. Part I* (P. J. Vinken and G. W. Bruyn, eds.), American Elsevier, New York, pp. 1–26.
286. Krebs, H. A., and Veech, R. L., 1969, *The Energy Level and Metabolic Control in Mitochondria* (S. Papa, J. M. Tager, E. Quagliariello, and E. C. Slater, eds.), Adriatica Editrice, Bari, pp. 329–382.
287. Lowry, O. H., Passonneau, J. V., Schulz, D. W., and Rock, M. K., 1961, *J. Biol. Chem.* **236**:2746–2755.
288. Burch, H. B., Bradley, M. E., and Lowry, O. H., 1967, *J. Biol. Chem.* **242**:4546–4554.
289. Vannucci, S. J., and Vannucci, R. C., 1980, *J. Neurochem.* **34**:1100–1105.
290. Alderman, J. L., and Shellenberger, M. K., 1974, *J. Neurochem.* **22**:937–940.
291. Balázs, R., 1970, *Handbook of Neurochemistry*, Volume 3 (A. Lajtha, ed.), pp. 1–36.
292. Sellinger, O. Z., and DeBalbian-Verster, F., 1962, *J. Biol. Chem.* **237**:2836–2844.

Oxidative Phosphorylation

Dennis W. Jung and Gerald P. Brierley

1. INTRODUCTION

Oxidative phosphorylation is the synthesis of ATP from ADP and inorganic phosphate (Pi) that accompanies the oxidation of NADH and other substrates by the mitochondrial electron transport chain. The overall process (shown for a $2e^-$ reaction in equation 1) can be considered to consist of the following components:

$$NADH + H^+ + 3ADP + 3Pi + \tfrac{1}{2}O_2 \rightarrow NAD^+ + 3ATP + H_2O \quad [1]$$

(1) the electron transport chain, (2) the ATP synthetase or F_0F_1 ATPase, and (3) a coupling mechanism that links the strongly exergonic redox reactions of electron transport to the endergonic synthesis of ATP.

Under normal conditions, the brain meets its relatively high energy demands exclusively by the oxidation of glucose. The glucose is degraded by the glycolytic enzymes of the cytosol, and the resulting pyruvate enters the mitochondrion via a specific membrane transporter. In the mitochondrial matrix, it is degraded to CO_2 by pyruvate dehydrogenase and the enzymes of the citric acid cycle. The resulting matrix NADH is oxidized by the NADH dehydrogenase, and reducing equivalents are passed sequentially through the components of the electron transport chain to molecular oxygen. The electron transport chain is localized in the inner membrane of the mitochondrion, and it is presently felt that the redox reactions of electron transport conserve energy as an electrochemical H^+ gradient across this membrane. The ATP synthetase, which is also localized in the inner membrane, appears to utilize this gradient or protonmotive force (Δp^*) to bring about the synthesis of ATP. Since the

---

* The abbreviations used are as follows: Δp, protonmotive force; DCCD, dicyclohexlycarbodiimide; OSCP, oligomycin-sensitivity-conferring protein; ΔpH, transmembrane pH gradients; $\Delta\psi$, membrane potential; R, the gas constant; T, temperature in kelvins; F, the Faraday; CoQ, ubiquinone or coenzyme Q; cyt, cytochrome.

Dennis W. Jung and Gerald P. Brierley • Department of Physiological Chemistry, College of Medicine, Ohio State University, Columbus, Ohio 43210.

synthesis of ATP occurs in the matrix compartment and most ATP-utilizing reactions of the cell are extramitochondrial, it is apparent that pathways for the transport of ATP, ADP, and Pi must be available in the inner membrane and contribute to the overall reaction.

As is developed in this chapter, all of these processes function as a highly integrated and closely regulated unit which responds to the cell's needs for ATP production. For the purposes of the present discussion, however, we consider each of the various segments of the reaction separately before attempting to deal with present concepts of how they are orchestrated in the functioning cell. It should be noted that a number of excellent reviews[1-9] are available which emphasize various aspects of oxidative phosphorylation and which should be used to supplement this rather brief survey.

2. THE ELECTRON TRANSPORT CHAIN

The electron transport chain consists of over 20 redox centers localized in the inner membrane of the mitochondrion. These centers include flavoproteins, iron–sulfur proteins, the cytochromes, and ubiquinone. The sequence of components shown in Fig. 1 is consistent with data available from several different experimental approaches. Although redox potentials vary with the energetic state of mitochondria, the redox centers appear to fall into four isopotential groups.[1,10] Such an arrangement tends to buffer the redox poise of the electron transport chain and permits components to interact sufficiently rapidly with each other to maintain a quasiequilibrium situation.

The electron transport chain can be resolved into four catalytically active complexes by fractionation with detergents and salt.[11-13] These are complex I, the NADH–ubiquinone (CoQ) reductase; complex II, succinate–ubiquinone reductase; complex III, ubiquinone–cyt c reductase; and complex IV, cyt c oxidase. Of the redox centers that have been implicated in electron transport, only CoQ and cyt c are not firmly associated with one of these complexes, and the four complexes, together with these two so-called mobile components, can be reconstituted to yield electron transport activity corresponding to that in the native membrane.

Complex I has a molecular weight of nearly 700,000, consists of 16–18 polypeptide subunits, has noncovalently bound FMN as a prosthetic group, and contains six or seven Fe–S centers in addition to CoQ and phospholipid.[11,14] The midpoint potentials of the Fe–S centers range over 350 mV. A CoQ-binding component may be present in order to stabilize ubiquinone radicals.[15] The enzyme NADH dehydrogenase, an iron–sulfur flavoprotein of molecular weight 75,000, has been isolated from complex I.[14] This enzyme represents the entering point for NADH reducing equivalents derived from Krebs cycle oxidations. The polypeptides of complex I are arranged asymmetrically across the inner membrane with the NADH dehydrogenase oriented toward the matrix side (see Fig. 2).[13-16] When complex I is incorporated into phospholipid vesicles, it catalyzes a translocation of H^+ coupled to the oxidation of NADH by CoQ.[17]

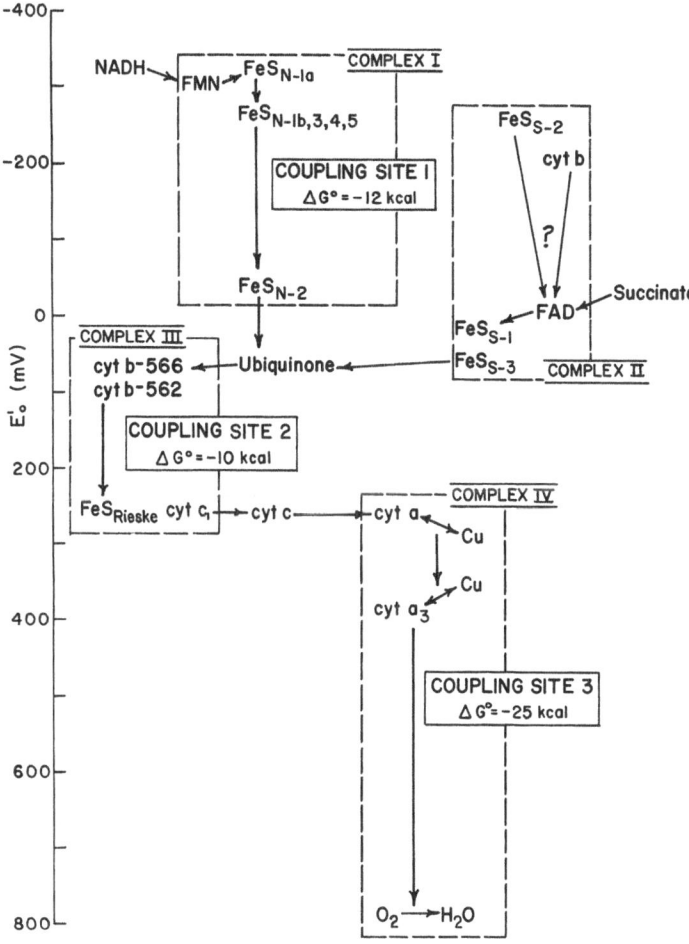

Fig. 1. Scheme of mitochondrial electron transport showing the midpoint potentials (E_0') of the carriers.[6,10,14,61] The actual redox potentials vary with the energetic and oxidation–reduction state of the components. This deviation is usually less than 100 mV but, in the case of cyt b-566, ranges from -30 to $+240$ mV on energization. This suggests that this component may play a direct role in energy conservation. Several redox centers operate at similar potentials and form "isopotential pools," with each separated by about 300 mV under state 4 conditions. These pools delineate the three so-called coupling sites as indicated. All of the components are electron carriers, but only the flavins and ubiquinone may carry hydrogens. The sequence of electron flow in complex III is probably cyclic in nature rather than linear as shown. Since the redox potentials of cyt a and a_3 and the associated coppers are all within 150 mV, it is suggested that energy conservation at site 3 may involve a cyt a_3–oxygen complex of high potential (600 mV) accepting electrons from cyt $a \cdot Cu$, thereby forming a 400-mV drop.

Complex II, which promotes the oxidation of succinate by CoQ, consists of four subunits.[11,14,18] It contains covalently bound FAD, heme b ($b_{557.5}$), and three iron–sulfur centers (see Fig. 1). Electron flow from succinate to CoQ does not result in H^+ translocation, and, in contrast to complex I, this segment of the chain is not associated with ATP formation. Succinate dehydrogenase

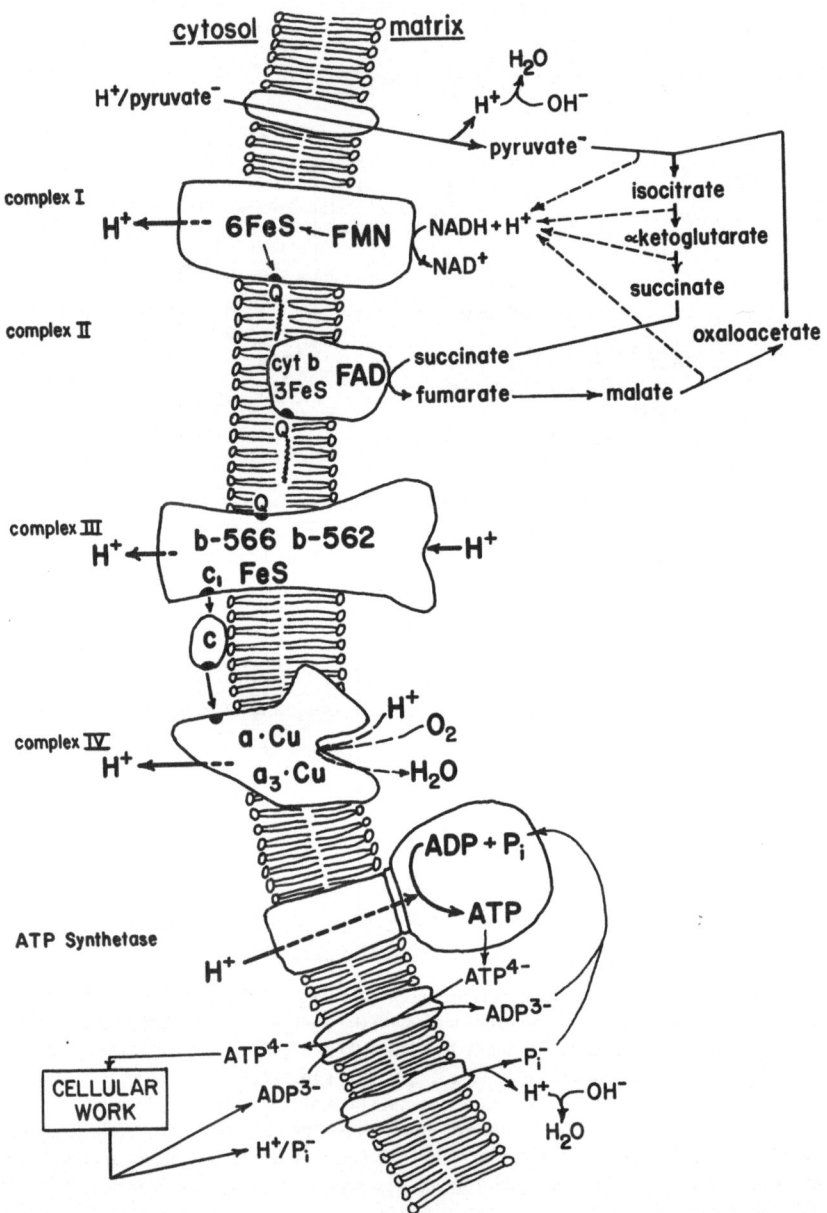

Fig. 2. A possible arrangement of the components of oxidative phosphorylation in the inner membrane of the mitochondrion. The four respiratory complexes, the ATP synthetase, and the pyruvate, Pi, and adenine nucleotide carriers are viewed as diffusible integral membrane protein complexes residing in a phospholipid bilayer. The lipid-soluble ubiquinone is approximately as long as the width of the bilayer and, because of its rigid structure, is probably positioned in the plane of the membrane and may be bound to proteins in complexes I, II, and III. The surface of cyt c is highly charged, and it interacts with the phospholipid head groups peripheral to the membrane. Reducing equivalents are shown entering the respiratory chain from NADH at complex I and from succinate at complex II. Electrons are thought to be shuttled between the complexes by lateral diffusion of ubiquinone and cyt c, although complexes I and III and II and III may form associations.

is oriented toward the matrix side of the membrane (Fig. 2).[14] One of the smaller subunits of complex II may be a specific CoQ-binding protein that stabilizes quinone radicals and is required for the linkage between complexes II and III.[19] The redox potentials of cyt b and of Fe–S center S-2 appear to be too low to accomodate their direct participation in the oxidation of succinate by ubiquinone (see Fig. 1).

Complex III, the bc_1 complex, transfers reducing equivalents from ubiquinone (reduced by complexes I or II) to cyt c, a peripheral protein located on the cytosol side of the inner mitochondrial membrane (Fig. 2).[11,13,18,20] It contains two cytochromes b, an iron–sulfur center (the Rieske iron–sulfur protein), and cyt c_1. The two b cytochromes, b-562 (b_k) and b-566 (b_T), are identified by differences in spectral properties, midpoint potentials, and kinetics of oxidation–reduction. Although they differ markedly from each other in physical properties and functional behavior in the membrane, the possibility remains that the two b cytochromes are identical but reside in different environments.[18,20] The exact sequence of electron transfer in complex III and the roles of the b cytochromes, ubiquinone, and the iron–sulfur protein are not clear. It is doubtful that a simple linear sequence of components operates in this complex, and several cyclic pathways have been proposed.[6,18,21,22]

The flow of electrons from ubiquinone to cytochrome c is accompanied by H^+ translocation from the matrix to the cytosol side of the inner membrane. Although the mechanism of this vectorial H^+ transport is not known, it has been proposed that an electrogenic H^+ pump linked to one of the b cytochromes is present.[23] The activity of this component might explain the shift in redox potential of cytochrome b-566 that is observed on energization.

Complex III is composed of eight or nine polypeptides with a probable functional dimer structure of 500,000 daltons.[6] Only 30% of the mass of complex III appears confined to the lipid bilayer, with the remainder extending above and below the surface. The b cytochromes seem to be located in the middle region, whereas cyt c_1 is on the cytosol side of the membrane, available for electron transfer to cyt c (Fig. 2). Isolated complex III contains considerable ubiquinone, and a ubiquinone-binding protein and stable ubiquinone radicals have been detected in this segment.[11,24]

Complex IV, cytochrome oxidase, represents the terminal segment of the electron transport chain. It passes electrons from cyt c to oxygen to produce water. Complex IV consists of seven polypeptides and four redox centers: two cytochromes, a and a_3, and low- and high-potential copper centers.[25–27] Cytochromes a and a_3 are distinguished by differences in spin state, oxidation-reduction midpoint potentials, and the fact that a_3 forms ligands with carbon monoxide and cyanide whereas cyt a does not. Like the other isolated respi-

---

During electron transport, H^+ is transported out of the matrix to the cytosol side at the three regions of the chain that constitute coupling sites. This establishes an electrochemical H^+ gradient (Δp). This gradient supplies energy for the uptake of Pi, pyruvate, and ADP via specific carriers and also for the synthesis of ATP by the ATP synthetase complex. The ATP is exported to the cytosol where it can be utilized for cellular work. Oxidative phosphorylation is thus dependent on the ability of the membrane to maintain Δp.

ratory complexes, cytochrome oxidase contains phospholipid which is essential for its enzymatic activity.

Ultracentrifugal and crystal studies suggest that cytochrome oxidase is a dimer containing four cytochromes *a* and four copper centers.[28] It is viewed as a Y-shaped structure spanning the inner membrane (Fig. 2), with the arms protruding a small distance into the matrix and the bottom section extending about 6 nm into the intermembrane space. Cytochrome *a* and cyt a_3 are located on the cytosol and matrix sides of the membrane, respectively; but have never been physically separated and do not appear to function independently. It is not certain where the reduction of oxygen occurs with respect to the membrane. Cytochrome oxidase reacts very quickly with oxygen with a K_m of 10^{-7}–10^{-8} M. The reduction of a molecule of oxygen to two molecules of water involves the transfer of four electrons and four H^+. A complex reaction mechanism is needed to prevent the production and release of intermediate radicals which could be destructive to cellular integrity. Recent studies have resulted in plausible suggestions as to how this might occur.[29,30]

Cytochrome oxidase preparations reconstituted into phospholipid vesicles have been shown to catalyze vectorial H^+ extrusion.[26,31] N',N'-Dicyclohexylcarbodiimide (DCCD), an inhibitor of H^+ translocation in ATP-linked systems, inhibits H^+ translocation in reconstituted cytochrome oxidase vesicles and mitochondria.[32] It binds specifically to one of the subunits which may form a membrane H^+ channel.[33] However, it has also been suggested that this DCCD-binding polypeptide in cytochrome oxidase may be a contaminant.[34]

Experiments with phospholipid-enriched mitoplasts (mitochondria whose outer membrane has been removed by hypotonic or detergent treatment[35]) suggest that the respiratory chain is not an ordered structured array of complexes but consists of structurally independent complexes diffusing randomly within the membrane.[36] Mitochondrial inner membranes increase in diameter in proportion to enrichment with exogenous phospholipid. This expansion produces a proportionate increase in the distance between integral proteins and a decrease in electron transfer activity between the respiratory chain complexes. Such observations suggest that ubiquinone acts as an independently diffusible, rate-limiting mobile electron carrier between the independently diffusible dehydrogenases and complex III. Cytochrome *c* is seen to diffuse laterally between complexes III and IV. These experiments suggest that the proper and rapid electron transport sequence between the components of the respiratory chain is mediated by their random diffusion and energetically favorable collisions in the plane of the inner membrane.[36]

Other evidence, however, suggests that multicomplex functional units exist in the membrane for complexes I and III and for complexes II and III. The resolution of these complexes is difficult and involves separation using detergents. Complex II may be a cleavage product of a functional succinate–cyt *c* reductase complex, since it is difficult to isolate complex III free of succinate dehydrogenase.[18] It also appears possible that the cyt *b* in complex II is derived from complex III. The discovery of protein-bound ubiquinone and the observation that fewer than three ubiquinone molecules per unit are needed for near-maximal activity of NADH–cyt *c* reductase[15] provide evidence that complexes I and III can combine to form a single functional unit. The above observations may be explained by rapid dissociation and reaggregation of the complexes in the membrane, with ubiquinone functioning as a reservoir of reducing power.[15,36]

Cytochrome c binds to both cyt c_1 and complex IV at a similar and probably identical site.[37] This can be taken as evidence that cyt c does not form a fixed link between complexes III and IV but diffuses laterally between them.[38] The fact that complex IV is readily separated from complexes I, II, and III also suggests that complexes III and IV are not tightly associated. The stoichiometry of $1:2:3:7:9:63$ for complexes I:II:III:IV:cyt c:ubiquinone[11,36] is consistent with functional complexes of I plus III and II plus III, but suggests a nonchain, random distribution of the other components.

3. THE ATP SYNTHETASE

The mitochondrial ATP synthetase (which is also called the F_0F_1 ATPase, the proton-translocating ATPase, and Complex V) is the complex that brings about the synthesis of ATP from ADP and Pi. When reconstituted into lipid membranes, this complex promotes an ATP-dependent ejection of H^+, and such systems are also able to utilize an artificially produced pH gradient to synthesize ATP. As is discussed in Section 5.1, the properties of this complex provide considerable support for the chemiosmotic model for energy coupling. Many excellent reviews on the chemistry and morphology of the complex are available.[2,13,39–42]

This multiprotein–lipid complex consists of at least 13 different polypeptides with a molecular weight of approximately 450,000. It can be separated into three parts: (1) F_1, which contains the active site of the enzyme; (2) a proteolipid complex (F_0) comprising a proton channel located in the lipid bilayer of the membrane; and (3) the oligomycin-sensitivity-conferring protein (OSCP) or stalk sector which connects F_1 to the membrane sector. High-resolution electron micrographs of negatively stained preparations of inner mitochondrial membrane reveal inner membrane spheres or headpieces extending 9–10 nm into the matrix and anchored to the membrane through a short stalk (Fig. 2). Biochemical evidence is consistent with F_1 being the headpiece and the oligomycin-sensitivity-conferring protein representing the stalk sector. The ATPase comprises about 15% of the total inner membrane protein of the mitochondria, but only approximately 17% of its mass is confined to the lipid bilayer.[43] Only the membrane sector contains lipid, and 20–30% phospholipid is needed for maximal activity of the reconstituted ATPase.[39] The stalk and F_1 sectors have been called "coupling factors," since they restore energy coupling when added to membranes depleted of these components.

F_1 is water soluble, contains no prosthetic groups, has a molecular weight of about 360,000, and consists of six different polypeptide subunits termed α, β, γ, δ, ϵ, and the inhibitor protein. The soluble F_1 catalyzes hydrolysis but not synthesis of ATP. It appears that the larger subunits, α and β, are present in equal amounts, and either $\alpha_3\beta_3\gamma_1\delta_1\epsilon_1$ or $\alpha_2\beta_2\gamma_2\delta_2\epsilon_2$ stoichiometry is suggested by several experimental approaches.[39,40] The β subunit appears to contain the active site of the enzyme, whereas the α subunit has high-affinity binding sites for adenine nucleotides that regulate activity. The δ and ε subunits function in binding the F_1 to the stalk sector.

The sixth subunit, the inhibitor protein, is a 10,000-dalton protein that specifically inhibits the ATPase activity of soluble or membrane bound F_1. It acts unidirectionally to block ATP hydrolysis but not ATP synthesis. It associates tightly with F_1 when the ATP + Mg level is high and dissociates at low ATP levels, at alkaline pH, and during electron transport.[39]

F_1 contains several binding sites for adenine nucleotides which appear to be involved in ATP synthesis and regulation. Tight binding sites for 3 mol of ATP and 2 of ADP per mol of F_1 have been reported after dialysis and gel filtration. A role for these bound nucleotides has been invoked in an alternate-site model for ATP synthesis.[39,41-43]

The stalk sector has no known intrinsic enzymatic activity but structurally links F_1 to the membrane sector. It probably functions in the proton-translocating activity of the ATPase by conducting protons or mediating conformational changes causing the release and uptake of protons. The sector consists of the OSCP and coupling factor 6 (F_6) of molecular weights 18,000 and 8,000, respectively. The OSCP does not contain inhibitor-binding sites but is needed for restoring sensitivity to oligomycin and other inhibitors that bind to the membrane sector. F_6 interacts with OSCP in binding F_1 to the membrane sector but may be confined to the membrane sector.

The membrane or F_0 sector combines with the stalk and F_1 to form the oligomycin-sensitive ATPase complex. It is composed of phospholipid and perhaps as many as four different polypeptides. At low concentrations, DCCD reacts covalently with acidic residues of a subunit of low molecular weight. This protein has received considerable attention, since it appears to form a proton channel that links ATP synthesis to the H^+ gradient. Because of its hydrophobic residues and solubility in organic solvents, this subunit is classed as a proteolipid. If F_1 is removed from the membrane, the ability to maintain a H^+ gradient is lost, and the proton channel appears to be "open." Oligomycin or DCCD plugs this leak and restores the ability to maintain a H^+ gradient. The binding of enzymatically inactive F_1 will also close the H^+ channel. The DCCD-binding protein probably occurs as a hexamer in the membrane.[45] Another membrane sector protein of 13,000 molecular weight also appears to participate in energy transfer catalyzed by the complete ATPase and is called factor B.

Radioactive photoaffinity labels show that mitochondria contain a specific uncoupler-binding site at a concentration comparable to the number of ATPase complexes.[46] This site fractionates with the ATPase complex and is absent from the other respiratory complexes. Two polypeptides are labeled by uncoupler, the α-subunit of F_1 and a 30,000-molecular-weight protein.

The ATP synthetase catalyzes several "partial" reactions of oxidative phosphorylation in the absence of electron transport that are believed to represent individual steps in the ATP-forming process. These include: (1) Pi–ATP exchange, the exchange of ^{32}Pi with the terminal phosphate group of ATP; (2)

Pi–H_2O exchange, the exchange of ^{18}O between $H_2^{18}O$ and HPO_4^{2-}; and (3) ADP–ATP exchange, the reversible transfer of the terminal phosphate group of ATP to ADP. These reactions have been useful in determining the sequence of steps and in formulating the catalytic mechanism of ATP formation.[42] The uncoupler-stimulated, oligomycin-sensitive ATP hydrolysis (ATPase activity) catalyzed by the ATP synthetase is believed to represent the reversal of the ATP-forming reaction.

4. ROLE OF MEMBRANE TRANSPORTERS IN OXIDATIVE PHOSPHORYLATION

The inner membrane of the mitochondrion has a high selective permeability. Most charged molecules are excluded by this membrane, and, as a consequence, the matrix compartment acts as an osmometer when mitochondria are suspended in sucrose, KCl, and many other solutes. However, as pointed out above (Section 3), the membrane contains transport pathways for substrates and the components of the phosphorylation reaction that provide the necessary communication for these reactants between the matrix and the cytosol compartments.[47–49]

The best characterized of these transport proteins is the adenine nucleotide translocator which shuttles ADP and ATP through the membrane.[50,51] This transporter is a hydrophobic, integral membrane protein (see Fig. 2) which can be isolated and reconstituted into artificial liposome systems. The complete transport system consists of two identical 32,000-dalton subunits. It is present in surprisingly large amounts, accounting for about 14% of the total protein of the inner membrane. The transporter catalyzes a 1:1 exchange of ADP^{3-} for ATP^{4-} so that its activity produces no net change in the size of the matrix adenine nucleotide pool.

Since respiring mitochondria maintain a negative interior membrane potential, the efflux of the more highly charged component ATP^{4-} in exchange for extramitochondrial ADP^{3-} is strongly favored (ADP^{3-} entry is favored by tenfold over that of ATP^{4-}). The reaction as it normally occurs during phosphorylation is therefore electrically unbalanced, with the loss of one e^- charge for each exchange event ($ADP_{in}^{3-}/ATP_{out}^{4-}$). This utilization of a portion of $\Delta\psi$ for the extrusion of ATP^{4-} must be accounted for in the overall stoichiometry of the reaction (i.e., the observed P/O ratio, ATP/$2e^-$, or other expression of efficiency of coupling). The adenine nucleotide transporter is highly specific for its substrates ADP and ATP, and the investigation of its properties has been greatly aided by the availability of the specific inhibitors atractylate, carboxyatractylate, and bongkrekate.

Inorganic phosphate (Pi) is transported across the inner membrane by a mechanism that is separate from that of the adenine nucleotides.[47–49,52] The transport pathway(s) for Pi are not completely clear. There is evidence for the presence of two pathways (1) a Pi/H^+ symport reaction on the so-called phosphate transporter (indistinguishable from Pi/OH^- antiport) which is sensitive to mercurials, N-ethylmaleimide, and other thiol reagents[53] and (2)

Pi/dicarboxylate exchange on the dicarboxylate transporter. This latter system exchanges various dicarboxylates (malate^{2-}, succinate^{2-}, malonate^{2-}) for each other as well as Pi and functions primarily in the uptake of substrates.

These two transporters can be coupled to allow net Pi and net dicarboxylate accumulation in the presence of a proton gradient. During oxidative phosphorylation, Pi enters the matrix on its transporter and exits from the mitochondria as the terminal pyrophosphate of ATP on the adenine nucleotide transporter. The evidence for the presence of two separate carriers for Pi is based primarily on differences in inhibitor sensitivity. The Pi carrier is highly sensitive to the abovementioned SH-group reagents, whereas the dicarboxylate carrier is inhibited by several nonphysiological dicarboxylates such as butylmalonate.[53] However, recent experiments with Pi analogues suggest that a single component could account for all of the observed Pi transport.[54] This issue should be resolved shortly with the availability of purified and reconstituted Pi transport components.[55]

The uptake of substrates, such as pyruvate, can also contribute to the overall stoichiometry of the phosphorylation process.[49] At high concentrations, pyruvate appears to enter the matrix by a nonspecific mechanism (involving the penetration of the free acid, which is present in equilibrium with the pyruvate anion at physiological pH), but the more usual entry pathway under physiological conditions involves a specific pyruvate/OH$^-$ antiporter (or H$^+$ symport as shown in Fig. 2). This mechanism therefore requires the expenditure of one H$^+$ equivalent of the Δp for each pyruvate taken up.

5. THE COUPLING MECHANISM

The exact mechanism by which the free energy of respiration is coupled to the synthesis of ATP in the mitochondrial membrane is not completely understood and retains several elements of controversy.[1] In the past, much experimental effort (and rhetoric) was expended on the premise that the two processes were linked by "high-energy" chemical intermediates analogous to those found in glycolytic phosphorylation (i.e., glyceraldehyde 3-phosphate dehydrogenase and 3-phosphoglycerol phosphate kinase). Interest in such chemical intermediates waned, however, as a number of potential candidates for such a role were examined and discarded on experimental grounds. Over the past decade or so, there has been increasing enthusiasm for Mitchell's chemiosmotic coupling model[56] until, at this point, this hypothesis has gained very wide (but not universal) acceptance. Enthusiasm for this model can be attributed to (1) the gradual realization that each of the basic chemiosmotic postulates can be verified experimentally and (2) the power of the model in explaining mitochondrial, chloroplast, and microbial membrane transport reactions in terms of electrical ($\Delta \psi$) and chemical (ΔpH) driving components.

5.1. Chemiosmotic Coupling

The chemiosmotic coupling model, first presented by Mitchell in 1961, links the respiratory chain to ATP formation indirectly through an electrochemical proton gradient, Δp.[4,56–59] The basic postulates of chemiosmotic coupling are as follows:

1. The respiratory chain consists of alternating hydrogen carriers (flavoproteins, ubiquinone) and electron carriers (cytochromes, Fe–S centers) arranged asymmetrically in the membrane in such a way that the redox activity of the electron transport chain moves H^+ from the matrix to the cytosol side of the membrane and returns electrons back to the matrix side. The net effect of these reactions is the electrogenic translocation of H^+ from the matrix to the cytosol.

2. The inner membrane of the mitochondrion (the coupling membrane) is relatively impermeable to H^+ and OH^- except through specific H^+-translocating pathways. As a result, an electrochemical H^+ potential is built up across the membrane as respiration continues. This H^+ potential is defined as the protonmotive force in mV, Δp, and consists of an electrical component (the membrane potential in mV, $\Delta\psi$) and a chemical component (the pH gradient, ΔpH) related by the expression given in equation 2.

$$\Delta p = \Delta\psi - \frac{2.3RT}{F}\,\Delta pH \qquad [2]$$

3. A highly asymmetric proton-translocating ATPase is also present in the coupling membrane. This component can utilize ATP hydrolysis to eject a pair of H^+ toward the cytosol side of the coupling membrane (H^+_c) but can also consume Δp to synthesize ATP from ADP and Pi in the matrix by reversal of this reaction (equation 3).

$$ATP \leftrightarrow ADP + Pi + 2H^+_c \qquad [3]$$

It is proposed that H^+ is carried through a H^+ channel in the membrane sector of the ATPase to the catalytic site on F_1, thus forming a field of high protonic potential.[4,56,58] It is postulated that the catalytic site faces the membrane at the bottom of a "proton well" and that this arrangement permits the direct interaction of ADP, Pi, and H^+ to form ATP.

4. The coupling membrane contains transport components that promote and regulate the influx and efflux of adenine nucleotides, Pi, substrates, and other components and provide the necessary communication between the matrix and cytosol compartments.[60]

The overall sequence of events in chemiosmotic coupling is shown diagramatically in Fig. 3. Respiratory activity produces the transmembrane H^+ gradient, Δp, which can be utilized by the ATPase to synthesize matrix ATP. The balance among this reaction, ATP-utilizing reactions of the matrix (such as the synthesis of citrulline in liver mitochondria), and the translocation of adenine nucleotides by its translocator is reflected in the matrix phosphorylation potential, $(\Delta G_{ATP})_m$ (equation 4).[61]

$$(\Delta G_{ATP})_m = \Delta G'_0 + RT\,\ln(ATP/ADP \cdot Pi)_m \qquad [4]$$

$\Delta G_0'$ is the standard free energy of ATP formation, and the components within the parentheses are the molar concentrations of ATP, ADP, and Pi in the matrix compartment. The activity of the adenine nucleotide translocator permits ATP to exit from the matrix and contribute to the cytosol phosphorylation potential $(\Delta G_{ATP})_c$ quantitated by an expression analogous to equation 4 involving concentrations of cytosol components. Back pressure from $(\Delta G_{ATP})_c$ and $(\Delta G_{ATP})_m$ limit ATP synthesis by the ATPase, and back pressure from Δp limits electron transport activity when the demand of the cell for ATP is low. The regulatory features of this sequence are discussed in Section 6.

These basic postulates of chemiosmotic coupling have now been verified experimentally to the satisfaction of many investigators, although, as we will discuss, basic issues with regard to the mechanism of H^+ ejection and utilization, stoichiometry, and regulation remain to be clarified. It has been established that mitochondrial electron transport results in electrogenic H^+ extrusion and that the coupling membrane is relatively impermeable to H^+ and OH^-.[62,63] The H^+-transporting ATPase is located asymmetrically in the membrane and is able to generate and utilize H^+ gradients as predicted by the chemiosmotic model.[2,7,40] The chemiosmotic predictions with regard to membrane transporters have also been confirmed, and good progress toward defining the molecular properties of these components has been made.[47–50] In addition, it is now reasonably clear that a membrane-enclosed two-compartment system is necessary for energy conservation and that uncouplers act by discharging the Δp by virtue of their ability to translocate H^+ across membranes (see Fig. 3).[64]

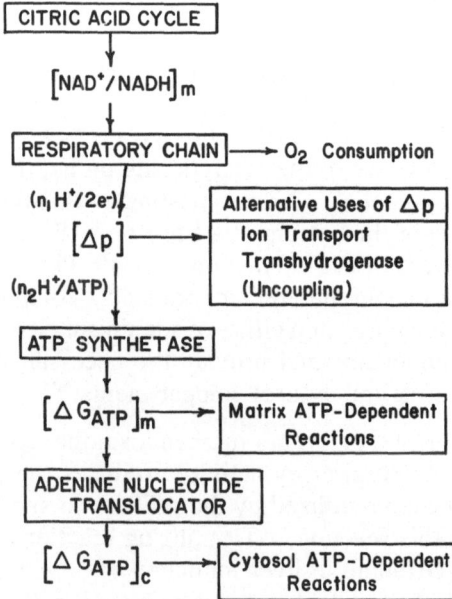

Fig. 3. Diagram showing the interrelationship among components of the oxidative phosphorylation reaction. See the text for details and explanation.

The magnitude of the Δp, $\Delta \psi$, and ΔpH can be estimated using ion distribution and membrane probe techniques, and a value of about 180 mV for the Δp in state 4 is indicated.[58,65] Direct confirmation of the existence of $\Delta \psi$ by microelectrode techniques is still lacking,[66] but many investigators question the reliability of such measurements when applied to mitochondria.[2,67]

Reconstitution experiments provide strong support for a chemiosmotic coupling mechanism. Each of the three respiratory chain complexes that contain a redox span sufficient to support ATP synthesis (complexes I, III, and IV) mediates vectorial H^+ translocation when incorporated into closed phospholipid vesicles.[62,63] The isolated F_0F_1 ATPase also supports H^+ translocation,[40] and when this component is incorporated in a vesicle with one of the H^+-translocating respiratory complexes, oxidative phosphorylation of ADP to ATP can be reconstituted. Especially persuasive evidence against more direct coupling models is provided by the reconstitution of mitochondrial ATPase and bacteriorhodopsin to produce a light-driven phosphorylation that does not involve an oxidation process.[63]

5.2. Molecular Mechanisms of Proton Gradient Formation

Although there is strong evidence that H^+ gradients may link the processes of respiration and ATP synthesis, the mechanism by which these gradients are formed and utilized remains uncertain. For example, it is not clear whether the proton ejection associated with respiration results directly from the redox reactions themselves (as postulated by Mitchell) or from the activity of some closely linked H^+-pumping mechanism. Direct production of Δp by electron transport would require that the redox carriers be organized vectorially in the membrane so as to form the postulated redox loops.[1,4,13,56,58] In Mitchell's proposed loop at site I, FMN is the hydrogen carrier, and Fe–S centers return electrons back across the membrane (see Fig. 4). Coupling sites II and III occur in the proposed protonmotive Q cycle.[21] This scheme requires vectorial movements of reduced ubiquinone across the membrane from the matrix to the cytosol side with $2H^+$ released to the cytosol phase. One electron is passed to cyt c_1, and the other is recycled to oxidized ubiquinone on the matrix side via the b cytochromes (Fig. 4). Thus, ubiquinone functions as the hydrogen carrier for both sites II and III by operating in a cycle. The Q cycle accounts for the known interdependence between cyt b reduction and cyt c_1 oxidation and for the anomalous behavior of b cytochromes in the presence of antimycin and satisfies the H^+-per-site stiochiometry of $2H^+/2e^-$ demanded by the chemiosmotic model. Discussions critical of this scheme have been presented.[22,68,69] Since no hydrogen carriers are present in the cyt c_1-to-oxygen segment, electron transport in this complex closes the final half loop and returns electrons to the matrix side of the membrane where they react with oxygen and matrix protons to form water (Fig. 4). It should be noted that redox loops cannot account for a $H^+/2e^-$ stoichiometry greater than 2 and that recent estimates of this ratio show values of 3 or even 4 for $H^+/2e^-$.[31,70]

As an alternative to this direct formation of Δp by electron transport, several investigators have proposed models in which the H^+ translocation is

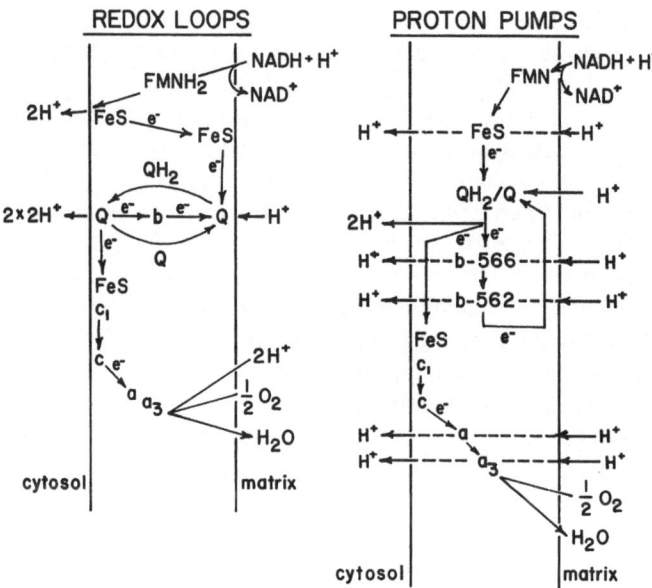

Fig. 4.   Comparison of redox loops as opposed to a proton pump mechanism for the ejection of H^+ by the respiratory chain. Redox loops consist of hydrogen and electron carriers alternately spanning the membrane to provide electrogenic H^+ transport.[4] A protonmotive ubiquinone cycle in which ubiquinol is oxidized with two H^+ released to the cytosol, one electron passed to the iron–sulfur center, and the other recycled via the b cytochromes to ubiquinone on the matrix side of the membrane comprises coupling sites 2 and 3.[21] The proton pump scheme is speculative and shows several redox components acting as proton pumps with a cyclic electron flow through the b cyt. The scheme assumes the H^+ released by the cyt a_3 pump is consumed in the formation of water. For details, see the text and refs. 4, 13, 21, 22, 31 from which the schemes were adapted.

closely associated with the redox components but is not a direct consequence of electron or H^+-transport activity.[1,13,22,23,68] These so-called H^+ pumps (Fig. 4) are usually postulated to result from conformational interaction between redox and H^+-translocating subunits. It is thought that acid or base side chains in a protein, if strategically placed near H^+ channels on the matrix and cytosol faces of the membrane, could transport H^+ as a result of conformational alterations in pK. The proton channels may be formed by α-helical structures arranged perpendicular to the plane of the membrane in analogy to those found in complex IV. Mechanisms for membrane H^+ translocation analogous to the Bohr effect in hemoglobin have been suggested. Here, the redox and H^+-transfer components are considered to be relatively far apart and to interact through protein structural changes.

   The pH dependence of the midpoint potentials of the b cytochromes and cyt a and a_3[10,22] suggests that they could function as redox-linked H^+ pumps. A H^+-pumping cyt b cycle with each of the b cytochromes acting as a H^+ pump and a cyclic pattern of electron transport has been proposed to account for the observed anomalies in this segment of the respiratory chain.[6,22,68,71] In contrast to the redox-loop model for H^+ gradient formation, proton H^+ pump models require a vectorial organization of the H^+-translocating regions but not necessarily of the redox centers.[13]

Topological investigations of the mitochondrial membrane indicate that the electron transport chain is indeed organized asymmetrically, but these studies also suggest that the hydrogen and electron carriers do not span the membrane as postulated by the chemiosmotic redox loops.[13] Although complex I contains both hydrogen (FMN) and electron (iron–sulfur centers) carriers, a loop arrangement across the membrane is not supported by experiments with ferricyanide, which show that this nonpenetrant electron acceptor does not react with the respiratory chain. Since no hydrogen carrier exists in the cyt *c* oxidase, no redox loop or energy conservation site is proposed for this segment, although a large free energy change (about 25 kcal) is available in the span from cyt *c* to oxygen. However, experiments with reconstituted cyt *c* oxidase indicate that H^+ is translocated across the membrane during electron flow.[85] Also, the complicated structure of the enzyme and the presence of a helical structure both suggest that mere electron translocation is not the only function of complex IV and lend further support to a H^+-pump-type mechanism.

There are also indications from studies using DCCD,[32] fluorescamine,[73] and other reagents that electron transport activity can be separated from H^+ translocation. Clearly, further exploration of this possibility will be helpful in resolving the issue of how these activities are interrelated.

5.3. Alternatives to Chemiosmotic Coupling

Chemiosmotic coupling requires that the membrane act only as a diffusion barrier so that Δp can be established between two osmotic compartments. Alternative coupling mechanisms have been proposed that are similar in concept to the chemiosmotic hypothesis but in which the proton flux is retained within the membrane[73a] or in phospholipid–water interphases along the membrane.[74]

There are also proposed coupling mechanisms that involve a more direct interaction between respiratory components and the synthesis of ATP than that postulated by the chemiosmotic model. One model for "conformational coupling" proposes that there is direct communication between respiratory carriers and the ATP synthetase by protein–protein interaction.[1,75] Compelling evidence for such direct interactions is not available at present, but variations on this theme would allow the generation and utilization of Δp by conformational changes such as those that have been postulated to produce Bohr protons. Along these lines, it has also been suggested that conformational changes at redox centers could be reflected in alternative activation of ATP synthetase or production of Δp (which in this case would be viewed as a device to support ion transport rather than an intermediate of oxidative phosphorylation).

A conformational model for ATP synthesis proposed by Boyer[42,43] is compatible with either conformational or chemiosmotic coupling mechanism but differs from the direct "H^+-well" ATPase mechanism proposed by Mitchell. In this model, a conformational change in the F_1 complex driven by a proton gradient or direct interaction with the respiratory chain leads to the release of preformed, tightly bound ATP. It is the release of ATP and not the actual formation from ADP and Pi that is postulated to require energy. This hypothesis is based partly on a differential sensitivity of coupling exchange reactions to uncouplers and partly on the presence of adenine nucleotide tightly bound to F_1. The model involves alternating binding sites for ATP and ADP + Pi. This is an indirect mechanism in that the interaction between the catalytic site and the H^+ gradient is mediated by a conformational change, and they do not occur at the same locus.

6. REGULATION OF OXIDATIVE PHOSPHORYLATION

The rate of respiration of isolated mitochondria depends on the availability of ADP as shown in Fig. 5. A suspension of mitochondria respires at a low rate in the presence of Pi and substrate until supplemented with ADP. A high rate of respiration, defined as state 3,[76] is then maintained until nearly all of the added ADP is phosphorylated to ATP. The production of ATP can be followed conveniently in such experiments by a sensitive pH recording (Fig. 5), since there is a net alkaline shift produced by the synthesis of ATP from ADP and Pi in this pH range.[77,78] When the phosphorylation reaction is complete, the rate of respiration returns to a low resting (or controlled) value which is defined as state 4. The ratio of the rate of state 3 to state 4 respiration is known as the respiratory control index and is widely used as a criterion for intact, functional preparations of mitochondria (see refs. 76 and 79–81 for more detailed exposition).

The efficiency of coupling can also be estimated from oxygen electrode records by determining the O_2 consumed in the phosphorylation of a known amount of ADP (cf. Fig. 5). Efficiencies are expressed in terms of the amount of Pi esterified (P/O ratio), ADP utilized (ADP:O ratio), or ATP formed (ATP:O ratio) and are most commonly given for a $2e^-$ oxidation reaction.[79,80]

For many years, the oxidation of a mole of NADH was thought to yield a limit of 3 mol of ATP (P/O = 3), since it was assumed that oxidation and phosphorylation were directly linked at specific "coupling sites." The indirect coupling mechanism provided by the chemiosmotic model would account for the fact that whole integer values for P/O ratios are seldom attained experimentally. In addition, since a portion of Δp is expended in substrate uptake and ATP^{4-} extrusion, it would seem likely that fractional P/O values less than 3 (or 2 for succinate) might be realistic. Unambiguous definition of theoretical P/O ratios awaits precise determination of the $H^+/2e^-$ ratio for various segments of the electron transport chain and the H^+/ATP ratio for the ATP synthetase. Since ΔG is not identical in the three segments of the chain that produce H^+ gradients (see Fig. 1), it is possible that $H^+/2e^-$ ratios (and presumably P/O ratios, since H^+/ATP is invariant) would vary from site to site.[22] Either 2^{82} or 3^{83} H^+/ATP appear to be consumed by the ATP synthetase, and 1 H^+/ATP is needed for translocation of ATP, ADP, and Pi,[6] for a total of 3 or 4 H^+ required per ATP synthesized and transported to the cytoplasm. Maxi-

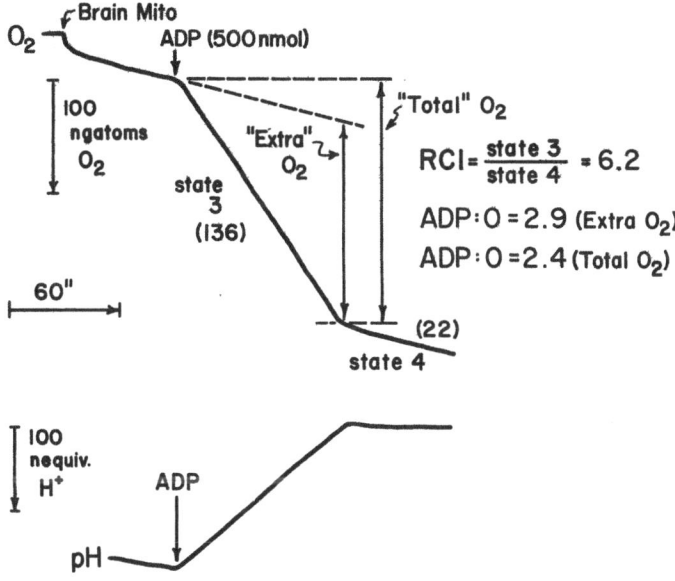

Fig. 5. Dependence of the rate of respiration on ADP (respiratory control). Isolated rat brain mitochondria (1 mg protein) were added to 1.5 ml of a medium consisting of KCl (120 mM), K phosphate (5 mM, pH 7.1), EGTA (1 mM), and glutamate (10 mM) held at 25° in a Gilson oxygraph. The oxygen content and the pH of the suspension were recorded simultaneously using a Clark electrode and a combination pH electrode. Addition of a known amount of ADP (0.5 μmol) results in an elevated rate of respiration (state 3) and an alkalinization of the medium which reflects the synthesis of ATP. The pH record will show the rate of ATP formation in mitochondria that have lost respiratory control because of the presence of free fatty acids or other conditions that produce "loose coupling" and can also be used to estimate ATPase activity.[78] The respiratory control index is calculated by dividing the rate of respiration in the presence of ADP (state 3) by the rate of controlled respiration which prevails after ATP synthesis is complete (state 4). The efficiency of phosphorylation (ADP:O) is estimated from the amount of ADP phosphorylated and the amount of O_2 consumed during state 3 (Total O_2).[79] Somewhat higher estimates of ADP:O are obtained if one uses the amount of O_2 taken up over and above the state 4 rate (the extra O_2).[81] Preparations of brain mitochondria[117] patterned after those of Clark and Niklas[114,115] averaged 136 ngatoms $O_2 \cdot min^{-1} \cdot mg \ protein^{-1}$ for state 3 respiration and showed an average ADP:O of 2.96 for glutamate (extra O_2) and a mean respiratory control index of over 6 for the conditions shown ($n =$ 7). Quite comparable results can be obtained using the brain mitochondria preparation of Bernard and Cockrell,[116] and both procedures can be applied to the preparation of mitochondria from dog and rat spinal cords (collaborative studies with C. Ansevin, J. N. Allen, N. Clendenon, S. Palayoor, and N. J. Bannon, unpublished data).

mum P/O ratios for matrix NADH could conceiveably be as low as 1.5 or as high as 4.0 depending on whether 6 or 12 H^+ are extruded by electron transport and whether 3 or 4 H^+ are consumed in ATP formation and transport. The recent determination of P/O values of 2.1–2.2 for β-hydroxybutyrate and malate oxidation[84] are consistent with ratios of H^+ extruded to H^+ consumed of either 9/4 or 6/3.

Isolated mitochondria maintain a highly reduced pool of NADH when respiring in state 4, and there is a marked shift of NADH toward a more oxidized state when state 3 (active phosphorylation) is inititated.[76] The other components of the respiratory chain are progressively more oxidized in state 4, with the terminal components, cyt $a + a_3$, almost completely in the oxidized form. When oxygen is not available to accept electrons, all of the components of the respiratory chain become reduced. In contrast to these results with isolated mitochondria, direct spectrophotometric estimates of the redox carriers of the brain *in vivo* show cyt aa_3 to be about 85% reduced.[85,86] The basis for this discrepancy is not understood at present.

The primary function of mitochondrial oxidative phosphorylation is the production of ATP to support the energy-requiring reactions of the cell. The process is closely regulated *in vivo* so that substrate utilization is efficiently matched to the ATP needs of the cell. The sequence of events that connect substrate utilization by citric acid cycle enzymes to the maintenance of extramitochondrial ATP levels is shown in Fig. 3. Increased utilization of ATP, as would follow from increased Na^+-K^+ ATPase activity accompanying the stimulation of a nerve, for example, produces a transient dislocation in $(\Delta G_{ATP})_c$ which is rapidly reflected in increased respiration and citric acid cycle activity.

There is presently a controversy as to exactly how such changes in the cytosolic adenine nucleotide pool are able to control respiration. Wilson and co-workers,[87–89] for example, postulate that the respiratory chain components up to the terminal oxidase are in a state of near equilibrium with the cytosol phosphorylation potential. In this model, a shift to lower $(\Delta G_{ATP})_c$ would shift the redox components toward more reduced cyt aa_3 and consequently toward higher rates of respiration.

In contrast to this thermodynamic regulatory model, another school of thought holds that the ratio of ATP/ADP exerts a kinetic control over the process at the level of the adenine nucleotide transporter.[90,91] In this model, a decrease in $(ATP/ADP)_c$ increases the exchange of ADP_{in} for ATP_{out}. This exchange tends to restore $(ATP)_c$ at the expense of matrix ATP, and the decreased $(\Delta G_{ATP})_m$ results in increased ATP synthetase activity and consumption of Δp. Since the respiratory chain is thought to be held in a low activity or controlled condition (state 4) by the back pressure exerted by high levels of Δp, this dissipation of the H^+ electrochemical gradient as a result of H^+ influx through the synthetase produces a stimulation of respiratory activity and a drop in the matrix NADH level. Since the enzymes of the citric acid cycle respond to decreased NADH with increased activity, there is a shift in overall cycle activity in the direction of restoration of matrix NADH. Such elevated levels of citric acid cycle, respiratory chain, and ATP synthetase activity would be maintained until the initiating dislocations in $(\Delta G_{ATP})_c$ are removed and the system returns to its initial state.

There is good experimental evidence that $(\Delta G_{ATP})_c > (\Delta G_{ATP})_m$, with typical values being about 14 and 10 kcal/mol, respectively.[92] The free energy required for synthesis of ATP in the cytosol should reflect the $(\Delta G_{ATP})_m$ plus the energy cost of translocating the ATP across the membrane. The dislocation in adenine nucleotide ratios between the two compartments has been shown to reflect the electrogenic nature of the adenine nucleotide transporter and $\Delta \psi$

by the expression given in equation 5.[50,61]

$$\Delta\psi = (RT/F) \ln[(ATP/ADP)_c/(ATP/ADP)_m]$$ [5]

Recent studies show that when cytosolic ATP/ADP ratios are varied at a nonlimiting Pi concentration, the ATP/ADP ratio, but not the ATP/ADP · Pi ratio, correlates with the respiratory rate.[90,91] Ratios of ATP/ADP between 5 and 100 control the rate of respiration, with rates being maximal and independent of the adenine nucleotides when this ratio falls below 5.

When respiration of rat liver mitochondria is stimulated to identical levels by (1) partial uncoupling (direct dissipation of Δp), (2) citrulline synthesis [direct utilization of $(\Delta G_{ATP})_m$], or (3) utilization of extramitochondrial ATP in the hexokinase reaction, identical $(ATP/ADP)_m$ ratios are found in the three systems.[93] The external ratio, $(ATP/ADP)_c$, varies considerably for comparable rates of respiration in these systems, and it is concluded that the respiratory rate is controlled directly by Δp, that ATP synthetase is at near equilibrium with Δp, and that there is no equilibrium at the step that transports ATP to the external compartment.[93]

In contrast, Holian and Wilson[87] have concluded from parallel estimates of the various parameters that there is no consistent correlation between Δp, $\Delta\psi$, or ΔpH and the respiratory rate and question whether Δp can serve as the primary intermediate in oxidative phosphorylation on thermodynamic grounds. They reason that the free energy of an intermediate component, such as Δp, should fall between that of the redox reactions and the resulting phosphorylation potential that is maintained. Their analysis suggests that three to seven H^+ would have to be transported per ATP synthesized for this condition to be met.

The stoichiometry of H^+ translocated as a result of the reactions of the respiratory chain ($n_1 H^+/2e^-$ in Fig. 3) as well as that for the number of H^+ consumed in ATP synthesis ($n_2 H^+/ATP$) is obviously an issue of considerable importance to the overall mechanism of oxidative phosphorylation. Estimates of the value of n_1 range from 2 to 4, and values of 3 for n_2 have been reported.[31,82,83,94] It should be noted that the original formulations of the chemiosmotic coupling model require that both n_1 and n_2 be equal to 2. Difficulties in estimating H^+ stoichiometries by oxidant and reductant pulse techniques have been discussed by Wikström and Krab.[22,31] They include underestimation caused by dissipating side reactions (Pi/H^+ symport) or particular reaction conditions and overestimation caused by slow-responding oxygen electrodes.

7. *ALTERNATIVE USES OF PROTONMOTIVE FORCE*

Mitochondria can utilize the Δp produced by electron transport to support reactions other than the synthesis of ATP (see Fig. 3). These include a wide variety of anion and cation transport reactions and the pyridine nucleotide transhydrogenase reaction. These reactions therefore compete with ATP synthesis and can affect the overall efficiency of oxidative phosphorylation under conditions in which the utilization of Δp by these pathways becomes significant.

7.1. Ion Transport

As has been summarized in recent reviews,[47–49,95–98] mitochondrial anion and cation transport reactions can be classified as (1) those that respond to ΔpH, (2) those that respond to $\Delta \psi$, and (3) exchange reactions that solely reflect the concentration gradients of the exchanged components. Since respiring mitochondria maintain a ΔpH (interior alkaline by about one pH unit), anion translocators in the first category (the phosphate, pyruvate, and glutamate transporters) promote net uptake of these components by H^+–anion symport or the equivalent $anion^-/OH^-$ antiport reactions. In contrast, anion transporters that respond to $\Delta \psi$ seem to provide pathways for the extrusion of more highly charged components from the mitochondrion at the expense of a portion of the $\Delta \psi$. The most prominent reaction in this category is the exchange of ATP_{out}^{4-} for ADP_{in}^{3-} on the adenine nucleotide translocator as discussed in Section 6. The electrogenic nature of the glutamate/aspartate transporter, which promotes the exchange of uncharged glutamate (in) for $aspartate^-$ (out), accounts for the unidirectional inward flow of cytosol reducing equivalents in the sequence known as the malate/aspartate shuttle.[47–49,99] The dicarboxylate and the α-ketoglutarate exchangers provide examples of mitochondrial anion translocators that respond to neither ΔpH nor $\Delta \psi$ and seem to depend on the concentration gradients of their substrates.

The uptake of cations by mitochondria occurs in response to the negative interior potential, and the most prominent of these reactions is the uptake of Ca^{2+} which is promoted by a high-affinity uniporter.[96–98] This highly active electrophoretic influx of Ca^{2+} is balanced by an electroneutral efflux reaction (Ca^{2+}/Na^+ in heart and brain mitochondria; Ca^{2+}/H^+ in liver). There is presently some question as to whether this balance between Ca^{2+} influx and efflux reactions functions to buffer the extramitochondrial free Ca^{2+} of the cell or to regulate Ca^{2+}-sensitive enzymes in the matrix.

Since mitochondria maintain a negative $\Delta \psi$ of about 200 mV and must function in an intracellular environment that contains roughly 140 mM K^+, it is clear that any significant uncontrolled permeability to K^+ would compromise both the osmotic integrity of the organelle and its ability to maintain Δp. The mitochondrion appears to cope with its K^+ environment first by maintaining a low electrophoretic permeability to monovalent cations and secondarily by utilizing ΔpH to extrude excess K^+ on a K^+/H^+ exchanger.[95] Recent evidence suggests that the low electrophoretic permeability to K^+ depends on the presence of NADPH[100] and that the K^+_{out}-for-H^+_{in} exchange is negatively regulated by divalent cations, so that it comes into play only when needed for osmotic control.[101] Mitochondria also contain a Na^+/H^+ exchange component which may contribute to net Ca^{2+} balance in mitochondria which extrude Ca^{2+} by Ca^{2+}/Na^+ exchange; this permits the extrusion of any Na^+ that might enter the mitochondrion under other circumstances.

7.2. Pyridine Nucleotide Transhydrogenase

Mitochondrial pyridine nucleotide transhydrogenase promotes the following reaction (equation 6):

$$NADH_m + NADP^+_m + nH^+_c \rightarrow NAD^+_m + NADPH_m + nH^+_m \quad [6]$$

The reaction in respiring mitochondria consumes a portion of the ΔpH component of Δp to convert $NADP^+$ to NADPH.[102,103] The transhydrogenase is an integral membrane protein consisting of two 120,000-dalton subunits[104] and does not appear to be structurally or functionally related to the NADH dehydrogenase of complex I. The stoichiometry of proton translocation (n in equation 6) has been reported as one[105] or two[106] H^+ per hydride ion transferred. The NADPH cannot traverse the membrane and is probably utilized in the matrix in the glutathione reductase reaction and in the maintenance of membrane permeability by thiol reduction.[107] The redox state of the $NADP^+$/NADPH couple remains relatively constant during wide swings in the redox state of the NAD^+/NADH pool.

8. INHIBITORS OF OXIDATIVE PHOSPHORYLATION

A large number of different reagents have been shown to interfere with the process of oxidative phosphorylation. As summarized in Table I, these can be grouped in four categories:

1. Respiratory chain inhibitors. These reagents block respiration at various points in the chain and eliminate the source of energy to support phosphorylation.[108,109]
2. ATP synthetase inhibitors. These react with either the F_1 (aurovertin) or the membrane sector (oligomycin, DCCD) of the ATPase to prevent its normal activity.[39–41,110]
3. Transport inhibitors. A number of compounds are available that block access of ADP (atractylate), Pi (mersalyl), or substrate (butylmalonate) to the matrix compartment.[47–49] Inhibitors of adenine nucleotide and Pi transport resemble the ATPase inhibitors in their observed effects in that they eliminate the stimulation of respiration by ADP (cf. Fig. 5). These reagents block the formation of ATP from extramitochondrial ADP but not from matrix ADP.
4. Uncouplers. These reagents, typified by 2,4-dinitrophenol, act to separate respiration from the synthesis of ATP.[46,64] Uncouplers stimulate respiration in the absence of ADP or in the presence of oligomycin, results that indicate that the rate of electron transport is no longer controlled by the components of the phosphorylation reaction. In the presence of an uncoupler, the free energy of respiration is dissipated as heat. Disruption of the coupling process by these reagents also permits the ATPase to run in reverse and hydrolyze ATP at an accelerated rate.

Table I
Some Inhibitors of Oxidative Phosphorylation[a]

Compound	Site of inhibition
Respiratory chain inhibitors	
Rotenone, amytal	NADH dehydrogenase
Carboxin, 2-thenoyltrifluoroacetone	Succinate dehydrogenase
Antimycin A, n-heptylquinoline N-oxide	Cyt b–c_1 region
Cyanide, carbon monoxide	Cyt c oxidase
ATP synthetase inhibitors	
Aurovertin, quercetin	F_1
Oligomycin, DCCD	Membrane sector
Transport inhibitors	
Atractylate, bongkrekate	ATP/ADP carrier
Mersalyl, N-ethylmaleimide	P_i carrier
Hydroxycinnamate, phenylpyruvate	Pyruvate carrier
Butylmalonate, mersalyl	P_i/dicarboxylate carrier
Uncouplers	
2,4-Dinitrophenol, dicumarol	Collapses H^+ gradient
Valinomycin + K^+	Induces futile cation cycling
Detergents, long-chain fatty acids	Alters membrane structure

[a] See text for references.

Classical uncouplers are lipid-soluble weak acids containing an aromatic ring structure. These reagents have been shown to promote H^+ passage through membranes and dissipation of proton gradients as predicted by the chemiosmotic hypothesis (see Fig. 3).

Uncoupling of oxidative phosphorylation is also accomplished by reagents that modify or disrupt membranes[64,111] so that the necessary ΔpH and membrane potential cannot be maintained. Ionophores, reagents that promote ion permeability,[112,113] often uncouple mitochondria by allowing futile cycling of ions to dissipate Δp. An example of this type of uncoupling is provided by ionophore A23187 which promotes the efflux of matrix Ca^{2+} by Ca^{2+}/H^+ exchange. This efflux, in combination with the rapid electrophoretic influx of Ca^{2+} on the endogenous Ca^{2+} uniporter, effectively discharges mitochondrial Δp and prevents the synthesis of ATP.

REFERENCES

1. Boyer, P. D., Chance, B., Ernster, L., Mitchell, P., Racker, E., and Slater, E. C., 1977, *Annu. Rev. Biochem.* **46**:955–1026.
2. Racker, E., 1976, *A New Look at Mechanisms in Bioenergetics,* Academic Press, New York.
3. Lee, C. P., Schatz, G., Ernster, L. (eds.), 1979, *Membrane Bioenergetics,* Addison-Wesley, London.
4. Mitchell, P., 1979, *Eur. J. Biochem.* **95**:1–20.
5. Schafer, G., and Klingenberg, M. (ed.), 1978, *Energy Conservation in Biological Membranes,* Springer-Verlag, New York.
6. von Jagow, G., and Engel, W. D., 1980, *Angew. Chem. [Engl.]* **19**:659–675.
7. Hinkle, P., and McCarty, R. E., 1978, *Sci. Am.* **238**(3):104–123.

8. Sanadi, D. R., and Wohlrab, H., 1976, *Chemical Mechanisms in Bioenergetics* (D. R. Sanadi, ed.), American Chemical Society, Washington, pp. 123–171.
9. Wainito, W. W., 1970, *The Mammalian Respiratory Chain,* Academic Press, New York.
10. Wilson, D. F., 1980, *Membrane Structure and Function* (E. E. Bittar, ed.), J. Wiley & Sons, New York, pp. 153–195.
11. Hatefi, Y., and Galante, Y. M., 1978, *Energy Conservation in Biological Membranes* (G. Schafer and M. Klingenberg, eds.), Springer-Verlag, New York, pp. 19–30.
12. Fleischer, S., and Packer, L. (eds.), 1978, *Methods in Enzymology,* Volume 53, Academic Press, New York.
13. DePierre, J. W., and Ernster, L., 1977, *Annu. Rev. Biochem.* **46:**201–262.
14. Ohnishi, T., 1979, *Membrane Proteins in Energy Transduction* (R. A. Capaldi, ed.), Marcel Dekker, New York, pp. 1–87.
15. Heron, C., Ragan, C. I., and Trumpower, B. L., 1978, *Biochem. J.* **174:**791–800.
16. Smith, S., and Ragan, I. C., 1980, *Biochem. J.* **185:**315–326.
17. Ragan, C. I., 1976, *Biochim. Biophys. Acta* **456:**249–290.
18. Trumpower, B. L., and Katki, A. G., 1979, *Membrane Proteins in Energy Transduction* (R. A. Capaldi, ed.), Marcel Dekker, New York, pp. 89–199.
19. Yu, C. A., Yu, L., and King, T. E., 1977, *Biochem. Biophys. Res. Commun.* **78:**259–265.
20. Rieske, J. S., 1976, *Biochim. Biophys. Acta* **456:**195–247.
21. Mitchell, P., 1976, *J. Theor. Biol.* **62:**327–367.
22. Wikström, M., and Krab, K., 1980, *Curr. Top. Bioenerget.* **10:**51–101.
23. Papa, S., 1976, *Biochim. Biophys. Acta* **456:**39–84.
24. Nagaoka, S., Yu, L., and King, T. E., 1981, *Arch. Biochem. Biophys.* **208:**334–343.
25. Malmstrom, B. G., 1979, *Biochim. Biophys. Acta* **549:**281–303.
26. Azzi, A., and Casey, R. P., 1979, *Mol. Cell. Biochem.* **28:**169–184.
27. Caughey, W. S., Wallace, W. J., Volpe, J. A., and Yoshikawa, S., 1976, *The Enzymes,* 3rd ed., Volume 13 (P. D. Boyer, ed.), Academic Press, New York, pp. 299–344.
28. Henderson, R., Capaldi, R. A., and Leigh, J. S., 1977, *J. Mol. Biol.* **112:**631–648.
29. Wikström, M., 1981, *Proc. Natl. Acad. Sci. U.S.A.* **78:**4051–4054.
30. Chance, B., Waring, A., and Saronio, C., 1978, *Energy Conservation in Biological Membranes* (G. Schafer and M. Klingenberg, eds.), Springer-Verlag, New York, pp. 56–73.
31. Wikström, M., and Krab, K., 1979, *Biochim. Biophys. Acta* **549:**177–222.
32. Casey, R. P., Thelen, M., and Azzi, A., 1980, *J. Biol. Chem.* **255:**3994–4000.
33. Azzi, A., 1980, *Biochim. Biophys. Acta* **594:**231–252.
34. Saraste, M., Penttila, T., Coggins, J. R., and Wikström, M., 1980, *FEBS Lett.* **114:**35–38.
35. Greenawalt, J. W., 1979, *Methods Enzymol.* **55:**88–98.
36. Hackenbrock, C. R., 1981, *Trends Biochem. Sci.* **6:**151–154.
37. Bosshard, H. R., Zurrer, M., Schagger, H., and von Jagow, G., 1979, *Biochem. Biophys. Res. Commun.* **89:**250–258.
38. Kawato, S., Sigel, E., Carafoli, E., and Cherry, R. J., 1981, *J. Biol. Chem.* **256:**7518–7527.
39. Senior, A. E., 1979, *Membrane Proteins in Energy Transduction* (R. A. Capaldi, ed.), Marcel Dekker, New York, pp. 233–278.
40. Futai, M., and Kanazawa, H., 1980, *Curr. Top. Bioenerget.* **10:**181–215.
41. Pedersen, P. L., Amzel, L. M., Soper, J. W., Cintron, N., and Hullhen, J., 1978, *Energy Conservation in Biological Membranes* (G. Schafer and M. Klingenberg, eds.), Springer-Verlag, New York, pp. 159–194.
42. Cross, R. L., 1981, *Annu. Rev. Biochem.* **50:**681–714.
43. Rosen, G., Gresser, M., Vinkler, C., and Boyer, P. D., 1979, *J. Biol. Chem.* **254:**10654–10661.
44. Soper, J. W., Decker, G. L., and Pedersen, P. L., 1979, *J. Biol. Chem.* **254:**11170–11176.
45. Graf, T., and Sebald, W., 1978, *FEBS Lett.* **94:**218–222.
46. Hanstein, W. G., 1976, *Biochim. Biophys. Acta* **456:**129–148.
47. Scarpa, A., 1979, *Membrane Transport in Biology,* Volume 2 (D. C. Tosteson, ed.), Springer-Verlag, New York, pp. 263–355.
48. Williamson, J. R., 1976, *Gluconeogenesis* (R. W. Hanson and M. A. Mehlman, eds.), John Wiley & Sons, New York, pp. 165–220.
49. LaNoue, K. F., and Schoolwerth, A. C., 1979, *Annu. Rev. Biochem.* **48:**871–922.

50. Klingenberg, M., 1980, *J. Membr. Biol.* **56:**97–105.
51. Vignais, P. V., 1976, *Biochim. Biophys. Acta* **456:**1–38.
52. Coty, W. A., and Pedersen, P. L., 1975, *Mol. Cell. Biochem.* **9:**109–124.
53. Fonyo, A., 1979, *Pharmacol. Ther.* **7:**627–645.
54. Freitag, H., and Kadenbach, B., 1978, *Eur. J. Biochem.* **83:**53–57.
55. Wohlrab, H., 1980, *J. Biol. Chem.* **255:**8170–8173.
56. Mitchell, P., 1976, *Biochem. Soc. Trans.* **4:**399–430.
57. Mitchell, P., 1961, *Nature* **191:**144–148.
58. Mitchell, P., 1966, *Biol. Rev.* **41:**445–502.
59. Mitchell, P., and Moyle, J., 1969, *Eur. J. Biochem.* **7:**471–484.
60. Mitchell, P., 1970, *Symp. Soc. Gen. Microbiol.* **20:**121–166.
61. Williamson, J. R., 1979, *Annu. Rev. Physiol.* **41:**485–506.
62. Kagawa, Y., 1978, *Biochim. Biophys. Acta* **505:**45–93.
63. Racker, E., 1979, *Acct. Chem. Res.* **12:**338–344.
64. Heytler, P. G., 1979, *Methods Enzymol.* **55:**462–472.
65. Rottenberg, H., 1979, *Methods Enzymol.* **55:**547–569.
66. Tedeschi, H., 1980, *Biol. Rev.* **55:**171–206.
67. Rottenberg, H., 1979, *Trends Biochem. Sci.* **4:**N182–N184.
68. Wikström, M., Krab, K., and Saraste, M., 1981, *Annu. Rev. Biochem.* **50:**623–55.
69. Storey, B. T., 1980, *Biochemistry of Plants,* Volume 2 (D. D. Davies, ed.), Academic Press, New York, pp. 125–195.
70. Reynafarje, B., and Lehninger, A. L., 1978, *J. Biol. Chem.* **253:**6331–6334.
71. Wikström, M., 1973, *Biochim. Biophys. Acta* **301:**155–193.
72. Sigel, E., and Carafoli, E., 1980, *Eur. J. Biochem.* **111:**299–306.
73. Tu, S., Lam, E., Ramirez, F., and Marecek, J. F., 1981, *Eur. J. Biochem.* **113:**391–396.
73a. Williams, R. J. P., 1978, *Biochim. Biophys. Acta* **505:**1–44.
74. Kell, D. B., 1979, *Biochim. Biophys. Acta* **549:**55–99.
75. Boyer, P. D., 1965, *Oxidases and Related Redox Systems,* Volume 2 (T. E. King, H. S. Mason, and M. Morrison, eds.), John Wiley & Sons, New York, pp. 994–1008.
76. Chance, B., and Williams, G. R., 1956, *Adv. Enzymol.* **17:**65–134.
77. Nishimura, M., Ito, T., and Chance, B., 1962, *Biochim. Biophys. Acta* **59:**177–182.
78. Jurkowitz, M., Scott, K. M., Altschuld, R. A., Merola, A. J., and Brierley, G. P., 1974, *Arch. Biochem. Biophys.* **165:**98–113.
79. Estabrook, R. W., 1967, *Methods Enzymol.* **10:**41–57.
80. Lessler, M. A., and Brierley, G. P., 1969, *Methods Biochem. Anal.* **17:**1–29.
81. Lehninger, A. L., 1975, *Biochemistry,* Worth Publications, New York.
82. Thayer, W. S., and Hinkle, P. C., 1973, *J. Biol. Chem.* **248:**5395–5402.
83. Azzone, G. F., and Massari, S., 1973, *Biochim. Biophys. Acta* **301:**195–226.
84. Hinkle, P. C., and Yu, M. L., 1979, *J. Biol. Chem.* **254:**2450–2455.
85. Jöbsis, F. F., and Rosenthal, M., 1978, *Cerebral Vascular Smooth Muscle and its Control, Ciba Foundation Symposium 56,* Elsevier, New York, pp. 129–148, 149–167.
86. Siesjö, B. K., 1978, *Brain Energy Metabolism,* John Wiley & Sons, New York.
87. Holian, A., and Wilson, D. F., 1980, *Biochemistry* **19:**4213–4221.
88. Sussman, I., Erecińska, M., and Wilson, D. F., 1980, *Biochim. Biophys. Acta* **591:**209–223.
89. Wilson, D. F., Stubbs, M., Veech, R. L., Erecińska, M., and Krebs, H. A., 1974, *Biochem. J.* **140:**57–64.
90. Kunz, W., Bohnensack, R., Böhme, G., Küster, U., Letko, G., and Schönfeld, P., 1981, *Arch. Biochem. Biophys.* **209:**219–229.
91. Davis, E. J., and Davis-van Thienen, W. I. A., 1978, *Biochem. Biophys. Res. Commun.* **83:**1260–1266.
92. Slater, E. C., 1979, *Methods Enzymol.* **55:**235–245.
93. Kuster, U., Letko, G., Kunz, W., Duszynsky, J., Bogucka, K., and Wojtczak, L., 1981, *Biochim. Biophys. Acta* **636:**32–38.
94. Mitchell, P., and Moyle, J., 1967, *Biochem. J.* **105:**1147–1162.
95. Brierley, G. P., 1978, *The Molecular Biology of Cell Membranes* (S. Fleischer, Y. Hatefi, D. H. MacLennan, and A. Tzagoloff, eds.), Plenum Press, New York, pp. 295–308.

96. Saris, N.-E., and Åkerman, K. E. O., 1980, *Curr. Top. Bioenerget.* **10**:103–179.
97. Fiskum, G., and Lehninger, A. L., 1980, *Fed. Proc.* **39**:2432–2436.
98. Carafoli, E., and Crompton, M., 1978, *Curr. Top. Membr. Trans.* **10**:151–216.
99. Dawson, A. G., 1979, *Trends Biochem. Sci.* **4**:171–176.
100. Jung, D. W., and Brierley, G. P., 1981, *J. Biol. Chem.* **256**:10490–10496.
101. Garlid, K. D., 1980, *J. Biol. Chem.* **255**:11273–11279.
102. Rydström, J., 1977, *Biochim. Biophys. Acta* **463**:155–184.
103. Earle, S. R., and Fisher, R. R., 1980, *Biochemistry* **19**:561–569.
104. Anderson, W. M., and Fisher, R. R., 1978, *Arch. Biochem. Biophys.* **187**:180–190.
105. Earle, S. R., and Fisher, R. R., 1980, *J. Biol. Chem.* **255**:827–830.
106. Moyle, J., and Mitchell, P., 1973, *Biochem. J.* **132**:571–585.
107. Chance, B., Sies, H., and Boveris, A., 1979, *Physiol. Rev.* **59**:527–605.
108. Singer, T. P., 1979, *Methods Enzymol.* **55**:454–462.
109. Slater, E. C., 1967, *Methods Enzymol.* **10**:48–57.
110. Linnett, P. E., and Beechey, R. B., 1979, *Methods Enzymol.* **55**:472–518.
111. Wojtczak, L., 1976, *J. Bioenerget. Biomembr.* **8**:293–311.
112. Gómez-Puyou, A., and Gómez-Lojero, C., 1977, *Curr. Top. Bioenerget.* **7**:221–257.
113. Reed, P. W., 1979, *Methods Enzymol.* **55**:435–454.
114. Clark, J. B., and Nicklas, W. J., 1970, *J. Biol. Chem.* **245**:4724–4731.
115. Lai, J. C. K., and Clark, J. B., 1979, *Methods Enzymol.* **55**:51–60.
116. Bernard, P. A., and Cockrell, R. S., 1979, *Biochim. Biophys. Acta* **548**:173–186.
117. Ansevin, C. F., 1980, *Neurology* (*N.Y.*) **30**:160–166.

Cerebral Metabolism in Vivo

William Sacks

1. INTRODUCTION

The brain is unique among the organs of the body in that it consumes about 2.2 mmol of oxygen per minute (a value that represents almost 20% of the resting human's total requirements) and in that it "extracts" from blood about 300 μmol of glucose per minute. Furthermore, experiments have demonstrated the brain's utter dependence on oxygen and glucose, with irreparable damage resulting from deprivation of either or both for only short periods of time.

Our knowledge of the functioning of the human brain under physiological conditions has been acquired primarily through the use of an arteriovenous method. Samples of blood going to and leaving the brain were drawn simultaneously from the femoral artery and the superior bulb of the internal jugular vein and then submitted to analyses. In this manner, it was found that there was an uptake by human brain of about 268 μmol O_2 and 56 μmol glucose per 100 ml blood and that about 268 μmol CO_2 per 100 ml blood was produced by brain. In reference to studies on the uptake of amino acids from blood, Waelsch[1] pointed out that

> Experiments in which the amino acid concentration in blood is raised to an unphysiological level may elicit an aspect of the blood–brain barrier not operative under physiological conditions. For determining the uptake of amino acids of the brain under physiological conditions, accurate measurements of the arteriovenous differences would be required.

Although brain slices *in vitro* and other brain preparations have proved of considerable value in biochemical research, it is true that, as Geiger[2] observed, "they obviously do not possess all the metabolic machinery which is involved in the physiological activity of the nerve cell."

The recent use of positron emission transverse tomography (PETT) with short-lived radioisotopes has permitted the investigation of regional cerebral metabolism *in vivo* with essentially noninvasive methods. Much of the background work leading to the PETT methodology was accomplished with long-

William Sacks • New York State Rockland Research Institute, Orangeburg, New York 10962.

lived radioisotopes combined with arteriovenous studies in human subjects or with brain perfusion studies in animals.

In this chapter those studies that give some insight into the biochemical environment of the intact brain and the effect of blood constituents on the physiological functioning of the brain are described. Although there are many *in vivo* techniques in the literature, the investigations cited herein are limited to those that use either arteriovenous, perfusion, or PETT methods.

2. ARTERIOVENOUS METHODS

2.1. Cerebral Blood Flow Determination

In 1943, Dumke and Schmidt,[3] using the rhesus monkey, obtained the first quantitative data on a brain *in vivo* under conditions that might be considered as approaching the normal physiological state. Two years later, Kety and Schmidt[4] published details of a procedure for the determination of cerebral blood flow in humans. The method, which made use of the Fick principle, consisted of the administration by inhalation of nitrous oxide for 10 min. During that time, arteriovenous blood samples were drawn at regular intervals. These were analyzed for nitrous oxide content, arterial and venous curves were drawn, and the integrated arteriovenous nitrous oxide differences was obtained. This value was divided into the amount of nitrous oxide taken up by the brain. The latter figure was obtained by multiplying the nitrous oxide content of the jugular venous blood at 10 min by the blood–brain partition coefficient (unity) for nitrous oxide. This method, which estimates cerebral blood flow, in terms of milliliters of whole blood/100 g brain per min, has been described by Kety.[5]

In the procedure summarized above, some blood samples were analyzed for O_2, CO_2, and glucose content. The product of the cerebral blood flow (CBF) multiplied by the arteriovenous (A–V) or venous–arterial (V–A) difference of these substrates gave an estimation of the consumption of oxygen and glucose and the production of carbon dioxide. For normal, healthy young men, average values have been reported[6] as 147 μmol O_2/100 g brain per min and 30 μmol glucose/100 g brain per min. The average CBF was 54 ml/100 g brain per min. Using an average weight of whole brain as 1400 g, the O_2 and glucose consumption for total brain were estimated as 2054 μmol O_2 and 422 μmol glucose per minute respectively.

More recently, local cerebral blood flow has been measured in humans[7] with methods employing ^{133}Xe inhalation and in animals employing [^{14}C]iodoantipyrine and autoradiography.[8] These methods are discussed elsewhere in this series.

2.2. Determination of Cerebral Metabolism in Human Subjects

In 1956, a report[9] from the author's laboratory gave details of a method for determining cerebral metabolism in humans *in vivo*. It made use of the arteriovenous technique of Kety and Schmidt and, in addition, employed the intravenous injection of a labeled substrate (such as [^{14}C]glucose).[12] It may be described briefly as follows. After the intravenous injection of a ^{14}C- and/or ^{3}H-labeled substrate, blood samples were drawn simultaneously over a period of 90–120 min. They were analyzed for radioactivity of the injected substrate, of the end product of metabolism of the substrate (i.e., $^{14}CO_2$ or ^{3}HOH), and of certain intermediates (i.e., lactate and pyruvate). From the data could be calculated the cerebral uptake of labeled substrate, the production of $^{14}CO_2$ (or ^{3}HOH) by brain, and the percent of labeled substrate taken up that was oxidized to $^{14}CO_2$ (or ^{3}HOH) by brain. In addition, by comparing the specific activity of venous blood (which was considered to represent brain) with the specific activity of brain $^{14}CO_2$* (or ^{3}HOH), the percent of brain carbon dioxide (or water) derived by oxidation of the substrate could be determined.

Methods employing PETT and short-lived radioisotopes make use of either [^{11}C]glucose[10] or [^{18}F]2-fluoro-2-deoxy-D-glucose (FDG)[11] as substrates injected intravenously. Radioactivities are measured in separate brain regions with PETT, and blood samples (arterial or "arterialized" venous) are drawn during the procedure in order to assay plasma glucose and [^{11}C]glucose or FDG. The data are used to determine local cerebral metabolic rates of glucose (LCMRGlc) and, with the [^{11}C]glucose procedure,[10] such parameters as glucose flux across the blood–brain barrier (BBB), the brain-to-blood glucose concentration ratio, the relative tissue free-glucose space, the brain free-glucose concentration, and the brain free-glucose turnover time. (These methods are discussed in greater detail in Volume 2 by J. D. Brodie, N. Volkow, and J. Rotrosen.)

2.3. Other Arteriovenous Methods

A technique similar to that of Sacks[12] was employed by Coxon and Robinson[13,14] to determine the venous–arterial specific activity ratios of carbon dioxide for different organs of dogs and monkeys after injection of [^{14}C]glucose. They examined the ability of various organs to produce $^{14}CO_2$ from blood [^{14}C]glucose and found that brain, liver, and kidney could raise the specific activity of CO_2 in blood passing through them but that resting muscle exerted a lowering effect. On the basis of their experimental evidence and model systems,[15,16] they suggested the use of continuous infusion (rather than a single injection) of the isotope to reduce the complexities of the problem of determining the rate of oxidation of a labeled substrate in an intact animal by keeping the specific activity of the injected isotope in the arterial blood relatively constant.

---

* Brain CO_2 specific activity as used in this chapter in reference to the author's studies refers to the "added increment" CO_2 specific activity,[12] i.e., the specific activity of the CO_2 produced by the brain and added to the venous blood leaving the brain. Specific activity equals the percent of injected activity per milligram of either CO_2 carbon or glucose carbon.

3. BRAIN PERFUSION METHODS

Brain perfusion methods attempt to isolate the cerebral circulation while allowing the brain to retain as many of its physiological functions as possible. Both partially isolated and completely isolated brain procedures have been used, and perfusates employed ranged from compatible whole blood of the same species to completely artificial blood. These studies are discussed only briefly since the subject is reviewed in detail by D. D. Gilboe in Volume 2 of this series.

3.1. Isolated Cat Brain Perfusion

The perfusion of the cat brain *in situ*, developed by Geiger and Magnes,[17] depended on the selective isolation of the cerebral venous outflow with simultaneous blocking of venous outlets of extracerebral tissues of the head. It also included a partial isolation of the arterial side of the brain circulation. A "simplified blood" consisting essentially of washed bovine erythrocytes suspended in Krebs–Ringer solution containing bovine serum albumin was used. Usually, it contained glucose at a concentration of 5.56 mmol/liter. Cat blood was ruled out because of the large volumes of blood needed, frequent occurrence of immunologic reactions, and the quick sedimentation of cat erythrocytes.

The apparatus included a flow meter, an oximeter, a pH meter, a heater, and equipment for automatically regulating CBF. The CBF was usually about 100 ml/100 g brain per min (that in man is about 54 ml/100 g brain per min). Geiger found it necessary to add cytidine and uridine[18] to "simplified blood" to maintain physiological activity of the brain for extended periods (4–5 hr). Allweis and Magnes,[19] on the other hand, apparently found the addition of these substances unnecessary. In Geiger's laboratory, a successful preparation exhibited the following physiological functions: normal corneal reflex, pupillary reactions to light, natural respiration, maintenance of systemic blood pressure and of vasomotor responses, normal electrocorticogram, spontaneous blinking, movements of facial structures, and easily elicited electric reactions in the brain to stimulation of one of the extremities.

The original apparatus[17] incorporated a heater, and presumably the perfusate was maintained at 37°C. In the report by Allweis and Magnes,[19] precautions were taken to minimize the metabolism of glucose by erythrocytes. The perfusate reservoir was immersed in ice and, prior to its entrance into the carotids, was heated rapidly to 37°C with a silver heating coil.

The high consumption of glucose and oxygen during the experiments indicates that the perfused cat brain may have been altered from its physiological state. Oxygen consumption, especially at the start of perfusion, was very high, sometimes reaching 268 μmol O_2/100 g brain per min. It declined within 5–10 min and then usually maintained a level of over 178 μmol as long as cerebral activity remained excellent. Thereafter, oxygen consumption declined slowly with progressive deterioration of cerebral activity. Although one could calculate an average value for glucose consumption of 32.8 μmol/100 g brain per min from their data, these investigators found that glucose uptake fluctuated considerably during an experiment and from one experiment to another (i.e., 22.8–48.3 μmol/100 g brain per min). Allweis and Magnes[19] gave somewhat higher values for these parameters, i.e., 51.1 μmol glucose/100 g brain per min and 308 μmol O_2/100 g brain per min. It was seen above that average values for normal human brain were 30.0 μmol glucose/100 g brain per min and 147 μmol O_2/100 g brain per min. In addition to the elevated uptake of O_2 and glucose, there was an excessive production of lactic acid by the perfused cat brain. Allweis and Magnes[19] found that lactic acid was liberated into the perfusate at a mean rate of 54.4 μmol/100 g brain per min and that the mean lactic acid content of brain cortex at the end of their experiment was 1967 μmol/100 g. As will be seen below, the normal human brain *in vivo* adds very little lactic acid to cerebral venous blood.[20–27]

3.2. Other Perfusion Techniques

A comparatively simple isolated, perfused rat brain preparation has been developed by Slotviter's group.[28] With a simpler surgical procedure than that of Geiger and Magnes, their preparation consisted of the skull and its contents, having the upper cervical vertebrae and small remnants of muscle attached. The rats were anesthetized by deep hypothermia, thus eliminating the problem of influence of chemical agents on cerebral metabolism and function. Early studies employed a perfusate consisting of canine erythrocytes in a solution of bovine albumin in Krebs–Ringer bicarbonate buffer. Spontaneous electroencephalographic (EEG) activity, which persisted for up to 5 h with open-circuit perfusion and about 2 h with recirculation, was used to indicate brain viability. The isolated rat brain and perfusate were allowed to reach room temperature before the start of the perfusion and remained there throughout the experiment. From Fig. 6 of ref. 28, one can approximate glucose consumption and lactate production of this preparation. The isolated, perfused rat brain consumed about 13.9 μmol glucose/100 g brain per min when the initial concentration of the perfusate was 1.11 mmol glucose/liter and about 40.6 μmol/100 g brain per min with an initial perfusate level of 1.68–2.22 mmol/liter. Values for lactate production were about 3.4 μmol/100 g brain per min with the 1.11 mmol/liter perfusate glucose and about 4.0 μmol/100 g brain per min with the 1.68–2.22 mmol/ liter perfusate glucose. These approximations indicate that in this perfusion preparation, done at lower than the animal's body temperature, the consumption of glucose varied considerably and did not differ widely from that in the perfused cat brain; however, production of lactic acid was less than in the Geiger–Magnes preparation.

Later studies by Sloviter *et al.*[29–31] made use of a liquid fluorocarbon with a high solubility for oxygen and carbon dioxide dispersed by ultrasonic treatment in a solution of bovine serum albumin in Krebs–Ringer bicarbonate buffer as perfusion medium. The EEG activity, glucose and oxygen consumption, as well as CO_2 and lactate production of the isolated perfused rat brain were similar when this fluorocarbon dispersion was employed as perfusate. Advantages of this preparation over the earlier "artificial blood" were listed as impossibility of hemolysis and absence of hemoglobin or other pigment to interfere with analytical procedures. Furthermore, it was claimed that the fluorocarbon perfusate could be used in conditions of pH and temperature and with toxic agents that the erythrocyte could not withstand.

The extracorporeal perfusion of the isolated dog head has been described by Gilboe et al.[32–34] With the brain left within the skull, perfusion was accomplished by heparinized, compatible donor blood (early studies) or diluted whole blood (later experiments) which entered through the common carotids and exited through a connector inserted into a small hole in the occipital bone to tap the confluent sinus. Electrocortical activity was used as an index of brain viability. A heat exchanger was incorporated into the system to maintain the brain at 37°C. Glucose consumption varied from 18.9 to 31.7 μmol/100 g brain per min and oxygen consumption averaged 161 μmol/100 g brain per min.

White, Albin, and Verdura[35] developed a method in which the brain of a monkey was surgically freed of all contiguous tissue except a small basal plate of bone and a thin bony strip over the sagittal sinus for supporting the brain and the electrodes. All extracranial vasculature was removed, vertebral arteries were ligated, and carotid arteries were cannulated, permitting autogenous cerebral perfusion until the start of extracorporeal circulation. The perfusion blood was freshly drawn, heparinized, compatible monkey blood diluted one-third with dextran. Intracerebral temperatures were maintained between 28 and 32°C to reduce cerebral metabolism. Therefore, (A–V) O_2 and (V–A) CO_2 differences were only 125 and 107 μmol per 100 ml blood, respectively. Glucose uptake after 2 h of closed-circuit perfusion was about 12.8 μmol/100 g brain per min. At that time, the perfusion blood had high levels of lactate (8.4 mmol/liter) and of pyruvate (1.0 mmol/liter). As perfusion progressed, there were elevations in free hemoglobin and a gradually developing acidosis.

The same laboratory[36] reported the successful transplantation, with survivals lasting from 6 hr to 2 days, of six isolated canine brains into the cervical vasculature of dogs. The brain was surgically isolated within the skull, and internalization of the brain graft was accomplished by connecting the carotid arteries of the isolated brain to the proximal carotid artery of recipient and by fixing the torcula cannula of the isolated brain within the jugular vein of the recipient. In addition to instrumentation for continuous monitoring of electrocortical activity, CBF, and temperature, catheters were implanted for sampling of arterial and venous blood from the transplanted brain. Average (A–V) O_2 and (V–A) CO_2 differences were 496 and 455 μmol per 100 ml blood, respectively, and the average (A–V) glucose difference was 134 μmol/100 ml blood. These values, though high, are consistent with the low average CBF of 24.2 ml/100 g brain per min. Oxygen and glucose consumption averaged 121 μmol O_2/100 g brain per min and 31 μmol glucose/100 g brain per min. Blood lactate determinations revealed no significant differences between arterial and venous samples.

A unique method was developed by Moss[37,38] for the benign, arterially isolated cerebral perfusion of the bull calf. This was accomplished by perfusion of a single carotid artery at 20 mm Hg pressure above systemic arterial pressure, which resulted in a pressure in the circle of Willis greater than aortic pressure, causing the collaterals to flow in a retrograde direction. The perfusate consisted of fresh donor-compatible blood. The procedure was apparently innocuous, since, after 6 hr of perfusion, no clinical, EEG, or histopathological evidence of cerebral dysfunction or damage could be detected. Unfortunately, no biochemical data were given in these reports, so the preparation could not be evaluated with the usual metabolic parameters.

3.3. Use of Filters in Perfusion

An inherent difficulty in perfusion techniques has been a steady increase in perfusion pressure and a build-up of lactic acid. Swank and Hissen[39] have suggested that this may be caused by a blockage of the microvasculature with platelet aggregates, since normal perfusion pressure could be maintained by the use of Dacron® wool or glass filters. Gilboe *et al.*[40] tested the hypothesis that occlusion of small blood vessels of the brain limits the access of brain tissue to blood and oxygen and thereby causes increased dependence on anaerobic metabolism. Two groups of isolated dog brains were used: one (1) perfused with blood filtered through Dacron® wool to remove platelet aggregates and one (2) perfused with blood filtered through a stainless steel screen to remove fibrin clots. No significant differences could be found in either glucose consumption (27.2 and 20.6 μmol/100 g brain per min for groups 1 and 2, respectively) or production of lactic acid (3.3 and 4.8 μmol/100 g brain per min for groups 1 and 2, respectively). Perfusion pressure increased more rapidly in the second group. The authors concluded that the excessive production of lactic acid observed by others with the isolated cat brain was probably caused by the use of "simplified blood."

The effects of filtering "simplified blood" through glass wool in cat brain perfusion have been reported by Allweis *et al.*[41] Use of the filter reduced to about 10% the incidence of progressive rise in perfusion pressure that occurred in 75% of experiments done without the glass wool filter. The pressure necessary to perfuse blood through the brain at a rate of 120 ml/100 g brain per min was about 30 mm Hg lower in experiments using the filter. It was found that CO_2 production by brain was significantly increased and that samples of cortex taken at the end of experiments showed greater glucose content and lower lactate content than in experiments employing the glass wool filter.

4. CEREBRAL METABOLISM OF CARBOHYDRATES IN VIVO

4.1. Glucose as the Primary Substrate

That carbohydrate provides the principal substrate of brain was suggested by evidence that the respiratory quotient of the brain is usually very close to unity. Further, that glucose *per se* is the primary foodstuff of the brain in humans is supported by studies showing that it alone will arouse patients in insulin hypoglycemic coma. Other substrates (i.e., glycerol and glutamic acid) reported to be capable of restoring contact during insulin hypoglycemia have been shown to cause an increase in blood glucose, so it is doubtful that they alone can serve to support brain metabolism. In certain disease states and under starvation conditions, substances other than glucose (i.e., ketone bodies) are metabolized by brain.[45]

4.2. Cerebral Uptake of Glucose Anomers

It has generally been assumed that glucose exists in blood as though its anomers were in mutarotational equilibrium (i.e., 36% α-glucose and 64% β-glucose) and that equilibrium glucose is taken up and metabolized by brain. To examine these assumptions, Sacks[42] studied the cerebral uptake of glucose anomers using the arteriovenous technique (with no injected isotope) in four volunteer subjects. Total blood glucose was analyzed by the Somogyi–Nelson method and β-glucose by an original procedure similar to the dialysis-coupled enzyme method employed by the same laboratory in the study of mutarotase in erythrocytes.[43,44] The apparent α-glucose was determined by difference. The overall average A–V differences for the four studies were 53.3 μmol/100 ml blood for total glucose, 41.1 μmol/100 ml blood for β-glucose and 12.2 μmol/100 ml blood for "apparent" α-glucose. These data indicate that about 77% of the glucose uptake by human brain *in vivo* may be in the form of β-glucose. This value exceeds that expected by the uptake of equilibrium glucose (which is 64% β-glucose).

4.3. Cerebral $^{14}CO_2$ Derived from $[^{14}C]$Glucose

Making use of the arteriovenous technique and intravenous injections of specifically labeled [^{14}C]glucose into normal human subjects, Sacks, in 1957, presented evidence suggesting that only part of cerebral CO_2 was derived from glucose.[12] Following a single injection of [U-^{14}C]glucose (uniformly labeled), it was found that brain $^{14}CO_2$ specific activity was relatively constant and that venous blood glucose specific activity (representing specific activity of glucose metabolized by brain) was considerably higher than brain $^{14}CO_2$ for most of the experiment.[12] Brain $^{14}CO_2$ specific activity curves were found following injections of [1-^{14}C]glucose, [2-^{14}C]glucose, and [6-^{14}C]glucose, and these values were used to derive the theoretical brain $^{14}CO_2$ specific activity curve for carbons 3 and 4 (C_3 and C_4) of [U-^{14}C]glucose. Since the derived curve was parallel to the venous glucose specific activity curve but was only about one-half as high, it was thought that only about 50% of brain $^{14}CO_2$ was derived from [^{14}C]glucose.[12] Also, it was shown that only about 40% of [^{14}C]glucose taken up by brain in 90 min came out as $^{14}CO_2$, and it was suggested that the ^{14}C remaining in brain became incorporated into the brain metabolic carbon pool consisting of glucose, glycogen, glutamine, glutamate, and N-acetylaspartate. Furthermore, comparison of cerebral $^{14}CO_2$ produced following injection of [1-^{14}C]glucose with that found after administration of [6-^{14}C]glucose led to the conclusion that practically all of the glucose-derived CO_2 arose via glycolysis.

The question of the extent of cerebral CO_2 derived from glucose has been examined by others. Allweis and Magnes[19] employed [U-^{14}C]glucose in their perfused cat brain studies and found that the radioactivity of CO_2 produced by brain reached an approximately constant value after about 30 min of perfusion. Their data suggested that only 22% of glucose taken up from blood was oxidized to CO_2. Most of the balance, they believed, could be accounted for as lactate accumulated in brain and liberated into blood. Similar results were found by Geiger *et al.*[46] After the addition of [U-^{14}C]glucose to "simplified blood," the concentration of ^{14}C in brain CO_2 increased steadily during the first 25–30 min to a concentration of 29–32% and remained at that level for the rest of the experiment. The results were interpreted to indicate that both glucose carbon and "cold" carbon from endogenous noncarbohydrate sources contributed to form respiratory CO_2. That brain glucose equilibrated quickly with glucose in the perfusing blood was established by showing that glucosazones prepared from blood and extracts of brain cortex pieces had similar specific activities.

Because of an investigation[47] that indicated that 30–40% of CO_2 produced by rat whole cerebrum slices was derived from endogenous sources, Gainer *et al.*[48] studied the intact anesthetized dog. The animals were given a continuous intravenous infusion of [U-^{14}C]glucose to maintain a constant plasma glucose specific activity. At intervals, arterial and venous blood from the superior saggital sinus were taken simultaneously for determination of specific activities of plasma glucose and CO_2. The specific activity of brain CO_2 rose during the first hour to a value about the same as that of blood glucose and then stayed level for the emaining 4.5 h of the experiment. The authors concluded that almost all of the metabolic CO_2 was derived from blood glucose.

In view of these results, Barkai and Allweis[49] thought that differences between experiments with the intact dog and perfused cat brain were caused by some peculiarities in the latter preparation (i.e., use of "simplified blood"). Therefore, they employed the continuous infusion technique with the narcotized cat. In these studies, at apparent isotopic equilibrium, it was observed that brain CO_2 specific activity was only 45% that of blood glucose specific activity. This value, which did not differ greatly from that found with the perfused cat brain (29–32%), was much lower than that obtained using the intact narcotized dog. The authors suggested that the latter dissimilarity may have resulted from either a species variation or from different origins of the cerebral venous blood sampled in the two experiments (i.e., in the dog, blood was obtained from the anterior superior sagittal sinus, which drains blood mainly from anterior parts of cortex). In similar experiments performed on intact narcotized cats, Gombos and co-workers[50] reported that brain CO_2 specific activity attained a value about 60% that of blood glucose after 4–5 h of perfusion.

Thus, there appears to be some controversy as to the extent of brain CO_2 derived from glucose. In the opinion of this author, differences between results found in humans and those obtained using the cat brain perfusion preparation resulted in part from use of "simplified blood." The findings of mutarotase (believed to be involved in glucose transport) in erythrocytes and of considerable species difference in activity of this enzyme[44] and of evidence that red blood cells may play some role in the transfer of glucose from blood into brain[51] suggest that it may not be desirable to use bovine cells in the cat preparation.

In addition, cat's normal plasma probably would be more likely to favor glucose transfer than the Krebs–Ringer solution plus bovine albumin used in "simplified blood."

Most likely, the main factor in determining the relationship of brain CO_2 specific activity to that of glucose is the dilution of ^{14}C (derived from [^{14}C]glucose) in traveling the Krebs tricarboxylic acid cycle (TCA). Following a series of experiments with [3-^{14}C]glucose as injected substrate (in humans), production of $^{14}CO_2$ almost kept pace with the net uptake of ^{14}C by brain.[52] In addition, brain CO_2 specific activities practically coincided with venous blood (brain) specific activities, suggesting little or no dilution of ^{14}C in traversing the Embden–Meyerhof–Parnas (EMP) pathway. Since intermediates of the TCA occur at concentrations in brain less than those of the glycolytic intermediates, brain CO_2 specific activity curves following injections of [2-^{14}C]glucose, [1-^{14}C]glucose, or [6-^{14}C]glucose indicated a significant dilution in traveling the TCA. From the extent of the estimated dilution (about 40-fold) and experimental data obtained in the author's laboratory and by others, a scheme for human cerebral metabolism (Fig. 1) was proposed.[52] It included a small, metabolically active pool of glutamate, γ-aminobutyrate, and succinic semialdehyde directly in the TCA, which was thought to be an open pool (i.e., with a constant outflow and a balancing inflow). The outflow was postulated to result from a slow equilibration of the small pool of glutamate with a large inactive pool of glutamate and from the formation of glutamine from the small glutamate pool. N-Acetylaspartate was considered a source of inflow into the TCA. It was further thought to supply substrate under emergency conditions (i.e., hypoglycemia).

As a result of this dilution of ^{14}C in traveling the TCA,[52] significant amounts of ^{14}C taken up by brain in the form of [^{14}C]glucose (when labeled on C_1, C_2, C_5, or C_6) remained in brain for long periods of time. Indeed, following a single intravenous injection of [2-^{14}C]glucose into a human subject, after 5 h about 21% of the [^{14}C]glucose uptake still remained in brain. Therefore, in cat brain perfusion studies, in which [U-^{14}C]glucose was always used, one could expect about two-thirds of the radioactivity to remain in brain for significant lengths of time. With incorporation of this ^{14}C into amino acids, proteins, fatty acids, and lipids (see below), one would not be surprised to find brain CO_2 specific activities only a fraction of blood glucose specific activities. In addition, it might well be that the extent of dilution varies in different structures of brain and in various animal species.

4.4. Incorporation of ^{14}C from Labeled Glucose into Brain Carbon Pool

Barkulis et al.[53] excised brain cortex slices at various time intervals following addition of [U-^{14}C]glucose to "simplified blood" in cat brain perfusion studies to determine specific activities of free amino acids. In the first samples (about 20 min), the combined amino acids had about 5–10% of the specific activity of the perfused glucose. Although the radioactivity of amino acids in the next samples (about 40 min) remained about the same, that of the third

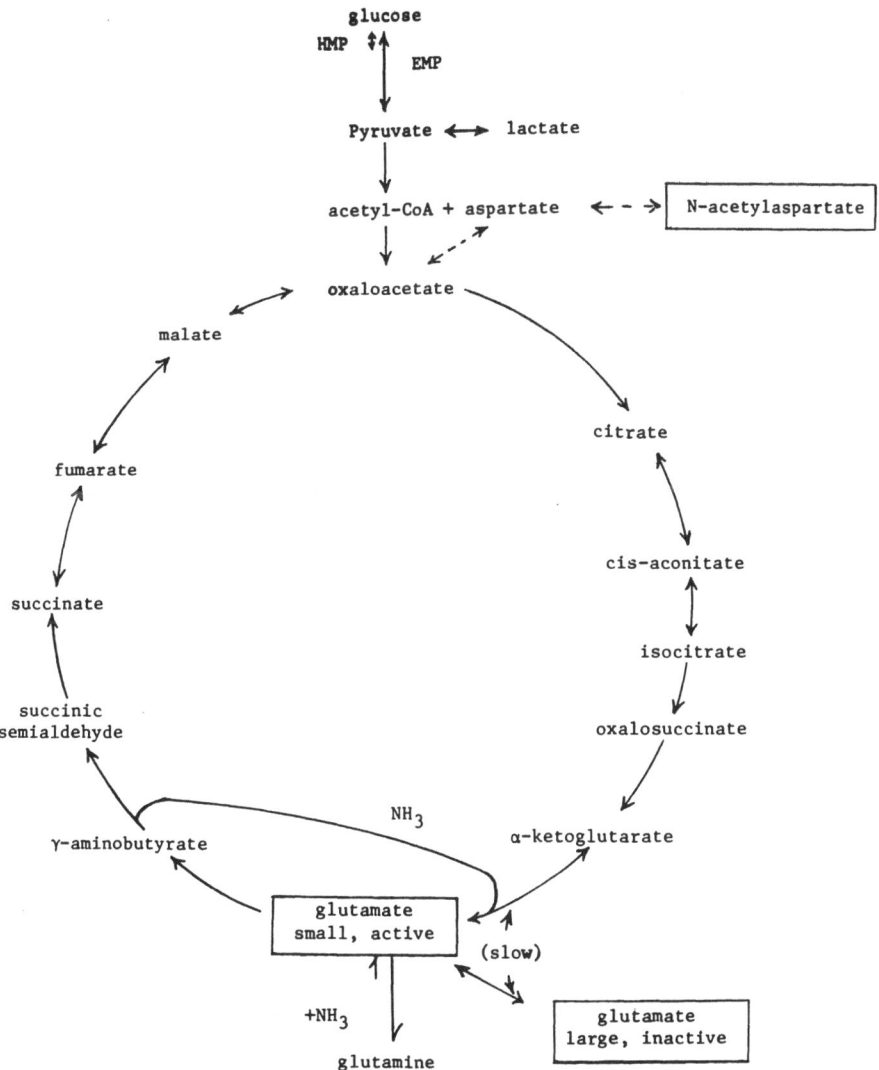

Fig. 1. Proposed scheme for the cerebral metabolism of glucose in humans *in vivo*. Substrates present in sufficiently high concentration in brain to be considered a pool are enclosed in rectangles. The N-acetylaspartate pool is thought to be a source of substrate in emergency conditions. The primary feature of the proposed pathway is the inclusion of glutamate, GABA, and succinic semialdehyde in the tricarboxylic acid cycle. The small metabolically active glutamate pool in the TCA cycle is considered to be in slow equilibrium with a large metabolically inactive glutamate pool in the brain. From Sacks.[52]

samples usually showed a pronounced increase (as much as fourfold) in radio-activity compared to the first samples and reached a specific activity 15–18% that of glucose; it thereafter remained constant. Glutamic and aspartic acids, which were further separated, showed the same trend in activities but leveled off at 23–24% of the glucose specific activity. In a subsequent paper,[54] it was found that ^{14}C derived from glucose was incorporated into proteins of brain cortex considerably more slowly than into free amino acids. It was further demonstrated[55] that ^{14}C (from glucose) was incorporated into N-acetyl-L-aspartic acid at a rate only about 1–5% that of its incorporation into aspartic acid in brain. The findings on incorporation of ^{14}C derived from glucose into various intermediary metabolites have been summarized by Gombos et al.[56] and are given in Table I.

4.5. Pyruvic and Lactic Acids

In the normal human brain, average arteriovenous (A–V) differences of oxygen and glucose are about 268–299 μmol O_2/100 ml blood and about 56 μmol glucose/100 ml blood. An uptake of 56 μmol glucose by brain would require about 335 μmol O_2; or, stated another way, an oxygen consumption of 268–299 μmol would be satisfied by 44–49 μmol glucose. This disparity between theoretical and actual values has usually been explained by the production and release by brain of 17.8 μmol lactic acid/100 ml blood[57] and of 2.5 μmol pyruvic acid/100 ml blood.[58]

Sacks has studied the human cerebral metabolism of [1-^{14}C]pyruvate[20] and specifically labeled [^{14}C]DL-lactic acid, D-[1-^{14}C]lactic acid, and L-[1-^{14}C]lactic acid.[20,27] With [1-^{14}C]pyruvate as injected substrate, brain specific activity curves resembled closely those found using [3-^{14}C]glucose (as would be expected, theoretically). With specifically labeled [^{14}C]DL-lactic acid, percentages of injected ^{14}C oxidized to $^{14}CO_2$ by brain varied in magnitude depending on the position of the isotope in the lactic acid injected, the order being as follows: [2-^{14}C]lactic > [1-^{14}C]lactic > [3-^{14}C]lactic > [U-^{14}C]lactic acid. With individual D- and L- isomers, L-[1-^{14}C]lactic acid had a value less than that for DL-[2-^{14}C]lactic but greater than the value for DL-[1-^{14}C]lactic acid. D-[1-^{14}C]Lactic acid gave percentages lower than any using DL[^{14}C]lactate. Sacks et al.[59] presented a convincing demonstration of the continuous uptake of DL-[2-^{14}C]lactate (A–V differences) by brain and the consistent cerebral oxidation of the [^{14}C]lactate to $^{14}CO_2$ (V–A differences) in an experiment in which the subject was given the isotope by constant infusion and blood samples were drawn continuously. Blood lactate and pyruvate determinations[20] indicated that both

Table I
Amounts of Radioactivity Derived from
[^{14}C]Glucose in the Various Metabolites in
Perfused Cat Brain, Expressed as Percent of
Their Total Carbon[a]

Free lactic acid	47–50
Respiratory CO_2	28–30
Free glutamic acid	31–35
Free aspartic acid	24–26
Free γ-aminobutyric acid	11–16

[a] Taken from Gombos et al.[56]

of these substances were added to blood as it flowed through the brain. For lactic acid, the average V–A difference for normal subjects was 6.1 μmol/100 ml blood; for pyruvate, the average V–A difference was 0.5 μmol/100 blood.

Other investigators have found that small quantities of lactic and pyruvic acids were added to cerebral venous blood,[21,23–26] although there was considerable variation in reported values. The study by Gottstein and co-workers[21] was particularly noteworthy since substrate-specific enzymic methods were used. These investigators reported that the cerebral production in normal humans was 2.7 μmol lactate/100 g brain per min (i.e., about 5.1 μmol/100 ml blood) and 0.6 μmol pyruvate/100 g brain per min (i.e., about 1.2 μmol/100 ml blood). A contradictory report was that by Scheinberg *et al.*,[22] in which no consistent differences were found in arterial and cerebral venous values for pyruvate and lactate in 25 normal resting subjects. Also, White and co-workers[36] observed no significant A–V or V–A blood lactate differences in transplanted canine brains (see above).

The apparently paradoxical situation in which the brain catabolizes lactate and pyruvate[20,27] while excreting both into the venous return blood suggests that compartmentalization of brain includes metabolism as well as function and that total production of lactic and pyruvic acids through glycolysis exceeds the amounts that are oxidized by way of the tricarboxylic acid cycle. Evidence obtained with the more specific methods available today suggests that earlier measurements of brain lactate and pyruvate production were probably too high and that the "extraction" of glucose does indeed exceed the production of CO_2, lactate, and pyruvate. Much data exist pointing to a significant incorporation of glucose carbon into lipids and proteins of brain (see above).

4.6. The Pentose Cycle

In the first report on the metabolism of [^{14}C]glucose by human brain *in vivo* from this laboratory, it was concluded that practically all of the glucose-derived CO_2 arose glycolytically.[12] In a later study, it was estimated that the pentose cycle (PC) accounted for about 1% of the cerebral metabolism of glucose *in vivo*.[52] This conclusion has been supported by others,[60,61] although some have estimated the activity as slightly higher. Thus, in experiments using the perfusion of the isolated monkey brain, the addition of [2-^{14}C]glucose to the perfusion blood allowed analyses indicating a 5 to 10% functioning of the PC in such specific areas of the monkey brain as the hypothalamus and brainstem.[61] A report by Moss[38] suggesting that the PC (or hexosemonophosphate shunt) was probably the major pathway *in vivo* for bovine cerebral glucose oxidation has been described in detail in a previous work,[62] and its experimental errors were pointed out. In a subsequent study, Hakim and Moss[63] estimated the extent of the PC by determining the turnover rate of 5-phosphogluconate. They reported that approximately 21% of the glucose consumed oxidatively by rat brain was metabolized through the PC. Again, in the opinion of this author, the experimental procedure employed and the data obtained make this conclusion very tenuous. This viewpoint is reenforced by our recent observation (see below) that a single electroconvulsive shock (ECS) or a series of nine ECSs can effect a 31% functioning of the PC in rat brain glucose metabolism.[64]

Sodium [1-^{14}C]gluconate was used as injected substrate in an attempt to demonstrate directly the functioning of the PC in human brain *in vivo*.[42] The results indicated no production of $^{14}CO_2$ by brain; however, in these studies, it could not be determined whether the [^{14}C]gluconate was taken up by brain.

4.7. The Tricarboxylic Acid Cycle

In the author's laboratory, experiments have been performed in which the following were employed as the injected substrates: [2-^{14}C]fumarate, [1,4-^{14}C]fumarate, [1,4-^{14}C]maleate, and [5-^{14}C]α-oxoglutarate. Maleic acid was included because of evidence of the existence of maleic hydratase, E.C. 4.2.1.31 (which converts maleic acid to D-malic acid) in humans.[65–68] When [2-^{14}C]fumarate was given intravenously to normal human subjects, it was readily oxidized to $^{14}CO_2$ by brain.[9] Studies with [1,4-^{14}C]fumarate[62] showed a delay of 6–30 min in production of $^{14}CO_2$ by brain and more fluctuations in V–A $^{14}CO_2$ differences than in [2-^{14}C]fumarate experiments. With [1,4-^{14}C]maleate, there was little or no production of $^{14}CO_2$ by brain, although $^{14}CO_2$ appeared in arterial blood very quickly after injection, indicating that the body readily oxidized this compound.[66] Following injection of [5-^{14}C]α-oxoglutarate, the data[62] showed a significant and consistent production of $^{14}CO_2$ by brain after an initial lag of 10–20 min. In these experiments with Krebs cycle intermediates, no significant ^{14}C labeling was found in glucose, amino acids, lactate, or pyruvate.

4.8. Cerebral Glucose Phosphatase Activity

In the author's studies on glucose metabolism,[12,42,52] a contradiction was found between actual glucose uptake and uptake in terms of [^{14}C]glucose. The data suggested that a labeled glucose intermediate was formed that had significant radioactivity. This led to the use of [^{14}C]glucose phosphate as injected substrate.[51] Following a single injection of [1-^{14}C]glucose-6-phosphate, it was found that the cerebral production of $^{14}CO_2$ was in excess of that expected from the brain uptake of [^{14}C]glucose (which was formed endogenously from the [1-^{14}C]glucose-6-phosphate). However, when the cerebral uptake of the [^{14}C]glucose phosphate was added to that of the endogenously formed [^{14}C]glucose, the net ^{14}C uptake by brain was of the order of that found in similar [1-^{14}C]glucose experiments having about the same cerebral $^{14}CO_2$ production.[51]

In studies in which the arterial blood levels of [^{14}C]glucose phosphate were elevated rapidly by constant infusion for 6–10 min, it was seen that there was a consistent uptake of [^{14}C]glucose phosphate (i.e., A–V [^{14}C]glucose phosphate differences) with a corresponding constant production of [^{14}C]glucose (as evidenced by V–A [^{14}C]glucose differences) while [^{14}C]glucose phosphate blood values were rising. In the latter studies (Fig. 2), addition of [^{3}H]glucose

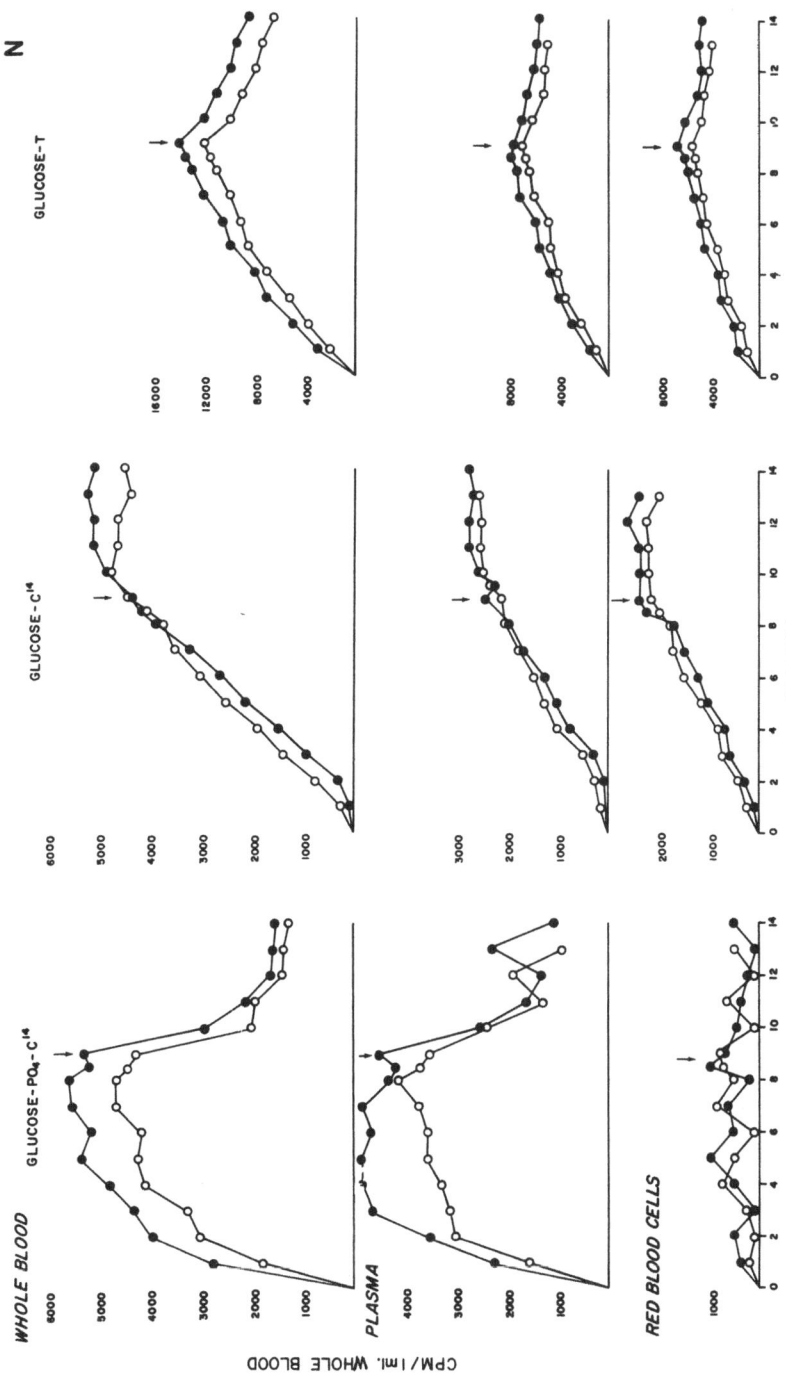

Fig. 2. Activity–time curves of arterial (femoral; solid circles) and venous (superior bulb of internal jugular; open circles) whole blood, plasma, and red blood cell (RBC) [14C]glucose phosphate, [14C]glucose, and [3H]glucose (glucose-T) during and shortly after constant intravenous infusion (8.5 min) of [1-14C]glucose-6-phosphate plus [6-3H]glucose (glucose-6-T). Samples were drawn simultaneously. Results for RBC were derived by subtracting values for plasma from those for whole blood, making use of hematocrit. The arrows indicate the end of constant infusion. From Sacks and Sacks.[51]

to the infusion demonstrated that isotopic glucose was taken up by brain in the usual manner while it was converting [^{14}C]glucose phosphate to [^{14}C]glucose. Thus, the evidence suggested that human brain *in vivo* exhibited glucose phosphatase activity. An experiment done with [U-^{14}C]glucose-1-phosphate showed similar results except that, as would be expected from [U-^{14}C]glucose studies, there was a much greater cerebral production of $^{14}CO_2$ than in the [1-^{14}C]glucose-6-phosphate experiments. From these studies, it could not be determined if the brain must convert glucose phosphate to glucose prior to oxidizing it to CO_2.

Some estimate of the extent of the continuous conversion of [1-^{14}C]glucose-6-phosphate to [^{14}C]glucose by human brain *in vivo* was obtained from a study (unpublished) in which [1-^{14}C]glucose-6-phosphate was given intravenously by constant infusion at a slow rate and blood samples were drawn continuously. During the first 30 min, at which time the production of $^{14}CO_2$ was negligible, the cumulative results indicated that about 34% of the [1-^{14}C]glucose-6-phosphate taken up by brain was converted continuously to [^{14}C]glucose.

Considerable evidence exists that glucose-6-phosphatase is normally present in mammalian brain.[69–80] Glucose 6-phosphate-hydrolyzing activity was demonstrated by Rosen[74] in the cerebellum of the mouse, rat, hamster, and marmoset. On the basis of a variety of histochemical criteria, he concluded that the enzyme activity appeared similar to glucose-6-phosphatase found in liver and kidney tissues. Though less active than in liver and kidney,[71] glucose-6-phosphatase activity in brain is nevertheless considerable (especially in humans *in vivo* as seen above). Its activity, demonstrated to be in neurons, was shown to vary greatly between regions.[77,78] Karnovsky *et al.*[79] have reported an increased glucose-6-phosphatase activity during slow-wave sleep.

In view of the evidence presented by Sacks and others, the concept that brain has no significant glucose-6-phosphatase activity[81,82] would seem to be unfounded.

4.9. Influence of Arterial Blood Glucose Level on Cerebral Uptake of Glucose

Using the perfused cat head, Geiger and co-workers[83] saw that when the "blood" glucose concentration was increased to 4.44 mmol/100 ml, the brain contained only 1.67 mmol/100 g and that this concentration was increased only slightly by doubling the "blood" glucose. Sata *et al.*[84] in studying cerebral glucose consumption by the Kety–Schmidt cerebral blood flow method, state that they could not demonstrate a parallelism between arterial glucose content and the A–V glucose difference observed in healthy subjects. Gottstein *et al.*[85] concluded that the cerebral uptake of glucose could not be improved in cerebral arteriosclerosis by elevating levels of arterial glucose. Intravenous infusions of 60 ml of 2.78 M glucose influenced neither CBF nor cerebral metabolism significantly in spite of elevated blood sugar level from 533 to 1472 μmol/100 ml blood.

In the author's laboratory, studies that combined a glucose tolerance test with our usual arteriovenous technique[62] indicated no correlation between cerebral uptake of glucose and arterial glucose concentration. In these experiments, there seemed to be no unusual results in either brain [^{14}C]glucose uptake or $^{14}CO_2$ production, although, in some cases, arterial glucose values were above 1111 μmol/100 ml. The reports cited fail to corroborate the findings of Rowe and co-workers[26] that larger quantities of glucose were taken up by brain when arterial glucose levels were rising. In the latter study, patients were fed a standard breakfast, which raised arterial blood glucose levels on the average from 583 to 833 μmol/100 ml.

With the exception of the report by Rowe *et al.*,[26] these investigations lead one to conclude that cerebral uptake of glucose was unaltered in hyperglycemia. What about the effect of hypoglycemia? In a study in which cerebral blood was sampled during therapeutic insulin treatment of schizophrenic patients, Himwich *et al.*[86] observed that oxygen uptake was reduced to an average value of 138.4 μmol/100 ml (from a control value of 299.1 μmol/100 ml blood) and that glucose uptake fell to an average of 23.3 μmol/100 ml (from a control value of 55.6 μmol/100 ml blood) during the ensuing hypoglycemia. Gottstein and Held[87] have expressed the belief that even a minor hypoglycemia with blood sugar values of 277.8 μmol/100 ml could result in a marked reduction of cerebral glucose uptake. It would seem, therefore, that although hyperglycemia probably had no effect, the cerebral glucose uptake could be significantly lowered by hypoglycemia. In regard to this matter, Butterfield *et al.*[88] proposed the concept of a brain glucose threshold similar to although slower acting than the peripheral glucose threshold they established previously.

4.10. Influence of Insulin on Uptake and Utilization of Glucose

It has been unequivocally stated for many years that the brain is insensitive to insulin; however, more recent studies *in vivo* have presented convincing evidence that insulin has some control on uptake and metabolism of glucose by brain. That the injection of insulin with the resultant hypoglycemia could reduce cerebral glucose uptake and metabolism was seen above. However, Gottstein and co-workers[85,87] were able to demonstrate a significant increase in cerebral glucose uptake (from 31.2 to 46.1 μmol/100 g brain per min) as a result of a combined infusion of glucose and insulin. A similar increase was found in diabetic patients given insulin. Butterfield *et al.*[88] have suggested that the brain glucose threshold (see above) is lowered by insulin. Not in agreement with these results found in humans *in vivo* is a report by Yamada and Sloviter[89] in which the authors saw no effect of insulin on glucose consumption of the isolated, perfused rat brain preparation.

4.11. Cerebral Uptake of Sugars Other than Glucose

After continuous administration of fructose for about 1 hr preceding perfusion and then perfusion with 5.56 mM fructose in "simplified blood," Geiger *et al.*[83] found an uptake of the sugar by cat brain; however, fructose apparently did not disappear from the brain during the experiment. Brain glucose content and oxygen consumption declined rapidly as in other glucose-free perfusion experiments. Allweis and Magnes[19] perfused cat brain with blood containing fructose (5.56 mM) and [U-^{14}C]fructose. Cerebral functional activity and oxygen consumption decreased rapidly, and only about 5% of brain CO_2 was derived from fructose. It was concluded that the isolated, perfused cat brain was able to oxidize fructose (in the absence of glucose) only to a very limited extent.

Ghosh et al.[31] found that mannose was a suitable substrate for maintaining metabolism of the isolated, perfused rat brain.

Using a simplified version of the Geiger–Magnes technique, Eidelberg et al.[90] found an uptake of arabinose as determined by measuring brain content. The data indicated a higher preference towards the (−)-arabinose (2:1) than the (+)-arabinose. The uptake could be inhibited by ouabain and by glucose. The authors concluded that the mechanism was of the carrier-mediated or of the active transport type. These data were of special interest in view of the author's demonstration of stereospecificity of erythrocyte mutarotase[44] in that the conversion of α-glucose to β-glucose was inhibited to a greater extent by L-arabinose than by D-arabinose and in view of the belief that the erythrocyte might play a role in the blood–brain barrier (BBB).

In the author's laboratory, [1-^{14}C]galactose, [2-^{14}C]galactose, and [1-^{14}C]ribose were used in studies in humans.[62] Although there was some $^{14}CO_2$ production by brain in all cases, the data indicated little or no cerebral uptake of either galactose or of ribose. It was felt that brain $^{14}CO_2$ arose from cerebral oxidation of [^{14}C]glucose that was synthesized by the body from the injected isotopes.

5. CEREBRAL METABOLISM OF LIPIDS IN VIVO

5.1. Fatty Acids

When [1-^{14}C]butyrate was injected intravenously into normal human subjects, there was a consistent and significant production of $^{14}CO_2$ by brain.[91] An average of 3.9% of injected ^{14}C was converted to $^{14}CO_2$ by brain in 90 min following injection. A comparison of brain CO_2 specific activity with venous blood butyrate specific activity gave an average of 9.8% of cerebral CO_2 derived from oxidation of butyrate. Under similar conditions, there was no significant production of $^{14}CO_2$ by brain with [1-^{14}C]octanoic acid[91] or with [1-^{14}C]acetate[62] as substrates.

Allweis et al.[92] demonstrated that [U-^{14}C]palmitic acid, which had been attached to purified bovine serum albumin and incorporated into the "simplified blood" used to perfuse the cat brain, was oxidized to $^{14}CO_2$. The relative specific activity of the $^{14}CO_2$ (compared to specific activity of palmitic acid in perfused blood) rose to an average of 2.8% and was increased slightly when glucose was omitted from the "simplified blood."

5.2. Glycerol

Glycerol is an integral part of all fats and of phosphoglycerides. Reportedly, its administration will alleviate the cerebral disorders produced by insulin hypoglycemia.[93,94] However, it was not clear that this effect was caused by glycerol *per se* and not by its rapid conversion to glucose by the body. The author has studied the cerebral oxidation of [U-^{14}C]glycerol and of [1,3-^{14}C]glycerol.[91] In estimating that 3.5% of cerebral $^{14}CO_2$ was derived from injected [^{14}C]glycerol, it was necessary to subtract from the brain CO_2 specific activity those values attributed to [^{14}C]glucose synthesized from [U-^{14}C]glycerol (which were calculated from venous blood glucose specific activities).

6. CEREBRAL METABOLISM OF AMINO ACIDS

It has been seen that after perfusion with "simplified blood" containing [U-^{14}C]glucose, ^{14}C was rapidly incorporated into free amino acids and, much more slowly, into proteins of brain (see above). However, there is a paucity of *in vivo* studies showing the oxidation of amino acids. A number of ^{14}C-labeled amino acids and derivatives have been used by the author as injected substrates with the usual arteriovenous procedure. With [U-^{14}C]aspartate, V–A $^{14}CO_2$ differences showed a small but significant metabolism by brain; however, little or no oxidation of DL-[4-^{14}C]aspartate, L-[U-^{14}C]lysine, DL-[1-^{14}C]glutamate, or L-[U-^{14}C]glutamate by brain was observed.[91] [1-^{14}C]GABA, [^{14}C]serine, and [^{14}C]alanine were readily oxidized to $^{14}CO_2$ by brain.[62] With [U-^{14}C]serine, V–A differences were not nearly as large as with DL-[1-^{14}C]serine, indicating that C_1 contributed most of the activity of the brain $^{14}CO_2$. [^{14}C]Phenylalanine, [^{14}C]tyrosine, L-[U-^{14}C]glutamine, and 3,4-dihydroxyphenyl [carboxy-^{14}C]alanine ([carboxy-^{14}C]DOPA) resulted in both A–V and V–A $^{14}CO_2$ differences, suggesting sporadic oxidation (or decarboxylation) or compartmentalization of brain with varying rates of oxidation (or decarboxylation) in compartments.

Maiolo and co-workers,[95] using the arteriovenous technique of Sacks, studied the cerebral metabolism of L-[U-^{14}C]glutamine and L-[U-^{14}C]glutamic acid in chronic mental patients. Under basal conditions, they found a lag (4–10 min) in production of cerebral $^{14}CO_2$ from both of these substrates. In one case, labeled GABA was found in cerebral venous blood after administration of L-[U-^{14}C]glutamine and in two cases after L-[U-^{14}C]glutamic acid injection. The data in this study corroborated experiments with [^{14}C]glutamic acid from the author's laboratory[91] and results with L-[U-^{14}C]glutamine (described above) indicating only sporadic production of $^{14}CO_2$ by brain.

In an investigation to determine the relative rate of pyruvic acid formation in brain by decarboxylation of oxaloacetate or malate, Gombos *et al.*[56] added ^{14}C-labeled aspartate to artificial blood in their cat brain perfusion experiments. It was seen that aspartic acid entered the brain very slowly, since the specific activity of free aspartate in brain after about 2 h of perfusion had only about 1% of the specific activity of aspartate in blood. By recalculating some of their data, one finds that after about 2 h of perfusion with L-[U-^{14}C]aspartate, brain $^{14}CO_2$ specific activity was about 31% that of brain free aspartate. Furthermore, the specific activity of the free glutamate of brain at that time was about 29% that of brain free aspartate. Thus, it becomes apparent that although aspartate was taken up by brain from blood only very slowly, an unusually large percentage (31%) of brain CO_2 was derived from the free aspartate of brain. That glutamate had nearly the same specific activity as brain respiratory CO_2 offers supporting evidence for the author's proposed scheme[52] for cerebral metabolism of glucose *in vivo* (Fig. 1). From the specific activity values found in cerebral lactate, it was concluded[56] that about 10% of pyruvate formed in brain was derived from decarboxylation of oxaloacetate or malate.

In order to test a hypothesis by Woolley[96] that a biochemical aberration in schizophrenia might be caused by a deficiency in the brain of the decarboxylating enzyme for 5-hydroxytryptophan (5-HTP), Sacks[97] employed DL-5-hydroxy[carboxy-^{14}C]tryptophan as injected substrate. It was believed that decarboxylation of this compound would be evidenced by V–A $^{14}CO_2$ differences permitting an *in vivo* assay of the decarboxylase in human subjects. The results were disappointing in that, following the intravenous injection of labeled 5-HTP in four control subjects and four chronic mental patients (schizophrenics), there were no significant V–A $^{14}CO_2$ differences. Levels of arterial blood $^{14}CO_2$ indicated that 5-HTP was decarboxylated readily by other body tissues. It was concluded that either 5-HTP did not enter the human brain from blood or that, if it did enter, it was not readily decarboxylated. This report, which contradicted current concepts, has impressive support in the literature. Gal and Marshall[98] reported that labeled serotonin could be found in the brain following the intracerebral injection of DL-[3-^{14}C]tryptophan but that no labeled serotonin could be found after the intraperitoneal injection of DL-[2-^{14}C]tryptophan, although as much as 10 to 20 times more of this labeled precursor was used. Other studies on tryptophan hydroxylase of rat brain have indicated fundamental differences between the biochemical and pharmacological properties of endogenous brain serotonin derived from exogenous precursors, and it was stated that the extracranial tryptophan or 5-HTP may not be the biological precursor of brain serotonin but that protein may uniquely provide the amino acids necessary.[99] In addition, there have been a number of reports of no or very low activities of aromatic amino acid decarboxylase, the enzyme that decarboxylates 5-HTP, DOPA, etc. in human brain samples.[100–104]

The uptake of individual amino acids by human brain *in vivo* has been studied by the author's laboratory,[105,106] by Knauff *et al.*,[107] and by Felig *et al.*[108] All three groups used similar analytical procedures, but, whereas Knauff *et al.* and Felig *et al.* assayed unlabeled amino acids in plasma samples, Sacks determined both unlabeled and ^{14}C-labeled amino acids in whole blood samples. Sacks summarized data of 29 A–V sets of whole blood samples from nine human subjects for 17 free amino acids determined using the amino acid analyzer. From one to five A–V sets were analyzed for each subject and results calculated using average values for each person. It was seen that these average values indicated A–V differences for nine of the unlabeled amino acids and V–A differences for eight unlabeled amino acids. With each amino acid, there were both A–V and V–A differences. This occurred both within the same study and from subject to subject, and in no case were there constant A–V or V–A differences for any of the 17 free amino acids assayed. Data from 11 human subjects given either a single injection or constant infusion of ^{14}C-labeled alanine, phenylalanine, glutamic acid, or methionine indicated that both A–V and V–A differences occurred in each experiment with the exception of the subject given L-[methyl-^{14}C]methionine by constant infusion. In that case, constant A–V differences were found.[106]

Knauff *et al.*[107] analyzed A–V differences of 19 free amino acids in plasma samples from 25 patients. They reported data similar to those of Sacks in that both A–V and V–A differences were found, and in no case were all 19 amino acids simultaneously consumed (A–V) or released (V–A). In contrast, Felig *et al.*,[108] who assayed duplicate plasma samples from eight subjects, reported that "consistently positive A–V differences were observed for virtually all amino acids, indicating a net uptake of essential and nonessential amino acids by the brain in intact man." Because of this discrepancy, Sacks reexamined experiments done with rhesus monkeys.[105] Those studies were important in two ways: (1) they allowed the evaluation of the accuracy of the arteriovenous technique, and (2) they provided indisputable evidence of the net uptake of amino acids by brain *in vivo*. Table II lists data from 12 monkeys given constant infusions of various ^{14}C-labeled amino acids. It is apparent that estimates of percentage of injected ^{14}C as estimated from A–V differences were in rather good agreement with the values actually found. On the other hand, the percentages to be expected based on the table published by Felig *et al.*[108] gave estimates manyfold greater than those found. It would thus seem that data of Sacks *et al.*[105,106]

Table II
Cerebral Uptake of ^{14}C-Labeled Amino Acids by Rhesus Monkeys in Vivo[a]

[^{14}C]Amino acid	Estimate from A–V ^{14}C differences[b] (% inj. ^{14}C)	Found in brain (% inj. ^{14}C)	Estimate from Felig *et al.*[108] data[b] (% inj. ^{14}C)
Methionine	2.8	1.8	10.3
Tryptophan	0.8	0.8	—
Alanine[c]	0.9	0.3	3.6
DOPA	−0.6	0.4	—
Glycine[c]	2.0	0.7	5.5
Aspartate	3.1	2.3	—
GABA	1.8	0.5	—
Glutamic acid[c]	1.1	0.2	—
Glutamine	1.9	0.8	—
Leucine	4.6	1.9	9.2
Valine	3.4	1.2	8.5
Threonine	1.1	0.6	7.0

[a] All experiments were performed by constant infusion.
[b] Corrected for ^{14}CO$_2$ produced by brain estimated from V–A ^{14}CO$_2$ differences.
[c] Uptake was shown to be, at least in part, of a labeled intermediate (e.g., pyruvate, lactate, glucose).

and Knauff *et al.*[107] approximate more closely the actual uptake of amino acids by human brain *in vivo*.

In most of the studies done in the author's laboratory, determinations of α-amino acid nitrogen were performed on whole blood samples. With 222 A–V sets of determinations done on 67 individuals, average values were 440 μmol/100 ml for arterial blood, 430 μmol/100 ml for cerebral venous blood, and 10 μmol for A–V difference. This gave an average value of about 2% for A–V free total α-amino acid nitrogen, a figure slightly lower than the 5% reported by Knauff *et al.*[107] but slightly higher than the average cerebral uptake (1.9%) of the various ^{14}C-labeled amino acids by rhesus monkeys *in vivo* (Table II).

Knauff *et al.*[107] concluded that, "There appears thus to be present an exchange mechanism going on in both directions, whereby significant quantities of free amino acids can be metabolized." Data from the author's laboratory showing a lack of consistent A–V or V–A differences for any of the 17 amino acids studied (with the possible exception of [14C]methionine) and experiments with monkeys indicating cerebral uptakes of 0.2 to 2.3% of injected ^{14}C-labeled amino acids lead to a similar conclusion.

7. CEREBRAL METABOLISM IN CHRONIC MENTAL PATIENTS IN VIVO

7.1. Reduced Cerebral Oxidation of [14C]Glucose in Chronic Mental Patients

Sacks[109] expanded the series of arteriovenous experiments with control subjects and performed similar studies with chronic mental patients. In Fig. 3 are shown average specific activity–time curves for these experiments. The percentage of cerebral CO_2 derived from oxidation of glucose (calculated from individual [U-14C]glucose experiments) shows a significant ($P < 0.02$) reduction of about one-third in chronic psychiatric patients. Composite curves of [U-14C]glucose specific activity and of brain $^{14}CO_2$ specific activity with variously labeled [14C]glucose are plotted in Fig. 4. Derived curves D (representing C_3 and C_4 of [U-14C]glucose) were quite different since that of the mental patients was about one-third lower (compared to corresponding G curves) than that of normal subjects.

In these studies comparing chronic mental patients with normal controls, the data could have been influenced by dietary and environmental differences associated with institutionalization. However, since the arterial curves, which are indicative of overall body metabolism, showed no real variation between the two groups, and there were striking similarities between experiments using the same substrate, these factors were probably of little importance.

That less of the brain CO_2 was derived from glucose in the psychotic subjects[109] was interpreted to indicate that (1) there probably was a decreased oxidation of carbohydrate by brain in these patients, since V–A CO_2 differences (in μmol/100 ml blood) were about the same as in normal subjects or that (2)

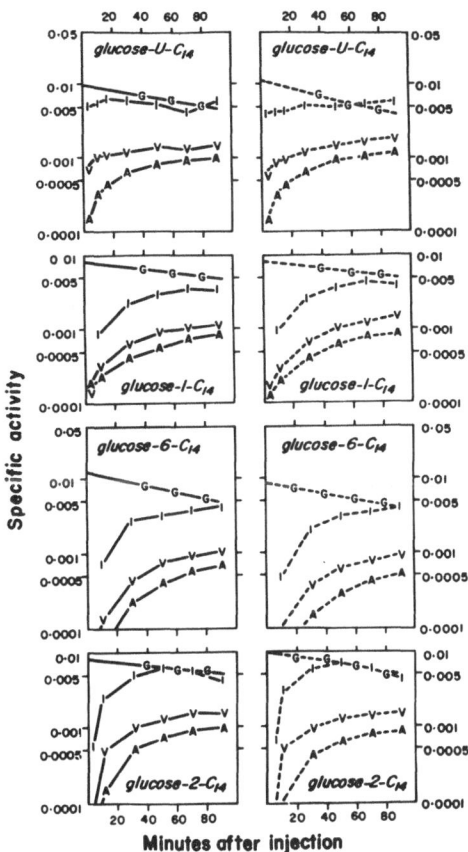

Fig. 3.   Average specific activity–time curves (plotted semilogarithmically) of blood glucose and CO_2. Curve G is venous glucose decay curve. Curves V and A represent specific activity of CO_2 in venous and arterial blood, respectively. Curve I represents specific activity of added increment (brain) $^{14}CO_2$. Solid lines represent normal subjects; broken lines, mental patients. From Sacks.[109]

there was a greater dilution of one or more of the carbohydrate intermediates by some protein and/or lipid intermediate.

7.2. Cerebral Oxidation of [3-^{14}C]Glucose in Chronic Mental Disease

With the availability of [3-^{14}C]glucose,* it was thought that a more direct demonstration of the difference in brain metabolism of mental patients could be accomplished. The initial experiments[20] indicated about 54% of brain CO_2 was derived from glucose in the normal subject and 35% in the mental patient. When the series was expanded, however, the differences were not as dramatic,

---

* Made by Dr. H. S. Isbell's laboratory[116] of the National Bureau of Standards with the aid of a grant from the National Institute of Mental Health (obtained through the efforts of Dr. C. Jelleff Carr of the Psychopharmacology Service Center).

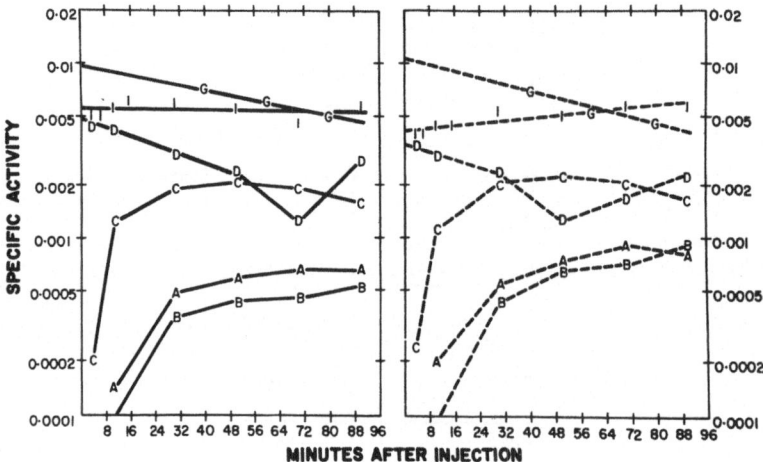

Fig. 4. Composite curves (plotted semilogarithmically) of [U-^{14}C]glucose specific activity (curve G) and specific activity of added increments of $^{14}CO_2$ with variously labeled glucose substrates. Curves A and B represent specific activity of added increment $^{14}CO_2$ following injection of [1-^{14}C]glucose and [6-^{14}C]glucose, respectively. One-sixth of the actual specific activity was plotted. Curve C represents specific activity of added increment $^{14}CO_2$ for [2-^{14}C]glucose plus [5-^{14}C]glucose. Since the curve for [5-^{14}C]glucose was taken to be the same as that for [2-^{14}C]glucose, one-third of actual specific activity was used in plotting curve C. Curve I (fitted by method of least squares) represents specific activity of added increment $^{14}CO_2$ when [U-^{14}C]glucose served as substrate. Curve D resulted when points on curves A, B, and C were added together and subtracted from corresponding experimental points of curve I. Curve D theoretically defines rate of decline of specific activity of added increments of $^{14}CO_2$ derived from carbon atoms 3 and 4 of [U-^{14}C]glucose. Solid lines represent normal subjects; broken lines, mental patients. From Sacks.[109]

although it was found that early values for brain $^{14}CO_2$ specific activity (curves I) were lower in mental patients.[52]

7.3. A Difference in the Cerebral Production of [1-^{14}C]Lactate from [3-^{14}C]Glucose in Chronic Mental Patients

In the 24 experiments done using [3-^{14}C]glucose, blood samples were assayed for [^{14}C]lactate formed endogenously from this substrate. These data, recently reexamined,[59] led to further elucidation of the mechanism whereby less $^{14}CO_2$ was produced from [^{14}C]glucose by the brain of the mental patient. It was seen that brain-produced lactate specific activities were considerably higher with mental patients than with control subjects. Ratios of brain-produced lactate to brain-uptake glucose specific activities varied about a theoretical value of 1 with control subjects but rose to about 4 with chronic mental patients. Of several possible explanations for this difference, the most likely was that involving a small lactate compartment(s) in some specific region(s) in which decarboxylation of the endogenously formed cerebral lactate was partially inhibited.[59] In this way, there could be a very rapid formation of high-specific-activity lactate shortly after injection which would "leak" into the large pool of brain lactate and/or enter the brain venous blood. Lactate data from two unusual studies with chronic psychiatric patients (exps. S_{13} and S_{14} of ref. 109) led to the speculation that these cases may have represented more extreme examples of the difference in brain-produced lactate than those of the 12 patients given [3-^{14}C]glucose. In those two studies, [^{14}C]lactate specific activities were 50–100 times that usually found in experiments with similarly labeled [^{14}C]glucose.[59]

Theoretically, the decarboxylation of lactate occurs after it is reconverted to pyruvate, which is then decarboxylated through the action of pyruvate dehydrogenase with thiamine pyrophosphate as a coenzyme. In an earlier study in which three control subjects and three mental patients were given [1-^{14}C]pyruvate,[20] it was evident that early brain [14]CO_2 specific activity values were lower with chronic mental patients, suggesting that [1-^{14}C]pyruvate decarboxylation was somewhat inhibited in those cases.

The demonstration of this difference in the production of [1-^{14}C]lactate from [3-^{14}C]glucose by the mental patient's brain and the proposed explanation offered further elucidation of the mechanism whereby less brain CO_2 was derived from glucose in psychotic subjects. Thus, it was postulated that a partial inhibition of the decarboxylation of endogenously formed [^{14}C]lactate and the subsequent "leaking" of a part of it into the brain's venous blood caused less [14]CO_2 to be produced from [^{14}C]glucose.[59]

8. AN ANIMAL MODEL FOR THE DETERMINATION OF CEREBRAL REGIONAL INTERMEDIARY GLUCOSE METABOLISM IN HUMANS IN VIVO USING SPECIFICALLY LABELED [^{11}C]GLUCOSE AND POSITRON EMISSION TRANSVERSE TOMOGRAPHY

8.1. Proposed Method Using [3,4-^{11}C]Glucose, [2,5-^{11}C]Glucose, and [1-^{11}C]Glucose

On the basis of data obtained with the arteriovenous technique, Sacks *et al.*[64] proposed a method for the determination of cerebral regional intermediary glucose metabolism in humans *in vivo* using specifically labeled [^{11}C]glucose and positron emission transverse tomography (PETT). (For a review of the principles and applications of PETT in neuroscience see Volume 2 of this series by J. D. Brodie, N. Volkow, and J. Rotrosen.) In it, the subject would be given successive intravenous injections of [3,4-^{11}C]glucose, [2,5-^{11}C]glucose, and [1-^{11}C]glucose. There would be a 30 min period of continuous PETT measurements following each injection and a 2-hr interval after the first and second injections. The data would be used with suitable equations and algorithms to estimate for each specific region of the subject's brain the dynamics of the EMP and the TCA metabolic pathways and the incorporation of glucose carbons into lactate and other carbon pools of the brain (i.e., glutamate, glutamine, GABA, aspartate, and alanine).

8.2. Use of ^{14}C as a Model for ^{11}C and Rat Brain Autoradiographs as a Model for PETT

In order to develop the rapid synthetic procedures that would be crucial to the proposed procedure, an animal model was employed.[64] Figure 5 shows some representative autoradiographs made with brain slices from rats given commercial, specifically labeled [^{14}C]glucose and sacrificed at time intervals selected to match those of the arteriovenous procedure with humans. As predicted, autoradiographs showed highest ^{14}C content from brain slices of [1-^{14}C]glucose experiments. Slices from [2-^{14}C]glucose studies had less ^{14}C, and those from [3,4-^{14}C]glucose experiments consistently had very little ^{14}C. Brain slices from rats given [6-^{14}C]glucose had ^{14}C activity similar to those given [1-^{14}C]glucose, indicating little PC pathway. The model was used to investigate effects of acute and chronic electroconvulsive shock (ECS) (Fig. 6).[64] One ECS produced an increase of 65% in dilution of glucose carbon in traversing the TCA and a 30% functioning of the PC. A series of nine ECSs (over 5 days) caused an increase of 88% in glucose carbon dilution in TCA and a 27% PC functioning.

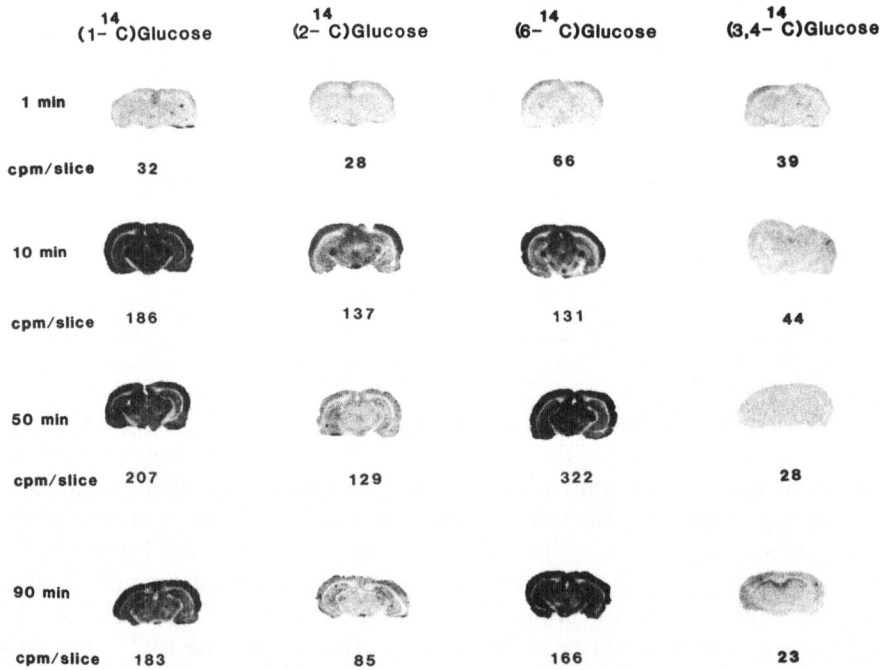

Fig. 5.   Autoradiographs made with brain slices from rats given specifically labeled [^{14}C]glucose intravenously. Note that ^{14}C content increased with time after injection up to 50 min and that C_2-labeled glucose slices had less than C_1 and that C_6 increased to values greater than C_1. Brain slices with [3,4-^{14}C]glucose had consistently little ^{14}C. Data were normalized to an average whole-blood glucose specific activity curve using values for glucose specific activity determined in each experiment. From Sacks *et al.*[64]

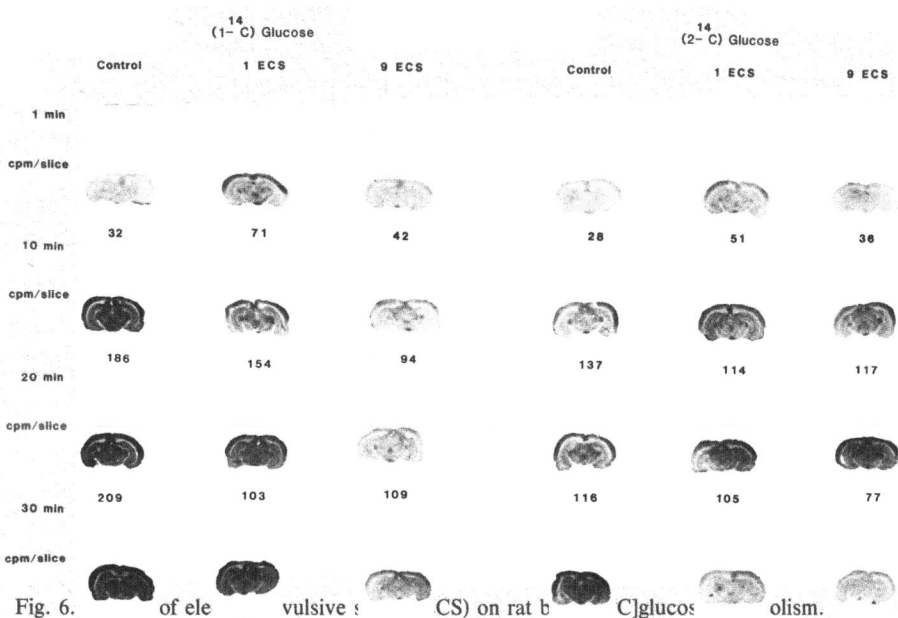

Fig. 6. ... of ele... vulsive ... CS) on rat b... C]glucos... olism. ... diographs show decreased ... C in brain slices even after one ... Following a series of nine... the decrease became more pronounced. From Sacks *et al.*[64]

Table III
Percentage of ^{14}C *Uptake Oxidized to* $^{14}CO_2$ *by Brain in Humans in Vivo*

Minutes after injection	[U-^{14}C]Glucose	[3-^{14}C]Glucose	[2-^{14}C]Glucose	[1-^{14}C]Glucose	[6-^{14}C]Glucose
		Normal Subjects			
2	10.6 (1)[a]	32.5 (2)	2.8 (5)	0.6 (5)	0.1 (4)
5	16.9	57.0	5.6	1.2	0.3
10	27.3	76.0	11.6	3.1	1.8
15	33.1	82.1	17.7	5.9	3.9
20	36.6	86.4	23.7	9.1	6.7
		Mental patients			
2	10.2 (1)	27.1 (2)	4.7 (5)	0.2 (6)	0 (4)
5	21.8	38.1	6.8	1.1	0.3
10	32.1	54.7	12.0	3.3	2.1
15	42.1	67.3	16.9	5.8	4.6
20	47.0	73.3	22.3	8.9	7.8

[a] Numbers in parentheses represent experiments averaged in obtaining values listed.

8.3. Advantages of Proposed Plan over Current PETT Methods

The procedure proposed by Sacks et al.[64] would not require many of the assumptions needed in methods presently used to estimate cerebral glucose metabolism in humans *in vivo* with short-lived isotopes and PETT. For example, nine kinetic constants are needed in the 2-deoxyglucose (2-DG) method.[82] More recently, as a result of increasing evidence showing the dephosphorylation of 2-deoxyglucose-6-phosphate (2-DG-6-P),[80,110,111] an additional constant has been employed in an attempt to account for the loss of 2-DG-6-P through the action of glucose-6-phosphatase in brain.[112] The justification for using an analogue of glucose rather than glucose *per se* because of ". . . significant losses of $^{14}CO_2$ from the brain within 2 min . . ."[113] would seem to be unfounded. As seen in Table III, with C_1-, C_6-, and even C_2-labeled glucose, such losses would be insignificant even 5 to 10 min after injection of the isotope.

The method employing [^{11}C]glucose and PETT as used by the St. Louis group[10] has two serious sources of error: (1) the [^{11}C]glucose produced photosynthetically is not only not uniformly labeled, but is contaminated with some [^{11}C]fructose[114,115] and, (2) because of a "washout effect" which occurs within the first minute following a single intravenous injection of isotopic glucose,[64] the short data collection time of 2–3 min could lead to significant errors (i.e., 150–250%)[64] in estimating glucose transport and metabolism.

In the method proposed by Sacks et al.,[64] there would be essentially three experiments with the same subject on the same day with only the position of the labeled carbon atom differing. Therefore, it could be reasonably assumed that the following factors would be relatively constant: (1) cerebral blood flow, (2) rate of uptake of glucose by brain from blood, (3) rate of production of CO_2, lactate, and pyruvate and their outflow from brain, and (4) rate of incorporation of C_1, C_2, C_5, and C_6 into brain carbon pools. The method would permit ongoing analyses of the EMP (and PC) and TCA pathways with resultant dilution in traveling the TCA in acute as well as chronic situations. The authors[64] emphasized that methods for determining cerebral glucose metabolism *in vivo* must be capable of assaying two practically equal (in humans, at least) components: (1) the oxidation to CO_2 and water and (2) the incorporation of C_1, C_2, C_5, and C_6 of glucose into the large, slow-turnover carbon pools (i.e., glutamate, glutamine, GABA, aspartate, alanine) and glucose carbons (primarily) 3 and 4 into the more rapidly turning over lactate pool.

REFERENCES

1. Waelsch, H., 1962, *Neurochemistry* (K. A. C. Elliott, I. H. Page, and J. H. Quastel, eds.), Charles C Thomas, Springfield, Illinois, p. 290.
2. Geiger, A., 1958, *Physiol. Rev.* **38**:1–20.
3. Dumke, P. R., and Schmidt, C. F., 1943, *Am. J. Physiol.* **138**:421–431.
4. Kety, S. S., and Schmidt, C. F., 1945, *Am. J. Physiol.* **143**:53–66.
5. Kety, S. S., 1948, *Methods in Medical Research*, Volume 1, Year Book Publishers, Chicago, pp. 204–217.

6. Kety, S. S., 1962, *Neurochemistry* (K. A. C. Elliott, I. H. Page, and J. H. Quastel, eds.), Charles C Thomas, Springfield, Illinois, p. 113.

7. Lassen, N. A., Ingvar, D. H., and Skinhøj, E., 1978, *Sci. Am.* **239(4)**:62–71.

8. Sakurda, O., Kennedy, C., Jehle, J. W., Brown, J. D., Carbin, G. L., and Sokoloff, L., 1978, *Am. J. Physiol.* **234**:H59–H66.

9. Sacks, W., 1956, *J. Appl. Physiol.* **9**:43–48.

10. Raichle, M. E., Larson, K. B., Phelps, M. E., Grubb, R. L., Jr., Welch, M. J., and Ter-Pogossiann, M. M., 1975, *Am. J. Physiol.* **228**:1936–1948.

11. Reivich, M., Kuhl, D., Wolf, A., Greenberg, J., Phelps, M., Ido, T., Casella, V., Fowler, J., Hoffman, E., Alavi, A., Som, P., and Sokoloff, L., 1979, *Circ. Res.* **44**:127–137.

12. Sacks, W., 1957, *J. Appl. Physiol.* **10**:37–44.

13. Coxon, R. V., and Robinson, R. J., 1956, *J. Physiol. (Lond.)* **132**:48P–49P.

14. Robinson, R. J., and Coxon, R. V., 1957, *Nature* **180**:1279–1281.

15. Coxon, R. V., and Robinson, R. J., 1959, *J. Physiol. (Lond.)* **147**:469–486.

16. Coxon, R. V., and Robinson, R. J., 1959, *J. Physiol. (Lond.)* **147**:487–510.

17. Geiger, A., and Magnes, J., 1947, *Am. J. Physiol.* **149**:517–537.

18. Geiger, A., and Yamasaki, S., 1956, *J. Neurochem.* **1**:93–100.

19. Allweis, C., and Magnes, J., 1958, *J. Neurochem.* **2**:326–336.

20. Sacks, W., 1961, *J. Appl. Physiol.* **16**:175–180.

21. Gottstein, U., Bernsmeirer, A., and Sedlmeyer, I., 1963, *Klin. Wochenschr.* **41**:943–948.

22. Scheinberg, P., Bourne, B., and Reinmuth, O. M., 1965, *Arch. Neurol.* **12**:246–250.

23. Sato, S., Tateyama, M., Sassamori, C., Kobayashi, S., Chiba, Y., and Takeda, Y., 1963, *Tohoku J. Exp. Med.* **81**:215–221.

24. Kneinerman, J., Sancetta, S. M., and Hackel, D. B., 1958, *J. Clin. Invest.* **37**:285–293.

25. Otani, H. I., 1963, *Jpn. Circ. J.* **27**:534–546.

26. Rowe, G. G., Maxwell, G. M., Castillo, C. A., Freeman, D. J., and Crumpton, C. W., 1959, *J. Clin. Invest.* **38**:2154–2158.

27. Sacks, W., 1965, *Ann. N.Y. Acad. Sci.* **119**:1091–1108.

28. Andjus, R. K., Suhara, K., and Sloviter, H. A., 1967, *J. Appl. Physiol.* **22**:1033–1039.

29. Sloviter, H. A., and Kamimoto, T., 1967, *Nature* **216**:458–460.

30. Sloviter, H. A., and Yamada, H., 1971, *J. Neurochem.* **18**:1269–1274.

31. Ghosh, A. K., Mukherji, B., and Sloviter, H. A., 1972, *J. Neurochem.* **19**:1279–1285.

32. Gilboe, D. D., Cotanch, W. W., and Glover, M. B., 1964, *Nature* **202**:399–400.

33. Gilboe, D. D., Cotanch, W. W., and Glover, M. B., 1965, *Nature* **206**:94–96.

34. Gilboe, D. D., Cotanch, W. W., Glover, M. B., and Levin, V. A., 1967, *Am. J. Physiol.* **212**:589–594.

35. White, R. J., Albin, M. S., and Verdura, J., 1964, *Nature* **202**:1082–1083.

36. White, R. J., Albin, M. S., Locke, G. E., and Davidson, E., 1965, *Science* **150**:779–781.

37. Moss, G., 1964, *J. Surg. Res.* **4**:170–177.

38. Moss, G., 1964, *Diabetes* **13**:585–591.

39. Swank, R. L., and Hissen, W., 1965, *Arch. Neurol.* **13**:93–100.

40. Gilboe, D. D., Glover, M. B., and Cotanch, W. W., 1967, *Am. J. Physiol.* **213**:11–15.

41. Allweis, C., Abeles, M., and Magnes, J., 1967, *Am. J. Physiol.* **213**:83–86.

42. Sacks, W., 1976, *Brain Metabolism and Cerebral Disorders,* 2nd ed. (H. E. Himwich, ed.), Spectrum, New York, pp. 89–127.

43. Sacks, W., 1967, *Science* **158**:498–499.

44. Sacks, W., 1968, *Arch. Biochem. Biophys.* **123**:507–513.

45. Cahill, G. F., and Aoki, T. T., 1980, *Cerebral Metabolism and Neural Function* (J. V. Passonneau, R. A. Hawkins, W. D. Lust, and F. A. Welsh, eds.), Williams & Wilkins, Baltimore, pp. 234–242.

46. Geiger, A., Kawakita, Y., and Barkulis, S. S., 1960, *J. Neurochem.* **5**:323–338.

47. Allweis, C. L., Gainer, H., and Chaikoff, I. L., 1960, *J. Appl. Physiol.* **15**:949–952.

48. Gainer, H., Allweis, C. L., and Chaikoff, I. L., 1963, *J. Neurochem.* **10**:903–908.

49. Barkai, A., and Allweis, C., 1966, *J. Neurochem.* **13**:23–33.

50. Gombos, G., Otsuki, S., Scruggs, W., Whitney, G., Schmolinske, A., and Geiger, A., 1963, *Fed. Proc.* **22**:633.

51. Sacks, W., and Sacks, S., 1968, *J. Appl. Physiol.* **24:**817–827.
52. Sacks, W., 1965, *J. Appl. Physiol.* **20:**117–130.
53. Barkulis, S. S., Geiger, A., Kawakita, Y., and Aguilar, V., *J. Neurochem.* **5:**339–348.
54. Geiger, A., Horvath, N., and Kawakita, Y., 1960, *J. Neurochem.* **5:**311–322.
55. Margolis, R. U., Barkulis, S. S., and Geiger, A., 1960, *J. Neurochem.* **5:**379–382.
56. Gombos, G., Geiger, A., and Otsuki, S., 1963, *J. Neurochem.* **10:**405–413.
57. Gibbs, W. L., Lennox, W. G., Nims, L. F., and Gibbs, F. A., 1942, *J. Biol. Chem.* **144:**325–332.
58. Himwich, W. A., and Himwich, H. E., 1946, *J. Neurophysiol.* **9:**133–136.
59. Sacks, W., Schechter, D. C., and Sacks, S., 1981, *J. Neurosci. Res.* **6:**225–236.
60. Coxon, R. V., 1957, *Metabolism of the Nervous System* (D. Richter, ed.), Pergamon Press, London, pp. 303–322.
61. Hostetler, K. Y., Landau, B. R., White, R. J., Albin, M. S., and Yoshon, D., *J. Neurochem.* **17:**33–39.
62. Sacks, W., 1969, *Handbook of Neurochemistry,* Volume 1 (A. Lajtha, ed.), Plenum Press, New York, pp. 301–324.
63. Hakim, A. H., and Moss, G., 1972, *Trans. N.Y. Acad. Sci.* **34:**473–484.
64. Sacks, W, Sacks, S., Badalamenti, A., and Fleischer, A., 1982, *J. Neurosci. Res.* **7:**57–69.
65. Sacks, W., and Jensen, C. O., 1951, *J. Biol. Chem.* **192:**231–236.
66. Sacks, W., 1958, *Science* **127:**594.
67. Angielski, S., 1963, *Acta Biol. Med.* (*Gdansk*) **7:**61–97.
68. Englard, S., Britten, J. S., and Listowsky, I., 1967, *J. Biol. Chem.* **242:**2255–2259.
69. Hers, H. G., 1957, *Le Metabolisme du Fructose,* Arscia, Bruxelles, p. 102.
70. Tewari, H. B., and Bourne, G. H., 1963, *J. Histochem. Cytochem.* **11:**121–122.
71. Nordlie, R. C., and Arion, W. J., 1965, *J. Biol. Chem.* **240:**2155–2164.
72. Sharma, N. N., 1967, *Acta Histochem.* **27:**165–171.
73. Prasannan, K. G., and Subrahamanyam, K., 1968, *Endocrinology* **82:**1–6.
74. Rosen, S. I., 1970, *Acta Histochem.* **36:**44–53.
75. Colilla, W., Jorgensen, R. A., and Nordlie, R. C., 1975, *Biochim. Biophys. Acta* **377:**117–125.
76. Dodd, P. R., Bradford, H. F., and Chain, E. B., 1971, *Biochem. J.* **125:**1027–1038.
77. Anchors, J. M., and Karnovsky, M. L., 1975, *J. Biol. Chem.* **250:**6408–6416.
78. Stephens, H. R., and Sandborn, E. B., 1976, *Brain Res.* **113:**127–146.
79. Karnovsky, M. L., Burrows. B. L., and Zoccoli, M. A., 1980, *Cerebral Metabolism and Neural Function* (J. V. Passonneau, R. A. Hawkins, W. D. Lust, and F. A. Welsh, eds.), Williams & Wilkins, Baltimore, pp. 359–366.
80. Hawkins, R. A., and Miller, A. L., 1978, *Neuroscience* **3:**251–258.
81. Scrutton, M. C., and Utter, M. F., 1968, *Annu. Rev. Biochem.* **37:**249–302.
82. Sokoloff, L., Reivich, M., Kennedy, C., Des Rosiers, M. H., Patlak, C. S., Pettigrew, K. D., Sakurada, O., and Shinohara, M., 1977, *J. Neurochem.* **28:**897–916.
83. Geiger, A., Magnes, J., Taylor, R. M., and Veralli, M., 1954, *Am. J. Physiol.* **177:**138–149.
84. Sato, S., Tateyama, M., Sasamori, C., Kobayashi, S., Chiba, Y., and Takeda, Y., 1963, *Tohoku J. Exp. Med.* **81:**207–214.
85. Gottstein, U., Held, K., Sebening, H., and Walpurger, G., 1965, *Klin. Wochenschr.* **43:**965–975.
86. Himwich, H., Bowman, K. M., Fazekas, J. F., and Goldfarb, W., *J. Nerv. Ment. Dis.* **89:**273–293.
87. Gottstein, U., and Held, K., 1967, *Klin. Wochenschr.* **45:**18–23.
88. Butterfield, W. J. H., Sells, R. A., Abrams, M. E., Sterky, G., and Whichelow, M. J., 1966, *Lancet* **1:**557–560.
89. Yamada, H., and Sloviter, H. A., 1969, *Fed. Proc.* **28:**574.
90. Eidelberg, E., Fishman, J., and Hams, M. L., 1967, *J. Physiol.* (*Lond.*) **191:**47–57.
91. Sacks, W., 1958, *J. Appl. Physiol.* **12:**311–318.
92. Allweis, C., Landau, T., Abeles, M., and Magnes, J., 1966, *J. Neurochem.* **13:**795–804.
93. Voegtlin, C., Dunn, E. R., and Thompson, J. W., 1925, *Am. J. Physiol.* **71:**574–582.

94. Sacks, W., 1973, *Biology of Brain Dysfunction,* Volume 1 (G. E. Gaull, ed.), Plenum Press, New York, pp. 143–189.
95. Maiolo, A. T., Bianchi Porro, G., Della Porta, P., Milani, R., Marchi, S., and Tagliabue, M., 1967, *I Comi Metabolici* (E. Polli, ed.), Edizioni L. Pozzi, Rome, pp. 463–487.
96. Woolley, D. W., 1958, *Chemical Concepts of Psychosis* (M. Rinkel and H. C. B. Denber, eds.), McDowell, Obolensky, New York, pp. 176–189.
97. Sacks, W., 1961, *J. Appl. Physiol.* **16**:1050–1054.
98. Gal, E. M., and Marshall, F. D., 1964, *Biogenic Amines,* Volume 8 (H. E. Himwich and W. A. Himwich, eds.), Elsevier, Amsterdam, pp. 56–60.
99. Green, H., and Sawyer, J. L., 1964, *Biogenic Amines,* Volume 8 (H. E. Himwich and W. A. Himwich, eds.), Elsevier, Amsterdam, pp. 150–167.
100. Ackermann, H., and Langemann, H., 1960, *Helv. Physiol. Pharmacol. Acta* **18**:C5–C6.
101. Håkanson, R., and Owman, C., 1966, *J. Neurochem.* **13**:597–605.
102. Robins, W., Robins, J. M., Croninger, A. B., Moses, S. G., Spencer, S. J., and Hudgens, R. W., 1967, *Biochem. Med.* **1**:240–251.
103. Vogel, W. H., Orfei, V., and Century, B., 1969, *J. Pharmacol. Exp. Ther.* **165**:196–203.
104. Sacks, W., Vogel, W. H., Nagatsu, T., Lloyd, K. G., and Sandler, M., 1979, *Catecholamines: Basic and Clinical Frontiers,* Volume 1 (E. Usdin, I. J. Kopin, and J. Barchas, eds.), Pergamon Press, New York, pp. 127–131.
105. Sacks, W., Sacks, S., Coxon, R. V., 1972, *Fed. Proc.* **31**:901.
106. Sacks, W., Coxon, R. V., and Sacks, S., 1979, *Abstracts of 7th International Society for Neurochemistry,* Jerusalem, p. 565.
107. Knauff, H. G., Gottstein, U., and Miller, B., 1964, *Klin. Wochenschr.* **42**:27–39.
108. Felig, P., Wahren, J., and Ahlborg, G., 1973, *Proc. Soc. Exp. Biol. Med.* **142**:230–231.
109. Sacks, W., 1959, *J. Appl. Physiol.* **14**:849–954.
110. Hawkins, R. A., 1980, *Cerebral Metabolism and Neural Function* (J. V. Passonneau, R. A. Hawkins, W. D. Lust, and F.A. Welsh, eds.), Williams & Wilkins, Baltimore, pp. 367–381.
111. Huang, M., Veech, R. L., and Passonneau, J. V., 1981, *Fed. Proc.* **40**:1621.
112. Huang, S. C., Phelps, M. E., Hoffman, E. J., Sideris, K., Selin, C. J., and Kuhl, D. E., 1980, *Am. J. Physiol.* **238**:E69–E82.
113. Sokoloff, L., 1981, *J. Cerebral Blood Flow Metab.* **1**:7–36.
114. Lifton, J. F., and Welch, M. J., 1971, *Radiat. Res.* **45**:35–40.
115. Straatmann, M., and Welch, M. J., 1973, *Int. J. Appl. Radiat. Isotopes* **24**:234–236.
116. Frush, H. L., Sniegoski, L. T., Holt, N. B., and Isbell, H. S., 1965, *J. Res. Natl. Bureau Stand. [A]* **69**:535–540.

Hypoglycemia

Bo K. Siesjö and Carl-David Agardh

1. INTRODUCTION

Hypoglycemia is associated with a great variety of medical disorders, e.g., liver failure, islet tumors of the pancreas, and premature birth. However, its clinical importance mainly resides in the fact that it frequently accompanies the treatment of diabetes mellitus.

Like hypoxia and ischemia, hypoglycemia leads to brain dysfunction and irreversible cell damage. From a neurochemical point of view, primary interest is usually focused on the metabolic perturbation induced by a deficient glucose supply and on the correlation between this perturbation and neuronal dysfunction. However, of equal importance are the biochemical mechanisms leading to irreversible cell damage. As the subsequent discussion will bear out, hypoglycemia offers a unique opportunity to study such mechanisms since it leads to cell damage even when oxygen supply is maintained, and it does so in the absence of cellular acidosis. It seems justified to begin, though, by a brief overview of the unique importance of glucose as a substrate for oxidative metabolism in the brain.

2. GLUCOSE AS A SUBSTRATE FOR BRAIN METABOLISM

It is now agreed that glucose is the main, under ordinary circumstances the sole, substrate for cerebral energy metabolism.[1,2] Several facts contribute to make brain tissues sensitive to even moderate reductions in plasma glucose concentrations. First, cerebral tissues have high energy requirements. In the whole human brain, cerebral metabolic rate for oxygen is about 1.6 $\mu mol \cdot g^{-1} \cdot min^{-1}$, corresponding to a glucose consumption of about 0.3 $\mu mol \cdot g^{-1} \cdot min^{-1}$. In the rat brain, the corresponding figures are about 3.0 and 0.6 $\mu mol \cdot g^{-1} \cdot min^{-1}$, respectively. However, some cortical and subcortical structures have much higher metabolic rates. For example, in the au-

Bo K. Siesjö and Carl-David Agardh • Laboratory for Experimental Brain Research, E-Bocket, University Hospital of Lund, S-221 85 Lund, Sweden.

ditory cortex of the rat, the local metabolic rate for glucose (l-CMR_{gl}) exceeds 1.5 $\mu mol \cdot g^{-1} \cdot min^{-1}$.[3,4] Second, the substrate stores are relatively slender in that glycogen and glucose concentrations amount to about 2–3 $\mu mol \cdot g^{-1}$ each. Thus, if the substrate supply from the blood is interrupted, the endogenous stores suffice to maintain normal cortical CMR_{gl} for 10–15 min in man and about 5–10 min in rats. Third, under normal circumstances other plasma constituents cannot substitute for glucose unless they are able to raise the blood glucose concentration (e.g., refs. 5,6).

An exception to this general rule is provided by conditions associated with increased plasma concentrations of ketone bodies, e.g., starvation and diabetes. In starvation, ketone bodies can support about 50% of the substrate requirements of the brain.[7] Animal experiments have verified this observation and shown that the extraction of ketone bodies by the brain is proportional to their plasma concentrations.[8,9] However, at least in insulin-induced hypoglycemia, ketone bodies do not seem to provide the missing substrates (see below).

3. CRITICAL GLUCOSE CONCENTRATIONS

When plasma glucose concentrations are progressively lowered, a series of clinical signs occur that reflect a progressive reduction in consciousness and in sensory and motor functions.[10] Early symptoms include somnolence and lassitude, progressing to stupor and coma. Motor function deteriorates in proportion to the reduction in consciousness and in sensory perception; however, overt coma may be preceded or accompanied by focal or generalized convulsions. When coma ensues, hypoglycemia may be accompanied by arterial hypoxia and by hypotension. In the clinical setting, it is frequently difficult to conclude whether any brain damage incurred was caused by the hypoglycemia *per se* or if secondary events (seizure activity and tissue hypoxia) contribute.

The deterioration of brain functions in hypoglycemia is paralleled by progressive impairment in electrophysiological functions. Thus, when the level of consciousness is reduced, there is generalized slowing of the EEG, progressing to complete cessation of electrical activity, and convulsive activity often shows up as polyspikes superimposed on the slow-wave activity.[11–14] Corresponding alterations occur in sensory evoked potentials. At moderate degrees of hypoglycemia, when the EEG shows a high-amplitude slow-wave activity, the amplitude of the sensory evoked response increases, and the evoked responses disappear when the EEG shows a suppression–burst activity,[15] reflecting deterioration of synaptic function.

Animal experiments allow a less precise documentation of neurological failure in hypoglycemia, but they have the virtue of permitting a close correlation between, on one hand, clinical behavior and electrical functions and, on the other hand, plasma and tissue glucose concentrations. For that reason, we use both clinical and experimental material to assess this correlation.

Observations in man demonstrate that when blood glucose concentrations are reduced towards 50% of control, i.e., to about 2.5 $\mu mol \cdot ml^{-1}$, signs of brain dysfunction such as restlessness and lassitude develop (e.g., ref. 16).

However, these as well as some other symptoms (sweating, anxiety, hunger, and palpitation) may at least partly be caused by increased epinephrine release.[17] With further reductions in blood glucose concentrations to 1.0–1.5 μmol $\cdot$ g^{-1}, coma ensues. In rats, behavioral signs of hypoglycemia (reduced mobility, posterior weakness) develop at blood glucose concentrations of below 2 μmol $\cdot$ g^{-1}; the animals are severely obtunded and weak at 1.0–1.5 μmol $\cdot$ g^{-1} and lose consciousness ("coma") at about 1.0 μmol $\cdot$ g^{-1}.[13,14] The corresponding EEG changes show a close correlation to blood glucose concentrations (Fig. 1). Such results allow the conclusion that the development of coma coincides with the cessation of spontaneous EEG activity ("isoelectric EEG") and that both occur at a blood glucose concentration of about 1.0 μmol $\cdot$ g^{-1}.

There has been some controversy regarding intracellular glucose concentrations, i.e., whether or not there are measurable amounts of free intracellular glucose under normal conditions (see refs. 2,18–20). In the opinion of the present reviewers, overwhelming evidence exists that free glucose is normally present. For example, in the ventilated rat under nitrous oxide analgesia and at a blood glucose concentration of 6–8 μmol $\cdot$ g^{-1}, one can derive an intracellular glucose concentration of about 3 μmol $\cdot$ g^{-1}.[21] Since CSF glucose concentrations in this species are about 50% of the whole-blood values, it seems permissible to use the extra- and intracellular compartments as a lumped space.[22]

Assuming that the extracellular fluid occupies 15% of brain volume and that the glucose concentration in cisternal CSF reflects the extracellular glucose concentration, the intracellular concentration approaches zero when plasma

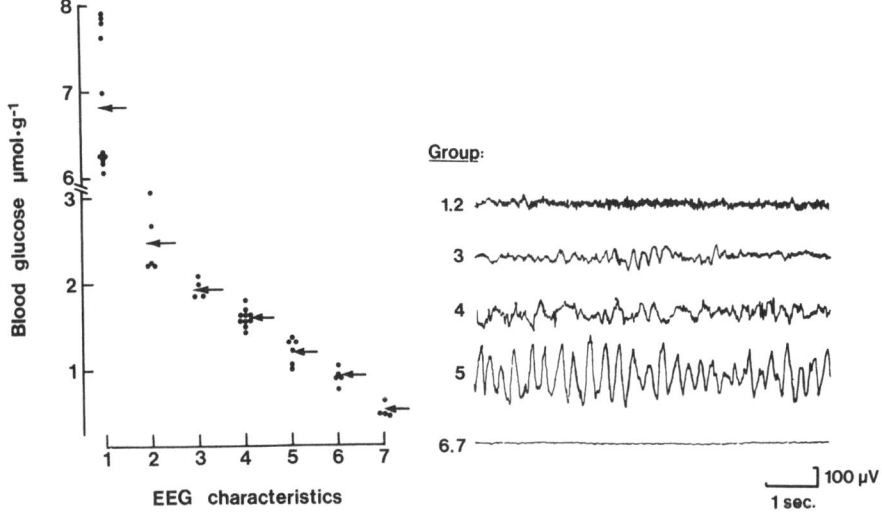

Fig. 1. Relationship between blood glucose concentrations and EEG characteristics during progressive, insulin-induced hypoglycemia. The right panel defines the progression of EEG changes from a control pattern in uninjected (1) and insulin-injected animals (2) to a slow-wave pattern (3, 4, 5) with interspersed epileptic spikes (4, 5) and to cessation of EEG activity (6, early; 7, late). The left panel correlates EEG patterns and blood glucose concentrations. Modified after Lewis *et al.*[13]

glucose is reduced below about 1.5 μmol $\cdot$ g^{-1}.[21] In other words, the development of symptoms of hypoglycemia seems to coincide with depletion of intracellular glucose. At that point, therefore, glucose transport from blood to tissue becomes rate limiting for cerebral glucose consumption.

4. CEREBRAL BLOOD FLOW, OXYGEN AND GLUCOSE CONSUMPTION

Information on cerebral oxygen consumption and on the rate of utilization of glucose and other exogenous substrates is crucial to a discussion of cellular metabolic changes during hypoglycemia. Thus, knowledge of the total oxygen consumption and of the part covered by glucose and other exogenous substrates will allow an assessment of oxidative breakdown of endogenous substrates.

4.1. Cerebral Metabolic Rate for Oxygen and Cerebral Blood Flow

Several studies have been devoted to measurements of CMRO$_2$ and CBF in man during moderate and severe hypoglycemia. All show that CMRO$_2$ is maintained close to control values at moderate degrees of hypoglycemia,[16,17,23] but results obtained in hypoglycemic coma differ. Thus, whereas Kety et al.[23] obtained results suggesting that CMRO$_2$ was reduced in hypoglycemic coma, Della Porta et al.[24] found no such decrease. Results obtained in rats support the latter view.[25] The controversy now seems resolved, since it has been found that CMRO$_2$ is maintained during the first 15 min of coma (defined as the period of ceased EEG activity) and falls to about 70% of control after 30 min.[26]

Results in man favor the view that CBF remains normal during hypoglycemia or increases but slightly.[16,23,24] In the rat, though, clear increases in overall and local CBF have been found both in the precomatose and the comatose stages.[25–27] We note that the increase in CBF is useful in the sense that it should increase glucose delivery to the tissue (cf. increase in CBF during hypoxia), but since the transport is mainly dependent on the capacity of the carrier, the effect is probably marginal. One finding provides a likely explanation for the somewhat varying CBF results reported. Thus, severe hypoglycemia leads to loss of cerebral vascular autoregulation; in other words, since blood flow varies directly with the cerebral perfusion pressure, an increase in CBF may be noted only if blood pressure is maintained.[28]

4.2. Cerebral Metabolic Rate for Glucose

A decrease in cerebral glucose utilization was reported to occur with only moderate decreases in blood glucose concentration (to about 2.5 μmol $\cdot$ g^{-1}, see ref. 16). It is less certain, though, that the figures reported really express true CMR$_{gl}$, since a steady state may not have been reached. Thus, if glucose equivalents were derived from the tissue stores of glucose and glycogen, the true glucose utilization must have exceeded that estimated from CBF and arteriovenous differences in glucose concentration (AVD$_{gl}$). It seems more

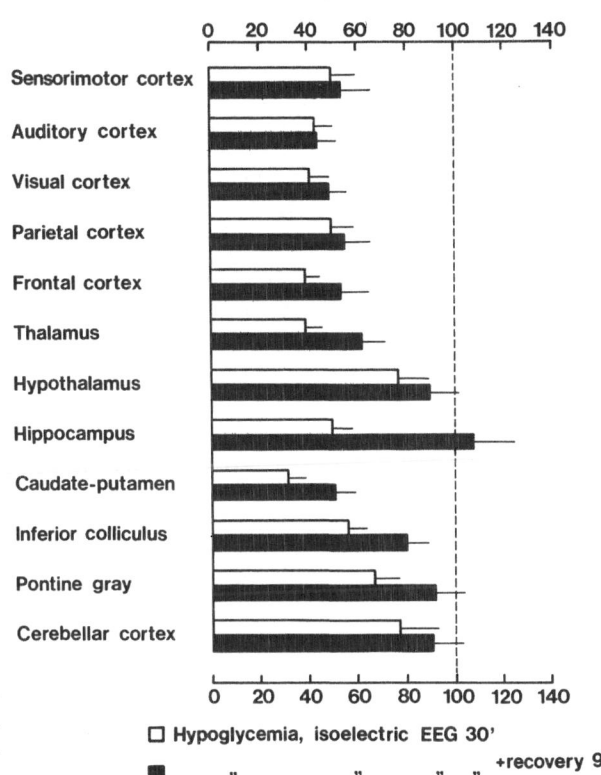

Local CMRgl in percent of control

Fig. 2. Local CMR_{gl} during and following severe hypoglycemia as estimated with the [^{14}C]deoxyglucose technique of Sokoloff *et al.*[3] Since the "lumped constant" is not known, the values (means ± S.E.M. in percent of control) are only semiquantitative. Reproduced with permission from Abdul-Rahman and Siesjö.[30]

clearly established that glucose utilization falls dramatically with more severe degrees of hypoglycemia, e.g., in coma. From data obtained in man[24] and experimental animals,[25,29] one can deduce that, in coma, glucose utilization falls to between one-third and one-half of control at an unchanged (or only moderately reduced) $CMRO_2$. Unless the balance can be accounted for by oxidation of other exogenous compounds, it provides information on the rate of consumption of endogenous substrates (see below).

It is now generally agreed that more than 90% of the glucose extracted by the brain is oxidized to CO_2 and water.[2] As a first approximation, therefore, we may assume that the ratio $6 \times AVD_{gl}/AVDO_2$ should be unity. In the rabbit, severe hypoglycemia was associated with a depression of this ratio to below 0.5 for up to 3 h, demonstrating oxidation of appreciable amounts of nonglucose substrates.[29] Comparable results have been reported for rats.[25,26] We can use the latter results to demonstrate that in the 30- and 60-min periods preceeding coma and during the first 30 min of coma, at least 25 μmol · g^{-1} (in terms of glucose equivalents) of endogenous substrates must have been consumed.[26] Since the normal stores of carbohydrate substrates (glucose plus glycogen) amount to only about 5 μmol · g^{-1}, it is clear that nonglucose substrates are mobilized.

At this point, it should be emphasized that the reduction in CMR_{gl} during hypoglycemia is not uniform. Abdul-Rahman and Siesjö[30] recently estimated local CMR_{gl} during severe hypoglycemia using the deoxyglucose technique of Sokoloff et al.[3] The results in Fig. 2 demonstrate that, in coma, estimated CMR_{gl} decreased less in some structures (e.g., cerebellum and hypothalamus). When these results are considered, it must be recalled that calculation of CMR_{gl} requires the use of a "lumped constant"[3] which changes precipitously in severe hypoglycemia.[31] Thus, CMR_{gl} values obtained are only semiquantitative estimates of the true rates. Nonetheless, it seems likely that the decrease in glucose delivery is inhomogeneous. For example, structures that display a pronounced lowering of "CMR_{gl}" during hypoglycemia remained metabolically depressed in the recovery period (see Fig. 2). Furthermore, the first category of structures shows other signs of a more pronounced substrate depletion, e.g., a marked hypoperfusion in the recovery period and histopathological signs of cell damage (see below).

The question must be raised of the mechanisms that lead to regional differences in the reduction of "CMR_{gl}" in hypoglycemia. Possibly, differences in CBF could contribute. For example, cerebellum and brainstem structures show a more pronounced increase in CBF than cerebral cortical structures.[27] However, it seems likely that local differences in glucose transport capacity are also involved. For example, if some structures have lower K_m values for glucose transport than others, or a higher V_{max}, they would maintain a higher "CMR_{gl}" during hypoglycemia. Since such differences may underlie the hitherto unexplained phenomenon of selective cell vulnerability, it seems highly justified that they be studied in more detail.

5. NEUROCHEMICAL CHANGES AT TISSUE LEVEL

Changes in the composition of cerebral tissues during hypoglycemia have been studied over a long period by many laboratories. No doubt, this interest stems from the facts that a deficient glucose supply drastically interferes with cerebral energy production and that oxidative metabolism continues at the expense of endogenous substrates, some of which are evidently derived from structural components. In 1955, Abood and Geiger[32] showed that when isolated cat brains were perfused with glucose-free blood, oxygen consumption continued at high rates, apparently at the expense of endogenous carbohydrate, lipid, and protein components. Early studies showed that hypoglycemia was associated with reduced tissue concentrations of glycogen, glucose, lactate, and pyruvate,[33,34] with changes in amino acid concentrations involving depletion of glutamate, glutamine, alanine, and GABA as well as accumulation of aspartate and ammonia,[35–37] and with decomposition of phospholipids.[38] In this period, attempts were also made to unravel the influence of hypoglycemia on tissue energy state, as this is reflected in the levels of phosphocreatine, ATP, ADP, and AMP.[33,34,39,40]

Although much of our knowledge of neurochemical changes in hypoglycemia was laid down by these early workers, studies performed during the last 10–15 years offer the advantages of increased specificity of methods employed, of improved techniques for tissue fixation, and of more comprehensive approaches involving a better correlation to neurological and electrophysiological functions, to blood and tissue glucose concentrations, to metabolic rate, and to histopathology. For these reasons, we focus our discussion on studies performed in that period.

5.1. Exogenous versus Endogenous Substrates

The disparity between oxygen and glucose utilization rates during severe hypoglycemia raises the question of whether other blood-borne compounds can provide some of the missing substrates. As stated, only glucose can reverse hypoglycemic coma. This is not because of an inability of brain tissues to metabolize other carbohydrate substrates than glucose, or amino acids, but resides in the fact that only glucose can be translocated across the barriers separating plasma and intracellular fluids by a carrier of sufficient capacity to support normal metabolic rate.[41–43]

Pertinent to this question are the blood concentrations, and the V_{max} values for the transport of substrates from blood to tissue. It would seem that the low V_{max} values for transport of amino acids (see refs. 42,43) exclude amino acids as candidates. Furthermore, there is no evidence that free fatty acids (FFAs) can provide the missing substrates.[7] The question remains whether or not ketone bodies enter as exogenous substrates. However, since the carboxylic acid carrier has a low capacity in mature animals,[42,43] and since the blood concentrations of β-hydroxybutyrate plus acetoacetate are below 1 μmol · g^{-1} in precoma and coma,[26] ketone bodies are unlikely to provide a quantitatively important exogenous substrate. We have to assume, therefore, that the missing substrates are derived from endogenous sources.

Assuming that exogenous noncarbohydrate substrates cannot make a significant contribution to oxidative reactions during hypoglycemia, we must consider to what extent the mismatch between oxygen and glucose uptake is covered by endogenous carbohydrate substrates. Available evidence indicates that the tissue stores of glucose approach zero when blood glucose concentrations fall below 1.5 μmol · g^{-1} and that the glycogen stores are depleted when coma ensues, i.e., at a blood glucose concentration of about 1 μmol · g^{-1}.[14,21,25,44,45] In other words, at the point at which EEG activity ceases, the missing substrates must be provided by carbohydrate intermediates (and lactate) and by noncarbohydrate compounds.

It has been clearly documented that hypoglycemia reduces the pool sizes of glycolytic metabolites and citric acid cycle intermediates.[21,25,34,40,45–47] Figure 3 illustrates changes observed both in the slow-wave–polyspike and the "isoelectric" periods (data from refs. 21,25). In the precomatose period, the tissue uses up not only the stores of glycogen and glucose (about 5 μmol · g^{-1} of glucose equivalents) but also part of the pools of glycolytic metabolites and citric acid cycle intermediates. The substrate contribution from the latter is small, being less than 2 μmol · g^{-1} for glycolytic metabolites and only 0.2 μmol · g^{-1} for citric acid cycle intermediates (in glucose equivalents). Furthermore, since no additional changes in these metabolites are observed after

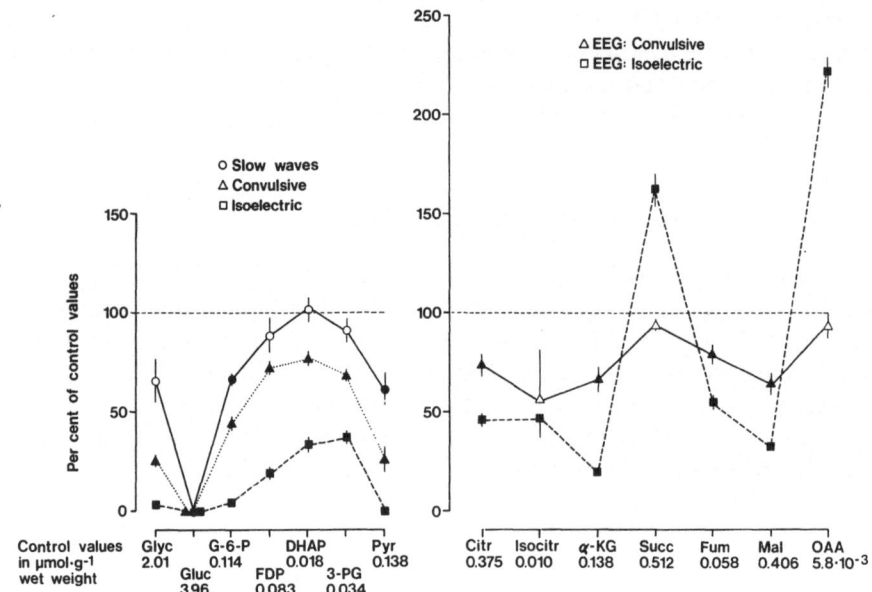

Fig. 3.   Influence of hypoglycemia on tissue concentrations of glycolytic metabolites and citric acid cycle intermediates. The values (means ± S.E.M. in percent of control) are from Lewis *et al.*[21] and Norberg and Siesjö.[25]

5 min of coma, further oxidation of "missing substrates" must occur at the expense of noncarbohydrate compounds.

In view of the fact that the amino acid pool is also reduced during hypoglycemia (see below), the reduction of the citric acid cycle pool most reflect a channeling of carbon skeletons from citric acid cycle intermediates to the glycolytic sequence. Probably, this occurs by reversal of one or more of the reactions that normally achieve CO_2 fixation for net synthesis of citric acid cycle intermediates.[25,40] As Fig. 3 shows, severe hypoglycemia leads to decreases in all citric acid cycle intermediates except succinate and oxaloacetate (OAA); in fact, the concentrations of the latter increased above normal in coma. Probably, the increase in succinate reflects inhibition of succinate dehydrogenase (E.C. 1.3.99.1) because of a shortage of reduced coenzyme Q_{10} and of ATP (see below). Two events probably contribute to the increase in OAA concentration: accumulation of OAA because of a shortage of acetyl-CoA for the condensation reaction yielding citrate[48] and a redox-dependent decrease of the malate/oxaloacetate ratio.[25]

This conclusion is supported by results suggesting that hypoglycemia is accompanied by an oxidation of cellular redox systems,[21,25] an event entirely in keeping with the redox change observed in isolated mitochondria in the substrate-depleted state.[49,50] Incidentally, such a redox change should exclude the possibility that cellular oxygen lack contributes to the deterioration of cerebral energy state (see below).

5.2. Amino Acids

As stated above, hypoglycemia leads to a clear perturbation of amino acid metabolism (Table I). Results obtained in animal experiments, in which electrical activity was assessed and blood (and tissue) glucose concentrations were measured,[21,25,34,45,51] demonstrate that carbon skeletons are mobilized from amino acids by both transamination and deamination reactions. The chief transamination reaction is that catalyzed by aspartate aminotransferase (E.C. 2.6.1.1),

$$\text{Glutamate} + \text{OAA} \rightleftharpoons \text{Aspartate} + \alpha\text{-KG} \qquad [1]$$

a reaction that allows oxidation of carbon skeletons when α-KG is oxidized to OAA in the citric acid cycle. It seems probable that this equilibrium reaction is shifted towards α-KG formation by the accumulation of OAA. The reaction also provides for oxidation of carbon skeletons from GABA, alanine, and glutamine via the reactions catalyzed by GABA aminotransferase (E.C. 2.6.1.19), alanine aminotransferase (E.C. 2.6.1.2), and glutaminase (E.C. 3.5.1.2).

$$\text{GABA} + \alpha\text{-KG} \rightleftharpoons \text{Succinate semialdehyde} + \text{Glutamate} \qquad [2]$$

$$\text{Alanine} + \alpha\text{-KG} \rightleftharpoons \text{Glutamate} + \text{Pyruvate} \qquad [3]$$

$$\text{Glutamine} + \text{H}_2\text{O} \rightleftharpoons \text{Glutamate} + \text{NH}_3 \qquad [4]$$

Some workers have noted that insulin hypoglycemia produces no decrease in the total content of free amino acids.[36,52] However, other studies have shown that at more severe degrees of hypoglycemia, a net loss of amino acids occurs, probably by deamination of glutamate via the glutamate dehydrogenase (E.C. 1.4.1.2) reaction (e.g., refs. 25,45).

Table I
Cerebral Cortical Concentrations of Amino Acids and Ammonia during Insulin-Induced Hypoglycemia[a]

		Isoelectric EEG (min)			
	Control ($n = 10$)	5 ($n = 6$)	10 ($n = 6$)	15 ($n = 6$)	30 ($n = 4$)
Glutamate	12.00 ± 0.16	6.72 ± 0.86	4.21 ± 0.34	4.47 ± 0.18	2.79 ± 0.16
Aspartate	3.52 ± 0.11	12.96 ± 0.85	15.28 ± 0.55	15.21 ± 0.44	13.74 ± 0.60
Glutamine	5.08 ± 0.12	1.57 ± 0.38	0.97 ± 0.16	0.91 ± 0.12	0.29 ± 0.09
Alanine	0.448 ± 0.019	$0.317 \pm 0.027^*$	0.251 ± 0.011	0.300 ± 0.028	0.253 ± 0.016
GABA	2.00 ± 0.04	1.78 ± 0.06	1.51 ± 0.09	1.50 ± 0.17	0.97 ± 0.03
	0.37 ± 0.04	3.94 ± 0.17	3.81 ± 0.15	3.70 ± 0.36	3.06 ± 0.16

[a] The values are given in $\mu\text{mol}\cdot\text{g}^{-1}$ wet weight and represent means $\pm$ S.E.M. $P < 0.01$ for all values except $^*P < 0.05$. Data from Agardh *et al.*[45]

$$\text{Glutamate} + NAD^+ + H_2O \rightleftharpoons \alpha\text{-KG} + NH_3 + NADH + H^+ \quad [5]$$

Probably, this reaction contributes significantly to accumulation of ammonia,[21,25] some of which is released into the venous effluent.[34] We note that reaction [5] can be shifted towards α-KG production by a redox change of the type discussed. In theory, carbon skeletons could be tapped off the aspartate pool via the reactions of the purine nucleotide cycle.[53] However, since these reactions, which lead to an equivalent production of ammonia, require ATP, they are less likely to be responsible for the decrease in amino acid pool size.

It should be mentioned that although brain tissues contain appreciable quantities of N-acetylaspartic acid, there is no evidence that this compound contributes to substrate supply during hypoglycemia.[52,54]

The glutamate dehydrogenase reaction is truly anaplerotic and must tend to increase the size of the citric acid pool. However, since the latter is reduced and not increased (see above), it is necessary to conclude that the carbon skeletons formed from the amino acids end up in the glycolytic chain via a reversal of the reactions normally leading to CO_2 fixation. Since the amino acid pool decreases by 5–6 $\mu mol \cdot g^{-1}$ after 30 min of hypoglycemic coma,[45] relatively large amounts of carbon skeletons must be shuttled along these pathways.

5.3. Energy Metabolism

Some years ago, results on the effect of hypoglycemia on cerebral energy metabolism were controversial. Thus, whereas some authors maintained that severe degrees of hypoglycemia were associated with a derangement of cerebral energy state,[12,34] others maintained that the tissue concentrations of high-energy phosphates remained unaltered.[39,44] It seems likely that the differences in results relate to the degree of deterioration of brain function. Thus, when unanesthetized animals are frozen during stupor (loss of righting reflex), the tissue concentrations of phosphocreatine and ATP are not measurably reduced.[46]

On the other hand, it has been repeatedly shown that whereas the concentrations of PCr, ATP, ADP, and AMP in ventilated animals are not measurably affected at precomatose levels of hypoglycemia corresponding to an EEG pattern of slow waves and polyspikes, they are precipitously altered when the EEG activity ceases at blood glucose concentrations of about 1 $\mu mol \cdot g^{-1}$ (Fig. 4).[13,14,25,45] It would thus seem that the deterioration of cerebral energy state is the cause of the sudden electrical failure. However, it should be emphasized that deterioration of cerebral energy state is less complete than in severe ischemia, since about one-third of the ATP concentration remains even if the period of hypoglycemic coma is prolonged to 60 min (see below).

Two additional findings should be recalled. First, hypoglycemia leads to a substantial decrease in the size of the adenine nucleotide pool (sum of ATP, ADP, and AMP),[13,14,25,45] and since adenosine, IMP, inosine, and hypoxanthine accumulate,[55] the results demonstrate that AMP is degraded via the 5'-nucleotidase (E.C. 3.1.3.5) and AMP deaminase (E.C. 3.5.4.6) reactions. Second,

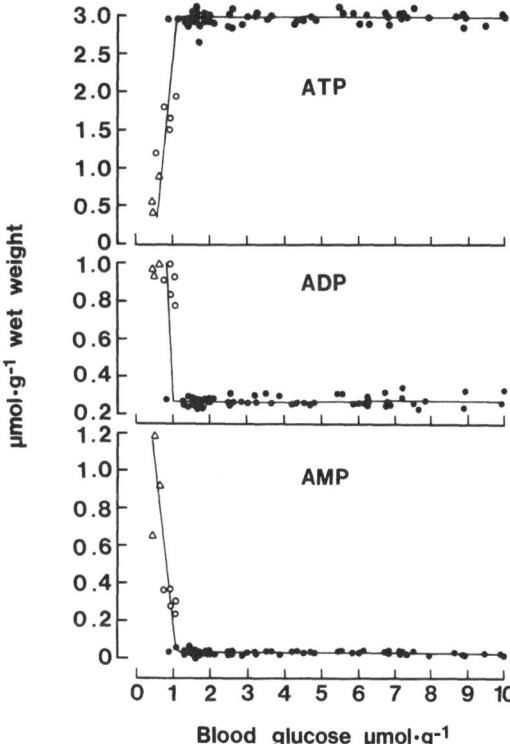

Fig. 4. Relationship between blood glucose concentration and tissue concentrations of ATP, ADP, and AMP during progressive insulin-induced hypoglycemia. Reproduced with permission from Lewis *et al.*[13]

as Fig. 5 shows, the reduction in ATP concentration is paralleled by corresponding decreases in the concentrations of GTP, UTP, and CTP.[55]

It is of considerable interest that hydrolysis of ATP and other nucleoside triphosphates occurs *pari passu* with the cessation of EEG activity, but that CMRO$_2$ remains unchanged for at least another 15 min (see above). These results do not support previous assumptions that cerebral metabolic rate is reduced in the precomatose phase of hypoglycemia.[44,46] We do not know why

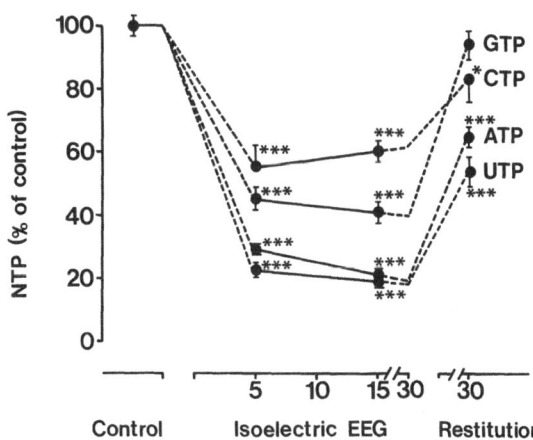

Fig. 5. Influence of severe insulin-induced hypoglycemia on brain tissue concentrations of ATP, CTP, GTP, and UTP. The values are means ± S.E.M. in percent of control. Reproduced with permission from Chapman *et al.*[55]

ATP falls in spite of an unchanged $CMRO_2$, but two possible explanations emerge: the energy requirements of the tissue are increased above normal, or the oxygen consumption is not accompanied by a corresponding production of ATP ("uncoupling of oxidative phosphorylation"). We return to this problem in Section 5.7.

5.4. Proteins, RNA, Phospholipids, and Free Fatty Acids

At present, the only experiments showing that hypoglycemia may lead to marked breakdown of protein constituents of the brain are those reported by Abood and Geiger.[32] Since these were obtained on isolated cat heads perfused with glucose-free erythrocyte suspensions, it is not known if the results apply to intact animals and to insulin-induced hypoglycemia. Available evidence is conflicting,[38,56] and it remains to be shown whether protein degradation occurs in hypoglycemia of sufficient severity to cause energy failure and breakdown of other macromolecular cell constituents. Recent studies have shown that at these degrees of hypoglycemia, the RNA content of the cerebral cortex is reduced by about 10%.[26]

Special attention must be paid to changes in tissue phospholipid composition, since several studies have shown that severe hypoglycemia causes substantial hydrolysis of phospholipids, and since these are normal constituents of cellular membrane structures. The results are far from unequivocal, though, since some workers have reported that tissue phospholipid content remains unchanged during hypoglycemia,[57,58] and others that a substantial loss of phospholipids occurs.[59,60] In one recent study,[26] both total phospholipid phosphorous and total fatty acid content decreased by about 10% in "coma" (Table II). That study also showed that this change occurred within 5 min after the EEG activity ceased, with no further changes in the subsequent 25-min period. Of interest in the present context is also the fact that the reduction in phospholipid content seemed to affect all individual phospholipid classes analyzed (see below).

From a pathophysiological point of view, the phospholipid degradation products that deserve interest are free fatty acids (FFA) and lysophospholipids.

Table II
Total Fatty Acid, Total Phospholipid Phosphorus, and Total Free Fatty Acid
Concentrations in Rat Cerebral Cortex during Hypoglycemia[a]

	Control	Isoelectric EEG (min)			
		5	15	30	60
Total fatty acids	119 ± 2.6	107.4 ± 2.8*	108.4 ± 2.3*	109.3 ± 1.9*	110.0 ± 3.4
Total phospholipid phosphorus	64.7 ± 1.5	59.5 ± 1.5*	60.2 ± 1.0*	60.1 ± 0.6*	60.0 ± 0.9*
Total free fatty acids	0.12 ± 0.01	0.75 ± 0.07**	0.74 ± 0.05**	0.75 ± 0.08**	0.96 ± 0.11**

[a] The values are given in $\mu mol \cdot g^{-1}$ cortex wet weight and represent means ± S.E.M.
* $P < 0.05$; ** $P < 0.001$ (Student's t-test). Data from Agardh et al.[26]

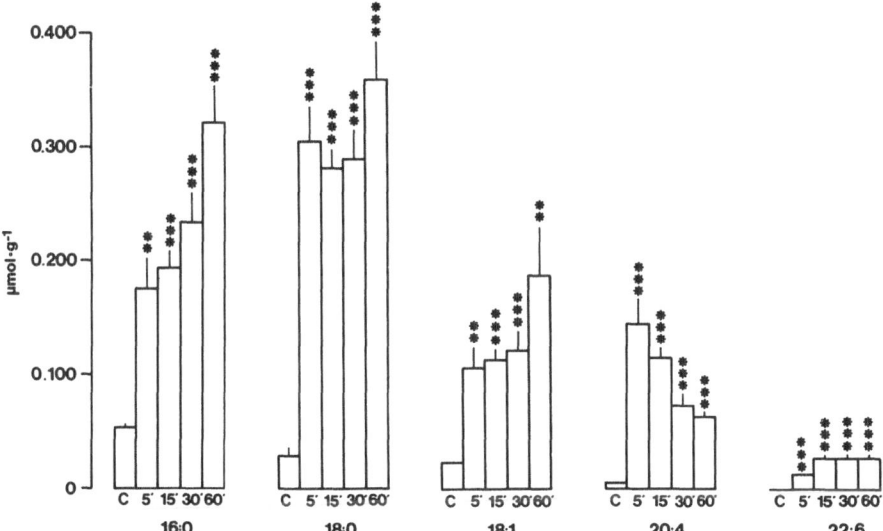

Fig. 6. Influence of hypoglycemic coma of 5 to 60 min duration on brain tissue concentrations of major free fatty acids. The values are means ± S.E.M. The black symbols represent statistical difference from controls (C): *$P < 0.01$; **$P < 0.001$. Modified after Agardh *et al.*[26]

This is because these compounds are toxic in several respects and because one of the FFAs (arachidonic acid) is a precursor for prostaglandins and related compounds, as well as for leukotrienes. The cascades of reactions elicited by arachidonic acid accumulation are catalyzed by fatty acid cyclooxygenase (E.C. 1.14.99.1) and lipoxygenase (E.C. 1.13.11.12), respectively. Results obtained on a variety of tissues indicate that many of the oxygenated products thus formed have adverse effects on vascular and parenchymal cells, some of which are caused by the free radical character of the degradation products.[61-64]

A recent study has demonstrated that in rats, the onset of hypoglycemic "coma" is accompanied by a six- to eightfold increase in cortical FFA content and that the FFAs remain elevated at that level for at least 1 h of continued coma.[26,65] In relative terms, the largest initial increase occurs in arachidonic acid concentration; however, prolongation of the hypoglycemia to 60 min is accompanied by a gradual decrease in the concentration of this polyenoic acid but not of the total FFA content (Fig. 6). It is tempting to speculate that the progressive decrease in the concentration of arachidonic acid is caused by its oxidative degradation along the cyclooxygenase and lipoxygenase pathways. We note that, in contrast to ischemia and hypoxia, hypoglycemia allows oxidative reactions to continue unabated, at least those that do not require an abundant supply of ATP.

5.5. The Balance of Payment: Oxygen Consumption versus Substrate Utilization

With the information at hand, it seems justified to attempt to relate oxygen consumption to changes in substrate levels. Since tissue glucose levels are already depleted at blood glucose concentrations of 1.5 μmol $\cdot$ g^{-1}, and since the glycogen levels must play only a minor role in supplying endogenous substrate during the slow-wave–polyspike period (which may last considerably longer than 1 h), it seems clear that other endogenous substrates must provide some of the missing substrates. No doubt, in this period glycolytic metabolites, citric acid cycle intermediates, and associated amino acids serve as fuels for oxidative metabolism. However, since the precomatose period is not accurately defined in terms of duration and utilization of endogenous substrates, it seems profitable to confine the discussion to the period of ceased electrical activity.

From results given above, it can be calculated that during the first 30 min of coma, the tissue uses about 100 μmol $\cdot$ g^{-1} of oxygen, corresponding to at least 15 μmol $\cdot$ g^{-1} of glucose equivalents. Since CMR$_{gl}$ is reduced to one-third to one-half of normal, other substrates must cover an oxygen consumption of, say, 10 μmol $\cdot$ g^{-1}. A rough estimate indicates that glycolytic metabolites, citric acid cycle intermediates, and amino acids should contribute about 1.0, 0.1, and 5.0 μmol $\cdot$ g^{-1}, respectively. The calculation suggests that at least 25% of the oxygen consumption remains unaccounted for. The decrease in RNA is, in this context, quantitatively of little importance. In theory, oxidation of 10% of the tissue phospholipid content would provide all the missing substrate equivalents. We lack sufficient information, however, to conclude that this is what really occurs, and two issues remain unresolved. First, the maximal reduction of phospholipid content occurred within the first 5 min and, since only about 10% of the loss of phospholipids could be accounted for in terms of FFAs, it is far from clear that lipid substrates were available or were utilized after that time. Second, sufficient information is not at hand to conclude that lipid substrates could efficiently support cerebral oxygen consumption. We must conclude, therefore, that the problem of defining the missing substrates has not been satisfactorily solved.

5.6. Ion Homeostasis

In recent years, information has been collected on the effect of hypoglycemia on extracellular potassium activity (K$_e^+$), on tissue impedance, on brain tissue electrolyte concentrations and water content, and on extra- and intracellular pH (pH$_e$ and pH$_i$, respectively). We lack data on changes in Ca^{2+} homeostasis, but other results allow at least some speculative deductions.

In ischemia, release of K$^+$ from cells occurs at a certain critical reduction in CBF, and, when the ischemia is severe, K$_e^+$ reaches levels of 50–60 μmol $\cdot$ ml^{-1}.[66] A corresponding efflux also occurs in hypoglycemia, and notably to a degree that abolishes spontaneous electrical activity.[67] In well-oxygenated animals with maintained blood pressure, K$_e^+$ stabilizes at 30–40 μmol $\cdot$ ml^{-1}, i.e., 10–20 μmol $\cdot$ ml^{-1} below the levels attained in severe (or complete) ischemia. In other words, although release of K$^+$ is extensive, some uptake must persist.

Information on overall brain tissue concentrations of electrolytes and on water content is controversial. Thus, Arieff et al.,[68] working on pentobarbital-anesthetized rats, concluded that insulin enhanced influx of K$^+$ into the brain and that hypoglycemia was associated with brain swelling secondary to expansion of both extra- and intracellular fluid volumes. However, results obtained in rats demonstrate little evidence of increased water content in the brain,[14] and recent results suggest that a shift of water occurs from extra- to intracellular fluids (see below). Possibly, the results of Arieff *et al.*[68] were influenced by changes in systemic variables. Thus, the arterial P_{O_2} values were low, and blood pressure was not given.

Recent results demonstrate that a reduction of blood glucose concentrations to levels that abolish EEG activity and cause energy failure leads to a sudden increase in tissue impedance, signalling a reduction of extracellular fluid volume to about 50% of control.[69] The impedance change is similar to that previously observed in asphyxia and ischemia.[70,71] In all probability, the impedance change reflects uptake of ions and water into glial cells, chiefly astrocytes (see below). This fluid shift seems related to the efflux of K^+ from neurons. Thus, evidence exists that glial cells accumulate K^+, Cl^-, and water when the extracellular K^+ activity rises above about 10 $\mu mol \cdot ml^{-1}$.[72–75]

It has been well documented that ischemia and epileptic seizures, two conditions leading to a deranged or strained brain energy metabolism, are associated with cellular uptake of Ca^{2+} from extracellular fluids.[76–78] Two events probably trigger this uptake: shortage of ATP for extrusion of Ca^{2+} and opening of voltage-dependent Ca^{2+} gates. It has been documented that in ischemia, Ca^{2+} influx occurs when the K_e^+ activity exceeds a value of about 13 $\mu mol \cdot ml^{-1}$.[78] By analogy, we must assume that a corresponding influx of Ca^{2+} occurs in hypoglycemia of sufficient severity to disrupt K^+ homeostasis. Furthermore, since Ca^{2+} may be the most important modulator of phospholipases, the precipitous rise in FFA concentrations suggests that the free intracellular Ca^{2+} activity increases.

Of all common conditions leading to brain cell damage, hypoglycemia is unique in that it does not lead to intracellular acidosis. Obviously, this is because the substrate for acid production, i.e., glucose, is lacking. The absence of cellular acidosis was documented some years ago,[21] but these results did not show the increase in pH that could be expected from the fact that metabolic acids are consumed in the glucose-deprived tissue. However, when it was realized that hypoglycemia is accompanied by an increased CBF (and thereby by a reduced P_{CO_2} difference between tissue and blood) and by a reduced extracellular fluid volume, it was possible to show that hypoglycemia leads to net intracellular alkalosis.[69,79] Since the extracellular pH also shifts somewhat in the alkaline direction, we can conclude that hypoglycemia neither leads to extracellular nor to intracellular acidosis. This fact must be taken into account when the mechanisms of hypoglycemic cell damage are considered (see below).

5.7. Metabolism of Isolated Mitochondria

We have already referred to the disparity between oxygen consumption and ATP production. It is tempting to conclude that this is caused by uncoupling of oxidative phosphorylation. First, it has been documented that accumulation of FFAs can uncouple mitochondria from a variety of tissues.[80,81] Second, it is now appreciated that certain conditions, notably influx of Na^+ into cells and accumulation of FFAs, will enhance release of Ca^{2+} from mitochondria, thereby inducing an energy-dependent, futile recycling of Ca^{2+} (and/or K^+) across the inner mitochondrial membranes.[82–84]

Only one study seems to have been published on mitochondrial metabolism in hypoglycemia.[85] In that, the results were disconcertingly negative. Thus, hypoglycemic coma of 30 or 60 min duration only reduced state 3 respiratory rates, and an increase in state 4 respiration was only noted in the recovery period following 60 min of coma. Furthermore, the phospholipid composition of the mitochondrial fraction remained unaltered. In view of the fact that mitochondria contain phospholipases,[86] this result was somewhat unexpected.

Obviously, the result quoted failed to demonstrate uncoupling of oxidative phosphorylation or structural damage to the mitochondrial population. In spite of this, it seems unwise to conclude that uncoupling does not occur *in vivo*; in fact, hypoglycemia is associated with those very conditions that would be expected to induce uncoupling. It is conceivable that the isolation procedure itself, i.e., the inclusion of a Ca^{2+} chelator (EDTA or EGTA) and an avid binder of FFAs (bovine serum albumin), removes the influence of uncouplers present *in vivo*. However, until more definitive data are at hand, any conclusion must be speculative.

5.8. Neurotransmitter Metabolism

The data presented above reveal that symptoms occur at degrees of hypoglycemia that do not measurably affect the cerebral energy state. We may conclude that, for example, cessation of spontaneous electrical activity and release of K^+ correlate with energy failure; however, we must explore other mechanisms in trying to explain the symptomatology of more moderate hypoglycemia. In analogy with conditions prevailing in, for example, arterial hypoxia (see ref. 87), it seems justified to consider "transmission failure," i.e., to look for signs of an aberrant transmitter metabolism.

Results presented above make it clear that severe hypoglycemia leads to a pronounced perturbation of cellular amino acid metabolism. The difficulty, though, is to equate changes in overall tissue content of excitatory and inhibitory amino acids to events occurring at synaptic junctions. It should suffice to say that the changes observed must reflect molecular havoc at synaptic terminals utilizing amino acids for signal transmission.

Comparable difficulties exist in interpreting changes affecting monamine transmitters and acetylcholine. However, for these transmitters, we stand a better chance of assessing not only tissue content but also synthesis and turnover rates. Results on acetylcholine appear clear-cut, since they show that insulin-induced hypoglycemia reduces the tissue content of acetylcholine.[88,89] Furthermore, even relatively moderate degrees of hypoglycemia reduce the rate of synthesis of acetylcholine even though the overall acetylcholine content does not decrease.[90] The sensitivity of acetylcholine synthesis to moderate hypoglycemia is best explained if it is assumed that newly formed acetylcholine arises from a small pool sensitive to even small reductions in the availability of the acetyl-CoA precursor. The situation is analogous to that in moderate hypoxia in which acetylcholine synthesis is affected at P_{O_2} levels that leave the cerebral energy state unaltered (see ref. 91).

Relatively few studies have been published on monoamine metabolism in insulin-induced hypoglycemia. It has been shown that insulin leads to increased tissue concentrations of tyrosine and tryptophan and that the latter is accompanied by an increased 5-HT turnover in the brain.[92–96] Working on unanesthetized rats, Gordon and Meldrum[96] noted that moderate insulin-induced hypoglycemia was associated with increased tissue concentrations of 5-HT and 5-HIAA and concluded that 5-HT turnover was enhanced.

Recently, cerebral metabolism of catechol- and indoleamines was studied in anesthetized and artificially ventilated rats during the slow-wave–polyspike period and in coma of 30 min duration.[97] As Table III shows, tissue concentrations of DA, NE, and 5-HT fell in precoma and reached very low levels in coma, whereas 5-HIAA concentrations rose. Results on tyrosine and tryptophan hydroxylation rates demonstrated increased NE and DA synthesis in the precomatose phase and during recovery. Obviously, hypoglycemia leads to a gross derangement of monoamine metabolism which includes an increased NE turnover during precoma and in the recovery phase.

5.9. Recovery Events

The metabolic changes associated with moderate degrees of hypoglycemia, i.e., those causing only depression of mental and electrical functions, are fully

Table III
Dopamine, Norepinephrine, 5-HT, and 5-HIAA Concentrations during Hypoglycemia[a]

	Control (saline)	Slow-wave polyspike	Iso-EEG (15 min)	Iso-EEG (30 min)
Dopamine (ng/g)				
Limbic	1176 ± 16	973 ± 29*	549 ± 23***	501 ± 19***
Striatum	3122 ± 156	2588 ± 152**	1316 ± 66***	843 ± 30***
Cortex	58 ± 13	21 ± 6**	19 ± 6**	6 ± 4***
Norepinephrine (ng/g)				
Limbic	554 ± 16	445 ± 9***	311 ± 17***	297 ± 7***
Striatum	187 ± 9	161 ± 13	129 ± 16**	115 ± 10**
Cortex	298 ± 7	219 ± 18**	157 ± 13***	125 ± 5***
5-HT (ng/g)				
Limbic	368 ± 3	309 ± 9*	196 ± 6***	158 ± 5***
Striatum	232 ± 37	252 ± 22	116 ± 10**	110 ± 0*
Cortex	255 ± 7	188 ± 3***	101 ± 6***	88 ± 7***
5-HIAA (ng/g)				
Limbic	399 ± 8	487 ± 8	728 ± 16***	765 ± 35***
Striatum	361 ± 55	449 ± 43	500 ± 53	471 ± 87
Cortex	282 ± 11	322 ± 21	450 ± 13***	424 ± 26***

[a] Shown are the means ± S.E.M., $n = 3$; * $P < 0.05$, ** $P < 0.01$, *** $P < 0.001$. Each experiment comprises pooled brain parts of two rats. Data from Agardh *et al.*[97]

and relatively quickly reversible.[34,46,47] It also seems probable that short periods of coma or of the equivalent "EEG silence" do not lead to irreversible metabolic and structural damage. For example, when the tissue is fixed by perfusion after 5 min of hypoglycemic coma, histopathological signs of cell damage cannot be observed by the light or the electron microscope (see Section 6.1). For that reason, we focus this discussion on coma periods of 30 or 60 min.

If recovery is induced after 30 min of hypoglycemic coma (by means of glucose administration), relatively extensive recovery of EEG and sensory evoked potentials is observed.[15] Corresponding to this functional recovery is a rapid normalization of tissue concentrations of PCr, ADP, and AMP as well as of glycolytic intermediates. Equally rapid is the reversal of the altered ratio between glutamate and aspartate. Some recovery events are much slower, though, e.g., the resynthesis of glycogen and the restoration of the adenine nucleotide and amino acid pools.

At first sight, it seems permissible to assume that recovery of these variables simply reflects the maximal capacity of rate-limiting synthesizing enzymes. For example, since the energy charge of the adenine nucleotide pool returns to control values, one may assume that the slow recovery of the pool size (and of the ATP concentration) simply reflects the limited capacity for *de novo* synthesis of purine bases and nucleosides.[55] However, several lines of evidence suggest that this is not necessarily the case. Rather, a hypoglycemic period of this duration probably induces (irreversible?) damage to a small population of cells. Some metabolic signs support this contention, e.g., the posthypoglycemic increases in lactate concentration and lactate/pyruvate ratio and the absence of any further resynthesis of adenine nucleotides in the 90- to 180-min period. However, the best index is provided by histopathological data (see Section 6.1).

After 60 min of hypoglycemia, the electrical activity of the brain recovers poorly if at all,[15] and the poor outcome is reflected in a late, secondary rise in extracellular K^+ activity.[85] In view of this and the histopathological results described below, there is a surprisingly extensive recovery of the cerebral energy state; for example, the charge of the remaining adenine nucleotide pool returns to more than 95% of control after 90 min of recovery.[45] Probably this disparity is a reflection of the resistance of mitochondria to the hypoglycemic insult (see above). However, the results also suggest that measurements of whole-tissue metabolite levels, however accurate they may be, constitute a blunt tool for assessing damage to part of the neuronal population.

6. HYPOGLYCEMIC CELL DAMAGE: OCCURRENCE, LOCALIZATION, AND UNDERLYING MECHANISMS

In this section, we explore the occurrence and localization of hypoglycemic cell damage, attempt a correlation of the histopathological and neurochemical alterations, and discuss the possible mechanisms that contribute to the cell damage.

6.1. Histopathology

Insulin-induced hypoglycemia can lead to widespread brain cell damage.[98,99] Much of the earlier data suffer from a number of shortcomings. For example, in virtually none of the clinical studies has it been possible to exclude the fact that the cellular alterations observed are influenced by complicating tissue hypoxia or epileptic seizures. Furthermore, in clinical material, it is seldom possible to assess both the severity and the duration of the hypoglycemic insult. Finally, in material fixed by immersion, it often becomes a matter of speculation if the changes observed are real or artifactual. For those reasons, we put some emphasis on experimental studies in which the tissue was fixed by perfusion for light and electron microscopy.

It must be recalled, however, that the identification of damaged or dead cells by such techniques is controversial. It is common knowledge that poor perfusion fixation can give rise to artifactual changes such as "dark cells" and "hydropic cell change."[100–102] One group has emphasized the difference between such artifactual cell alterations and "ischemic cell change"[99,103] and pointed out that the latter involves a series of transition events starting with microvacuolation, proceeding to ischemic cell changes, eventually with "incrustations," and ending with "homogenizing cell change" and dissolution of the cell (cf. ref. 104). Typically, these events are observed in anoxia–ischemia. In these conditions, microvacuolation at the light microscopic level was shown to consist of high-amplitude swelling of mitochondria in the electron microscope.[103] However, the identification of "ischemic cell change" as nonartifactual is not accepted by all.[105]

In a study on spontaneously breathing, anesthetized baboons that were monitored with respect to physiological variables, EEG, and somatosensory evoked potentials and were subjected to hypoglycemia of such severity that evoked potentials were suppressed for 42–90 min, Brierley *et al.*[106] fixed the brains by perfusion with formalin–acetic acid–methanol for light microscopy after 30–240 min of recovery. The authors identified three transition stages, microvacuolation, ischemic cell change, and ischemic cell change with incrustations, in the brains of the hypoglycemic animals and concluded that the structural alterations are similar in anoxia–ischemia and in hypoglycemia. The similarity was emphasized by the fact that much the same "selectively vulnerable regions" were affected (neocortex and hippocampus, with scattered neuronal lesions in striatum and thalamus). No ultrastructural study was performed in the hypoglycemic animals, however, and the interpretation of microvacuolation in terms of mitochondrial swelling was by analogy to changes observed in anoxia–ischemia.

A subsequent study on rats in which the brains were fixed by perfusion with glutaraldehyde after 30 and 60 min of hypoglycemic coma or after 30 or 180 min of recovery gave somewhat different results.[51,107] Thus, although a number of neurons in layers 3 and 5–6 of the cerebral cortex were condensed and dark staining (about 14% after 30 min and about 30% after 60 min), other neurons were slightly swollen with subplasmalemmal clearing. Furthermore, although the condensed dark-staining cells showed vacuoles, electron microscopy suggested that these must mainly have represented expanded cisternae of rough endoplasmic reticulum and of Golgi complexes. There was little evidence of mitochondrial swelling; if anything, the mitochondria appeared contracted, i.e., they showed ultrastructural characteristics similar to those described by Hackenbrock[50] for substrate-depleted mitochondria *in vitro*. Finally, in the glutaraldehyde-fixed tissue, astrocytic swelling was more conspicuous than what emerges from the description of Brierley *et al.*[106]

The inevitable conclusion from these studies was that the structural alterations observed in hypoglycemia are not identical to those observed in anoxia–ischemia. We recognize that astrocytic swelling is the expected result of the efflux of potassium and the impedance change described above. It is of interest that some cells showed rarefaction of the ribosomal units. Possibly this alteration forms the ultrastructural counterpart to the reduction in RNA content observed by biochemical analysis (see above).

When recovery was induced in rats after 30 and 60 min of hypoglycemia, the slightly swollen cells with subplasmalemmal clearing were no longer discernable, indicating that this type of cell alteration is reversible. The number of condensed, dark-staining cells decreased, but those remaining probably represent the proportion that suffered irreversible damage. Some of these showed moderate mitochondrial swelling. Possibly this is an ultrastructural counterpart of the posthypoglycemic increase in state 4 respiration (see above).

In view of the difficulty of properly assessing cell damage and cell death at a time when dissolution of the cells has not yet occurred, the final solution to the problem must await the development of more specific techniques. At present, it suffices to recall that in the baboon, the changes appeared to follow a time course[106] (see also ref. 103 for an account of similar changes observed in anoxia–ischemia). Furthermore, in the rat, condensed, dark-staining cells were not observed in the control animals or after 5 min of hypoglycemia; they increased in number when the hypoglycemic period was prolonged from 30 to 60 min, and some of them remained in the recovery period.

6.2. Localization of Cell Damage: Metabolic and Circulatory Correlates

It has already been mentioned that hypoglycemic cell damage tends to be concentrated in pyramidal cells in the neocortex and hippocampus, two classical vulnerable regions. However, it is noteworthy that damage is not inflicted on the cerebellar Purkinje cells, traditionally assumed to be vulnerable.[108,109] It seems probable that at least part of this selectivity in response results from the difference among brain structures with regard to glucose availability (see above). In other words, in the hypoglycemic animals, the damage incurred by any given structure may be inversely proportional to the efficiency of the (endothelial) glucose carrier and/or to the increase in CBF. This is reflected in the fact that a "resistant" structure, the cerebellum, develops signs of energy failure more slowly than a vulnerable one, the cerebral cortex.[108]

The regional differences in vulnerability seem to be paralleled by corresponding differences in posthypoglycemic blood flow rates.[27] It has been shown that following long periods of ischemia, an initial postischemic ("reactive") hyperemia is soon followed by a pronounced secondary hypoperfusion (see ref. 109). The mechanisms responsible are not known, nor has it been established that this hypoperfusion may jeopardize recovery events. Nonetheless, there are reasons to believe that the degree of hypoperfusion parallels the insult inflicted during the ischemia. It is, therefore, of interest that hypoglycemia is also followed by a secondary hypoperfusion (Fig. 7),[27] and that the secondary hypoperfusion seems most pronounced in areas that show a dense accumulation of damaged cells as defined by histopathological criteria (e.g., the cerebral cortex and the hippocampus) but that a secondary hypoperfusion is not observed in areas of apparently less vulnerability (e.g., the cerebellum). Thus, the damage inflicted seems directly correlated both to the severity of the decrease in glucose availability during hypoglycemia and to the degree of posthypoglycemic hypoperfusion. It is not known why certain cells in the selectively vulnerable areas are preferentially affected. As described elsewhere,[109] this could be a function of the calcium conductance of such cells (see below).

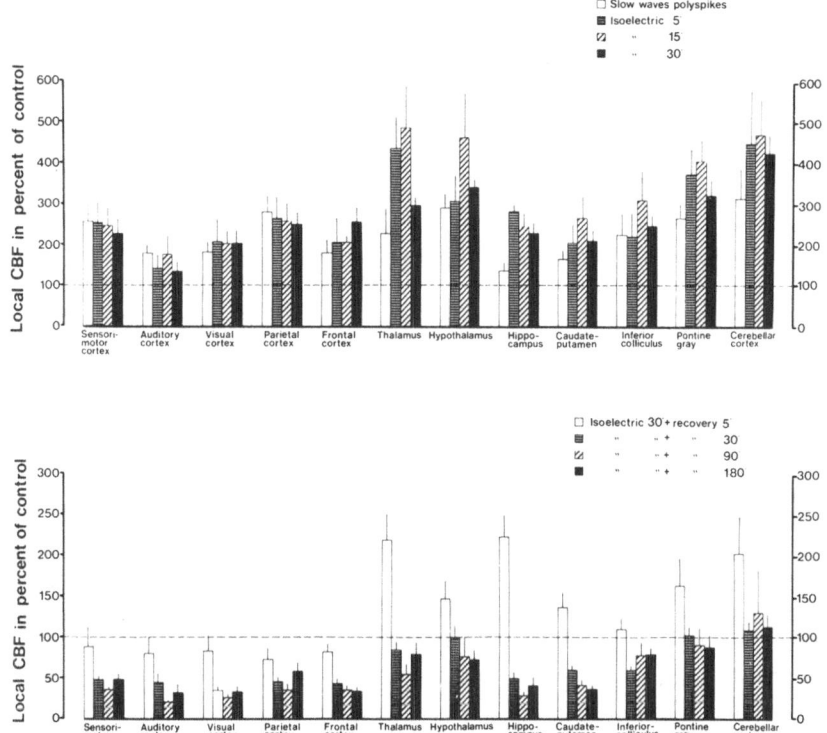

Fig. 7. Local CBF during insulin-induced hypoglycemia (upper panel) as well as in the recovery period following glucose administration (lower panel). The values are means ± S.E.M. in percent of controls. Reproduced with permission from Abdul-Rahman *et al.*[27]

6.3. Mechanisms of Cell Damage in Hypoglycemia

Several adverse conditions lead to neuronal cell damage that often has a preferential localization to certain selectively vulnerable regions. In some of these conditions, cerebral energy production is either seriously hindered (hypoxia and ischemia) or pathologically enhanced (status epilepticus), and common to these conditions is a marked or excessive lactic acidosis.[109] In ischemia and hypoxia, the presence of excessive acidosis markedly influences the outcome[110–112] (for further literature, see ref. 109).

It has been clearly documented that hypoglycemic cell damage develops in the absence of intra- or extracellular acidosis (see above). At first sight, it would seem that energy failure with a drastic reduction in ATP concentrations provides a mechanism that is common to hypoglycemia and to hypoxia/ischemia. However, it then remains to be explained why cell damage is incurred in status epilepticus in spite of the fact that the cerebral energy state is only marginally affected.[113–115]

Furthermore, two findings indicate that the relationship between hypoglycemic cell damage and tissue energy failure is far from tight. One of these relates to the fact that most cerebellar cells, including the Purkinje cells, are spared even if the period of hypoglycemic coma is extended to 60 min,[108] i.e., after a hypoglycemic insult that causes widespread damage to cells in the cerebral cortex and hippocampus. This difference in sensitivity cannot be explained by the fact that the cerebellar energy failure is less severe. Thus, although energy failure takes longer to develop in the cerebellum (see above), it is marked after 60 min. The second finding concerns cell damage in the cerebral cortex. Recent results demonstrate that pretreatment of animals with phenobarbital ameliorates the neuronal and astrocytic alterations even though extensive energy failure occurs (Kalimo, H., Agardh, A., Olsson, Y., and Siesjö, B. K., unpublished data).

At present, identification of mechanisms contributing to cell damage in hypoglycemia and other adverse conditions is hampered by the lack of suitable techniques for assessment of biochemical changes at the level of individual cells. As described elsewhere, though, it seems justified to focus interest on a derangement of phospholipid metabolism and on changes affecting ion homeostasis, particularly those affecting Ca^{2+}.[109]

A tentative synthesis of these events is as follows. First, there are reasons to believe that efflux of K^+ from neurons is what triggers both uptake of osmotic equivalents and water into glial cells and translocation of Ca^{2+} from extra- to intracellular fluids. Perineuronal and perivascular swelling of astrocytic processes could well impede oxygen transport to neighboring neurons and further compromise glucose transport to these cells (see ref. 73).

Furthermore, several mechanisms could contribute to a precipitous rise in intracellular Ca^{2+} activity: voltage-dependent opening of Ca^{2+} gates in plasma membranes, release of Ca^{2+} sequestered in endoplasmic reticulum because of ATP shortage, and enhancement of Ca^{2+} release from mitochondria, partly as a result of influx of Na^+ into cells. Once the Ca^{2+} activity rises and phospholipases are activated, the accumulation of FFAs can further enhance the release of Ca^{2+} from mitochondria. It is possible that futile recycling of Ca^{2+} (and/or K^+) then occurs, diverting respiratory energy away from ATP synthesis. Finally, accumulation of arachidonic acid could lead to both cell swelling (cf. ref. 116) and enhanced production of prostaglandins and related substances as well as leukotrienes. The potential toxicity of such compounds is now being realized, and future research will no doubt clarify the molecular mechanisms involved. At present, it suffices to recall that accumulation of FFA occurs in all injury conditions discussed (see ref. 109) and that the extent and localization of hypoglycemic cell damage seem to correlate with the accumulation of FFA.[117]

7. SUMMARY AND CONCLUSIONS

Hypoglycemia, especially that accompanying insulin treatment of diabetes, is a common cause of brain dysfunction and is often accompanied by irreversible brain cell damage. Hypoglycemia does this even when an adequate cerebral oxygen supply is maintained, demonstrating that the cellular effects are secondary to shortage of substrate. In severe hypoglycemia, the CMR_{gl}-to-$CMRO_2$ ratio is reduced for extended periods of time. Since no other exogenous substrate has been identified that could provide the missing substrate, the balance must be made up of endogenous substrates. At moderate degrees of hypoglycemic ischemia, these are derived from endogenous stores of glycogen, glycolytic metabolites, and citric acid cycle intermediates as well as from transamination reactions. When hypoglycemia becomes severe, phospholipids and RNA are broken down, and net oxidation of amino acids leads to a reduction of the free amino acid pool and to ammonia formation. The consumption of endogenous substrates, many of which exist as anions of relatively strong acids, leads to extra- and intracellular alkalosis. Even if the hypoglycemia is severe, cerebral oxygen supply is upheld. Thus, cerebral blood flow is increased, and cellular redox systems appear to shift towards oxidation.

When hypoglycemia is severe enough to abolish spontaneous EEG activity, the cerebral energy state is deranged. At that point, K^+ is released from cells, and tissue impedance increases. The first of these events probably triggers both swelling of astrocytic processes and influx of Ca^{2+} into cells, whereas the second provides the physiological counterpart to the appearance of *status spongiosus* in light and electron micrographs. Probably, the influx of Ca^{2+} into cells and its release from intracellular sequestration sites are what triggers breakdown of phospholipids and accumulation of free fatty acids. Influx of Na^+ and Ca^{2+} and accumulation of FFAs can be assumed to induce futile cycling of Ca^{2+} (and/or K^+) across mitochondrial membranes. This would further disturb intracellular Ca^{2+} homeostasis and may explain why the cerebral energy state deteriorates in spite of a maintained $CMRO_2$.

Histopathological results suggest that irreversible neuronal damage is incurred after 30 min of hypoglycemic coma and exaggerated after 60 min. The data demonstrate that cell damage is localized mainly to the neocortex and the hippocampus but absent in the cerebellum. This selective vulnerability of regions correlates directly to the degree of reduction in glucose delivery during hypoglycemia and to the reduction in local blood flow in the recovery phase. The mechanisms leading to cell damage have not been defined. Results obtained on cerebral cortex and the cerebellum suggest that cell damage correlates poorly to deterioration of tissue energy state. It is suggested that accumulation of FFAs and oxidative breakdown of arachidonic acid could trigger a series of events the end result of which is cell damage. There is no unequivocal explanation for the selective vulnerability of cells within regions, but variations in cellular Ca^{2+} conductance could contribute.

ACKNOWLEDGMENTS. Work from the author's own laboratory was supported by grants from the Swedish Medical Research Council, the United States Public Health Service via the NIH, and from the Medical Faculty, University of Lund. Expert secreterial assistance was given by Yvonne Hansson and Gillian Sjödahl.

REFERENCES

1. McIlwain, H., and Bachelard, H. S., 1971, *Biochemistry and the Central Nervous System,* Churchill Livingstone, Edinburgh, London.
2. Siesjö, B. K., 1978, *Brain Energy Metabolism,* John Wiley & Sons, New York, Chichester.
3. Sokoloff, L., Reivich, M., Kennedy, C., Des Rosiers, M. H., Patlak, C. S., Pettigrew, K. D., Sakurada, O., and Shinohara, M., 1977, *J. Neurochem.* **28:**897–916.
4. Sokoloff, L., 1981, *J. Cereb. Blood Flow Metab.* **1:**7–36.
5. Kety, S. S., 1950, *Am. J. Med.* **8:**205–217.
6. Sokoloff, L., Fitzgerald, G. G., and Kaufman, E. E., 1977, *Nutrition and the Brain,* Volume 1 (R. J. Wurtman and J. J. Wurtman, eds.), Raven Press, New York, pp. 87–139.
7. Owen, O. E., Morgan, A. P., Kemp, H. G., Sullivan, J. M., Herrera, M. G., and Cahill, G. F., Jr., 1967, *J. Clin. Invest.* **46:**1589–1595.
8. Hawkins, R. A., Williamson, D. H., and Krebs, H. A., 1971, *Biochem. J.* **122:**13–18.
9. Ruderman, N. B., Ross, P. S., Berger, M., and Goodman, M. N., 1974, *Biochem. J.* **138:**1–10.
10. Ghajar, J. B. G., Plum, F., Duffy, T. E., 1982, *J. Neurochem.,* **38:**397–409.
11. Creutzfeldt, O. D., and Meisch, J. J., 1963, *Electroencephalogr. Clin. Neurophysiol. [Suppl.]* **24:**158–171.
12. Hinzen, D. H., and Müller, U., 1971, *Pfluegers Arch.* **322:**47–59.
13. Lewis, L. D., Ljunggren, B., Ratcheson, R. A., and Siesjö, B. K., 1974, *J. Neurochem.* **23:**673–679.
14. Feise, G., Kogure, K., Busto, R., Scheinberg, P., and Reinmuth, O. M., 1976, *Brain Res.* **126:**263–280.
15. Agardh, C.-D., and Rosén, I., 1979, *Neurosci. Lett. [Suppl.]* **3:**45.
16. Gottstein, U., and Held, K., 1967, *Klin. Wochenschr.* **45:**18–23.
17. Eisenberg, S., and Seltzer, H. S., 1962, *Metabolism* **11:**1162–1168.
18. Bachelard, H. S., 1969, *Handbook of Neurochemistry,* Volume 1 (A. Lajtha, ed.), Plenum Press, New York, pp. 25–31.
19. Bachelard, H. S., 1980, *Cerebral Metabolism and Neural Function* (J. V. Passonneau, R. A. Hawkins, D. W. Lust, and F. A. Welsh, eds.), Williams & Wilkins, Baltimore, pp. 106–119.
20. Lund-Andersen, H., 1979, *Physiol. Rev.* **59:**305–352.
21. Lewis, L. D., Ljunggren, B., Norberg, K., and Siesjö, B. K., 1974, *J. Neurochem.* **23:**659–671.
22. Buschiazzo, P. M., Terrell, E. B., and Regen, D. M., 1970, *Am. J. Physiol.* **219:**1505–1513.

23. Kety, S. S., Woodford, R. B., Harmel, M. H., Freyhan, F. A., Appel, K. E., and Schmidt, C. F., 1948, *Am. J. Psychiatry* **104**:765–770.
24. Della Porta, P., Maiolo, A. T., Negri, V. U., and Rosella, E., 1964, *Metabolism* **13**:131–140.
25. Norberg, K., and Siesjö, B. K., 1976, *J. Neurochem.* **26**:345–352.
26. Agardh, C.-D., Chapman, A. G., Nilsson, B., and Siesjö, B. K., 1981, *J. Neurochem.* **36**:490–500.
27. Abdul-Rahman, A., Agardh, C.-D., and Siesjö, B. K., 1980, *Acta Physiol. Scand.* **109**:307–314.
28. Nilsson, B., Agardh, C.-D., Ingvar, M., and Siesjö, B. K., 1981, *Acta Physiol. Scand.* **111**:455–463.
29. Pappenheimer, J. R., and Setchell, B. P., 1973, *J. Physiol. (Lond.)* **233**:529–551.
30. Abdul-Rahman, A., and Siesjö, B. K., 1980, *Acta Physiol. Scand.* **110**:149–159.
31. Crane, P. D., Pardridge, W. M., Braun, L. D., Nyerges, A. M., and Oldendorf, W. H., 1981, *J. Neurochem.* **36**:1601–1604.
32. Abood, L. G., and Geiger, A., 1955, *Am. J. Physiol.* **182**:557–560.
33. Olsen, N. S., and Klein, R. J., 1947, *Arch. Biochem.* **13**:343–347.
34. Tews, J. K., Carter, S. H., and Stone, W. E., 1965, *J. Neurochem.* **12**:679–693.
35. Dawson, R. M. C., 1950, *Biochem. J.* **47**:386–391.
36. Cravioto, R. O., Massieu, G., and Izquierdo, J. J., 1951, *Proc. Soc. Exp. Biol. Med.* **78**:856–858.
37. De Ropp, R. S., and Snedeker, E. H., 1961, *J. Neurochem.* **7**:128–134.
38. Knauff, H. G., Mark, D., and Mayer, G., 1961, *Hoppe Seylers Z. Physiol. Chem.* **326**:227–234.
39. Tarr, M., Brada, D., and Samson, F. E., Jr., 1962, *Am. J. Physiol.* **203**:690–692.
40. Goldberg, N. D., Passonneau, J. V., and Lowry, O. H., 1966, *J. Biol. Chem.* **241**:3997–4003.
41. Oldendorf, W. H., 1971, *Am. J. Physiol.* **221**:1629–1639.
42. Oldendorf, W. H., 1981, *Res. Methods Neurochem.* **5**:91–112.
43. Pardridge, W. M., and Oldendorf, W. H., 1977, *J. Neurochem.* **28**:5–12.
44. Ferrendelli, J. A., and Chang, M. M., 1973, *Arch. Neurol.* **28**:173–177.
45. Agardh, C.-D., Folbergrová, J., and Siesjö, B. K., 1978, *J. Neurochem.* **31**:1135–1142.
46. Gorell, J. M., Dolkart, P. H., and Ferrendelli, J. A., 1976, *J. Neurochem.* **27**:1043–1049.
47. Gorell, J. M., Law, M. M., Lowry, O. H., and Ferrendelli, J. A., 1977, *J. Neurochem.* **29**:187–191.
48. Dawson, R. M. C., 1953, *Biochim. Biophys. Acta* **11**:548–552.
49. Chance, B., and Williams, G. R., 1955, *J. Biol. Chem.* **217**:383–427.
50. Hackenbrock, C., 1966, *J. Cell Biol.* **30**:269–297.
51. Agardh, C.-D., Kalimo, H., Olsson, Y., and Siesjö, B. K., 1980, *Acta Neuropathol. (Berl.)* **50**:31–41.
52. Mukherji, B., Turinsky, J., and Sloviter, H. A., 1971, *J. Neurochem.* **18**:1783–1785.
53. Lowenstein, J. M., 1972, *Physiol. Rev.* **52**:382–414.
54. Jacobson, K. B., 1959, *J. Gen. Physiol.* **43**:323–333.
55. Chapman, A. G., Westerberg, E., and Siesjö, B. K., 1981, *J. Neurochem.* **36**:179–189.
56. Yoshino, Y., and Elliott, K. A. C., 1970, *Can. J. Biochem.* **48**:236–243.
57. Page, I. H., Pasternak, L., and Burt, M. L., 1931, *Biochem. Z.* **231**:113–122.
58. Petersen, V. P., and Schou, M., 1955, *Acta Physiol. Scand.* **33**:309–315.
59. Hinzen, D. H., Becker, P., and Muller, U., 1970, *Pfluegers Arch.* **321**:1–14.
60. Stone, W. E., Tews, J. K., Whistler, K. E., and Brown, D. J., 1972, *J. Neurochem.* **19**:321–332.
61. Jackschik, B. A., Falkenhein, S., and Parker, C. W., 1977, *Proc. Natl. Acad. Sci. U.S.A.* **74**:4577–4581.
62. Flower, J. R., 1979, *Ciba Found. Symp.* **65**:123–139.
63. Lands, W. E. M., 1979, *Annu. Rev. Physiol.* **41**:633–652.
64. Murphy, R., Hammarström, S., and Samuelsson, B., 1979, *Proc. Natl. Acad. Sci. U.S.A.* **76**:4275–4279.
65. Agardh, C.-D., Westerberg, E., and Siesjö, B. K., 1980, *Acta Physiol. Scand.* **109**:115–116.
66. Branston, N. M., Strong, A. J., and Symon, L., 1977, *J. Neurol. Sci.* **32**:305–321.

67. Astrup, J., and Norberg, K., 1976, *Brain Res.* **103**:418–423.
68. Arieff, A. J., Doerner, T., Zelig, H., and Massry, S. G., 1974, *J. Clin. Invest.* **54**:654–663.
69. Pelligrino, D., Almquist, L.-O., and Siesjö, B. K., 1981, *Brain Res.* **221**:129–147.
70. Van Harreveld, A., 1965, *Brain Tissue Electrolytes,* Butterworths, London.
71. Hossmann, K.-A., Sakaki, S., and Zimmerman, V., 1977, *Stroke* **8**:77–81.
72. Bourke, R. S., Kimelberg, H. K., West, C. R., and Bremer, A. M., 1975, *J. Neurochem.* **25**:323–328.
73. Bourke, R. S., Daze, M. A., and Kimelberg, H. K., 1978, *Dynamic Properties of Glia Cells* (E. Scoffeniels, G. Franck, L. Hertz, and D. B. Tower, eds.), Pergamon Press, Oxford, New York, pp. 337–346.
74. Kimelberg, H. K., Narumi, S., Biddlecome, S., and Bourke, R. S., 1978, *Dynamic Properties of Glia Cells* (E. Schoffeniels, G. Franck, L. Hertz, and D. B. Tower, eds.), Pergamon Press, Oxford, New York, pp. 347–357.
75. Hertz, L., 1981, *J. Cereb. Blood Flow Metab.* **1**:143–153.
76. Nicholson, C., Ten Bruggencate, G., Steinberg, R., and Stöckle, H., 1977, *Proc. Natl. Acad. Sci. U.S.A.* **74**:1287–1290.
77. Nicholson, C., 1979, *The Neurosciences: Fourth Study Program* (F. O. Schmitt and F. G. Worden, eds.), MIT Press, Cambridge, Massachusetts, pp. 457–476.
78. Harris, R. J., Symon, L., Branston, N. M., and Bayhan, M., 1981, *J. Cereb. Blood Flow Metab.* **1**:203–209.
79. Pelligrino, D., and Siesjö, B. K., 1981, *J. Cereb. Blood Flow Metab.* **1**:85–96.
80. Vázquez-Colón, L., Ziegler, F. D., and Elliott, W. B., 1966, *Biochemistry* **5**:1134–1139.
81. Wojtczak, L., 1976, *J. Bioenerget. Biomembr.* **8**:293–311.
82. Harris, E. J., 1977, *Biochem. J.* **168**:447–456.
83. Crompton, M., Moser, R., Lüdi, H., and Carafoli, E., 1978, *Eur. J. Biochem.* **82**:25–31.
84. Crompton, M., and Heid, I., 1978, *Eur. J. Biochem.* **91**:599–608.
85. Agardh, C.-D., Chapman, A. G., Pelligrino, D., and Siesjö, B. K., 1981, *J. Neurochem.* **38**:662–668.
86. Woelk, H., and Porcelatti, G., 1973, *Hoppe Seylers Z. Physiol. Chem.* **354**:90–100.
87. Siesjö, B. K., and Plum, F., 1973, *Biology of Cerebral Dysfunction,* Volume 1 (G. E. Gaull, ed.), Plenum Press, New York, pp. 319–372.
88. Welsh, J. H., 1943, *J. Neurophysiol.* **6**:329–336.
89. Crossland, J., Elliott, K. A. C., and Pappius, H. M., 1955, *Am. J. Physiol.* **183**:32–34.
90. Gibson, G. E., and Blass, J. P., 1976, *J. Neurochem.* **27**:37–42.
91. Gibson, G. E., and Duffy, T. E., 1981, *J. Neurochem.* **36**:28–33.
92. Fernstrom, J. D., and Wurtman, R. J., 1971, *Science* **174**:1023–1025.
93. Tagliamonte, A., Demontis, M. G., Olianas, M., Onali, P. L., and Gessa, G. L., 1975, *Pharmacol. Res. Commun.* **7**:493–499.
94. Fernando, J. C. R., Knott, P. J., and Curzon, G., 1976, *J. Neurochem.* **27**:343–345.
95. MacKenzie, R. G., and Trulson, M. E., 1978, *J. Neurochem.* **30**:1205–1208.
96. Gordon, A. E., and Meldrum, B. S., 1970, *Biochem. Pharmacol.* **19**:3042–3044.
97. Agardh, C.-D., Carlsson, A., Lindqvist, M., and Siesjö, B. K., 1979, *Diabetes* **28**:804–809.
98. Meyer, A., 1963, *Greenfield's Neuropathology* (W. Blackwood, W. H. McMenemy, A. Meyer, R. M. Norman, D. S. Russel, eds.), Arnold, London, pp. 235–287.
99. Brierley, J. B., 1976, *Greenfield's Neuropathology* (W. Blackwood and J. A. N. Corsellis, eds.), Arnold, London, pp. 41–85.
100. Cammermeyer, J., 1961, *Acta Neuropathol. (Berl.)* **1**:245–270.
101. Cammermeyer, J., 1978, *Histochemistry* **56**:97–115.
102. Cohen, E. B., and Pappas, G. D., 1965, *J. Comp. Neurol.* **136**:375–395.
103. Brown, A. W., and Brierley, J. B., 1973, *Acta Neuropathol. (Berl.)* **23**:9–22.
104. Spielmeyer, W., 1922, *Histopathologie des Nervensystems,* Springer-Verlag, Berlin.
105. Cammermeyer, J., 1973, *Arch. Neurol.* **29**:391–393.
106. Brierley, J. B., Brown, A. W., and Meldrum, B. S., 1971, *Brain Res.* **25**:483–499.
107. Kalimo, H., Agardh, C.-D., Olsson, Y., and Siesjö, B. K., 1980, *Acta Neuropathol. (Berl.)* **50**:43–52.

108. Agardh, C.-D., Kalimo, H., Olsson, Y., and Siesjö, B. K., 1981, *J. Cereb. Blood Flow Metab.* **1:**71–84.
109. Siesjö, B. K., 1981, *J. Cereb. Blood Flow Metab.* **1:**155–185.
110. Myers, R. E., 1979, *Advances in Perinatal Neurology* (R. Korobkin and G. Guilleminault, eds.), Spectrum, New York, pp. 85–114.
111. Myers, R. E., 1979, *Adv. Neurol.* **26:**195–213.
112. Rehncrona, S., Rosen, I., and Siesjö, B. K., 1980, *Acta Physiol. Scand.* **110:**435–437.
113. Chapman, A. G., Meldrum, B. S., and Siesjö, B. K., 1977, *J. Neurochem.* **28:**1025–1035.
114. Blennow, G., Brierley, J. B., Meldrum, B. S., and Siesjö, B. K., 1978, *Brain* **101:**687–700.
115. Söderfeldt, B., Kalimo, H., Olsson, Y., and Siesjö, B. K., 1981, *Acta Neuropathol. (Berl.)* **54:**219–231.
116. Chan, P. H., and Fishman, R. A., 1978, *Science* **201:**358–360.
117. Agardh, C.-D., and Siesjö, B. K., 1981, *J. Cereb. Blood Flow Metab.* **1:**267–275.

Glutamate

Richard P. Shank and Graham LeM. Campbell

1. DISTINCTIVE BIOCHEMICAL PROPERTIES OF GLUTAMATE IN NEURAL TISSUES

For nearly 50 years glutamate has been among the most intensely investigated substances known to occur in CNS tissues. Probably no other substance has been investigated for so many different reasons. Interest in glutamate was motivated initially by the recognition that its concentration in the brain is appreciably higher than that in most other tissues. Quastel and Wheatley[1] and Krebs[2] provided the first evidence that glutamate is metabolically active in CNS tissues and suggested that glutamate may serve a role in intermediary metabolism. Weil-Malherbe devoted much of his career to investigating the function of glutamate in neural tissues, focusing primarily on the role of glutamate in ammonia detoxification.[3,4] The early work of these biochemists stimulated clinical interest in glutamate, and in the 1940s a number of reports were published that indicated that glutamate might have a beneficial effect on several neurological disorders, including hypoglycemic coma, epilepsy, and mental retardation.[4,5] Rigorous follow-up studies demonstrated that the beneficial effects were only marginal.[4,5] Therefore, the therapeutic use of glutamate was discontinued.

In the 1950s, a number of distinctive properties of glutamate were revealed. These observations raised many questions regarding the metabolic and physiological significance of glutamate in neural tissues and served as the impetus for the accelerating interest in glutamate that has ensued during the past two decades. Included among these distinctive properties of glutamate are the unusual metabolic compartmentation described by Waelsch and his colleagues,[6] the potent excitatory action on vertebrate CNS neurons[7] and arthropod muscle fibers,[8] and the finding that glutamate can initiate neuronal cell death in the retina.[9] It was also during the 1950s that glutamate was shown to be the immediate metabolic precursor of GABA,[10] and the exceptionally high degree to which glucose carbon is incorporated into glutamate as part of oxidative me-

Richard P. Shank and Graham LeM. Campbell • Department of Neurology, The Graduate Hospital, Philadelphia, Pennsylvania 19146, and Department of Physiology, Temple University School of Medicine, Philadelphia, Pennsylvania 19140.

tabolism in CNS tissues was first observed.[11] Additional distinctive properties of glutamate that have been identified more recently include its exceptionally low rate of penetration across the blood–brain barrier,[12] its low concentration in cerebrospinal fluid,[13] and its ability to initiate neuronal cell death in various regions of the CNS.[14] Other distinctive properties that involve neural tissues include the ability of glutamate in its monosodium salt form to enhance the flavor of foods[15] and the involvement of glutamate as a causal factor in the Chinese restaurant syndrome.[16]

In writing this review, our principal objective has been to relate these distinctive properties of glutamate to specific metabolic and physiological roles served by this amino acid. We have been mindful of the numerous recent reviews pertaining to the putative neurotransmitter function of glutamate and the metabolic compartmentation phenomenon. Accordingly, we emphasize conceptual issues that we believe have either been neglected or not properly addressed.

2. SIGNIFICANCE OF GLUTAMATE IN BRAIN FUNCTION

Current knowledge indicates that glutamate serves a number of functions in neural tissues. Some of these are common to all animal tissues, whereas some are either unique to the nervous system or are more manifest in neural tissues than in other tissues.

2.1. Glutamate as an Intermediate in Energy Metabolism

It is well established that when ^{14}C-labeled glucose is metabolized by CNS tissues, a high percentage of the label passes through glutamate prior to the formation of $^{14}CO_2$.[17–19] Typically, within a few minutes after the administration of labeled glucose, approximately 40% of the total label in the tissue is located in glutamate, a percentage considerably higher than that found in other tissues. Such a rapid and extensive accumulation of label in glutamate does not necessarily mean that glutamate is an obligatory intermediate in energy metabolism, but it is evident from such data that a large portion of the glutamate in CNS tissues is rapidly synthesized from a metabolically active pool of α-ketoglutarate. Since the total amount of glutamate is essentially constant, the glutamate so formed must be quickly metabolized. Some of the glutamate is metabolized to GABA and glutamine, but most of it is converted back to α-ketoglutarate.[17,19]

This exceptionally rapid interconversion of α-ketoglutarate and glutamate in CNS tissues has occasionally been described as an isotopic exchange.[17] Although this interconversion would not necessarily have to serve any useful metabolic purpose, a functional metabolic pathway has been identified that in fact requires an interconversion of α-ketoglutarate and glutamate. This pathway has been termed the malate–aspartate shuttle[20] and serves the purpose of transferring reducing equivalents from the cytosol into mitochondria (Fig. 1). In order for the malate–aspartate shuttle to operate, α-ketoglutarate must be trans-

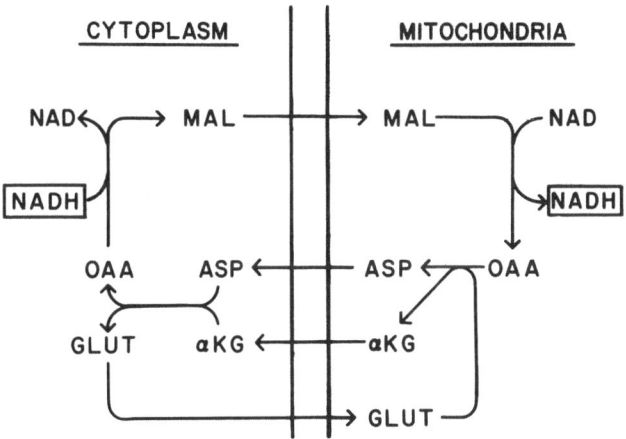

Fig. 1. The metabolic components of the malate–aspartate shuttle. Although it is not evident from the diagram, the flux of aspartate (ASP) across the mitochondrial membrane is coupled to the flux of glutamate (GLUT) in the reverse direction. The fluxes of malate (MAL) and α-ketoglutarate (αKG) may be similarly coupled.

located from mitochondria into the cytosol and then be converted to glutamate by the catalytic action of aspartate aminotransferase. Under steady-state conditions, an equimolar amount of glutamate is transported from the cytosol into mitochondria, whereupon the glutamate is converted back into α-ketoglutarate. Again, aspartate aminotransferase serves as the enzyme mediating this reaction.

Because aerobic glycolysis is especially vigorous in CNS tissues, the reduction of cytosolic NAD^+ to NADH occurs at a high rate. Assuming that the oxidation of cytosolic NADH is mediated predominantly by the malate–aspartate shuttle, it would be expected that each of the component reactions of this metabolic process would be quite active. This expectation is supported experimentally by the observations that the enzymes that catalyze the various reactions of the malate–aspartate shuttle are prevalent in CNS tissues[17–19] and the observations that isolated CNS mitochondria can rapidly take up glutamate and malate and, under appropriate conditions, release stoichiometric amounts of aspartate and α-ketoglutarate.[21–24]

Another means by which reducing equivalents can be transferred across the mitochondrial membrane is via a cyclic metabolic process referred to as the glycerol phosphate shuttle. In this process, cytosolic NADH is oxidized by a reaction in which dihydroxyacetone phosphate is reduced to glycerol phopshate, which is then translocated into mitochondria and reoxidized to dihydroxyacetone phosphate with a concomitant reduction of NAD^+. The cycle is completed by the translocation of the oxidized product back into the cytosol. At least one of the enzymes needed for this metabolic cycle to operate, glycerolphosphate dehydrogenase, is much more active in oligodendrocytes than in neurons and astrocytes.[25] Consequently, the glycerol phosphate shuttle may not be operative in neurons, leaving the malate–aspartate shuttle as the sole mechanism for transferring equivalents into mitochondria. Immunocytochemical analysis of cytosolic aspartate aminotransferase indicate that this enzyme is especially prevalent in certain types of neurons,[26] an observation consistent with the hypothesis that the malate–aspartate shuttle is especially active in these cells.

2.2. The Putative Neurotransmitter Function

From the time its neuroexcitatory effect was first demonstrated, the potent excitatory action of glutamate on neurons has never been seriously disputed. Yet the concept that glutamate may function as a neurotransmitter was met with considerable skepticism. Even today, this putative function is shrouded with controversy. Initially, the idea that glutamate may be a transmitter was resisted in part because the prevailing dogma held that a neurotransmitter should be restricted in its occurrence to those neurons in which it serves the transmitter function. Other reasons for resisting a transmitter role for glutamate included the observation that glutamate excited virtually all neurons examined and that its excitatory action did not appear to be terminated by enzymatic degradation of the glutamate molecule.[27]

More recently, the controversy surrounding glutamate has resulted from an inability to establish that glutamate satisfies all the accepted criteria needed to prove rigorously that a substance functions as a neurotransmitter. In this context, it should be emphasized that formidable technical problems are encountered experimentally when these criteria are used to evaluate the transmitter function of glutamate.[27-29]

Despite this, the evidence that glutamate does function as a major excitatory neurotransmitter is very compelling, and it appears that most neuroscientists actively engaged in research on neurotransmitters in the CNS now include glutamate on their list of "accepted neurotransmitters." At the present time, neurochemical research on glutamate is dominated by studies relevant to the neurotransmitter function. This is in part because of the apparent prominence of glutamate as a neurotransmitter and partly because of the widespread interest of neuroscientists in neurotransmitters. A major thrust at the present time is to identify the neurons within the CNS that utilize glutamate as a neurotransmitter.

The first studies designed to identify neurons that utilize glutamate as a neurotransmitter were undertaken by Aprison and his colleagues.[30,31] As a theoretical basis for their studies, it was assumed that glutamate would be more concentrated in glutamatergic neurons than in other types of neurons. A comparison of the content of glutamate in discrete regions of the spinal cord and in the dorsal and ventral roots revealed a distribution consistent with the concept that glutamate is the transmitter released by primary afferent neurons.[30] A selective destruction of interneurons in the spinal cord resulted in a significant reduction in glutamate content in this tissue, suggesting that some spinal interneurons might be glutamatergic.[30] More recently, an analysis of the content of glutamate in normal tissue and in tissue subjected to conditions inducing neuronal cell death has provided evidence for glutamatergic neurons in the cerebellum, hippocampal formation, cerebral cortex, and retina (Table I).

Table I
Neurons in Which Glutamate May Function as a Neurotransmitter

Type of neuron	Experimental evidence[a] and references
Neocortex: projection neurons to striatum, thalamus, amygdala, nucleus accumbens, lateral geniculate, superior colliculus, red nucleus, spinal cord	A: 32,33 B: 32–37 C: 35,43,44
Hippocampal formation: perforant afferents, commisural fibers,[b] Schaffer collaterals, mossy fibers, projection neurons to the septum	A: 40,42,47 B: 38,39,41,42,48 C: 38,42
Olfactory system: projection neurons from bulb to the cortex (mitral cells)[b]	A: 45 C: 46,48
Cerebellum: granule cells (parallel fibers)	A: 50,51 B: 50,52 C: 52
Retina: some photoreceptor cells and some ganglion and amacrine cells	A: 59 B: 53,54
Spinal cord and brainstem: some interneurons,[b] some dorsal root fibers, afferents to the cochlear nucleus, baroreceptor sensory neurons	A: 30,31,56,57 B: 55,56 C: 55

[a] Types of experimental evidence: A, selectively high concentration of glutamate based on distribution of content and/or decrease in content after selective lesions; B, high uptake rate and reduced uptake after selective lesions; C, stimulus-evoked release and reduced release after selective lesions.

[b] Aspartate may be a transmitter in addition to, or in place of, glutamate.

An analysis of the distribution of glutamate is not an ideal way of attempting to identify glutamatergic neurons because the data are not definitive. Although an ideal marker of glutamatergic neurons has yet to be found, a method that has become popular in recent years is the use of autoradiography to determine the cellular localization of ^{3}H-labeled glutamate or aspartate, particularly D-aspartate.[34,41,54,60] The theoretical basis for this procedure is the assumption that the membrane transport carriers for glutamate (and aspartate) are selectively expressed in glutamatergic (and aspartatergic) neurons. In fact, there is a wealth of evidence indicating that glutamate transport carriers are prevalent in the membrane of astrocytes (see Section 3.3.4). Despite this problem, the autoradiographic procedure used in several laboratories does appear to be useful in identifying glutamatergic neurons. A possible explanation for the selective labeling of presumed glutamatergic neurons is that the labeled substrate may be metabolized much more rapidly in astrocytes, thereby resulting in a selective retention of label in the neurons.

Another procedure occasionally used to identify glutamatergic neurons is stimulus-evoked release of glutamate (either endogenous or preloaded exogenous) from tissue slices.[35,42,46,61] As with the previous two procedures, the effect of selective destruction of the presumed glutamatergic neurons is sometimes used as an adjunct of this procedure.

In evaluating the role of glutamate as a neurotransmitter, the mounting evidence that some neurons may release two or more synaptically active compounds must be kept in mind. Furthermore, it is possible that not all neurons of a particular histological type release the same neurotransmitter agent and that the transmitter released by a particular type of neuron may not be the same in all species of animals.

2.3. Glutamate as a Metabolic Precursor of GABA

GABA now appears to be universally recognized as a major inhibitory neurotransmitter in the CNS of vertebrates. Glutamate has been established as the immediate precursor of GABA,[62] although some GABA may be derived from putrescine.[63] Immunocytochemical studies have shown that glutamate decarboxylase occurs predominantly, if not exclusively, in GABAergic neurons and is most prevalent in the terminals of these neurons.[64] An unresolved and somewhat controversial issue is the rate at which GABA is synthesized. GABA is labeled quite rapidly when [U-^{14}C]glucose is supplied as a metabolic substrate.[65–67] Assuming a simple precursor–product relationship between the entire tissue pool of glutamate and GABA, it would appear that the rate of GABA synthesis is about one-half the rate at which acetyl moieties are fed into the cycle.[66–68] This value is unrealistically high because it is unlikely that even the total metabolic activity in GABAergic neurons could account for nearly half of the oxidative metabolism in CNS tissues.[68] On the basis of various metabolic considerations, it is evident that in these metabolic studies the pool of glutamate that is metabolized to GABA must have specific activity about four times that in the tissue as a whole.[66–68] With this taken into account, the calculated rate of GABA synthesis is about one-tenth the rate at which acetyl moieties enter the citric acid cycle (in rats, this would be about 0.2 μmol/min per g tissue). Even this figure may be high because in experiments utilizing inhibitors to selectively prevent GABA degradation, the net accumulation of GABA was less than 0.1 μmol/min per g tissue.[68] However, even if GABA is synthesized at a rate only 5% of the entry of acetyl-CoA into the cycle, it is possible that in GABAergic neurons perhaps 50% or more of the α-ketoglutarate formed from citrate is metabolized through the GABA shunt.

The degradative enzyme (GABA aminotransferase) is relatively low in nerve terminals as compared to other cellular entities such as astrocytes and dendritic processes.[69] This situation had led to speculation that most if not all of the GABA synthesized in GABAergic neurons is released prior to metabolic conversion to succinate.[68] If this is true, the rate of release from GABAergic terminals must be quite high. This would necessitate a correspondingly high net input of carbon moieties into the citric acid cycle (perhaps via pyruvate carboxylation) and a net input of amino moieties into glutamate (perhaps via glutamine uptake and conversion to glutamate).

Recent immunocytochemical observations, in addition to metabolic data, indicate that pyruvate carboxylase is much more prevalent in astrocytes than in neurons,[70,71] and if glutamine were used in stoichiometric amounts to replenish both the carbon and amino moieties of glutamate, then GABA should be more readily labeled from glutamine than from glucose. This is not the case, at least in vivo.[72] This appears to be inconsistent with the concept that GABA must be released from nerve terminals before being metabolized to succinate. As an alternative, it is possible that despite the low level of GABA transaminase in GABAergic nerve terminals, the activity of the enzyme may be sufficient to convert most of the GABA to succinate prior to release when consideration is made for the very high content of GABA in this compartment. The metabolic significance of the low level of GABA transaminase may be that it contributes to the maintenance of a high content of GABA in GABAergic terminals, whereas the high amount of GABA transaminase in other compartments insures that GABA will be maintained at a low level in these locations.

2.4. Glutamate and the Detoxification of Ammonia

Free ammonia is normally present in the CNS at a concentration of about 0.1 to 0.2 mM,[4] but an accumulation to an appreciably higher level has a deleterious effect on the functional state of neurons.[4] The biochemical basis for this deleterious effect on neuronal activity is not well understood but is probably multifactorial.[68,73-75] Ammonia is generated in neural tissues as the result of a number of deamination reactions. Consequently, it must be constantly removed to prevent an accumulation to toxic levels. The principal mechanism by which this is achieved appears to be the conjugation of ammonia with glutamate by the catalytic action of glutamine synthetase, a microsomal enzyme that is located predominantly in astrocytes.[76,77] Some free ammonia is also removed by the coupling of ammonia with α-ketoglutarate by glutamate dehydrogenase.[18]

In quantitative terms, the participation of glutamate in the detoxification of ammonia is probably quite small in comparison to the interconversion of glutamate with α-ketoglutarate or the rate of metabolic conversion to GABA. However, if the detoxification of ammonia does result in a net synthesis of glutamine from glutamate, then assuming that the glutamine is subsequently released into the bloodstream, this process could result in a net loss of glutamate from the CNS parenchyma. To prevent a depletion of glutamate, some restorative mechanism must exist. At the present time, such a mechanism can only be conjectured. One possibility is that the synthesis of glutamate from α-ketoglutarate and ammonia is matched with the utilization of glutamate for ammonia detoxification purposes, but this appears to be unlikely because $^{15}NH_4^+$ is incorporated much more extensively into the amide moiety than the amino moiety of glutamine.[18] Other possibilities include a net uptake from the bloodstream of amino acids that can be metabolized directly to glutamate (e.g., arginine and ornithine) or that can be used to supply the amino moiety for the synthesis of glutamate from α-ketoglutarate via transamination reactions (e.g., valine and leucine). Present evidence regarding the net flux of amino acids from blood into the CNS parenchyma is conflicting, although it does appear that there is not a high rate of net influx or efflux for any amino acid.[75,78,79] An accurate measurement of small net fluxes is difficult because the high rate of blood flow minimizes arterial–venous differences.

2.5. Glutamate as a Constituent of Protein

The amount of glutamate present in hydrolysates of protein extracted from brain indicates that glutamate and glutamine together account for 8 to 10% of the total amino acid residues.[80,81] These two amino acid moieties are present in about equal amounts in the total protein pool[81]; however, this may vary depending on the metabolic and physiological state of the tissue. In this context, Wherrett and Tower[81] found that 16% of the glutaminyl moieties were deamidated in the cerebral cortex of cats within a few minutes after the onset of complete ischemia. The amount of deamidation was found to be remarkably constant, suggesting that only certain proteins are susceptible to this type of hydrolytic activity. This deamidation process was prevented by methionine sulfoximine but not by barbiturates. The physiological significance of this deamidation of proteins in CNS tissues remains to be established. Presumably, if such a metabolic process were to serve some physiological role, a mechanism must be present that converts the glutamyl moiety back to the glutaminyl form. Attempts to demonstrate the existence of a specific amidation reaction were not successful.[81]

The deamidation of glutaminyl moieties may be one of the consequences of ischemia that contributes significantly to the short-term and long-term adverse effects of this condition on CNS tissues. The release of ammonia could be a causal factor in the swelling of cells that occurs with ischemia, and if reamidation does not occur, it could contribute to the irreversible cellular damage associated with profound ischemia.

Glutamate is particularly high in membrane proteins and probably contributes appreciably to the net negative charge exhibited by many of these protein molecules.[82] Two of the known brain-specific proteins, S-100 and 14-3-2, contain a high portion of glutamyl residues.[83] The function of S-100 has yet to be established, but 14-3-2 is now known to be a neuron-specific enolase.[84]

2.6. Glutamate as a Constituent of Small Peptides

In addition to glutathione, a number of γ-glutamyl dipeptides occur in small amounts in CNS tissues.[85] These dipeptides are intermediates in a metabolic process referred to as the γ-glutamyl cycle, which may serve as a Na^+-independent membrane transport system for a number of amino acids.[86]

A number of neuroactive peptides contain the cyclic glutamate derivative pyroglutamate (5-oxoproline, 2-pyrrolidone-5-carboxylate) as the N-terminal residue. These peptides include luteinizing hormone-releasing hormone, thyrotropin-releasing hormone, neurotensin, bombesin, and several others.

N-Acetylglutamate is another derivative of glutamate that occurs in CNS tissues. This compound serves as a regulator of urea synthesis in the liver and has an activating effect on glutaminase.[87] The function of N-acetylglutamate in CNS tissues is not known, but it has been found to inhibit α-ketoglutarate transport.[88]

2.7. The Contribution of Glutamate to Intracellular Osmotic and Ionic Activity

The average content of glutamate in the gray matter of CNS generally ranges between 6 and 12 μmol per g wet tissue.[73] The average concentration in intracellular fluid is presumably about 20 to 40% higher. Therefore, if most of the glutamate molecules are osmotically active, this amino acid would contribute about 2 to 5% of the total intracellular osmolarity and a substantially greater portion of the osmotically active anionic activity. Although glutamate contributes a small percentage of the average total osmolarity in CNS tissues, the contribution may reach 10 to 15% in the terminals of neurons in which glutamate serves as a neurotransmitter. By contrast, the contribution in GABAergic neurons and glia cells may be considerably less.[66,70] Because all cells must maintain essentially the same intracellular osmotic activity, the cells possessing low levels of glutamate must compensate by having a high level of some other osmotically active anionic substance. In GABAergic neurons, the presence of GABA would serve as partial compensation, but since GABA does not have a net negative charge at physiological pH, some other substance must make up the anionic deficit. A compound that may serve this function is N-acetylaspartate[89] which is present in the CNS at concentrations much higher than in other tissues.[90]

3. BIOCHEMICAL BASIS FOR THE NEUROTRANSMITTER FUNCTION OF GLUTAMATE

Assuming that glutamate does function as a major excitatory neurotransmitter in the vertebrate CNS, an understanding of the biochemical processes that make this function possible should be a subject of high priority in the field of neurochemistry. In recent years, a number of neurochemists have become engaged in research projects relevant to this issue. Despite these efforts, the present information provides only a rudimentary understanding. Because of the complexity of glutamate metabolism, it is likely that a definitive understanding of the biochemical processes underlying the transmitter function will not be achieved in the near future. Here we try to put into perspective some of the current literature. It should be evident from our discussion that much of the reported data must be interpreted cautiously.

3.1. Metabolic Origin of the Neurotransmitter Pool

It is generally assumed that for any neurotransmitter there is a rather continuous drain from the reservoir of molecules contained in the "transmitter pool" within the nerve terminal. This drain reflects the release evoked by action potentials and also the spontaneous release that elicits miniature postsynaptic responses. At present, there is no reason to believe that this condition does not apply to the transmitter pool of glutamate. If this condition does hold for glutamate, the transmitter pool must be constantly replenished, presumably by a net synthesis of new molecules within the nerve terminal. Although axonal transport could conceivably supply new glutamate molecules to the nerve terminal, a more likely possibility is that some precursor is transported from the interstitial fluid into the nerve terminal and is therein metabolized to glutamate. A number of compounds are now being investigated as possible precursors of the glutamate transmitter pool. These include glucose, α-ketoglutarate, glutamine, ornithine, and arginine.

3.1.1. Glucose as a Precursor of the Glutamate Transmitter Pool

Glucose has been regarded as a possible precursor of the glutamate transmitter pool primarily because of the extensive incorporation of glucose carbon into glutamate during oxidative metabolism. Furthermore, it has been demonstrated that a large portion of the glutamate molecules released from brain slices by electrical stimulation are derived metabolically from glucose.[91,92] Despite this relationship between glucose metabolism and the transmitter pool of glutamate, it is not valid to conclude that glucose carbon is utilized for replenishing the transmitter. Assuming that the transmitter pool is derived from or even exchanges with the cytosol of the nerve terminal, the rapid metabolic turnover of glutamate because of its involvement in the malate–aspartate shuttle could account for the labeling of the transmitter pool from glucose.

In order for glucose to serve as a precursor, it must first be metabolized to α-ketoglutarate. The utilization of α-ketoglutarate to replenish the transmitter pool would cause a depletion of the content of citric acid cycle intermediates, which would then have to be restored by an anaplerotic process. Hence, the utilization of α-ketoglutarate derived metabolically within the nerve terminal from glucose to replenish the glutamate transmitter pool would require a stoichiometric replenishment of the citric acid cycle. Although the level of anaplerotic activity of CNS tissues is comparatively high,[93,94] an analysis of metabolic compartmentation data indicates that anaplerotic activity appears to occur predominantly, if not exclusively, in astrocytes.[70] Furthermore, as noted previously, the enzyme that mediates anaplerotic activity in CNS tissues (pyruvate carboxylase[94]) appears to be located predominantly, if not exclusively, in astrocytes.[71] This suggests that glucose oxidatively metabolized within nerve terminals cannot be utilized for a net restoration of the transmitter pool of glutamate, even though carbon derived from glucose may enter the transmitter pool extensively.

3.1.2. α-Ketoglutarate as a Precursor of the Glutamate Transmitter Pool

The evidence that the anaplerotic enzyme pyruvate carboxylase is selectively expressed in astrocytes suggests that neurons are dependent on these glia cells for a supply of one or more citric acid cycle intermediates in order to prevent a depletion of these metabolites. Support for such a metabolic dependence of nerve terminals on astrocytes has been provided by the observation that nerve-terminal-enriched preparations vigorously import α-ketoglutarate by as many as three Na^+-dependent, high-affinity transport systems.[88,95] On the basis of the apparent localization of anaplerotic activity in astrocytes and the rapid uptake of α-ketoglutarate by nerve-terminal-enriched preparations, it has been hypothesized that the transmitter pool of glutamate may be replenished in part by a net transfer of α-ketoglutarate from astrocytes to glutamatergic nerve terminals.[88] This hypothesis is further supported by metabolic data indicating that the α-ketoglutarate imported by nerve terminals is rapidly metabolized to glutamate and is associated with a net increase in the content of glutamate.[88,96,97]

Because the synthesis of glutamate from α-ketoglutarate requires a source for the amino moiety, another precursor must contribute to this metabolic process. Several possibilities exist. These include aspartate, glutamine, orninthine, arginine, and free ammonia.

3.1.3. Glutamine as a Precursor of the Glutamate Transmitter Pool

In recent years, glutamine has received considerable attention as a possible precursor of the transmitter pool of glutamate.[61,70,98–105] Present evidence strongly favors a precursor role for glutamine, but the conclusion that this amino acid is the sole or even principal metabolic precursor of transmitter glutamate is not yet justified. In thin tissue slices and synaptosomal preparations, glutamine is rapidly metabolized to glutamate, some of which can be released by membrane depolarization in a Ca^{2+}-dependent manner.[61,101–105] In more intact CNS preparations such as the isolated toad brain[98,100] and the rat brain *in situ*,[72] the relative conversion of glutamine to glutamate is reduced considerably. This reduced activity probably reflects in part the greater degree to which glutaminase is inhibited in these more physiological preparations, but the higher rate at which other metabolic processes such as the malate–aspartate shuttle operate in the more normal tissue also may be a factor in the reduced contribution of glutamine to the synthesis of glutamate.

Astrocytes presumably serve as the principal origin of glutamine. These cells almost certainly represent the principal location of glutamine synthetase,[77] and current metabolic data suggest that there is a net synthesis of glutamine in these cells (see Chapter 15, this volume). Glutamine appears to be transported into nerve terminals by three carriers: the alanine-preferring system, which probably serves as the dominant carrier, the large neutral amino acid carrier, and the basic amino acid carrier.[97] Glutamine is also readily transported across the membrane of astrocytes.[106] Glutaminase, the enzyme that converts glutamine to glutamate, is discussed in detail by E. Kvamme in Volume 4 of this series. The phosphate-dependent form of this enzyme is prevalent in nerve terminals,[101] but neuronal lesion experiments indicate that it is not concentrated in the terminals of glutamatergic neurons.[107] Phosphate-dependent glutaminase is a mitochondrial enzyme and is partially regulated by glutamate.[108]

3.1.4. Ornithine and Arginine as Precursors of Transmitter Glutamate

These basic amino acids have only recently been proposed as possible precursors of glutamate.[109,110] Metabolic studies have demonstrated that glutamate can be formed from these amino acids in CNS tissues via a common pathway in which glutamate semialdehyde is an intermediate.[109,111] In synaptosomal preparations, the rate of glutamate synthesis from ornithine is slow in comparison to that from glutamine and α-ketoglutarate; furthermore, ornithine is more readily metabolized to glutamate in cell bodies than in nerve-terminal-enriched preparations.[111]

The presumed source of arginine and ornithine is the bloodstream. These amino acids cross the blood–brain barrier fairly readily,[12] a process presumably mediated by the basic amino acid carrier in the membrane of endothelial cells. Present evidence provides no indication that there is a marked net transfer of these amino acids from blood into the CNS parenchyma[78,79]; however, the information relevant to this issue is inconclusive. It should be emphasized that the first step in the conversion of ornithine to glutamate involves a transamination reaction in which α-ketoglutarate is converted to glutamate.

3.2. The Mechanism of Glutamate Release into the Synaptic Cleft

The release of glutamate from CNS tissues has been the subject of many studies in recent years. Most of these studies have been oriented toward establishing the cellular and subcellular origin of release rather than the mechanism of release. Present evidence indicates that there is both a Ca^{2+}-dependent and a Ca^{2+}-independent release from nerve terminals.[112]

Based on the evidence that most other known neurotransmitters and hormones are released by exocytosis, it would be expected that glutamate is released in this manner. Neurotransmitter release at arthropod excitatory neuromuscular junctions, where glutamate is virtually certain to be the transmitter, is known to be quantal.[113] But quantal release may not necessarily be mediated by exocytosis. It does appear that glutamate can be released by a carrier-mediated process, but the physiological significance of carrier-mediated release is questionable.[114] Bradford and his colleagues[115] have attempted to ascertain the intracellular origin of stimulus-evoked release by prelabeling glutamate and then comparing the specific activity of glutamate released from synaptosomes to that in the nerve terminal cytosol and isolated synaptic vesicles. They found that the specific activity of the glutamate released by a depolarization stimulus was nearer that in the cytosol than in the vesicles.

Although these results suggest that glutamate is released from the cytosol, the data must be interpreted cautiously. The procedures that are presently used to isolate nerve terminals and synaptic vesicles probably result in an appreciable loss of glutamate from both the cytosol and the vesicles,[100] and the residual glutamate in these compartments may not be representative of that prior to the isolation process. Furthermore, metabolism of glutamate during the isolation process may result in a marked change in the specific radioactivity of glutamate in the different subcellular compartments.

Another concept promoted recently is that glutamate newly synthesized from glutamine[61,112] or derived by uptake into the nerve terminals[42] is selectively released during membrane depolarization. The evidence for this is based on a comparison of the specific radioactivity of glutamate released from brain slices by high K^+ with the specific activity of glutamate remaining in the tissue slices. In these studies, it was found that when slices were incubated with either [^{14}C]glutamine or [^{14}C]glutamate, the specific radioactivity of the glutamate subsequently released from the slices by a depolarization stimulus was considerably higher than the specific activity of glutamate remaining in the slice. An important consideration in this type of experiment that was ignored in these studies is the likelihood that glutamate will be much more highly labeled in the superficial portions of the slices than in the interior portions. The glutamate released by a depolarization stimulus may come predominantly from the superficial portions and consequently have a higher specific activity than that remaining in the tissue as a whole.

3.3. Biochemical Characterization of the Postsynaptic Glutamate Receptors

An analysis of the membrane responses indicates that glutamate causes a selective increase in the permeability of cations through the membrane.[27-29,113] The membrane response is very quick in onset and terminates abruptly, indicating that glutamate directly activates cation-selective channels in the membrane. A prominent biochemical alteration evoked by glutamate is the elevation of cyclic GMP.[116] This raises the possibility that this cyclic nucleotide may be a second messenger for glutamate; however, it appears more likely that the elevation occurs as a result of the depolarization process.[116]

Glutamate has been reported to elicit detectable membrane depolarization responses at concentrations as low as 1 to 10 μM.[117,118] This ability of glutamate to induce membrane responses at low concentrations, coupled with a pharmacological analysis of the action of numerous analogues of glutamate, leaves no doubt that glutamate interacts with specific receptors located on the external surface of neuronal cell membranes.[118]

Recognition of this has provided impetus for the development of experiments intended to characterize these receptors biochemically. The usual approach has been to prepare a cellular fraction enriched in nerve terminals, which is extensively washed with a hypotonic medium to disrupt the integrity of cell membranes, thereby promoting the removal of endogenous glutamate. The preparation is then incubated in the presence of radiolabeled glutamate, and the amount of glutamate bound to saturable binding sites determined. Usually Na^+ is omitted from the incubation medium to prevent binding of glutamate to transport sites, although it has not yet been conclusively determined that the absence of Na^+ does indeed prevent binding to these sites. The information most commonly obtained is the dissociation constant (K_D) and the number of binding sites in a given amount of protein. Reported K_D values have ranged from 10^{-8} to 10^{-4} M, although the reported values most frequently lie within 5×10^{-7} and 5×10^{-6} M.[119-124]

It is likely that the extracellular concentration of glutamate is never appreciably less than 10^{-6} M; hence, the physiological significance of binding sites with K_D values less than 10^{-6} M is questioned, since the sites would be nearly saturated under *in vivo* conditions. In this context, it is of interest that the presence of Na^+ greatly reduces the affinity of glutamate for those sites reported to have K_D values of approximately 10^{-8} M.[123] Therefore, the reported dissociation constants may be quite different from the actual dissociation constants occurring *in vivo*. Since it is not yet possible to assay these binding sites in any meaningful way, their physiological significance cannot be verified directly. Competitive inhibition between glutamate and various active analogues of glutamate has been used as a means of correlating binding activity to pharmacological activity. Inferences drawn from such comparisons indicate that there are at least four types of glutamate receptors in mammalian CNS tissues.[118] At least one type of receptor is concentrated in postsynaptic membranes.[124]

A glutamate-binding glycoprotein derived from synaptic membranes has been isolated recently by Michaelis and his colleagues.[125] This protein has a molecular weight of approximately 14,000 and contains an iron atom. Whether it represents a type of glutamate postsynaptic receptor remains to be established. The production of an antibody to this protein should help to establish its physiological role.

3.4. Membrane Transport as the Mechanism of Transmitter Inactivation

In order for glutamate to function effectively as a neurotransmitter, its extracellular steady-state concentration presumably must be maintained at a level below that causing membrane depolarization. The concentration of glutamate in interstitial fluid of CNS tissues is not known, but based on the concentration in CSF[13] and steady-state extracellular concentrations achieved by isolated toad brains,[73,98] it would appear to be between 1 and 10 μM. A wealth of evidence indicates that the low extracellular concentration of glutamate is maintained by transport into neurons and glial cells.[126–129] Another factor contributing to the low extracellular level is the low rate of penetration into the CNS parenchyma from blood.[12] In addition, some glutamate may be transported out of the CNS parenchyma into the blood.[130]

During the brief instant after the release of glutamate into the synaptic cleft, the concentration within the cleft must increase considerably, perhaps to about 1 mM. Because of the very short distance from the synaptic cleft to the adjacent interstitial space, a high portion of the glutamate may diffuse out of cleft. In this case, the return of the concentration in the cleft to the resting "steady-state" level, and the consequent termination of the postsynaptic response, would be mediated predominantly by diffusion rather than reuptake into the nerve terminal.

A number of reported studies indicate that uptake by neurons and astrocytes is mediated primarily by Na^+-dependent, high-affinity carriers.[131] In the membrane of neurons, these carriers may be particularly prevalent in nerve terminals.[132] In most studies, a kinetic analysis of initial uptake revealed only one apparent high-affinity carrier, and usually the reported affinity constant (K_t) was between 10 and 30 μM, although occasionally affinity constants of 2 μM or less were observed.[126,127] In a few studies, two high-affinity systems were observed, which exhibited apparent K_t values of approximately 2 and 20 μM.[133,134] In one of these studies, the two transport systems were found to be present in both glial-enriched and granule-cell-enriched fractions prepared from the mouse cerebellum.[133] A more recent study indicated that astrocytes may possess a carrier with an exceptionally high-affinity $(K_t \sim 0.2\mu M)$.[132]

A wide range of V_{max} values has been observed.[126,127] The reported values on the average are higher in astrocyte-enriched preparations than in neuronal-enriched material including synaptosomal preparations. These observations, coupled with the metabolic evidence that exogenous glutamate is selectively taken up by astrocytes in intact CNS tissue preparations, has led to the suggestion that glutamate is inactivated (removed from the interstitial fluid) primarily by uptake into astrocytes.[126,127] Present information probably does not justify such a conclusion. The rate of uptake by glutamatergic nerve terminals may be considerably greater than that observed for neuronal cell bodies or for the entire population of nerve terminals (and gliosomes) present in synaptosomal preparations. This possibility is supported by lesion experiments in which a loss of glutamatergic terminals is usually associated with an appreciable decline in uptake.[112] In addition, in intact tissue, exogenous glutamate may be more accessible to astrocyte transport sites than to neuronal sites, whereas for glutamate released from nerve terminals, the situation would be the reverse.

It has been questioned whether the Na^+-dependent, high-affinity uptake represents a net uptake of just a homoexchange process.[135] Recent studies indicate that although some homoexchange may occur, there can be an appreciable Na^+-dependent net uptake of glutamate.[136,137] An impressive example of the ability of CNS tissues to effect a net uptake of glutamate is the isolated toad brain which, when incubated in a volume of medium approximately 50 times the tissue mass, removed glutamate from the medium until a steady-state concentration of less than 2 μM was achieved.[137] The initial concentration in the medium was varied from 0 to 100 μM, and the final extracellular concentration achieved was the same regardless of the initial concentration. The replacement of NaCl with sucrose resulted in a marked flux of glutamate from the tissue into the medium,[100] indicating that the net uptake was Na^+ dependent.

3.5. Regulation of the Glutamate Transmitter Pool

For any given species of animal and specified region of CNS tissue, the content of glutamate shows little variation among a defined population. This suggests that the content of glutamate is tightly regulated. The ability of CNS tissues to maintain glutamate at a precise concentration is further illustrated by results obtained with the isolated toad brain. In an experiment in which the net uptake of glutamate was theoretically sufficient to increase the tissue content by one-third, the concentration was actually increased only marginally and eventually returned to the same level as that in tissue incubated without glutamate in the medium.[98,137]

Obviously, glutamate can function effectively as a neurotransmitter only if the transmitter pool is tightly regulated. At the present time, there is little definitive information pertaining to the mechanism(s) by which the transmitter pool is regulated. In order to meaningfully investigate the factors regulating the concentration of glutamate, the metabolic precursor(s) that serve to replenish the transmitter pool must be established, and it must be determined if the transmitter pool is metabolically distinct from other pools of glutamate within glutamatergic neurons.

With regard to the latter issue, the observation that the glutamate released from brain slices by an electrical stimulus is readily labeled from [^{14}C]glucose[91,92] suggests that the transmitter pool is not metabolically distinct from the pool contributing to the operation of the malate–aspartate shuttle. Therefore, it may be that the process regulating the transmitter pool is the same as that regulating the total content of glutamate inside the terminals of glutamatergic neurons. A related consideration is the regulation of the glutamate content in other neurons and in glial cells. Conceivably, a perturbation in the content of glutamate in any of these compartments could influence the extracellular steady-state level and thereby interfere with the ability of glutamate to function effectively as a transmitter.

Since aspartate aminotransferase appears to be extremely active in the cytosol of neurons in general and glutamatergic neurons in particular, it is unlikely that glutamate is regulated totally independently of aspartate, α-ketoglutarate, and oxaloacetate. Because cytosolic malic dehydrogenase is also quite active, it may be that malate must be included in this group of metabolic intermediates. Assuming that the content of glutamate is not regulated totally independently of these four compounds, it follows that any biochemical process that mediates the net production or net reduction of any of the five compounds can contribute to the regulation of glutamate.

Three prominent enzyme-mediated reactions fit into this category: (1) carboxylation of pyruvate, resulting in a net synthesis of oxaloacetate, (2) conversion of malate to pyruvate by malate enzyme, and (3) the formation of glutamate from glutamine by phosphate-dependent glutaminase (or another glutaminase enzyme). Since pyruvate carboxylation appears to occur predominantly, if not exclusively, in astrocytes, this reaction would presumably be excluded as a factor contributing directly to the regulation of the transmitter pool of glutamate. However, in place of this reaction, a transport system mediating a net uptake of α-ketoglutarate into glutamatergic nerve terminals could function as a regulatory mechanism. Indeed, recent observations suggest that the transport of α-ketoglutarate into nerve terminals is a highly regulated process.[95]

Another factor to be considered is the regulation of the amino nitrogen content present in glutamate plus aspartate. The synthesis of glutamate from glutamine is one way in which the content of amino nitrogen can be regulated. An additional reaction that may participate in this regulatory process is the interconversion of glutamate and α-ketoglutarate catalyzed by glutamate dehydrogenase. This reaction would make it possible for the total content of glutamate plus aspartate to be altered without an equivalent change occurring in the total content of α-ketoglutarate, oxaloacetate, and malate. Glutamate dehydrogenase is prevalent in CNS tissues[138] but appears to be much more active in astrocytes than in nerve terminals.[139] The characteristics of this enzyme are discussed in detail by E. Kvamme in Volume 4 of this series.

Still another consideration is the participation of these five compounds in energy metabolism. Because of this metabolic role, the content of glutamate will be influenced by the state of energy metabolism. In this context, it is of interest that changes in the state of energy metabolism frequently cause the content of glutamate to shift opposite to that of aspartate.[75]

Membrane transport processes may also contribute to the regulation of the transmitter pool of glutamate. For example, the transport of glutamate across the cell membrane could contribute to the regulation of glutamate,[140] although it would seem that the primary role of this process is to insure that glutamate is rapidly removed from the interstitial fluid and maintained at low extracellular steady-state levels. Also, the translocation of glutamate, α-ketoglutarate, aspartate, malate, and oxaloacetate (if it occurs) across the mitochondrial membrane could contribute to the regulation of glutamate in the cytosol. Clearly, the regulation of glutamate in CNS tissues must be a complex and multifactorial process.

4. BIOCHEMICAL BASIS FOR THE METABOLIC COMPARTMENTATION OF GLUTAMATE

In the intervening time between the last reviews on this subject[141,142] and the present, the most significant development has been the experimental evidence reported by Norenberg and his colleagues that indicates that glutamine synthetase is selectively expressed in astrocytes.[77] These investigators used immunocytochemical procedures to examine the cellular localization of glutamine synthetase in several regions of rat CNS tissues and invariably found specific reaction product only in astrocytes and ependymal cells. Although these results leave virtually no doubt that glutamine synthetase is much more prevalent in astrocytes than in neurons, it would be premature to conclude that neurons are totally devoid of this enzyme. Several biochemical studies using populations of cell bodies enriched in neurons suggest that glutamine synthetase is present at least in neuronal cell bodies.[132,143-145]

The demonstration that glutamine synthetase is much more prevalent in astrocytes than in neurons provides at least a partial explanation for the selective conversion of so many metabolites to glutamine relative to glutamate formation. However, it can not account for the low rate at which most carbon atoms derived from glucose, pyruvate, glycerol, lactate, and β-hydroxybutyrate are incorporated into glutamine.[18,141,142]

In this context, pyruvate provides a striking example of the selective incorporation of only certain carbon moieties into glutamine. The carboxyl carbon of pyruvate is selectively incorporated into glutamine, whereas the carbon atoms in the 2 and 3 positions are not.[70] Since pyruvate carboxylation would be required in order for the carboxyl carbon to enter the citric acid cycle and to be incorporated subsequently into glutamine, this suggests that pyruvate carboxylase must be selectively expressed either in the same cellular compartment as glutamine synthetase or in one that is metabolically associated with this compartment. In contrast, the carbons in the 2 and 3 positions enter the citric acid cycle primarily via the pathway catalyzed by the pyruvate de-hydrogenase complex.[146] Since the incorporation of these two carbons into glutamine is low, the activity of this enzyme complex must be low in the glu-tamine-forming compartment.

Therefore, based on the metabolic results obtained with labeled pyruvate, it would appear that pyruvate carboxylase is located predominantly in astro-cytes, whereas the pyruvate dehydrogenase complex is much more active in neurons than in astrocytes. A selective localization of the pyruvate dehydrog-enase complex in neurons could also account for the low incorporation of car-bon derived from glucose, lactate, and glycerol into glutamine.

Other biochemical differences between neurons and astrocytes almost cer-tainly contribute to the metabolic compartmentation of glutamate in CNS tis-sues. For example, the pool of glutamate associated with glutamine synthesis accounts for only a small percentage of the total content of gluta-mate.[18,142,143] Assuming that this pool resides in astrocytes, which account for about 25% of the total tissue volume, the concentration of glutamate in these cells must be appreciably lower than in CNS tissue as a whole. It is likely that differences in the rate at which carbon exchanges between glutamate and citric acid cycle intermediates also contribute to metabolic compartmentation.

Biochemical differences among different types of neurons may also con-tribute to the metabolic compartmentation of glutamate. In this context, we have discussed previously the evidence that the concentration of glutamate in GABAergic neurons is much less than that in other compartments in which glutamate is readily labeled from glucose (Section 2.3).

5. NEUROTOXIC EFFECTS OF GLUTAMATE

When applied to CNS tissues in sufficient amounts, glutamate can cause cells, particularly neurons, to swell.[147] Associated with this swelling is a grad-ually spreading depression of neuronal activity. Morphological changes in neu-rons can be observed within a few minutes after the application of glutamate, and within a few hours, irreversible changes characteristic of cell death can occur.[14] Actual cell death initiated by glutamate is limited to neurons, and some types of neurons are much more susceptible than other types.

When injected intravenously in large amounts, glutamate can induce neu-ronal cell death only in some regions of the CNS; particularly susceptible are neurons within the circumventricular organs which do not have a highly de-veloped blood–brain barrier.[14] As might be expected, neurons in immature animals which do not have a well-developed blood–brain barrier are more sus-ceptible than neurons in adult animals. When it is administered systemically, the amount of glutamate required to induce detectable cell death is sufficient to elevate the concentration in blood several orders of magnitude above the normal level.[14] Because normal homeostatic mechanisms are quite effective in preventing glutamate absorbed from the intestinal lumen from causing an ap-preciable elevation in the systemic circulation,[148] it is highly improbable that glutamate ingested in the diet could have a direct neurotoxic effect on CNS neurons, except perhaps in certain pathological conditions. Since glutamate is the principal causative factor in the Chinese restaurant syndrome, it is evident that dietary glutamate can have temporary adverse neurological effects in some people. Whether this is because of some peripheral action of glutamate or a direct effect on CNS neurons is not known.[149]

A number of glutamate analogues are also neurotoxic.[14,150–154] Olney and co-workers have examined the effects of a large number of compounds and found a strong correlation between the neuroexcitatory and neurotoxic potencies of these compounds.[14] Based on this correlation, Olney developed an "excitotoxic" hypothesis of neurotoxicity.[152] This hypothesis holds that the neuroexcitatory compounds can induce a sustained increase in Na^+ permeability which causes a depletion of energy reserves of sufficient magnitude to prevent the cell from performing metabolic processes needed to maintain cellular viability. Recent evidence that the neurotoxicity of kainate in the caudate nucleus of rats is dependent on an intact corticostriatal pathway (a presumed glutamatergic pathway) indicates that the neurotoxic effect of at least some excitatory compounds involves more than just a sustained depolarization of the cell membrane.[153] In this context, it is noteworthy that several analogues of glutamate that exhibit little or no neuroexcitatory action but rather antagonize the excitatory action of glutamate are selectively toxic to astrocytes.[14] The possibility that these analogues depolarize astrocyte membranes remains to be investigated.

The concept that glutamate and its analogues induce cell death merely as a consequence of a sustained membrane depolarization must be regarded only as a working hypothesis. It is possible that at the concentrations that induce cell death these compounds have deleterious effects on specific biochemical systems, particularly those involved in glutamate metabolism. An investigation of the mechanism(s) by which glutamate and its analogues induce cell death is of value because the information derived from such experimentation may provide new insights into the physiological roles of glutamate and the important phenomenon of neuronal and astrocyte cell death. In addition, the neurotoxicity of these compounds can serve as a useful tool for studying interactions between different types of neurons and between neurons and astrocytes.[154]

6. CONCLUDING REMARKS

In the future, neurochemical research on glutamate will likely continue to be dominated by studies pertaining to some aspect of the neurotransmitter function. The strong interest in identifying the neurons that use glutamate as a neurotransmitter should continue for many years, as should the interest in the characterization of glutamate receptors. Research will be greatly facilitated in these areas when specific markers of glutamatergic neurons are found and when specific antibodies to glutamate receptors are prepared and made available.

The role served by astrocytes in mediating the inactivation of glutamate and in supplying metabolic precursors to the nerve terminals should continue to be of interest. Research in this area will be facilitated by advances in cell separation techniques that make possible rapid and complete separation of astrocyte and neuronal cellular material. The formidable problem of determining how the neurotransmitter pool is regulated should attract considerable interest. Research in this area could be facilitated by procedures that make possible the separation of glutamatergic nerve terminals from other types of nerve terminals. Research pertaining to the neurotoxicity of glutamate analogues should continue to develop, particularly for analogues other than kainate which now dominates this area.

Because glutamate is an intermediate in the malate–aspartate shuttle, attention should continue to be given to its role in energy metabolism. Particular attention should be given to the uptake and metabolism of glutamate by mitochondria, and some attention should be focused on the regulation of the translocation across the membrane. The observation by Wherrett and Tower[81] that an appreciable and constant percentage of glutaminyl moieties in cerebral cortical protein is rapidly deamidated by complete ischemia should be confirmed, and the metabolic, physiological, and pathological significance of this investigated.

ACKNOWLEDGMENTS. The authors are grateful to Dr. C. J. Van den Berg for providing us with copies of two unpublished manuscripts, and to Dr. John J. O'Neil for constructive criticisms of our manuscript.

REFERENCES

1. Quastel, J. H., and Wheatley, A. H. M., 1932, *Biochem. J.* **26**:725–744.
2. Krebs, H. A., 1935, *Biochem. J.* **29**:1951–1969.
3. Sourkes, T. L., 1976, *Neuropharmacology* **15**:443–448.
4. Weil-Malherbe, H., 1950, *Physiol. Rev.* **30**:549–568.
5. Waelsch, H., 1951, *Adv. Protein Chem.* **6**:299–341.
6. Berl, S., Lajtha, A., and Waelsch, H., 1961, *J. Neurochem.* **7**:186–197.
7. Curtis, D. R., Phillis, J. W., and Watkins, J. C., 1960, *J. Physiol. (Lond.)* **150**:656–682.
8. Van Harreveld, A., 1960, *Inhibition in the Nervous System and Gamma Aminobutyric Acid* (E. Roberts, C. F. Baxter, A. Van Harreveld, C. A. G. Wiersma, W. R. Adey, and K. F. Killam, eds.), Pergamon Press, New York, pp. 454–459.
9. Lucas, D. R., and Newhouse, J. P., 1957, *Arch. Ophthalmol.* **58**:193.
10. Roberts, E., and Frankel, S., 1950, *J. Biol. Chem.* **187**:55–63.
11. Beloff-Chain, A., Cantanzaro, R., Chain, E. B., Masi, I., and Pocchiara, F., 1955, *Proc. R. Soc. Lond. [Biol.]* **144**:22–28.
12. Oldendorf, W., 1971, *Am. J. Physiol.* **221**:1629–1639.
13. Gjessing, L. R., Gjeshahl, P., and Sjaastad, O., 1972, *J. Neurochem.* **19**:1807–1808.
14. Olney, J. W., 1978, *Kainic Acid as a Tool in Neurobiology* (E. G. McGeer, J. W. Olney, and P. L. McGeer, eds.), Raven Press, New York, pp. 95–121.
15. Yamaguchi, S., and Kimizuka, A., 1979, *Glutamic Acid: Advances in Biochemistry and Physiology* (L. J. Filer, Jr., S. Garattini, M. R. Kare, W. A. Reynolds, and R. J. Wurtman, eds.), Raven Press, New York, pp. 35–54.
16. Schaumburg, H. A., Byck, R., Gerstl, R., and Mashman, J. H., 1969, *Science* **163**:826–828.
17. Balázs, R., and Haslam, R. J., 1965, *Biochem. J.* **94**:131–141.
18. Berl, S., and Clarke, D. D., 1969, *Handbook of Neurochemistry*, Volume 2 (A. Lajtha, ed.), Plenum Press, New York, pp. 447–472.
19. Van den Berg, C. J., 1970, *Handbook of Neurochemistry*, Volume 3 (A. Lajtha, ed.), Plenum Press, New York, pp. 514–544.

20. Brand, M. D., and Chappel, J. B., 1974, *Biochem. J.* **140**:205–210.
21. Dennis, S. C., and Clark, J. B., 1977, *Biochem. J.* **168**:521–527.
22. Dennis, S. C., and Clark, J. B., 1977, *Biochem. J.* **172**:155–162.
23. Kerpel-Fronius, S., Beck, D. P., and Von Korff, R. W., 1977, *J. Neurochem.* **28**:871–875.
24. Minn, A., and Gayet, J., 1977, *J. Neurochem.* **29**:873–881.
25. McCarthy, K. D., and DeVellis, J., 1980, *J. Cell Biol.* **85**:890–902.
26. Altschular, R., Harmison, G., Neisen, G., Wenthold, R. J., and Fex, J., 1981, *Proc. Natl. Acad. Sci. U.S.A.* **78**:6553–6557.
27. Curtis, D. R., 1979, *Glutamic Acid: Advances in Biochemistry and Physiology* (L. J. Filer, Jr., S. Garattini, M. R. Kare, W. A. Reynolds, and R. J. Wurtman, eds.), Raven Press, New York, pp. 163–175.
28. Watkins, J. C., 1978, *Kainic Acid as a Tool in Neurobiology* (E. G. McGeer, J. W. Olney, and P. L. McGeer, eds.), Raven Press, New York, pp. 37–70.
29. Puil, E., 1981, *Brain Res. Rev.* **3**:229–322.
30. Graham, L. T., Jr., Shank, R. P., Werman, R., and Aprison, M. A., 1967, *J. Neurochem.* **14**:465–472.
31. Davidoff, R. A., Graham, L. T., Jr., Shank, R. P., Werman, R., and Aprison, M. A., 1967, *J. Neurochem.* **14**:1025–1031.
32. Divac, I., Fonnum, F., and Storm-Mathisen, J., 1977, *Nature* **266**:377–378.
33. McGeer, P. L., McGeer, E. G., Sherer, D., and Singh, K., 1977, *Brain Res.* **128**:369–373.
34. Fonnum, F., Soreide, A., Kvale, I., Walker, J., and Walaas, I., 1981, *Glutamate as a Neurotransmitter* (G. DiChiara and G. L. Gessa, eds.), Raven Press, New York, pp. 29–41.
35. Baughman, R. W., and Gilbert, C. D., 1981, *J. Neurosci.* **1**:427–439.
36. Young, A. B., Bromberg, M. D., and Penney, J. B., Jr., 1981, *J. Neurosci.* **1**:241–249.
37. Fonnum, F., Storm-Mathisen, J., and Divac, I., 1981, *Neuroscience* **5**:863–873.
38. Nadler, J. V., Vaca, K. W., White, W. F., Lynch, G. S., and Cotman, C. W., 1976, *Nature* **260**:538–540.
39. Storm-Mathisen, J., 1977, *Brain Res.* **120**:379–386.
40. Crawford, I. L., and Connor, J. D., 1973, *Nature* **244**:442–443.
41. Storm-Mathisen, J., 1981, *Glutamate as a Neurotransmitter* (G. DiChiara and G. L. Gessa, eds.), Raven Press, New York, pp. 43–55.
42. Nadler, J. V., White, W. F., Vaca, K. W., Perry, B. W., and Cotman, C. W., 1978, *J. Neurochem.* **31**:147–155.
43. Moroni, F., Biamehi, C., Tanganelli, S., Moneti, G., and Beani, L., 1981, *J. Neurochem.* **36**:1691–1699.
44. Rowlands, G. J., and Roberts, P. J., 1980, *Exp. Brain Res.* **39**:239–240.
45. Harvey, J. A. Scholfield, C. N., Graham, L. T., Sr., and Aprison, M. A., 1975, *J. Neurochem.* **24**:445–449.
46. Bradford, H. F., and Richards, C. D., 1976, *Brain Res.* **105**:168–172.
47. Nitsch, C., 1981, *Adv. Biochem. Psychopharmacol.* **29**:97–104.
48. Zaczek, R., Hedreen, J. C., and Coyle, J. T., 1979, *Exp. Neurol.* **65**:145–156.
49. Collins, G. G., and G. A. Probett, 1981, *Brain Res.* **23**:231–234.
50. Young, A. B., Oster-Granite, M. L., Herndon, R. M., and Snyder, S. H., 1974, *Brain Res.* **73**:1–13.
51. McBride, W. J., Aprison, M. H., and Kusano, K., 1976, *J. Neurochem.* **26**:867–870.
52. Rea, M. A., McBride, W. J., and Rohde, B. H., 1981, *Neurochem. Res.* **6**:33–39.
53. Redburn, D. A., 1981, *Glutamate as a Neurotransmitter* (G. DiChiara and G. L. Gessa, eds.), Raven Press, New York, pp. 79–89.
54. Cuenod, M., Beaudet, A., Canzek, V., Streit, P., and Reubi, J. C., 1981, *Glutamate as a Neurotransmitter* (G. DiChiara and G. L. Gessa, eds.), Raven Press, New York, pp. 57–68.
55. Wenthold, R. J., 1981, *Glutamate as a Neurotransmitter* (G. DiChiara and G. L. Gessa, eds.), Raven Press, New York, pp. 69–78.
56. Reis, D. J., Granata, A. R., Perrone, M. H., and Talman, W. T., 1981, *J. Auto. Nerv. Syst.* **3**:321–324.
57. Johnson, J. L., and Aprison, M. A., 1971, *Brain Res.* **26**:141–148.
58. Berger, S. J., Carter, J. C., and Lowry, O. H., 1977, *J. Neurochem.* **28**:149–158.
59. Kennedy, A. J., Neal, M. J., and Lolley, R. N., 1977, *J. Neurochem.* **29**:157–159.

60. Fonnum, F., and Malthe-Sorenssen, D., 1981, *Glutamate: Transmitter in the Central Nervous System* (P. J. Roberts, J. Storm-Mathisen, Jr., and G. A. R. Johnston, eds.), John Wiley & Sons, New York, pp. 205–222.

61. Cotman, C. W., and Nadler, J. V., 1981, *Glutamate: Transmitter in the Central Nervous System* (P. J. Roberts, J. Storm-Mathisen, Jr., and G. A. R. Johnston, eds.), John Wiley & Sons, New York, pp. 117–154.

62. Baxter, C. F., 1970, *Handbook of Neurochemistry*, Volume 3 (A. Lajtha, ed.), Plenum Press, New York, pp. 289–353.

63. Seiler, N., and Eichentopf, B., 1975, *Biochem. J.* **152**:201–210.

64. Barber, R., Vaughn, J. E., Saito, K., McLaughlin, B. J., and Roberts, E., 1978, *Brain Res.* **141**:35–55.

65. Gaitonde, M. M., Dahl, D. R., and Elliott, K. A. C., 1965, *Biochem. J.* **94**:345–352.

66. Balázs, R., Machiyama, Y., and Patel, A. J., 1973, *Metabolic Compartmentation in the Brain* (R. Balázs and J. E. Cremer, eds.), Macmillan, London, pp. 57–70.

67. Shank, R. P., and Aprison, M. H., 1971, *J. Neurobiol.* **2**:145–151.

68. Van den Berg, C. J., Matheson, D. F., Nijenmanting, M. C., and Beuntink, R., 1982, *Neurotransmitter Interaction and Compartmentation* (H. Bradford, ed.), Plenum Press, New York (in press).

69. Hyde, J. C., and Robinson, N., 1976, *Histochemistry* **49**:51–65.

70. Shank, R. P., and Aprison, M. H., 1981, *Life Sci.* **28**:837–842.

71. Shank, R. P., Campbell, G. LeM., Freytag, S. O., and Utter, M. F., 1981, *Soc. Neurosci. Abstr.* **7**:936.

72. Costa, E., Guidotti, A., Moroni, A., and Peralta, E., 1979, *Glutamic Acid: Advances in Biochemistry and Physiology* (L. J. Filer, Jr., S. Garattini, M. R. Kare, W. A. Reynolds, and R. J. Wurtman, eds.), Raven Press, New York, pp. 151–161.

73. Shank, R. P., and Graham, L. T., Jr., 1978, *Advances in Neurochemistry*, Volume 3 (B. Agranoff and M. H. Aprison, eds.), Plenum Press, New York, pp. 165–201.

74. Lux, H. D., Loracher, C., and Neher, E., 1970, *Exp. Brain Res.* **11**:431–447.

75. Siesjo, B. K., 1978, *Brain Energy Metabolism*, John Wiley & Sons, New York.

76. Utley, J. D., 1964, *Biochem. Pharmacol.* **13**:1383–1392.

77. Norenberg, M. D., 1979, *J. Histochem. Cytochem.* **27**:756–762.

78. Hills, A. G., Reid, E. L., and Kerr, W. D., 1967, *Am. J. Physiol.* **223**:1470–1476.

79. Betz, A. L. and Gilboe, D. D., 1973, *Am. J. Physiol.* **224**:580–587.

80. Minard, F. N., and Richter, D., 1968, *J. Neurochem.* **15**:1463–1468.

81. Wherrett, J. R., and Tower, D. B., 1971, *J. Neurochem.* **18**:1027–1042.

82. Tower, D. B., and Wherret, J. R., 1971, *J. Neurochem.* **18**:1043–1051.

83. Moore, B. W., 1965, *Biochem Biophys. Res. Commun.* **19**:739–744.

84. Kato, K., Suzuki, F., and Umeda, Y., 1981, *J. Neurochem.* **36**:793–797.

85. Reichelt, K. L., 1970, *J. Neurochem.* **17**:19–25.

86. Meister, A., 1979, *Glutamic acid: Advances in Biochemistry and Physiology* (L. J. Filer, Jr., S. Garattini, M. R. Kare, W. A. Reynolds, and R. J. Wurtman, eds.), Raven Press, New York, pp. 69–84.

87. Katunuma, N., Katsunuma, T., Towatari, T., Tomino, I., 1973, *The Enzymes of Glutamine Metabolism* (S. Prusiner and E. R. Stadtman, eds.), Academic Press, New York, pp. 227–258.

88. Shank, R. P., and Campbell, G. LeM., 1981, *Life Sci.* **28**:843–850.

89. McIntosh, J. C., and Cooper, J. R., 1965, *J. Neurochem.* **12**:825–835.

90. D'Adamo, A. F., Jr., and Yatsu, F. M., 1966, *J. Neurochem.* **13**:961–965.

91. Potashner, S. J., 1978, *J. Neurochem.* **31**:177–186.

92. Potashner, S. J., and Lake, N., 1981, *Glutamate as a Neurotransmitter* (G. DiChiara and G. L. Gessa, eds.), Raven Press, New York, pp. 139–145.

93. Cheng, S. C., 1971, *Int. Rev. Neurobiol.* **14**:125–157.

94. Patel, M. S., 1974, *J. Neurochem.* **11**:887–898.

95. Shank, R. P., and Campbell, G. LeM., 1982, *Trans. Am. Soc. Neurochem.* **13**:264.

96. Shank, R. P., and Campbell, G. LeM., 1981, *Trans. Am. Soc. Neurochem.* **12**:376.

97. Shank, R. P., and Campbell, G. LeM., 1982, *Neurochem. Res.* **7**:601–615.

98. Shank, R. P., and Aprison, M. H., 1979, *Glutamic Acid: Advances in Biochemistry and Physiology* (L. J. Filer, Jr., S. Garattini, M. R. Kare, W. A. Reynolds, and R. J. Wurtman, eds.), Raven Press, New York, pp. 139–150.
99. Quastel, J. H., 1974, *Biochem. Soc. Trans.* 2:765–780.
100. Shank, R. P., and Aprison, M. H., 1977, *J. Neurochem.* 28:1189–1196.
101. Bradford, H. F., Ward, H. K., and Thomas, A. J., 1978, *J. Neurochem.* 30:1453–1459.
102. Hamberger, A. C., Chiang, G. H., Nylen, E. S., Scheff, S. W., and Cotman, C. W., 1979, *Brain Res.* 168:513–530.
103. Reubi, J. C., Van den Berg, C. J., and Cuenod, M., 1978, *Neurosci. Lett.* 10:171–174.
104. Tapia, R., and Gonzalez, R. M., 1978, *Neurosci. Lett.* 10:165–169.
105. Kemel, J. L., Gauchy, C., Glowinski, J., and Besson, M. J., 1979, *Life Sci.* 24:2139–2150.
106. Hertz, L., 1979, *Prog. Neurobiol.* 13:277–323.
107. McGeer, E. G., McGeer, P. L., and Hattori, T., 1979, *Glutamic Acid: Advances in Biochemistry and Physiology* (L. J. Filer, Jr., ed.), Raven Press, New York, pp. 187–201.
108. Benjamin, A. M., 1981, *Brain Res.* 208:363–377.
109. Roberts, E., 1981, *Glutamate as a Neurotransmitter* (G. DiChiara and G. L. Gessa, eds.), Raven Press, New York, pp. 91–102.
110. Wong, P. T., and McGeer, E. G., 1981, *Brain Res.* 227:519–529.
111. Shank, R. P., and Campbell, G. LeM., 1983, *J. Neurosci. Res.* 9:(in press).
112. Cotman, G. W., Foster, A., and Lanthon, T., 1981, *Glutamate as a Neurotransmitter* (G. DiChiara and G. L. Gessa, eds.), Raven Press, New York, pp. 1–27.
113. Usherwood, P. N. R., 1978, *Adv. Comp. Physiol. Biochem.* 7:227–310.
114. Cutler, R. W. P., and Dudzinski, D. S., 1975, *Brain Res.* 8:415–423.
115. DeBelleroche, J. S., and Bradford, H. F., 1977, *J. Neurochem.* 29:335–343.
116. Garthwaite, J., and Balázs, R., 1981, *Glutamate as a Neurotransmitter* (G. DiChiara and G. L. Gessa, eds.), Raven Press, New York, pp. 317–326.
117. Hosli, L., Andres, P. E., and Hosli, E., 1976, *Pfluegers Arch.* 363:43–48.
118. Johnston, G. A. R., 1979, *Glutamic Acid: Advances in Biochemistry and Physiology* (L. J. Filer, Jr., S. Garattini, M. R. Kare, W. A. Reynolds, and R. J. Wurtman, eds.), Raven Press, New York, pp. 177–186.
119. Michaelis, E. K., 1975, *Biochem. Biophys. Res. Commun.* 65:1004–1012.
120. Foster, A. C., and Roberts, P. J., 1978, *J. Neurochem.* 31:1467–1477.
121. Biziere, K., Thompson, H., and Coyle, J. T., 1980, *Brain Res.* 183:421–433.
122. Baudry, M., and Lynch, G., 1980, *Proc. Natl. Acad. Sci. U.S.A.* 77:2298–2302.
123. Vargas, F., and Costa, E., 1981, *Glutamate as a Neurotransmitter* (G. DiChiara and G. L. Gessa, eds.), Raven Press, New York, pp. 307–316.
124. Foster, A. C., Mena, E. E., Fagg, G. E., and Cotman, C. W., 1981, *J. Neurosci.* 1:620–625.
125. Michaelis, E. K., 1979, *Biochem. Biophys. Res. Commun.* 87:106–112.
126. Hertz, L., 1979, *Prog. Neurobiol.* 13:277–323.
127. Schousboe, A., *Int. Rev. Neurobiol.* 22:1–45.
128. Schousboe, A., and Hertz, L., 1981, *Glutamate as a Neurotransmitter* (G. DiChiara and G. L. Gessa, eds.), Raven Press, New York, pp. 103–113.
129. Johnson, J. L., 1978, *Prog. Neurobiol.* 10:155–202.
130. Pardridge, W., 1979, *Glutamic Acid: Advances in Biochemistry and Physiology* (L. J. Filer, Jr., S. Garattini, M. R. Kare, W. A. Reynolds, and R. J. Wurtman, eds.), Raven Press, New York, pp. 125–138.
131. Fagg, C. E., and Lane, J. D., 1980, *Neuroscience* 4:1015–1036.
132. Shank, R. P., and Campbell, G. LeM. *J. Neurosci.*, submitted.
133. Campbell, G. LeM., and Shank, R. P., 1978, *Brain Res.* 153:618–622.
134. Fagg, C. E., Jones, I. M., and Jordan, C. C., 1978, *Neurosci. Lett.* 9:71–75.
135. Raiteri, M., Federico, R., Coletti, A., and Levi, G., 1975, *J. Neurochem.* 24:1243–1250.
136. Johnston, G. A. R., 1976, *Adv. Biochem. Psychopharmacol.* 15:175–184.
137. Shank, R. P., and Baxter, C. F., 1975, *J. Neurochem.* 24:641–646.
138. Chee, P. Y., Dahl, J. L., and Fahien, L. A., 1979, *J. Neurochem.* 33:53–60.
139. Patel, A. J., Hunt, A., Tahourdin, C. S. M., Gordon, R. D., and Bunn, S., 1981, *Soc. Neurosci. Abstr.* 7:935.

140. Levi, G., Kandera, J., and Lajtha, A., 1967, *Arch. Biochem. Biophys.* **119**:303–311.
141. Van den Berg, C. J., Reignierse, G. L. A., Blochuis, G. G. C., Kron, M. C., Ronda, G., Clarke, D. D., and Garfinkel, D., 1976, *Metabolic Compartmentation and Neurotransmission—Relation to Brain Structure and Function* (S. Berl, D. D. Clarke, and S. Schneider, eds.), Plenum Press, New York, pp. 515–544.
142. Van den Berg, C. J., Matheson, D. F., and Nijenmanting, W. C., 1978, *Amino Acids as Chemical Transmitters* (F. Fonnum, ed.), Plenum Press, New York, pp. 709–723.
143. Piddington, R., 1977, *Brain Res.* **128**:505–514.
144. Ward, H. K., and Bradford, H. F., 1979, *J. Neurochem.* **33**:339–342.
145. Weiler, C. T., Nystrom, B., and Hamberger, A., 1979, *Brain Res.* **160**:539–543.
146. Cheng, S. C., 1973, *Metabolic Compartmentation in the Brain* (R. Balázs and J. E. Cremer, eds.), Macmillan, London, pp. 107–118.
147. Van Harreveld, A., and Fifkova, E., 1970, *J. Neurobiol.* **2**:213–229.
148. Munro, H., 1979, *Glutamic Acid: Advances in Biochemistry and Physiology* (L. J. Filer, Jr., S. Garattini, M. R. Kare, W. A. Reynolds, and R. J. Wurtman, eds.), Raven Press, New York, pp. 55–68.
149. Kenny, R. A., 1979, *Glutamic Acid: Advances in Biochemistry and Physiology* (L. J. Filer, Jr., S. Garattini, M. R. Kare, W. A. Reynolds, and R. J. Wurtman, eds.), Raven Press, New York. pp. 363–374.
150. Herndon, R. M., and Coyle, J. T., 1978, *Kainic Acid as a Tool in Neurobiology* (E. G. McGeer, J. W. Olney, and P. L. McGeer, eds.), Raven Press, New York, pp. 189–200.
151. Hampton, C. K., Garcia, C., and Redburn, D. A., 1981, *J. Neurosci. Res.* **6**:99–112.
152. Olney, J. W., Rhee, V., and de Gubareff, T., 1977, *Brain Res.* **120**:151–157.
153. McGeer, E. G., McGeer, P. L., and Singh, K., 1978, *Brain Res.* **139**:381–383.
154. McGeer, P. L., McGeer, E. G., and Hattori, T., 1978, *Kainic Acid as a Tool in Neurobiology* (E. G. McGeer, J. W. Olney, and P. L. McGeer, eds.), Raven Press, New York, pp. 123–138.

Glutamine

Elling Kvamme

1. CONTENT IN CNS

1.1. Content in Various Brain Areas

Glutamine (Gln) is one of the dominant amino acids in the brain, and the Gln family, consisting of Gln, Glu, Asp, and GABA, constitutes 70–80% of the free amino acid nitrogen in mammalian brain.[1] The content of Gln, Glu, and GABA in various brain areas is shown in Table I. The Gln concentration is always lower than that of Glu and higher than that of GABA. The concentration varies greatly from one animal species to another, being highest in the dog and lowest in the rat. It should, however, be kept in mind that the measurements were done by different investigators, which may partly explain the variation. Thus, Bradford and Thomas[2] found 3.0, 8.6, and 1.2 μmol/g wet wt. of rat cerebral cortex for Gln, Glu, and GABA, respectively, which are higher than the figures of Gründig and Hanbauer.[3] Similar values for rat brain cortex to those reported by Bradford and Thomas[2] were found by others,[4,5] and they did not differ very much from those of guinea pig brain.[6] In human cortex cerebri, Robinson and Williams[7] found values within the same range for Gln and Glu as Perry *et al.*[8] As shown in Table I, the concentration of Gln, Glu, and GABA appears to be remarkably constant for different brain areas within the same species, as is the sum of these amino acids. This might indicate that the amino acids in question are interrelated. Wiechert and Göllnitz[9] found, however, lower values for the three amino acids in dog medulla than in brain. In the dorsal root, Gln tends to be higher than Glu, which has been explained by low glutaminase activity.[11–13]

In summary, the members of the Gln family, Gln, Glu, Asp, and GABA, are the dominant amino acids in brain, and their contents are rather constant from one brain area to another, suggesting functional similarities for each of these amino acids throughout the brain.

Elling Kvamme • Neurochemical Laboratory, Preclinical Medicine, Oslo University, Oslo, Norway.

Chapter 15

Table I

Glutamine, Glu, and GABA in Various CNS Areas and Animal Species ($\mu mol \cdot g$ wet wt.$^{-1}$)

	Cerebral cortex				Cerebellum				Nucleus caudatus				Thalamus				Medulla			
	Gln	Glu	GABA	Sum	Gln	Glu	GABA	Sum	Gln	Glu	GABA	Sum	Gln	Glu	GABA	Sum	Gln	Glu	GABA	Sum
Human[8]	4.4	10.8	2.1	17.3[a]	5.2	9.6	2.3	17.1	3.2	10.5	3.0	16.7	3.7	8.4	2.5	14.6				
Dog[9]	11.8	17.8	1.8	31.4	13.3	16.0	1.8	31.1[c]	16.0	18.7	3.4	38.1	8.7	15.1	2.7	26.5	5.7	8.8	0.8	14.5
Bovine[10d]	4.5	7.9	2.2	14.6[b]	4.5	7.0	1.4	12.9[c]	5.6	7.5	3.6	14.9	3.4	6.5	2.4	12.3	3.9	6.3	1.1	11.3
Rat[3d]	1.7	4.9	1.3	7.9	1.8	4.3	1.5	7.6					1.9	4.2	1.9	8.0				

[a] Temporal.
[b] Parietal.
[c] Cerebellar cortex.
[d] Recalculated assuming that the brain protein concentration is 10%.

1.2. Subcellular Distribution

The subcellular distribution of Gln is similar to that of Glu, Asp, and GABA. Following precipitation of the microsomes, the high-speed supernatant contained about 75%, the synaptosomes 8–10%, and the myelin fraction 7–9% of most amino acids, including Gln, Glu, and GABA.[6] The most abundant amino acid in synaptosomes is Glu, followed by Gln, Asp, GABA, and Tau.[2,10] Recent work indicates that the Gln content in synaptosomes is difficult to estimate,[14] since this amino acid, in contrast to Glu,[14,15] is easily lost by leakage during the preparation. Most authors have found the synaptosomal Gln content in cerebral cortex to be on the order of 3–6 μmol/g protein, whereas that of Glu is 10–20 μmol/g protein and GABA 2–4 μmol/g.[10] The synaptosomal vesicles contain about the same amounts of these amino acids,[10,16,17] but it is remarkable that Tau is the most abundant amino acid in the vesicles.[10,17] Taurine is the only amino acid highly enriched in the vesicle fraction of all brain areas.[10] Van Gelder[18] reported that the fluctuation of Tau in cerebral cortex was related to changes in the content of Gln and Glu. He also suggested that the release of Glu in epilepsy was mediated by Tau.

Thus, the amino acids of the Gln family are the most abundant in synaptosomes, but in the synaptosomal vesicle fraction, only Tau appears to be enriched.

1.3. Developmental Changes

A relatively high Gln value is reported in the brain of newborn rat, rabbit,[19,20] mouse,[21] cat,[22] and dog,[23] and the level decreases during the first 7 days after birth. Thereafter, the levels increase to the adult age. In newborn mouse, brain Gln is 78%, Glu 37%, GABA 54%, and Asp 55% of the levels found in the mature brain. At 7 days of age, the values are Gln 58%, Glu 57%, GABA 69%, and Asp 68% of the adult values.[21] It is of interest that these values for Glu and its derivatives are similar to the percentages of adult levels at 7 days of age for brain weight, proteins and phosphatides.[24] It is also of interest that only members of the Gln family increase with maturation, whereas other amino acids decrease. However, it may have some bearing on the fact that myelination begins on the ninth day in the mouse,[24] and the deposition of myelin continues for the following 40 days.

1.4. Other Factors Affecting the Content

Glutamine is the only amino acid that shows an age-related change in mouse brain.[25] The Gln content increases approximately 50% in the senescent mouse brain, and the ammonia concentration increases about 60%. The increase in Gln has been related to increased synthesis in an attempt to detoxify the higher levels of ammonia. Thus, the increase in Gln could not be explained by a change in the activity of glutaminase and γ-glutamyltransferase. The age dependence of the amino acid content found in mouse brain deserves to be studied in other species as well.

Acute hypercapnic conditions cause increased levels of Gln in rat brain with a reciprocal fall in the Glu level. In hypocapnia, Gln is reduced and Glu increased.[26] Since the ammonia content is increased in hypercapnic conditions, a stimulated synthesis of Gln from Glu and ammonia has been suggested.

Intrauterine malnutrition affects amino acids and enzymes of the Gln family. Thus, if pregnant rats are fed with wheat and Bengal gram (black chick peas), the brains of 1-day-old rats show decreased concentrations of Gln, Glu, Ala, and GABA and decreased activities of Gln synthetase, Gln transferase, phosphate-activated glutaminase, Gln–oxoacid aminotransferase, and Glu decarboxylase.[27] The concentrations of DNA, RNA, and protein also decreased, but that of Asp increased. It is possible that the reduced levels of amino acids were caused by decreased enzyme activities. The changes observed in the brain were restored when the diet was fortified with Lys and with Met, Cys, or Trp and also when 20% casein was included in the diet.[27]

In another study, thiamine deficiency reduced the levels of Glu and Asp in certain brain areas, particularly in medulla oblongata, cerebellum and midbrain regions, but left Gln unchanged.[28] This was explained by a diminished entry of pyruvate into the tricarboxylic acid cycle.

The levels of Gln and GABA in brain are influenced by ethanol, but the effects appear to be very complex.[5] Significant amounts of ethanol-derived acetate are incorporated into brain Gln and GABA following intraperitoneal injection of [14C]ethanol.[29] During alcohol intoxication, glucose utilization is partially inhibited, and the brain might adapt its metabolism to utilize ethanol-derived acetate obtained from Gln. It is also of interest that treatment with Asn or Gln caused substantial increases in the Gln concentration in cerebellum and medulla oblongata.[30] Since cerebral Glu concentration has been found to be diminished in neurological disorders such as Friedreich's ataxia, treatment with Asn or Gln has been suggested.

In summary, the brain levels of Gln and Glu are affected by age, hypercapnia, hypocapnia, malnutrition, and ethanol, and treatment with Gln or Asn is reported to increase the Gln content in cerebellum and medulla oblongata.

1.5. Compartmentalization

The basis for this concept is the so-called Waelsch effect.[31] Thus, on injection of isotopically labeled Glu or ammonia, bicarbonate, acetate, butyrate, citrate, or some amino acids other than Glu, the specific activity of Gln was higher than that of the precursor isolated shortly after the injection. The effect was, however, not found by injecting labeled glucose, pyruvate, lactate, or glycerol. Since other possibilities could be ruled out, the Waelsch effect was explained by assuming that there are at least two metabolic pools of Glu. Thus, a small Glu pool, which does not mix with another large Glu pool, has a rapid turnover into a large Gln pool. The Waelsch effect appears to be a specific property of the nervous system, and it has been demonstrated both *in vivo* and *in vitro*.

Since Gln synthetase appears to be predominantly localized in glial cells, and glutaminase has been assumed to be most active in neurons, the concept of the Gln cycle has been proposed. According to this hypothesis, Glu is taken up by glial cells and converted to Gln by the synthetase reaction, and this Gln enters neurons to form Glu and GABA.[32–34].

2. TRANSPORT AND UPTAKE

2.1. Transport across the Blood–Brain Barrier

The mechanism of the net transport of Gln into the brain from blood has been subject to different views. Following the administration of large doses of Gln, Schwerin[35] reported in 1950 that it readily penetrates into brain. However, other workers later found that the administration of 50 mg Gln per kg body weight to rat failed to elevate the brain content,[36,37] and Gln has been claimed not to enter brain.[37,38] Studies on the efflux of Gln and various amino acids from brain indicated that a carrier-mediated process is operating rather than passive diffusion.[39,40] Abdul-Ghani *et al.*[41] also studied the release of Gln and Glu from the cat's brain into the cerebral venous blood following afferent electrical stimulation of the contralateral brachial plexus. They concluded that the kinetic behavior of the Gln transport was compatible with a carrier-mediated process and not with passive diffusion.

2.2. Glutamine Uptake and Release in CNS Cellular Constituents

Glutamine generally enters eukaryotic cells by a low-affinity uptake ($K_m > 0.05$ mM) that is either sodium independent or sodium dependent. Three different sodium-dependent uptake systems are observed in Ehrlich cells, pigeon erythrocytes, rabbit reticulocytes, and rat hepatocytes.[42] In kidney mitochondria, it appears that Gln enters by a carrier mechanism[43] that is sodium dependent and inhibited by proline.[44] Most investigators believe, however, that the Gln uptake in CNS is characterized by low affinity and sodium independence. This applies to Gln uptake in brain slices,[45,46] neuroblastoma, gliablastoma,[47] astrocytes,[48–50] and synaptosomes[46,48,51] (Table II) and is in striking contrast to the sodium-dependent high-affinity systems for Glu, Asp, and GABA found in neuronal constituents and astrocytes. High-affinity sodium-dependent uptake has, however, been described in brain prisms,[52] dorsal roots,[53] and dorsal root ganglia.[54] Since the V_{max} also was low, as seen in Table II, the significance of this finding is doubtful. Roberts[54] suggested that the high-affinity uptake was an artifact derived from homoexchange of extracellular labeled Gln with the large Gln pool within the satellite glial cells. It has also been suggested that the high-affinity uptake might be caused by contamination of the labeled Gln with labeled Glu.[55]

The extracellular Gln content is high in many species, as reflected by the CSF concentration. In humans, CSF[56,57] and plasma[58] levels of Gln are approximately 0.5 mM, in dog 0.02 and 0.03,[59] and in cat 0.02 and 0.3, respectively.[41,60] It is of interest that a high V_{max} for uptake of Gln has been reported

Table II
The Kinetics of Gln Uptake

	Ref.	V_{max} (μmol·min^{-1}·g wet wt.$^{-1}$)	K_m (mM)
Brain slices	45	0.3	2.5
	46	0.2	0.66, 2.25
Brain prisms	52	0.2	0.05
Dorsal roots	53	0.01	0.05
Dorsal root ganglia	54	0.01	0.05
Neuroblastoma (C-1300)	47	3–7	0.70
138 MG glioma cell line	47	2.9	0.49
Bulk-prepared astrocytes	48	0.2	0.63
Astrocytes in primary cell cultures	49	0.2	0.15
	50	5.0	3.30
Synaptosomes	51	1.0	0.26
	48	0.3	0.24
	46	5.5	0.25

in synaptosomes[46,51] and neuroblastoma,[47] which indicates a substantial accumulation of this amino acid. The uptake of Gln at 0.25 mM concentration into nerve terminals also exceeds that into brain cortex slices.[46] The astrocytes, however, appear to have a rather low V_{max} in relation to the relatively high K_m.[49,50] Thus, a higher Gln uptake in neuronal cell constituents compared to glial cells is suggested in support of the concept of the Gln cycle (See Section 4.2). However, the high V_{max} for the 138 MG glioma cell line may not warrant this general conclusion. Experiments with rabbit retinas also failed to support the hypothesis that the GABA skeleton is rapidly transported between glial cells and GABAergic neurons in the form of Gln or Glu.[61]

A possible regulation of branched-chain L-amino acids and of Phe and Met on Gln uptake into rat brain cortex slices has been reported.[62] These amino acids suppressed the uptake of Gln, but the inhibition was blocked by high potassium, which itself had little or no effect on Gln uptake. The inhibitory effect was found in immature rat brains and in crude synaptosomal preparations derived from adult rat brain. Relatively low concentrations (e.g., 1 mM) of branched-chain L-amino acids (i.e., Val, Leu, and Ile) or of L-Phe or L-Met were required to inhibit the uptake of L-Gln (2 mM) into brain cells approximately 40%. There was, however, no inhibition of the uptake of Glu and GABA. It is of interest that these inhibitory amino acids are elevated in patients with certain inborn errors of metabolism such as maple syrup urine disease, phenylketonuria, and homocystinuria. This may explain the Gln depletion found in some cases of phenylketonuria and maple syrup urine disease.

Glutamine is rapidly released from whole brain, brain slices,[63] hemisections,[64] and astrocytes in primary cultures.[49] Whether Gln is released in a potassium-dependent,[51,65] calcium-independent[66] manner, as has been reported, is a matter of debate. Thus, no potassium dependence has been found by other workers for the efflux of Gln from brain slices,[67,68] synaptosomes,[69] and astrocytes in primary cultures.[55] By a countertransport technique, which makes use of high concentrations of unlabeled compounds, it was indicated that Gln was released by a facilitated diffusion transport system.[70] The Gln release from brain slices was, however, enhanced by branched-chain L-amino acids (i.e., Val, Leu, and Ile) and L-Phe and L-Met in 1 mM concentrations, and this release appeared to be diminished by high potassium.[62]

To conclude, the Gln uptake in CNS appears generally to occur by a low-affinity, sodium-independent mechanism. In spite of that, the uptake may be substantial because the extracellular Gln concentration is high (0.5 mM). The Gln release appears to occur by a facilitated diffusion transport system. Both the uptake and release of Gln are probably regulated by branched-chain L-amino acids, L-Phe, and L-Met.

3. METABOLISM

3.1. General Metabolic Functions

Glutamine is not an essential amino acid in the diet of mammals. Ammonia is transported by Gln in the form of amine and amide nitrogen. Similarly, Gln is a vehicle for transporting Glu, which otherwise must depend on rather restrictive carrier-mediated systems, e.g., penetration of the inner mitochondrial membrane in order to serve as fuel for respiration. The human plasma concentration of Gln was measured to be 0.55 mM.[58] This accounts for about 20% of the total amino acid content in plasma and is thus the most significant amino acid. The plasma Gln level appears to parallel the P_{CO_2}.[58] In chronic acidosis, excess acid is excreted in urine in the form of ammonium salts. This ammonia is derived from blood Gln[71] that is hydrolyzed by the phosphate-activated glutaminase in kidney.

Skeletal muscle is considered to be the most important source of endogenous Gln,[72] and the small intestine is the main Gln-utilizing tissue. Glutamine is more important as respiratory fuel in the small intestine than is glucose.[73] Glutamine is also a major source of energy in other rapidly growing cells such as reticulocytes,[74] fibroblasts,[75] and various malignant cells.[76,77] Glutamine is a precursor of Ala, citrulline, and Pro. About 60% of the urea nitrogen excreted is derived from ammonia and Ala that is formed from Gln[73] (for review see ref. 78). Moreover, the Gln amide nitrogen is a precursor for His, Trp, Asn, amino sugars, purines, and also carbamyl phosphate, which is used for the synthesis of the pyrimidine ring. However, these reactions have been little studied in CNS. Of great importance to CNS is the precursor relationship of Gln to transmitter Glu and GABA, which is discussed below.

Brain mitochondria oxidize Gln in the presence of malate at about the same rate as GABA but at a lower rate than Glu.[79] The nonsynaptic mitochondria show higher rates of oxygen uptake with either pyruvate or Glu plus malate as substrate than do the synaptic mitochondria. This difference has, however, not been found for Gln plus malate. Oxygen uptake in the presence of Gln plus malate follows single-phase kinetics with an apparent K_m for Gln of 0.28 mM, whereas the oxygen uptake in the presence of Glu plus malate appears to be biphasic with apparent K_m values of 0.25 mM and 2.4 mM.[80] Brain mitochondria, which oxidize Gln, export Glu to the surrounding medium.[81] The Gln oxidation is dependent on several enzymes such as glutaminases, aminotransferases, and possibly Glu translocases.

Two Glu translocases have been suggested in brain mitochondria, namely, Glu–Asp translocase and a relatively slow Glu–OH⁻ translocase. Nonsynaptic mitochondria appear to possess both translocases, but synaptic mitochondria have only the Glu–Asp translocase.[82] A Gln–Glu translocase, which also has been suggested, is highly controversial and probably nonexistent.[81] Therefore, the lower oxygen uptake with Gln than with Glu as substrate cannot be explained by an effect of a translocase, because Gln is not itself subject to the restraints of a mitochondrial translocase and acts as an intramitochondrial source of Glu.[80]

Glutamine is a much less efficient fuel for synaptosomal respiration and maintenance of tissue potassium level than is an equivalent amount of glucose. Thus, the contribution of Gln to synaptosomal energy metabolism appears to be minimal.[83] However, the presence of Gln is necessary in incubation media containing glucose to maintain the amino acid pools of the synaptosomes. Therefore, Krebs–Ringer–Gln is suggested as the best medium for synaptosomes.[84]

On the other hand, Glu serves as a fuel for brain respiration both in the absence and in the presence of glucose.[85] Endogenous Glu is largely oxidized by an initial reaction with Glu dehydrogenase.[86] Exogenous Glu, however, undergoes initial transamination to Asp and 2-oxoglutarate. Thus, 49% is converted to Asp, 37% to Gln, and the rest is oxidized through Glu dehydrogenase. There is no obvious reason why exogenous Glu is more easily oxidized than Gln which penetrates the mitochondrial membranes with no difficulty. The distinction between endogenous and exogenous Glu is also puzzling and indicates that the endogenous Glu is compartmentalized. This is supported by recent work showing that endogenous Glu, in contrast to exogenous Glu, exerts very little inhibition of phosphate-activated glutaminase and is therefore largely unavailable to this enzyme.[14]

In conclusion, Gln has an important transport function in CNS for Glu and ammonia, and it serves as metabolic precursors for, e.g., other amino acids, amino sugars, proteins, purines, and pyrimidines. Glutamine, in contrast to Glu, does not seem to be generally important for energy metabolism in CNS.

Table III
Enzymes Synthesizing and Degrading Gln

Synthesis
  Gln synthetase (E.C. 6.3.1.2)
Degradation
  Glutaminases (Gln amidotransferases)
    Phosphate-activated glutaminase (phosphate-dependent glutaminase, E.C. 3.5.1.2)
    Maleate-activated glutaminase (phosphate-independent glutaminase, γ-glutamyltransferase,
      γ-glutamyl transpeptidase, E.C. 2.3.2.2)
    Transglutaminase (E.C. 2.3.2.13)
    Other enzymes with Gln amidotransferase activity
  Glutamine aminotransferase (glutaminase II, E.C. 2.6.1.15) and ω-amidase (E.C. 3.5.1.3)

3.2. Synthesis and Degradation

Ammonia is formed by brain slices in the absence of glucose. Following addition of glucose or Glu, ammonia production is suppressed, and Gln is formed,[87] whereas labeled Glu and Asp are formed on incubating brain slices with [14C]Gln.[88] These simple experiments suggest the involvement of a number of enzymes listed in Table III, which contains the most important enzyme systems synthesizing and degrading Gln. Otherwise, the Gln-degrading enzymes are dealt with in Volume 4, Chapter 4 of this *Handbook*.

3.2.1. Glutamine Synthetase (E.C. 6.3.1.2)

3.2.1a. Purification and Molecular Properties. The enzyme has been studied in rat, ox, sheep, pig, and human brain and also particularly in liver and *E. coli*. The purification of the brain enzyme[89] usually proceeds through four steps: the enzyme is precipitated by acid from an acetone powder extract and is further subject to column chromatography on hydroxylapatite and DEAE-cellulose, whereby a purification of about 200-fold is obtained. The yield is 15–30%.

Glutamine synthetase from brain and liver is composed of eight subunits which are identical and show a cubelike appearance on electron microscopy. It is considered to possess D_4 symmetry in the model for subunit structure proposed by Haschemeyer[90] and is formed by isologous association of two heterologously bonded tetramers. The brain enzyme has a molecular weight of 400,000. It is of interest that all Gln synthetases have subunits with mol. wt. 44,000–50,000 and that the mammalian enzymes show a rather similar amino acid composition.[91]

3.2.1b. Reactions. In addition to the important synthetic reaction (No. 1) the enzyme catalyzes several other reactions[92]:

$$\text{Glu} + \text{NH}_3 + \text{ATP} \rightleftharpoons \text{Gln} + \text{ADP} + \text{Pi} \qquad [1]$$

$$\text{L-Gln} + \text{NH}_2\text{OH} \rightarrow \text{L-}\gamma\text{-Glutamylhydroxamate} + \text{NH}_3 \qquad [2]$$

$$\text{L-Gln} + \text{H}_2\text{O} \rightarrow \text{L-Glu} + \text{NH}_3 \qquad [3]$$

$$\text{Glu} + \text{ATP} \rightarrow \text{Pyroglutamate} + \text{ADP} + \text{Pi} \qquad [4]$$

$$\text{L-Met-S-sulfoximine} + \text{ATP} \rightarrow \text{L-Met-S-sulfoximine phosphate} + \text{ADP} \qquad [5]$$

$$\beta\text{-Glutamyl phosphate} + \text{ADP} \rightarrow \text{ATP} + \beta\text{-Glu} \qquad [6]$$

$$\text{Carbamyl phosphate} + \text{ADP} \rightarrow \text{ATP} + \text{CO}_2 + \text{NH}_3 \qquad [7]$$

$$\text{CycloGlu} + \text{ATP} \rightarrow \text{Cycloglutamylphosphate} + \text{ADP} \qquad [8]$$

The mammalian Gln synthetases are irreversibly inhibited by Met sulfoximine. ATP and magnesium are necessary for the binding of Glu to the enzyme, whereby it becomes activated. The activated intermediate is probably enzyme-bound γ-glutamyl phosphate. The enzyme reacts in a sequential manner with metal nucleotide, Glu, and ammonia (reaction 1). Certain anions, particularly bicarbonate and chloride, activate brain and liver synthetases when nonsaturating levels of L-Glu are used. Liver Gln synthetase is, in addition, significantly activated by 2-oxoglutarate, but the brain synthetase is much less affected by this compound. The mammalian enzymes are inhibited by inorganic phosphate, which, at least in part, is ascribed to the reversibility of the reaction. Brain and liver Gln synthetases are inhibited by carbamyl phosphate, whereby the cells are provided with a means for supply of Gln for pyrimidine biosynthesis. Thus, Gln-dependent carbamyl phosphate synthetases specific for pyrimidine synthesis are found in mammalian tissues, but the activity in brain is low.[93] It is of interest that brain Gln synthetase can also catalyze ATP synthesis from ADP and carbamyl phosphate.

3.2.1c. Distribution and Induction. The enzymes Gln synthetase and glutaminase appear to be unevenly distributed in CNS. Berl[94] determined the distribution of Gln synthetase in 16 brain areas of the adult rat. Neocortex showed the highest enzyme activity, about three times greater than corpus callosum and two times greater than the caudate nucleus, hippocampus, thalamus, hypothalamus, cerebellum, pons, and medulla. In developing kitten brains, the neocortex had the strongest increase in enzyme activity and achieved the highest values. In this area, the rate of enzyme development coincided with the pattern of development of compartmentatlization of Glu. Both occur mainly during the third to the sixth postnatal weeks. The activity of Gln synthetase, glutaminase, Glu dehydrogenase, GABA aminotransferase, Asp aminotransferase, and Glu decarboxylase were all found to be higher in gray matter than in white matter of the cat spinal cord.[95] The distribution correlated to that of the Gln concentration.[96]

In order to obtain tissue consisting chiefly of astrocytic glia, the technique of retrograde degeneration was used in the cat CNS.[97] Glutamine synthetase was found only in the glial capillary wall portions of CNS, whereas glutaminase appeared to be distributed equally between the neurons and the glial capillary wall. Glutamine synthetase increased from the periphery to the cerebral cortex in the visual and somatic systems. Glutaminase was highest in the dorsal thalamus in the visual system and in the cuneate nucleus in the somatic sensory system.

Normal glial cells obtained from the brain hemispheres of newborn mice were cultured for 3 weeks by Schousboe *et al.*[98] During this period of time, the activity of the Glu-metabolizing enzymes, Gln synthetase, Glu dehydrogenase, and Asp-2-oxoacid aminotransferase rose from low neonatal values toward the levels in adult brain (i.e., 25.9, 12.3, and 206 nmol $\cdot$ min^{-1} $\cdot$ mg^{-1} cell protein for the three enzymes, respectively). The activities of the Glu-metabolizing enzymes in the astrocytes corresponded with values obtained for homogenates of adult brain.[99-102] The high activity of Gln synthetase was also in agreement with the suggestion of a glial localization of this enzyme.[33,86]

By subcellular fractionation, Gln synthetase has been found to be preferentially localized in microsomes or the soluble fration,[103,104] and most workers agree that Gln synthetase has mainly a glial localization[103-106] and is low in synaptosomes.[2,104] However, experiments using cesium chloride gradients for isolating the synaptosomes from pig brain suggested that Gln synthetase was predominantly localized in the nerve endings.[109] Recently, this was also found by Dennis *et al.*[110] in synaptosomes from rat brain isolated by Ficoll® gradients. These authors suggested that a dissociation of Gln synthetase from membranes might be caused by the hyperosmotic sucrose used for isolation of synaptosomes by the previous authors.

Thus, the very important question of the localization of the enzyme is unsettled, but the glial localization found by immunohistochemical technique appears to be rather convincing.[107,108] In addition, Nicklas *et al.*[111,112] found that a kainic-acid-induced lesion of the striatum caused increased Gln synthetase activity, which correlated well with the relative increase in glial cells and loss of neurons following kainic acid lesioning. This is consistent with a glial localization of Gln synthetase.

Glutamine synthetase is an inducible enzyme. Thus, the enzyme in chick retina,[113-115] in chick cerebral hemispheres,[116] in cat hypothalamic cultures,[117] and in mouse primary astrocyte cultures[118] was induced by hydrocortisone. The induction in astrocyte cultures was prevented by actinomycin D, indicating its dependence on RNA and protein synthesis.

To conclude, brain Gln synthetase has been highly purifed from various animal species, and molecular and kinetic properties have been studied in great detail. The enzyme is unevenly distributed in brain, being highest in neocortex and the gray matter of medulla. Most workers believe that Gln synthetase has predominantly a glial localization. It is an inducible enzyme.

4. FUNCTION

4.1. Stimulus Release of Transmitter Glu and GABA from Gln

The amino acids that are the highest ranking as possible transmitter candidates in the CNS are Glu, GABA, and Gly. In establishing a neurotransmitter role for a compound, it is essential that it can be released and collected from a well-defined synapse as a result of neuronal firing. A commonly used method has been to measure the release of transmitter candidates from tissue slices, synaptosmes, etc. following depolarizing stimulation. Thus, Osborne *et al.*[119] prepared synaptosome beds from rat olfactory lobes and found that endogenous Glu, Asp, and GABA were specifically retained in the synaptosome beds on incubation in Krebs–bicarbonate medium, possibly because of continuous reuptake, and that the same amino acids were selectively released on electrical stimulation. This suggests that they have a neurotransmitter function in the olfactory lobe. Another approach has been to allow tissue slices to take up an exogenous labeled transmitter candidate and to measure the efflux by depolarizing stimulation. The results are, however, often difficult to interpret because the compound is taken up by other tissue elements than the nerve terminals and released on stimulation.

Berl et al.[120] showed in 1961 that Gln is a general precursor for Glu and GABA in brain. In recent years, several release studies have been published that indicate that Gln is also a precursor for transmitter Glu and GABA. Indeed, this may be a major functional role for Gln in CNS. Thus, Reubi et al.[121] reported a potassium-induced, calcium-dependent release of [³H]Glu and [³H]GABA from slices of pigeon optic tectum that were incubated with [³H]Gln. This is of particular interest because there is evidence that Glu and GABA have a transmitter role in pigeon optic tectum.

A similar potassium-induced release of [¹⁴C]Glu and [¹⁴C]GABA derived from [¹⁴C]Gln has been reported[122] in mouse brain cortex slices. Moreover, double-label experiments using [¹⁴C]glucose with [³H]Gln, compared to single-labeled experiments with these substances, revealed that the Glu released by stimulation with veratrine or high potassium from rat brain synaptosomes was derived principally (80%) from Gln.[83] The contribution of Gln relative to glucose to the total Glu, Asp, and GABA was in the range 50–70%. Shank and Aprison also found that [¹⁴C]Gln is more rapidly converted to GABA than [¹⁴C]glucose in the toad brain.[64]

Hamberger et al.[123] used the rabbit dentate gyrus of the hippocampal formation for release studies of labeled Glu because this system appears to be glutamergic. The slices were incubated with ¹⁴C-labeled glucose, pyruvate, and Gln. They found potassium- and calcium-stimulated release of labeled Glu from these precursors, but Gln was the preferred precursor. Following unilateral lesion in the entorhinal cortex, the calcium-dependent release of Glu derived from glucose or Gln was markedly reduced, suggesting that the transmitter pool of Glu is in perforant path terminals and can be synthesized from glucose or Gln. The synthesis from Gln appeared to be regulated both by the Gln uptake and the activity of glutaminase.[124]

[³H]Glutamine was also found to be incorporated into Glu, Asp, and GABA in retina cells from rat, cat, frog, pigeon, and guinea pig. Glutamine appeared particularly to be a source of GABA that might be located in amacrine or ganglion cells.[125] However, the labeling of Glu, Asp, and GABA derived from Gln decreased on light stimulation in the rat retina, whereas no such decrease was found in the labeling of these amino acids derived from glucose. Glutamate and Asp are assumed to be photoreceptor neurotransmitters, and as such, they are expected to be released in the dark.[126]

In summary, it is suggested that Gln is a precursor for transmitter Glu and GABA in brain, and experiments supporting this have been performed using mouse brain cortex, pigeon optic tectum, rabbit dentate gyrus, rat brain synaptosomes, and retina from various mammals. Direct proof for this function of Gln appears, however, to be lacking.

4.2. Glutamine as Glial–Neuronal Glu Transporter

As discussed above, compartmentalization is a key word in understanding the interrelation among Gln, Glu, and GABA. (see ref. 151). The concept of the Gln cycle is thus based on cellular compartmentalization of Gln and Glu caused by an assumed predominant glial localization of Gln synthetase and synaptosomal localization for glutaminase.

There are, however, certain obstacles to this concept. Although glutaminase may be higher in nerve endings than in glial cells,[84,104] astrocytes in primary cultures contain an active glutaminase capable of degrading the same amount of Gln as homogenate of neonatal and adult mouse brain.[50] Furthermore, no high-affinity uptake of Gln has been demonstrated with certainty in neurons, even if it is possible that Gln at the actual extracellular concentrations preferentially enters the neuronal cellular elements.[46,51] However, it should be remembered that the synaptosomes do not appear to be devoid of Gln synthetase and that very little Glu or GABA is needed for the transmitter function. It is thus still possible that the Gln serving as precursor for the transmitters Glu and GABA is formed within the nerve endings itself.

4.3. Detoxification of Ammonia

In addition to the function of transporting Glu, e.g., through membranes, Gln also serves as a carrier for ammonia. The toxic effect of ammonia in brain is not fully elucidated. In hepatic coma lasting longer than 24 h, oxygen consumption and glucose utilization are reduced in brain by 50%, and brain Gln and 2-oxoglutaramate are increased. Glutamine and 2-oxoglutaramte itself appear, however, to be nontoxic.[127] The toxic effect of ammonia is probably not caused by depletion of 2-oxoglutarate or ATP, as might be expected.[128] An increased activity of Glu dehydrogenase,[129-131] particularly in astrocytes,[131] has been reported in hepatic coma, but others have found normal activities for this enzyme and also for glutaminase.[132] Ammonia appears to exert a direct postsynaptic effect on cerebellar slices,[133] since spontaneous action potentials are abolished by ammonia after a short period of spike activity. In addition, a presynaptic effect has been found. Thus, the evoked release of Glu following depolarization with potassium was greatly reduced by ammonia.[134]

Since the CNS lacks a complete urea cycle, a means of binding toxic concentrations of ammonia is essential. Glutamate dehydrogenase and Gln synthetase perform this function, and ammonia nitrogen is recycled into the valuable amino acids Glu and Gln and further into proteins, etc. Evidently, brain cannot waste nitrogen by forming urea.

4.4. Other Functions

As previously stated, Gln is a precursor of the Asp needed in both purine and pyrimidine synthesis. It has been suggested that Gln also is the precursor for transmitter Asp,[84] but Asn appears to be a more likely candidate.[135] Glucosamine-6-phosphate, which is a key intermediate for the synthesis of glycolipids, glycoproteins, etc., is produced from L-Gln and fructose-6-phosphate by an amidotransferase. Moreover, it is of interest that amino sugars formed from Gln appear to be essential for intercellular adhesion of mouse teratoma cells.[136]

Glutamine may also exert an unknown effect on protein synthesis. Thus, Roux *et al.*[137,138] showed that radioactive Gln, Asp, and Asn were incorporated into calf brain tRNA in the presence of homologous aminoacyl tRNA synthetases. When the glutaminyl and asparaginyl tRNAs were deaminoacylated, the original amino acids were recovered. In addition, two new compounds were identified as the cyclic compounds α-aminoglutarimide and α-aminosuccinimide, respectively. Thus, considerable fractions of glutaminyl and asparaginyl tRNAs were deesterified in forming these cyclic compounds. It is possible that deesterification of glutamyl and asparaginyl tRNA has a role in the regulation of protein synthesis by controlling the concentration of these compounds. The newly identified imides may also serve as intermediates in other metabolic reactions. The authors point out that several derivatives of the imides possess neuropharmacological properties; e.g., 2-ethyl-2-phenylglutarimide (glutethimide) is a hypnotic sedative, and cycloheximide and streptovitacin A, both derivatives of glutarimide, are known to inhibit protein synthesis in eukaryotes. Moreover, 2-ethyl-2-methylsuccinimide (ethosuximide) is an antiepileptic drug, and N-2-dimethyl-2-phenylsuccinimide (methsuximide) is an anticonvulsant.

4.5. Glutamine in Pathological Conditions

The possible inhibition of Gln uptake in certain inborn errors of metabolism has been discussed in Section 2.2, and the role of Gln in detoxification of ammonia in Section 4.3. Phenylacetic acid has recently been used successfully in the treatment of hyperammonemia.[139,140] Thus, phenylacetic acid conjugates with Gln to form phenylacetyl-Gln, which is excreted practically quantitatively in urine. In this way, a 45% increase in urinary nitrogen excretion has been described. Glutamine itself appears to be largely nontoxic, but changes in its level in brain may alter the levels of physiologically potent compounds derived from Gln and thereby produce clinical symptoms. This has been observed, e.g., in convulsions. Convulsions produced in rats by pentylenetetrazole caused increased ammonia level and glutaminase activity and decreased Gln level and Gln synthetase activity.[141,142] These convulsions have been attributed to an imbalance between Glu and GABA.[143]

In vitro studies with guinea pig brain slices indicated that lithium, which is used in the treatment of manic–depressive disorders, affects the metabolism of the Gln–Glu system.[144] Thus, leakage of Glu, Gln, Asp, and GABA into the media increased markedly, whereas the labeling of Gln from [1-^{14}C]acetate decreased in the presence of Li$^+$. It was suggested that depletion of transmitter Glu is associated with the therapeutic effects of the lithium treatment. However, the concentrations of Li$^+$ used (40–100 mM) were very much higher than the therapeutically effective serum concentrations.

It is also indicated that impaired metabolism of Gln may be involved in Huntington's disease. Glutamine levels are thus reduced in postmortem frontal cortical tissue samples from patients with this disease,[145] and Gln protects against the toxic effects of Glu on Huntington's fibroblasts *in vitro*.[146] Moreover, the neurotoxin kainic acid produces a lesion similar to that seen in Huntington's disease when injected into the striatum of animals,[147] and the actions of kainic acid are related to Gln. The Gln levels are reduced by kainic acid in rat cerebellar slices.[148] Glutamine also has a protective action against the effects of intrastriatal kainic acid administration in rats.[149] Carter[150] has reported a postmortem study of Gln synthetase in brain from patients with Huntington's disease. Glutamine synthetase activity was reduced in choreic tissue to 79% of the mean control level in the frontal cortex (Brodmann area 11), to 70% in area 38 of temporal cortex, to 74% in the putamen, and to 87% in the cerebellum. This reduction could not be ascribed to neuronal or glial cell loss or to differences in treatment before or after death and appeared to be related to the disease.

5. CONCLUDING REMARKS

Glutamine is an amino acid with multiple functions. Metabolically, it has a key role in many organs for energy production and synthesis of proteins, purines, pyrimidines, etc. In brain, it is one of the dominant amino acids, being a general precursor for Glu and GABA. It is also indicated that it is precursor for transmitter Glu and GABA. Glutamine generally enters the cells of the CNS by a low-affinity sodium-independent uptake. It penetrates most membranes easily, although the transport may be carrier mediated, but Gln is assumed to be compartmentalized on the cellular level. The most important enzymes regulating the CNS level of Gln are Gln synthetase, phosphate-activated glutaminase, and Gln aminotransferase. Most workers assume that Gln synthetase is predominantly localized in glial cells and glutaminase in neurons, forming the basis for the concept of the Gln cycle, but this is a matter of debate. Brain lacks a complete urea cycle, and ammonia is detoxified by Gln formation. In this way, nitrogen is also saved for synthesis of important compounds such as proteins. Although Gln itself is nontoxic, Gln metabolism may be altered by convulsions and other pathological conditions and by certain inborn errors of metabolism, e.g., Huntington's disease.

REFERENCES

1. Timiras, P. S., Hudson, D. B., and Oklund, S., 1973, *Prog. Brain Res.* **40**:267–275.
2. Bradford, H. F., and Thomas, A. J., 1969, *J. Neurochem.* **16**:1495–1504.
3. Gründig, E., and Hanbauer, I., 1970, *J. Neurochem.* **17**:215–220.
4. Kandera, J., Levi, G., and Lajtha, A., 1968, *Arch. Biochem. Biophys.* **126**:249–260.
5. Sytinsky, I. A., Guzikov, B. M., Gomanko, M. V., Eremin, V. P., and Konovalova, N. N., 1975, *J. Neurochem.* **25**:43–48.
6. Mangan, J. L., and Whittaker, V. P., 1966, *Biochem. J.* **98**:128–137.
7. Robinson, N., and Williams, C. B., 1965, *Clin. Chim. Acta* **12**:311–317.
8. Perry, T. L., Berry, K., Hansen, S., Diamond, S., and Mok, C., 1971, *J. Neurochem.* **18**:513–519.
9. Wiechert, P., and Göllnitz, G., 1970, *J. Neurochem.* **17**:137–147.
10. Kontro, P., Marnela, K.-M., and Oja, S. S., 1980, *Brain Res.* **184**:129–141.
11. Duggan, A. W., Johnston, G. A. R., 1970, *J. Neurochem.* **17**:1205–1208.
12. Johnson, J. L., and Aprison, M. H., 1970, *Brain Res.* **24**:285–292.

13. Johnson, J. L., 1974, *Brain Res.* **69**:366–369.
14. Kvamme, E., and Lenda, K., 1981, *Neurosci. Lett.* **25**:193–198.
15. Geddes, J. W., Newstead, J. D., and Wood, J. D., 1980, *Neurochem. Res.* **5**:1107–1116.
16. De Belleroche, J. S., and Bradford, H. F., 1973, *J. Neurochem.* **21**:441–451.
17. Lähdesmäki, P., Karppinen, A., Saarni, H., and Winter, R., 1977, *Brain Res.* **138**:295–308.
18. Van Gelder, N. M, 1978, *Can. J. Physiol. Pharmacol.* **56**:362–374.
19. Agrawal, H. C., Davis, J. M., and Himwich, W. A., 1966, *J. Neurochem.* **13**:607–615.
20. Agrawal, H. C., Davis, J. M., and Himwich, W. A., 1967, *Brain Res.* **3**:374–380.
21. Agrawal, H. C., Davis, J. M., and Himwich, W. A., 1968, *J. Neurochem.* **15**:917–923.
22. Berl, S., and Purpura, D. P., 1966, *J. Neurochem.* **13**:293–304.
23. Dravid, A. R., Himwich, W. A., and Davis, J. M., 1965, *J. Neurochem.* **12**:901–906.
24. Folch-Pi, J., 1955, *Biochemistry of the Nervous System* (H. Waelsch, ed.), Academic Press, New York, pp. 121–136.
25. Kirzinger, S. S., and Fonda, M. L., 1978, *Exp. Gerontol.* **13**:255–261.
26. Weyne, J., Van Leuven, F., and Leusen, I., 1973, *Life Sci.* **12**:211–218.
27. Prasad, C., and Agarwal, K. N., 1980, *J. Neurochem.* **34**:1270–1273.
28. Hamel, E., Butterworth, R. F., and Barbeau, A., 1979, *J. Neurochem.* **33**:575–577.
29. Roach, M. K., and Reese, W. N., Jr., 1972, *Biochem. Pharmacol.* **21**:2013–2019.
30. Butterworth, R. F., Landreville, F., Hamel, E., Merkel, A., Giguere, F., and Barbeau, A., 1980, *J. Can. Sci. Neurol.* **7**:447–450.
31. Waelsch, H., 1960, *Structure and Function of the Cerebral Cortex* (D. B. Tower, and J. P. Schadé, eds.), Elsevier, Amsterdam, pp. 313–326.
32. Balázs, R., Machiyama, Y., Hammond, B. J., Julian, T., and Richter, D., 1970, *Biochem. J.* **116**:445–467.
33. Benjamin, A. M., and Quastel, J. H., 1972, *Biochem. J.* **128**:631–646.
34. Van den Berg, C. F., Matheson, D. F., Ronda, G., Reijnierse, G. L. A., Blokhuis, G. G. D., Kroon, M. C., Clarke, D. D., and Garfinkel, D., 1975, *Metabolic Compartmentation and Neurotransmission* (S. Berl, D. D. Clarke, and D. Schneider, eds.), Plenum Press, New York, pp. 515–543.
35. Schwerin, P., Bessman, S. P., and Waelsch, H., 1950, *J. Biol. Chem.* **184**:37–44.
36. Dobkin, J., 1972, *J. Neurochem.* **19**:1195–1202.
37. O'Neal, R. M., and Koeppe, R. E., 1966, *J. Neurochem.* **13**:835–847.
38. Berl, S., Takagaki, G., Clarke, D. D., and Waelsch, H., 1962, *J. Biol. Chem.* **237**:2562–2569.
39. Lajtha, A., and Toth, J., 1962, *J. Neurochem.* **9**:199–212.
40. Levi, G., Blasberg, R., and Lajtha, A., 1966, *Arch. Biochem. Biophys.* **114**:339–351.
41. Abdul-Ghani, A.-S., Marton, M., and Dobkin, J., 1978, *J. Neurochem.* **31**:541–546.
42. Kilberg, M. S., Handlogten, M. E., and Christensen, H. N., 1980, *J. Biol. Chem.* **255**:4011–4019.
43. Simpson, D. P., 1980, *J. Biol. Chem.* **255**:7123–7128.
44. Alleyne, G. A. O., McFarlane Anderson, N., and Scott, B., 1980, *Int. J. Biochem.* **12**:99–102.
45. Cohen, S. R., and Lajtha, A., 1972, *Handbook of Neurochemistry*, Volume 7 (A. Lajtha, ed.), Plenum Press, New York, pp. 543–572.
46. Benjamin, A. M., Verjee, Z. H., and Quastel, J. H., 1980, *J. Neurochem.* **35**:67–77.
47. Walum, E., and Weiler, C., 1978, *Proc. Eur. Soc. Neurochem.* **1**:499.
48. Weiler, C. T., Nyström, B., and Hamberger, A., 1979, *J. Neurochem.* **32**:559–565.
49. Balcar, V. J., and Hauser, K. L., 1978, *Proc. Eur. Soc. Neurochem.* **1**:498.
50. Schousboe, A., Hertz, L., Svenneby, G., and Kvamme, E., 1979, *J. Neurochem.* **32**:943–950.
51. Baldessarini, R. J., and Yorke, C., 1974, *J. Neurochem.* **23**:839–848.
52. Balcar, V. J., and Johnston, G. A. R., 1975, *J. Neurochem.* **24**:875–879.
53. Roberts, P. J., and Keen, P., 1974, *Brain Res.* **76**:352–357.
54. Roberts, P. J., 1976, *Adv. Exp. Med. Biol.* **69**:165–178.
55. Hertz, L., 1979, *Prog. Neurobiol.* **13**:277–323.
56. Gjessing, L. R., Gjesdahl, P., and Sjaastad, O., 1972, *J. Neurochem.* **19**:1807–1808.
57. Johnson, J. L., 1978, *Prog. Neurobiol.* **10**:155–202.

58. Welbourne, T. C., 1980, *Eur. J. Appl. Physiol.* **45**:185–188.
59. Bito, L., Davison, H., Levin, E., Murray, M., and Snider, N., 1966, *J. Neurochem.* **13**:1057–1067.
60. Crowshaw, K., Jessup, S. F., and Ramwell, P. W., 1967, *Biochem. J.* **103**:79–85.
61. Ehinger, B., 1977, *Exp. Eye Res.* **25**:221–234.
62. Benjamin, A. M., Verjee, Z. H., and Quastel, J. H., 1980, *J. Neurochem.* **35**:78–87.
63. Machiyama, Y., Balázs, R., Hammond, B. J., Julian, T., and Richter, D., 1970, *Biochem. J.* **116**:469–481.
64. Shank, R. P., and Aprison, M. H., 1977, *J. Neurochem.* **28**:1189–1196.
65. Clark, R. M., and Collins, G. G. S., 1976, *J. Physiol. (Lond.)* **262**:383–400.
66. Benjamin, A. M., and Quastel, J. H., 1977, *Can. J. Physiol. Pharmacol.* **55**:347–355.
67. Arnfred, T., and Hertz, L., 1971, *J. Neurochem.* **18**:259–265.
68. Tapia, R., and Gonzales, R. M., 1978, *Neurosci. Lett.* **10**:165–169.
69. De Belleroche, J. S., and Bradford, H. F., 1972, *J. Neurochem.* **19**:585–602.
70. Walum, E., 1979, *Biochem. Biophys. Res. Commun.* **88**:1271–1274.
71. Pitts, R. F., Pilkington, L. A., MacLeod, M. B., and Leal-Pinto, E., 1972, *J. Clin. Invest.* **51**:557–565.
72. Marliss, E. B., Aoki, T. T., Pozefsky, T., Most, A. S., and Cahill, G. F., Jr., 1971, *J. Clin. Invest.* **50**:814–817.
73. Windmueller, H. G., and Spaeth, A. E., 1980, *J. Biol. Chem.* **255**:107–112.
74. Rapoport, S., Rost, J., and Schultze, M., 1971, *Eur. J. Biochem.* **23**:166–170.
75. Donnelly, M., and Scheffler, I. E., 1976, *J. Cell. Physiol.* **89**:39–51.
76. Lavietes, B. B., Regan, D. H., and Demopoulos, H. B., 1974, *Proc. Natl. Acad. Sci. USA.* **71**:3993–3997.
77. Reitzer, L. J., Wice, B. M., and Kennell, D., 1979, *J. Biol. Chem.* **254**:2669–2676.
78. Lund, P., 1980, *FEBS Lett.* **117**:K86–K92.
79. Lai, J. C. K., and Clark, J. B., 1976, *Biochem. J.* **154**:423–432.
80. Dennis, S. C., Lai, J. C. K., and Clark, J. B., 1977, *Biochem. J.* **164**:727–736.
81. Dennis, S. C., Land, J. M., and Clark, J. B., 1976, *Biochem. J.* **156**:323–331.
82. Brand, M. D., and Chappell, J. B., 1974, *Biochem. J.* **140**:205–210.
83. Bradford, H. F., Ward, H. K., and Thomas, A. J., 1978, *J. Neurochem.* **30**:1453–1459.
84. Bradford, H. F., and Ward, H. K., 1976, *Brain Res.* **110**:115–125.
85. Gonda, O., and Quastel, J. H., 1962, *Biochem. J.* **84**:394–406.
86. Benjamin, A. M., and Quastel, J. H., 1974, *J. Neurochem.* **23**:457–464.
87. Krebs, H. A., 1935, *Biochem. J.* **29**:1951–1969.
88. Reichelt, K. L., and Kvamme, E., 1967, *J. Neurochem.* **14**:987–996.
89. Ronzio, R. A., Rowe, W. B., Wilk, S., and Meister, A., 1969, *Biochemistry* **8**:2670–2674.
90. Haschemeyer, R. H., 1970, *Adv. Enzymol.* **33**:71–118.
91. Meister, A., 1974, *The Enzymes*, (Boyer, P. D., ed.) Volume 10, 3rd ed., Academic Press, New York, pp. 699–754.
92. Tate, S. S., and Meister, A., 1973, *The Enzymes of Glutamine Metabolism* (S. Prusiner and E. R. Stadtman, eds.), Academic Press, New York, London, pp. 77–127.
93. Tatibana, M., and Ito, K., 1969, *J. Biol. Chem.* **244**:5403–5413.
94. Berl, S., 1966, *Biochemistry* **5**:916–922.
95. Graham, L. T., Jr., and Aprison, M. H., 1969, *J. Neurochem.* **16**:559–566.
96. Graham, L. T., Jr., Shank, R. P., Werman, R., and Aprison, M. H., 1967, *J. Neurochem.* **14**:465–472.
97. Utley, J. D., 1964, *Biochem. Pharmacol.* **13**:1383–1392.
98. Schousboe, A., Svenneby, G., and Hertz, L., 1977, *J. Neurochem.* **29**:999–1005.
99. Awapara, J., and Seale, B., 1952, *J. Biol. Chem.* **194**:497–502.
100. Strominger, J. L., and Lowry, O. H., 1955, *J. Biol. Chem.* **213**:635–646.
101. Wu, C., 1963, *Comp. Biochem. Physiol.* **8**:335–351.
102. Van Gelder, N. M., 1974, *Can. J. Physiol. Pharmacol.* **52**:952–959.
103. Waelsch, H., 1959, *Proceedings of the Fourth International Congress of Biochemistry*, Volume 3, Pergamon Press, London, pp. 36–45.
104. Salganicoff, L., and De Robertis, E., 1965, *J. Neurochem.* **12**:287–309.

105. Benjamin, A. M., and Quastel, J. H., 1975, *J. Neurochem.* **25**:197–206.
106. Martinez-Hernandez, A., Bell, K. P., and Norenberg, M. D., 1977, *Science* **195**:1356–1358.
107. Norenberg, M. D., and Martinez-Hernandez, A., 1979, *Brain Res.* **161**:303–310.
108. Norenberg, M., 1979, *J. Histochem. Cytochem.* **27**:756–762.
109. Kornguth, S. E., Flangas, A. L., Siegel, F. L., Geison, R. L., O'Brien, J. F., Lamar, C., Jr., and Scott, G., 1971, *J. Biol. Chem.* **246**:1177–1184.
110. Dennis, S. C., Lai, J. C. K., and Clark, J. B., 1980, *Brain Res.* **197**:469–475.
111. Nicklas, W. J., Duvoisin, R. C., and Berl, S., 1979, *Brain Res.* **167**:107–117.
112. Nicklas, W. J., Nunez, R., Berl, S., and Duvoisin, R., 1979, *J. Neurochem.* **33**:839–844.
113. Moscona, A. A., and Piddington, R., 1966, *Biochim. Biophys. Acta* **121**:409–411.
114. Reif-Lehrer, L., and Amos, H., 1968, *Biochem. J.* **106**:425–430.
115. Kovacs, S. H., 1977, *In Vitro* **13**:24–30.
116. Piddington, R., 1971, *J. Exp. Zool.* **177**:219–228.
117. Vaccaro, D. E., Leeman, S. E., and Reif-Lehrer, L., 1979, *J. Neurochem.* **33**:953–957.
118. Juurlink, B. H. J., Schousboe, A., Jørgensen, O. S., and Hertz, L., 1981, *J. Neurochem.* **36**:136–142.
119. Osborne, R. H., Duce, I. R., and Keen, P., 1976, *J. Neurochem.* **27**:1483–1488.
120. Berl, S., Lajtha, A., and Waelsch, H., 1961, *J. Neurochem.* **7**:186–197.
121. Reubi, J.-C., Van den Berg, C., and Cúenod, M., 1978, *Neurosci. Lett.* **10**:171–174.
122. Tapia, R., and González, R. M., 1978, *Neurosci. Lett.* **10**:165–169.
123. Hamberger, A. C., Chiang, G. H., Nylén, E. S., Scheff, S. W., and Cotman, C. W., 1979, *Brain Res.* **168**:513–530.
124. Hamberger, A., Chiang, G. H., Sandoval, E., and Cotman, C. W., 1979, *Brain Res.* **168**:531–541.
125. Voaden, M. J., Lake, N., Marshall, J., and Morjaria, B., 1978, *J. Neurochem.* **31**:1069–1076.
126. Voaden, M. J., and Morjaria, B., 1980, *J. Neurochem.* **35**:95–99.
127. Zieve, L., and Nicoloff, D. M., 1975, *Annu. Rev. Med.* **26**:143–157.
128. Hindfelt, B., and Siesjø, B. K., 1970, *Life Sci.* **9**:1021–1028.
129. Colombo, J. P., Bachmann, C., Peheim, E., and Berüter, J., 1977, *Enzyme* **22**:399–406.
130. Sadasivudu, B., Indira Rao, T., and Radhnakrishna Murthy, C., 1977, *Neurochem. Res.* **2**:639–655.
131. Norenberg, M. D., 1976, *Arch. Neurol.* **33**:265–269.
132. O'Neill, B. B., and O'Donovan, D. J., 1979, *Biochem. Soc. Trans.* **7**:35–36.
133. Benjamin, A. M., Okamoto, K., and Quastel, J. H., 1978, *J. Neurochem.* **30**:131–143.
134. Hamberger, A., Hedquist, B., and Nyström, B., 1979, *J. Neurochem.* **33**:1295–1302.
135. Reubi, J. C., Toggenburger, G., and Cuénod, M., 1980, *J. Neurochem.* **35**:1015–1017.
136. Oppenheimer, S. B., Edidin, M., Orr, C. W., and Roseman, S., 1969, *Proc. Natl. Acad. Sci. USA.* **63**:1395–1402.
137. Roux, H., and Murthy, M. R. V., 1975, *J. Neurochem.* **24**:1163–1172.
138. Murthy, M. R. V., Thénot, J. P., and Roux, H., 1975, *J. Neurochem.* **24**:1173–1180.
139. Brusilow, S., Valle, D. L., Batshaw, M. L., 1979, *Lancet* **2**:452–454.
140. Brusilow, S., Tinker, J., and Batshaw, M. L., 1980, *Science* **207**:659–661.
141. Dobkin, J., 1970, *J. Neurochem.* **17**:237–246.
142. Wiechert, P., and Göllnitz, G., 1969, *J. Neurochem.* **16**:1007–1016.
143. Bhaskar Rao, A., Joseph, P. K., Ramakrishna Rao, P., Rajan, R., and Ramakrishnan, S., 1978, *Ind. J. Biochem. Biophys.* **15**:308–310.
144. Berl, S., and Clarke, D. D., 1972, *Brain Res.* **36**:203–213.
145. Kremzner, L. T., Berl, S., Stellar, S., and Coke, L. J., 1979, *Adv. Neurol.* **23**:537–547.
146. Gray, P. N., May, P. C., Mundy, L., and Elkins, J., 1980, *Biochem. Biophys. Res. Commun.* **95**:707–714.
147. McGeer, E. G., and McGeer, P. L., 1976, *Nature* **263**:517–519.
148. Nicklas, W. J., Krespan, B., and Berl, S., 1980, *Eur. J. Pharmacol.* **62**:209–213.
149. McGeer, E. G., and McGeer, P. L., 1978, *Neurochem. Res.* **3**:501–517.
150. Carter, C. J., 1981, *Lancet* **24**:782–783.
151. Shank, R. P., and Aprison, M. H., 1981, *Life Sci.* **28**:837–842.

16

γ-Aminobutyric Acid
Metabolism and Biochemistry of Synaptic Transmission

Ricardo Tapia

1. INTRODUCTION

The interest in γ-aminobutyric acid (GABA) has been increasing steadily in the last few years. After the initial GABA symposium in 1959,[1] hundreds of publications on its role in the CNS have appeared, and several other symposia devoted entirely to GABA[2-5] or to more general subjects but including GABA to a great extent[6-11] have been held. In addition, many general reviews on the biochemistry, pharmacology, and physiology of GABA have been published, including an extensive chapter in the first edition of this *Handbook*.[12-23] During these years, evidence has accumulated to the extent that GABA is accepted as the most widely distributed inhibitory neurotransmitter in the mammalian central nervous system as well as in the nervous system of some invertebrates.

The present chapter focuses on the biochemical mechanisms related to the neurotransmitter role of GABA in central synapses. GABA receptors are not considered, since there is a special chapter on this topic in Volume 6 of this *Handbook*. Thus, mainly presynaptic mechanisms are reviewed. The properties of the enzymes directly involved in GABA metabolism are first discussed, as well as the turnover of GABA, the compartmentation of GABA metabolism, and the possible participation of GABA in cerebral protein synthesis. Subsequently, data on the uptake and release of GABA are covered. This metabolic and transport information is then integrated into a model of the GABAergic terminal, together with the effects of drugs *in vivo* on the function of GABA synapses, in Section 4. The concluding section refers briefly to the importance of the localization of GABAergic neuronal pathways in the CNS and to the possible involvement of GABA in the control of the hypothalamic–pituitary function.

Ricardo Tapia • Department of Neurosciences, Center for Investigations in Cellular Physiology, Universidad Nacional Autónoma de México, 04510 Mexico, D. F., Mexico.

Whenever possible, references to the most recent published reviews covering each specific topic are made at the beginning of each section. Specific references to articles that appeared after the most recent reviews are given, and the attempt is made to integrate their results in the best possible way. Mention of publications appearing prior to the reviews is made only when it was considered necessary for this integration or to emphasize specific points.

2. METABOLISM

The metabolism of GABA depends basically on two enzymes, glutamate decarboxylase (GAD), which decarboxylates the 1-carboxyl of L-glutamic acid to yield GABA, and GABA transaminase (GABA-T), which transfers the amino group of GABA to α-ketoglutarate to yield glutamate and succinic semialdehyde. This section first reviews the biochemical properties of these two enzymes. The problem of GABA turnover and the precursor pools of GABA (compartmentation) are subsequently discussed.

2.1. Glutamate Decarboxylase

Although some alternate pathways of GABA synthesis have been proposed,[12,24] there is general agreement that GABA is synthesized mainly, if not only, by glutamate decarboxylase (GAD, L-glutamate-1-carboxylyase, E.C. 4.1.1.15). The possibility of the existence of a second type of GAD in neurons (now called nonneuronal GAD[25,26]) with different kinetic properties is now considered very low.[25–30] The only alternate pathway that has received some further support in recent years is that from putrescine, formed via ornithine decarboxylation, a pathway suggested by data obtained in neuroblastoma cells,[31] chick embryo retina[32] and brain,[33] and adult rat brain.[34] However, the role of putrescine and other polyamines in brain is still far from clear and might be unrelated to the synthesis of GABA.[34–37]

The importance of GAD as the rate-limiting step in the synthesis of GABA was evident since the initial studies on the relationship between GAD inhibition *in vivo* and the appearance of seizures (see refs. 1,12–16,38–40 for reviews). During the subsequent years, this enzyme has attracted increasing interest because it appears to be closely linked to the function of GABA as an inhibitory neurotransmitter and therefore is a marker of GABAergic synapses much more reliable than GABA-T, particularly since the latter enzyme does not seem to participate directly in the synaptic GABA mechanisms.

Glutamate decarboxylase has been purified to homogeneity from mouse brain, and several physicochemical, kinetic, and immunochemical properties of the pure enzyme have been studied.[41,42] Glutamate decarboxylase from bovine and catfish brain has also been purified to yield a single electrophoretic band.[43,44] Several other attempts to purify GAD from *Drosophila*,[45] bovine cerebellum,[46] human brain,[47] rat brain,[48,49] and mouse brain[50,51] have been published. Conventional biochemical techniques were employed for these purifications, including the use of affinity chromatography columns with α-

methylglutamate[49] or with some 5'-phosphopyridoxyl derivatives, mainly the N-(5'-phosphopyridoxyl)-2,4-diaminobutyrate,[50,51] as ligands. Antibodies have been raised against some of these GAD preparations,[43,44,46,48,51] and they have been used to localize immunohistochemically central GABAergic synapses in several regions of the brain (see Section 5). By applying immunocytochemical techniques, it has been possible to locate GAD-containing terminals, but GAD in neuronal soma or dendrites has been detected in only a few cases[51-53] (see reviews of the development of GAD immunocytochemistry in refs. 17 and 54). However, after treatment with the axonal transport blocker colchicine, GAD could be visualized in somata and dendrites of the cerebellum, Ammon's horn,[55] corpus striatum,[56] nucleus reticularis thalamis,[57] and retina.[58] Since this effect of colchicine does not seem to be caused by an unspecific or damaging effect, these observations suggest that GAD is synthesized in the neuronal soma but is so rapidly transported along the axon that the enzyme concentration in the soma is too low to be detected by the usual immunocytochemical procedures.[55]

The molecular characterization of the purified GAD and many of its kinetic properties, including the effect of a great variety of inhibitors, as well as its immunochemical properties, have been comprehensively reviewed by Wu[41] and Saito,[42] and there is no need to repeat them here. Rather, some other kinetic and molecular properties, reported after the date of those reviews, are considered. These properties, although determined in less pure preparations of GAD, may have important implications for the regulation of GAD activity and for the physiological role of this enzyme as the rate-limiting step in the synthesis of a neurotransmitter pool of GABA (see also Section 4).

2.1.1. Mechanism, Regulation, and Inhibition Studies of GAD Activity

Although some of the endogenous compounds that behave as inhibitors of GAD activity studied by Roberts and colleagues,[41,59] for example, Zn^{2+} or α-ketoglutarate, might participate in the regulation of GAD activity, there is much evidence that the coenzyme of GAD, pyridoxal 5'-phosphate (PLP), plays a fundamental role in this regulation (see refs. 14,16,24,39,40,60 for reviews). Several reports indicate that the dependence of GAD activity on PLP is a phenomenon that occurs not only in vitro but also in vivo, as expected for a physiological regulator. In fact, when the concentration of PLP is decreased in brain pharmacologically through the inhibition of pyridoxal kinase or as a consequence of a dietary deficiency of pyridoxine, GAD activity is decreased, and this decrease is reversed both by the administration of pyridoxine in vivo and by the addition of PLP to the incubation mixtures.[61-69] A similar phenomenon has been described when GAD was inhibited in vivo by PLP analogues capable of antagonizing the coenzymatic function of PLP.[70-72] Furthermore, it has been shown that this dependence of GAD activity on PLP concentration in vivo occurs in nerve endings.[73] Finally, as has been stressed by Tower,[40,74] a human counterpart of the convulsions observed experimentally as a consequence of GAD inhibition secondary to PLP deficiency (see Section 4.1) was observed in humans in 1954, when many children fed with a B_6-deficient liquid formula had seizures that rapidly disappeared after the intravenous administration of pyridoxine.[75,76] In the following 15 years, several cases of pyridoxine-dependent children who also showed convulsions unless their diet was supplemented with pyridoxine were reported.[77-81]

From a series of kinetic studies of GAD in mouse brain carried out in our laboratory, both in the absence and in the presence of specially designed inhibitors,[82-85] we have reached the following conclusions which are very relevant for the regulation of GAD activity: (1) two forms of GAD activity exist in brain, one that is dependent on free PLP and the other independent of the free coenzyme; (2) the K_ms for glutamate (K_s) of both GAD activities are identical; (3) the free PLP-independent GAD possesses tightly bound coenzyme not displaced by PLP analogues or by dialysis[82-84]; (4) the ratio free PLP-dependent/free PLP-independent GAD is about 2.8, calculated from three different types of kinetic analysis,[84] which means that from the total GAD activity about 25% corresponds to the free PLP-independent activity (i.e., it is completely saturated with the coenzyme), and 75% to the free PLP-dependent activity; (5) the PLP–glutamate Schiff base cannot be a substrate for either form of the enzyme[84]; (6) the two types of GAD activity are independent and cannot be interconverted by changes in the availability of substrate or coenzyme, although the kinetic analysis does not give information on the physical relationship between the two activities—the two corresponding catalytic sites might be in the same molecule, in different subunits, or in different molecules; (7) the free PLP-dependent GAD activity seems to follow a random bireactant mechanism with regard to the coenzyme and the substrate[85]; (8) as a consequence of the above kinetic properties, only the free PLP-dependent activity would be susceptible to regulation by the available PLP, and all endogenous compounds capable of forming Schiff bases with PLP, including GABA, would be potential regulators of this activity provided that they exist in concentrations similar to or higher than the dissociation constant of their corresponding PLP–Schiff base.[84,85]

The postulated mechanistic model accounting for all of the kinetic observations, which was reproduced by a computer using the experimentally obtained kinetic constants,[84,85] is the following:

$$\text{Ea} \underset{K_{s1}=K_{s2}}{\rightleftharpoons} \text{EaS} \xrightarrow{K_p} \text{Ea} + \text{P}$$

$$\text{S} + \text{A} \xrightarrow{K_0} \text{SA}$$

$$\text{E} \xrightarrow{K_{s2}=K_{s1}} \text{ES}$$

$$k_a \updownarrow \qquad \updownarrow k_a$$

$$\text{AE} \xrightarrow{K_{s2}=K_{s1}} \text{AES} \xrightarrow{K_p'} \text{E} + \text{P}$$

where Ea is the free PLP-independent catalytic site of GAD ("a" represents the tightly bound PLP); E is the free PLP-dependent catalytic site; S is glutamate; A is PLP, P represents reaction products, and SA the glutamate–PLP Schiff base. The corresponding equilibrium constants (K) are indicated by the appropriate subscripts, and K_p and K_p' are the rate constants for product formation for the two types of enzyme activity.

Another series of articles on the regulation of GAD activity by PLP and other compounds has been published by the group of Miller and Martin.[86–93] They have concluded that *in vivo* the saturation of GAD by PLP may not be greater than 35% and that both glutamate and ATP dissociate the coenzyme from the apoenzyme, whereas Pi seems capable of inhibiting this dissociation and consequently activating the enzyme.[88,89,91] Since these effects are observed at concentrations lower than the endogenous brain concentrations, these authors have suggested that the balance of the levels of the nucleotide, Pi, and glutamate might participate in the regulation of GABA synthesis *in vivo* mainly through the regulation of the degree of binding of the coenzyme to the apoenzyme. Similar results were recently obtained with a partially purified GAD from hog brain.[92] In this report, Martin *et al.* suggest that the two types of PLP-binding sites described above[84] might be present in only one molecular species, since they found no evidence of subunit dissociation in their studies of activation by PLP and Pi and inhibition by ATP and glutamate. However, it has been reported that the inhibitory effect of ATP on GAD activity is slightly greater when tested in a low-molecular-weight GAD than in a high-molecular-weight GAD, as separated in Sephadex G-200, whereas no difference was observed in the inhibitory effect of chloride ions.[94] In contrast to the above results with ATP, activation of GAD by this nucleotide in the absence of PLP was observed in a hypotonically disrupted crude synaptosomal fraction.[95]

The inactivation of GAD by glutamate can be explained by the impossibility of the glutamate–Schiff base to behave as a substrate. In fact, as already mentioned, any endogenous compound capable of forming a Schiff base with PLP and present in the nerve endings at concentrations higher than the corresponding Schiff base dissociation constant is a potential regulator of the free PLP-dependent GAD activity. It is clear, however, that neither PLP nor any of the compounds that seem to facilitate the dissociation of the coenzyme could participate in the regulation of the GAD activity possessing tightly bound PLP.

Other studies on the regulation of GAD activity have been focused on the possibility that depolarization of the nerve ending may modify it. Gold *et al.* have reported that GAD is activated in striatal slices on depolarization by high K^+ concentration or by veratridine,[96,97] and the activation by K^+ depolarization seems to be dependent on external calcium.[97] Glutamate decarboxylase was also found to be activated by protoveratrine-A in synaptosomes, and this activation was blocked by tetrodotoxin, whereas that produced by high K^+ concentration was not.[98] Although there are some problems in the interpretation of these data, since GAD activity was measured in most cases under adverse conditions (presence of O_2, pH 7.4, high chloride concentrations), the fact that the net GABA formation also increases by depolarization is an indication that GABA synthesis is increased as a consequence of GAD activation.[97] The GABA thus synthesized appears to remain in the tissue, although one would expect it to be released (see Section 4).

On the grounds of the results of similar depolarization experiments in synaptosomes from the substantia nigra, it has been suggested that the activation of GAD by veratridine or high K^+ concentrations is related to an increased saturation of GAD by PLP.[90,93] Although this increase requires a relatively long period (20 min), it might be linked to the stimulation of GABA release induced by depolarization.[93]

In 1976, Baxter[24] challenged the view that PLP is the most important factor regulating the activity of GAD *in vivo*, mainly on the basis of work by Gey and Georgi[99] who could not find a correlation between the changes produced by some drugs on PLP levels and on GAD activity. In the light of the overwhelming evidence reviewed above (see also Section 4.1), and since the drugs failing to give a good correlation (amphetamine, theophylline, atropine)[99] possess many effects on other neurotransmitter systems which could secondarily affect GAD activity through different mechanisms, one can safely conclude that PLP availability and its interaction with the apoenzyme (at least with one of its forms) is the fundamental regulatory mechanism of GAD activity. This conclusion does not mean that it is the only cerebral enzyme that may be regulated by PLP. Since 1971, it was shown that 3,4-dihydroxyphenylalanine (DOPA) decarboxylase is also very much affected by a decrease in PLP concentration *in vivo*.[67] However, in the same article, it was demonstrated that the convulsions produced by drugs that decrease PLP levels are not produced by inhibition of DOPA decarboxylase. This conclusion was supported by more recent publications[71] (see also Section 4.1).

Besides the inhibitors used in the abovementioned publications, which include stable phosphopyridoxyl derivatives,[82,83] ATP, Pi,[88–91,94] α-ketoglutarate and GABA,[85] other recent reports have studied the inhibitory effect of other compounds on GAD activity. Certain anesthetics, such as ketamine and γ-hydroxybutyric acid, were weak GAD inhibitors ($K_i > 5$ mM).[100] Several structural analogues of glutamate, such as the hydroxamates of L-aspartic acid, L-glutamic acid, and DL-methionine as well as malic and glutaric acids, were reversible inhibitors of GAD with K_i values higher than 1 mM, whereas 3-mercaptopropionic acid and thiomalic acid were much more potent inhibitors ($K_i = 12$ μM).[101] L-α-Hydroxyglutaric acid, but not its D- isomer, as well as D- and L-etomidate behaved as irreversible GAD inhibitors.[101] 2-Keto-4-pentenoic acid (most probably the active compound derived from allylglycine[102–104]) behaves as a competitive inhibitor of GAD, but when it is preincubated with the enzyme, the kinetics corresponds to a noncompetitive inhibition. Thus, this drug seems to act through two different mechanisms, a reversible one on the binding of substrate to the active site and an irreversible one inhibiting subsequent catalytic steps.[102,104] 1-Hydroxy-3-aminopyrrolidone-2,4-diaminobutyrate, 4-N-hydroxy-2,4-diaminobutyrate, and *p*-hydroxybenzaldehyde behave as inhibitors of GAD at relatively high concentrations (0.5–5 mM) in the absence of exogenous PLP.[105]

(2RS,3E)-2-Methyl-3,4-didehydroglutamic acid, designed as a "suicide" inhibitor of GAD, was tested in a partially purified GAD preparation from chick embryo brain.[106] The results of the experiments appear to fulfill the expectations of the drug design, but in view of the possible differences between the immature GAD and the adult enzyme (see the following section), these results should be regarded with caution. A very potent competitive GAD inhibitor (IC_{50} 0.61 μM) has been isolated from *Streptomyces* and identified as 4,5-dihydroxyisophthalic acid.[107]

The stereochemistry of glutamate decarboxylation by GAD has been studied recently. Whereas in one report the results indicate that the decarboxylation occurs with retention of configuration,[108] other authors observed an at least partial inversion of configuration.[109]

With respect to the amino acid residues involved in GAD activity, in addition to the probable involvement of sulfhydryl groups in GAD activity discussed by Wu,[41] some studies have recently been carried out on the participation of arginine residues. Using phenylglyoxal, a compound that reacts specifically with arginine residues, it has been found that such residues near or at the binding site of glutamate are probably involved in GAD activity.[110]

2.1.2. Developmental Studies on GAD

In mouse brain, it has been reported that some properties of GAD, such as its sensitivity to hypotonicity, Triton X-100, centrifugation, and heating at 37°C, are different in the newborn GAD than in the adult enzyme: the newborn enzyme is inactivated by these procedures, whereas the adult enzyme is practically unaffected.[111] This sensitivity is progressively lost during development, suggesting that the immature enzyme protein changes during development or is replaced by the mature form.[112] Furthermore, the activation of the newborn GAD by saturating PLP concentration is less than that of the adult enzyme, suggesting that the abovementioned differences in sensitivity are somehow related to the degree of saturation by the coenzyme.[68,111]

The inhibitory effect of Triton X-100 on the immature GAD has also been reported in the brain and retina of chick embryos.[113] Other developmental studies on GAD activity, which have not considered the possibility of different GAD properties and consequently have been carried out under identical experimental conditions at all ages, in most cases in the presence of Triton-X-100, have been published recently. These include studies on chick retina,[32] chick spinal cord,[114,115] toad brain,[116] spinal cord,[116] and retina,[117,118] rabbit retina,[119] and rat cerebellum.[120] With the exception of toad brain and spinal cord, where only modest changes were observed, notable increases of at least fourfold were detected during development in the regions and species mentioned. These increases in general correlate well with the known maturational changes of other parameters of GABAergic transmission such as GABA uptake and GABA release.[118,119]

In one study,[121] GAD from newborn and from adult mice were partially purified. In both cases, two peaks of GAD activity were separated by gel filtration and by gel electrophoresis, one peak with a molecular weight higher than 200,000 and the other of about 85,000. These four peaks were immunochemically indistinguishable and showed similar K_m and optimum pH. However, since the starting GAD preparations were supernatants of homogenates prepared in hypotonic medium, it seems possible that part of the newborn GAD activity was destroyed by this treatment.[111,112]

2.2. GABA Transaminase

It is well established that the metabolic degradation of GABA is accomplished through its transamination with α-ketoglutarate, catalyzed by GABA transaminase (GABA-T, E.C. 2.6.1.19), to yield glutamate and succinic semialdehyde, which is oxidized to succinate by succinic semialdehyde dehydrogenase.[12,13,24] GABA transaminase has been purified from the brain of the mouse,[41] human,[122] pig,[123] rat,[124] and rabbit.[125] The kinetic, immunochemical, and physicochemical properties of the purified GABA-T have been reviewed by Wu,[41] Saito,[42] and Schousboe et al.[126] Although this is also a PLP-dependent enzyme, in contrast to GAD, the coenzyme seems to be tightly bound to GABA-T under physiological conditions, since a decrease in the concentration of free PLP in vivo does not affect its activity.[12,14,67] In vitro, the coenzyme has been dissociated, and the mechanisms of reactivation of the enzyme have been studied by fluorescence spectroscopy.[127] Also in contrast to GAD, GABA-T has been found to be present in relatively high concentrations in other mammalian tissues different from the brain, particularly in liver and kidney.[128] It has been shown by kinetic studies[129] and by immunologic techniques[130] that the enzyme from the nervous system is indistinguishable from that of peripheral organs.

2.2.1. Mechanism, Regulation, and Inhibition Studies of GABA-T Activity

The kinetic mechanism of GABA-T has been postulated to be of the type ping–pong bi–bi.[124,129,131,132] Not much is known of the regulation of its activity. Several inhibitors that exist in brain have been described,[41] and some of them might participate in GABA-T regulation under physiological conditions. For example, the availability of α-ketoglutarate inside the mitochondria might play such a role.[133] Tunnicliff et al.[132] have found that GABA-T in the intrasynaptosomal mitochondria has eight times less affinity for GABA than the enzyme in the cytoplasmic mitochondria and that the former is more susceptible to inhibition by aminooxyacetic acid and by 2,4-diaminobutyric acid than the latter. These differences could be important for the physiological metabolic degradation of the pool of GABA taken up by the nerve endings after GABA synaptic action.

Many inhibitors of GABA-T have been described in recent years, most of them with the purpose of inhibiting the enzyme in vivo and as a consequence increasing GABA concentration in brain. Metcalf[134] has recently reviewed the mechanism of action of several types of inhibitors, including carbonyl-trapping agents, hydrazinopropionic acid, dipropylacetate, cycloserine, ethanol-amine-O-sulfate, γ-acetylenic-GABA, γ-vinyl-GABA, and gabaculine.

The latter five compounds seem to be irreversible inhibitors of GABA-T by a mechanism known as "suicide" inactivation. This type of inhibition requires that the inhibitor be accepted by the active site of the enzyme, and then that an initial catalytic step, which results in the irreversible inhibition of the subsequent steps and of the binding of substrates, be carried out.[135,136] Because of this mechanism, such suicide inhibitors are in theory highly specific, an important property, since many of the drugs used to block GABA-T activity in vivo also inhibit GAD.[14] Thus, ethanolamine-O-sulfate,[137] 4-aminohex-5-ynoic acid (γ-acetylenic-GABA),[138,139] 4-aminohex-5-enoic acid (γ-vinyl-GABA),[136,139,140] 5-amino-1,3-cyclohexadiene carboxylic acid (gabaculine),[141–143] and 3-amino-1,5-cyclohexadiene carboxylic acid (isogabaculine)[144] have been found to be potent irreversible inhibitors of cerebral GABA-T, both *in vivo* and *in vitro*. Furthermore, the diamine corresponding to γ-acetylenic GABA (5-hexyene-1,4-diamine) seems to be oxidized *in vivo* by brain mitochondrial monoamine oxidase to γ-acetylenic-GABA and as a consequence inhibits GABA-T.[145] In the last year, a group of new irreversible suicide GABA-T inactivators, the 4-amino-5-halopentanoic acids with Cl, F, or Br in position 5, was described.[146,147] Table I shows the structure of the suicide-type inhibitors of GABA-T that have been studied.

Table I
Irreversible Suicide Inhibitors of GABA-T

GABA	CH₂—CH₂ chain: NH₂—CH₂—CH₂—CH₂—COOH
γ-Acetylenic-GABA	CH≡C—CH(NH₂)—CH₂—CH₂—COOH
γ-Vinyl-GABA	CH₂=CH—CH(NH₂)—CH₂—CH₂—COOH
4-Amino-5-halopentanoic acids (X = Cl, F or Br)	X—CH₂—CH(NH₂)—CH₂—CH₂—COOH
Gabaculine	5-amino-1,3-cyclohexadiene carboxylic acid ring structure
Isogabaculine	3-amino-1,5-cyclohexadiene carboxylic acid ring structure

Other compounds with inhibitory effect on GABA-T are 4-N-hydroxy-2,4-diaminobutyrate and *p*-hydroxybenzaldehyde, the latter being much more potent than the former,[105] and two noncompetitive and relatively weak inhibitors, 6-azauracil[148] and thiopental.[149]

Most of the GABA-T inhibitors mentioned above block GABA metabolism *in vivo*, elevate GABA levels in brain, and possess some anticonvulsant action (although some of them also inhibit GAD activity). Their pharmacological actions, including their anticonvulsant effects, have been recently reviewed.[150] These effects on GABA metabolism and on convulsions have been shown for gabaculine and isogabaculine,[151–153] for γ-acetylenic-GABA and γ-vinyl-GABA,[138–140,154–160] for ethanolamine-O-sulfate,[137,159,161–164] and for dipropylacetate.[160,161,164–167] In the case of the latter compound, it has been claimed that its effects on GABA metabolism may be mediated by inhibition of succinic semialdehyde dehydrogenase rather than by inhibition of GABA-T.[161,168]

The inhibitory effect *in vivo* on GABA degradation of all the above compounds, with the consequent elevation in GABA levels, has been frequently correlated with their anticonvulsant action. The effects of aminooxyacetic acid or other GABA-T inhibitors *in vivo*, such as L-glutamyl-γ-hydrazide, have been studied in relation to their anticonvulsant effects (see for example refs. 60,167,169–174). The picture of the relationships between changes in GABA metabolism in whole brain or in certain regions and the decrease in excitability (seizure protection) resulting from increased GABAergic synaptic function is not clear.

With regard to the structure–activity relationships in GABA-T, using phenylglyoxal as a specific reagent, it has been found that arginine residues participate at or near the PLP-binding site of GABA-T.[175] A direct participation of lysine residues at the catalytic site has also been shown using O-phthalaldehyde to block such residues.[176]

The turnover number of GABA-T has recently been estimated in different rat brain regions using a radioimmunoassay of GABA-T. The values obtained vary from 3.5 nmol of GABA transaminated per second per nanomole of GABA-T in the cerebellum to 30.6 in the internal capsule.[177] Regarding the half-life of GABA-T, on the basis of the recovery of its activity after the irreversible inhibition produced by γ-vinyl-GABA, it has been estimated to be about 3.5 days.[140]

GABA has also been considered a precursor of other neuroactive compounds. For example, there is evidence of its conversion *in vivo* to γ-hydroxybutyric acid through succinic semialdehyde.[178] 2-Pyrrolidinone, which might be formed by cyclization of GABA, has been detected in mouse brain by gas chromatography–mass spectrometry.[179]

2.3. GABA Turnover

In recent publications,[15,16,170,180] previous reports as well as the difficulties in the estimation of GABA turnover have been discussed. One of the important problems is the presence of different metabolic pools of GABA which originate from different metabolic precursors and are probably related to specific synaptic roles (see Sections 2.4 and 4). This problem is a major one when labeled precursors are used, whereas the methods employing GABA-T inhibitors *in vivo* also have a number of limitations.[16] From experiments with inhibitors of GABA-T, turnover rates of 4.0 μmol/g per h and 6.5 μmol/g per h, with γ-acetylenic-GABA and γ-vinyl-GABA, respectively, have been reported in whole mouse brain.[138,140] With another method, based on the continuous intravenous infusion of glucose labeled with ^{13}C and measurement of the percent incorporation of the label into glutamate and GABA, turnover rates varying from 384 to 562 nmol/mg protein per hour have been obtained in four rat brain regions.[181] If the former value of 6.5 μmol/g per h is converted to nanomoles per milligram protein per hour on the basis of a 14% protein content, a value ten times smaller than that observed with [^{13}C]glucose is obtained.

2.4. Compartmentation of GABA Metabolism

Since the discovery of the compartmentation of glutamate metabolism in brain,[182,183] which has been reviewed extensively in the first edition of this *Handbook*[184] and elsewhere[24,185–189] and is also the subject of a chapter in this volume, many publications have appeared on the way this compartmentation is related to GABA metabolism and GABA synaptic function.[23] These studies have shown that glutamine seems to be an important precursor of GABA in the "large" glutamate compartment and that GABA is converted to glutamine in the "small" glutamate compartment.[186,188,190,191] Since the large glutamate compartment seems to be located in nerve endings,[188,192–195] several studies have attempted to demonstrate that glutamine is a precursor of the metabolic pool of GABA directly related to its synaptic function (releasable pool).

Thus, it has been shown that labeled GABA derived from labeled glutamine is released in a calcium-dependent manner on depolarization by veratrine[195] or by high potassium concentrations,[195–199] These results were obtained in rat cortical synaptosomes[195] and in slices of mouse cortex,[196] pigeon optic tectum, paleostriatum, hippocampus, and cerebellum,[197,198] and rat striatum, hippocampus, cerebellum, substantia nigra, and cochlear nucleus.[198] Furthermore, it has been shown that intrastriatal kainic acid or hemitransection resulted in parallel decreases of GAD activity and of the release of labeled GABA synthesized from labeled glutamine in the substantia nigra, thus indicating that this phenomenon occurs in the GABAergic nerve endings.[199] Although a pool of GABA synthesized from glutamine seems to be preferentially released, it has been found that GABA derived from pyruvate,[200] glutamate,[196,201,202] and glucose[195,203–205] can also be released from brain preparations *in vitro*. In recent reports, it has been demonstrated that GABA synthesized from labeled glutamine is released *in vivo* from the substantia nigra and from the pallido–entopeduncular nuclei and that this release is increased by K$^+$ depolarization or by stimulation of the caudate nucleus.[206] Spontaneous release of GABA synthesized from glucose has been also reported *in vivo* in rat hypothalamus.[207] The properties of the different GABA pools at the synaptic level are discussed further in Sections 3 and 4.

The metabolism of GABA seems to be compartmentalized in retina similarly to brain, although there are some species variations. Thus, GABA is a precursor of the small glutamate pool,[208,209] and glutamine is a good precursor of GABA.[208,210,211]

2.4.1. Role of Glial Cells in GABA Metabolism

The metabolic compartmentation of GABA has been related to structural compartmentation. Since the small glutamate pool, which is closely associated with glutamine synthesis, is located mainly in glial cells,[186,188,190,192,193,212–214] it has been postulated that the GABA taken up by glia is converted there to glutamine, which is then transported into nerve endings where it is deaminated to glutamate by glutaminase.[194,215] Subsequent decarboxylation of glutamate would then produce GABA in a pool available for release, as described above. This hypothesis is supported by the following data:

1. A high-affinity (K_m 0.2–40 μM) sodium-dependent uptake of GABA has been reported in several glial preparations, including peripheral ganglia,[216,217] bulk-isolated glial cells,[218,219] glioma cell lines,[220–222] primary cultures from rat cerebellum,[223] mouse[224,225] and rat[226] brain hemispheres, and the filum terminale of the frog.[227] In the primary cultures, the V_{max} values are much greater than those of the other preparations.
2. Although GABA-T is present in both neurons and glial cells, it seems to be concentrated in glial cells.[228,229]
3. Glutamate decarboxylase is mainly located in the nerve endings,[73,230,231] and it is practically absent in glial cells.[228,229,232]

Most of the data on the transport and metabolic properties of glial cells related to GABA have been reviewed recently.[233–236] Comparative kinetic and enzymatic data in different glial preparations and among glia, brain slices, synaptosomes, and neuronal perikarya are summarized in these reviews (see also Section 3).

2.5. GABA, GAD, and Protein Synthesis

An interesting possible metabolic role of GABA is that related to protein synthesis. In 1976, Baxter[237] reviewed the data indicating that GABA added to *in vitro* cell-free systems from brain tissue stimulates protein synthesis as well as the efforts made to elucidate its possible mechanisms of action and the failure to confirm the stimulatory effect of GABA on protein synthesis *in vitro*. After this review, to our knowledge, only two publications have appeared supporting the possibility that a certain pool of GABA is linked to the rate of protein synthesis.[238,239]In these reports, it was found, with the use of drugs, that both *in vivo* and in brain cortex slices the incorporation of labeled leucine into brain protein was decreased in parallel with the activity of GAD, rather than with GABA levels, and that this effect was brain specific.[238] Subsequently, it was shown that the decrease of protein synthesis produced by pharmacological GAD inhibition affected proteins from all brain subcellular fractions as well as all proteins separated by gel electrophoresis.[239]

These results have been reviewed in comparison with the effects on protein synthesis of other neurotransmitter suspects such as dopamine and serotonin.[240] The interpretation of the available data on the relationship between GABA and protein synthesis is obviously difficult because of the poor reproducibility of the results in cell-free systems and the possible multiple effects of the drugs used in the pharmacological approach. In spite of these difficulties, it is clear that the eventual demonstration of a causal relationship between GABA and protein synthesis would be of great interest for its possible implications in the plasticity of GABAergic synapses.[240]

According to the immunocytochemical findings, GAD is probably synthesized in neuronal soma and transported to the terminals by axoplasmic flow.[55] This view is also supported by measurements of GAD activity in the substantia nigra and the striatum after hemisection interrupting the axons of the striatal neurons.[241] In crayfish peripheral nerve, there is also evidence that GAD is transported from the neuronal soma,[242] but other findings with the same preparation suggest that GAD may be synthesized locally in the nerve.[243]

3. UPTAKE AND RELEASE

Both *in vitro*, in brain slices or synaptosomes, and *in vivo* with cups or push–pull cannulas, it has been possible to demonstrate that GABA has a low spontaneous rate of release and that its release is stimulated by depolarizing conditions in a calcium-dependent manner. There are different pools of GABA, originated from different metabolic pools of glutamate, which seems to be differentiated by the release mechanisms.

Since GABA-T is a mitochondrial enzyme, the termination of the synaptic action of GABA is not by enzymatic breakdown but appears to occur through an active, Na^+- and Cl^--dependent uptake of the amino acid into the nerve endings or the glial cells. To what extent the high-affinity uptake represents a net uptake or a hemoexchange process and what the physiological role of the homoexchange is are questions not yet completely answered.

3.1. Uptake Studies

3.1.1. Uptake Mechanisms

Several reviews listing the kinetic constants for the high- and low-affinity uptake of GABA by synaptosomes, brain slices, glial preparations, ganglia and retina, have been published.[22,234–236,244,245] The K_m values for the high-affinity uptake vary from less than 0.5 μM in rabbit synaptosomes and glia,[218,222] cerebellar glia,[223] and sympathetic ganglia[246] to more than 100 μM in chick embryo brain,[247] but most values are in the 10–40 μM range.[244] The low-affinity K_m values are well over 300 μM. The high-affinity transport system, which is dependent on Na^+ and has recently been shown also to depend on Cl^-,[248] seems to be specific for amino acids with a neurotransmitter function, with the possible exception of proline.[244,249,250]

Martin[244] has discussed lucidly and extensively the kinetics and the possible mechanisms of the participation of Na^+ in the high-affinity uptake of GABA, taking into account the data of similar Na^+-dependent transport of amino acids in other tissues outside the nervous system. Such mechanisms include the possibility that the energy for the transport is obtained from the sodium concentration gradient and the electrical membrane gradient, but another factor that might play a role is the electrogenic Na^+–K^+ ATPase activity. The earlier kinetic analyses in synaptosomes indicate the participation of two or three Na^+ ions per GABA molecule transported (see ref. 244). More recent kinetic studies support the view that there is a coupling of at least two Na^+ per one GABA both in synaptosomes[251–258] and in cultured[254] or bulk-prepared astrocytes.[255]

Wheeler[252,253] has developed a model fitting the kinetic equations in cortical and hypothalamic synaptosomes. According to this model, an effective translocation of GABA requires that two sodium ions bind sequentially to the carrier before GABA can bind to it; although GABA can bind to the carrier without sodium, this binding would result only in a negligible translocation. Furthermore, both V_{max} and the K_m for GABA depend on Na^+ concentration. Qualitatively, hypothalamic synaptosomes behaved as cortical synaptosomes, although some quantitative differences have been found.[252,253]

A recent report by Kanner,[256] using membrane vesicles presumably derived from synaptic plasma membranes, indicates that the active transport of GABA occurs by a process driven only by ionic gradients and is not affected by ouabain. In this preparation, the kinetic constants are similar to those of synaptosomes, and the GABA influx is strictly dependent on both a Na^+ gradient (out > in) and a gradient of certain small anions (out < in), chloride being the most effective anion. The transport of GABA in the presence of a K^+ gradient (in > out) is notably stimulated by the K^+ ionophore valinomycin which creates or enhances a membrane potential (interior negative). From these observations, it is concluded that the uptake of GABA is driven by the Na^+ concentration gradient and additionally, in agreement with other studies in depolarized or hyperpolarized synaptosomes,[251] by the membrane potential. The role of chloride ions could be related to the creation of membrane potential (interior negative) as a consequence of its transport into the terminals together with Na^+ and GABA or to the activation of the GABA carrier by Cl^- binding to an external site. The stoichiometry for the Na^+–Cl^-–GABA cotransport would be 2 or $3:1:1$.[256]

In a subsequent publication,[257] the GABA carrier was solubilized and reconstituted in phospholipid membranes. The reconstituted system had all the properties of the native system regarding the requirements of ionic gradients and the effects of cation ionophores. In these experiments, the chloride dependence was observed in the presence of valinomycin, suggesting that its effect is not related only to the membrane potential.[257] These findings on the ionic gradient requirements for the transport of GABA in membrane vesicles derived from synaptosomes were confirmed more recently.[258] Furthermore, in the latter report, it was shown that the uptake of GABA measured chemically has identical properties to that of radioactive GABA.

Some kinetic studies on the low-affinity transport of GABA (at concentrations above 200 μM) and its energy requirements in brain slices have also been published. It has been found that in the presence of glucose this uptake is slower, although the rate equations are similar to those in its absence.[259,260]

3.1.2. Uptake and Homoexchange

The demonstration that the high-affinity, Na^+-dependent uptake of GABA in synaptosomes could primarily represent a 1:1 exchange with intrasynaptosomal GABA[244,261-263] cast some doubts on the hypothesis that the removal of GABA from the synapse was by a transport mechanism.[244] Subsequently, it has been shown that under certain conditions, for example, depletion of endogenous GABA[263-265] or increased intrasynaptosomal K^+,[266] there is a net uptake of GABA by the high-affinity process. In these reports, it was also demonstrated that a small external K^+ concentration (1 mM) was required for a net uptake of GABA in GABA-deficient synaptosomes.[263,264] Other authors[267] have proposed that under *in vivo* conditions the high-affinity transport may be responsible for a net uptake of GABA, assuming that the membrane potential in synaptosomes is less than that *in vivo*. Under these conditions, the cation fluxes under depolarizing conditions might be sufficient to shift the concentrative uptake from about 10^{-4} M GABA (low-affinity range) to 10^{-6} M.[267] In this report, however, the actual results *in vitro* agree with the absence of net uptake at GABA concentrations below 10^{-4} M.

Levi and co-workers have comprehensively discussed the above results,[266,268] and they have proposed,[266,269] on the basis of a number of homoexchange experiments in which the fluxes of Na^+ and Ca^{2+} were modified by ionophores or by preincubations and subsequent superfusions of the synaptosomes with media containing different Na^+ concentrations, that GABA transport in nerve endings is modulated both by the extracellular GABA concentration and by the changes in Na^+ and Ca^{2+} fluxes occurring during the depolarization–repolarization physiological cycle. According to this proposed model, under resting conditions, there would be a 1:1 exchange between the synaptic and the intraterminal GABA. On depolarization, the accompanying Na^+ and Ca^{2+} inward fluxes would favor a carrier-mediated stoichiometry of efflux/influx greater than unity, which would increase the concentration of GABA in the synaptic cleft. This increase would become greater with the GABA released through the carrier-independent, Ca^{2+}-dependent mechanism. In a subsequent phase, the inverse cationic fluxes characteristic of the repolarization of the terminal, plus the possible low-affinity GABA uptake (depending on the GABA concentration reached in the synaptic space), would invert the stoichiometry of efflux/influx, and GABA would be taken up by the terminal.[266,269]

Very recently, a report was published in which net high-affinity uptake was observed in synaptosomes, and only a small proportion (about 10% of the uptake) represented homoexchange.[270] This is the first study that has failed to reproduce the initial homoexchange findings. The authors attribute their different findings to possible differences in the quality of the synaptosomes preparation used or to possible synaptosomal damage resulting from filtration (which they avoid using). However, besides the numerous control data supporting a real homoexchange process (Na$^+$ and temperature dependence, for example), the other reports mentioned that reproduced the lack of net uptake under "resting" conditions[264,267] used centrifugation instead of filtration, and the synaptosomal preparations used were obtained from Ficoll®[264] or sucrose[267] gradients. The discrepancy, therefore, has to be attributed to a different factor, perhaps the tenfold dilution of the synaptosomes when the unlabeled GABA was added, in the abovementioned report.[270]

In cortical slices[271] incubated in small volumes of medium to facilitate the study of bidirectional transport of GABA by measuring changes in the medium, it has been shown that no net high-affinity uptake of GABA occurred at concentrations below 25 μM, and at 25 μM, only 50% of the uptake could be net uptake, the difference being accounted for by exchange. In spinal cord slices, all of the uptake was found to be through homoexchange.[272]

3.1.3. Inhibitors of GABA Uptake

If uptake is involved in the termination of GABA synaptic action, it is obvious that the development of specific inhibitors would be of great interest to study GABAergic synapses. A list of many compounds with inhibitory action, and of others that had been tested with negative results, was published in 1975[14] (see also ref. 15), and structural and kinetic information on some of them was reviewed in 1976.[273] In recent years, this list has increased considerably. In 1975, a series of isoxazoles structurally related to the potent GABA-receptor agonist muscimol were tested as inhibitors of GABA uptake in brain slices. This study led to the finding that nipecotic acid (piperidine-3-carboxylic acid) behaves as a potent noncompetitive inhibitor of GABA uptake in brain slices, with K_i of 11 μM, whereas weak inhibitory effects were found for some isoxazoles and amino acid analogues.[274] It was found subsequently that the (−)isomer of nipecotic acid was more potent than the (+)isomer and that both isomers behaved as competitive inhibitors when added simultaneously to GABA and as noncompetitive when preincubated before GABA addition.[275]

It was found later that the analogues of nipecotic acid isolated from the betel nut, guvacine and arecaidine, were also potent inhibitors of GABA uptake in brain slices and were devoid of effects on the GABA receptor.[276,277] The structure and kinetics of the inhibition by this class of compounds in brain slices have been reviewed by Krogsgard-Larsen[278] and in glial preparations by Schousboe.[233,234,]

In vivo, both isomers of nipecotic acid and (+)2,4-diaminobutyric acid, applied iontophoretically, enhance the depressant action of GABA applied through another barrel of the micropipette, in neurons of different regions of the CNS. However, these compounds failed to affect the inhibitory action of synaptically released GABA, as judged from experiments in which cerebellar Purkinje cells were inhibited by stimulation of the basket cells.[279] The latter result might be explained by the capacity of the tissue to take up both uptake inhibitors,[22,280,281] because this would decrease the probability of attaining their necessary concentration at sufficient synapses. With this consideration in mind, the results of this publication[279] agree with the postulate that GABA transport is important for the physiological removal of GABA in the synaptic space. This conclusion was also drawn from similar *in vivo* experiments in the spinal cord using muscimol to inhibit neuronal firing.[282]

From cortical slices and frog retina used as model preparations for neuronal uptake, and sympathetic ganglia and rat retina as glial uptake models, it has been shown that *cis*-3-aminocyclohexanecarboxylic acid, a compound know to be an inhibitor of GABA uptake in cortical slices,[14,283] seems to inhibit selectively the uptake into neuronal elements.[284,285] It had previously been reported that L-2,4-diaminobutyric acid is also a more potent inhibitor of neuronal GABA uptake than of glial uptake, whereas the opposite is true for β-alanine[286,287] (see refs. 15,22,288,289 for review on this point). This specific effect of β-alanine on glial GABA uptake has been confirmed in sympathetic ganglia.[290] Furthermore, in isolated cerebellar Purkinje cell perikarya[291] and in a general population of isolated cerebellar neuronal perikarya,[292] a high-affinity GABA uptake has been recently described, and in these neuronal preparations, β-alanine showed markedly less inhibitory effect than 2,4-diaminobutyric acid[291] or *cis*-1,3-aminocyclohexanecarboxylic acid.[292] Similar differences in the potency of inhibition of GABA uptake between 2,4-diaminobutyric acid and β-alanine have been observed in isolated cerebellar glomeruli.[293] In these structures, the high-affinity uptake of GABA, and also that of diaminobutyric acid, occurs specifically into the inhibitory terminals of the Golgi axons.[294,295]

An interesting report was recently published[296] demonstrating kinetically and with inhibition experiments that in a glioma cell line a single transport system is responsible for the high-affinity uptake of both β-alanine and taurine and for the low-affinity uptake of GABA. Therefore, it seems possible that the low-affinity uptake of GABA observed in thick slices is carried out in glial cells by the β-alanine and taurine high-affinity system.[296] Whether this explanation can also account for the low-affinity GABA transport in synaptosomes remains to be explored.

Recently, three-carbon-ring cyclic analogues of GABA, *cis*- and *trans*-2(aminomethyl)cyclopropanecarboxylic acids, have been synthesized and tested. The *trans* acid was a weak inhibitor of GABA uptake, but it behaved as a depressant of neuronal firing more potent than GABA, and this effect was bicuculline sensitive.[297] Other analogues of GABA that are inhibitors of GABA-T activity, such as gabaculine, γ-acetylenic-GABA, and γ-vinyl-GABA, are also relatively weak inhibitors of GABA uptake in synaptosomal fractions (IC$_{50}$ 81 μM to 560 μM), whereas ethanolamine-O-sulfate or *n*-dipropylacetate have negligible effects.[298]

3.2. Release Studies

The release of GABA from nerve endings is obviously one of the crucial steps in its physiological action. As with other neurotransmitters, Ca^{2+} seems to play a fundamental role in this process, since it has been found that in different preparations, *in vitro* and *in vivo*, it is required for the release of GABA stimulated by depolarization. Besides this mechanism of depolarization–secretion coupling, there seems to exist another mechanism of GABA release which is dependent on the metabolic GABA-synthesizing system and is possibly linked to Ca^{2+}, although in a different way. The relationship between the metabolism and the release of GABA constitutes, therefore, a subject of great interest. Different GABA pools, fed preferentially by different metabolic precursors and with specific properties regarding their release, have been postulated to exist in nerve endings. In this respect, the metabolic compartmentation of GABA, discussed above, is clearly connected with the mechanisms of its release.

Diverse methods have been designed to study the release of GABA both *in vitro* and *in vivo*. *In vitro*, brain slices or synaptosomes incubated or superfused with appropriate media have been used. *In vivo*, cups placed on the cerebral surface or push–pull cannulas located in specific brain regions have been employed. Both *in vivo* and *in vitro*, the release has been studied by preloading the tissue with labeled GABA by the high-affinity uptake mechanism and measuring the release of radioactivity in the presence of aminooxyacetic acid to inhibit GABA metabolism. Alternatively, the release of endogenous GABA or of labeled GABA synthesized from different labeled metabolic precursors loaded previously into the tissue has been measured. The latter type of experiment has permitted the analysis of the relationship between different metabolic pools of GABA and their release.

3.2.1. Calcium and the Depolarization-Induced Release of GABA

Many studies have been published in recent years with the methodological approaches mentioned above on the action of Ca^{2+} on the release of GABA stimulated by depolarization. Reviews of most of these publications have recently been published.[299,300] Some methodological and interpretation problems, as well as discussions of the advantages and disadvantages of the depolarization produced by electrical pulses, by high K^+ concentrations, and by drugs promoting ion transport, can be found in these reviews. In this regard, one of the findings that seems to be reproducible is as follows: in slices from several brain regions, the release of exogenous GABA stimulated by electrical pulses is less dependent on Ca^{2+} than the stimulation produced by depolarizing K^+ concentrations,[301–306] whereas this difference is small when endogenous GABA release is studied.[307] The form and polarity of the electrical stimulatory pulses are important parameters in this type of study.[300,303] When the depolarization of slices was induced by veratridine, no consistent Ca^{2+} dependence of GABA release was found, and the veratridine-stimulated release was even enhanced in the absence of Ca^{2+} as compared to 1 or 2 mM Ca^{2+}.[303,308–310]

In synaptosomes, depolarizing K^+ concentrations stimulate the release of labeled GABA in a Ca^{2+}-dependent manner, although the dependence has been in general only partial,[263] probably because there is a relatively important release from nonsynaptosomal structures which is stimulated by K^+ in a Ca^{2+}-independent way.[311–314] Cotman and co-workers have published a series of experiments on the relationship between the external Ca^{2+} concentration

($[Ca^{2+}]_o$) and the release of GABA stimulated by high K^+ concentrations in synaptosomal fractions.[311–315] They have demonstrated that divalent cations such as magnesium and manganese inhibited this release and that barium and strontium were effective when replacing Ca^{2+} in the medium. These authors have also shown that the Ca^{2+} ionophore A23187 induces the release of GABA when Ca^{2+} is present in the medium.[311–313]

In kinetic studies, they found that the relationship between $[Ca^{2+}]_o$ and efflux was sigmoidal and not hyperbolic and calculated a K_m of about 0.2 mM, which did not differ among several brain regions.[315] The Ca^{2+}-dependent release stimulated by veratridine-induced depolarization observed in the latter publication could not be analyzed kinetically because of the shape of the curves.[315] This anomaly can probably be accounted for by the already mentioned finding that the effect of veratridine on GABA release in slices is greater in the absence than in the presence of Ca^{2+}.[303,308–310] This phenomenon has also been reported in synaptosomes,[266] and it has been explained by the possibility that the veratridine-stimulated release is caused not only by the increased external Ca^{2+} transport produced by depolarization but also by the increase in the intraterminal Na^+ concentration that this alkaloid produces.[266,308,309]

The possibility that this Ca^{2+}-independent release may be via the Na^+-dependent carrier-mediated transport is improbable, since L-2,4-diaminobutyrate, which, as previously mentioned, is an inhibitor of this process in neurons, does not affect it.[266,309,316] Therefore, it has been postulated that the stimulatory effect of veratridine on GABA release in the absence of external Ca^{2+} results from efflux of Ca^{2+} from intraterminal mitochondria mediated by Na^+–Ca^{2+} exchange.[309,310,316] This view is supported by the observations that the efflux of GABA from synaptosomes in the absence of external Ca^{2+} is stimulated by uncouplers of oxidative phosphorylation such as dinitrophenol and carbonyl cyanide-p-trifluormethoxyphenylhydrazone, and by quinidine,[317] as well as by other procedures that increase intracellular Na^+ concentration by different independent mechanisms such as ouabain, gramicidin D, and K^+-free medium.[316,318] The enhancement of GABA efflux produced by these treatments was dependent on the presence of external Na^+, but it was additive to the effect of external GABA (homoexchange) and was unaffected by L-2,4-diaminobutyric acid, thus ruling out the participation of an outward carrier-mediated transport.[316]

The dependence of the K^+-stimulated release of GABA on external Ca^{2+} has also been studied using drugs that block the transport of Ca^{2+}. Thus, the inorganic dye ruthenium red[319,320] and lanthanum[321] block the Ca^{2+}-dependent release of GABA in synaptosomal fractions. The inhibitory effect of ruthenium red was later also observed in brain slices.[309] In this respect, it is of interest that La^{3+} enhances the spontaneous release of GABA,[321,322] although this increase is certainly much less than that of acetylcholine previously demonstrated to occur in neuromuscular junctions treated with lanthanum.[323] In fact, the spontaneous release of labeled acetylcholine synthesized from labeled choline in synaptosomes was not affected by lanthanum,[321] suggesting that the release mechanisms of acetylcholine in central synapses are not identical to those in peripheral synapses.

A release of GABA stimulated by depolarization in a Ca^{2+}-dependent manner has also been observed in several retinal preparations from chick,[324] toad,[118] and rabbit[119,325] but not from rat.[326] The Ca^{2+} ionophore A23187 also induces a Ca^{2+}-dependent release of GABA from chick retina in the absence of depolarization.[327]

3.2.1a. Glial Cells and Ca^{2+}-Dependent Release. The discovery that glial cells were able to take up GABA by a high-affinity transport process similar to that of nerve endings led to studies of the possibility that GABA might also be released from such glial cells. When the release of labeled GABA was first studied in dorsal root ganglia as an experimental model for glial transport, a stimulation of GABA release by high K^+ concentrations was found, and a partial Ca^{2+} dependence of this phenomenon was observed.[328,329] Subsequently evidence has been accumulating in favor of the idea that, although glia can release the captured exogenous GABA in the presence of high K^+ concentrations, this release is entirely independent of Ca^{2+}. Thus, the Ca^{2+} antagonists D-600[330] and La^{3+}[328] failed to affect such GABA efflux in dorsal root ganglia. Furthermore, in bulk-prepared glial cells, this efflux was completely unaffected by the absence of external Ca^{2+}, whereas in neuronal perikarya and synaptosomes studied comparatively, the release was notably decreased.[331]

In agreement with the above results, Bowery, Neal, and co-workers[332–334] observed a complete Ca^{2+} independence of the K^+-induced release in three different preparations that accumulate exogenous GABA mainly, if not only, in glial cells, namely, rat retina and spinal and sympathetic ganglia. In contrast, an almost total Ca^{2+} dependence was observed in cortical slices and frog retina, which accumulate GABA preferentially in neuronal elements. Furthermore, the release of GABA in the glial preparations was not affected by veratridine, a finding previously reported by Minchin,[330] whereas in the neuronal models, a notable stimulation was observed.[332] In sympathetic ganglia,[333] it was observed that electrical stimulation produced an enhanced release of GABA, which diminished progressively with repeated stimulation and was not affected by tetrodotoxin or high Mg^{2+} with zero Ca^{2+}. An important finding in this work[334] was that electrical recordings from glial cells at different $[K^+]_o$ and as a response to orthodromic stimulation of the ganglion showed that the extracellular K^+ quantities accumulated during repetitive nerve stimulation were below those necessary to stimulate the release of GABA from glial cells. Thus, the authors concluded that the release of GABA from glial is unlikely to be physiological and may be considered only a consequence of the high intracellular concentrations of GABA generated by the carrier-mediated process.[333,334]

In cortical slices in which the uptake of GABA into neurons and glia was selectively inhibited by 2,4-diaminobutyrate and β-alanine, respectively (see Section 3.1), the K^+-stimulated release of GABA in the presence of Ca^{2+} was nearly abolished only after incubation with diaminobutyrate, suggesting that in this preparation the origin of the released GABA was neuronal.[335]

In cultures enriched in neurons or glia, starting from dissociated cerebellar cells, it was also found that a K^+-stimulated, Ca^{2+}-dependent release of GABA occurred from neurons, whereas negligible release was observed in glial cells.[336]

From all of the above observations, it may be safely concluded that the Ca^{2+} dependence of the release of GABA induced by K^+ depolarization is a reasonable criterion for establishing its neuronal origin.

3.2.2. Metabolic Pools of Releasable GABA

Some implications of the cerebral compartmentation of glutamate and GABA metabolism on GABA release have been already mentioned in Section 2.4. The Ca^{2+}-dependent, depolarization-stimulated release of GABA derived from glutamine has been demonstrated in several preparations *in vitro*,[196–199] and in some reports it has been shown that glutamine is in this respect a better precursor than glutamate.[196,337] *In vivo*, glutamine is also a good precursor of GABA released both synaptically (as a response to afferent stimulation) and by high-K^+ depolarization.[206] Other metabolic precursors of the GABA released by depolarization are pyruvate,[200] glutamate,[196,201,202,337] and glucose.[195,203–205] In some of these reports,[200–202] it has been observed that the recently synthesized GABA is preferentially released, suggesting a relationship between the synthesis of GABA and its release.

In accord with this possibility, it had been previously postulated, on the basis of the relationships between the inhibition of GAD activity and the appearance of convulsions, that under resting conditions GAD activity is coupled to the release of GABA.[69,338] More direct evidence supporting this hypothesis was obtained more recently in cortical slices by the finding that the inhibition of GAD activity was well correlated with a decrease of the spontaneous release of GABA synthesized from glutamate.[201] Since in cortex slices this spontaneous release of GABA derived from glutamate was higher than that synthesized from glutamine,[196] it seems possible that exogenous glutamate feeds a pool of GABA more directly connected with the synthesis–release coupling, whereas glutamate derived from glutamine feeds a GABA pool released by depolarization. Further support for the coupling of GABA synthesis with its spontaneous release was provided by the experiments *in vivo* of Van der Heyden *et al.* who, with the aid of a push–pull cannula implanted in the substantia nigra[339] in the striatum,[340] found that the spontaneous release of endogenous GABA decreased 50–60% immediately after superfusion with the GAD inhibitor 3-mercaptopropionic acid. This inhibition correlates nicely with the abovementioned inhibition observed in brain cortex slices.[169,201]

Also in agreement with a close association between GABA synthesis and its release are the results of De Belleroche and Bradford.[341] These authors used radioactive glucose as precursor and found, on the basis of the specific radioactivity of the GABA released as compared to that contained in different subcellular fractions, that the cytoplasm was the most probable site of origin of the GABA released by depolarization. To date, this is probably the more direct study negating the possibility that GABA stored in synaptic vesicles is released by exocytosis.

3.2.3. In Vivo Release Studies

Some studies in which the release of GABA *in vivo* was measured using a perfusion technique through a push–pull cannula have been already mentioned.[206,207,339,340] The release of radioactive GABA synthesized from labeled glutamine in the substantia nigra and in the pallido–entopeduncular nuclei was stimulated by K^+ depolarization, and that in the substantia nigra by electrical stimulation of the caudate nucleus.[206] During perfusion of several hypothalamic regions, spontaneous release of GABA derived from labeled glucose was detected mainly from the lateral edge of the ventromedial nucleus.[207]

The release of endogenous GABA from both the substantia nigra and the striatum was stimulated by K^+ depolarization in a Ca^{2+}-dependent manner and also by electrical stimulation of the striatum.[339,340] The release from the striatum was stimulated by K^+ depolarization, and 3-mercaptopropionic acid and tetrodotoxin partially inhibited this stimulation.[340] Endogenous GABA was spontaneously released with about 30-min rhythmicity from the cat posterior hypothalamus, and the release was stimulated by high K^+ concentrations in a Ca^{2+}-dependent manner and by electrical stimulation of the hypothalamus itself or of the locus coeruleus.[342]

All of the above results were obtained in perfusion experiments of deep nuclei using push–pull cannulas. Other recent publications have employed superficial perfusion of brain regions by means of cups[343,344] or cannulas.[345–347] One of these superfusion systems[345,346] makes it possible to carry out experiments in the unanesthetized, freely moving animal. With this technique, the release of endogenous GABA from the sensorimotor and the visual cortices could be detected only after treatment with the GABA-T inhibitors γ-vinyl-GABA or γ-acetylenic-GABA.[345] Under these experimental conditions, the release of endogenous GABA was stimulated by contralateral brachial plexus stimulation or by superfusion with a scorpion venom toxin (tityustoxin). Most of these results have been collected in a single publication.[347]

The procedure of fixing a cylinder on the surface of the brain to superfuse it has been used in recent years to study the release of endogenous GABA from the visual cortex after K^+ depolarization and after the addition of Ca^{2+} ionophores to the medium. A Ca^{2+}-dependent K^+-stimulated release was observed,[348] and the ionophore A23187 notably stimulated the release in the presence of Ca^{2+} and in the absence of depolarization.[349] In one superfusion study using a cup fixed onto the surface of the dorsal medulla, no stimulation of [^{3}H]GABA release by K^+ depolarization was observed,[343] and the same negative result was obtained in slices of the same region. However, electrical stimulation and veratridine did stimulate the release in the slices, and high K^+ also enhanced the release when tested in synaptosomal preparations. It is therefore possible that some swelling effect of raised K^+ may account for these differences.[343] In another study using cups placed on control or denervated slabs of motor cortex, it was found that electrical stimulation increased the release of labeled GABA from both control and denervated cortex. Seizure activity induced by the application of methacholine resulted in a decrease of GABA release, but again, no difference between the control and denervated cortex was observed.[344]

In experiments using a method for perfusing the fourth ventricle similar to that described by Obata,[350] it has recently been observed that the release of labeled GABA was stimulated by high K^+ concentrations in the presence of Ca^{2+} and also by β-alanine, probably because of its inhibitory action on GABA uptake.[351]

The release of endogenous amino acids into the vitreous of the rat eye *in vivo* has recently been reported.[352] No spontaneous GABA release could be detected, and, in agreement with a previous result *in vitro* in this species,[326] neither high K^+ nor light stimulation induced any GABA release.

3.2.4. *Pharmacological Modification of GABA Release*

In contrast to the inhibition of GABA uptake, probably because of the superficial present knowledge of GABA release mechanisms, few studies exist on the specific inhibition or stimulation of the release of GABA. As mentioned in Section 3.2.2, the spontaneous release of GABA is diminished as a consequence of inhibition of GAD activity.[201,339,340] Also mentioned in Section 3.2.2 is the finding that the Ca^{2+} antagonists D-600[302] ruthenium red,[309,319,320] and lanthanum[321] inhibit the K^+-stimulated, Ca^{2+}-dependent release of GABA. Several drugs with actions on the CNS have been studied regarding their effect on GABA release.[14,19] Pentobarbital at 200–500 μM inhibits the Ca^{2+}-dependent release of exogenous GABA in synaptosomes[353] and of endogenous GABA in midbrain slices,[354] suggesting that this drug acts through modifications of Ca^{2+} transport. In contrast, it was reported that similar concentrations of this drug stimulate the electrically evoked release of GABA from olfactory cortex slices[355] and the spontaneous release from rabbit retina.[356] Olsen *et al.*[357,358] have studied the effect of several drugs on the calcium-dependent release of GABA in synaptosomes. In this study, 100 μM pentobarbital had no effects; neither did 100 μM phenytoin, pentylenetetrazole, bicuculline, picrotoxinin, bicucine methyl ester, *d*-tubocurarine, colchicine, β-*p*-chlorophenylGABA, muscimol, or benzylpenicillin. Haloperidol, imipramine, chlorpromazine, and diazepam inhibited the Ca^{2+}-dependent release with 4–80 μM IC_{50} values.[358] Furthermore, diazepam at concentrations above 50 μM stimulated the spontaneous GABA release, and this effect was enhanced by sodium.[357] Since diazepam also notably inhibited GABA uptake, these results suggest an involvement of the Na^+-dependent carrier mechanism in diazepam action.

In other reports, penicillin at 3.4–34 mM concentrations produced a dose-dependent inhibition of the Ca^{2+}-dependent release of GABA in cortex slices,[359,360] whereas 4 μM β-*p*-chlorphenylGABA had no significant effect.[361] However, in synaptosomes, this drug stimulated the spontaneous GABA release at 50 μM concentration and had no effect on GABA uptake.[362]

Certain venom toxins have been shown to enhance the release of GABA and to block its uptake, although their effect in general is not specific for GABA. A neurotoxin from the black widow spider venom was purified using as criterion its ability to stimulate the spontaneous release of GABA from synaptosomal fractions.[363] This stimulation was also observed in K^+-depolarized synaptosomes and was not affected by L-2,4-diaminobutyrate, tetrodotoxin, dithiotreitol, absence of Ca^{2+}, and replacement of Ca^{2+} by Sr^{2+} but was decreased in Na^+-free medium and abolished in Na^+- and Ca^{2+}- free medium. The uptake of GABA by synaptosomes was greatly inhibited by low concentrations of the toxin.[363] In cortex slices, a purified toxin from the same black widow venom, α-latrotoxin, also stimulated the spontaneous release of GABA and depleted the synaptic vesicle population in nerve endings in the slices.[364] It is not clear whether α-latrotoxin and the toxin purified by Grasso and Senni[363] correspond to the same molecule.

β-Bungarotoxin, a new toxin purified from *Bungarus* snakes, enhances the spontaneous release of GABA in synaptosomes,[365,366] and this effect is decreased in Ca^{2+}-free medium or in medium in which Ca^{2+} was replaced by Sr^{2+}.[365] This toxin also inhibited the high-affinity GABA uptake.[365,366] The possible participation of the phospholipase activity of the toxin in this releasing effect is controversial. For example, whereas in the presence of Ca^{2+} the toxin releases lactate dehydrogenase from the synaptosomes, and this action is abolished by substituting Sr^{2+} for calcium,[365] the release of nontransmitter amino acids is not affected by the toxin.[366] This specificity of action, however, had not been found in a previous study in which GABA and 2-deoxyglucose release was observed.[367]

Tityustoxin, a neurotoxin purified from the scorpion *Tityus serrulatus*, devoid of phospholipase activity, also enhances the spontaneous and the K^+-stimulated release of both endogenous and exogenous GABA from synaptosomes.[365,368] This effect is only partially dependent on external Ca^{2+}, but it is completely suppressed by tetrodotoxin. A similar action of tityustoxin was observed *in vivo* in superfused sensorimotor cortex of unanesthetized animals, and it is not restricted to GABA but similarly affects other amino acid transmitters.[368] These results have been interpreted as implying a depolarizing action of tityustoxin probably related to an effect on Na^+ channels. This action would also cause the inhibition of GABA uptake observed in synaptosomes.[366,368]

It should be mentioned that the above-described observations with neurotoxins from different animal venoms are not specific for GABA release but, because of their mechanisms of action, affect indiscriminately the release of all neurotransmitters, including that of acetylcholine in the neuromuscular junction. A more specific effect on GABA release has been reported with tetanus toxin.[369] When animals were injected *in vivo* with the toxin in the substantia nigra and the striatum, the Ca^{2+}-dependent release of labeled GABA stimulated by K^+ depolarization in slices of those regions was about 40% inhibited. In the striatal slices, a smaller inhibitory effect was observed on dopamine release, whereas that of acetylcholine and serotonin was not affected.[369]

A macrocyclic antiparasitic agent, avermectin B_1a, produces a long-lasting stimulation of radioactive GABA release independently of the presence of external Ca^{2+} in synaptosomal fractions, but no effect was observed on glutamate release.[370]

3.3. Developmental Studies of Uptake and Release

Since the parameters of uptake and release of GABA are intimately related to the synaptic function of GABA, they have been used in several articles as markers of the establishment and development of GABAergic synapses. Results of this type of experiment have not been consistent, and a number of discrepancies exist. In rat brain cortex homogenates, high-affinity uptake of GABA notably increased from 5 to 15 days of age and then declined slowly to the adult value.[371] In contrast, other studies reported a decrease in V_{max} in the "light" synaptosomal fraction from rat cortex from 7 days of age to adult.[372] No changes in V_{max} were observed in a crude rat striatal synaptosomal fraction from 14 days of age to the adult, with zero values at 12 days.[373]

In other studies with rat synaptosomes, the uptake of GABA was maximal at 7 days of age and decreased progressively to the adult value,[374] and a slightly higher uptake was found at 8 days than in the adult.[375] A decrease in the high-affinity GABA uptake was also observed in crude synaptosomal fraction from mouse brain from 5 or 10 days of age to the adult.[376] In contrast, in homogenates of chick embryo optic lobes and cerebral hemispheres, notable progressive increases were observed between 6 days of age and hatching, and there was a decrease at the eighth post-hatch day.[377]

No better agreement exists regarding the GABA release data during development. In slices from rat brain hemispheres or spinal cord, a similar partial Ca^{2+}-dependent release of GABA stimulated by K^+ depolarization was observed at 1 day, 10 days, or 13 weeks of age.[378] In contrast, no Ca^{2+}-dependent release was observed in synaptosomes from 7-day-old rats, only a small release was found at 14 days of age, and even at 21 days, only half the release value of that of the adult was observed.[374] Several conditions of GABA release have been comparatively studied in synaptosomal fractions from 8-day-old and adult rats.[375] In this detailed work, it was found that the homoexchange process and the carrier-mediated GABA release, which are dependent on monovalent cationic fluxes (see ref. 269), are already present in synaptosomes from 8-day-old rats, whereas the Ca^{2+}-dependent release stimulated by depolarization is not observed in these immature synaptosomes unless they are depleted of Na^+. Furthermore, in these synaptosomes, the uptake of GABA was inhibited by β-alanine, whereas in the adult synaptosomes it was not, thus suggesting greater gliosomal contamination in the immature synaptosomal fractions than in the adult synaptosomes.[375]

From the available data, it is difficult to conclude to what extent the release of GABA is an adequate parameter to assess the development of GABAergic synapses. It seems, however, that the uptake does not represent a good marker for such development. The possible damage of immature nerve endings during isolation, as compared to the adult terminals, and the different participation of glial structures at different ages are clearly factors that complicate the correct interpretation of these developmental studies.

4. THE BIOCHEMICAL DOMAIN OF THE GABAERGIC TERMINAL

Of course, the neurochemical studies of GABA in the CNS are aimed ultimately at an understanding of its physiological role at GABAergic synapses.

In this section, an attempt is made to integrate in a coherent way the results discussed in the previous sections and others not mentioned yet at the GABAergic ending, as if we were seated on top of a single GABA terminal and looked down at it. Admittedly, there are many results in the GABA literature that do no fit precisely in the integration attempted, and also not everything in it has been unambiguously demonstrated. However, it is my belief that this integration is reasonable in view of the available literature to date and that it may be valuable as a framework for future studies. After this integration is made, the pertinent data on the experimental alterations of GABAergic synaptic function and their consequences, mainly convulsive activity, are discussed in a subsection on pharmacological effects *in vivo*. These effects will, of course, be related to the integrative model.

Table II summarizes the fundamental neurochemical events that have been discussed in previous sections. These events are localized at the nerve ending shown in Fig. 1. The existence of two types of GAD activity, one independent of and the other dependent on free PLP (events I and III), the importance of pyridoxal kinase in the synthesis of PLP (event II), and the possible regulation

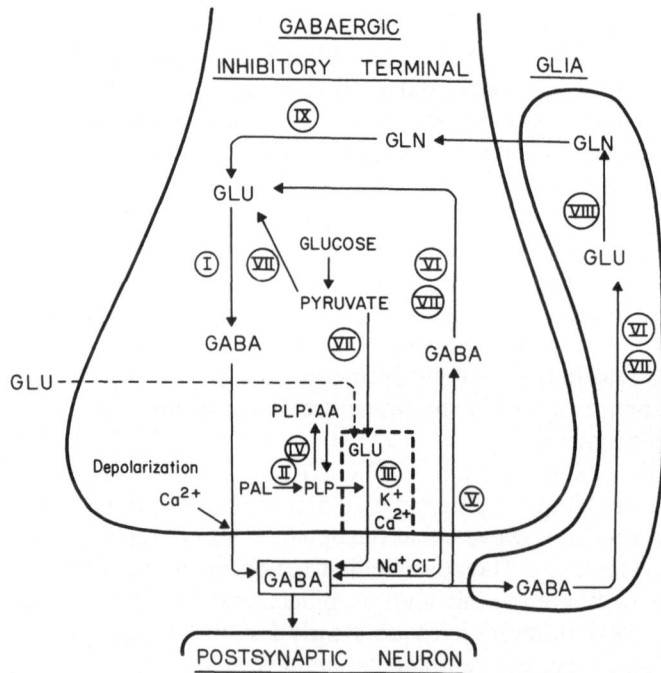

Fig. 1. The biochemical domain of the GABAergic synaptic terminal. This figure shows the location and relationships of the biochemical events listed in Table II (roman numbers) in the nerve ending and its surroundings. The thick discontinuous line indicates that event III occurs in the presynaptic membrane in the presence of K^+ and Ca^{2+}. The arrow from PLP to event III indicates the participation of this coenzyme in its regulation. PAL, pyridoxal. The thin discontinuous line refers to the uptake of glutamate into the terminal, a phenomenon that probably occurs only under experimental conditions. Discussion and justification for the postulates included in this integrated view can be found in the text.

Table II
Biochemical Events of GABAergic Transmission at the Presynaptic Ending (see Fig.
1 and Text for Explanation and Discussion)[a]

I. Free PLP-independent GAD activity and its probable relationship to release:

$$\text{GAD--PLP} \xleftrightarrow{\text{GLU}} \text{GAD--PLP·GLU} \xrightarrow{\text{CO}_2} \text{GABA} \xrightarrow{\text{Ca}^{2+}} \text{GABA}$$

II. Pyridoxal kinase activity:

$$\text{PYRIDOXAL} \xrightarrow{\text{ATP} \quad \text{ADP}} \text{PLP}$$

III. Free PLP-dependent GAD activity and its probable relationship to release:

$$\text{GAD} \xrightarrow{\text{PLP (GLU)}} \left\{ \begin{array}{c} \text{GAD·PLP} \\ \text{GAD·GLU} \end{array} \right\} \xrightarrow{\text{GLU (PLP)}} \text{GAD·PLP·GLU} \xrightarrow{\substack{\text{K+} \\ \text{Ca2+}}} \text{GABA}$$

IV. Trapping of PLP by amino groups (AA) (or, otherwise, dissociation of PLP from the free PLP-dependent GAD):

$$\text{PLP} + \text{AA} \leftrightarrow \text{PLP·AA}$$

V. Carrier-mediated uptake or release of GABA:

$$\text{GABA} \xleftarrow{\text{Na}^+, \text{Cl}^-} \text{GABA}$$

VI. GABA-T activity:

$$\text{GABA} \xleftrightarrow{\alpha\text{-KG}} \text{SSA} + \text{GLU}$$

VII. Glutamate synthesis through Krebs cycle reactions and glutamic–oxalacetic transaminase (GOT):

$$\left. \begin{array}{c} \text{Pyruvate} \\ \text{SSA} \end{array} \right\} \rightarrow \alpha\text{-KG} \leftrightarrow \text{GLU}$$

VIII. Glutamine synthetase activity:

$$\text{GLU} + \text{NH}_3 \xrightarrow{\text{ATP}} \text{GLN}$$

IX. Glutaminase activity:

$$\text{GLN} \rightarrow \text{GLU} + \text{NH}_3$$

[a] The vertical black bars in I, III, and V represent the presynaptic membrane. In I, III, and IV, the dots between names indicate high dissociation constants, whereas the dash represents a low dissociation constant. Abbreviations not defined previously: SSA, succinic semialdehyde; α-KG, α-ketoglutarate.

of PLP availability by its trapping in the form of Schiff bases or by its dissociation from the apoenzyme (event IV) are discussed in Section 2.1.1. The role
of Ca^{2+} and the relationships between GABA metabolism and its release
(events I, III, VIII, and IX) are reviewed in Sections 2.4, 3.2.1, and 3.2.2,
whereas the transport of GABA (uptake and release) by a Ca^{2+}-independent,
Na^+- and Cl^--dependent, carrier-mediated process (event V) is analyzed in
Sections 3.1.1 and 3.1.2. The degradation of GABA (event VI) is discussed in
Section 2.2.

The topographical location of events I–IX in the nerve ending and its
surroundings, shown in Fig. 1, requires some justification. Whereas no further
discussion will be made at this point about the metabolic compartmentation
scheme (the GABA–GLU–GLN–GLU–GABA cycle and the possible role of
glial cells), since it has been previously discussed at length (Sections 2.4 and
3.2.2 and refs. 23,24,185–190), the distinction between the location of two different GAD populations and its implications for GABA release have not been
adequately justified.

The subcellular location of GAD is in general accepted as synaptoplasmic,
since biochemical determinations of the enzyme activity have shown that most
of it is in the supernatant after osmotic disruption of the synaptosomes.[73,230,231]
However, electron microscopic immunocytochemical studies of GAD have
shown that at least part of the enzyme is associated with the presynaptic membrane and other membranous structures inside the terminal.[53] Furthermore, it
has been demonstrated that in the presence of Ca^{2+}, GAD binds strongly to
the nerve ending membrane.[230,231]

On these bases, it seems possible that at least part of the total GAD population of molecules may be bound to the presynaptic membrane under physiological conditions, a possibility that is particularly relevant for the hypothesis
of a synthesis–secretion coupling of GABA (Section 3.2.2, refs. 69,200–
202,338). The mechanism of this Ca^{2+}-dependent binding of GAD has recently
been studied in our laboratory using phospholipid vesicles as a membrane
model.[379,380] It was found that the Ca^{2+}-dependent binding of GAD to these
membranes, as well as the lack of binding of other soluble synaptosomal enzymes such as cholineacetyltransferase and lactate dehydrogenase, was very
similar, both qualitatively and quantitatively, to that previously reported with
synaptic plasma membranes.[60,230,231,379,380] It was shown that this binding was
primarily of ionic nature and, accordingly, that a concentration of potassium
ions similar to its physiological intracellular concentration (100 mM) was almost
as effective as 2 mM Ca^{2+} in inducing the binding of GAD. Furthermore, it
was observed that only part of the population of GAD molecules is capable of
binding and that this part was mostly the free PLP-dependent enzyme.[379,380]

All of the above results have been taken into account to differentiate the
GAD activities I and III in Fig. 1. Whereas the free PLP-independent activity
is cytoplasmic, the free PLP-dependent one is bound to the presynaptic membrane by the high intraterminal K^+ concentration and possibly also by Ca^{2+}.
The role of the latter divalent cation at GABAergic terminals, therefore, might
be double: to participate in the depolarization–secretion coupling and in the
binding of GAD to the presynaptic membrane for the synthesis–secretion

coupling[69] As already mentioned, direct evidence for the existence of the latter mechanism of release has been obtained by inhibiting GAD activity and demonstrating an immediate decrease in spontaneous GABA release, both *in vitro*[201] and *in vivo*, in the substantia nigra[339] and in the striatum.[340]

According to the foregoing discussion, two mechanisms of GABA release in which Ca^{2+} participates are postulated in Fig. 1: (1) a phasic release by the depolarization–secretion coupling discussed in Section 3.2.1, which utilizes a pool of GABA previously synthesized in the cytoplasm by the free PLP-independent GAD system I, and (2) a tonic, continuous release of the GABA synthesized at the membrane level by the GAD system III, which would be also promoted by the high intracellular K^+ concentration. This GAD system is regulated by the free PLP concentration which, in turn, is controlled by the activity of pyridoxal kinase (II) and by the possibility of being displaced by other intracellular compounds (IV). Besides these two release mechanisms in which Ca^{2+} participates, the Na^+-dependent carrier-mediated mechanism (V) discussed in Section 3.1.2 would also function for the removal of excess GABA after its phasic release or, under resting conditions, for the maintenance of a certain amount of GABA at the synapse, which would contribute to the tonic inhibition exerted by GABA on the postsynaptic neurons.

The metabolic origin of the glutamate substrates for GAD systems I and III is discussed in Section 3.2.2. Glutamine seems to be a better precursor for the glutamate used in the depolarization-stimulated GABA release than the exogenously supplied glutamate,[196,198,337] although glucose and pyruvate can also feed this glutamate pool.[200,203–205] In contrast, exogenous glutamate, presumably mixing with the glutamate derived from glucose metabolism (VII), seems to feed preferentially the GAD system (III) for the spontaneous, synthesis-dependent release of GABA.[196] These metabolic relationships of the releasable pools of GABA are not, of course, completely independent from one another. All that is postulated in Fig. 1 is the predominant equilibrium situation.

4.1. Pharmacological Studies of GABA Presynaptic Mechanisms in Vivo

There is one behavioral alteration that seems to be closely related to a decrease or inhibition of the functioning of GABA synapses *in vivo*: the occurrence of convulsive activity. Therefore, it is reasonable to use the criterion of appearance of convulsions after selective inhibition of the events summarized in Table II and Fig. 1 to confirm their involvement in GABAergic transmission *in vivo*. Thus, as has been reviewed before,[14,38–40,169,338] the inhibition of GAD activity *in vivo* to a certain level results almost invariable in seizure activity. This effect is independent of the concentration of GABA in brain[69] and, as discussed in Section 2.1.1, may be produced as a consequence of decreasing PLP concentration through the inhibition of pyridoxal kinase, by carbonyl-trapping drugs, or by vitamin B_6-deficient diets.[62–74]

A series of publications has appeared in recent years on the effects of some convulsant drugs on GAD activity *in vivo*. The drugs used include the D- and L-isomers of allylglycine (the effects of this drug *in vivo* seem to result from its metabolite, α-keto-4-pentenoic acid[102–104]),[71,102–104,381–385] some PLP antagonists such as methyldithiocarbazinate, thiosemicarbazide, and 4-deoxypyridoxine,[71] or methoxpyridoxine,[72,386] and 3-mercaptopropionic acid.[387] The results of all of these experiments agree in general with the existence of a causal relationship between GAD inhibition and occurrence of convulsions. That other PLP-dependent enzymes, such as dihydroxyphenylalanine decarboxylase, may be inhibited by a deficiency in the PLP coenzymatic function had been shown some years ago, but it has also been demonstrated that the inhibition of enzymes other than GAD is not related to the production of convulsions.[68]

These findings of the specific relationship between GAD inhibition and seizures have been confirmed recently using other drugs.[71,388] It is also important to emphasize that with some of the convulsant drugs mentioned, it has been demonstrated that the inhibition of GABA synthesis occurs in the nerve endings,[73] a finding that has also been confirmed recently.[389] All of the above reviewed reports clearly support the events II, III and IV of Fig. 1. Regarding in particular event III, it is worth recalling that the inhibition of GAD by locally applied 3-mercaptopropionic acid results in an immediate decrease of the spontaneous release of GABA in vivo.[339,340]

The elevation of GABA levels in vivo produced by the administration of GABA-T inhibitors was reviewed in Section 2.2.1, as was the possible relationship of this elevation to anticonvulsant effects. As mentioned in that section, the problems of interpretation of these results are caused mainly by the different structural and metabolic compartments that might be specifically or preferentially affected by the drugs (see Sections 2.4 and 3.2.2).

Because of this situation, and in order to be able to establish these relationships at a synaptic level, several attempts have been made recently to correlate the concentration of GABA in synaptosomes with the convulsant or anticonvulsant action of drugs that modify the metabolism or uptake of GABA. Wood et al.[390,391] first showed that it was possible to measure the effect of drugs administered in vivo on GABA levels in synaptosomes without serious limitations of postmortem changes or redistribution during subcellular fractionation. The elevation of GABA in total brain produced by an anticonvulsant dose of aminooxyacetic acid[14] was also observed in synaptosomal fractions, where GABA-T activity was inhibited.[390] However, the synaptosomal elevation was less than that in the soluble fraction, and, as expected from other findings,[169,338] the time of maximal elevation after treatment was not correlated with the maximal protection against convulsions.[390]

More recently, the same group[392] reported that synaptosomal GABA also increased after treatment with cycloserine or gabaculine and that the change in synaptosomal GABA levels was correlated with the anticonvulsant effect of these compounds against convulsions produced by some GAD inhibitors (isoniazid, asymmetric dimethylhydrazine, and high doses of aminooxyacetic acid).[14] However, the anticonvulsant activity was manifested not by suppression but only by a delay in the appearance of convulsions, and, interestingly, a similar percent decrease in synaptosomal GABA concentration (about 25%) was observed at the onset of convulsions produced by the three convulsants used.[392] When the three convulsants mentioned were administered without anticonvulsant, a decrease in synaptosomal GABA content was observed in spite of the fact that GABA levels in total brain and in the soluble fraction were increased.[391] These results, therefore, support the postulate that GAD system III is more important for the maintenance of the normal excitability level than the elevation of GABA as a result of blocking event VI.

A similar conclusion was reached by Sarhan and Seiler[393] on the basis of (1) the changes in synaptosomal GABA produced by γ-acetylenic-GABA and γ-vinyl-GABA, (2) the inhibitory effect of the former on GAD activity, and (3) the anticonvulsant action of these drugs against seizures induced by two GAD inhibitors, 3-mercaptopropionic acid and pyridoxal phosphate-glutamyl hydrazone (although these authors used a crude synaptosomal fraction containing free mitochondria). The latter drug had previously been shown to inhibit GAD activity *in vivo* in synaptosomal fractions.[73]

Löscher[394] has reported that the main GABA-T inhibitors used recently, aminooxyacetic acid, gabaculine, ethanolamine-O-sulfate, γ-acetylenic-GABA, γ-vinyl-GABA, and dipropylacetate, administered at anticonvulsant doses inhibited GABA-T and increased GABA levels in synaptosomes. The most effective drugs on these two parameters were aminooxyacetic acid, γ-vinyl-GABA, and γ-acetylenic-GABA, whereas the effects of dipropylacetate were very small. A surprising finding was that all of the drugs studied, with the exception of γ-acetylenic-GABA which inhibited it, increased the activity of GAD in synaptosomes. This effect was particularly notable in the case of ethanolamine-O-sulfate.[394] This observation is probably the only one in the literature in which GAD activity is increased *in vivo* by pharmacological means, and, if confirmed, might be at least partially responsible for the anticonvulsant effects of the drugs mentioned.

A reasonable conclusion from the above-described experiments is that an elevation of the intraterminal GABA level, rather than that in extraterminal compartments, is probably related to a decrease of neuronal excitability because the release by mechanism I will be facilitated. In fact, in their experiments *in vivo*, Van der Heyden et al.[339] found a delayed increase in endogenous GABA release from the substantia nigra after local treatment with aminooxyacetic acid, and Abdul-Ghani et al.[395] reported increased GABA release from the sensorimotor cortex after systemic administration of γ-vinyl-GABA and γ-acetylenic-GABA (but no effect was observed with dipropylacetate). However, in spite of this facilitation, if GAD activity is inhibited,[391,392] the relative contribution of system III would decrease, the synaptic extracellular GABA would be reduced below the necessary concentration to maintain normal excitability, and convulsions will occur.

Similar conclusions on the correlation between increased intraterminal GABA concentration and protection against convulsions were reached by Gale and Iadarola in a series of experiments using a completely different approach.[396–399] These authors studied the effect of anticonvulsant doses of aminooxyacetic acid, γ-vinyl-GABA, and dipropylacetate on GABA content in the substantia nigra after surgically destroying its afferent pathways on one side by hemitransection. Since this lesion results in the loss of GABAergic terminals, the drug-induced increases in GABA levels in the denervated substantia nigra have to be ascribed to an extraterminal GABA pool. The results of these experiments indicate that the effect of aminooxyacetic acid is more notable in the non-nerve-terminal compartment, whereas that of dipropylacetate, and of γ-vinyl-GABA after 60 h, was mostly on the nerve terminal pool. Although the elevation of GABA observed was not quantitatively very impressive (about 30%), it correlated well with the anticonvulsant action of the drugs mentioned.[396-399]

The postulate that the pool of GABA released by depolarization via a Ca^{2+}-mediated process is also important for the physiological control of excitability is supported by the observation that the intracranial administration of ruthenium red, a drug that inhibits the Ca^{2+}-dependent release of GABA,[319,320] produces long-lasting convulsive states.[400]

If the uptake mechanism (V) is physiologically functioning, it would be expected that its inhibition *in vivo* would result in protection against convulsions. This expectation has been confirmed recently by two groups.[401,402] It was observed that several inhibitors of the high-affinity GABA uptake, of which nipecotic acid and especially its ethyl ester were the most effective, increased the thresholds of convulsions produced by electroshock or pentylenetetrazole[401] or diminished the percentage of convulsive symptoms of audiogenic seizures in DBA/2 mice.[402] The mechanism of such anticonvulsant action, however, is not clear, because it has been reported that nipecotic ethyl ester administered *in vivo* produces an elevation of synaptosomal GABA levels.[403] Since this increase is higher than that observed in whole brain, whereas the opposite occurs with GABA-T inhibitors, it has been proposed that the anticonvulsant effect of GABA uptake inhibitors might be caused by a diminished uptake into glial cells or neuronal perikarya.[403] This effect would result in an increase in the extracellular GABA, and possibly this would be responsible for the anticonvulsant effect. If this were true, the increased synaptosomal GABA would be a consequence of increased extracellular GABA rather than a mechanism for the anticonvulsant action. However, this possibility implies that nipecotic acid would not be inhibiting GABA uptake into nerve endings. Obviously, it is difficult at the present time to have a clear picture of the action *in vivo* of GABA uptake inhibitors.

5. CONCLUDING REMARKS: GABAERGIC PATHWAYS IN THE CNS

The four previous sections in this chapter have focused on the metabolic and transport mechanisms of GABA as related to the synaptic phenomena of GABAergic transmission. The knowledge of these phenomena is essential for the understanding of the role of GABA-dependent inhibition in the functions of the CNS and of the biochemical alterations involved in neuropsychiatric diseases and their rational pharmacological treatments. It is clear, however, that besides those mechanisms, knowledge of the neuronal pathways using GABA as a transmitter in the brain is also necessary.

Several experimental approaches toward the acquisition of this knowledge have been followed. These include the binding of specific GABA receptor agonists in different brain zones, the autoradiographic localization of labeled GABA taken up through the high-affinity uptake previously discussed, the Ca^{2+}-dependent, depolarization-stimulated release of GABA from specific regions, the measurement of GABA concentration and GAD activity in different brain regions, the changes in these five parameters after surgical or chemical lesions of afferent pathways, and the immunohistochemical localization of GAD by means of specific antibodies. The results of these studies have already given valuable information on the mapping of GABAergic pathways in the CNS, although a complete picture is not yet available.

Many reviews have been published in recent years on the localization of GABA neurons and pathways with the biochemical and autoradiographic approaches mentioned above.[404–419] In these reviews, tables may be found with the values of GABA levels and GAD activity in a large number of CNS areas, including the retina, in various species as well as summaries of the results of lesion experiments. An analysis of these reviews will reveal that the concentrations of GABA and GAD vary much less between regions than the content and enzymes of other neurotransmitters, such as catecholamines and serotonin (see, for example, ref. 405 for compared values). This can be taken as an indication of the ubiquity of GABAergic synapses in the CNS and therefore of the importance of GABAergic inhibitory mechanisms in its functions.

The possibilities of mapping the GABAergic pathways in the CNS increased enormously with the production of specific antibodies against GAD and their use for immunocytochemical studies. Some of the publications in which GABAergic terminals and neurons have been located with this immunologic technique have been already mentioned.[17,51–58] Two of these[17,54] review the development of the immunocytochemical methods for GABA neurons as well as the main observations made in the cerebellum, spinal cord, olfactory bulb, and basal ganglia.[17,54] Results obtained after these reviews appeared include the demonstration of GAD-containing neuronal somata, dendrites, and axon terminals in the visual cortex,[420] in the neostriatum and the globus pallidus,[421] in neuronal somata of the nucleus reticularis thalami,[57] and in terminals presynaptic to primary afferent nerve endings in the substantia gelatinosa of the spinal cord.[422] In other articles, it has been shown that GAD immunocytochemically visualized decreases in the substantia nigra after lesioning the striatonigral and the pallidonigral pathways[423] and in the lateral habenula after lesion of the stria medullaris.[424] Besides the already mentioned work on the GAD immunocytochemistry in rabbit retina,[58] this technique has been used also to locate GAD-containing cells in the retina of the goldfish,[425] frog,[426] and rat.[427,428]

Many of the previous findings on the location of GAD-containing terminals in the cerebellum, spinal cord, and substantia nigra were recently confirmed.[51] In this publication, two types of terminals were identified immunohistochemically, differing in size, intensity, and location, and it was suggested that the large, strongly immunoreactive terminals correspond to Golgi type I neurons, whereas the small, weakly to moderately immunoreactive, and finely distributed terminals belong to Golgi type II neurons.[51]

The distribution of GABA terminals in the hypothalamus, and particularly in the median eminence,[51,414,429] is suggestive of a role for GABA in the control of hormones from the anterior pituitary. Such a possibility, particularly regarding prolactin secretion, is supported by several experimental findings, which have been reviewed recently.[430–435] This possible role of GABA increases the already enormous range of CNS functions in which GABA seems to participate as an inhibitory neurotransmitter.

ACKNOWLEDGMENTS. Part of the work from the author's laboratory mentioned in this chapter was supported by a grant from the Consejo Nacional de Ciencia y Tecnología, México, D. F. (Project PCCBNAL 790214).

REFERENCES

1. Roberts, E., Baxter, C. F., Van Harreveld, A., Wiersma, C. A. G., Adey, W. R., and Killam, K. F. (eds.), 1960, *Inhibition in the Nervous System and Gamma-Aminobutyric Acid*, Pergamon Press, Oxford.
2. Roberts, E., Chase, T. N., and Tower, D. B. (eds.), 1976, *GABA in Nervous System Function*, Raven Press, New York.
3. Krogsgaard-Larsen, P., Scheel-Krüger, J., and Kofod, H. (eds.), 1978, *GABA-Neurotransmitters: Pharmacochemical, Biochemical and Pharmacological Aspects*, Munksgaard, Copenhagen.
4. Mandel, P., and DeFeudis, F. V. (eds.), 1979, *GABA—Biochemistry and CNS Functions*, Plenum Press, New York.
5. Lal, H. (ed.), 1980, *GABA Neurotransmission: Current Developments in Physiology and Neurochemistry*, Brain Res. Bull. 5(Suppl. 2).
6. Berl, S., Clarke, D. D., and Schneider, D. (eds.), 1975, *Metabolic Compartmentation and Neurotransmission*, Plenum Press, New York.
7. Bradford, H. F., and Marsden, C. D. (eds.), 1976, *Biochemistry and Neurology*, Academic Press, London.
8. Levi, G., Battistin, L., and Lajtha, A. (eds.), 1976, *Transport Phenomena in the Nervous System*, Plenum Press, New York.
9. Garattini, S., Pujol, J. F., and Samanin, R. (eds.), 1978, *Interactions between Putative Neurotransmitters in the Brain*, Raven Press, New York.
10. Fonnum, F. (ed.), 1978, *Amino Acids as Chemical Transmitters*, Plenum Press, New York.
11. Tapia, R., and Cotman, C. W. (eds.), 1981, *Regulatory Mechanisms of Synaptic Transmission*, Plenum Press, New York.
12. Baxter, C. F., 1970, *Handbook of Neurochemistry*, Volume 3 (A. Lajtha, ed.), Plenum Press, New York, pp. 289–353.
13. Iversen, L. L., 1972, *Perspectives in Neuropharmacology* (S. H. Snyder, ed.), Oxford University Press, London, pp. 75–111.
14. Tapia, R., 1975, *Handbook of Psychopharmacology*, Volume 4 (L. L. Iversen, S. D. Iversen, and S. H. Snyder, eds.), Plenum Press, New York, pp. 1–58.
15. Iversen, L. L., 1978, *Psychopharmacology: A Generation of Progress* (M. A. Lipton, A. DiMascio, and K. F. Killam, eds.), Raven Press, New York, pp. 25–38.
16. Mao, C. C., and Costa, E., 1978, *Psychopharmacology: A Generation of Progress* (M. A. Lipton, A. DiMascio, and K. F. Killam, eds.), Raven Press, New York, pp. 307–318.
17. Roberts, E., 1978, *Psychopharmacology: A Generation of Progress* (M. A. Lipton, A. DiMascio, and K. F. Killam, eds.), Raven Press, New York, pp. 95–102.
18. DeFeudis, F. V., 1975, *Annu. Rev. Pharmacol.* 15:105–130.
19. Johnston, G. A. R., 1978, *Annu. Rev. Pharmacol. Toxicol.* 18:269–289.

20. Curtis, D. R., and Johnston, G. A. R., 1974, *Ergeb. Physiol.* **69**:97–188.
21. Krnjević, K., 1974, *Physiol. Rev.* **54**:418–540.
22. Iversen, L. L., and Kelly, J. S., 1975, *Biochem. Pharmacol.* **24**:933–938.
23. Tapia, R., 1980, *Glutamine: Metabolism, Enzymology and Regulation* (J. Mora and R. Palacios, eds.), Academic Press, New York, pp. 285–297.
24. Baxter, C. F., 1976, *GABA in Nervous System Function* (E. Roberts, T. N. Chase, and D. B. Tower, eds.), Raven Press, New York, pp. 61–87.
25. Wu, J.-Y., 1980, *Brain Res. Bull.* **5**(Suppl. 2):31–36.
26. Wu, J.-Y., 1977, *J. Neurochem.* **28**:1359–1367.
27. Miller, L. P., and Martin, D. L., 1973, *Life Sci.* **13**:1023–1032.
28. Martin, D. L., and Miller, L. P., 1976, *GABA in Nervous System Function* (E. Roberts, T. N. Chase, and D. B. Tower, eds.), Raven Press, New York, pp. 57–58.
29. Walsh, J. M., and Clark, J. B., 1976, *J. Neurochem.* **26**:1307–1309.
30. Kanazawa, I., Iversen, L. L., and Kelly, J. S., 1976, *J. Neurochem.* **27**:1267–1269.
31. Kremzner, L. T., Hiller, J. M., and Simon, E. J., 1975, *J. Neurochem.* **25**:889–894.
32. De Mello, F. G., Bachrach, U., and Nirenberg, M., 1976, *J. Neurochem.* **27**:847–851.
33. Sobue, K., and Nakajima, T., 1978, *J. Neurochem.* **30**:277–279.
34. Murrin, L. C., 1980, *J. Neurochem.* **34**:1779–1781.
35. Seiler, N., Bink, G., and Grove, J., 1980, *Neuropharmacology* **19**:251–258.
36. Nistico, G., Ientile, R., Rotiroti, D., and DiGiorgio, R. M., 1980, *Biochem. Pharmacol.* **29**:954–957.
37. Pajunen, A. E. I., Hietala, O. A., Baruch-Virransalo, E. L., and Piha, R. S., 1979, *J. Neurochem.* **32**:1401–1408.
38. Wood, J. D., 1975, *Prog. Neurobiol.* **5**:77–95.
39. Meldrum, B. S., 1975, *Int. Rev. Neurobiol.* **17**:1–36.
40. Tower, D. B., 1976, *GABA in Nervous System Function* (E. Roberts, T. N. Chase, and D. B. Tower, eds.), Raven Press, New York, pp. 461–478.
41. Wu, J.-Y., 1976, *GABA in Nervous System Function* (E. Roberts, T. N. Chase, and D. B. Tower, eds.), Raven Press, New York, pp. 7–55.
42. Saito, K., 1976, *GABA in Nervous System Function* (E. Roberts, T. N. Chase, and D. B. Tower, eds.), Raven Press, New York, pp. 103–111.
43. Wu, J.-Y., Su, Y. Y. T., Lam, D. M. K., Brandon, C., and Denner, L., 1980, *Brain Res. Bull.* **5**(Suppl. 2):63–70.
44. Su, Y. Y. T., Wu, J.-Y., and Lam, D. M. K., 1979, *J. Neurochem.* **33**:169–179.
45. Chude, O., Roberts, E., and Wu, J.-Y., 1979, *J. Neurochem.* **32**:1409–1415.
46. Hadjian, R. A., and Stewart, J. A., 1977, *J. Neurochem.* **28**:1249–1257.
47. Blindermann, J.-M., Maitre, M., Ossola, L., and Mandel, P., 1978, *Eur. J. Biochem.* **86**:143–152.
48. Oertel, W. H., Tappaz, M. L., Kopin, I. J., Ransom, D. H., and Schmechel, D. E., 1980, *Brain Res. Bull.* **5**(Suppl. 2):713–719.
49. Yamaguchi, T., and Matsumura, Y., 1977, *Biochim. Biophys. Acta* **481**:706–711.
50. Possani, L. D., Bayón, A., and Tapia, R., 1977, *Neurochem. Res.* **2**:51–57.
51. Pérez de la Mora, M., Possani, L. D., Tapia, R., Terán, L., Palacios, R., Fuxe, K., Hökfelt, T., and Ljungdahl, Å., 1981, *Neuroscience* **6**:875–895.
52. Barber, R., and Saito, K., 1976, *GABA in Nervous System Function* (E. Roberts, T. N. Chase, and D. B. Tower, eds.), Raven Press, New York, pp. 113–132.
53. Wood, J. G., McLaughlin, B. J., and Vaughn, J. E., 1976, *GABA in Nervous System Function* (E. Roberts, T. N. Chase, and D. B. Tower, eds.), Raven Press, New York, pp. 133–148.
54. Roberts, E., 1978, *Interactions between Putative Neurotransmitters* (S. Garattini, J. F. Pujol, and R. Samanin, eds.), Raven Press, New York, pp. 89–105.
55. Ribak, C. E., Vaughn, J. E., and Saito, K., 1978, *Brain Res.* **140**:315–332.
56. Ribak, C. E., Vaughn, J. E., and Roberts, E., 1979, *J. Comp. Neurol.* **187**:261–284.
57. Houser, C. R., Vaughn, J. E., Barber, R. P., and Roberts, E., 1980, *Brain Res.* **200**:341–354.
58. Brandon, C., Lam, D. M. K., and Wu, J.-Y., 1979, *Proc. Natl. Acad. Sci. U.S.A.* **76**:3557–3561.

59. Wu, J.-Y., and Roberts, E., 1974, *J. Neurochem.* **23:**759–767.
60. Tapia, R., and Covarrubias, M., 1978, *Amino Acids as Chemical Transmitters* (F. Fonnum, ed.), Plenum Press, New York, pp. 431–438.
61. Roberts, E., Younger, F., and Frankel, S., 1951, *J. Biol. Chem.* **191:**277–285.
62. Minard, F. N., 1967, *J. Neurochem.* **14:**681–692.
63. Dakshinamurti, K., and Stephens, M. C., 1969, *J. Neurochem.* **16:**1515–1522.
64. Bayoumi, R. A., and Smith, W. R. D., 1972, *J. Neurochem.* **19:**1883–1897.
65. Tapia, R., Pérez de la Mora, M., and Massieu, G. H., 1969, *Ann. N.Y. Acad. Sci.* **166:**257–266.
66. Tapia, R., and Awapara, J., 1969, *Biochem. Pharmacol.* **18:**145–152.
67. Tapia, R., and Pasantes, H., 1971, *Brain Res.* **29:**111–122.
68. Tapia, R., Pasantes-Morales, H., Taborda, E., and Pérez de la Mora, M., 1975, *J. Neurobiol.* **6:**159–170.
69. Tapia, R., Sandoval, M. E., and Contreras, P., *J. Neurochem.* **24:**1283–1285.
70. Killam, K. F., Dasgupta, S. R., and Killam, E. K., 1960, *Inhibition in the Nervous System and Gamma-Aminobutyric Acid* (E. Roberts, C. F. Baxter, A. Van Harreveld, C. A. G. Wiersma, W. R. Adey, and K. F. Killam, eds.), Pergamon Press, Oxford, pp. 302–316.
71. Sawaya, C., Horton, R., and Meldrum, B., 1978, *Biochem. Pharmacol.* **27:**475–481.
72. Nitsch, C., 1980, *J. Neurochem.* **34:**822–830.
73. Pérez de la Mora, M., Feria-Velasco, A., and Tapia, R., 1973, *J. Neurochem.* **20:**1575–1587.
74. Tower, D. B., 1969, *Basic Mechanisms of the Epilepsies* (H. H. Jasper, A. A. Ward, Jr., and A. Pope, eds.), Little, Brown, Boston, pp. 611–638.
75. Coursin, D. B., 1954, J.A.M.A. **154:**406–408.
76. Coursin, D. B., 1955, *Am. J. Dis. Child.* **90:**344–348.
77. Hunt, A. D., Stokes, J., McCrory, W. W., and Stroud, H. H., 1954, *Pediatrics* **13:**140–145.
78. Sokoloff, L., Lassen, N. A., McKhann, G. M., Tower, D. B., and Albers, W., 1959, *Nature* **173:**751–753.
79. Coursin, D. B., 1964, *Vitam. Horm.* **22:**755–786.
80. Coursin, D. B., 1969, *Ann. N.Y. Acad. Sci.* **166:**7–15.
81. Scriver, C. R., and Whelan, D. T., 1969, *Ann. N.Y. Acad. Sci.* **166:**83–96.
82. Tapia, R., and Sandoval, M. E., 1971, *J. Neurochem.* **18:**2051–2059.
83. Bayón, A., Possani, L. D., and Tapia, R., 1977, *J. Neurochem.* **29:**513–517.
84. Bayón, A., Possani, L. D., Tapia, M., and Tapia R., 1977, *J. Neurochem.* **29:**519–525.
85. Bayón, A., Possani, L. D., Rode, G., and Tapia, R., 1978, *J. Neurochem.* **30:**1629–1631.
86. Miller, L. P., and Martin, D. L., 1976, *Life Sci.* **19:**281–288.
87. Miller, L. P., Walters, J. R., and Martin, D. L., 1977, *Nature* **266:**847–848.
88. Miller, L. P., Martin, D. L., Mazumder, A., and Walters, J. R., 1978, *J. Neurochem.* **30:**361–369.
89. Seligman, B., Miller, L. P., Brockman, D. E., and Martin, D. L., 1978, *J. Neurochem.* **30:**371–376.
90. Miller, L. P., and Walters, J. R., 1979, *J. Neurochem.* **33:**533–539.
91. Martin, S. B., and Martin, D. L., 1979, *J. Neurochem.* **33:**1275–1283.
92. Martin, D. L., Meeley, M. P., Martin, S. B., and Pedersen, S., 1980, *Brain Res. Bull.* **5**(Suppl. 2):57–61.
93. Miller, L. P., Walters, J. R., Eng, N., and Martin, D. L., 1980, *Brain Res. Bull.* **5**(Suppl. 2):89–94.
94. Turský, T., and Laššánová, M., 1978, *J. Neurochem.* **30:**903–905.
95. Tunnicliff, G., and Ngo, T. T., 1979, *Can. J. Physiol. Pharmacol.* **57:**873–877.
96. Gold, B. I., Simon, J. R., and Roth, R. H., 1978, *Life Sci.* **22:**187–194.
97. Gold, B. I., and Roth, R. H., 1979, *J. Neurochem.* **32:**883–888.
98. Huger, F. P., and Gold, B. I., 1980, *Biochem. Pharmacol.* **29:**3034–3036.
99. Gey, K. F., and Georgi, H., 1974, *J. Neurochem.* **23:**725–738.
100. Dye, D. J., and Taberner, P. V., 1975, *J. Neurochem.* **24:**977–1001.
101. Taberner, P. V., Pearce, M. J., and Watkins J. C., 1977, *Biochem. Pharmacol.* **26:**345–349.
102. Horton, R. W., 1978, *Biochem. Pharmacol.* **27:**1471–1477.

103. Orlowski, M., Reingold, D. F., and Stanley, M. E., 1977, *J. Neurochem.* **28**:349–353.
104. Reingold, D. F., and Orlowski, M., 1979, *J. Neurochem.* **32**:907–913.
105. De Boer, T., Bruinvels, J., and Bonta, I. L., 1979, *J. Neurochem.* **33**:597–601.
106. Chrystal, E., Bey, P., and Rando, R. R., 1979, *J. Neurochem.* **32**:1501–1507.
107. Endo, A., Kitahara, N., Oka, H., Miguchi-Fukazawa Y., and Terahara, A., 1978, *Eur. J. Biochem.* **82**:257–259.
108. Bouclier, M., Jung, M. J., and Lippert, B., 1979, *Eur. J. Biochem.* **98**:363–368.
109. Galli-Kienle, M., Bosisio, E., Manzocchi, A., and Santaniello, E., 1979, *J. Neurochem.* **32**:1361–1363.
110. Tunnicliff, G., and Ngo, T. T., 1978, *Experientia* **34**:989–990.
111. Tapia, R., and Meza-Ruiz, G., 1975, *J. Neurobiol.* **6**:171–180.
112. Tapia, R., and Meza-Ruiz, G., 1976, *Neurochem. Res.* **1**:133–140.
113. López-Colomé, A. M., Salceda, R., and Tapia, R., 1979, *Neurochem. Res.* **4**:567–573.
114. Maderdrut, J. L., 1979, *Neuroscience* **4**:995–1005.
115. Reitzel, J. L., Maderdrut, J. L., and Oppenheim, R. W., 1979, *Brain Res.* **172**:487–504.
116. Loh, Y. P., 1976, *J. Neurochem.* **26**:1303–1305.
117. Ma, P. M., and Grant, P., 1978, *Brain Res.* **140**:368–373.
118. Hollyfield, J. G., Rayborn, M. E., Sarthy, P. V., and Lam, D. M. K., 1979, *J. Comp. Neurol.* **188**:587–598.
119. Lam, D. M. K., Fung, S.-C., and Kong, Y. C., 1980, *J. Comp. Neurol.* **193**:89–102.
120. Gilad, G. M., and Kopin, I. J., 1979, *J. Neurochem.* **33**:1195–1204.
121. Wu, J.-Y., Wong, E., Saito, K., Roberts, E., and Schousboe, A., 1976, *J. Neurochem.* **27**:653–659.
122. Cash, C., Maitre, M., Ciesielski, L., and Mandel, P., 1974, *FEBS Lett.* **47**:199–203.
123. Bloch-Tardy, M., Rolland, B., and Gonnard, P., 1974, *Biochimie* **56**:823–832.
124. Maitre, M., Ciesielski, C., Cash, C., and Mandel, P., 1975, *Eur. J. Biochem.* **52**:157–169.
125. John, R. A., and Fowler, L. J., 1976, *Biochem. J.* **155**:645–651.
126. Schousboe, A., Saito, K., and Wu, J.-Y., 1980, *Brain Res. Bull.* **5**(Suppl. 2):71–76.
127. Beeler, T., and Churchich, J. E., 1978, *Eur. J. Biochem.* **85**:365–371.
128. Tews, J. K., Riegel, E. A., and Harper, A. E., 1980, *Brain Res. Bull.* **5**(Suppl. 2):245–251.
129. White, H. L., and Sato, T. K., 1978, *J. Neurochem.* **31**:41–47.
130. Wu, J.-Y., Moss, L. G., and Chude, O., 1978, *Neurochem. Res.* **3**:207–219.
131. Schousboe, A., Wu, J.-Y., and Roberts, E., 1973, *Biochemistry* **12**:2868–2873.
132. Tunnicliff, G., Ngo, T. T., Rojo-Ortega, J. M., and Barbeau, A., 1977, *Can. J. Biochem.* **55**:479–484.
133. Walsh, J. M., and Clark, J. B., 1976, *Biochem. J.* **160**:147–157.
134. Metcalf, B. W., 1979, *Biochem. Pharmacol.* **28**:1705–1712.
135. Rando, R. R., 1977, *Biochemistry* **16**:4604–4610.
136. Lippert, B., Metcalf, B. W., Jung, M. J., and Casara, P., 1977, *Eur. J. Biochem.* **74**:441–445.
137. Fowler, L. J., and John, R. A., 1972, *Biochem. J.* **130**:569–573.
138. Jung, M. J., Lippert, B., Metcalf, B. W., Schechter, P. J., Böhlen, P., and Sjoerdsma, A., 1977, *J. Neurochem.* **28**:717–723.
139. Lippert, B., Jung, M. J., and Metcalf, B. W., 1980, *Brain Res. Bull.* **5**(Suppl. 2):375–379.
140. Jung, M. J., Lippert, B., Metcalf, B. W., Böhlen, P., and Schechter, P. J., 1977, *J. Neurochem.* **29**:797–802.
141. Rando, R. R., and Bangerter, F. W., 1976, *J. Am. Chem. Soc.* **98**:6762–6764.
142. Kobayashi, K., and Miyazawa, S., 1977, *FEBS Lett.* **76**:207–210.
143. Allan, R. D., Johnston, G. A. R., and Twitchin, B., 1977, *Neurosci. Lett.* **4**:51–54.
144. Metcalf, B. W., and Jung, M. J., 1979, *Mol. Pharmacol.* **16**:539–545.
145. Danzin, C., Jung, M. J., Seiler, N., and Metcalf, B. W., 1979, *Biochem. Pharmacol.* **28**:633–639.
146. Silverman, R. B., and Levy, M. A., 1980, *Biochem. Biophys. Res. Commun.* **95**:250–255.
147. Silverman, R. B., and Levy, M. A., 1981, *Biochemistry* **20**:1197–1203.
148. Krooth, R. S., Hsiao, W.-L., and Lam, G. E. M., 1979, *Biochem. Pharmacol.* **28**:1071–1076.

149. Chang, S.-C., and Brunner, E. A., 1979, *Biochem. Pharmacol.* **28:**105–109.
150. Palfreyman, M. G., Schechter, P. J., Buckett, W. R., Tell, G. P., and Koch-Weser, J., 1981, *Biochem. Pharmacol.* **30:**817–824.
151. Rando, R. R., and Bangerter, F. W., 1977, *Biochem. Biophys. Res. Commun.* **76:**1276–1281.
152. Matsui, Y., and Deguchi, T., 1977, *Life Sci.* **20:**1291–1295.
153. Schechter, P. J., Tranier, Y., and Grove, J., 1979, *Life Sci.* **24:**1173–1182.
154. Schechter, P. J., Tranier, Y., Jung, M. J., and Böhlen P., 1977, *Eur. J. Pharmacol.* **45:**319–328.
155. Jobe, P. C., Graham, L. T., Jr., Ray, T. B., and Mims, M. E., 1980, *Life Sci.* **27:**2005–2011.
156. Kendall, D. A., Fox, D. A., and Enna, S. J., 1981, *Neuropharmacology* **20:**351–355.
157. Smialowski, A., 1980, *Neurosci. Lett.* **19:**331–335.
158. Myslobodsky, M. S., and Valenstein, E. S., 1980, *Epilepsia* **21:**163–175.
159. Nisticó, G., Di Giorgio, R. M., De Luca, G., and Macaione, S., 1979, *J. Neurochem.* **33:**343–346.
160. Wood, J. D., Durhan, J. S., and Peesker, S. J., 1977, *Neurochem. Res.* **2:**707–715.
161. Anlezark, G., Horton, R. W., Meldrum, B. S., and Sawaya, M. C. B., 1976, *Biochem. Pharmacol.* **25:**413–417.
162. Fletcher, A., and Fowler, L. J., 1980, *Biochem. Pharmacol.* **29:**1451–1454.
163. Leach, M. J., and Walker, J. M. G., 1977, *Biochem. Pharmacol.* **26:**1569–1572.
164. Horton, R. W., Anlezark, G. M., Sawaya, M. B. C., and Meldrum, B. S., 1977, *Eur. J. Pharmacol.* **41:**387–397.
165. Schechter, P. J., Tranier, Y., and Grove, J., 1978, *J. Neurochem.* **31:**1325–1327.
166. Patsalos, P. N., and Lascelles, P. T., 1981, *J. Neurochem.* **36:**688–695.
167. Iadarola, M. J., Raines, A., and Gale, K., 1979, *J. Neurochem.* **33:**1119–1123.
168. Whittle, S. R., and Turner, A. J., 1978, *J. Neurochem.* **31:**1453–1459.
169. Tapia, R., 1980, *Neurochemistry and Clinical Neurology* (L. Battistin, G. Hashim, and A. Lajtha, eds.), Alan R. Liss, New York, pp. 123–131.
170. Walters, J. R., Eng, N., Pericić, D., and Miller, L. P., 1978, *J. Neurochem.* **30:**759–766.
171. Wood, J. D., and Peesker, S. J., 1976, *Can. J. Physiol. Pharmacol.* **54:**534–540.
172. Löscher, W., and Frey, H.-H., 1978, *Biochem. Pharmacol.* **27:**103–108.
173. Löscher, W., 1979, *Biochem. Pharmacol.* **28:**1397–1407.
174. Carmona, E., Gomes, C., and Trolin, G., 1980, *Naunyn Schmiedebergs Arch. Pharmacol.* **312:**51–55.
175. Tunnicliff, G., 1980, *Biochem. Biophys. Res. Commun.* **97:**160–165.
176. Kim, D. S., and Churchich, J. E., 1981, *Biochem. Biophys. Res. Commun.* **99:**1333–1340.
177. Ossola, L., Maitre, M., Blinderman, J.-M., and Mandel, P., 1980, *J. Neurochem.* **34:**293–296.
178. Gold, B. I., and Roth, R. H., 1977, *J. Neurochem.* **28:**1069–1073.
179. Callery, P. S., Geelhaar, L. A., and Stogniew, M., 1978, *Biochem. Pharmacol.* **27:**2061–2063.
180. Pérez de la Mora, M., Fuxe, K., Hökfelt, T., and Ljungdahl, A., 1975, *Neurosci. Lett.* **1:**109–114.
181. Bertilsson, L., Mao, C. C., and Costa, E., 1977, *J. Pharmacol. Exp. Ther.* **200:**277–284.
182. Waelsch, H., 1962, *Amino Acid Pools* (J. T. Holden, ed.), Elsevier, Amsterdam, pp. 722–730.
183. Berl, S., Lajtha, A., and Waelsch, H., 1961, *J. Neurochem.* **7:**186–197.
184. Berl, S., and Clarke, D. D., 1969, *Handbook of Neurochemistry*, Volume 2 (A. Lajtha, ed.), Plenum Press, New York, pp. 447–472.
185. Quastel, J. H., 1975, *Metabolic Compartmentation and Neurotransmission* (S. Berl, D. D. Clarke, and D. Schneider, eds.), Plenum Press, New York, pp. 337–361.
186. Van den Berg, C. J., Matheson, D. F., Ronda, G., Reijnierse, G. L. A., Blokhuis, G. G. D., Kroon, M. C., Clarke, D. D., and Garfinkel, D., 1975, *Metabolic Compartmentation and Neurotransmission* (S. Berl, D. D. Clarke, and D. Schneider, eds.), Plenum Press, New York, pp. 515–543.
187. Berl, S., and Clarke, D. D., 1978, *Amino Acids as Chemical Transmitters* (F. Fonnum, ed.), Plenum Press, New York, pp. 691–708.

188. Van den Berg, C. J., Matheson, D. F., and Nijenmanting, W. C., 1978, *Amino Acids as Chemical Transmitters* (F. Fonnum, ed.), Plenum Press, New York, pp. 709–723.
189. Van Gelder, N. M., and Drujan, B., 1980, *Brain Res.* **200**:443–455.
190. Quastel, J. H., 1974, *Biochem. Soc. Trans.* **2**:765–780.
191. Shank, R. P., and Aprison, M. H., 1977, *J. Neurochem.* **28**:1189–1196.
192. Balázs, R., Machiyama, Y., Hammond, Y., Julian, B. J., and Richter, D., 1970, *Biochem. J.* **116**:445–467.
193. Minchin, M. C. W., and Beart, P. M., 1975, *Brain Res.* **83**:437–449.
194. Bradford, H. F., and Ward, H. K., 1976, *Brain Res.* **110**:115–125.
195. Bradford, H. F., Ward, H. K., and Thomas, A. J., 1978, *J. Neurochem.* **30**:1453–1459.
196. Tapia, R., and González, R. M., 1978, *Neurosci. Lett.* **10**:165–169.
197. Reubi, J. C., Van den Berg, C., and Cuénod, M., 1978, *Neurosci. Lett.* **10**:171–174.
198. Reubi, J. C., 1980, *Neuroscience* **5**:2145–2150.
199. Kemel, M. L., Gauchy, C., Glowinski, J., and Besson, M. J., 1979, *Life Sci.* **24**:2139–2150.
200. Gauchy, C. M., Iversen, L. L., and Jessell, T. M., 1977, *Brain Res.* **138**:374–379.
201. Tapia, R., 1976, *Transport Phenomena in the Nervous System* (G. Levi, L. Battistin, and A. Lajtha, eds.), Plenum Press, New York, pp. 385–394.
202. Ryan, L. D., and Roskoski, R., Jr., 1975, *Nature* **258**:254–256.
203. Minchin, C. M. W., 1977, *Exp. Brain Res.* **29**:515–526.
204. Potashner, S. J., 1978, *J. Neurochem.* **31**:177–186.
205. Potashner, S. J., 1978, *J. Neurochem.* **31**:187–195.
206. Gauchy, C., Kemel, M. L., Glowinski, J., and Besson, M. J., 1980, *Brain Res.* **193**:129–141.
207. Meeker, R., and Myers, R. D., 1979, *Neuroscience* **4**:495–506.
208. Starr, M. S., 1975, *Biochem. Pharmacol.* **24**:1193–1197.
209. Voaden, M. J., Lake, N., and Nathwani, B., 1977, *J. Neurochem.* **28**:457–459.
210. Voaden, M. J., Lake, N., Marshall, J., and Morjaria, B., 1978, *J. Neurochem.* **31**:1069–1076.
211. Morjaria, B., and Voaden, M. J., 1979, *J. Neurochem.* **33**:541–551.
212. Martínez-Hernandez, A., Bell, K. P., and Norenberg, M. D., 1977, *Science* **195**:1356–1358.
213. Norenberg, M. D., and Martínez-Hernandez, A., 1979, *Brain Res.* **161**:303–310.
214. Ward, H. K., and Bradford, H. F., 1979, *J. Neurochem.* **33**:339–342.
215. Nicklas, W. J., Nunez, R., Berl, S., and Duvoisin, R., 1979, *J. Neurochem.* **33**:839–844.
216. Schon, R., and Kelly, J. S., 1974, *Brain Res.* **66**:289–300.
217. Roberts, P. J., 1976, *Brain Res.* **113**:206–209.
218. Henn, F. A., and Hamberger, A., 1971, *Proc. Natl. Acad. Sci. U.S.A.* **68**:2686–2690.
219. Sellström, Å., and Hamberger, A., 1975, *J. Neurochem.* **24**:847–852.
220. Schrier, B. K., and Thompson, E. J., 1974, *J. Biol. Chem.* **249**:1769–1780.
221. Schubert, D., 1975, *Brain Res.* **84**:87–98.
222. Hutchison, H. T., Werrbach, K., Vauce, C., and Haber, B., 1974, *Brain Res.* **66**:265–274.
223. Lasher, R. S., 1975, *J. Neurobiol.* **6**:597–608.
224. Schousboe, A., Hertz, L., and Svenneby, G., 1977, *Neurochem. Res.* **2**:217–229.
225. Hertz, L., Wu, P. H., and Schousboe, A., 1978, *Neurochem. Res.* **3**:313–323.
226. Balcar, V. J., Mark, J., Borg, J., and Mandel, P., 1979, *Neurochem. Res.* **4**:339–354.
227. Glusman, S., Pacheco, M., González-Robles, A., and Haber, B., 1979, *Brain Res.* **172**:259–276.
228. Sellström, Å., Sjöberg, L. B., and Hamberger, A., 1975, *J. Neurochem.* **25**:393–398.
229. Hauser, K., and Bernasconi, R., 1980, *Tissue Culture in Neurobiology* (E. Giacobini, A. Vernadakis, and A. Shahar, eds.), Raven Press, New York, pp. 205–219.
230. Salganicoff, L., and DeRobertis, E., 1965, *J. Neurochem.* **12**:287–309.
231. Fonnum, F., 1968, *Biochem. J.* **106**:401–412.
232. Schousboe, A., Svenneby, G., and Hertz, L., 1977, *J. Neurochem.* **29**:999–1005.
233. Schousboe, A., 1979, *GABA-Biochemistry and CNS Functions* (P. Mandel and F. V. De-Feudis, eds.), Plenum Press, New York, pp. 219–237.
234. Schousboe, A., 1978, *GABA-Neurotransmitters: Pharmacochemical, Biochemical and Pharmacological Aspects* (P. Krogsgaard-Larsen, J. Scheel-Krüger, and H. Kofod, eds.), Munksgaard, Copenhagen, pp. 263–280.

235. Hertz, L., 1979, *Prog. Neurobiol.* **13**:277–323.
236. Hertz, L., and Schousboe, A., 1980, *Brain Res. Bull.* **5**(Suppl. 2):389–395.
237. Baxter, C. F., 1976, *GABA in Nervous System Function* (E. Roberts, T. N. Chase, and D. B. Tower, eds.), Raven Press, New York, pp. 89–102.
238. Sandoval, M. E., and Tapia, R., 1975, *Brain Res.* **96**:279–286.
239. Sandoval, M. E., Palacios, R., and Tapia, R., 1976, *J. Neurochem.* **27**:667–672.
240. Tapia, R., and Sandoval, M. E., 1977, *The Neurobiology of Sleep and Memory* (R. R. Drucker-Colín and J. L. McGaugh, eds.), Academic Press, New York, pp. 17–32.
241. Storm-Mathisen, J., 1975, *Brain Res.* **87**:107–109.
242. Hariri, M., and Hammerschlag, R., 1977, *Neurosci. Lett.* **7**:319–325.
243. Sarne, Y., Schrier, B. K., and Gainer, H., 1976, *Brain Res.* **110**:91–97.
244. Martin, D. L., 1976, *GABA in Nervous System Function* (E. Roberts, T. N. Chase, and D. B. Tower, eds.), Raven Press, New York, pp. 347–386.
245. Roberts, P. J., 1980, *Brain Res. Bull.* **5**(Suppl. 2):83–88.
246. Bowery, N. G., and Brown, D. A., 1972, *Nature [New Biol.]* **238**:89–91.
247. Levi, G., 1970, *Arch. Biochem. Biophys.* **138**:347–349.
248. Kuhar, M. J., and Zarbin, M. A., 1978, *J. Neurochem.* **31**:251–256.
249. Snyder, S. H., Logan, W. J., Bennett, J. P., and Arregui, A., 1973, *Neurosci. Res.* **5**:131–157.
250. Bennett, J. P., Mulder, A. H., and Snyder, S. H., 1974, *Life Sci.* **15**:1045–1056.
251. Blaustein, M. P., and King, C., 1976, *J. Membr. Biol.* **30**:153–173.
252. Wheeler, D. D., and Hollingsworth, R. G., 1979, *J. Neurosci. Res.* **4**:265–289.
253. Wheeler, D. D., 1980, *J. Neurosci. Res.* **5**:323–337.
254. Larsson, O. M., Hertz, L., and Schousboe, A., 1980, *J. Neurosci. Res.* **5**:469–477.
255. Sellström, Å., Henn, F., Estborn, L., Hansson, E., and Hamberger, A., 1980, *Brain Res. Bull.* **5**(Suppl. 2):95–99.
256. Kanner, B. I., 1978, *Biochemistry* **17**:1207–1211.
257. Kanner B. I., 1978, *FEBS Lett.* **89**:47–50.
258. Roskoski, R., Jr., 1981, *J. Neurochem.* **36**:544–550.
259. Cohen, S. R., 1980, *J. Neurochem.* **35**:1008–1012.
260. Cohen, S. R., 1981, *Brain Res.* **205**:157–168.
261. Levi, G., and Raiteri, M., 1974, *Nature* **250**:735–737.
262. Simon, J. R., Martin, D. L., and Kroll, M., 1974, *J. Neurochem.* **23**:981–991.
263. Raiteri, M., Federico, R., Coletti, A., and Levi, G., 1975, *J. Neurochem.* **24**:1243–1250.
264. Ryan, L. D., and Roskoski, R., Jr., 1977, *J. Pharmacol. Exp. Ther.* **200**:285–291.
265. Roskoski, R., Jr., 1978, *J. Neurochem.* **31**:493–498.
266. Levi, G., Banay-Schwartz, M., and Raiteri, M., 1978, *Amino Acids as Chemical Transmitters* (F. Fonnum, ed.), Plenum Press, New York, pp. 327–350.
267. Sellström, Å., Venema, R., and Henn, F., 1976, *Nature* **264**:652–653.
268. Levi, G, Poce, U., and Raiteri, M., 1976, *Transport Phenomena in the Nervous System* (G. Levi, L. Battistin, and A. Lajtha, eds.), Plenum Press, New York, pp. 273–289.
269. Levi, G., and Raiteri, M., 1978, *Proc. Natl. Acad. Sci. U.S.A.* **75**:2981–2985.
270. Pastuszko, A., Wilson, D. F., and Erecińska, M., 1981, *Proc. Natl. Acad. Sci. U.S.A.* **78**:1242–1244.
271. Cutler, R. W. P., and Young, J., 1979, *Brain Res.* **165**:261–270.
272. Moscowitz, J. A., and Cutler, R. W. P., 1980, *J. Neurochem.* **35**:1394–1399.
273. Johnston, G. A. R., 1976, *GABA in Nervous System Function* (E. Roberts, T. N. Chase, and D. B. Tower, eds.), Raven Press, New York, pp. 395–411.
274. Krogsgaard-Larsen, P., and Johnston, G. A. R., 1975, *J. Neurochem.* **25**:797–802.
275. Johnston, G. A. R., Krogsgaard-Larsen, P., Stephanson, A. L., and Twitchin, B., 1976, *J. Neurochem.* **26**:1029–1032.
276. Johnston, G. A. R., Krogsgaard-Larsen, P., and Stephanson, A. L., 1975, *Nature* **258**:627–628.
277. Lodge, D., Johnston, G. A. R., Curtis, D. R., and Brand, S. J., 1977, *Brain Res.* **136**:513–522.

278. Krogsgaard-Larsen, P., 1978, *Amino Acids as Chemical Transmitters* (F. Fonnum, ed.), Plenum Press, New York, pp. 305–321.
279. Curtis, D. R., Game, C. J. A., and Lodge, D., 1976, *Exp. Brain Res.* **25**:413–428.
280. Kelly, J. S., Dick, F., and Schon, F., 1975, *Brain Res.* **85**:255–259.
281. Johnston, G. A. R., Stephanson, A. L., and Twitchin, B., 1976, *J. Neurochem.* **26**:83–87.
282. Lodge, D., Curtis, D. R., and Johnston, G. A. R., 1978, *J. Neurochem.* **31**:1525–1528.
283. Beart, P. M., Johnston, G. A. R., and Uhr, M. L., 1972, *J. Neurochem.* **19**:1855–1861.
284. Bowery, N. G., Jones, G. P., and Neal, M. J., 1976, *Nature* **264**:281–284.
285. Neal, M. J., and Bowery, N. G., 1977, *Brain Res.* **138**:169–174.
286. Schon, F., and Kelly, J. S., 1975, *Brain Res.* **86**:243–257.
287. Kelly, J. S., and Dick, F., 1976, *Cold Spring Harbor Symp. Quant. Biol.* **40**:93–106.
288. Iversen, L. L., Dick, F., Kelly, J. S., and Schon, F., 1975, *Metabolic Compartmentation and Neurotransmission* (S. Berl, D. D. Clarke, and D. Schneider, eds.), Plenum Press, New York, pp. 65–89.
289. Storm-Mathisen, J., Fonnum, F., and Malthe-Sorenssen, D., 1976, *GABA in Nervous System Function* (E. Roberts, T. N. Chase, and D. B. Tower, eds.), Raven Press, New York, pp. 387–394.
290. Bowery, N. G., Brown, D. A., White, R. D., and Yamini, G., 1979, *J. Physiol.* (Lond.) **293**:51–74.
291. Henn, F. A., Venema, R., Anderson, D., and Sellström, Å., 1980, *J. Neurochem.* **34**:1671–1677.
292. East, J. M., Dutton, G. R., and Currie, D. N., 1980, *J. Neurochem.* **34**:523–530.
293. Wilson, J. E., Wilkin, G. P., and Balázs, R., 1976, *J. Neurochem.* **26**:957–965.
294. Wilkin, G. P., Balázs, R., Schon, F., and Kelly, J. S., 1974, *Nature* **252**:397–399.
295. Kelly, J. S., Dick, F., and Schon, F., 1975, *Brain Res.* **85**:255–259.
296. Martin, D. L., and Shain, W., 1979, *J. Biol. Chem.* **254**:7076–7084.
297. Allan, R. D., Curtis, D. R., Headley, P. M., Johnston, G. A. R., Lodge, D., and Twitchin, B., 1980, *J. Neurochem.* **34**:652–654.
298. Löscher, W., 1980, *J. Neurochem.* **34**:1603–1608.
299. Fagg, G. E., and Lane, J. D., 1979, *Neuroscience* **4**:1015–1036.
300. Orrego, F., 1979, *Neuroscience* **4**:1037–1057.
301. Orrego, F., and Miranda, R., 1976, *J. Neurochem.* **26**:1033–1038.
302. Vargas, O., Doria de Lorenzo, M. C., Saldate, M. C., and Orrego, F., 1977, *J. Neurochem.* **28**:165–170.
303. Szerb, J. C., 1979, *J. Neurochem.* **32**:1565–1573.
304. Mulder, A. H., and Snyder, S. H., 1974, *Brain Res.* **76**:297–308.
305. López-Colomé, A. M., Tapia, R., Salceda, R., and Pasantes-Morales, H., 1978, *Neuroscience* **3**:1069–1074.
306. Johnston, G. A. R., 1977, *Brain Res.* **121**:179–181.
307. Valdés, F., and Orrego, F., 1978, *Brain Res.* **141**:357–363.
308. Minchin, M. C. W., 1980, *Biochem. J.* **190**:333–339.
309. Cunningham, J., and Neal, M. J., 1981, *Br. J. Pharmacol.* **73**:655–667.
310. Neal, M. J., and Bowery, N. G., 1979, *Brain Res.* **167**:337–343.
311. Redburn, D. A., Shelton, D., and Cotman, C. W., 1976, *J. Neurochem.* **26**:297–303.
312. Haycock, J. W., Levy, W. B., Denner, L. A., and Cotman, C. W., 1978, *J. Neurochem.* **30**:1113–1125.
313. Cotman, C. W., Haycock, J. W., and White, W. F., 1976, *J. Physiol.* (*Lond.*) **254**:475–505.
314. Haycock, J. W., Levy, W. B., and Cotman, C. W., 1978, *Brain Res.* **155**:192–195.
315. Haycock, J. W., Levy, W. B., Denner, L. A., and Cotman, C. W., 1979, *Neuroscience* **4**:1341–1346.
316. Sandoval, M. E., 1980, *J. Neurochem.* **35**:915–921.
317. Sandoval, M. E., 1980, *Brain Res.* **181**:357–367.
318. Sandoval, M. E., 1981, *Regulatory Mechanisms of Synaptic Transmission* (R. Tapia and C. W. Cotman, eds.), Plenum Press, New York, pp. 187–203.
319. Tapia, R., and Meza-Ruiz, G., 1977, *Brain Res.* **126**:160–166.

320. Meza-Ruiz, G., and Tapia, R., 1978, *Brain Res.* **154**:163–166.
321. Tapia, R., and Arias, C., 1981, *Regulatory Mechanisms of Synaptic Transmission* (R. Tapia and C. W. Cotman, eds.), Plenum Press, New York, pp. 169–186.
322. Osborne, R. H., and Bradford, H. F., 1975, *J. Neurochem.* **25**:35–41.
323. Miledi, R., Molenaar, P. C., and Polak, R. L., 1980, *J. Physiol. (Lond.)* **309**:199–214.
324. López-Colomé, A. M., Salceda, R., and Pasantes-Morales, H., 1978, *Neurochem. Res.* **3**:431–441.
325. Redburn, D. A., 1977, *Exp. Eye Res.* **25**:265–275.
326. Kennedy, A. J., and Neal, M. J., 1978, *Exp. Eye Res.* **26**:71–75.
327. Pasantes-Morales, H., and Quesada, O., 1980, *Neurochem. Res.* **5**:1077–1088.
328. Minchin, M. C. W., and Iversen, L. L., 1974, *J. Neurochem.* **23**:533–540.
329. Roberts, P. J., 1974, *Brain Res.* **74**:327–332.
330. Minchin, M. C. W., 1975, *J. Neurochem.* **24**:571–577.
331. Sellström, Å., and Hamberger, A., 1977, *Brain Res.* **119**:189–198.
332. Neal, M. J., and Bowery, N. G., 1979, *Brain Res.* **167**:337–343.
333. Bowery, N. G., Brown, D. A., and Marsh, S., 1979, *J. Physiol. (Lond.)* **293**:75–101.
334. Adams, P. R., and Brown, D. A., 1979, *J. Physiol. (Lond.)* **293**:95–101.
335. Hammerstad, J. P., and Lytle, C. R., 1976, *Brain Res.* **27**:399–403.
336. Pearce, B. R., Currie, D. N., Beale, R., and Dutton, G. R., 1981, *Brain Res.* **206**:485–489.
337. Čanžek, V., and Reubi, J. C., 1980, *Exp. Brain Res.* **38**:437–441.
338. Tapia, R., 1974, *Neurohumoral Coding of Brain Function* (R. D. Myers and R. R. Drucker-Colín, eds.), Plenum Press, New York, pp. 3–26.
339. Van der Heyden, J. A. M., Venema, K., and Korf, J., 1979, *J. Neurochem.* **32**:469–476.
340. Van der Heyden, J. A. M., Venema, K., and Korf, J., 1980, *J. Neurochem.* **34**:1648–1653.
341. De Belleroche, J. S., and Bradford, H. F., 1977, *J. Neurochem.* **29**:335–343.
342. Dietl, H., and Philippu, A., 1980, *Naunyn Schmiedebergs Arch. Pharmacol.* **308**:143–147.
343. Roberts, F., Hill, R. G., Osborne, R. H., and Mitchell, J. F., 1979, *Brain Res.* **178**:467–477
344. Reiffenstein, R. J., 1979, *Can. J. Physiol. Pharmacol.* **57**:798–803.
345. Abdul-Ghani, A. S., Coutinho-Netto, J., and Bradford, H. F., 1980, *Brain Res.* **191**:471–478.
346. Abdul-Ghani, A. S., Bradford, H. F., Cox, D. W. G., and Dodd, P. R., 1979, *Brain Res.* **171**:55–66.
347. Bradford, H. F., 1981, *Regulatory Mechanisms of Synaptic Transmission* (R. Tapia and C. W. Cotman, eds.), Plenum Press, New York, pp. 103–140.
348. Clark, R. M., and Collins, G. G.S., 1976, *J. Physiol. (Lond.)* **262**:383–400.
349. Collins, G. G. S., 1977, *J. Neurochem.* **28**:461–463.
350. Obata, K., 1976, *GABA in Nervous System Function* (E. Roberts, T. N. Chase, and D. B. Tower, eds.), Raven Press, New York, pp. 251–254.
351. Vellucci, S. V., and Webster, R. A., 1980, *Neuropharmacology* **19**:1099–1104.
352. Coull, B. M., Owens, D. K., and Cutler, R. W. P., 1981, *Brain Res.* **210**:301–309.
353. Haycock, J. W., Levy, W. B., and Cotman, C. W., 1977, *Biochem. Pharmacol.* **26**:159–161.
354. Waller, M. B., and Richter, J. A., 1980, *Biochem. Pharmacol.* **29**:2189–2198.
355. Collins, G. G. S., 1980, *Brain Res.* **190**:517–528.
356. Bauer, B., 1979, *Brain Res.* **163**:307–317.
357. Olsen, R. W., Lamar, E. E., and Bayless, J. D., 1977, *J. Neurochem.* **28**:299–305.
358. Olsen, R. W., Ticku, M. K., Van Ness, P. C., and Greenlee, D., 1978, *Brain Res.* **139**:277–294.
359. Cutler, R. W. P., and Young, J., 1979, *Brain Res.* **170**:157–163.
360. De Boer, T., Stoof, J. C., and van Duyn, H., 1980, *Brain Res.* **192**:296–300.
361. Potashner, S. J., 1979, *J. Neurochem.* **32**:103–109.
362. Roberts, P. J., Gupta, H. K., and Shargill, N. S., 1978, *Brain Res.* **155**:209–212.
363. Grasso, A., and Senni, M. I., 1979, *Eur. J. Biochem.* **102**:337–344.
364. Tzeng, M.-C., Cohen, R. S., and Siekevitz, P., 1978, *Proc. Natl. Acad. Sci. U.S.A.* **75**:4016–4020.
365. Tse, C. K., Dolly, J. O., and Diniz, C. R., 1980, *Neuroscience* **5**:135–143.
366. Smith, C. C. T., Bradford, H. F., Thompson, E. J., and MacDermont, J. J., 1980, *J. Neurochem.* **34**:487–494.

367. Wernicke, J. F., Vanker, A. D., and Howard, B. D., 1975, *J. Neurochem.* **25**:483–496.
368. Coutinho-Netto, J., Abdul-Ghani, A.-S., Norris, P. J., Thomas, A. J., and Bradford, H. F., 1980, *J. Neurochem.* **35**:558–565.
369. Collingridge, G. L., Collins, G. G. S., Davis, J., James, T. A., Neal, M. J., and Tongroach, P., 1980, *J. Neurochem.* **34**:540–547.
370. Pong, S.-S., Wang, C. C., and Fritz, L. C., 1980, *J. Neurochem.* **34**:351–358.
371. Johnston, M. V., and Coyle, J. T., 1980, *J. Neurochem.* **34**:1429–1441.
372. Hitzemann, R. J., and Loh, H. H., 1978, *Brain Res.* **159**:29–40.
373. Ramsay, P. B., Krigman, M. R., and Morrell, P., 1980, *Brain Res.* **187**:383–402.
374. Redburn, D. A., Broome, D., Ferkany, J., and Enna, S. J., 1978, *Brain Res.* **152**:511–519.
375. Levi, G., Gallo, V., Ciotti, T., and Raiteri, M., 1979, *J. Neurochem.* **33**:1043–1053.
376. Meza-Ruiz, G., 1980, *Brain Res. Bull.* **5**(Suppl. 2):121–125.
377. Bondy, S. C., and Purdy, J. L., 1977, *Brain Res.* **119**:403–416.
378. Davies, L. P., Johnston, G. A. R., and Stephanson, A. L., 1975, *J. Neurochem.* **25**:387–392.
379. Covarrubias, M., and Tapia, R., 1978, *J. Neurochem.* **31**:1209–1214.
380. Covarrubias, M., and Tapia, R., 1980, *J. Neurochem.* **34**:1682–1688.
381. Fisher, S. K., and Davis, W. E., 1976, *Biochem. Pharmacol.* **25**:1881–1885.
382. Taberner, P. V., and Keen, P., 1977, *J. Neurochem.* **29**:595–597.
383. Horton, R. W., Chapman, A. G., and Meldrum, B. S., 1978, *J. Neurochem.* **30**:1501–1504.
384. Marigold, J., and Taberner, P. V., 1978, *Biochem. Pharmacol.* **27**:1109–1112.
385. Beart, P. M., and Bilal, K., 1979, *Biochem. Pharmacol.* **28**:449–454.
386. Nitsch, C., and Okada, Y., 1976, *Brain Res.* **105**:173–178.
387. Adcock, T., and Taberner, P. V., 1978, *Biochem. Pharmacol.* **27**:246–248.
388. Sawaya, C., Edmonson, F., Horton, R., and Meldrum, B., 1979, *Biochem. Pharmacol.* **28**:2854–2856.
389. Matsuda, M., Abe, M., Hoshino, M., and Sakurai, T., 1979, *Biochem. Pharmacol.* **28**:2785–2789.
390. Wood, J. D., Kurylo, E., and Newstead, J. D., 1978, *Can. J. Biochem.* **56**:667–672.
391. Wood, J. D., Russell, M. P., Kurylo, E., and Newstead, J. D., 1979, *J. Neurochem.* **33**:61–68.
392. Wood, J. D., Russell, M. P., and Kurylo, E., 1980, *J. Neurochem.* **35**:125–130.
393. Sarhan, S., and Seiler, N., 1979, *J. Neurosci. Res.* **4**:399–421.
394. Löscher, W., 1981, *J. Neurochem.* **36**:1521–1527.
395. Abdul-Ghani, A.-S., Coutinho-Netto, J., Druce, D., and Bradford, H. F., 1981, *Biochem. Pharmacol.* **30**:363–368.
396. Iadarola, M. J., and Gale, K., 1979, *Eur. J. Pharmacol.* **59**:125–129.
397. Gale, K., and Iadarola, M. J., 1980, *Brain Res.* **183**:217–223.
398. Gale, K., and Iadarola, M. J., 1980, *Science* **208**:288–291.
399. Iadarola, M. J., and Gale, K., 1980, *Brain Res. Bull.* **5**(Suppl. 2):13–19.
400. Tapia, R., Meza-Ruiz, G., Durán, L., and Drucker-Colín, R. R., 1976, *Brain Res.* **116**:101–109.
401. Frey, H.-H., Popp, C., and Löscher, W., 1979, *Neuropharmacology* **18**:581–590.
402. Horton, R. W., Collins, J. F., Anlezark, G. M., and Meldrum, B. S., 1979, *Eur. J. Pharmacol.* **59**:75–83.
403. Wood, J. D., Schousboe, A., and Krogsgaard-Larsen, P., 1980, *Neuropharmacology* **19**:1149–1152.
404. Storm-Mathisen, J., 1976, *GABA in Nervous System Function* (E. Roberts, T. N. Chase, and D. B. Roberts, eds.), Raven Press, New York, pp. 149–168.
405. Fahn, S., 1976, *GABA in Nervous System Function* (E. Roberts, T. N. Chase, and D. B. Roberts, eds.), Raven Press, New York, pp. 169–186.
406. Otsuka, M., and Konishi, S., 1976, *GABA in Nervous System Function* (E. Roberts, T. N. Chase, and D. B. Roberts, eds.), Raven Press, New York, pp. 197–202.
407. Kuriyama, K., and Kimura, H., 1976, *GABA in Nervous System Function* (E. Roberts, T. N. Chase, and D. B. Roberts, eds.), Raven Press, New York, pp. 203–216.
408. Okada, Y., and Shimada, C., 1976, *GABA in Nervous System Function* (E. Roberts, T. N. Chase, and D. B. Roberts, eds.) Raven Press, New York, pp. 223–228.

409. Okada, Y., 1976, *GABA in Nervous System Function* (E. Roberts, T. N. Chase, and D. B. Roberts, eds.), Raven Press, New York, pp. 229–233.
410. Okada, Y., 1976, *GABA in Nervous System Function* (E. Roberts, T. N. Chase, and D. B. Roberts, eds.), Raven Press, New York, pp. 235–243.
411. Fonnum, F., 1978, *Amino Acids as Chemical Transmitters* (F. Fonnum, ed.), Plenum Press, New York, pp. 143–153.
412. Storm-Mathisen, J., 1978, *Amino Acids as Chemical Transmitters* (F. Fonnum, ed.), Plenum Press, New York, pp. 155–171.
413. Emson, P. C., 1978, *Amino Acids as Chemical Transmitters* (F. Fonnum, ed.), Plenum Press, New York, pp. 181–192.
414. Tappaz, M. L., 1978, *Amino Acids as Chemical Transmitters* (F. Fonnum, ed.), Plenum Press, New York, pp. 193–212.
415. Karlsen, R. L., 1978, *Amino Acids as Chemical Transmitters* (F. Fonnum, ed.), Plenum Press, New York, pp. 241–256.
416. Voaden, M. J., 1978, *Amino Acids as Chemical Transmitters* (F. Fonnum, ed.), Plenum Press, New York, pp. 257–274.
417. Gottesfeld, Z., and Jacobowitz, D. M., 1978, *Interactions between Putative Neurotransmitters* (S. Garattini, J. F. Pujol, and R. Samanin, eds.), Raven Press, New York, pp. 109–126.
418. Gottesfeld, Z., Brandon, C., Jacobowitz, D. M., and Wu, J.-Y., 1980, *Brain Res. Bull.* **5**(Suppl 2):1–6.
419. Emson, P. C., and Lindvall, O., 1979, *Neuroscience* **4**:1–30.
420. Ribak, C. E., 1978, *J. Neurocytol.* **7**:461–478.
421. Ribak, C. E., Vaughn, J. E., and Roberts, E., 1979, *J. Comp. Neurol.* **187**:261–284.
422. Barber, R. P., Vaughn, J. E., Saito, K., McLaughlin, B. J., and Roberts, E., 1978, *Brain Res.* **141**:35–55.
423. Ribak, C. E., Vaughn, J. E., and Roberts, E., 1980, *Brain Res.* **192**:413–420.
424. Gottesfeld, Z., Brandon, C., and Wu, J.-Y., 1981, *Brain Res.* **208**:181–186.
425. Lam, D. M. K., Su, Y. Y. T., Swain, L., Marc, R. E., Brandon, C., and Wu, J.-Y., 1979, *Nature* **278**:565–567.
426. Brandon, C., Lam, D. M. K., Su, Y. Y. T., and Wu, J.-Y., 1980, *Brain Res. Bull.* **5**(Suppl. 2):21–29.
427. Vaughn, J. E., Famiglietti, E. V., Jr., Barber, R. P., Saito, K., Roberts, E., and Ribak, C. E., 1981, *J. Comp. Neurol.* **197**:113–127.
428. Famiglietti, E. V., Jr., and Vaughn, J. E., 1981, *J. Comp. Neurol.* **197**:129–139.
429. Tappaz, M., Aguera, M., Belin, M. F., and Pujol, J. F., 1980, *Brain Res.* **186**:379–391.
430. Jones, M. T., Hillhouse, E. W., and Cole, R. S., 1978, *Interactions between Putative Neurotransmitters* (S. Garattini, J. F. Pujol, and R. Samanin, eds.), Raven Press, New York, pp. 245–261.
431. Makara, G. B., and Stark, E., 1978, *Interactions between Putative Neurotransmitters* (S. Garattini, J. F. Pujol, and R. Samanin, eds.), Raven Press, New York, pp. 263–283.
432. Vale, W., Rivier, C., Rivier, J., and Brown, M., 1978, *Psychopharmacology: A Generation of Progress* (M. A. Lipton, A. DiMascio, and K. F. Killam, eds.), Raven Press, New York, pp. 403–421.
433. Kizer, J. S., and Youngblood, W. W., 1978, *Psychopharmacology: A Generation of Progress* (M. A. Lipton, A. DiMascio, and K. F. Killam, eds.), Raven Press, New York, pp. 465–486.
434. Grandison, L., 1980, *Brain Res. Bull.* **5**(Suppl. 2):701–704.
435. Tapia, R., 1980, *Front. Horm. Res.* **6**:86–103.

Glycine

E. C. Daly and M. H. Aprison

1. INTRODUCTION

Glycine is structurally the simplest amino acid and is found in all body fluids and proteinaceous tissue in relatively large amounts. Although glycine is a nonessential amino acid, it is an essential intermediate in the metabolism of protein, peptides, one-carbon fragments, nucleic acids, porphyrins, and bile salts. In addition, within the last 20 years a neurotransmitter role for glycine has been recognized within the CNS.[1] Subsequent neurochemical investigations have attempted to separate the "general metabolic" and "transmitter" functions of glycine within the CNS.

Facts known in the early 1960s led to combined neurochemical and neurophysiological search for the spinal postsynaptic inhibitory transmitter.[2,3] These facts were: (1) a detailed knowledge of spinal cord anatomy and physiology[4]; (2) recognition of the inhibitory effects of certain amino acids on spinal neurons[5]; and (3) the strychnine resistance of GABA-induced inhibitions.[6] The finding of a marked intrasegmental variation in the content of glycine within the spinal cord led Aprison and Werman[1] to postulate that glycine was the elusive compound. Subsequent studies have confirmed glycine's role in the spinal cord and have suggested a similar role in more rostral portions of the CNS.

Although the "general metabolic" roles of glycine are required for CNS function, the present chapter will mainly address the specific neurochemical role of glycine in CNS neurotransmission. Previous reviews have been published.[2,3,7–10]

2. DETERMINATION OF GLYCINE IN BIOLOGICAL FLUIDS AND NERVOUS TISSUE

Ion-exchange column chromatography has largely replaced the classical techniques of paper and thin-layer chromatograpy for analysis of amino acids.

E. C. Daly and M. H. Aprison • The Institute of Psychiatric Research and Departments of Psychiatry, Neurology, and Biochemistry, Indiana University School of Medicine, Indianapolis, Indiana 46223.

Microprocessor-controlled amino acid analyzers with computerized data reduction capabilities are becoming common. The dedicated nature of these instruments makes them economically unfeasible unless there is a continuous
requirement for the analyses of a large number of samples. Gas chromatographic methods have the advantage of a less-dedicated instrument. A sensitive
mass fragmentographic method for glycine exists,[11] but the sophisticated instrumentation would make it unlikely that such an approach will become com-
monplace.

The rapidly evolving field of high-performance liquid chromatography
(HPLC) offers the best chance of a precise, rapid, automated, and affordable
analysis of amino acids for the generalized laboratory in the future. However,
even constantly improved HPLC technology has not been able to obtain the
resolution of established amino acid analyzer technology.[12] The initial and
maintenance expenses for these sophisticated instruments make thin-layer
chromatography of amino acid derivates (dansyl chloride and dinitrofluoro-
benzene) alternatives in selected cases.[13,14]

There are several procedures for the determination of glycine based on
spectrophotometric[15–18] and/or fluorometric methods.[16,19] An extremely sensitive (approximately 50 fmol) enzymatic method for glycine has been described
employing enzymatic recycling of NAD.[20]

Except for a possible decline in the cerebellar cortex, the content of glycine
is relatively constant during maturation in various regions of the rat CNS[21,22];
in other species, glycine decreases during maturation.[23,24] There appears to be
a circadian rhythm in the rat hindbrain and spinal cord[25,26] and possibly more
rostral areas of the CNS[27]; however, the contents of glycine in this latter study
are much higher than those usually reported for the rat CNS.[28] There is little
diurnal variation in plasma glycine levels, but there tends to be an increase
during darkness hours.[29] The content of glycine is stable immediately postmortem but increases significantly within an hour in rat, mouse, and human
brain.[30–32] The postmortem increase in glycine content may be more rapid in
guinea pig brain.[31]

3. DISTRIBUTION OF GLYCINE IN NERVOUS TISSUE

3.1. Vertebrates

Aprison and Werman[1] first suggested that glycine may function as inhibitory neurotransmitter in the spinal cord on the basis of its markedly increased
content in lumbosacral gray matter compared to spinal roots, approximately
7 and 1.2 µmol/g, respectively. Further studies from their laboratory demonstrated that the content of glycine in the ventral gray matter was higher than
three other amino acid transmitter candidates (Glu, Asp, GABA) and that the
spinal cord contained the highest content of glycine within the CNS in six of
seven species studied.[33,34] It was of interest that in animals without limbs, the
glycine content was evenly distributed in the rostro–caudal axis of the spinal

cord, whereas in animals with limbs, the content was higher in the cervical and lumbar enlargements.[34]

The glycine content of the spinal gray matter was correlated with the number of interneurons remaining after ischemic insult.[35,36] After severe ischemic insult to the dorsal horn of the cat, the glycine content decreased 27% in the dorsal white matter, but the change was statistically insignificant, probably on the basis of the small sample size.[36] The high spinal content of glycine was confirmed by other laboratories.[37–40] The content of glycine has been found to be as high as 9.6 μmol/g in discrete regions of the ventral horn of the rabbit.[40] These studies strongly suggest that glycine is present in interneurons within spinal gray matter and that the high levels found in dorsolateral and ventral white matter reflect glycinergic propriospinal interneurons.[34]

The distribution of glycine is also high in the brainstem—medulla, pons, midbrain, and tectum. Specific nuclei within the brainstem with a relatively high glycine level include dorsal column nuclei,[41] inferior colliculus,[42] substantia nigra,[43] and cochlear nucleus.[44] The cerebrum and cerebellum have lower levels.[28,34,45] The concentration of glycine is high in the inner layers of the retina,[46] especially the amacrine[47,48] and horizontal[48] cell layers. It is of interest to note that the glycine content of proteins is not increased in the CNS regions where the free content is high.[49]

Subcellular fractionation studies indicate that glycine and other amino acids are not specifically localized in either synaptosomes[50–52] or synaptic vesicles.[52,53] However, glycine has been shown to be enriched relative to other amino acids in a subpopulation of synaptosomes isolated from the medulla and spinal cord.[50] The content of glycine in crude synaptosome preparations (P_2) from various brain regions reflects its level in homogenates.[54] These relationships were not confirmed in bovine brain.[52]

It is of interest to note that GABA has been estimated to be present in nerve terminals at a concentration of 50–100 nM, although one cannot measure such high concentrations after subcellular fractionation.[55] An equilibrium between cytoplasmic and vesicular pools may be disrupted during homogenization and centrifugation procedures.[56] Although other transmitters (ACh, NE, and 5-HT) are localized in vesicles, the general inability to find glycine (and other amino acids) so localized may be a reflection of large metabolic pools. Even though the amino acid content of synaptic vesicles is small, it is larger than its ACh content.[53] On the other hand, there is no *a priori* requirement that amino acids be released from synaptic vesicles. There is evidence that GABA, Glu, and Asp may be released from the cytoplasmic pool of synaptosomes,[57] and this may be also true for glycine.

3.2. Invertebrates

The abdominal ganglia of the *Aplysia* contain certain neurons the glycine content of is much higher than that of their neighbors.[58] These neurons have specific uptake and axonal transport systems for glycine.[59,60] Electrical stimulation of the cells and glycine perfusion of arterial muscle fibers produced identical effects.[61] After incubation of the cell bodies in [^{3}H]glycine, an association was found between the axonally transported label and vesicles (and mitochondria).[62] These studies suggest that glycine acts as an intracellular messenger in *Aplysia*, and studies of its release after physiological stimulation are awaited. Extensive investigations have not been done in other invertebrates.

4. METABOLISM OF GLYCINE IN NERVOUS TISSUE

4.1. Carbon Source for Glycine

The influx of glycine into the CNS from the blood is slow and less than the influx of serine and essential amino acids.[28,63] Although the influx of most amino acids is carrier mediated, glycine appears to enter the CNS by simple diffusion from the blood.[64] The content of glycine in the brain increases very little within an hour after systemically administered glycine has produced a manyfold increase in the plasma level.[65] However, there is a significant increase after chronic (2 day) elevation of plasma glycine.[66] Although still slow, the influx of glycine is quicker in the immature animal.[67,68]

Since the CNS glycine content is relatively high and its influx is relatively low, glycine probably is synthesized *de novo* in brain. Glucose and/or serine may be the sources of glycine. Although glycine received label at a slow rate after systemic [^{14}C]glucose, calculated flux rates indicated that glucose might serve as a main source of carbon units via serine.[28] However, serine is taken up by the brain as rapidly as essential amino acids, which may indicate that it is essential for the brain although not essential for the body as a whole.[63] Glyoxylate uptake by the brain is slow, apparently by the monocarboxylic acid carrier system.[69]

There are a number of possible pathways leading from glucose to glycine (Fig. 1). In the first two pathways (I, II), serine is the immediate precursor of glycine via serine hydroxymethyltransferase (SHMT, E.C. 2.1.2.1). In the latter two pathways (IV, V), glyoxylate is the immediate precursor of glycine via glycine:2-oxoglutarate aminotranserase (glycine transaminase, GT, E.C. 2.6.1.4). In pathway III, glycine is formed by the condensation of NH_3,CO_2, and N^5, N^{10}-methylenetetrahydrofolate by the reverse reaction (glycine synthetase, E.C. 2.1.2.10.) of the glycine cleavage system (GCS). The activated one-carbon unit may be formed from the third carbon of serine during the SHMT reaction.

Serine can be formed by the phosphorylated pathway (I)[70] or the non-phosphorylated pathway (II).[71] The evidence is best for the former within the CNS.[70,72] Glyoxylate can be formed either by the nonoxidative decarboxylation of hydroxypyruvate to glycoaldehyde (IV) or by the action of isocitrate lyase (V). The activity of the nonoxidative decarboxylation is relatively high within the CNS.[73] Although isocitrate lyase is generally considered a microbial enzyme, it may be present in animals.[74]

The catabolism of glycine may be via the GCS and/or SHMT (see below). The GT is usually considered irreversible. D-Amino acid oxidase, converting

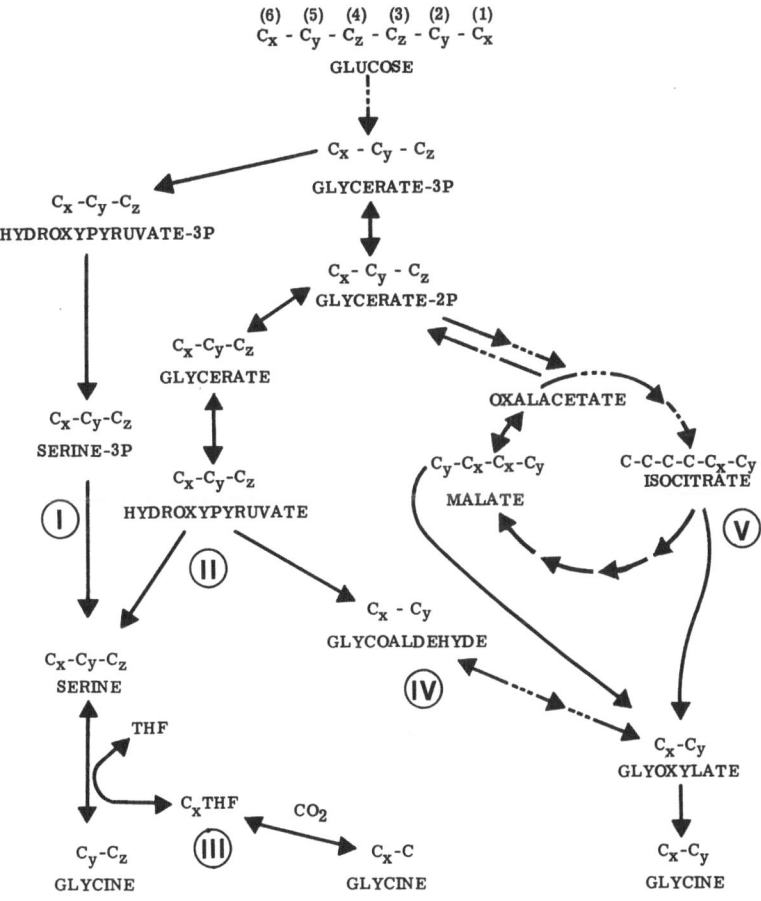

Fig. 1. Pathways illustrating carbon flow from glucose to glycine. Reprinted with permission from Aprison *et al.*[56] and Aprison and Daly.[9]

glycine to glyoxylate, is present in the CNS, but glycine is a very poor substrate for the enzyme.[75] In addition, the activity is very low in the rostral regions of the CNS and appears to be mainly localized in lysosomes or atypical microperoxisomes within astrocytes.[75-77]

4.2. Interconversion of Serine and Glycine

4.2.1. In Vivo Studies

From the labeling pattern of glycine and serine after the intraperitoneal or intracisternal injection of [^{14}C]glucose or [^{14}C]serine, it was concluded that glycine is derived from glucose via serine.[9,28,78] Although glyoxylate did label glycine after intracisternal injection, an estimate of the flux of label could not be calculated because of the inability to measure the low endogenous levels of glyoxylate in the CNS.[28] However, intracisternal injection of glucose, citrate, glycerate, and ethanolamine revealed little evidence for flow of carbon through glyoxylate (pathways IV and V, Fig. 1).[9,56] After intracisternal injection of [3-^{14}C]serine, little evidence was found for glycine synthetase (pathway III, Fig. 1).[9,28]

From the increased content of glycine found in the spinal cord 1 h after systemic threonine injection, it was suggested that threonine may serve as a precursor of glycine within the CNS.[79] However, labeled glycine could not be found within the medulla after the intracisternal injection of [U-^{14}C]threonine.[80] The significance of threonine as a precursor of glycine within the CNS is uncertain. Glycine can be formed by the action of threonine dehydrogenase and aminoacetone synthetase in rat liver mitochondria.[81]

4.2.2. In Vitro Studies

The labeling pattern of the glycine incorporated into protein during incubation of brain slices with [^{14}C]glucose indicated that glycine was synthesized from serine.[82] Unfortified crude synaptosomal (P_2) fractions of various brain regions actively interconverted [^{14}C]glycine and [^{14}C]serine.[54] The most interesting finding was that the cerebellum fractions converted serine to glycine faster, but converted glycine to serine slower, than the other regions. The relatively high rate of glycine formation suggested that the activity may be important for the generation of one-carbon units for the general metabolism of cells[54] (see Section 4.4).

Preliminary studies demonstrated that [^{14}C]glyoxylate could be rapidly converted to [^{14}C]glycine in those P_2 fractions.[54] Once again, the small endogenous content of glyoxylate could not be measured. Consequently, the rate of conversion could not be estimated.

4.3. Enzyme Studies

4.3.1. Serine Hydroxymethyltransferase

Bridgers[83] originally reported the activity of SHMT in whole mouse brain. The enzymatic activity has been measured in dialyzed soluble phosphate buffer extracts of rat brain[84] and purified from such extracts of fresh-frozen bovine brain.[85] Since the SHMT within the CNS is largely localized in particulate fractions after homogenization in 0.32 M sucrose (see below), the efficiency of extraction is uncertain in the latter two studies.

Two laboratories have studied the regional distribution of SHMT within the CNS.[86,87] Davies and Johnston[86] found no correlation between glycine content and SHMT activity in five regions of the rat CNS (cerebral cortex, cerebellum pons–medulla, diencephalon-mesencephalon, and spinal cord). However, they did find significantly higher levels of SHMT in cat spinal gray matter than in white matter. In addition, the ventral gray matter had significantly higher levels than the dorsal gray matter.[86] These findings are at least qualitatively correlated to the content of glycine in the spinal cord.[33]

Table I

Distribution of Glycine, Serine Hydroxymethyltransferase, Succinate Dehydrogenase, Glycine Transaminase, and the Glycine Cleavage System in Five Regions of the Rat CNS

	Telencephalon	Midbrain	Cerebellum	Medulla–pons	Spinal cord
Glycine[a] (μmol/g tissue)	0.83	1.50	0.68	4.15	4.20
SHMT[a] (μmol/h per g tissue)	3.49	3.74	4.80	5.08	4.35
SHMT[a] (nmol/h per mg protein)	28.4	30.9	40.6	45.9	44.6
SDH[a] (μmol/h per mg protein)	2.70	2.63	3.27	2.30	1.86
GT[a] (nmol/h per mg protein)	162	162	158	166	133
GCS[b] (nmol/h per mg protein)	4.35	0.88	2.06	0.39	0.07
DNA[c] (mg/g tissue)	0.86	0.94	4.57	0.83	0.48

[a] Values from Daly and Aprison.[87]
[b] Values from Daly et al.[112]
[c] Values from Aprison and Daly.[9]

Daly and Aprison[87] have also studied the regional distribution of SHMT in five similar regions of the rat CNS (Table I). Values varied between 28 and 46 μmol/h per mg protein, and, except for the level found in the cerebellum, the SHMT appeared to be correlated with the glycine content in the various areas.[87] When the SDH levels in the five areas studied (Table I) are used to adjust the SHMT activity per relative number of mitochondria (the SHMT within the CNS appears to be a mitochondrial enzyme—see below), there was an excellent correlation between the SHMT activity and the glycine content in the five regions.[87]

Our laboratory has reported the nearly exclusive localization of SHMT in particulate fractions.[87] The activity was highest in mitochondrial fractions. The activity found in synaptosomal fractions is probably a reflection of synaptosomal mitochondria, but indirect evidence[87–89] suggests that SHMT may be associated with a subpopulation of synaptosomes that band below 1.2 M sucrose during subcellular fractionation—the so-called "noncholinergic" synaptosomes.[90] It is interesting that there is direct evidence that glycine appears concentrated in "heavy" synaptosomes isolated from the P_1 fractions of the medulla and spinal cord.[50] Experiments are needed to verify the possibility of an association between glycine and SHMT in "heavy" synaptosomes.

Although Davies and Johnston[86] reported a bimodal distribution of SHMT between soluble and particulate matter in subcellular fractions of rat CNS, the prominent localization of SHMT in particulate fractions within the CNS[87] has been confirmed by two separate laboratories.[89,91] Yoshida and Kikuchi[92] also found the brain to contain extremely small amounts of soluble SHMT activity. In fact, the soluble SHMT activity in brain was the lowest of the nine tissues examined. (The monkey brain appears to contain more soluble SHMT.[93]) The relative lack of a soluble SHMT activity may be unique to the CNS, since approximately 50% of the SHMT in rat liver is soluble.[94] In the liver, there is good evidence for two separate isozymes.[94,95] It would appear that the CNS contains little of the cystolic isozyme, and the compartmentation of SHMT exclusively in mitochondria may imply important and unique regulatory functions for glycine and/or one-carbon metabolism within the brain. The correlation of SHMT activity with glycine levels suggests the possibility that the enzyme might be rate limiting in the synthesis of glycine. The studies of Davies and Johnston[86] and Daly and Aprison[87] also do not agree on the correlation of SHMT levels and glycine content or the subcellular distribution of SHMT within the rat CNS. Possible explanations for these discrepancies have been discussed.[9,87]

The high level of SHMT activity in the cerebellum was also suggested by the studies using P_2 fractions discussed above.[54] From the data in Table I, it can be seen that the cerebellum has a greater relative number of mitochondria than the other areas. More dramatically, the cerebellum contains approximately five to ten times the amount of DNA of the other regions. This increased content of DNA is most likely a reflection of the greater number of cells per gram of tissue in the cerebellum. Subsequently, a greater demand for one-carbon units may exist. Support for this idea is found in maturation studies in which there appears to be a correlation between SHMT levels and growth of the CNS.[22,83,96]

It is of interest that a similar suggestion has recently been made to explain the peak of SHMT in the rat liver during the perinatal period.[97] In addition, even in the adult brain, the activity of thymidylate synthetase is highest in the cerebellum,[98] suggesting greater one-carbon metabolism in this region.

We have speculated that the SHMT activity in the "metabolic pool" is relatively constant from cell to cell in the CNS and that the increased levels of the enzyme in the medulla–pons and spinal cord reflect its role in the "transmitter pool" in these areas. Consequently, if one accepts the idea that the "extra" glycine found in the high-glycine regions of the CNS represents a "transmitter pool," then serine appears to be a precursor of glycine in both the metabolic and functional compartments.[9,56,87]

4.3.2. Glycine Transaminase

The suggestion that glyoxylate may serve as a precursor of glycine within the CNS was first made by Johnston and Vitali[99] when they noted the presence of a transaminating activity between glutamate and glyoxylate in acetone powder extracts prepared from rat spinal cord. The activity was localized in soluble and mitochondrial fractions prepared from sucrose homogenates.[100–102] The GT activity was higher in spinal gray matter than in white matter.[101] At least 11 amino acids are capable of participating in a transamination reaction with glyoxylate in dialyzed homogenates of rat whole brain. Glutamate, glutamine and ornithine are the best substrates for the reaction.[100,102] Whether these activities represent separate enzymes or multiple reactions of a few enzymes has not been studied. However, the possibility of at least different isoenzymes is suggested by the finding that the glutamine:glyoxylate transaminase activity appears to be localized primarily in the mitochondrial fraction, whereas the glutamate:glyoxylate transaminase activity appears to be present in both the mitochondrial and soluble fractions.[100,102]

The activities of the transaminase were the same in the telencephalon, midbrain, cerebellum, and medulla–pons of the rat, whereas the spinal cord contained a significantly lower amount of activity (Table I).[87] This distribution of GT among the five regions does not correlate with the glycine content and is distinctly different from that found for SHMT (Table I). Approximately two-fold higher GT activity has been reported in the retina.[104] Maturation studies indicate a tendency for the transaminase activity to increase during postnatal development.[22,103]

Factors that favor a role for glyoxylate in glycine metabolism within the CNS are (1) the large increase of glyoxylate and glycine in experimental thiamine deficiency[105] (however, see ref. 106), (2) the presence of GT in CNS tissues, and (3) the high levels of hydroxypyruvate decarboxylase activity (pathway IV, Fig. 1) found in the CNS.[73] Factors that argue against a major role for glyoxylate in glycine metabolism within the CNS are (1) the low levels of glyoxylate in the CNS—approximately 10 nmol/g,[78,101,105,107] (2) the toxic effect of glyoxylate on nerve cells,[107] (3) the possible nonspecificity of the glycine transaminase,[28] (4) the apparent lack of a readily available precursor of glyoxylate within the CNS,[56,78] and (5) the very slow influx of glyoxylate from the blood into the CNS.[69]

4.3.3. Glycine Cleavage System

The GCS appears to be the major degradative enzyme in peripheral tissues.[108] In liver, the GCS has been found to be a complex of four proteins[108] and, together with a portion of the SHMT, is located within the inner membrane fraction of mitochondria.[109] The properties of the GCS within the CNS[110–112] appear to be similar to those described for the system in liver.[108] The enzymatic activity is exclusively located in mitochondria and is not a typical amino acid decarboxylase in that it is dependent on both NAD^+ and tetrahydrofolate. The liberation of the carboxyl carbon of glycine is closely associated with the formation of serine via the condensation reaction between the α carbon (attached to tetrahydrofolate) and a second molecule of glycine catalyzed by SHMT. The detailed experiments with liver mitochondria describing the membrane-bound nature of the four-protein complexes have not been performed with CNS tissues, but Triton X-100 completely abolished the GCS activity in CNS homogenates. This fact suggested that the detergent causes disruption of membrane associations that are essential for enzymic activity.[110–112]

Bruin and associates[110] found the total GCS activity to be the highest in the cerebrum of most animals. Uhr[111] found higher enzyme levels in homogenates of the rostral portions of the CNS than in the medulla–pons and spinal cord of the sheep and cat. Our laboratory confirmed the lower apparent activity in the medulla–pons and spinal cord of the rat (Table I).[112]

The 25-fold difference in activity within the CNS suggested the possibility of the presence of an inhibitor in those regions where the activity was low.[112] Evidence for such inhibition was found when CNS homogenates were incubated with liver homogenates. The combined GCS activity of the liver and any region of the CNS was significantly less than the sum of the activities measured separately in each homogenate. This nonadditivity appeared greater in regions where the GCS activity was low (spinal cord and medulla–pons) than in regions where the GCS activity was high (telencephalon and cerebellum).[112] This apparent "inhibition" has recently been confirmed for the GCS within the CNS.[113]

It remains to be seen whether this endogenous "inhibition" has any significance in the metabolic and/or transmitter roles of glycine *in vivo* or whether it is merely a serendipitous finding caused by tissue homogenization and the *in vitro* assay. The potency of the inhibition, the fact that the GCS activity appears inversely related (exponentially) to the concentration of the inhibitor within the five regions, and the apparent importance of the GCS in the liver suggest that the apparent activity *in vitro* may not reflect either the amounts of the enzyme system *in vivo* or the possible significance of the GCS in the metabolism of glycine within the CNS. The experiments studying the interconversion of serine and glycine in P_2 fractions[54] (see Section 4.2.2) found the conversion of glycine to serine greater in the medulla–pons and spinal cord preparations than in cerebellar preparations. The portion of this conversion mediated by the GCS is unknown.

4.4. Interrelationships among Serine, Glycine, and the One-Carbon Pool in the CNS

The CNS is unique with regard to folate metabolism in several ways. The concentration of folate in CSF is approximately threefold higher than in plasma,[114] and folates have the highest CSF/plasma ratio among numerous endogenous compounds investigated.[115] There appears to be a saturable concentrative uptake of N^5-methyltetrahydrofolate (methylTHF) from serum into the CSF,[116] and this uptake may explain the sparing of CNS folate levels during folic acid deficiency.[117] The CNS folates are distributed nearly equally between particulate and soluble fractions,[118] in sharp distinction to the distribution in liver tissue where greater than 90% of the folates are in the soluble fraction.[119] The particulate activity appears to be associated with mitochondria and synaptosomes.[88,120] Approximately 50% of folates are in the pentaglutamate form, with significant amounts of the hexa- and heptaglutamate derivates present. Less than 20% of the CNS folates contained three or fewer glutamate residues, a distribution similar to that found in peripheral tissue.[120] Tetrahydrofolates predominate in the CNS in distinction to the predominance of methylated derivatives in the liver.[121] The turnover of brain folates is very slow, and that of synaptosomal particulate fraction appears almost inert.[122]

The formation of one-carbon units and glycine within the mitochondria, their utilization within the cytosol,* and the necessity for strict control of glycine levels in the transmitter pool emphasize the need for regulation in both the metabolism and transport of these compounds within the mitochondrion. If the N^5,N^{10}-methyleneTHF generated by the GCS condenses with another molecule of glycine, a molecule of serine will be formed via the SHMT. Therefore, a "coupled" GCS and SHMT would convert two molecules of glycine to one molecule each of serine and CO_2 without net alteration of the one-carbon pool. (The other reaction products are disregarded in this discussion.) An "uncoupled" GCS would degrade glycine with generation of one-carbon units, whereas an "uncoupled" SHMT would consume one-carbon units during the degradation of glycine. *In vivo* studies[28] suggest that little "coupling" of SHMT and the GCS occurs in glycine formation (pathway III, Fig. 1). However, there appears to be a significant "coupling" of the two activities in the degradation of glycine both *in vitro*[110,112] and *in vivo*.[9] One may envision a cycle in which an "uncoupled" SHMT (eq. 1) would generate one-carbon units and glycine. A "coupled" GCS (eq. 2) and SHMT (eq. 3) would prevent buildup of glycine during one-carbon generation. The net reaction would be the generation of two

---

* MethyleneTHF dehydrogenase is located in both soluble and particulate fractions,[88] whereas the methylTHF methyltransferase and methyleneTHF reductase are limited to the soluble fraction in the CNS.[89,91]

one-carbon units and a molecule of CO_2 for each molecule of serine consumed (eq. 4).

$$2\,\text{Serine} + 2\,\text{THF} \rightarrow 2\,\text{Glycine} + 2\,\text{CTHF} \qquad [1]$$

$$\text{Glycine} + \quad \text{THF} \rightarrow \text{C-THF} \quad + \quad CO_2 \qquad [2]$$

$$\text{Glycine} + \text{C-THF} \rightarrow \text{Serine} \quad + \quad \text{THF} \qquad [3]$$

$$\text{Serine} + 2\,\text{THF} \rightarrow 2\,\text{C-THF} + \quad CO_2 \qquad [4]$$

The two isoenzymes of SHMT in the liver have been proposed to function in a one-carbon shuttle system linking folate metabolism in the mitochondrial and cytolic compartments.[123] Such a shuttle is necessitated by the finding of transport of glycine and serine across the rat liver mitochondrial membrane but the lack of transport of N^5,N^{10}-methyleneTHF and THF. In addition, the methyleneTHF reductase is generally considered irreversible *in vivo*.[123-125] In this model, methyleneTHF formed within the mitochondrion would condense with glycine to form serine via the mitochondrial SHMT; serine would be transported to the cytoplasm; and the one-carbon unit would be released into the cytoplasmic pool by the conversion of serine to glycine via the soluble SHMT.[123,125] There is a concentrative uptake mechanism for serine in reticulocyte mitochondria,[126] whereas glycine is transported by a non-energy-dependent, passive carrier system in both rat liver and brain mitochondria.[127]

The need for active one-carbon units may be less in the adult CNS than in liver. The SHMT,[22,83,96] GCS,[113] methyltransferase,[84,96,128] methyleneTHF reductase,[128] and dihydrofolate reductase[129] all decrease with maturation, as do the levels of total folates[130] and thymidylate synthetase activity[98] within the brain. Consequently, slow transport of folate across the mitochondrial membrane may be sufficient to meet the metabolic needs of adult CNS tissue. In addition, there is indirect evidence for the formation of methyleneTHF from methylTHF within the CNS.[128,131,132] By generating one-carbon units in the cytoplasm (reverse of the methyleneTHF reductase), this reaction would tend to lessen the need for the interconversion of serine and glycine in generating the one-carbon units for maintenance of CNS structures.

5. PHYSIOLOGICAL EFFECTS OF GLYCINE ON THE CNS

The depressant action of glycine on the firing rate of neurons was recognized in the early 1960s but was considered nonspecific.[5] After the intrasegmental distribution of glycine was discovered,[1] the inhibitory action of glycine on spinal neurons was reinvestigated. Glycine hyperpolarized the postsynaptic membrane and increased its conductance. Inhibitory synaptic potentials evoked in lumbar motorneurons by dorsal root stimulation and those produced by iontophoresis of glycine were shown to be equivalent.[133] This potent inhibition of spinal motoneurons was quickly confirmed and found to be similar on spinal interneurons.[134-136]

In addition to motoneurons, a strong depressant action of glycine has been found on cat spinal interneurons,[135] Renshaw cells,[135] sacral parasympathetic preganglionic neurons,[137] and dorsal horn cells.[138,139] Glycine is also a strong depressant of rat spinal neurons[140] and produces a hyperpolarization and conductance increase of the postsynaptic membrane in cultured cells of rat spinal cord.[141] It is of interest that the action of glycine on chick embryo spinal cord–muscle cultures may be either facilitatory or inhibitory depending on the degree of maturation.[142]

Within the brainstem, glycine is inhibitory on cells in the cuneate nucleus,[143,144] red nucleus,[145] hypoglossus nucleus,[146] medullary reticular formation,[147] medullary respiratory neurons,[148] and Deiter's nucleus.[149] Two regions on the ventral medulla regulating vasomotor tone are sensitive to topically applied glycine.[150]

Glycine inhibits the firing of neurons in the substantia nigra.[151] Unilateral intranigral injection of glycine leads to a bilateral decrease of dopamine release in the caudate nucleus[152,153] and an increase in caudate neuronal activity.[154] There is a K^+-stimulated release of [^{3}H]glycine from striatal slices which is blocked by GABA,[155] but the presence of a high-affinity uptake system within the substantia nigra is controversial.[43,156] These facts, together with the relatively high endogenous content[9] and [^{3}H]strychnine binding[157] within the midbrain, lend support to the hypothesis that there is a glycinergic inhibitory interneuron interspaced between the GABAergic striatonigral and dopaminergic nigrostriatal pathways.[152] Glycine inhibits the firing of cells in the entopeduncular nucleus, a homologue of the primate medial pallidal segment.[158] Similar glycinergic interneurons interacting with dopaminergic neurons have been suggested within the striatum[10] and ventral tegmentum.[159]

In most other areas of the CNS, the depressant action of glycine is weak. Areas in which this weak activity is observed include thalamus,[160] hypothalamus,[161] olfactory bulb,[162] cerebellum,[160,163,164] and cerebral cortex.[136,160,165] The reversal potential for the weak hyperpolarization by glycine was much more positive than the cortical IPSP,[165] and the postsynaptic action of glycine was less than 8% the action of GABA in cat cerebral cortex.[136] In comparison, the action of glycine was five times more potent than that of GABA in the spinal cord.[136] Recent studies, however, suggest that there may be a subpopulation of synapses in the rat cerebral cortex that use glycine as a transmitter.[166] In addition, in the rat medial hypothalamus, a subpopulation of neurons (10% of neurons tested) was found more sensitive to glycine than to GABA.[167]

Glycine has depressant effects on retinal neurons,[168] and antagonists have provided evidence for separate neurons using GABA and glycine.[169] Glycine is inhibitory to bipolar, amacrine, and ganglion cells but does not appear to have an effect on horizontal cells.[170] A model for the postulated separate roles of GABA- and glycine-utilizing amacrine cells has been presented.[171] Taurine has similar effects on retinal neurons, and strychnine blocks the action of both glycine and taurine.[172] However, the intraretinal distribution of taurine does not support a role as a transmitter within the inner layers of the retina.[46,47]

In addition to these studies in mammals, glycine has a strong depressant effect on the Mauthner cell in the medulla of the goldfish[173] and in the reticular formation and spinal cord of the lamprey.[174,175] However, the action of glycine is variable on spinal motorneurons of the frog.[176–178]

The interaction of glycine with the postsynaptic membrane has been studied in feline spinal motoneurons and lamprey medullary neurons. The sigmoidal nature of dose (iontophoretic current)–response (conductive change) curves suggests that glycine molecules interact in a cooperative manner with their receptor. When the data are plotted on log–log coordinates, a slope of approximately 2 is found, consistent with the hypothesis that two or more molecules of glycine attach to the receptor before activation occurs.[136,173,175] There appears to be positive cooperativity among four molecules of glycine in activating the postsynaptic receptor of the Mauthner cell in the medulla of the goldfish.[173] In addition to its action on its own receptor, GABA appears to interact with the glycine receptor to modulate its affinity for glycine, probably by lowering the energy barrier for binding of the first of the four glycine molecules required.[173] It is of interest that bicuculline inhibits strychnine binding to spinal cord membranes[179] and that there is some evidence that glycine and GABA may share the same ionic conductance channel in spinal cord.[180]

Glycine and GABA (and probably other transmitters) may have a modulating function on glial cells. Glial cell membranes are depolarized indirectly by the efflux of K^+ from neighboring neurons into the extracellular fluid. The efflux of K^+ is assumed to arise from the interaction of the transmitter with the neuronal receptor site.[181]

The "somatic" region of the postsynaptic membrane of cat motoneurons demonstrates a preferential topographic sensitivity to glycine compared to the dendritic region.[182] This physiological study is in good agreement with the electron microscopic visualization of the nerve terminals of Ia inhibitory interneurons (presumed glycinergic) which form axosomatic and proximal axo-dendritic synapses.[183] Data collected in the hypoglossus nucleus suggest that glycine was more potent than GABA on the soma, whereas GABA was more potent on the dendrites of the neurons.[146] Studies using cerebellar slices have demonstrated marked variation in the topographic sensitivity of putative transmitters and antagonists on Purkinje cell dendrites.[164] The effects of glycine are hyperpolarizing when it is applied to the cell body but are multiphasic when it is applied at peripheral sites of the postsynaptic membrane of cultured mouse spinal neurons.[184]

All of these studies suggest that there is a selective topographic density of glycine receptors in the perisomatic region of the postsynaptic neuron. They also suggest that unless a putative transmitter or drug is applied on the same topographic region of the postsynaptic membrane as the natural transmitter, the electrical events recorded by an intracellular electrode may not be equivalent.[182] In addition, these topographic relationships emphasize the need for caution in comparing the relative potencies of different compounds.

6. RELEASE OF GLYCINE FROM NERVOUS TISSUE

Evidence for the release of endogenous glycine *in vivo* after a physiological stimulus has been obtained.[185–187] The effluxes of GABA, alanine, and glycine were significantly increased in perfusates of the dorsal column nuclei of rat medulla after stimulation of the medial lemiscus in the thalamus.[185] The *in vivo* release of endogenous glycine into the perfused rat spinal subarachnoid space occurred after stimulation of the sciatic nerve.[186] The effux of glycine could only be measured if *p*-chloromercuribenzoate, a nonspecific uptake inhibitor, was present in the perfusate or if strychnine was given intravenously prior to stimulation.[186] In other experiments, K^+-evoked release into the subarachnoid perfusate (without strychnine or *p*-chloromercuribenzoate) depended on whether endogenous or preloaded labeled glycine was measured. Although there was no release of labeled glycine, there was a transient evoked release of endogenous glycine.[186] Release of endogenous glycine into the perfused central canal of the feline spinal cord has been reported after stimulation of descending tracts in the presence of uptake inhibitors.[187] *In vivo* photic stimulation released endogenous glycine from pigmented[188] but not albino rat retina.[189] However, superfusion of the vitreal space with an elevated K^+ concentration in the latter species did evoke release.[189]

Exogenous as well as endogenous glycine can be released in the region of the pigeon optic tectum after electrical stimulation of the nucleus isthmi, pars parvocellularis (IPC).[190,191] This release, the depressent action of glycine on tectal neurons, a high-affinity uptake system, and glycine-sensitive [³H]strychnine binding in the optic tectum provide strong evidence for glycine being a transmitter in the IPC–tectal pathway in the pigeon.[192–194] Further support is provided by the retrograde labeling of this pathway after [³H]glycine injection into the tectum.[195] The endogenous levels of glycine in this pathway before and after IPC ablation would be of interest.

Endogenous glycine was not released from the visual cortex spontaneously or after K^+ stimulation *in vivo*,[196] whereas a small, nonconsistent release of glycine from the cerebral cortex was found after stimulation of the reticular formation.[197] Endogenous glycine was not released into the perfused fourth ventricule after cerebellar stimulation.[198]

The only other data for the release of endogenous glycine come from experiments *in vitro*. Two experiments *in situ* have measured the release of exogenous [¹⁴C]glycine from the hemisected toad spinal cord after stimulation of the dorsal roots[9,56] or the rostral cord.[199] The correlation of release with stimulation frequency, the dependence of release on calcium, and the failure to find release of either nontransmitter amino acids or intracellular and extracellular water markers suggest that the evoked release of the labeled glycine may have been from the transmitter pool. The stimulation of dorsal roots activated inhibitory interneurons mainly within the stimulated segments, whereas stimulation of the rostral cord activated descending pathways that would (1) directly release glycine, (2) activate inhibitory interneurons in a number of segments, or (3) activate motoneuron collaterals, thereby activating Renshaw cells (if present). Simultaneous stimulation of dorsal and ventral roots has also evoked release of exogenous [¹⁴C]glycine from the perfused feline lumbar spinal cord *in vivo*.[200] However, this release could only be demonstrated in the presence of *p*-hydroxymercuribenzoate, a weak and nonspecific inhibitor of glycine uptake.[201]

Exogenous [¹⁴C]glycine was released from slices of rat spinal cord after electrical[202,203] or K^+-evoked depolarization.[202,204] This evoked release of glycine appeared to be specific in that release of nontransmitter amino acids or urea was not enhanced. The release was dependent on the presence of calcium and preloading via the high-affinity uptake system.[204] Potassium-stimulated release of exogenous glycine has also been reported in slices of frog medulla and spinal cord.[205] The release of glycine from CNS tissue slices may be via a rate-limiting carrier-mediated process.[186,202]

The regional specificity of the evoked release of exogenous glycine is controversial. Some investigations have found a calcium-dependent release from slices of cerebral cortex,[205–207] whereas others have not.[208,209] Analysis of the time course of release after K^+ stimulation suggested that the preloaded glycine might have been released from glial cells.[206] Conflicting reports also exist for the cerebellar release of glycine.[207,209,210] Calcium-dependent release has been reported in slices from basal ganglia plus thalamus.[207] Release from slices of striatum, hypothalamus, superior colliculi, and inferior colliculi has been reported without investigation of the calcium dependence.[209] Potassium-evoked release of both exogenous and endogenous glycine was found in isolated rat dorsal root ganglia.[211]

The K^+-induced release of endogenous glycine from tissue slices of various regions of the CNS has been studied. Release was reported from rat spinal cord[212] and pigeon cerebellum[213] but not rat cerebellum.[214] Release was found in rat midbrain slices that was not clearly calcium dependent.[215] Glycine was one of many amino acids released from slices of the guinea pig cochlear nucleus.[216]

Release of preloaded [^{3}H]glycine from retina has been reported *in vivo* (cat) and *in vitro* (rabbit).[217] Similar to studies *in vivo* (see above), K^+ stimulation produced release of endogenous glycine in isolated retina from albino rats, but photic stimulation did not.[218]

Bradford and co-workers have shown release of endogenous amino acids from preparations of synaptosomes after electrical or K^+ stimulation. There was a significant calcium-dependent release of endogenous glycine from synaptosomes isolated from the medulla and spinal cord which could be blocked by treatment of the rats *in vivo* with tetanus toxin.[50] Release of exogenous glycine from spinal cord "particles" was also blocked after *in vitro* incubation with tetanus toxin.[219] In contrast to spinal cord–medullary synaptosomes, Bradford and associates[51] have not found release of endogenous glycine from synaptosomes isolated from hypothatamus or cerebral cortex. However, Levi and associates[220] have reported the release of endogenous glycine from cortical synaptosomes.

7. CLEARANCE AND UPTAKE OF GLYCINE

7.1. CSF Clearance

Although the concentration of glycine in the blood is high, its concentration in the CSF is low.[1,221] The plasma–CSF ratio for glycine is one of the highest among common amino acids.[221] The glycine concentration has been reported lower in lumbar CSF than in cisternal CSF or ventricular CSF.[221,222]

One reason for the low levels of glycine in the CSF and for its low net uptake into the CNS from blood could be the efficient, carrier-mediated system transporting glycine from the CSF to the blood.[223] Both the choroid plexus and the spinal subarachnoid membrane may be important sites for this transport.[223-225] The spinal subarachnoid membrane may be quantitatively more efficient, since glycine is cleared from the CSF at a greater rate during ventriculolumbar perfusions than during ventriculocisternal perfusions.[224] The high-affinity uptake systems described below for transporting glycine into CNS tissues may also function to keep the concentration of glycine low in the CSF. Until a specific inhibitor is found, the rapid clearance of glycine from the subarachnoid space will hinder attempts to collect glycine after release *in vivo*.

7.2. Uptake into Nervous Tissue

Numerous studies have provided evidence for the suggestion of Aprison and Werman[2] that the synaptic action of glycine might be terminated by uptake mechanisms. The discovery of a high-affinity system in slices of spinal cord was the first direct evidence for such a mechanism.[201] Early studies demonstrated that the spinal cord had both high- and low-affinity uptake systems (K_m < 50 μM and > 100 μM, respectively), whereas the cerebral cortex contained only a low-affinity system.[226,227] The high-affinity uptake system was shown to be sodium dependent.[228]

Labeled glycine appeared to be taken up by a subpopulation of synaptosomes within the spinal cord more dense than those accumulating GABA.[229,230] The identification of this subpopulation of synaptosomes as well as the K^+-evoked release of exogenous glycine from spinal cord slices were dependent on the presence of sodium during preloading of the tissue with radioactive glycine.[204,230] From these experiments, it was postulated that the high-affinity, sodium-dependent uptake of exogenous glycine labels the transmitter pool within the nerve terminal.[204]

The possibility that high-affinity uptake systems are the result of homo-exchange between exogenous and endogenous amino acid is controversial.[231,232] All of the experiments cited above measured the accumulation of label after incubation with radioactive glycine. Even the one study[233] showing net uptake of glycine into a crude synaptosomal preparation has been critized.[231]

The high-affinity uptake of glycine into glia cells is hard to reconcile with the suggestion that exogenous radioactive glycine labels the transmitter pool. High-affinity uptake of glycine has been measured in glia cells, and the accumulation of label was greater in glial preparations from spinal cord than from cerebral cortex.[234,235] Autoradiography has also shown labeled glia cells after high-affinity uptake (see below). Our position is that the high-affinity uptake of glycine may be an efficient method for the termination of its synaptic action, whether it occurs into glia or presynaptic or postsynaptic neurons. However, the presence of such a system is neither necessary nor sufficient to postulate a transmitter function in a particular region of the CNS. Nevertheless, high-affinity uptake of glycine has been found in cerebral cortex,[220,236,237] cerebellar Golgi terminals,[238] hypothalamus,[239] and retina.[240-242]

7.3. Autoradiographic Studies

After slices of rat spinal cord are incubated with [³H]glycine, label appears in gray matter associated with nerve terminals to a greater extent than in white matter.[243,244] Especially prominent labeling occurred in the central gray matter. Whether there is a small amount of labeling over glial elements or thin myelinated axons is uncertain. Between 28 and 40% of synapses in the spinal cord accumulate [³H]glycine. Within the gray matter there was an association of glycine accumulation with terminals containing flattened vesicles.[243,244] Such an association was also found *in vivo* after perfusion of the central canal of the spinal cord or after direct intraspinal injection of [³H]glycine.[245,246] These findings are of interest in view of the hypothesis that terminals of inhibitory synapses contain "flattened" vesicles.[247] Direct electron microscopic visualization of spinal cord nerve terminals revealed that the axodendritic terminals of Ia inhibitory interneurons (presumably glycinergic) contained slightly more "flattened" vesicle profiles than did excitatory terminals.[183] Cervical subarachnoid injection of [³H]glycine in the rat labeled glial and neuronal elements in the dorsal horn of the cervical spinal cord (substantia gelatinosa Rolandi). Highest grain densities were seen over neuronal perikarya in lamina III.[248]

Neuronal and glial elements in cultures of spinal cord, medulla, and cerebellum accumulate label after incubation with [³H]glycine, but the latter area only weakly.[249] The uptake appears to be greater in cerebellar slices[250] than in cerebellar cultures. There is diffuse accumulation of label in slices of the dorsal cochlear nucleus.[251] Extensive investigations have localized the high-affinity uptake of glycine into inner retinal layers, especially the amacrine and ganglion cell layers.[252–254] Three distinct subpopulations of cat amacrine cells accumulate label 4 hr after intravitreal injection of glycine, but no identification of the label was reported.[253] Recent evidence in goldfish retina suggests that [³H]glycine may also be accumulated by interplexiform cells.[254]

8. INTERACTIONS WITH DRUGS AND TOXINS

8.1. Strychnine

The studies by Eccles and colleagues and their extension and refinement by Curtis explained the well-known convulsant properties of strychnine by a blockage of spinal postsynaptic inhibition. Most spinal postsynaptic inhibitions of short latency and duration are blocked by strychnine, whereas strychnine does not affect presynaptic inhibitions or excitatory postsynaptic potentials.[6,8] The case for glycine's role in inhibitory transmission was greatly strengthened by the finding that strychnine blocked the depressant effect of glycine but not the effects of GABA on spinal neurons.[134,135] With the assumption that glycine is the transmitter at these strychnine-sensitive synapses, the experiments can be most easily explained by a reversible and specific interaction between strychnine and the spinal postsynaptic receptor for glycine.[255,256]

It is of interest that glycine exerts a depolarizing action on immature spinal neurons, and this effect is also antagonized by strychnine.[257,258] Strychnine antagonizes the descending postsynaptic inhibition of neck motoneurons from the medial vestibular nucleus.[259] Some segmental inhibitions within the dorsal horn of the spinal cord are strychnine sensitive, whereas, the effect of strychnine on descending inhibitions of dorsal horn neurons is uncertain.[250,251,260–262] A controversy also exists on whether mutual inhibition (inhibition of Renshaw cells by Renshaw cells) is strychnine sensitive.[263–266] Some investigators postulate the existence of two glycine receptors, one strychnine sensitive and the other strychnine resistant.[263,264] Others have postulated two pools of Renshaw cells, one glycinergic and the other GABAergic.[266] Finally, others have found mutual inhibition to be strychnine sensitive, thus making the postulate of two glycine receptors or two pools of Renshaw cells unnecessary.[265]

Although the primary effects of strychnine are postsynaptic, higher concentrations can result in nonspecific alterations of membrane.[8,9,174] Strychnine is also known to antagonize the postsynaptic effects of several amino acids—glycine, taurine, β-alanine, and proline.[8,178] However, based on its relative specificity in spinal inhibitions, strychnine sensitivity has become an important technique for screening of transmitter candidates at inhibitory synapses. In the context of designing experiments in which the goal is to identify a neurotransmitter, the usefulness and validity of strychnine sensitivity as a screening procedure have been challenged.[3,9,267,268] Nevertheless, strychnine sensitivity of the routine transmitter can at times be the only evidence for the identification of the transmitter if the conductance changes of GABA and glycine are the same.[174,269] The antagonism of glycine by strychnine may be more specific than the antagonism of GABA by bicuculline or picrotoxin.[270]

Earlier studies have shown both glycine and the IPSP to be sensitive to strychnine in various regions of the brainstem including the reticular formation, hypoglossal nucleus, trigeminal (motor) nucleus, and perhaps cuneate nucleus.[8,9] Strychnine antagonizes inhibition of vestibular neurons,[271] cochlear neurons,[272] medullary descending trigeminal neurons,[273] respiratory driving neurons,[148] and amacrine cells of the retina.[169–171] Generally speaking, strychnine sensitivity is not characteristic of postsynaptic inhibitions in more rostral areas of the CNS including the cerebral cortex, hippocampus, olfactory bulb, cerebellar cortex, thalamus, and hypothalamus.[8] However, some postsynaptic inhibitions in the intralaminar nuclei of the feline thalamus are strychnine sensitive.[274,275]

8.2. Tetanus Toxin

Tetanus toxin blocks both postsynaptic and presynaptic inhibitions in the spinal cord.[276] With the assumption that glycine and GABA are transmitters in spinal cord, pharmacological studies have postulated a presynaptic site of action of tetanus toxin in blocking neurotransmission.[277,278] The electrically induced release of glycine, GABA, and aspartate from synaptosomes isolated from the spinal cord and medulla is significantly decreased in animals poisoned with tetanus toxin.[50] Ultrastructural studies have shown an increase in synaptic vesicles in the presynaptic nerve terminal at both the neuromuscular junction and axosomatic synapses in the spinal cord. In the latter, flattened synaptic vesicles appeared to be selectively increased.[279]

A number of studies have looked at the levels of amino acids in spinal cord from animals with experimentally induced tetanus. Glycine has been found to be decreased,[280] unchanged,[281,282] or increased[283] within the affected segment of spinal cord in experimental local tetanus. These latter carefully controlled experiments[283] provide support for the proposed presynaptic blockade of glycine release by tetanus toxin. Unfortunately, the results of intraspinal GABA levels were not reported.[283]

The fact that tetanus toxin binds to nervous tissue has been known since the beginning of the century, and there is general agreement that the toxin reaches the CNS by ascent of peripheral nerves.[276] The postulated presynaptic mechanism of action of tetanus toxin and its retrograde ascent of peripheral nerves requires migration of the molecule across at least two synapses (the peripheral neuromuscular junction and at least one central synapse). Electron microscopic autoradiography of the spinal cord after intramuscular injection of labeled toxin showed accumulation of label in the presynaptic elements surrounding motoneurons. Moreover, there was good temporal correlation between the appearance of the presynaptic label and the clinical signs of tetanus intoxication.[284] After intraspinal injection of labeled toxin, the radioactivty appears to localize in the presynaptic elements of axosomatic synapses of spinal gray matter.[285,286] This is the same compartment that accumulates [^{3}H]glycine after intraspinal injection.[246]

9. THE "GLYCINE RECEPTOR"

Strychnine binds to synaptic membrane fractions in a specific manner.[157] The fact that the regional distribution of binding paralleled the endogenous glycine content and physiological sensitivity of neurons to glycine suggested that [^{3}H]strychnine binding occurred at the postsynaptic receptor for glycine. The binding of strychnine could be reversed by glycine and those amino acids (taurine, serine, L-alanine, β-alanine) whose iontophoretic effects are also antagonized by strychnine.[157] The appearance of this strychnine-binding activity in the spinal cord of the chick embryo was well correlated with the biochemical, physiological, and anatomic evidence for the time of emergence of postsynaptic inhibition.[287] Further experiments suggested that strychnine and glycine bind to the receptor in a cooperative fashion; glycine binds at the "glycine recognition site," and strychnine binds at a distinct site which may represent the chloride ionic gate mechanism.[288]

Zarbin, Wamsley, and Kuhar[289] studied the autoradiography of [^{3}H]strychnine binding in the greatest detail to date. The highest binding occurred in the spinal cord; the dorsal horn and central gray matter had the highest binding in the cervical and thoracic segments, whereas in the cervical and lumbar enlargements, there was prominent labeling in the ventral gray matter in addition to the above regions. In the brainstem, there appeared to be localized labeling of certain nuclear groups. Especially prominent were the hypoglossal, intercalatus, trigeminal (motor), facial, and cuneiform nuclei. The entire reticular formation had significant, but lower, binding.[289] In the thalamus, the parafascicular nucleus and zona incerta had the highest binding, whereas in the cerebellum and cerebral cortex, little if any binding was observed.[289] These authors have thoroughly reviewed the correlation of their autoradiographic binding data with the symptoms and signs of strychnine toxicity. It is of interest that binding of [^{3}H]muscimol (GABA receptor) and [^{3}H]strychnine are generally the inverse of each other within the CNS.[290] Cultures of rat spinal cord, brainstem, and cerebellum bound radioactive β-alanine, glycine, and strychnine in similar manner, whereas GABA binding was distinctly different.[291] [^{3}H]Strychnine is also bound to retinal membranes.[292,293] However, the high levels of taurine and its potent antagonism of [^{3}H]strychnine binding make any conclusions about the glycine receptor in the retina difficult. The strychnine-binding receptor (glycine receptor?) has recently been solubilized from rat spinal cord and appears to be a polypeptide (mol. wt. 48,000).[294,295]

It is of interest that the clinical potency of strychnine analogues correlated well with their ability to inhibit strychnine binding.[296,297] Also, there was a marked reduction in strychnine binding in the anterior gray matter but not in the dorsal gray matter or white matter in postmortem studies of the thoracic spinal cord of patients with amyotrophic lateral sclerosis, a disease known to result in degeneration of the spinal anterior horn cells (and consequently postsynaptic receptors). There was no evidence for loss of dopaminergic, muscarinic, β-adrenergic, or GABAergic receptors.[298]

Benzodiazepines are inhibitors of [^{3}H]strychnine binding, and their clinical efficacy paralleled the potency of antagonism of *in vitro* strychnine binding.[288] It was suggested that these drugs may exert their anxiolytic and muscle relaxant properties via an ability to mimic glycine at the postsynaptic receptors. However, no interactions were found between benziodiazepines and glycine or strychnine when studied iontophoretically.[299,300] The relatively high *in vitro* concentrations needed to displace the [^{3}H]strychnine binding compared to the clinically effective doses of the benzodiazepines also cast doubt on such an interaction.[301] Moreover, glycine or strychnine did not affect the binding of [^{3}H]diazepam to rat brain preparations.[302] Nonspecific lipid solubility properties of benzodiazepines may be responsible for the *in vitro* displacement of [^{3}H]strychnine from membranes.[303]

The sodium-independent (nontransport?) binding of [^{3}H]glycine has been investigated in the rat CNS.[304] The affinity of binding was positively correlated to the glycine content in the five regions examined, but the apparent number of binding sites was negatively correlated. These authors suggest that the number of binding sites may include both postsynaptic and nonpostsynaptic receptors and that the affinity of glycine binding appears to be a better indicator of the glycinergic system. The results of strychnine[289] and glycine[304] binding differ significantly in the rostral portions of the rat CNS. Although there is little evidence for a large number of glycinergic synapses in the cerebral cortex or cerebellum, recent evidence suggests there may be a subpopulation of glycinergic neurons in these areas.[220,238] Quantitation and characteristics of these neurons will help clarify the discrepancies in glycine and strychnine binding studies.

10. GLYCINE IN PATHOLOGICAL STATES OF THE CNS

10.1 Hyperglycinemia Syndromes

There are at least three hereditary disorders of glycine metabolism—glycine encephalopathy, nonketotic hyperglycinemia, and ketotic hyperglycinemia. All three disorders are believed to have an autosomal recessive mode of inheritance. The first two disorders have only recently been distinguished clinically and chemically.[305–307]Glycine encephalopathy is characterized by a relatively normal neurological status at birth which quickly deteriorates in the first few days of postnatal life to hypotonia, lethargy, coma, respiratory failure, and seizures. Those who survive appear to improve somewhat but are left with spasticity, seizures, and severe psychomotor retardation. Nonketotic hyperglycinemia is characterized by the onset of relatively mild neuropsychiatric symptoms in childhood or adolescence.[308,309] At least one asymptomatic child has been reported to have the apparent biochemical defect.[309] The term ketotic hyperglycinemia was formerly used to describe the disorder characterized by hyperglycinemia, severe neonatal illness, but with organic aciduria, ketosis, neutropenia, and thrombocytopenia. Further elucidation of the biochemical defect suggested that the hyperglycinemia was secondary to the primary defect in the metabolism of an organic acid.[310]

The biochemical distinction among the hyperglycinemias is not unequivocal,[311] but a classification based on a proposal by Dalla-Bernardina and associates[307] is presented in Table II. The distinction between glycine encephalopathy and nonketotic hyperglycinemia cannot be made biochemically and relies on the severe clinical course of the former for diagnosis. The metabolic defect in ketotic hyperglycinemia is thought to involve the inhibition of the activity or synthesis of the GCS by various accumulated products of defective branched-chain amino acid catabolism.[314,315] Methylmalonate and propionate have been shown to inhibit the uptake of glycine by liver and brain mitochondria.[316] Methylmalonate also inhibits uptake of glycine by spinal cord synaptosomes.[317] The lesion in the "nonketotic" syndromes may be a primary defect in the GCS.[312,318,319] Glycine levels within the brain are greatly elevated in glycine encephalopathy but are not increased in nonspecific and ketotic hyperglycinemia.[305,306] The GCS was greatly reduced or not detectable in the brain in the former disorder but was mildly or not decreased in the latter syndromes.[305,306,319] Although the activity of the GCS was decreased in the liver in all three disorders, the activity in glycine encephalopathy tended to be higher.[305,306] The role of serine hydroxymethyltransferase has not been fully investigated in glycine encephalopathy.[9]

The neonate with glycine encephalopathy has greatly elevated levels of CSF glycine and a severely abnormal EEG immediately after birth, even before symptoms develop.[320] Neither the plasma concentration nor the in vivo peripheral metabolism of glycine is normal in obligate heterozygotes (parents).[321] Although generally asymptomatic, heterozygotes frequently have soft neurological signs on detailed investigation.[322] No effective therapy has been found for glycine encephalopathy. This is probably a reflection of the neurotoxicity

Table II

Biochemical Findings in the Hyperglycinemia Syndromes

Hyperglycinemia	Glycine			CNS	Organic acids	Typical cases[a]
	Plasma	CSF	CSF/plasma			
Glycine encephalopathy	N or ↑	↑	↑	↑	—	306,307
Nonketotic	↑	↑	↑	?	—	308,309
Nonspecific	↑	N	N	N	—	305,312
Ketotic	↑	N	N	N	+[b]	310
D-Glyceric acidemia	↑	N or ↑	N or ↑	?	+[c]	313

[a] References with case reports.
[b] Organic acid.
[c] Glyceric acid.

of glycine in the immature nervous system.[323] Although there may be a disruption of the transmitter pool in hyperglycinemia,[324,325] the clinical symptoms and widespread pathological changes seen in glycine encephalopathy suggest that a disruption of the metabolic functions of glycine may be more important in pathogenesis.

10.2. Hyperphenylalaninemia

The classical biochemical lesion in phenylketonuria (PKU) involves a defective phenylalanine hydroxylase resulting in a deficiency of tyrosine and the accumulation of phenylalanine and its metabolites.[326] Recent studies in animals have shown that chronic hyperphenylalaninemia results in elevated levels of glycine in the immature brain.[327–329] The etiology and pathological significance (if any) are not known. Elevated glycine levels are not seen after acute inhibition of phenylalanine hydroxylase[330] in the spinal cord and liver during chronic hyperphenylalinemia[329] or in the CSF of newborns with PKU.[331]

10.3 Spasticity

Interruption of descending tracts produces spastic paralysis in the muscles innervated by spinal segments caudal to the lesion. After a variable period of spinal shock characterized by flaccid paralysis and areflexia, spasticity occurs in experimental animals whose spinal cords have been transected. The pathophysiological basis of spasticity is not known, although decreased postsynaptic and recurrent inhibition may play some role.[332]

A number of studies have suggested that glycine may play a role in clinical and experimental spasticity. Temporary aortic occlusion produces spinal interneuronal loss and a spastic paraplegia in cats. The glycine content of the lumbar spinal cord is reduced in proportion to cell loss.[35,36] With nearly complete interneuronal loss (accompanied by marked anterior horn cell loss and glial proliferation), there is an approximate 80% decrease in the glycine content in gray matter.[36]

Three weeks after transection of T_{10}, the glycine content was decreased by 18% in the dorsolateral white matter of the cat but unchanged in the gray and ventral white matter.[333] Glycine was decreased by as much as 40% in the central and ventral gray matter of the dog spinal cord to 8 weeks after midthoracic transection.[332] Systemic administration of glycine seemed to ameliorate some components of spasticity in dogs after transection[334] and in rats after aortic occlusion.[335] Preliminary studies of oral glycine therapy (1–4 g/day) in humans with spasticity have shown some benefit without evidence of toxicity.[336,337] Although its mode of action is believed to be a blockade of excitatory transmitter release, baclofen, a clinically useful antispastic drug, increased the flux of label from [^{14}C]glucose into [^{14}C]glycine in slices of guinea pig cerebral cortex.[338]

10.4. Seizure Disorders

A biochemical lesion in human epilepsy is not known. Glycine has been found to be elevated in human epileptogenic foci removed at surgery for intractable seizures.[339,340] Glycine is also elevated in the chronic epileptogenic focus in animals with cobalt-induced seizures.[341] The elevation of glycine levels appeared to be proportional to the severity of the induced seizure disorder.[341] Systemic, topical, and intracisternal administration of glycine attentuates seizure activity in acute models of epilepsy in rodents[342,343] but not in cats.[344]

It is of interest that patients treated with dipropylacetic acid (valproic acid), a clinically useful anticonvulsant, have elevated levels of glycine in the plasma, urine, and CSF,[345] and the degree of glycinuria is proportional to the dose of medication.[346] *In vivo* administration of dipropylacetate to rats results in decreased activity of the liver GCS.[346,347] The mechanism of action of valproic acid is unknown. Little is known of its effect on glycine metabolism or levels within the CNS. Whole-brain levels of glycine were not altered acutely after systemic valproic acid in mice, and seizure protection appeared to be temporally related to elevated GABA levels.[348]

10.5. Psychiatric Disorders

Very few studies have investigated glycine metabolism in disorders of higher cortical function. This is in a large part because of evidence that few glycinergic synapses are present in rostral areas of the CNS. Among the numerous biochemical theories of schizophrenia, the transmethylation theory and the methyl-activation theory are two of the most controversial.[349] The interconversion of serine and glycine plays a prominent role in each theory. Indirect evidence for the latter theory is provided by the decreased activity of SHMT found in the erythrocytes of schizophrenic patients.[349] The decreased urinary excretion of serine during acute psychotic episodes in four patients is consistent with the transmethylation theory. Serine administration was able to reprecipitate the thought disorder in all four patients.[350]

The glycine content of erythrocytes was reported elevated in patients with bipolar affective disorders.[351] However, further experiments suggested that this elevation was secondary to lithium therapy.[352,353] The elevated glycine content in the cerebral hemispheres of rats receiving chronic lithium administration is of interest as a possible factor in its therapeutic effects.[354]

REFERENCES

1. Aprison, M. H., and Werman, R., 1965, *Life Sci.* **4**:2075–2083.
2. Aprison, M. H., and Werman, R., 1968, *Neurosci. Res.* **1**:143–174.
3. Werman, R., and Aprison, M. H., 1968, *Structure and Function of Inhibitory Neuronal Mechanisms* (C. von Euler, S. Skogland, and U. Soderberg, eds.), Pergamon Press, New York, pp. 473–486.
4. Eccles, J. C., 1964, *The Physiology of Synapses*, Springer-Verlag, Berlin.
5. Curtis, D. R., and Watkins, J. C., 1960, *J. Neurochem.* **6**:117–141.
6. Curtis, D. R., 1963, *Pharmacol. Rev.* **15**:333–364.
7. Krnjević, K., 1974, *Physiol. Rev.* **54**:418–540.
8. Curtis, D. R., and Johnston, G. A. R., 1974, *Ergeb. Physiol. Biol. Chem. Pharmakol.* **69**:97–188.
9. Aprison, M. H., and Daly, E. C., 1978, *Adv. Neurochem.* **3**:203–294.
10. Pycock, C. J., and Kerwin, R. W., 1981, *Life Sci.* **28**:2679–2686.

11. Lapin, A., and Karobath, M., 1980, *J. Chromatogr.* **193:**95–99.
12. Fernstrom, M. H., and Fernstrom, J. D., 1981, *Life Sci.* **29:**2119–2130.
13. Shank, R. P., and Aprison, M. H., 1970, *Anal. Biochem.* **35:**136–145.
14. Briel, G., and Neuhoff, V., 1972, *Hoppe Seylers Z. Physiol. Chem.* **353:**540–553.
15. Aprison, M. H., and Shank, R. P., 1970, *Exp. Physiol. Biochem.* **3:**31–38.
16. Sardesai, V. M., and Provido, H. S., 1970, *Clin. Chim. Acta* **29:**67–71.
17. Goodwin, J. F., and Stampwala, S., 1973, *Clin. Chem.* **19:**1010–1015.
18. Ohmori, S., Ikeda, M., Watanabe, Y., and Hirota, K., 1970, *Anal. Biochem.* **90:**662–670.
19. Frazier, P. D., and Summer, G. K., 1971, *Anal. Biochem.* **44:**66–76.
20. Berger, S. J., Carter, J., G., and Lowry, O. H., 1975, *Anal. Biochem.* **65:**232–240.
21. Cutler, R. W. P., and Dudzinski, D. S., 1974, *J. Neurochem.* **23:**1005–1009.
22. Davies, L. P., and Johnston, G. A. R., 1974, *J. Neurochem.* **22:**107–112.
23. Lajtha, A., and Toth, J., 1973, *Brain Res.* **55:**238–241.
24. Agrawal, H. C., and Himwich, W. A., 1970, *Developmental Neurobiology* (W. A. Himwich, ed.), Charles C Thomas, Springfield, Illinois, pp. 287–310.
25. Piepho, R. W., and Freidman, A. H., 1971, *Life Sci.* **10:**1355–1362.
26. Ross, F. H., Sermons, A. L., and Walker, C. W., 1980, *Life Sci.* **26:**1019–1022.
27. Ross, F. H., Sermons, A. L., Owasoyo, J. O., and Walker, C. W., 1980, *Pharmacol. Res. Commun.* **12:**891–897.
28. Shank, R. P., and Aprison, M. H., 1970, *J. Neurochem.* **17:**1461–1475.
29. Fernstrom, J. D., Larin, F., and Wurtman, R. J., 1971, *Life Sci.* **10:**813–819.
30. Shank, R. P., and Aprison, M. H., 1971, *J. Neurobiol.* **2:**145–151.
31. Lajtha, A., and Toth, J., 1974, *Brain Res.* **76:**546–551.
32. Perry, T. L., Hansen, S., and Gandham, S. S., 1981, *J. Neurochem.* **36:**406–412.
33. Graham, L. T., Jr., Shank, R. P., Werman, R., and Aprison, M. H., 1967, *J. Neurochem.* **14:**465–572.
34. Aprison, M. H., Shank, R. P., and Davidoff, R. A., 1969, *Comp. Biochem. Physiol.* **28:**1345–1355.
35. Davidoff, R. A., Graham, L. T., Jr., Shank, R. P., Werman, R., and Aprison, M. H., 1967, *J. Neurochem.* **14:**1025–1031.
36. Homma, S., Suzuki, T., Murayama, S., and Otsuka, M., 1979, *J. Neurochem.* **32:**691–698.
37. Miyata, Y., and Otsuka, M., 1975, *J. Neurochem.* **25:**239–244.
38. Johnston, G. A. R., 1968, *J. Neurochem.* **15:**1013–1017.
39. Boehme, D. H., Fordice, M. W., Marks, N., and Vogel, W., 1973, *Brain Res.* **50:**353–359.
40. Berger, S. J., Carter, J. G., and Lowry, O. H., 1977, *J. Neurochem.* **28:**149–158.
41. Roberts, F., and Hill, R. G., 1978, *J. Neurochem.* **31:**1549–1551.
42. Adams, J. C., and Wenthold, R. J., 1979, *Neuroscience* **4:**1947–1951.
43. James, T. A., and Starr, M. S., 1979, *Eur. J. Pharmacol.* **57:**115–125.
44. Godfrey, D. A., Carter, J. G., Lowry, O. H., and Matschinsky, F., 1978, *J. Histochem. Cytochem.* **26:**118–126.
45. Shaw, R. K., and Heine, J. D., 1965, *J. Neurochem.* **12:**151–155.
46. Cohen, A. I., McDaniel, M., and Orr, H., 1973, *Invest. Ophthalmol.* **12:**686–693.
47. Kennedy, A. J., Neal, M. J., and Lolley, R. N., 1977, *J. Neurochem.* **29:**157–159.
48. Berger, S. J., McDaniel, M. L., Carter, J. G., and Lowry, O. H., 1977, *J. Neurochem.* **28:**159–163.
49. Serra, S., Grynbaum, A., Lajtha, A., and Marks, N., 1972, *Brain Res.* **44:**579–592.
50. Osborne, R. H., Bradford, H. F., and Jones, D. G., 1973, *J. Neurochem.* **21:**407–419.
51. Bradford, H. F., Bennett, G. W., and Thomas, A. J., 1973, *J. Neurochem.* **21:**495–505.
52. Kontro, P., Marnela, K.-M., and Oja, S. S., 1980, *Brain Res.* **184:**129–141.
53. DeBelleroche, J. S., and Bradford, H. F., 1973, *J. Neurochem.* **21:**441–451.
54. McBride, W. J., Daly, E., and Aprison, M. H., 1973, *J. Neurobiol.* **4:**557–566.
55. Fonnum, F., and Walberg, F., 1973, *Brain Res.* **62:**577–579.
56. Aprison, M. H., Daly, E. C., Shank, R. P., and McBride, W. J., 1975, *Metabolic Compartmentation and Neurotransmission* (S. Berl, D. D. Clarke, and D. Schneider, eds.), Plenum Press, New York, pp. 37–63.
57. DeBelleroche, J. S., and Bradford, H. F., 1977, *J. Neurochem.* **29:**335–343.

58. Iliffe, T. M. McAdoo, D. J., Beyer, C. B., and Haber, B., 1977, *J. Neurochem.* **28:**1037–1042.
59. Price, C. H., Goggeshall, R. E., and McAdoo, D. J., 1978, *Brain Res.* **154:**25–40.
60. Price, C. H., McAdoo, D. J., Farr, W., and Okura, E. R., 1979, *J. Neurobiol.* **10:**551–571.
61. Sawadra, M., McAdoo, D. J., Blankovship, J. E., and Price, C. H., 1981, *Brain Res.* **207:**486–490.
62. Price, C. H., and McAdoo, D. J., 1981, *Brain Res.* **219:**307–315.
63. Banos, G., Daniel, P. M., Moorhouse, S. R., and Pratt, O. E., 1975, *J. Physiol. (Lond.)* **246:**539–548.
64. Oldendorf, W. H., and Szabo, J., 1976, *Am. J. Physiol.* **230:**94–98.
65. Battistin, L., Grynbaum, A., and Lajtha, A., 1971, *Brain Res.* **29:**85–99.
66. Toth, E., and Lajtha, A., 1981, *Trans. Am. Soc. Neurochem.* **12:**158.
67. Banos, G., Daniel, P. M., Moorhouse, S. R., and Pratt, O. E., 1971, *J. Physiol. (Lond.)* **213:**45P–46P.
68. Seta, K., Sershen, H., and Lajtha, A., 1972, *Brain Res.* **47:**415–425.
69. O'Fallon, J. V., and Brosemer, R. W., 1978, *J. Neurochem.* **31:**365–366.
70. Bridgers, W. F., 1965, *J. Biol. Chem.* **240:**4591–4597.
71. Uhr, M. L., and Sneddon, M. K., 1972, *J. Neurochem.* **19:**1495–1500.
72. Feld, R. D., and Sallach, H. J., 1974, *Brain Res.* **73:**558–562.
73. Hedrick, J. L., and Sallach, H. J., 1964, *Arch. Biochem. Biophys.* **105:**261–269.
74. Kondrashova, M. N., and Rodlonova, M. A., 1971, *Dokl. Akad. Nauk SSSR* **196:**1225–1227.
75. DeMarchi, W. J., and Johnston, G. A. R., 1969, *J. Neurochem.* **16:**355–361.
76. Gaunt, G. L., and DeDuve, C., 1976, *J. Neurochem.* **26:**749–759.
77. Arnold, G., Liscum, L., and Holtzman, E., 1979, *J. Histochem. Cytochem.* **27:**735–745.
78. Shank, R. P., Aprison, M. H., and Baxter, C. F., 1973, *Brain Res.* **52:**301–308.
79. Maher, T. J., and Wurtman, R. J., 1980, *Life Sci.* **26:**1283–1286.
80. Siemers, E. R., Daly, E. C., and Aprison, M. H., 1980, *Trans. Am. Soc. Neurochem.* **11:**69.
81. Bird, M. I., and Nunn, P. B., 1979, *Biochem. Soc. Trans.* **7:**1276–1277.
82. Sky-Peck, H. H., Rosenbloom, C., and Winzler, R. J., 1966, *J. Neurochem.* **13:**223–228.
83. Bridgers, W. F., 1968, *J. Neurochem.* **15:**1325–1328.
84. Ordoñez, L. A., and Wurtman, R. J., 1973, *J. Neurochem.* **21:**1447–1455.
85. Broderick, D. S., Candland, K. L., North, J. A., and Mangum, J. H., 1972, *Arch. Biochem. Biophys.* **148:**196–198.
86. Davies, L. P., and Johnston, G. A. R., 1973, *Brain Res.* **54:**149–156.
87. Daly, E. C., and Aprison, M. H., 1974, *J. Neurochem.* **22:**877–885.
88. McClain, L. D., Carl, G. F., and Bridgers, W. F., 1975, *J. Neurochem.* **24:**719–722.
89. Rassin, D. K., and Gaull, G. E., 1975, *J. Neurochem.* **24:**969–978.
90. DeRobertis, E., Pellegrino DeIraldi, A., Rodriquez DeLores Arnaiz, G., and Salganicoff, L., 1962, *J. Neurochem.* **9:**23–35.
91. Burton, E. G., and Sallach, H. J., 1975, *Arch. Biochem. Biophys.* **166:**483–494.
92. Yoshida, T., and Kikuchi, G., 1973, *J. Biochem.* **73:**1013–1022.
93. Rassin, D. K., Sturman, J. A., and Gaull, G. E., 1981, *J. Neurochem.* **36:**1263–1271.
94. Nakano, Y., Fujioka, M., and Wada, H., 1968, *Biochim. Biophys. Acta* **159:**19–26.
95. Ogawa, H., and Fujioka, M., 1981, *J. Biochem.* **90:**381–390.
96. Gaull, G. E., VonBerg, W., Raiha, N. C. R., and Sturman, J. A., 1973, *Pediatr. Res.* **7:**527–533.
97. Snell, K., 1980, *Biochem. J.* **190:**451–455.
98. Suleiman, S. A., and Spector, R., 1982, *J. Neurochem.* **38:**392–396.
99. Johnston, G. A. R., and Vitali, M. V., 1969, *Brain Res.* **12:**471–472.
100. Johnston, G. A. R., and Vitali, M. V., 1969, *Brain Res.* **15:**201–208.
101. Johnston, G. A. R., Vitali, M. V., and Alexander, H. M., 1970, *Brain Res.* **20:**361–367.
102. Benuck, M., Stern, F., and Lajtha, A., 1972, *J. Neurochem.* **19:**949–957.
103. Benuck, M., Stern, F., and Lajtha, A., 1971, *J. Neurochem.* **18:**1555–1567.
104. Dasgupta, P., and Narayanaswami, A., 1981, *J. Neurochem.* **37:**1603–1606.
105. Liang, C.-C., 1962, *Biochem. J.* **82:**429–434.
106. Steward, C. B., Grammer, J., and Brosemer, R. W., 1981, *Ann. Nutr. Metab.* **25:**289–298.

107. Dennis, M. J., and Clarke, J. T. R., 1979, *J. Neurochem.* **33**:383–385.
108. Kikuchi, G., 1973, *Mol. Cell Biochem.* **1**:169–187.
109. Motokawa, Y., and Kikuchi, G., 1971, *Arch. Biochem. Biophys.* **146**:461–466.
110. Bruin, W. J., Frantz, B. M., and Sallach, H. J., 1973, *J. Neurochem.* **20**:1649–1658.
111. Uhr, M. L., 1973, *J. Neurochem.* **20**:1005–1009.
112. Daly, E. C., Nadi, N. S., and Aprison, M. H., 1976, *J. Neurochem.* **26**:179–185.
113. Lahoya, J. L., Benavides, J., and Ugarte, M., 1980, *Dev. Neurosci.* **3**:75–80.
114. Herbert, V., and Zalusky, R., 1961, *Fed. Proc.* **20**:453.
115. Reynolds, E. H., 1976, *Clin. Haematol.* **5**:661–696.
116. Levitt, M., Nixon, P. F., Pincus, J. H., and Bertino, J. R., 1971, *J. Clin. Invest.* **50**:1301–1308.
117. Fehling, C., Jagerstad, M., Lindstrand, K., and Elmquist, D., 1976, *Z. Ernahrungswiss.* **15**:1–8.
118. Bridgers, W. F., and McClain, L. D., 1972, *Adv. Biochem. Psychopharmacol.* **4**:81–92.
119. Wang, F. K., Koch, J., and Stokstad, E. L. R., 1967, *Biochem. Z.* **346**:458–466.
120. Brody, T., Shin, Y. S., and Stokstad, E. L. R., 1976, *J. Neurochem.* **27**:409–413.
121. Shin, Y. S., Williams, M. A., and Stokstad, E. L. R., 1972, *Biochem. Biophys. Res. Commun.* **47**:35–43.
122. Carl, G. F., Peterson, J. D., McClain, L. D., and Bridgers, W. F., 1980, *J. Neurochem.* **34**:1442–1452.
123. Cybulski, R. L., and Fisher, R. R., 1976, *Biochemistry* **15**:3183–3187.
124. Cybulski, R. L., and Fisher, R. R., 1977, *Biochemistry* **16**:5116–5120.
125. Cybulski, R. L., and Fisher, R. R., 1981, *Biochim. Biophys. Acta* **646**:329–333.
126. Rapoport, S., Mueller, M., Dumdey, R., and Rashman, J., 1980, *Eur. J. Biochem.* **108**:449–455.
127. Benavides, J., Garcia, M. L., Lopez-Lahoya, J., Ugarte, M., and Valdiviejo, F., 1980, *Biochim. Biophys. Acta* **598**:588–594.
128. Ordonez, L. A., and Villarroel, H., 1976, *J. Neurochem.* **27**:305–307.
129. Spector, R., Levy, P., and Abelson, H. T., 1977, *J. Neurochem.* **29**:919–921.
130. McClain, L. D., and Bridgers, W. F., 1970, *J. Neurochem.* **17**:763–766.
131. Fuller, R. W., 1976, *Life Sci.* **19**:625–628.
132. Turner, A. J., 1977, *Biochem. Pharmacol.* **26**:1009–1014.
133. Werman, R., Davidoff, R. A., and Aprison, M. H., 1968, *J. Neurophysiol.* **31**:81–95.
134. Curtis, D. R., Hösli, L. Johnston, G. A. R., and Johnston, I. H., 1968, *Exp. Brain Res.* **5**:235–258.
135. Curtis, D. R., Hosli, L., and Johnston, G. A. R., 1968, *Exp. Brain Res.* **6**:1–18.
136. Krnjević, K., Puil, E., and Werman, R., 1977, *Can. J. Physiol. Pharmacol.* **55**:658–669.
137. Ryall, R. W., and DeGroat, W. C., 1972, *Brain Res.* **37**:345–347.
138. Game, C. J. A., and Lodge, D., 1975, *Exp. Brain Res.* **23**:75–84.
139. Johnston, S. E., and Davies, J., 1981, *Neurosci. Lett.* **26**:43–47.
140. Biscoe, T. J., Duggan, A. W., and Lodge, D., 1972, *Comp. Gen. Pharmacol.* **3**:423–433.
141. Hösli, L., Hösli, E., and Andres, P. F., 1973, *Brain Res.* **62**:597–602.
142. Obata, K., Oide, M., and Tanaka, H., 1978, *Brain Res.* **144**:179–184.
143. Kelly, J. S., and Renaud, L. P., 1973, *Br. J. Pharmacol.* **48**:369–386.
144. Kelly, J. S., and Renaud, L. P., 1973, *Br. J. Pharmacol.* **48**:387–395.
145. Altmann, H., Bruggencate, G. Ten, Pickelman, P., and Steinberg, R., 1976, *Brain Res.* **111**:337–345.
146. Altmann, H., Bruggencate, G. Ten, and Sonnhöf, U., 1972, *Pfluegers Arch.* **331**:90–94.
147. Tebĕcis, A. K., 1973, *Exp. Neurol.* **40**:297–308.
148. Champagnat, J., Denavit-Saubie, M., Moyanova, S., Rondouin, G., and Boudinot, E., 1981, *Adv. Biochem. Psychopharmacol.* **29**:89–95.
149. Bruggencate, G. Ten, and Enberg, I., 1971, *Brain Res.* **25**:431–448.
150. Feldberg, W., and Rocha E. Silva, Jr., M., 1981, *Br. J. Pharmacol.* **72**:17–24.
151. Dray, A., Gonye, T. J., and Oakley, N. R., 1976, *J. Physiol.* (*Lond.*) **259**:825–849.
152. Cheramy, A., Nieoullon, A., and Glowinski, J., 1978, *Eur. J. Pharmacol.* **47**:141–147.
153. Leviel, V., Cheramy, A., Nieoullon, A., and Glowinski, J., 1979, *Brain Res.* **175**:259–270.

154. Finnerty, E. P., and Chan, S. H. H., 1980, *Eur. J. Pharmacol.* **67**:477–479.
155. Kerwin, R. W., and Pycock, C. J., 1980, *Eur. J. Pharmacol.* **64**:169–172.
156. Kerwin, R. W., and Pycock, C. J., 1979, *Eur. J. Pharmacol.* **54**:93–98.
157. Young, A. B., and Snyder, S. H., 1973, *Proc. Natl. Acad. Sci. U.S.A.* **70**:2832–2836.
158. Obata, K., and Yoshida, M., 1973, *Brain Res.* **64**:455–459.
159. Grundlach, A. L., and Beart, P. M., 1982, *J. Neurochem.* **38**:574–581.
160. Curtis, D. R., Duggan, A. W., Felix, D., Johnston, G. A. R., and McLennan, H., 1971, *Brain Res.* **33**:57–73.
161. Dreifuss, J. J., and Matthews, E. K., 1972, *Brain Res.* **45**:599–603.
162. Felix, D., and McLennan, H., 1971, *Brain Res.* **25**:661–664.
163. Kawamura, H., and Provini, L., 1970, *Brain Res.* **24**:293–304.
164. Okamoto, K., and Sakai, Y., 1980, *Br. J. Pharmacol.* **69**:407–413.
165. Kelly, J. S., and Krnjević, K., 1969, *Exp. Brain Res.* **9**:155–163.
166. Marciani, M. G., Stanzione, P., Cherubini, E., and Benardi, G., 1980, *Neurosci. Lett.* **18**:169–172.
167. Blume, H. W., Pittman, Q. J., and Renaud, L. P., 1981, *Brain Res.* **209**:145–158.
168. Ames, A., and Pollen, D. A., 1969, *J. Neurophysiol.* **32**:424–442.
169. Wyatt, H. J., and Daw, N. N., 1976, *Science* **191**:204–205.
170. Miller, R. F., Frumkes, T. E., Slaughter, M., and Dacheux, R. F., 1981, *J. Neurophysiol.* **45**:764–782.
171. Frumkes, T. E., Miller, R. F., Slaughter, M., and Dacheux, R. F., 1981, *J. Neurophysiol.* **45**:783–804.
172. Cunningham, R., and Miller, R. F., 1980, *Brain Res.* **197**:123–138.
173. Werman, R., 1979, *Adv. Exp. Med. Biol.* **123**:287–301.
174. Matthews, G., and Wickelgren, W. O., 1979, *J. Physiol. (Lond.)* **293**:393–415.
175. Homma, S., and Rooainen, C. M., 1978, *J. Physiol. (Lond.)* **279**:231–252.
176. Barker, J. L., Nicoll, R. A., and Padjen, A., 1975, *J. Physiol. (Lond.)* **245**:521–536.
177. Sonnhöf, U., Grafe, P., Krumnakl, J., Linder, M., and Schindler, L., 1975, *Brain Res.* **100**:327–341.
178. Nistri, A., 1981, *Adv. Biochem. Psychopharmacol.* **29**:263–269.
179. Goldinger, A., and Muller, W. E., 1980, *Neurosci. Lett.* **16**:91–95.
180. Barker, J. L., and McBurney, R. N., 1979, *Nature* **277**:234–236.
181. Hösli, L., Hösli, E., Andres, P. F., and Landolt, H., 1981, *Exp. Brain Res.* **42**:43–48.
182. Zieglgansberger, W., and Champagnat, J., 1979, *Brain Res.* **160**:95–104.
183. Rostad, J., 1981, *Brain Res.* **223**:397–401.
184. Barker, J. L., and Ransom, B. R., 1978, *J. Physiol. (Lond.)* **280**:331–354.
185. Roberts, P. J., 1974, *Brain Res.* **67**:419–428.
186. Cutler, R. W. P., 1976, *Adv. Exp. Biol. Med.* **69**:435–446.
187. Fagg, G. E., Jordan, C. C., and Webster, R. A., 1978, *Brain Res.* **158**:159–170.
188. Coull, B. M., and Cutler, R. W. P., 1978, *Invest. Ophthalmol.* **17**:682–684.
189. Coull, B. M., Owens, D. K., and Cutler, R. W. P., 1981, *Brain Res.* **210**:301–309.
190. Reubi, J.-C., and Cúenod, M., 1976, *Brain Res.* **112**:347–361.
191. Wolfensberger, M., Reubi, J. C., Canzek, V., Redweik, U., Curtius, H., and Cúenod, M., 1981, *Brain Res.* **224**:327–336.
192. Barth, R., and Felix, D., 1974, *Brain Res.* **80**:532–537.
193. Henke, H., and Cúenod, M., 1978, *Brain Res.* **152**:105–119.
194. LeFort, D., Henke, H., and Cúenod, M., 1978, *J. Neurochem.* **30**:1287–1291.
195. Streit, P., Knecht, E., and Cúenod, M., 1980, *Brain Res.* **187**:59–67.
196. Clark, R. M., and Collins, G. G. S., 1976, *J. Physiol. (Lond.)* **262**:383–400.
197. Jasper, H. H., and Koyama, I., 1969, *Can. J. Physiol. Pharmacol.* **47**:889–905.
198. Obata, K., and Takeda, K., 1969, *J. Neurochem.* **16**:1043–1047.
199. Roberts, P. J., and Mitchell, J. F., 1972, *J. Neurochem.* **19**:2473–2481.
200. Jordan, C. C., and Webster, R. A., 1971, *Br. J. Pharmacol.* **43**:441p–442p.
201. Neal, M. J., 1971, *J. Neurophysiol.* **215**:103–117.
202. Hopkin, J., and Neal, M. J., 1971, *Br. J. Pharmacol.* **42**:215–223.
203. Cutler, R. W. P., Murray J. E., and Hammerstad, J. P., 1972, *J. Neurochem.* **19**:539–542.

204. Mulder, A. H., and Snyder, S. H., 1974, *Brain Res.* **76:**297–308.
205. Davidoff, R. A., and Adair, R., 1976, *Brain Res.* **118:**403–415.
206. Vargas, O., deLorenzo, M., and Oriego, F., 1977, *Neuroscience* **2:**383–390.
207. Bondy, S. C., Burks, J. S., and Harrington, M. E., 1979, *Arch. Neurol.* **36:**540–543.
208. Mitchell, J. F., Neal, M. J., and Srinivasan, V., 1969, *Br. J. Pharmacol.* **36:**201P–202P.
209. Lopez-Colome, A. M., Tapia, R., Salcedaj, R., and Pasantes-Morales, H., 1978, *Neuroscience* **3:**1069–1074.
210. Okamoro, K., and Namina, M., 1978, *J. Neurochem.* **31:**1393–1402.
211. Roberts, P. J., 1974, *Brain Res.* **74:**327–332.
212. Beart, P. M., and Bilal, K. B., 1976, *Neurosci. Abstr.* **2:**594.
213. Toggenburger, G., and Henke, H., 1981, *Neurosci. Lett.* [*Suppl.*] **7:**567.
214. Foster, A. C., and Roberts, P. J., 1980, *J. Neurochem.* **35:**517–519.
215. Waller, M. B., and Richter, J. A., 1980, *Biochem Pharmacol.* **29:**2189–2198.
216. Wenthold, R. J., 1979, *Brain Res.* **162:**338–343.
217. Ehinger, B., and Lindberg-Bauer, B., 1976, *Brain Res.* **113:**535–549.
218. Kennedy, A. J., and Neal, M. J., 1978, *Exp. Eye Res.* **26:**71–75.
219. Bigalke, H., Heller, I., Bizzini, B., and Haberman, E.,1981,*Naunyn Schmiedebergs Arch. Pharmacol.* **316:**244–251.
220. Levi, G., Bernardi, G., Cherubini, E., Gallo, V., Marciani, M. G., and Stanzione, P., 1982, *Brain Res.* **236:**121–131.
221. McGale, E. H. F., Pye, I. F., Stonier, C., Hutchinson, E. C., and Aber, G. M. 1977, *J. Neurochem.* **29:**291–297.
222. Franklin, G. M., Dudzinski, D. S., and Cutler, R. W. P., 1975, *J. Neurochem.* **24:**367–372.
223. Dudzinski, D. S., and Cutler, R. W. P., 1974, *J. Neurochem.* **22:**355–361.
224. Murray, J. E., and Cutler, R. W. P., 1970, *J. Neurochem.* **17:**703–704.
225. Wright, E. M., 1974, *Brain Res.* **74:**354–358.
226. Johnston, G. A. R., and Iversen, L. L., 1971, *J. Neurochem.* **18:**1951–1961.
227. Logan, W. J., and Snyder, S. H., 1972, *Brain Res.* **42:**413–431.
228. Bennett, J. P., Logan, W. J., and Snyder, S. H., 1972, *Science* **178:**997–999.
229. Iversen, L. L., and Johnston, G. A. R., 1971, *J. Neurochem.* **18:**1939–1950.
230. Arregui, A., Logan, W. J., Bennett, J. P., and Snyder, S. H., 1972, *Proc. Natl. Acad. Sci. U.S.A.* **69:**3485–3489.
231. Levi, G., and Raiteri, M., 1974, *Nature* **250:**735–737.
232. Iversen, L. L., 1975, *Nature* **253:**481.
233. Aprison, M. H., and McBride, W. J., 1973, *Life Sci.* **12:**449–458.
234. Henn, F. A., 1975, *Metabolic Compartmentation and Neurotransmission* (S. Berl, D. D. Clarke, and D. Schneider, eds.), Plenum Press, New York, pp. 91–97.
235. Hannuniemi, R., and Oja, S. S., 1981, *Neurochem. Res.* **6:**873–884.
236. Peterson, N. A., and Raghupathy, E., 1973, *J. Neurochem.* **21:**97–110.
237. Muller, W. E., and Snyder, S. H., 1978, *Brain Res.* **143:**487–498.
238. Wilkin, G. P., Csillag, A. Balaźs, R., Kingsbury, A. E., Wilson, J. E., and Johnson, A. L., 1981, *Brain Res.* **216:**11–33.
239. McGeer, E. G., and Singh, E. A., 1980, *Neurosci. Lett.* **17:**85–87.
240. Bruun, A., and Ehinger, B., 1972, *Invest. Ophthalmol.* **11:**191–198.
241. Neal, M. J., 1976, *Gen. Pharmacol.* **7:**321–332.
242. Lund Karlsen, R., 1978, *J. Neurochem.* **31:**1055–1061.
243. Matus, A. I., and Dennison, M. E., 1972, *J. Neurocytol.* **1:**27–34.
244. Ljungdahl, A., and Hökfelt, T., 1973, *Brain Res.* **62:**587–595.
245. Dennison, M. E., Jordan, C. C., and Webster, R. A., 1976, *J. Physiol.* (*Lond.*) **258:**55P–56P.
246. Price, D. P., Stocks, A., Griffin, J. W., Young, A., and Peck, K., 1976, *J. Cell Biol.* **68:**389–395.
247. Uchizono, K., 1965, *Nature* **207:**642–643.
248. Riberio-da-Silva, A., and Coimbra, A., 1980, *Brain Res.* **188:**449–464.
249. Hösli, L., and Hösli, E., 1978, *Rev. Physiol. Biochem. Pharmacol.* **81:**135–188.
250. Hökfelt, T., and Ljungdahl, A., 1972, *Adv. Biochem. Psychopharmacol.* **6:**1–36.
251. Schwartz, I. R., 1981, *Exp. Neurol.* **73:**601–617.

252. Voaden, M. J., Marshall, J., and Murani, M., 1974, *Brain Res.* **67:**116–132.
253. Pourcho, R. G., 1980, *Brain Res.* **198:**333–346.
254. Marc, R. E., and Lam, D. M. K., 1981, *J. Neurosci.* **1:**152–165.
255. Larson, M. D., 1969, *Brain Res.* **15:**185–200.
256. Curtis, D. R., Duggan, A. W., and Johnston, G. A. R., 1971, *Exp. Brain Res.* **12:**547–565.
257. Evans, R. H., 1978, *Br. J. Pharmacol.* **62:**431P–432P.
258. Obata, K., Oide, M., and Tanaka, H., 1978, *Brain Res.* **144:**179–184.
259. Felpel, L. P., 1972, *Exp. Brain Res.* **14:**494–502.
260. Belcher, G., Ryall, R. W., and Schaffner, R., 1978, *Brain Res.* **151:**307–321.
261. Bagust, J., Green, K. A., and Kerkut, G. A., 1981, *Brain Res.* **217:**425–429.
262. Duggan, A. W., Griersmith, B. T., and Johnson, S. M., 1981, *Brain Res.* **210:**231–241.
263. Ryall, R. W., Piercey, M. F., and Polosa, C., 1972, *Brain Res.* **41:**119–129.
264. Belcher, G., Davis, J., and Ryall, R. W., 1976, *J. Physiol. (Lond.)* **256:**651–662.
265. Curtis, D. R., Game, C. J. A., Lodge, D., and McCulloch, R. M., 1976, *J. Physiol. (Lond.)* **258:**227–242.
266. Cullheim, S., and Kellerth, J.-O., 1981, *J. Physiol. (Lond.)* **312:**209–224.
267. Werman, R., 1966, *Comp. Biochem. Physiol.* **18:**745–766.
268. Werman, R., 1972, *Annu. Rev. Physiol.* **34:**337–374.
269. Straughn, D. W., 1974, *Neuropharmacology* **13:**495–508.
270. Hill, R. G., Simmons, M. A., and Straughn, D. W., 1976, *Br. J. Pharmacol.* **56:**9–19.
271. Precht, W., Schwindt, P. C., and Baker, R., 1973, *Brain Res.* **62:**222–226.
272. Caspary, D. M., Havey, D. C., and Faingold, C. L., 1979, *Brain Res.* **172:**179–185.
273. Yokota, T., Nishikawa, N., and Nishikawa, Y., 1979, *Brain Res.* **168:**430–434.
274. Kitano, K., 1976, *Neuropharmacology* **15:**777–783.
275. Ishida, Y., 1977, *Neuropharmacology* **16:**163–170.
276. Mellanby, J., and Green, J., 1981, *Neuroscience* **6:**281–300.
277. Curtis, D. R., and DeGroat, W. C., 1968, *Brain Res.* **10:**208–212.
278. Curtis, D. R., Felix, D., Game, C. J. A., and McCulloch, R. M., 1973, *Brain Res.* **51:**358–362.
279. Kryzhanovsky, G. N., 1973, *Naunyn Schmiedebergs Arch. Pharmacol.* **276:**247–270.
280. Semba, T., and Kano, M., 1969, *Science* **64:**571–572.
281. Johnston, G. A. R., DeGroat, W. C., and Curtis, D. R., 1969, *J. Neurochem.* **16:**797–800.
282. Fedineć, A. A., and Shank, R. P., 1971, *J. Neurochem.* **18:**2229–2234.
283. Hilbig, G., Räker, K. O., and Wellhöhoner, H. H., 1979, *Naunyn Schmiedebergs Arch. Pharmacol.* **307:**287–290.
284. Schwab, M. E., and Thoenen, H., 1976, *Brain Res.* **105:**213–227.
285. Price, D. L., Griffin, J. W., and Peck, K., 1977, *Brain Res.* **121:**379–384.
286. Price, D. L., and Griffin, J. W., 1981, *Neurosci. Lett.* **23:**149–155.
287. Zukin, S. R., Young, A. B., and Snyder, S. H., 1975, *Brain Res.* **83:**525–530.
288. Snyder, S. H., 1975, *Br. J. Pharmacol.* **53:**473–484.
289. Zarbin, M. A., Wamsley, J. K., and Kuhar, M. J., 1981, *J. Neurosci.* **1:**532–547.
290. Palacios, J. M., Wamsley, J. K., Zarbin, M. A., and Kuhar, M. J., 1981, *Adv. Biochem. Psychopharmacol.* **29:**445–451.
291. Hösli, E., and Hösli, L., 1981, *Brain Res.* **213:**242–245.
292. Schaffer, J. M. and Anderson, S. M., 1981, *J. Neurochem.* **36:**1597–1600.
293. Borbe, H. O., Müller, W. E., and Wohlert, U., 1981, *Brain Res.* **205:**131–139.
294. Graham, D., Pfeiffer, F., and Betz, H., 1981, *Biochem. Biophys. Res. Commun.* **102:**1330–1335.
295. Pfeifer, F., and Betz, H., 1981, *Brain Res.* **226:**273–279.
296. Mackerer, C. R., Kochman, R. L., Shen, T. F., and Hershenson, F. M., 1977, *J. Pharmacol. Exp. Ther.* **201:**326–331.
297. Hershenson, F. M., Prodan, K. A., Kochman, R. L., Bloss, J. L., and Mackerer, C. R., 1977, *J. Med. Chem.* **20:**11–14.
298. Hayashi, H., Suga, M., Satake, M., and Tsubaki, T., 1981, *Ann. Neurol.* **9:**292–294.
299. Curtis, D. R., Lodge, D., Johnston, G. A. R., and Brand, S. J., 1976, *Brain Res.* **118:**344–347.

300. Dray, A., and Straughn, D. W., 1976, *J. Pharm. Pharmacol.* **28**:314–315.
301. Costa, E., Guidotti, A., Mao, C. C., and Suria, A., 1975, *Life Sci.* **17**:167–186.
302. Squires, R. F., and Braestrup, C., 1977, *Nature* **266**:732–734.
303. Hunt, P., and Raynaud, J.-P., 1977, *J. Pharm. Pharmacol.* **29**:442–444.
304. Kishimoto, H., Simon, J. R., and Aprison, M. H., 1981, *J. Neurochem.* **37**:1015–1024.
305. Perry, T. L., Urquhart, N., MacLean, J., Evans, M. E., Hansen, S., Davidson, A. G. F., Applegarth, D. A., MacLeod, P. J., and Lock, J. E., 1975, *N. Engl. J. Med.* **292**:1269–1272.
306. Perry, T. L., Urquhart, N., Hansen, S., and Mamer, O. A., 1977, *Pediatr. Res.* **11**:1192–1197.
307. Dalla Bernardina, B., Aicardi, J., Goutières, F., and Plouin, P., 1979, *Neuropaediatrie* **10**:209–225.
308. Frazier, D. M., Summer, G. K., and Chamberlain, H. R., 1978, *Am. J. Dis. Child.* **132**:777–781.
309. Holmgren, G., and Blomquist, H. K., 1977, *Neuropaediatrie* **8**:67–72.
310. Rosenberg, L. E., 1978, *The Metabolic Basis of Inherited Disease*, 4th ed. (J. B. Stanbury, J. B. Wyngaarden, and D. S. Frederickson, eds.), McGraw-Hill, New York, pp. 411–429.
311. Harris, D. J., Thompson, R. M., Wolf, B., and Yang, B. I.-Y., 1981, *J. Med. Genet.* **18**:156–157.
312. Bank, W. J., Pizer, L., and Pfender, W., 1978, *Adv. Neurol.* **21**:267–278.
313. Brandt, N. J., Rasmussen, K., Brandt, S., Kolvraa, S., and Shonheyder, F., 1976, *Acta Paediatr. Scand.* **65**:17–22.
314. O'Brien, W. E., 1978, *Arch. Biochem. Biophys.* **189**:291–297.
315. Kolvraa, S., 1979, *Pediatr. Res.* **13**:889–893.
316. Ugarte, M., Lopez-Lahoya, J., Garcia, M. L., Benavides, J., and Valdivieso, F., 1979, *J. Inherit. Metab. Dis.* **2**:93–96.
317. Lopez-Lahoya, J., Garcia, M. L., Benavides, J., and Ugarte, M., 1981, *J. Neurochem.* **36**:325–327.
318. Ando, T., Nyhan, W. L., Gerritsen, T., Gong, L., Heiner, D. C., and Bray, P. F., 1968, *Pediatr. Res.* **2**:254–263.
319. Hiraga, K., Kochi, H., Hayasaka, K., Kikuchi, G., and Nyhan, W., 1981, *J. Clin. Invest.* **68**:525–534.
320. Wendt, L. von, Similä, S., Saukkonen, A.-L., Koivisto, M., and Kouvalainen, K., 1981, *Am. J. Dis. Child.* **135**:1072.
321. Garcia-Castro, J. M., Isales-Forsythe, C. M., Levy, H. L., Shih, V. E., Lao Velez, C. R., Gonzalez-Rios, M. D. C., and DeTorres, L. C. R., 1982, *N. Engl. J. Med.* **306**:79–81.
322. Wendt, L. V., Alanko, H., Sorri, M., Toivakka, E., Saukkonen, A.-L., and Similä, S., 1981, *Clin. Genet.* **19**:94–100.
323. DeGroot, C. J., Boelieverts, V., Touwen, B. C. L., and Hommes, F. A., 1978, *Prog. Brain Res.* **48**:199–205.
324. Ransom, B. R., and Nelson, P. G., 1976, *N. Engl. J. Med.* **294**:1295–1296.
325. Benavides, J., Lopez-Lahoya, J., Valdivieso, F., and Ugarte, M., 1981, *J. Neurochem.* **37**:315–320.
326. Wellner, D., and Meister, A., 1981, *Annu. Rev. Biochem.* **50**:911–968.
327. Lane, J. D., Schone, B., Langenbeck, U., and Neuhoff, V., 1980, *Biochim. Biophys. Acta* **627**:144–156.
328. Isaacs, C. E., and Greengard, O., 1980, *Biochem. J.* **192**:441–448.
329. Dienel, G. A., 1981, *J. Neurochem.* **36**:34–43.
330. Valdivieso, F., Ugarte, M., Maties, M., Gimenez, C., and Mayor, F., 1977, *J. Ment. Defic. Res.* **21**:95–102.
331. Snyderman, S. E., Sansaricq, C., Norton, P. M., and Castro, J. V., 1981, *J. Pediatr.* **99**:63–67.
332. Hall, P. V., Smith, J. E., Campbell, R. L., Felten, D. L., and Aprison, M. H., 1976, *Life Sci.* **18**:1467–1472.
333. Rizzoli, A. A., 1968, *Brain Res.* **11**:11–18.
334. Smith, J. E., Hall, P. V., Galvin, M. R., Jones, A. R., and Campbell, R. L., 1979, *Neurosurgery* **4**:152–156.

335. Stern, P., and Hadžović, J., 1970, *Life Sci.* **9**:955–959.
336. Barbeau, A., 1974, *Neurology (Minneap.)* **24**:392.
337. Stern, P., and Bokonjić, R., 1974, *Pharmacology* **12**:117–119.
338. Potashner, S. J., 1979, *J. Neurochem.* **32**:103–109.
339. vanGelder, N. M., Sherwin, A. L., and Rasmussen, T., 1972, *Brain Res.* **40**:385–393.
340. Perry, T. L., and Hansen, S., 1981, *Neurology (Minneap.)* **31**:872–876.
341. vanGelder, N. M., and Courtois, A., 1972, *Brain Res.* **43**:477–484.
342. Lapin, I. P., 1981, *Eur. J. Pharmacol.* **71**:495–498.
343. Cherubini, E., Bernardi, G., Stanzione, P., Marciani, M. G., and Mercuri, N., 1981, *Neurosci. Lett.* **21**:93–97.
344. Fariello, R. G., 1979, *Exp. Neurol.* **66**:55–63.
345. Similä, S., Wendt, L. von, and Linna, S. L., 1980, *J. Neurol. Sci.* **45**:83–86.
346. Mortensen, P. B., Kolvraa, S., and Christensen, E., 1980, *Epilepsia* **21**:563–569.
347. Hayasaka, K., Kochi, H., Hiraga, K., and Kikuchi, G., 1981, *Biochem. Int.* **1**:410–416.
348. Schecter, P. J., Tranier, Y., and Grove, J., 1978, *J. Neurochem.* **31**:1325–1327.
349. Carl, G. F., Crews, E. L., Carmichael, S. M., Benesh, F. C., and Smythies, J. R., 1978, *Biol. Psychiatry* **13**:773–776.
350. Pepplinkhuizen, L., Blom, W., Bruinvels, J., and Moleman, P., 1980, *Lancet* **1**:454–456.
351. Rosenblatt, S., Gaull, G. E., Chanley, J. D., Rosenthal, J. S., Smith, H., and Sarkozi, L., 1979, *Am. J. Psychiatry* **136**:672–674.
352. Shea, P. A., Small, J. G., and Hendrie, H. C., 1981, *Biol. Psychiatry* **16**:825–830.
353. Deutsch, S. I., Peselow, E. D., Banay-Schwartz, M., Gershon, S., Virgilio, J., Fieve, R. R., and Rotrosen, J., 1981, *Am. J. Psychiatry* **138**:683–684.
354. Deutsch, S. I., Stanley, M., Banay-Schwartz, M., Peselow, E. D., Virgilio, J., Geisler, S., and Mohs, R. C., 1981, *Eur. J. Pharmacol.* **75**:75–76.

Taurine

S. S. Oja and Pirjo Kontro

1. INTRODUCTION

Taurine (2-aminoethanesulfonic acid) was first discovered in ox bile and subsequently in virtually all animals, many higher plants, molds, and bacteria.[1] In the liver, the conjugation of taurine with cholic acid has long been known, but its other possible physiological functions are less clear. It has been considered an osmoregulator in marine animals,[2] a stabilizer of muscle membranes,[3] a regulator of cardiac excitability,[4] an inhibitory neurotransmitter or -modulator in the brain and retina,[5] and has been assigned a number of related regulatory functions in the CNS.[6] Such intriguing hypotheses have lately prompted extensive investigation. Four recent international symposia have focused solely on taurine,[7-10] and a number of review articles have appeared on the subject.[5,11-16]

2. TAURINE IN NERVOUS TISSUES

There are millimolar quantities of taurine in mammalian tissues. The organ most enriched is often the heart, 2–40 mmol/kg wet weight in various species, together with other muscular tissue.[17,18] Relatively high levels have been noted in the spleen, adrenal glands, kidneys,[18] and all sex organs.[19,20] Taurine content is also remarkably high in the pituitary and pineal glands as well as in the neurohypophysis.[21-23] Throughout the nervous system, taurine is one of the most abundant amino acids.[24,25] Marked species differences also obtain in the brain, the content being high, for instance (3–8 mmol/kg), in the mouse and rat, whereas the levels in the cat and guinea pig are considerably lower.[15] Many reported taurine concentrations are misleadingly too high, however, because of unsatisfactory separation of taurine from the phosphates of ethanolamine and glyceroethanolamine in single-column amino-acid analyzers.[25,26]

S. S. Oja and Pirjo Kontro • Department of Biomedical Sciences, University of Tampere, Tampere, Finland.

2.1. Regional Distribution

The regional distribution of taurine in the CNS is somewhat heterogeneous. The cerebral cortex, cerebellum, olfactory bulbs, and striatum contain more taurine than the pons–medulla and spinal cord in the rat.[24,27–29] Of the individual nuclei, the lateral geniculate and inferior colliculus contain much taurine.[21,30] In the inferior colliculus, the posterior region has the highest and the anterior region the lowest taurine concentration in the cat.[30] In the rat spinal cord and thalamus, taurine is fairly evenly distributed.[31,32] However, the canine lumbar spinal cord shows segmental differences, and selective decrements occur in taurine content after thoracic transection.[33] In the rat striatum, some taurine may be preferentially associated with kainic acid-sensitive[34] inhibitory interneurons.[35]

The intraregional distribution of taurine has been most thoroughly studied in the rat cerebellum, where it has been considered an inhibitory transmitter released from stellate cells.[36] Consistent with this, the taurine levels seem to be lower in the excitatory granular cells but higher in the inhibitory stellate cells[37,38] than in the other cerebellar cortical cells. At variance with this, no decrement in taurine levels was observed in the molecular layer or synaptosomal fractions in the cerebellar cortex in adult neuron-deficient rats.[39] Moreover, in an autoradiographic study, only Purkinje cell somata and dendrites concentrated exogenous taurine.[40] The distribution of taurine in cerebellar peduncles and four medullary nuclei[41] and the depressant action of taurine on Purkinje cell activity[42] do not support a specific role for taurine in the cerebellar excitatory climbing fibers in spite of high taurine levels in these.[43–45] More information is needed before taurine can be assigned a definite transmitter role in any specific tracts, even in the cerebellum.

Taurine is the most abundant retinal amino acid in several animal species.[46–48] In the retina, it is concentrated within the photoreceptor cell layer, particularly in the inner segments, the outer nuclear layer, and synaptic terminals.[48–51] Exogenous [^{3}H]taurine accumulates in the photoreceptor cells and Müller fibers.[48,52] The retinal taurine content is markedly reduced in genetically photoreceptorless mice,[53] in rats afflicted with retinitis pigmentosa,[54,55] and in cats in which the photoreceptors have degenerated because of a taurine-free diet.[56] Furthermore, a degeneration of the inner and outer cell layers of rat retinas is accompanied by changes in retinal taurine levels.[57] Considerable evidence thus supports the conclusion that most taurine in the retina is associated with the photoreceptor cells, it being essential for their viability.[58,59]

2.2. Cellular Distribution

Both glial and neuronal cells contain taurine.[60,61] The major part of brain taurine has been recovered in soluble subcellular fractions.[62] In the bovine brain, only 3–21% of total tissue taurine is sequestered in nerve endings, the largest proportions being encountered in lenticular, caudate, and medullary nuclei.[25] In the rat brain, the synaptosomal fractions of striatum, cerebral cortex, and cerebellum contain the highest amounts of taurine.[29] The subcellular distribution in the retina is similar to that in the brain.[63]

Taurine has been found to be the most abundant amino acid in synaptic vesicles isolated from the rat cerebral cortex.[64] It is the only amino acid distinctly enriched in synaptic vesicles in all the various bovine brain areas studied, and although it represents only about 1% of total tissue taurine,[25] that is still enough for a possible transmitter function. It has been suggested that taurine is a constituent of a great number of functionally differing or even of all synaptic vesicles, being most probably bound to vesicular or other synaptic membranes.[25] In keeping with this concept, taurine has also been shown to be enriched in the synaptic membrane subfractions of the bovine brain.[65]

2.3. Developmental Changes

The taurine concentration is high in the fetal brain,[66,67] with a subsequent gradual decrease during postnatal development.[67–69] The magnitude of this postnatal decrease varies from area to area,[16,70,71] being several times greater in the spinal cord and medulla than, for example, in the cerebellar or cerebral cortex.[72] In contrast to taurine, the concentrations of the other functional amino acids generally increase during development.[66] It has therefore been suggested that taurine may have a role in brain development *per se*[67,73] in addition to its functional role in the mature brain. Synaptic taurine also increases during development in relation to total taurine in the rat brain.[62] On the other hand, the taurine content decreases uniformly in all subcellular fractions of the brain in taurine-deficient kittens.[74] In contrast to all brain regions, including the central components of the visual system (optic nerves, optic tracts, lateral geniculate bodies, and superior colliculi), the taurine content in the retina increases during development in the rabbit and rat.[75–77]

2.4. Factors Affecting Taurine Concentrations

The brain taurine levels are generally fairly stable. They show in all brain regions[78] and the pineal gland[79] a circadian rhythm. Ambient illumination influences taurine levels in the chick retina.[80] Drugs such as hydrazine, reserpine, pentobarbital,[81] and haloperidol[82] have failed to alter brain taurine concentrations. Amphetamine and chlorpromazine increase and diethylether, chlordiazepoxide, and reserpine reduce taurine concentrations in the mouse pituitary gland and hypothalamus.[83] In other brain regions, chlorpromazine has variably affected total tissue taurine,[24] exerting no effect on the synaptosomal taurine content.[84] In ethanol-dependent rats, the brain taurine decreases,[85] and in barbiturate-dependent rats, the cerebellar taurine content increases.[86] 2-Guanidinoethanesulfonate has been proposed as a rapidly acting and convenient tool for depletion of tissue taurine contents,[87] but its efficacy is less marked in the brain than in other tissues.

3. TAURINE BIOSYNTHESIS

Radioactive taurine is formed in animals from labeled carbohydrate precursors.[88–91] [^{35}S]Methionine is converted to cysteine, cysteinesulfinic acid (CSA), cysteic acid (CA), hypotaurine, and taurine (Fig. 1) in the rat brain[92] and cystine and cysteine to taurine.[28] Also, rat pineals incubated with [^{14}C]cysteine can synthetize taurine,[93] and the major metabolite of cysteine in squid nerve is hypotaurine, with CSA and taurine formed in smaller amounts.[94] The rat brain has been found to synthetize taurine from labeled cystine *in vivo*, other products identified being glutathione, CA, CSA, hypotaurine, and cystathionine.[95] Taurine formation was, however, more rapid in the liver than in the brain. All of the above results indicate that L-cysteine is a precursor of taurine. In the rat liver, 31% of cysteine is metabolized via the pyruvate pathway, and the remaining 69% via the taurine pathway.[96]

Taurine may be formed by several theoretically possible routes,[1] of which the pathways shown in Fig. 2 may alone be of significance. The multiplicity of metabolic pathways leading to taurine creates qualitative and quantitative differences among different tissues and animal species. For instance, the rat liver is capable of synthetizing taurine from CSA, whereas the cat liver is not.[97]

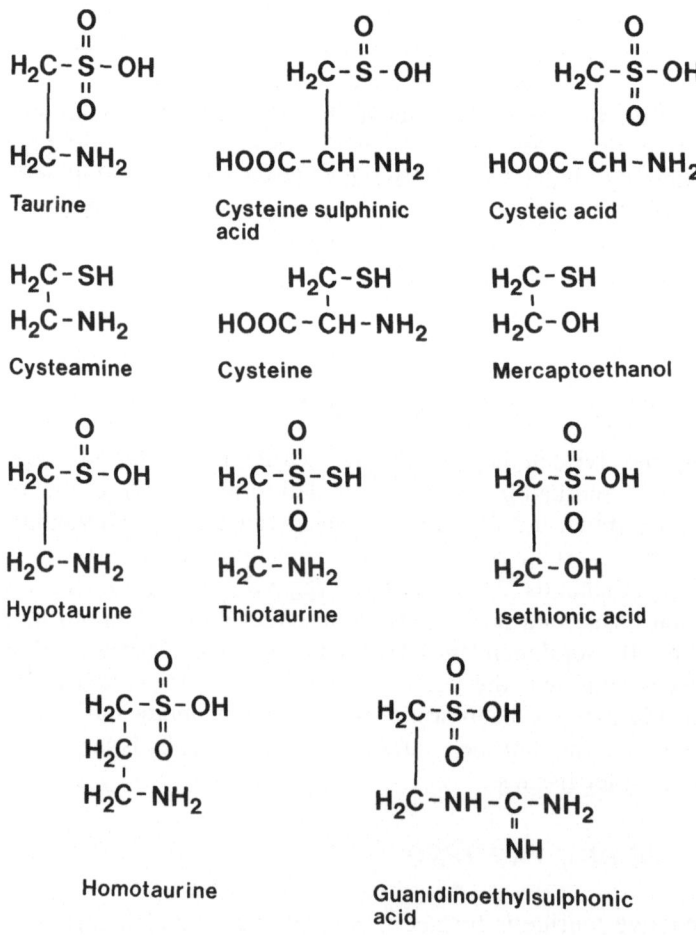

Fig. 1.   Structural formulas of some taurine analogues.

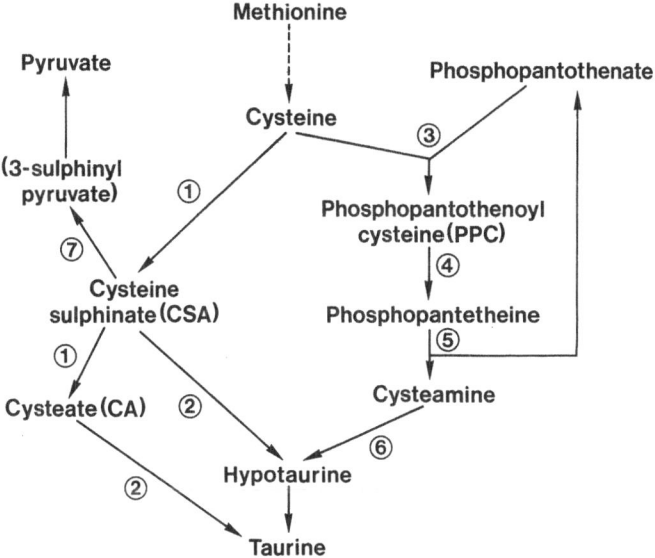

Fig. 2. The main pathways of taurine biosynthesis. The enzymes in the cysteinesulfinate (CSA) pathway are (1) cysteine dioxygenase (E.C. 1.13.11.20) and (2) CSA decarboxylase (E.C. 4.1.1.29), and in the cysteamine pathway, (3) phosphopantothenoyl-cysteine (PPC) synthetase (E.C. 6.3.2.5), (4) PPC decarboxylase (E.C. 4.1.1.36) (5), pantetheinase (E.C. 3.5.1.-), and (6) cysteamine dioxygenase (E.C. 1.13.11.19). The CSA can be also broken down by (7) CSA aminotransferase (E.C. 2.6.1.-). The exact definition of the enzyme(s) oxidizing hypotaurine to taurine is still open.

Therefore, inferences generalized from one single animal experiment can be misleading when an integrated picture of taurine biosynthesis is sought.

3.1. The Cysteinesulfinate Pathway

One of the major pathways leading to taurine starts with oxidation of cysteine to CSA, which can be either decarboxylated to hypotaurine with subsequent oxidation to taurine or further oxidized to CA followed by direct decarboxylation to taurine. This pathway is often considered to predominate in nervous tissue.

3.1.1. Cysteine Dioxygenase

Cysteine is oxidized to CSA by cysteine dioxygenase, "cysteine oxidase" (L-cysteine:oxygen oxidoreductase, E.C. 1.13.11.20) in the liver,[98] brain,[99] and retina.[100,101] The difficult purification of cysteine dioxygenase—it is extremely unstable in standard procedures—has recently been accomplished from rat liver.[102] Aerobic inactivation of the enzyme was prevented by anaerobic incubation with L-cysteine and by addition of cytoplasmic activating protein.[103] The liver enzyme has been better characterized than the brain enzyme, although the enzyme activities are of the same order of magnitude in both organs.[96,104] The hepatic enzyme has a molecular weight of about 22,500, is dependent on

NAD^+, contains one atom of iron per mole of enzyme, and is markedly induced by L-cysteine and corticosteroids.[103,105] The optimum pH of the enzyme reaction is 8.5–9.5.[102]

Brain cysteine dioxygenase also needs NAD^+ as cofactor but has an optimum pH of 6.8.[99] Enzyme activities are highest in the hypothalamus, colliculi, and pons–medulla, with a relatively low activity in the cerebral cortex.[99,106] A considerable proportion of the enzyme activity is in the crude mitochondrial fraction and distributes on subfractionation in both synaptosomal (about 15%) and mitochondrial fractions.[106,107] The activity of rat brain and liver cysteine dioxygenase increases during fetal and postnatal development.[99,108] In the brain, one-third of the reaction product is CA,[99] whereas in the retina, the product is almost exclusively CSA.[100] Cysteine dioxygenase has been localized in crude mitochondrial and supernatant fractions in ox retina[100] or in particulate components, including synaptosomes.[101,109] Cysteine dioxygenase is also predominantly localized in inner segments and photoreceptor nerve endings, although it is absent from outer segments.[101] In the retina, the activity increases with age as in the brain.[100]

3.1.2. Cysteinesulfinate Decarboxylase

Cysteinesulfinic acid and CA are obviously decarboxylated in the liver and brain by the same enzyme, cysteinesulfinate decarboxylase (CSD) or cysteate decarboxylase (CAD) (L-cysteinesulphinate carboxy-lyase, E.C. 4.1.1.29).[1,110,111] However, the brain decarboxylase appears to be different from the liver enzyme.[1,110] The CSD activity displays enormous species differences, particularly in the liver.[112] No activity of CSD or CAD has been detected in the hearts of rats, cats, dogs, rabbits, guinea pigs, or mice.[112–114] Glutamate decarboxylase (L-glutamate 1-carboxy-lyase, E.C. 4.1.1.15, GAD), the enzyme responsible for the biosynthesis of 4-aminobutyric acid (GABA), accepts both CSA and CA as substrates.[115]

The CSD has been purified about 500-fold from rat liver.[110] The enzyme has a molecular weight of about 100,000 and consists of two identical subunits. Partially purified CSD from calf brain has a molecular weight of about 65,000.[111] Recently, CSD has also been purified from soluble and mitochondrial fractions of rat brain.[116] The CSD activity paralleled that of GAD in all other purification steps, but under special conditions of Sepharose gel elution, a labile component emerged containing only CSD activity which comprised about 20% of the GAD/CSD activity recovered. Cysteinesulfinate decarboxylase requires pyridoxal phosphate as coenzyme, which highly activates brain CSD in assays *in vitro*.[117,118] In pyridoxine deficiency, no taurine is synthetized in the rat brain.[117] Convulsant doses of pyridoxal phosphate antagonistic drugs can also decrease cerebral CSD activity in mice.[119]

Most brain CSD (over 60%) is in particulate fractions, particularly in synaptosomes.[106,107,118] The CSD activity associated with particles and that remaining in the supernatant fraction are found to differ with respect to optimum pH and apparent kinetic constants.[118] Both CSD and cysteine dioxygenase are released into the soluble fraction after hypoosmotic resuspension of the crude mitochondrial fraction, indicating that these enzymes are located within the synaptoplasm.[120] Rat brain CSD shows a heterogeneous regional distribution, with highest activities in colliculi, olfactory bulbs, and hypothalamus and lowest in spinal cord and pons.[106] In the calf, cerebellum and mesencephalon possess the highest activities, the olfactory bulbs a moderately high, and the cerebral cortex the lowest.[121] The CAD activity is similarly distributed in bovine brain to CSD in rat brain.[114] In adult chicks, the enzyme activity is high in the cerebral cortex, cerebellum, and along the visual pathways.[122] In the rat spinal cord, CSD activity is higher in the dorsal half than in the ventral half of the cord and highest in the dorsal part of the dorsal horn.[32] The regional distribution of cysteine dioxygenase and CSD in the brains of several animal species does not show correlation with the regional distribution of taurine[27–29,32,113,122] with the possible exception of the different areas of the rat olfactory bulb and nucleus.[123]

The CSD activity is very low in the fetal liver but increases during gestation.[108] During postnatal development, CSD activity increases in the brains of rat and calf.[118,121,124,125] The activity associated with the particulate fraction increases more than tenfold, whereas soluble CSD increases only twofold.[118] Electrical stimulation of rat brain slices significantly enhances the decarboxylation of CSA in both adult and newborn animals.[124]

Cysteinesulfinate decarboxylase activity has been detected in the retinas of several animal species.[63,126] In the chick retina, the enzyme is recovered in the soluble and crude mitochondrial fractions, with a considerable portion in synaptosomes.[101] Furthermore, in the adult chick retina, the activity of CSD is higher in the inner plexiform and nuclear layers than in the respective outer layers.[122]

3.1.3. Cysteinesulfinate Aminotransferase

Cysteinesulfinate can also be further metabolized to pyruvate with the aid of L-cysteinesulfinate:2-oxoglutarate aminotransferase (CSA-T, E.C. 2.6.1.-), as recently demonstrated with mouse and rat nervous tissues.[127,128] Isoenzymes of CSA-T (of supernatant and mitochondrial origin) have been purified from rat brain, and their properties have been characterized and compared to those of aspartate aminotransferase (L-aspartate:2-oxoglutarate aminotransferase, E.C. 2.6.1.1).[129,130] The existence of a specific CSA-T has not been established, since CSA-T and aspartate aminotransferase may be a single protein. It has been proposed that CSA-T regulates taurine biosynthesis in brain and retina by regulating the levels of CSA, one of the key intermediates.[123,127,131,132]

3.2. The Cysteamine Pathway

Cysteamine is an intermediate in the other main pathway from cysteine to taurine (Fig. 2). Radioactive cystine and cysteine are converted to cysteamine in rat tissues,[133] various animal tissues contain endogenous cysteamine,[133,134] and taurine and hypotaurine are cysteamine metabolites both *in vivo* and *in vitro*.[135–139] Taurine can obviously be formed via cysteamine, but no enzyme has been found that could decarboxylate cysteine directly. It has therefore been suggested that cysteamine is produced in the course of the biosynthesis of CoA and phosphopantetheine and that the following enzymes are involved:

1. Phosphopantothenoyl-cysteine synthetase (E.C. 6.3.2.5) joins 4'-phosphopantothenate derived from pantothenic acid with cysteine. The enzyme is well characterized,[140,141] being active in most cells.
2. Phosphopantothenoyl-cysteine decarboxylase (E.C. 4.1.1.36) decarboxylates phosphopantothenoyl-cysteine to 4'-phosphopantetheine, which is further split into panthetheine and phosphate by a phosphatase. The decarboxylase has been purified from the horse liver.[142]
3. Pantetheinase (E.C. 3.5.1.-) cleaves pantetheine to pantothenic acid and cysteamine. The enzyme has been purified from the horse kidney.[143,144] It has been demonstrated *in vitro* that 4'-phosphopantothenoyl-cysteine is indeed converted to cysteamine through the combined actions of partially purified phosphopantothenoyl-cysteine decarboxylase and pantetheinase.[145,146]
4. Cysteamine dioxygenase (E.C. 1.13.11.19) produces hypotaurine by adding an oxygen molecule to the sulfhydryl group of cysteamine in the presence of Na_2S. The enzyme is a nonheme iron protein with a molecular weight of about 83,000.[147] It has been identified in the liver, muscle, heart, and kidney of the rat, ox, horse, pig, and sheep[133,148] and partially purified from horse kidney.[147,149]

The relative importance of the cysteamine pathway in taurine biosynthesis *in vivo* is not yet established, since no intermediates have been detected in rat brain or liver *in vivo*.[95] Most animal tissues are, however, provided with the necessary enzymatic sequence. Moreover, in most tissues, this route has been preferred over the CSA pathway in *in vitro* experiments.[150] It is very likely that the cysteamine pathway is the major biosynthetic source of cardiac taurine. It has been argued that more taurine is formed in the rat brain through the cysteamine than the CSA pathway, whereas in the liver the latter dominates.[151,152]

3.3. Oxidation of Hypotaurine to Taurine

In both the CSA and cysteamine pathways, hypotaurine occupies a key position as the last intermediate. The oxidation of hypotaurine to taurine has long been a puzzle. Some workers have been unable to detect any oxidation by animal tissues *in vitro*,[153,154] whereas others have done so.[124,155,156] *In vivo*, hypotaurine is readily converted to taurine in mice.[157] It has been thought that because of its chemical lability, hypotaurine could be nonenzymatically oxidized to taurine by the trace amounts of H_2O_2 produced in cellular metabolism[154] or, for instance, by ultraviolet irradiation.[158] The Enzyme Commission has, however, prematurely coined and coded an enzyme, hypotaurine dehydrogenase (hypotaurine:NAD^+ oxidoreductase, E.C. 1.8.1.3), on the basis of a poorly checked short communication.[159]

We have recently thoroughly characterized hypotaurine oxidation in the mouse liver and brain *in vitro*.[157,160] Hypotaurine oxidation exhibited properties characteristic of an enzyme-catalyzed reaction, being pH and temperature dependent, enhanced by oxygenation, NAD^+, Cu^{2+}, and Fe^{2+} ions, and obeying Michaelis–Menten kinetics. The activity was higher in both organs in developing than in adult mice. Most oxidation was recovered in brain samples in the soluble fraction, whereas in the liver the activity was more evenly distributed among subcellular fractions. Hypotaurine is also oxidized by several retinal subcellular fractions.[155,156] It is not yet established whether the oxidation is effected by a specific new enzyme or by some already known oxidoreductase as a side reaction. Moreover, the properties of the liver and brain enzymes do not appear to be identical.[157,160]

3.4. An Inorganic Pathway

A third pathway for taurine production proposed by Martin and co-workers starts from inorganic sulfate and serine.[161,162] Inorganic sulfate reacts first with ATP to form adenosine 5'-phosphosulfate and 3'-phosphoadenosine 5'-phosphosulfate, which in turn reacts with 2-aminoacrylic acid derived from serine to form cysteic acid, this being finally decarboxylated to taurine.[161,163–165] The enzymes involved are sulfate adenylyltransferase (ATP:sulfate adenylyltransferase, E.C. 2.7.7.4), adenylylsulfate kinase (ATP:adenylylsulfate 3'-phosphotransferase, E.C. 2.7.1.25), L-serine dehydratase (L-serine hydro-lyase, deaminating, E.C. 4.2.1.13), and cysteate decarboxylase (CAD, see above).

This route has been investigated mainly in the liver but has also been claimed to operate in other animal tissues, including the brain.[166,167] These concepts have been disputed by other investigators, however. No transfer of radioactivity from labeled 3'-phosphoadenosine 5'-phosphosulfate to taurine has been detected in the rat heart,[168] and injected radioactive sulfate has not been incorporated *in vivo* into taurine in cats, not even in those kept on a taurine-deficient diet.[169] It has also been calculated[150,170] from Martin's own data that only one of 40,000 taurine molecules is labeled after administration of 740 kBq of $^{35}SO_4^{2-}$. The inorganic pathway, if it exists at all, may be a constitutive enzyme system of significance only when other pathways are impaired or absent.[167,171,172]

4. TAURINE CATABOLISM

Only inorganic sulfate and isethionic acid (2-hydroxyethanesulfonic acid) have seriously been considered possible breakdown products of taurine.[1] Minimal amounts of $^{35}SO_4^{2-}$ formed from $[^{35}S]$taurine *in vivo*[173] are apparently produced by gut microbes and subsequently reabsorbed,[174] since no sulfate production from taurine has been observed in germ-free mice.[175] The slow deamination of taurine to isethionic acid[92,176,177] may similarly be effected by microorganisms,[178,179] since, again, no conversion occurs in germ-free mice.[175] Another metabolic route from cysteine bypassing taurine has been suggested in which the precursor of isethionic acid is mercaptoethanol.[180,181]

5. TRANSPORT OF TAURINE

5.1. Influx in Vivo

Taurine is readily absorbed from the gut, and plasma taurine levels can be substantially increased,[174,182–184] but accumulation in the brain is rather minimal.[185–187] There occurs some blood–brain exchange of taurine,[92,182,185] but the exchange rates are considerably lower than elsewhere in the organism.[113,173] The highly polar character of taurine molecules apparently limits penetration. The uptake of taurine by different brain areas does not vary greatly and shows no clear correlation to the regional distribution of endogenous taurine.[27,29,188] Developmental changes in the blood–brain exchange of taurine are likewise not great,[189] the computed transport rates being about two times greater in adult than in 7-day-old mice.[173] On the other hand, exogenous taurine accumulates in the rat brain more readily in fetuses[190] and during the neonatal period[70] than in adults.

5.2. Influx in Vitro

Taurine exchange between incubation medium and nervous tissue preparations is very slow,[191–193] and the tissue/medium ratios generated are not very high except in the retina.[194–196] Taurine uptake is at least partly energy dependent and temperature sensitive.[54,193–195,197–205] Two different types of taurine transport, saturable and nonsaturable, have been observed in various preparations of nervous tissue. Two components, low and high affinity, have been generally detected in the saturable transport.[54,192–197,199,200,202–204,206–213] The transport constants (K_m) for the high-affinity uptake vary from 3 to 100 μmol/1, the lowest values generally applying to cultivated cells. The K_m values of the low-affinity uptake vary within a range of 0.2–11 mmol/1. The published kinetic parameters have been itemized in other recent review articles.[5,15,214,215] The constants for the high-affinity transport are of the same order of magnitude as those of GABA and glycine.[5] The maximal velocity of the high-affinity uptake of taurine is considerably lower than that of glutamate, GABA, and hypotaurine, for instance.[193,213] Doubts have arisen concerning the capacity of the high-affinity transport to eliminate taurine efficiently enough from synaptic clefts.[203]

Taurine uptake has been sodium dependent in all nervous tissue preparations studied.[192,193,195,198–202,212,216–219] The high-affinity transport component is totally abolished in the absence of sodium ions, whereas the low-affinity one is affected less.[217] The Na$^+$ dependence curves of taurine uptake have been reported as sigmoidal,[193,212,217,219] linear,[202] or hyperbolic.[218] The majority of studies thus indicate that more than one sodium ion is involved in the transfer of a taurine molecule across brain cell membranes. Both the absence and depolarizing concentrations of K$^+$ also reduce taurine uptake.[193,201,202,217,218,220,221] These ionic dependencies together with ouabain sensitivity[193,202,219] show the involvement of Na$^+$–K$^+$ ATPase in sustained transport. The reported Ca^{2+} and Mg^{2+} effects[193,201,202,217,219] are too inconsistent to warrant any conclusions.

The high-affinity uptake has been found in all brain areas studied.[28,210] It has been argued that the uptake occurs predominantly into nerve endings[198] different from GABA-accumulating[222] and glutamate-accumulating[29] synaptosomes. Results of kainic acid lesions have suggested that the high-affinity taurine uptake is confined to interneurons in the rat cerebellum and striatum.[223,224] Also, glial cells possess a high-affinity uptake system for taurine.[199,200,209,218] It has been suggested that glial cells account for the major part of the taurine transport in brain slices.[225,226] Furthermore, in rat brainstem and spinal cord cultures, labeled taurine is accumulated by both neurons and glial cells, but in cerebellar cultures mostly by glial cells.[205] The picture of cellular localization of taurine uptake in the brain seems as yet incomplete.

Radioactive taurine concentrates in the inner plexiform layer and the innermost part of the inner nuclear layer of the chicken retina.[227] In the rabbit retina, the glial Müller cells appear to take up most of taurine,[207] whereas in the retinas of the rat, mouse, guinea pig, baboon, pigeon, cat, and frog, taurine accumulates mostly in photoreceptor cells and pigment epithelium.[228] Accordingly, the high-affinity uptake of taurine has been associated with viable photoreceptor cells and nonsaturable influx with the inner retina.[54,204] Active taurine transport into rat and frog retinal pigment epithelium[201,229,230] suggests that the cells of pigment epithelium accumulate taurine *in vivo* from the blood and gradually pass it into the retina.[52,201,229] A specific taurine transport system is also present in the retina in the early stages of development.[219]

Chlorpromazine, haloperidol, imipramine, diazepam, and *p*-chloromercuribenzoate inhibit taurine uptake.[198,199,213] The convulsants strychnine, pentylenetetrazole, oxotremorine, and picrotoxin do not affect and bicuculline moderately inhibits taurine uptake in crude synaptosomal fractions.[202] In glial cells, strychnine has proved a potent inhibitor and *p*-chloromercuriphenylsulfonate more efficient on the neuronal clones.[218] Cyclic nucleotide derivatives are ineffective in both neuroblastoma and glial cells.[211] Of various structural analogues tested, hypotaurine and β-alanine, followed by GABA, are the most potent competitive inhibitors of taurine uptake.[193,195,197–200,202,212,218,219] Furthermore, 3-guanidinopropionic acid, 2-guanidinoethanesulfonic acid[202] (Fig. 1), and L-2,4-diaminobutyric acid (L-DABA)[193] reduce taurine uptake. It is inferred from inhibition studies that "the hypothetical carrier at the cell membrane recognizes equally strongly ionized electropositive and electronegative ends of an acceptable molecule separated by two or three carbon atoms."[197] That conclusion has been confirmed by others.[199,202,218]

Taurine uptake in brain slices has been shown to be higher[231] and lower[197] in newborn than adult rats, to increase during fetal and early postnatal periods, to reach a maximum between postnatal 2 and 3 weeks, and then to decline to adult levels.[232] Furthermore, taurine uptake progressively increases into glia and decreases into neurons with cell culture time.[233]

The influx mechanisms of taurine have been extensively investigated *in vitro* in an attempt to prove whether or not they qualify for termination of the synaptic action of a neurotransmitter candidate. Taurine seems to fulfill the criteria for the most part satisfactorily enough.

5.3. Efflux in Vivo

The strikingly slow efflux of taurine from the brain *in vivo* is complex, probably originating from different intracellular pools.[186,234] The average half-life of brain taurine is several days.[28,29,184] Spontaneous efflux of taurine from developing rhesus monkey brains is slightly faster just before the rapid growth spurt of the brain,[70,77] but the half-life of brain taurine is about five times longer in 7-day-old than in adult mice.[173] Electrical stimulation, high K^+ concentrations, and veratrum alkaloids enhance taurine efflux from cerebral and cerebellar cortex.[234-239] The stimulated release of taurine has been reported to be Ca^{2+} dependent in the cerebellar cortex[239] but not in the visual cerebral cortex,[236] although calcium ionophores release taurine there also.[240] An increased Ca^{2+} concentration *per se* enhances taurine efflux from the superfused cat cerebral cortex.[235] In the cerebellar cortex, taurine efflux thus behaves as might be expected if it were from nerve endings.

5.4. Efflux in Vitro

Taurine efflux is also very slow *in vitro*[191,241] The spontaneous efflux from synaptosomes contains two components with widely differing half-lives, the slower component probably representing release from intrasynaptosomal compartments.[242] The more complex efflux from brain slices cannot be similarly resolved into few components.[241,243] Taurine efflux is not greatly modified by homoexchange or heteroexchange with GABA.[241,242,244] Muscimol and pentobarbital depress, and picrotoxin increases, taurine release.[245]

Different authors report surprisingly inconsistent properties of the stimulated released of taurine from brain preparations. Electrical and potassium stimulation have not brought about any change in taurine release in slices from various brain regions,[246-249] or caused a major enhancement.[221,249-252] Synaptosomes have generally responded to both types of stimuli.[213,242,244,253] Potassium ions induce in purified synaptosomes a new efflux component which may originate from emptying of synaptic vesicles.[242] The calcium dependence of the stimulated taurine efflux has also not yet been satisfactorily established. The absence of Ca^{2+} ions has attenuated K^+-stimulated[192,244,253] and electrically induced[247,254] release from cerebral and cerebellar slices and crude synaptosomes, has not affected the potassium-evoked release from cerebral cortex slices,[247] or even has increased the electrically stimulated release[255] and spontaneous efflux[248] in slices from rat cerebral cortex and hypothalamus. Only a part of the stimulated efflux of taurine may be calcium dependent, possibly the component associated with emptying of the content of synaptic vesicles into medium.[242]

The slow efflux of taurine from the retina comprises two phases.[194] In contrast to nervous tissue, it is significantly enhanced by homoexchange.[195,256] Electrical stimulation[195,256] and high potassium ion concentrations[257-259] markedly increase taurine efflux from retinas in a calcium-dependent manner. The potassium-induced release is sensitive to Mg^{2+} ions, verapamil, and ruthenium red.[257] Ionophores that mimic depolarizations release taurine from retinas,[260,261] indicating that calcium movements across membranes are the main driving force to efflux. It has also been suggested that taurine has a role in the regulation of calcium fluxes in the retina.[262] High K^+ concentrations do not stimulate taurine release from chick retinal synaptosomes, indicating that the observed release from whole retinas originates from other cell structures.[258] Also, the taurine release induced by ionophores is not characteristic of a neurotransmitter.[263] Taurine release may rather be related to a process involving contractile proteins, since colchicine and cytochalasin B diminish the stimulated efflux of taurine from the chick retina.[264]

Taurine efflux increases severalfold after a single light flash,[54,256,265] but the light-induced efflux may occur by a mechanism separate from the potassium-stimulated release.[257] Retinal perfusion with Ca^{2+}- or Na^+-free media or glucose omission almost abolish this light effect.[266] The light-induced release originates in the frog from the outer rod segments[267] and in the chick from cells spared by kainic acid treatment, most probably from the photoreceptor cells.[268]

5.5. Axonal Transport

Taurine—unlike other amino acids including GABA—is axonally transported in the goldfish visual system, the rate of migration being similar to fast axonal protein transport.[269,270] In mammalian optic axons, taurine transport is not so directly linked to the axonal transport of proteins, reaching a maximum prior to and during the period of major synapse formation.[77,271] The axonally transported taurine may possibly regulate electrical activity and facilitate the development of axons and the formation of synaptic connections.[16]

6. MEMBRANE EFFECTS OF TAURINE

6.1. Actions on Membranes and Ionic Effects

Like some other short-chain ω-amino acids (glycine, β-alanine, GABA), taurine often inhibits neuronal firing by producing hyperpolarization via altered membrane permeability to ions.[5,272] Taurine has been proposed as a transmitter in the cerebral cortex,[273] since electrophoretically administered taurine hyperpolarizes neuronal membranes in the cat sensorimotor cortex.[274] It also modifies the direct cortical responses[275] and depresses the firing of single neurons in the pericruciate[276] and postcruciate[277] cortex in cats. The depression in the postcruciate cortex is blocked by both bicuculline and strychnine.[278] This suggests that taurine does not act as transmitter at cortical inhibitory synapses that should be selectively sensitive to either antagonist. The depressant effect of taurine on thalamic neurons is also blocked by both.[279]

Taurine and GABA suppress spontaneous discharge dose dependently in the rat cerebellum *in vivo*[42] and guinea pig cerebellar slices *in vitro*.[280,281] The first suppressive effect was antagonized by picrotoxin and bicuculline but not by strychnine.[282,283] The synapses of the stellate neurons on the Purkinje cell dendrites have been assumed to be the most taurine-sensitive site.[282] Taurine is indeed more effective in the dendritic zones than near the somata of the Purkinje cells.[42] Taurine increases Cl^- and K^+ permeabilities but also affects Na^+ permeability by interacting with strychnine-sensitive sites at the neuronal membranes.[280] In view of the electrophysiological data and the differential distribution of taurine in the cerebellar cortex, the possibility that taurine might be the transmitter of the cerebellar stellate neurons cannot be ruled out.

In the brainstem, taurine depresses medullar and bulbar reticular neurons,[284,285] causing hyperpolarization and increasing membrane conductance.[286] The neurons in rat and human brainstem cultures respond similarly to taurine and glycine.[287] Since strychnine, but not bicuculline, reversibly blocks taurine effects, taurine appears to act on receptors identical or similar to glycinelike receptors.

Taurine also depresses the firing of several types of spinal neurons (motoneurons, Renshaw cells, dorsal horn interneurons).[288-290] The inhibitory action requires one acidic and one basic group in the molecule, separated optimally by two or three carbon atoms. Taurine causes Cl^--dependent hyperpolarization in motoneurons with a decrease in excitability and membrane resistance.[291] Taurine and β-alanine hyperpolarize the ventral roots of the frog spinal cord, which effect is antagonized by strychnine.[292] Taurine may also evoke depolarizations in the amphibian spinal cord.[293-296] Taurine and β-alanine also strongly depolarize the primary afferent terminals in the isolated frog spinal cord,[293] which effect is blocked by picrotoxin, bicuculline, and strychnine, suggesting the existence of a taurine/β-alanine receptor.[294] Picrotoxin and strychnine also antagonize the taurine-induced dorsal root depolarization.[296]

Taurine also depresses invertebrate neurons by increasing Cl^- conductance.[297-299] In lobster giant axons and isolated rat muscle preparations, taurine increases membrane permeability to K^+ and Cl^- but not to Na^+ ions.[300,301] This causes the membrane potential to stabilize near the equilibrium potentials of the first two ions, which action has been thought to explain the overall inhibitory effects of taurine in nervous and muscular tissues. It appears that taurine fairly generally inhibits neuronal activity, mimicking, alternatively, GABA or glycine and β-alanine. The specificity of the depressant action of taurine must remain undefined until a taurine-specific antagonist has been found.

Perfusion of isolated frog retinas with taurine or intravitreal taurine injections in the chicken diminish the b-wave amplitude in the electroretinograms.[302,303] In the chicken, strychnine abolishes the depressant action of taurine.[304] Externally applied taurine dissociates the on and off activities at the ganglion cell level in perfused rabbit retina–eyecup preparations.[302] Strychnine blocks this taurine effect. Taurine and glycine totally suppress off-discharges of transient ganglion cells without modifying their on-discharges, strychnine again showing antagonism.[305,306] Taurine or glycine or both have been proposed as the amacrine-released neurotransmitter in the mudpuppy retina,[306] being selectively involved in off-channel activity, whereas GABA may subserve a similar role for the on channel.[306,307]

6.2. Binding to Membranes

The majority of stipulated roles of taurine in nervous tissue presuppose interaction with neural membranes. Taurine-sensitive receptors have been demonstrated on antennular filaments of the spiny lobster (*Panulirus argus*).[308] In the presence of sodium, taurine binds to feline synaptosomal fractions but also to liver and spleen preparations.[309] Synaptic membranes from chicken brain[310] and retina[311] apparently bind taurine in a sodium-dependent manner. No sodium-independent taurine binding, which could represent interaction with possible postsynaptic receptor sites, has been found in membrane fractions isolated from the mouse brain,[312] rat cerebral cortex and spinal cord,[313] and chick retina.[311] In the same conditions, characteristic postsynaptic GABA binding was clearly evident. A sodium-independent, calcium-dependent, high-affinity taurine binding to calf brain synaptic membranes has been reported,[314] but this binding was stated to be largely mixed with transport into empty membrane pouches. However, two protein fractions have been prepared from calf brain synaptic membranes which bind small amounts of [^{35}S]taurine.[315,316] The demarcation of pre- and postsynaptic taurine binding has not yet been entirely successful, and the characterization of the postulated postsynaptic receptors for taurine must probably also await the discovery of a specific taurine antagonist.

6.3. Interactions with Calcium

Taurine may modulate intracellular calcium concentrations, which in turn regulate neuronal excitability. It does not mobilize synaptic membrane-bound calcium[317] but could modulate the intraneuronal calcium concentration by interacting with calcium retention mechanisms in mitochondria.[318] Taurine increases the association of calcium with cardiac mitochondria, possibly by reacting with calcium on the mitochondrial outer membrane and/or intermembrane spaces and thus increasing the mitochondrial calcium-binding capacity.[319] Taurine suppresses calcium release from brain mitochondria, which may diminish the cytoplasmic calcium concentrations.[318] Taurine inhibits the uptake and release of calcium in rat brain synaptosomal preparations.[320–322] Furthermore, taurine inhibits calcium binding to brain microsomes in conditions simulating depolarization but not in the resting state.[323]

Calcium fluxes have been similarly modified by taurine in the retina. Taurine specifically stimulates the ATP-dependent calcium uptake in the disk membranes from frog photoreceptor cell outer segments, which suggests that it regulates translocation of calcium within the rod outer segments.[324] On the other hand, taurine inhibits calcium transport in subcellular fractions from the chick retina, most in the fraction containing the outer segments and pigment epithelium.[262] Taurine also interferes with the binding and uptake of calcium in the endoplasmic reticulum and sarcolemma of the heart and striated muscle[3,325–328] but not in microsomal preparations.[329]

The molecular mechanisms of the calcium–taurine interactions are not known. Taurine can chelate calcium,[330] but that is not the likely mode of action in the cells.[331,332] Moreover, in most aforementioned experiments, the test concentrations of taurine have been very high, at the highest possible limit of the physiological level.

7. PHYSIOLOGICAL AND PHARMACOLOGICAL ACTIONS OF TAURINE

7.1. Interactions with Neurotransmitters

Taurine is a weak β-adrenergic agonist in cultured pineal glands.[333] It suppresses potassium-stimulated release of norepinephrine from the rat superior cervical ganglia and cerebral cortex slices without affecting uptake and spontaneous release.[334] Intraventricularly injected taurine increases the synthesis of dopamine and norepinephrine in all rat brain regions studied, inhibits firing of dopamine neurons, and excites norepinephrine neurons.[335] The interference of taurine with motor behavior and temperature regulation has been suggested to be mediated via such effects on catecholaminergic systems.[335] There is also some evidence of taurine effects on cholinergic systems. In rats, taurine protects cholinergic neurons against the neurotoxic actions of kainic acid[336] and inhibits potassium-evoked release of acetylcholine from the superior cervical ganglia and cerebral cortical slices.[334] Taurine apparently blocks a second muscarinic receptor in dog ganglia.[337] Taurine potentiates stimulus-evoked release of GABA from the rat brain cerebral cortex without inhibiting GABA reuptake.[338] It also interferes with membrane benzodiazepine and GABA receptors.[339]

7.2. Thermoregulation

Intraventricularly injected taurine disrupts temperature regulation in various animal species, producing hypothermia in a cool environment.[335,340–342] In the sheep, taurine inhibits all autonomic effectors regulating heat production, heat loss, and peripheral vasomotor tone.[343] Neuronal activity in the efferent pathways controlling the peripheral vasomotor tone and heat production is inhibited, and the arousal level depressed.[344] The hypothermic effect may be mediated by the central serotonergic systems (rats),[345] but an increase in the hypothalamic acetylcholine may also contribute (mice).[346] At variance with the above studies, taurine injected unilaterally into the lateral ventricle has been found to increase the core temperature in rats.[347] Bilateral injections of low taurine doses into the preoptic region of the anterior hypothalamus also induced hyperthemia, but high doses induced hypothermia. In rabbits, intracerebroventricular taurine reduces the hyperthermia caused by prostaglandin E_1 (PGE_1).

It also inhibits the onset but extends the duration of fever induced by a bacterial endotoxin[344] and intravenous leukocytic pyrogen.[348]

7.3. Anticonvulsive Effects

Among other possible causes, insufficient amounts of inhibitory amino acids are believed to lead to excessive neuronal discharges in epilepsies. In the majority of experimental animal models of epilepsy, the concentrations of the inhibitory amino acids GABA and taurine tend to diminish in the primary epileptogenic foci,[349–354] but not constantly in all investigations.[355–359] Any decrement in taurine has been less pronounced in the secondary focus.[352] Taurine has been the only amino acid to show a significant decrement in the focal area—preceding similar changes in GABA—before the manifestation of epilepsy after topical application of penicillin or cobalt on the cerebral cortex in cats.[360] The decrement in taurine could thus possibly increase the overall excitability of a neuronal population and contribute to the initiation of seizures. Analyses on autopsy or biopsy samples from human epileptic brains have, however, yielded strikingly discrepant data with respect to amino acid concentrations, that of taurine included,[361–363] and no conclusions are warranted.

Taurine, administered as a drug, has been effective in the control of seizures in a variety of experimental models of epilepsy,[11,349,353,364–383] but negative results have also been reported.[352,368,374,377,382,384,385] Some success has been met in taurine trials on epileptic patients.[366,386–391] The response to taurine has greatly varied from patient to patient; not all of them show significant improvement despite comparable taurine dosage.[389–392] Abnormalities in brain electrical activity have generally disappeared more slowly and less completely than the clinical signs[389,393,394]; even a clear aggravation of the EEG pathology, albeit associated with a clinical improvement, has been reported.[390] One cannot infer from the small patient materials published so far whether any specific type of epilepsy is more amenable to taurine treatment than the others.

The failure of taurine to protect against convulsions may result from its poor penetration into the brain.[173] In animal epilepsy models, intraventricularly given taurine has been most often effective, but intraperitoneal, intravenous, or oral taurine has sometimes been entirely ineffective.[368,377] Systemic taurine may have easier access to the focal areas in epilepsies caused by local brain damage than in genuine forms of the disease. A taurine derivative that possesses the anticonvulsant properties of taurine but penetrates better into the brain would appear an excellent therapeutic agent. The efficacy of taurine in some epileptics but not in others may also depend on individual differences in genetically regulated taurine transport capacities.[395] It must be also noted that only intractable cases have so far been subjected to taurine medication.

Several hypotheses have been put forward on the molecular mechanisms of taurine action in epilepsies. Taurine seems to reverse the abnormal amino acid patterns in epilepsy.[349,353,396–398] In particular, the high glutamine levels in the cerebrospinal fluid tend to diminish.[391] Taurine may act by maintaining the proper balance between the concentrations of glutamine and glutamate in neuronal and glial cell compartments.[399] Taurine may also chelate those trace metals, Zn^{2+} ions in particular, that are inhibitory to the cerebral ATPase.[387] The increase in membrane Cl^- conductance[280,291] is a likely contributing factor, and a complexation of Ca^{2+} ions with taurine has also been considered.[373]

7.4. Other Actions

Taurine administration generally inhibits behavioral responses in animal tests,[400–403] possibly by modulating the central motor systems[404] or by interfering with learning and memory.[405] It also blocks the mouse-killing behavior of aggressive rats.[406] The contralateral circling behavior induced by a taurine injection into the substantia nigra possibly indicates some role in the nigral functions.[407] The reduced activity of CSD in putamen in Huntington's chorea[408] apparently reflects a similar change in the GAD activity. Both uptake and content of taurine in platelets are decreased in retinitis pigmentosa,[409,410] but even large doses of taurine for 1 year have not slowed down the progress of the disease.[411]

Taurine may also be involved in the hypothalamic control of thirst and appetite.[412] Intraventricularly injected taurine and hypotaurine increase prolactin secretion in rats[413] but do not affect the secretion of luteinizing and follicle-stimulating hormones.[414] Taurine antagonizes stress-induced elevation in blood sugar by reducing the epinephrine output from the adrenal glands in rats.[415] The increased output of growth hormone induced by morphine administration in rats is also completely blocked by an intraventricular injection of taurine.[416] Taurine may also promote recovery from akinesia and analgesia induced by D-Ala^2-enkephalinamide.[417] Taurine significantly delays the onset of sleep, shortens its duration, and helps in conserving the righting reflex in mice during acute intoxication with ethanol.[418] The other depressant effects of ethanol are also markedly reduced by taurine.[419] Taurine may be useful in prevention and treatment of alcohol-withdrawal symptoms in man.[420]

The neuropharmacology of taurine has also been discussed elsewhere[11] together with its extensively studied cardiac actions.[4,421]

8. NUTRITIONAL ASPECTS

The high tissue concentrations of taurine, in developing brain and retina in particular, combined with low rates of biosynthesis suggest that a considerable amount of taurine must be supplied by food. Rats fed on a diet deficient in pyridoxine[186,422] or thiamine[81] maintain their levels of taurine unaltered. The activities of cysteine dioxygenase and CSD and the concentration of taurine in the brain have also remained unaltered in rats kept on diets containing 18% or 60% casein.[423] On the other hand, rats undernourished for a prolonged period[424] or eating a normal diet fortified with taurine[186] also show no change. Taurine added to the diet of growing rats previously fed with a diet low in sulfur compounds restores the lowered brain taurine levels.[425] Considerable amounts of taurine are indeed normally transferred from the mother via milk into the rat pup.[426] Also, long-term administration of taurine in a liquid diet increases the brain taurine content.[427]

In cats, deficiency can be produced with diets devoid of taurine. Here, organ taurine levels drop markedly, and retinal photoreceptor cells degenerate,[56,59,428–430] although the retina and olfactory bulb resist taurine depletion more effectively than other brain areas.[74,429] All subcellular pools of taurine are uniformly depleted.[431] The retinal degeneration in the cat can be prevented by feeding taurine but not with methionine, cysteine, inorganic sulfate, pyridoxine, or pyridoxine together with cysteine.[59,169,432] In taurine-deficient cats, the lattice arrangement of tapetal rods is severely disorganized.[433] The morphological changes become visible only when the retinal taurine concentration has diminished to below 50% of normal after several months on a taurine-free diet.[432] On the other hand, the electroretinogram responses are directly proportional to the retinal concentration decrement, being manifested before any structural changes are ophthalmoscopically visible.[434] It has been suggested that taurine deficiency results in abnormal retinal ionic concentrations, but the mechanism by which photoreceptor cells die is not known.[204] The cat conjugates bile acids with taurine only,[435] and since endogenous taurine biosynthesis is limited, cats fed on taurine-deficient diets are unable to maintain sufficient body concentrations. Taurine has thus been considered an essential nutrient for the cat,[169] not only for the prevention of retinal degeneration but also for the general maintenance of normal amino acid concentrations in tissues.[429]

In the monkey, taurine depletion causes a significant growth depression but no retinal degeneration.[436] Accordingly, no CSD activity has been found in the monkey liver and brain. Also, human infants fed with synthetic milk formulas become taurine depleted because these formulas contain very little taurine.[437,438] Human infants and beagle pups also become taurine deficient with solutions used in total parenteral nutrition because these do not contain taurine.[438,439] Taurine is the second most abundant free amino acid in human milk.[440] Moreover, a preterm human infant has only a limited capacity to convert methionine to cystine and cystine to taurine.[437] It has therefore been suggested that taurine may also be an essential nutrient for the human infant,[437] though no obvious clinical signs have resulted from taurine depletion.

9. TAURINE DERIVATIVES AND TAURINE-CONTAINING PEPTIDES

9.1. Taurine Derivatives

β-Alanine, the carboxylic acid analogue of taurine, has been considered a neurotransmitter on its own[205] and a specific marker of glial GABA pools.[441] Iontophoretically applied β-alanine inhibits firing and hyperpolarizes single neurons,[272,284] which actions are antagonized by strychnine. High-affinity uptake[192] and potassium-stimulated release[442] have been detected for β-alanine in brain slices.

Hypotaurine, (Fig. 1) the sulfinic acid analogue of taurine, has been suggested to have a synaptic function in the CNS.[443] It depresses dose-dependently the spiking of Purkinje cells in guinea pig cerebellum.[283] The action is competitively blocked by picrotoxin and strychnine. The most hypotaurine-sensitive sites are located in the dendritic region of the cerebellar molecular layer. Hypotaurine reduces the b-wave amplitude in the frog retina, though to a lesser degree than taurine.[303] Hypotaurine is also released from brain slices by potassium stimulation.[241,243] It possesses an effective energy-, temperature-, and sodium-dependent high-affinity transport system in brain slices, similar to that of neurotransmitter amino acids.[444–447] The uptake is inhibited by GABA and its analogues but not by taurine.[193,448] In other ways as well, hypotaurine transport resembles GABA transport more than taurine transport.[449] In the male and female reproductive organs, the very high hypotaurine concentrations may have an essential role in fertility.[19,20]

Homotaurine (3-aminopropanesulfonic acid) is the most potent inhibitor of cerebral cortical neurons[276] and cerebellar Purkinje cells.[283] The most homotaurine-sensitive site lies near the Purkinje cell somata. The anticonvulsant potency of homotaurine is greater than that of taurine.[368] Homotaurine causes hypothermia and reduces skeletal muscle tonus.[450] Homotaurine has been suggested to be a GABA agonist.[451]

Taurocyamine (guanidinotaurine, 2-(or β-)guanidinoethanesulfonic acid) has a six times stronger inhibitory action on cerebellar Purkinje cells than taurine.[283] It is synthetized in both vertebrates and invertebrates by oxidation of hypotaurocyamine.[452] The substance has been identified in the brains of rat, ox, and guinea pig,[453] but at very low concentrations.[454] Taurocyamine has no effect on the frog electroretinogram[303] but is a convulsant *in vivo*.[455]

Isethionic acid (2-hydroxyethanesulfonic acid) is probably formed from mercaptoethanol, an alternative substrate for cysteamine dioxygenase.[180] It is the major anion in the axoplasm of squid giant nerve cells,[456] but its concentration in the mammalian brain is very low.[179] Isethionate efflux from the rat CNS is multiphasic and faster than that of taurine.[457]

Cysteine sulfinic acid (CSA) and cysteic acid (CA) have strong excitatory actions in the spinal cord.[458] The concentration of CSA is higher in the cerebral cortex and cerebellum and lower in the midbrain and olfactory bulbs than elsewhere in the brain.[459] A sodium-dependent high-affinity uptake in crude synaptosomal fractions and potassium-stimulated, calcium-dependent release from brain slices have been reported for CSA.[460] It is bound specifically and sodium-independently to crude synaptic membrane fractions. Intracerebroventricularly injected CSA and CA are convulsants in the rat.[460]

9.2. Taurine-Containing Peptides

Taurine is a constituent of small N-acetylaspartyl peptides formed *in vitro* by homogenates of mouse and monkey brains.[461,462] Similar peptides have been characterized in calf brain nerve terminals.[463] High potassium and calcium concentrations and electrical stimulation release these peptides from synaptic vesicles. A dipeptide, γ-L-glutamyltaurine (glutaurine, Litoralon®), with hormonelike properties[464,465] has recently been isolated from protein-free aqueous extracts of bovine parathyroid powder[466] and is assumed also to exist in brain tissue. In cerebellar preparations, taurine is active in the assay of γ-glutamyltransferase as an acceptor of γ-glutamyl residues.[467]

10. CONCLUSIONS

Taurine is a ubiquitous constituent of animal tissues, abundant in electrically excitable organs and particularly enriched in synaptic vesicles in the brain and photoreceptor cells in the retina. Taurine biosynthesis probably proceeds either via cysteinesulfinate or cysteamine intermediaries, but most taurine may be derived from food. It appears to be an essential nutrient for kittens, which are blinded in taurine deficiency, but no clinical deficiency signs are known in human infants. The neurobiological functions of taurine are not yet defined. It has been considered a neurotransmitter or -modulator or stabilizer of membrane functions. Taurine is released from brain and retinal cells by specific stimuli, but the calcium dependence of the stimulated release has not been proved. Taurine generally inhibits neuronal firing and hyperpolarizes neurons by increasing the membrane permeability for chloride ions. No specific taurine antagonist is known. As a drug, taurine is a potent anticonvulsant, although its efficacy is limited by poor penetration into brain tissue. Since taurine is apparently not broken down in animal tissues, only its efficient high-affinity uptake could terminate the possible synaptic actions. Taurine certainly affects neuronal excitability, but there is no conclusive evidence for it having a role as natural transmitter in any brain area. Several other hypotheses have recently been put forward in an attempt to explain the observed actions in excitable tissues.

ACKNOWLEDGMENTS. We are grateful to the personnel of the Medical Section of the Library of the University of Tampere for their generous help in the literature search and to Miss Riitta Mero and Miss Eija Kyrölä for skillful secretarial assistance. Dr. Pirjo Kontro is a Junior Research Fellow of the Finnish Academy.

REFERENCES

1. Jacobsen, J. G., and Smith, L. H., Jr., 1968, *Physiol. Rev.* **48**:424–511.
2. Simpson, J., Allen, K., and Awapara, J., 1959, *Biol. Bull.* **117**:371–381.
3. Huxtable, R., and Bressler, R., 1973, *Biochim. Biophys. Acta* **323**:573–583.
4. Baskin, S. I., and Finney, C. M., 1979, *Sulfur-Containing Amino Acids* **2**:1–18.
5. Oja, S. S., Kontro, P., and Lähdesmäki, P., 1977, *Prog. Pharmacol.* **1**(3):1–119.
6. Baskin, S. I., Leibman, A. J., and Cohn, E. M., 1976, *Adv. Biochem. Psychopharmacol.* **15**:153–164.
7. Huxtable, R., and Barbeau, A. (eds.), 1976, *Taurine*, Raven Press, New York.
8. Barbeau, A., and Huxtable, R. J. (eds.), 1978, *Taurine and Neurological Disorders*, Raven Press, New York.
9. Schaffer, S. W. Baskin, S. I., and Kocsis, J. J., (eds.), 1981, *The Action of Taurine on Excitable Tissues*, MTP Press, Lancaster.
10. Huxtable, R. J., and Pasantes-Morales, H. (eds.), 1982, *Taurine in Nutrition and Neurology*, Plenum Press, New York.
11. Barbeau, A., Inoue, N., Tsukada, Y., and Butterworth, R. F., 1975, *Life Sci.* **17**:669–678.
12. Mandel, P., Pasantes-Morales, H., and Urban, P. F., 1976, *Transmitters in the Visual Process* (S. L. Bonting, ed.), Pergamon Press, New York, pp. 89–105.
13. Collins, G. G. S., 1977, *Essays in Neurochemistry and Neuropharmacology*, Volume 1 (M.

B. H. Youdim, W. Lowenberg, D. F. Sharman, and J. R. Lagnodo, eds.), John Wiley & Sons, New York, pp. 43–72.

14. Sturman, J. A., Rassin, D. K., and Gaull, G. E., 1977, *Life Sci.* **21**:1–22.
15. Mandel, P., and Pasantes-Morales, H., 1978, *Reviews of Neuroscience*, Volume 3 (S. Ehrenpreis and I. Kopin, eds.), Raven Press, New York, pp. 157–193.
16. Sturman, J. A., and Hayes, K. C., 1980, *Advances in Nutritional Research*, Volume 3 (H. H. Draper, ed.), Plenum Press, New York, pp. 231–299.
17. Kocsis, J. J., Kostos, V. J., and Baskin, S. I., 1976, *Taurine* (R. Huxtable and A. Barbeau, eds.), Raven Press, New York, pp. 145–153.
18. Yoshikawa, K., and Kuriyama, K., 1976, *Jpn. J. Pharmacol.* **26**:649–654.
19. Kochakian, C. D., 1976, *Taurine* (R. Huxtable and A. Barbeau, eds.), Raven Press, New York, pp. 327–334.
20. Kochakian, C. D., 1980, *Natural Sulfur Compounds: Novel Biochemical and Structural Aspects* (D. Cavallini, G. E. Gaull, and V. Zappia, eds.), Plenum Press, New York, pp. 213–224.
21. Guidotti, A., Badiani, G., and Pepeu, G., 1972, *J. Neurochem.* **19**:431–435.
22. Crabai, F., Sitzia, A., and Pepeu, G., 1974, *J. Neurochem.* **23**:1091–1092.
23. Krusz, J. C., Dix, R. K., and Baskin, S. I., 1978, *Fed. Proc.* **37**:907.
24. Piha, R. S., Oja, S. S., and Uusitalo, A. J., 1962, *Ann. Med. Exp. Biol. Fenn.* **40**(Suppl. 5):1–28.
25. Kontro, P., Marnela, K.-M., and Oja, S. S., 1980, *Brain Res.* **184**:129–141.
26. Tachiki, K. H., and Baxter, C. F., 1979, *J. Neurochem.* **33**:1125–1129.
27. Kandera, J., Levi, G., and Lajtha, A., 1968, *Biochem. J.* **126**:249–260.
28. Collins, G. G. S., 1974, *Brain Res.* **76**:447–459.
29. Lombardini, J. B., 1976, *Taurine* (R. Huxtable and A. Barbeau, eds.), Raven Press, New York, pp. 311–326.
30. Adams, J. C., and Wenthold, R. J., 1979, *Neuroscience* **4**:1947–1951.
31. Yoneda, Y., Takashima, S., Hirai, K., Kurihara, E., Yukawa, Y., Tokunaga, H., and Kuriyama, K., 1977, *Jpn. J. Pharmacol.* **27**:881–888.
32. Yoneda, Y., and Kuriyama, K., 1978, *J. Neurochem.* **30**:821–825.
33. Lane, J. D., Smith, J. E., Hall, P. V., and Campbell, R. L., 1978, *Brain Res.* **152**:386–390.
34. Nicklas, W. J., Duvoisin, R. C., and Berl, S., 1979, *Brain Res.* **167**:107–117.
35. Placheta, P., Singer, E., Schönbeck, G., Heckl, K., and Karobath, M., 1979, *Neuropharmacology* **18**:399–402.
36. McBride, W. J., and Frederickson, R. C. A., 1980, *Fed. Proc.* **39**:2701–2705.
37. McBride, W. J., Nadi, N. S., Altman, J., and Aprison, M. H., 1976, *Neurochem. Res.* **1**:141–152.
38. Nadi, N. S., McBride, W. J., and Aprison, M. H., 1977, *J. Neurochem.* **28**:453–455.
39. Rea, M. A., McBride, W. J., and Rohde, B. H., 1981, *Neurochem. Res.* **6**:33–39.
40. Assumpção, J. A., Bernardi, N., Dacke, C. G., Davidson, N., and Eichelberger, H. G., 1980, *J. Physiol. (Lond.)* **298**:35P.
41. McBride, W. J., Rea, M. A., Felten, D. L., Sinisi, N., and Rohde, B. H., 1980, *Neurochem. Res.* **5**:337–344.
42. Frederickson, R. C. A., Neuss, M., Morzorati, S. L., and McBride, W. J., 1978, *Brain Res.* **145**:117–126.
43. McBride, W. J., Rea, M. A., and Nadi, N. S., 1978, *Neurochem. Res.***3**:793–801.
44. Rohde, B. H., Rea, M. A., and McBride, W. J., 1978, *Brain Res.* **156**:202–205.
45. Rea, M. A., McBride, W. J., and Rohde, B. H., 1980, *J. Neurochem.* **34**:1106–1108.
46. Pasantes-Morales, H., Klethi, J., Ledig, M., and Mandel, P., 1972, *Brain Res.* **41**:494–497.
47. Starr, M. S., 1973, *Brain Res.* **59**:331–338.
48. Voaden, M. J., Lake, N., Marshall, J., and Morjaria, B., 1977, *Exp. Eye Res.* **25**:249–257.
49. Kennedy, A. J., and Voaden, M. J., 1974, *J. Neurochem.* **23**:1093–1095.
50. Orr, H. T., Cohen, A. I., and Lowry, O. H., 1976, *J. Neurochem.* **26**:609–611.
51. Kennedy, A. J., Neal, M. J., and Lolley, R. N., 1977, *J. Neurochem.* **29**:157–159.
52. Pourcho, R. G., 1977, *Exp. Eye Res.* **25**:119–127.
53. Cohen, A. I., McDaniel, M., and Orr, H., 1973, *Invest. Ophthalmol.* **12**:686–693.
54. Schmidt, S. Y., 1978, *Exp. Eye Res.* **26**:529–535.

55. Schmidt, S. Y., and Berson, E. L., 1978, *Taurine and Neurological Disorders* (A. Barbeau and R. J. Huxtable, eds.), Raven Press, New York, pp. 281–287.
56. Hayes, K. C., Carey, R. E., and Schmidt, S. Y., 1975, *Science* **188**:949–951.
57. Salceda, R., Cárabez, A., Pacheco, P., and Pasantes-Morales, H., 1979, *Exp. Eye Res.* **28**:137–146.
58. Schmidt, S. Y., Berson, E. L., and Hayes, K. C., 1976, *Invest. Ophthalmol.* **15**:47–52.
59. Berson, E. L., Hayes, K. C., Rabin, A. R., Schmidt, S. Y., and Watson, G., 1976, *Invest. Ophthalmol.* **15**:52–58.
60. Schubert, D., Carlisle, W., and Look, C., 1975, *Nature* **254**:341–343.
61. Sellström, Å., Sjöberg, L.-B., and Hamberger, A., 1975, *J. Neurochem.* **25**:393–398.
62. Rassin, D. K., Sturman, J. A., and Gaull, G. E., 1977, *J. Neurochem.* **28**:41–50.
63. Macaione, S., Tucci, G., De Luca, G., and Di Giorgio, R. M., 1976, *J. Neurochem.* **27**:1411–1415.
64. De Belleroche, J. S., and Bradford, H. F., 1973, *J. Neurochem.* **21**:441–451.
65. Marnela, K.-M., Kontro, P., Pitkänen, R. I., and Oja, S. S., 1980, *Acta Univ. Oul. A 97* [*Biochem.*] **29**:11–16.
66. Oja, S. S., Uusitalo, A. J., Vahvelainen, M.-L., and Piha, R. S., 1968, *Brain Res.* **11**:655–661.
67. Sturman, J. A., and Gaull, G. E., 1975, *J. Neurochem.* **25**:831–835.
68. Oja, S. S., and Piha, R. S., 1966, *Life Sci.* **5**:865–870.
69. Sturman, J. A., Rassin, D. K., and Gaull, G. E., 1978, *Taurine and Neurological Disorders* (A. Barbeau and R. J. Huxtable, eds.), Raven Press, New York, pp. 49–71.
70. Sturman, J. A., 1979, *J. Neurochem.* **32**:811–816.
71. Sturman, J. A., Rassin, D. K., Gaull, G. E., and Cote, L. J., 1980, *J. Neurochem.* **35**:304–310.
72. Cutler, R. W. P., and Dudzinski, D. S., 1974, *J. Neurochem.* **23**:1005–1009.
73. Oja, S. S., 1966, *Ann. Acad. Sci. Fenn.* [*Med.*] **125**:1–69.
74. Sturman, J. A., Rassin, D. K., Hayes, K. C., and Gaull, G. E., 1978, *J. Nutr.* **108**:1462–1476.
75. Macaione, S., Ruggeri, P., De Luca, G., and Tucci, G., 1974, *J. Neurochem.* **22**:887–891.
76. Macaione, S., Tucci, G., and Di Giorgio, R. M., 1975, *Ital. J. Biochem.* **24**:163–174.
77. Sturman, J. A., 1979, *J. Neurobiol.* **10**:221–237.
78. Iwata, H., Matsuda, T., Yamagami, S., Tsukamoto, T., and Baba, A., 1978, *Brain Res.* **143**:383–386.
79. Grosso, D. S., Bressler, R., and Benson, B., 1978, *Life Sci.* **22**:1789–1798.
80. Pasantes-Morales, H., Klethi, J., Ledig, M., and Mandel, P., 1973, *Brain Res.* **57**:59–65.
81. Iwata, H., Baba, A., and Yoneda, Y., 1976, *Taurine* (R. Huxtable and A. Barbeau, eds.), Raven Press, New York, pp. 85–90.
82. Perry, T. L., Hansen, S., and Kish, S., 1979, *Life Sci.* **24**:283–288.
83. Neuhoff, V., and Tonge, S. R., 1973, *J. Pharm. Pharmacol.* **25**:138P–139P.
84. Lähdesmäki, P., and Oja, S. S., 1976, *Selected Topics in Environmental Biology* (B. Bhatia, G. S. Chhina, and B. Singh, eds.), Interprint Publications, New Delhi, pp. 173–177.
85. Iwata, H., Matsuda, T., Lee, E., Yamagami, S., and Baba, A., 1980, *Experientia* **36**:332–333.
86. Iwata, H., Matsuda, T., Yamagami, S., Hirata, Y., and Baba, A., 1978, *Biochem. Pharmacol.* **27**:1955–1960.
87. Huxtable, R. J., Laird, H. E., II, and Lippincott, S. E., 1979, *J. Pharmacol. Exp. Ther.* **211**:465–471.
88. Gilles, R., and Schoffeniels, E., 1968, *Arch. Int. Physiol. Biochim.* **76**:441–451.
89. Gilles, R., and Schoffeniels, E., 1969, *Comp. Biochem. Physiol.* **28**:417–423.
90. Minchin, M. C. W., and Beart, P. M., 1975, *J. Neurochem.* **24**:881–884.
91. Finney, C. M., 1978, *Comp. Biochem. Physiol.* **61**:409–414.
92. Peck, E. J., Jr., and Awapara, J., 1967, *Biochim. Biophys. Acta* **141**:499–506.
93. Ebels, I., Benson, B., and Larsen, B. R., 1980, *J. Neural Transm.* **48**:101–117.
94. Hoskin, F. C. G., Pollock, M. L., and Prusch, R. D., 1975, *J. Neurochem.* **25**:445–449.
95. Pasantes-Morales, H., Chatagner, F., and Mandel, P., 1980, *Neurochem. Res.* **5**:441–451.

96. Yamaguchi, K., Sakakibara, S., Asamizu, J., and Ueda, I., 1973, *Biochim. Biophys. Acta* **297**:48–59.

97. Hardison, W. G. M., Wood, C. A., and Proffitt, J. H., 1977, *Proc. Soc. Exp. Biol. Med.* **155**:55–58.

98. Yamaguchi, K., 1980, *Natural Sulfur Compounds: Novel Biochemical and Structural Aspects* (D. Cavallini, G. E. Gaull, and V. Zappia, eds.), Plenum Press, New York, pp. 175–186.

99. Misra, C. H., and Olney, J. W., 1975, *Brain Res.* **97**:117–126.

100. Di Giorgio, R. M., Tucci, G., and Macaione, S., 1975, *Life Sci* **16**:429–436.

101. Macaione, S., Di Giorgio, R. M., and De Luca, G., 1980, *Natural Sulfur Compounds: Novel Biochemical and Structural Aspects* (D. Cavallini, G. E. Gaull, and V. Zappia, eds.), Plenum Press, New York, pp. 265–276.

102. Sakakibara, S., Yamaguchi, K., Hosokawa, Y., Kohashi, N., Ueda, I., and Sakamoto, Y., 1976, *Biochim. Biophys. Acta* **422**:273–279.

103. Yamaguchi, K., Hosokawa, Y., Kohashi, N., Kori, Y., Sakakibara, S., and Ueda, I., 1978, *J. Biochem (Tokyo)* **83**:479–491.

104. Misra, C. H., 1979, *Int. J. Biochem.* **10**:201–204.

105. Yamaguchi, K., Sakakibara, S., Koga, K., and Ueda, I., 1971, *Biochim. Biophys. Acta* **237**:502–512.

106. Pasantes-Morales, H., Loriette, C., and Chatagner, F., 1977, *Neurochem. Res.* **2**:671–680.

107. Rassin, D. K., and Gaull, G. E., 1975, *J. Neurochem.* **24**:969–978.

108. Loriette, C., and Chatagner, F., 1978, *Experientia* **34**:981–982.

109. Macaione, S., and Di Giorgio, R. M., 1977, *Life Sci.* **20**:617–622.

110. Guion-Rain, M. C., Portemer, C., and Chatagner, F., 1975, *Biochim. Biophys. Acta* **384**:265–276.

111. Heinämäki, A. A., and Piha, R. S., 1980, *Acta Chem. Scand.* [*B*] **34**:363–367.

112. Jacobsen, J. G., Thomas, L. L., and Smith, L. H., Jr., 1964, *Biochim. Biophys. Acta* **85**:103–116.

113. Spaeth, D. G., and Schneider, D. L., 1974, *Proc. Soc. Exp. Biol. Med.* **147**:855–858.

114. Wu, J.-Y., Moss, L. G., and Chen, M.-S., 1979, *Neurochem. Res.* **4**:201–212.

115. Wu, J.-Y., Matsuda, T., and Roberts, E., 1973, *J. Biol. Chem.* **248**:3029–3034.

116. Urban, P. F., Reichelt, P., and Mandel, P., 1981, *Amino Acid Neurotransmitters* (F. V. DeFeudis and P. Mandel, eds.), Raven Press, New York, pp. 537–544.

117. Rassin, D. K., and Sturman, J. A., 1975, *Life Sci.* **16**:875–882.

118. Pasantes-Morales, H., Mapes, C., Tapia, R., and Mandel, P., 1976, *Brain Res.* **107**:575–589.

119. Sawaya, C., Edmondson, F., Horton, R., and Meldrum, B., 1979, *Biochem. Pharmacol.* **28**:2854–2856.

120. Rassin, D. K., 1975, *Metabolic Compartmentation and Neurotransmission. Relation to Brain Structure and Function* (S. Berl, D. D. Clarke, and D. Schneider, eds.), Plenum Press, New York, pp. 559–565.

121. Piha, R. S., and Saukkonen, H., 1966, *Suomen Kemistilehti* [*B*] **39**:112–114.

122. Mathur, R. L., Klethi, J., Ledig, M., and Mandel, P., 1976, *Life Sci.* **18**:75–80.

123. Austin, L., Recasens, M., Mathur, R. L., and Mandel, P., 1978, *Neurosci. Lett.* **9**:59–63.

124. Oja, S. S., Karvonen, M.-L., and Lähdesmäki, P., 1973, *Brain Res.* **55**:173–178.

125. Di Giorgio, R. M., Macaione, S., and De Luca, G., 1978, *Ital. J. Biochem.* **27**:83–93.

126. Pasantes-Morales, H., López-Colomé, A. M., Salceda, R., and Mandel, P., 1976, *J. Neurochem.* **27**:1103–1106.

127. Gabellec, M. M., Recasens, M., and Mandel, P., 1978, *Life Sci.* **23**:1263–1270.

128. Recasens, M., Gabellec, M. M., Austin, L., and Mandel, P., 1978, *Biochem. Biophys. Res. Commun.* **83**:449–456.

129. Recasens, M., Benezra, R., Gabellec, M. M., Delaunoy, J.-P., and Mandel, P., 1979, *FEBS Lett.* **99**:51–54.

130. Recasens, M., Benezra, R., Basset, P., and Mandel, P., 1980, *Biochemistry* **19**:4583–4589.

131. Recasens, M., Gabellec, M. M., Mack, G., and Mandel, P., 1978, *Neurochem. Res.* **3**:27–35.

132. Recasens, M., and Benezra, R., 1981, *Amino Acid Neurotransmitters* (F. V. DeFeudis and P. Mandel, eds.), Raven Press, New York, pp. 545–550.

133. Huxtable, R., and Bressler, R., 1976, *Taurine* (R. Huxtable and A. Barbeau, eds.), Raven Press, New York, pp. 45–57.
134. Cavallini, D., De Marco, C., and Mondovi, B., 1955, *Ric. Sci.* **25:**2901–2903.
135. Cavallini, D., Mondovi, B., and De Marco, C., 1954, *Ric. Sci.* **24:**2649–2651.
136. Cavallini, D., De Marco, C., and Mondovi, B., 1961, *Enzymologia* **23:**101–110.
137. Cavallini, D., Federici, G., Ricci, G., Dupré, S., Antonucci, A., and De Marco, C., 1975, *FEBS Lett.* **56:**348–351.
138. Eldjarn, L., 1954, *J. Biol. Chem.* **206:**483–490.
139. Eldjarn, L., Pihl, A., and Sverdrup, A., 1956, *J. Biol. Chem.* **223:**353–358.
140. Brown, G. M., 1959, *J. Biol. Chem.* **234:**370–377.
141. Abiko, Y., Tomikawa, M., and Shimizu, M., 1968, *J. Biochem. (Tokyo)* **64:**115–117.
142. Scandurra, R., Barboni, E., Granata, F., Pensa, B., and Costa, M., 1974, *Eur. J. Biochem.* **49:**1–9.
143. Cavallini, D., Dupré, S., Graziani, M. T., and Tinti, M. G., 1968, *FEBS Lett.* **1:**119–121.
144. Dupré, S., Graziani, M. T., Rosei, M. A., Fabi, A., and Del Grosso, E., 1970, *Eur. J. Biochem.* **16:**571–578.
145. Dupré, S., Granata, F., Santoro, L., Scandurra, R., Federici, G., and Cavallini, D., 1975, *Ital. J. Biochem.* **24:**369–376.
146. Cavallini, D., Scandurra, R., Dupré, S., Federici, G., Santoro, L., Ricci, G., and Barra, D., 1976, *Taurine* (R. Huxtable and A. Barbeau, eds.), Raven Press, New York, pp. 59–66.
147. Cavallini, D., De Marco, C., Scandurra, R., Dupré, S., and Graziani, M. T., 1966, *J. Biol. Chem.* **241:**3189–3196.
148. Dupré, S., and De Marco, C., 1964, *Ital. J. Biochem.* **13:**386–390.
149. Cavallini, D., Scandurra, R., and Dupré, S., 1971, *Methods Enzymol.* **17B:**479–483.
150. Scandurra, R., Federici, G., Dupré, S., and Cavallini, D., 1978, *Bull. Mol. Biol. Med.* **3:**141–147.
151. Cavallini, D., Scandurra, R., Dupré, S., Santoro, L., and Barra, D., 1976, *Physiol. Chem. Phys.* **8:**157–160.
152. Scandurra, R., Politi, L., Dupré, S., Moriggi, M., Barra, D., and Cavallini, D., 1977, *Bull. Mol. Biol. Med.* **2:**172–177.
153. Cavallini, D., De Marco, C., Mondovi, B., and Stirpe, F., 1954, *Biochim. Biophys. Acta* **15:**301–302.
154. Fiori, A., and Costa, M., 1969, *Acta Vitaminol. (Milano)* **23:**204–207.
155. Di Giorgio, R. M., Macaione, S., and De Luca, G., 1977, *Life Sci.* **20:**1657–1662.
156. Di Giorgio, R. M., De Luca, G., and Macaione, S., 1978, *Bull. Mol. Biol. Med.* **3:**115–120.
157. Kontro, P., and Oja, S. S., 1980, *Natural Sulfur Compounds: Novel Biochemical and Structural Aspects* (D. Cavallini, G. E. Gaull, and V. Zappia, eds.), Plenum Press, pp. 201–212.
158. Ricci, G., Dupré, S., Federici, G., Spoto, G., Matarese, R. M., and Cavallini, D., 1978, *Physiol. Chem. Phys.* **10:**435–442.
159. Sumizu, K., 1962, *Biochim. Biophys. Acta* **63:**210–212.
160. Oja, S. S., and Kontro, P., 1981, *Biochim. Biophys. Acta* **667:**350–357.
161. Martin, W. G., Sass, N. L., Hill, L., Tarka, S., and Truex, R., 1972, *Proc. Soc. Exp. Biol. Med.* **141:**632–633.
162. Gorby, W. G., and Martin, W. G., 1975, *Proc. Soc. Exp. Biol. Med.* **148:**544–549.
163. Martin, W. G., Miraglia, R. J., Spaeth, D. G., and Patrick, H., 1966, *Proc. Soc. Exp. Biol. Med.* **122:**841–844.
164. Miraglia, R. J., and Martin, W. G., 1969, *Proc. Soc. Exp. Biol. Med.* **132:**640–644.
165. Sass, N. L., and Martin, W. G., 1972, *Proc. Soc. Exp. Biol. Med.* **139:**755–761.
166. Miraglia, R. J., and Martin, W. G., 1967, *Proc. West Va. Acad. Sci.* **39:**158–163.
167. Martin, W. G., Truex, R. C., Tarka, S. M., Hill, L. J., and Gorby, W. G., 1974, *Proc. Soc. Exp. Biol. Med.* **147:**563–565.
168. Huxtable, R. J., 1978, *Taurine and Neurological Disorders* (A. Barbeau and R. J. Huxtable, eds.), Raven Press, New York, pp. 5–17.
169. Knopf, S., Sturman, J. A., Armstrong, M., and Hayes, K. C., 1978, *J. Nutr.* **108:**773–778.
170. Federici, G., Ricci, G., Santoro, L., Antonucci, A., and Cavallini, D., 1980, *Natural Sulfur*

Compounds: Novel Biochemical Structural Aspects (D. Cavallini, G. E. Gaull, and V. Zappia, eds.), Plenum Press, New York, pp. 187–194.

171. Yamaguchi, K., 1975, *J. Jpn. Biochem. Soc.* **47**:241–263.

172. Robeson, B. L., and Martin, W. G., 1978, *Fed. Proc.* **37**:537.

173. Oja, S. S., Lehtinen, I., and Lähdesmäki, P., 1976, *Q. J. Exp. Physiol.* **61**:133–143.

174. Sturman, J. A., Hepner, G. W., Hofmann, A. F., and Thomas, P. J., 1975, *J. Nutr.* **105**:1206–1214.

175. Fellman, J. H., Roth, E. S., Avedovech, N. A., and McCarthy, K. D., 1980, *Arch. Biochem. Biophys.* **204**:560–567.

176. Read, W. O., and Welty, J. D., 1962, *J. Biol. Chem.* **237**:1521–1522.

177. Lähdesmäki, P., and Korhonen, K., 1978, *J. Neurochem.* **30**:705–711.

178. Fellman, J. H., Roth, E. S., and Fujita, T. S., 1978, *Taurine and Neurological Disorders* (A. Barbeau and R. J. Huxtable, eds.), Raven Press, New York, pp. 19–24.

179. Remtulla, M. A., Applegarth, D. A., Clark, D. G., and Williams, I. H., 1977, *Life Sci.* **20**:2029–2036.

180. Cavallini, D., Dupré, S., Federici, G., Solinas, S., Ricci, G., Antonucci, A., Spoto, G., and Matarese, M., 1978, *Taurine and Neurological Disorders* (A. Barbeau and R. J. Huxtable, eds.), Raven Press, New York, pp. 29–34.

181. Dupré, S., Federici, G., Ricci, G., Spoto, G., Antonucci, A., and Cavallini, D., 1978, *Enzyme* **23**:307–313.

182. Urquhart, N., Perry, T. L., Hansen, S., and Kennedy, J., 1974, *J. Neurochem.* **22**:871–872.

183. Wheler, G. H. T., Osborne, R. H., Bradford, H. F., and Davison, A. N., 1974, *Biochem. Soc. Trans.* **2**:285–286.

184. Lefauconnier, J.-M., Urban, F., and Mandel, P., 1978, *Biochimie* **60**:381–387.

185. Minato, A., Hirose, S., Ogiso, T., Uda, K., Takigawa, Y., and Fujihira, E., 1969, *Chem. Pharm. Bull. (Tokyo)* **17**:1498–1505.

186. Sturman, J. A., 1973, *J. Nutr.* **103**:1566–1580.

187. Wheler, G. H. T., Osborne, R. H., Bradford, H. F., and Davison, A. N., 1977, *Brain Res.* **136**:535–542.

188. Levi, G., 1968, *Prog. Brain Res.* **29**:219–228.

189. Baños, G., Daniel, P. M., Moorhouse, S. R., and Pratt, O. E., 1971, *J. Physiol. (Lond.)* **213**:45P–46P.

190. Sturman, J. A., Rassin, D. K., and Gaull, G. E., 1977, *J. Neurochem.* **28**:31–39.

191. Oja, S. S., 1971, *J. Neurochem.* **18**:1847–1852.

192. Okamoto, K., and Namima, N., 1978, *J. Neurochem.* **31**:1393–1402.

193. Kontro, P., 1981, *Amino Acid Neurotransmitters* (F. V. DeFeudis and P. Mandel, eds.), Raven Press, pp. 161–167.

194. Starr, M. S., and Voaden, M. J., 1972, *Vision Res.* **12**:1261–1269.

195. Kennedy, A. J., and Voaden, M. J., 1976,, *J. Neurochem.* **27**:131–137.

196. Starr, M. S., 1978, *Brain Res.* **151**:604–608.

197. Lähdesmäki, P., and Oja, S. S., 1973, *J. Neurochem.* **20**:1411–1417.

198. Schmid, R., Sieghart, W., and Karobath, M., 1975, *J. Neurochem.* **25**:5–9.

199. Schousboe, A., Fosmark, H., and Svenneby, G., 1976, *Brain Res.* **116**:158–164.

200. Sieghart, W., and Karobath, M., 1976, *J. Neurochem.* **26**:981–986.

201. Edwards, R. B., 1977, *Invest. Ophthalmol. Vis. Sci.* **16**:201–208.

202. Hruska, R. E., Padjen, A., Bressler, R., and Yamamura, H. I., 1978, *Mol. Pharmacol.* **14**:77–85.

203. Kontro, P., and Oja, S. S., 1978, *J. Neurochem.* **30**:1297–1304.

204. Schmidt, S. Y., 1980, *Exp. Eye Res.* **31**:373–379.

205. Hösli, E., and Hösli, L., 1980, *Neuroscience* **5**:145–152.

206. Pasantes-Morales, H., Klethi, J., Urban, P. F., and Mandel, P., 1972, *Physiol. Chem. Phys.* **4**:339–348.

207. Ehinger, B., 1973, *Brain Res.* **60**:512–516.

208. Henn, F. A., 1975, *Metabolic Compartmentation and Neurotransmission. Relation to Brain Structure and Function* (S. Berl, D. D. Clarke, and D. Schneider, eds.), Plenum Press, New York, pp. 91–97.

209. Borg, J., Balcar, V. J., and Mandel, P., 1976, *Brain Res.* **118**:514–516.
210. Lombardini, J. B., 1977, *J. Neurochem.* **29**:305–312.
211. Borg, J., Balcar, V. J., and Mandel, P., 1979, *Brain Res.* **166**:113–120.
212. Martin, D. L., and Shain, W., 1979, *J. Biol. Chem.* **254**:7076–7084.
213. Lähdesmäki, P., Pasula, M., and Oja, S. S., 1975, *J. Neurochem.* **25**:675–680.
214. Oja, S. S., Kontro, P., and Lähdesmäki, P., 1976, *Adv. Exp. Med. Biol.* **69**:237–251.
215. Oja, S. S., and Kontro, P., 1978, *Taurine and Neurological Disorders* (A. Barbeau and R. J. Huxtable, eds.), Raven Press, New York, pp. 181–200.
216. Lajtha, A., and Sershen, H., 1975, *J. Neurochem.* **24**:667–672.
217. Kontro, P., and Oja, S. S., 1978, *Neuroscience* **3**:761–765.
218. Borg, J., Balcar, V. J., Mark, J., and Mandel, P., 1979, *J. Neurochem.* **32**:1801–1806.
219. Salceda, R., 1980, *Neurochem. Res.* **5**:561–572.
220. Starr, M. S., 1973, *Biochem. Pharmacol.* **22**:1693–1700.
221. Benjamin, A. M., and Quastel, J. H., 1977, *Can. J. Physiol. Pharmacol.* **55**:356–362.
222. Sieghart, W., and Karobath, M., 1974, *J. Neurochem.* **23**:911–915.
223. Meiners, B. A., Speth, R. C., Bresolin, N., Huxtable, R. J., and Yamamura, H. I., 1980, *Fed. Proc.* **39**:2695–2700.
224. Singer, E., and Placheta, P., 1979, *Wien. Klin. Wochenschr.* **91**:359.
225. Riddall, D. R., Leach, M. J., and Davison, A. N., 1976, *J. Neurochem.* **27**:835–839.
226. Schousboe, A., 1978, *Dynamic Properties of Glia Cells* (E. Schoffeniels, G. Franck, L. Hertz, and D. B. Tower, eds.), Pergamon Press, Oxford, pp. 173–182.
227. Pasantes-Morales, H., Bonaventure, N., Wioland, N., and Mandel, P., 1973, *Int. J. Neurosci.* **5**:235–241.
228. Lake, N., Marshall, J., and Voaden, M. J., 1978, *Exp. Eye Res.* **27**:713–718.
229. Lake, N., Marshall, J., and Voaden, M. J., 1977, *Brain Res.* **128**:497–503.
230. Miller, S., and Steinberg, R. H., 1976, *Exp. Eye Res.* **23**:177–189.
231. Bleecker, M., and Gfeller, E., 1971, *Trans. Am. Soc. Neurochem.* **2**:57.
232. Piccoli, F., Grynbaum, A., and Lajtha, A., 1971, *J. Neurochem.* **18**:1135–1148.
233. Borg, J., Ramaharobandro, N., Mark, J., and Mandel, P., 1980, *J. Neurochem.* **34**:1113–1122.
234. Assumpção, J. A., Bernardi, N., Dacke, C. G., and Davidson, N., 1977, *J. Physiol. (Lond.)* **270**:48P–49P.
235. Kaczmarek, L. K., and Adey, W. R., 1974, *Brain Res.* **76**:83–94.
236. Clark, R. M., and Collins, G. G. S., 1976, *J. Physiol. (Lond.)* **262**:383–400.
237. Bernardi, N., Assumpção, J. A., Dacke, C. G., and Davidson, N., 1977, *Pfluegers Arch.* **372**:203–205.
238. Bernardi, N., Assumpção, J. A., Davidson, N., and Dacke, C. G., 1977, *Experientia* **33**:914–915.
239. Davidson, N., 1979, *J. Physiol. (Paris)* **75**:673–676.
240. Collins, G. G. S., 1977, *J. Neurochem.* **28**:461–463.
241. Oja, S. S., Korpi, E. R., and Kontro, P., 1981, *Amino Acid Neurotransmitters* (F. V. DeFeudis, and P. Mandel, eds.), Raven Press, New York, pp. 175–181.
242. Kontro, P., 1979, *Neuroscience* **4**:1745–1749.
243. Korpi, E. R., Kontro, P., Marnela, K.-M., Nieminen, K., and Oja, S. S., 1981, *Life Sci.* **29**:811–816.
244. Sieghart, W., and Heckl, K., 1976, *Brain Res.* **116**:538–543.
245. Collins, G. G. S., 1980, *Brain Res.* **190**:517–528.
246. Lähdesmäki, P., and Oja, S. S., 1972, *Exp. Brain Res.* **15**:430–438.
247. Collins, G. G. S., and Topiwala, S. H., 1974, *Br. J. Pharmacol.* **50**:451P–452P.
248. López-Colomé, A. M., Tapia, R., Salceda, R., and Pasantes-Morales, H., 1978, *Neuroscience* **3**:1069–1074.
249. Collins, G. G. S., Anson, J., and Probett, G. A., 1981, *Brain Res.* **204**:103–120.
250. Kaczmarek, L. K., and Davison, A. N., 1972, *J. Neurochem.* **19**:2355–2362.
251. Vargas, O., Del Carmen Doria De Lorenzo, M., and Orrego, F., 1977, *Neuroscience* **2**:383–390.
252. Collins, G. G. S., 1979, *Br. J. Pharmacol.* **66**:109P–110P.

253. Placheta, P., Singer, E., Sieghart, W., and Karobath, M., 1979, *Neurochem. Res.* **4**:703–712.
254. Wheler, G. H. T., Bradford, H. F., Davison, A. N., and Thompson, E. J., 1979, *J. Neurochem.* **33**:331–338.
255. Orrego, F., Miranda, R., and Saldate, C., 1976, *Neuroscience* **1**:325–332.
256. Pasantes-Morales, H., Klethi, J., Urban, P. F., and Mandel, P., 1974, *Exp. Brain Res.* **19**:131–141.
257. López-Colomé, A., Erlij, D., and Pasantes-Morales, H., 1976, *Brain Res.* **113**:527–534.
258. López-Colomé, A. M., Salceda, R., and Pasantes-Morales, H., 1978, *Neurochem. Res.* **3**:431–441.
259. Pycock, C. J., and Smith, L. F. P., 1980, *Br. J. Pharmacol.* **70**:55P.
260. Pasantes-Morales, H., Salceda, R., and Gómez-Puyou, A., 1974, *Biochem. Biophys. Res. Commun.* **58**:847–853.
261. Salceda, R., and Pasantes-Morales, H., 1975, *Brain Res.* **96**:206–211.
262. Pasantes-Morales, H., Ademe, R. M., and López-Golomé, A. M., 1979, *Brain Res.* **172**:131–138.
263. Pasantes-Morales, H., and Quesada, O., 1980, *Neurochem. Res.* **5**:1077–1088.
264. Pasantes-Morales, H., Salceda, R., and López-Colomé, A. M., 1980, *J. Neurochem.* **34**:172–177.
265. Pasantes-Morales, H., Urban, P. F., Klethi, J., and Mandel, P., 1973, *Brain Res.* **51**:375–378.
266. Pasantes-Morales, H., Salceda, R., and López-Colomé, A. M., 1976, *Taurine* (R. Huxtable and A. Barbeau, eds.), Raven Press, New York, pp. 191–200.
267. Salceda, R., López-Colomé, A. M., and Pasantes-Morales, H., 1977, *Brain Res.* **135**:186–191.
268. Pasantes-Morales, H., Quesada, O., and Cáradez, A., 1981, *J. Neurochem.* **36**:1583–1586.
269. Ingoglia, N. A., Sturman, J. A., Lindquist, T. D., and Gaull, G. E., 1976, *Brain Res.* **115**:535–539.
270. Ingoglia, N. A., Sturman, J. A., Rassin, D. K., and Lindquist, T. D., 1978, *J. Neurochem.* **31**:161–170.
271. Politis, M. J., and Ingoglia, N. A., 1979, *Brain Res.* **166**:221–231.
272. Curtis, D. R., and Johnston, G. A. R., 1974, *Rev. Physiol.* **69**:97–188.
273. Aleksandrov, A. A., 1978, *Fiziol. Zh. S.S.S.R.* **64**:1057–1065.
274. Alexandrov, A. A., and Batuev, A. S., 1979, *J. Neurosci. Res.* **4**:59–64.
275. Kaczmarek, L. K., and Adey, W. R., 1975, *Electroencephalogr. Clin. Neurophysiol.* **39**:292–294.
276. Crawford, J. M., and Curtis, D. R., 1964, *Br. J. Pharmacol.* **23**:313–329.
277. Curtis, D. R., Game, C. J. A., and Lodge, D., 1976, *Exp. Brain Res.* **25**:413–428.
278. Curtis, D. R., Duggan, A. W., Felix, D., Johnston, G. A. R., and McLennan, H., 1971, *Brain Res.* **33**:57–73.
279. Curtis, D. R., and Tebēcis, A. K., 1972, *Exp. Brain Res.* **16**:210–218.
280. Okamoto, K., Quastel, D. M. J., and Quastel, J. H., 1976, *Brain Res.* **113**:147–158.
281. Okamoto, K., and Sakai, Y., 1979, *Br. J. Pharmacol.* **65**:277–285.
282. Okamoto, K., and Sakai, Y., 1980, *Br. J. Pharmacol.* **69**:407–413.
283. Okamoto, K., and Sakai, Y., 1981, *Brain Res.* **206**:371–386.
284. Haas, H. L., and Hösli, L., 1973, *Brain Res.* **52**:399–402.
285. Haas, H. L., and Hösli, L., 1973, *Experientia* **29**:542–544.
286. Hösli, L., Hösli, E., Andrés, P. F., and Wolff, J. R., 1975, *Golgi Centennial Symposium: Perspectives in Neurobiology* (M. Santini, ed.), Raven Press, New York, pp. 473–488.
287. Hösli, L., Haas, H. L., and Hösli, E., 1973, *Experientia* **29**:743–744.
288. Curtis, D. R., Duggan, A. W., Felix, D., and Johnston, G. A. R., 1971, *Brain Res.* **32**:69–96.
289. Sonnhof, U., Grafe, P., Krumnikl, J., Linder, M., and Schindler, L., 1975, *Brain Res.* **100**:327–341.
290. Krnjević, K., and Puil, E., 1976, *Taurine* (R. Huxtable and A. Barbeau, eds.), Raven Press, New York, pp. 179–189.
291. Nicoll, R. A., Padjen, A., and Barker, J. L., 1976, *Neuropharmacology* **15**:45–53.

292. Evans, R. H., and Watkins, J. C., 1975, *Br. J. Pharmacol.* **55**:519–526.
293. Barker, J. L., Nicoll, R. A., and Padjen, A., 1975, *J. Physiol. (Lond.)* **245**:521–536.
294. Barker, J. L., Nicoll, R. A., and Padjen, A., 1975, *J. Physiol. (Lond.)* **245**:537–548.
295. Davidoff, R. A., 1972, *Exp. Neurol.* **35**:179–193.
296. Nistri, A., and Constanti, A., 1976, *Neuropharmacology* **15**:635–641.
297. Koidl, B., and Florey, E., 1975, *Comp. Biochem. Physiol. [C]* **51**:13–23.
298. Homma, S., 1979, *Brain Res.* **173**:287–293.
299. Hue, B., Pelhate, M., and Chanelet, J., 1979, *Can. J. Neurol. Sci.* **6**:243–250.
300. Gruener, R., and Bryant, H. J., 1975, *J. Pharmacol. Exp. Ther.* **194**:514–521.
301. Gruener, R., Bryant, H., Markovitz, D., Huxtable, R., and Bressler, R., 1976, *Taurine* (R. Huxtable and A. Barbeau, eds.), Raven Press, New York, pp. 225–242.
302. Cunningham, R., and Miller, R. F., 1976, *Brain Res.* **117**:341–345.
303. Urban, P. F., Dreyfus, H., and Mandel, P., 1976, *Life Sci.* **18**:473–480.
304. Bonaventure, N., Wioland, N., and Mandel, P., 1974, *Brain Res.* **80**:281–289.
305. Bonaventure, N., Wioland, N., and Roussell, G., 1980, *Pfluegers Arch.* **385**:51–64.
306. Cunningham, R. A., and Miller, R. F., 1980, *Brain Res.* **197**:123–138.
307. Cunningham, R. A., and Miller, R. F., 1980, *Brain Res.* **197**:139–151.
308. Fuzessery, Z. M., Carr, W. E. S., and Ache, B. W., 1978, *Biol. Bull.* **154**:226–240.
309. Gadea-Ciria, M., García-Gracia, M., Camacho, J. G., Somoza, G., Balfagón, G., and De-Feudis, F. V., 1975, *J. Neurosci. Res.* **1**:393–397.
310. Kumpulainen, E., Jokisalo, V. J., and Lähdesmäki, P., 1978, *Int. J. Neurosci.* **8**:123–128.
311. López-Colomé, A. M., and Pasantes-Morales, H., 1980, *J. Neurochem.* **34**:1047–1052.
312. Kontro, P., 1981, *Abstracts of the Eighth Meeting of the International Society for Neurochemistry*, Nottingham, p. 35.
313. Pasantes-Morales, H., and Gamboa, A., 1980, *Natural Sulfur Compounds: Novel Biochemical and Structural Aspects* (D. Cavallini, G. E. Gaull, and V. Zappia, eds.), Plenum Press, New York, pp. 307–318.
314. Lähdesmäki, P., Kumpulainen, E., Raasakka, O., and Kyrki, P., 1977, *J. Neurochem.* **29**:819–826.
315. Kumpulainen, E., Olkinuora, M., and Lähdesmäki, P., 1979, *Neurosci. Lett.* **11**:215–218.
316. Kumpulainen, E., 1980, *Acta Univ. Oul. A 97 [Biochem]* **29**:129–133.
317. Tan, A. T., 1975, *J. Neurochem.* **24**:127–134.
318. Nakagawa, K., and Kuriyama, K., 1979, *Jpn. J. Pharmacol.* **29**:309–312.
319. Dolara, P., Agresti, A., Giotti, A., and Pasquini, G., 1973, *Eur. J. Pharmacol.* **24**:352–358.
320. Kuriyama, K., Muramatsu, M., Nakagawa, K., and Kakita, K., 1978, *Taurine and Neurological Disorders* (A. Barbeau and R. J. Huxtable, eds.), Raven Press, New York, pp. 201–216.
321. Remtulla, M. A., Katz, S., and Applegarth, D. A., 1979, *Life Sci.* **24**:1885–1892.
322. Pasantes-Morales, H., and Gamboa, A., 1980, *J. Neurochem.* **34**:244–246.
323. Izumi, K., Ngo, T. T., and Barbeau, A., 1978, *Taurine and Neurological Disorders* (A. Barbeau and R. J. Huxtable, eds.), Raven Press, New York, pp. 137–149.
324. Kuo, C.-H., and Miki, N., 1980, *Biochem. Biophys. Res. Commun.* **94**:646–651.
325. Dolara, P., Agresti, A., Giotti, A., and Sorace, E., 1976, *Can. J. Physiol. Pharmacol.* **54**:529–533.
326. Chovan, J. P., Kulakowski, E. C., Benson, B. W., and Schaffer, S. W., 1979, *Biochim. Biophys. Acta* **551**:129–136.
327. Chovan, J. P., Kulakowski, E. C., Sheakowski, S., and Schaffer, S. W., 1980, *Mol. Pharmacol.* **17**:295–300.
328. Read, W. O., Jaqua, M. J., and Steffen, R. P., 1980, *Proc. Soc. Exp. Biol. Med.* **164**:576–582.
329. Remtulla, M. A., Katz, S., and Applegarth, D. A., 1978, *Life Sci.* **23**:383–390.
330. Dolara, P., Ledda, F., Mugelli, A., Mantelli, L., Zilletti, L., Franconi, F., and Giotti, A., 1978, *Taurine and Neurological Disorders* (A. Barbeau and R. J. Huxtable, eds.), Raven Press, New York, pp. 151–159.
331. Igisu, H., Izumi, K., Goto, I., and Kina, K., 1976, *Pharmacology* **14**:362–366.
332. Irving, C. S., Hammer, B. E., Danyluk, S. S., and Klein, P. D., 1980, *J. Inorg. Biochem.* **13**:137–150.

333. Wheler, G. H. T., Weller, J. L., and Klein, D. C., 1979, *Brain Res.* **166**:65–74.
334. Muramatsu, M., Kakita, K., Nakagawa, K., and Kuriyama, K., 1978, *Jpn. J. Pharmacol.* **28**:259–268.
335. Garcia de Yebenes Prous, J., Carlsson, A., and Mena Gomez, M. A., 1978, *Naunyn Schmiedebergs Arch. Pharmacol.* **304**:95–99.
336. Sandberg, P. R., Staines, W., and McGeer, E. G., 1979, *Brain Res.* **161**:367–370.
337. Hilton, J., 1977, *J. Pharmacol. Exp. Ther.* **203**:426–434.
338. Leach, M. J., 1979, *J. Pharm. Pharmacol.* **31**:533–535.
339. Williams, M., Risley, E. A., and Totaro, J. A., 1980, *Life Sci.* **26**:557–560.
340. Sgaragli, G., and Pavàn, F., 1972, *Neuropharmacology* **11**:45–56.
341. Sgaragli, G. P., Magnani, M., Carlà, V., and Giotti, A., 1976, *Naunyn Schmiedebergs Arch. Pharmacol.* **295**:95–97.
342. Clark, S. M., and Lipton, J. M., 1981, *Exp. Aging Res.* **7**:17–24.
343. Bligh, J., Silver, A., Smith, C. A., and Bacon, M. J., 1979, *J. Therm. Biol.* **4**:9–14.
344. Harris, W. S., and Lipton, J. M., 1977, *J. Physiol. (Lond.)* **266**:397–410.
345. Sgaragli, G. P., Pavàn, F., and Galli, A., 1975, *Naunyn Schmiedebergs Arch. Pharmacol.* **288**:179–184.
346. Hruska, R. E., Thut, P. D., Huxtable, R., and Bressler, R., 1976, *Taurine* (R. Huxtable and A. Barbeau, eds.), Raven Press, New York, pp. 347–356.
347. Kerwin, R. W., and Pycock, C. J., 1979, *J. Pharm. Pharmacol.* **31**:466–470.
348. Lipton, J. M., and Ticknor, C. B., 1979, *J. Physiol. (Lond.)* **287**:535–543.
349. van Gelder, N. M., 1972, *Brain Res.* **47**:157–165.
350. Koyama, I., 1972, *Can. J. Physiol. Pharmacol.* **50**:740–752.
351. Craig, C., and Hartman, E. R., 1973, *Epilepsia* **14**:409–414.
352. Joseph, M. H., and Emson, P. C., 1976, *J. Neurochem.* **27**:1495–1501.
353. Carruthers-Jones, D. I., and van Gelder, N. M., 1978, *Neurochem. Res.* **3**:115–123.
354. Emson, P. C., 1978, *Taurine and Neurological Disorders* (A. Barbeau and R. J. Huxtable, eds.), Raven Press, New York, pp. 319–338.
355. Hansen, S., Perry, T. L., Wada, J. A., and Sokol, M., 1973, *Brain Res.* **50**:480–483.
356. Frigyesi, T. L., and Lombardini, J. B., 1979, *Life Sci.* **24**:1251–1260.
357. Iwata, H., Yamagami, S., Lee, E., Matsuda, T., and Baba, A., 1979, *Jpn. J. Pharmacol.* **29**:503–507.
358. Battistin, L., Varotto, M., Tezzon, F., and Pistollato, L., 1979, *Neurochem. Res.* **4**:457–464.
359. van Gelder, N. M., Edmonds, H. L., Jr., Hegreberg, G. A., Chatburn, C. C., Clemmons, R. M., and Sylvester, D. M., 1980, *J. Neurochem.* **35**:1087–1091.
360. Mutani, R., Durelli, L., Mazzarino, M., Valentini, C., Monaco, F., Fumero, S., and Mondino, A., 1977, *Brain Res.* **122**:513–521.
361. van Gelder, N. M., Sherwin, A. L., and Rasmussen, T., 1972, *Brain Res.* **40**:385–393.
362. Perry, T. L., Hansen, S., Kennedy, J., Wada, J. A., and Thompson, G. B., 1975, *Arch. Neurol.* **32**:752–754.
363. van Gelder, N. M., 1976, *Taurine* (A. Barbeau and R. Huxtable, eds.), Raven Press, New York, pp. 293–302.
364. Durelli, L., Quattrocolo, G., Buffa, C., Valentini, C., and Mutani, R., 1975, *Acta Neurol. (Napoli)* **30**:540–545.
365. Durelli, L., Quattrocolo, G., Buffa, C., Valentini, C., and Mutani, R., 1976, *Riv. Neurol.* **46**:254–261.
366. Bergamini, L., and Mutani, R., 1973, *Riv. Sper. Freniat.* **97**:738–747.
367. Derouaux, M., Puil, E., and Naquet, R., 1973, *Electroencephalogr. Clin. Neurophysiol.* **34**:770.
368. Adembri, G., Bartolini, A., Bartolini, R., Giotti, A., and Zilletti, L., 1974, *Br. J. Pharmacol.* **52**:439P–440P.
369. Mutani, R., Bergamini, L., Fariello, R., and Delsedime, M., 1974, *Brain Res.* **70**:170–173.
370. Mutani, R., Bergamini, L., Delsedime, M., and Durelli, L., 1974, *Brain Res.* **70**:330–332.
371. Izumi, K., Donaldson, J., Minnich, J. L., and Barbeau, A., 1973, *Can. J. Physiol. Pharmacol.* **51**:885–889.
372. Izumi, K., Igisu, H., and Fukuda, T., 1974, *Brain Res.* **76**:171–173.

373. Izumi, K., Igisu, H., and Fukuda, T., 1975, *Brain Res.* **88**:576–579.
374. Wada, J. A., Osawa, T., Wake, A., and Corcoran, M. E., 1975, *Epilepsia* **16**:229–234.
375. Barbeau, A., Tsukada, Y., and Inoue, N., 1976, *Taurine* (R. Huxtable and A. Barbeau, eds.), Raven Press, New York, pp. 253–266.
376. Gaito, J., 1976, *Bull. Psychon. Soc.* **7**:397–400.
377. Laird, H. E., and Huxtable, R., 1976, *Taurine* (R. Huxtable and Barbeau, A., eds.), Raven Press, New York, pp. 267–274.
378. Savoldi, F., Tartara, A., and Bo, P., 1976, *Farmaco* [*Prat.*] **31**:27–34.
379. van Gelder, N. M., Koyama, I., and Jasper, H. H., 1977, *Epilepsia* **18**:45–54.
380. Huxtable, R., and Laird, H., 1978, *Can. J. Neurol. Sci.* **5**:215–221.
381. Roches, J.-C., Zumstein, H. R., Fässler, A., Scollo-Lavizzari, G., and Hösli, L., 1979, *Eur. Neurol.* **18**:26–32.
382. Lapin, I. P., 1980, *J. Neural Transm.* **48**:311–316.
383. Sanberg, P. R., and Willow, M., 1980, *Neurosci. Lett.* **16**:297–300.
384. Burnham, W. M., Albright, P., and Racine, R. J., 1978, *Can. J. Physiol. Pharmacol.* **56**:497–500.
385. Ramabadran, K., Bansinath, M., Iyer, H. S., Karanth, S., and Guruswami, M. N., 1980, *Indian J. Exp. Biol.* **18**:1035–1037.
386. Barbeau, A., and Donaldson, J., 1973, *Lancet* **2**:387.
387. Barbeau, A., and Donaldson, J., 1974, *Arch. Neurol.* **30**:52–58.
388. Bergamini, L., Mutani, R., Delsedime, M., and Durelli, L., 1974, *Eur. Neurol.* **11**:261–269.
389. Pennetta, R., Masi, G., Perniola, T., and Ferrannini, E., 1977, *Acta Neurol.* (*Napoli*) **32**:316–322.
390. Mongiovi, A., 1978, *Riv. Neurol.* **48**:305–325.
391. Airaksinen, E. M., Oja, S. S., Marnela, K.-M., Leino, E., and Pääkkönen, L., 1980, *Progress in Clinical and Biological Research*, Volume 39 (L. Battistin, G. Hashim, and A. Lajtha, eds.), Alan R. Liss, New York, 157–166.
392. Mantovani, J., and DeVivo, D. C., 1979, *Arch. Neurol.* **36**:672–674.
393. König, P., Kriechbaum, G., Presslich, O., Schubert, H., Schuster, P., and Sieghart, W., 1977, *Wien. Klin. Wochenschr.* **89**:111–113.
394. Rumpl, E., Gerstenbrand, F., Hengl, W., and Binder, H., 1977, *EEG EMG* **8**:77–81.
395. Goodman, H. O., Connolly, B. M., McLean, W., and Resnick, M., 1980, *Clin. Chem.* **26**:414–419.
396. van Gelder, N. M., and Courtois, A., 1972, *Brain Res.* **43**:477–484.
397. Mutani, R., Monaco, F., Durelli, L., and Delsedime, M., 1975, *Epilepsia* **16**:765–769.
398. van Gelder, N. M., Sherwin, A. L., Sacks, C., and Andermann, F., 1975, *Brain Res.* **94**:297–306.
399. van Gelder, N. M., 1978, *Can. J. Physiol. Pharmacol.* **56**:362–374.
400. Persinger, M. A., Valliant, P. M., and Falter, H., 1976, *Dev. Psychobiol.* **9**:131–136.
401. Baskin, S. I., Hinkamp, D. L., Marquis, W. J., and Tilson, H. A., 1974, *Neuropharmacology* **13**;591–594.
402. Sanberg, P. R., and Fibiger, H. C., 1979, *Psychopharmacology* **62**:97–99.
403. Valliant, P. M., Persinger, M. A., and Satinder, K. P., 1978, *Dev. Psychobiol.* **12**:515–518.
404. Sanberg, P. R., and Ossenkopp, K.-P., 1977, *Psychopharmacology* **53**:207–209.
405. Persinger, M. A., Lafreniére, G. F., and Falter, H., 1976, *Psychopharmacology* **49**:249–252.
406. Mack, G., and Mandel, P., 1976, *C. R. Acad. Sci.* [*D*] (*Paris*) **283**:361–362.
407. Kaakkola, S., and Kääriäinen, I., 1980, *Acta Pharmacol. Toxicol.* (*Kbh.*) **46**:293–298.
408. Wu, J.-Y., Bird, E. D., Chen, M. S., and Huang, W. M., 1979, *Neurochem. Res.* **4**:575–586.
409. Airaksinen, E. M., Sihvola, P., Airaksinen, M. M., and Tuovinen, E., 1979, *Lancet* **1**:474–475.
410. Airaksinen, E. M., Airaksinen, M. M., Sihvola, P., and Marnela, K.-M., 1981, *Metab. Pediatr. Ophthalmol.* **5**:45–48.
411. Reccia, R., Pignalosa, B., Grasso, A., and Campanella, G., 1980, *Acta Neurol.* (*Napoli*) **35**:132–136.
412. Thut, P. D., Hruska, R. E., Huxtable, R., and Bressler, R., 1976, *Taurine* (R. Huxtable and A. Barbeau, eds.), Raven Press, New York, pp. 357–364.

413. Scheibel, J., Elsasser, T., and Ondo, J. G., 1980, *Brain Res.* **201**:99–106.
414. Scheibel, J., Elsasser, T., and Ondo, J. G., 1980, *Neuroendocrinology* **30**:350–354.
415. Nakagawa, K., and Kuriyama, K., 1975, *Jpn. J. Pharmacol.* **25**:737–746.
416. Collu, R., Charpenet, G., and Clermont, M. J., 1978, *Can. J. Neurol. Sci.* **5**:139–142.
417. Izumi, K., Munekata, E., Yamamoto, H., Nakanishi, T., and Barbeau, A., 1980, *Peptides* **1**:139–146.
418. Iida, S., and Hisada, Y., 1974, *Jpn. J. Pharmacol.* **24**(Suppl.):91.
419. Iida, S., and Hikichi, M., 1976, *J. Stud. Alcohol* **37**:19–26.
420. Ikeda, H., 1977, *Lancet* **2**:509.
421. Grosso, D. S., and Bressler, R., 1976, *Biochem. Pharmacol.* **25**:2227–2232.
422. Hope, D. B., 1957, *J. Neurochem.* **1**:364–369.
423. Loriette, C., Pasantes-Morales, H., Portemer, C., and Chatagner, F., 1979, *Nutr. Metab.* **23**:467–475.
424. Awapara, J., 1956, *J. Biol. Chem.* **218**:571–576.
425. Lombardini, J. B., and Medina, E. V., 1978, *J. Nutr.* **108**:428–433.
426. Sturman, J. A., Rassin, D. K., and Gaull, G. E., 1977, *Pediatr. Res.* **11**:28–33.
427. Toth, E., and Lajtha, A., 1981, *Trans. Am. Soc. Neurochem.* **158**:177.
428. Berson, E. L., Schmidt, S. Y., and Rabin, A. R., 1976, *Br. J. Ophthalmol.* **60**:142–147.
429. Anderson, P. A., Baker, D. H., Corbin, J. E., and Helper, L. C., 1979, *J. Anim. Sci.* **49**:1227–1234.
430. Barnett, K. C., and Burger, I. H., 1980, *J. Small Anim. Pract.* **21**:521–534.
431. Rassin, D. K., Sturman, J. A., Hayes, K. C., and Gaull, G. E., 1978, *Neurochem. Res.* **3**:401–410.
432. Schmidt, S. Y., Berson, E. L., Watson, G., and Huang, C., 1977, *Invest. Ophthalmol. Vis. Sci.* **16**:673–678.
433. Wen, G. Y., Sturman, J. A., Wisniewski, H. M., Lidsky, A. A., Cornwell, A. C., and Hayes, K. C., 1979, *Invest. Ophthalmol. Vis. Sci.* **18**:1201–1206.
434. Schmidt, S. Y., Berson, E. L., and Hayes, K. C., 1976, *Trans. Am. Acad. Ophthalmol. Otolaryngol.* **81**:687–693.
435. Rabin, B., Nicolosi, R. J., and Hayes, K. C., 1976, *J. Nutr.* **106**:1241–1246.
436. Hayes, K. C., Stephan, Z. F., and Sturman, J. A., 1980, *J. Nutr.* **110**:2058–2064.
437. Gaull, G. E., Rassin, D. K., Räihä, N. C. R., and Heinonen, K., 1977, *J. Pediatr.* **90**:348–355.
438. Rigo, J., and Senterre, J., 1977, *Biol. Neonate* **32**:73–76.
439. Malloy, M. H., Gaull, G. E., Heird, W. C., and Rassin, D. K., 1980, *Natural Sulfur Compound: Novel Biochemical and Structural Aspects* (D. Cavallini, G. E. Gaull, and V. Zappia, eds.), Plenum Press, New York, pp. 245–252.
440. Rassin, D. K., Sturman, J. A., and Gaull, G. E., 1978, *Early Hum. Dev.* **2**:1–13.
441. Schon, F., and Kelly, J. S., 1975, *Brain Res.* **86**:243–257.
442. Johnston, G. A. R., 1977, *Brain Res.* **121**:179–181.
443. Kontro, P., 1980, *Acta Univ. Oul. A 88 [Biochem.]* **27**:1–52.
444. Oja, S. S., and Kontro, P., 1980, *J. Neurochem.* **35**:1303–1308.
445. Kontro, P., and Oja, S. S., 1981, *The Action of Taurine on Excitable Tissues* (S. W. Schaffer, S. I. Baskin, and J. J. Kocsis, eds.), NTP Press, Lancaster, pp. 49–57.
446. Kontro, P., and Oja, S. S., 1981, *J. Neurochem.* **37**:297–304.
447. Oja, S. S., and Kontro, P., 1981, *Taurine in Nutrition and Neurology* (R. J. Huxtable and H. Pasantes-Morales, eds.), Plenum Press, New York, pp. 115–126.
448. Kontro, P., 1980, *Acta Univ. Oul. A 97 [Biochem.]* **29**:111–116.
449. Kontro, P., and Oja, S. S., 1981, *Neurochem. Res.* **6**:1173–1185.
450. Sgaragli, G. P., Carlà, V., Magnani, M., and Giotti, A., 1978, *Naunyn Schmiedebergs Arch. Pharmacol.* **305**:155–158.
451. Nistri, A., and Constanti, A., 1979, *Prog. Neurobiol.* **13**:117–235.
452. Thoai, N. V., Zappacosta, S., and Robin, Y., 1963, *Comp. Biochem. Physiol.* **10**:209–225.
453. Blass, J. P., 1960, *Biochem. J.* **77**:484–489.
454. Matsumoto, M., Kobayashi, K., and Mori, A., 1979, *J. Neurochem.* **32**:645–646.

455. Mizuno, A., Mukawa, J., Kobayashi, K., and Mori, A., 1975, *IRCS Med. Sci. Neurobiol. Neurophysiol. Pharmacol.* **3**:385.
456. Koechlin, B. A., 1955, *J. Biophys. Biochem. Cytol.* **1**:511–529.
457. Lombardini, J. B., and Homan, J. A., 1979, *Neurochem. Res.* **4**:449–455.
458. Curtis, D. R., and Watkins, J. C., 1960, *J. Neurochem.* **6**:117–141.
459. Baba, A., Yamagami, S., Mizuo, H., and Iwata, H., 1980, *Anal. Biochem.* **101**:288–293.
460. Iwata, H., and Baba, A., 1981, *Taurine in Nutrition and Neurology* (R. J. Huxtable and H. Pasantes-Morales, eds.), Plenum Press, New York, pp. 211–219.
461. Reichelt, K. L., and Kvamme, E., 1973, *J. Neurochem.* **21**:849–859.
462. Reichelt, K. L., and Edminson, P. D., 1974, *FEBS Lett.* **47**:185–189.
463. Lähdesmäki, P., Airaksinen, K., Vartiainen, M., and Halonen, P., 1980, *Acta Chem. Scand.* [B] **34**:343–348.
464. Feuer, L., Török, L., and Csaba, G., 1979, *Endokrinologie* **73**:367–369.
465. Török, L. J., Feuer, L., and Csaba, G., 1979, *Gen. Comp. Endocrinol.* **38**:285–289.
466. Furka, A., Sebestylén, F., Feuer, L., Horváth, A., Hercsel, J., Ormai, S., and Bánayai, B., 1980, *Acta Biochim. Biophys. Acad. Sci. Hung.* **15**:39–47.
467. Lisý, V., Dutton, G. R., and Currie, D. N., 1980, *Life Sci.* **27**:2615–2620.

Methionine Metabolism in the Brain

H. H. Tallan, D. K. Rassin, J. A. Sturman, and G. E. Gaull

1. INTRODUCTION

Special interest in the role of sulfur amino acids in brain began with the observations that cystathionine was present in large concentrations in human brain[1] but was absent from the brain of patients with homocystinuria caused by cystathionine β-synthase (E.C. 4.2.1.22) deficiency.[2] The contrast between the amount of cystathionine in brain versus liver and our later observation of the low concentration of cystathionine in fetal human brain led to a more detailed analysis of the special adaptations of sulfur amino acid metabolism in brain. What follows is a review of the metabolism of methionine in the central nervous system as seen by a group of us who worked together on this problem for over a decade. The point of view is, broadly speaking, developmental and nutritional. Questions related to inborn errors of methionine metabolism will be discussed in a chapter in Volume 10. In the present chapter, we follow the pathway of sulfur metabolism only as far as cyst(e)ine. The synthesis and metabolism of taurine and of glutathione have been reviewed in Chapters 18 and 22 (this volume). Reference to earlier literature not cited here may be found in the comprehensive 1970 review by Gaitonde.[3]

The metabolism of methionine to cyst(e)ine and the enzymes involved in these conversions were first well understood in liver. From these studies, it was clear that the metabolism of the sulfur-containing amino acids was inti-

H. H. Tallan and G. E. Gaull • Department of Human Development and Nutrition, New York State Institute for Basic Research in Developmental Disabilities, Staten Island, New York 10314; and Division of Human Genetics, Department of Pediatrics, Mount Sinai School of Medicine of the City University of New York, New York, New York 10029. *D. K. Rassin* • Division of Developmental Nutrition and Metabolism, Department of Pediatrics, University of Texas Medical Branch, Galveston, Texas 77550. *J. A. Sturman* • Developmental Neurochemistry Laboratory, Department of Pathological Neurobiology, New York State Institute for Basic Research in Developmental Disabilities, Staten Island, New York 10314.

mately linked with that of the B_6 vitamers.[4] It is of particular interest, therefore, that pyridoxal kinase (E.C. 2.7.1.35), the enzyme that activates pyridoxal to pyridoxal phosphate, has greater activity in brain than in any other tissue (for review see ref. 4).

2. SULFUR AMINO ACIDS IN BRAIN

2.1. Methionine

Methionine in the circulation, derived from the diet or from other tissues, crosses the blood–brain barrier and enters the brain via the L uptake system for large neutral amino acids.[5,6] The concentration of methionine in whole brain is relatively low, 10 to 100 nmol/g wet weight in various species.[7] Differences among brain regions are not great.[8-10] Although the concentration in normal rat serum follows a diurnal rhythm with a peak at 4 a.m. and a dip at 4 p.m., the concentration in brain does not vary.[11,12] The effect of diet on brain methionine concentration is small because of the competition between methionine and the other large neutral amino acids for transport into the brain (see ref. 6); however, administration of methionine alone does increase the brain methionine concentration.[11,13] Brain methionine is affected by administration of L-dihydroxyphenylalanine (DOPA), but both increases[14] and decreases[6,15] as well as no effect[16] have been reported, perhaps depending on dosage and timing.

Methionine in the free amino acid pool of brain appears to be utilized mainly for protein synthesis. Thus, 3 h after intracisternal injection of [^{35}S]methionine into rats, 80% of the ^{35}S in the brain was in the protein fraction, three-quarters as [^{35}S]methionine and one-quarter as [^{35}S]cysteine.[17] This may be an overestimate (see Section 3.7).

Metabolism of free methionine to cysteine is initiated by conversion to S-adenosylmethionine (AdoMet) in a reaction catalyzed by methionine adenosyltransferase (E.C. 2.5.1.6) (MAT). Transamination of methionine to 4-methylthio-2-oxobutyrate also takes place in brain.[18]

2.2. S-Adenosylmethionine

S-Adenosylmethionine occurs in rat brain at a concentration of about 25 nmol/g,[19-23] in mouse brain at 25[24] or 75[25] nmol/g, and in rabbit brain at 22 nmol/g.[23] It is evenly distributed throughout the brain.[20,21,26] The concentration in rat brain follows the diurnal rhythm of serum methionine; there is a rise in brain AdoMet to 32 nmol/g at 4 a.m. and a decline to 22 nmol/g at 4 p.m.[11,27] Administration of methionine generally results in modest increases in brain AdoMet concentration.[11,19,20,28,29] The concentration of brain AdoMet is affected by a number of compounds. Methionine sulfoximine causes a decrease, particularly in striatum and cerebellum, by an undetermined mechanism.[20] Methyl acceptors, e.g., pyrogallol and purpurogallin,[28] DOPA,[15,16,25,27] and α-methyl-DOPA,[30] decrease brain AdoMet, presumably by increasing utilization. Cycloleucine[25] and L-2-amino-4-hexynoic acid[31] decrease AdoMet by inhibition of MAT. Other compounds reported to decrease brain AdoMet are pargyline

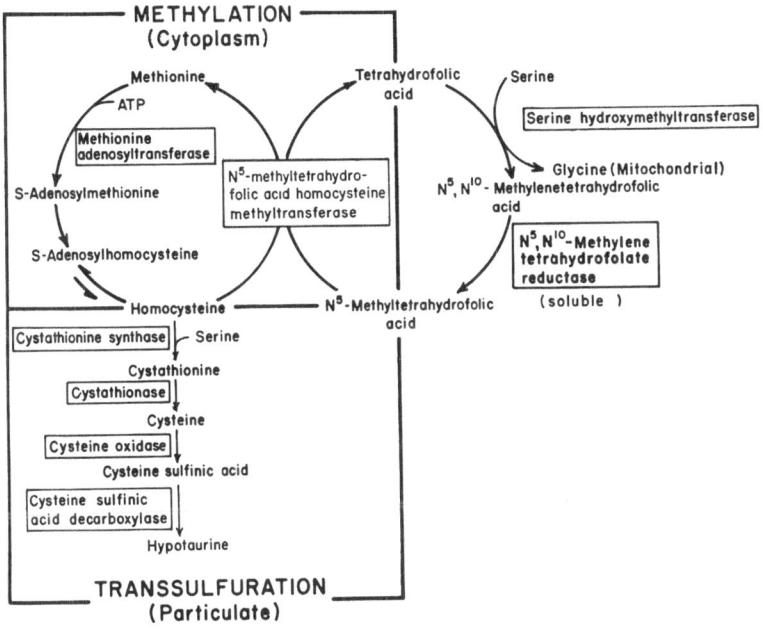

Fig. 1. Pathways of methionine metabolism.

and imipramine,[25,28] *d*-amphetamine, desmethylimipramine, and chlorimipramine.[25] There is a decrease in brain AdoMet in folic-acid-deficient rats[16]; this would appear to have a more complex cause than diminished folate-dependent remethylation of homocysteine to methionine (see below), for the brain methionine concentration is not altered significantly.[16]

S-Adenosylmethionine serves a number of important functions in the body (Fig. 1) (for review, see ref. 32). First, it is the active methyl donor for most transmethylation reactions, which in brain include the 3-O-methylation of catecholamines, the formation of epinephrine, methylation of histamine and of carnosine, methylation of RNA and DNA, methylation of histones, and formation of phosphatidylcholine.[22,33] Second, the coproduct of methylation, S-adenosylhomocysteine (AdoHcy), is hydrolyzed to homocysteine, which may then either be remethylated to methionine or may enter the transsulfuration pathway to provide cysteine. Third, AdoMet is decarboxylated to S-adenosyl-(5')-3-methylthiopropylamine, which provides the propylamine moiety required for synthesis of the polyamines, spermidine from putrescine and spermine from spermidine (see refs. 34,35). Fourth, AdoMet regulates the activity of several enzymes (see ref. 32). It has both stimulatory effects on MAT at low concentrations and inhibitory effects at high concentrations[36-38]; it inhibits spermine synthase,[39] spermidine synthase,[39] and methylenetetrahydrofolate reductase (E.C. 1.1.1.171)[40-42]; it activates methyltetrahydrofolate : homocysteine methyltransferase (E.C. 2.1.1.13)[43,44] and cystathionine β-synthase.[45] How these varied actions interact *in vivo* to help regulate sulfur amino acid metabolism remains to be worked out.

Decarboxylated AdoMet, which is utilized as the propylamine donor for the synthesis of the polyamines, is found in brain at about half the concentration of AdoMet: 15 nmol/g in the rat, 10 in the rabbit.[23] A smaller value, 0.9 nmol/g in rat brain, is also reported.[34] Decarboxy-AdoMet has also been shown to inhibit the AdoMet hydrolase (E.C. 3.3.1.2) of rat liver,[23] brain methylenetetrahydrofolate reductase,[42] and the methylation of histamine and acetylserotonin.[46]

2.3. S-Adenosylhomocysteine

S-Adenosylhomocysteine, the demethylated coproduct of the transmethylation reactions, has been found in rat brain at concentrations of 0.5 to 6.9 nmol/g.[19,22,24] The concentration in brain is increased 100-fold, without increase in AdoMet, by intraperitoneal administration of large amounts of adenosine and homocysteine thiolactone, evidently by action of AdoHcy hydrolase (E.C. 3.3.1.1) in the direction of synthesis.[24] Binding of AdoHcy to a membrane fraction from rat cerebral cortex has been described.[47]

S-Adenosylhomocysteine is a competitive inhibitor of the AdoMet transmethylation reactions, and it has been proposed that the extent of these methylations is regulated by the ratio of AdoHcy concentration to that of AdoMet (see ref. 48). Decreased activity of a number of transmethylases in brain following elevation of AdoHcy content has been reported.[24] In addition to inhibiting most AdoMet-dependent transmethylases, AdoHcy inhibits spermine synthase,[34] betaine:homocysteine methyltransferase (E.C. 2.1.1.5),[49] and methyltetrahydrofolate:homocysteine methyltransferase[50]; it activates cystathionine β-synthase.[49]

2.4. Homocyst(e)ine

Homocysteine is formed on hydrolysis of AdoHcy by AdoHcy hydrolase. The reaction equilibrium favors synthesis but readily proceeds in the direction of hydrolysis if the products are removed. Adenosine is rapidly deaminated to inosine; homocysteine is remethylated to methionine or condensed with serine to form cystathionine. These and possibly other mechanisms of removal are very efficient, for free homocysteine is not detected in normal tissues. Free homocysteine has been reported present[51,52] and absent[53,54] in liver from patients with homocystinuria; it has not been detected in brain.[2,54-56] Another possible route for removal of free homocysteine is by formation of mixed disulfides with cysteine or with proteins[53,57] Cysteine–homocysteine mixed disulfide is found in plasma and urine of patients with homocystinuria[58] and has been shown to occur in the plasma of fasting normal subjects at a concentration of about 0.3 μmol/dl[59,60] (see ref. 61). It was not found in liver.[53] Homocysteine in disulfide linkage with protein has been found in the plasma of both patients with homocystinuria[57,62] and normal subjects (0.14 μmol/dl for the latter) and in tissues, including brain (0.4 μmol/g protein), from a patient with homocystinuria.[57] The concentration of the protein disulfide in brain was five- to sevenfold greater than that in liver and kidney.[57] These findings raise the possibility that the presumed toxicity of homocysteine to the patient with homocystinuria is exerted through formation of mixed disulfides with proteins and consequent conformational changes in these proteins.

2.5. Cystathionine

Transsulfuration, i.e., transfer of the sulfur atom of methionine to the carbon chain of serine to produce cysteine, is initiated by the condensation of homocysteine and serine to form cystathionine (Fig. 1). This reaction, catalyzed by cystathionine β-synthase, is irreversible under physiological conditions, but *in vitro* it can proceed in the reverse direction of β-hydrolysis in the presence of 5,5'-dithiobis-(2-nitrobenzoic acid), which removes the homocysteine formed.[63] Transsulfuration is completed by the action of γ-cystathionase (E.C. 4.4.1.1), which splits cystathionine to cysteine, 2-oxobutyrate, and ammonia.

Cystathionine, the asymmetrical thioether formed in this process, has been the subject of much investigation since the realization that it is present in larger amounts in human and monkey brain than in other human tissues or in the brain of other species.[1] The concentration in human brain is of the order of 1–2 μmol/g but varies considerably in different brain regions. Occipital cortex, corpus callosum, and thalamus have high concentrations[2,10,64]; white matter has more cystathionine than the corresponding gray matter.[64-66] A similar regional distribution with a wide range of values is found in monkey brain.[67-69] The concentration in whole brain of other mammals is 0.01–0.23 μmol/g[7]; the region in rat brain with the greatest concentration is cerebellum, with 0.2–0.5 μmol/g,[8,70-73] whereas in cat brain it is caudate nucleus, with 0.5 μmol/g.[9] Autopsy specimens of human cortex and cerebellum have been found to contain less cystathionine than biopsy specimens, arguing for continued action of γ-cystathionase post-mortem,[74] but such differences have not been observed in other studies.[64,75]

Cystathionine is absent from the brain of patients with homocystinuria, who lack or have reduced cystathionine β-synthase activity.[2,55] The concentration in brain of pyridoxine-deficient rats is increased,[72,76,77] presumably as a result of the greater diminution of γ-cystathionase activity than of β-synthase activity because of the loss of the cofactor pyridoxal phosphate. Administration of DOPA to rats results in an increase in brain cystathionine, perhaps indicative of increased flow through the transsulfuration pathway to remove the homocysteine formed after methylation of the DOPA.[78] The concentration of cystathionine in the brain (and other organs) of mice is increased by administration of cysteine plus homoserine, either by "reverse" γ-cystathionase action or by inhibition of γ-cystathionase by the cysteine.[79,80] A dipeptide, γ-aminobutyryl-cystathionine, has been found in human and baboon brain but not in brain of other species; its distribution in human brain generally parallels that of cystathionine, with a concentration in occipital cortex reaching 90 nmol/g.[81]

In addition to its role in the transfer of sulfur, a neurotransmitter function for cystathionine has been proposed (see Section 5). Whether the high concentration of cystathionine in primate brain serves some special function remains an unanswered but intriguing question. What determines the concentration of cystathionine in brain also is uncertain. The relative amounts of cystathionine β-synthase and of γ-cystathionase in brain, brain regions, and other tissues of various species do not correlate with the content of cystathionine.[67,82–84] However, the activity of γ-cystathionase in brain generally is rather low, and in certain conditions in which γ-cystathionase activity is particularly decreased or low, as in vitamin B_6 deficiency in rats[76] or in fetal monkey liver,[68] cystathionine accumulates. During development of the monkey, the amount of cystathionine in brain does correlate with the amount of cystathionine β-synthase. In the rat, however, this correlation is found only for the first 16 days of postnatal life, after which the concentration of cystathionine decreases in the face of constant activity of cystathionine β-synthase (see below).

2.6. Cyst(e)ine

Hydrolysis of cystathionine is carried out by γ-cystathionase, yielding cysteine and, nominally, homoserine. The latter compound is at once broken down to 2-oxobutyrate and ammonia by the γ-cystathionase, which has intrinsic homoserine dehydratase activity. In addition, diet-derived cysteine and cysteine formed by transsulfuration in other tissues enter the brain from the general circulation. Studies with the intact rat indicate that uptake of cysteine is by the leucine-preferring (L) transport system[85]; studies with isolated rat brain capillaries, however, suggest uptake of cystine by the alanine-serine-cysteine-preferring (ASC) transport system.[86] Further work is needed to resolve this difference.

The concentration of free cyst(e)ine in the brain is very low, ranging from barely detectable to about 0.07 μmol/g in biopsy or fresh specimens from man, cat, and rat; there is little variation with region.[8,9,74,77] The content of human brain specimens taken at autopsy is 10- to 15-fold greater,[10,74] probably the result of postmortem autolysis.

Traces of the mixed disulfide of cysteine and glutathione are detectable in biopsied human brain specimens,[74] and considerable amounts (up to 1.25 μmol/g) are found in autopsy specimens.[10,74] In plasma, half of the cysteine is bound to protein by disulfide linkage,[87] and it is possible that this type of mixed disulfide occurs in tissues as well. Cysteine is a component of the tripeptide glutathione, formation of which may utilize a large part of the available cysteine. Concentrations of glutathionine in brain regions are very variable but can reach totals (reduced plus oxidized, expressed as reduced) of 1–3 μmol/g.[8–10,74] Cysteine is a component of proteins and reenters the free amino acid pool as protein turns over. Further metabolism of cysteine leads to taurine and to sulfate (see Chapter 18).

An indication of the distribution of cysteine into these pathways was achieved by intracisternal injection of L-[^{35}S]-cystathionine into rats.[70] At 1 h, with half the radioactivity remaining as cystathionine, 4% of the radioactivity was in protein, 7% in glutathione, and 8% about equally distributed in cysteic plus cysteine sulfinic acids, taurine, and glutathione–cysteine mixed disulfide. By 9 h, the brain retained 60% of the radioactivity that had been present at 1 h; 20% of that amount remained as cystathionine, 22% was in protein, 18% in glutathione, and 18% in the other metabolites. Little radioactivity was found in the cystine fraction at any time.

3. ENZYMES OF SULFUR AMINO ACID METABOLISM IN BRAIN

3.1. Methionine Adenosyltransferase

Methionine adenosyltransferase is found in extracts of rat brain at a specific activity of about 3 nmol product/h per mg protein[19,29,42,83,88-92] (other studies used different units[21,69,93]), in extracts of monkey brain at 4 nmol/h per mg,[83] and of human brain at 1.6.[83] The distribution of enzyme activity in parts of rat brain is fairly even.[91,93] The variation of enzyme activity among brain regions is greater in monkey brain than in rat brain,[33,67] but these reports differ on the pattern of distribution. Recent studies in the rat indicate that the MAT of brain, in common with other extrahepatic MATs, is a different molecular species from the MAT of liver.[94] There is also evidence that the MAT of rat cerebellum differs in certain physical and kinetic properties from the MAT of rat cerebral cortex.[95] The significance of these findings remains to be elucidated.

3.2. S-Adenosylhomocysteine Hydrolase

The specific activity of AdoHcy hydrolase in brain is greater than that of MAT, but a precise figure cannot be stated, published values not being in agreement. Values reported for rat brain range from about 20 nmol product/h per mg protein to 700 nmol/h per mg[19,22,96-99]; the distribution is fairly uniform.[99] Mouse brain[100] and rabbit and chicken brain[97] have values comparable to rat brain, but only traces are reported for dog brain.[97] Assay in the direction of synthesis gives a much higher specific activity than assay in the direction of hydrolysis.[97] The preparation of highly purified AdoHcy hydrolase from rat brain[96] will permit delineation of the complex kinetics[97,98] of this enzyme.

3.3. S-Adenosylmethionine Decarboxylase

S-Adenosylmethionine decarboxylase (E.C. 4.1.1.50) is present in rat brain at a specific activity of about 0.1–0.7 nmol product/h per mg protein,[101-104] in monkey brain at 0.45 nmol/h per mg,[105] and in human brain at 0.35 nmol/h per mg.[106] It has also been determined in mouse brain.[107] The regional distribution in rat brain shows the greatest activity in parietal gray matter, one-third as much in corpus striatum.[104]

3.4. Cystathionine β-Synthase

This enzyme is found in rat brain at a specific activity of about 40 nmol product/h per mg according to most studies,[73,76,83,84,89,108] although lower[90] and higher[109] values are reported. Comparable values were found in rabbit[84] and monkey[67,68,83] brain, higher in mouse brain,[84] and lower in guinea pig[84] and human[83] brain. The distribution of the enzyme in rat[73,108] and monkey[67,69] brain regions showed no marked differences, cerebellum tending to be high; there was no difference between gray and white matter of human brain.[64] In the vitamin B_6-deficient rat, brain cystathionine β-synthase is reported to be unaffected,[76] only slightly decreased,[82] or decreased by half,[109] perhaps depending on the severity of the deficiency. Retention of activity in the face of vitamin B_6 deficiency indicates very tight binding of cofactor. Studies carried out with only the soluble enzyme may require reevaluation (see Section 5), because rat brain β-synthase activity is about 50% in the particulate fraction.[109]

3.5. γ-Cystathionase

The specific activity of γ-cystathionase in brain has been reported to be very low or undetectable[64,82] when assayed by some older procedures. Similarly, the assay of Mudd et al.,[83] which is rate limited by substrate concentration, has given values for rat brain of 0.15–0.22 nmol product/h per mg.[83,89,90] However, application of the method of Gaull et al.,[110] which is not rate limited by substrate, has revealed the presence of considerably larger amounts of γ-cystathionase, 5–10 nmol/h per mg in rat brain.[73,76,84,111] There are comparable activities in guinea pig, rabbit, and mouse brain,[84] and higher activity in human[112] and monkey brain.[67] The distribution in regions of monkey brain ranges from 72 nmol/h per mg in medulla to 22 in parietal gray matter, with most regions having specific activities of 30–50 nmol/h per mg.[67] In the rat, vitamin B_6 deficiency results in loss of cofactor (pyridoxal phosphate) from γ-cystathionase in liver, pancreas, and kidney; however, the enzyme in brain is not affected.[76]

3.6. Methyltetrahydrofolate:Homocysteine Methyltransferase

Remethylation of homocysteine to methionine in brain is effected primarily by methyltetrahydrofolate:homocysteine methyltransferase. Reported specific activities in rat brain vary from 0.06 to 8.4 nmol product/h per mg protein, but most values range from 0.7 to 4.6.[16,29,42,113–120] The distribution in rat[115] and rabbit[120] brain is relatively uniform. In human brain, the specific activity is 1.4 nmol/h per mg,[112] in monkey brain 2.4.[121] Bovine brain contains exceptionally large amounts of the enzyme (36 nmol/h per mg), more than in any other tissue but pancreas.[122] The enzyme was partially purified from bovine brain and, interestingly, at one stage showed two activity peaks on DEAE-Sephadex.[122] Activity of a purified preparation was higher with 5-methyltetrahydropteroyl polyglutamates than with the monoglutamate.[123]

Exposure of rats to nitrous oxide leads to loss of methyltetrahydrofolate:homocysteine methyltransferase activity in both brain and liver as a result of oxidation of the vitamin B_{12} cofactor from the cob(I)alamin form to the inactive cob(III)alamin form.[119,124] Cycloleucine, by inhibiting MAT and the synthesis of AdoMet, may reduce the transferase activity, for which AdoMet is a cofactor.[125] Folic acid deficiency in the rat results in no[124] or slight[16] loss of activity in the brain and larger losses in liver and kidney.[16]

Betaine:homocysteine methyltransferase is generally considered to be lacking in brain. It has not been detected in rat[113] or monkey[121] brain but is measurable in human brain at a specific activity of 0.37 nmol product/h per mg protein.[112]

3.7. Metabolic Studies

Although the enzymatic apparatus for carrying out the metabolism of the sulfur amino acids has been demonstrated to exist in brain, and the products of the reactions have been detected in brain tissue, there are few direct measurements of the fluxes of metabolites through these pathways in brain. Studies in which labeled compounds are administered peripherally must be discounted, for metabolism undoubtedly occurs in the liver, with subsequent transport of products to the brain. Even when a test compound is injected directly into the CNS, there may be rapid exit to the peripheral circulation[82] and return of metabolites to the brain. These difficulties are avoided by *in vitro* studies, with the caveat, however, that some processes may be disrupted in homogenates or slices of brain.

When [^{35}S]methionine was injected intracisternally in rats, 80% of the ^{35}S still in the brain after 3 h was found incorporated into protein.[17,126] This may not be a true estimate of the amount of methionine utilized for protein synthesis relative to the amount metabolized, for the turnover of protein in the brain would be slower than the loss of free labeled compound to the periphery. The ^{35}S in the soluble fraction was as [^{35}S]methionine (14%), [^{35}S]cystine plus glutathione (67%), and [^{35}S]taurine (19%); however, as noted by the authors, the labeled metabolites could have come from the liver.[126] [^{35}S]Cystathionine was found in rat brain after subarachnoid injection of [^{35}S]methionine[73] and undoubtedly was formed *in situ*, for cystathionine does not enter the brain from the periphery.[82,90] In *in vitro* experiments with rat brain slices or cell suspensions incubated with [^{35}S]methionine, 1.8% of the original ^{35}S was found in cystine plus glutathione (mostly the latter), 4.5% in taurine, and 0.16% in sulfate after 3 h.[126] In other experiments, the rate of formation of [^{35}S]cysteine from [^{35}S]methionine by rat brain slices was found to be 5.6%/h per g tissue; the intermediates [^{35}S]homocysteine and [^{35}S]cystathionine were detected but not identified conclusively.[127] These studies indicate that transsulfuration, methionine → homocysteine → cystathionine → cysteine, does take place in brain but is rather slow as compared with liver.

The metabolism of cystathionine also has been investigated. After intracerebral injection of [^{35}S]cystathionine, [^{35}S]cysteine was found in the brain protein and in the soluble fraction; the latter also contained labeled taurine and sulfate.[90] Other workers found that after intracisternal injection of [^{35}S]cystathionine into rats, at 9 h, 20% of the residual ^{35}S remained as cystathionine, 22% was in protein, 18% in glutathione, and 18% as a mixture of cysteic acid, cysteine sulfinic acid, taurine, and glutathione–cysteine mixed disulfide.[70] The rapid exit of [^{35}S]cystathionine to the peripheral plasma,[82] however, leaves open the possibility that metabolism occurred in the liver.

Recently, interest has focused on the question of how much of the methionine in the brain is newly derived from the diet and how much has been recycled from homocysteine (see ref. 6). Such information would provide insight into the main role of methionine in the brain (aside from protein synthesis): is it as part of a cycle that provides methyl groups (derived from methyltetrahydrofolate or, ultimately, from serine) for AdoMet-mediated methylations, or is it as a sulfur donor to convert serine to cysteine? Functioning of the remethylation cycle *in vivo* has been studied by Spector *et al.*,[128] who infused [^{3}H-CH$_3$]methionine and [^{35}S]methionine of high specific activity into rats intravenously. After 5 h, the ratio of ^{3}H to ^{35}S, relative to the ratio in the methionine infused, was 0.55–0.66 in methionine isolated from brain, 0.43 in that from liver, and 0.75 in plasma methionine. The decrease in ratio represents loss of ^{3}H-labeled CH$_3$ groups and remethylation of [^{35}S]homocysteine with unlabeled CH$_3$. The higher ratio in plasma than in either organ would suggest that little if any remethylated methionine was transported from liver to brain. However, injection of doubly labeled methionine intraventricularly into rabbits gave, after 2.5 h, only a slight decrease in ^{3}H/^{35}S ratio, and there was no decrease in the methionine incubated with rabbit brain slices.[128] It seems evident that remethylation does take place in the brain, but its extent remains to be determined.

4. DEVELOPMENT OF SULFUR AMINO ACID METABOLISM IN BRAIN

The pathways of methionine metabolism discussed in the previous sections refer to the situation that exists in mature tissue. Prior to maturity, however, major changes occur in the activities of some of the enzymes and in the concentrations of some of the intermediates and end products. These changes occur at different stages of development among animals.

4.1. Man

The chief metabolic pathway of methionine metabolism in brain is the remethylation cycle via methyltetrahydrofolate:homocysteine methyltransferase (see above). The activity of this enzyme is considerably greater in early fetal brain than in mature brain.[112] There is only a small activity of betaine:homocysteine methyltransferase in mature brain and even less in fetal brain. Brain has apparently limited transsulfuration capability at maturity and none detectable in early stages of development (*in utero*, 3 to 6 months after conception). Cystathionine β-synthase activity is present in fetal human brain, although less is present than in mature human brain.[129,130] The concentration of cystathionine in fetal human brain is small (40 nmol/g) compared with that in mature human brain (approx. 1–2 μmol/g). This large difference probably results from the greater activity of the remethylation cycle in fetal brain, which removes homocysteine more rapidly than it does in mature brain, as well as from the smaller activity of cystathionine β-synthase in fetal brain. In rat, the K_m for methyltetrahydrofolate:homocysteine methyltransferase is about 10^{-5} M, whereas those for cystathionine β-synthase and for S-adenosylhomocysteine hydrolase are each about 10^{-3} M. Thus, the remethylation of homocysteine to methionine rather than the accumulation of cystathionine would tend to be favored.[112]

Transsulfuration results in a loss of the methionine carbon skeleton from the remethylation cycle. The only other loss of this carbon skeleton occurs when AdoMet is decarboxylated. S-Adenosylmethione decarboxylase, the first step in the biosynthesis of the polyamines spermidine and spermine, is present in fetal brain at a smaller specific activity than in mature brain.[106] The net result of these adaptations of methionine metabolism in human fetal brain, as compared with mature brain, is that less of the carbon skeleton of methionine is lost from the remethylation cycle, and the cycle itself is more active in fetal brain than in mature brain. No information on methionine metabolism in human brain at intermediate stages, i.e., late fetal and neonatal, is available.

4.2. Rhesus Monkey

As in human brain, the remethylation cycle of methionine metabolism via methyltetrahydrofolate:homocysteine methyltransferase is prominent in monkey brain, although there are no notable differences in activity during development.[121] However, only the last half of gestation was studied. Since the monkey infant is more precocious at birth than the human infant,[131] it is likely that, if there is a period of greater activity of this enzyme in the monkey, as there is in man, it took place prior to the earliest time studied in the monkey. Activity of betaine:homocysteine methyltransferase is present in trace amounts at most at all stages of brain development. The changes in the transsulfuration pathway during development of monkey brain resemble those that take place in human brain, at least with respect to the findings during the periods available for study. Thus, the concentration of cystathionine and the activity of cystathionine β-synthase are low during fetal development and increase slowly during neonatal life, reaching values found in the adult by 3 months after birth (Fig. 2A,B).[68]

As in man, activity of AdoMet decarboxylase is low during fetal development and increases slowly during neonatal life, reaching values found in the adult by 3 months after birth.[105] This adaptation results in greater synthesis of spermidine in mature brain than in fetal brain (Fig. 3A,B).

4.3. Rat

Rat brain resembles primate brain in having the remethylation cycle of methionine metabolism via methyltetrahydrofolate:homocysteine methyltransferase as the major pathway.[113] Activity of this enzyme clearly decreases during development, which would be expected in that the rat pup is even less

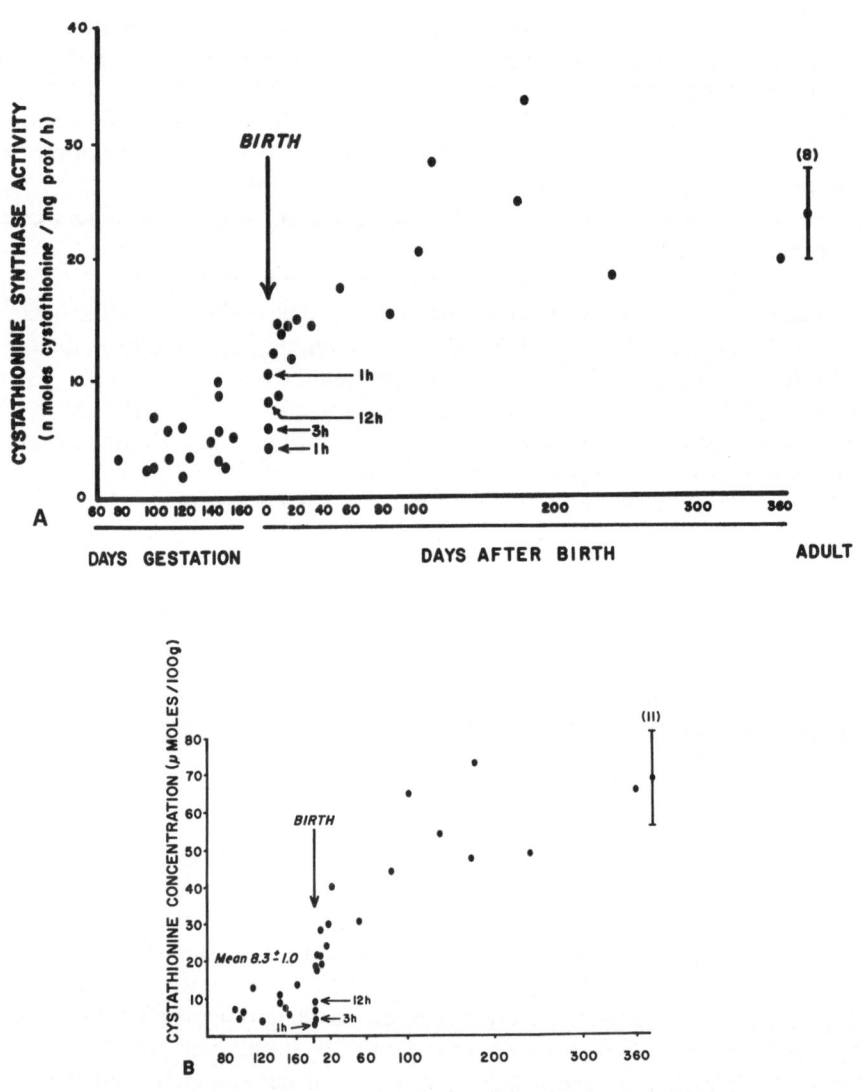

Fig. 2.   Specific activity of cystathionine synthase (A) and concentration of cystathionine (B) in rhesus monkey brain during development. From Sturman *et al.*[68] with permission.

precocious at birth than the human infant.[131] The changes that occur in the transsulfuration pathway during development of rat brain in some respects resemble those in primate brain: cystathionine β-synthase activity increases during development, and γ-cystathionase activity is said to increase slightly, although even at maturity it is low.[89,109,111] There are, however, differences: the concentration of cystathionine increases to a peak value 15 to 21 days after birth, then decreases to 10–25% of that concentration in adult brain.[73,132–134] This pattern is found both in whole brain and in cerebellum.

The activity of AdoMet decarboxylase is low in newborn rat brain and increases slowly during development, reaching values found in the adult some time after weaning.[102] The concentration of spermidine, however, decreases with development of the rat brain,[104] again unlike the situation in primate brain in which the concentration of spermidine increases in parallel with the activity of AdoMet decarboxylase.[105]

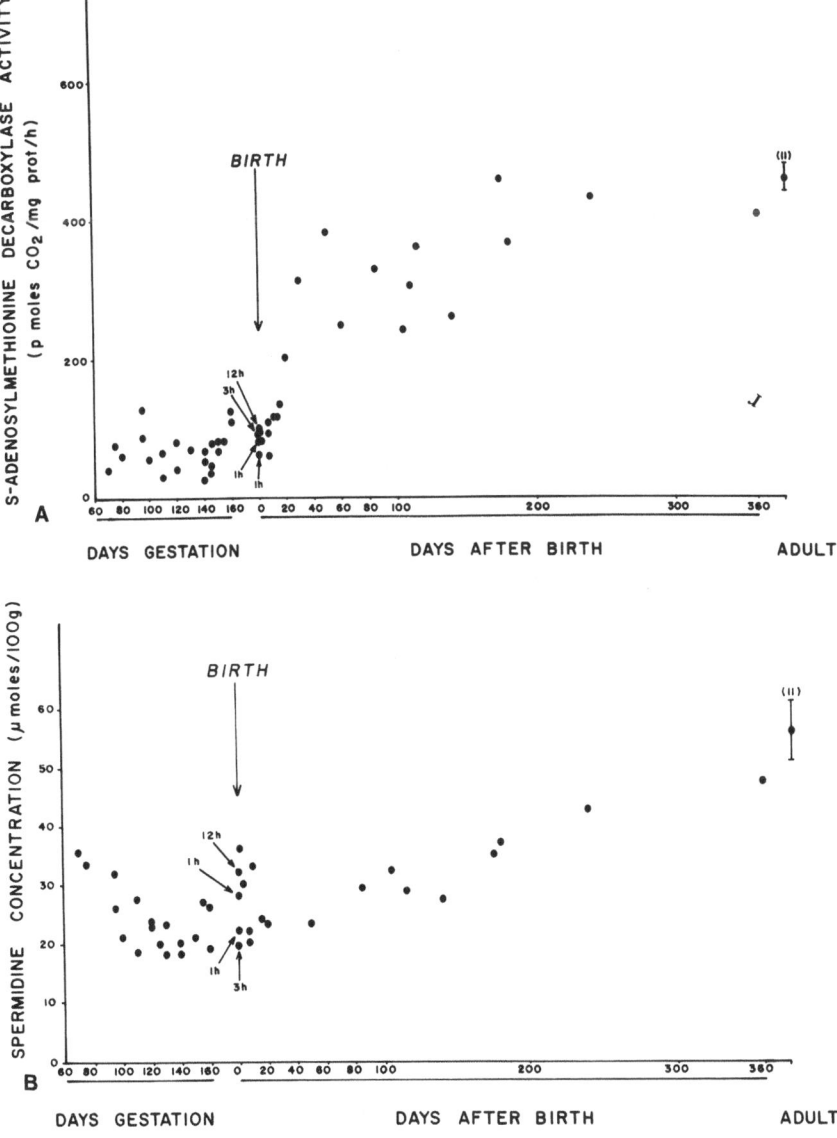

Fig. 3. Specific activity of S-adenosylmethionine decarboxylase (A) and concentration of spermidine (B) in rhesus monkey brain during development. From Sturman and Gaull[105] with permission.

4.4. Nutritional and Clinical Implications

The changes in methionine metabolism in developing brain are similar (except for the behavior of cystathionine and spermidine) among the three species that have been studied: man, rhesus monkey, and rat. Young brain has an active remethylation cycle via methyltetrahydrofolate:homocysteine methyltransferase, and minimal losses from the cycle occur by removal of the carbon skeleton of methionine for polyamine biosynthesis. There is evidence of a relatively small conversion of the sulfur of methionine to that of cyst(e)ine in brain any time during development. As the remethylation cycle activity is decreased, polyamine biosynthesis is increased with concomitant increases in the concentration of spermidine in the primate but not in the rat (see above). It is difficult precisely to correlate these developmental changes in brain with the rates of brain growth in the three species studied. Perhaps the correlation is with the development of function; however, the specific functions of cystathionine and spermidine in mature brain remain to be elucidated. It is of interest that Seiler et al.[135] have shown that spermidine synthesis increases strikingly in dissociated cell cultures of 8-day-old chick cerebrum. These changes occur after cell proliferation ceases and continue after spermine and putrescine concentrations have reached a plateau. They suggest a role for spermidine in the synthesis of the specialized membranes of the dendritic and synaptic regions. The role of methionine metabolism in polyamine synthesis and function in brain has been discussed by us in greater detail.[68]

At this point, it is worth describing the changes that occur in methionine metabolism in liver during development. These changes are more varied and extreme than those that occur in brain and have an effect on the overall nutrient requirements of the animal. As found in brain, the remethylation cycle of methionine metabolism via methyltetrahydrofolate:homocysteine methyltransferase is greater in early developing liver than in mature liver.[89,112,121] Mature liver also has an active remethylation cycle via betaine:homocysteine methyltransferase. The activity of this pathway increases during development in man and the monkey.[112,121] Finkelstein et al.[89] report this enzyme activity to be greater in developing rat liver than in mature rat liver. This apparent difference between rat liver and primate liver requires clarification.

The activity of the transsulfuration pathway in liver increases during development. In contrast to the brain, however, the change in liver is a dramatic one. Human fetal liver has virtually no transsulfuration capability,[129,130] for it has no activity of γ-cystathionase. Rhesus monkey fetal liver has detectable but low γ-cystathionase activity which increases approximately tenfold during development.[68] Rat fetal liver has easily measurable γ-cystathionase activity which also increases more than tenfold during development.[111]

The activity of AdoMet decarboxylase in fetal liver is considerably higher than that in mature liver,[136] the opposite of what occurs in brain (see above). Thus, mature brain resembles fetal liver in the primate (but not in the rat) in the sense that the concentrations of cystathionine and spermidine are relatively high. In both cases, the higher concentration of spermidine is associated with a greater activity of AdoMet decarboxylase.

In the case of fetal liver, the increased concentration of cystathionine is associated with an absence of γ-cystathionase. In the case of brain, activity of γ-cystathionase is never as high as in mature liver, and the concentration of cystathionine may be a function of cystathionine β-synthase rather than of γ-cystathionase.

These developmental changes are of considerable potential nutritional and clinical significance. Zlotkin *et al.*[137a] have recently extended our early and incomplete findings on the postnatal development of human hepatic γ-cystathionase. In a complete and thoroughgoing study, they have now shown that the development of this enzyme in man resembles that of the rhesus monkey. Zlotkin *et al.*[137] and Rigo and Senterre[138] have shown that plasma concentrations of free cystine are not maintained at normal levels when human infants are nourished with a total parenteral solution of amino acids that is low in cyst(e)ine. Zlotkin *et al.* did not find any decrease in rate of growth or in overall nitrogen balance as a result of this deficiency. Their study is inconclusive, however, because such solutions are also low in tyrosine, and plasma concentrations of tyrosine fall well below normal. It is not likely, therefore, that one could demonstrate cyst(e)ine to be limiting when tyrosine may also be limiting. Furthermore, the balance studies were only for periods of 3 days. It is unlikely that such a short balance period could define "essentiality," especially with an amino acid that has a considerable reservoir as protein-bound cysteine[87] and as cysteine incorporated into glutathione.[139]

Our recent studies in the beagle pup model for total parenteral nutrition (TPN) bear on this matter and provide evidence to suggest that the developing brain depends on liver as a major source of cysteine.[140] During TPN employing an amino acid solution deficient in cyst(e)ine, cerebral methionine and cystathionine concentrations increase, whereas the cerebral cyst(e)ine concentration decreases. In contrast, both methionine and cystathionine are readily metabolized by the liver, and the hepatic cyst(e)ine content was maintained. Plasma concentrations of methionine, cystathionine, and cyst(e)ine remain unaffected. Thus, the liver was able to maintain its own cyst(e)ine content but not that of the brain. In the same model, although total body weight remained stable under these conditions, cerebral weight and protein content were decreased.

Human infants are treated with TPN solutions similar to those employed in the beagle pup studies. The fact that in the pup the brain cysteine concentrations fell in the face of plasma cysteine concentrations that are maintained gives one cause for concern about the human infant, in whom plasma cysteine concentrations actually fall. Whether or not brain cysteine concentrations in the human infant fall as well is not known.

5. SUBCELLULAR COMPARTMENTATION OF SULFUR AMINO ACID METABOLISM IN BRAIN

The development by Gray and Whittaker[141] and DeRobertis and co-workers[142] of subcellular fractionation techniques that could be applied to brain tissue provided a potent new technique for the investigation of the varied functions of brain metabolites. Assignment of a metabolite and/or its associated enzymes to a subcellular fraction could be used to infer the role(s) of the components of various metabolic pathways. These techniques have been widely applied to the investigation of the neurochemical properties of acetylcholine, the catecholamines, serotonin, histamine, brain peptides, and amino acids. Particularly important was the introduction of a method that allowed nerve-ending particles or synaptosomes[143] to be isolated. The availability of these organelles for biochemical study made it possible for neurochemists to implement investigations that could define many of the properties of proposed neurotransmitter and neuromodulator compounds. This section addresses the application of these techniques to the sulfur amino acid metabolic pathway as a means to gain further understanding of the function of these compounds in brain.

The sulfur-containing amino acids (Fig. 1) appear to be important to the brain for a number of functional roles.[144] The production of AdoMet from methionine serves to supply the major methyl donor used in brain metabolism, particularly for methylation of the catecholamines and histamine. The reconversion of homocysteine to methionine involves methylation utilizing the donor N^5-methyltetrahydrofolate. The remethylation of tetrahydrofolate to N^5-methyltetrahydrofolate involves the enzyme serine hydroxymethyltransferase (E.C. 2.1.2.1) which is primarily responsible for the synthesis of the neurotransmitter glycine in the brain.[145]

Whereas the first portion of the pathway of methionine metabolism appears to be primarily involved in methylation functions, the portion after homocysteine (especially the transsulfuration reactions and the synthesis of taurine) appears to be associated mainly with neurotransmitter and/or neuromodulator functions. Cystathionine has been proposed as a putative inhibitory neurotransmitter,[146] although this function has been questioned[147] because of the uniformity of its distribution in spinal cord. Cysteine and its metabolites cysteine sulfinic acid and cysteic acid have also been implicated in brain function in a manner compatible with neurotransmitter properties.[144,148–150] The last products of the pathway, taurine and hypotaurine, will be discussed elsewhere but have also been suggested to be neurotransmitters or neuromodulators (see refs. 144,151,152 for other reviews). Thus, the pathway of methionine metabolism functionally subserves both a methylation role and a putative neurotransmitter role. These dual functions plus the requirement to subserve protein and spermidine synthesis suggested the need for some form of compartmentation to separate the various functional pools of these amino acids from one another.

Application of subcellular fractionation techniques to the distribution of amino acids has consistently shown approximately 20% of the brain content of these compounds to be associated with synaptosomes. Mangan and Whittaker[153] first showed an association of some amino acids with synaptosomes. Other investigators[154] demonstrated that GABA could be stored and released by synaptosomes when the latter were subjected to osmotic shock. Bradford[155] found that a number of amino acids were released from synaptosomes subjected to electrical stimulation, particularly those amino acids implicated in neurotransmission. Investigations designed to demonstrate that amino acids were stored in synaptic vesicles in a manner analogous to ace-tylcholine were generally unsuccessful,[153,156,157] with the possible exception of two compounds, taurine and glutamate.[157-159] With the exception of taurine (to be discussed elsewhere in this volume), the sulfur-containing amino acids are not present in sufficiently high concentrations in brain to permit reliable subcellular distribution studies to be made by conventional amino acid analysis.

The studies discussed above were concerned primarily with the major amino acid constituents of brain: glutamate, glutamine, aspartate, γ-amino-butyrate, glycine, serine, taurine, alanine, and threonine. However, using tech-niques more sensitive than conventional amino acid analysis, approximately 30% of rat brain AdoMet has been identified with a synaptosomal fraction isolated from rat brain,[26] and some cystathionine has been identified within synaptosomes.[160] Thus, subcellular studies of sulfur-containing amino acids have been unable to define distribution patterns that could be used to design a model of compartmentation for the entire pathway.

Studies of the transport of sulfur-containing amino acids by synaptosomal fractions support a neurotransmitter role for those compounds that have been studied. Cysteic acid and cysteine sulfinic acid are potent inhibitors of the high-affinity transport system that serves to inactivate glutamic acid.[161,162] Indeed, the potency of their action is such that they could be the primary substrates for such a system. In addition, high-affinity transport mechanisms have been found in synaptosomes for both cystine[163] and cysteine.[164] Despite the low concentrations of sulfur-containing amino acids in synaptosomes, their trans-port characteristics give evidence for a functional role in these organelles.

The next step was to examine the distribution of the enzymes associated with these amino acids, on the premise that synthesis would occur within the organelles responsible for the appropriate functions. Early experiments had provided information about subcellular distribution of enzymes in brain that was sometimes contradictory. Cystathionine β-synthase was found to be mi-tochondrial in origin[165] after preliminary experiments had demonstrated a par-ticulate component for this enzyme.[109] Other investigators found cystathionine β-synthase to be entirely soluble in distribution.[133] Cysteine sulfinate decar-boxylase (E.C. 4.1.1.29) was described as being a synaptosomal component,[166] and serine hydroxymethyltransferase as being a mitochondrial enzyme.[167,168]

On the basis of simultaneous study of the subcellular distribution of many of the enzymes involved in methionine metabolism, we then postulated a com-partmentation model[169] similar to that suggested for the enzymes of glutamate metabolism.[170] The model consisted of three components: (1) the transmeth-ylation function, which appeared to be located in the soluble fractions; (2) the neurotransmitter function, including the transsulfuration reactions and the syn-thesis of taurine, which appeared to be located in synaptosomes or nerve end-ings; and (3) the synthesis of glycine, which appeared to be located in mito-chondria (Fig. 1).

In order to test the model, we examined in rat brain the subcellular dis-tribution of a number of the enzymes involved in the metabolism of methio-nine.[169] Methyltetrahydrofolate:homocysteine methyltransferase and MAT were both found to be primarily located in the soluble compartment, even to a greater extent than the cytoplasmic marker lactate dehydrogenase (E.C. 1.1.1.27), some of which is occluded within synaptosomes. Cystathionine β-synthase, cysteine dioxygenase (E.C. 1.13.11.20), and cysteine sulfinate decarboxylase were all found primarily in the synaptosomal fractions isolated from rat brain. Activity of γ-cystathionase is present in brain in amounts too small to allow detailed study by techniques of subcellular fractionation.[67] Nonetheless, preliminary fractionation studies did show that brain γ-cystathionase had a particulate component, which would be compatible with the other compartmentation findings.[169]

Serine hydroxymethyltransferase was found in the mitochondria and, interestingly, was present in mitochondria that were not of synaptosomal origin. This enzyme was only minimally present in isolated synaptosomes, whereas much more of the mitochondrial marker fumarase (E.C. 4.2.1.2) was associated with the mitochondria occluded within the synaptosomes. This location of fumarase has been found by others.[171] On the basis of comparison of the distribution of fumarase and serine hydroxmethyltransferase, we suggested that glycine synthesis might occur either in the glia or the neuronal cell body.[169] A model for the control of glycine release at the synapse based on the mitochondrial location of its synthesizing enzyme has been suggested.[172] It is not yet clear, however, whether such a model should be based on a neuronal or glial localization for glycine synthesis. In addition, serine hydroxymethyltransferase might be useful as a marker for glial cells if it is primarily derived from glial mitochondria. Cultured glial cells do have the capacity to synthesize glycine,[173] a finding that supports this possibility.

Other findings that bear on our basic hypothesis include those in which cystathionine β-synthase was located in mitochondria.[165] This finding has now been confirmed, and further experiments in our laboratory using rhesus monkey brain[174] substantiate our previous findings in rat brain of the synaptosomal localization of this enzyme. A report by Misra and Olney[175] that cysteine dioxygenase was microsomal in origin was later corrected by these workers on reexamination of their subcellular fractionation techniques.[176] The synaptosomal location of cysteine dioxygenase also has been confirmed by other investigators.[177,178]

Methyltetrahydrofolate : homocysteine methyltransferase has been found to be in the soluble compartment of rat brain,[114] as has methylenetetrahydrofolate reductase.[114] Both of these findings are compatible with the soluble nature of the methylation functions of methionine pathways and confirm our findings and hypothesis. Further evidence in support of this hypothesis is the fact that some of the enzymes that use methyl groups, such as catechol O-methyltransferase (E.C. 2.1.1.6), are also soluble.[171] Others of this class of enzymes, such as histamine methyltransferase (E.C. 2.1.1.8), appear to be soluble in some brain areas and particulate in others.[179] More recently, we demonstrated the soluble nature of the enzyme methylenetetrahydrofolate reductase in rhesus monkey brain.[174] S-Adenosylhomocysteine hydrolase, which catalyzes the conversion of AdoHcy to homocysteine, has been found to be primarily soluble when measured in fractions prepared from rat brain.[99] Less of this enzyme was associated with the synaptosomal fractions (25%) than was lactate dehydrogenase, the cytoplasmic marker (51%).[99]

The synaptosomal location of cysteine sulfinate decarboxylase has been a consistent finding by a number of laboratories with various animal models.[166,169,180,181] It is not yet clear, however, whether cysteine sulfinate decarboxylase in brain is a distinct enzyme or is a secondary activity of glutamate decarboxylase (E.C. 4.1.1.15),[182–184] also an enzyme located in synaptosomes.

In addition, we have studied the brain subcellular distribution of some of the enzymes and amino acids of the sulfur amino acid metabolic pathway during development of the rhesus monkey. Methyltetrahydrofolate:homocysteine methyltransferase is present in relatively greater amounts in the crude mitochondrial and synaptosomal fractions of fetal monkey brain than it is in the adult animal.[174] Methylenetetrahydrofolate reductase also undergoes a reduction in the activity that is associated with the crude mitochondrial and synaptosomal fraction during the transition from fetal to adult life.[174] Serine hydroxymethyltransferase, as well as the associated amino acids glycine and serine, and cystathionine β-synthase maintain a fairly constant subcellular distribution during development.[174] Cysteine sulfinate decarboxylase is not specifically associated with any one subcellular fraction during fetal development but becomes primarily a synaptosomal constituent in the neonatal and adult brain subcellular fractions.[183]

Patterns of subcellular distribution of the enzymes of methionine metabolism are consistent with a compartmentation of methylation and neurotransmitter functions. Glycine synthesis occurs in mitochondria, possibly of glial origin. Such a compartmentation of metabolites provides further evidence for the importance of the sulfur-containing amino acids in the function of the brain.

6. CONCLUSIONS

The intermediates and enzymes of methionine metabolism are all found in brain. Their presence and functions in brain are adapted in ways that suggest that methylation and polyamine synthesis are of particular importance and that the transsulfuration of methionine to cysteine and its metabolites is very carefully controlled. Studies of the subcellular location of these intermediates, their associated enzymes, and their uptake systems suggest that what transsulfuration does take place subserves a neurotransmitter or neuromodulator function. It is this last consideration that may require that the transsulfuration function be limited, since cysteine sulfinate and cysteic acid are potent neuroexcitatory compounds, and taurine is a potent neuroinhibitory conpound. Finally, the pattern of development of these metabolic pathways gives further evidence of the special function of these pathways during brain development. Indeed, current clinical and nutritional studies suggest these adaptations to be of considerable significance.

ACKNOWLEDGMENTS. Studies from the authors' laboratories were supported by the New York State Office of Mental Retardation and Developmental Disabilities.

REFERENCES

1. Tallan, H. H., Moore, S., and Stein, W. H., 1958, *J. Biol. Chem.* **230**:707–716.
2. Brenton, D. P., Cusworth, D. C., and Gaull, G. E., 1965, *Pediatrics* **35**:50–56.
3. Gaitonde, M. K., 1970, *Handbook of Neurochemistry*, Volume 3 (A. Lajtha, ed.), Plenum Press, New York, pp. 225–287.
4. Sturman, J. A., and Rivlin, R. S., 1975, *Biology of Brain Dysfunction*, Volume 3 (G. E. Gaull, ed.), Plenum Press, New York, pp. 425–475.
5. Sershen, H., and Lajtha, A., 1979, *J. Neurochem.* **32**:719–726.
6. Zeisel, S. H., and Wurtman, R. J., 1979, *Transmethylation* (E. Usdin, R. T. Borchardt, and C. R. Creveling, eds.), Elsevier/North Holland, New York, pp. 59–68.
7. Gaull, G. E., Tallan, H. H., Lajtha, A., and Rassin, D. K., 1975, *Biology of Brain Dysfunction*, Volume 3 (G. E. Gaull, ed.), Plenum Press, New York, pp. 47–143.
8. Shaw, R. K., and Heine, J. D., 1965, *J. Neurochem.* **12**:151–155.
9. Perry, T. L., Sanders, H. D., Hansen, S., Lesk, D., Kloster, M., and Gravlin, L., 1972, *J. Neurochem.* **19**:2651–2656.
10. Perry, T. L., Berry, K., Hansen, S., Diamond, S., and Mok, C., 1971, *J. Neurochem.* **18**:513–519.
11. Rubin, R. A., Ordonez, L. A., and Wurtman, R. J., 1974, *J. Neurochem.* **23**:227–231.
12. Burnet, F. R., 1979, *J. Neurochem.* **33**:603–605.
13. Daniel, R. G., and Waisman, H. A., 1969, *J. Neurochem.* **16**:787–795.
14. Liu, Y. P., Ambani, L. M., and Van Woert, M. H., 1972, *J. Neurochem.* **19**:2237–2239.
15. Ordonez, L. A., and Wurtman, R. J., 1973, *Biochem. Pharmacol.* **22**:134–137.
16. Ordonez, L. A., and Wurtman, R. J., 1974, *Arch. Biochem. Biophys.* **160**:372–376.
17. Gaitonde, M. K., and Richter, D., 1956, *Proc. R. Soc. Lond.* [*Biol.*] **145**:83–99.
18. Mitchell, A. D., and Benevenga, N. J., 1978, *J. Nutr.* **108**:67–78.
19. Eloranta, T. O., 1977, *Biochem. J.* **166**:521–529.
20. Schatz, R. A., and Sellinger, O. Z., 1975, *J. Neurochem.* **24**:63–66.
21. Baldessarini, R. J., and Kopin, I. J., 1966, *J. Neurochem.* **13**:769–777.
22. Hoffman, D. R., Cornatzer, W. E., and Duerre, J. A., 1979, *Can. J. Biochem.* **57**:56–65.
23. Zappia, V., Cartení-Farina, M., and Porchelli, M., 1979, *Transmethylation* (E. Usdin, R. T. Borchardt, and C. R. Creveling, eds.), Elsevier/North Holland, New York, pp. 95–104.
24. Schatz, R. A., Wilens, T. E., and Sellinger, O. Z., 1981, *J. Neurochem.* **36**:1739–1748.
25. Taylor, K. M., and Randall, P. K., 1975, *J. Pharmacol. Exp. Ther.* **194**:303–310.
26. Yu, P. H., 1978, *Anal. Biochem.* **86**:498–504.
27. Wurtman, R. J., Rose, C. M., Matthysse, S., Stephenson, J., and Baldessarini, R., 1970, *Science* **169**:395–397.
28. Baldessarini, R. J., 1966, *Biochem. Pharmacol.* **15**:741–748.
29. Carl, G. F., Benesh, F. C., and Hudson, J. L., 1978, *Biol. Psychiatry* **13**:661–669.
30. Lo, C.-M., Kwok, M.-L., and Wurtman, R. J., 1976, *Neuropharmacology* **15**:395–402.
31. Lombardini, J. B., and Talalay, P., 1973, *Mol. Pharmacol.* **9**:542–560.
32. Lombardini, J. B., and Talalay, P., 1971, *Adv. Enzyme Regul.* **9**:349–384.
33. Volpe, J. J., and Laster, L., 1970, *J. Neurochem.* **17**:413–424.
34. Pegg, A. E., and Hibasami, H., 1979, *Transmethylation* (E. Usdin, R. T. Borchardt, and C. R. Creveling, eds.), Elsevier/North-Holland, New York, pp. 105–116.
35. Sturman, J. A., and Gaull, G. E., 1978, *Advances in Polyamine Research*, Volume 2 (R. A. Campbell, D. R. Morris, D. Bartos, G. D. Daves, Jr., and F. Bartos, eds.), Raven Press, New York, pp. 213–240.
36. Mudd, S. H., 1962, *J. Biol. Chem.* **237**:PC1372–PC1375.
37. Lombardini, J. B., Chou, T.-C., and Talalay, P., 1973, *Biochem. J.* **135**:43–57.

38. Chou, T.-C., and Talalay, P., 1972, *Biochemistry* **11**:1065–1073.
39. Hibasami, H., Tanaka, M., Nagai, J., Ikeda, T., and Pegg, A. E., 1980, *FEBS Lett.* **110**:323–326.
40. Kutzbach, C., and Stokstad, E. L. R., 1967, *Biochim. Biophys. Acta* **139**:217–220.
41. Kutzbach, C., and Stokstad, E. L. R., 1971, *Biochim. Biophys. Acta* **250**:459–477.
42. Turner, A. J., 1979, *Transmethylation* (E. Usdin, R. T. Borchardt, and C. R. Creveling, eds.), Elsevier/North-Holland, New York, pp. 69–76.
43. Taylor, R. T., and Weissbach, H., 1969, *Arch. Biochem. Biophys.* **129**:728–744.
44. Taylor, R. T., and Weissbach, H., 1969, *Arch. Biochem. Biophys.* **129**:745–766.
45. Finkelstein, J. D., Kyle, W. E., Martin, J. J., and Pick, A.-M., 1975, *Biochem. Biophys. Res. Commun.* **66**:81–87.
46. Zappia, V., Zydek-Cwick, C. R., and Schlenk, F., 1969, *J. Biol. Chem.* **244**:4499–4509.
47. Fonlupt, P., Rey, C., and Pacheco, H., 1981, *J. Neurochem.* **36**:165–170.
48. Borchardt, R. T., 1977, *The Biochemistry of Adenosylmethionine* (F. Salvatore, E. Borek, V. Zappia, H. G. Williams-Ashman, and F. Schlenk, eds.), Columbia University Press, New York, pp. 151–171.
49. Finkelstein, J. D., Kyle, W. E., and Harris, B. J., 1974, *Arch. Biochem. Biophys.* **165**:774–779.
50. Burke, G. T., Mangum, J. H., and Brodie, J. D., 1971, *Biochemistry* **10**:3079–3085.
51. Tada, K., Yoshida, T., Hirono, H., and Arakawa, T., 1967, *Tohoku J. Exp. Med.* **92**:325–332.
52. Tada, K., Yoshida, Y., and Arakawa, T., 1970, *Tohoku J. Exp. Med.* **101**:223–226.
53. Rassin, D. K., Longhi, R. C., and Gaull, G. E., 1977, *J. Pediatr.* **91**:574–577.
54. Kanwar, Y. S., Manaligod, J. R., and Wong, P. W. K., 1976, *Pediatr. Res.* **10**:598–609.
55. Gerritsen, T., and Waisman, H. A., 1964, *Science* **145**:588.
56. Wong, P. W. K., Justice, P., Hruby, M., Weiss, E. B., and Diamond, E., 1977, *Pediatrics* **59**:749–756.
57. Kang, S.-S., Wong, P. W. K., and Becker, N., 1979, *Pediatr. Res.* **13**:1141–1143.
58. Brenton, D. P., Cusworth, D. C., and Gaull, G. E., 1965, *J. Pediatr.* **67**:58–68.
59. Gupta, V. J., and Wilcken, D. E. L., 1978, *Eur. J. Clin. Invest.* **8**:205–207.
60. Wilcken, D. E. L., and Gupta, V. J., 1979, *Clin. Sci.* **57**:211–215.
61. Bremer, H. J., Duran, M., Kamerling, J. P., Przyrembel, H., and Wadman, S. K., 1981, *Disturbances of Amino Acid Metabolism: Clinical Chemistry and Diagnosis*, Urban & Schwarzenberg, Baltimore, Munich, p. 52.
62. Malloy, M. H., Rassin, D. K., and Gaull, G. E., 1981, *Am. J. Clin. Nutr.* **34**:2619–2621.
63. Brown, F. C., and Gordon, P. H., 1971, *Can. J. Biochem.* **49**:484–491.
64. Shimizu, H., Kakimoto, Y., and Sano, I., 1966, *J. Neurochem.* **13**:65–73.
65. Gjessing, L. R., and Torvik, A., 1966, *Scand. J. Clin. Lab. Invest.* **18**:565.
66. DeFeudis, F. V., Delgado, J. M. R., and Roth, R. H., 1970, *Brain Res.* **18**:15–23.
67. Sturman, J. A., Rassin, D. K., and Gaull, G. E., 1970, *J. Neurochem.* **17**:1117–1119.
68. Sturman, J. A., Gaull, G. E., and Niemann, W. H., 1976, *J. Neurochem.* **26**:457–463.
69. Volpe, J. J., and Laster, L., 1970, *J. Neurochem.* **17**:425–437.
70. Lefauconnier, J.-M., Portemer, C., and Chatagner, F., 1978, *Neurochem. Res.* **3**:345–356.
71. Hudson, D. B., Vernadakis, A., and Timiras, P. S., 1970, *Brain Res.* **23**:213–222.
72. Hope, D. B., 1964, *J. Neurochem.* **11**:327–337.
73. Heinonen, K., 1975, *Acta Endocrinol. (Kbh.)* **80**:487–500.
74. Perry, T. L., Hansen, S., Berry, K., Mok, C., and Lesk, D., 1971, *J. Neurochem.* **18**:521–528.
75. Perry, T. L., Hansen, S., and Gandham, S. S., 1981, *J. Neurochem.* **36**:406–412.
76. Sturman, J. A., Cohen, P. A., and Gaull, G. E., 1969, *Biochem. Med.* **3**:244–251.
77. Kurtz, D. J., Levy, H., and Kanfer, J. N., 1972, *J. Nutr.* **102**:291–298.
78. Brown, F. C., and DeFoor, M., 1974, *Biochem. Pharmacol.* **23**:1135–1137.
79. Wong, P. W. K., and Fresco, R., 1972, *Pediatr. Res.* **6**:172–181.
80. Sturman, J. A., Schneidman, K., and Gaull, G. E., 1971, *Biochem. Med.* **5**:404–411.
81. Perry, T. L., Hansen, S., Schier, G. M., and Halpern, B., 1977, *J. Neurochem.* **29**:791–795.
82. Brown, F. C., and Gordon, P. H., 1971, *Biochim. Biophys. Acta* **230**:434–445.

83. Mudd, S. H., Finkelstein, J. D., Irreverre, F., and Laster, L., 1965, *J. Biol. Chem.* **240**:4382–4392.
84. Sturman, J. A., Rassin, D. K., and Gaull, G. E., 1970, *Int. J. Biochem.* **1**:251–253.
85. Wade, L. A., and Brady, H. M., 1981, *J. Neurochem.* **37**:730–734.
86. Hwang, S. M., Weiss, S., and Segal, S., 1980, *J. Neurochem.* **35**:417–424.
87. Malloy, M. H., Rassin, D. K., and Gaull, G. E., 1981, *Anal. Biochem.* **113**:407–415.
88. Guchhait, R. B., and Grau, J. E., Jr., 1978, *J. Neurochem.* **31**:921–925.
89. Finkelstein, J. D., 1967, *Arch. Biochem. Biophys.* **122**:583–590.
90. Griffiths, R., and Tudball, N., 1976, *Life Sci.* **19**:1217–1224.
91. Hiemke, C., and Ghraf, R., 1981, *J. Neurochem.* **37**:613–618.
92. Raina, A., Eloranta, T., and Kajander, O., 1976, *Biochem. Soc. Trans.* **4**:968–971.
93. Chase, H. P., Volpe, J. J., and Laster, L., 1968, *J. Clin. Invest.* **47**:2099–2108.
94. Okada, G., Teraoka, H., and Tsukada, K., 1981, *Biochemistry* **20**:934–940.
95. Diez Altares, M. C., and Sellinger, O. Z., 1976, *Enzyme* **21**:53–65.
96. Schatz, R. A., Vunnam, C. R., and Sellinger, O. Z., 1979, *Transmethylation* (E. Usdin, R. T. Borchardt, and C. R. Creveling, eds.), Elsevier/North-Holland, New York, pp. 143–153.
97. Walker, R. D., and Duerre, J. A., 1975, *Can. J. Biochem.* **53**:312–319.
98. Finkelstein, J. D., and Harris, B., 1973, *Arch. Biochem. Biophys.* **159**:160–165.
99. Broch, O. J., and Ueland, P. M., 1980, *J. Neurochem.* **35**:484–488.
100. Schatz, R. A., Vunnam, C. R., and Sellinger, O. Z., 1977, *Life Sci.* **20**:375–383.
101. Pegg, A. E., and Williams-Ashman, H. G., 1969, *J. Biol. Chem.* **244**:682–693.
102. Schmidt, G. L., and Cantoni, G. L., 1973, *J. Neurochem.* **20**:1373–1385.
103. Russell, D. H., and Lombardini, J. B., 1971, *Biochim. Biophys. Acta* **240**:273–286.
104. Shaskan, E. G., Haraszti, J. H., and Snyder, S. H., 1973, *J. Neurochem.* **20**:1443–1452.
105. Sturman, J. A., and Gaull, G. E., 1975, *J. Neurochem.* **25**:267–272.
106. Sturman, J. A., and Gaull, G. E., 1974, *Pediatr. Res.* **8**:231–237.
107. Russell, D. H., and Meier, H., 1975, *J. Neurobiol.* **6**:267–275.
108. Kohl, R. L., and Quay, W. B., 1979, *J. Neurosci. Res.* **4**:189–196.
109. Kashiwamata, S., 1971, *Brain Res.* **30**:185–192.
110. Gaull, G. E., Rassin, D. K., and Sturman, J. A., 1969, *Neuropaediatrie* **1**:199–226.
111. Heinonen, K., 1973, *Biochem. J.* **136**:1011–1015.
112. Gaull, G. E., von Berg, W., Räihä, N. C. R., and Sturman, J. A., 1973, *Pediatr. Res.* **7**:527–533.
113. Finkelstein, J. D., Kyle, W. E., and Harris, B. J., 1971, *Arch. Biochem. Biophys.* **146**:84–92.
114. Burton, E. G., and Sallach, H. J., 1975, *Arch. Biochem. Biophys.* **166**:483–494.
115. Ordóñez, L. A., and Wurtman, R. J., 1973, *J. Neurochem.* **21**:1447–1455.
116. Finkelstein, J. D., Martin, J. J., Kyle, W. E., and Harris, B. J., 1978, *Arch. Biochem. Biophys.* **191**:153–160.
117. Ordóñez, L. A., and Villarroel H., O. A., 1976, *J. Neurochem.* **27**:305–307.
118. Turner, A. J., Pearson, A. G. M., and Mason, R. J., 1979, *Biochemical and Pharmacological Roles of Adenosylmethionine and the Central Nervous System* (V. Zappia, ed.), Pergamon Press, New York, pp. 51–69.
119. Deacon, R., Lumb, M., Perry, J., Chanarin, I., Minty, B., Halsey, M., and Nunn, J., 1980, *Eur. J. Biochem.* **104**:419–422.
120. Suleiman, S. A., and Spector, R., 1980, *J. Neurochem.* **35**:1250–1252.
121. Sturman, J. A., Gaull, G. E., and Niemann, W. H., 1976, *J. Neurochem.* **27**:425–431.
122. Mangum, J. H., Steuart, B. W., and North, J. A., 1972, *Arch. Biochem. Biophys.* **148**:63–69.
123. Coward, J. K., Chello, P. L., Cashmore, A. R., Parameswaran, K. N., DeAngelis, L. M., and Bertino, J. R., 1975, *Biochemistry* **14**:1548–1552.
124. Suleiman, S. A., and Spector, R., 1980, *Life Sci.* **27**:2427–2432.
125. Greco, C. M., Powell, H. C., Garrett, R. S., and Lampert, P. W., 1980, *Neuropathol. Appl. Neurobiol.* **6**:349–360.
126. Gaitonde, M. K., and Richter, D., 1957, *Metabolism of the Nervous System* (D. Richter, ed.), Pergamon Press, London, pp. 449–455.
127. Peck, E. J., Jr., and Awapara, J., 1967, *Biochim. Biophys. Acta* **141**:499–506.

128. Spector, R., Coakley, G., and Blakely, R., 1980, *J. Neurochem.* **34**:132–137.
129. Sturman, J. A., Gaull, G., and Räihä, N. C. R., 1970, *Science* **169**:74–76.
130. Gaull, G., Sturman, J. A., and Räihä, N. C. R., 1972, *Pediatr. Res.* **6**:538–547.
131. Dobbing, J., and Sands, J., 1979, *Early Hum. Dev.* **3**:79–83.
132. Agrawal, H. C., Davis, J. M., and Himwich, W. A., 1966, *J. Neurochem.* **13**:607–615.
133. Volpe, J. J., and Laster, L., 1972, *Biol. Neonate* **20**:385–403.
134. Timiras, P. S., Hudson, D. B., and Oklund, S., 1973, *Neurobiological Aspects of Maturation and Aging* (D. H. Ford, ed.), Elsevier, Amsterdam, pp. 267–275.
135. Seiler, N., Sarhan, S., and Roth-Schechter, B. F., 1981, *Dev. Neurosci.* **4**:181–187.
136. Sturman, J. A., and Gaull, G. E., 1976, *Biochim. Biophys. Acta* **428**:70–77.
137. Zlotkin, S. H., Bryan, M. H., and Anderson, G. H., 1981, *Am. J. Clin. Nutr.* **34**:914–923.
137a. Zlotkin, S. H., and Anderson, G. H., 1982, *Pediatr. Res.* **16**:65–68.
138. Rigo, J., and Senterre, J., 1977, *Biol. Neonate* **32**:73–76.
139. Tate, S. S., Grau, E. M., and Meister, A., 1979, *Proc. Natl. Acad. Sci. U.S.A.* **76**:2715–2719.
140. Malloy, M. H., Rassin, D. K., Heird, W. C., and Gaull, G. E., 1981, *Am. J. Clin. Nutr.* **34**:1520–1525.
141. Gray, E. G., and Whittaker, V. P., 1962, *J. Anat.* **96**:79–88.
142. DeRobertis, E., Pellegrino de Iraldi, A., Rodriguez de Lores Arnaiz, G., and Gomez, C. J., 1961, *J. Biophys. Biochem. Cytol.* **9**:229–235.
143. Whittaker, V. P., Michaelson, I. A., and Kirkland, R. J. A., 1964, *Biochem. J.* **90**:293–303.
144. Rassin, D. K., and Gaull, G. E., 1978, *Amino Acids as Chemical Transmitters* (F. Fonnum, ed.), Plenum Press, New York, pp. 571–597.
145. Shank, R. P., and Aprison, M. H., 1970, *J. Neurochem.* **17**:1461–1475.
146. Werman, R., Davidoff, R. A., and Aprison, M. H., 1966, *Life Sci.* **5**:1431–1440.
147. Johnston, G. A. R., 1968, *J. Neurochem.* **15**:1013–1017.
148. Rassin, D. K., 1975, *Metabolic Compartmentation and Neurotransmission* (S. Berl, D. D. Clarke, and D. Schneider, eds.), Plenum Press, New York, pp. 559–565.
149. Key, B. J., and White, R. P., 1970, *Neuropharmacology* **9**:349–357.
150. Olney, J. W., Ho, O. L., and Rhee, V., 1971, *Exp. Brain Res.* **14**:61–76.
151. Gaull, G. E., and Rassin, D. K., 1979, *Neural Growth and Differentiation* (E. Meisami and M. A. B. Brazier, eds.), Raven Press, New York, pp. 461–477.
152. Rassin, D. K., 1981, *Amino Acid Neurotransmitters* (F. V. DeFeudis and P. Mandel, eds.), Raven Press, New York, pp. 127–134.
153. Mangan, J. L., and Whittaker, V. P., 1966, *Biochem. J.* **98**:128–138.
154. Neal, M. J., and Iversen, L. L., 1969, *J. Neurochem.* **16**:1245–1252.
155. Bradford, H. F., 1970, *Brain Res.* **19**:239–247.
156. Rassin, D. K., 1972, *J. Neurochem.* **19**:139–148.
157. DeBelleroche, J. S., and Bradford, H. F., 1973, *J. Neurochem.* **21**:441–451.
158. Rassin, D. K., Sturman, J. A., and Gaull, G. E., 1977, *J. Neurochem.* **28**:41–50.
159. Rassin, D. K., Sturman, J. A., Hayes, K. C., and Gaull, G. E., 1978, *Neurochem. Res.* **3**:401–410.
160. Tudball, N., and Beaumont, A., 1979, *Biochim. Biophys. Acta* **588**:285–293.
161. Balcar, V. J., and Johnston, G. A. R., 1972, *J. Neurochem.* **19**:2657–2666.
162. Balcar, V. J., and Johnston, G. A. R., 1972, *J. Neurobiol.* **3**:295–301.
163. Segal, S., and Hwang, S. M., 1979, *J. Neurochem.* **33**:697–704.
164. Hwang, S. M., and Segal, S., 1979, *J. Neurochem.* **33**:1303–1308.
165. Kashiwamata, S., 1971, *FEBS Lett.* **19**:69–71.
166. Agrawal, H. C., Davison, A. N., and Kaczmarek, L. K., 1971, *Biochem. J.* **122**:759–763.
167. Davies, L. P., and Johnston, G. A. R., 1973, *Brain Res.* **54**:149–156.
168. Daly, E. C., and Aprison, M. H., 1974, *J. Neurochem.* **22**:877–885.
169. Rassin, D. K., and Gaull, G. E., 1975, *J. Neurochem.* **24**:969–978.
170. Salganicoff, L., and DeRobertis, E., 1965, *J. Neurochem.* **12**:287–309.
171. Broch, O. J., Jr., and Fonnum, F., 1972, *J. Neurochem.* **19**:2049–2055.
172. Aprison, M. H., and Daly, E. C., 1978, *Adv. Neurochem.* **3**:203–294.
173. Schrier, B. K., and Thompson, E. J., 1974, *J. Biol. Chem.* **249**:1769–1780.

174. Rassin, D. K., Sturman, J. A., and Gaull, G. E., 1981, *J. Neurochem.* **36:**1263–1271.
175. Misra, C. H., and Olney, J. W., 1975, *Brain Res.* **97:**117–126.
176. Misra, C. H., Mena, E. E., Rhee, V., and Olney, J. W., 1977, *Fed. Proc.* **36:**751.
177. Byrne, M. C., and Salganicoff, L., 1977, *Fed. Proc.* **36:**1007.
178. Pasantes-Morales, H., Loriette, C., and Chatagner, F., 1977, *Neurochem. Res.* **2:**671–680.
179. Snyder, S. H., and Taylor, K. M., 1972, *Perspectives in Neuropharmacology* (S. H. Snyder, ed.), Oxford University Press, New York, pp. 43–73.
180. Pasantes-Morales, H., Mapes, C., Tapia, R., and Mandel, P., 1976, *Brain Res.* **107:**579–589.
181. Macaione, S., Tucci, G., De Luca, G., and Di Giorgio, R. M., 1976, *J. Neurochem.* **27:**1411–1415.
182. Spears, R. M., and Martin, D. L., 1980, *Soc. Neurosci. Abstr.* **6:**443.
183. Rassin, D. K., Sturman, J. A., and Gaull, G. E., 1981, *J. Neurochem.* **37:**740–748.
184. Rassin, D. K., 1981, *Taurine in Nutrition and Neurology* (R. J. Huxtable and H. Pasantes-Morales, eds.), Plenum Press, New York, pp. 257–268.

The Significance of Tryptophan, Phenylalanine, Tyrosine, and Their Metabolites in the Nervous System

Simon N. Young

1. INTRODUCTION

The aromatic amino acids tryptophan, phenylalanine, and tyrosine are important as precursors of 5-hydroxytryptamine (5-HT) and the catecholamines. The extent to which availability of these amino acids can influence biogenic amine synthesis and function is still a matter of active research. Other pathways of metabolism lead to the trace amines (discussed in Volume 1), the pineal indoles, and products of the kynurenine type in which the indole ring is cleaved. Until recently, methodological limitations have precluded adequate study of the metabolism of many of these compounds. For some, their significance in brain is a matter purely for speculation. For others, the data available point to some function, although exactly what is often not clear. The aromatic amino acids lead to such diversity of metabolites that they cannot all be discussed here. For example, condensation products, such as tetrahydro-β-carboline, have recently been demonstrated in brain using mass spectrometric methods[1] and can influence neuronal activity in a variety of ways.[2] Such condensation products, which can be formed from all three amino acids in the course of their metabolism, are not within the scope of this chapter. The purpose of this chapter is to follow tryptophan, phenylalanine, and tyrosine from the diet through the periphery to the brain, to review some of the metabolic processes that occur there, and to indicate to some extent what the significance of these processes may be.

Simon N. Young • Department of Psychiatry, McGill University, Montreal, Quebec H3A 1A1, Canada.

2. FACTORS INFLUENCING THE CONTENTS OF TRYPTOPHAN, PHENYLALANINE, AND TYROSINE IN BRAIN

2.1. Dietary Intake

The effects of diets low in tryptophan have been studied using artificial diets deficient in tryptophan,[3] diets low in protein,[4] and a diet of corn, which has low levels of tryptophan.[5] With all these diets, in a few days plasma and brain tryptophan and hence brain 5-hydroxytryptamine (5-HT) are depressed. The low brain 5-HT seen after chronic corn consumption may be responsible for the increased response to a painful stimulus seen in this condition and may, thus, be functionally significant.[6] The catecholamine precursors have been studied less in chronic experiments. Tyrosine is, of course, not an essential amino acid as are the other two, being formed from phenylalanine. The important protective effects of diets low in phenylalanine are well established in phenylketonuria.

Changes in brain aromatic amino acid concentrations are also seen after acute dietary manipulations. When rats that have been deprived of food overnight are given carbohydrate, the content of all three amino acids in brain increases.[7] This is because carbohydrate stimulates release of insulin, which enhances uptake of the branched-chain amino acids into muscle. As discussed in more detail below (Section 2.4), the branched-chain amino acids share the same transport system as the aromatic amino acids for uptake into brain. Thus, a lower level of the branched-chain amino acids decreases competition for the transport system and increases uptake of the aromatic amino acids into brain.[8] When protein is added to the carbohydrate, the levels in brain of the essential amino acids tryptophan and phenylalanine decline, whereas tyrosine increases.[7] In each case, these changes are explained by the changes in the amino acid pattern, specifically by changes in the serum concentration ratio of each amino acid to the sum of the other neutral amino acids which compete with it for uptake into brain. Such changes may be of clinical significance, as ingestion of a balanced meal caused a decrease in CSF tryptophan and 5-hydroxyindoleacetic acid in humans.[9]

In the rat, shifts in plasma amino acid patterns that affect brain amine synthesis appear to be important for feeding regulatory mechanisms.[10] For example, protein intake in rats able to select their diet was inversely correlated with the ratio of tryptophan to the other neutral amino acids in plasma and may thus be dependent on brain 5-HT. Energy intake appears to be influenced by brain catecholamine activity. Humans often find that food ingestion alters their mood or behavior, and one challenge for the future is to determine to what extent such changes can be explained by altered levels of the aromatic amino acids in brain and hence altered amine synthesis and function.

2.2. Peripheral Protein Metabolism

Although an adult mammal in nitrogen balance maintains a corresponding balance with respect to all the essential amino acids, over short periods of time the free amino acid content will fluctuate depending on the rates of dietary intake, excretion, and metabolism. The amount of aromatic amino acids in the urine is small compared with dietary intake. Therefore, catabolic routes must eventually account for most of the amino acids ingested. If intake and catabolism were the only factors controlling free amino acid levels, they would fluctuate considerably throughout the day. However, protein metabolism diminishes these fluctuations. After a protein meal, there is rapid uptake of amino acids by liver and other tissues. The liver disposes of these amino acids fairly rapidly by catabolism and by protein synthesis. There is also protein synthesis in muscle, but in this tissue, the amino acids tend to remain in the free form longer.[11]

The rapid incorporation of the aromatic amino acids into protein in the liver after ingestion of food limits the rise in their concentration in plasma, whereas their release as a consequence of protein catabolism in the postabsorptive state prevents a large decline in plasma amino acid levels.[12] Tryptophan is released from the liver into the blood as early as 2 h after food is removed from rats.[13] After 24 h of starvation, there is a larger release of tryptophan, accompanied by a fall in the liver tryptophan content, possibly mediated by altered transport at the cell membrane.[13] Starvation is accompanied by an increase in brain tryptophan and 5-hydroxyindoles,[7] and in this situation, the extra tryptophan in the brain is supplied by breakdown of protein in the liver and other tissues.

The stores of free tryptophan, phenylalanine, and tyrosine in a rat are equivalent to between 2 and 4% of the rat's daily requirement for these amino acids.[12] Thus, small differences in the net rate of protein synthesis or catabolism as a result of differences in the dietary or hormonal state of the animal could cause relatively large changes in free tryptophan. However, although some of the dietary and hormonal factors that influence protein metabolism have been elucidated,[14] there is no clear understanding of what controls the balance between free amino acid pools and labile protein stores. Increased knowledge in this area should increase understanding of the physiological variations in brain aromatic amino acid levels.

Tryptophan seems to play a special role in the regulation of protein synthesis. An increase in the tryptophan content of the liver causes an increase in ribosome aggregation,[15] and force-feeding of tryptophan will increase the rate of RNA and protein synthesis, even in the livers of well-fed rats.[16] This role of tryptophan may stem from the fact that it is the least abundant amino acid in the pool available for protein synthesis.[12] Because of the small pool size of free tryptophan, an increase in the rate of incorporation of tryptophan into protein can deplete free tryptophan stores. Thus, if protein synthesis is stimulated by growth hormone in the hypophysectomized rat, brain tryptophan and 5-hydroxyindole levels decline.[17] When rats that have been trained to eat their normal daily meal in 2 h are given a single tryptophan-deficient meal, there is a marked fall in serum and brain tryptophan and in brain 5-HT which persists for more then 24 h.[18]

2.3. *Peripheral Amino Acid Metabolism*

Quantitatively the most important irreversible route of tryptophan metabolism is the kynurenine pathway which is initiated by tryptophan pyrrolase (L-tryptophan-2,3-dioxygenase), primarily a liver enzyme. Other pathways such as 5-HT or tryptamine synthesis account for less than 5% of ingested tryptophan.[19,20] Thus, as tryptophan pyrrolase must catabolize most of the dietary tryptophan, and as it is induced by both glucocorticoids and its substrate tryptophan, it can influence availability of tryptophan to the brain.

Induction of tryptophan pyrrolase by hydrocortisone lowers rat brain tryptophan and 5-hydroxyindoles.[21] This effect of hydrocortisone is mediated by the increase in the liver enzyme, as no effect on brain metabolism is seen in circumstances in which hydrocortisone does not induce tryptophan pyrrolase. This is so in rats over 100 days old[22] and in the mongolian gerbil (*Meriones unguiculatus*).[21] The decline in tissue tryptophan content in these circumstances ranges from 1.5 to 4 µg/g.[23] In rats of 125 g body weight, the rate of tryptophan catabolism in the liver, when tryptophan pyrrolase is induced by hydrocortisone, is about 20 µg/min.[24] Degradation of tryptophan in the liver at this rate for less than 1 h will account for the depletion of free tryptophan stores in the whole animal, and yet tryptophan pyrrolase remains elevated for several hours after hydrocortisone. Presumably, as tissue tryptophan falls, there is net protein catabolism and release of more tryptophan. A much larger and longer-lasting induction of tryptophan pyrrolase occurs with the substrate analogue DL-α-methyltryptophan. When this compound is injected into rats, there is severe depletion of tryptophan in the liver, blood, and brain accompanied by considerable net protein catabolism and loss of body weight.[25] These data indicate how protein metabolism interacts with the catabolic pathway to prevent any very large effect on brain tryptophan and 5-HT in normal circumstances.

After a tryptophan load, protein synthesis will play no important part in controlling tryptophan levels, and in these circumstances, induction of tryptophan pyrrolase either by hydrocortisone or by immobilization stress sharply decreases the rise in brain tryptophan that occurs after the load.[26] Induction of tryptophan pyrrolase by a tryptophan load will also limit the rise in brain tryptophan. Thus, growth hormone, which antagonizes induction of tryptophan pyrrolase by tryptophan, potentiates the increase in brain tryptophan and 5-hydroxyindoles when rats are given the amino acid.[27] These data suggest that tryptophan pyrrolase acts in some ways like an overflow system to remove excess tryptophan from the system. Obviously, when tryptophan is given clinically, the liver will attempt to minimize any change in the brain tryptophan content, a fact that may limit the efficacy of this psychopharmacological approach.

For phenylalanine, the most important irreversible pathway is hydroxylation to tyrosine. In phenylketonuria, this enzyme does not function, and plasma phenylalanine levels are greatly elevated. In heterozygotes, base-line plasma phenylalanine tends to be high, and after a phenylalanine load, the increase of phenylalanine tends to be greater and that of tyrosine less than in controls. However, there is overlap between the groups.[28] Normally, transamination of phenylalanine is relatively insignificant, but it becomes more important as the amino acid level rises.[29]

The main catabolic route of tyrosine is initiated by tyrosine transaminase.

In the liver, this enzyme is induced by tryptophan (not tyrosine)[30] and glucocorticoids.[31] Unlike tryptophan pyrrolase, it is also induced by insulin and glucagon.[32] The diurnal variation of tyrosine transaminase in the liver probably results mainly from variation in protein intake, of which tryptophan may be the active component.[33] Liver tyrosine transaminase and brain tyrosine have an inverse diurnal relationship,[34] but it is not possible to say whether this relationship implies cause and effect. Tyrosine transaminase is present in brain as well as liver, although at lower levels, and is induced by glucocorticoids in both tissues.[35] When rats are treated with hydrocortisone with or without glucagon, there is a decline in plasma and brain tyrosine, although there is no agreement on whether catecholamine levels fall also.[35,36]

2.4. Transport into Brain

Transport of amino acids from plasma into brain cells occurs in two stages. The first is at the blood–brain barrier, which is localized in the endothelium of brain capillaries, and the second is from the extracellular fluid of the brain across the cell membrane into the cell. Both of these processes have been studied using a variety of methods. Uptake at the blood–brain barrier has been studied most extensively using the Oldendorf technique in which a solution containing labeled amino acid and [³H]water (as a freely diffusible standard) is injected into the carotidartery. The uptake of the amino acid is measured, relative to that of water, after a single passage of the solution through the brain microcirculation. An extensive study of self-inhibition and cross inhibition using this technique has demonstrated a carrier system for neutral amino acids which transports tryptophan, phenylalanine, tyrosine, leucine, isoleucine, valine, threonine, histidine, cysteine, serine, glutamine, and asparagine.[37] The affinity of the last four amino acids for the transport system is very low. The K_m of the first eight amino acids for the system transporting them across the blood–brain barrier is close to their plasma amino acid levels. This is unlike transport into the liver, intestine, kidney, and erythrocyte where the K_m is tenfold greater than the plasma levels.[38] Thus, at the blood–brain barrier, but not at the other tissues, there is competition among the amino acids for entry into tissue. Experimental results indicate that the inhibition is simple competitive inhibition, as K_i values approximate K_m values.[39,40]

Because of this competition, the contents in brain of the aromatic amino acids do not follow their plasma concentrations in any simple way but also depend on the concentrations of the other amino acids. Because the aromatic amino acids are metabolized primarily in the liver and the branched-chain amino acids mainly in muscle, the two groups are susceptible to different influences. Thus, as discussed in Section 2.1, glucose stimulates release of insulin, which enhances uptake of the branched-chain amino acids into muscle. As the concentrations of the branched-chain amino acids in plasma fall, there is less competition for the uptake system, and the level of the aromatic amino acids in brain increase.[7] In hepatic cirrhosis also, the plasma level of the branched-chain amino acids is lowered, whereas that of the aromatics is raised, thus explaining in part the elevated levels of the aromatic amino acids in brain.[41]

Changes in brain amino acid levels after protein ingestion depend on the pattern of increase of all the large neutral amino acids in plasma.[7]

Although the other amino acids in plasma are in free solution, most of the tryptophan is bound to albumin.[42] The proportion of tryptophan in plasma that is in free solution (i.e., non-albumin bound) depends in part on the nonesterified fatty acids. These fatty acids bind to albumin at the same site as tryptophan and displace it.[43] Many factors that influence plasma free tryptophan affect rat brain tryptophan in the same way.[44,45] The free plasma tryptophan is increased by drugs such as salicylate, clofibrate, and probenecid, which bind to albumin and displace tryptophan, and also by starvation and lipolytic drugs, such as heparin, caffeine, and isoproterenol, which increase fatty acids in plasma. All these treatments raise brain tryptophan, and these data have been taken to indicate that it is free plasma tryptophan that controls the brain level. However, this view has been challenged. In a long series of acute dietary experiments, changes in rat brain tryptophan were related to the ratio of total plasma tryptophan to the sum of the competing amino acids and not to a similar ratio involving free plasma tryptophan.[8,46] To confuse matters further, there is controversy about the validity of some of the methods used to measure free plasma tryptophan.[47,48]

The types of experiments described above provide insufficient data to make definite conclusions about the role of albumin binding. Thus, an increase in brain tryptophan cannot be attributed to an increase in free plasma tryptophan unless it is known that there is no decrease in the plasma concentration of the competing amino acids. In most cases, these have not been measured. The results from the dietary experiments are also not conclusive. The difficulty in interpreting results from this type of experiment is illustrated by the study, described in Section 2.1., that looked at human CSF tryptophan after ingestion of food.[9] The changes in CSF tryptophan corresponded to changes in the ratio of free plasma tryptophan to the sum of the competing amino acids when this was calculated using concentrations in micrograms per milliliter plasma.[9] However, when the ratio was recalculated using molar concentration, the CSF changes corresponded better to a ratio employing total and not free plasma tryptophan.[8] Neither method of calculation is entirely valid, as they do not take into account the different affinities of the different amino acids for the transport system.

An experimental approach that gives results more amenable to interpretation is to look at the effect of albumin on tryptophan uptake. When albumin is added to the injected bolus in the Oldendorf method, there is a decrease in the uptake of tryptophan.[49] However, the decrease is not as large as would be expected from measurements of the free tryptophan fraction in the injected bolus. The decrease in uptake associated with albumin lessens as tryptophan is bound less tightly to albumin, and the uptake is unchanged by albumin when about half the tryptophan is in free solution.[50] The data indicate that, depending on the conditions, between 50% and 100% of the tryptophan bound to albumin is available for uptake by the brain. Although the effect of albumin binding is small, it is magnified to a certain extent by the inhibition of the amino acids.

Thus, albumin reduces the effective tryptophan concentration, and at a lower tryptophan concentration, inhibition by the other amino acids will be greater. In one situation, then, albumin reduces tryptophan uptake 26% in the absence of competing amino acids and 33% in their presence.[50] Depending on the physiological conditions. (i.e., free fatty acids and competing amino acids in plasma), the apparent free fraction of tryptophan *in vivo* may approximate anything from the free plasma tryptophan measured *in vitro* to the total plasma tryptophan.[51] These data indicate that the views that attributed control of brain tryptophan either to the free or to the total plasma tryptophan were too highly polarized, although both may be useful approximations in different circumstances.

Another technique that has been used to study transport of amino acids into brain involves infusion of labeled amino acid into the blood over a period of several minutes.[40] When a constant specific activity of the amino acid is maintained in the circulation, label accumulates linearly in the brain for up to 5 min. This technique,[40] like the Oldendorf technique,[49] shows that uptake occurs by a nonsaturable process (presumably diffusion) as well as by active transport. However, the diffusion component is small under normal circumstances. The infusion technique reveals a high rate of influx. For example, at normal plasma tryptophan concentrations, the influx is around 2 nmol/min per g tissue.[52] The tryptophan content of brain is of the order of 20 nmol/g tissue, so the free tryptophan pool in brain must have a high turnover rate. Experimental studies suggest that the half-life of free amino acids in brain varies from 3 to 30 min for the essential amino acids.[53]

Once the amino acids have been transported across the blood–brain barrier into the extracellular fluid of the brain, they must be transported into the cell. This second process has been studied with a variety of tissue preparations including slices, synaptosomes, and cultured cells. Transport into slices is similar to that across the blood–brain barrier, as it consists of two components, an active transport and passive diffusion, and is a system shared by all large neutral amino acids.[54]

Several groups have addressed the problem of whether transport of tryptophan or tyrosine into neurons has special properties related to the role of these amino acids as neurotransmitter precursors. In synaptosomes prepared from brain (which, of course, will only contain a very small proportion of synaptosomes from 5-HT or catecholamine neurons), the active uptake of tyrosine and tryptophan is enhanced by conditions similar to those observed during neuronal depolarization; i.e., it is stimulated by Ca^{2+} and K^+ and inhibited by Na^+.[55,56] Although it is not possible to isolate specific neuronal systems, uptake of tryptophan, phenylalanine, and tyrosine has been studied in three neuroblastoma clones that synthesize norepinephrine, 5-HT, and acetylcholine, respectively.[57] There were no substantial differences among the three clones in their ability to take up the three aromatic amino acids. In the enteric nervous system, 5-HT will stimulate uptake of tryptophan,[58] but no such effect has been observed in the central nervous system. Although catecholamine and 5-HT neurons will use much more tyrosine or tryptophan than other brain cells, the lack of any evidence for special uptake systems in these neurons may indicate that the additional requirement for amino acids is satisfied by decreased efflux rather than enhanced uptake.

3. THE AROMATIC AMINO ACIDS AND PROTEIN SYNTHESIS

Estimates of the rate of turnover of protein in brain indicate that the amount of amino acids that are incorporated and released from protein within 30 min equals the total cerebral pool of most free amino acids.[59] Thus, the free amino acid pool in brain exchanges rapidly with both free amino acids in plasma and cerebral protein. For brain as a whole, protein synthesis is the most active (although reversible) pathway of aromatic amino acid metabolism. However, this is probably not so within 5-HT or catecholamine neurons. Although such neurons cannot be studied in isolation, the pineal gland, which synthesizes 5-HT very actively as an intermediate on the pathway to melatonin, can. In cultured rat pineals, labeled tryptophan is converted via 5-HT at 40 times the rate it is incorporated into protein.[60] Suggestions that the requirement for an amino acid as a biogenic amine precursor can influence, in some circumstances, protein synthesis within the neuron remain speculation.[61]

As discussed in Section 2.2, tryptophan availability may influence protein synthesis in the periphery because it is the least abundant amino acid in the pool available for protein synthesis. However, tryptophan probably does not limit protein synthesis in the adult brain. Thus, tRNA species including tryptophyl-tRNA are present in immature mouse brain at about the same level as in adult brain[62] even though the rate of cerebral protein synthesis is much greater in the immature animals.[63] Also, the tryptophan-activating enzyme has a particularly low K_m for tryptophan, as studied in the brain of the adult water buffalo.[64] However, even if tryptophan is not limiting for protein synthesis in brain, tryptophan administration has been reported to stimulate incorporation of amino acids into rat brain proteins.[65]

Apart from the possible stimulatory effect of tryptophan on brain protein synthesis, the aromatic amino acids may influence protein synthesis indirectly through their metabolites, 5-HT and dopamine. Both of these amines can cause disaggregation of brain polysomes and suppress brain protein synthesis.[66] Studies in vitro indicate that 5-HT may act by inhibiting initiation of mRNA-directed protein synthesis.[67]

4. TRYPTOPHAN AND 5-HYDROXYTRYPTAMINE SYNTHESIS

Tryptophan hydroxylase, the rate-limiting enzyme in 5-HT synthesis, is localized solely within 5-HT neurons in the CNS.[68] Much evidence indicates that tryptophan hydroxylase is not normally saturated with its substrate and that changes in the content of tryptophan in brain affect the rate of 5-HT synthesis. Many other factors can influence 5-HT synthesis, but they are not within the scope of this chapter. The first indication that tryptophan hydroxylase may not be fully saturated with tryptophan came from an experiment in which rats pretreated with a monoamine oxidase inhibitor were given a tryptophan load, resulting in an increase in brain 5-HT.[69] Later studies showed that large loads of tryptophan, given alone, increased both 5-HT and 5-hydroxyindoleacetic acid concentrations in brain, although 5-hydroxytryptophan remained undetectable.[70] This may be of importance physiologically, as doses of tryptophan as low as 12.5 mg/kg (less than 5% of the normal daily dietary intake of tryptophan) can raise brain 5-HT significantly.[71] Investigations on the accumulation of 5-HT in rat brain after inhibition of monoamine oxidase[72] or on the accumulation of 5-hydroxytryptophan after inhibition of aromatic amino acid decarboxylase[73] indicate that tryptophan hydroxylase is about half saturated under normal circumstances and that tryptophan loading can double the rate of 5-HT synthesis. This is in reasonable agreement with data obtained *in vitro*. The K_m for rabbit hindbrain tryptophan hydroxylase is 50 μM with tetrahydrobiopterin as cofactor,[74] whereas the normal brain tryptophan content is equivalent to about 30 μM. Thus, data both *in vivo* and *in vitro* indicate that the K_m for tryptophan hydroxylase is about the same as the brain tryptophan concentration.

Although it is commonly assumed that tetrahydrobiopterin is the natural cofactor for tryptophan hydroxylase, the evidence for this is not conclusive. The K_m of the particulate enzyme (presumably in synaptosomes), which can be determined without exogenous pteridine, is the same as that found with tetrahydrobiopterin.[75] However, with tetrahydrobiopterin, there is substrate inhibition at concentrations of tryptophan above 0.2 mM.[74] There is one report of substrate inhibition *in vivo*.[76] Doses of tryptophan as low as 25 mg/kg were reported to increase brain 5-HT maximally, whereas 50 mg/kg produced control levels of 5-HT. However, numerous other studies have not reported substrate inhibition even at much higher doses (e.g., refs. 72 and 73). Although high tryptophan levels may increase readings after a large tryptophan load when 5-HT is determined fluorometrically,[76] no substrate inhibition is seen, even after large tryptophan loads, when 5-HT is determined by high-performance liquid chromatography with electrochemical detection,[77] a method not subject to interference by tryptophan. Thus, the balance of evidence indicates that substrate inhibition does not occur *in vivo*, although at present, there is no indication why there is this discrepancy between the data *in vivo* and *in vitro*.

In humans, as in rodents, tryptophan hydroxylase is probably only about half saturated under normal circumstances. Thus, the concentrations of tryptophan and 5-hydroxyindoleacetic acid in human ventricular,[78,79] cisternal, and lumbar[80] cerebrospinal fluid are positively correlated. This suggests that physiological variations in brain tryptophan may influence brain 5-HT synthesis in humans. When humans are given a tryptophan load, CSF 5-hydroxyindoleacetic acid rises.[81] A load of 3 g (about two times the normal daily dietary intake) doubles it, but 6 g causes no further increase,[82] indicating that tryptophan hydroxylase is already saturated.

If tryptophan availability influences brain 5-HT synthesis, an obvious question is whether it also affects 5-HT release or function. Administration of tryptophan to animals inhibits many types of behaviors,[83-85] and the time course of the behavioral alteration is related to the increase in brain 5-HT.[83] Tryptophan is also capable of reversing some of the behavioral effects of *para*-chlorophenylalanine, an inhibitor of 5-HT synthesis.[86] However, although some of the effects of tryptophan administration can be reversed by 5-HT blockers, they are not necessarily mediated by 5-HT, as they can be blocked by peripheral decarboxylase inhibitors.[85,87] Possible candidates as mediators of some of tryptophan's behavioral effects are tryptamine, tetrahydronorharmane, tryptophol, and kynurenine.[85] Also, in many circumstances, an increase in brain 5-HT caused by tryptophan administration does not bring about the behavioral effects that might be expected.[88,89]

An alteration in 5-HT release in response to tryptophan loading would presumably imply an increase in 5-HT release each time a 5-HT neuron fires. Single-unit recording studies on identified cells in the raphe nuclei, the site of the cell bodies of 5-HT neurons, have revealed a slow rhythmic discharge rate[90] that shows a strong positive correlation with the level of behavioral arousal.[91] Experimental manipulations of neuronal firing indicate that the release of labeled 5-HT endogenously formed from labeled tryptophan is dependent on nerve activity.[92] During sleep, when firing rates of raphe neurons are lowered, there is diminished release of endogenous 5-HT,[93] suggesting that the relationship between release of 5-HT and neuronal activity is of physiological as well as pharmacological significance. When animals are given tryptophan, there is a dose-related decline in raphe unit activity[94] which is dependent on an increase in local (raphe) 5-HT content.[90]

If tryptophan administration increases the amount of 5-HT released each time a neuron fires and decreases the rate of firing, the net result may be little or no change in 5-HT release; experimental results support this idea. Release of 5-HT into the perfused ventricles of the rat was examined.[95] Acute injection of 5-hydroxytryptophan caused an immediate, important, and long-lasting increase in 5-HT release, whereas tryptophan led to a transient and moderate elevation of 5-HT release that was only detected during the second hour after tryptophan administration. When 5-HT release was examined in the brain of the freely moving unanesthetized rat using *in vivo* voltammetry, tryptophan administration was found not to alter 5-HT release at all.[96]

The data above indicate that tryptophan administration can alter brain function, but the extent to which these alterations are mediated by changes in 5-HT release is doubtful.

5. PHENYLALANINE, TYROSINE, AND CATECHOLAMINE SYNTHESIS

As with 5-HT, there are numerous factors that influence catecholamine synthesis, but only precursor availability comes within the scope of this chapter.

The rate-limiting step in the synthesis of dopamine, norepinephrine, or epinephrine in brain is tyrosine hydroxylase. This enzyme converts tyrosine to 3,4-dihydroxyphenylalanine (DOPA) in one step or phenylalanine to DOPA in two steps via tyrosine. In synaptosomes, which contain the natural cofactor for tyrosine hydroxylase (probably tetrahydrobiopterin), formation of catechol compounds from phenylalanine occurs at about half the rate from tyrosine.[97] Tyrosine formed from phenylalanine within synaptosomes does not seem to equilibrate with endogenous tyrosine.[98–100] There is competition between the two precursors for the hydroxylating system, although of an uncompetitive type. Tyrosine inhibits catecholamine synthesis from phenylalanine more than phenylalanine inhibits catecholamine synthesis from tyrosine.[100] As phenylalanine is rapidly converted to tyrosine in the liver, this may limit the extent to which phenylalanine acts as a precursor of catecholamine synthesis even after phenylalanine loading. However, phenylalanine can act efficiently as an amine precursor, and indeed, phenylalanine administration can increase DOPA synthesis in rats given an inhibitor of aromatic amino acid decarboxylase more than tyrosine administration can in the striatum and limbic forebrain, although the increases are similar in the rest of the hemispheres.[101] Large doses of phenylalanine decrease catecholamine synthesis, presumably because hydroxylation of phenylalanine occurs to the exclusion of tyrosine.[77,101] The implications of the relatively high rate of catecholamine synthesis from phenylalanine, and its compartmentation from catecholamine synthesis from endogenous tyrosine, are not fully understood.

Although it has been known for over 10 years that tyrosine administration has functional effects that could be mediated by catecholamines, such as a decrease in blood pressure in rats made hypertensive,[102] metabolic effects of tyrosine have only been studied recently. In metabolic studies, the situation is complicated by the fact that tyrosine is the precursor of both dopamine and norepinephrine, with a minor contribution going to epinephrine. Thus, studies that looked at DOPA accumulation after inhibition of aromatic amino acid decarboxylase were studying a heterogeneous population of neurons. In studies of this type, tyrosine increased DOPA to a small extent in whole rat brain[103] and to a larger extent in the hemispheres in which norepinephrine is the predominant catecholamine.[101] Treatments that lowered brain tyrosine decreased DOPA accumulation[103] and indicated that tyrosine hydroxylase is about 75% saturated with tyrosine under normal circumstances.[101]

Dopamine and norepinephrine have been studied separately with respect to tyrosine availability, but epinephrine has not been looked at. Variations in brain tyrosine affect rat brain 4-hydroxy-3-methoxyphenylethylene glycol sulfate in the same way, thus indicating that tyrosine availability controls norepinephrine synthesis under normal conditions.[104] This is not so for dopamine. Tyrosine loading does not increase homovanillic acid (HVA) under normal circumstances, but it does when animals are given haloperidol[105] or reserpine[106] or when their nigrostriatal pathway is lesioned.[107] Under these circumstances, impulse flow in the dopamine neurons and dopamine synthesis are increased. Possibly, increased dopamine synthesis will lower levels of tyrosine within the dopaminergic neurons, making tyrosine hydroxylase less saturated with tyrosine and thus more responsive to tyrosine loading.

Tyrosine loads probably do increase dopamine release as well as dopamine synthesis when dopamine impulse flow is increased, as in the situations described above. However, under more normal circumstances, the situation may be similar to that concerning tryptophan and 5-HT release. As with 5-HT, treatments that increase availability of dopamine or norepinephrine tend to decrease firing of dopaminergic or noradrenergic neurons.[108] Although this has not been looked at after tyrosine administration, it may well be that increased catecholamine synthesis and decreased firing of catecholamine neurons after tyrosine will balance each other out. Although tyrosine loading definitely has functional effects,[77,102,109] these are not necessarily mediated by changes in catecholamine release in all circumstances. As with tryptophan, other minor pathways may be important. This is probably true of phenylalanine, which can be converted both to phenylethylamine and phenylethanolamine.[110] In this respect, it should be noted that the behavioral effects of phenylalanine and tyrosine are not always the same.[77,109]

6. MINOR TRYPTOPHAN METABOLITES

6.1. Tryptamine and N-Methylated Tryptamines

Tryptamine is present in rat brain at a concentration that is two orders of magnitude less than that of 5-HT. At one time, in common with some other compounds that are the product of minor pathways of aromatic amino acid metabolism, it was considered a metabolic accident. There are enough data now to indicate that it may have functional significance in brain, but unfortunately, some of the data in the literature are unreliable. Because of the low levels of tryptamine in brain, the analytical methodology used to detect it has not always been reliable, a recurring problem in studies on minor metabolites. Thus, the content of tryptamine in brain has been reported to be of the order of 20–70 ng/g using radiochemical derivative,[111] spectrophotofluorometric,[112] and radioenzymatic[113] assays, whereas three separate mass spectrometric assays indicate that the correct value is no more than 0.5 ng/g.[114–116] It is distributed heterogeneously in the brain, with highest levels in the caudate nucleus and hypothalamus.[117]

Tryptamine is formed by the action of aromatic amino acid decarboxylase on tryptophan. The K_m for this enzyme is high (14 mM),[75] and thus, tryptamine synthesis should be dependent on tryptophan concentrations over a wide range. Experimental data support this idea.[118,119] Unlike 5-HT, tryptamine is sufficiently lipophilic to diffuse across the blood–brain barrier,[120] and perhaps as much as half of the tryptamine in brain is synthesized peripherally.[121] Tryptamine is not stored in vesicles,[119] and its release is not dependent on nerve activity.[92] It depresses the firing of most neurons in rat cortex, and at subthreshold doses, it can markedly modulate the effect of 5-HT on neuronal firing.[122] Some of the behavioral effects of tryptophan can be modified by peripheral decarboxylase inhibitors[85,87] which would inhibit tryptamine synthesis peripherally and reduce the level of this amine in brain.[121] Thus, some of the behavioral effects of tryptophan may be mediated in part by tryptamine.

The level of the tryptamine metabolite indoleacetic acid in rat CSF is about 6% of the level of 5-HIAA.[121] The relative rates of metabolism of the parent amines in rat brain are about the same.[118] In human CSF, the equivalent value is 20%,[121] indicating that CNS tryptamine metabolism may be relatively more important in humans than in rats. Human CSF indoleacetic acid is significantly and positively correlated with CSF tryptophan,[80] and it increases in proportion to the size of a tryptophan load on tryptophan administration.[82] Thus, tryptophan availability is involved in the control of tryptamine metabolism in human CNS.

Although the main pathway of tryptamine metabolism is by monoamine oxidase to indoleacetic acid, N-methylation can also occur, giving N-methyltryptamine and N,N-dimethyltryptamine. Similar transformations can occur with 5-HT, giving 5-hydroxy,-N,N-dimethyltryptamine (bufotenine) and 5-methoxy-N,N-dimethyltryptamine. The interest in these compounds depends on the fact that they are hallucinogenic and thus may play some role in mental illness.[123,124] They are formed by the action of indoleamine-N-methyltransferase, with S-adenosylmethionine acting as methyl donor. Suggestions that methyltetrahydrofolate could act as methyl donor arose because of a methodological artifact.[125] The enzyme has been demonstrated in human brain and lung.[123.] Reliable methods employing mass spectrometry have shown the presence of very low levels of N-methyl- and N,N-dimethyltryptamine and bufotenine in human blood or urine.[126–128] Methylated tryptamines are more lipophilic than tryptamine and should cross the blood–brain barrier with relative ease. Thus, their presence in blood or urine implies their existence in brain. However, it is not possible to say whether they are present in brain at a concentration sufficient to influence neuronal function in any perceptible way.

6.2. Melatonin, N-Acetyl-5-Hydroxytryptamine, and 5-Methoxytryptamine

5-Hyroxytryptamine can be converted to a number of important products. Melatonin is formed from 5-HT in the pineal by the action of 5-HT N-acetyltransferase, giving N-acetyl-5-HT, and hydroxyindole-0-methyltranserase. The former enzyme is the rate-limiting step in this process, and its activity is controlled by a β-adrenergic receptor. The adrenergic stimulation of the pineal is turned off by light, which stimulates the retina. The signal passes from there through a variety of neuronal pathways and reaches the pineal via the superior cervical ganglion. Thus, during dark, the N-acetyltransferase activity and melatonin levels are high, whereas light diminishes both greatly.[129] Melatonin is released from the pineal into blood and exhibits a similar rhythm in rat pineal and serum.[130] In human plasma, rhythms are seen that depend not only on the time of day but also on the time of year.[131]

The main target organ for melatonin is the brain and, in particular, the hypothalamus.[132] When labeled melatonin is given to rats, it concentrates preferentially in the hypothalamus,[133] and binding sites for melatonin, which have been reported recently in animal brains, are present at a high concentration in the hypothalamus.[134,135] The antigonadotrophic actions of melatonin are presumably mediated by its action on the hypothalamus and thus on release of pituitary hormones.

The introduction of radioimmunoassays for the measurement of melatonin has greatly increased knowledge about this compound and has led to the realization that it is not purely a pineal compound. Thus, melatonin is synthesized in the retina, where it shows a diurnal rhythm,[136] and also in the digestive tract of the rat, where no rhythm is seen.[137]

The immediate precursor of melatonin is N-acetyl-5-HT. An enzyme, different from pineal N-acetyltransferase, that is capable of acetylating 5-HT has been reported in rat brain.[138] By use of an antiserum to N-acetyl-5-HT, this compound has been detected in the cerebellum, the spinal tract of the trigeminal roots, and the pontal and spinal reticular formation, all tissues devoid of melatonin.[139] Thus, N-acetyl-5-HT may have functional significance in its own right as well as being a precursor of melatonin.

Although 5-HT in the pineal is usually acetylated before being methylated, there is evidence that, under some circumstances, it can be methylated directly to give 5-methoxytryptamine. Levels of this compound in pineal are low and vary from species to species. For example, rat pineals did not contain measurable 5-methoxytryptamine, although it could be detected in cow pineals.[140] This is consistent with data obtained in vitro that showed that hydroxyindole-0-methyltransferase from bovine pineals can methylate 5-HT but that the enzyme from rat can not.[141] The significance of 5-methoxytryptamine in pineals is not known.

Reports that 5-methoxytryptamine is measurable in rat hypothalamus resulted from an artifact.[142] Melatonin in the hypothalamus was deacetylated during the assay procedure. As the artifactual report used mass spectrometry, this illustrates the importance of assessing all types of methodology critically when dealing with very low levels of compounds.

6.3. 5-Hydroxytryptophol, 5-Methoxytryptophol, and Tryptophol

Monoamine oxidase converts 5-HT to an aldehyde that can then be oxidized to 5-hydroxyindoleacetic acid by aldehyde dehydrogenase or reduced to 5-hydroxytryptophol by aldehyde reductase. Because the K_m for the reductase is very much higher than that for the dehydrogenase, most of the aldehyde is oxidized.[143,144] Thus, the concentration of 5-hydroxytryptophol in rat brain is only about 1% of the 5-hydroxyindoleacetic acid level.[145] In pineal, the relevant fraction is about 20%.[146] The pineal contains very high levels of 5-HT and presumably also of the aldehyde derived from 5-HT by the action of monoamine oxidase. Higher levels of the aldehyde tend to favor more production of the reduced meabolite,[143,144] thereby possibly explaining the relatively greater production of 5-hydroxytryptophol in the pineal. Ethanol will also favor production of the reduced metabolites. Ethanol increases the $NADH:NAD^+$ ratio. As nicotinic cofactors are used for both aldehyde dehydrogenase and reductase, the presence of ethanol will alter the relative enzyme activities.[143,144] Thus, 5-

hydroxytryptophol levels are elevated in human CSF during ethanol intoxication.[147]

Hydroxyindole-0-methyltransferase is capable of methylating 5-hydroxytryptophol to 5-methoxytryptophol,[141] and the latter compound has been detected in rat pineals at about the same level as melatonin using gas chromatography–mass spectrometry.[148]

5-Hydroxy- and 5-methoxytryptophol are somewhat more hydrophilic than tryptophol, which is highly lipophilic and capable of diffusing across the blood–brain barrier more easily than water.[149] However, both compounds are capable of diffusing into brain and can influence brain function when administered peripherally to animals. 5-Methoxy- and 5-hydroxytryptophol, like melatonin, affect pituitary and ovarian function,[150] and the hydroxy compound, like tryptophol itself, can induce a sleeplike state.[151]

6.4. Products of the Action of Indoleamine-2,3-Dioxygenase

It has been known for many years that brain homogenates are capable of converting tryptophan to kynurenine.[152] The enzyme responsible for the formation of kynurenine in brain is different from the liver tryptophan-2,3-dioxygenase discussed in Section 2.3. The enzyme in brain has a much lower activity than the liver enzyme[153] and a broader range of specificity, for it is active towards 5-hydroxytryptophan,[154] 5-HT, D-tryptophan, and melatonin[155] as well as L-tryptophan. Thus, it is called indoleamine-2,3-dioxygenase. Within rabbit brain, highest activity is found in the pineal and choroid plexus. The activity within the pineal may be important, as the main metabolite formed from labeled tryptophan in cultured pineals is kynurenine.[156] Also, when labeled melatonin was injected intracisternally in the rat, the major metabolite was the product formed by ring cleavage, N-acetyl-5-methoxykynurenamine.[155] 5-Hydroxykynurenine, which would be formed by the action of indoleamine-2,3-dioxygenase on 5-HT, has been detected in mouse brain[157] and can produce contractions of dog and human cerebral arteries through 5-HT receptors.[158] Thus, the little that is known about the functional significance of indoleamine-2,3-dioxygenase indicates that more research would be useful in this relatively neglected area.

7. MINOR PHENYLALANINE AND TYROSINE METABOLITES

7.1. Phenylethylamine and Phenylethanolamine

Phenylethylamine is present in brain in small amounts. The actual content is a matter of controversy, and some of the methods used for its quantitation may be unreliable. Unfortunately, some of the data on phenylethylamine in the literature must be discounted because of these methodological problems. This matter is dealt with in detail in Chapter 8 of Volume 1.

Phenylethylamine is formed by the action of aromatic amino acid decarboxylase on phenylalanine. The K_m of the enzyme for phenylalanine is high

$(10^{-2}$ M),[159] suggesting that the rate of phenylethylamine synthesis should vary sensitively with changes in phenylalanine availability, and indeed, phenylalanine loading is known to increase levels of the amine in brain.[160] This may be important in phenylketonuria, as the phenylethylamine metabolite phenylacetic acid is one of the compounds suggested to be responsible for the behavioral deficits in this disease.[161]

After inhibition of monoamine oxidase, phenylethylamine accumulates in brain at about the same rate as the catecholamines.[162] As its concentration in brain is much less than that of the catecholamines, its half-life must be short, probably less than 1 min.[162] However, the data on accumulation of the amines in brain do not necessarily imply that phenylethylamine is synthesized in brain at the same rate as the catecholamines. Phenylethylamine is highly lipophilic and crosses the blood–brain barrier with ease, indeed, more easily than tryptamine.[163] Most tissues of the rat contain higher concentrations of phenylethylamine than whole brain,[164] and the contribution of peripheral sources to brain phenylethylamine is probably appreciable. In human CNS, as in rat brain, the rate of metabolism (if not synthesis) of phenylethylamine is probably comparable to the rate of catecholamine metabolism, as the concentration in human CSF of phenylacetic acid is as great as that of the dopamine metabolite homovanillic acid.[165]

The role of phenylethylamine in brain function is not known, but much interest has been stimulated by the structural and behavioral similarities between phenylethylamine and amphetamine.[166]

Studies with labeled phenylalanine indicate that phenylethylamine can be metabolized further to phenylethanolamine,[110] but specific methodology for this compound suggests that it is present in rat brain at very low levels and is detectable only after inhibition of monoamine oxidase.[167]

7.2. Tyramines and Octopamines

The three isomers of tyramine (o, m, and p) are excreted in rat urine in the approximate ratio $1:3:7$.[168] The m and p isomers have been detected reliably in rat brain at about 0.3 and 1 ng/g, although, as with other trace amines, there have been numerous unreliable reports of larger amounts (see Chapter 8 of Volume 1). The quantities of each tyramine found increase with the number of pathways by which it can be synthesized. p-Tyramine can be formed by decarboxylation of tyrosine, although aromatic amino acid decarboxylase is not very active towards the natural p isomer of tyrosine, and this pathway is not quantitatively important.[169] The other pathways for the formation of p-tyramine are hydroxylation of phenylethylamine[168] and dehydroxylation of dopamine.[170] Decarboxylation is not a possibility for the formation of m-tyramine, as m-tyrosine does not occur naturally, but the other two pathways are active.[168,170] Only hydroxylation of phenylethylamine is a possibility for the formation of o-tyrosine. The enzymes responsible for the hydroxylation and dehydroxylation reactions have not been characterized.

Octopamine has, like tyramine, been misidentified and finally identified in low concentrations in rat brain as the m and p isomers (see Chapter 8 of Volume

1). It can be formed by the action of dopamine-β-hydroxylase on tyramine or from DOPA, presumably by dehydroxylation of norepinephrine.[171]

The acid metabolites of *p*-tyramine and *p*-octopamine have been detected in human CSF. Both compounds increase on administration of probenecid, indicating turnover of the parent amines in human CNS.[172] However, the roles of the tyramines and octopamines in brain function are not known.

8. THE D AMINO ACIDS

The D isomers of the aromatic amino acids are of interest because they may have effects on the brain different from those of the L isomers. Although the effects of D-phenylalanine on the brain have been studied the most, more information is known about the metabolism of D-tryptophan. D-Tryptophan is transported actively across barriers such as those in the intestine[173] and the blood–brain interface,[174] but to a smaller extent than the L isomer. At the blood–brain barrier, it is transported at about 10% of the rate of the natural amino acid.[174] There are important differences in the metabolism of D-tryptophan between species. The rat can convert D-tryptophan to L-tryptophan in two steps.[175] The first is a transamination to indolepyruvic acid which is then reaminated to the L amino acid. Dogs are capable of doing this, but to a lesser extent,[176] and humans even less so.[177] Thus, when rats are given D-tryptophan, brain indoleamine synthesis increases as a result of the rise of L-tryptophan in brain,[174] but in dogs, the rise of brain tryptophan is caused by the D isomer, and brain indoleamine synthesis is not elevated.[176]

D-Phenylalanine is, according to one report, capable of elevating rat brain phenylethylamine by a mechanism independent of its interconversion to L-phenylalanine.[178] Although the method used for determining phenylethylamine in this study is of doubtful specificity (see Chapter 8 of Volume 1), this alleged effect has been used clinically in an attempt to treat depressed patients with D-phenylalanine, although with disappointing results.[179] D-Phenylalanine has also been reported to inhibit potassium-evoked release of norepinephrine from slices of rat hypothalamus[180] and to inhibit degradation of enkephalin by peptidases and thus prolong enkephalin action.[181] More information is needed on the metabolism and behavioral effects of this interesting compound. However, its clinical usefulness seems doubtful, as D-phenylalanine may be one of the few D amino acids that humans can convert to the L isomer.[182]

Little is known about D-tyrosine, but it can lower blood pressure in hypertensive rats at a lower dose than L-tyrosine,[183] an action that may possibly be mediated by conversion to a trace amine such as tyramine.

9. CONCLUSIONS

The factors that control the level of aromatic amino acids in brain are complex. They include peripheral metabolism of the amino acids, both into protein and through catabolic pathways, and competition among the various large neutral amino acids for uptake into brain. However, our present level of understanding does permit us to explain variations of brain aromatic amino acid levels in most physiological (e.g., dietary intake) and pharmacological circumstances. The extent to which variations in brain amino acid levels can influence biogenic amine synthesis still needs further elucidation for phenylalanine, tyrosine, and the catecholamines. One of the challenges for the future will be to determine how much such changes in synthesis lead to changes in release and function, especially in physiological circumstances.

Subtle changes in mood and behavior are a common experience after food ingestion, and some of these changes might result from altered aromatic amino acid levels. However, even in circumstances in which changes in the levels of amino acids in brain (e.g., after amino acid administration) are known to produce behavioral changes, such changes can not necessarily be attributed to changes in biogenic amine function. The aromatic amino acids are metabolized to a wide variety of compounds many of which are known to have psychoactive properties and may influence brain function in spite of their low levels in brain. Careful work will be needed to determine which of these minor metabolites are important in physiological circumstances and which are important for an understanding of the neuropsychopharmacology of the aromatic amino acids.

One essential requirement for work in this area is specific and reliable analytical methodology. Unfortunately, this requirement has often not been met in the past. Further research on aromatic amino acid metabolism and on the function of aromatic amino acid metabolites should lead to a greater understanding of those diseases in which such metabolism is altered and may lead in some circumstances to relatively simple treatments through pharmacological approaches designed to manipulate brain amino acid levels. Such an approach is not new but has recently been gaining in popularity and should continue to do so with increased understanding of the implications of altered brain amino acid levels.

REFERENCES

1. Honecker, H., and Rommelspacher, H., 1978, *Naunyn Schmiedebergs Arch. Pharmacol.* **305**:135–141.
2. Buckholtz, N. S., 1980, *Life Sci.* **27**:893–903.
3. Culley, W. J., Saunders, R. N., Mertz, E. T., and Jolly, D. H., 1963, *Proc. Soc. Exp. Biol. Med.* **113**:645–648.
4. Dickerson, J. W. T., and Pao, S. K., 1975, *J. Neurochem.* **25**:559–564.
5. Fernstrom, J. D., and Wurtman, R. J., 1971, *Nature [New Biol.]* **234**:62–64.
6. Messing, R. B., Fisher, L. A., Phebus, L., and Lytle, L. D., 1976, *Life Sci.* **18**:707–714.
7. Fernstrom, J. D., and Faller, D. V., 1978, *J. Neurochem.* **30**:1531–1538.
8. Fernstrom, J. D., 1976, *Metabolism* **26**:207–223.
9. Perez-Cruet, J., Chase, T. N., and Murphy, D. L., 1979, *Nature* **248**:694–695.
10. Anderson, G. H., 1979, *Can. J. Physiol. Pharmacol.* **57**:1043–1057.
11. Christensen, H. N., 1964, *Mammalian Protein Metabolism*, Volume 1 (H. N. Munro and J. B. Allison, eds.), Academic Press, New York, pp. 564–566.
12. Munro, H. N., 1970, *Mammalian Protein Metabolism*, Volume 4 (H. N. Munro, ed.), Academic Press, New York, pp. 299–386.
13. Bloxam, D. L., Warren, W. H., and White, P. J., 1974, *Life Sci.* **15**:1443–1455.

14. Munro, H. N., 1964, *Mammalian Protein Metabolism*, Volume 1 (H. N. Munro and J. B. Allison, eds.), Academic Press, New York, pp. 381–481.

15. Sidransky, H., Sarma, D. S. R., Bongiorno, M., and Verney, E., 1968, *J. Biol. Chem.* 243:1123–1132.

16. Majumdar, A. P. N., and Jørgensen, A. J. F., 1976, *Biochem. Med.* 16:266–276.

17. Cocchi, D., di Giulio, A., Groppetti, A., Mantegazza, P., Muller, E. E., and Spano, P. F., 1975, *Experientia* 31:384–386.

18. Gessa, G. L., Biggio, G., Fadda, F., Corsini, G. V., and Tagliamonte, A., 1974, *J. Neurochem.* 22:869–870.

19. Sjoerdsma, A., Weissbach, H., and Udenfriend, S., 1956, *Am. J. Med.* 20:520–532.

20. Young, S. N., St-Arnaud-McKenzie, D., and Sourkes, T. L., 1978, *Biochem. Pharmacol.* 27:763–767.

21. Green, A. R., Sourkes, T. L., and Young, S. N., 1975, *Br. J. Pharmacol.* 53:287–292.

22. Green, A. R., and Curzon, G., 1975, *Biochem. Pharmacol.* 24:713–716.

23. Green, A. R., Woods, H. F., Knott, P. J., and Curzon, G., 1975, *Nature* 255:170.

24. Young, S. N., and Sourkes, T. L., 1975, *J. Biol. Chem.* 250:5009–5014.

25. Sourkes, T. L., 1971, *Fed. Proc.* 30:897–903.

26. Joseph, M. H., Young, S. N., and Curzon, G., 1976, *Biochem. Pharmacol.* 25:2599–2604.

27. Young, S. N., and Oravec, M., 1979, *Can. J. Biochem.* 57:517–522.

28. Jervis, G. A., 1960, *Clin. Chim. Acta* 5:471–475.

29. Kaufman, S., 1977, *Advances in Neurochemistry*, Volume 2 (B. W. Agranoff, and M. H. Aprison, eds.), Plenum Press, New York, pp.1–132.

30. Cihak, A., Lamar, C., and Pitot, H. C., 1973, *Arch. Biochem. Biophys.* 156:188–194.

31. Knox, W. E., 1963, *Trans. N.Y. Acad. Sci.* 25:503–512.

32. Hager, C. B., and Kenney, F. T., 1968, *J. Biol. Chem.* 243:3296–3300.

33. Wurtman, R. J., 1974, *Life Sci.* 15:827–847.

34. Wurtman, R. J., and Fernstrom, J. D., 1972, *Perspectives in Neuropharmacology* (S. H. Snyder, ed.), Oxford University Press, Oxford, pp. 143–193.

35. Laborit, H., and Thuret, F., 1977, *Res. Commun. Chem. Pathol. Pharmacol.* 17:77–85.

36. Benkert, O., and Matussek, N., 1970, *Nature* 228:73–75.

37. Oldendorf, W. H., and Szabo, J., 1976, *Am. J. Physiol.* 230:94–98.

38. Pardridge, W. M., and Oldendorf, W. H., 1977, *J. Neurochem.* 28:5–12.

39. Pardridge, W. M., 1977, *J. Neurochem.* 28:103–108.

40. Pratt, O. E., 1979, *J. Neural Transm.* [*Suppl.*] 15:29–42.

41. Soeters, P. B., and Fischer, J. E., 1976, *Lancet* 2:880–882.

42. McMenamy, R. H., Lund, C. C., and Oncley, J. L., 1957, *J. Clin. Invest.* 36:1672–1679.

43. McMenamy, R. H., 1964, *J. Biol. Chem.* 239:2835–2841.

44. Curzon, G., and Knott, P. J., 1974, *Br. J. Pharmacol.* 50:197–204.

45. Gessa, G. L., and Tagliamonte, A., 1974, *Serotonin: New Vistas: Biochemistry and Behavioral and Clinical Studies* (E. Costa, G. L. Gessa, and M. Sandler, eds.), Raven Press, New York, pp. 119–131.

46. Fernstrom, J. D., Hirsch, M. J., and Faller, D. V., 1976, *Biochem. J.* 106:589–595.

47. Baumann, P., and Perey, M., 1977, *Clin. Chim. Acta* 76:223–231.

48. Fernstrom, J. D., Munro, H. N., and Wurtman, R. J., 1977, *Nature* 265:277.

49. Etienne, P., Young, S. N., and Sourkes, T. L., 1976, *Nature* 262:144–145.

50. Yuwiler, A., Oldendorf, W. H., Geller, E., and Braun, L., 1977, *J. Neurochem.* 28:1015–1023.

51. Pardridge, W. M., 1979, *Life Sci.* 25:1519–1528.

52. Daniel, P. M., Moorhouse, S. R., and Pratt, O. E., 1976, *Psychol. Med.* 6:277–286.

53. Toth, J., and Lajtha, A., 1977, *Neurochem. Res.* 2:149–160.

54. Sourkes, T. L., 1979, *J. Neural Transm.* [*Suppl.*] 15:107–114.

55. Bruinvels, J., and Moleman, P., 1980, *J. Neurochem.* 34:1065–1070.

56. Morre, M. C., and Wurtman, R. J., 1981, *Life Sci.* 28:65–75.

57. Archer, E. G., Breakfield, X. O., and Sharata, M. N., 1977, *J. Neurochem.* 28:127–135.

58. Gershon, M. D., and Dreyfus, C. F., 1980, *Brain Res.* 184:229–233.

59. Lajtha, A., 1974, *Aromatic Amino Acids in the Brain* (G. E. W. Wolstenholme and D. W. Fitzsimons, eds.), Elsevier, Amsterdam, pp. 25–41.

60. Wurtman, R. J., Shein, H. M., Axelrod, J., and Laren, F., 1969, *Proc. Natl. Acad. Sci. U.S.A.* **62**:749–755.

61. Shaw, D. M., Riley, G., Tidmarsh, S., and Blazek, R., 1976, *Lancet* **1**:363–364.

62. Johnson, T. C., and Chou, L., 1973, *J. Neurochem.* **20**:405–414.

63. Johnson, T. C., and Luttges, M. W., 1966, *J. Neurochem.* **13**:545–552.

64. Lin, C. C., Chung, C. H., and Lee, M. L., 1973, *Biochem. J.* **135**:367–373.

65. Jørgensen, A. J. F., and Majumdar, A. P. N., 1976, *Biochem. Med.* **16**:37–46.

66. Weiss, B. F., Liebschutz, J. L., Wurtman, R. J., and Munro, H. N., 1975, *J. Neurochem.* **24**:1191–1195.

67. Majumdar, A. P. N., 1980, *Res. Commun. Chem. Pathol. Pharmacol.* **28**:435–445.

68. Joh, T. H., Shikimi, T., Pickel, V. M., and Reis, D. J., 1975, *Proc. Natl. Acad. Sci. U.S.A.* **72**:3575–3579.

69. Hess, S. M., and Doepfner, W., 1961, *Arch. Int. Pharmacodyn.* **134**:89–99.

70. Ashcroft, G. W., Eccleston, D., and Crawford, T. B. B., 1965, *J. Neurochem.* **12**:489–492.

71. Fernstrom, J. D., and Wurtman, R. J., 1971, *Science* **173**:149–152.

72. Grahame-Smith, D. G., 1971, *J. Neurochem.* **18**:1053–1066.

73. Carlsson, A., and Lindqvist, M., 1978, *Naunyn Schmiedebergs Arch. Pharmacol.* **303**:157–164.

74. Friedman, P. A., Kappelman, A. H., and Kaufman, S., 1972, *J. Biol. Chem.* **247**:4165–4173.

75. Ichiyama, A., Nakamura, S., Nishizuka, Y., and Hayaishi, O., 1970, *J. Biol. Chem.* **245**:1699–1709.

76. Gál, E. M., Young, R. B., and Sherman, A. D., 1978, *J. Neurochem.* **31**:237–244.

77. Gibson, C. J., Deikel, S. M., Young, S. N., and Binik, Y. M., 1981, *Psychopharmacology* **76**:118–121.

78. Curzon, G., Kantamaneni, B. D., Bartlett, J. R., and Bridges, P. K., 1976, *J. Neurochem.* **26**:613–615.

79. Vapalaliti, M., Hyyppä, M. T., Nieminen, V., and Rinne, U. K., 1978, *J. Neurosurg.* **48**:58–63.

80. Young, S. N., Gauthier, S., Anderson, G. M., and Purdy, W. C., 1980, *J. Neurol. Neurosurg, Psychiatry* **43**:438–445.

81. Eccleston, D., Ashcroft, G. W., Crawford, T. B. B., Stanton, J. B., Wood, D., and McTurk, P. H., 1970, *J. Neurol. Neurosurg. Psychiatry* **33**:269–272.

82. Young, S. N., and Gauthier, S., 1981, *J. Neurol. Neurosurg. Psychiatry* **44**:323–328.

83. Smith, J. E., Hingten, J. N., Lane, J. D., and Aprison, M. H., 1976, *J. Neurochem.* **26**:537–541.

84. Raleigh, M. J., Brammer, G. L., Yuwiler, A., Flannery, J. W., McGuire, M. T., and Geller, E., 1980, *Exp. Neurol.* **68**:322–334.

85. Tricklebank, M. D., Drewitt, P. N., and Curzon, G., 1980, *Psychopharmacology* **69**:173–177.

86. Marsden, C. A., and Curzon, G., 1976, *Neuropharmacology* **15**:165–171.

87. Patkina, N. A., and Lapin, I. P., 1976, *Pharmacol. Biochem. Behav.* **5**:241–245.

88. Modigh, K., 1973, *Psychopharmacology* **30**:123–134.

89. Fernando, J. C. R., and Curzon, G., 1981, *Neuropharmacology* **20**:115–122.

90. Gallager, D. W., and Aghajanian, G. K., 1976, *Neuropharmacology* **15**:149–156.

91. Trulson, M. E., and Jacobs, B. L., 1979, *Brain Res.* **163**:135–150.

92. Héry, F., Simonnet, G., Bourgoin, S., Soubrié, P., Artaud, F., Hamon, M., and Glowinski, J., 1979, *Brain Res.* **169**:317–334.

93. Puizillout, J. J., Gaudin-Chazal, G., Daszuta, A., Seyfritz, N., and Ternaux, J. P., 1979, *J. Physiol [Paris]* **75**:531–537.

94. Trulson, M. E., and Jacobs, B. L., 1976, *Neuropharmacology* **15**:339–344.

95. Ternaux, J. P., Boireau, A., Bourgoin, S., Hamon, M., Héry, F., and Glowinski, J., 1976, *Brain Res.* **101**:533–548.

96. Marsden, C. A., Conti, J., Strope, E., Curzon, G., and Adams, R. N., 1979, *Brain Res.* **171**:85–99.

97. Karobath, M., and Baldessarini, R. J., 1972, *Nature* **236**:206–208.

98. Bagchi, S. P., and Zarycki, E. P., 1975, *Biochem. Pharmacol.* **24**:1381–1390.

99. Katz, I., Lloyd, T., and Kaufman, S., 1976, *Biochim. Biophys. Acta* **445**:567–578.

100. Kapatos, G., and Zigmond, M., 1977, *J. Neurochem.* **28:**1109–1119.
101. Carlsson, A., and Lindqvist, M., 1978, *Naunyn Schmiedebergs Arch. Pharmacol.* **303:**157–164.
102. Laborit, H., London, A., and Weber, B., 1970, *Agressologie* **11:**139–151.
103. Wurtman, R. J., Larin, F., Mostafapour, S., and Fernstrom, J. D., 1974, *Science* **185:**183–184.
104. Gibson, C. J., and Wurtman, R. J., 1978, *Life Sci.* **22:**1399–1406.
105. Scally, M. C., Ulus, I., and Wurtman, R. J., 1977, *J. Neural. Trans.* **41:**1–6.
106. Sved, A. F., Fernstrom, J. D., and Wurtman, R. J., 1979, *Life Sci.* **25:**1293–1300.
107. Melamed, E., Hefti, F., and Wurtman, R. J., 1980, *Proc. Natl. Acad. Sci. U.S.A.* **77:**4305–4309.
108. Aghajanian, G. K., 1978, *Essays in Neurochemistry and Neuropharmacology,* Volume 3 (M. B. H. Youdim, W. Lovenberg, D. F. Sharman, and J. R. Lagnado, eds.), John Wiley & Sons, New York, Chichester, pp. 1–32.
109. Thurmond, J. B., Lasley, S. M., Conkin, A. L., and Brown, J. W., 1977, *Pharmacol. Biochem. Behav.* **6:**475–478.
110. Snodgrass, S. R., 1974, *J. Pharm. Pharmacol.* **26:**931–936.
111. Snodgrass, S. R., and Horn, A. S., 1973, *J. Neurochem.* **21:**687–696.
112. Sloan, J. W., Martin, W. R., Clements, T. H., Buchwald, W. F., and Bridges, S. R., 1975, *J. Neurochem.* **24:**523–532.
113. Saavedra, J. M., and Axelrod, J., 1973, *J. Pharmacol. Exp. Ther.* **185:**523–529.
114. Philips, S. R., Durden, D. A., and Boulton, A. A., 1974, *Can. J. Biochem.* **52:**447–451.
115. Warsh, J. J., Godse, D. D., Stancer, H. C., Chan, P. W., and Coscina, D. V., 1977, *Biochem. Med.* **18:**10–20.
116. Artigas, F., and Gelpí, E., 1979, *Anal. Biochem.* **92:**233–242.
117. Philips, S. R., Durden, D. A., and Boulton, A. A., 1974, *Can. J. Biochem.* **52:**447–451.
118. Warsh, J. J., Coscina, D. V., Godse, D. D., and Chan, P. W., 1979, *J. Neurchem.* **32:**1191–1196.
119. Young, S. N., Anderson, G. M., and Purdy, W. C., 1980, *J. Neurochem.* **34:**309–315.
120. Oldendorf, W. H., and Braun, L. D., 1976, *Brain Res.* **113:**219–224.
121. Young, S. N., Anderson, G. M., Gauthier, S., and Purdy, W. C., 1980, *J. Neurochem.* **34:**1087–1092.
122. Jones, R. S. G., and Boulton, A. A., 1980, *Life Sci.* **27:**1849–1856.
123. Gillin, J. G., Kaplan, J., Stillman, R., and Wyatt, R. J., 1976, *Am. J. Psychiatry* **133:**203–208.
124. Luchins, D., Ban, T. A., and Lehmann, H. E., 1978, *Int. Pharmacopsychiatry* **13:**16–33.
125. Fuller, R. W., 1976, *Life Sci.* **19:**625–628.
126. Oon, M. C. H., Murray, R. M., Rodnight, R., Murphy, M. P., and Birley, J. L. T., 1977, *Psychopharmacology* **54:**171–175.
127. Räisänen, M., and Kärkkäinen, J., 1979, *J. Chromatogr.* **162:**579–584.
128. Walker, R. W., Mandel, L. R., Kleinman, J. E., Gillin, J. C., Wyatt, R. J., and Vandenheuvel, W. J. A., 1979, *J. Chromatogr.* **162:**539–546.
129. Reiter, R. J., 1980, *The Pineal,* Volume 5, Eden Press, Westmount, Quebec.
130. Illnerova, H., Backstrom, M., Sääf, J., Wetterberg, L., and Vangbo, B., 1978, *Neurosci. Lett.* **9:**189–193.
131. Arendt, J., Wirz-Justice, A., Bradtke, K., and Kornemark, M., 1979, *Ann. Clin. Biochem.* **16:**307–312.
132. Shaw, K. M., 1979, *Adv. Drug Res.* **11:**75–96.
133. Anton-Tay, F., and Wurtman, R. J., 1969, *Nature* **221:**474–475.
134. Niles, L. P., Wong, Y. W., Mishra, R. K., and Brown, G. M., 1979, *Eur. J. Pharmacol.* **55:**219–220.
135. Cardinali, D. P., Vacas, M. I., and Boyer, E. E., 1979, *Endocrinology* **105:**437–441.
136. Pang, S. F., Yu, H. S., Suen, H. C., and Brown, G. M., 1980, *J. Endocrinol.* **87:**89–93.
137. Bubenik, G. A., 1980, *Horm. Res.* **12:**313–323.
138. Friedhoff, A. J., and Miller, J. C., 1977, *Res. Commun. Chem. Pathol. Pharmacol.* **16:**225–244.

139. Bubenik, G. A., Brown, G. M., and Grota, L. G., 1976, *Brain Res.* **118**:417–427.
140. Bosin, T. R., Jonsson, G., and Beck, O., 1979, *Brain Res.* **173**:79–88.
141. Quay, W. B., 1974, *Pineal Chemistry,* Charles C. Thomas, Springifeld, Illinois.
142. Narasimhachari, N., Kempster, E., and Anbar, M., 1980, *Biomed. Mass Spectrom.* **7**:231–235.
143. Duncan, R. J. S., and Sourkes, T. L., 1974, *J. Neurochem.* **22**:663–669.
144. Turner, A. J., Illingsworth, J. A., and Tipton, K. J., 1974, *Biochem. J.* **144**:353–360.
145. Cheifetz, S., and Warsh, J. J., 1980, *J. Neurochem.* **34**:1093–1099.
146. Anderson, G. M., Young, J. G., Young, S. N., Batter, D. K., Cohen, D. J., and Shaywitz, B. A., 1981, *J. Chromatogr.* **224**:315–320.
147. Beck, O., Borg, S., Holmstedt, B., and Stibler, H., 1980, *Biochem. Pharmacol.* **29**:693–696.
148. Wilson, B. W., Lynch, H. J., and Ozaki, Y., 1978, *Life Sci.* **23**:1019–1024.
149. Cornford, E. M., Bocash, W. D., Braun, L. D., Crane, P. D., Oldendorf, W. H., and MacInnis, A. J., 1979, *J. Clin. Invest.* **63**:1241–1248.
150. Reiter, R. J., and Vaughan, M. K., 1977, *Life Sci.* **21**:159–172.
151. Seed, J. R., and Sechelski, J., 1977, *Life Sci.* **21**:1603–1610.
152. Gál, E. M., Armstrong, J. C., and Ginsberg, B., 1966, *J. Neurochem.* **13**:643–654.
153. Yoshida, R., and Hayaishi, O., 1978, *Proc. Natl. Acad. Sci. U.S.A.* **75**:3998–4000.
154. Tsuda, H., Noguchi, T., and Kido, R., 1972, *J. Neurochem.***19**:887–890.
155. Hirata, F., Hayaishi, O., Tokuyama, T., and Senoh, S., 1974, *J. Biol. Chem.* **249**:1311–1313.
156. Fujiwara, M., Shibata, M., Watanabe, Y., Nukiwa, T., Hirata, F., Mizuno, N., and Hayaishi, O., 1978, *J. Biol. Chem.* **253**:6081–6085.
157. Makino, K., Joh, Y., and Hasegauva, F., 1962, *Biochem. Biophys. Res. Commun.* **6**:432–437.
158. Toda, N., 1975, *J. Pharmacol. Exp. Ther.* **193**:385–392.
159. Christenson, J. G., Dairman, W., amd Udenfriend, S., 1970, *Arch. Biochem. Biophys.* **141**:356–367.
160. Saavedra, J. M., 1974, *J. Neurochem.* **22**:211–216.
161. Fulton, T. R., Triano, T., Rabe, A., and Loo, Y. H., 1980, *Life Sci.* **27**:1271–1281.
162. Durden, D. A., and Philips, S. R., 1980, *J. Neurochem.* **34**:1725–1732.
163. Oldendorf, W. H., 1971, *Am. J. Physiol.* **221**:1629–1639.
164. Durden, D. A., Philips, S. R., and Boulton, A. A., 1973, *Can. J. Biochem.* **51**:995–1002.
165. Sandler, M., Ruthven, C. R. J., Goodwin, B. L., and Coppen, A., 1979, *Clin. Chim. Acta* **93**:169–171.
166. Sabelli, H. C., Borison, R. L., Diamond, B. I., Havdala, H. S., and Narasimhachari, N., 1978, *Biochem. Pharmacol.* **27**:1708–1711.
167. Danielson, T. J., Boulton, A. A., and Robertson, H. A., 1977, *J. Neurochem.* **29**:1131–1135.
168. Boulton, A. A., Dyck, L. E., and Durden, D. A., 1974, *Life Sci.* **14**:1673–1683.
169. Fellman, J. H., Roth, E. S., and Fujita, T. S., 1976, *Arch. Biochem. Biophys.* **174**:562–567.
170. Boulton, A. A., and Dyck, L. E., 1974, *Life Sci.* **14**:2497–2506.
171. Brandau, K., and Axelrod, J., 1972, *Naunyn Schmiedebergs Arch. Pharmacol.* **273**:123–133.
172. Karoum, F., Bunney, W., Gillin, J. C., Jimerson, D., Van Kammen, D., and Wyatt, R. J., 1977, *Biochem. Pharmacol.* **26**:629–632.
173. Bosin, T. R., Hathaway, D. R., and Maickel, R. P., 1974, *Arch. Int. Pharmacodyn. Ther.* **212**:32–35.
174. Yuwiler, A., 1973, *J. Neurochem.* **20**:1099–1109.
175. Loh, H. H., and Berg, C. P., 1971, *J. Nutr.* **101**:1351–1358.
176. Budny, J., Dow, R. C., Eccleston, D., Hill, A. G., and Ritchie, I. M., 1976, *Br. J. Pharmacol.* **58**:3–7.
177. Hankes, L. V., Brown, R. R., Leklem, J., Schmaeler, M., and Jesseph, J., 1972, *J. Invest. Dermatol.***58**:85–95.
178. Borison, R. L., Maple, P. J., Havdala, H. S., and Diamond, B. I., 1978, *Res. Commun. Chem. Pathol. Pharmacol.* **21**:363–366.
179. Mann, J., Peselow, E. D., Snyderman, S., and Gershon, S., 1980, *Am. J. Psychiatry* **137**:1611–1612.

180. Fernández Pardal, J., 1980, *J. Neurosci. Res.* **5**:155–161.
181. Bodnar, R. J., Lattner, M., and Wallace, M. M., 1980, *Pharmacol. Biochem. Behav.* **13**:829–833.
182. Tokuhisa, S., Saisu, K., Naruse, K., Yoshikawa, H., and Baba, S., 1981, *Chem. Pharm. Bull.* **29**:514–518.
183. Shalita, B., and Dikstein, S., 1979, *Pfluegers Arch.* **379**:245–250.

Imino Acids of the Brain

Ezio Giacobini

1. INTRODUCTION

Extensive and detailed review articles and chapters on brain amino acids can easily be found in the literature, including handbooks of neurochemistry. To my knowledge, this is the first chapter to discuss separately data on imino acids (cyclic secondary imino acids) present in the nervous tissue. Thirty-two substances are listed by Himwhich and Agrawal in Volume 1 of the *Handbook of Neurochemistry*[1] as amino acids and analogues present in the brain of five mammalian species (mouse, rabbit, guinea pig, cat, and dog). The concentrations reported in brain vary widely from a few nanomoles (half-cystine) to several micromoles (glutamic acid) per gram fresh tissue. So far, only three imino acids have been related to brain function: proline (PRO), hydroxyproline (HYP), and pipecolic acid (PA) (Fig. 1). Their concentrations in whole brain are relatively low compared to other cerebral amino acids: PRO (30–80 nmol/ g), HYP (40–80 nmol/g), and PA (18 $\pm$ 4 nmol/g[1a]). Cat brain has ten times lower concentrations of PRO than GABA.[2] However, this is the same range shown by several amino acids such as methionine, leucine, isoleucine, tyrosine, phenylalanine, and ornithine.[3] It is interesting to note that several of these "trace amino acids" are present at relatively high levels (proline, valine, isoleucine, tyrosine, ornithine, and phenylalanine) in the early postnatal life and fall to significantly lower levels at 2–3 weeks after birth. These changes may result from (1) qualitative changes in protein synthesis after birth, (2) high concentrations in the mother's blood prior to delivery, or (3) slower accumulation from circulation into the brain in the adult because of changes in the blood–brain barrier (BBB) after birth, or (4) postnatal activation of breakdown and secretion.

For simplicity, in this chapter, PRO and HYP on one hand and PA on the other will be discussed in two separate sections.

Ezio Giacobini • Laboratory of Neuropsychopharmacology, Department of Biobehavioral Sciences, University of Connecticut, Storrs, Connecticut 06268. *Present address*: Department of Pharmacology, Southern Illinois University, School of Medicine, Springfield, Illinois, 62708.

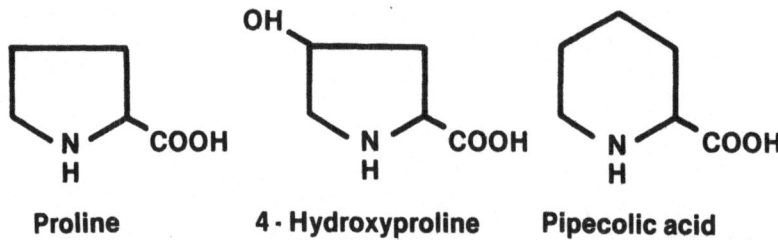

Proline   4 · Hydroxyproline   Pipecolic acid

Fig. 1.   Structures of the three imino acids isolated in mammalian brain.

2. PROLINE AND HYDROXYPROLINE

2.1. Synthesis and Metabolism in Brain

The metabolism of these imino acids has been extensively studied in bacteria and in many organs in mammals, as reported in the recent comprehensive review by Adams and Frank.[4] In brain, our knowledge about PRO and HYP is much more limited, and the formation and function of HYP are still matters of discussion.[5-7] However, hydroxyprolinemia associated with mental retardation has been reported[8,9]; therefore, the importance of this imino acid for the nervous system cannot be dismissed. In contrast to the invertebrate nervous system, PRO is found in only relatively low concentration in the mammalian brain.[10-13] Because of their common rigid pyrrolidine ring, PRO and HYP share unique chemical and biochemical properties. The presence of a secondary amino group excludes them from the pyridoxal-linked reactions of many amino acids such as transamination, decarboxylation, and racemization.

Hydroxyproline originates only from PRO through a special route that acts on peptide-bound PRO residues as substrate.[4] The degradative metabolisms of PRO and HYP are similar and may share some enzymes. The HYP group includes several isomers such as 4-HYP, 3-HYP, and 3,4-diHYP, each showing some metabolic peculiarity.[4] Both 4-HYP and 3-HYP are found mainly in connective tissue and may not be a component of any specific brain protein.

Nutritional tracers have established L-glutamic acid and L-ornithine as precursors of PRO (Fig. 2).[4] In addition, proline can be synthesized from arginine (Fig. 2). The metabolic relationship among glutamate, ornithine, arginine, and the PRO–HYP group has been reviewed.[4,14] It has been reported that in brain

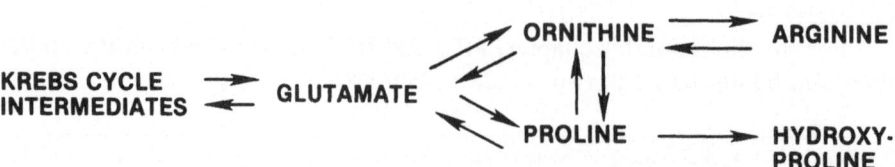

Fig. 2.   Pathways of proline and hydroxyproline biosynthesis.[4]

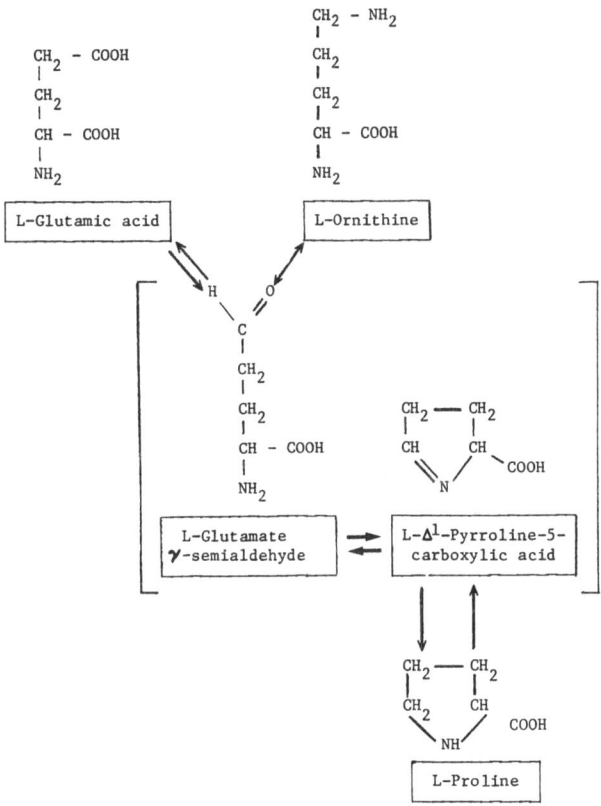

Fig. 3. Pathways of proline biosynthesis and degradation in mammalian cells.[18]

PRO can be formed from glucose,[15] probably through the glutamate step (Fig. 3). In one study, HYP was not oxidized by brain preparations that were capable of oxidizing PRO.[7] Indirect evidence for the synthesis of PRO from glutamate in brain is the finding that carbon from labeled glucose is incorporated into proline.[15] The alternative pathway from arginine is probably present in brain, as both arginine and ornithine-S-transaminase are found to be active.[16,17] The enzymatic conversion of glutamic acid to pyrroline-5-carboxylic acid, the immediate precursor of proline, has been recently demonstrated in homogenates of mammalian cells but not in nerve cells (Fig. 3.)[18] Proline can be utilized for synthesis of both arginine and HYP, and a further option is oxidation to glutamate (Figs. 2,3). The first step in this pathway, which is common with arginine, is the reaction catalyzed by PRO-oxidase resulting in the formation of Δ^1-pyrroline-5 carboxylate.[4,19] This metabolite can be oxidized by a DPN (TPN)-dependent dehydrogenase to glutamate.[4,20] This reaction has not yet been demonstrated in nervous tissue; however, a high yield of labeled glutamate can be obtained from labeled proline.[21] In addition, it has been suggested that proline can be converted to other amino acids by brain *in vivo*, but the details of these reactions have not been studied.[21]

An interesting finding in brain is the report of Yamanishi et al.[22] of the in vitro production of pyrrolidine from decarboxylation of PRO in rat cerebral tissue. Pyrrolidine, an alicyclic amine, has been reported to be present in mammalian brain and to cause sedation in rats.[23–26] The function of this brain amine, as well as of the structurally related piperidine, is unknown.[26].

2.2. Physiological Role

Regional differences in the distribution of PRO in CNS may be a useful indication of its possible role. Histochemical autoradiographic data have shown a differential accumulation pattern of PRO uptake following i.p. injection into the cat cerebellar cortex[27,28] and in the cerebral and cerebellar cortex[29–31] and the subcommissural organ of the mouse.[32] Significantly higher concentrations of PRO are present at the spinal level in dorsal roots as compared to ventral roots,[33] and excitatory fibers show a higher content of PRO than inhibitory ones.[11] In contrast with ventral spinal cord, in dorsal sensory neurons, PRO metabolism has been shown to be related to the glutamate compartment.[34] The relationship of PRO to glutamate metabolism raises the possibility that PRO may serve, as reported for the lobster neuromuscular junction, as an inactive form of glutamate precursor.[11] Other results suggest instead that PRO may function as an antagonist to the release of glutamate.[35]

During the last few years, increasing neurophysiological and biochemical evidence has been accumulated for the role of PRO as a CNS neurotransmitter. First, its uptake and release characteristics both in brain slices and synaptosomes[36–39] are similar to those described for known amino acid transmitter candidates. The interaction between PRO and GABA uptake is weak,[40] and both allylglycine and piperidine derivatives strongly interfere with it.[41.] Second, several important neurophysiological findings point to a specific role of PRO in CNS. They have been summarized in the review by Felix and Kunzle.[28] These authors tested and compared the action of PRO with GABA, glycine, and their specific antagonists on single neurons of the cat cerebellar cortex and pigeon tectal neurons. In the pigeon, on all cells tested, PRO, similarly to GABA or glycine, depressed the neuronal discharge. The inhibitory action of PRO was not blocked by known GABA and glycine antagonists, indicating that PRO is not competing for glycine or GABA receptors and may act on a different type of receptor.

Zarzecki et al.[42] observed that microiontophoretically applied PRO acts as an antagonist of glutamate in selected areas of the CNS of cats. However, as pointed out by other authors, PRO is clearly not a universal blocker of the effect of glutamate. Nistri and Morelli[43] reported that PRO, similarly to GABA and taurine, produced Cl^- dependent depolarization or hyperpolarization of the ventral root potentials of the frog spinal cord in vitro. They concluded that PRO in low concentrations is a fairly potent compound that might have a physiological role in certain synaptic mechanisms at the spinal level.

Neurophysiological microiontophoretic studies summarized in the papers of Felix and Kuntzle,[28] Nistri and Morelli,[43] and Zarzecki et al.[42] strongly indicate that PRO may in fact have a specific function in the CNS, probably acting as an inhibitory neurotransmitter. The question remains open whether the PRO action is direct or glutamate mediated. Pharmacological knowledge of PRO and HYP is limited. Intracerebral injection of L-PRO induced retrograde amnesia (after 24 h) in chicks.[44] This finding was interpreted as being related to swelling of dendritic spines which serve as substrate of short-term memory and, on stabilization, as the basis for long-term memory. This phenomenon may be related to the block of intracellular glutamate release exerted by PRO.[35,45] Van Harreveld and Fifkova[45] postulated a central role for glutamate, independent of its function as a putative neurotransmitter, that is blocked by PRO in a reversible and apparently competitive manner.

2.3. Relationship between Proline and Pipecolic Acid Metabolism

Autoradiographic observations revealed great regional differences in the utilization of D-PRO in neuronal protein synthesis by mouse brain.[30,31,46,47] It was shown that the incorporated radioactivity originated exclusively from L-PRO. In the absence of amino acid-racemizing enzymes in mammalian tissues, this finding was explained by postulating a two-step enzyme pathway consisting of (1) oxidation of D-PRO by D-amino acid oxidase and (2) reduction of the resulting Δ^1-pyrroline-2-carboxylate by Δ^1-pyrroline-2-carboxylate reductase. The latter is the same enzyme that catalizes the reduction of Δ^1-piperidine-2-carboxylate, an intermediate in the brain-specific metabolism of lysine leading to the formation of pipecolic acid. This relationship was found to be true in homogenates of mouse forebrain by Garweg *et al.*[48] The relationships between PA and PRO transport and uptake in brain are discussed in the following sections.

3. PIPECOLIC ACID

3.1. Origin, Biosynthesis, and Metabolism

The imino acid pipecolic acid (piperidine-2-carboxylic acid) was first identified as a constituent of plants by Zacharias *et al.*,[49] Grobbelaar and Stewart,[50] and Morrison.[51] It is widely distributed in the plant kingdom[52] where it represents a precursor in the biosynthesis of piperidine alkaloids (nicotiana alkaloids) such as nicotine and anabasine. In addition, it is found in *Neurospora* as a product of lysine degradation.[53]

The first evidence of PA formation from lysine in vertebrates was presented by Rothstein and Miller.[54,55] They demonstrated that PA is a major metabolite of lysine in the intact rat *in vivo*.[54,55] The *in vitro* conversion of lysine to PA was first shown to take place in the rat liver.[56] However, it was later reported[57] that PA is not formed in the intact mouse or in mouse liver *in vitro*. Our laboratory was the first to demonstrate *in vitro* formation of PA from lysine in the mouse brain and various other organs.[1a,58] Subsequently, *in vitro* synthesis of L-[^{14}C]PA from L-[U-^{14}C]lysine was also demonstrated in rat tissues, including brain, by Hernandez and Chang.[59] An acetyl donor was not

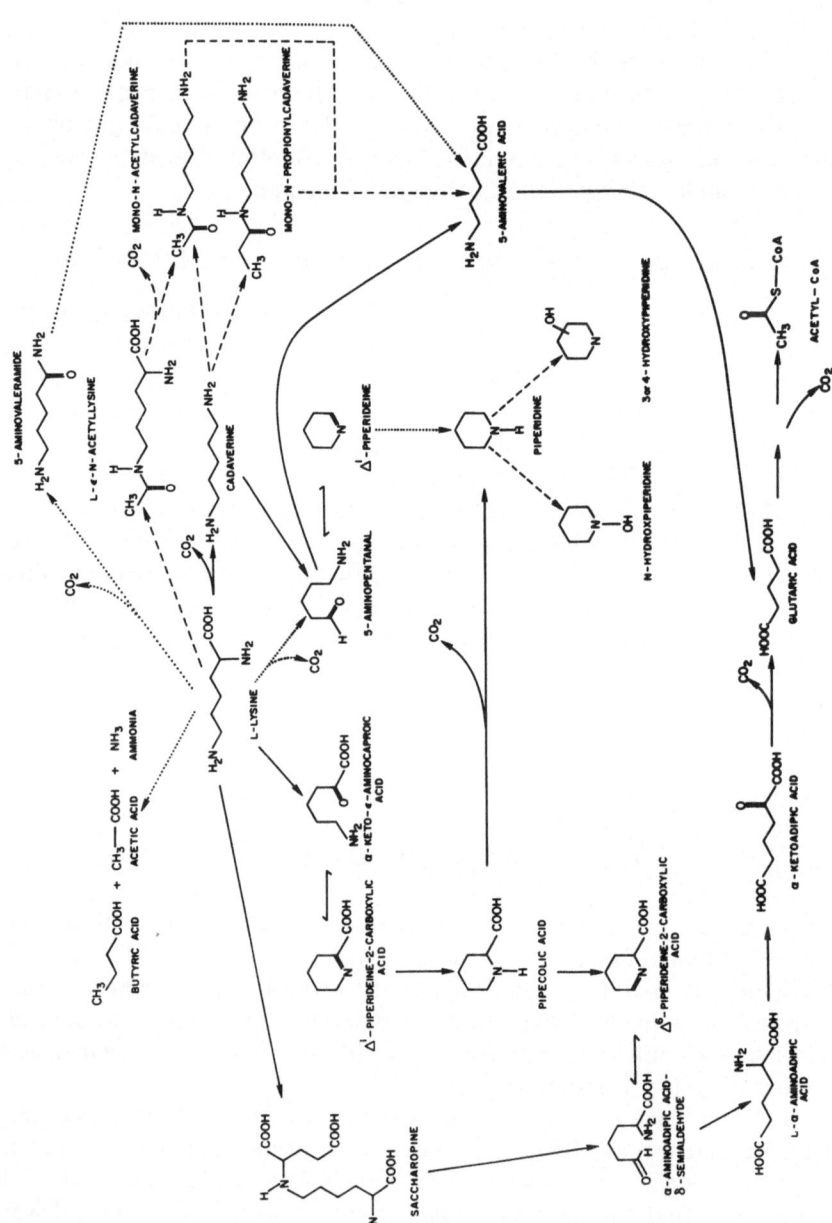

Fig. 4. Pathways of pipecolic acid formation and metabolism.[58]

required for this synthesis, and no labeling of ε-N-acetyl-L-lysine was detected. We have for the first time identified and measured mass spectrometrically this imino acid in the mouse brain. Its concentration in whole brain is 18 ± 4 nmol/g.[1a,58]

Meister *et al.*[60] and Meister and Buckley[61] described a pyridine nucleotide-dependent reduction of the α-keto acid analogue of lysine to L-PA in mammalian liver, its formation from Δ^1-piperidine-2-carboxylic acid, as well as the formation of L-PRO from Δ'-pyrroline-2-carboxylic acid. In rat liver mitochondria, PA forms α-aminoadipic acid and glutaric acid, establishing its position on the main pathway of lysine metabolism[62] (Fig. 4). It has been demonstrated that PA is degraded first to Δ^6-piperidine-6-carboxylic acid and further via α-aminoadipic acid semialdehyde to α-aminoadipic acid (Fig. 4).[63,64]

Experiments performed with rat liver mitochondria indicated saccharopine [ε-N-(glutaryl-2)-L-lysine] as a major intermediate (Fig. 4) of lysine catabolism in mammalian tissues.[65] Chang[66–68] showed that after intraventricular injection of L- and D-[^{14}C]lysine, labeled PA is found in the rat brain although no saccharopine is detectable. Thus, the mammalian brain differs in this respect from other organs. In mouse brain, formation of α-aminoadipic acid has been demonstrated following intraventricular injections of PA.[69] These results, in two species, strongly suggest that mammalian brain metabolizes lysine mainly through the L-pipecolic acid pathway.

Piperidine, an alicyclic amine that is present in the vertebrate and invertebrate brain,[26] was first suggested to be a decarboxylation product of PA by Kasé *et al.*[70] Piperidine has a selective cholinomimetic effect on characterized neurons of the molluscan brain,[71] shows a hypnogenic action in the cat,[72,73] and has been suggested to be involved in mechanisms of sleep and hibernation.[74,75] According to Kasé *et al.*[76,77] piperidine can be produced *in vitro* and *in vivo* in the rat brain. We were unable to confirm these results.[69,78] The presence of piperidine in brain is well established; however, the conversion of PA into piperidine in the brain is still doubtful and at most does not represent a major metabolic pathway for PA.[1a,26]

There are only few data concerning the endogenous levels of PA in brain. Kasé *et al.*[79] investigated its regional distribution in the dog brain and found values from 1 to 10 nmol/g, which are in the same range as we found in mouse brain (18 ± 4 nmol/g[1a]). A value of 200 μmol/g PA has been reported by Gatfield *et al.*[80] in human brain and in an individual suffering from hyperpipecolatemia. Pipecolate formation shows vast regional differences in the CNS of the mouse, dog, and monkey as reflected by Δ^1-pyrroline-2-carboxylate reductase activity.[48] The rate of reduction is highest in certain telencephalic and diencephalic regions, lower in brainstem, and not measurable in the cerebellum and spinal cord. This enzymatic distribution compares well with both accumulation and levels of PA found by us in the mouse brain.[69]

3.2. Relationship of Pipecolic Acid Metabolism to Piperidine and Cadaverine Formation

Cadaverine and pipecolic acid metabolism was investigated *in vitro* in several organs of the mouse by measuring $^{14}CO_2$ formation from labeled precursors.[58] The liver showed the highest formation of $^{14}CO_2$ from [1,5-^{14}C]cadaverine, whereas brain demonstrated a much lower rate of formation. Anaerobiosis and monoamine oxidase (MAO) inhibition significantly reduced $^{14}CO_2$ formation in every organ, whereas diamine oxidase (DAO) inhibition showed no inhibitory effect in brain and kidney. Piperidine was formed *in vitro* only in the large intestine and its contents. This formation is probably of bacterial origin. Under a variety of experimental conditions, we were unable to demonstrate any formation of piperidine in brain from cadaverine. Some reduction in [1,5-^{14}C]cadaverine levels following a 1-h incubation at 37°C was detectable in all tissues examined except brain and heart. Biosynthesis *in vitro* of [^{3}H]piperidine from D,L-[^{3}H]pipecolic acid was very low in brain and kidney. Furthermore, following a 1-h incubation at 37°C under aerobic conditions, the levels of [^{14}C]PA and [^{3}H]piperidine recovered from mouse brain homogenate did not indicate a high level of degradation of these two substances.

These results suggest that under *in vitro* conditions cadaverine is not a precursor of piperidine in brain, liver, heart, and kidney and that only very low levels of piperidine may be formed from PA in brain. Outside the brain, formation of piperidine from PA is only detectable in kidney and in the contents of the large intestine. The latter is probably of bacterial origin. Our results do not support previous findings from other authors[70,76] of an endogenous origin of piperidine in brain from cadaverine and PA and suggest that (1) cadaverine is not a precursor of piperidine in brain; (2) the conversion of PA into piperidine in the brain does not constitute a major metabolic pathway; and (3) the main source of piperidine in CNS may be brain-exogenous.

3.3. Brain Uptake of Pipecolic Acid

Plasma/brain ratios of lysine following i.v. injections of labeled lysine, the precursor of PA, demonstrated that in both mouse and rat, the BBB for this amino acid is not particularly effective.[68,81] In the same investigation,[68] labeled PA was detected in significant amounts in the rat brain within 30 s following i.v. injections of lysine, indicating a high rate of formation. We modified and applied the method described by Oldendorf[82] in the rat to assess the BBB permeability by a rapid injection of a radiolabeled substance into the common carotid artery of the mouse.[83] Using this technique, we compared the brain uptake of PA in the mouse to that of several amino acids and amines in the rat. In general, brain uptake index (BUI) values were lower in the mouse than those previously reported in the rat (Table I). The only exception was PRO. The BUI of D,L-[^{3}H]PA was 3.38% (at 0.114 mM). This value places PA in the same category as acidic amino acids and amines, i.e., substances with a low transport rate.[82,84] Lysine showed a higher BUI than PA.

The PA uptake was saturable between the concentrations of 0.114 and 3.44 mM and unsaturable above this value. The kinetic analysis suggested the presence of two kinds of transport systems. Substances structurally related to PA, such as nipecotic acid, isonipecotic acid, L-proline, and piperidine (Fig. 5) showed a significant inhibitory effect. Among the amino acids tested, only GABA showed an inhibitory effect. Our results demonstrate that an appreciable

Table I
Uptake of ^{14}C-labeled Substances in the Mouse and Rat Brain

	Concn. (mM)	Mouse brain uptake index[a]	Concn. (mM)	Rat brain uptake index[a,b]
^{3}HOH reference		100		100
Inulin	0.082	1.62 ± 0.03 (3)	0.100	1.95 ± 0.38 (3)
L-Leucine	0.0084	27.3 ± 2.1 (4)	0.008	54 ± 5 (3)
	0.0168	24.5 ± 3.5 (4)		
L-Tyrosine	0.0047	25.0 ± 3.3 (5)	0.006	50 ± 2 (3)
	0.0094	21.0 ± 0.7 (4)		
L-Tryptophan	0.039	15.2 ± 2.2 (3)	0.022	36 ± 1 (3)
	0.098	12.9 ± 0.3 (3)		
L-Lysine	0.031	8.67 ± 0.39 (4)	0.008	16 ± 3.2 (3)
L-Proline	0.040	6.04 ± 1.62 (4)	0.010	3.3 ± 0.35 (3)
Tryptamine	0.065	9.80 ± 0.51 (6)	0.140	12.5 ± 3.2 (3)
	0.097	8.20 ± 0.47 (3)		
5-OH-Tryptamine	0.288	2.35 ± 0.44 (4)	0.022	2.63 ± 1.8 (3)
Cadaverine	0.316	2.83 ± 0.71 (4)	0.500	3.8 (8)[c]
D,L-Pipecolic acid	0.114	3.38 ± 0.81 (4)		

[a] Brain uptake index values are means ± S.D. Number of animals is given in parentheses.
[b] Data from Oldendorf.[82]
[c] Data from Sershen and Lajtha.[111]

amount of PA can cross the BBB and enter the brain. Chang[67,68] however, suggested the existence of a complete BBB for PA in the rat.

Additional evidence for a brain uptake of PA is provided by our experiments with i.p. and i.v. injections of [^{3}H]PA.[85] Approximately 1% of the total injected PA is recovered in the mouse brain 1 h after i.p. injection. This PA is then slowly eliminated via the kidney. There is also indication of PA metabolism in both brain and kidney.[85] The brain is not the only organ that is able to synthesize PA[58,68]; in addition, a dietary origin of PA should be considered. The active brain uptake mechanisms for PA demonstrated by us could contribute to the buildup via circulation of increased cerebral levels of this substance under pathological conditions.[80,86,87]

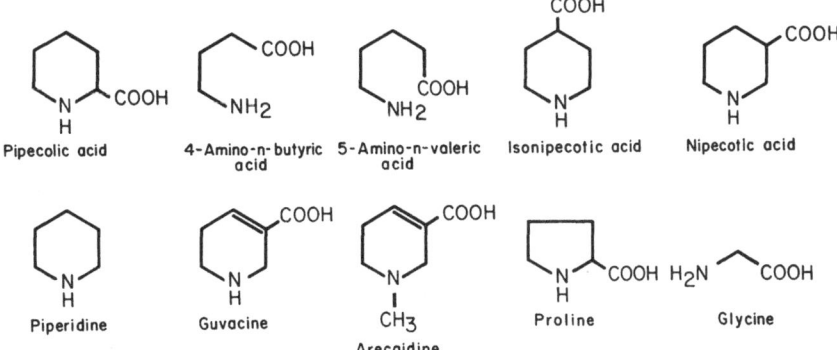

Fig. 5. Pipecolic acid and structurally related substances.[1a]

3.4. Accumulation, Metabolism, and Secretion of Exogenous Pipecolic Acid

Pipecolic acid that is derived from the diet or from synthesis in other organs such as large intestine, liver, and kidney[1a,58] can be transported into and accumulated in the brain. It is therefore important to determine the mechanisms of transport and secretion that might be responsible for building up and retaining cerebral levels of this substance. A subsequent, but not secondary, question is whether PA is metabolized in brain. The long-term accumulation of PA, as well as its disappearance following exogenous administration, was studied in brain and other organs of mice pulse-injected i.p. or i.v. with D,L-[^{3}H]PA.[85] Three features of the pattern of retention of PA were most salient: first, a rapid accumulation in brain; second, a rapid secretion of this compound in the urine; and third, long-lasting steady levels of radioactivity maintained in brain (Figs. 6a,b).

Sixty minutes after i.v. injection, the brain/plasma ratio of PA is approximately 0.2 and approaches unity at 5 h. The percent recovered as PA in brain is 78 at 30 min and 71 at 120 min, suggesting slow metabolic activity. The kidney shows a rapid secretion of radioactivity which correlates well with the exponential decrease in plasma and urine (Fig. 6b). Administration of probenecid (200 mg/kg) significantly increases radioactivity caused by PA in brain, liver, and urine. The formation of α-aminoadipic acid, a known metabolite of PA, could be demonstrated in kidney 30 min after i.p. injections. These results indicate that most of the PA taken up by the brain is rapidly removed through circulation and secreted in the urine; however, a small part is retained and metabolized by brain and kidney. Following intracerebral ventricular (i.c.v.) injections of D,L-[^{3}H]PA in the mouse, this is reabsorbed through the ventricles and redistributed to various brain regions.[69] The highest accumulation is found in four brain regions homolateral to the injection site: right hippocampus, right neocortex, diencephalon, and right striatum (Fig. 7). This correlates well with the distribution of the activity of Δ^1-pyrroline-2-carboxylate reductase in different brain regions.[48] Following i.c.v. D,L-[^{3}H]PA preloading, the radioactivity is released from hippocampus slices in the perfusion medium after depolarization induced by high K^+. During perfusion with a Ca^{2+}-free medium containing EGTA, a significant reduction of release is observed.

The recovery of D,L-[^{3}H]PA in the brain shows a more rapid phase of disappearance from 0 to 2 h and a slower phase from 2 to 5 h (Fig. 8). At 5 h, only 28% radioactivity, represented mainly by PA, is left in the brain. Kidney secretion represents the major route of elimination of the injected PA. The presence of α-aminoadipic acid both in brain and urine was observed.[85] This substance induces striking gliotoxic and neurotoxic changes when administered subcutaneously to infant mice.[88] The synthesis of L-α-aminoadipic acid is of interest in view of its known inhibitory effect on the neuroexcitatory properties of glutamate and aspartate on CNS neurons. Probenecid (200 mg/kg) significantly increases the accumulation of i.c.v. injected D,L-[^{3}H]PA in brain and kidney. The presence of a selective regional accumulation of PA in certain brain regions, its metabolism in brain, its enhanced retention following prob-

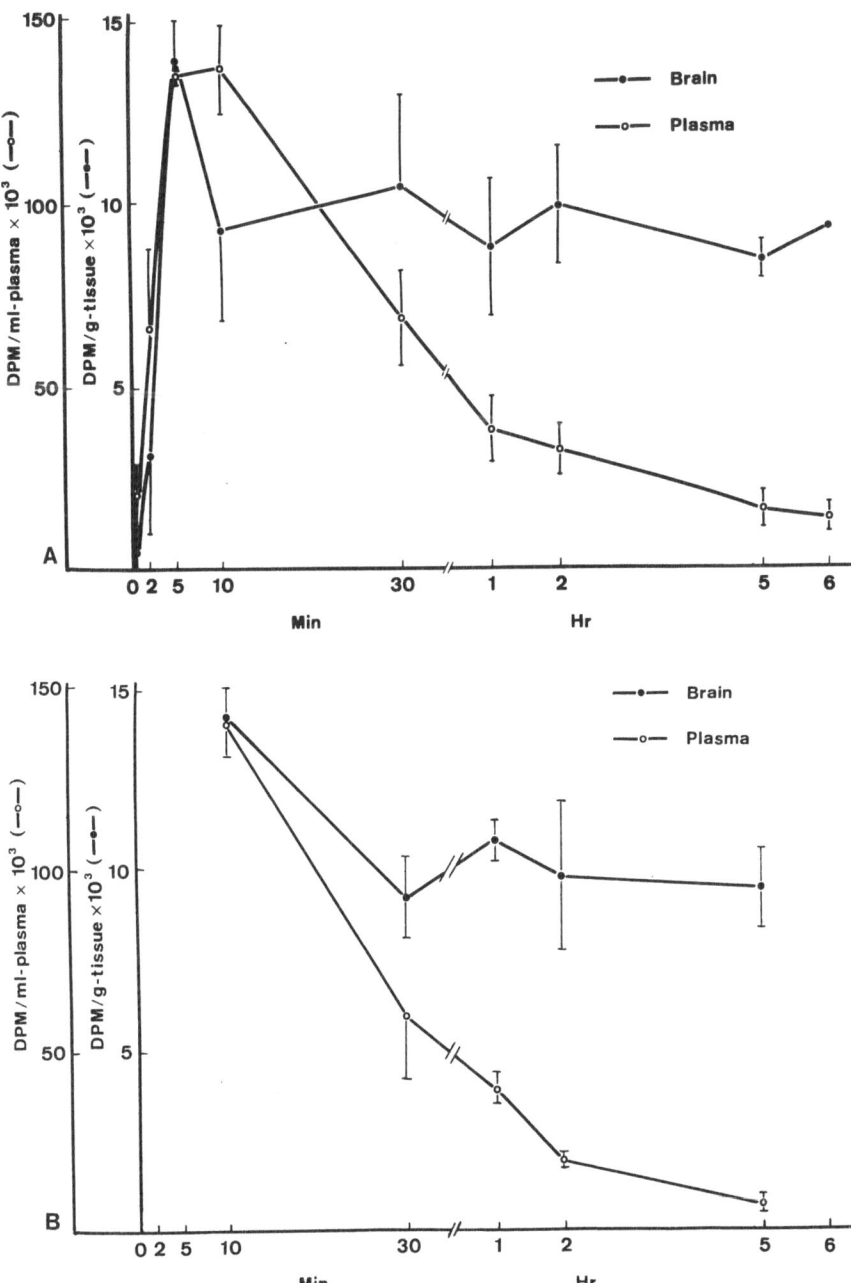

Fig. 6. (a) Changes in total radioactivity in mouse brain and plasma following i.p. injections (6.9 nmol/mouse) of D,L-[³H]pipecolic acid.[85] (b) Changes in total radioactivity in mouse brain and plasma following i.v. injections (6.9 nmol/mouse) of D,L-[³H]pipecolic acid.[85]

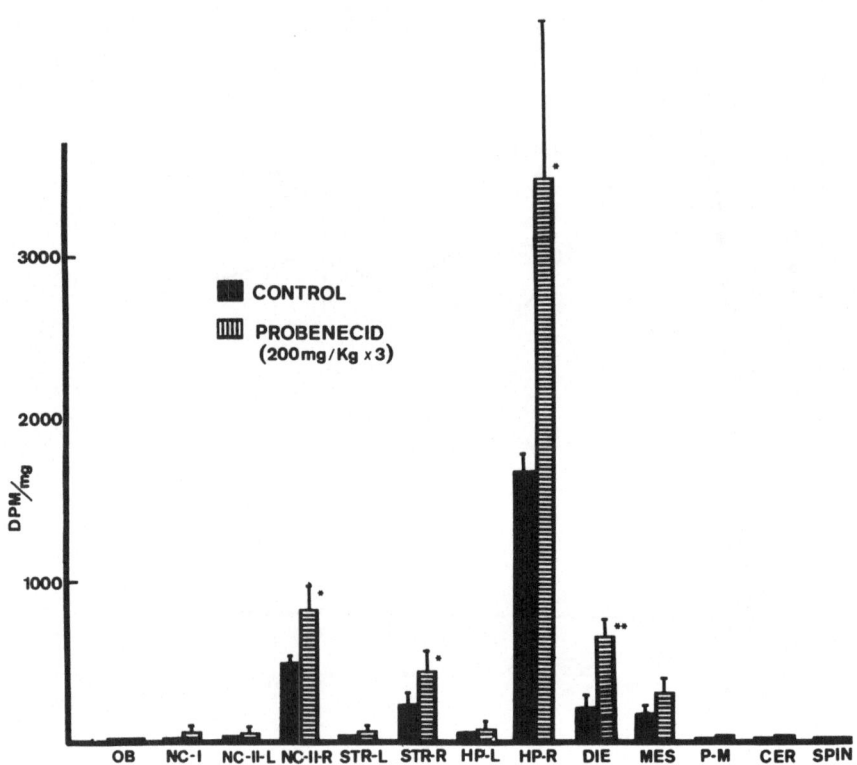

Fig. 7.   Accumulation of D,L-[³H]pipecolic acid 5 h after i.c.v. injection (right lateral ventricle) in 13 brain regions of the mouse.[69]

enecid administration, and its Ca^{2+}-dependent release following high-K^+ stimulation all constitute indirect evidence for a neuronal localization of this brain endogenous imino acid.

3.5. *Uptake of [³H]PA in Brain Slices, Synaptosomal Preparations, and Glial Cells*

We have demonstrated that [³H]PA (4×10^{-7} M) is taken up by mouse synaptosomal preparations by means of a Na^+-dependent, temperature-sensitive mechanism.[89] The uptake is strongly affected by ouabain (10^{-4} M). Kinetic studies indicated that the uptake is saturable at higher substrate concentration. A two-component system is present with a K_{m_1} of 3.9×10^{-6} M and a K_{m_2} of 90.2×10^{-6} M, which suggests that PA is transported by a high- and a low-affinity transport system. Compounds structurally related to PA (Fig. 5), such as glycine, L-proline, 4-amino-*n*-butyric acid (GABA), and 5-aminovaleric acid, show an inhibitory effect on uptake at a concentration of 10^{-4} M or less. A mutual inhibition of PA and L-PRO was shown, favoring the hypothesis of a common transport system for these two imino acids. It should be noted that L-PRO also inhibits the transport of PA *in vivo*.[83] Pipecolic acid significantly inhibits the initial GABA uptake in synaptosomal fractions in the rat[90] and increases the high-K^+-induced release of GABA.[91]

To gain further knowledge about a possible relationship between GABA and PA, we investigated and compared the uptake and release of both substances in synaptosomal fractions, brain slices, and glial-cell-enriched fractions of rat brain.[92] Our results show that the synaptosomal uptake of [^{3}H]PA is both shorter in duration and weaker than GABA uptake. The V_{max} value of the high-affinity uptake for PA is 100-fold lower than that for GABA. The K_m value is almost the same for both substances. Pipecolic acid and GABA mutually inhibit their uptake in synaptosomes; however, PA, GABA, and nipecotic acid produce a greater inhibition of synaptosomal [^{14}C]GABA uptake than of synaptosomal [^{3}H]PA uptake (Table II). The fact that nipecotic acid, a potent inhibitor of GABA uptake,[93] does not inhibit PA uptake as strongly as GABA uptake (Table II) may indicate that these substances are not taken up into synaptosomal fractions via the same carrier-mediated transport system. It is possible that PA may be taken up into sites different from GABA-containing neurons in the CNS. In addition, since GABA (10^{-4} M) affected neither spontaneous nor K^+-induced release of PA (Table II), the latter may not be dis-

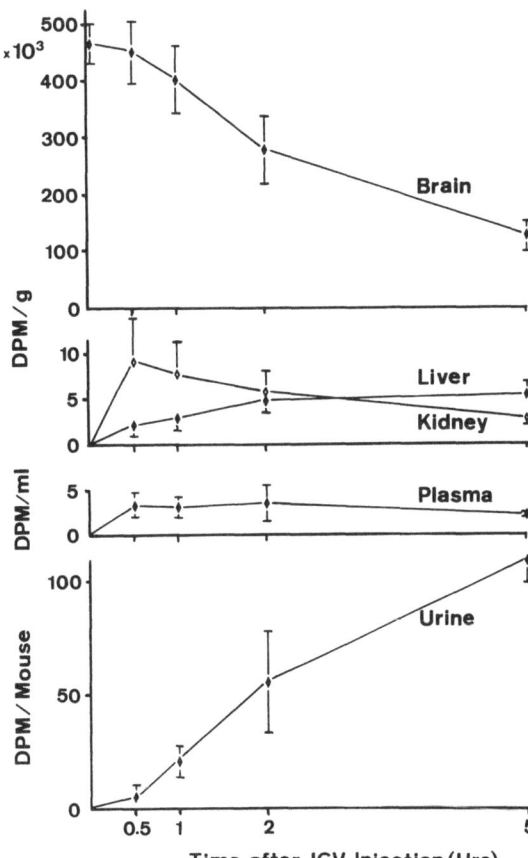

Fig. 8. Changes in total radioactivity in mouse brain, liver, kidney, plasma, and urine following i.c.v. injections of D,L-[^{3}H]pipecolic acid (0.69 nmol/mouse).[69]

Table II
Effect of PA, GABA, and Nipecotic Acid on GABA and PA Uptake and Release

	Brain slices	Synaptosomes	Glial cells
Pipecolic acid (10^{-4} M)	GABA release ↑	PA uptake ↓	
		GABA uptake ↓	GABA uptake ↓ ↓ ↓
GABA (10^{-4} M)	PA release 0	GABA uptake ↓ ↓ ↓	GABA uptake ↓ ↓ ↓
		PA uptake ↓	
Nipecotic acid (10^{-4} M)		GABA uptake ↓ ↓ ↓ ↓	GABA uptake ↓ ↓ ↓ ↓
		PA uptake ↓	

placed by GABA at its storage site. These results, taken together with the evidence for the presence, localization, and metabolism of PA in the brain,[1a] support the possibility that PA-containing neurons exist in the brain.

With regard to a glial localization of PA, a small amount of [^{3}H]PA was detectable in glial cell fractions. This contrasted with the marked uptake of [^{14}C]GABA and its different time course into glial cell fractions. In addition, PA inhibited the initial [^{14}C]GABA uptake into glial cell fractions and enhanced GABA release from brain slices (Table II). It is possible that PA physiologically plays a regulatory role on GABAergic neurotransmission by controlling GABA concentration at the synaptic cleft either by inhibiting its reuptake in both glial cells and endings or by increasing its release as demonstrated in brain slices (Fig. 9).

3.6. Release of Pipecolic Acid from Rat and Mouse Brain Slices and Synaptosomal Preparations

In our study of a release mechanism for PA, we investigated [94] whether or not a high concentration of K$^+$ could induce the release of PA from brain slices preloaded with radioactive PA. The K$^+$-induced depolarization of brain slices was accompanied by the release of radioactive PA from rat brain slices.

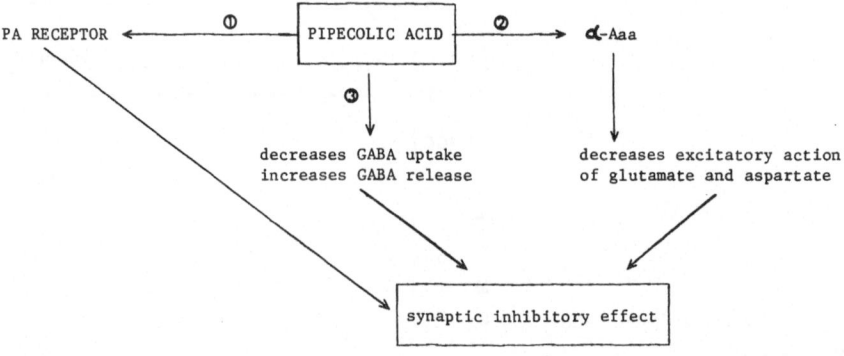

Fig. 9.   Proposed alternatives of neuromodulatory action of pipecolic acid in the CNS.

Since the K^+-induced release of this imino acid was significantly inhibited when the preparation was perfused with Ca^{2+}-free medium in the presence of EGTA for 15 min, it seems probable that this release would be Ca^{2+} dependent as reported for other putative neurotransmitters. Verapamil at a concentration of 10^{-5} M, which is known to inhibit Ca^{2+} influx, significantly inhibited the high-K^+-induced release of labeled PA. This can be taken as evidence that the influx of Ca^{2+} into nerve terminals containing PA is a necessary event in the release mechanism. In the mouse, following D,L-[^{3}H]PA preloading with i.c.v. injections, radioactivity was released from hippocampus slices in the perfusion medium after high-K^+-induced depolarization.[69] This release was Ca^{2+} dependent. GABA (10^{-4} M) affected neither spontaneous nor high-K^+-induced release of [^{3}H]PA from rat brain slices.[92] On the other hand, it has been demonstrated that PA enhances a high-K^+-induced release of [^{14}C]GABA from brain slices.[91]

We investigated the uptake and release of PA in the crude mitochondrial P_2 fractions and glial-enriched fractions of the rat brain.[92] The K^+-induced depolarization evoked a release of preloaded [^{3}H]PA from the fraction. The above results, together with our data on brain uptake *in vivo*, indicate that PA physiologically may play a regulatory role on GABAergic neurotransmission by controlling GABA concentration at the synaptic cleft (Fig. 9).

3.7. Pipecolic Acid Uptake and Biosynthesis during Development

In vitro formation of PA from L-lysine was studied in whole mouse embryo and brain.[95] The whole mouse embryo is able to form PA from L-lysine at 13 days of gestational age, suggesting that this animal can actively metabolize lysine at this stage of development.

In vitro formation of PA from L-lysine could be detected in mouse brain as early as the 17th day of gestation. A slight increase in formation occurs up to the first days after birth; however, at 90 days, the concentration is back to perinatal values. The brain uptake following i.p. injection of [^{3}H]PA in the 1-day-old mouse shows a different pattern than that in the adult (H. Nishio and E. Giacobini, unpublished data) (Fig. 10). First, PA is more strongly accumulated; secondly, higher levels of [^{3}H]PA are maintained for at least 24 h following a single i.p. injection. In a parallel investigation, the biosynthesis of PA was studied in the chick embryo. Such a preparation offers the advantage that formation of PA can be investigated in the whole animal without contribution from bacteria.[95]

All of the chick embryo organs (including brain) examined at different developmental stages showed PA formation from L-lysine. This formation showed a slight tendency to increase throughout development. It has been previously shown that L-lysine is metabolized in the chicken by two alternate routes through pipecolate or saccharopine, leading to the formation of α-aminoadipate[96,97] (Fig. 4). The observations that formation of PA in the chick embryo brain occurs at early stages of development and that its transport and accumulation are increased in the newborn mouse are important in view of the

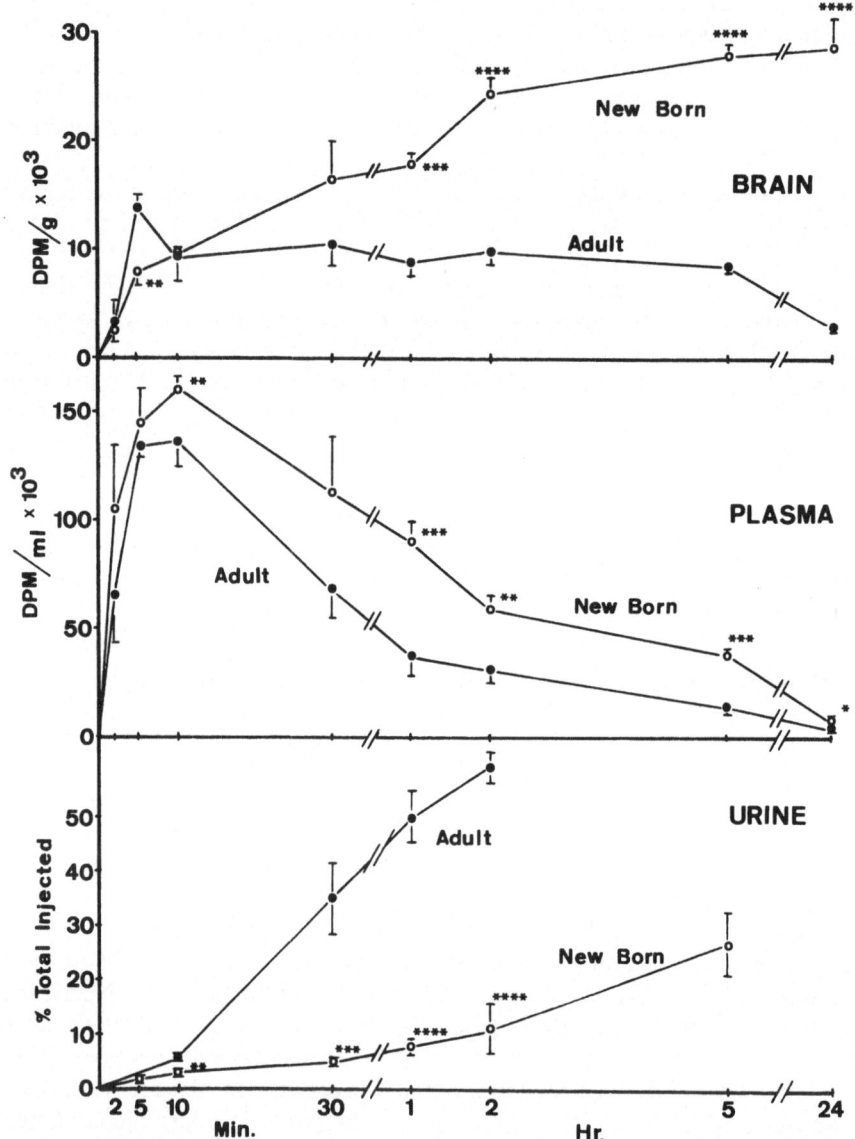

Fig. 10.   Brain uptake of D,L-[³H]PA following i.p. injections in 1-day-old and adult mice.

abnormally elevated levels of PA found in brain of hyperpipecolic infants affected by severe mental retardation[80,87] (Table III).

3.8. Physiological Role of Pipecolic Acid in Brain

Neurophysiological and neuropharmacological results supporting a specific physiological role of PA in mammalian brain are scanty. This contrasts with the large body of biochemical evidence summarized in the previous sec-

Table III
Disorders of PA Metabolism

Authors	No. of patients	Age at death	Symptoms	Pathology	Brain localization	Pipecolic acid			
						Serum[a]	Urine[b]	CSF	Brain[c]
Gatfield et al.[80]	1	104 wk	Hypothonia Paralysis Retardation	Hepatomegaly Demyelination, gliosis Neuronal degeneration (CNS)	Pons–medulla Cerebellum	●●●	(●)	●	●●
Woody and Popene[105]	4	4–12 yr	Hyperlysinemia				●● (3–6)		
	4	6–23 days	Premature				●● (1–8)		
Danks et al.[107]	8	2–34 wk	Retardation Hypothonia Icterus Cer. dysf.	Renal cysts Liver fibrosis CNS: Cerebral gliosis dem. gyral folding Subependymal cysts	Paravent. Cerebellum Brainstem	●●	●●● (33–218)		
Thomas et al.[87]	1	129 wk	Hypothonia Hepatomegaly Retardation	Retinal and optic nerve degen. Hepatomegaly	CNS general Retina Optic nerve	●●●	●?		
Brun et al.[109]	2	14 wk	Retardation Hypothonia Hepatomegaly Icterus	Hepatocytic necrosis Micropolygyria Heterotypia, gliosis Hypoplasia Scant myelination	Pons–medulla Cerebellum		(●)		

[a] In serum: ●, 5–20; ●●, 20–50; ●●●, 50–350 μM.
[b] In urine: ●, ≤1; ●●, 1–10; ●●●, 10–200 μg/mg creatinine.
[c] In CSF: ●, 1–2 μM.
[d] In brain: ●●●, 200 μmol/kg.

tions. Intraventricular administration of PA in the mouse produces ptosis, hypotonia, and sedation accompanied by suppression of fighting behavior.[98] In the cat, intracerebral administration with chronically implanted cannulae and electrodes causes depression in both EEG activity and behavior following injections into the cerebellum and hippocampus.[98]

More recently, Kasé et al.[99] investigated the effect of PA on the activity of cortical and pyramidal neurons in the rat brain by using single-unit recording and microelectrophoretic techniques. Electrophoretically applied PA exerted in most cases a significant depressant effect on both cerebral cortical and hippocampal pyramidal neurons. These findings seem to support the results obtained by the same authors with intraventricular and intracerebral administration of PA. The favored interpretation of these electrophysiological responses was either an action on GABA receptors or a direct effect on PA receptor sites. The former could be produced by affecting GABAergic transmission, either by inhibiting uptake of endogenous GABA as demonstrated by Nomura et al.[89,92] or by facilitating its release as shown by Nomura et al.[94] and Okuma et al.[91] In addition, the inhibitory affect of L-α-aminoadipic acid, a major metabolite of PA in brain, on the excitatory properties of glutamate and aspartate should not be neglected. The possibility of a direct interaction with a PA-specific receptor site is supported by our recent findings of a [^{3}H]PA high-affinity binding site.[100] It remains to be demonstrated whether specific PA-receptive sites exist in the cerebral cortex and hippocampus structures examined electrophysiologically.

4. THE IMINOACIDURIAS

A primary iminoacidopathy is a hereditary defect in an enzymatic step of the metabolic pathway of one or more imino acids or in a carrier-mediated mechanism necessary for transport of a certain imino acid into or out of cells. Two main groups are presently known.

4.1. Disorders of Proline and Hydroxyproline Metabolism

Hyperprolinemia (type I), which has been reported in a few families, is related to mental retardation, abnormal EEG, photosensitivity, nerve deafness, and defective renal development.[6,101,102] Diffuse neuronal loss and delayed myelination have been reported in this familial disease characterized by the absence or reduction of PRO oxidase activity. Fasting plasma PRO levels are 7.5–10 mg/dl. In hyperprolinemia, type II, PRO and pyrroline-5-carboxylate are excreted. The disease is associated with convulsions and mental retardation. In both type I and type II, PRO, HYP, and glycine are found in the urine. A competition occurs in the iminoglycine transport system,[103] suggesting that PRO, HYP, glycine, and perhaps serine share a common tubular transport. In neonatal iminoglycinuria, PRO, HYP, and glycine metabolism are affected with a low-capacity system for transport of all three imino acids.[103] Iminoaciduria persists about 3 months; glycinuria persists 5–6 months.[103] There are too few known clinical cases to provide a clear-cut differentiation between hydroxy-prolinemia and hyperprolinemia. However, in some cases, there are fewer symptoms in the former type of disease than in the latter.

4.2. Disorders of Pipecolic Acid Metabolism

Abnormally elevated levels of PA have been detected in serum of young patients with hyperthyroidism,[104] hyperlysinemia,[105] and kwarshiorkor.[106] Mental retardation is a common symptom of these patients.

Directly related to PA metabolism is the newly discovered metabolic disorder of hyperpipecolatemia (hyper-PA acidemia). This syndrome was described for the first time in an 18-month-old infant by Gatfield *et al.*[80] in Canada and then confirmed in a second case by Thomas *et al.*[87] (Table III). This inborn disease is characterized by high levels of PA in brain serum and urine. The level of serum amino acids is normal. The clinical picture is characterized by a severe and progressive mental retardation which relates to a progressive malformation and degeneration of the CNS, hypotonia, and hepatomegalia (Table III).

In a report by Danks *et al.*[107] eight cases of a cerebro–hepato–renal syndrome showing a fault in PA metabolism were identified. This disturbance has been designated Zellweger syndrome. Since its first description in 1962[108] in the United States, further cases have been recognized in other parts of the world, including a first Scandinavian case reported by Brun *et al.*[109] These various hyperpipecolatemic syndromes occur in infants of both sexes. Most cases die within weeks or months, the oldest case reaching the age of 12 years[105] (Table III). The frequency of the disease has been estimated to be of the order of 1/100,000 births. Directly related to disturbances of PA metabolism is α-aminoadipic aciduria (α-AAA-uria), which has been described in several cases of mental retardation in children.[110] In some cases, it is associated with α-ketoadipic aciduria (α-KAA-uria), whereas in other patients, α-KAA levels are normal. In these patients, oral loading with lysine results in an increase of α-AAA levels in serum. A defect in the degradation pathway of L-lysine between α-AAA and glutaryl-CoA has been suggested.[110] It is not clear from studies performed on fibrolasts and liver homogenates isolated from these patients whether α-AAA transaminase and α-KAA decarboxylase activities are normal or defective.[110]

Etiology and pathogenesis of diseases related to abnormal PA metabolism leading to severe mental retardation and CNS degeneration are unknown. Equally obscure is the relationship among the various PA syndromes described. Our ignorance stems mainly from the fact that our knowledge about the role and metabolism of PA in the organism, particularly in brain, is still poor.

The devastating effect of hyperpipecolatemia on the human CNS suggests that (1) PA or some of its metabolites (α-aminoadipic acid?) are accumulated in the brain and allowed to exert pronounced neurotoxic effects or that (2) normal PA metabolism, catabolism, and secretion are essential for the physiological function and development of the brain or that (3) an inborn malfunction in the process of uptake and transport of brain-exogenous PA or a breakdown in the mechanism of efflux of PA (or of its metabolites) from the brain could be present.

5. CONCLUDING REMARKS

Biochemical and electrophysiological evidence strongly suggests that PRO may have a specific function in mammalian CNS, probably acting as an inhibitory neuromodulator. Similar evidence is lacking for HYP. The question remains open whether PRO action is direct or glutamate mediated. With regard to PA, results from our laboratory,[1a] as well as from other authors, demonstrate that this imino acid not only can be transported and accumulated in the brain from exogenous sources but also can be synthesized and metabolized in the CNS.

Evidence for selective brain areas specifically accumulating, synthesizing, and releasing PA is now available.[1a] Taken together, these results strongly suggest that PA may be selectively localized in neuronal and synaptic structures where it could act as physiological neuromodulator. On the basis of the present findings, three major hypotheses can be formulated with regard to the mechanism of action of PA: (1) direct modulation of a PA receptor site, (2) modulation through GABA neurotransmission, and (3) modulation through a decreased excitatory action of glutamate and aspartate by α-aminoadipic acid, a major metabolite of PA (Fig. 9).

The present evidence proposes that imino acids may actually constitute a new group of neuromodulators in mammalian CNS. This group should be considered together with the numerous cerebral amino acids, the neurotransmitter action of which has already been demonstrated.

ACKNOWLEDGMENT. The investigations carried out in the author's laboratory were supported by United States Public Health grants NS 11430 and NS 14086 and by grants from the University of Connecticut Research Foundation.

REFERENCES

1. Himwich, W. A., and Agrawal, H. C., 1970, *Handbook of Neurochemistry*, Volume 1 (A. Lajtha, ed.), Plenum Press, New York, pp. 33–52.
1a. Giacobini, E., Nomura, Y., and Schmidt-Glenewinkel, T., 1980, *Cell. Mol. Biol.* **26:**135–146.
2. Perry, T. L., Sanders, H. D., Hansen, S., Lesk, D., Kloster, M., and Gravlin, L., 1972, *J. Neurochem.* **19:**2651–2656.
3. Lajtha, A., and Toth, J., 1973, *Brain Res.* **55:**238–241.
4. Adams, E., and Frank, L., 1980, *Annu. Rev. Biochem.* **49:**1005–1061.
5. Strecker, H. J., 1970, *Handbook of Neurochemistry*, Volume 3 (A. Lajtha, ed.), Plenum Press, New York, p. 184.
6. Guroff, G., 1980, *Molecular Neurobiology* (G. Guroff, ed.), Marcel Dekker, New York, pp. 116–117.

7. Taggart, J. V., and Krakaur, R. B., 1949, *J. Biol. Chem.* **177**:641.
8. Efron, M. L., Bixby, E. M., and Pryles, C. V., 1965, *N. Engl. J. Med.* **272**:1299–1308.
9. Efron, M. L., 1965, *N. Engl. J. Med.* **273**:1243.
10. Evans, P. D., 1973, *J. Neurochem.* **21**:11–17.
11. McBride, W. J., Shank, R. P., Freeman, A. R., and Aprison, M. H., 1974, *Life Sci.* **14**:1109–1120.
12. Osborne, N. N., and Neuhoff, V., 1974, *Brain Res.* **80**:251–264.
13. Sorenson, M. M., 1973, *J. Neurochem.* **20**:1231–1245.
14. Adams, E., 1970, *Int. Rev. Connect. Tissue Res.* **5**:1–91.
15. Lindsay, J., and Bachelard, H. S., 1966, *Biochem. Pharmacol.* **15**:1045–1052.
16. Ratner, S., Morell, H., and Corvalho, E., 1960, *Arch. Biochem. Biophys.* **91**:280–289.
17. Peraino, C., and Pitot, H. C., 1963, *Biochim. Biophys. Acta* **73**:222–231.
18. Smith, R. J., Downing, S. J., Phang, J. P., Lodato, R. F., and Aoki, T. T., 1980, *Proc. Natl. Acad. Sci. U.S.A.* **77**:5221–5225.
19. Johnson, A. B., and Strecker, H. J., 1962, *J. Biol. Chem.* **237**:1876–1882.
20. Strecker, H. J., 1960, *J. Biol. Chem.* **235**:3218–3223.
21. Sporn, M. B., Dingman, W., and De Falco, A., 1959, *J. Neurochem.* **4**:141–147.
22. Yamanishi, Y., Kasé, Y., Miyata, T., and Kataoka, M., 1970, *Life Sci.* **9**:409–414.
23. Honegger, C. G., and Honegger, R., 1960, *Nature* **185**:530.
24. Perry, T. L., Shaw, K. N. F., Walker, D., and Realich, D., 1962, *Pediatrics* **30**:576.
25. Kasé, Y., Miyata, T., Kamikawa, Y., and Kataoka, M., 1969, *Jpn. J. Pharmacol.* **19**:300.
26. Giacobini, E., 1976, *Advances in Biochemical Psychopharmacology*, Volume 15 (E. Costa, E. Giacobini, and R. Paoletti, eds.), Raven Press, New York, pp. 17–56.
27. Kunzle, H., and Felix, D., 1974, *Experientia* **30**:680.
28. Felix, D., and Kunzle, H., 1976, *Advances in Biochemical Psychopharmacology*, Volume 15 (E. Costa, E. Giacobini, and R. Paoletti, eds.), Raven Press, New York, pp. 165–173.
29. Garweg, G., 1969, *Naturwissenschaften* **56**:463–464.
30. Garweg, T., 1970, *Experientia* **26**:1348–1349.
31. Garweg, G., and Schneider, E. J., 1971, *Experientia* **27**:377–378.
32. Garweg, G., and Kinsky, I., 1970, *Naturwissenschaften* **57**:253.
33. Roberts, P. J., Keen, P., and Mitchell, J. F., 1973, *J. Neurochem.* **21**:199–209.
34. Johnson, J. L., 1975, *Brain Res.* **96**:192–196.
35. Van Harreveld, A., and Fifkova, E., 1973, *J. Neurochem.* **20**:947–962.
36. Balcar, V. J., and Johnston, G. A. R., 1974, *Biochem. Pharmacol.* **23**:821–827.
37. Bennett, J. P., Mulder, A. H., and Snyder, S. H., 1974, *Life Sci.* **15**:1045–1056.
38. Peterson, N. A., and Raghupathy, E., 1972, *J. Neurochem.* **19**:1423–1438.
39. Raghupathy, E., and Peterson, N. A., 1977, *J. Neurochem.* **29**:859–863.
40. Krogsgaard-Larsen, P., and Johnston, G. A. R., 1975, *J. Neurochem.* **25**:797–802.
41. Balcar, V. J., Johnston, G. A. R., and Stephanson, A. L., 1976, *Brain Res.* **102**:143–151.
42. Zarzecki P., Blum, P. S., Cordingley, G. E., and Somjen, G. G., 1975, *Brain Res.* **89**:187–191.
43. Nistri, A., and Morelli, P., 1978, *Neuropharmacology* **17**:21–27.
44. Cherkin, A., Eckardt, M. J., and Gerbrandt, L. K., 1976, *Science* **193**:242–244.
45. Van Harreveld, A., and Fifkova, E., 1974, *Brain Res.* **81**:455.
46. Garweg, G., and Dahnke, H. G., 1973, *Anat. Anz.* **134**:297–303.
47. Garweg, G., and Dahnke, H. G., 1973, *Naturwissenschaften* **60**:201–202.
48. Garweg, G., von Rehren, D., and Hintze, U., 1980, *J. Neurochem.* **35**:616–621.
49. Zacharias, R. M., Thompson, J. F., and Stewart, F. C., 1952, *J. Am. Chem. Soc.* **74**:2949–2950.
50. Grobbelaar, N., and Stewart, F. C., 1953, *J. Am. Chem. Soc.* **75**:4341.
51. Morrison, R. I., 1953, *Biochem. J.* **53**:474.
52. Greenberg, J. P., and Winitz, M., 1961, *Chemistry of the Aminoacids*, Volume 3, John Wiley & Sons, New York, pp. 2529–2558.
53. Schweet, R. S., Holden, J. T., and Lowy, P. H., 1954, *J. Biol. Chem.* **211**:517.
54. Rothstein, M., and Miller, L. L., 1953, *J. Am. Chem. Soc.* **75**:4371.
55. Rothstein, M., and Miller, L. L., 1954, *J. Biol. Chem.* **211**:851–858.

56. Rothstein, M., and Miller, L. L., 1954, *J. Biol. Chem.* **206:**243.
57. Higashino, K., Fujioka, M., and Yamamura, Y., 1971, *Arch. Biochem. Biophys.* **142:**606–614.
58. Schmidt-Glenewinkel, T., Nomura, Y., and Giacobini, E., 1977, *Neurochem. Res.* **2:**619–637.
59. Hernandez, M. F., and Chang, Y. F., 1980, *Biochem. Biophys. Res. Commun.* **93:**762–769.
60. Meister, A., Radhakrishnan, A. N., and Buckley, S. D., 1957, *J. Biol. Chem.* **229:**789–800.
61. Meister, A., and Buckley, S. D., 1957, *Biochim. Biophys. Acta* **23:**202.
62. Rothstein, M., and Greenberg, D. M., 1960, *J. Biol. Chem.* **235:**714–718.
63. Rothstein, M., Cooksey, K. E., and Greenberg, D. M., 1962, *J. Biol. Chem.* **237:**2828–2830.
64. Meister, A., 1965, *Biochemistry of the Amino Acids* Volume 2, Academic Press, New York, pp. 941–951.
65. Higashino, K., Tsukada, K., and Lieberman, I., 1965, *Biochem. Biophys. Res. Commun.* **20:**285.
66. Chang, Y. F., 1976, *Biochem. Biophys. Res. Commun.* **69:**174–180.
67. Chang, Y. F., 1978, *J. Neurochem.* **30:**347–354.
68. Chang, Y. F., 1978, *J. Neurochem.* **30:**355–360.
69. Nishio, H., Giacobini, E., and Ortiz, J., 1982, *Neurochem. Res.* **7:**373–385.
70. Kasé, Y., Kataoka, M., and Miyata, T., 1967, *Life Sci.* **6:**2427–2431.
71. Stepita-Klauco, N., Dolezalova, H., and Giacobini, E., 1973, *Brain Res.* **63:**141–152.
72. Miyata, T., Kamata, K., Nishikibe, M., Kasé, Y., Takehama, K., and Okano, Y., 1974, *Life Sci.* **15:**1135–1152.
73. Drucker-Colin, R. R., and Giacobini, E., 1975, *Brain Res.* **88:**186–189.
74. Giacobini, E., 1975, *Sleep, 1974* (P. Levin and P. Koella, eds.), S. Karger, Basel, pp. 59–65.
75. Dolezalova, H., Stepita-Klauco, M., and Fairweather, R., 1974, *Brain Res.***72:**115–122.
76. Kasé, Y., Okano, Y., Yamanishi, Y., Kataoka, M., Kitahara, K., and Miyata, T., 1970, *Life Sci.* **9:**1381–1387.
77. Kasé, Y., Okano, Y., Miyata, T., Kataoka, M., and Yonehara, N., 1974, *Life Sci.* **14:**785–791.
78. Schmidt-Glenewinkel, T., Giacobini, E., Nomura, Y., Okuma, Y., and Segawa, T., 1978, *2nd Meeting European Society for Neurochemistry* (*Gottingen*) *Abstracts*, Verlag Chemie Weinheim, New York, p. 618.
79. Kasé, Y., Kataoka, M., Miyata, T., and Okano, Y., 1973, *Life Sci.* **13:**867–873.
80. Gatfield, P. D., Taller, E., Hinton, G. G., Wallace, A. C., Abdelnour, G. M., and Haust, M. D., 1968, *Can. Med. Assoc. J.* **99:**1215–1233.
81. Lajtha, A., 1958, *J. Neurochem.* **2:**209–215.
82. Oldendorf, W. H., 1971, *Am. J. Physiol.* **221:**1629–1639.
83. Nishio, H., and Giacobini, E., 1981, *Neurochem. Res.* **6:**835–845.
84. Oldendorf, W. H., and Szabo, J., 1976, *Am. J. Physiol.* **230:**94–98.
85. Nishio, H., Ortiz, J., and Giacobini, E., 1981, *Neurochem. Res.* **6:**1241–1252.
86. Gatfield, P. D., and Taller, E., 1971, *Brain Res.* **29:**170–174.
87. Thomas, G. H., Haslam, R. H., Batsahw, M. L., Capute, A. J., Neidengard, L., and Ransom, J. L., 1975, *Clin. Genet.* **8:**376–382.
88. Olney, J. W., de Gubareff, T., and Collins, J. F., 1980, *Neurosci. Lett.* **19:**277–282.
89. Nomura, Y., Schmidt-Glenewinkel, T., and Giacobini, E., 1980, *Neurochem. Res.* **5:**1163–1173.
90. Nomura, Y., Okuma, Y., and Segawa, T., 1978, *J. Pharmacodyn.* **1:**251–255.
91. Okuma, Y., Nomura, Y., and Segawa, T., 1979, *J. Pharmacodyn.* **2:**261–265.
92. Nomura, Y., Okuma, Y., Segawa, T., Schmidt-Glenewinkel, T., and Giacobini, E., 1981, *Neurochem. Res.* **6:**391–400.
93. Johnston, G. A. R., Stephanson, A. L., and Twitchen, B., 1976, *J. Neurochem.* **26:**83–87.
94. Nomura, Y., Okuma, Y., Segawa, T., Schmidt-Glenewinkel, T., and Giacobini, E., 1979, *J. Neurochem.* **33:**803–805.
95. Nomura, Y., Schmidt-Glenewinkel, T., and Giacobini, E., 1978, *Dev. Neurosci.* **1:**239–249.
96. Grove, J. A., and Roghair, H. G., 1971, *Arch. Biochem. Biophys.* **144:**230–236.
97. Wang, S. H., and Nesheim, M. C., 1975, *J. Nutr.* **102:**583–596.

98. Miyata, T., Kamata, K., Noguchi, M., Okano, Y., and Kasé, Y., 1973, *Jpn. J. Pharmacol.* **23**(Suppl.):81.

99. Kasé, Y., Takahama, K., Hashimoto, T., Kaisaku, J., Okano, Y., and Miyata, T., 1980, *Brain Res.* **193**:608–613.

100. Giacobini, E., Nishio, H., Ortiz, J., and Gutierrez, M., 1981, *Soc. Neurosci. Abstr.* **7**:322.

101. Efron, M. L., and Ampola, M. G., 1967, *Pediatr. Clin. North Am.* **14**:881–899.

102. Waisman, H. A., 1970, *Biochemistry of Brain and Behavior*, (R. Bowman and S. P. Datta, eds.), Plenum, New York, p. 223.

103. Scriver, C. R., Clow, C. L., and Lamm, P., 1973, *Clin. Biochem.* **6**:142–188.

104. Sonoda, Y., 1957, *Proc. Jpn. Acad.* **33**:162.

105. Woody, N. C., and Popene, M. B., 1970, *Pediatr. Res.* **4**:89–95.

106. Whitehead, R. G., 1964, *Nature* **204**:389.

107. Danks, D. M., Tippett, P., Adams, C., and Campbell, P., 1975, *J. Pediatr.* **86**:382–407.

108. Bowen, P., Lee, C. C. N., Zellweger, H., and Lindenberg, R., 1962, *Bull. Johns Hopkins Hosp.* **114**:402–414.

109. Brun, A., Gilboa, M., Meeuwisse, G. W., and Nordgren, H., 1978, *Eur. J. Pediatr.* **127**:229–245.

110. Manders, A. J., Oostrom, C. G. v., Trijbels, J. M. F., Rutten, F. J., and Kleijer, W. J., 1981, *Eur. J. Pediatr.* **136**:51–55.

111. Sershen, H., and Lajtha, A., 1976, *Exp. Neurol.* **53**:465–474.

Glutathione

Howard S. Maker

1. INTRODUCTION

Glutathione (GSH) is a tripeptide formed from glutamic acid, cysteine, and glycine:

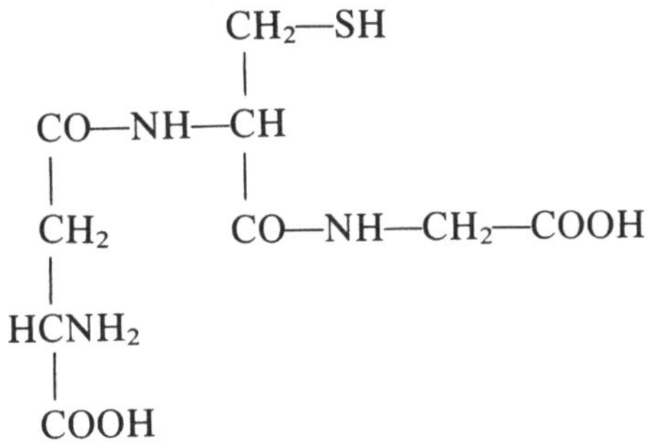

$$CH_2\text{—}SH$$
$$|$$
$$CO\text{—}NH\text{—}CH$$
$$|\qquad\qquad|$$
$$CH_2\qquad CO\text{—}NH\text{—}CH_2\text{—}COOH$$
$$|$$
$$HCNH_2$$
$$|$$
$$COOH$$

Although its functions in brain are still uncertain, it has continued to draw attention because its concentration (2–3 μmol/g tissue)[1] exceeds that of any other amino acid or peptide except glutamate and glutamine. Two major roles have been postulated for GSH: reduction or conjugation of oxidative or electrophilic endogenous or exogenous toxins and participation in amino acid transport across membranes. Several reviews have been published.[2–5] This chapter serves to update that of Orlowski and Karkowski[3] which should be consulted for a history of the subject and a complete bibliography of earlier work.

2. GLUTATHIONE IN BRAIN

Almost all (97%) of the acid-soluble thiol groups available for reaction in rat cerebral cortex are in GSH.[6] As in other tissues, glutathione is found in

Howard S. Maker • Department of Neurology, Mount Sinai School of Medicine of the City University of New York, New York, New York 10029, and The Bronx Veterans Administration Medical Center, Bronx, New York 10468.

reduced (GSH) and oxidized (GSSG, glutathione disulfide) forms. The greater the care that is taken to prevent the oxidation of GSH during tissue preparation, the lower is the concentration of GSSG found. The most reliable studies indicate a GSSG concentration in rat brain less than 0.7% that of GSH.[7] In rat brain, 62% of GSH was soluble (cytoplasmic), with a particulate distribution in mitochondrial (14%), synaptosomal (8.4%), and myelin (8.2%) fractions.[8] At birth, cat neocortex GSH levels were 80% of adult levels and attained adult concentrations by the end of the second postnatal week.[9] Glutathione in other regions of the cat brain also increased postnatally, but the midbrain slightly lagged the neocortex. Brain contains other peptides besides GSH containing the γ-glutamyl moiety.[1]

In addition to free GSH, mixed disulfides formed by the conjugation of the thiol groups of GSH and tissue proteins are present in brain as elsewhere in the body.[3] Using the best available techniques, Folbergrova *et al.*[7] found a concentration of GSH in rat cerebral cortex of 2.05 ± 0.2 μmol/g with that of GSSG 0.014 ± 0.002 μmol/g. Similar results were found by Cooper *et al.*[10] after freeze-blowing the entire rat brain, but the GSSG concentration was 0.0017 ± 0.007 μmol/g. Glutathione in cerebellum was 2.01 μmol/g, and that in brainstem 1.37 μmol/g. Folbergrova[7] *et al.* estimated the expected level of brain GSSG based on the apparent equilibrium constant of glutathione reductase and the ratio of $NADP^+/NADPH$ calculated from the levels of substrates and products of enzymes utilizing $NADP^+$ as a cofactor. Assuming equilibrium in the glutathione reductase reaction, the expected GSSG concentration was 0.1 pmol/g in contrast to the measured 14 nmol/g.[7] The investigators attributed part of the difference to the oxidation of GSH during tissue preparation.

3. ASSAY OF GSH AND GSSG

Difficulties encountered in the assay of GSH and GSSG have resulted from the lack of stability of tissue GSH and the lack of selectivity of earlier methods. Freshly prepared GSH is fairly stable in deionized water, although commercial GSH may contain 1 to 2% GSSG; GSH is less stable in tissue containing such prooxidants as ferric ion and lipid peroxides, and it oxidizes during tissue preparation. This oxidation has relatively little effect on GSH levels, but the very low GSSG concentrations may be considerably increased. Indeed, it is probable that the true level of brain GSSG has never been determined.[7] The best present methods for GSSG measurement freeze the brain *in situ* and immediately add to the tissue homogenate a substance that can covalently bind to and prevent the oxidation of the thiol. N-Ethylmaleimide[7] and 2-vinylpyridine[11] are suitable thiol reagents. The mixed disulfides of glutathione have usually not been measured or assayed together with GSSG. A significant proportion of the GSH may be in the form of such disulfides.

There are now sensitive and reliable methods for the assay of GSH and GSSG, but the evaluation of past studies requires assessment of the methods used. Determinations using ion-exchange chromatography yielded concentrations well below those now known to be present.[3] The Ellman reagent (5,5′-dithiobis-2-nitrobenzoic acid) is still in active use after two decades. It reacts with thiol groups to form a product with high molar absorbance for spectrophotometric measurement. Eighty percent of the thiol groups of brain soluble in acid after precipitation of protein and reacting with the reagent are in GSH.[6,12] The assay, therefore, overestimates tissue GSH. O-Phthalaldehyde reacts with GSH at pH 8.0 and with GSSG at pH 12.0 to form a fluorescent product.[13] The assay has a sensitivity (twice the blank) of 0.2 nmol but lacks specificity for GSSG.[14] Enzymatic assays utilize the inherent specificity of the enzyme and are more accurate than those with thiol reagents. Yeast glyoxylase can conjugate methylglyoxal to GSH, forming S-lactosyl-GSH, which is measured spectrophotometrically,[6] but most modern assays use glutathione reductase to transfer hydrogen from NADPH to GSSG. The decrease in molar absorbance of NADPH can be followed at 340 nm, or the NADP$^+$ formed can be induced to fluoresce in strong base for a more sensitive assay (0.3 nmol).[15] The GSH is assayed as GSSG after oxidation.[16]

Several investigators have developed systems based on enzymatic cycling. A substance that is easily measured in its reduced form is added in its oxidized form to a reaction mixture containing GSH which nonenzymatically reduces the indicator agent. The GSSG formed in the nonenzymatic reaction is then reduced back to GSH by NADPH and glutathione reductase. The GSH is thus regenerated, and one molecule of the indicator substance is reduced for each turn of the cycle. The rate of the reaction is proportional to the amount of GSH or GSSG originally present, and the greater the number of turns of the cycle, the greater the amplification in the form of the amount of the indicator substance reduced and measured. In the frequently used method of Tietze,[17] the Ellman reagent is the indicator, and the practical amplification allows a sensitivity of 0.1 nmol of GSH or GSSG. The reaction mixture contains excess NADPH, and the 2-nitro-5-thiobenzoate formed is measured at 412 nm. For added sensitivity, the NADP$^+$ formed can be converted back to NADPH which is naturally fluorescent (sensitivity 2 pmol[18]) or amplified still further by the enzymatic cycling procedure of Lowry and Passonneau.[19] To measure GSSG alone in the original procedure, GSH was masked with N-ethylmaleimide, but this substance is a potent inhibitor of the reductase, and excess must be removed by ether extraction or column chromatography prior to the cycling. 2-Vinylpyridine, which also reacts with GSH, is not a glutathione reductase inhibitor and should facilitate the assay.[11]

Glutathione can be determined accurately on an amino acid analyzer, but 10 nmol are required.[20] High-pressure liquid chromatography, however, is accurate and has a sensitivity in the nanomole range. Moreover, this is the only technique that allows separate determination of selected mixed disulfides.

4. SYNTHESIS AND DEGRADATION OF GLUTATHIONE

4.1. Biosynthesis of Glutathione

4.1.1. γ-Glutamylcysteine Synthetase (E.C. 6.3.2.2)

$$\text{L-glutamate} + \text{L-cysteine} + \text{ATP} \xrightarrow{\text{Mg}^{2+}} \text{L-}\gamma\text{-glutamyl-L-cysteine} + \text{ADP} + \text{Pi}$$

γ-Glutamylcysteine synthetase (GCS) is a soluble enzyme requiring energy

Table I

Enzymes of Glutathione Synthesis and Degradation in Brain (nmol/per min)

Enzyme	Substrate	Rabbit[a]		Bovine[b]		Mouse[c] brain
		Cerebral cortex	Choroid plexus	Cerebral cortex	Choroid plexus	
γ-Glutamylcysteine synthetase	L-α-aminobutyrate	309	656	286	495	310
Glutathione synthetase	γ-Glutamyl-L-α-aminobutyrate	—	—	—	—	72
γ-Glutamyl transpeptidase	L-α-glutamyl-p-nitroanilide	430	40,130	390	4,638	118
γ-Glutamyl cyclotransferase	L-γ-Glutamyl-methionine	766	1,490	768	1,520	4,950
5-Oxoprolinase	L-5-Oxoproline	2.19	4.38	—	—	—

[a] Okonkwo et al.[63]
[b] Orlowski and Wilk.[29]
[c] Orlowski and Wilk.[27]

in the form of ATP and magnesium ion for activation. Its activity is lower in mouse and rabbit kidney than in rat kidney, but activity in brain is similar in the three species (Table I). The GCS activity in choroid plexus is approximately twice that in brain. The K_m of renal GCS is 0.35 mM.[22] The concentrations of L-glutamic acid and L-cysteine in rat brain are, respectively, 9.70 μmol/g and 55.9 nmol/g.[23] Cysteine (approx. 60 μM) is therefore present at a level below that at which the enzyme is half saturated, and cysteine availability will limit the rate of reaction. L-Aminobutyrate is a useful substitute for cysteine in assaying the enzyme, reacting with a similar rate but with a higher K_m.[22] Rat renal GCS has a molecular weight of approximately 92,000.[22] Activity of the enzyme is dependent on a free thiol group, and it is strongly inhibited by relatively low concentrations of thiol-reactive agents.[22] Activity of GCS is irreversibly inhibited by L-methionine-S-sulfoximine following phosphorylation of the inhibitor in tissue to methionine sulfoximine phosphate.[24] Phosphorylated methionine sulfoximine is also an irreversible inhibitor of glutamine synthetase.[25] At a dose of 2.5 μmol/g body weight, a racemic mixture of the inhibitor produces recurrent seizures in mice.[26,27] The mechanism of the seizure induction is not known.

4.1.2. Glutathione Synthetase (E.C. 6.3.2.3)

$$\text{L-}\gamma\text{-glutamyl-L-cysteine} + \text{glycine} + \text{ATP} \xrightarrow{\text{Mg}^{2+}}$$
$$\text{L-4-glutamyl-L-cysteine} + \text{ADP} + \text{Pi}$$

Glutathione synthetase (GS) resembles GCS in requiring energy and magnesium. Its optimal activity is at pH 8.0. The GS of bovine erythrocytes has a molecular weight of 123,000, possibly as a dimer of two equal subunits.[28] The substrate γ-glutamyl cysteine is also a substrate of γ-glutamyl cyclotransferase, and it is postulated that GSH synthesis is separated in cells from sites of the degradative enzyme.[29] The GS activity in brain and other tissues is two orders of magnitude less than that of GCS (Table I). However, the major regulation of GSH synthesis may be the supply of cysteine limiting the synthesis of γ-glutamyl cysteine. There is little data on the concentration of γ-glutamyl cysteine, but its level in erythrocytes is approximately 20 μM.[3] Physiological concentrations of GSH inhibit GCS, and feedback inhibition of the initial enzyme in the synthesis sequence may be an additional control.[3]

4.2. Degradation of Glutathione

4.2.1. γ-Glutamyl Transpeptidase (E.C. 2.3.2.2)

GSH + amino acid (or peptide) →
γ-glutamyl amino acid (or peptide) + cysteinylglycine

γ-Glutamyl transpeptidase (GTP) is a membrane-bound (particulate) enzyme catalyzing the transfer of the γ-glutamyl group of GSH to amino acids or peptides and releasing glycine. It also hydrolyzes the γ-glutamyl bond of

GSH and catalyzes the transfer of the γ-glutamyl moiety from one molecule of GSH to another, forming γ-glutamyl GSH and cysteinylglycine.[3] In addition, GTP may function as an oxidase of GSH, forming GSSG.[30] Amino substrates include all neutral amino acids except proline and many peptides. The highest affinities of serum GTP are for glutamine and cysteine, and the lowest for basic and acidic amino acids.[31] The enzyme is a glycoprotein containing up to 36% carbohydrate.[32] This may contribute to its relatively high resistance to proteolytic enzymes.[33] Sheep renal GTP (molecular weight 90,000) consists of two polypeptide chains of 65,000 and 27,000 molecular weight.[34,35] Four different proteins with GTP activity were isolated from rat brain having molecular weights from 74,000 to 234,000 and K_ms ranging from 0.07 mM to 8.6 mM.[36] The greatest proportion of the total enzyme activity was associated with the plasma membrane and endoplasmic-reticulum-rich fractions.[36] Bulk-isolated glial cells also contained four isoenzymes, but the molecular weights and affinity constants differed from those of the total homogenate.[37] γ-Glutamyl transpeptidase is strongly activated by divalent cations (Ca^{2+} and Mg^{2+}) and to a lesser degree by sodium and potassium.[34,35] It is inactivated by thiol-reactive agents such as N-ethylmaleimide.[38,39] However, the degree of enzyme inhibition differs among enzymes prepared from different sources, and it is unlikely that a thiol group is essential to the catalytic process. Borate, serine,[40] and analogues of the substrate[41] also inhibit GTP.

γ-Glutamyl transpeptidase can be solubilized using detergents.[32] Assay is facilitated by synthetic substrates in which a γ-glutamyl acid is bound to easily measured substances such as p-nitroanilide or 2-naphthylamine.[32,38] The highest GTP activities were found in kidney; in mouse kidney, the activity was almost four orders of magnitude greater than that in liver.[29] Brain activity (mouse) (Table I) was 20-fold that of liver. The GTP activity in choroid plexus was among the highest of any tissue, and in some species, activity equal to that in kidney was found. In sheep brain, GTP activity was highest in the thalamus (1.10 μmol/g per min) and lowest in the cerebral cortex (0.01 μmol/g per min).[42] Activities in the head of the caudate nucleus and cerebellum were, respectively, 0.27 and 0.06 μmol/g per min.[42] Most of the particulate enzyme activity was in a cellular debris fraction (probably containing capillaries), but a synaptosomal-enriched microsomal fraction had the highest specific activity.[42] There was little regional variation of GTP in mouse brain, although activities were slightly higher in the medulla and thalamus and lower in the frontal cortex and hippocampus than in other regions.[43] The GTP activities in rabbit brain regions were much greater than those of mouse brain, but like the mouse brain, there was little regional variation.[43]

γ-Glutamyl transpeptidase can be localized in tissue sections by trapping the naphthylamine released by the enzyme with the formation of a colored derivative.[44-46] Histochemical studies have shown a widespread distribution including the brush order of proximal renal tubules,[45-47] the apical portion of the cells of the intestinal epithelium,[48] and the epithelial cells of the ciliary body and the lens capsule.[49,50] In rodent brain, histochemically demonstrated activity was most abundant in the brush border of the choroid plexus epithelial cells, capillary endothelial cells throughout the brain, and the ependymal cells lining the central canal of the spinal cord.[51] Enzyme activity was also detected in some neuronal groups in the diencephalon,[51] cells of the cerebral cortex,[36] cerebellar Purkinje cells, and anterior horn cells. Biochemical studies of a capillary-rich fraction of bovine brain homogenate confirmed the association of GTP with mammalian endothelial cells,[53] and the enzyme is currently used as a marker for such cells in isolating them from whole-brain preparations.

The GTP of isolated mouse brain endothelial cells decreased during proliferation in cell culture, but activity was restored by cocultivating them with irradiated rat glioma cells.[54] However, GTP was not found in chick brain endothelial cells.[55] The distribution of the enzyme is, therefore, more favorable for a role in amino acid transport in choroid plexus than in brain endothelium (Section 8.3.). The activity of GTP in serum is a sensitive indicator of liver disease.[38,56] Elevations in blood or CSF have also been reported in neurological and muscle diseases,[57] but the relationship of the elevations to the neurological conditions was not clear, since the effects of drugs used clinically and affecting the liver were not evaluated.

4.2.2 γ-Glutamyl Cyclotransferase (E.C. 2.3.2.4)

γ-glutamyl amino acid → pyrrolidone carboxylate + amino acid

γ-Glutamyl cyclotransferase (γ-glutamyl lactamase) catalyzes the irreversible release of the amino acid from a γ-glutamyl amino acid and the cyclization of the γ-glutamyl moiety to pyrrolidone carboxylate. The multiple names of the latter product confuse the clinical literature particularly. Pyrrolidone carboxylate is also called 5-oxoproline, pyroglutamic acid, 5-oxopyrrolidone-2-carboxylic acid, and 2-pyrrolidone-5-carboxylic acid.[2] Glutamate is not formed. The soluble enzyme obtained from pig liver had a molecular weight of approximately 22,000 (and from rat liver 27,000).[58]

Although two apparent isoenzymes differing in isoelectric point were found in several species and tissues, the other properties of the two forms including their amino acid compositions were sufficiently similar to indicate that they are variants of the same protein.[58] High activities of γ-glutamyl cyclotransferase were found in kidney, liver, testis, and brain (Table I). The activity in choroid plexus was approximately twice that in cerebral cortex. In the tissues of a single species, the substrate specificities of the enzyme for several different amino acids combined with glutamic were similar, but there were marked differences in substrate specificities among different species.[3] Among the naturally occurring substrates tested, L-γ-glutamyl-L-glutamine was the best substrate for mouse, rat, and human brain.[27,59,60] γ-Glutamyl derivatives of branched-chain and aromatic amino acids were poor substrates.[27,60] Apparently, a free carboxyl group in the amino acid moiety is necessary for activity. Pyrrolidone carboxylate (pyrrolidone-5-carboxylic acid) has been located at the N-terminus of some proteins such as the λ-type light chain of mouse immunoglobulin. Glutamine is a precursor of the terminal acid, and GTP and γ-glutamyl cyclotransferase may participate in its formation.

4.2.3. Pyrrolidone Carboxylate Hydrolase

L-pyrrolidone carboxylate + ATP + $2H_2O \rightarrow$ L-glutamate + ADP + Pi

This enzyme is active in kidney, liver, lungs, and in rat, mouse, and rabbit brain.[29,62,63] The activity is low (Table I), but like other GSH degradative enzymes, activity in choroid plexus is twice that in cerebral cortex. Pyrrolidone carboxylate hydrolase requires ATP and both a divalent (Mg^{2+} or Mn^{2+}) and a monovalent (K^+ or NH_4^+ but not Na^+) cation for activity.[64] The enzyme from rat kidney appears to consist of four subunits, each with a molecular weight of 115,000.[65] It is strongly inhibited by thiol-reactive agents and competitively inhibited by L-2-imidozolidone-4-carboxylate.[66]

4.2.4. Cysteinyl Dipeptidase (E.C. 3.4.3.5)

L-cysteinylglycine + $H_2O \rightarrow$ L-cysteine + glycine

Although it is present in all tissues, little is known about this enzyme.[3,67] Its activity, together with those of the enzymes described above, restores the three original amino acids of the GSH tripeptide.

5. TURNOVER AND COMPARTMENTATION OF GLUTATHIONE

5.1. Glutathione Turnover

The turnover rates of GSH in several tissues have been estimated after the administration of isotopically labeled glycine or glutamate.[3,5] In liver, the turnover was 650 nmol/g per h, and in skeletal muscle 33–49 nmol/g per h.[68] The GSH half-lives were: rat kidney, 22 min[69]; rat liver, 4.7 h[70]; human erythrocyte, 96 h[71]; and rabbit skeletal muscle, 103 h.[68] The first-order rate constants (related to half-life by the formula $t_{1/2} = 0.693/\text{rate constant}$) were: rat kidney 24×10^{-3} min^{-1},[69] rat liver 4.77×10^{-3} min^{-1},[70] human erythrocyte 0.12×10^{-3} min^{-1},[71] and rabbit skeletal muscle 0.11×10^{-3} min^{-1}.[68] Assuming a homogeneous pool of precursor and product, the half-life of GSH in rat brain following the intraperitoneal or intracisternal injection of labeled glycine was 71 h[72] for a first-order rate constant of 0.17×10^{-3} min^{-1}, intermediate between liver and erythrocyte. A more rapid but not specified synthetic rate was found following the intracisternal injection of glycine,[72] and cerebral cortical slices actually formed GSH more rapidly than did liver slices.[73] There is thus uncertainty as to the rate of synthesis of GSH in brain, but it is certainly fairly rapid.

5.2. Glutathione Compartmentation and Development

Although the intercellular and intracellular compartmentation of enzymes and metabolites influences incorporation studies in all tissues, the presence in brain of major compartments differing in metabolic rates and with highly specialized and active metabolic pathways makes it particularly difficult to assess the results of such studies in brain.[74,75] Studies of glutamate metabolism in brain indicate that GSH metabolism is also compartmentalized.[72,76] The smaller of the two major glutamate compartments contains glutamine synthetase and responds to NH_3 stimulation by increased formation and release of glutamine.[76] Glutamine synthetase is a glial enzyme, and it appears that the small compartment in which glutamine is formed from glutamate is glial.[76,77] Glutamic acid derived from glutamine is incorporated into brain GSH more rapidly than is glutamic acid of the total brain glutamate pool,[72,75] and it is probable that GSH in the glial compartment turns over at a more rapid rate than that in the neuronal compartment. The activity of the GSH-utilizing enzyme glutathione peroxidase is higher in bulk-isolated glial[78] than neuronal perikarya and in a malignant glial than a neuronal cell line.[53] However, GSH is synthesized in both compartments and is probably present in all cells.

The specific activity of the enzyme responsible for keeping GSH in the reduced form, glutathione reductase, is present in synaptosomes at a level approximately one-half that in whole-brain homogenate (H. S. Maker and C. Weiss, unpublished data) and was actually of higher activity in a neuronal than a glial cell line.[15] Following the intracisternal injection of labeled glutamate, the regional specific activity of GSH reflected the rate of diffusion of the substrate from the site of injection without any clear regional difference in synthetic rate.[72] During the development of cat brain, evidence of compartmentation of GSH synthesis appeared at 3 to 6 weeks in the cerebral cortex but somewhat earlier in the hippocampus and later in the mesodiencephalon.[9] The GSH concentration increased from 1.6 µmol/g at birth to an adult value of 2.0 µmol/g by 3 weeks of age.[9] Compartmentation appeared at a time of both neuronal maturation and the proliferation and maturation of glial cells.

6. GLUTATHIONE-ASSOCIATED ENZYMES

6.1. Glutathione Reductase (E.C. 1.6.4.2.)

$$GSSG + NADPH + H^+ \rightarrow 2GSH + NADP^+$$

Oxidized glutathione (GSSG) is formed when GSH is oxidized nonenzymatically or enzymatically by glutathione peroxidase, glutathione sulfotransferase, or dehydroascorbate reductase.[5] At higher than normal concentrations, GSSG may leak from cells,[79] but its usual fate is to be rapidly reduced by the cytosolic enzyme glutathione reductase (GSR). The ratio of GSSG to GSH in most tissues approaches that found at the equilibrium of the GSR reaction (Section 2). Although NADH is also a substrate, the affinity of the enzyme for NADPH is greater, and in tissue this is the major or only pyridine nucleotide substrate.[80] Oxidized glutathione is by far the most effective disulfide substrate, although other disulfides such as various protein–GSH conjugates and dihydrolipoic acid react at much lower rates.[81–83]

The gene for the enzyme has been localized to human chromosone 8.[84] Glutathione reductase is found as a monomer (molecular weight 44,000–60,000) or a dimer (102,000–120,000)[86,87] and contains one flavin adenine dinucleotide (FAD) molecule, not covalently bound, in each subunit.[88–90] A dietary deficiency of riboflavin can and has in the past caused erroneously low estimates of GSR activity, but this problem can be overcome by including FMN or FAD in the assay reagent.[90] Enzyme activity does not follow classical enzyme kinetics. There may be a branched-type reaction with simultaneous ping–pong and sequestered mechanisms.[87] The K_m of hepatic GSR for GSSG is 50 μM, and that for NADPH is 3 μM.[91] The chemical constituents of the active site of human erythrocyte GSR include FAD and a redox-active disulfide derived from a cystine moiety of the enzyme protein.[92] Some investigators have found that NADPH concentrations greater than 100 μM inhibited the hepatic enzyme activity, but we did not find inhibition of GSR activity in mouse brain homogenates with concentrations as high as 0.6 mM in the presence of GSSG.[15]

The oxidation of NADPH by GSR facilitates the assay of the enzyme by spectrophotometric or fluorometric assay of the pyridine nucleotide. Alternatively, the disappearance of thiol groups can be followed with the Ellman reagent (Section 3). Glutathione reductase has been found in all rat tissues examined. Activity in mouse tissue homogenates (nmol NADPH oxidized/mg protein per min at 22°C) was: kidney, 1214; liver, 1183; brain, 804; and erythrocyte, 540 (C. Weiss and H. S. Maker, unpublished data). Brain activity was higher in acetone-dried powders than in fresh homogenates, and a paradoxical increase on prolonged standing was noted.[93] The maximal rate was 240 μmol/ g per h in brain powder.[93] Half-maximal activity rates were obtained with 80 μM GSSG.[93]

Glutathione reductase is unstable post-mortem, declining 40% by 24 h in mouse brain under conditions similar to that to which human brain is exposed (T. S. Brannan and H. S. Maker, unpublished data).

The regional distribution of GSR[94] in rat brain is shown in Table II. Of the regions examined, the highest activity was found in the caudate–putamen. Activity (per mg protein) in a synaptosomal preparation made from rat brain homogenate was approximately half that in the whole homogenate (H. S. Maker and C. Weiss, unpublished data). During development (birth to adult) of mouse brain, activity of GSR increased at a slightly greater rate than did brain protein, from 8.0 to 11.0 nmol/min per mg protein (0.45 to 1.17 nmol/min per g).[95] Activity in a malignant cell line with neuronal markers (neuroblastoma NIE 115) was 46.2 nmol/min per mg protein, and that in a line with glial cell markers (C6) was 17.2 nmol/min per mg protein.[15] Several enzyme reactions could supply the NADPH for GSSG reduction including NADP-linked isocitrate dehydrogenase, malic enzyme, and the hexose monophosphate pathway enzymes, glucose-6-phosphate and 6-phosphogluconate dehydrogenases. In erythrocytes[96] and liver,[97] the activity of the hexose monophosphate pathway appears to be regulated by the ratio of GSSG to GSH, and we have found evidence of such a regulation in rat brain (T. S. Brannan and H. S. Maker, unpublished data), supporting the studies of Hotta.[97a] The coupling of NADP$^+$ reduction to carbohydrate metabolism allows GSH to be the major final cellular

Table II

Regional Distribution of Glutathione Peroxidase and Glutathione Reductase in Rat Brain (S.E.M.)

	Glutathione peroxidase[a]	Glutathione reductase[b]
Corpus callosum	11.5 (1.08)	3.34 (0.17)
Interpeduncular nucleus	12.1 (1.92)	—
Occipital cortex	17.2 (0.78)	—
Hippocampus	—	5.24 (0.28)
Frontal cortex	17.7 (0.98)	7.70 (0.30)
Cerebellar cortex	19.3 (0.78)	7.43 (0.27)
Septum	18.9 (1.40)	6.82 (0.26)
Hypothalamus	—	7.01 (0.23)
Olfactory tubercle	20.1 (1.34)	5.67 (0.34)
Nucleus accumbens	21.3 (2.06)	5.80 (0.39)
Substantia nigra	23.8 (2.00)	4.32 (0.31)
Caudate–putamen	28.9 (1.32)	8.87 (0.43)

[a] Expressed in nmol GSH oxidized/mg protein per min in homogenates at 22°C. Brannan *et al.*[110]
[b] Expressed in nmol GSSG reduced/mg protein per min in homogenates at 24°C. Brannan *et al.*[94]

reductant, whereas its own reduction depends on the readily available glucose supply.

6.2. Glutathione Peroxidase (Glutathione : H_2O_2 Oxidoreductase, E.C. 1.11.1.9)

$$2GSH + ROOH \rightarrow GSSG + ROH$$

Glutathione peroxidase (GPx) plays a major role in the destruction of cytoplasmic and mitochondrial H_2O_2 and in the reduction of organic peroxides. There are two major enzyme activities reducing organic peroxides, GPx and glutathione transferase(s). The latter enzymes are described in Section 7.1. Only GPx can reduce H_2O_2 using GSH as the reducing agent.[98] In brain, 74% of the enzymatic activity reducing organic peroxides is in GPx, as compared to 35% in liver, 69% in kidney, and 100% in heart, lung, spleen, thymus, and intestinal mucosa.[98] Glutathione peroxidase is a selenium-requiring enzyme found in both the soluble and mitochondrial cell fractions. Bovine blood GPx consists of four subunits, each with a molecular weight of 21,000 and containing one atom of selenium per subunit.[99,100] Evidence of an additional membrane-bound non-selenium-requiring GPx has been found in heart and liver mitochondria.[101]

6.2.1. Assay of Glutathione Peroxidase

Assay methods for GPx have been reviewed by Ganther *et al.*,[102] and only those aspects related to the selection of an assay and the interpretation of results are discussed here. In brain, which has comparatively low activity, the most useful assays have been those based on the coupling of the production of GSSG produced by GPx to the oxidation of NADPH by added GSR.[103] The decrease in absorbance of NADPH or the increase of fluorescence of NADP$^+$ induced to fluoresce at alkaline pH can be followed.[15] The sensitivity of the latter method is sufficient to assay GPx in less than 100 μg of brain tissue, and greater sensitivities are possible.[15] The activity of GPx in blood (almost entirely in the cellular elements) is high (rat blood, 407 nmol/min per μl),[104] so it is best to perfuse animals with saline to remove blood. Full activity of the partially bound enzyme is released by detergent[105] or freeze–thawing a homogenate.[15]

Enzyme activity is initiated by the enzyme in its reduced form, and it is activated by thiol-containing agents. This is most easily accomplished by preincubating the enzyme in the assay reagent containing GSH before adding the peroxide. The optimal pH is 8.8,[106] although the assay is usually carried out at pH 7.0–7.4. Reduced glutathione is not stable in the presence of H_2O_2, particularly at 37°C, and lower than physiological temperatures are more practical for the assay. The use of the coupled reaction has been criticized because of a reported inhibition of GPx by NADPH,[107] but we found no inhibition of the brain enzyme to a concentration of 2.5 mM NADPH.[15] Organic peroxides as substrates for brain GPx yield a sum of the activities of selenium-requiring GPx and glutathione tranferase.[98,108] Hydrogen peroxide as substrate in the assay detects only the former. Glutathione peroxidase activity has been reported both as moles of GSH oxidized and as moles of GSSG or NADPH formed; the latter means of expression giving a value half that obtained as GSH oxidized.

Glutathione peroxidase does not follow Michaelis–Menten kinetics but, rather, a ter–uni ping–pong mechanism with two different intermediates formed rather than a classical enzyme–substrate complex.[106] There are no true maximal velocities or affinity constants for H_2O_2 or GSH. Both are linear functions of the GSH concentration. The overall reaction rate also varies with the ionic conditions of the assay.[106] These facts must be considered when comparing GPx activities determined under different assay conditions. For practical use, a concentration of GSH can be selected equal to that of the tissue being studied, and an apparently pseudo-zero-order reaction linear with time and tissue concentration obtained. Glutathione peroxidase activities can then be compared among different tissues and tissue regions. At a concentration of 3.0 mM GSH and 0.5 mM H_2O_2 (pH 7.0, 22°C), the "V_{max}" of a rat brain homogenate was 45.9 nmol GSSG formed/min per mg protein.[15]

6.2.2. Glutathione Peroxidase Activity in Brain

The highest activities of GPx were found in rat blood cells, spleen, liver (269 nmol NADPH/min per mg protein), and heart, intermediate levels in lung and kidney (146 nmol/min per mg protein), and lower levels in fat and brain (33 nmol/min per mg protein).[104] Initially, no GPx activity was found in rat brain,[109] but use of detergent[105] or freeze–thawing[15] made possible the study of GPx in brain (Table II). At concentrations of 3.0 mM GSH and 0.5 mM H_2O_2 (pH 7.0, 22°C), the highest activities of GPx were in the rat caudate–putamen and substantia nigra and the lowest activity in the corpus callosum.[110]

The regional distribution of selenium was similar to that of GPx, with the highest concentrations in the putamen and substantia nigra.[111] Activity in bulk-isolated glial perikarya was reported to be tenfold higher than that in neuronal peri-karya.[78] However, the regional distribution of the enzyme and the finding that 25% of rat brain GPx was in the synaptosomal fraction[105] indicate that the glial/neuronal ratio cannot be as high as in the bulk isolates. Glutathione peroxidase activity in a malignant cell line carrying neuronal markers (neuroblastoma NIE 115) was 1.57 nmol/min per mg protein, and that in a cell line with glial marker proteins (C6) was 9.7 nmol/min per mg protein.[15] During development of rat brain, GPx activity increased as did brain protein, increasing 70% from birth to adulthood per milligram tissue.[95] Glutathione peroxidase is stable post-mortem (T. S. Brannan and H. S. Maker, unpublished data).

6.3. Glutathione-S-Transferase(s) (E.C. 2.5.1.18)

See Section 7.1 for a discussion of this enzyme.

7. POSSIBLE FUNCTIONS OF GLUTATHIONE

7.1. Detoxification

The GSH thiol group can react nonenzymatically or enzymatically with many substances. Possibly because of the solubility of GSH and the exposure of its thiol, it is much more reactive than are cellular proteins with "hidden" thiol groups.[5] A major role of nucleophilic GSH in liver is the conjugation of potentially toxic endogenous and exogenous electrophilic substances, thus blocking their toxic actions, rendering them more soluble, and facilitating their excretion.[113] Following conjugation, the GSH compounds may be converted to N-acetylcysteine conjugates called mercapturic acids by the sequential enzymatic removal of γ-glutamyl and glycine from the GSH moiety and the N-acetylation of the cysteine remaining attached to the electrophilic substance.[114]

Although conjugation may occur nonenzymatically, rat liver contains a group of enzymes called glutathione-S-transferases (GST). They are specific for GSH, but each can react with several different substances including alkyls, aryls, alkylaryls, alkenes, epoxides, quinones, and prostaglandins.[113] Individual GSTs may react more readily with one or another of such substances, but they are not selective. These substrates are characterized by having an electrophilic carbon within an essentially hydrophobic structure which can react with the nucleophilic thiol of GSH to yield a thioether.[113] In addition to catalytic properties, at least one of the GSTs of rat liver has a high binding affinity for bilirubin and some other substances and has been called ligandin.[113] In contrast to rat liver, the GST activity of human liver appears to be a single protein which can become modified *in vivo* by deamidation to yield several different bands on electrophoresis.[114] The human enzyme is a dimer of units of 22,000–25,000 molecular weight.[114,115]

The GSTs of liver would be important to neurological well-being if their only function were to remove from circulation hydrophobic electrophilic substances such as bilirubin or exogenous toxins before they reached the nervous system. However, they may also play a more direct role protecting the brain from toxins escaping systemic destruction or produced within the nervous system (Section 7.2.). Earlier work had indicated little GST activity in brain,[116] but more recent studies have reported brain activities of 323 nmol of 1-chloro-2,4-dinitrobenzene conjugated/min per mg protein (rabbit),[117] 182 nmol/min per mg protein (mouse),[117] 162 nmol/min per mg protein (rat).[118] These values were one-half to one-sixth the activities in the livers of these species.[117,118]

There was a uniform regional distribution of GST in rat brain.[118] During development, rat brain activity increased 3.2-fold from 48 nmol/min per mg protein at birth.[118] The brain activities are well above the non-selenium-dependent GPx activity (presumably GST) of brain, suggesting that GST functions better as a transferase than as a peroxidase. A role for GSH as a protectant of neuronal thiol groups from toxins has been postulated.[119] The GST of rat brain (and liver) can conjugate GSH to the neurotoxin acrylamide,[120] but further work is necessary to evaluate the importance of this system in the brain.

7.2. Glutathione as an Intracellular Reductant

The oxidation–reduction states of enzymes,[121] neurotransmitter receptors,[122] microtubules,[123,124] and other macromolecules are critical to their functions. Glutathione is a major source of reducing equivalents for regulating the oxidation status within cells. Ascorbate is the other major cytosolic reductant, and it is present in brain at concentrations equal to those of GSH. The oxidation of ascorbate or its functioning as a reductant produces dehydroascorbate which is a potent cytotoxin.[125] The nonenzymatic or enzymatic reduction of dehydroascorbate by GSH links the two reductants and maintains the level of reduced ascorbate.[126,127] Oxidized glutathione (GSSG) is also a potential toxin. *In vivo*, GSSG can irreversibly inhibit erythrocyte hexokinase,[128] brain caudate nucleus adenylate cyclase,[121] hepatic phosphorylase phosphatase,[129] hepatic glycogen synthetase,[130] and brain creatine kinase (H. S. Maker and C. Weiss, unpublished data). However, GSSG concentrations of 0.1 to 1.0 mM are required for these effects, far above those found *in vivo*, but indicative of the importance of the rapid reduction of GSSG by GSR to cell function.

7.2.1. Glutathione in the Prevention of Oxidant Damage

During the course of normal cellular aerobic metabolism, several products formed in the reduction sequence of O_2 to H_2O are known cytotoxins. These include the superoxide anion radical, H_2O_2, and the hydroxyl radical.[131,132] Even in minute amounts, these powerful oxidants can damage nucleic acids,[133] proteins, and unsaturated fatty acids such as those in the β position of membrane phospholipids and glycolipids.[131,132,134] The rapid removal of superoxide by superoxide dismutase (see Volume 4) and of H_2O_2 prevents the formation of the hydroxyl radical. Hydrogen peroxide is reduced either by GPx or catalase. It is probable that the selection of which of these two enzymes actually functions to destroy a given molecule of H_2O_2 is dependent on its location. In brain as in other tissues, catalase is confined to microbodies called peroxisomes in which H_2O_2 is both produced and destroyed.[124,135] In contrast, GPx is available to destroy the peroxide formed in the cytosol and in mitochondria.[132] In addition, GPx but not catalase can reduce organic peroxides formed from DNA[133] and fatty acids.[136,137] The GSSG formed by GPx is in turn reduced by GSR using NADPH as the reductant. This series of linked reactions reducing H_2O_2 and maintaining GSH has been referred to as the GPx complex and serves to couple the reduction of toxic oxidants to glucose metabolism by the hexose monophosphate pathway.[132,137–140]

Lipid peroxidation occurs in brain tissue and has been implicated in certain pathological conditions.[141,142] Locally injected polyunsaturated fatty acids (PUFA) produce brain edema, apparently by acting on capillary endothelia.[134,143] In brain slices, the production of swelling paralleled an increase in superoxide formation and lipid peroxidation.[134] Edema produced by cold injury was accompanied by lipid peroxidation,[144] and the oxidation of the PUFA released during ischemia could contribute to vasogenic brain edema. Indeed, some investigators have postulated that free radicals play a role in the structural changes of ischemia.[145] In apparent support of this hypothesis, it was reported that hypoxia (8.7% O_2) decreased both GSR and acid-soluble thiols in rat brain.[146] However, oxidant damage would not be expected under hypoxic conditions, and later investigators found that the GSH/GSSG ratio of rat brain was unaffected by hypoxia or brief ischemia. Even when GSH was decreased (by 30 min of complete ischemia), the GSSG concentration did not increase.[10,147] Following a 24-h period of reperfusion, the GSH concentration was still low without a change in GSSG.[10] These findings indicate good GSR function even during ischemia but do not exclude the possiblity that GSH functions in protection of tissue against oxidant damage during the period of oxygen restoration (reperfusion).

The destruction of cells by irradiation and several environmental and pharmacological agents depends on the production within cells of oxidative radicals.[148,149] The glutathione system functions in the protection of cells against radiation damage,[133,148,150] the chemotherapeutic agent daunomycin, and toxic concentrations of oxygen and ozone.[137]

7.2.2. Glutathione in Protection against Amine-Related Toxins

The enzymatic and nonenzymatic oxidation of biogenic amines (dopamine, norepinephrine, epinephrine, serotonin) produces several strong cytotoxins. The monoamine oxidase reaction produces H_2O_2 in amounts equal to those of the amine aldehyde product.[151] Stored intracellular amines are protected from the action of mitochondrial monoamine oxidase, but amines taken up again after their release at a terminal are rapidly oxidized to the aldehyde and H_2O_2. The biogenic amine aldehyde is potentially toxic but is rapidly reduced to the glycol (norepinephrine) or oxidized to the corresponding acid (dopamine).[152] Hydrogen peroxide must also be rapidly reduced. Although catalase is present in neurons, it is confined to scattered peroxisomes,[135] and it is probable that GPx is primarily responsible for the elimination of H_2O_2 in nerve terminals. Brain tissue can couple the oxidation of biogenic amines by monoamine oxidase to the oxidation of GSH to GSSG by the action of GPx on H_2O_2.[151] There was sufficient GPx, GSR, and glucose-6-phosphate dehydrogenase in brain to link the destruction of H_2O_2 produced during the oxidation of dopamine to glucose-6-phosphate oxidation and the hexose monophosphate pathway (T. S. Brannan and H. S. Maker, unpublished data).

In addition to reduction products of oxygen, the nonenzymatic oxidation of amines yields the toxic aminoquinones and aminochromes.[153] These arise as artifacts in amine preparations, particularly at neutral or alkaline pH, and there is indirect evidence of their production in brain, albeit at very low rates. Brain lacks the melanin-forming enzyme tyrosinase, and it is probable that the slow accumulation of neuromelanin in the substantia nigra, locus coeruleus, and other pigmented neurons is a result of the polymerization of amine oxidation products that have escaped reduction. Glutathione can remove aminoquinones by direct or indirect (through ascorbate) reduction or by conjugation, forming a GSH-amine adduct.[154] The GSH conjugation product of 6-hydroxydopamine (formed via the quinone) has been detected in rat brain after the local injection of the amine.[154] Any of these highly toxic substances derived from oxygen or amines might lead to cell loss through excess production, failure to sequester the toxins, or deficiencies of the protective mechanisms linked to GSH, and several investigators have postulated that the aminergic cell loss that occurs in Parkinson's disease and in normal aging could be caused by such defects.[153,155,156] Glutathione reduction and conjugation are key mechanisms in preventing such damage.

7.3 The γ-Glutamyl Cycle and Amino Acid Transport

The enzymatic reactions for the synthesis and degradation of GSH have been linked in a hypothetical cycle called the γ-glutamyl cycle (Fig. 1). During the cycle's operation, some, though not all, amino acids are incorporated and

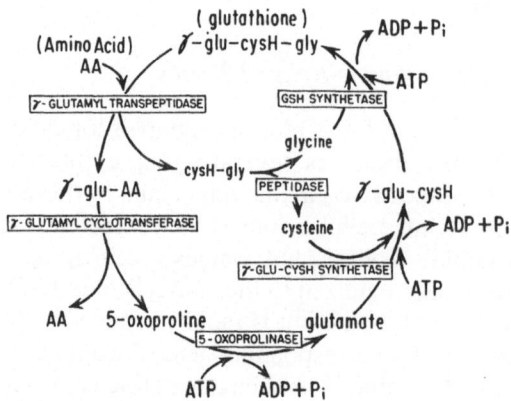

Fig. 1. The δ-glutamyl cycle. From Meister.[157]

released, leading Meister and Orlowski to propose that it could function in amino acid transport. The possibility that the cycle functions in brain has been reviewed by its progenitors.[3,157] In operation, the cycle would have to be spatially oriented so that an amino acid became bound to GSH by membrane-bound GTP at one location, was transported intracellularly, and was released at another location by γ-glutamyl cyclotransferase. Evidence compatible with the operation of the cycle in transport in the brush border of the proximal renal tubule[69] has been found, but its operation in other tissues remains unproven.[3]

8. EXPERIMENTAL MANIPULATION OF GLUTATHIONE

8.1. Manipulation of Glutathione Concentration

Agents that oxidize or conjugate GSH have been used to decrease GSH concentrations primarily in the liver of intact animals or in cultured cells. Irreversible alkylation with thiol reagents such as N-ethylmaleimide is not specific for GSH, and these agents are general cytotoxins. Diethylmaleate conjugates GSH more readily than other tissue thiols, and it has been used to decrease hepatic GSH levels in animal studies of the effect of such depletion on hepatic detoxification.[158] The effect of diethymaleate on brain GSH has not been reported, and it may not cross the blood–brain barrier. Diazene derivatives (e.g., N,N-dimethylamide, diamide) oxidize GSH intracellularly to GSSG, and GSH is much more sensitive than other tissue thiols.[5] Diazene penetrates cell membranes, and the related reagent, diazene dicarboxylic bis-CN'-methyl-piperazimide, which does not readily penetrate cell membranes, can be used to differentiate between intracellular and extracellular GSH depletion.

Inhibitors of the several steps of the γ-glutamyl cycle that include the GSH-synthesizing enzymes have been found. Such inhibitors have reduced mouse hepatic GSH concentration by two-thirds but have had little effect on brain.[41] Methionine sulfoximine inhibits GCS and lowers GSH concentrations in mouse liver and kidney (70% decrease after 2.5 μmol/g, i.p.).[41] Brain GCS was inhibited, but no decline in brain GSH levels was found, although methionine sulfoximine crosses the blood–brain barrier (Section 4.1.1). The apparent resistance of brain GSH to the GCS inhibitor may have resulted from the slower rate of turnover of brain GSH as compared to liver and kidney (Section 5.1).

8.2. Manipulation of Glutathione-Metabolizing Enzymes

The glutathione peroxidase activity that reduces H_2O_2 is a selenium-requiring enzyme. Selenium depletion will deplete isolated cells of GPx, and a selenium-poor diet depletes the enzyme from most tissues.[102] However, despite repeated attempts, depleting selenium from animals by deprivation from gestation onwards, no loss of brain GPx activity could be demonstrated.[102] Brain is somehow protected, and no satisfactory method for inhibiting brain GPx is available.

The chemotherapeutic agent 1,3-bis(chloroethyl)-1-nitrosourea (BCNU) rapidly decreased human erythrocyte GSR after intravenous administration.[159] Of 20 erythrocyte enzymes examined, only GSR was significantly reduced; FAD failed to restore the activity. Glutathione reductase was also relatively specifically inhibited *in vitro*, but the loss could be prevented by 10 mM mercaptoethanol.[160] More recently, an active product of the drug has been shown also to inhibit chymotrypsin and transglutaminase.[161] Tissue GSR including that of mouse brain was inhibited by up to 80% by 200 μm BCNU. We have confirmed the *in vivo* inhibition of brain GSR with BCNU but have not found the chemically related 1-(2-chloroethyl)-3-cyclohexyl-1-nitrosourea (CCNU) inhibitory (H. S. Maker and C. Weiss, unpublished data).

Another substance that inhibits GSR activity but is much less specific than BCNU is 6-aminonicotinamide (6-AN).[162] This compound is incorporated into 6-amino analogues of NAD(H) and NADP(H) by tissue NAD^+ glycohydrolase, but enzymes requiring NADP(H) as cofactor are much more irreversibly inhibited than those requiring NAD(H).[162] The most sensitive enzyme appears to be 6-phosphogluconate dehydrogenase, and large amounts of its substrate accumulate in brain and other tissues as the hexose monophosphate pathway is blocked.[163] Thus, both GSR and its NADPH cofactor are diminished by 6-AN. Although dopamine depletion and an L-DOPA-remediable "parkinson-like" rigidity was produced in mice by 5–10 mg/kg 6-AN, damage was not confined to the basal ganglia; both neurons and glia of other regions were destroyed.[165] 3-Acetyl pyridine also inhibits enzymes requiring a pyridine nucleotide cofactor, with GSR the most vulnerable.[166]

9. DISEASES AFFECTING GLUTATHIONE METABOLISM

Because of the uncertainties about the function of GSH, attempts have been made to infer possible functions from "experiments of nature" in which defects in the synthesis or metabolism of GSH have been found, usually in patients studied because of hematological, systemic metabolic, or neurological abnormalities. As in any such investigation, it may be difficult to define the relationship between the enzyme defect and the observed structural or functional abnormalities. Even when the enzyme abnormality has been located in a specific cell, its role may be uncertain. Indeed, in at least one family, the abnormality in erythrocyte morphology that had prompted the enzyme study was found to genetically segregate independently of a reduction of erythrocyte GSR activity that was also present.[167] The enzyme abnormality was apparently unrelated to the erythrocyte defect. The defects in human GSH metabolism that have been described may be assymptomatic, cause only increased fragility of erythrocytes under specific metabolic stress, or be associated with defects of other cell types including those of nervous tissue.

9.1. Glutathione Synthetase Deficiency

Two major clinical syndromes of GSS deficiency have been described.[168] The first type presents in the newborn with hemolytic anemia (induced by prooxidant substances) and with a metabolic acidosis caused by 5-oxoproli-nemia. These patients excrete large amounts of oxoproline in their urine and have deficiencies of GSH and GS in those cells (erythrocytes, leucocytes, and fibroblasts) that have been examined. The accumulation of oxoproline, which causes the metabolic acidosis, is not a result of decreased destruction of this metabolite but is caused by its increased synthesis. γ-Glutamyl cysteine is a substrate for both GS and γ-glutamyl cyclotransferase. With decreased GS, more γ-glutamyl cysteine becomes available for oxoproline formation by the latter enzyme. In addition, increased synthesis may occur because of low GSH concentrations, since GSH is normally a feedback inhibitor of GS.[169]

The second type of GS deficiency presents at various ages with an isolated hemolytic anemia.[170] The defect of GSH formation in the latter form is confined to the erythrocyte and appears to result from a greater than normal instability of the enzyme which is lost at a more rapid rate than normal during erythrocyte aging. Erythrocytes lack the apparatus for protein synthesis (nuclei and ribosomes), and any premature enzyme loss may be manifested in these "cells" and not elsewhere in the body. In contrast, nucleated cells can replace lost enzymes. In both the systemic and localized forms, cell destruction is caused by the failure of mechanisms protecting cells against oxidative toxins. When leukocytes are involved as in type one, leukopenia may develop as white cells are lost through the normal generation of oxidative toxins such as H_2O_2 during phagocytosis.[171] Loss of both red and white cells can be prevented by administration of the antioxidant α-tocopherol.[172]

The neurological interest in this enzyme arises from reports of three patients with GSH synthesis defects and neurological symptoms. Because they have been cited so often as neurochemical illnesses, the cases will be summarized in detail. A single patient was the subject of many reports over a period of 18 years.[173,174] He was the product of a normal gestation and birth but had retarded psychomotor development. At age 17, he began to suffer frequent episodes of vomiting which were found at age 19 to be caused by excess production of oxoproline secondary to a GS deficiency. Subsequent episodes were treated with alkylizing agents. At his initial neurological examination at age 19, he was retarded and had spastic tetraparesis, greater in the lower extremities, intention tremor, dystaxic gait, and dysarthria. At 23, generalized seizures began, his mental dysfunction deteriorated into dementia, and he died at age 28. The major brain findings at postmortem were a diffuse loss of cerebellar granular cells with relative sparing of Purkinje cells, small infarcts of the right centrum and left thalamus, and focal neuronal loss in the visual cortex.[174] The reported pathology thus does not account for the many clinical findings (such as dementia), nor is it clear if the cerebellar changes were related to the enzyme defect, the repeated episodes of acidosis, the seizures, the seizure treatment, or to a still unrecognized process.

9.2. γ-Glutamylcysteine Synthetase Deficiency

Two siblings with GCS deficiency and neurological abnormalities have been reported.[175] The sister presented at age 27 with hemolytic anemia and low erythrocyte GSH and GS. Absent deep tendon reflexes in her lower extremities were noted. At age 29, she suffered a severe hemolytic crisis and "psychotic behavior" after receiving sulfa drugs for a urinary tract infection. At 35, she had only mildly impaired coordination, four-limb dystaxia, and impaired thinking. Her brother was first seen at age 29 with a hemolytic anemia, slurred speech, and pes cavus. By age 36, he had dystaxia, dysmetria, dysdiadochokinesis, "muscle" weakness, absent deep tendon reflexes in the lower and decreased reflexes in the upper extremities, and decreased position and vibratory sensation in all four extremities. At 36, he developed irregular staccato speech, mental changes, and myoclonic spasms of the right lower leg. Clinical and electromyographic studies were said to indicate a "myopathy," but peripheral nerve conduction velocities were normal. Both patients had severe loss of erythrocyte GSH and GCS but lesser decreases in muscle and leukocytes. Both suffered episodes of hemolytic anemia, but neither had metabolic acidosis because of the failure to form the oxoproline precursor γ-glutamyl cysteine. These patients thus suffered symptoms similar to some cases of hereditary ataxia but also had unaccounted for, atypical features such as myoclonus and signs of a myopathy. It is inappropriate to use these cases to support a specific relationship of GSH metabolism to cerebellar function.[157]

9.3. γ-Glutamyl Transpeptidase Deficiency

A single case of GTP deficiency has been reported,[176] a 33-year-old man who was mentally retarded and who secreted excessive amounts of GSH in his urine. The case supports a role for GSH and GTP in renal tubular transport. Evidently, his urine was screened because of the retardation, but there is no evidence that the mental abnormality had any relation to the enzyme defect.

9.4. Defects in Glutathione Antioxidation and Detoxification

Defects in those enzymes that participate in the reduction of GSSG (GSR),[164] the supply of NADPH for GSSG reduction (glucose-6-phosphate dehydrogenase[177]), or the destruction of hydrogen or organic peroxides (GPx[178]) have been described in humans either with no clinical abnormalities or suffering episodic hemolysis precipitated by the ingestion of prooxidative substances. No neurological symptoms have been reported. It is possible, however, that prooxidative damage could be caused in other tissues, including the CNS, during amine or drug metabolism if local antioxidant mechanisms were disturbed. Such defects might be manifested independently of those in the red cell. Too little is known at present about possible differences in glutathione metabolism in different tissues to evaluate this possibility.

10. SUMMARY

In summary, brain tissue contains relatively high concentrations of GSH and the enzymes needed for its synthesis, degradation, and utilization. Glutathione turnover in brain is intermediate between that of liver and erythrocyte. Although GSH is probably more actively synthesized in glia than in neurons, enzymes utilizing it are active in synaptosomes. In other tissues, GSH provides reducing equivalents to neutralize oxidative toxins, removes electrophilic toxins by conjugation, and may participate in amino acid transport. Some or all of these mechanisms could function in brain, but it remains for future investigators to evaluate the importance of GSH in the detoxification of endogenous and exogenous toxins in brain. The clinical consequences to the nervous system of defects in GSH metabolism are also uncertain. The evidence at hand indicates that GSH is important to brain metabolism, but its precise roles remain to be determined.

ACKNOWLEDGMENT. Supported by the Clinical Center for Research in Parkinson's and Allied Diseases, Grant NS-11631 from the United States Public Health Service. I should like to thank Dr. Marian Orlowski of Mount Sinai School of Medicine on whose earlier review this chapter is based.

Note Added in Proof
Since the acceptance of this chapter, a new case of hyperpipecolic acidemia has been reported by Arneson *et al.*[179] This is an infant with increased plasma (.01 μmole/ml) and urinary (.06 μmole/ml) concentrations of PA presenting the clinical features of the cerebrohepatorenal Syndrome of Zellweger. The patient was followed up from 2 days of age to her death at 10.5 months.

REFERENCES

1. Pisano, J. J., 1969, *Handbook of Neurochemistry,* Volume 1 (A. Lajtha, ed.), Plenum Press, New York, pp. 53–74.
2. Floché, L., Benöhr, H. C., Sies, H., Waller, H. D., and Wendel, A. (eds.), 1974, *Glutathione. Proceedings of the 16th Conference of the German Society of Biological Chemistry, Tubingen, March, 1973,* Georg Thieme, Stuttgart.
3. Orlowski, M., and Karkowsky, A., 1976, *Int. Rev. Neurobiol.* **19:**75–121.
4. Arias, I. M., and Jakoby, W. B. (eds.), 1976, *Glutathione: Metabolism and Function,* Raven Press, New York.
5. Kosower, N. S., and Kosower, E. M., 1978, *Int. Rev. Cytol.* **54:**109–160.
6. Martin, H., and McIlwain, H., 1959, *Biochem. J.* **71:**275–280.
7. Folbergrova, J., Rehncrona, S., and Siesjo, B. K. 1979, *J. Neurochem.* **32:**1621–1627.
8. Reichelt, L. L., and Fonnum, F., 1969, *J. Neurochem.* **16:**1409–1416.
9. Berl, S., and Purpura, D. P., 1966, *J. Neurochem.* **13:**293–304.
10. Cooper, A. J. C., Pulsinelli, W. A., and Duffy, T. E., 1980, *J. Neurochem.* **35:**1242–1245.
11. Griffith, O. W., 1980, *Anal. Biochem.* **106:**207–212.
12. Boyne, A. F., and Ellman, G. L., 1972, *Anal. Biochem.* **46:**639–653.
13. Hissin, P. J., and Hilf, R., 1976, *Anal. Biochem.* **74:**214–226.
14. Beutler, E., and West, C., 1977, *Anal. Biochem.* **81:**458–460.
15. Weiss, C., Maker, H. S., and Lehrer, G. M., 1980, *Anal. Biochem.* **106:**512–516.
16. Hadley, W. M., Bousquet, W. F., and Miya, T. S., 1974, *J. Pharmacol. Sci.* **63:**57–59.
17. Tietze, F., 1969, *Anal. Biochem.* **27:**502–522.
18. Brehe, J. E., and Burch, H. B., *Anal. Biochem.* **74:**189–197.
19. Lowry, O. H., and Passonneau, J. V., 1972, *A Flexible System of Enzymatic Analysis,* Academic Press, New York, pp. 130–146.

20. Tabor, C. H., and Tabor, H., 1977, *Anal. Biochem.* **78:**543–553.
21. Reed, D. J., Babson, J. R., Beatty, P. W., Brodie, A. E., Ellis, W. W., and Potter, D. W., 1980, *Anal. Biochem.* **106:**55–62.
22. Orlowski, M., and Meister, A., 1971, *J. Biol. Chem.* **246:**7095–8105.
23. Gaitonde, M. K., 1970, *Handbook of Neurochemistry*, Volume 3 (A. Lajtha, ed.), Plenum Press, New York, pp. 225–287.
24. Richman, P. G., Orlowski, M., and Meister, A., 1973, *J. Biol. Chem.* **248:**6684–6690.
25. Ronzio, R. A., Rowe, B. W., and Meister, A., 1969, *Biochemistry* **8:**1066–1075.
26. Mellanby, E., 1946, *Br. Med. J.* **2:**885–887.
27. Orlowski, M., and Wilk, S., 1975, *J. Neurochem.* **25:**601–606.
28. Wendel, A., Schaich, E., Weber, U., and Flohé, L., 1972, *Hoppe Seylers Z. Physiol. Chem.* **353:**514–522.
29. Orlowski, M., and Wilk, S., 1975, *Eur. J. Biochem.* **53:**581–590.
30. Tate, S. S., and Orlando, J., 1979, *J. Biol. Chem.* **254:**5573–5575.
31. Karkowski, A. M., and Orlowski, M., 1978, *J. Biol. Chem.* **253:**1574–1581.
32. Orlowski, M., and Meister, A., 1965, *J. Biol. Chem.* **240:**338–347.
33. Taniguchi, N., 1974, *J. Biochem. (Tokyo)* **75:**473–480.
34. Orlowski, M., Okonkwo, P. D., and Green, J. P., 1973, *FEBS Lett.* **31:**237–240.
35. Zelazo, P., and Orlowski, M, 1976, *Eur. J. Biochem.* **61:**147–155.
36. Reyes, E., and Barela, T. D., 1980, *Neurochem. Res.* **5:**159–170.
37. Reyes, E., Lewis, L., and Saland, L., 1980, *Proc. West. Pharmacol. Soc.* **23:**381–383.
38. Orlowski, M., 1963, *Arch. Immunol. Ther. Exp. (Warsz.)* **11:**1–61.
39. Richter, R., 1969, *Arch. Immunol. Ther. Exp. (Warsz.)* **17:**476–495.
40. Revel, J. P., and Ball, E. G., 1959, *J. Biol. Chem.* **234:**577–582.
41. Griffith, O. W., and Miester, A., 1977, *Proc. Natl. Acad. Sci. USA* **74:**3330–3334.
42. Reyes, E., and Palmer, G. C., 1976, *Res. Commun. Chem. Pathol. Pharmacol.* **14:**759–762.
43. Liśy, V., Šťastny, F., and Lodin, Z. 1979, *Neurochem. Res.* **4:**747–753.
44. Albert, Z., Orlowski, M., and Szewczuk, A., 1961, *Nature* **191:**767–768.
45. Albert, Z., Orlowski, J., Orlowski, M., and Szewczuk, A., 1964, *Acta Histochem.* **18:**78–89.
46. Glenner, G. G., Folk, J. E., and McMillan, P. J., 1962, *J. Histochem. Cytochem.* **10:**481–489.
47. Rutenburg, A. M., Kim, H., Fischbein, J. W., Hanker, J. S., Wasserkrug, H. L., and Seligman, A. M., 1969, *J. Histochem. Cytochem.* **17:**517–526.
48. Greenberg, E., Wollaeger, E. E., Fleisher, G. A., and Engstrom, G. V., 1967, *Clin. Chim. Acta* **16:**79–89.
49. Reddy, V., and Unaker, J. M., 1975, *Exp. Eye Res.* **17:**405–408.
50. Ross, L. L., Barber, L., Tate, S. S., and Meister, A., 1973, *Proc. Natl. Acad. Sci. U.S.A.* **70:**2211–2214.
51. Albert, Z., Orlowski, M., Rzucidlo, Z., and Orlowski, J., 1966, *Acta Histochem.* **25:**312–320.
52. Meister, A., Tate, S. S., and Ross, L. L., 1976, *Membrane-Bound Enzymes*, Volume 3 (A. Martinosi, ed.), Plenum Press, New York, pp. 315–317.
53. Orlowski, M., Sessa, G., and Green, J. P., 1974, *Science* **184:**66–68.
54. DeBault, L. E., and Pasquale, C. A., 1980, *Science* **207:**653–655.
55. Stewart, P. A., 1980, *Exp. Neurol.* **67:**442–446.
56. Rosalki, S. B., 1975, *Adv. Clin. Chem.* **17:**53–107.
57. Ewen, L. M., and Griffiths, J., 1973, *Am. J. Clin. Pathol.* **59:**2–9.
58. Adamson, E. D., Szewczuk, A., and Connell, G. E., 1971, *Can. J. Biochem.* **49:**218–226.
59. Orlowski, M., Richman, P. G., and Meister, A., 1969, *Biochemistry* **8:**1048–1055.
60. Orlowski, M., and Meister, A., 1973, *J. Biol. Chem.* **248:**2836–2844.
61. Burstein, Y., and Schecter, I., 1977, *Biochem. J.* **165:**347–354.
62. Tate, S. S., Ross, L. L., and Meister, A., 1973, *Proc. Natl. Acad. Sci. U.S.A.* **70:**1447–1449.
63. Okonkwo, P. O., Orlowski, M., and Green, J. P., 1974, *J. Neurochem.* **22:**1053–1058.
64. Van Der Werf, P., Orlowski, M., and Meister, A., 1971, *Proc. Natl. Acad. Sci. U.S.A.* **68:**2982–2985.
65. Wendel, A., Flugge, U. I., and Jenke, H. S., 1975, *Hoppe Seylers Z. Physiol. Chem.* **356:**881–885.

66. Van Der Werf, P., Stephani, R. A., Orlowski, M., and Meister, A., 1973, *Proc. Natl. Acad. Sci. U.S.A.* **70:**759–761.

67. Marks, N., 1970, *Handbook of Neurochemistry*, Volume 3 (A. Lajtha, ed.), Plenum Press, New York, pp. 133–171.

68. Henriques, O. B., Henriques, S. B., and Neuberger, A., 1955, *Biochem. J.* **60:**409–424.

69. Sekura, R., and Meister, A., 1974, *Proc. Natl. Acad. Sci. U.S.A.* **71:**2969–2972.

70. Douglas, G. W., and Mortenson, R. A., 1956, *J. Biol. Chem.* **222:**581–585.

71. Dimant, E., Landsberg, E., and London, I. M., 1955, *J. Biol. Chem.* **213:**769–776.

72. Berl, S., Lajtha, A., and Waelsch, H., 1961, *J. Neurochem.* **7:**186–197.

73. Takahashi, Y., and Akabane, Y., 1961, *J. Neurochem.* **7:**89–96.

74. Berl, S., 1973, *Metabolic Compartmentation in the Brain* (R. Balaźs and J. C. Cremer, eds.), MacMillan, London, pp. 3–17.

75. Lajtha, A., Berl, S., and Waelsch, H., 1959, *J. Neurochem.* **3:**322–332.

76. Garfinkle, D., 1966, *J. Biol. Chem.* **241:**3918–3929.

77. Lajtha, A., Clarke, D. D., and Maker, H. S., 1981, *Basic Neurochemistry*, 3rd ed. (G. J. Siegel, R. W. Albers, R. Katzman, R., and B. W. Agranoff, eds.), Little, Brown, Boston, pp. 329–353.

78. Savolainen, H., 1978, *Res. Commun. Chem. Pathol. Pharmacol.* **21:**173–176.

79. Sies, H., Gerstenecker, C., Menzel, H., and Flohé, L., 1972, *FEBS Lett.* **27:**171–175.

80. Beutler, E., and Yeh, M. K. Y., 1963, *Blood* **21:**573–585.

81. Srivastava, S. K., and Beutler, E., 1970, *Biochem. J.* **119:**353–377.

82. Smith, J. E., 1971, *Biochim. Biophys. Acta* **242:**36–38.

83. Mannervik, B., and Erikson, S. A., 1974, *Glutathione. Proceedings of the 16th Conference of the German Society of Biological Chemistry* (L. Flohé, H. C. Benohr, H. Sies, H. D. Waller, and A. Wendel, eds.), George Thieme, Stuttgart, pp. 120–132.

84. Kucherlapati, R. S., Nichols, E. A., Greagan, R. P., Chen, S., Borgaonokad, D. S., and Ruddle, F. H., 1974, *Am. J. Hum. Genet.* **26:**51A.

85. Ray, L. E., and Prescott, J. M., 1975, *Proc. Soc. Exp. Biol. Med.* **148:**402–409.

86. Ichio, I. I., and Sakai, H., 1974, *Biochim. Biophys. Acta* **350:**141–150.

87. Flohé, L., and Gunzler, W. A., 1976, *Glutathione: Metabolism and Function* (I. M. Arias and W. B. Jakoby, eds.), Raven Press, New York, pp. 17–34.

88. Worthington, D. J., and Rosemeyer, M. A., 1975, *Eur. J. Biochem.* **60:**455–466.

89. Nakashima, K., Miwa, S., and Yamauchi, K., 1976, *Biochem. Biophys. Acta* **445:**309–323.

90. Beutler, E., 1969, *J. Clin. Invest.* **48:**1957–1966.

91. Mize, C. E., and Langdon, R. G., 1962, *J. Biol. Chem.* **237:**1589–1595.

92. Boggaram, V., and Mannervik, B., 1979, *Acta Chim. Scand. [B]* **33:**593–594.

93. McIlwain, H., and Treisze, M. A., 1957, *Biochem. J.* **65:**288–296.

94. Brannan, T. S., Maker, H. S., Raes, I., and Weiss, C., 1980, *Brain Res.* **200:**474–477.

95. Brannan, T. S., Maker, H. S., and Weiss, C., 1981, *Neurochem. Res.* **6:**39–43.

96. Metz, S., Balcerzak, P., and Sagone, A. L., Jr., 1974, *Blood* **44:**691–697.

97. Eggleston, L. V., and Krebs, H. A., 1974, *Biochem. J.* **138:**425–435.

97a. Hotta, A. S., 1962, *J. Neurochem.* **9:**43–51.

98. Lawrence, R. A., and Burk, R. F., 1978, *J. Nutr.* **108:**211–215.

99. Flohé, L., Eisle, B., and Wendel, A., 1971, *Hoppe Seylers Z. Physiol. Chem.* **352:**151–158.

100. Flohé, L., Gunzler, W. A., and Schock, H. H., 1973, *FEBS Lett.* **32:**132–134.

101. Katki, A. G., and Meyers, C. E., 1980, *Biochem. Biophys. Res. Commun.* **96:**85–91.

102. Ganther, H. E., Hafeman, D. G., Lawrence, R. A., Serfass, R. E., and Hoekstra, W. G., 1976, *Trace Elements in Human Health and Disease, Volume II: Essential and Toxic Elements* (A. S. Prasad and D. Oberleas, eds.), Academic Press, New York, pp. 165–234.

103. Paglia, D. E., and Valentine, W. V., 1967, *J. Lab. Clin. Med.* **70:**158–169.

104. Mills, G. C., 1960, *Arch. Biochim. Biophys.* **86:**1–5.

105. Prohaska, J. R., and Ganther, H. E., 1976, *J. Neurochem.* **27:**1379–1387.

106. Flohé, L., Loschen, G., Gunzler, W. A., and Eichele, E., 1972, *Hoppe Seylers Z. Physiol. Chem.* **353:**987–999.

107. Little, C., and O'Brien, P. J., 1968, *Biochem. Biophys. Res. Commun.* **31:**145–150.

108. Burk, R. F., Nishiki, K., Lawrence, R. A., and Chance, B., 1978, *J. Biol. Chem.* **253**:43–46.
109. DeMarchena, O., Guarnieri, M., and McKhann, G., 1974, *J. Neurochem.* **22**:773–776.
110. Brannan, T. S., Maker, H. S., Weiss, C., and Cohen, G., 1980, *J. Neurochem.* **35**:1013–1014.
111. Larsen, N. A., Pakkenberg, H., Damsgaard, E., and Heydorn, K., 1979, *J. Neurol. Sci.* **42**:407–416.
112. Chasseaud, F. L., 1974, *Glutathione. Proceedings of the 16th Conference of the German Society of Biological Chemistry* (L. Flohé, H. C. Benöhr, H. Sies, H. D. Waller, and A. Wendel, eds.), George Thieme, Stuttgart, pp. 90–113.
113. Jakoby, W. B., 1978, *Adv. Enzymol.* **46**:383–414.
114. Habig, W. H., Pabst, M. J., and Jakoby, W. B., 1974, *J. Biol. Chem.* **249**:7130–7139.
115. Scully, N. C., and Mantle, J. J., 1981, *Biochem. J.* **193**:367–370.
116. Johnson, M. K., 1966, *Biochem. J.* **98**:44–56.
117. Dixit, R., Mukhtar, H., Seth, P. K., and Krishna Murti, C. R., 1980, *Neurotoxicology* **2**:193–196.
118. Das, M., Dixit, R., Seth, P. K., and Mukhtar, H., 1981, *J. Neurochem.* **36**:1439–1442.
119. Makinen, A., Savolainen, H., Lehtonen, E., and Vainio, H., 1977, *Res. Commun. Chem. Pathol. Pharmacol.* **16**:577–580.
120. Dixit, R., Seth, P. K., and Mukhtar, H., 1980, *Biochem. Int.* **1**:547–552.
121. Baba, A., Lee, E., Matsuda, T., Kihara, T., and Iwata, H., 1978, *Biochem. Biophys. Res. Commun.* **85**:1204–1210.
122. Sven, E. T., Stefanini, E., and Clement-Cormier, Y. C., 1980, *Biochem. Biophys. Res. Commun.* **96**:953–960.
123. Beck, W. T., 1980, *Biochem. Pharmacol.* **29**:2333–2337.
124. Oliver, J. M., Albertini, D. F., and Berlin, R. D., 1976, *J. Cell Biol.* **71**:921–932.
125. Patterson, J. W., 1950, *J. Biol. Chem.* **183**:81–88.
126. Grimble, R. F., and Hughes, R. E., 1967, *Experientia* **23**:362.
127. Bigley, R. H., and Stankova, L., 1974, *J. Exp. Med.* **139**:1084–1092.
128. Magnani, M., Stocchi, V., Ninfali, P., Dacha, M., and Fornaini, G., 1980, *Biochim. Biophys. Acta* **615**:113–120.
129. Shimazu, T., Tototake, S., and Usami, M., 1978, *J. Biol. Chem.* **253**:7376–7382.
130. Ernest, M. J., and Kim, K. H., 1973, *J. Biol. Chem.* **248**:1550–1555.
131. Fridovich, I., 1978, *Science* **201**:875–880.
132. Chance, B., Sies, H., and Boveris, A., 1980, *Physiol. Rev.* **59**:527–605.
133. Christophersen, B. O., 1969, *Biochim. Biophys. Acta* **186**:387–389.
134. Chan, P. H., and Fishman, R. A., 1980, *J. Neurochem.* **35**:1004–1007.
135. McKenna, O., Arnold, G., and Holtzman, E., 1976, *Brain Res.* **117**:181–194.
136. Chow, C. K., and Tappel, A. L., 1972, *Lipids* **7**:158–524.
137. Chow, C. K., and Tappel, A. L., 1974, *J. Nutr.* **104**:444–451.
138. Sies, H., Gerstenecker, C., Summer, K. H., Menzel, H., and Flohé, L., 1974, *Glutathione. Proceedings of the 16th Conference of the German Society of Biological Chemistry* (L. Flohé, H. C. Benöhr, H. Sies, H. D. Waller, and A. Wendel, eds.), George Thieme, Stuttgart, pp. 261–276.
139. Videla, L. A., Fernandez, V., Ugarle, G., and Valenzuela, A., 1980, *FEBS Lett.* **111**:6–10.
140. Gibson, D. D., Hoornbook, K. R., and McKay, P. B., 1980, *Biochim. Biophys. Acta* **630**:572–582.
141. Konat, G., 1973, *J. Neurochem.* **20**:1247–1251.
142. Bishayee, S., and Balasubramanian, A. S., 1971, *J. Neurochem.* **18**:909–920.
143. Priouleau, G. R., Fishman, R. A., and Chan, P. H., 1979, *Ann. Neurol.* **6**:156.
144. Suzuki, O., and Yagi, K., 1974, *Experientia* **30**:248.
145. Flamm, E. S., Demopoulos, H. B., Seligman, M. L., Poser, R. G., and Ransohoff, J., 1978, *Stroke* **9**:445–447.
146. Wideman, J., and Domańska-Jansk, K., 1974, *Resuscitation* **3**:27–36.
147. Rehncrona, S., Folbergrova, J., Smith, D. S., and Siesjö, B. K., 1980, *J. Neurochem.* **34**:477–486.

148. Bacq, Z. M. (ed.), *Sulfur-Containing Radioprotective Agents. International Encyclopedia of Pharmacology and Therapeutics,* Section 79, Pergamon Press, New York, pp. 161–301.

149. Bozzi, A., Mavelli, I., Mondovi, B., Strom, R., and Rotilio, G., 1981, *Biochem. J.* **194**:369–372.

150. Morse, M. L., and Dahl, R. H., 1978, *Nature* **271**:660–662.

151. Brannan, T. S., Maker, H. S., and Raes, I. P., 1981, *J. Neurochem.* **36**:307–309.

152. Cooper, J. R., Bloom, F. E., and Roth, R. H., 1978, *The Biochemical Basis of Neuropharmacology,* 3rd ed., Oxford University Press, New York, London, pp. 102–195.

153. Tse, D. C. S., McCreery, R. L., and Adams, R. N., 1976, *J. Med. Chem.* **19**:37–40.

154. Liang, Y. O., Plotsky, P. M., and Adams, R. M., 1977, *J. Med. Chem.* **20**:581–583.

155. Cohen, G., Dembiec, D., Mytilineou, C., and Heikkila, R. E., 1976, *Advances in Parkinsonism* (W. Birkmayer and O. Horneykiewicz, eds.), Editiones Roche, Basel, pp. 251–257.

156. Ambani, L. M., Van Woert, M. H., and Murphy, S., 1975, *Arch. Neurol.* **32**:114–118.

157. Meister, A., 1978, *Adv. Neurol.* **21**:289–302.

158. Wendel, A., and Feurstein, S., 1981, *Biochem. Pharmacol.* **30**:2513–2520.

159. Frischer, H., and Ahmad, T., 1977, *J. Lab. Clin. Med.* **89**:1080–1090.

160. Shinohara, K., 1979, *Clin. Chim. Acta* **92**:147–152.

161. Laki, K., Sipka, S., and Csak, F., 1978, *Fed. Proc.* **37**:1544.

162. Cooper, H. U., and Neubert, D., 1964, *Biochim. Biophys. Acta* **89**:23–32.

163. Kauffman, F. C., and Johnson, E. V., 1974, *J. Neurobiol.* **5**:379–382.

164. Loos, D., Halbhübner, K., Kehr, W., and Herken, H., 1979, *Neuroscience* **4**:667–676.

165. Horita, N., Ishii, T., and Izumyama, Y., 1980, *Acta Neuropathol. (Berl.)* **49**:19–27.

166. Herken, H., 1968, *Z. Klin. Chem.* **6**:357–367.

167. Nakashima, K., Yamauchi, K., Miwa, S., Fujimura, K., Mizutani, A., and Kuramoto, A., 1978, *Am. J. Hematol.* **4**:141–150.

168. Spielberg, S. P., Garrick, M. D., Corash, L. M., Butler, J. DeB., Tietze, F., Rogers, L., and Schulman, J. D., 1978, *J. Clin. Invest.* **61**:1417–1420.

169. Richman, P. G., and Meister, A., 1975, *J. Biol. Chem.* **250**:1422–1426.

170. Mohler, D. N., Majerus, P. W., Minnich, V., Hess, C. E., and Garrick, M. D., 1970, *N. Engl. J. Med.* **283**:1253–1257.

171. Spielberg, S. P., Boxer, L. A., Oliver, J. M., Butler, E. J., and Schulman, J. D., 1979, *Br. J. Haematol.* **42**:215–223.

172. Boxer, L. A., Oliver, J. M., Spielberg, S. P., Allen, J. M., and Schulman, J. D., 1979, *N. Engl. J. Med.* **301**:901–905.

173. Jellum, E., Kluge, T., Borresen, H. C., Stokke, O., and Eldjarn, L., 1970, *Scand. J. Clin. Lab. Invest.* **26**:327–335.

174. Skellerud, K., Marstein, S., Schrader, H., Brundelet, P. J. and Jellum, E., 1980, *Acta Neuropathol. (Berl).* **52**:235–238.

175. Richards, F., Cooper, M. R., Pearce, L. A., Cowan, R. J., and Spurr, C. L., 1974, *Arch. Intern. Med.* **134**:534–537.

176. Goodman, S. I., Mace, J. W., and Pollak, S., 1971, *Lancet* **1**:234–235.

177. Burka, E. R., Weaver, Z. III, and Marks, P. A., 1966, *Ann. Intern. Med.* **64**:817–825.

178. Nicheles, T. F., 1974, *Glutathione. Proceedings of the 16th Conference of the German Society of Biological Chemistry* (L. Flohé, H. C. Benöhr, H. Sies, H. D. Waller, and A. Wendel, eds.), Geog Thieme, Stuttgart, pp. 173–179.

179. Arneson, D. W., Tipton, R. E., and Ward, J. C., 1982, *Arch. Neurol.* **39**:713–716.

Metabolism and Neurotransmission

Gary E. Gibson and John P. Blass

1. INTRODUCTION

Oxidative metabolism is essential for normal neuronal function. The brain consumes 20% of the oxygen that is used by the body even though it represents only 2% of the total body mass. Most of this oxygen is utilized for the catabolism of glucose and the production of ATP. However, mild to moderate decreases in the availability or utilization of oxygen or glucose impair brain function without reducing the levels of energy metabolites (i.e., ATP).[1-4] The cerebral dysfunction that accompanies impaired oxidative metabolism is associated with changes in neurotransmitter metabolism. Thus, an understanding of the pathophysiological basis of altered neuronal function requires knowledge of the relationship of metabolism to neurotransmission.

A productive approach to this problem has been to examine the effects of interrupted oxidative metabolism on neurotransmitter synthesis. Although a variety of experimental models have been used to do this, this chapter is primarily concentrated on the effects of hypoxia or low oxygen on brain function. The interaction of neurotransmitters and oxidative metabolism during hypoxia is likely to be representative of many other conditions in which altered metabolism disrupts brain function. Although acetylcholine is probably not the only neurotransmitter that is closely related to oxidative metabolism, we primarily discuss it because we are familiar with a relatively larger amount of data for acetylcholine.

Atkinson[5] provided a valuable parallel to the interaction of oxidative metabolism and neurotransmitter synthesis with studies of yeast cells. In response to metabolic insults, the yeast cells maintained ATP levels and the adenylate energy charge [i.e., (ATP + $\frac{1}{2}$ ADP)/(ATP + ADP + AMP)] at the expense of biosynthetic pathways. Thus, biosynthetic activities (i.e., protein synthesis)

Gary E. Gibson and John P. Blass • Department of Neurology, Cornell University Medical College, Burke Rehabilitation Center, White Plains, New York 10605.

may be better indicators of altered metabolism than are the concentrations of ATP or the adenylate energy charge. In the brain, neurotransmitter synthesis is one of those biosynthetic processes. Maintenance of brain ATP levels during severe metabolic insults to the brain has been well documented in hypoxia[1-4] and hypoglycemia.[4] However, unchanged ATP concentrations do not necessarily indicate that energy metabolism was unaffected, since severe hypoxia alters the turnover of the high-energy phosphates[3] and the oxidation–reduction state of the brain[4] without changing ATP levels. Nevertheless, as metabolism shifts to maintain the levels of energy metabolites (i.e., ATP), there is a marked decrease in biosynthetic activities which include diminished neurotransmitter formation. The incorporation of glucose,[4,6,7] pyruvate,[8] or 3-hydroxybutyrate[9] into lipids, nucleic acids and proteins declines in parallel with the reduction in oxidation caused by a variety of inhibitors.

Severe acute insults can produce long-lasting alterations in biosynthetic activity. For example, brief periods of total anoxia *in vivo* impaired subsequent *in vitro* protein synthesis.[10] The metabolic shift to maintain ATP concentrations may be a protective mechanism to preserve cellular integrity, since the cell may not be able to recover from a decrease in ATP concentrations. Yatsu *et al.*[11] postulated that ischemic tissue damage only occurs when the brain can no longer maintain the levels of energy metabolites. This response of the cell to preserve ATP concentrations at the expense of biosynthetic pathways may explain the vulnerability of neurotransmitter synthesis to acute metabolic insults. However, the interaction of brain metabolism with neurotransmitter formation is difficult to interpret, because there are many primary and secondary events that accompany the shifts in various pathways to maintain ATP levels.

Before discussing how the various neurotransmitters respond to interruption of metabolism in five disorders—hypoxia, hypoglycemia, thiamine deficiency, and two inborn errors of metabolism—the normal relationship of the neurotransmitters to metabolism is considered.

2. ACETYLCHOLINE

2.1. Linkage

Acetylcholine and oxidative metabolism are tightly coupled. The glucose requirement for acetylcholine synthesis was first demonstrated by Mann *et al.*[12,13] in the 1930s. However, the closeness of this linkage was not realized until quite recently. Glucose and pyruvate are the best precursors of the acetyl group for acetylcholine synthesis both *in vitro*[14] and *in vivo*.[15] *In vitro*, the specific activity (dpm/nmol) of acetylcholine quickly equilibrates with the specific activity of its radioactive carbohydrate precursor.[14,16,17] Detailed evidence of this close link between acetylcholine synthesis and carbohydrate metabolism was obtained in studies that examined the effects of inhibitors of pyruvate,[8] glucose,[4,6,7] and 3-hydroxybutyrate[9] oxidation. Although less than 1% of the oxidized pyruvate was converted to acetylcholine, its catabolism (i.e., $^{14}CO_2$ production from either $[1-^{14}C]$- or $[2-^{14}C]$pyruvate) and acetylcholine synthesis declined in parallel whether metabolism was impaired by noncompetitive (3-bromopyruvate) or competitive (2-ketobutyrate and other 2-oxoacids) inhibitors of pyruvate dehydrogenase as well as by inhibitors of the electron transport chain (e.g., barbiturate or KCN).[8] Similarly, when glucose oxidation was reduced by low oxygen tensions,[6,18] cyanide, or the 2-oxoacids,[7] there was a proportional decline in acetylcholine synthesis, even though only about 1% of the glucose oxidized was incorporated into acetylcholine.

The decline in [^{14}C]acetylcholine synthesis correlated highly ($r^2 = 0.97$) with $^{14}CO_2$ production from [U-^{14}C]glucose by synaptosomes or slices with various oxygen tensions in high- or low-K$^+$ media.[18] In infant brains, the ketone bodies, acetoacetate[19] and 3-hydroxybutyrate,[9] were used preferentially to glucose for acetylcholine synthesis. Inhibition of 3-hydroxybutrate oxidation by various concentrations of methylmalonic acid led to a proportional decline in acetylcholine synthesis, even though less than 1% of the oxidized ketone was eventually converted to acetylcholine.[9]

Although most of the above *in vitro* studies were performed with brain slices, the tight coupling between acetylcholine synthesis and carbohydrate oxidation also exists in synaptosomes.[18,20–23]

In vivo, impaired carbohydrate oxidation also reduced acetylcholine synthesis, and the linkage appeared to be as tight as it was *in vitro.* Even mild hypoxia reduced acetylcholine synthesis markedly without altering ATP levels.[4,24] Similarly, acetylcholine synthesis fell in animals that were made hypoglycemic with insulin.[4] Pharmacological treatments directed toward improving cholinergic function ameliorated the behavioral effects of these conditions.[4] These data are discussed in more detail in the second half of the chapter.

Another indication of this close linkage between acetylcholine synthesis and pyruvate metabolism is a parallel distribution of pyruvate dehydrogenase and choline acetyltransferase in cats[25] and in man.[26,27] Furthermore, lesions of the habenular interpeduncular pathway reduce choline acetyltransferase (-90%) and pyruvate dehydrogenase (-50%) activities even though other mitochondrial enzymes remain unaffected.[28]

2.2. Compartmentation

Although the precise mechanism of the tight coupling of carbohydrate or ketone body oxidation to acetylcholine synthesis is unknown, much of it may be explained by metabolic compartmentation. Citrate is a good source of acetyl groups for lipids but not for acetylcholine *in vivo*[15] or *in vitro.*[29,30] The rapidly released pool of acetylcholine was preferentially labeled by exogenous glucose but not by exogenous pyruvate.[31] The glucose that enters the brain and subsequently labels acetylcholine does not mix uniformly with other metabolic pools either *in vivo* or *in vitro.*[32] These findings and those of Lefresne *et al.*[22,23] demonstrate that glucose or pyruvate can be preferentially oxidized or used for acetylcholine synthesis under various conditions. After incubation of slices with or *in vivo* injection of [U-^{14}C]glucose, the specific activity (dpm/nmol) of acetylcholine in brain exceeds that of its obligate precursor pyruvate. This result suggests that there is a rapidly turning over metabolic pool of pyruvate that labels acetylcholine and that inhibition of oxidation decreases metabolism in this rapidly turning over pool as well as in the more slowly turning over pools. This hypothesis is consistent with the observation that flux through the tricarboxylic acid cycle is more rapid in nerve endings than in the glial compartment.[33]

2.3. Metabolism of the Acetyl Moiety

The close linkage between acetylcholine synthesis and oxidative metabolism may result from a limitation in the availability of the acetyl-CoA moiety, but this hypothesis cannot be tested until the precursor of the acetyl group is known. Although it has been documented for over 40 years that glucose and pyruvate are the best acetyl group precursors, the metabolic pathway from pyruvate to the acetyl-CoA that is utilized by choline acetyltransferase is unknown. Discovery of this acetyl-CoA precursor and regulation of its availability would add a new dimension to the study of the interactions of neurotransmitter and metabolism.

By analogy with studies on liver,[34,35] citrate has generally been regarded as the transport form of acetyl groups out of the mitochondria, but the evidence for this speculation is controversial. When brain slices were incubated with [6-^{3}H-^{14}C]glucose, the ^{3}H/^{14}C ratios of both citrate and acetylcholine were two-thirds of that in the glucose precursor.[36,37] Such a labeling pattern is consistent with citrate as a precursor, but that observation only shows that citrate and acetylcholine come from a similar metabolic pool. Citrate is a poor precursor of acetylcholine *in vivo*[15] and *in vitro*,[30] although the activity of citrate lyase appears adequate to maintain acetylcholine synthesis.[38,39] Furthermore, interruption of citrate metabolism does not alter acetylcholine metabolism except under extreme conditions. High concentrations of hydroxycitrate, an inhibitor of citrate lyase, or inhibitors of mitochondrial citrate transport do not proportionally suppress acetylcholine synthesis.[30,40] These results suggest that citrate may be a source of acetyl groups but that it cannot be the exclusive donor.

Other potential sources for the acetyl group have been investigated. Acetate is a good precursor of the acetyl group in nonmammalian species and at the neuromuscular junction, but repeated efforts failed to demonstrate its effectiveness in mammalian brain.[41] Acetylcarnitine stimulated synthesis in acetone powders.[17] In brain slices, acetylcarnitine stimulated acetylcholine formation, but it did not appear to be the direct donor of acetyl groups.[30] Mitochondria that were incubated with carnitine released acetylcarnitine, but this acetylcarnitine was not coupled to acetylcholine synthesis.[42,43] Carnitine decreased acetate incorporation into acetylcholine *in vivo*.[15] When the habenular interpeduncular pathway was lesioned, carnitine acetyltransferase activities declined, but to a lesser extent than choline acetyltransferase activities.[28] Other potential acetyl donors that have been studied are acetylphosphate, acetylaspartate, acetylglycine, acetylmethionine, and acetylglutamate. Acetyl-phosphate and acetylaspartate can be donors in brain slices,[30] but their role in normal acetylcholine metabolism has not been studied in detail.

2.4. Importance of Mitochondria

The close link of acetylcholine synthesis and metabolism appears to depend on the integrity of the mitochondrial membrane. It may limit synthesis by restricting pyruvate entry into the tricarboxylic acid cycle. When pyruvate influx was decreased by α-hydroxycinnamate, acetylcholine synthesis was inhibited.[44] The membrane may also limit efflux of active acetyl groups out of the mitochondria. Solubilization of the synaptosomal membrane with increasing concentrations of the detergent Triton X-100 decreased acetylcholine synthesis. However, once the concentration of Triton X-100 was sufficient to release mitochondrial enzymes, acetylcholine synthesis increased. This finding is consistent with the hypothesis that the mitochondrial membrane limits efflux of acetyl groups. Mitochondria are quite impermeable to efflux of acetyl CoA[42] unless the membranes are damaged (e.g., by treatment with ether).[45] Membrane permeability during normal metabolism may be regulated by alterations in mitochondrial phospholipase activity.[46]

Since the transport form of the acetyl group from the mitochondria is unknown, an indirect approach has been used to demonstrate the link between acetylcholine synthesis and the efflux of the acetyl group. The total nanomoles of glucose that enter the tricarboxylic acid cycle (i.e., flux through pyruvate dehydrogenase) were estimated by collection of $^{14}CO_2$ from [3,4-^{14}C]glucose. The $^{14}CO_2$ from [2-^{14}C]glucose was assumed to represent the nanomoles oxidized by the tricarboxylic acid cycle. The difference in nanomoles between these estimates the efflux of carbons from the cycle. A decrease in acetylcholine synthesis in slices or synaptosomes by a variety of inhibitors mirrored the decline in efflux.[47] Altered efflux may result from changes in the redox state as discussed below, or it may merely reflect neuronal metabolism responding to a metabolic insult.

2.5. Role of Pyruvate Dehydrogenase

The activation state of the pyruvate dehydrogenase complex, which produces acetyl CoA by oxidation of pyruvate, may play an important role in the coupling of acetylcholine synthesis, carbohydrate oxidation, and neuronal function. This enzyme complex exists in active (dephosphorylated) and inactive (phosphorylated) forms. Brief bursts of high-frequency synaptic stimulation that produced an extremely long-lasting increase in synaptic efficacy (long-term potentiation) increased the phosphorylation of the 41,000-dalton α-subunit of pyruvate dehydrogenase, one measure of its activation state.[48–50] Learning has been reported to have similar effects.[51]

Modulation of pyruvate dehydrogenase activity might influence synaptic physiology through mitochondrial alterations that lead to fluctuations in the synthesis of acetyl groups for acetylcholine. An alternative explanation that does not exclude an interaction with the availability of acetyl groups is that the activation of pyruvate dehydrogenase might be linked to the calcium-sequestering capability of the mitochondria and thereby to neuronal plasticity.[48] Mitochondrial calcium accumulation parallels pyruvate dehydrogenase activity in a variety of conditions; when the pyruvate concentration that was added to isolated brain mitochondria was increased from 0.025 to 0.100 mM, $^{45}Ca^{2+}$ uptake and pyruvate dehydrogenase activities approximately doubled.[48] This relationship was maintained in the presence or absence of dichloroacetate, an inhibitor of pyruvate dehydrogenase kinase.[48] In striatal slices, a fourfold increase in pyruvate concentrations (1.25 to 5 mM) doubled acetylcholine synthesis.[16]

Since the transport form of the acetyl donor has been difficult to determine, the hypothesis of Lefresne et al.[22,23,52] that a cytoplasmic pyruvate dehydrogenase furnishes acetyl moieties for acetylcholine synthesis seemed quite attractive. Their evidence is indirect and, as they suggest, is equally compatible with metabolic compartmentation. A cytoplasmic pyruvate dehydrogenase is difficult to reconcile with the following observations: (1) the inability to detect pyruvate dehydrogenase activity in nonmitochondrial fractions of rat brain with even very sensitive assays,[39,53] (2) the close coupling of acetylcholine synthesis and mitochondrial metabolism, (3) the inability of synaptoplasmic fractions to synthesize acetylcholine,[42] (4) the increase in acetylcholine synthesis with solubilization of mitochondrial membranes,[17] and (5) the utilization of ketone bodies for acetylcholine synthesis.[9,19]

2.6. Choline Availability and Metabolism

The availability of the choline for acetylcholine synthesis may also be altered by interruption of metabolism. Inhibition of the high-affinity choline uptake system that has been demonstrated to be coupled to synthesis in mammalian brain can impair acetylcholine synthesis.[54] This high-affinity choline uptake system appears to be energy dependent, since it requires glucose and is inhibited by 2,4-dinitrophenol and other metabolic inhibitors.[55–57] In spite of this energy dependence, acetylcholine synthesis is more sensitive to inhibition of energy metabolism than is the high-affinity choline uptake system. Acetylcholine synthesis was inhibited more than choline uptake by pentobarbital (21% greater inhibition), NaCN (45%), and bromopyruvate (38%).[21] Thus, impaired metabolism may alter acetylcholine synthesis, in part, by reducing choline availability.

2.7. Metabolism and Vesicular Loading

Another possible effect of altered metabolism on the cholinergic system is interference with the loading of acetylcholine into synaptic vesicles. This possibility has not been directly pursued in mammalian systems, but MgATP and HCO_3^- influence the loading of "energized" vesicles from electroplaques.[58,59] Although MgATP alone suppressed uptake, the combination of MgATP and HCO_3^- stimulated uptake. These results suggested that uptake may involve the acidification of vesicles by an ATPase-driven proton pump. The presence of ATPase activity in cholinergic synaptic vesicles is consistent with this hypothesis.[60,61]

2.8. Regulation by Cellular Redox State

Although ATP and the adenylate energy charge are maintained during a variety of insults, the redox state ($[NAD^+]/[NADH]$ ratio) appears to be a sensitive indicator of changes that are related to altered neurotransmitter metabolism. Most estimates of the cellular redox state are calculated from the concentrations of oxidized and reduced metabolites of suitable pyridine nucleotide-linked dehydrogenases. The $[NAD^+]/[NADH]$ ratio is calculated from the following general equation in which K_{eq} is the equilibrium constant for the reaction:

$$[NAD^+]/[NADH] = [H^+][\text{oxidized substrate}]/[K_{eq}][\text{reduced substrate}]$$

This method was originally devised by Krebs and Veech[62] for liver cells, but many of the assumptions that were valid in liver have proven controversial in brain.[63] Furthermore, in conditions that might alter intracellular pH, the inability to accurately estimate $[H^+]$ limits the applicability of the method; a constant pH is generally presumed.

There is little disagreement that the lactate dehydrogenase equilibrium can be used to estimate the cytosolic redox state in both brain and liver:

$$[NAD^+]/[NADH] = [H^+][\text{pyruvate}]/[\text{lactate}][1.11 \times 10^{-4} M]$$

However, the calculation of the mitochondrial redox state in brain is controversial. In liver, the ratio can be determined from either the glutamate dehydrogenase or the 3-hydroxybutyrate dehydrogenase equilibrium, which seem to be in reasonable agreement.[62,64,65] The equation for the calculation with the glutamate dehydrogenase equilibrium follows:

$[NAD^+]/[NADH]_{\text{mitochondrial}}$

$$= [H^+][\text{2-oxoglutarate}][NH_4^+]/[\text{glutamate}][3.87 \times 10^{-6} M]$$

The validity of this equilibrium as a measure of brain redox state is questioned because during hypoxia, the $[NADH]/[NAD^+]$ ratio that was calculated from the glutamate dehydrogenase equilibrium decreased rather than increased as expected.[66,67] Estimates from the cytochromes[68] and 3-hydroxybutyrate dehydrogenase[63] demonstrated the expected increases. The cytochrome determinations are complicated because they represent only relative measurements. In the studies that utilized the 3-hydroxybutyrate dehydrogenase equilibrium, the animals had to be starved for 24 h, and measurements were made on nitrous-oxide-anesthetized animals. The use of the glutamate dehydrogenase equilibrium has been validated in other experimental paradigms.[69] Although there are limitations to the use of the glutamate dehydrogenase equilibrium, there is no ideal method for estimation of the mitochondrial $[NAD^+]/[NADH]$ potential.

Both the mitochondrial and cytosolic derived measures of redox can be converted to potentials by use of the Nernst equation ($E_0 = 0.337\ V$)[62]:

$$E = E_0 + (RT/nF) \log ([NAD^+]/[NADH])$$

A transmitochondrial potential can be estimated by subtracting the E(mitochondrial) from the E(cytoplasmic). The magnitude of the transmitochondrial redox potential calculated in this way agrees with estimates of the transmitochondrial hydrogen ion potential, which may be an integral part of oxidative phosphorylation.[70] Although the calculation of this potential with measurements from several laboratories showed that it tends to decrease during metabolic insults (e.g., hypoxia), the values vary considerably and should still be used at best as relative and not absolutes.[71]

The redox state and the mitochondrial membrane potential may play an integral role in neuronal metabolism either by interacting with mitochondrial calcium metabolism or by control of efflux or influx of other metabolites. Although calcium efflux is reduced as the mitochondrial redox state declines,[72] the role of this potential in the control of the efflux of metabolites (e.g., neurotransmitters or acetyl CoA units for acetylcholine synthesis) from the mitochondria has not been well studied.

The importance of the redox state to acetylcholine metabolism was suggested by early studies which demonstrated that the addition of nicotinamide, an inhibitor of endogenous NADase, to brain acetone powders stimulated acetylcholine production.[73] Recently, Benjamin and Quastel[46] demonstrated that high concentrations of NAD^+ stimulate barbiturate-depressed acetylcholine synthesis. When brain slices were incubated with KCN, amobarbital, low glucose concentrations, or under nitrogen, there was a reduction in the transmitochondrial membrane potential that was highly correlated ($r = 0.96$) with a decrease in acetylcholine formation and $^{14}CO_2$ production from [3,4-^{14}C]glucose (i.e., an estimate of flux through pyruvate dehydrogenase; $r = 0.82$).[6] These changes were more highly correlated with the transmitochondrial redox potential than with the mitochondrial or cytosolic redox potential or any metabolite measurement. The importance of this potential or the cellular redox state in the control of the synthesis of the other glucose-derived neurotransmitters requires further investigation.

3. CATECHOLAMINES AND INDOLEAMINES

The availability of molecular oxygen may play a critical role in the biosynthesis of the neurotransmitters dopamine, norepinephrine, and serotonin. Oxygen is required in the rate-limiting steps of their synthesis (dopamine-β-hydroxylase, tyrosine hydroxylase, and tryptophan hydroxylase, respectively). Since the K_m for oxygen (7 mm Hg = 12 μM)[74] is approximately the same as normal brain oxygen tensions,[75] their biosynthesis is one possible site of interaction with metabolism.

Once the neurotransmitter is synthesized, its uptake into synaptic vesicles may be altered by the metabolic state of the cytosol. Although *in vitro* uptake into chromaffin granules is increased severalfold by ATP and magnesium, there is no simple quantitative correlation between Mg^{2+}-dependent ATPase activities and catecholamine transport. However, this ATPase may help establish a hydrogen ion electrochemical gradient, and vesicular loading of catecholamines was highly correlated to the gradient in the absence of ATP and Mg^{2+}. In their presence, catecholamine uptake was better correlated to the membrane potential than to ATPase activity. When the hydrogen potential was reduced by uncouplers, vesicular uptake of catecholamines declined. It is plausible that the vesicular membrane potential may be influenced by both the mitochondrial and cytosolic redox potentials. Further evidence that vesicular loading may be closely related to energy metabolism is that ATP is released with norepinephrine under normal conditions.[78]

Interruption of energy metabolism may alter several other aspects of neurotransmitter metabolism. It might affect the release of neurotransmitters, since inhibition of Na^+-K^+ ATPase by ouabain increases their release. Inactivation of these neurotransmitters may also depend on metabolism, since both reuptake and the degradative enzyme monoamine oxidase are oxygen dependent. Determining which of these steps is primarily or secondarily altered by impaired oxidative metabolism is quite difficult.[78]

4. AMINO ACIDS

Many of the links between metabolism and acetylcholine synthesis are also true of the glucose-derived amino acid neurotransmitters [i.e., serine, glycine, glutamate, γ-aminobutyrate (GABA), and aspartate]. However, they are much more difficult to define because of metabolic compartmentation.[33] Current methods do not distinguish the metabolic pools of the amino acids from their neurotransmitter pools. The glucose that enters the brain and is not lost as CO_2 is largely in glutamate. At 30 min after injection of [U-^{14}C]glucose, 37% is in glutamate, and 9% in glutamine.[79] Inhibitors of the tricarboxylic acid cycle such as fluoroacetate decrease the formation of the amino acids derived from the tricarboxylic acid cycle.[80–82] As discussed in the next section, the biosynthesis of these putative neurotransmitter amino acids is as sensitive to hypoxic insults as acetylcholine formation.[83]

5. IMPLICATIONS FOR DISEASE

The interaction of metabolism with neurotransmitters may be important in the pathophysiology of the cerebral dysfunction associated with several different diseases, particularly the metabolic encephalopathies. These are a group of disorders in which systemic changes alter brain function (e.g., hypoxia, hypoglycemia, hyperammonemia, inborn errors of metabolism, heavy metal intoxication, and nutritional deficits). Plum and Posner[84] suggested that these metabolic encephalopathies share common clinical signs and symptoms: loss of interest and attentiveness that lead to diminished orientation, cognition, memory, and perception; obtundation that is followed by stupor, coma, and finally death. The observation of a stereotyped clinical pattern despite diverse etiologies suggests that these disorders may have a common pathophysiological mechanism. The hypothesis that different metabolic encephalopathies have similar molecular mechanisms was proposed as early as 1937 by Stevens: "The characteristic picture of fatigue and depression in human beri-beri may occur solely by virtue of histotoxic anoxia."[85] Thus, the search for a common biochemical lesion guides many of the studies that use animal models as well as biochemical measurements on human tissues. The interaction of neurotransmitters and oxidative metabolism may also be important in aging and age-related neurological deficits (e.g., the dementias).

The focus of many studies has been on the cholinergic system because of its relationship to carbohydrate metabolism which was demonstrated by Quastel and his colleagues 40 years ago.[86] In addition to the close link of oxidative metabolism and acetylcholine synthesis, several other observations support the hypothesis that the cholinergic system may be important in the production of the neurological symptoms of the metabolic encephalopathies. Acetylcholine is an important neurotransmitter in both the thalamic and extrathalamic divisions of the reticular activating systems, which are essential for attention, learning, memory, and, ultimately, consciousness. Cholinergic antagonists such as scopolamine produce symptoms that resemble those of the metabolic encephalopathies in man.[87] Furthermore, the acetylcholinesterase inhibitor physostigmine delays the time until death and seizures in chemical[4] or hypoxic hypoxia.[88] Conditions that produce metabolic encephalopathies (i.e., hypoxia,[89] hypoglycemia,[89] or thiamine deficiency[90]) impair cholinergic synaptic transmission without diminishing axonal conduction. The discussion below focuses on some of the more standardized metabolic encephalopathies.

5.1. Hypoxia

Perhaps the best studied metabolic encephalopathy is hypoxia (i.e., reduced oxygen availability). The effects of acute hypoxia (i.e., 15-min exposure) on mental function in man were documented in studies of aviation medicine and pulmonary physiology that have been repeatedly summarized.[2,91,92] When the alveolar oxygen (P_{AO_2}) was reduced from 95 to 85 mm Hg (i.e., equivalent to an altitude of 5000 feet), the ability to dark adapt declined. A P_{AO_2} of 60–45 mm Hg (i.e., 8,000 to 10,000 feet) reduced short-term memory and the ability to perform complex tasks. At a P_{AO_2} of 35–45 mm Hg (i.e., 15,000 to 20,000 feet), lethargy, euphoria, impaired critical judgment, and muscular incoordination occurred. A P_{AO_2} of 35 mm Hg (i.e., above 20,000 feet) was compatible with consciousness for only a short time. On the other hand, severe hypoxia is required to alter even gross measures of brain metabolism; the cerebral metabolic rate for oxygen in man is unchanged even at a P_{AO_2} of less than 40 mm Hg, although there is a loss of consciousness and electroencephalographic slowing.[93] In animal models of hypoxia, a flat electroencephalogram occurs with normal ATP concentrations and adenylate energy charge.[2] Thus, large changes in cerebral function accompany minimal changes in brain energy utilization, which indicates that the effects of hypoxia must occur through interference with some other process.

Experiments to assess the role of acetylcholine metabolism during hypoxia preceded those on energy metabolism. After studies in the late 1930s demonstrated that glucose and oxygen were required for acetylcholine synthesis in brain slices[12,13] and the superior cervical ganglion,[94] the hypothesis that the changes in mental function that accompany hypoxia are caused by alterations in acetylcholine metabolism was widely tested. Unfortunately, in those early studies, one was limited to measuring concentrations of acetylcholine, which have proven to be relatively insensitive to alterations in cholinergic function. Despite these limitations, Welsh[95] concluded in 1943 that impaired formation of acetylcholine was important in the pathophysiology of hypoxia since it decreased brain acetylcholine content. Other early studies in brain[96] or in ganglia[97] did not find changes in acetylcholine levels. The development of reliable methods for assessing turnover made direct testing of this hypothesis possible.

Mice were made chemically hypoxic by injection of $NaNO_2$ (anemic hypoxia) which converts hemoglobin to methemoglobin and thus reduces the oxygen-carrying capacity of the blood or by injection of KCN (histotoxic hypoxia) which combines with cytochrome oxidase to prevent oxygen utilization. The synthesis of acetylcholine from [U-^{14}C]glucose and [$^2H_4$]choline declined at even mild degrees of both forms of chemical hypoxia.[4,83] However, the tissue concentrations of acetylcholine, ATP, or the adenylate energy charge did not change except with the most severe insults. The decline in synthesis either paralleled or preceded increases in brain lactate,[4,98,99] a relatively early indicator of hypoxia. Since chemical hypoxia may also produce alterations that are distinct from tissue hypoxia (e.g., high concentrations of $NaNO_2$ activate guanylate cyclase), the effects of low oxygen on acetylcholine synthesis were also studied.[22] If arterial oxygen tensions were lowered to 60 mm Hg (i.e., an elevation of about 13,500 feet), incorporation of [U-^{14}C]glucose and [$^2H_4$]choline into acetylcholine declined about 45%. If the arterial oxygen tension was lowered to 45 mm Hg (i.e., an elevation of about 24,000 feet), synthesis was reduced another 40%. Thus, the decline in synthesis of acetylcholine at low oxygen tensions coincided with decreased mental function in man.

Several lines of evidence suggest that these deficits in the cholinergic system are physiologically important. An anoxic superior cervical ganglion can maintain axonal conduction and postsynaptic firing for several hours, but cholinergic transmission across the synapse is lost relatively rapidly.[89] Pretreatment of chemical[4] or hypoxic hypoxic[88] mice with the acetylcholinesterase inhibitor physostigmine prolongs the time to seizures and death. Furthermore, pretreatment of hypoxic mice with alkaline tris partially reversed the symptoms of hypoxia and the reduction in the turnover of acetylcholine.[99]

Although the changes in the cholinergic system during hypoxia appeared physiologically important, acetylcholine was not the only neurotransmitter that hypoxia altered. The syntheses of the other glucose-derived neurotransmitters (i.e., the amino acids) were also diminished by hypoxia.[83] The physiological importance of changes in the synthesis of the amino acids is more difficult to evaluate than that for acetylcholine, because no method distinguishes between the metabolic and neurotransmitter pools. During moderately severe hypoxia (i.e., 5% O_2), [U-^{14}C]glucose incorporation into alanine, aspartate, glutamate, glutamine, and GABA declined about 55%, but that into serine was unaffected.[100] In milder forms of hypoxia, when the percent oxygen in the inspired air was reduced from 20 to 15 or 10%, the syntheses of acetylcholine and of the amino acids from [U-^{14}C]glucose declined similarly.[83] Analogous decreases in neurotransmitter metabolism and increases in lactate concentrations were observed during anemic hypoxia (37.5 or 75 mg/kg of $NaNO_2$). In more severe forms of chemical hypoxia, the synthesis of acetylcholine and those amino acids derived from the tricarboxylic acid cycle (i.e., glutamate, aspartate, and GABA) declined in parallel, whereas incorporation into the glycolytic amino acids (i.e., serine, glycine, and alanine) began to rise. The physiological effects of altered amino acid metabolism during hypoxia have not been assessed.

Catecholamine and serotonin syntheses are also reduced by hypoxia. When the oxygen content of the inspired air decreased from 20 to 10%, dopamine (-20%) and serotonin (-26%) synthesis declined,[101,102] but their concentrations were unchanged. However, when catecholamine synthesis was stimulated by stressing the animal, hypoxia did not diminish formation.[101,102]

An alternative way to examine neurotransmitter metabolism is to measure their second messengers, the cyclic nucleotides. During mild anemic or histotoxic hypoxia, cyclic GMP levels increased, but cyclic AMP concentrations were unchanged in even severe hypoxia.[98] In hypoxic hypoxic rats that were anesthetized with nitrous oxide, only severe hypoxia increased cyclic GMP, but, again, no changes in cyclic AMP were observed.[103] Both of these studies have their limitations. In the first study, the hypoxic agents (KCN or $NaNO_2$) may have activated guanylate cyclase, although this occurs only at concentrations several orders of magnitudes higher than those calculated to have been present in the brains of the hypoxic animals. The latter study was done during nitrous oxide anesthesia, which is known to alter the cholinergic system[24] and perhaps the response of these second messengers.

5.2. Hypoglycemia

Acetylcholine metabolism is also sensitive to glucose availability. The early studies of Mann et al.[12,13] demonstrated that glucose was required for acetylcholine synthesis in vitro, but the closeness of this relationship has only recently been appreciated. As the glucose concentration in the incubation medium was reduced, acetylcholine synthesis and oxidative metabolism declined in parallel even though less than 1% of the glucose was converted to acetylcholine.[6,7] This relationship was true with high or low K^+, with synaptosomes or slices, and in the presence of various metabolic inhibitors.[18] In 1943, Welsh[95] found that hypoglycemia decreased the in vivo concentrations of acetylcholine and concluded that the cholinergic system was important in the pathophysiology of this disorder. The decreased incorporation of [$^2H_4$]choline into acetylcholine during insulin-induced hypoglycemia further supports this hypothesis.[4] These changes occurred with normal levels of acetylcholine, although choline levels increased two- to fourfold in whole brain[4] and in five brain regions.[104] The alterations in the cholinergic system with hypoglycemia appear to be physiologically important. Ganglia that were incubated without glucose lost synaptic transmission within minutes, although axonal conduction was maintained for several hours.[89] After a large dosage of insulin, the time until death can be delayed by injection of the acetylcholinesterase inhibitor physostigmine.[4]

5.3. Thiamine Deficiency

In 1929, Kinnersley and Peters[105] demonstrated that thiamine deficiency impaired brain pyruvate oxidation and suggested that there was "an intimate relation between the symptoms and some phase of carbohydrate metabolism." Mann *et al.*, in 1938[12] and 1939,[13] showed that acetylcholine synthesis depended on glucose oxidation, which suggested that the link Kinnersley and Peters had alluded to might be acetylcholine. Thus, efforts to test the hypothesis that thiamine depletion decreases acetylcholine metabolism because of thiamine's role as a cofactor in the pyruvate and 2-oxoglutarate dehydrogenase steps began immediately after the studies of Mann *et al.*[12,13] and still continue.

The early *in vivo* studies on the role of acetylcholine in thiamine deficiency were rather equivocal. Technical limitations restricted the investigations to measurement of acetylcholine concentrations, which have proven to be inadequate indicators of cholinergic function. In 1939, MacIntosh[97] found no change in acetylcholine content of thiamine-deficient brains from pigeons, rats, or mice or rat ganglia, but in 1943, Lissak *et al.*[106] found decreased levels in rat brain. Even with the development of better assays and tissue fixation methods, the effects of thiamine deficiency on brain acetylcholine content are controversial.[107,108] The contradictory measurements of acetylcholine concentrations in these many studies of thiamine deficiency have been summarized elsewhere.[109] There has been uniform agreement that severe thiamine deficiency depresses acetylcholine synthesis *in vivo*, whether formation is measured from [3-^{14}C]pyruvate,[110] [U-^{14}C]glucose, or [^{2}H$_4$]choline[108] or estimated by the hemicholinium method.[107,111]

Decreased *in vitro* acetylcholine synthesis has been observed consistently during severe thiamine deficiency. In 1940, Mann and Quastel[112] demonstrated that thiamine deficiency reduced acetylcholine synthesis (-40%) from pigeon brain slices regardless of whether or not the slices were depolarized with K^+. These deficits in formation could be ameliorated by addition of thiamine to the incubation media. This decreased *in vitro* synthesis of acetylcholine after *in vivo* deprivation of thiamine has been confirmed in brain acetone powders,[113] brain homogenates,[110] and superior cervical ganglia.[114,115]

These deficits in acetylcholine synthesis in severe thiamine deficiency appear to be caused by a decline in oxidative metabolism rather than by a direct effect on other aspects of the cholinergic system. Neither the activities of choline acetyltransferase[113–116] or acetylcholinesterase[114–116] nor the high-affinity choline uptake system[117] changed. On the other hand, the activities of the thiamine-requiring oxidative enzymes declined at degrees of thiamine deficiency that reduced acetylcholine synthesis. Both 2-oxoglutarate and pyruvate dehydrogenase[118–123] declined about 50–75% in late stages. Although this may[116] or may not[124] alter the levels of acetyl-CoA, acetylcholine formation and the enzymes of carbohydrate oxidation are inhibited similarly.

Deficits in oxidative metabolism and acetylcholine synthesis during thiamine deficiency appear to be physiologically important. The synaptic response in thiamine-deficient animals is lost at rapid stimulation rates, although the amplitude and conduction velocity of the compound action potential of the sympathetic trunk is indistinguishable from that in controls.[90,114,115] The persistence of normal function at low rates of stimulation and its loss at higher rates imply that acetylcholine synthesis in thiamine-deficient animals cannot meet an increased physiological demand.

Another approach to test the importance of the cholinergic lesion is to determine the efficacy of cholinergic drugs on thiamine-induced behavioral deficits. Cheney *et al.*[110] showed that the acetylcholinesterase inhibitor physostigmine prolonged the time (56%) from the beginning of seizures until death if it was given before the seizures began. More extensive pharmacological manipulations of thiamine-induced behavioral deficits have been performed during early stages of thiamine deficiency with a tight-rope test.[108,109,125] After only 2 days of treatment with the antimetabolite pyrithiamine and a thiamine-deficient diet, about 40% of the rats lost their ability to perform on the tight-rope test. This deficit could be partially ameliorated by treatment with physostigmine as effectively as by a single injection of a large dosage of thiamine. However, the peripherally acting acetylcholinesterase inhibitor neostigmine was ineffective. The beneficial action of physostigmine was blocked by atropine but not by methatropine or by the nicotinic blocker mecamylamine. The muscarinic agonist arecoline was also effective, whereas nicotine was not. These data suggested that thiamine deficiency produced a central muscarinic deficit of the cholinergic system.

Thiamine deficiency may also decrease the biosynthesis of the amino acid neurotransmitters, but currently, one cannot distinguish the metabolic and neurotransmitter pools of these compounds. Gaitonde and his colleagues[126,127] found that thiamine deficiency decreased incorporation of intraperitoneally injected [U-^{14}C]glucose into brain amino acids by 37–53%. When Gubler *et al.*[128] injected [U-^{14}C]glucose intraventricularly, they found a decline in the synthesis of aspartate and glutamate but not of GABA. Koeppe *et al.*[129] did not find a reduced incorporation into glutamate, but they used anesthetized animals. The physiological importance of these changes in amino acid synthesis has not been examined, in part because no good pharmacological agents are available for these neurotransmitter systems.

Although serotonin and catecholamine metabolism are impaired by decreased oxidative metabolism, their relationship to thiamine deficiency and to the deficits in oxidative metabolism is not as obvious as for the glucose-derived neurotransmitters. Thiamine deficiency reduced synaptosomal uptake and brain concentrations of 5-hydroxytryptamine.[117,130,131] The reduced catecholamine and indoleamine synthesis in severely thiamine-deficient mice[132] might be caused by an altered redox state of the pyridine cofactors that are required for their formation, but this possibility has not been explored.

5.4. Inborn Errors of Metabolism

Since acetylcholine synthesis is tightly coupled to oxidative metabolism, a genetic defect in any energy-related enzyme may lead to a deficit in cholinergic function. For example, the brain does not have a large excess of the enzyme pyruvate dehydrogenase (i.e., a pyruvate dehydrogenase activity of 5 μmol/min per g tissue[53] versus a pyruvate flux of 4.1 μmol/min per g tissue during seizures[133]). Thus, a decrease in pyruvate dehydrogenase activity would likely impair carbohydrate metabolism and subsequently reduce acetylcholine synthesis. Although a direct demonstration of deficits in brain acetylcholine metabolism in man are not possible, pharmacological studies, which are at best equivocal, suggest that they may be important. Neurological deficits such as ataxias and mental retardation are associated with decreased pyruvate dehydrogenase activity.[134] Some patients with documented deficiencies in pyruvate metabolism (e.g., spinocerebellar degenerations) may be improved by treatment with an acetylcholinesterase inhibitor (physostigmine)[135] or with choline[137] or lecithin.[138]

A second type of inborn error in which these relationships may be important is when an abnormality in liver metabolism leads to an accumulation of toxic metabolites in the blood. Maple syrup urine disease and phenylketonuria are inherited disorders of amino acid metabolism that are associated with severe neurological deficits.[139] The 2-oxo acids that are metabolites of the branched-chain amino acids accumulate in maple syrup urine disease, whereas phenylalanine and 2-oxo-3-phenylpropionate increase in phenylketonuria. The increased 2-oxo acids in maple syrup urine diseases inhibited $^{14}CO_2$ production from [U-^{14}C]glucose or [1-^{14}C]- or [2-^{14}C]pyruvate as well as acetylcholine production in rat brain slices.[7] This occurred at concentrations within the range reported in the serum of patients, although these determinations have been quite variable.[139] In contrast to these results, the metabolites that are elevated in phenylketonuria did not inhibit carbohydrate utilization or acetylcholine synthesis.[7]

6. CONCLUSION

The metabolic encephalopathies are common disorders, and study of their mechanisms is likely to have widespread clinical implications. Both theoretical and practical considerations indicate the need for further basic neurochemical studies of the interrelations and regulatory mechanisms linking carbohydrate and energy metabolism to neurotransmission.

ACKNOWLEDGMENTS. Supported in part by grants number NS16997, NS03346, NS15125, the Winifred Masterson Burke Relief Foundation, the Will Rogers Institute, and the Brown and Williamson Company. The authors thank Christine Peterson for careful criticism of this chapter.

REFERENCES

1. Gurdjian, E. S., Stone, W. E., and Webster, J. E., 1944, *Arch. Neurol. Psychiatry* **54**:472–477.
2. Siesjö, B., Johannsson, H., Ljunggren, B., and Norberg, K., 1974, *Brain Dysfunction in Metabolic Disorders* (F. Plum, ed.), Raven Press, New York, pp. 75–112.
3. Duffy, T. E., Nelson, S. R., and Lowry, O. H., 1972, *J. Neurochem.* **19**:1043–1048.
4. Gibson, G. E., and Blass, J. P., 1976, *J. Neurochem.* **27**:37–42.
5. Atkinson, D. E., 1977, *Cellular Energy Metabolism and Its Regulation,* Academic Press, New York.
6. Gibson, G. E., and Blass, J. P., 1976, *J. Biol. Chem.* **251**:4127–4130.
7. Gibson, G. E., and Blass, J. P., 1976, *J. Neurochem.* **26**:1073–1078.
8. Gibson, G. E., Jope, R., and Blass, J. P., 1975, *Biochem. J.* **148**:17–23.
9. Gibson, G. E., and Blass, J. P., 1979, *Biochem. Pharmacol.* **28**:133–140.
10. Yanagahira, T., 1974, *J. Neurochem.* **22**:113–117.
11. Yatsu, F. M., Lee, L. W., and Liao, C. L., 1975, *Stroke* **6**:678–683.
12. Mann, P. J. G., Tennenbaum, M., and Quastel, J. H., 1938, *Biochem. J.* **32**:243–261.
13. Mann, P. J. G., Tennenbaum, M., and Quastel, J. H., 1939, *Biochem. J.* **33**:822–835.
14. Browning, E. T., 1976, *Biology of Cholinergic Function* (A. M. Goldberg and I. Hanin, eds.), Raven Press, New York, pp. 193–196.
15. Tucek, S., and Cheng, S. C., 1974, *J. Neurochem.* **22**:893–914.
16. Lefresne, P., Guyenet, P., and Glowinski, J., 1973, *J. Neurochem.* **20**:1083–1097.
17. Lefresne, P., Hamon, M., Beaujouan, J. C., and Glowinski, J., 1975, *J. Neurochem.* **25**:415–422.
18. Ksiezak, H., and Gibson, G. E., 1981, *J. Neurochem.* **37**:305–314.
19. Itoh, T., and Quastel, J. H., 1970, *Biochem. J.* **116**:641–655.
20. Barker, L. A., Mittag, T. W., and Krespan, B., 1977, *Cholinergic Mechanisms and Psychopharmacology* (D. J. Jenden, ed.), Plenum Press, New York, pp. 465–480.
21. Jope, R. S., 1978, *Cholinergic Mechanisms and Psychopharmacology* (D. J. Jenden, ed.), Plenum Press, New York, pp. 497–510.
22. Lefresne, P. J., Beaujouan, C., and Glowinski, J., 1978, *Biochimie* **60**:479–487.
23. Lefresne, P. J., Beaujouan, C., and Glowinski, J., 1978, *Nature* **274**:497–500.
24. Gibson, G. E., and Duffy, T. E., 1981, *J. Neurochem.* **36**:28–33.
25. Reynolds, S. F., and Blass, J. P., 1976, *Neurology (Minneap.)* **26**:625–628.
26. Sorbi, S., Amaducci, L., and Blass, J. P., 1982, *The Aging Brain: Cellular and Molecular Mechanisms of Aging in the Nervous System* (E. Giacobini, G. Giacobini, and A. Vernadakis, eds.), Raven Press, New York, pp. 223–230.
27. Perry, E. K., Perry, R. H., Tomlinson, B. E., Blessed, G., and Gibson, P. H., 1980, *Neurosci. Lett.* **18**:105–110.
28. Sterri, S. H., and Fonnum, F., 1980, *J. Neurochem.* **35**:249–254.
29. Nakamura, R., Cheng, S.-C., and Naruse, H., 1970, *Biochem. J.* **118**:443–450.
30. Gibson, G. E., and Shimada, M., 1981, *Biochem. Pharmacol.* **29**:167–174.
31. Molenaar, P. C., and Polak, R. L., 1976, *J. Neurochem.* **26**:95–99.
32. Gibson, G. E., Blass, J. P., and Jenden, D. J., 1978, *J. Neurochem.* **30**:71–76.
33. Berl, S., Clarke, D. D., and Schneider, D. (eds.), 1975, *Metabolic Compartmentation and Neurotransmission,* Raven Press, New York.
34. Watson, J. A., and Lowenstein, J. M., 1970, *J. Biol. Chem.* **245**:5993–6002.
35. Spencer, A. F., and Lowenstein, J. M., 1962, *J. Biol. Chem.* **237**:3640–3648.
36. Sterling, G. H., and O'Neill, J. J., 1978, *J. Neurochem.* **31**:525–530.
37. Sollenberg, J., and Sorbo, B., 1970, *J. Neurochem.* **17**:201–207.
38. Tucek, S., 1967, *J. Neurochem.* **14**:531–545.
39. Szutowicz, A., Stepien, M., Lysiak, W., and Angielski, S., 1976, *Acta Biochim. (Pol.)* **23**:227–234.
40. Tucek, S., Dolezal, V., and Sullivan, A. C., 1981, *J. Neurochem.* **36**:1331–1337.
41. Tucek, S., and Cheng, S. C., 1970, *Biochem. Biophys. Acta* **208**:538–540.

42. Polak, R. L., Molenaar, P. C., and Braggaar-Schaap, P., 1978, *Cholinergic Mechanisms and Psychopharmacology* (D. J. Jenden, ed.), Plenum Press, New York, pp. 511–524.
43. Robinson, B. H., Williams, G. R., Halperin, M. L., and Leznoff, C. C., 1971, *Eur. J. Biochem.* **20**:65–71.
44. Jope, R. S., Weiler, M. H., and Jenden, D. J., 1978, *J. Neurochem.* **30**:949–954.
45. Tucek, S., 1967, *Biochem. J.* **104**:749–756.
46. Benjamin, A. M., and Quastel, J. H., 1981, *Science* **213**:1495–1497.
47. Ksiezak, H. J., and Gibson, G. E., 1981, *J. Neurochem.* **37**:88–94.
48. Browning, M., Baudry, M., Bennett, W. F., and Lynch, G., 1981, *J. Neurochem.* **36**:1932–1940.
49. Browning, M., Bennett, W., Kelly, P., and Lynch, G., 1981, *Brain Res.* **218**:255–266.
50. Browning, M., Dunwiddie, T., Bennett, W., Gispen, W., and Lynch, G., 1981, *Science* **203**:60–62.
51. Morgan, D., and Routtenberg, A., 1981, *Science* **214**:470–471.
52. Lefresne, P., Hamon, M., Beaujouan, S. C., and Glowinski, I., 1977, *Biochemie* **59**:197–215.
53. Ksiezak-Reding, H., Blass, J. P., and Gibson, G. E., 1982, *J. Neurochem.* **38**:1627–1636.
54. Kuhar, M. J., 1978, *Cholinergic Mechanisms and Psychopharmacology* (D. J. Jenden, ed.), Plenum Press, New York, pp. 447–456.
55. Yamamura, H. I., and Snyder, S. H., 1973, *J. Neurochem.* **21**:1355–1374.
56. Simon, J. R., and Kuhar, M. J., 1976, *J. Neurochem.* **27**:93–99.
57. Dowdall, M. J., and Simon, E. J., 1973, *J. Neurochem.* **21**:969–982.
58. Parsons, S. M., and Koenigsberger, R., 1980, *Proc. Natl. Acad. Sci. U.S.A.* **77**:6234–6238.
59. Koenigsberger, R., and Parsons, S., 1980, *Biochem. Biophys. Res. Commun.* **94**:305–312.
60. Breer, H., Morris, S. J., and Whittaker, V. P., 1977, *Eur. J. Biochem.* **80**:313–318.
61. Rothlein, J. E., and Parsons, S. M., 1979, *Biochem. Biophys. Res. Commun.* **88**:1069–1076.
62. Krebs, H. A., and Veech, R. L., 1969, *Advances in Enzyme Regulation*, Volume 7 (G. Weber, ed.), Pergamon Press, Oxford, pp. 397–413.
63. Siesjö, B. K., and Berntman, L., 1979, *Adv. Neurol.* **26**:319–323.
64. Greenbaum, A. L., Gumma, K. A., and McLean, P., 1971, *Arch. Biochem. Biophys.* **143**:617–663.
65. Williamson, D. H., Lund, P., and Krebs, H. A., 1967, *Biochem. J.* **103**:514–527.
66. Lai, F. M., and Miller, A. T., 1973, *Comp. Biochem. Physiol. [B.]* **44**:829–835.
67. Siesjö, B. K., and Nilsson, L., 1971, *Scand. J. Clin. Lab. Invest.* **27**:83–96.
68. Rosenthal, M., LaManna, J. C., Jobsis, F. F., Levassew, J. E., Kontos, H. A., and Patterson, J. L., 1976, *Brain Res.* **108**:143–154.
69. Miller, A. L., Hawkins, R. A., and Veech, R. L., 1973, *J. Neurochem.* **20**:1393–1400.
70. Mitchell, P., 1968, *Chemiosmotic Coupling and Energy Transduction*, Glynn Research, Bodmin, United Kingdom.
71. Blass, J. P., and Gibson, G. E., 1979, *Adv. Neurol.* **26**:229–254.
72. Fishkum, G., and Lehninger, A. L., 1980, *Fed. Proc.* **39**:2432–2436.
73. Harpur, R. P., and Quastel, J. H., 1949, *Nature* **164**:779–782.
74. Fisher, D. B., and Kaurman, S., 1972, *J. Neurochem.* **19**:1359–1366.
75. Lubbers, D. W., 1968, *Oxygen Transport In Blood and Tissue* (D. W. Lubbers, U. P. Luft, G. C. Thews, and E. Witzleb, eds.), Georg Thieme, Stuttgart, pp. 124–139.
76. Johnson, R. G., Carlson, N. J., and Scarpa, A., 1978, *J. Biol. Chem.* **253**:1512–1521.
77. Holz, R. W., 1978, *Proc. Natl. Acad. Sci. U.S.A.* **75**:5190–5194.
78. Smith, A. D., 1979, *The Release of Catecholamines from Adrenergic Neurons* (D. M. Paton, ed.), Pergamon Press, New York, pp. 1–17.
79. Gaitonde, M. K., Dahl, D. R., and Elliot, K. A. C., 1965, *Biochem. J.* **94**:345–352.
80. Clarke, D. D., Nicklas, W. J., and Berl, S., 1970, *Biochem. J.* **120**:345–351.
81. Patel, A., and Koenig, H., 1971, *J. Neurochem.* **18**:621–628.
82. Cheng, S. C., Kumar, S., and Casella, G. A., 1972, *Brain Res.* **42**:112–128.
83. Gibson, G. E., Peterson, C., and Sansone, J., 1981, *J. Neurochem.* **37**:192–201.
84. Plum, F., and Posner, J. B., 1980, *The Diagnosis of Stupor and Coma*, 3rd ed., F. A. Davis, Philadelphia.

85. Stevens, H., 1937, *J. Comp. Psychol.* **24**:441–458.
86. Quastel, J. H., 1978, *Cholinergic Mechanisms and Psychopharmacology* (D. J. Jenden, ed.), Plenum Press, New York, pp. 411–430.
87. Drachman, D. A., 1978, *Psychopharmacology: A Generation of Progress* (M. A. Lipton, A. DiMascio, and K. F. Killam, eds.), Raven Press, New York, pp. 651–662.
88. Scremin, A. M. E., and Scremin, O. U., 1979, *Stroke* **10**:142–143.
89. Dolivo, M., 1974, *Fed. Proc.* **33**:1043–1048.
90. Perri, V., Sacchi, O., and Casella, C., 1970, *Q. J. Exp. Physiol.* **55**:25–35.
91. Luft, U. C., 1965, *Handbook of Physiology,* Volume II, Section 3: *Respiration* (W. O. Fenn and H. Rahn, eds.), American Physiological Society, Washington, pp. 1099–1145.
92. Gibson, G. E., Pulsinelli, W. A., Blass, J. P., and Duffy, T. E., 1981, *Am. J. Med.* **70**:1247–1253.
93. Cohen, P., Alexander, S., Smith, R., Reivich, M., and Wollman, H., 1967, *J. Appl. Physiol.* **23**:183–189.
94. Kahlson, G., and MacIntosh, F. C., 1939, *J. Physiol. (Lond.)* **96**:277–292.
95. Welsh, J. H., 1943, *J. Neurophysiol.* **6**:329–336.
96. Cortell, R., Feldman, J., and Gellhorn, E., 1941, *Am. J. Physiol.* **132**:588–593.
97. MacIntosh, F. C., 1939, *J. Physiol. (Lond.)* **96**:16P.
98. Gibson, G. E., Shimada, M., and Blass, J. P., 1978, *J. Neurochem.* **31**:757–760.
99. Gibson, G. E., Shimada, M., and Blass, J. P., 1979, *Biochem. Pharmacol.* **28**:747–750.
100. Yoshino, Y., and Elliot, K. A. C., 1970, *Can. J. Biochem.* **48**:228–235.
101. Davis, J. N., and Carlsson, A., 1973, *J. Neurochem.* **20**:913–915.
102. Davis, J. N., Giron, L. T., Staton, E., and Maury, W., 1979, *Adv. Neurol.* **26**:319–323.
103. Folbergrova, J., Nilsson, B., Sakabe, T., and Siesjo, B., 1981, *J. Neurochem.* **36**:1670–1674.
104. Gorell, J. M., Navarro, C. P., and Schwendner, S. P. W., 1981, *J. Neurochem.* **36**:321–324.
105. Kinnersley, H. W., and Peters, P. A., 1929, *Biochem. J.* **23**:1126–1136.
106. Lissak, K., Kovacs, T., and Nagy, E. K., 1943, *Arch. Ges. Physiol.* **247**:124–131.
107. Vorhees, C. V., Schmidt, D. E., and Barrett, R. J., 1978, *Brain Res. Bull.* **3**:493–496.
108. Barclay, L. L., Gibson, G. E., and Blass, J. P., 1981, *J. Pharmacol. Exp. Ther.* **217**:537–543.
109. Gibson, G. E., Barclay, L. L., and Blass, J. P., 1982, *Ann. N.Y. Acad. Sci.* **378**:382–403.
110. Cheney, D. L., Gubler, C. J., and Jaussi, A. W., 1969, *J. Neurochem.* **16**:1283–1291.
111. Vorhees, C. V., Schmidt, D. E., Barrett, R. J., and Schenker, S., 1977, *J. Nutr.* **107**:1902–1908.
112. Mann, P. J. G., and Quastel, J. H., 1940, *Nature* **145**:856–857.
113. Bhagat, B., and Lockett, M. F., 1962, *J. Pharm. Pharmacol.* **14**:37–40.
114. Sacchi, O., Consolo, S., Peri, I., Prigioni, S., Ladinsky, H., and Perri, V., 1978, *Brain Res.* **151**:443–456.
115. Sacchi, O., Ladinsky, H., Prigione, I., Consolo, S., Peri, G., and Perri, V., 1978, *Brain Res.* **151**:609–614.
116. Heinrich, C. P., Stadler, H., and Weiser, H., 1973, *J. Neurochem.* **21**:1273–1281.
117. Plaitakis, A., Nicklas, W., and Berl, S., 1978, *Neurology (Minneap.)* **28**:691–698.
118. Dreyfus, P. M., and Hauser, G., 1965, *Biochim. Biophys. Acta* **104**:78–84.
119. Reinauer, H., Frassow, G., and Hollman, S., 1968, *Hoppe Seylers Z. Physiol. Chem.* **349**:969–978.
120. Takahashi, K., Nakamura, A., and Nose, Y., 1971, *J. Vitaminol.* **17**:207–274.
121. Gubler, C. J., 1961, *J. Biol. Chem.* **236**:3112–3120.
122. Bennett, C. D., Jones, J. H., and Nelson, J., 1966, *J. Neurochem.* **13**:449–459.
123. Holowach, J., Kauffman, F., Ikossi, M. G., Thomas, C., and McDougal, D. B., 1968, *J. Neurochem.* **15**:621–631.
124. Reynolds, S. F., and Blass, J. P., 1975, *J. Neurochem.* **24**:185–186.
125. Barclay, L. L., Gibson, G. E., and Blass, J. P., 1981, *Pharmacol. Biochem. Behav.* **14**:153–157.
126. Gaitonde, M. K., and Nixey, R. W. K., 1974, *J. Neurochem.* **22**:53–61.
127. Gaitonde, M. K., Fayein, N. A., and Johnson, A. L., 1975, *J. Neurochem.* **24**:1215–1223.

128. Gubler, C. J., Adams, B. L., Hammond, B., Yaun, E. C., Guo, S. M., and Bennion, M., 1974, *J. Neurochem.* **22**:831–836.
129. Koeppe, R. E., O'Neal, R. M., and Han, C. H., 1964, *J. Neurochem.* **11**:695–699.
130. Chan-Palay, V., 1977, *J. Comp. Neurol.* **176**:463–494.
131. Chan-Palay, V., Plaitakis, A., Nicklas, W., and Berl, S., 1977, *Brain Res.* **138**:380–384.
132. Iwata, H., 1976, *J. Nutr. Sci. Vitaminol.* **22**(Suppl):25–27.
133. Siesjö, B. K., 1978, *Brain Energy Metabolism,* John Wiley & Sons, New York.
134. Blass, J. P., 1982, *The Metabolic Basis of Inherited Disease* (J. B. Stanbury, J. B. Weingaarden, D. S. Fredrickson, J. Goldstein, and M. Brown, eds.), McGraw-Hill, New York (in press).
135. Kark, R. A. P., Blass, J. P., and Spence, M. A., 1977, *Neurology (Minneap.)* **27**:70–72.
136. Livingstone, J. R., and Mastaglia, F. L., 1979, *Br. Med. J.* **2**:939.
137. Barbeau, A., 1978, *Can. J. Neurol. Sci.* **5**:157–160.
138. Stanbury, J. B., Wyngaarden, J. B., and Frederickson, D., 1972, *The Metabolic Basis of Inherited Disease,* 3rd ed., McGraw-Hill, New York.

24

Blood Flow

Bo K. Siesjö and Martin Ingvar

1. INTRODUCTION

The circulation of the brain and spinal cord is of interest to those who deal with both the theoretical and practical aspects of neurochemistry, and to neurologists and neurosurgeons, the subject is of crucial interest. This is because a number of clinical conditions are primary disorders of circulation, and many others are complicated by decreases in nutritional blood flow. As a result of this, continuous efforts have been made to devise suitable techniques for measuring CBF in both experimental animals and man. In many applications, one main purpose has been to use CBF data for deriving the cerebral metabolic rate for oxygen ($CMRO_2$) and/or glucose (CMR_{gl}).

At the time of the Second World War, knowledge of the physiology and pathophysiology of cerebral circulation was limited, mostly because of lack of quantitative CBF techniques. The work of Kety and associates set the stage for an almost explosive progress. Thus, whereas the development of the Kety–Schmidt technique[1,2] allowed measurements of overall CBF and $CMRO_2$, the work of Kety[3] led to the development of two other techniques allowing derivation of regional (or local) CBF. One of these, which involves calculation of CBF from the externally recorded clearance of a radioactive isotope following its intracarotid administration,[4,5] forms the basis of most of our present knowledge of regional CBF in man. The other technique is based on autoradiographic measurements of tracer uptake in animal experiments[6] (see also ref. 7). The latter technique, which allows assessment of local blood flow with high spatial resolution, has recently been supplemented with an ingenious method for measuring local glucose consumption (l-CMR_{gl}).[8]

In animal experiments, therefore, it is now possible to measure both CBF and CMR_{gl} in circumscribed brain structures. A corresponding development has now occurred with techniques applicable to man. Thus, the ^{15}O techniques developed by Ter-Pogossian and associates[9,10] (see also ref. 11) made it possible

B. K. Siesjö and Martin Ingvar • Laboratory for Experimental Brain Research, University Hospital of Lund, S-221 85 Lund, Sweden.

to measure both regional CBF and $CMRO_2$. Related techniques utilizing positron emission tomography now permit three-dimensional measurements of CBF, CMR_{gl}, and $CMRO_2$ in man.[12-14]

The literature on CBF is at present overwhelmingly large. This chapter represents an attempt to summarize the most important features of CBF methodology, physiology, and pathophysiology. Since the chapter is directed towards a group of readers whose main interest is neurochemistry, emphasis is laid on information that relates to cerebral metabolism. Unfortunately, the literature on the subject is scattered over a large number of journals, but a number of textbooks and review articles have been published.[15-23] Furthermore, a new scientific journal (*Journal of Cerebral Circulation and Metabolism*) is solely devoted to the subject. A recent supplement to that journal provides a valuable source to the literature on the subject.[24]

2. METHODS FOR MEASURING OVERALL AND REGIONAL CEREBRAL BLOOD FLOW

2.1. Anatomic Considerations

Two features of cerebrovascular anatomy complicate measurements of CBF.[19,22,25] First, the brain is supplied with blood by as many as four main arteries (the two carotid and the two vertebral arteries), and it drains its venous blood by two main veins (the internal jugular veins). Second, extensive communications exist between these major supply and drainage vessels and other vessels. For example, mixing of blood occurs between the internal and external carotid artery territories and between vertebral and spinal arteries. These communications differ among species, and among individuals within a species. As a result of these communications, and of a normally extensive mixing of blood in the circle of Willis, several species can tolerate obstruction of both carotid arteries or of both vertebral arteries, and some even obstruction of three of the four major arteries.[26-28] The gerbil, though, lacks a full set of communicating arteries, and a large proportion of animals develop unilateral infarcts if one carotid artery is obstructed.[26,29]

The communications among arteries make it a formidable, if not an impossible, task to isolate arteries for accurate measurements of arterial inflow to the brain, and the quantitative significance of, e.g., electromagnetic measurements of internal carotid flow is obscure.[30,31] Similar difficulties arise when attempts are made to isolate cerebral outflow vessels for measurements of CBF. However, these problems are easier to overcome, and several groups have developed useful and widely exploited techniques by cannulating cerebral venous sinuses.[32-34] Although at least one of these techniques has been successfully used for measurements of $CMRO_2$ as well[34] (see also refs. 35,36), we discuss it in connection with continuous CBF measurements.

Two additional anatomic features should be recalled. First, there are few arterial communications at tissue level, and, at least as a first approximation, the cerebral arteries have the character of "end arteries." As a result, obstruction of a proximal artery (e.g., the middle cerebral artery) leads to distal ischemia and infarction. Second, most results favor the view that arteriovenous anastomoses are infrequent. Thus, microspheres of a certain minimal size will be trapped in the tissue (see below).

2.2. Cerebral Blood Flow Techniques

When describing the major CBF methods applicable to animal experiments and/or man, we include only a brief description of the principles on which they are based and put some emphasis on the advantages and limitations of the techniques.

2.2.1. The Kety–Schmidt Technique

This method is based on the Fick principle applicable to an inert, i.e., nonmetabolizable substance whose concentration in arterial blood is allowed to rise as a function of time.[2,3] At any given time, the following equation applies

$$dQ_i/dt = F(C_a - C_{\bar{v}})$$ [1]

i.e., the amount taken up is a function of blood flow (F) and of the arterial (C_a) and mean venous ($C_{\bar{v}}$) concentrations. Since C_a and $C_{\bar{v}}$ change with time, the equation must be integrated. This can be done if F remains constant during the time considered (T). Furthermore, if T is made sufficiently long, the tissue concentration can be replaced by the venous concentration times the partition coefficient (λ). We then obtain

$$\text{CBF} = C_{\bar{v}}(T){\cdot}\lambda / \int_0^T (C_a - C_{\bar{v}})\, dt$$ [2]

Quite naturally, the equation is equally applicable to the desaturation phase. For the sake of simplicity, we consider desaturation only.

The Kety–Schmidt procedure gives valid estimates of CBF only if certain requirements are fulfilled (see ref. 22). The most important of these are: (1) CBF must remain constant during the measurement; (2) all parts of the tissue must be cleared from tracer after time T (i.e., the time chosen for sampling of arterial and venous blood); and (3) the cerebral venous blood sampled should not be contaminated with blood from extracerebral sources. In this context, it should be recalled that the method is valid even if only a part of the cerebral blood is drained through the vein used for sampling.

Let us illustrate the problems by considering a situation in which a subject (or animal) has been allowed to inhale an inert tracer (e.g., [133]xenon) until the brain appears to be saturated with tracer. When tracer supply is interrupted, the arterial and cerebral venous "washout" curves may be those depicted in Fig. 1 (a and b). In case *a*, problems in calculating CBF arise, since there is a lingering arteriovenous difference in tracer activity. This can be caused by two things: brain tissues contain poorly perfused masses that are only slowly de-

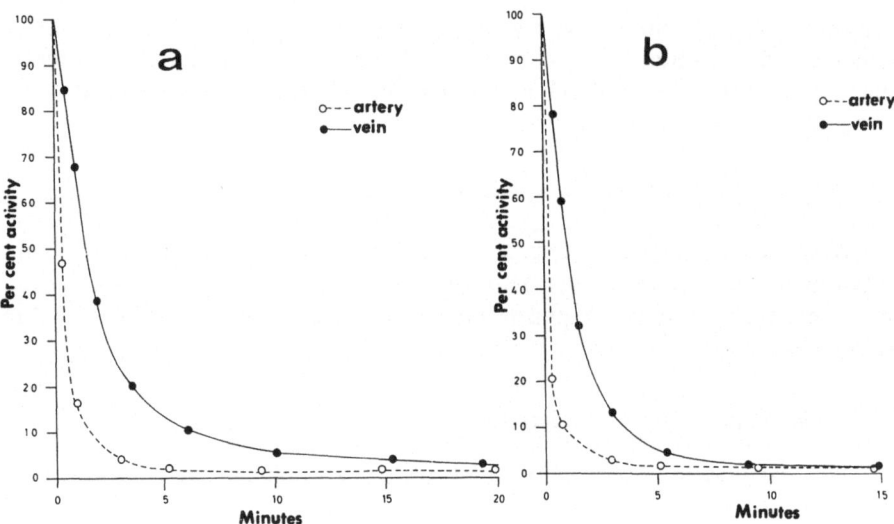

Fig. 1. Arterial (open circles) and cerebrovenous (filled circles) ^{133}Xe activities during desaturation, used for calculating CBF according to the Kety–Schmidt principle. The diagram, which is slightly modified after Nilsson and Siesjö,[47] illustrates one difficulty in calculating CBF. Thus, when a lingering arteriovenous difference in ^{133}Xe activity is encountered (case a), this may either be because of the presence of slowly perfused tissue masses or because of extracerebral contamination of venous blood. These errors can only be excluded if the arteriovenous difference approaches zero at a time when all parts of the brain have been desaturated (case b).

saturated; alternatively, slowly perfused (and thereby slowly cleared) extracerebral tissues deliver blood to the venous sampling site. With the first of these alternatives, calculated CBF will overestimate true CBF[37]; with the second, CBF will be underestimated.

Curves of the *a* type are regularly obtained in man when venous blood is sampled from the jugular bulb during a 10-min saturation or desaturation period,[2,38] and the delay in equilibration is exaggerated if flow rates are reduced.[39] Working on the assumption that the lack of equilibration within 10 min is caused by the presence within the brain of slowly perfused tissue masses, some workers sampled blood for somewhat longer periods and extrapolated the arteriovenous difference in tracer activity to infinity.[40,41] When this is done, calculated CBF and $CMRO_2$ are 10–20% lower than those reported by Kety and Schmidt.[2] It should be emphasized, though, that such a procedure cannot be adopted if the lack of equilibration is caused by extracerebral contamination of venous blood. Although such contamination is small under normal conditions in man,[42] it may increase in pathological states. There is some uncertainty, therefore, regarding absolute CBF values in man. We can only conclude that mean CBF in the human brain is somewhere between 0.45 and 0.65 ml·g^{-1}·min^{-1}, and $CMRO_2$ between 1.35 and 1.65 μmol·g^{-1}·min^{-1}.

When the Kety–Schmidt procedure is applied to rats with sampling of venous blood from the superior sagittal sinus, the curves obtained are those of Fig. 1b, and the arteriovenous difference in tracer activity approaches zero at a time when all parts of the tissue are desaturated.[43,44] Probably, the CBF and $CMRO_2$ values obtained pertain to cerebral cortical tissue.[34,45] In the rat, sampling of blood from the retroglenoid vein gives information on whole–brain CBF and $CMRO_2$[46–48] (see also ref. 49). Such experiments have given the interesting information that small animals have higher CBF and $CMRO_2$ values than larger ones or man (see refs. 22,47).

The advantages of the Kety–Schmidt technique are that the method is based on the Fick principle, and that it yields quantitative values for $CMRO_2$ (and CMR_{gl}). The method is often used, therefore, to validate other techniques. However, the method is exacting, and it gives only mean values for the whole brain or the cerebral cortex. For many applications, therefore, it is inferior to those giving information on regional or local flow rates.

2.2.2. Clearance Techniques for Measurements of Regional CBF

Starting from equation 1, Kety[3] derived an equation describing the clearance of an inert substance from a homogeneous tissue. If diffusion equilibration between blood and tissue is instantaneous, we obtain an equation expressing tissue activity at any given time $[C_i(T)]$ as a function of the initial activity (C_i) and blood flow (K_i)

$$C_i(T) = C_i \cdot \exp[-K_i(T)] \qquad [3]$$

The equation applies if the tissue can be quickly "saturated" and the tracer is effectively cleared when blood passes the lungs. A practically useful method was worked out by Scandinavian workers[4,5] who injected radioactive tracer into the internal carotid artery and monitored clearance from the tissue by external detectors. Originally, β-emitting isotopes were employed, but the subsequent use of γ-emitting ones (mainly ^{133}Xe) allowed the method to be used in man.[50–53] Furthermore, when an increasing number of (collimated) detectors were employed, regional measurements became feasible. However, problems of interpretation arose from the fact that clearance was not monoexponential. This was solved by exponential stripping of the curves in two main components, a fast one and a slow one, corresponding to CBF values of about 0.8 and 0.2 $ml \cdot g^{-1} \cdot min^{-1}$, respectively. These were for a while assumed to mainly reflect blood flow in gray and white matter, respectively. However, with the continued development of autoradiographic techniques, it became clear that a spectrum of blood flow rates exists in the brain (see below). Furthermore, the fast and slow flow rates thus calculated do not correlate well with direct estimates of blood flow in gray and white matter.[54]

In clinical material, it is now customary to calculate CBF from the initial slope of the clearance curves,[55] a procedure that gives a practical index of CBF changes but not values that can be referred to any given anatomic structure in the brain. From a quantitative point of view, another application of the clearance technique is of greater interest. The derivation is according to Zierler[56] (see also refs. 53,57) who showed that the mean transit time $(\bar{t})$ of an inert substance through a tissue is related to total flow (F) via the equation

$$\bar{t} = V/F \qquad [4]$$

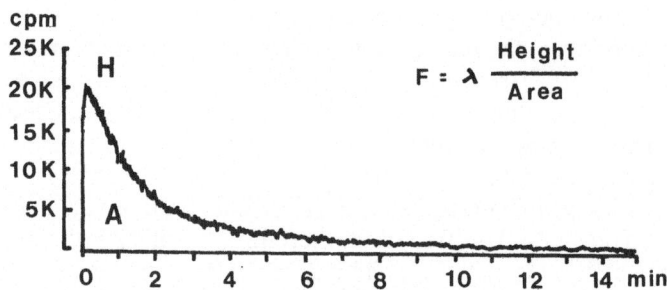

Fig. 2.   Clearance of a radioactive tracer from the brain as measured by external detectors. If a "slug" injection of the tracer is given into one carotid artery, mean CBF can be calculated from the initial height of the curve (H), the area under the curve (A), and the tissue/blood partition coefficient for the tracer.

where V is the volume of distribution. It can be shown that if a "slug" injection of tracer is given, CBF can be calculated from the simple equation

$$CBF = \lambda \cdot (\text{height})/(\text{area}) \qquad [5]$$

i.e., CBF is a function of the initial tracer concentration divided by the area under the clearance curve (Fig. 2).

The importance of the Zierler[56] derivation is that it follows the Fick principle (law of conservation of matter) and makes no predictions about speed of equilibration of tracer or homogeneity of tissue flow rates. It is similar in principal, therefore, to the Kety–Schmidt technique, but, unless $CMRO_2$ is measured as well, it requires no cannulation or puncture of cerebral veins. The method has been much used in man, and its careful adaptation to experiments in baboons has given much useful information on cerebrovascular physiology[58–60] (see below). It should be emphasized, though, that at high flow rates, the method gives lower CBF values than those derived by microsphere techniques.[54]

During recent years, the clearance technique has been modified so as to allow administration of tracer (^{133}Xe) by inhalation[61] or by i.v. infusion.[62] However, although these modifications have greatly facilitated CBF measurements in man, recirculation of tracer and extracerebral contamination reduce the application of the technique for measurements of cerebral metabolic rates.[20] Furthermore, the quantitative significance of compartmental ^{133}Xe data has been questioned.[54]

Invasive techniques based on similar principles have allowed measurements of regional CBF in animal experiments. Undoubtedly, the most useful of these is that based on measurements of hydrogen clearance by electrodes inserted in the tissue.[63,64] For example, this method has been successfully used to determine critical levels of regional CBF in baboons following clipping of the middle cerebral artery.[65,66]

Hydrogen clearance techniques have also been developed in which the hydrogen gas is not inhaled but is generated by the electrode assembly itself.[67]

When such electrodes are placed on the surface of the brain, and electrodes of small diameters are used, it is possible to measure "microflow" in circumscribed surface areas. However, the problem then becomes similar to those encountered when sophisticated techniques are used to measure pial arterial diameter.[68,69] Thus, it becomes a matter of speculation if pial arterial flow (or diameter) reflects CBF changes in the tissue proper.

Hydrogen clearance techniques have the advantage of allowing repeated measurements in the same animal, and the animal does not have to be sacrificed once a series of measurements has been performed. The chief disadvantages are that CBF can be measured in only a limited number of structures and that some tissue damage could occur. Furthermore, several potential errors have been recognized (see ref. 64).

2.2.3. Methods Based on Tissue Uptake of CBF Markers

Two such methods are at present widely used. One of these, which is based on the indicator fractionation principle,[70] employs the intracardiac injection of labeled microspheres which are trapped in the tissues in proportion to tissue blood flow rates, whereas the other involves measurement of the tissue uptake of freely diffusible radioactive tracer substances.

The microsphere technique[71-73] is based on the principle that if a known amount of radioactively labeled microspheres is injected into the left atrium, the amount trapped in any given tissue is proportional to the fraction of cardiac output that perfuses that tissue. If cardiac output is known, tissue blood flow can be calculated. Since the microspheres can be labeled by as much as five different γ-emitting isotopes, sequential blood flow rates can be determined. When the last injection has been given, the animal is killed, and tissue samples are dissected for γ counting together with a series of arterial samples or an "integrated" reference blood sample. If an integrated reference sample is used, CBF (in $ml \cdot g^{-1} \cdot min^{-1}$) can be calculated from the formula

$$CBF = C_B \times RBF/C_R \qquad [6]$$

where C_B is counts per gram of tissue, RBF is the rate of withdrawal of the reference sample, and C_R is the total counts in the reference sample.[72,73]

The microsphere technique has been adapted to several species. It has the virtue of allowing repeated measurements of regional blood flow. However, care must be taken in achieving adequate mixing of the microsphere suspension, and such phenomena as plasma skimming and axial streaming can distort the results; furthermore, the results are not quantitative if the spheres are not trapped in the tissue, and obstruction of capillary flow could create some degree of ischemia, especially with repeated injections (see refs. 23,30). In order to circumvent the latter problem, the number of spheres injected must be restricted. Since the tissue counts then become low, the spatial resolution of the technique is limited. As a result, the technique is less suited for use in small animals like the rat. However, workers who have used 15-μm spheres and who have carefully applied the technique to larger animals have reported reproducible measurements of regional brain blood flow that are probably quantitatively valid.[4,72] Such measurements have also demonstrated that virtually no shunting of such spheres occurs.[54]

The dominant tissue uptake technique in small animals is that based on autoradiographic measurements of freely diffusible radioactive tracers. Kety[3] rewrote equation 1 for regional CBF as

$$dC_i/dt = F_i (C_a - C_v)/V_i \qquad [7]$$

in which the subscript i denotes regional values. If one assumes that the tracer comes into instantaneous diffusion equilibrium between blood and tissue, C_v can be replaced by the ratio C_i/λ. Furthermore, if F is homogeneous, the resulting equation can be integrated and solved for C_i at any time (T):

$$C_i(T) = \lambda \cdot K_i \int_0^T C_a \cdot \exp[-K_i(T)] \, dt \qquad [8]$$

In other words, regional flow (which is proportional to K_i) can be derived from C_i/λ and from the integral of the arterial tracer activity during time T. In the original application,[6] a radioactive tracer (trifluoro[^{131}I]iodomethane) was infused i.v. during 60 sec, repeated arterial samples were drawn, and the animal was then decapitated for the subsequent measurements of C_i by means of quantitative autoradiography. A more practical tracer was later employed ([^{14}C]antipyrine, see ref. 74), but since this proved diffusion limited,[75,76] most subsequent measurements have been carried out with [^{14}C]iodoantipyrine.[77]

The unique importance of the autoradiographic technique is that it allows measurements of CBF with a resolution that is only limited by that of ^{14}C autoradiography (about 0.2 mm). It is possible, therefore, to measure CBF in a great many structures of the rat brain (Table I). The method can be used in unanesthetized animals that are only partly restrained[77] or virtually unrestrained.[78] With some precautions (sampling of arterial blood from a short brachial catheter, infusion of isotope within 20 sec), even very high flow rates can be reproducibly assessed. Since the deoxyglucose technique of Sokoloff et al.[8] utilizes similar autoradiographic measurements, it is possible to correlate blood flow and metabolic rate in the same structures. Recent modifications of these techniques allow such correlations to be made in one and the same animal.[79]

Three drawbacks with the autoradiographic techniques can be listed. First, repeated measurements of CBF in the same animal are not feasible. Second, any tracer used must show some degree of diffusion limitation, and, at high flow rates, CBF is probably underestimated. Third, because of isotope costs, the method is less suitable for large animals.

For certain experimental situations, it has proved advantageous to use the principle of indicator fractionation to assess the rate of uptake of diffusible tracers in the brain, thereby allowing calculation of CBF. Thus, if an artificial organ is created by withdrawing blood into a syringe at a constant rate, CBF can be derived even if the i.v. infusion of diffusible indicator is as short as 5

Table I
Local CBF[a] in Unparalyzed, Air-Breathing Rats

Structure	Sakurada et al.[77]	Dahlgren et al.[217]
Frontal cortex	1.39 ± 0.09	1.20 ± 0.05
Striatum	1.37 ± 0.07	1.38 ± 0.04
Substantia nigra	1.17 ± 0.07	1.11 ± 0.05
Cerebellum	1.00 ± 0.08	1.44 ± 0.07
Sensory–motor cortex	1.99 ± 0.22	1.46 ± 0.06
Parietal cortex	1.69 ± 0.13	1.52 ± 0.06
Visual cortex	1.50 ± 0.12	1.26 ± 0.04
Lateral geniculate body	1.39 ± 0.10	1.44 ± 0.07
Superior colliculus	1.34 ± 0.14	1.75 ± 0.05
Auditory cortex	2.05 ± 0.10	2.18 ± 0.08
Medial geniculate body	2.02 ± 0.15	1.93 ± 0.05
Inferior colliculus	2.46 ± 0.21	3.16 ± 0.07
Superior olive	1.73 ± 0.12	2.31 ± 0.10
Hippocampus	1.00 ± 0.09	1.00 ± 0.03
Amygdala	0.88 ± 0.05	1.15 ± 0.09
Septal nuclei	0.86 ± 0.04	1.27 ± 0.07
Hypothalamus	0.86 ± 0.04	1.27 ± 0.06

[a] Values are expressed in $ml \cdot g^{-1} \cdot min^{-1}$ as means ± S.E.

sec.[80] Even more importantly, the method can be used to measure simultaneously the undirectional flux of substances across the blood–brain barrier.[81,82] This makes it possible to correct for any diffusion limitation of the CBF marker at high flow rates.

2.2.4. Techniques for Continuous Measurements of CBF

In some experimental settings, it is of considerable interest to obtain information on rapid changes in CBF. These are most easily assessed by continuous venous outflow techniques. Such methods were developed by cannulation of the superior sagittal sinus in the cat[32] and of the confluens sinum in the dog.[33] However, the most widely used technique is that developed by Theye and Michenfelder[34] who cannulated the superior sagittal sinus in the dog after elimination of ethmoidal anastomoses. By weighing the tissue supplying blood to the outflow catheter, the authors could express their flow rates per unit weight of tissue, and a comparison with the Kety–Schmidt technique suggested that calculated CBF and $CMRO_2$ values were quantitatively valid.[35] The method has been extensively used to delineate changes in CBF and $CMRO_2$ caused by a variety of sedatives and anesthetics (see Table 32 in ref. 22).

Recently, a venous outflow technique has been worked out for the rat in which the rate of flow through one retroglenoid vein is assessed during compression of the contralateral vein.[46,47] By measurements of arteriovenous differences in O_2 content, deductions can be made of any extracerebral contamination. The method is only semiquantitative. However, as Fig. 3 shows, it allows assessment of second-to-second changes in CBF.

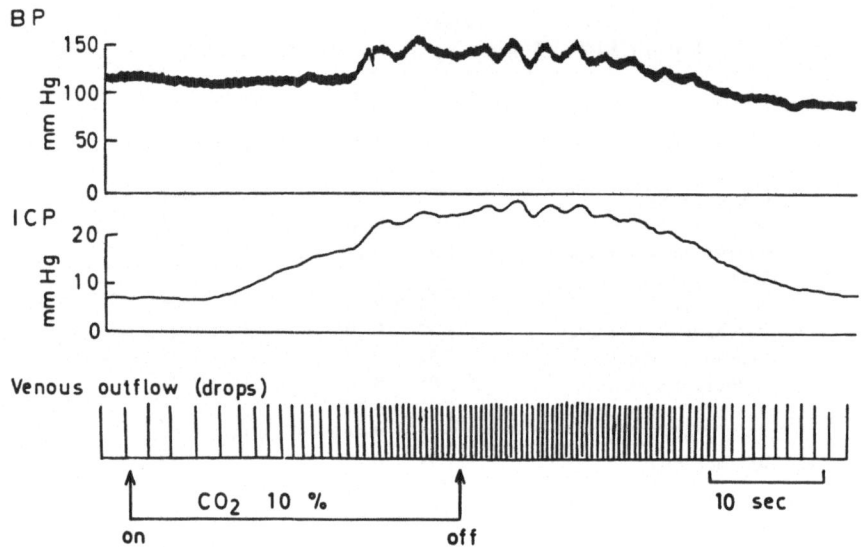

Fig. 3. Illustration of time resolution of changes in CBF as measured by a continuous venous outflow technique in the rat. The on/off marks denote the period during which 10% CO_2 was added to the gas mixture delivered to the respirator. Since the respirator gives a 5 to 10-sec delay in altering arterial P_{CO_2}, the CBF response is virtually instantaneous. BP, mean arterial blood pressure; ICP, intracranial CSF pressure. Reproduced with permission from Nilsson *et al.*[182]

2.2.5. The ^{15}O Technique of Ter-Pogossian

When Ter-Pogossian and his collaborators[9,10] developed the ^{15}O technique, it represented the only method for measuring regional CBF and $CMRO_2$ in man and experimental animals. The method is based on external detection of ^{15}O washout following the intracarotid injection of ^{15}O-labeled water to measure CBF (height-over-area method) and ^{15}O-labeled erythrocytes to measure $CMRO_2$. When $CMRO_2$ is derived, the fractional oxygen extraction is first calculated from the time–activity curve produced by the injection of [^{15}O]oxyhemoglobin. The $CMRO_2$ is then calculated as the product of fractional oxygen utilization times arterial oxygen content times CBF.

Values obtained for CBF and $CMRO_2$ have been found to correlate favorably with these obtained with the Kety–Schmidt technique.[11] The method has given much useful information on regional CBF and $CMRO_2$, but it suffers from some drawbacks. First, it requires expensive equipment including a cyclotron to produce the isotope. Second, since the tissue is viewed by external detectors, the tissue structures studied are less well defined. The method has great potential value, though, especially when it is combined with equipment yielding results in three dimensions (see below).

2.2.6. Methods Based on Positron Emission Tomography

With the advent of refined techniques for positron emission tomography,[83] it became possible to adopt the [^{14}C]deoxyglucose technique of Sokoloff *et al.*[8]

to measurements of l-CMR$_{gl}$ in man.[12,13,84] Furthermore, steady-state ^{15}O techniques were soon developed allowing measurements of local CMRO$_2$ as well.[14] Finally, application of CBF methods originally developed for animal experiments to positron emission tomography has yielded useful techniques for measurements of l-CBF as well[14,85,86] (for further literature, see ref. 24, pp. S1–S59). As a result of this rapid technocological development, it is now possible to obtain three-dimensional measurements of metabolic rate and blood flow in the human brain in health and disease.

3. PRINCIPAL FEATURES OF CEREBRAL BLOOD FLOW PHYSIOLOGY

We begin by summarizing the principal features of CBF regulation. Figure 4 illustrates four important relationships.

3.1. Autoregulation

The cerebral circulation shows what is denoted as autoregulation. Thus, when the cerebral perfusion pressure is varied over a relatively large range (about 60–160 mm Hg), CBF remains constant. At the lower limit of the autoregulatory range, a further reduction in perfusion pressure leads to a fall in

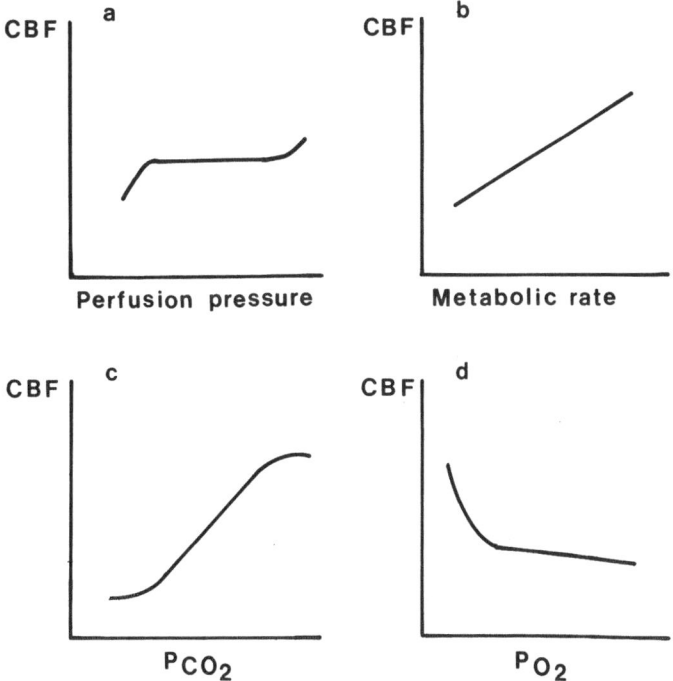

Fig. 4. Schematic diagram illustrating the relationship between, on one hand, cerebral perfusion pressure (a), metabolic rate (b), Pco$_2$ (c), and Po$_2$ (d) and, on the other hand, CBF.

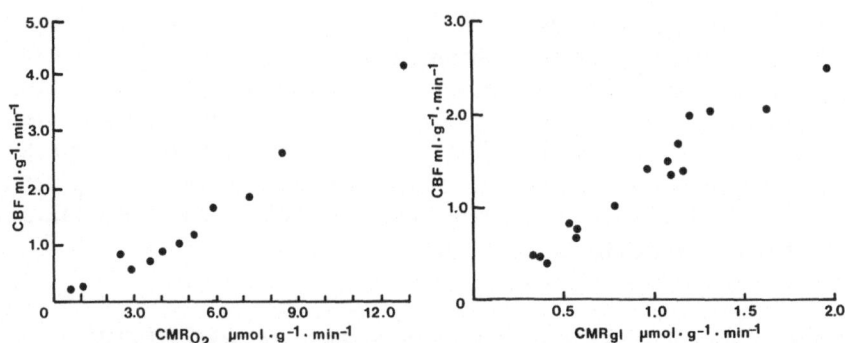

Fig. 5. Relationship between metabolic rate and CBF. The left panel shows the relationship between $CMRO_2$ and CBF. The points represent mean values for groups of animals in which $CMRO_2$ was either reduced by anesthesia or hypothermia or elevated by epileptic seizures. For further details, see Nilsson et al.[182] The right panel shows the relationship between local CMR_{gl} and local CBF. The data were obtained from Sokoloff et al.[8] and Sakurada et al.,[77] respectively.

CBF,[16,33,87–90] and if the pressure acutely increases to above about 160 mm Hg, CBF rises ("breakthrough of autoregulation," see refs. 91,92). Such a "breakthrough" is associated with an increased permeability of the blood–brain barrier to protein-bound tracers.[93–95]

Three additional features of the relationship between cerebral perfusion pressure and CBF should be recalled. First, CBF is maintained at a somewhat lower perfusion pressure when the latter is caused by a raised intracranial pressure than by a reduced blood pressure.[96,97] Second, with the definition of perfusion pressure used, CBF does not cease altogether when the perfusion pressure is zero, since the pulse pressure may propel blood through the tissue. Third, a number of adverse conditions impair or abolish autoregulation (see below). Failure of autoregulation in normal animals is often a sign of maltreatment of the tissue or the cerebral vasculature.

3.2. Coupling of Metabolic Rate and Blood Flow

Usually, the perfusion of a tissue is tightly correlated to its metabolic rate. That this principle applies to the brain is amply illustrated by Fig. 5 which relates $CMRO_2$ in rats under a variety of conditions to the corresponding CBF values as well as CMR_{gl} in a variety of structures of the same brain to the corresponding local CBF values. These results illustrate that reversible alterations in metabolic rate of the rat brain are accompanied by blood flow rates ranging from about 0.15 to about 7 $ml \cdot g^{-1} \cdot min^{-1}$ and that in the normal brain, local perfusion rates vary between about 0.1 (white matter) and 2.5 (inferior colliculus) $ml \cdot g^{-1} \cdot min^{-1}$. Quite naturally, the tight relationship between metabolic rate and blood flow raises the question about the coupling factors involved. These are discussed below.

3.3. Carbon Dioxide Responsiveness of the Cerebral Circulation

One of the most conspicuous features of the cerebral circulation is its extreme sensitivity to changes in arterial (and tissue) CO_2 tensions. In several species, e.g., the baboon and the rat, the relationship between P_{CO_2} and CBF forms an S-shaped curve (see below) and is relatively linear around normal carbon dioxide tensions (where CBF changes with 4–6% per mm Hg change in arterial P_{CO_2}; see refs. 59,82,98,99). At CO_2 tensions of 60–80 mm Hg, overall CBF in the baboon reaches maximal values of about 1.5 $ml \cdot g^{-1} \cdot min^{-1}$ (control value about 0.6 $ml \cdot g^{-1} \cdot min^{-1}$); in the rat, blood flow in cerebral cortical structures may exceed 5 $ml \cdot g^{-1} \cdot min^{-1}$ (control values about 1.5 $ml \cdot g^{-1} \cdot min^{-1}$). Since autoregulation is lost during hypercapnia, the CBF values attained are directly related to the existing perfusion pressure.

As is discussed in greater detail below, the CO_2 responsiveness is modified by a number of factors such as the type and depth of anesthesia and by drugs. When P_{CO_2} is decreased, CBF is never reduced below about 40% of control. In all probability, this is because at very low CO_2 tensions, CBF becomes limiting for cerebral oxygen supply. In other words, in this situation, the vasoconstrictor influence of hypocapnia is counteracted by factors related to impending tissue hypoxia. Results on changes in $CMRO_2$ during hypercapnia are controversial (see ref. 100). However, all data agree in showing that CBF changes out of proportion to any change in $CMRO_2$.

3.4. Influence of Hypoxia (and Hypoglycemia) on CBF

A relatively steep increase in CBF occurs at arterial P_{O_2} values below 50 mm Hg.[101–103] However, this represents an exaggerated response rather than a step function, since an increase in arterial P_{O_2} above normal reduces CBF; for example, in hyperbaric oxygenation, CBF may fall by as much as 25%.[101,104] Animal experiments have also shown that CBF changes inversely with P_{O_2} in the physiological P_{O_2} range.[105]

Increases in CBF also occur in anemic hypoxia, i.e., in situations in which P_{O_2} is normal but the actual or effective hemoglobin concentration is reduced.[106–109] The increases observed are quantitatively similar to those recorded in hypoxic hypoxia. For example, MacMillan[110] found that exposure of rats to 1.5% CO increased cerebral cortical flow rates about threefold. In such instances, the hyperemia is clearly related to tissue hypoxia. However, in normovolemic anemia, a reduced blood viscosity undoubtedly contributes to the increased blood flow rates.[108,111,112]

The increase in cerebral blood flow in hypoxia represents a purposeful homeostatic mechanism for securing an adequate oxygen supply to the tissue. It is of interest that a similar homeostatic mechanism seems to operate in hypoglycemia.[113,114] Thus, whenever the tissue is threatened by oxygen or substrate lack, a compensatory increase in CBF occurs.

4. MECHANISMS REGULATING CEREBRAL BLOOD FLOW

We proceed to discuss mechanisms that regulate CBF in health and disease and consider in turn neurogenic influences, circulating monoamines, metabolic feedback signals, and prostaglandins. Whenever possible, the question is raised whether the mechanisms discussed qualify as general coupling factors adjusting

blood flow to metabolic rate or whether they enter as modulators of the relationships illustrated in Fig. 5 (see above).

4.1. Extrinsic Innervation of Cerebral Vessels

Cerebral vessels are innervated by sympathetic fibers originating in the
superior cervical ganglion and by parasympathetic fibers purported to be contained in the greater superficial petrosal nerve (for review of the older literature,
see refs. 16,17,25). Many investigators attempted to delineate the physiological
role of this innervation, but the results were largely negative.[87] Thus, section
or stimulation of the cervical sympathetic chain induced no or only small
changes in CBF (of the order 5–15%), and manipulation of the parasympathetic
innervation gave equally modest changes in flow.

An attractive explanation of the negative findings reported was proposed
by Harper *et al.*[115] According to the series resistance hypothesis proposed by
these authors, sympathetic stimulation constricts innervated arteries, but CBF
remains largely unchanged because of dilatation of more distal resistance ar-

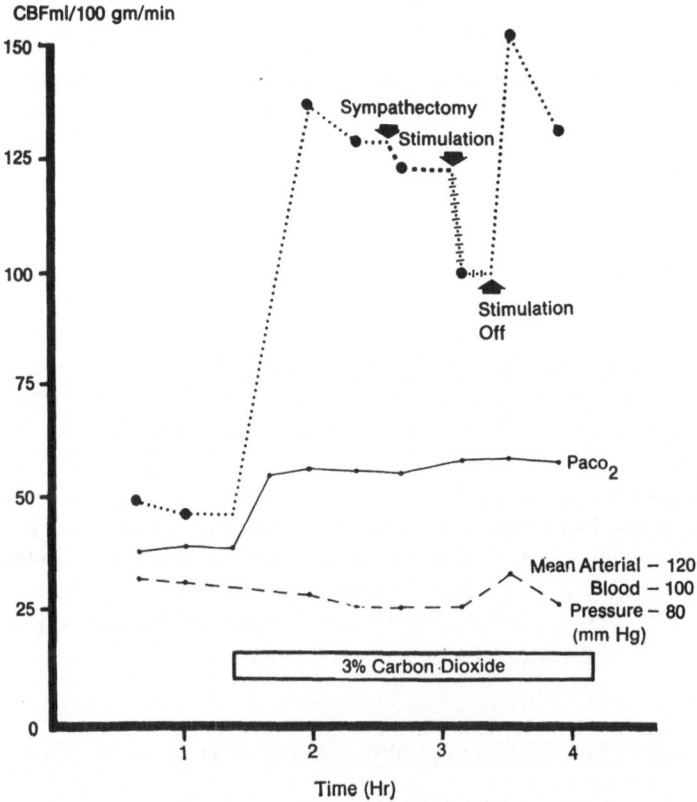

Fig. 6.   Influence of stimulation of the cervical sympathetic chain on CBF during hypercapnia in
a baboon maintained artificially ventilated on phencyclidine–N$_2$O. Reproduced with permission
from Harper *et al.*[115]

teries purported to be under metabolic control. The hypothesis predicts that if the latter vessels are maximially dilated, sympathetic stimulation or systemic administration of sympathomimetic drugs will reduce CBF.

As seen in Fig. 6, this was also shown to be the case in hypercapnia.[115] Subsequent studies showed a smaller or no effect of sympathetic stimulation during hypercapnia (see ref. 31). Probably, the differences in results result from the species used and the anesthesia employed. For example, the negative results reported by the Iowa City group were obtained on animals anesthetized with chloralose–urethane, an anesthetic mixture that seems to reduce CBF by about 50% (see differences in CBF values between awake and anesthetized animals at similar P_{CO_2} values in references 116,117). Subsequent results from the same group also showed that during bicuculline-induced seizures, a condition associated with increased cervical sympathetic nerve activity, unilateral sympathectomy gave rise to increased CBF values on the ipsilateral side.[118]

Subsequent studies have greatly improved our knowledge of the physiology and pharmacology of the neurogenic control of cerebral vessels (for reviews, see refs. 30,31,119). First, studies with modern histochemical fluorescence techniques have shown a dense noradrenergic innervation of the cerebral vasculature also involving intraparenchymal vessels.[30,120] Second, it has been clearly demonstrated that cerebral vessels are provided with α-receptors, and that they respond to α-adrenoceptor agonists with vasoconstriction both *in vitro* and *in vivo*.[30,119] although cerebral vessels are less sensitive and less discriminating in their response to α-receptor agonists.[31] Many studies have documented that this innervation modulates the autoregulatory ability of the cerebral circulation, notably at the lower and upper limits of the autoregulatory range (Fig. 7).

Working on baboons, Fitch *et al.*[121] noted that CBF fell when the cerebral perfusion pressure was reduced (by bleeding) to about 65 mm Hg. However, acute sympathectomy or administration of an α-adrenoceptor blocker (phenoxybenzamine) extended the autoregulatory range down to a pressure of about 35 mm Hg, indicating that sympathetic discharge during hemorrhage contributes to the reduction in CBF. We note that this result conforms to the "series resistance" hypothesis. Nonconfirmatory results were obtained in dogs under chloralose–urethane anesthesia,[118] but a subsequent study by the same authors on cats anesthetized with methohexital–chloralose–urethane showed that sympathetic discharge during hemorrhagic hypotension did somewhat reduce CBF.[122] Thus, differences in results may be related to species and type of anesthesia.

The original publication demonstrating that the sympathetic nervous system affects the upper limit of autoregulation was published by Bill and Linder[123] who noted that sympathetic stimulation minimized or prevented the increase in CBF following an acute rise in blood pressure. Subsequent studies have confirmed this finding and demonstrated that sympathetic stimulation also prevents or minimizes extravasation of protein-bound tracers.[122,124–126] Thus, an increased sympathetic tone may serve to prevent a potentially harmful "breakthrough of autoregulation" in stressful situations.

Less is known about the physiological role of the parasympathetic inner-

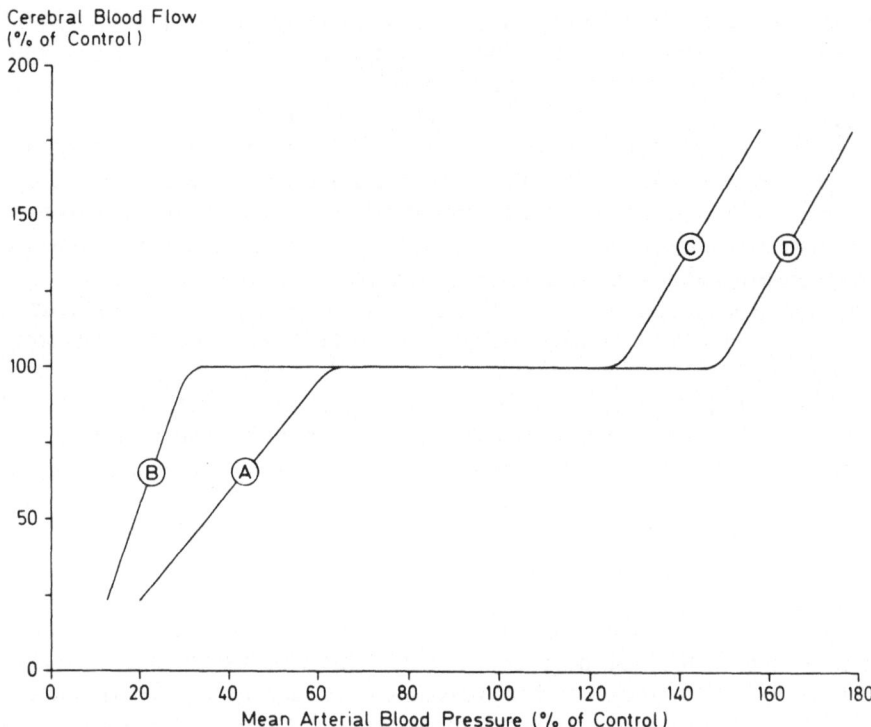

Fig. 7.   Diagram illustrating autoregulation of CBF and the modifying influence of sympathetic nervous activity. Curves A and C denote lower and upper limits of autoregulation, respectively, under control conditions. Curve B illustrates extension of autoregulation to lower pressures when the sympathetic activity is depressed with either α-adrenergic blockade or cervical sympathectomy. Curve D illustrates extension of upper limit of autoregulation to higher pressures with stimulation of the cervical sympathetic chain. Reproduced with permission from Edvinsson and MacKenzie.[30]

vation.[30,119] Thus, although it seems established that a muscarinic innervation of vessels mediates vasodilatation[127] and that a corresponding nicotinic innervation may modulate norepinephrine release at vascular nerve terminals via axoaxonal interaction, neither section nor stimulation of the parasympathetic innervation nor administration of parasympathomimetics or muscarinic antagonists have unequivocal effects on CBF under control or pathophysiological situations.[128] Evidently, it will remain for future research to unravel the physiological role of the parasympathetic innervation.

In this context, it should be recalled that James *et al.*[129] produced evidence to show that the autoregulatory ability of the cerebral circulation was controlled from carotid sinus baroreceptors via a reflex arc whose efferent components were the cervical sympathetic chain and the seventh cranial nerve. Furthermore, a later study from that group purported to show that the response of the cerebral circulation to hypoxia (and hypercapnia) was at least in part mediated by reflexes from the carotid artery chemoreceptors.[130] Several subsequent studies have failed to confirm these studies, however (see refs. 131–133), and it seems unlikely that baroreceptor or chemoreceptor reflexes play any significant role in the control of cerebral circulation.

In summary, although cerebral vessels receive a fairly dense innervation with α-adrenergic vasoconstrictory and with muscarinic vasodilatory fibers, this innervation neither contributes significantly to resting cerebrovascular tone nor does it modify CBF changes in all but some extreme situations (hypercapnia, epileptic seizures, hemorrhagic hypotension, acute blood pressure rises).

4.2. Circulating Monoamines

In 1950, Kety proposed that grave apprehension and anxiety may increase CBF and $CMRO_2$. A subsequent study[134] showed that large i.v. doses of epinephrine increased CBF and $CMRO_2$ by about 20%, whereas norepinephrine in doses that induced similar increase in blood pressure had no such effects. Several subsequent studies failed to confirm that catecholamines influence CBF and $CMRO_2$. Rather, these results suggested that norepinephrine induced an expected (but small) decrease in CBF (for literature, see refs. 22,30,119). It was also argued that the properties of the blood–brain barrier would prevent circulating catecholamines from acting on intracerebral receptors. Thus, the barrier permeability to catecholamines is low,[94,135–137] and even if some net transfer occurs, the vascular endothelium contains enzymes that degrade monoamines by oxidation and methylation.[138,139]

Results from two laboratories provide an explanation for the discrepant results mentioned and define the cerebral circulatory and metabolic effects of catechol- and indoleamines. MacKenzie *et al.*[59,60] circumvented the barrier by injecting norepinephrine intraventricularly, by releasing it from cerebral storage sites by means of reserpine, and by infusing the amine into the carotid artery following osmotic opening of the barrier. In all instances, CBF rose, and, whenever measured, so did $CMRO_2$ (Fig. 8). Subsequent results from the same laboratory showed that intracarotid infusion of a dopamine agonist had similar effects,[140] whereas the effects of 5-HT were the opposite: CBF and $CMRO_2$ fell.[141]

Induction of "immobilization stress" (withdrawal of N_2O supply in ventilated rats that were given local anesthesia and protected from external stimuli) was shown by our own laboratory to virtually double $CMRO_2$ with a two- to threefold increase in CBF.[142,143] Since these effects were prevented by prior adrenalectomy or by administration of propranolol, it seemed likely that they were caused by circulating catecholamines acting on β-adrenoceptors in the brain. Subsequent results also showed that equally large increases in the CBF/$CMRO_2$ couple could be provoked by i.v. epinephrine in a dose of 8 $\mu g \cdot kg^{-1} \cdot min^{-1}$, the effects of norepinephrine being somewhat less marked[144] (see also ref. 145). Furthermore, it could be shown that epinephrine was without effect on CBF and $CMRO_2$ if the blood pressure was kept constant during infusion and that an increase in pressure *per se* (i.e., in the absence of epinephrine infusion) gave equally negative effects.[146] Thus, the results hint that

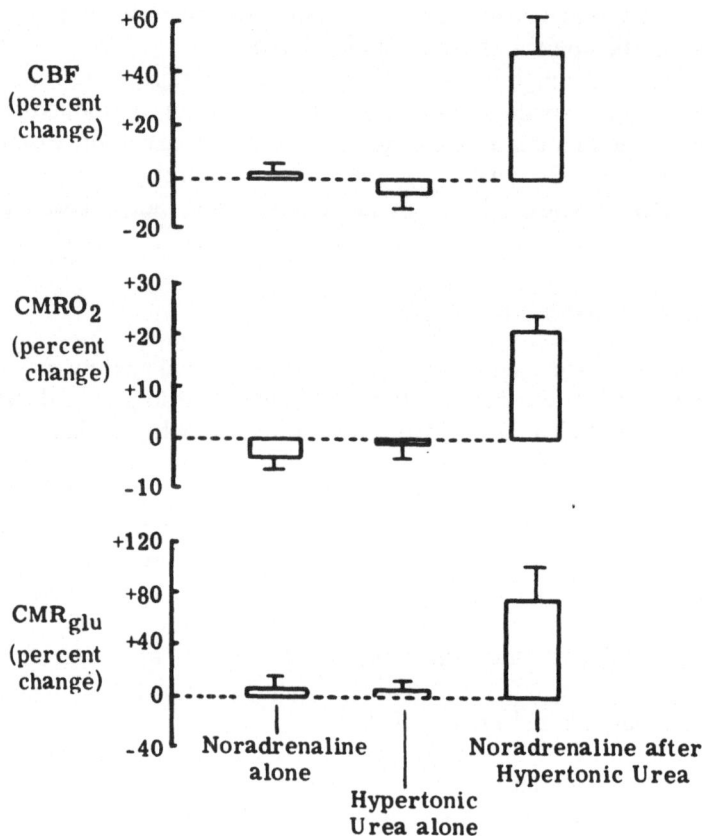

Fig. 8. Influence of intracarotid administration of norepinephrine (50 ng·kg^{-1}·min^{-1}) with or without disruption of the blood–brain barrier by hypertonic urea on CBF, CMRO$_2$, and CMR$_{gl}$ in ventilated baboons maintained on phencyclidine–N$_2$O. The bars denote percent changes from baseline values (means ± S.E.M.). Reproduced with permission from MacKenzie *et al.*[57]

catecholamines exert their dramatic effects on CBF and CMRO$_2$ only when the associated increase in blood pressure allows the amines to pass the blood–brain barrier.

4.3. Intrinsic Innervation

The results quoted demonstrate that exogenous or endogenous catecholamines exert effects on CBF that are different from these resulting from norepinephrine release at vascular nerve terminals. In all instances in which monoamines were found to influence CBF, CMRO$_2$ (or CMRl$_{gl}$) changed simultaneously. It is tempting to conclude, therefore, that the CBF changes recorded were secondary to alterations in metabolic rate. Whatever the primary change, it seems justified to discuss the possibility that the intrinsic monoaminergic systems influence CBF.

The diffuse noradrenergic innervation of the forebrain, brainstem, and cerebellum mainly arises from the nucleus locus coeruleus.[147,148] It is generally accepted that the activity of this system is enhanced in stressful situations and that stimulation of the locus coeruleus inhibits the firing rate of neurons in, e.g., the cerebellum and hippocampus.[149,150] There is now evidence for a central noradrenergic innervation of cerebral microvessels as well.[151-153] Results presented a few years ago hinted that noradrenergic fibers from the nucleus locus coeruleus exert a resting, vasoconstrictory tone on cerebral vessels. Thus, Raichle *et al.*[154] reported that stimulation of the locus coeruleus in monkeys could reduce CBF and that intraventricular administration of an α-adrenoceptor blocker (phentolamine) increased CBF. Similar results were reported in cats by Bates *et al.*[155] who observed that bilateral destruction of the locus coeruleus increased resting CBF and attenuated the CBF response to hypercapnia.

Subsequent results have cast doubts on the conclusions drawn from these studies. For example, recent data demonstrate that cerebral microvessels contain adrenoceptors of the β rather than the α type.[156-158] Furthermore, two sets of data have failed to confirm that the intrinsic noradrenergic system modulates resting cerebrovascular tone or the response to hypercapnia. First, neither electrolytic lesions of the locus coeruleus system nor 6-hydroxydopamine lesions of the ascending noradrenergic bundle in the rat influence resting or hypercapnic vessel tone (Table II).[159] Second, administration of propranolol, a potent β-adrenoceptor blocker, does not reduce CBF under resting conditions.[59,160-162] It is still unsettled whether or not the drug attenuates the CO_2 response, but recent evidence suggests that any effect present is not necessarily specific (i.e., mediated by blockade of β-adrenoceptors; see ref. 162).

As stated, experiments with administration of a dopamine agonist indicate that dopamine increases CBF and $CMRO_2$. However, there is no evidence that the intrinsic dopaminergic system makes "synaptic" contacts with brain vessels, and since administration of pimozide (a dopamine antagonist) neither alters resting CBF nor influences the CBF response to hypercapnia,[140] the dopamine system does not seem directly involved in cerebrovascular control. Results obtained following unilateral 6-hydroxydopamine lesions of the nigrostriatal dopamine pathway allow similar conclusions.[163] However, since the hypercapnic CBF response was slightly enhanced in the denervated caudate–putamen, the data suggest that dopamine released from the nigrostriatal system may exert a slight vasoconstrictory effect, possibly secondary to a metabolic change. In rats subjected to a unilateral lesion of the dopaminergic system and injected with amphetamine, a much larger CBF increase was seen on the damaged side than on the unlesioned side,[164] supporting the conclusion of Dahlgren *et al.*[163]

An innervation of cerebral microvessels by 5-HT neurons has been reported,[165] but its physiological role remains elusive. Thus, in the experiments in which 5-HT was found to reduce CBF, $CMRO_2$ was depressed as well.[141] Furthermore, recent results with chemical lesioning of the 5-HT systems (by intraventricular 5,7-dihydroxytryptamine) have failed to show any influence on the CBF response to hypercapnia,[141,163] and they indicate that if normocapnic CBF is increased by the procedure, this increase is coupled to a corresponding rise in $CMRO_2$.

Table II

Body Temperature, Mean Arterial Blood Pressure (MABP), Arterial Blood Gases, pH, Total Oxygen Content in Arterial Blood (Tao_2) Arteriovenous Difference for Oxygen ($AVDo_2$), Cerebral Blood Flow (CBF), and Metabolic Rate for Oxygen ($CMRo_2$) in Rats with Bilateral 6-Hydroxydopamine Lesions of the Ascending Noradrenergic Fiber Bundle from the Locus Coeruleus at Normoxia, Normocapnia, Hypoxia, and Hypercapnia

Experimental group	Temp (°C)	MABP (mm Hg)	Pco_2 (mm Hg)	Po_2 (mm Hg)	pH	Tao_2 (μmol·ml^{-1})	$AVDo_2$ (μmol·ml^{-1})	CBF (ml·g^{-1}·min^{-1})	$CMRO_2$ (μmol·g^{-1}·min)
Sham operated ($n = 4$)	37.2 ± 0.2	149 ± 5	36.8 ± 1.0	100 ± 3	7.40 ± 0.01	10.51 ± 0.2	2.92 ± 0.43	1.51 ± 0.44	3.81 ± 0.37
Lesioned ($n = 4$)	37.3 ± 0.2	160 ± 4	39.0 ± 0.7	109 ± 3	7.36 ± 0.01	10.40 ± 0.23	3.08 ± 0.19	1.36 ± 0.15	4.13 ± 0.34
Hypoxia, control ($n = 6$)	36.6 ± 0.1	122 ± 4	39.4 ± 0.8	26 ± 1	7.18 ± 0.01	2.20 ± 0.07	1.12 ± 0.05	5.69 ± 0.44	6.40 ± 0.28
Lesioned, hypoxia ($n = 4$)	36.6 ± 0.3	119 ± 7	39.7 ± 1.1	23 ± 1	7.15 ± 0.03	1.99 ± 0.21	1.07 ± 0.07	6.07 ± 0.55	6.64 ± 0.08
Hypercapnia, control ($n = 7$)	36.6 ± 0.1	156 ± 6	84 ± 2	123 ± 2	7.13 ± 0.01	9.97 ± 0.38	0.82 ± 0.07	5.60 ± 0.55	4.55 ± 0.41
Lesioned, hypercapnia ($n = 6$)	36.9 ± 0.2	153 ± 4	80 ± 2	123 ± 4	7.15 ± 0.02	9.82 ± 0.22	0.85 ± 0.03	6.17 ± 0.68	5.23 ± 0.53

In conclusion, although the catecholamines epinephrine, norepinephrine, and dopamine and the indoleamine 5-HT modulate cerebrovascular tone, and although at least cerebral noradrenergic and 5-HT neurons appear to innervate cerebral microvessels, it has not been conclusively shown that the intrinsic monoaminergic systems regulate cerebrovascular tone. It seems more likely that the effects described, e.g., those following administration of the appropriate agonists or antagonists, are secondary to an altered neuronal metabolism. However, since published results have been mainly concerned with normal vascular tone and with CO_2 responses, we still lack information on the influence of monoaminergic systems, if any, on CBF in a variety of pathological situations as well as on the adjustment of microcirculation to various functional needs. Thus, it still remains a possibility that these systems are cybernetic in the sense that they help to modulate the nutritional environment of other neuronal systems subserving information transmittal in the CNS (see ref. 166).

4.4. Regulation of CBF by Metabolic Signals

Most workers in the field have been inclined to conclude that overall or local blood flow is adjusted to the metabolic needs by products of an altered metabolism, and some have advanced the hypothesis that such metabolic "error signals" are responsible for changes in cerebrovascular resistance in pathological situations as well (for reviews, see refs. 18,119,167). The rationale behind these proposals is best understood if we consider the increase in blood flow that accompanies an enhanced metabolic rate. All experience shows that such an enhancement may be quite localized in the brain, as exemplified by the local increases in CBF during differential sensory stimulation or focal activation of motor areas (see ref.168). It follows from such results that regulation of CBF must be equally localized, excluding blood-borne factors or activation of diffusely projecting neuronal systems.

It seems justified to recall the metabolic perturbation induced on activation of a cluster of neurons. Depolarization is followed by efflux of K^+ from cells and influx of Na^+ and Ca^{2+}, ionic shifts that will lead to activation of Na^+–K^+ (and Ca^{2+})-dependent ATPases. The ensuing hydrolysis of ATP is reversed by oxidative phosphorylation, causing an increased consumption of O_2 and production of CO_2. Hydrolysis of ATP will, via the associated rise in ADP, induce a shift in the adenylate kinase equilibrium and thereby lead to accumulation of AMP. This, in turn, can be expected to trigger accumulation of degradation products of AMP, including adenosine. Production of CO_2 is bound to lower pH; however, with intense activation, stimulation of aerobic glycolysis may exceed the needs, and lactic acid production will contribute to the acidotic shift. Stimulation of neuronal activity will also enhance turnover of membrane-bound phospholipids,[169,170] possibly induced by influx of Ca^{2+} through voltage- or agonist-dependent gates. Conceivably, influx or release of Ca^{2+} leads to activation of phospholipases with accumulation of FFA, notably arachidonic acid, the latter leading to enhanced production of prostaglandins and leukotrienes.[171]

All of the stimulus-coupled events discussed have been assumed to be involved in regulation of CBF. We now discuss them in historical order.

4.4.1. Extracellular pH

Several findings led to the proposal that extracellular pH is an important, if not the main determinant of CBF (for reviews, see refs. 17,18,119,167,172,173). First, early results showed that whereas hypercapnia increased CBF, i.v. infusion of "fixed" acid did not, suggesting that the decisive change was the decrease in pH on the tissue side of the blood–brain barrier. This conclusion was strengthened by the increases in CBF found in hypoxia and epileptic seizures, conditions associated with enhanced lactic acid production in the brain. Furthermore, one study indicated that when induced hypoglycemia depleted the brain of glucose, the precursor of lactic acid, hypoxia did not lead to vasodilatation.[103] Second, compelling experimental results from Betz's laboratory (summarized in refs. 18,173; see also ref. 174) as well as data obtained in man[175] showed a close (inverse) correlation between extracellular pH and CBF.[176] It should also be recalled that ventriculocisternal perfusion with acid solutions has been found to increase regional CBF. Third, microapplication of solutions to pial vessels unequivocally demonstrates that an acid pH dilates, and an alkaline pH constricts, arteries.[119,177–179]

The results quoted leave no doubt that extracellular pH is an important determinant of CBF. However, the question arises as to its physiological importance, and several sets of results question the general validity of the "pH hypothesis." For example, it has now been demonstrated that maximal vasodilatation on induction of hypoxia and epileptic seizures may occur before the extracellular fluids are acidified; in fact, dilatation also occurs when the pH shows a transient alkaline shift as a result of a fall in P_{CO_2} (see refs. 167,180–182). It should also be recalled that CBF varies inversely with P_{O_2} over a range of arterial P_{O_2} values that do not seem to be associated with alterations in tissue lactate concentration. Furthermore, in hypoglycemia, CBF increases[113,114] in spite of an increase in extracellular pH,[183] a comparable dissociation between pH and CBF being observed in amphetamine intoxication.[181] Finally, changes in cerebrovascular tone over the autoregulatory range are not accompanied by alterations in extracellular pH,[184] and, in anesthesia, CBF may fall dramatically in the absence of an alkaline shift in pH.[185] Obviously, although extracellular pH undoubtedly influences cerebrovascular resistance, changes in pH cannot adequately explain why CBF changes in a number of circumstances.

4.4.2. Extracellular K^+ and Ca^{2+}

There is considerable evidence that cerebral vessels dilate when extracellular K^+ activity increases and when Ca^{2+} activity decreases.[18,119,173,186,187] Available data indicate that K^+ is vasodilatory up to a concentration of about $10 \mu mol \cdot ml^{-1}$ but that concentrations exceeding about 20 $\mu mol \cdot ml^{-1}$ are vasoconstrictory. Furthermore, whereas decreased Ca^{2+} activities dilate vessels, increased activities lead to vasoconstriction. There is also an interaction among

K^+, Ca^{2+}, and H^+. For example, at acid pH values, K^+ has virtually no vasodilatory effect, and the effect of Ca^{2+} is dependent on both K^+ and H^+ activities (see refs. 119,173,187).

Again, we must raise the question whether changes in K^+ or Ca^{2+} activities enter as important general modulators of CBF. In all probability, they do not. To take only two examples, in hypoxia CBF can increase three to fourfold without any detectable increase in extracellular K^+ activity,[181,182] and in ischemia, changes in extracellular K^+ and Ca^{2+} activities occur first when the ischemia is severe enough to abolish spontaneous or evoked electrical activity.[188-190] We want to emphasize, though, that neuronal activity is bound to alter the activities of all three ions (H^+, K^+, and Ca^{2+}) in a direction favoring vasodilatation. It is conceivable, therefore, that the combined influences of these changes provide a stronger stimulus than what is apparent when each is studied in isolation.[18,173]

4.4.3. Adenosine

Adenosine, a potent dilator of some extracerebral vessels, has recently been suggested to modulate cerebrovascular resistance in several conditions. The original evidence stems from two sets of observations. First, application of adenosine to pial vessels elicits marked arterial dilatation.[191-194] Second, brain tissue concentrations of adenosine were found to be increased in hypoxia and on electrical stimulation of the brain, two conditions associated with an increase in CBF.[195]

Results obtained from our own laboratory did not corroborate those reported by Rubio *et al.*[195] which were based on a potentially traumatic technique for tissue freezing. Thus, when the tissue was frozen *in situ*, no increases in cerebral cortical adenosine concentration were found at moderate or severe degrees of hypoxia.[196] Furthermore, in one series, the adenosine concentration was unchanged when the tissue was frozen 2–3 sec or 20 min following the induction of bicuculline-induced seizures,[196] but, in another, a fivefold increase was noted when the freezing was begun 1 min following seizure induction.[197]

The results quoted left little support for the hypothesis that adenosine enters as a major modulator of cerebrovascular resistance in hypoxia and epileptic seizures. However, a recent series of articles from the Charlottesville group provides circumstantial but nevertheless compelling evidence that adenosine is in fact an important coupling factor (for summary, see ref. 198). Using freeze-blowing for rapid inactivation of enzymes in the rat brain, these authors have recorded increases in brain adenosine concentration not only in hypoxia[199] and epileptic seizures[200] (see also ref. 201) but also during marked hypocapnia.[199] Furthermore, their data indicate that a reduction of mean arterial blood pressure to values that are within the normal autoregulatory range lead to increases in tissue adenosine concentration.[202]

Winn *et al.*[198,199] tentatively concluded that the discrepant results obtained in hypoxia by the Charlottesville and Lund groups can be explained by the fact that the slow *in situ* freezing allows deamination of the adenosine accumulated (cf. also differences in results obtained during seizures). Since this seems a reasonable suggestion, it must be seriously considered that adenosine enters as an important modulator of CBF. This also evolves from the fact that application of adenosine deaminase to the surface of the brain seems to markedly blunt the circulatory response to hypoxia.[203] Thus, although it remains to be shown that adenosine accumulates at more moderate degrees of hypoxia, that the adenosine accumulated in hypoxia, hypocapnia, hypotension, and seizures occurs at extracellular sites close to resistance vessels, and that the vascular effects of adenosine are not blunted by concomitant changes in K^+ or H^+ activities (see ref. 193), the evidence presented is compelling.

Thus, it remains a clear possibility that at least part of the cerebrovascular dilatation during hypoxia and epileptic seizures is mediated by adenosine and that adenosine accumulation prevents further vasoconstriction during marked hypocapnia. It should also be recalled that hypoglycemia, another condition associated with cerebrovascular dilatation (see above), leads to marked accumulation of adenosine.[204] Finally, although most workers in the field have been inclined to conclude that autoregulation is caused by a myogenic mechanism, the results of Winn *et al.*[202] suggest that metabolic factors cannot be excluded.

Two groups have reported that intracarotid infusion of adenosine leads to cerebrovascular dilatation.[205,206] In one of these studies,[205] infusion of ATP had an even more pronounced effect. Although that effect may well have been one secondary to degradation of ATP to adenosine by appropriate ectoenzymes, the results hint the interesting possibility that circulatory effects of ATP and/or adenosine could be elicited by activity in purinergic nerve endings.[207] Clearly, the results discussed provide promising hints to the nature of the coupling factors involved in adjusting CBF to changes in brain metabolism.

4.4.4. Prostaglandins

In 1973, Pickard and MacKenzie reported results indicating that prostaglandins enter as important modulators of CBF.[208] These authors, working on baboons under phencyclidine–N_2O anesthesia, found that indomethacin (10 mg·kg^{-1} i.v. or 0.04–0.2 mg·kg^{-1}· min^{-1} by intracarotid administration) reduced resting CBF by an average of 38% with no effect on $CMRO_2$ and markedly attenuated the CBF response to hypercapnia. The results therefore suggested that some undefined fatty acid cyclooxygenase products are instrumental in maintaining normal cerebrovascular tone and in mediating the CBF response to hypercapnia. However, a subsequent study from the same group showed that indomethacin did not affect autoregulation.[209]

The results of Pickard and MacKenzie[208] were subsequently confirmed by Sakabe and Siesjö[210] working on rats ventilated on 70% N_2O. In that species, resting CBF was reduced by about 50% with no significant change in $CMRO_2$, and the CO_2 response was markedly attenuated. If allowance is made for the difference in resting CBF values, the circulatory response to indomethacin administration was strikingly similar in baboons and rats (Fig. 9). Sakabe and Siesjö[210] also showed that indomethacin failed to affect the circulatory response to hypoxia. Subsequent studies from the same laboratory demonstrated that

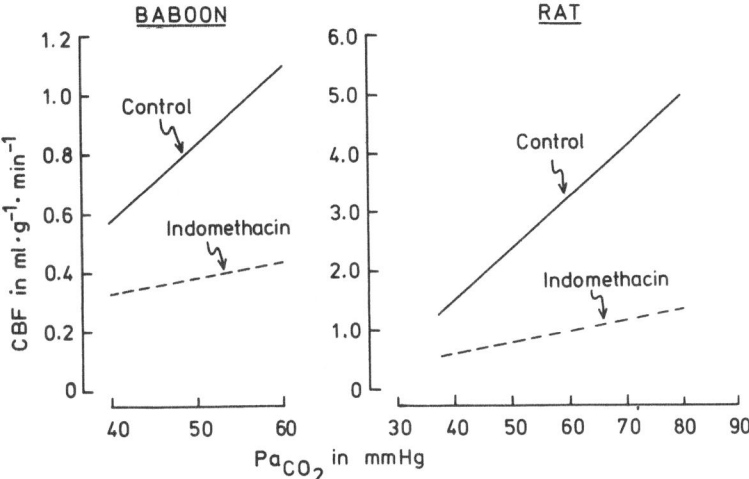

Fig. 9. The influence of indomethacin (10 mg·kg^{-1}) on CBF under normocapnic and hypercapnic conditions. Results on baboons are from Pickard and MacKenzie,[208] and those on rats from Sakabe and Siesjö.[210]

indomethacin administration left the CBF response during hypoglycemia[211] and epileptic seizures[212] unchanged. Thus, apart from affecting normal cerebro-vascular tone, the drug seemed to have a "preferential" effect on the CO_2 response.

Three findings provided hints to the mechanisms whereby indomethacin alters CBF. First, since cerebral circulatory effects were observed with only 1 mg·kg^{-1} of indomethacin, with maximal vasodilatation being achieved at 3–5 mg·kg^{-1}, it seemed likely that the effects were secondary to inhibition of fatty acid cyclooxygenase.[213] In support, dose–response curves (effects on CBF versus indomethacin dose) showed a striking similarity to those previously reported by Abdel-Halim *et al.*[214] who measured prostaglandin formation in brain tissue *in vitro* following prior *in vivo* administration of indomethacin (see ref. 213). Second, results reported by Pickard and co-workers suggest that the circulatory effects were secondary to decreased production of prostacyclin rather than of, e.g., PGE_2 or $PGF_{2\alpha}$. Thus, intracarotid administration of PGE_2 and $PGF_{2\alpha}$ with or without prior opening of the blood–brain barrier reduced CBF and $CMRO_2$.[215] Third, in a subsequent study, these authors could show that intracarotid infusion of prostacyclin increased CBF and reversed the vasoconstriction induced by indomethacin.[216] It should also be mentioned that indomethacin reduces CBF within a few seconds following its i.v. injection.[213]

The effects of indomethacin on CBF in rats ventilated on 70% N_2O were verified by measurements of local CBF in awake rats with [^{14}C]iodoantipyrine autoradiography.[213] The results verified those obtained with a ^{133}Xe modification of the Kety–Schmidt technique in showing reductions in l-CBF to 40–60% of control. Since some areas, notably cerebral cortical structures, showed reductions to 40% of control, the question arose if such pronounced reductions in flow rates occur in the awake state. Experiments designed to answer this

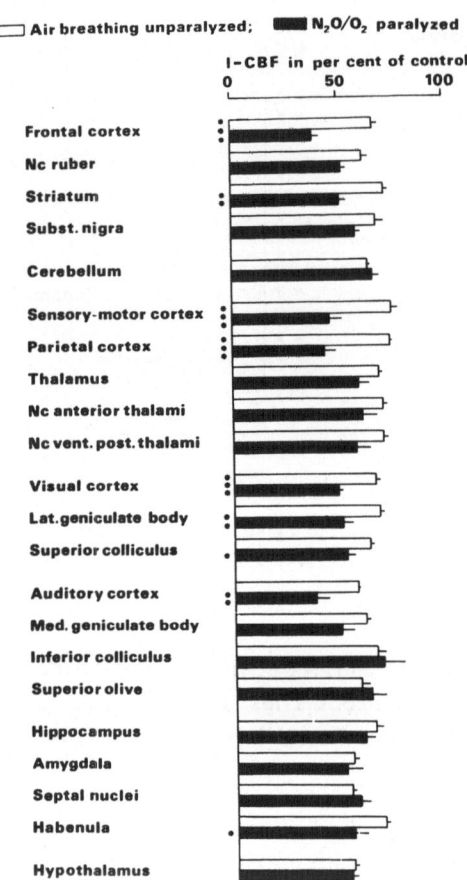

Fig. 10. The influence of indomethacin (10 mg·kg⁻¹) on local CBF in 22 brain structures as measured in air-breathing, unanesthetized rats (open bars) or in ventilated rats maintained on 70% N_2O (filled bars). The values are given as percent of controls (means ± S.E.M.). *$P < 0.05$; **$P < 0.01$; ***$P < 0.001$. Reproduced with permission from Dahlgren et al.[217]

question showed that the indomethacin response was exaggerated by the anesthetic procedure.[217] However, as Fig. 10 shows, marked reductions in l-CBF (to 55–75% of control) were also observed in awake, minimally restrained rats (cf. ref. 218).

Two groups of workers have failed to find that indomethacin reduces control CBF or attenuates the CO_2 response.[219,220] Since these experiments were carried out on animals (rabbits and cats, respectively) maintained on barbiturate anesthesia, the question arose if depression of metabolic rate had blunted the circulatory response to indomethacin. Thus, previous studies have shown that barbiturates and other depressant drugs lower CBF and attenuate the circulatory response to hypercapnia (ref. 30). However, a subsequent study showed that although barbiturate anesthesia blunted the CBF response to indomethacin at normal P_{CO_2}, the drug nevertheless reduced the CO_2 responsiveness four- to fivefold.[221]

Since the effects of the drug observed in baboons and rats were subsequently confirmed in man,[222] it seemed that the methods used by Cuypers *et al.*[219] and Wei *et al.*[220] were not well suited to measure tissue blood flow. However, results reported during the Fifth International Conference on Prostaglandins in Florence (May 18–21, 1982) brought new light on the problem. Thus, Busija and Heistad reported that indomethacin (10 mg.kg^{-1}) failed to alter normocapnic or hypercapnic CBF in anesthetized cats and unanesthetized rabbits, Wennmalm *et al.* reported that although indomethacin reduced normocapnic and hypercapnic CBF in man, neither aspirin nor naprosyn had any effect, and Siesjö and Wieloch reported that, in the rat, diclofenac (or piroxicam) failed to mimic the circulatory effects of indomethacin.[223,224] These new results suggest that the effects of indomethacin are species-dependent, and that, in the baboon, rat, and man, the drug reduces CBF by mechanisms that seem unrelated to inhibition of cyclooxygenase. Thus, although the results fail to show that cyclooxygenase products enter as important regulators of CBF they suggest that further work on the mechanisms of action of indomethacin may shed light on CBF regulation.

5. PATHOLOGY OF CEREBRAL CIRCULATION

Needless to say, the most important cerebrovascular pathology is the decrease in overall or regional CBF during ischemia secondary to a reduction in cerebral perfusion pressure. To a first approximation, the degree of reduction in CBF is proportional to the fall in cerebral perfusion pressure, whether this is caused by reduced blood pressure or by raised intracranial pressure. It should be recalled, though, that the reduction in local CBF is inhomogeneous. This is because certain areas, notably those that are situated inbetween the distribution territories of the major cerebral arteries and thereby form "watershed regions," suffer greater damage than others[225,226] (for further literature, see ref. 227). Many clinical conditions are assoicated with an altered reactivity of cerebral vessels that may well have pathophysiological importance. In fact, such changes in vascular reactivity seem to aggravate the impact of an ischemic insult, and the potentially harmful effects of postinsult hypoperfusion states are now under exploration. We discuss such adverse vascular reactions under three headings.

5.1. Loss of CO_2 Responsiveness

Several pathological conditions are accompanied by a reduced or abolished CBF response to hypercapnia (and/or hypocapnia), e.g., trauma, transient hypoxia, and ischemia.[228–231] In some of these conditions, notably severe hypoxia (and in epileptic seizures), the vessels are already maximally dilated, and no further increase in CBF can be expected when the P_{CO_2} is increased. However, hypocapnia then fails to reduce CBF, probably because the mechanisms causing vasodilation have an overriding influence on cerebrovascular resistance.

Probably, loss of CO_2 responsiveness has no pathophysiological importance *per se* but rather reflects damage to the cerebral vasculature. The mechanisms have not been defined. However, in view of the results discussed below, it is tempting to speculate that loss of CO_2 responsiveness results at least in part from a perturbed fatty acid and Ca^{2+} metabolism.

5.2. Disturbances of Autoregulation

Autoregulation, a property of the cerebral vasculature of all species examined, is impaired or lost in a variety of clinical disorders. For example, hypertensive patients who suffer decreases in CBF may show symptoms of brain ischemia when blood pressure is reduced to normotensive levels.[232–234] Thus, although hypertensive patients show autoregulation around their increased pressures, their autoregulatory curve is shifted to the right along the pressure axis.

A disturbed or abolished capacity for autoregulation is also observed whenever the vessels are dilated, e.g., during hypercapnia, hypoxia, and epileptic seizures (see refs. 89,235–237). In the first of these conditions, the tissue has not suffered mechanical or chemical damage. In the latter two, and in hypoglycemia,[211] there is either shortage of oxygen or substrate (hypoxia and hypoglycemia) or increased metabolic demands (epileptic seizures). Loss of autoregulation persists for some time in the posthypoxic period[238] and is also commonly observed in focal ischemia (and in trauma).[230,239,240] In all of these circumstances, loss of autoregulation can hazard the tissue in two ways. First, since CBF is pressure passive, even moderate reductions in blood pressure can further curtail oxygen supply, e.g., in hypoxia[241] (see also ref. 22). Second, because of the dilated cerebrovascular bed, rapid increases in blood pressure can lead to extravasation of proteins and other blood-borne macromolecules, and, probably, it predisposes to edema formation.[94,95,242] It should also be emphasized that loss of autoregulation is often synonomous with vascular damage, implying that the tissue has lost its capacity to purposefully adjust local circulation to the functional and metabolic needs.

5.3. Vascular Spasm and Postinsult Hypoperfusion Syndromes

In neurosurgery, spasm of cerebral vessels represents an urgent problem.[243,244] Angiographically detectable spasm of major cerebral arteries is often observed about 1 week after subarachnoidal hemorrhage (SAH). Many patients with SAH develop serious cerebral ischemia but, since this does not correlate well with spasm as observed with angiography, the major problem may be constriction of smaller (resistance) vessels.

In spite of almost staggering attempts to elucidate the underlying mechanisms, they still remain elusive. However, it seems likely that vasoconstrictory substances are released from blood constituents following bleeding into the CSF spaces.[244–246] A majority of workers agree that monoamines such as norepinephrine or 5-HT are not responsible. Possibly, spasm develops by mechanisms that involve damage to the vascular endothelium, partly by inactivation of the prostacyclin synthesis and partly by free radical reactions. Thus, it has been suggested that lipid hydroperoxides are released from blood constituents, causing both inactivation of prostacyclin synthesis and structural damage to vessels.[247] At present, the supportive evidence is only circumstantial, but the data available help to focus interest on products formed during the oxidative conversion of polyenoic fatty acids such as arachidonic acid.

Another adverse vascular reaction, possibly related to the vasoconstriction observed after SAH, is the secondary hypoperfusion that is observed in the recovery periods following ischemia[248-250] and hypoglycemia.[114] This reduction seems out of proportion to any reduction in metabolic rate, and it cannot be excluded that a "mismatch" between blood flow and metabolic rate leads to a secondary ischemic insult to the tissue.[251-253] At least under some circumstances, this secondary hypoperfusion can be ameliorated by pretreatment with indomethacin or by postischemic infusion of indomethacin plus prostacyclin.[254] The beneficial effect of indomethacin is best explained if one assumes that a postischemic arachidonic acid cascade upsets the delicate balance between the production of the vasodilatory prostacyclin and the vasoconstrictory (and proaggregating) thromboxane A_2.

In support of this idea, gross accumulation of prostaglandins has been shown to occur in the immediate postischemic period[255] during which accumulated arachidonic acid is only slowly metabolized.[256-258] Possibly, enhanced arachidonic acid oxidation inactivates prostacyclin (but not thromboxane) synthesis, explaining why prostacyclin must be administered with indomethacin in the postischemic period.

As a final point, we recall that paradoxical vascular reactions may contribute to ischemic cell damage by preventing purposeful dilatation of vessels in ischemic tissue. Thus, although experimental clipping of a middle cerebral artery leads to an initial dilatation of arteries in the affected tissue, and although some vessels remain dilated, others subsequently show intense vasoconstriction.[259] In this case, though, the vasoconstriction may be caused by gross extracellular accumulation of K^+ ions and possibly also by influx of Ca^{2+} through voltage-dependent gates; the vasoconstriction has been shown to be at least transiently reversed by Ca^{2+} antagonists.[260]

5.4. Therapeutic Aspects

In spite of considerable experimental efforts and numerous clinical trials, there is no established clinical drug therapy that improves cerebral circulation by an action on cerebral vessel resistance. Instead, successful interventions are those that ameliorate the cause of a reduced blood flow, e.g., by alleviating a reduced blood pressure, an increased intracranial pressure, or a manifest edema. In fact, beneficial effects may be achieved by drugs that normally reduce CBF. Such effects have been noted when barbiturates have been administered to patients with increased intracranial pressure[261] or to animals with induced regional[262-264] or global[265] (however, see also ref. 266) ischemia. In all probability, such drugs improve the blood-flow-to-metabolism ratio by an action on cellular metabolism.

Until quite recently, the basis for a successful treatment directed towards cerebral circulation seemed meager since the diseases by themselves appeared to give "maximal" dilatation of resistance vessels. However, the recognition that several conditions lead to aberrant vascular reactions ("spasm") has prepared the ground for further work. In this respect, results suggesting that pathological vasoconstriction may be Ca^{2+} mediated and that a disturbed balance between prostacyclin and thromboxane synthesis may complicate cerebrovascular disorders offer potentially useful hints. However, future research may well reveal that other neurohormones and neuromodulators enter as important determinants of cerebrovascular resistance in health and disease. At present, available results have demonstrated that certain peptides are markedly vasoactive when tested in *in vitro* systems, applied topically to pial vessels, or infused into the carotid arteries.[267–270] No doubt, the physiological significance of these systems will soon become established.

ACKNOWLEDGMENTS: Work from the authors' own laboratory was supported by grants from the Swedish Medical Research Council (14x-263), the United States Public Health Service via the National Institutes of Health (2R01 NS-07838), and from the Medical Faculty, University of Lund. Expert secretarial assistance was given by Yvonne Hansson and Gillian Sjödahl.

REFERENCES

1. Kety, S. S., and Schmidt, C. F., 1945, *Am. J. Physiol.* **143:**53–66.
2. Kety, S. S., and Schmidt, C. F., 1948, *J. Clin. Invest.* **27:**476–483.
3. Kety, S. S., 1960, *Methods in Medical Research*, Volume 8 (H. D. Bruner, ed.), Year Book Medical Publishers, Chicago, pp. 223–227.
4. Lassen, N. A., and Ingvar, D. H., 1961, *Experientia* **17:**42–45.
5. Ingvar, D. H., and Lassen, N. A., 1962, *Acta Physiol. Scand.* **54:**325–338.
6. Landau, W. M., Freygang, W. H., Rowland, L. P., Sokoloff, L., and Kety, S. S., 1955, *Trans. Am. Neurol. Assoc.* **80:**125–129.
7. Freygang, W. H., and Sokoloff, L., 1958, *Advances in Biological and Medical Physics*, Volume 6 (C. A. Tobias and J. H. Lawrence, eds.), Academic Press, New York, pp. 263–279.
8. Sokoloff, L., Reivich, M., Kennedy, C., Des Rosiers, M. H., Patlak, C. S., Pettigrew, K. D., Sakurada, O., and Shinohara, M., 1977, *J. Neurochem.* **28:**897–916.
9. Ter-Pogossian, M. M., Eichling, J. O., Davis, D. O., and Welch, M. J., 1970, *J. Clin. Invest.* **49:**381–391.
10. Ter-Pogossian, M. M., Davis, D. O., Eichling, J. O., and Carter, C. C., 1971, *Seventh Conference on Cerebral Vascular Diseases* (J. T. Toole, J. Moossy, and R. Janeway, eds.), Grune & Stratton, New York, pp. 103–108.
11. Raichle, M. E., Grubb, R. L., Eichling, J. O., and Ter-Pogossian, M. M., 1976, *J. Appl. Physiol.* **40:**638–640.
12. Reivich, M., Kuhl, D., Wolf, A., Greenberg, J., Phelps, M., Ido, T., Casella, E., Hoffman, A., Lavi, A., and Sokoloff, L., 1979, *Circ. Res.* **44:**127–137.
13. Phelps, M. E., Huang, S. C., Hoffman, E. J., Selin, C., Sokoloff, L., and Kuhl, D. E., 1979, *Ann. Neurol.* **6:**371–388.
14. Frackowiak, R. S. J., Lenzi, G. L., Jones, T., and Heather, J. D., 1980, *J. Comput. Assist. Tomogr.* **4:**727–736.
15. Kety, S. S., 1950, *Am. J. Med.* **8:**205–217.
16. Lassen, N. A., 1959, *Physiol. Rev.* **39:**183–238.
17. Sokoloff, L., 1959, *Pharmacol. Rev.* **11:**1–85.
18. Betz, E., 1972, *Physiol. Rev.* **52:**595–630.
19. Purves, M. J., 1972, *The Physiology of the Cerebral Circulation*, Cambridge University Press, Cambridge.

20. Lassen, N. A., and Ingvar, D. H., 1972, *Prog. Nucl. Med.* **1**:376–409.
21. Lassen, N. A., and Christensen, M. S., 1976, *Br. J. Anesth.* **8**:719–841.
22. Siesjö, B. K., 1978, *Brain Energy Metabolism*, John Wiley & Sons, New York.
23. Lacombe, P., Meric, P., and Seylaz, J., 1980, *Brain Res. Rev.* **2**:105–169.
24. Raichle, M. E., Grubb, R. L., and Ter-Pogossian, M. M. (eds.), 1981, *J. Cereb. Blood Flow Metab.* **1**(Suppl. 1):1–586.
25. Schmidt, C. F., 1950, *The Cerebral Circulation in Health and Disease*, Charles C Thomas, Springfield, Illinois.
26. Levine, S., and Payan, H., 1966, *Exp. Neurol.* **16**:255–262.
27. Eklöf, B., and Schwartz, S., 1969, *Arch. Surg.* **99**:695–701.
28. Pulsinelli, W. A., and Brierley, J. B., 1979, *Stroke* **10**:267–272.
29. Kahn, K., 1972, *Neurology (Minneap.)* **22**:510–515.
30. Edvinsson, L., and MacKenzie, E. T., 1977, *Pharmacol. Rev.* **28**:275–347.
31. Heistad, D. D., and Marcus, M. L., 1978, *Circ. Res.* **42**:295–302.
32. Ingvar, D. H., and Söderberg, U., 1958, *Acta Physiol. Scand.* **42**:130–143.
33. Rapela, C. E., and Green, H. D., 1964, *Circ. Res.* **14, 15**(Suppl. 1):205–212.
34. Theye, R. R., and Michenfelder, J. D., 1968, *Anesthesiology* **29**:1119–1124.
35. Michenfelder, J. D., Messick J. M., Jr., and Theye, R. A., 1968, *J. Surg. Res.* **8**:475–481.
36. Takeshita, H., Michenfelder, J. D., and Theye, R. A., 1972, *Anesthesiology* **37**:605–612.
37. Sapirstein, L. A., and Ogden, E., 1956, *Circ. Res.* **4**:245–249.
38. McHenry, L. C., Jr., 1964, *Neurology (Minneap.)* **14**:785–793.
39. Lassen, N. A., and Klee, A., 1965, *Circ. Res.* **16**:26–32.
40. Lassen, N. A., and Munck, O., 1955, *Acta Physiol. Scand.* **33**:30–49.
41. Alexander, S. C., Wollman, H., Cohen, P. J., Chase, P. E., Melman, E., and Behar, M., 1964, *Anesthesiology* **25**:37–42.
42. Shenkin, H. A., Harmel, M. H., and Kety, S. S., 1948, *Arch. Neurol. Psychiatry* **60**:240–252.
43. Norberg, K., and Siesjö, B. K., 1974, *Acta Physiol. Scand.* **91**:154–164.
44. Berntman, L., Dahlgren, N., and Siesjö, B. K., 1978, *Acta Physiol. Scand.* **104**:101–108.
45. Homburger, E., Hinwich, W. A., Etsten, B., York, G., Maresca, R., and Himwich, H. E., 1946, *Am. J. Physiol.* **147**:343–345.
46. Nilsson, B., 1974, *Acta Physiol. Scand.* **92**:142–144.
47. Nilsson, B., and Siesjö, B. K., 1976, *Acta Physiol. Scand* **96**:72–82.
48. Nilsson, B., and Siesjö, B. K., 1982, *Stroke* (in press).
49. Gjedde, A., Caronna, J. J., Hindfelt, B., and Plum, F., 1975, *Am. J. Physiol.* **229**:113–118.
50. Glass, H. I., and Harper, A. M., 1963, *Br. Med. J.* **1**:593.
51. Lassen, N. A., Hoedt-Rasmussen, K., Sorensen, S. C., Skinhoj, E., Cronquist, S., Bodforss, B., and Ingvar, D. H., 1963, *Neurology (Minneap.)* **13**:719–727.
52. Ingvar, D. H., Cronqvist, S., Ekberg, R., Risberg, J., and Hoedt-Rasmussen, K., 1965, *Acta Neurol. Scand. [Suppl.]* **14**:72–78.
53. Hoedt-Rasmussen, K., Sveinsdottir, E., and Lassen, N. A., 1966, *Circ. Res.* **18**:237–247.
54. Marcus, M. L., Bischof, C. J., and Heistad, D. D., 1981, *Cir. Res.* **48**:748–761.
55. Olesen, J., Paulson, O. B., and Lassen, N. A., 1971, *Stroke* **2**:519–540.
56. Zierler, K. L., 1965, *Circ. Res.* **16**:309–321.
57. Meier, P., and Zierler, K. L., 1954, *J. Appl. Physiol.* **6**:731–744.
58. Harper, A. M., Deshmukh, V. D., Sengupta, D., Rowan, J. O., and Jennett, W. B., 1972, *Neuroradiology* **3**:134–136.
59. MacKenzie, E. T., McCulloch, J., and Harper, A. M., 1976, *Am. J. Physiol.* **231**:489–494.
60. MacKenzie, E. T., McCulloch, J. O'Keane, M., Pickard, J. D., and Harper, A. M., 1976, *Am. J. Physiol.* **231**:483–488.
61. Mallet, B. L., and Veall, N., 1963, *Lancet* **1**:1081–1082.
62. Austin, G., Horn, N., Rouhe, S., and Hayward, W., 1972, *Eur. Neurol.* **8**:43–51.
63. Aukland, K., Bower, B. F., and Berliner, R. W., 1964, *Circ. Res.* **14**:164–187.
64. Young, W., 1980, *Stroke* **11**:552–564.
65. Pasztor, E., Symon, L., Dorsch, N. W. C., and Branston, N. M., 1973, *Stroke* **4**:556–557.
66. Symon, L., Pasztor, E., and Branston, N. M., 1974, *Stroke* **5**:355–364.

67. Stosseck, K., Lübbers, D. W., and Cottin, N., 1974, *Pfluegers Arch.* **348**:225–238.
68. Wahl, M., Kuschinsky, W., Bosse, O., and Thurau, K., 1973, *Circ. Res.* **32**:162–169.
69. Harper, A. M., and MacKenzie, E. T., 1977, *J. Physiol. (Lond.)* **271**:735–741.
70. Sapirstein, L. A., 1958, *Am. J. Physiol.* **193**:161–168.
71. Roth, J. A., Greenfield, A. J., Kaihara, S., and Wagner, H. N., Jr., 1970, *Am. J. Physiol.* **219**:96–101.
72. Marcus, M. L., Heistad, D. D., Ehrhardt, J. C., and Abboud, F. M., 1976, *J. Appl. Physiol.* **40**:501–507.
73. Heyman, M. A., Payne, B. D., Hoffman, J. J. E., and Rudolph, A. M., 1977, *Prog. Cardiovasc. Dis.* **20**:55–79.
74. Reivich, M., Jehle, J., Sokoloff, L., and Kety, S. S., 1969, *J. Appl. Physiol.* **27**:296–300.
75. Eklöf, B., Lassen, N. A., Nilsson, L., Norberg, K., Siesjö, B. K., and Torlöf, P., 1974, *Acta Physiol. Scand.* **91**:1–10.
76. Eckman, W. W., Phair, R. D., Fenstermacher, J. D., Patlak, C. S., Kennedy, C., and Sokoloff, L., 1975, *Am. J. Physiol.* **223**:215–221.
77. Sakurada, O., Kennedy, C., Jehle, J., Brown, J. D., Carbin, C. L., and Sokoloff, L., 1978, *Am. J. Physiol.* **234**:H59–H66.
78. Dahlgren, N., Ingvar, M., Yokoyama, H., and Siesjö, B. K., 1981, *J. Cereb. Blood Flow Metab.* **2**:211–218.
79. Jones, S. C., Lear, J. L., Greenberg, J. H., and Reivich, M., 1979, *Acta Neurol. Scand.* [*Suppl.*] **72**:202–203.
80. Van Uitert, R. L., and Levy, D. E., 1978, *Stroke* **9**:67–72.
81. Gjedde, A., Hansen, A. J., and Siemkowicz, E., 1980, *Acta Physiol. Scand.* **108**:321–330.
82. Sage, J. I., Van Uitert, R. L., and Duffy, T. E., 1981, *J. Neurochem.* **36**:1731–1738.
83. Ter-Pogossian, M. M., Raichle, M. E., and Sobel, B. E., 1980, *Sci. Am.* **243**(4):141–155.
84. Huang, S. C., Phelps, M. E., Hoffman, E. J., Sideris, K., Selin, C. J., and Kuhl, D. E., 1980, *Am. J. Physiol.* **238**:E69–E82.
85. Kuhl, D. E., Phelps, M. E., Kowell, A. P., Metter, E. J., Selin, C., and Winter, J., 1980, *Ann. Neurol.* **8**:47–60.
86. Raichle, M. E., 1979, *Brain Res. Rev.* **1**:47–68.
87. Lassen, N. A., 1974, *Circ. Res.* **34**:749–760.
88. Häggendal, E., and Johansson, B., 1965, *Acta Physiol. Scand.* [*Suppl.,*] **258**(66):27–53.
89. Harper, A. M., 1966, *J. Neurol. Neurosurg. Psychiatry* **29**:398–403.
90. Zwetnow, N. N., 1970, *Acta Physiol. Scand.* [*Suppl.*] **339**:1–31.
91. Strandgaard, S., MacKenzie, E. T., Sengupta, D., Rowan, J. O., Lassen, N. A., and Harper, A. M., 1974, *Circ. Res.* **34**:435–440.
92. Strandgaard, S., 1978, *Acta Neurol. Scand.* [*Suppl.*] **66**:1–82.
93. Johansson, B. B., Li, C.-L., Olsson, Y., and Klatzo, I., 1970, *Acta Neuropathol. (Berl.)* **16**:117–124.
94. Rapoport, S. I., 1976, *Blood–Brain Barrier in Physiology and Medicine*, Raven Press, New York.
95. Johansson, B. B., 1980, *Adv. Exp. Med. Biol.* **131**:211–226.
96. Miller, J. D., Stanek, A., and Langfitt, T. W., 1971, *Prog. Brain Res.* **35**:411–432.
97. Grubb, R. L., Jr., Raichle, M. E., Phelps, M. E., and Ratcheson, R. A., 1975, *J. Neurosurg.* **43**:385–398.
98. Harper, A. M., and Glass, H. I., 1965, *J. Neurol. Neurosurg. Psychiatry* **28**:449–452.
99. Norberg, K., and Siesjö, B. K., 1974, *Acta Physiol. Scand.* **91**:154–164.
100. Siesjö, B. K., 1980, *Anesthesiology* **52**:461–465.
101. Kety, S. S., and Schmidt, C. F., 1948, *J. Clin. Invest.* **27**:484–491.
102. Cohen, P. J. S., Alexander, S. C., Smith, F. C., Reivich, M., and Wollman, H., 1967, *J. Appl. Physiol.* **23**:183–189.
103. Kogure, K., Scheinberg, P., Reinmuth, O. M., Fujishima, M., and Busto, R., 1970, *J. Appl. Physiol.* **29**:223–229.
104. Lambertsen, C. J., Kough, R. H., Cooper, D. Y., Emmel, G. L., Loeschcke, H.H., and Schmidt, C. F., 1953, *J. Appl. Physiol.* **5**:471–486.
105. Borgström, L., Johannsson, H., and Siesjö, B. K., 1975, *Acta Physiol. Scand.* **93**:423–432.

106. Häggendal, E., and Norbäck, B., 1966, *Acta Chir. Scand.* [*Suppl.*] **364**:13–21.
107. Häggendal, E., Nilsson, N. J., and Norbäck, B., 1966, *Acta Chir. Scand.* [*Suppl.*] **364**:3–12.
108. Paulson, O. B., Parving, H.-H., Olesen, J., and Skinhoj, E., 1973, *J. Appl. Physiol.* **35**:111–116.
109. Michenfelder, J. D., and Theye, R. A., 1969, *Anesthesiology* **31**:449–457.
110. MacMillan, V., 1975, *Can. J. Physiol. Pharmacol.* **53**:644–650.
111. Johannsson, H., and Siesjö, B. K., 1974, *Acta Physiol. Scand.* **91**:136–138.
112. Borgström, L., Johannsson, H., and Siesjö, B. K., 1975, *Acta Physiol. Scand.* **93**:505–514.
113. Norberg, K., and Siesjö, B. K., 1976, *J. Neurochem.* **26**:345–352.
114. Abdul-Rahman, A., Agardh, C.-D., and Siesjö, B. K., 1980, *Acta Physiol. Scand.* **109**:307–314.
115. Harper, A. M., Deshmukh, V. D., Rowan, J. O., and Jennet, W. B., 1972, *Arch. Neurol.* **27**:1–6.
116. Mueller, S. M., Heistad, D. D., and Marcus, M. L., 1977, *Circ. Res.* **41**:350–356.
117. Gross, P. M., Marcus, M. L., and Heistad, D. D., 1980, *J. Appl. Physiol.* **48**(2):213–217.
118. Mueller, S. M., Heistad, D. D., and Marcus, M. L., 1979, *Am. J. Physiol.* **237**(2):H178–H184.
119. Kuschinsky, W., and Wahl, M., 1978, *Physiol. Rev.* **58**:656–689.
120. Owman, C., and Edvinsson, L., 1977, *Neurogenic Control of the Brain Circulation* (C. Owman and L. Edvinsson, eds.), Pergamon Press, Oxford, pp. 15–38.
121. Fitch, W., MacKenzie, E. T., and Harper, A. M., 1975, *Circ. Res.* **37**:550–557.
122. Gross, P. M., Heistad, D. D., Strait, M. R., Marcus, M. L., and Brody, M. J., 1979, *Circ. Res.* **44**:288–294.
123. Bill, A., and Linder, J., 1976, *Acta Physiol. Scand.* **96**:114–121.
124. Edvinsson, L., Owman, C., and Siesjö, B. K., 1976, *Brain Res.* **117**:518–523.
125. McKenzie, E. T., Strandgaard, S., Graham, D. J., Jones, J. V., Harper, A. M., and Farrar, J. K., 1976, *Circ. Res.* **38**:33–41.
126. Heistad, D. D., Marcus, M. L., and Gross, P. M., 1978, *Am. J. Physiol.* **235**(5):H544–H552.
127. Kuschinsky, W., Wahl, M., and Neiss, A., 1974, *Pflüegers Arch.* **347**:199–208.
128. Busija, D. W., and Heistad, D. D., 1981, *Circ. Res.* **48**:62–69.
129. James, I. M., Millar, R. A., and Purves, M. J., 1969, *Circ. Res.* **25**:77–93.
130. Ponte, J., and Purves, M. J., 1974, *J. Physiol.* (*Lond.*) **237**:315–340.
131. Heistad, D. D., Marcus, M. L., Ehrhardt, J. C., and Abboud, F. M., 1976, *Circ. Res.* **38**:20–25.
132. Heistad, D. D., and Marcus, M. L., 1976, *Stroke* **7**:239–243.
133. Rapela, C. E., Green, H. D., and Denison, A. B., Jr., 1967, *Circ. Res.* **21**:559–568.
134. King, B. D., Sokoloff, L., and Wechsler, R. L., 1952, *J. Clin. Invest.* **31**:273–279.
135. Weil-Malherbe, H., Whitby, L. G., and Axelrod, J., 1961, *Regional Neurochemistry* (S. S. Kety and J. Eldes, eds.), Pergamon Press, Oxford, New York, pp. 284–292.
136. Oldendorf, W. H., 1971, *Am. J. Physiol.* **221**:1629–1639.
137. Oldendorf, W. H., 1981, *Research Methods in Neurochemistry*, Volume 5 (N. Marks and R. Rodnight, eds.), Plenum Press, New York, pp. 91–112.
138. Bertler, A., Falck, B., Owman, C., and Rosengren, E., 1966, *Pharmacol. Rev.* **18**:369–385.
139. Hardebo, J. E., and Owman, C., 1980, *Ann. Neurol.* **8**:1–11.
140. McCulloch, J., and Harper, A. M., 1977, *Am. J. Physiol.* **233**:H222–H227.
141. Harper, A. M., and MacKenzie, E. T., 1977, *J. Physiol.* (*Lond.*) **271**:721–733.
142. Carlsson, C., Hägerdal, M., and Siesjö, B. K., 1976, *Acta Physiol. Scand.* **95**:206–208.
143. Carlsson, C., Hägerdal, M., Kaasik, A. E., and Siesjö, B. K., 1977, *Brain Res.* **119**:223–231.
144. Berntman, L., Dahlgren, N., and Siesjö, B. K., 1978, *Acta Physiol. Scand.* **104**:101–108.
145. Abdul-Rahman, A., Dahlgren, N., Johansson, B. B., and Siesjö, B. K., 1979, *Acta Physiol. Scand.* **107**:227–232.
146. Dahlgren, N., Rosen, I., Sakabe, T., and Siesjö, B. K., 1980, *Brain Res.* **184**:143–152.
147. Amaral, D. G., and Sinnamon, H. M., 1977, *Prog. Neurobiol.* **9**:147–196.
148. Lindvall, O., and Björklund, A., 1978, *Handbook of Psychopharmacology*, Volume 9 (L. Iversen, S. D. Iversen, and S. H. Snyder, eds.), Plenum Press, New York, pp.139–231.
149. Korf, J., Aghajanian, G. K., and Roth, R. H., 1973, *Neuropharmacology* **12**:933–938.
150. Bloom, F. E., 1975, *Rev. Physiol. Biochem. Pharmacol.* **74**:1–103.

151. Hartman, B. K., Zide, D., and Udenfriend, S., 1972, *Proc. Natl. Acad. Sci. U.S.A.* **69:**2722–2726.

152. Lai, F. M., Udenfriend, S., and Spector, S., 1975, *Proc. Natl. Acad. Sci. U.S.A.* **72:**4622–4625.

153. Swanson, L. W., Connelly, M. A., and Hartman, B. K., 1977, *Brain Res.* **136:**166–173.

154. Raichle, M. E., Hartman, B. K., Eichling, J. O., and Sharpe, L. G., 1975, *Proc. Natl. Acad. Sci. U.S.A.* **72:**3726–3730.

155. Bates, D., Weinshilboum, R. M., Campbell, R. J., and Sundt, T. M., Jr., 1977, *Brain Res.* **136:**431–443.

156. Herbst, T. J., Raichle, M. E., and Ferrendelli, J. A., 1979, *Science* **204:**330–332.

157. Nathanson, J. A., and Glaser, G. H., 1979, *Nature* **278:**567–569.

158. Harik, S. J., Sharma, V. K., Wetherbee, J. R., Warren, R. H., and Banarjee, S. P., 1980, *Eur. J. Pharmacol.* **61:**207–208.

159. Dahlgren, N., Lindvall, O., Sakabe, T., and Siesjö, B. K., 1981, *Brain Res.* **209:**11–23.

160. Berntman, L., Carlsson, C., and Siesjö, B. K., 1978, *Brain Res.* **151:**220–224.

161. Olesen, J., Hougaard, K., and Hertz, M., 1978, *Stroke* **9:**344–349.

162. Dahlgren, N., Ingvar, M., and Siesjö, B. K., 1981, *J. Cereb. Blood Flow Metab.* **1:**429–436.

163. Dahlgren, N., Lindvall, O., Nobin, A., and Stenevi, U., 1981, *Brain Res.* **230:**221–233.

164. Lindvall, O., Ingvar, M., and Stenevi, U., 1981, *Brain Res.* **211:**211–216.

165. Reinhard, J. F., Jr., Liebmann, J. E., and Schlosberg, A. J., 1979, *Science* **206:**85–87.

166. Roberts, E., 1980, *Antiepileptic Drugs: Mechanisms of Action* (G. H. Glaser, J. K. Penry, and D. M. Woodbury, eds.), Raven Press, New York, pp. 667–713.

167. Siesjö, B. K., Berntman, L., and Nilsson, B., 1980, *Microvasc. Res.* **19:**158–170.

168. Lassen, N. A., Ingvar, D. H., and Skinhoj, E., 1978, *Sci. Am.* **239**(4):62–71.

169. Michell, R. H., 1975, *Biochim. Biophys. Acta* **415:**81–147.

170. Hawthorne, J. N., and Pickard, M. R., 1979, *J. Neurochem.* **32:**5–14.

171. Hirata, F., and Axelrod, J., 1980, *Science* **209:**1082–1090.

172. Lassen, N. A., 1968, *Scand. J. Clin. Lab. Invest.* **22:**247–251.

173. Betz, E., and Csornai, M., 1978, *Pfluegers Arch.* **374:**67–72.

174. Pannier, J. L., and Leusen, I., 1973, *Pfluegers Arch.* **338:**347–359.

175. Fencl, V., Vale, J. R., and Broch, J. A., 1969, *J. Appl. Physiol.* **27:**67–76.

176. Pannier, J. L., Weyne, J., Demeester, G., and Leusen, I., 1972, *Pfluegers Arch.* **333:**337–351.

177. Kuchinsky, W., Wahl, M., Bosse, O., and Thurau, K., 1972, *Circ. Res.* **16:**240–247.

178. Kontos, H. A., Raper, A. J., and Patterson, J. L., Jr., 1977, *Stroke* **8:**358–360.

179. Kontos, H. A., Wei, E. P., Raper, A. J., and Patterson, J. L., 1977, *Stroke* **8:**226–229.

180. Nilsson, B., and Siesjö, B. K., 1977, *Neurogenic Control of the Brain Circulation* (C. Owman and L. Edvinsson, eds.), Pergamon Press, Oxford, pp. 295–300.

181. Astrup, J., Heuser, D., Lassen, N. A., Nilsson, B., Norberg, K., and Siesjö, B. K., 1978, *Cerebral Vascular Smooth Muscle and Its Control. CIBA Foundation Symposium 56* (M. Purves, ed.), Elsevier/Excerpta Medica/North-Holland, Amsterdam, Oxford, New York, pp. 313–332.

182. Nilsson, B., Rehncrona, S., and Siesjö, B. K., 1978, *Cerebral Vascular Smooth Muscle and Its Control. CIBA Foundation Symposium 56* (M. Purves, ed.), Elsevier/Excerpta Medica/North-Holland, Amsterdam, Oxford, New York, pp. 199–214.

183. Pelligrino, D., and Siesjö, B. K., 1981, *J. Cereb. Blood Flow Metab.* **1:**85–96.

184. Wahl, M., and Kuschinsky, W., 1979, *Pfluegers Arch.* **382:**203–208.

185. McDowall, D. G., Okuda, Y., Heuser, D., and Keaney, N. P., 1978, *Cerebral Vascular Smooth Muscle and Its Control. CIBA Foundation Symposium 56* (M. Purves, ed.), Elsevier/Excerpta Medica/North-Holland, Amsterdam, Oxford, New York, pp. 257–267.

186. Cameron, I. R., and Caronna, J., 1976, *J. Physiol. (Lond.)* **262:**415–430.

187. Betz, E., Enzenross, H. G., and Vlahov, V., 1973, *Pfluegers Arch.* **343:**79–88.

188. Branston, N. M., Strong, A. J., and Symon, L., 1977, *J. Neurol. Sci.* **32:**305–321.

189. Astrup, J., Symon, L., Branston, N. M., and Lassen, N. A., 1977, *Stroke* **8:**51–57.

190. Harris, R. J., Symon, L., Branston, N. M., and Bayhan, M., 1981, *J. Cereb. Blood Flow Metab.* **1:**203–209.

191. Berne, R. M., Rubio, R., and Curnish, R. R., 1974, *Circ. Res.* **35**:262–271.
192. Wahl, M., and Kuchinsky, W., 1976, *Pfluegers Arch.* **362**:55–59.
193. Wahl, M., and Kuschinsky, W., 1977, *Blood Vessels* **14**:285–293.
194. Gregory, P. C., Boisvert, D. P. J., and Harper A. M., 1980, *Pfluegers Arch.* **386**:187–192.
195. Rubio, R., Berne, R. M., Bockman, E. L., and Curnish, R. R., 1975, *Am. J. Physiol.* **228**:1896–1902.
196. Rehncrona, S., Siesjö, B. K., and Westerberg, E., 1978, *Acta Physiol. Scand.* **104**:453–463.
197. Siesjö, B. K., Ingvar, M., Folbergrova, J., and Chapman, A., 1981, *Status Epilepticus: Mechanisms of Brain Damage and Treatment*, Raven Press, New York (in press).
198. Winn, H. R., Rubio, R., and Berne, R. M., 1981, *J. Cereb. Blood Flow Metab.* **1**:239–244.
199. Winn, H. R., Rubio, R., and Berne, R. M., 1981, *Am. J. Physiol.* **241**:H235–H242.
200. Winn, H. R., Welsh, J., Rubio, R., and Berne, R. M., 1980, *Circ. Res.* **47**:481–491.
201. Schrader, J., Wahl, M., Kuschinsky, W., and Kreutzberg, G. W., 1980, *Pfluegers Arch.* **387**:245–251.
202. Winn, H. R., Welsh, J., Rubio, R., and Berne, R. M., 1980, *Am. J. Physiol.* **239**:H644–H651.
203. Wei, E. P., and Kontos, H. A., 1981, *J. Cereb. Blood Flow Metab.* **1**(Suppl. 1):395–396.
204. Chapman, A. G., Westerberg, E., and Siesjö, B. K., 1981, *J. Neurochem.* **36**:179–189.
205. Forrester, T., Harper, A. M., McKenzie, E. T., and Thomson, E. M., 1979, *J. Physiol. (Lond.)* **296**:343–355.
206. Heistad, D. D., Marcus, M. L., Gourley, J. K., and Busija, D. W., 1981, *Am. J. Physiol.* **240**:H775–H780.
207. Burnstock, G., 1975, *J. Exp. Zool.* **194**:103–134.
208. Pickard, J. D., and MacKenzie, E. T., 1973, *Nature [New Biol.]* **245**:187–188.
209. Pickard, J. D., MacDonell, L. A., MacKenzie, E. T., and Harper, A. M., 1977, *Circ. Res.* **40**:198–203.
210. Sakabe, T., and Siesjö, B. K., 1979, *Acta Physiol. Scand.* **107**:283–284.
211. Nilsson, B., Agardh, C.-D., Ingvar, M., and Siesjö, B. K., 1981, *Acta Physiol. Scand.* **111**:455–463.
212. Ingvar, M., Nilsson, B., and Siesjö, B. K., 1981, *Acta Physiol. Scand.* **111**:205–212.
213. Dahlgren, N., Nilsson, B., Sakabe, T., and Siesjö, B. K., 1981, *Acta Physiol. Scand.* **111**:475–485.
214. Abdel-Halim, M. S., Sjöquist, B., and Änggård, E., 1978, *Acta Pharmacol. Toxicol.(Kbh.)* **43**:266–272.
215. Pickard, J. D., MacDonnel, L. A., MacKenzie, E. T., and Harper, A. M., 1977, *Eur. J. Pharmacol.* **43**:343–351.
216. Pickard, J., Tamura, A., Stewart, M., McGeorge, A., and Fitch, W., 1980, *Brain Res.* **197**:425–431.
217. Dahlgren, N., Ingvar, M., Yokoyama, H., and Siesjö, B. K., 1981, *J. Cereb. Blood Flow Metab.* **1**:233–236.
218. Bill, A., 1979, *Acta Physiol. Scand.* **105**:437–442.
219. Cuypers, J., Cuevas, A., and Duisburg, R., 1978, *Neurochirurgie* **21**:62–66.
220. Wei, E. P., Ellis, E. F., and Kontos, H. A., 1980, *Am. J. Physiol.* **238**:H226–H230.
221. Dahlgren, N., and Siesjö, B. K., 1981, *J. Cereb. Blood Flow Metab.* **1**:109–115.
222. Wennmalm, A., Eriksson, S., and Wahren, J., 1981, *Clin. Physiol.* **1**:227–234.
223. Siesjö, B. K., Wieloch, T., 1983, *Advances in Prostaglandin, Thromboxane, Leukotriene Research*, Volume 12 (B. Samuelsson, R. Paoletti, and P. Ramwell, eds.), pp. 339–344.
224. Hougaard, K., Nilsson, B., and Wieloch, T., 1983, *Acta Physiol. Scand.*, in press.
225. Zülch, K. J., 1953, *Zentralbl. Allg. Pathol.* **90**:402.
226. Lindenberg, R., 1959, *J. Neuropathol. Exp. Neurol.* **18**:348–349.
227. Brierley, J. B., 1976, *Greenfield's Neuropathology* (W. Blackwood and J. A. N. Corsellis, eds.), Edward Arnold, London, pp. 44–85.
228. Waltz, A. G., 1970, *Stroke* **1**:27–37.
229. Fieschi, C., 1971, *Seventh Conference on Cerebral Vascular Diseases* (J. F. Toole, J. Moossy, and R. Janeway, eds.), Grune & Stratton, New York, pp. 130–139.
230. Paulson, O. B., 1971, *Stroke* **2**:327–360.
231. Olesen, J., 1974, *Cerebral Blood Flow: Methods for measurement, regulation, effects of drugs,*

and changes in disease, Thesis, University of Copenhagen, FADLs Forlag, Copenhagen-Århus-Odense.

232. Kety, S. S., King, B. D., Horvath, S. M., Feffers, W. A., and Hafkenschiel, J. H., 1950, *J. Clin. Invest.* **29:**402–407.
233. Finnerty, F. A., Witkin, L., and Fazekas, J. F., 1954, *J. Clin. Invest.* **33:**1227–1232.
234. Strandgaard, S., Olesen, J., Skinhoj, E., and Lassen, N. A., 1973, *Br. Med. J.* **1:**507–510.
235. Harper, A. M., 1965, *Br. J. Anaesth.* **37:**225–235.
236. Häggendal, E., and Johansson, B., 1965, *Acta Physiol. Scand.* [*Suppl.*] **258:**27–53.
237. Plum, F., Posner, J. B., and Troy, B., 1968, *Arch. Neurol.* **18:**1–13.
238. Freeman, J., and Ingvar, D. H., 1968, *Exp. Brain Res.* **5:**61–71.
239. Waltz, A. G., 1968, *Neurology (Minneap.)* **18:**613–621.
240. Reivich, M., Marshall, W. J. S., and Kassell, N., 1971, *Seventh Princeton Conference on Cerebral Vascular Diseases* (J. F. Toole, J. Moossy, and R. Janeway, eds.), Grune & Stratton, New York, pp. 66–69.
241. Siesjö, B. K., and Nilsson, L., 1971, *Scand. J. Clin. Lab. Invest.* **27:**83–96.
242. Klatzo, I., 1967, *J. Neuropathol. Exp. Neurol.* **26:**1–14.
243. Sundt, T. M., and Davis, D. H., 1980, *Cerebral Arterial Spasm* (R. H. Wilkins, ed.), Williams & Wilkins, Baltimore, pp. 244–250.
244. Brandt, L., 1981, *Aspects on Cerebral Vasospasm*, Thesis, University of Lund, Lund.
245. Alksne, J. F., and Branson, P. J., 1980, *Neurol. Res.* **2:**274–282.
246. Boullin, D. J., (ed.), 1980, *Cerebral Vasospasm*, John Wiley & Sons, New York, Chichester.
247. Sano, K., Asano, T., Tanishima, T., Sasaki, T., 1980, *Neurol. Res.* **2:**253–271.
248. Hossmann, K.-A., and Kleihues, P., 1973, *Arch. Neurol.* **29:**375–382.
249. Hossmann, K.-A., 1977, *Brain and Heart Infarct* (K. J. Zülch, W. Kaufmann, K.-A. Hossmann, and V. Hossmann, eds.), Springer Verlag, Berlin, Heidelberg, New York, pp. 107–122.
250. Levy, D. E., Van Uitert, R. L., and Pike, C. L., 1979, *Neurology* (N.Y.) **29:**1245–1252.
251. Ginsberg, M. D., Budd, W. W., and Welsh, F. A., 1978, *Ann. Neurol.* **3:**482–492.
252. Kofke, W. A., Nemoto, E. M., Hossmann, K.-A., Taylor, F., Kessler, P. D., and Stezoski, S. W., 1979, *Stroke* **10:**554–560.
253. Siesjö, B. K., 1981, *J. Cereb. Blood Flow Metab.* **1:**155–185.
254. Hallenbeck, J. M., and Furlow, T. W., Jr., 1979, *Stroke* **10:**629–637.
255. Gaudet, R. J., and Levine, L., 1979, *Biochem. Biophys. Res. Commun.* **86:**893–901.
256. Siesjö, B. K., and Nilsson, B., 1981, *Adv. Prostaglandin Thromboxane Res.* (in press).
257. Rehncrona, S., Siesjö, B. K., and Smith, D. S., 1980, *Acta. Physiol. Scand.* **492:**135–140.
258. Yoshida, S., Inoh, S., Asano, T., Sano, K., Kubota, M., Shimazaki, H., and Ueta, N., 1980, *J. Neurosurg.* **53:**323–331.
259. Tamura, A., Graham, D., McCulloch, J., and Teasdale, G. M., 1981, *J. Cereb. Blood Flow Metab.* **1:**61–69.
260. Brandt, L., Andersson, K.-E., Edvinsson, L., Ljunggren, B., 1981, *J. Cereb. Blood Flow Metab.* **1:**339–347.
261. Marshall, L. F., Durity, F., Lounsbury, R., Graham, D. I., Welsh, F., and Langfitt, T. W., 1975, *J. Neurosurg.* **43:**308–317.
262. Smith, A. L., Hoff, J. T., Nielsen, S. L., and Larson, C. P., 1974, *Stroke* **5:**1–7.
263. Hoff, J. T., Smith, L., Hankinson, H. L., and Nielsen, S. L., 1975, *Stroke* **6:**28–33.
264. Michenfelder, J. D., and Milde, J. H., 1975, *Stroke* **6:**405–410.
265. Bleyaert, A. L., Nemoto, E. M., Safar, P., Stezoski, S. W., Mickell, J. J., Moossy J., and Gutti, R. R., 1978, *Anesthesiology* **49:**390–398.
266. Steen, P. A., and Michenfelder, J. D., 1979, *Anesthesiology* **50:**404–408.
267. Heistad, D. D., Marcus, M. L., Said, S. J., and Gross, P. M., 1980, *Am. J. Physiol.* **239:**H73–H80.
268. Wahl, M., 1981, *J. Cereb. Blood Flow Metab.* **1**(Suppl. 1:323–324.
269. Edvinsson, L., McCulloch, J., and Uddman, R., 1981, *J. Cereb. Blood Flow Metab.* **1**(Suppl. 1):319–321.
270. Auer, L. M., Johansson, B. B., Kuschinsky, W., and Edvinsson, L., 1981, *J. Cereb. Blood Flow Metab.* **1**(Suppl. 1):311–312.

Index